A Textbook of
ELECTRICAL ENGINEERING

For

First Year B.Tech./B.E. Students

By

J.B. Gupta

B.Sc. (Elec. Engg.); C. Eng. (I);
F.I.E. (India); M.I.S.T.E.

S.K. KATARIA & SONS®

Publisher of Engineering Books

4885/109, Prakash Mahal, Dr. Subhash Bhargav Lane,
Opposite Delhi Medical Association, Daryaganj, New Delhi–110002 (India)
Phone: +91-11-43551243; Mobile: +91-9871775858, 7678425209
e-mail: katariabook@gmail.com, katariabooks@yahoo.com
Website: www.skkatariaandsons.com
(e-books, subscriptions and bulk buying also available on our website)

Published by:

S.K. Kataria & Sons®

4885/109, Prakash Mahal, Dr. Subhash Bhargav Lane,
Opposite Delhi Medical Association, Daryaganj, New Delhi-110002
Phone: +91-11-43551243; Mobile: +91-9871775858, 7678425209
e-mail: katariabook@gmail.com, katariabooks@yahoo.com

Edited By: **Rajeev Manglik**

B.E. (Elec.) Pb. University

and

Rohit Manglik

B.E. (Elec.) University of Roorkee

First Edition: March, 2001; Thirteenth Edition: August, 2014; Reprint: 2026

Laser Type Setting by:
SARA ASSIGNMENTS, Delhi

Printed at:
A.S. OFFSET PRINTERS, Delhi

PREFACE TO THE THIRTEENTH EDITION

The author feels great pleasure in presenting the Thirteenth Edition of the book titled "ELECTRICAL ENGINEERING" with marked improvements over the previous one. The book has been thoroughly revised and rewritten so as to meet the needs of First Year students of Degree Engineering (all branches) of U.P. Technical Univ. (U.P.T.U.) Lucknow and Uttrakhand Technical Univ. (U.T.U.) Dehradun. The book has been redesigned keeping in view the latest syllabi prescribed by U.P.T.U. Lucknow and U.T.U. Dehradun. The subject matter has been reorganised so as to facilitate the students as well as teachers. Chapter 10 of "Introduction To Power System" has been replaced by Chapter 18 of "Introduction To Earthing and Electrical Safety". Few topics such as "multimeter", "megger", "ac excitation in magnetic circuits" have been deleted as these have not been included in the latest syllabi prescribed by U.P.T.U. Lucknow and U.T.U. Dehradun.

It is earnestly hoped that with the extensive additions, revisions, modifications and improvement, the present edition will facilitate the students in preparing for their regular examination and prove more useful to the students than the earlier edition.

Every care has been taken to eliminate misprints, omissions and errors but it is too much to expect that no inaccuracy/error/misprint has crept in and the author would be grateful to the readers for bringing to notice any of such errors/misprints/omissions they may come across while going through the book.

The author will like to thank the readers (both teachers and students) who sent their valuable suggestions that became the basis for the revision of the book.

The author shall feel satisfied if the book meets the needs of the students for whom it is meant. Suggestions and criticism for the further improvement of the book shall be welcomed.

The author will like to appreciate the sincere efforts and contribution made by his sons Rajeev Manglik and Rohit Manglik in presentation of the new edition in this form.

The author will like to thank all the scientists, engineers, authors and publishers whose works and text have been the source of enlightenment, inspiration and guidance in presenting this new edition.

Lastly, but not the least, the author is thankful to his publisher, Sanjeev Kumar Kataria for bringing out the book in such a short time and for pricing it moderately despite heavy cost of paper and printing.

August 11, 2014 **J.B. GUPTA**

PREFACE TO THE FIRST EDITION

The primary aim of presenting this book is to meet the needs of first year students of degree engineering (all branches) studying in Engineering Colleges affiliated to U.P. Technical University. The subject matter has been presented in an easy comprehensive manner, logical sequence using simple language and to give a detailed exposition of fundamentals.

The book consists of 14 chapters providing a comprehensive coverage of course as prescribed by U.P. Technical University for first year students. A large number of worked-out examples, theoretical exercises and numerical problems, all of degree standard, have been included in the text. The solution to illustrative examples covering almost all topics and subtopics makes the entire presentation easy to follow. Exercises and numerical problems with answers given at the end of each chapter helps the students in evaluating their understanding of the subject.

20 Practicals, prescribed by U.P. Technical University for 1st year degree engineering students have also been included in the book.

Every care has been taken to eliminate misprints, omissions and errors but it is too much to expect that no inaccuracy/ misprint/ error has crept in and the author would be grateful to the readers for bringing to his notice any such errors/ misprints/ omissions they may come across while going through the book.

The author shall feel satisfied if the book meets the needs of the students for whom it is meant. Suggestions and criticism for the improvement of the book shall be welcomed.

The author will like to thank to all the Scientists, Engineers, Authors and Publishers whose works and text have been the source of enlightenment, inspiration and guidance to me in the presentation.

Lastly, but not the least, the author is thankful to his publisher for bringing out the book in such a short time and for pricing it moderately despite heavy cost of paper and printing.

January 1, 2001 **J.B. GUPTA**

CONTENTS

CHAPTER 1

Introductory Concepts

1.1. NATURE OF ELECTRICITY

It is comparatively easy to describe electricity by its action than by its nature. Electricity is convenient form of energy and have got numerous applications, such as for lighting, transportation, communication, heating, elevators, and for driving of various types of machine tools owing to its inherent numerous advantages.

Although no one knows precisely what electricity is, it has been possible to develop theories about electricity through experiments and observations of its behaviour. As a result, it is now believed that all matters are essentially electrical in nature.

In order to appreciate the nature of electricity, it is necessary to consider the structure of matter. Matter has been defined any thing that occupies space and has weight such as copper, iron, water, air, smoke, aluminium, silver etc. Matter is made up of extremely small particles known as *molecules*. A molecule is the smallest particle into which a given substance may be divided without changing its identity. Molecules can further be subdivided chemically into still minute particles known as *atoms*. An atom may be regarded as the smallest particle of an element that can exist. The substances, whose molecules consist of similar atoms, are called the *elements* whereas the substances whose molecules consist of dissimilar atoms are called the *compounds*.

According to modern electron theory, the atoms of all elements are composed of two parts, namely, the central *nucleus* and the surrounding or *orbital electrons* (Fig. 1.1).

The electrons each have a negative electric charge of 1.602×10^{-19} coulomb and some particles within the nucleus have a positive charge of the same magnitude. Since opposite charges attract each other, a force of attraction exists between the oppositely charged electrons and nucleus. Compared to the mass of nucleus, electrons are relatively tiny particles of almost negligible mass (each electron weighing 9.107×10^{-31} kg).

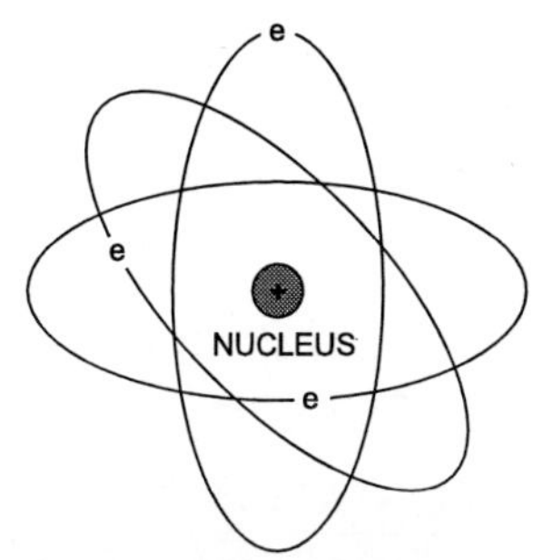

Fig. 1.1 *Rutherford's Nuclear Model of Atomic Structure*

The nucleus of an atom is largely a cluster of two types of particles, called the *protons* and *neutrons*. Protons each have a positive electrical charge, equal in magnitude (but opposite in polarity) to the negative charge of an electron. A neutron has no charge at all. Protons and neutrons, each have mass about 1,800 times the mass of an electron. For a given atom, the number of protons in the nucleus normally is equal to the number of orbiting electrons. Thus, an atom is electrically neutral. If an atom loses an electron, it loses the negative charge and becomes a *positive ion*. Similarly if an atom gains an additional electron, it becomes a *negative ion*. However, nearly all atoms have some electrons which are loosely bound to their nuclei. These electrons are known as *free electrons* and may be dislodged by one means or another and transferred from one atom to another. The body, that contains unequal number of electrons and protons is said to be *electrically charged.* If a body contains electrons more than its normal number then the body is said to be – *vely charged.* Similarly a body containing electrons less than its normal number is said to be + *vely charged.*

The atoms of some materials such as silver, copper, aluminium and zinc, known as *conductors,* have many free electrons. These free electrons of a metal move about haphazardly in all directions from atom to atom but when a certain electrical pressure or potential is applied to such metals at the two ends, the electrons move only in one direction. The drift of electrons in a conductor in one direction is known as the *electric current i.e.* the flow of electric current takes place by the movement of the electrons in a conductor. Since electrons are negatively charged, the direction of their motion is opposite to the direction of conventional current which is from higher potential to the lower potential. In non-metallic materials, such as glass, mica, slate and porcelain, the electrons are very closely bound to the nucleus and it is difficult to remove the electrons from the atoms. Such materials are known as *non-conductors or insulators.*

1.2. ELECTRIC CURRENT

Electric current may be defined as the time rate of net motion of an electric charge across a cross-sectional boundary (Fig. 1.2). A random motion of electrons in a metal does not constitute a current unless there is a net transfer of charge with time

i.e. electric current, i = Rate of transfer of electric charge

$$= \frac{\text{Quantity of electric charge transferred during a given time duration}}{\text{Time duration}}$$

$$= \frac{dQ}{dt} \qquad \text{...(1.1)}$$

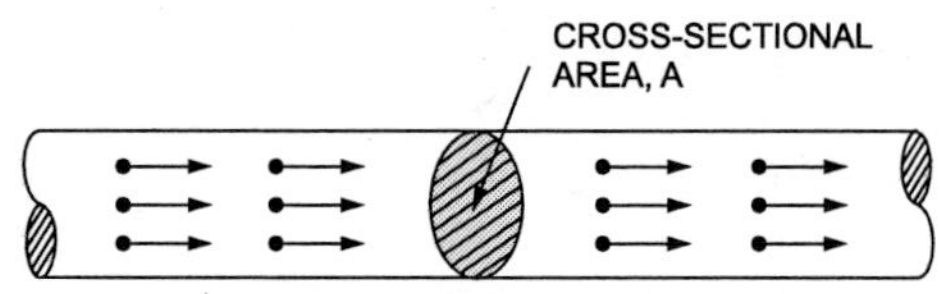

Fig. 1.2

Coulomb is the practical as well as SI unit for measurement of electric charge. One coulomb is approximately equal to 624×10^{16} electrons.

Since current is the rate of flow of electric charge through a conductor and coulomb is the unit of electric charge, the current may be specified in coulombs per second. In practice, the term coulomb per second is seldom used, a shorter term, *ampere* is used instead.

1.3. ELECTROMOTIVE FORCE AND POTENTIAL DIFFERENCE

Electromotive force (emf) is the force that causes an electric current to flow in an electric circuit while the *potential difference* (pd) between two points in an electric circuit is that difference in their electrical state which tends to cause flow of electric current between them.

Volt is a unit of electromotive force as well as potential difference in practical as well as in SI system of units.

The volt is defined as that potential difference between two points of a conductor carrying a current of one ampere when the power dissipated between these points is equal to one watt.

1.4. RESISTANCE

Resistance may be defined as that property of a substance which opposes (or restricts) the flow of an electric current (or electrons) through it.

The practical as well as MKS (or SI) unit of resistance is ohm (Ω), which is defined as that resistance between two points of a conductor when a potential difference of one volt, applied between these points, produces in this conductor a current of one ampere, the conductor not being a source of any emf.

For insulators having high resistance, much bigger units kilo ohm or kΩ (10^3 ohm) and mega ohm or MΩ (10^6 ohm) are used. In case of very small resistances smaller units like milli ohm (10^{-3} ohm) or micro ohm (10^{-6} ohm) are employed.

1.5. OHM'S LAW

As the rate of flow of water through a pipe is directly proportional to the effective pressure (*i.e.* difference of pressure at two ends) and inversely proportional to the frictional resistance, similarly the current flowing through a conductor is directly proportional to the potential difference across the ends of the conductor and inversely proportional to the conductor resistance. This relation was discovered by German Physicist Georg Simon Ohm and so it is known as Ohm's law.

If I is the current flowing through a conductor of resistance R across which a potential difference V is applied then according to Ohm's law

$$I \propto V \quad \text{and} \quad I \propto \frac{1}{R} \quad \text{or} \quad I \propto \frac{V}{R} \quad \text{or} \quad I = \frac{V}{R} \qquad ...(1.2)$$

where V is in volts, R is in ohms and I is in amperes.

Ohm's law may be defined as follows:

Physical state i.e. temperature etc. remaining the same, the current flowing through a conductor is directly proportional to the potential difference applied across its ends.

Or

The ratio of potential difference applied across a conductor and current flowing through it remains constant provided physical state i.e. temperature etc. of the conductor remains unchanged.

$$i.e. \quad \frac{V}{I} = \text{Constant} = R \qquad ...(1.3)$$

where R is known as the resistance of the conductor.

Ohm's law may be alternatively expressed as

$$V = IR \qquad ...(1.4)$$

Equations (1.2), (1.3) and (1.4) give Ohm's law in three forms with which the student should be familiar.

Ohm's law cannot be applied to circuits consisting of electronic tubes or transistors because such elements are not bilateral *i.e.* they behave in different way when the direction of flow of current is reversed as in case of a diode. Ohm's law also cannot be applied to circuits consisting of nonlinear elements such as powdered carbon, thyrite, electric arc etc. For example, for silicon carbide, the relationship between applied voltage (or potential difference) V and current flowing I is given as $V = K I^m$ where K and m are constants and m is less than unity.

1.6. LAWS OF RESISTANCE

The resistance of a wire depends upon its length, area of x-section, type of material, purity and hardness of material of which it is made of and the operating temperature.

Resistance of a wire is

(*a*) directly proportional to its length, l *i.e.* $R \propto l$

(*b*) inversely proportional to its area of x-section, a *i.e.* $R \propto 1/a$

Combining above two facts we have $R \propto l/a$

$$\text{or} \quad R = \rho \frac{l}{a} \qquad ...(1.5)$$

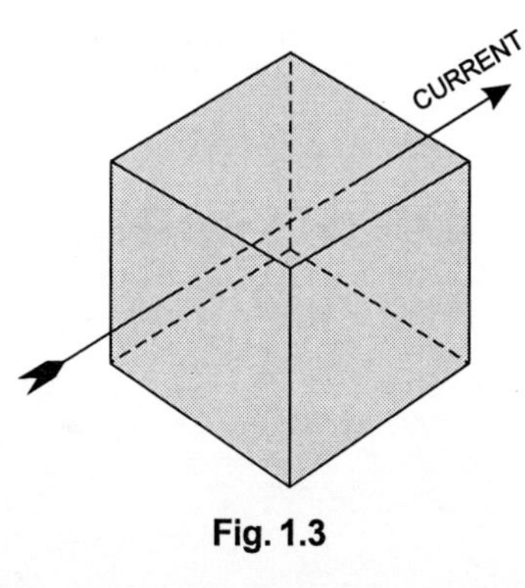

Fig. 1.3

where ρ (rho) is a constant depending upon the nature of the material and is known as the *specific resistance or resistivity of* the material of the wire.

To determine the nature of the constant ρ let us imagine a conductor of unit length and unit cross-sectional area, for example a cube whose edges are each of length one unit, and let the current flow into the cube at right angles to one face and out at the other face. Then putting $l = 1$ and $a = 1$ in Eq. (1.5) we have R = ρ. *Hence resistance of a material of unit length having unit cross-sectional area is defined as the resistivity or specific resistance of the material.*

Specific resistance or resistivity of a material is also defined as the resistance between opposite faces of a unit cube of that material.

Resistivity is measured in ohm-metres (Ω-*m*) or ohms per metre cube in MKS (or SI) system and ohm-cm (Ω-cm) or ohm per cm cube in CGS system.

$$1\ \Omega\text{-m} = 100\ \Omega\text{-cm}$$

1.7. CONDUCTANCE AND CONDUCTIVITY

The reciprocal of resistance *i.e.* 1/R is called the *conductance* and is denoted by English letter G. It is defined as the inducement offered by the conductor to the flow of current and is measured in siemens (S)*. Earlier, the unit of conductance was mho (℧).

$$1 \text{ siemen} = 1 \text{ mho}$$

From Eq. (1.5)

$$G = \frac{1}{R} = \frac{1}{\rho \frac{l}{a}} = \frac{1}{\rho} \cdot \frac{a}{l} = \sigma \frac{a}{l} \qquad \text{...(1.6)}$$

where $\sigma = \frac{1}{\rho}$ and is known as *specific conductance* or *conductivity* of the material. Hence *conductivity* is the reciprocal of the resistivity and is defined as the conductance between the two opposite faces of a unit cube. The unit of conductivity is siemens/metre (S/m).

Example 1.1 **A coil consists of 2,000 turns of copper wire having a cross-sectional area of 0.8 mm². The mean length per turn is 80 cm, and the resistivity of copper is 0.02 μ Ω-m. Find the resistance of the coil.**

Solution:

Length of the coil, l = Number of turns × mean length per turn = 2,000 × 0.8 = 1,600 m

Cross-sectional area of wire, $a = 0.8 \text{ mm}^2 = 0.8 \times 10^{-6} \text{ m}^2$

Resistivity of copper, $\rho = 0.02\ \mu\Omega\text{-m} = 2 \times 10^{-2} \times 10^{-6}\ \Omega\text{-m} = 2 \times 10^{-8}\ \Omega\text{-m}$

Resistance of the coil, $R = \rho \frac{l}{a} = 2 \times 10^{-8} \times \frac{1{,}600}{0.8 \times 10^{-6}} = 40\ \Omega$ **Ans.**

Example 1.2 **The resistance of a conductor 1 mm² in cross section and 20 m long is 0.346 Ω. Determine the specific resistance of the conductor material.** [*Bangalore Univ. Elec. Circuits-I, 1991*]

Solution:

Resistance of conductor, $R = 0.346\ \Omega$

Length of conductor, $l = 20$ m

Cross-sectional area of conductor, $a = 1 \text{ mm}^2 = 1 \times 10^{-6} \text{ m}^2$

Specific resistance of conductor material, $\rho = \frac{R \times a}{l} = \frac{0.346 \times 1 \times 10^{-6}}{20} = 1.73 \times 10^{-8}\ \Omega\text{-m}$ **Ans.**

Example 1.3 **A wire of length 1 m has a resistance of 2 Ω. Obtain the resistance if specific resistance is doubled, diameter is doubled and the length is made three times of the first.**

[*Pb. Technical Univ. Basic Electrical and Electronics Engineering Second Semester 2004-05*]

Solution:

New specific resistance, $\rho_2 = 2\,\rho_1$

New length of wire, $l_2 = 3\, l_1$

New cross-sectional area, $a_2 = \frac{\pi}{4}\left(d_2^2\right) = \frac{\pi}{4}(2d_1)^2 = 4\, a_1$

New resistance, $R_2 = \frac{\rho_2 l_2}{a_2} = \frac{2\rho_1 \times 3 l_1}{4 a_1} = 1.5 \frac{\rho_1 l_1}{a_1} = 1.5\, R_1 = 1.5 \times 2 = 3\ \Omega$ **Ans.**

* Named after German electrical engineer Werner von Siemens

1.8. TYPES OF SUPPLY

There are two types of supply viz dc and ac.

The voltage or current available from batteries or solar cells is *direct* in the sense that the polarity remains the same. Such sources are called direct current sources. A plot of output of such sources with respect to time is a straight line parallel to time axis as illustrated in Fig. 1.4. This shows that the voltage / current output of a dc source is constant with respect to time, unless the chemicals in the battery are exhausted or the light incident on solar cell varies.

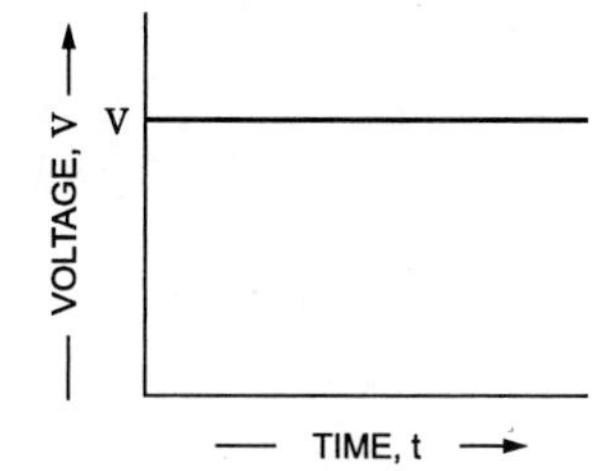

(*a*) *Output Characteristic of Battery*

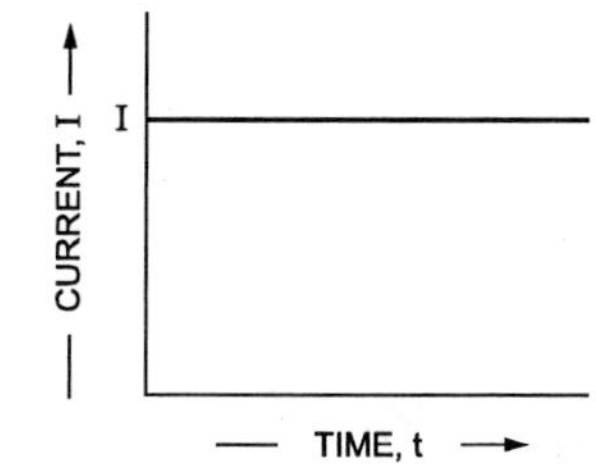

(*b*) *Output Characteristic of Solar Cell*

Fig. 1.4

Power obtainable from batteries or solar cells is very small as compared to the total power we need. The dc power in bulk can be generated by using dc generators, and will be discussed in detail later on. Standard dc voltages are 230 and 460 V.

A current (or voltage) is called *alternating* if it periodically changes its direction and magnitude whereas direct currents are steady and in one direction. In more restricted sense, alternating current is a periodically varying current, the average value of which, over a period, is zero.

At present a large percentage of electrical energy (nearly all) being used for domestic, commercial and industrial purposes is generated as alternating current because of technical and economical reasons (refer to Art 3.1).

For industrial applications of direct current like electrolytic processes, welding processes and variable speed motor drives, the present trend is to generate ac and then convert it into dc by converters. For conversion of ac into dc the dc generator as a part of motor (ac)- generator (dc) set has to compete the SCR rectifiers and other various power controlled devices which usually are cheaper in cost, compact in size, relatively noise-free in operation and require minimum maintenance, though they suffer from the disadvantages of having poor power factor, harmonic generation, poor braking etc.

In our country the standard voltage for low tension (*lt*) distribution is 240 / 415 V (240 V between phase and neutral wires and 415 V from line-to-line). This is a reasonably safe voltage. Higher the voltage used, lower is the installation cost i.e. economy is achieved but increase of voltage means more hazard to human life and property. Hence, a compromise is made. All over the world the standard voltage used for *lt* distribution is in the range of 200 / 350 V to 250 / 440 V except in American continent where 110 /190 V is the distribution voltage. Generation, transmission and *ht* distribution voltages are governed by economy and operation.

1.9. RESISTANCE VARIATION WITH TEMPERATURE

The resistance of all pure metallic conductors increases with the increase in temperature but the resistance of the insulators and non-metallic materials generally decreases with the increase in temperature.

If the resistance of any pure metal is plotted on a temperature base, it is found that over the range of temperature from 0 to 100 °C the graph is practically a straight line, as illustrated in Fig. 1.5. If this straight line is extended, it cuts the temperature axis at some temperature, $-t_0$ °C, known as *inferred zero resistance temperature.* This does not mean that the resistance of the metal is actually zero at that temperature, but $-t_0$ °C is the temperature at which the resistance would

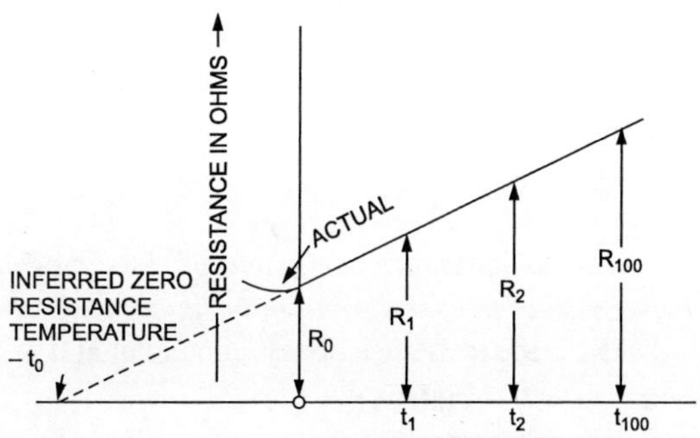

Fig. 1.5

be zero if the rate of decrease between 100 and 0 °C were maintained constant at all temperatures. From the similarity of the triangles in Fig. 1.5

$$\frac{R_2}{R_1} = \frac{t_0 + t_2}{t_0 + t_1} \quad ...(1.7)$$

where R_1 and R_2 are the resistances at temperatures t_1 °C and t_2 °C respectively. Thus, if the resistance R_1 for any temperature t_1 °C is known, then resistance for any other temperature t_2 can be computed from above equation provided that t_0 for that particular material is known.

Note: It *has been found that the resistivity of the metallic conductors is significantly affected by variations in temperature but it is also affected by other factors like tension and pressure to some extent. For most metals except lithium and calcium resistivity decreases with the increase in pressure. However, resistivity increases with increase in tension.*

The variation of resistance with temperature is often utilized in determining temperature variations. For example, in testing of an electric machine, the resistance of the coil is measured both before and after the test run, and the increase in resistance is a measure of the rise in temperature. For computation of temperature rise Eq. (1.7) may be transposed to the following form

$$t_2 - t_1 = \frac{R_2 - R_1}{R_1}(t_0 + t_1) \quad ...(1.8)$$

A temperature of 20 °C has been adopted as the *standard reference temperature* for measurement of resistance, and the handbooks give the resistance of the various materials at that temperature. Consequently, when a designer is computing the resistance of any conductor from its dimensions, the initial temperature *t*, at which the resistance is known, is generally 20 °C.

1.10. TEMPERATURE COEFFICIENT OF RESISTANCE

Let a metallic conductor having a resistance of R_0 at 0 °C be heated to *t* °C and let its resistance at this temperature be R_t. From Eq. (1.7)

$$\frac{R_t}{R_0} = \frac{t_0 + t}{t_0 + 0}$$

$$\text{or } R_t = R_0 + \frac{1}{t_0} R_0 t$$

$$\text{or change in resistance, } \Delta R = R_t - R_0 = \frac{1}{t_0} R_0 t = \alpha_0 R_0 t \quad ...(1.9)$$

where $\alpha_0 = \frac{1}{t_0}$ and is called the temperature coefficient of resistance of the material at 0 °C

From Eq. (1.9) it may be concluded that change in resistance due to change in temperature

(*a*) varies directly as its initial resistance,

(*b*) varies directly as rise in temperature and

(*c*) depends on the nature of the material of the conductor.

The Eq. (1.9) may be rewritten as

$$\alpha_0 = \frac{\Delta R}{R_0 t} \quad ...(1.10)$$

So the temperature coefficient of resistance may be defined as the ratio of increase in resistance per degree rise of temperature to the original resistance.

If R_0 is the resistance of any conductor at 0 °C and α_0 is the temperature coefficient of resistance at 0 °C, then resistance at *t* °C is given as

$$R_t = \text{Original resistance} + \text{increase in resistance} = R_0 + R_0 \alpha_0 t = R_0(1 + \alpha_0 t) \quad ...(1.11)$$

The above expression holds good for both increase as well as decrease in temperature.

It is to be noted that

(*i*) temperature coefficient of resistance for all pure metallic conductors is positive *i.e.* the resistance of all pure metallic conductors increase with the increase in temperature, that of non-metallic materials such as of carbon is negative i.e. the resistance of non-metallic materials such as of carbon decreases with the increase in temperature. The temperature coefficient of resistance of alloys like constantan and manganin is negligible.

(*ii*) temperature coefficient of resistance is not constant but depends on the initial temperature on which the increment in resistance is based. When the increment is based on the resistance measured at 0 °C, then the temperature coefficient of resistance has the value of α_0. At any other temperature t °C, value of temperature coefficient of resistance is α_t and so on. For any material the temperature coefficient of resistance at 0 °C *i.e.* α_0 has the maximum value.

The temperature coefficient of resistance at any temperature t_1 is given as

$$\alpha_1 = \frac{1}{\frac{1}{\alpha_0} + t_1} \qquad ...(1.12)$$

The temperature coefficient of resistance at temperature t_2 in terms of temperature coefficient of resistance at temperature t_1 is given as

$$\alpha_2 = \frac{1}{\frac{1}{\alpha_1} + (t_2 - t_1)} \qquad ...(1.13)$$

If R_1 is the resistance of any conductor at t_1 °C and α_1 is the temperature coefficient of resistance at t_1 °C, then resistance of the conductor at t_2 °C is given

$$R_2 = R_1\,[1 + \alpha_1\,(t_2 - t_1)] \qquad ...(1.14)$$

(*iii*) As the resistance of the material changes with the change in temperature so it is obvious that resistivity of the material depends on temperature.

Knowledge of temperature coefficient of resistance is utilized in determining the temperature rise of electrical machines, as already described in Art 1.9.

1.11. SI SYSTEM OF UNITS

SI is the latest form of metric system and absorbs in it the rationalized MKSA system. SI stands for "Systeme International d' Unites" in French. This abbreviation is now adopted by the International Standardising Organisation as the abbreviated name of this new system of units in all languages.

The SI system is, in fact, simply the RMKSA* system expanded by adding the degree Kelvin, Candela and mole as basic units of temperature, luminous intensity and amount of substance respectively. The SI system is a comprehensive, logical and coherent system, designed for use in all branches of science, engineering and technology.

This system derives all the units from the following seven base units.

Quantity	*Unit*	*Symbol*
Length	metre	m
Mass	kilogram	kg
Time	second	s
Intensity of electric current	ampere	A
Thermodynamic temperature	Kelvin	K
Luminous intensity	candela	cd
Amount of substance	mole	mol

* RMKSA stands for rationalized metre, kilogram, second, ampere

The SI system, besides seven base units, has following supplementary units.

Quantity	*Unit*	*Symbol*
Plane angle	radian	rad
Solid angle	steradian	sr

Recommended prefixes for formation of multiples and submultiples of units are given below:

Multiple	Prefix	Symbol	Fraction	Prefix	Symbol
10	deca	*da**	0.1	deci	*d**
100	hecto	*h**	0.01	centi	*c**
1000	kilo	k	0.001	milli	*m*
1000 000	mega	M	0.000 001	micro	μ
1000 000 000	giga	G	0.000 000 001	nano	*n*
1000 000 000 000	tera	T	0.000 000 000 001	pico	*p*
			0.000 000 000 000 001	fento	*f*
			0.000 000 000 000 000 001	atto	*a*

*Should be restricted as much as possible.

1.12. WORK, POWER AND ENERGY

Work is said to be done by or against a force, when its point of application moves in or opposite to the direction of the force and is measured by the product of the force and the displacement of the point of application in the direction of force.

$$i.e. \text{ Work done, } W = \text{Force } [F] \times \text{distance } [d]$$

The SI or MKS unit of work is the joule, which is defined as the work done when a force of one newton acts through a distance of one metre in the direction of the force. Hence, if a force F acts through distance d in its own direction,

$$W = F \text{ [newtons]} \times d \text{ [metres]} = Fd \text{ joules}$$

Power is defined as the rate of doing work or the amount of work done in unit time.

The MKS or SI unit of power is the joule/second or watt. In practice, the watt is often found to be inconveniently small and so a bigger unit, the *kilowatt* is frequently used.

$$1 \text{ kilowatt} = 1{,}000 \text{ watts}$$

The bigger unit of power, most commonly used in engineering practice (not at all in SI system) is horse power defined as below :

Metric Horse Power : It is the practical unit of power in MKS system (not in SI system) which according to ISI specifications is equal to 75 kgf-m of work done per second.

Energy is defined as the capacity of doing work. Its units are same as those of work, mentioned above.

If a body having mass m, in kg, is moving with velocity v, in metres/second,

$$\text{Kinetic energy} = \frac{1}{2} m\text{v}^2 \text{ joules}$$

If a body having mass m, in kg, is lifted vertically through height h, in metres, and if g is the gravitational acceleration, in metres/second^2 in that region, potential energy acquired by the body

$$= \text{Work done in lifting the body} = mgh \text{ joules} = 9.81\, mh \text{ joules}$$

As already stated, in SI system the unit of energy of all forms is joule. Bigger unit of energy is mega joules (MJ) where 1 MJ = 10^6 J.

The thermal units of energy, calorie (gm calorie) and kilo calorie (kilogram calorie), are defined below:

Calorie : It is the amount of heat required to raise the temperature of one gram of water through 1 °C.

$$1 \text{ calorie} = 4.18 \text{ J} \simeq 4.2 \text{ J}$$

Kilo Calorie : It is the amount of heat required to raise the temperature of 1 kg of water through 1 °C.

$$1 \text{ k. calorie} = 1{,}000 \text{ calories} = 4{,}180 \text{ joules} \simeq 4{,}200 \text{ J}$$

1.13. ELECTRICAL UNITS OF WORK, POWER AND ENERGY

The unit of work done and of energy expended is joule. It is equal to the energy expended in passing 1 coulomb of charge through a resistance of 1 ohm *i.e.* the energy expended in passing one ampere current for 1 second through a resistance of one ohm is taken as one joule. It may also be expressed as 1 watt-second *i.e.* one watt of power consumed for one second.

i.e. 1 joule = 1 watt-second

The unit of energy, joule or watt-second is too small for practical purposes, so a bigger unit Mega joule (MJ) or kilowatt-hour (kWh) is used in electrical engineering.

$$1 \text{ kWh} = 1{,}000 \text{ watt-hours} = 1{,}000 \times 3{,}600 \text{ watt-seconds or joules} = 3.6 \text{ MJ}$$

The kWh, also called the *Board of Trade* (BOT) *unit,* is the energy absorbed by supplying a load of 1 kW or 1,000 watts for the period of one hour. This is legal unit on which charges for electrical energy are made, and, therefore, it is called the Board of Trade (BOT) unit.

Watt: It is defined as the power expended when there is an unvarying current of one ampere between two points having a potential difference of one volt. As already stated the bigger unit of power is kW or Mega watt.

$$1 \text{ kW} = 1{,}000 \text{ watts}$$
$$1 \text{ MW} = 1{,}000 \text{ kW} = 1 \times 10^6 \text{ watts}$$

1.14. CONVERSION OF ELECTRICAL UNITS INTO MECHANICAL AND THERMAL UNITS OR VICE VERSA

$$1 \text{ watt} = 1 \text{ joule/second} = 1 \text{ N-m/s}$$

$$1 \text{ kW} = 1{,}000 \text{ watts or J/s or N-m/s} = \frac{1{,}000}{735.5} \text{ } i.e. \text{ } 1.36 \text{ hp (metric)}$$

$$1 \text{ kWh} = 1{,}000 \text{ watt-hours} = 3{,}600{,}000 \text{ watt-seconds or joules} = 1.36 \text{ hp-hour (metric)}$$

$$1 \text{ calorie} = 4.18 \text{ J or watt-seconds}$$

$$1 \text{ k cal} = 4{,}180 \text{ J or watt-seconds} = \frac{4{,}180}{3{,}600{,}000} \text{ } i.e. \text{ } \frac{1}{860} \text{ kWh}$$

$$1 \text{ kWh} = 36 \times 10^5 \text{ watt-seconds} = \frac{36 \times 10^5}{4{,}180} \text{ i.e. } 860 \text{ k cals}$$

1.15. HEATING EFFECT OF ELECTRIC CURRENT

When an electric current flows through a conductor, electrical energy is expended in overcoming the frictional resistance between the electrons and the molecules of the wire.

If potential difference of V volts is applied across a conductor and current of I amperes flows through it for time of t seconds, then energy expended will be equal to VIt watt-seconds or joules.

If R is the resistance of the conductor through which a current of I amperes flows and V is the potential difference applied across its ends then by Ohm's law

$$V = IR$$

$$\text{and energy expended, } W = VIt = IR \times I \times t = I^2 R t \text{ joules} \qquad ...(1.15)$$

$$\text{or W also} = V \times \frac{V}{R} t = \frac{V^2}{R} t \text{ joules} \qquad ...(1.16)$$

According to the *law of conservation of energy* this electrical energy expended must be converted in some other form of energy and that other form is heat *i.e.* electrical energy expended is converted into heat energy and conversion of electrical energy into heat energy is called the heating or *thermal effect of electric current.*

1.16. JOULE'S LAW OF ELECTRIC HEATING

From Eq. (1.15) the energy expended or heat generated in joules when a current of I amperes flows through a resistance of R ohms for t seconds is given as

$$H = I^2Rt \text{ joules}$$

The above expression is known as *Joule's law,* which states that the amount of heat produced in an electric circuit is

(*i*) proportional to the square of the current *i.e.* $H \propto I^2$

(*ii*) proportional to the resistance of the circuit *i.e.* $H \propto R$, and

(*iii*) proportional to the time duration for which the current flows through the circuit *i.e.* $H \propto t$

Heat produced in k cals,

$$H = \frac{I^2Rt}{4,180} \simeq \frac{I^2Rt}{4,200} \quad [\because 1 \text{ kcal} = 4,180 \text{ J} \simeq 4,200 \text{ J}]$$

1.17. SERIES CIRCUITS

When the resistors are connected end to end, so that they form only one path for the flow of current, then resistors are said to be connected in series and such circuits are known as *series circuits.*

Let resistors R_1, R_2 and R_3 be connected in series, as shown in Fig. 1.6, and the potential difference of V volts be applied between extreme ends A and D to cause flow of current of I amperes through all the resistors R_1, R_2 and R_3.

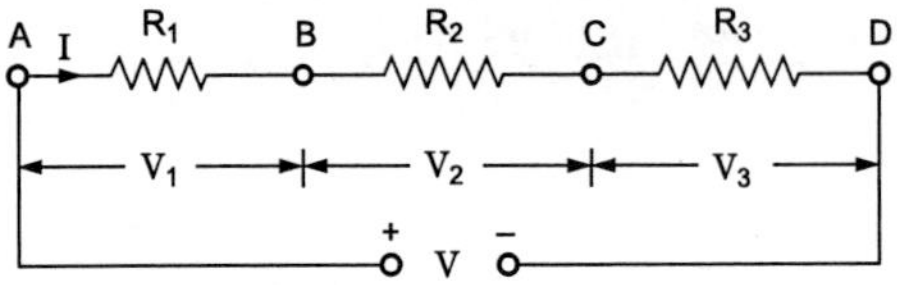

Fig. 1.6 *Series Circuit*

Now according to Ohm's law

Voltage drop across resistor R_1, $V_1 = IR_1$

Voltage drop across resistor R_2, $V_2 = IR_2$

Voltage drop across resistor R_3, $V_3 = IR_3$

Voltage drop across whole circuit,

V = Voltage drop across resistor R_1 + voltage drop across resistor R_2 + voltage drop across resistor R_3

$$i.e. \; V = IR_1 + IR_2 + IR_3 = I(R_1 + R_2 + R_3)$$

$$\text{or} \quad \frac{V}{I} = R_1 + R_2 + R_3 \quad \text{...(1.17)}$$

And according to Ohm's law $\frac{V}{I}$ gives the whole circuit resistance, say R

∴ Effective resistance of the series circuit,

$$R = R_1 + R_2 + R_3 \quad \text{...(1.18)}$$

Thus, when a number of resistors are connected in series, the equivalent resistance is given by the arithmetic sum of their individual resistances.

$$i.e. \; R = R_1 + R_2 + R_3 + \ldots R_n \quad \text{...(1.19)}$$

From the above discussions for a series circuit we conclude that

1. same current flows through all parts of the circuit,
2. applied voltage is equal to the sum of voltage drops across the different parts of the circuit,
3. different resistors have their individual voltage drops,
4. voltage drop across individual resistor is directly proportional to its resistance, current being the same in each resistor,
5. voltage drops are additive,
6. resistances are additive,
7. powers are additive.

Series circuits are common in electrical equipment. The tube filaments in small radios are usually in series. Current controlling devices are wired in series with the controlled equipment. Fuses are in series with the equipment they protect. A thermostat switch is in series with the heating element in an electric iron.

Automatic house-heating equipment has a thermostat, electromagnet coils, and safety cut-outs in series with a voltage source. Rheostats are placed in series with the coils in large motors for motor current control.

Example 1.4 **Three resistors are connected in series across a 12 V battery. The first resistor has the value of 1 ohm, second has a voltage drop of 4 V and third has a power dissipation of 12 W. Calculate the value of each resistance and circuit current.**

Solution: Let the three resistors be of R_1 (= 1 Ω), R_2 and R_3 ohms, current flowing through these three resistors R_1, R_2 and R_3 be of I amperes and voltage drops across resistors R_1, R_2 and R_3 be of V_1, V_2 and V_3 volts respectively. The circuit is shown in Fig. 1.7.

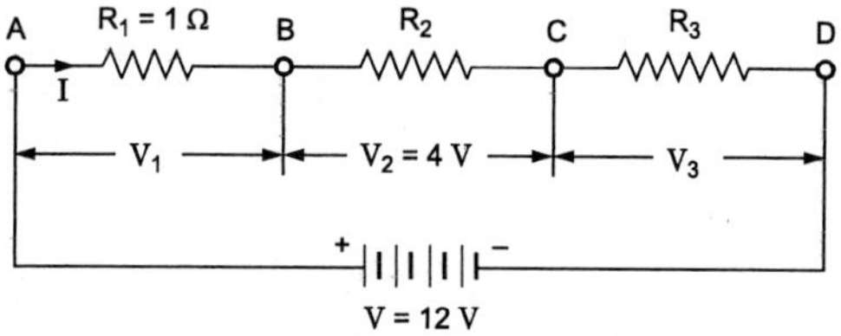

Fig. 1.7

$$\text{Now } V_1 = I R_1 = I \text{ volts}$$

$$V_2 = I R_2 = 4 \text{ volts}$$

$$V_3 = \frac{\text{Power dissipation}}{I} = \frac{12}{I} \text{ volts}$$

$$\text{Since } V = V_1 + V_2 + V_3$$

$$\text{So } 12 = I + 4 + \frac{12}{I}$$

$$\text{or } I^2 - 8I + 12 = 0$$

$$\text{or } (I - 6)(I - 2) = 0$$

$$\text{or } I = 6 \text{ or } 2 \text{ A} \quad \textbf{Ans.}$$

When I = 6 A;

$$R_2 = \frac{V_2}{I} = \frac{4}{6} = \frac{2}{3}\ \Omega$$

$$\text{and } R_3 = \frac{P}{I^2} = \frac{12}{6^2} = \frac{1}{3}\ \Omega$$

$$\text{i.e. } R_1 = 1\ \Omega;\ R_2 = \frac{2}{3}\ \Omega \text{ and } R_3 = \frac{1}{3}\ \Omega \quad \textbf{Ans.}$$

When I = 2 A;

$$R_2 = \frac{V_2}{I} = \frac{4}{2} = 2\ \Omega$$

$$\text{and } R_3 = \frac{P}{I^2} = \frac{12}{2^2} = 3\ \Omega$$

$$\text{i.e. } R_1 = 1\ \Omega;\ R_2 = 2\ \Omega \text{ and } R_3 = 3\ \Omega \quad \textbf{Ans.}$$

1.18. PARALLEL CIRCUITS

When a number of resistors are connected in such a way that one end of each of them is joined to a common point and the other ends being joined to another common point, as shown in Fig. 1.8, then resistors are said to be connected in parallel and such circuits are known as *parallel circuits*. In these circuits current is divided into as many paths as the number of resistances.

Let the resistors R_1, R_2 and R_3 be connected in parallel, as shown in Fig. 1.8, and the potential difference of V volts be applied across the circuit.

Since potential difference across each resistor is same and equal to potential difference applied to the circuit *i.e.* V

∴ According to Ohm's law

$$\text{Current in resistor } R_1,\ I_1 = \frac{V}{R_1} \quad \text{....(1.20)}$$

$$\text{Current in resistor } R_2,\ I_2 = \frac{V}{R_2} \quad \text{....(1.21)}$$

$$\text{Current in resistor } R_3,\ I_3 = \frac{V}{R_3} \quad \text{...(1.22)}$$

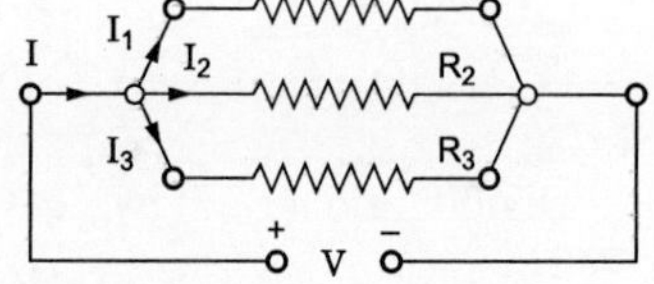

Fig. 1.8 *Parallel Circuit*

Adding Eqs. (1.20), (1.21) and (1.22) we have

$$I_1 + I_2 + I_3 = \frac{V}{R_1} + \frac{V}{R_2} + \frac{V}{R_3} = V\left(\frac{1}{R_1} + \frac{1}{R_2} + \frac{1}{R_3}\right) \qquad ...(1.23)$$

and since $I_1 + I_2 + I_3 = I$, the total current flowing through the circuit

$$\text{so } I = I_1 + I_2 + I_3 = V\left(\frac{1}{R_1} + \frac{1}{R_2} + \frac{1}{R_3}\right)$$

$$\text{or } \frac{I}{V} = \frac{1}{R_1} + \frac{1}{R_2} + \frac{1}{R_3}$$

and since $\frac{I}{V} = \frac{1}{R}$ where R is the equivalent resistance of the whole circuit.

$$\therefore \frac{1}{R} = \frac{1}{R_1} + \frac{1}{R_2} + \frac{1}{R_3} \qquad ...(1.24)$$

Thus, when a number of resistors are connected in parallel, the reciprocal of the equivalent resistance is given by the arithmetic sum of the reciprocals of their individual resistances.

In general if n resistors of resistances R_1, R_2, R_3,.......R_n are connected in parallel, then equivalent resistance R of the circuit is given by the expression

$$\frac{1}{R} = \frac{1}{R_1} + \frac{1}{R_2} + \frac{1}{R_3} + \ldots\ldots\ldots\ldots + \frac{1}{R_n} \qquad ...(1.25)$$

$$\text{Also } G = G_1 + G_2 + G_3 +, \ldots\ldots G_n \qquad ...(1.26)$$

where $G = \frac{1}{R}$, $G_1 = \frac{1}{R_1}$, $G_2 = \frac{1}{R_2}$, $G_3 = \frac{1}{R_3}$ and so on.

From the above discussions for a parallel circuit we conclude that

1. same voltage acts across all branches of the circuit,
2. different resistors (or branches) have their individual currents,
3. total circuit current is equal to the sum of individual currents through the various resistors (or branches),
4. branch currents are additive,
5. conductances are additive,
6. powers are additive,
7. the reciprocal of the equivalent or combined resistance is equal to the sum of the reciprocals of the resistances of the individual branches.

Parallel circuits are very common in use. Various lamps and appliances in a house are connected in parallel, so that each one can be operated independently. A series circuit is an "all or none" circuit, in which either every thing operates or nothing operates. For individual control, devices are wired in parallel.

1.19. CURRENT DISTRIBUTION IN PARALLEL CIRCUITS

Let two resistors of resistances R_1 and R_2 be connected in parallel across a pd of V volts, According to Ohm's law

Current flowing through resistor R_1, $I_1 = \frac{V}{R_1}$ ...(1.27)

Current flowing through resistor R_2, $I_2 = \frac{V}{R_2}$ ...(1.28)

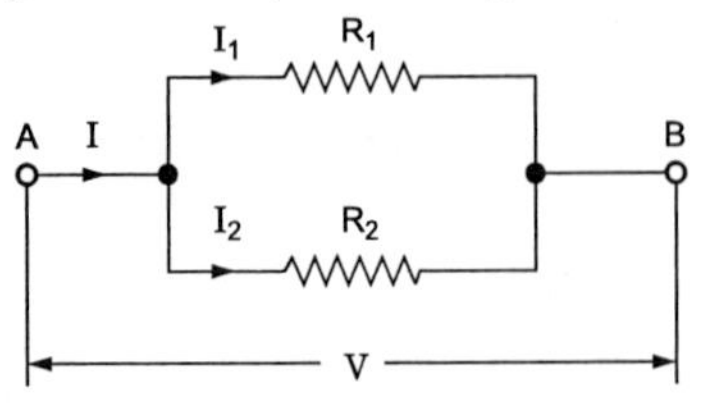

Fig. 1.9

Dividing Eq. (1.27) by Eq. (1.28) we have

$$\frac{I_1}{I_2} = \frac{V/R_1}{V/R_2} = \frac{R_2}{R_1} \qquad ...(1.29)$$

Hence, current flowing through each resistor, when connected in parallel, is inversely proportional to their respective resistances.

Since conductance is reciprocal of resistance and if G_1 and G_2 are the respective conductances of resistors R_1 and R_2 then

$$\frac{1}{G_1} = R_1 \text{ and } \frac{1}{G_2} = R_2$$

Substituting the above values of R_1 and R_2 in Eq. (1.29) we have

$$\frac{I_1}{I_2} = \frac{1/G_2}{1/G_1} = \frac{G_1}{G_2} \quad ...(1.30)$$

Adding 1 on both sides of the above expression we have

$$\frac{I_1}{I_2} + 1 = \frac{G_1}{G_2} + 1$$

$$\text{or} \quad \frac{I_1 + I_2}{I_2} = \frac{G_1 + G_2}{G_2}$$

$$\text{or} \quad I_2 = \frac{I_1 + I_2}{G_1 + G_2} \times G_2$$

$$\text{Since} \quad I_1 + I_2 = I \text{ and } G_1 + G_2 = \frac{1}{R_1} + \frac{1}{R_2} = \frac{1}{R} = G$$

$$\therefore \quad I_2 = I \times \frac{G_2}{G} \quad ...(1.31)$$

$$\text{and} \quad I_1 = I \times \frac{G_1}{G} \quad ...(1.32)$$

Hence, current in any branch of a parallel circuit is directly proportional to its respective conductance and is equal to the total current flowing through the circuit multiplied by the ratio of the conductance of the branch to that of the circuit.

The same relation holds good for parallel circuit consisting of more than two resistors and is very useful for its solution.

Let us consider a circuit consisting of resistances R_1, R_2, R_3 and R_4 ohms respectively connected in parallel across a potential difference of V volts, as shown in Fig. 1.10.

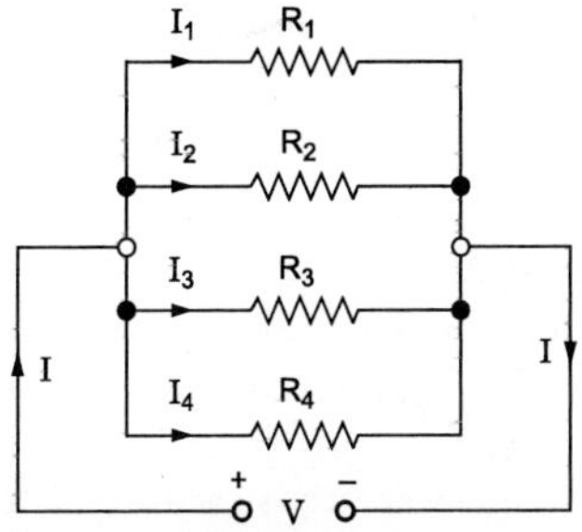

Fig. 1.10

$$\text{Conductance of first resistor, } G_1 = \frac{1}{R_1}$$

$$\text{Conductance of second resistor, } G_2 = \frac{1}{R_2}$$

$$\text{Conductance of third resistor, } G_3 = \frac{1}{R_3}$$

$$\text{Conductance of fourth resistor, } G_4 = \frac{1}{R_4}$$

Total conductance of the whole circuit is given by

$$G = \frac{1}{R} = \frac{1}{R_1} + \frac{1}{R_2} + \frac{1}{R_3} + \frac{1}{R_4}$$

$$= G_1 + G_2 + G_3 + G_4$$

$$\text{Total current flowing through the circuit, } I = \frac{V}{R} = V \times G$$

$$\text{or } V = \frac{I}{G} \qquad ...(1.33)$$

$$\text{Current flowing through resistance } R_1, I_1 = \frac{V}{R_1} = \frac{I}{G} \times G_1 = \frac{G_1}{G} I$$

$$\text{Similarly current flowing through resistance } R_2, I_2 = V \times G_2 = I \times \frac{G_2}{G}$$

$$\text{Current flowing through resistance } R_3, I_3 = V \times G_3 = I \times \frac{G_3}{G}$$

$$\text{and current flowing through resistance } R_4, I_4 = V \times G_4 = I \times \frac{G_4}{G}.$$

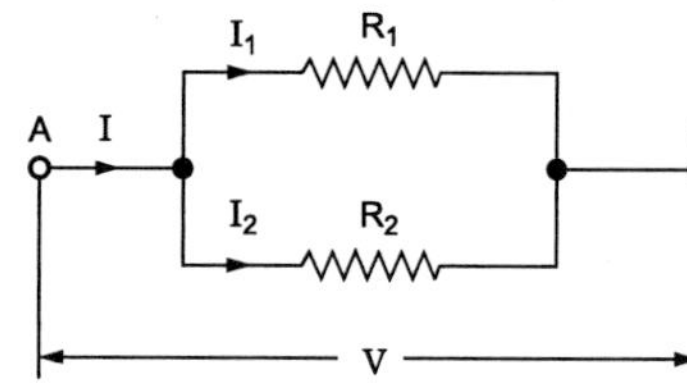

Fig. 1.11

Example 1.5 Two resistances, one of 30 Ω and another of unknown value are connected in parallel, the total power dissipated in the circuit is 450 watts when the applied voltage is 90 volts. Find the value of unknown resistance.

Solution: Power dissipated in the circuit, P = 450 watts
Voltage applied across the circuit, V = 90 volts

$$\text{Power dissipated in 30 Ω resistance, } P_1 = \frac{V^2}{R_1} = \frac{(90)^2}{30} = 270 \text{ watts}$$

$$\text{Power dissipated in the unknown resistance, } P_2 = P - P_1 = 450 - 270 = 180 \text{ watts}$$

$$\text{The value of unknown resistance, } R_2 = \frac{V^2}{P_2} = \frac{(90)^2}{180} = 45\ \Omega \textbf{ Ans.}$$

1.20. SERIES-PARALLEL CIRCUITS

So far, only simple series and simple parallel circuits have been considered. Practical electric circuits very often consist of combinations of series and parallel resistances. Such circuits may be solved by the proper application of Ohm's law and the rules for series and parallel circuits to the various parts of the complex circuit. There is no definite procedure to be followed in solving complex circuits, the solution depends on the known facts concerning the circuit and the quantities which one desires to find. One simple rule may usually be followed, however — reduce the parallel branches to an equivalent series branch and then solve the circuit as a simple series circuit.

For example consider a series-parallel circuit shown in Fig. 1.12 for solution.

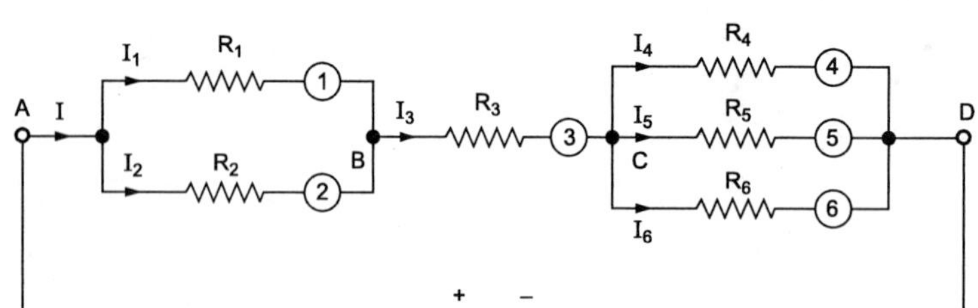

Fig. 1.12

First of all equivalent resistances of all parallel branches are determined separately *e.g.* of branches AB and CD by the law of parallel circuits discussed in Art 1.18.

Equivalent resistance of parallel branch AB,

$$R_{AB} = \frac{1}{\frac{1}{R_1} + \frac{1}{R_2}} = \frac{R_1 R_2}{R_1 + R_2}$$

and equivalent resistance of parallel branch CD,

$$R_{CD} = \frac{1}{\frac{1}{R_4} + \frac{1}{R_5} + \frac{1}{R_6}} = \frac{R_4 R_5 R_6}{R_5 R_6 + R_4 R_6 + R_4 R_5}$$

Now the circuit shown in Fig. 1.12 gets reduced to a simple series circuit shown in Fig. 1.13 consisting of three resistances,

$$R_{AB} = \frac{R_1 R_2}{R_1 + R_2}, \quad R_{BC} = R_3$$

and $$R_{CD} = \frac{R_4 R_5 R_6}{R_5 R_6 + R_4 R_6 + R_4 R_5}$$

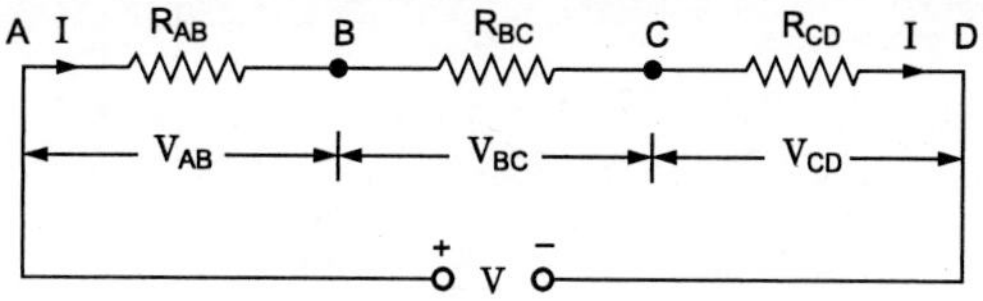

Fig. 1.13

Total resistance of circuit, $R_T = R_{AB} + R_{BC} + R_{CD}$

Now circuit current may be determined from the relation

$$I = \frac{V}{R_T}$$

After knowing I, potential differences across branches AB, BC and CD are determined from the relations

PD across branch AB, $V_{AB} = I.R_{AB} = \frac{V}{R_T} R_{AB}$

PD across branch BC, $V_{BC} = I.R_{BC} = \frac{V}{R_T} R_{BC}$

and PD across branch CD, $V_{CD} = IR_{CD} = \frac{V}{R_T} R_{CD}$.

After determination of potential difference across each parallel branch, the currents in the various resistances are determined from the relations

Current in resistance $R_1 = I_1 = \frac{V_{AB}}{R_1}$

Current in resistance $R_2 = I_2 = \frac{V_{AB}}{R_2}$

Current in resistance $R_3 = I_3 = I$

Current in resistance $R_4 = I_4 = \frac{V_{CD}}{R_4}$

Current in resistance $R_5 = I_5 = \frac{V_{CD}}{R_5}$

Current in resistance $R_6 = I_6 = \frac{V_{CD}}{R_6}$

Thus, equivalent resistance of the whole circuit, voltage drop across each branch and currents in the various resistors may be determined.

Example 1.6 **Two resistances of 20 Ω and 30 Ω respectively are connected in parallel. These two parallel resistances are further connected in series with a resistance of 15 Ω. If the current through the 15 Ω resistance is 3 A find *(a)* the currents through the 20 Ω and 30 Ω resistances respectively *(b)* the voltage across the whole circuit *(c)* the total power consumed.**

Solution: Equivalent resistance of branch AB, $R_{AB} = \frac{1}{\frac{1}{20} + \frac{1}{30}} = 12\ \Omega$

Effective resistance of the circuit, $R_{eff} = R_{AB} + R_{BC} = 12 + 15 = 27\ \Omega$

Circuit current, I = Current through 15 Ω resistance = 3 A

(b) Voltage across the whole circuit, $V = I R_{eff} = 3 \times 27 = 81$ V **Ans.**

(*c*) Total power consumed, $P = VI = 81 \times 3 = 243$ watts **Ans.**

(*a*) Voltage drop across branch AB, $V_{AB} = I R_{AB} = 3 \times 12 = 36$ V

Current through 20 Ω resistance, $I_1 = \dfrac{V_{AB}}{R_1} = \dfrac{36}{20} = 1.8$ A **Ans.**

Current through 30 Ω resistance, $I_2 = \dfrac{V_{AB}}{R_2} = \dfrac{36}{30} = 1.2$ A **Ans.**

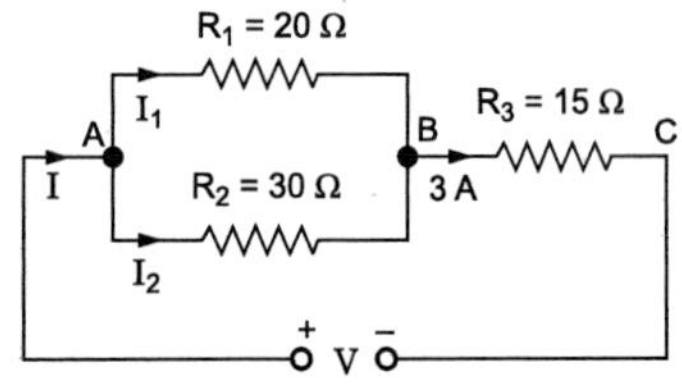

Fig. 1.14

Example 1.7 **For the circuit shown in Fig. 1.15, using the method of series-parallel combination, find V_1 and I_2.**

Solution: Total resistance of the circuit,

R_{eq} = 25 + 2.5 + equivalent resistance of parallel combination of resistors of 10, 50 and 20 Ω

$$= 25 + 2.5 + \frac{1}{\frac{1}{10}+\frac{1}{50}+\frac{1}{20}}$$

$$= 25 + 2.5 + \frac{100}{10+2+5} = \frac{1,135}{34}\ \Omega$$

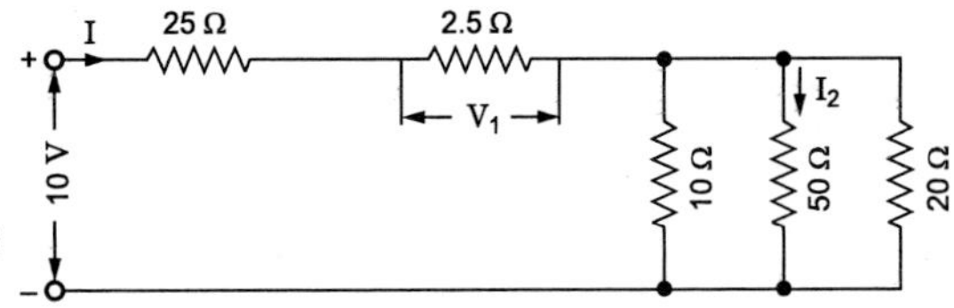

Fig. 1.15

Applied voltage, V = 10 V

Current drawn from the supply, $I = \dfrac{V}{R_{eq}} = \dfrac{10}{1,135/34} = \dfrac{10 \times 34}{1,135} = \dfrac{68}{227}$ A

Voltage drop across resistor of 2.5 Ω, $V_1 = I \times 2.5 = \dfrac{68}{227} \times 2.5 = \dfrac{170}{227}$ V **Ans.**

Voltage drop across parallel combination, $V_2 = \dfrac{68}{227} \times \dfrac{100}{17} = \dfrac{400}{227}$ V

Current through 50 Ω resistor, $I_2 = \dfrac{V_2}{50} = \dfrac{400}{227 \times 50} = \dfrac{8}{227}$ A **Ans.**

Example 1.8 **In the given circuit shown in Fig. 1.16 find the resistance between the points B and C.**

[*G.B. Technical Univ. Electrical Engineering Even Semester 2009-10*]

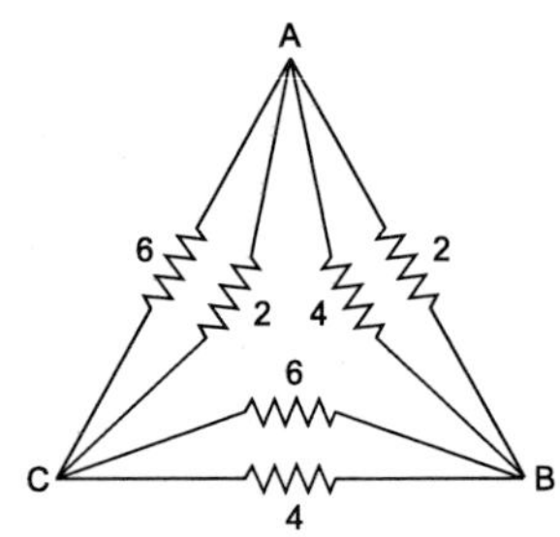

Fig. 1.16

Solution: Resistance in branch AB, $R_{AB} = 4 \| 2 = \dfrac{4}{1+2} = 1.333\ \Omega$

Resistance in branch, BC, $R_{BC} = 6 \| 4 = \dfrac{6 \times 4}{4+6} = 2.4\ \Omega$

Resistance is branch CA, $R_{CA} = 6 \| 2 = \dfrac{6 \times 2}{6+2} = 1.5\ \Omega$

Resistance between points B and C $= R_{BC} \| (R_{AB} + R_{CA}) = 2.4 \| (1.333 + 1.5) = 2.4 \| 2.833$

$$= \frac{2.4 \times 2.833}{2.4 + 2.833} = 1.3\ \Omega \quad \textbf{Ans.}$$

Example 1.9 **Find the equivalent resistance for the following circuit and hence calculate the current supplied by the source.**

[*G.B. Technical Univ. Electrical Engineering Second Semester 2011-12*]

Solution: The given circuit can be solved simply by simplification as shown in Fig. 1.18(*a*), (*b*), (*c*), (*d*), (*e*), (*f*) and (*g*).

So equivalent resistance of the circuit,

$$R = 4 + 1 = 5\ \Omega$$

Current supplied by the source, $I = \dfrac{V}{R} = \dfrac{120}{5} = 24$ A **Ans.**

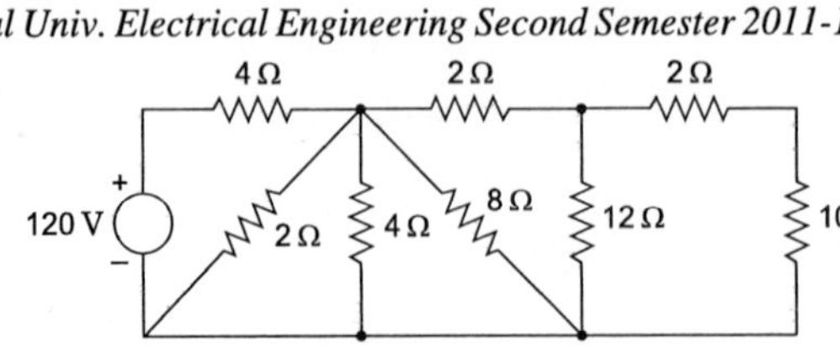

Fig. 1.17

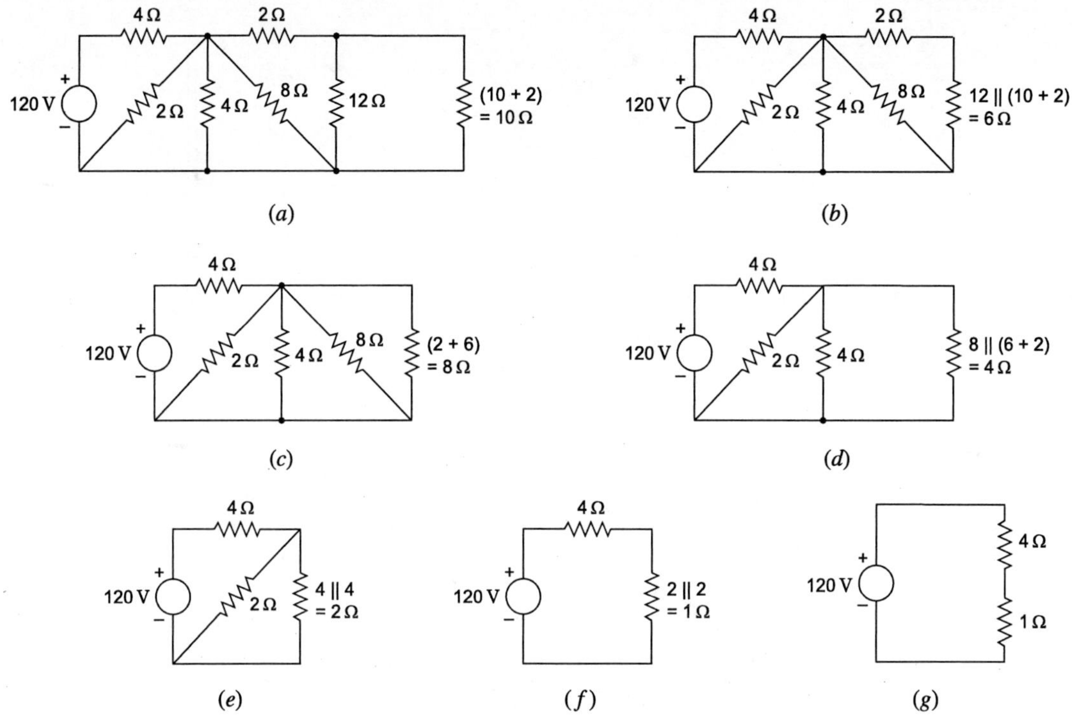

Fig. 1.18

1.21. MAGNETIC FIELD DUE TO A CURRENT CARRYING CONDUCTOR

In 1819 it was discovered by a Danish Physicist, Hans Christian Oersted that an electric current is always accomplished by certain magnetic effects.

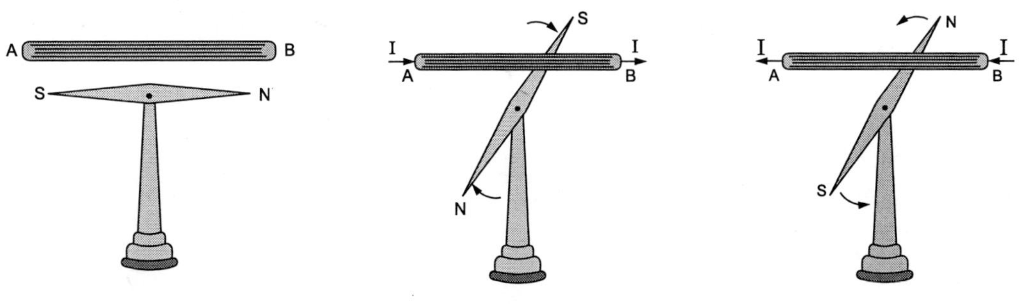

(*a*) *No Current Flow* (*b*) *Current Flowing From A to B* (*c*) *Current Flowing From B to A*

Fig. 1.19

Oersted found that when current is passed through a conductor placed above the magnetic needle, the needle turns in a certain direction, as shown in Fig. 1.19. He also found that when the direction of flow of current is reversed the magnetic needle also deflects in opposite direction.

Further investigation showed that the field around the current carrying conductor consists of lines of force, which encircles the conductor. It can be proved experimentally by passing a current carrying conductor AB in the card board and plotting the field with the help of magnetic needle on it, as shown in Figs. 1.20 (*a*) and 1.20 (*b*).

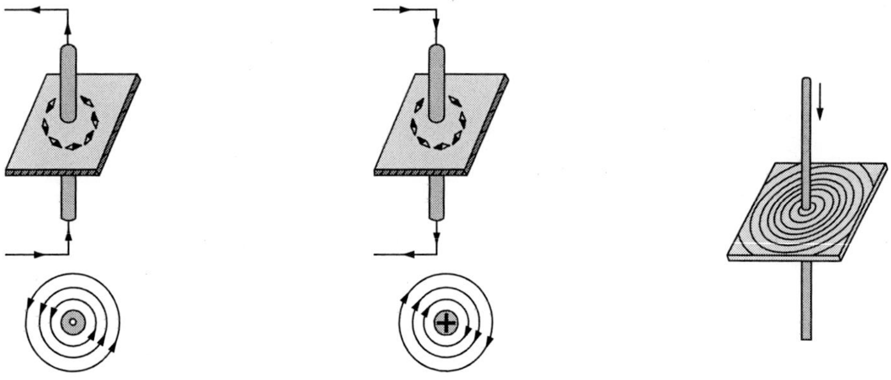

(*a*) (*b*) (*c*) *Magnetic Field Around A Current Carrying Conductor*

Fig. 1.20

From Figs. 1.20 (*a*) and 1.20 (*b*) it is observed that when the current is passed through conductor in upward direction, the direction of lines of force is counter-clockwise direction (observed from the top of the conductor) and when the current is passed through the conductor in downward direction, the direction of lines of force is clockwise (observed from the top of the conductor).

The properties of the lines of magnetic induction around a current carrying conductor are summarized as below:

(*i*) Lines of magnetic induction are circles, symmetrical about, and concentric with, the axis of the conductor.

(*ii*) The spacing between the lines of induction decreases as we move closer to the conductor [Fig. 1.20 (*c*)]

(*iii*) The direction of lines of magnetic induction depends on the direction of flow of current through the conductor.

(*iv*) Magnetic induction or flux density depends upon the strength (or magnitude) of the current flowing through the conductor.

1.22. DETERMINATION OF DIRECTION OF MAGNETIC FIELD AROUND A CURRENT CARRYING CONDUCTOR

The direction of lines of force (magnetic field) around a straight current carrying conductor may be determined by any of the following rules:

1. Corkscrew Rule. If the right handed corkscrew is held with its axis parallel to the conductor pointing the direction of flow of current and the head of the screw is rotated in such a direction that the screw moves in the direction of flow of current then the direction in which the head of screw is rotated, will be the direction of magnetic lines of force.

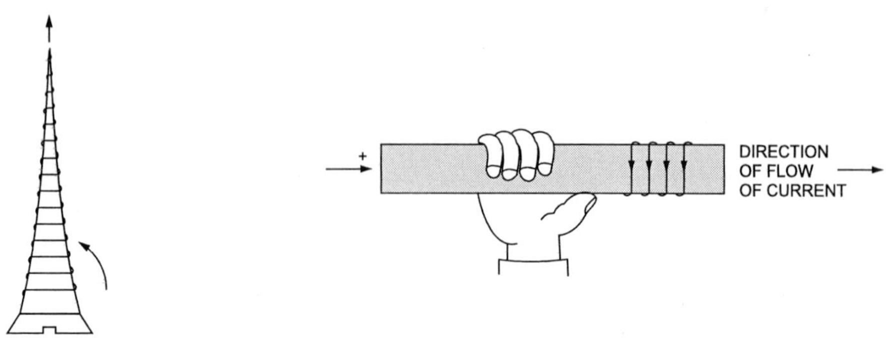

(*a*) *Cork Screw Rule* (*b*) *Right Hand Rule*

Fig. 1.21

2. Right Hand Rule. If the current carrying conductor is held in right hand by the observer so that it is encircled by fingers stretching the thumb at right angle to the fingers in the direction of flow of current then finger tips will point the direction of magnetic lines of force, as shown in Fig. 1.21 (*b*).

1.23. MAGNETIC FIELD DUE TO A CIRCULAR LOOP

If a single turn wire carrying current is bent in the form of a loop (or ring), as shown in Fig. 1.22, the lines of magnetic induction around it will be concentric circles, leaving the plane of the loop (or ring) on one side and entering on the other. The loop acts as the true magnet having north and south poles.

The direction of magnetic field may be determined by applying either of the two rules namely (*i*) right hand rule or (*ii*) corkscrew rule, already explained in Art 1.22.

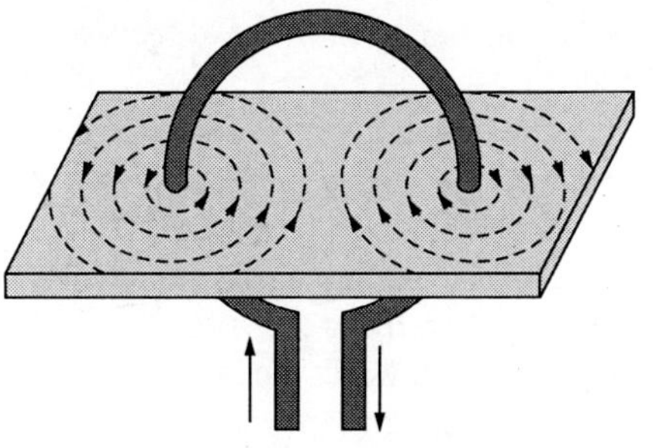

Fig. 1.22 *Magnetic Field Due To a Circular Loop*

1.24. SOLENOID

The current carrying wire wound spirally in the form of helix about an axis, as shown in Fig. 1.23, is known as *solenoid* or *coil.* Magnetic field produced due to current carrying solenoid is fairly uniform over a small region in the middle of the coil. It acts just like a bar magnet having north and south poles.

There are several methods used to determine the polarity of the solenoid.

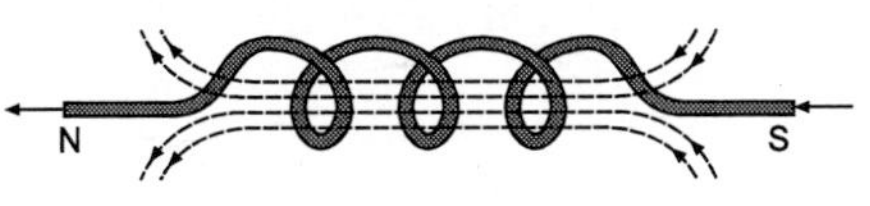

Fig. 1.23 *Solenoid*

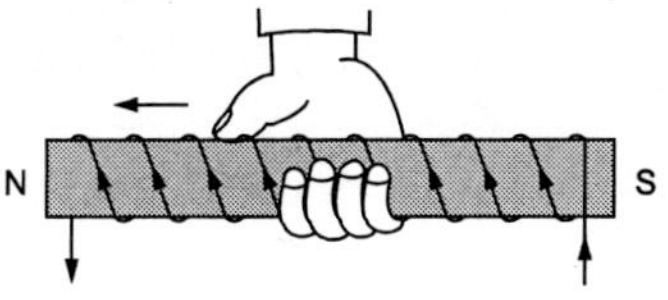

Fig. 1.24 *Helix Rule*

1. By Use of Compass Needle. If one of the poles (say north pole) of the compass needle be brought into close proximity to one of the poles of the current carrying solenoid of unknown polarity the action of the compass needle will immediately classify the pole as north or south depending upon whether the needle is repelled or attracted.

2. Helix Rule. If the helix is held in right hand in such a manner that the finger tips point in the direction of flow of current and thumb is outstretched longitudinally along the coil, it will point towards north pole (Fig. 1.24).

1.25. FORCE ON A CURRENT CARRYING CONDUCTOR LYING IN THE MAGNETIC FIELD

In Fig. 1.25 (*a*) a uniform magnetic field between the two opposite poles is shown.

In Fig. 1.25 (*b*) the cross section of a conductor carrying current in inward direction placed between the two magnets, the field being temporarily removed, is shown. By applying the right hand thumb rule, the direction of the field around the conductor is found to be clockwise.

If the current carrying conductor shown in Fig. 1.25 (*b*) is placed in the magnetic field shown in Fig. 1.25 (*a*), the resultant magnetic field would be similar to that shown in Fig. 1.25 (*c*).

The lines of force above the conductor are strengthened, since they are in same direction but the lines of force below the conductor are weakened because the two fields below the conductor are opposite in directions and hence tends to destroy each other. Magnetic lines like rubber bands have a tendency to strengthen out and, therefore, a force is experienced on the conductor in the downward direction, as shown in Fig. 1.25 (*c*).

If the direction of current is reversed in the conductor, as shown in Fig. 1.25 (*d*), the direction of force experienced is reversed. In this case the lines of force above the conductor are weakened while those below the conductor are strengthened.

Hence, it is observed that when a current carrying conductor is placed at right angle to the direction of magnetic field, a mechanical force is experienced on the conductor in a direction perpendicular to both the direction of magnetic field and flow of current.

The direction of this force can be determined by applying *Fleming's Left Hand Rule* which states that if the thumb, forefinger and middle finger of the left hand are stretched in such a way that they are at right angles to each other mutually and forefinger points towards the direction of the magnetic field, middle finger towards the direction of the flow of current then thumb will point the direction of force acting on the conductor [Fig. 1.26 (*a*)].

If the current in the conductor is reversed, keeping the direction of magnetic field unchanged the direction of force will reverse.

Similarly, if the direction of the magnetic field is reversed, keeping the direction of flow of current in the conductor unchanged, the direction of force will reverse.

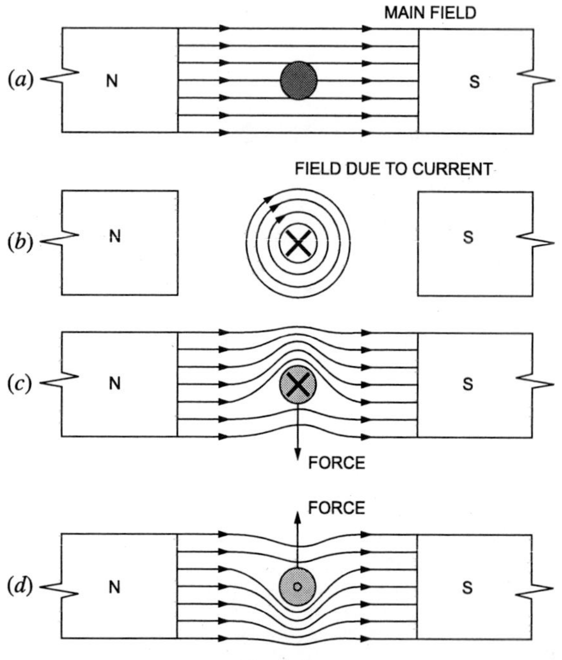

Fig. 1.25

It should be noted that no force is exerted on a conductor when it lies parallel to the magnetic field.

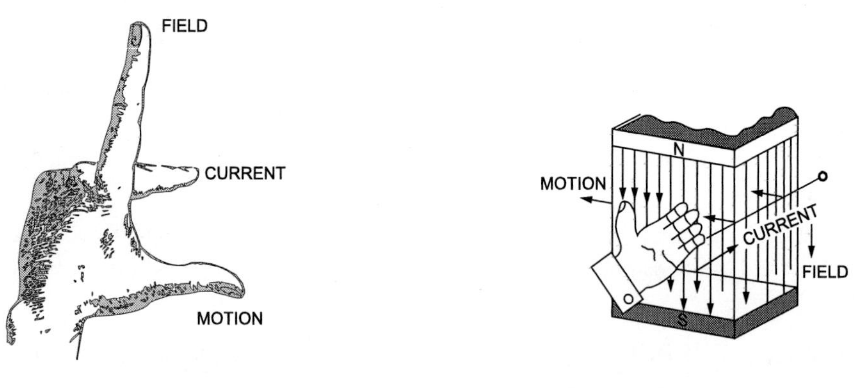

(*a*) *Fleming's Left Hand Rule* (*b*)

Fig. 1.26

Fig. 1.26 (*b*) shows another method of finding the direction of force acting on a current carrying conductor placed in a magnetic field. It is known as *Flat left hand rule*. If the palm of the left hand is so placed that the lines of induction leaving the N-pole enter the palm and the four fingers are in the direction of flow of current in the conductor, then the thumb will indicate the direction of force acting on the conductor.

The force experienced on the conductor is directly proportional to

(1) Flux density (field strength), B

(2) The current flowing through the conductor, I and

(3) The length of the conductor, l

$\therefore$ The magnitude of the force is given by, $F = B\,I\,l$ (1.34)

where F is the force in newtons, B is in teslas (Wb/m^2), I is in amperes and l is in metres.

In general, if the conductor lies at an angle θ with a magnetic field of flux density B webers/metre^2, then mechanical force experienced on a current carrying conductor is given by

$$F = B\,I\,l \sin\theta \text{ newtons} \tag{1.35}$$

where l is the length of the conductor in metres and I is the current carried by the conductor in amperes.

Example 1.10 **A straight long conductor of 1 metre length carrying a current of 60 amperes is placed at right angle to a uniform magnetic field of strength 2.5 T. Determine the mechanical force acting on the conductor.**

Solution: Flux density, B = 2.5 T

Current flowing through the conductor, I = 60 A

Length of conductor, l = 1 metre

Force acting on the conductor, $F = B\,I\,l = 2.5 \times 60 \times 1 = 150$ N **Ans.**

HIGHLIGHTS

1. All matters are made of electrons, protons and neutrons.
2. The controlled movement of electrons (or drift) through a substance is called the electric current. Current is expressed in amperes.
3. Electromotive force (emf) or potential difference is that which causes the current to flow in a closed circuit. The unit of potential difference in SI system is volt.
4. Resistance may be defined as that property of substance which opposes (or restricts) the flow of an electric current (or electrons) through it. The unit of resistance in SI system is ohm (Ω).
5. Ohm's law may be given as

   $$\frac{V}{I} = \text{Constant} = R \qquad \text{for no change in physical state}$$

6. Laws of resistance are summed up as

   $$R = \rho\,\frac{l}{a}$$

   where ρ is the specific resistance or resistivity of the material of wire, which may be defined as the resistance between opposite faces of a unit cube of that material.

   Resistivity is measured in ohm-metres (Ω-m) or ohms per metre cube.
7. Conductance G is reciprocal of resistance i.e. $G = \frac{1}{R}$. The unit of conductance is siemens (S).
8. Conductivity σ is reciprocal of resistivity ρ i.e. $\sigma = \frac{1}{\rho}$. The unit of conductivity is siemens per metre.
9. A current (or voltage) is called alternating if it periodically changes its direction and magnitude whereas direct currents are steady and flow in one direction only.
10. In case of good conductors, especially metals, the variations in resistance with temperature (within normal range) are given by

    $$R_t = R_0\,(1 + \alpha_0 t)$$

    Similarly, resistivity of given material varies as

    $$\rho_t = \rho_0\,(1 + \alpha_0 t)$$

11. Temperature coefficient of resistance is not constant but depends upon the initial temperature on which the increment in resistance is based. The values of temperature coefficient for different temperatures can be had from the relations

    $$\alpha_1 = \frac{1}{\frac{1}{\alpha_0} + t_1} \quad \text{and} \quad \alpha_2 = \frac{1}{\frac{1}{\alpha_1} + (t_2 - t_1)}$$

12. *Power* is defined as the rating of doing work or the amount of work done in unit time. It is expressed in watts or kilowatts.

$$1 \text{ kW} = 1{,}000 \text{ watts}$$

Power dissipated in a resistor of R ohms is given as

$$P = VI = I^2R = \frac{V^2}{R} \text{ watts}$$

13. *Energy* is defined as the capacity of doing work. Its units are joules, newton-metres, watt-seconds or kilowatt-hours (kWh)

$$1 \text{ kWh} = 3{,}600{,}000 \text{ watt-seconds or joules}$$

Energy expended or heat generated in a conductor of resistance R ohms when a current of I amperes flows through it for t seconds is given as

$$H = I^2Rt = VIt = \frac{V^2}{R}t \text{ watt-seconds or joules}$$

14. When a number of resistors are connected in series, the equivalent resistance is given by the arithmetic sum of their individual resistances

$$i.e. \; R = R_1 + R_2 + R_2 + \dots \dots + R_n$$

15. When a number of resistors are connected in parallel, the reciprocal of the equivalent resistance is given by the arithmetic sum of the reciprocals of their individual resistances

$$i.e. \; \frac{1}{R} = \frac{1}{R_1} + \frac{1}{R_2} + \frac{1}{R_3} + \dots \dots + \frac{1}{R_n}$$

16. When two resistances R_1 and R_2 are in parallel and carry a total current I, then

$$\text{current through resistance } R_1, I_1 = \frac{R_2}{R_1 + R_2} I$$

$$\text{and current through resistance } R_2, I_2 = \frac{R_1}{R_1 + R_2} I$$

17. When n resistors, each of R ohms, are connected in parallel, then total resistance is given by $R_T = \frac{R}{n}$ ohms.

18. Current in any branch of a parallel circuit is directly proportional to its conductance and is equal to the total current flowing through the circuit multiplied by the ratio of the conductance of the branch to that of the circuit.

19. The current carrying wire wound spirally in the form of helix is known as solenoid or coil.

20. When a current carrying conductor is placed at right angle to the direction of magnetic field, a mechanical force is experienced on the conductor in a direction perpendicular to both the direction of magnetic field and flow of current. The direction of this force can be determined by Fleming's left hand rule.

21. According to Fleming's left hand rule if the thumb, forefinger and middle finger of the left hand are stretched in such a way that they are at right angles to each other mutually and forefinger points towards the direction of the magnetic field, middle finger towards the direction of the flow of current then thumb will point the direction of force acting on the conductor (Fig. 1).

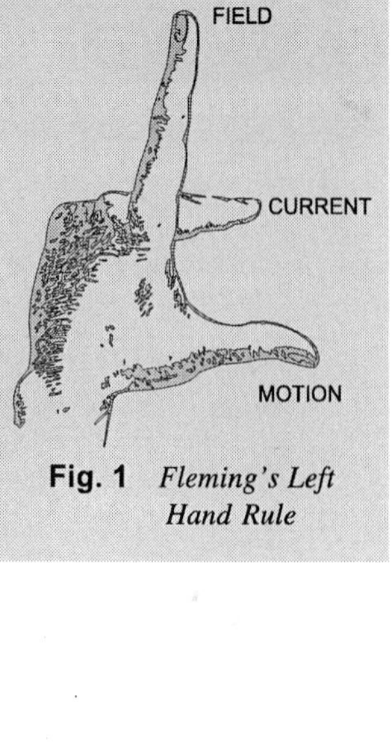

Fig. 1 *Fleming's Left Hand Rule*

EXERCISES

1. What is ohm's law ? State its limitations.

[Pb. Technical Univ. Basic Electrical and Electronics Engineering December 2006]

2. State laws of resistance and hence define specific resistance.
3. On what factors does the resistance offered by a conductor depend ?
4. What is the effect of temperature on the resistance of copper, aluminium, porcelain and carbon ? Define temperature coefficient of resistance. How does its value change with temperature ?
5. How is energy related to power ? State the SI units of work, power and energy. What are the practical units of electrical energy and mechanical power ?
6. Explain the terms watt, joule, watt-hour and kWh.
7. State and explain Joule's law of electric heating.
8. Compute equivalent resistance of three resistances R_1, R_2 and R_3 connected in (*i*) series (*ii*) parallel.
9. Derive an expression for the electrodynamic force on a conductor carrying current and lying in the magnetic field.
10. State and explain Fleming's left hand rule.

SHORT ANSWER TYPE QUESTIONS WITH ANSWERS

Q. 1. Do conventional current and electron current flow in the same direction in an electric circuit ?

Ans. No, the direction of flow of electrons is opposite to that of conventional current.

Q. 2. What is drift current ?

Ans. The steady flow of electrons in one direction caused by the applied electric field constitutes electric current, called the drift current.

Q. 3. What is difference between emf and potential difference ?

Ans. EMF (electromotive force) is the force that causes a current to flow in an electric circuit while the potential difference between any two points in an electric circuit is that difference in their electrical state which tends to cause flow of electricity from one point to the other.

Q. 4. What is relation between voltage, current and resistance of an electric circuit ?

Ans. $\frac{V}{I} = R$

Q. 5. Enlist the limitations of Ohm's law ? **[U.P. Technical Univ. Even Semester 2013-14]**

Ans. Ohm's law cannot be applied to the circuits consisting of (*i*) electronic tubes or transistors (because such elements are not bilateral) and (*ii*) nonlinear elements such as powdered iron, electric arc etc.

Q. 6. On what factors does the resistance of a conductor depend ? **[G.B. Technical Univ. Odd Semester 2012-13]**

Ans. The resistance of a conductor depends upon its length, area of x-section and type of material, purity and hardness of material of which it is made of and the operating temperature.

Q. 7. How does the resistance of a homogeneous material having constant cross section vary with its length ?

Ans. Resistance of a homogeneous material having constant cross section varies linearly with its length i.e. $R \propto l$.

Q. 8. How does the resistance of a homogeneous material having constant length vary with the changing cross-sectional area ?

Ans. Resistance of a homogeneous material having constant length is inversely proportional to its area of x-section i.e. $R \alpha \frac{1}{a}$.

Q. 9. If the length of a wire of resistance R is uniformly stretched to n times its original value, what will be its new resistance ?

Ans. New resistance

$$= \text{Old resistance} \times \frac{\text{new length}}{\text{old length}} \times \frac{\text{old cross section}}{\text{new cross section}} = R \times \frac{nl}{l} \times \frac{a}{a/n} = n^2 R$$

Q. 10. What is meant by conductivity and how is it related with resistivity ?

Ans. Conductivity is defined as the conductance between the two opposite faces of a unit cube. It is reciprocal of resistivity.

Q. 11. How does increase in temperature affect the resistance of (*i*) conductors and (*ii*) insulators ?

Ans. The resistance of the conductors (except of carbon) increases with the increase in temperature, therefore, on high temperature good conductors become bad conductors even.

The insulators lose their insulating power with the increase in temperature and some may even become good conductors if they are strongly heated.

Q. 12. Write the practical units of electrical energy and mechanical power. **[U.P. Technical Univ. Even Semester 2013-14]**

Ans. The practical units of electrical energy and mechanical power are kWh and horse power respectively.

Q. 13. Give relation between kWh, MJ and kilocalorie.

Ans. 1 kWh = 3.6 MJ = 860 kcals.

Q. 14. Express Joule's law of electric heating.

Ans. According to Joule's law of electric heating, the energy expended or heat generated in joules is given as

$$\text{Work done or heat generated} = I^2Rt = \frac{V^2}{R}t = VIt = Wt \text{ joules.}$$

Q. 15. How will you find the direction of magnetic field around a current carrying conductor ?

Ans. By Corkscrew rule or right hand rule.

Q. 16. How will you determine the polarity of a solenoid ?

Ans. By using compass needle or by using helix rule.

Q. 17. On what factors does the force experienced by a current carrying conductor placed in a magnetic field depend ?

Ans. The force experienced on a current carrying conductor placed in a magnetic field is directly proportional to (*i*) flux density (field strength), B (*ii*) the magnitude of current flowing through the conductor (*iii*) the length of the conductor and (*iv*) sine of the angle at which the conductor lies with the magnetic field *i.e.* $F = BIl \sin \theta$ newtons.

Q. 18. What is Fleming's left hand rule ?

Ans. Fleming's left hand rule is used to determine the direction of mechanical force experienced by a current carrying conductor placed in a magnetic field.

Fleming's left hand rule states that if the thumb, forefinger and middle finger of the left hand are stretched in such a way that they are at right angles to each other mutually and forefinger points towards the direction of the magnetic field, middle finger towards the direction of flow of current then thumb will point in the direction of force acting on the conductor [Fig. 1.26(*a*)].

PROBLEMS

1. A piece of silver wire has a resistance of 1 Ω. What will be the resistance of the manganin wire of one-third length and one-third the diameter of silver wire, if the specific resistance of manganin is 30 times of silver ?
   [Ans. 90 Ω**]**
2. The resistivity of a ferric-chromium-aluminium alloy is 51×10^{-8} Ω-m. A sheet of material is 15 cm long, 6 cm wide and 0.014 cm thick. Determine the resistance between (*a*) opposite ends and (*b*) opposite sides.
   [Allahabad Univ. Elec. Circuits 1983]
   [Ans. (*a*) 0.0091 Ω (*b*) 79.3×10^{-10} Ω**]**
3. The equivalent resistance of four resistors joined in parallel is 30 Ω. The currents flowing through them are 4, 2, $\frac{4}{3}$ and $\frac{2}{3}$ amperes. Determine the value of each resistor. **[Ans.** 60 Ω ; 120 Ω ; 180 Ω and 360 Ω**]**
4. A current of 78 A is divided into three branches. The lengths of the wire in the three branches are in the ratio of 8, 12 and 16. The wires are of same material and cross-sectional area. Calculate the current in each branch.
   [Ans. 36 A, 24 A and 18 A**]**
5. How many 60 W lamps may be safely run on a 230 V circuit fitted with a 5 A fuse ? **[Ans.** 19**]**
6. Twenty lamps of 60 watts each are run for 5 hours from 240 V supply. Calculate (a) the units consumed. (b) the current through the mains, and (c) resistance of each lamp. **[Ans.** (*a*) 6 units; (*b*) 5 A; (*c*) 960 Ω**]**

UNIT-I

DC Circuit Analysis and Network Theorems

CONTENTS IN THIS UNIT

DC Circuit Analysis and Network Theorems:

Circuit Concepts: Concept of network, Active and passive elements, Voltage and current sources, Concept of linearity and linear network, Unilateral and bilateral elements, R-L and C as linear elements, Source transformation.

Kirchhoff's laws, Loop and nodal methods of analysis, Star-delta transformation.

Network Theorems: Superposition theorem, Thevenin's theorem, Norton's theorem, Maximum power transfer theorem.

2

CHAPTER

DC Circuit Analysis and Network Theorems

2.1. CONCEPT OF NETWORK

An electric circuit (or network) is an interconnection of physical electrical devices. The purpose of electric networks is to distribute and convert electrical energy into some other forms. Accordingly, the basic circuit components are an energy source (or sources), an energy converter (or converters), and conductors connecting them.

An *energy source* (or source), such as a primary or secondary cell, a generator, and the like, is a device that converts chemical, mechanical, thermal or some other form of energy into electrical energy.

An *energy convertor,* also called the *load*, (such as lamp, heating appliance, or an electric motor) converts electrical energy into light, heat, mechanical work and so on.

Events in an electrical circuit may be defined in terms of emf (or voltage) and current.

When electrical energy is generated, transmitted and converted under conditions such that the currents and voltages involved remain constant with time, the electric circuit is identified as *direct current* (dc) *circuit.* If the currents and voltages do change with time, the circuit is defined as *alternating current (*ac*) circuit.*

A graphic representation of an electric circuit is called a *circuit diagram* (Fig. 2.1). Such a diagram consists of interconnected symbols called *circuit elements or circuit parameters.* Two elements are necessary to represent processes in a dc circuit. These are source of emf E_S and of internal (or source) resistance R_S and the load resistance (which includes the resistance of the conductors) R.

In any electric circuit the energy convertor (or load) and the conductors connecting it to the source make up the *external circuit* in which current flows from the +ve side to the –ve side of the source whereas inside the source, current flows in the opposite direction, *i.e.* from the –ve side to the +ve side. The source emf is directed from the terminal at a lower potential to that at a higher one. In diagrams this is shown by arrows.

The *source emf* (or open-circuit voltage) is the voltage that appears across the source when no load is connected across it.

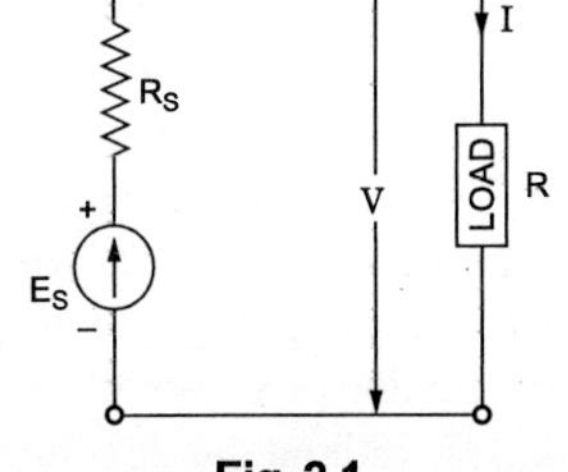

Fig. 2.1

When a load is connected to the source terminals and the circuit is closed, an electric current starts flowing through the circuit. Now voltage across source terminals (called the *terminal voltage*) is not equal to source emf. It is due to voltage drop inside the source, *i.e.* across the source resistance.

Voltage drop inside the source $= I\,R_S$

The relationship between the current through a resistance and the voltage across the same resistance is called its *volt-ampere (or voltage-current) characteristic.* When represented graphically, voltages are laid off as abscissae and currents as ordinates.

There are two types of volt-ampere characteristics–straight line and nonlinear (curve), as shown in Figs. 2.2 (a) and 2.2 (b) respectively.

Resistive elements for which the volt-ampere characteristic is a straight line [Fig. 2.2 (a)] are called *linear*, and the electric circuits containing only linear resistances are called *linear circuits*.

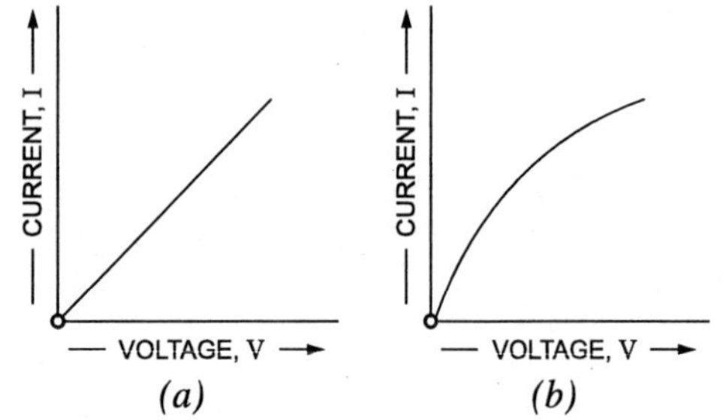

Fig. 2.2

Resistive elements for which the volt-ampere characteristic is other than a straight line are termed *non-linear,* and so the electric circuits containing them are called *nonlinear circuits.* Examples of nonlinear elements are tungsten lamps, vacuum tubes and transistors, etc.

An electric circuit, whose characteristics or properties are same in either direction (e.g. a distribution or transmission line), is called the *bilateral circuit.* The distribution or transmission line can be made to perform its function equally well in either direction.

An electric circuit, whose characteristics or properties change with the direction of its operation (*e.g.* a diode rectifier), is called the *unilateral circuit.* A diode rectifier cannot perform rectification in both directions.

A network is said to be *passive* if it contains no source of emf in it. The *equivalent resistance* between any two terminals of a passive network is the ratio of potential difference across the two terminals to the current flowing into (or out of) the network. When a network contains one or more sources of emf and/or current, it is said to be *active.*

A circuit consisting of a limited (finite) number of circuit elements is referred to as *'lumped parameter circuit'* while a circuit containing unlimited (infinite) number of circuit elements is called a *distributed parameter circuit.* Transmission line is an example of distributed parameter network.

In case, a branch is removed from an electric network, the remainder of the network is left with a pair of terminals. The part of the network, which is considered with respect to the removed branch or terminal pair or port is termed as *one-port network.* When two branches are removed so that the network is left with four terminals or two pairs of terminals, the remainder network is called the *two-port network.* Usually one port accepts a source and the other port is coupled to a load, so that there is an input port and an output port in any two-port system.

2.2. SINGLE- AND MULTI-MESH NETWORK

Electric network may provide a single closed path (known as a *mesh or loop)* or several closed paths for the flow of current. An elementary *single-mesh* network, in which all the elements carry the same current, is shown in Fig. 2.1. An elementary multi-mesh network is shown in Fig. 2.3. It has 6 nodes, seven branches, three loops and two meshes. A *junction* (*or node**) is a point in a network where two or more branches meet. A *branch* is any section of a network which joins two nodes directly, that is, without passing through a third node. A *loop* is a closed path in a network formed by a number of connected branches. Mesh is a loop that contains no other loop within it.

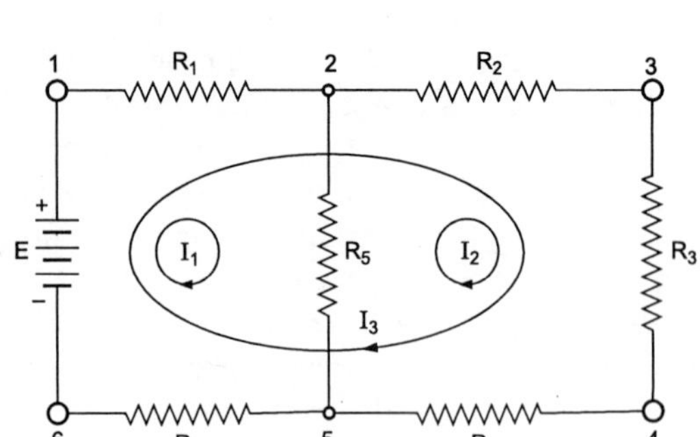

Fig. 2.3

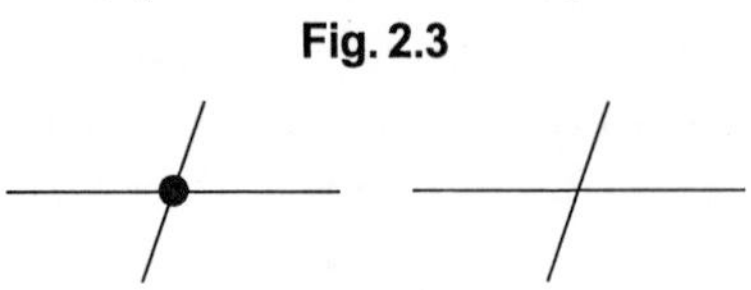

Fig. 2.4

If in the circuit diagram, as in Fig. 2.4 (*a*), there is a bold dot at the intersection of two branches, these

* Nodes are considered to be resistanceless and infinitesimal.

branches are electrically connected and have a common node. Otherwise they simply cross, as in Fig. 2.4. (*b*), and are not connected electrically.

2.3. ACTIVE AND PASSIVE ELEMENTS

Network elements may be classified into two categories viz. active elements and passive elements.

The elements which supply energy to the network are known as *active elements*. The voltage sources like batteries, dc generators, ac generators and current sources like photoelectric cells, metadyne generators fall under the category of active elements. Most of the semiconductor devices like transistors are treated as current sources.

The components which dissipate or store energy are known as *passive components*. Resistors, inductors and capacitors fall under the category of passive elements. The resistor is the only component which dissipates electrical energy. The inductor and capacitors are the components which store energy, the inductor stores energy by virtue of a current passing through it whereas the capacitor stores energy by virtue of potential difference across it.

2.4. RESISTANCE

Resistance is a dissipative element, which converts electrical energy into heat, when the current flows through it in any direction. The process of energy conversion is irreversible.

The circuit element used to represent energy dissipation is most commonly described by requiring the voltage across the element be directly proportional to the current through it. Mathematically, the voltage is

$$v = Ri \text{ volts} \qquad \text{...(2.1)}$$

where i is the current in amperes. The constant of proportionality R is the resistance of the element and is measured in ohms (abbreviated Ω). The voltage-current relation expressed by Eq. (2.1) is known as Ohm's law. A physical device whose principal electrical characteristic is resistance is known as a *resistor.*

Since an electric charge gives up energy when passing through a resistor, the voltage v in Eq. (2.1) is a voltage drop in the direction of current. Alternatively, v is a voltage rise in the direction opposite to the current. The conventional diagrammatic representation of a resistance, together with designations of the current direction and voltage polarity, is shown in Fig. 2.5. The plus and minus signs denote decrease of potential, and hence a voltage drop, from left to right (or plus to minus). The element has two terminals (also called nodes).

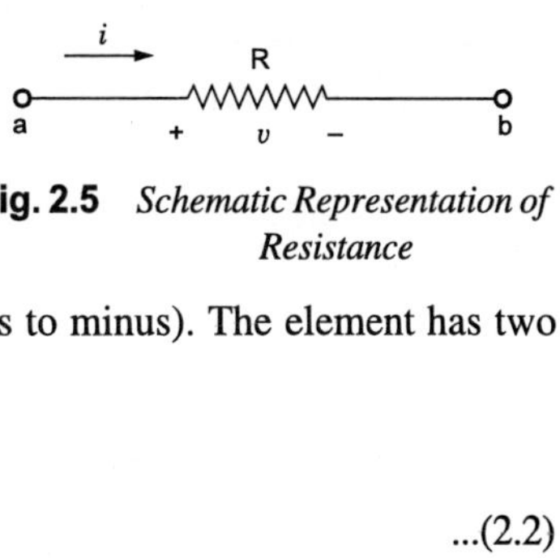

Fig. 2.5 *Schematic Representation of Resistance*

The power dissipated by resistance may be given by expression

$$p = vi = i^2R = \frac{v^2}{R} \text{ watts} \qquad \text{...(2.2)}$$

Eq. (2.1) gives the voltage across a resistor in terms of its current. A reciprocal relationship providing the current in terms of voltage is often of equal or greater value in a particular case. As a result, Ohm's law is often expressed as

$$i = Gv \text{ amperes} \qquad \text{...(2.3)}$$

where

$$G = \frac{1}{R} \qquad \text{...(2.4)}$$

Reciprocal of resistance R i.e. G is called *conductance* and is measured in mhos or siemens (SI unit of conductance is siemens, but mho is more frequently used).

Power dissipated can then be expressed in the alternative form

$$p = vi = v\,(Gv) = v^2G = i \times \frac{i}{G} = \frac{i^2}{G} \text{ watts} \qquad \text{...(2.5)}$$

2.5. CAPACITANCE

Capacitance is a two-terminal element that has the capability of charge storage and, consequently, energy storage. The stored energy can be fully retrieved.

The current through the capacitor is proportional to the derivative of voltage across it and is given by expression

$$i = C\frac{dv}{dt} \quad \text{...(2.6)}$$

where C has the unit of farads, the practical unit being a microfarad (μ F) because a farad is physically a large unit. Integrating above Eq. (2.6) we have

$$v = \frac{1}{C}\int_0^t i\, dt + v_c(0) \quad \text{...(2.7)}$$

where $v_c(0)$ = Capacitance voltage at $t = 0$

For an initially uncharged capacitor $v_c(0) = 0$, so that

$$v = \frac{1}{C}\int_0^t i\, dt = \frac{q}{C} \quad \text{or} \quad C = \frac{q}{v} \quad \text{or} \quad q = Cv \quad \text{...(2.8)}$$

The proportionality constant C expresses the charge-storing property of the element and is called the *capacitance*. With q in coulombs and v in volts, the capacitance C is in farads (abbreviated F). A capacitor is a physical element which exhibits the property of capacitance.

The schematic representation of capacitance, in which current and voltage reference directions are indicated, is depicted in Fig. 2.6. In this figure and in Eqs. (2.6) and (2.8), a voltage drop exists in the direction of flow of current. Charge flow from a higher potential to a lower potential i.e. from plus to minus, signifies that energy can be removed from the circuit and stored. The capacitive effect may be thought of as opposing a change of voltage.

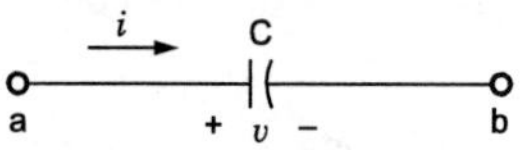

Fig. 2.6 *Schematic Representation of Capacitance*

The power associated with a capacitance is

$$p = vi = Cv\frac{dv}{dt} \text{ watts} \quad \text{...(2.9)}$$

Energy stored in the capacitance may be had by integrating above Eq. (2.9) as

$$W = \int p\, dt = \int Cv\frac{dv}{dt} = \frac{1}{2}Cv^2 \text{ joules} \quad \text{...(2.10)}$$

A capacitor offers low impedance to ac but very high impedance to dc. So capacitors are used to couple alternating voltage from one circuit to another circuit and at the same time to block dc voltage from reaching the next circuit. It is also employed as a *bypass capacitor* where it passes the ac through it without letting the dc to go through the circuit across which it is connected. A capacitor forms a *tuned circuit* in series or in parallel with an inductor.

2.6. INDUCTANCE

The circuit element used to represent the energy stored in a magnetic field is defined by the relation

$$v = L\frac{di}{dt} \quad \text{...(2.11)}$$

The above expression describes a situation in which the voltage across the element is proportional to the time rate of change of current through it. The constant of proportionality L is the *self-inductance* or simply the *inductance* of the element, and is measured in *henrys* (abbreviated H).

The voltage v in Eq. (2.11) is a voltage drop in the direction of current and can be considered to oppose an increase in current. Fig. 2.7 depicts the schematic representation of an inductance and its associated reference direction for current and voltage polarity.

Integrating Eq. (2.11) we have

$$i = \frac{1}{L}\int_0^t v\,dt + i(0) \qquad ...(2.12)$$

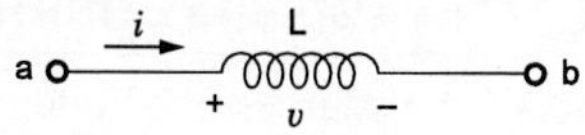

Fig. 2.7 *Schematic Representation of Inductance*

where $i(0)$ = Inductance current at $t = 0$.

According to Eq. (2.12) *current through an inductance* cannot change instantly (compared with capacitance voltage) as it would require infinite voltage.

Because the effect of inductance is to oppose the change in the magnitude of current, inductance is analogous to mass or inertia in a mechanical system and to the mass of liquid in hydraulics. Inductance prevents the current from changing instantly as it requires infinite voltage to cause an instantaneous change in current, just as the mass of an automobile prevents it from stopping or starting instantaneously.

The power associated with the inductive effect in a circuit is

$$p = vi = Li\frac{di}{dt} \text{ watts} \qquad ...(2.13)$$

and the energy stored is

$$W = \int p\,dt = \int Li\frac{di}{dt}dt = \int Li\,di = \frac{1}{2}Li^2 \text{ joules} \qquad ...(2.14)$$

unlike the resistive energy, which is transformed into heat, the inductive energy is stored in the same sense that kinetic energy is stored in a moving mass. Eq. (2.14) reveals that the magnitude of stored energy depends on the magnitude of current and not in the manner of attaining that magnitude. The stored inductive energy reappears in the circuit as the current is reduced to zero. For example, if a switch is opened in a current carrying inductive circuit, the current decays rapidly, but not instantaneously. In accordance with Eq. (2.11), a relatively high voltage appears across the separating contacts of the switch, and an arc may form. The arc makes it possible for the stored energy to be dissipated as heat in the arc and the circuit resistances.

In case of an inductor current does not change instantaneously. It offers high impedance to ac but very low impedance to dc *i.e.* it blocks ac signal but passes dc signal.

A piece of wire, or a conductor of any type, has inductance *i.e.* a property of opposing the change of current through it. By coiling the wire the inductance is increased as the square of the number of turns. The inductance is represented by English capital letter L and measured in henrys.

2.7. SUMMARY OF RELATIONSHIPS FOR THE PARAMETERS

Some of the relationships discussed so far in this chapter are summarised in tabular form (Table 2.1). These equations are encountered so frequently in the study of electrical engineering that they should be memorized.

Table 2.1 Summary of Relationships For The Parameters

Parameter	*Basic Relationship*	*Voltage-Current Relationships*		*Energy*
Resistance, R	$v = Ri$	$v_R = Ri_R$	$i_R = Gv_R$	$W_R = \int_{-\infty}^{t} v_R i_R\,dt$
$G = \frac{1}{R}$				
L (or M)	$\psi = Li$	$v_L = L\frac{di_L}{dt}$	$i_L = \frac{1}{L}\int_{-\infty}^{t} v_L dt$	$W_L = \frac{1}{2}LI^2$
C	$q = Cv$	$v_C = \frac{1}{C}\int_{-\infty}^{t} i_C\,dt$	$i_C = C\frac{dv_C}{dt}$	$W_C = \frac{1}{2}CV^2$

2.8. VOLTAGE AND CURRENT SOURCES

Most of the sources encountered in everyday life (such as batteries, dynamos, alternators etc.) are voltage sources but some current sources do exist. Some examples are; photoelectric cells as used in light meters, metadyne generators as used in military gun controls. Other devices may be regarded as current sources, such as the collector circuits of transistors and the anode circuits of pentode thermionic tubes.

2.8.1. Independent and Dependent Sources. The source (voltage or current) may be independent or dependent. A source is said to be independent when it does not depend on any other quantity in the circuit. Fig. 2.8 (*a*) shows an independent dc voltage source whereas Fig. 2.8 (*b*) depicts a time varying voltage source. The positive sign indicates that terminal A is positive with respect to terminal B *i.e.* the potential of terminal A is V volts higher than that of terminal B.

Similarly an ideal constant current source is shown in Fig. 2.8 (*c*) whereas time varying current source is shown in Fig. 2.8 (*d*). The arrow indicates the direction of current flow at any moment under consideration.

A *dependent source* is one which depends on some other quantity in the circuit which may be either a voltage or a current. In electronic circuits, we very often find that the current through an element, (say collector current through a bipolar junction transistor) is dependent on a current through some other element or in a MOSFET it is dependent on the voltage across some other element. Such a source is called a *dependent source*. In a dependent source the output voltage (or current) depends on another voltage (or current). The relationship may be linear or nonlinear. There are four possible dependent sources as are represented in Fig. 2.9.

Such sources can also be either constant sources or time varying sources.

Independent sources actually exist as physical entities such as an accumulator, a dc generator and an alternator. But dependent sources are parts of *models* that are used to represent electrical properties of electronic devices such as operational amplifiers and transistors etc.

Another major difference is that four terminals are required to define a controlled source; whereas only two are required for an independent source. Of the four dependent source terminals, one pair provides the control and the second pair exhibits the properties of the source.

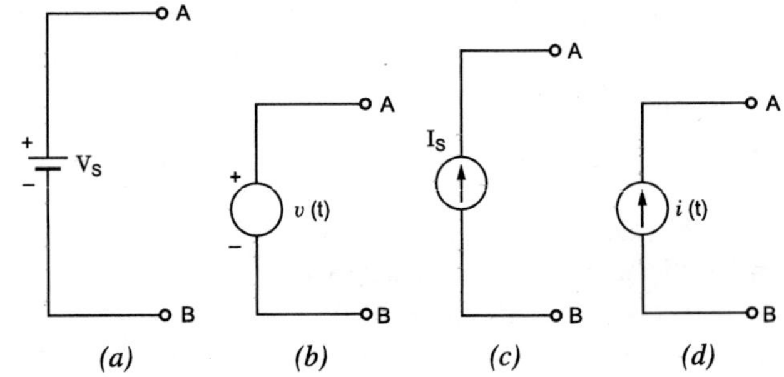

Fig. 2.8

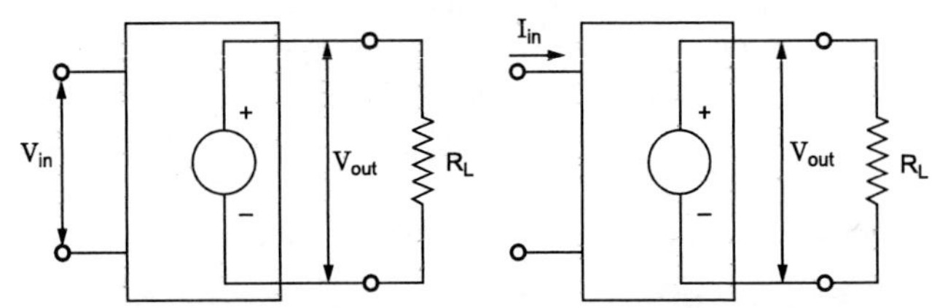

$V_{out} = k_1 V_{in}$; k_1 dimensionless

(a) Voltage-Dependent Voltage Source

$V_{out} = k_2 I_{in}$; k_2 in ohms

(b) Current-Dependent Voltage Source

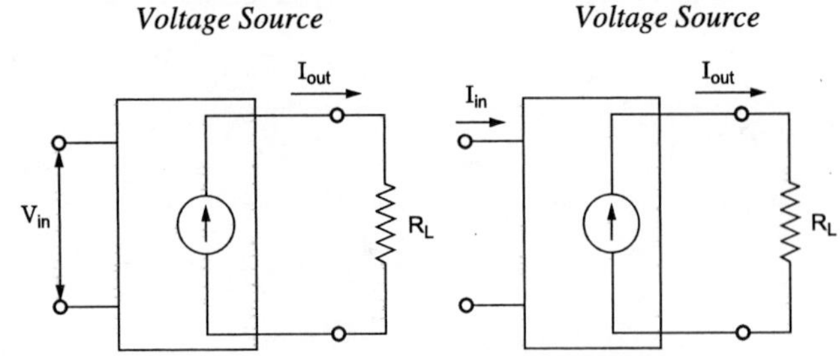

$I_{out} = k_3 V_{in}$; k_3 in mhos

(c) Voltage-Dependent Current Source

$I_{out} = k_4 I_{in}$: k_4 dimensionless

(d) Current Dependent Current Source

Fig. 2.9 *Dependent Sources*

2.8.2. Ideal Voltage Source. A constant voltage source is an ideal source element capable of supplying any current at a given voltage. If the internal resistance of a voltage source is zero, the terminal voltage (voltage across the load) is equal to the voltage across the source (the source emf), and is independent of the amount of load current, or in other words the voltage of an ideal voltage source is independent of load current supplied by it. For example, if the terminals are connected together the source will supply an infinite current. The symbolic representations of dc and ac ideal voltage sources are given in Figs. 2.10 (*a*) and 2.10 (*b*) respectively.

There are two noteworthy points regarding ideal voltage sources. First, ideal voltage source cannot be short circuited (because this will be contrary to the definition of the ideal voltage source itself). Secondly (and for the same reason) two ideal voltage sources of unequal output voltages cannot be placed in parallel.

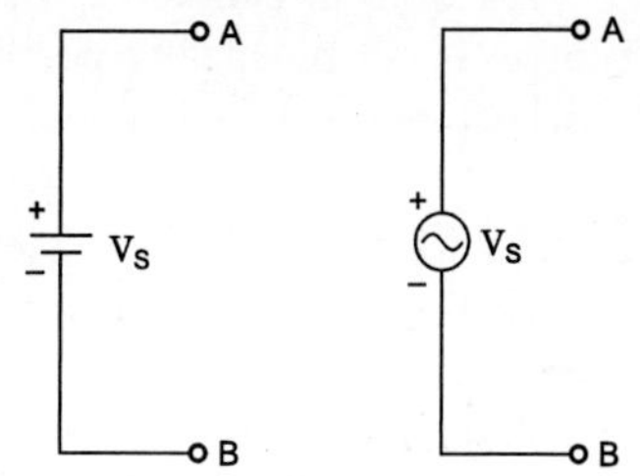

(*a*) (*b*)

Fig. 2.10

An ideal voltage source is not practically possible. There is no voltage source which can maintain its terminal voltage constant even when its terminals are short circuited.

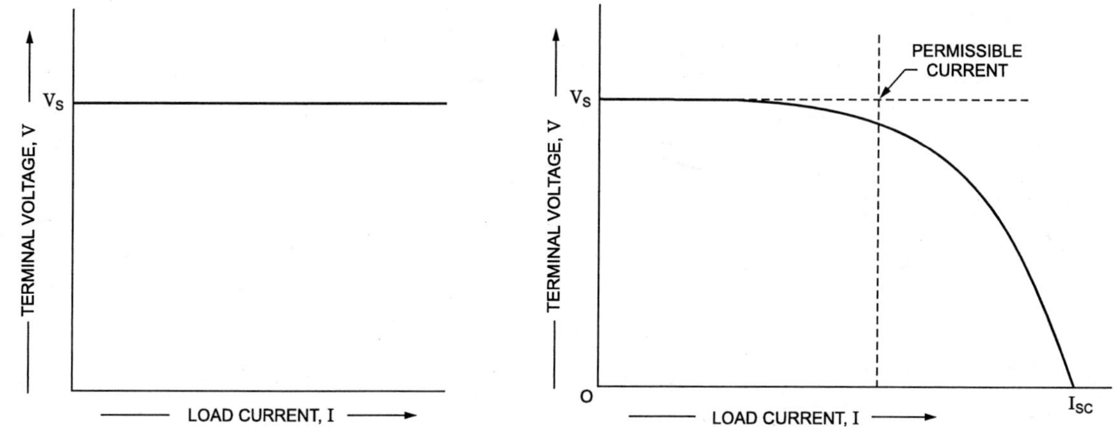

(*a*) *For Ideal Voltage Source* (*b*) *For a Practical Voltage Source*

Fig. 2.11 *V-I Characteristics of Voltage Sources*

A lead-acid battery or a dry cell are examples of a constant voltage source when the current drawn is below a certain limit. However, a practical source always shows a drop in its terminal voltage which increases with load current. A dc generator or a rectifier operating on mains supply is a voltage source (the output voltage is contaminated with ripples) and exhibit a voltage drop which is load dependent.

The volt-ampere (V-I) characteristics of an ideal voltage source are depicted in Fig. 2.11 (*a*) in comparison to that of a practical voltage source shown in Fig. 2.11 (*b*). Dotted line is that of an ideal voltage source for differentiation. Within a permissible range of current a practical dc voltage source maintains the terminal (output) voltage within a narrow range of its nominal voltage. Beyond this value of current, called the *rated current*; the voltage drops rapidly till the short-circuit current I_{SC}, where the terminal voltage drops to zero. In contrast, an ideal dc voltage source would have maintained the voltage even up to infinitely large current.

2.8.3. Ideal Current Source. Like a constant voltage source, there may be a constant current source—a source that supplies a constant current to a load even if its impedance varies. Ideally, the current supplied by such a source should remain constant irrespective of the load impedance. A symbolic representation of such an ideal constant current source is shown in Fig. 2.8 (*c*). The arrow inside the circle indicates the direction of flow of current in the circuit, when a load is connected to the source.

There are two noteworthy points regarding ideal current sources. First, an ideal current source cannot be open circuited (because this will be contrary to the definition of the constant current source itself). Again, for the same reason, two ideal current sources of different output currents cannot be placed in series.

An ideal current source, like an ideal voltage source, is not practically possible. There is no current source which can maintain current supplied by it constant even when its terminals are open circuited. An ideal current source does not exist in practice. A practical current source can be represented as shown in Fig. 2.12 (*a*).

A solar cell is an example of *current source*. It provides constant current to a resistance within a specified range of output voltage. The value of the current delivered by a current source depends on the flux incident on the cell. An ideal current source provides a constant current to a resistor of resistance R ohms ($0 < R < \infty$). V-I characteristics of a practical current source is compared with that of an ideal current source (dotted vertical lines) in Fig. 2.12 (*b*). The limit V_R up to which a current source maintains the current is drawn by horizontal dotted line. Beyond this value of output voltage (hence load resistance) the current falls. A current source can also be had by connecting a high resistance R_{se} in series with a high voltage dc source. It will behave as a current source in a range R_L is less than 5 to 10% of series resistance R_{se}. In that case, the current is primarily decided by R_{se}. The current through load resistance R_L is nearly constant unless R_L is significant in comparison to R_{se}.

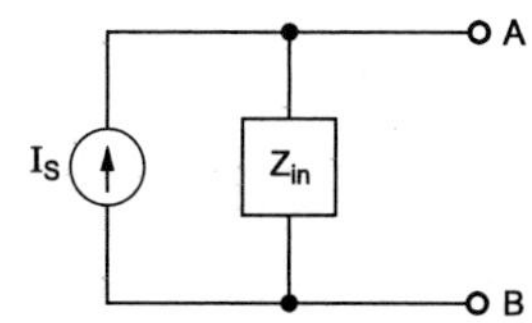

(*a*) *Practical Current Source*

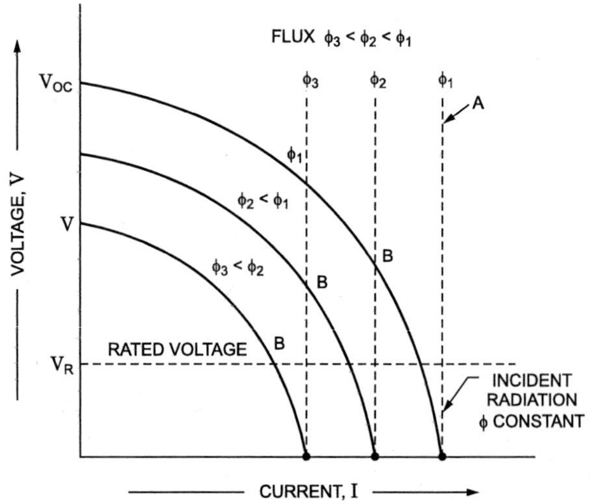

(*b*) *Comparison of V-I Characteristics of an Ideal Current Source (A) With That of a Practical Current Source (B) For Different Values of Radiation Intensity*

Fig. 2.12

2.8.4. Practical Sources. In most applications, ideal sources are approximation. The internal resistance R_{in} of a voltage source, which is responsible for a drop in terminal voltage of a source on load is quite small in comparison to the load resistance of a network connected across the voltage source. Hence voltage drop is considered negligible. Internal resistance needs to be taken into account when it is significant in comparison to load resistance R_L.

A practical dc voltage source is represented in Fig. 2.13 (*a*). V_S is the internal voltage of the voltage source and R_{in} is the internal resistance of the voltage source. When we measure the voltage across the source terminals without any load (no resistance connected to the output terminals of the source) the terminal voltage is $V = V_S$ as there is no current through R_{in} to cause a voltage drop.

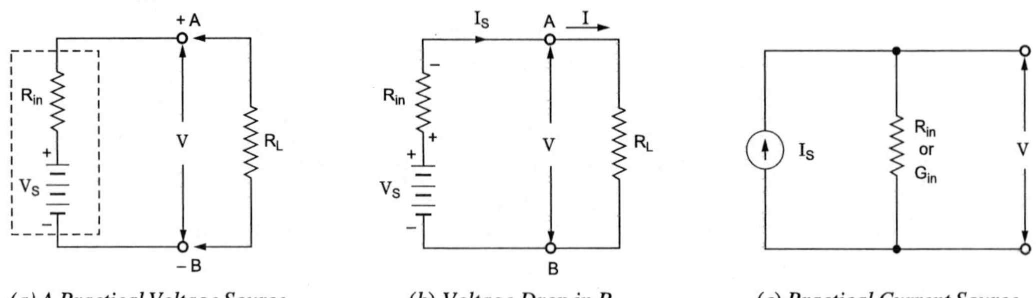

(*a*) *A Practical Voltage Source* (*b*) *Voltage Drop in R_{in}* (*c*) *Practical Current Source*

Fig. 2.13 *Practical Sources*

Assuming V_S to be constant, the terminal voltage falls on loading by an external or load resistance R_L. The resulting current,

$$I = \frac{V_S}{R_{in} + R_L} \quad ...(2.15)$$

and the terminal voltage is then $V = IR_L = \frac{V_S}{R_{in} + R_L} \times R_L$ ...(2.16)

As the load resistance R_L reduces (load on the source increases), terminal voltage V falls linearly, if R_{in} and V_S are assumed constant.

When the terminals of a practical voltage source are short circuited by a thick wire of zero resistance resulting in a short circuit, source current I_S is given as

$$I_S = I_{SC} = \frac{V_S}{R_{in}} \quad ...(2.17)$$ In Fig. 2.13 (*b*), R_L being zero.

$$\text{Hence, } R_{in} = \frac{V_S}{I_{SC}} = \frac{\text{Open-circuit voltage of the source, } V_{OC}}{\text{Short-circuit current, } I_{SC}} \quad ...(2.18)$$

Similarly an ideal current source must produce infinite voltage on open circuit. A practical current source will have a finite output voltage. A practical current source is represented as shown in Fig. 2.13 (*c*). In this case short-circuit current,

$$I_{SC} = I_S \quad ...(2.19)$$

and the open-circuit voltage, $V_{OC} = I_S R_{in}$ ...(2.20)

$$\text{Hence, } R_{in} = \frac{V_{OC}}{I_{SC}} = \frac{\text{Open-circuit voltage}}{\text{Short-circuit current}} \quad ...(2.21)$$

$$\text{or } G_{in} = \frac{1}{R_{in}} = \frac{I_{SC}}{V_{OC}} = \frac{\text{Short-circuit current}}{\text{Open-circuit voltage}} \quad ...(2.22)$$

2.9. SOURCE EQUIVALENCE OR TRANSFORMATION

Practically, a voltage source is not different from a current source. In fact, a source can either operate as a current source or as a voltage source. It merely depends upon its operating conditions. If load impedance is very large in comparison to internal impedance of the source, it will be advantageous to treat the source as a voltage source. On the other hand, if the load impedance is very small in comparison to the internal impedance of the source, it is better to represent the source as a current source. From the circuit point of view it does not matter at all whether the source is treated as a voltage source or a current source. In fact, it is possible to convert a voltage source into a current source and vice versa.

Consider a voltage source of voltage V_S and internal resistance R_{in} shown in Fig. 2.13 (*a*) for conversion into an equivalent current source. The current supplied by this voltage source, when a short circuit is put across terminals A and B, will be equal to V_S/R_{in}. A current source supplying this current $I_S = V_S/R_{in}$ and having the same resistance across it will represent the equivalent current source [Fig. 2.13 (*c*)].

Similarly a current source of output current I_S in parallel with resistance R_{in} can be converted into an equivalent voltage source of voltage $V_S = I_S R_{in}$ and a resistance R_{in} in series with it [Fig. 2.13 (*a*)].

It should be noted that a voltage source-series resistance combination is equivalent to a current source–parallel resistance combination if, and only if their respective open-circuit voltages are equal, and their respective short-circuit currents are equal.

For example, a voltage source branch consisting of a 10 V source in series with a resistance of 2.5 Ω may be replaced by a current source branch consisting of a 4 A source in parallel with a 2.5 Ω resistance and vice versa, as shown in Figs. 2.14 (*a*) and 2.14 (*b*) respectively.

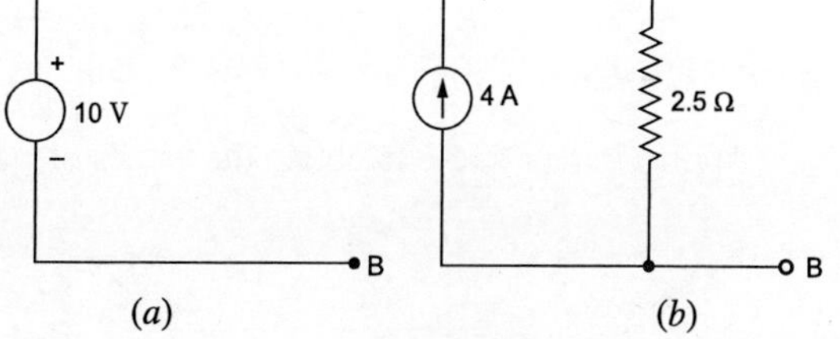

Fig. 2.14

Example 2.1 **Determine the current in 28 Ω resistor in the circuit shown in Fig. 2.15.**

Solution: The two current sources cannot be combined together because of the 28 Ω resistance presence between points A and C. However, this hurdle can be removed by converting 10 A current source into the equivalent voltage source of 40 V, as shown in Fig. 2.16 (*a*). Similarly 5 A current source can be converted into the equivalent voltage source of 40 V, as shown in Fig. 2.16 (*b*).

The circuit shown in Fig. 2.16 (*b*) is a simple series circuit having total voltage of 80 V and total resistance of 4 + 28 + 8 *i.e.* 40 Ω. So the current flowing through resistance of 28 Ω is 2 A **Ans.**

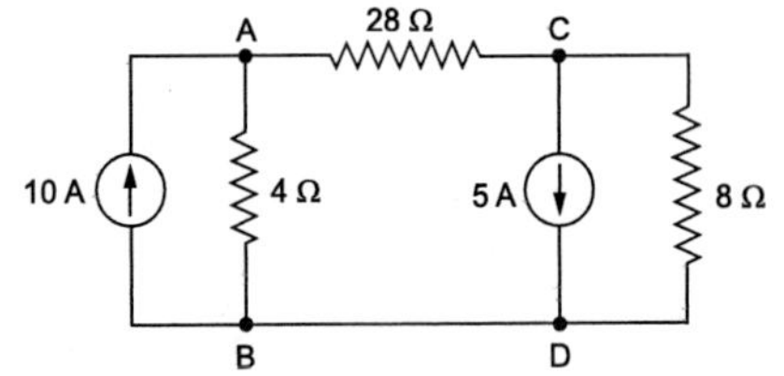

Fig. 2.15

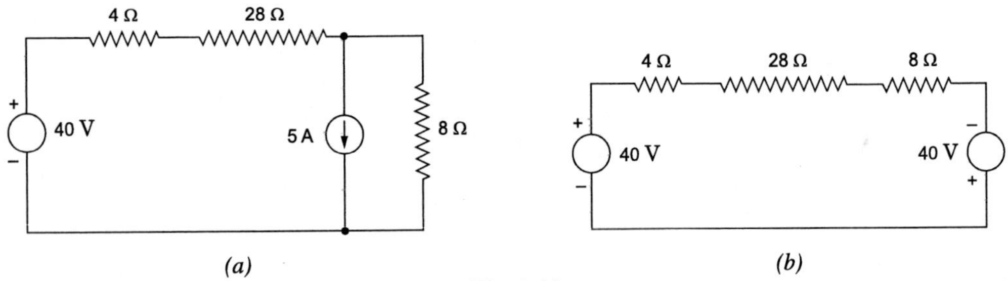

Fig. 2.16

2.10. SOLUTION OF SIMULTANEOUS EQUATIONS

Let us consider two simultaneous equations

$$a_1x + b_1y = m \qquad ...(i)$$

$$a_2x + b_2y = n \qquad ...(ii)$$

Here, the two unknown are x and y and a_1, b_1, a_2, b_2 are coefficients of these unknown whereas m and n are constants. These equations may be solved, by use of determinants as follows:

Write the two equations in matrix form as

$$\begin{bmatrix} a_1 & b_1 \\ a_2 & b_2 \end{bmatrix} \begin{bmatrix} x \\ y \end{bmatrix} = \begin{bmatrix} m \\ n \end{bmatrix}$$

The common determinant is given as

$$\Delta = \begin{vmatrix} a_1 & b_1 \\ a_2 & b_2 \end{vmatrix} = a_1\, b_2 - b_1\, a_2$$

For finding the determinant for x, replace the coefficients of x in the original matrix by the constants so that we get determinant

$$\Delta_1 = \begin{vmatrix} m & b_1 \\ n & b_2 \end{vmatrix} = (m\, b_2 - b_1\, n)$$

For finding the determinant for y, replace the coefficients of y in the original matrix by the constants, so we get determinant

$$\Delta_2 = \begin{vmatrix} a_1 & m \\ a_2 & n \end{vmatrix} = (a_1\, n - m\, a_2)$$

Apply Cramer's rule to obtain the values of x and y

$$x = \frac{\Delta_1}{\Delta} = \frac{m\, b_2 - b_1\, n}{a_1\, b_2 - b_1\, a_2}$$

$$\text{and } y = \frac{\Delta_2}{\Delta} = \frac{a_1\, n - m\, a_2}{a_1\, b_2 - b_1\, a_2}$$

Now let us consider three simultaneous equations in three unknown

$$a_1 x + b_1 y + c_1 z = m \quad ...(i)$$

$$a_2 x + b_2 y + c_2 z = n \quad ...(ii)$$

$$a_3 x + b_3 y + c_3 z = p \quad ...(iii)$$

The above equations can be put in the matrix form as

$$\begin{bmatrix} a_1 & b_1 & c_1 \\ a_2 & b_2 & c_2 \\ a_3 & b_3 & c_3 \end{bmatrix} \begin{bmatrix} x \\ y \\ z \end{bmatrix} = \begin{bmatrix} m \\ n \\ p \end{bmatrix}$$

The common determinant is given as

$$\Delta = \begin{vmatrix} a_1 & b_1 & c_1 \\ a_2 & b_2 & c_2 \\ a_3 & b_3 & c_3 \end{vmatrix} = a_1 (b_2 c_3 - c_2 b_3) - a_2 (b_1 c_3 - c_1 b_3) + a_3 (b_1 c_2 - c_1 b_2)$$

The determinant for x can be found by replacing coefficients of x in the original matrix by constants

$$\Delta_1 = \begin{vmatrix} m & b_1 & c_1 \\ n & b_2 & c_2 \\ p & b_3 & c_3 \end{vmatrix} = m (b_2 c_3 - c_2 b_3) - n (b_1 c_3 - b_3 c_1) + p (b_1 c_2 - c_1 b_2)$$

$$\text{Similarly determinant for } y, \Delta_2 = \begin{vmatrix} a_1 & m & c_1 \\ a_2 & n & c_2 \\ a_3 & p & c_3 \end{vmatrix} = a_1 (n c_3 - c_2 p) - a_2 (m c_3 - c_1 p) + a_3 (m c_2 - c_1 n)$$

$$\text{and determinant for } z, \Delta_3 = \begin{vmatrix} a_1 & b_1 & m \\ a_2 & b_2 & n \\ a_3 & b_3 & p \end{vmatrix} = a_1 (b_2 p - n b_3) - a_2 (b_1 p - m b_3) + a_3 (b_1 n - m b_2)$$

Now according to Cramer's rule

$$x = \frac{\Delta_1}{\Delta}; y = \frac{\Delta_2}{\Delta} \text{ and } z = \frac{\Delta_3}{\Delta}.$$

2.11. KIRCHHOFF'S LAWS

The basic laws, that electric circuits follow rationally from the nature of the electrical quantities, have already been defined. They lead directly to methods for the systematic analysis of electric circuits. These laws are known as *Kirchhoff's laws,* and they describe the relationships among circuit voltages and circuit currents that must be satisfied. These laws are very helpful in determining the equivalent resistance of a complex network and the current flowing in the various branches of the network.

Gustov Robert Kirchhoff derived two basic laws governing networks, one commonly known as *first law* or *current law* (KCL) or *point law,* whereas the second is called *second law* or *voltage law* (KVL) or *mesh law.*

1. Kirchhoff's First Law or Current Law (KCL) or Point Law. According to this law in any network of wires carrying currents, the algebraic sum of all currents meeting at a point (or junction) is zero or the sum of incoming currents towards any point is equal to the sum of outgoing currents away from that point. If I_1, I_2, I_3, I_4, I_5, and I_6 are the currents meeting at junction O, flowing in the directions of arrowheads marked on them (Fig. 2.17), taking incoming currents as positive and outgoing currents as negative, according to Kirchhoff's first law (KCL)

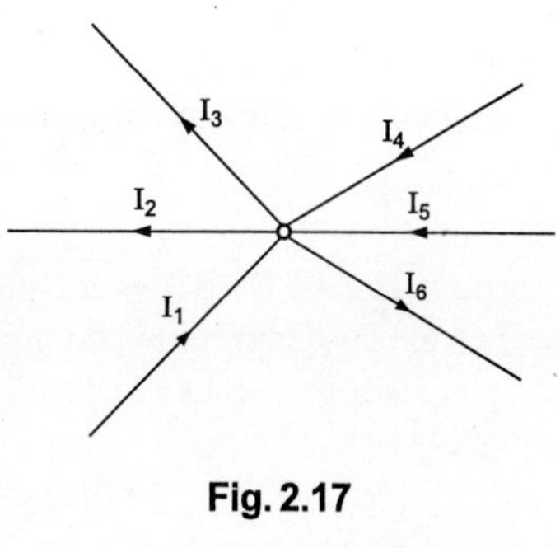

Fig. 2.17

$$I_1 - I_2 - I_3 + I_4 + I_5 - I_6 = 0$$
$$\text{or } I_1 + I_4 + I_5 = I_2 + I_3 + I_6$$

2. Kirchhoff's Second Law or Voltage Law (KVL) or Mesh Law. According to this law in any closed circuit or mesh the algebraic sum of emfs acting in that circuit or mesh is equal to the algebraic sum of the products of the currents and resistances of each part of the circuit.

If the circuit shown in Fig. 2.18 is considered, then according to Kirchhoff's second law (or KVL)

In mesh AFCBA,

$$E_1 = R_1 (I_1 + I_2) + R_2 (I_1 + I_2) + R_5 I_1$$
$$\text{or } E_1 = (R_1 + R_2)(I_1 + I_2) + R_5 I_1$$

In mesh FEDCF,

$$- E_2 = - R_3 I_2 - R_4 I_2 - R_2 (I_1 + I_2) - R_1 (I_1 + I_2)$$
$$\text{or } E_2 = R_3 I_2 + R_4 I_2 + (R_1 + R_2)(I_1 + I_2)$$

and in mesh AFEDCBA,

$$E_1 - E_2 = - R_3 I_2 - R_4 I_2 + R_5 I_1$$
$$\text{or } E_1 - E_2 = R_5 I_1 - (R_3 + R_4) I_2$$

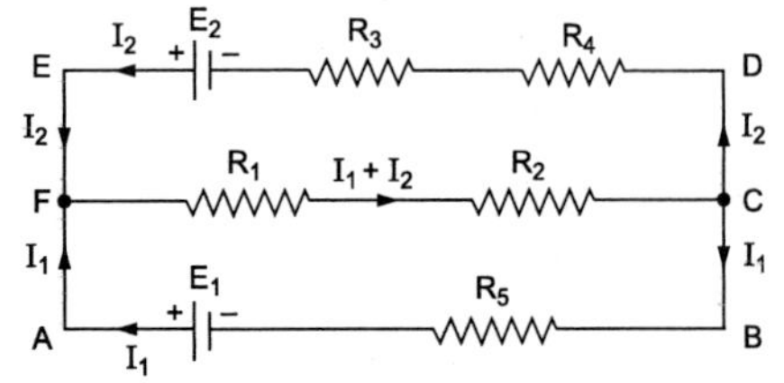

Fig. 2.18

2.11.1. Application of Kirchhoff's Laws To Circuits. First of all the current distribution in various branches of the circuit is made with directions of their flow complying with first law of Kirchhoff. Then Kirchhoff's second law is applied to each mesh (one by one) separately and algebraic equations are obtained by equating the algebraic sum of emfs acting in a mesh equal to the algebraic sum of respective drops in the same mesh. By solving the equations so obtained unknown quantities can be determined. While applying Kirchhoff's second law, the question of algebraic signs may be troublesome and is a frequent source of error. If, however, the following rules are kept in mind, no difficulty should occur.

The resistive drops in a mesh due to current flowing in clockwise direction must be taken positive drops.

The resistive drops in a mesh due to current flowing in counter-clockwise direction must be taken as negative drops.

Similarly the battery emf causing current to flow in clockwise direction in a mesh must be taken as positive emf and the battery emf causing current to flow in counter-clockwise direction in a mesh must be taken as negative emf.

For example, for the circuit shown in Fig. 2.18 let the current distribution be made as shown, which satisfy Kirchhoff's first law fully.

Taking first, mesh AFCBA for the application of Kirchhoff's second law, we see that there is only one emf acting in the mesh (E_1) and since it tries to send current in clockwise direction so E_1 be taken as positive, similarly all the resistive drops *i.e.* $R_1 (I_1 + I_2)$, $R_2 (I_1 + I_2)$ and $R_5 I_1$ are clockwise, so these must be taken as positive.

∴ According to Kirchhoff's second law in mesh AFCBA

$$E_1 = R_1 (I_1 + I_2) + R_2 (I_1 + I_2) + R_5 I_1$$

In mesh FEDCF, there is only one emf acting in the mesh (E_2) and since it tries to send current in counter-clockwise direction through the mesh under consideration, it may be taken as negative. Since all of the resistive drops $R_3 I_2$, $R_4 I_2$, $R_2 (I_1 + I_2)$ and $R_1 (I_1 + I_2)$ are counter-clockwise, these may be taken as negative.

Hence according to Kirchhoff's second law in mesh FEDCF we get,

$$- E_2 = - R_3 I_2 - R_4 I_2 - R_2 (I_1 + I_2) - R_1 (I_1 + I_2)$$
$$\text{or } E_2 = (R_3 + R_4) I_2 + (R_1 + R_2)(I_1 + I_2)$$

In mesh AFEDCBA, emf E_1 tries to cause current in clockwise direction so be taken as positive and emf E_2 tries to cause current in counter-clockwise direction so be taken as negative. Similarly resistive drop $R_5 I_1$ being clockwise be taken as positive and resistive drops $R_3 I_2$ and $R_4 I_2$ being counter - clockwise be taken as negative.

$$\text{Hence } E_1 - E_2 = - R_3 I_2 - R_4 I_2 + R_5 I_1 = R_5 I_1 - (R_3 + R_4) I_2$$

Example 2.2 **Find the voltage drops across R_1 and R_2 (Fig. 2.19). The resistance R_3 is not specified.**
[G.B. Technical Univ. Electrical Engineering Odd Semester 2010-11]

Solution: Voltage drop across unknown resistor R_3 = Voltage across 80 Ω resistor = 0.5 × 80 = 40 V

Applying in kVL to LHS loop we have

$$I R_1 + I R_2 + \text{voltage drop across } R_3 = 100 \text{ V}$$

where I is current supplied by 100 V battery

$$\text{or} \quad 10\,I + 10\,I + 40 = 100$$

$$\text{or} \quad I = \frac{100 - 40}{20} = 3 \text{ A}$$

So voltage drop across $R_1 = 3 \times 10 = 30$ V **Ans.**

Voltage drop across $R_2 = 3 \times 10 = 30$ V **Ans.**

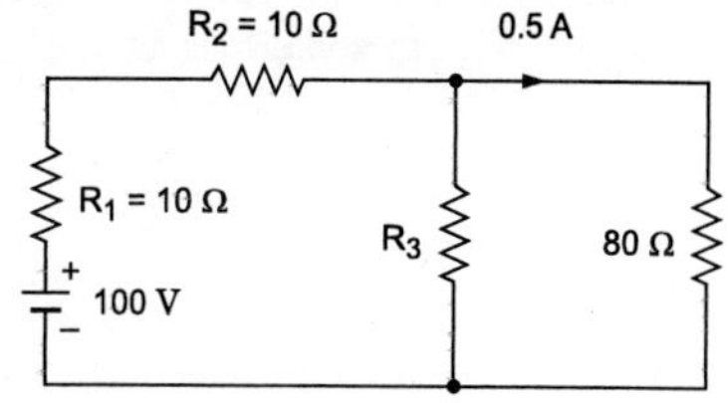

Fig. 2.19

Example 2.3 **Find the current I in the circuit given in Fig. 2.20 (*a*).**

[U.P. Technical Univ. Electrical Engineering Second Semester 2006-07]

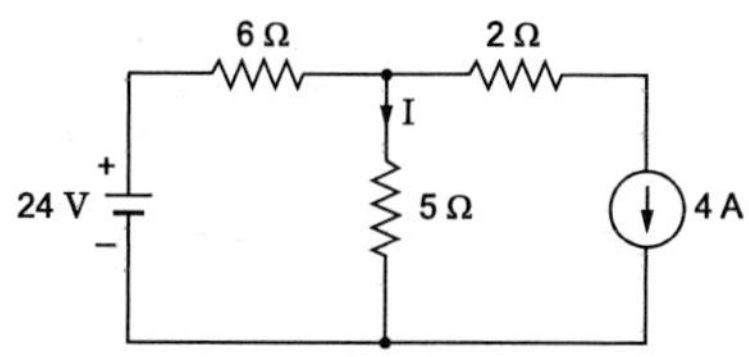

Fig. 2.20 (*a*)

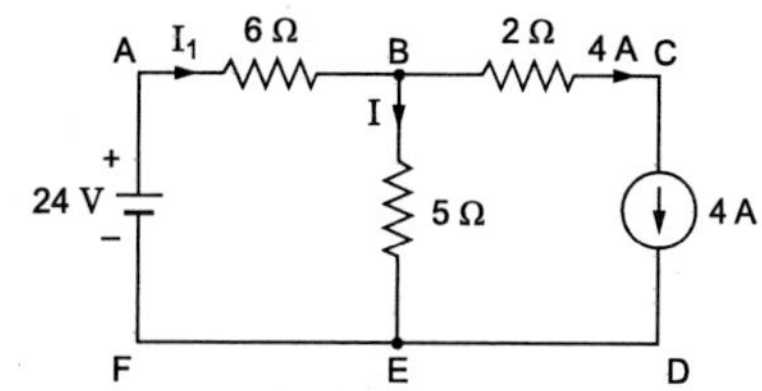

Fig. 2.20 (*b*)

Solution: Applying Kirchhoff's current law at junction B we have

Current supplied by 24 V battery, $I_1 = (I + 4)$ amperes

Applying Kirchhoff's voltage law to mesh ABEFA we have

$$6\,I_1 + 5\,I = 24$$

$$\text{or} \quad 6\,(I + 4) + 5\,I = 24$$

$$\text{or} \quad 11\,I + 24 = 24$$

$$\text{or} \quad I = 0 \textbf{ Ans.}$$

Example 2.4 **Find current in 2 Ω resistor.**
[U.P. Technical Univ. Basic Electrical Engineering Odd Semester 2013-14]

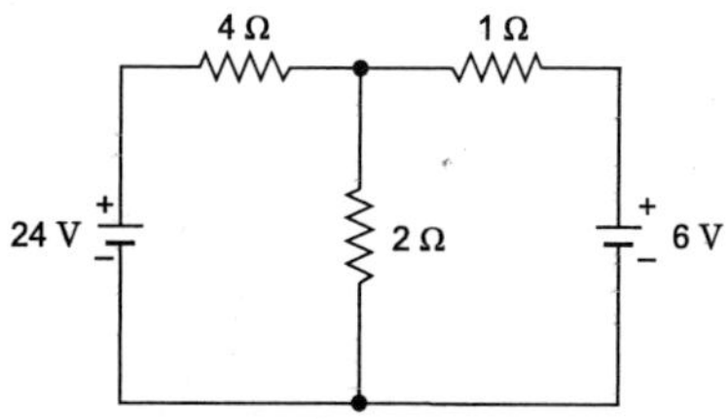

Fig. 2.21

Solution: Let the current distribution in the network be as shown in Fig. 2.22.
Applying KVL to meshes ABEFA and BCDEB we have respectively

$$4\,I_1 + 2\,(I_1 + I_2) = 24 \qquad ...(i)$$

$$\text{and } I_2 \times 1 + 2\,(I_1 + I_2) = 6 \qquad ...(ii)$$

$$\text{or } 3\,I_1 + I_2 = 12 \qquad ...(iii)$$

$$\text{and } 2\,I_1 + 3\,I_2 = 6 \qquad ...(iv)$$

Solving Eqs (*iii*) and (*iv*) we have

$$I_1 = \frac{30}{7} \text{ A}$$

$$\text{and } I_2 = \frac{-6}{7} \text{ A}$$

$$\text{Current in 2 Ω resistor } = I_1 + I_2 = \frac{30}{7} + \left(\frac{-6}{7}\right) = \frac{24}{7} \text{ A} \quad \textbf{Ans.}$$

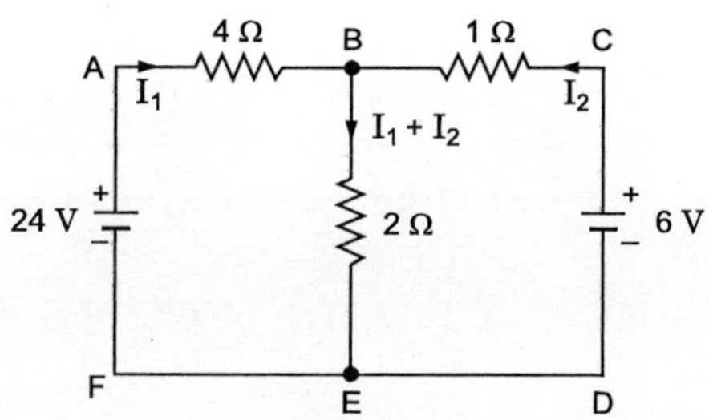

Fig. 2.22

Example 2.5 **Find the current in, and voltage across, the 2 Ω resistance in the following figure (Fig. 2.23).**

[*U.P. Technical Univ. Electrical Engineering First Semester 2008-09*]

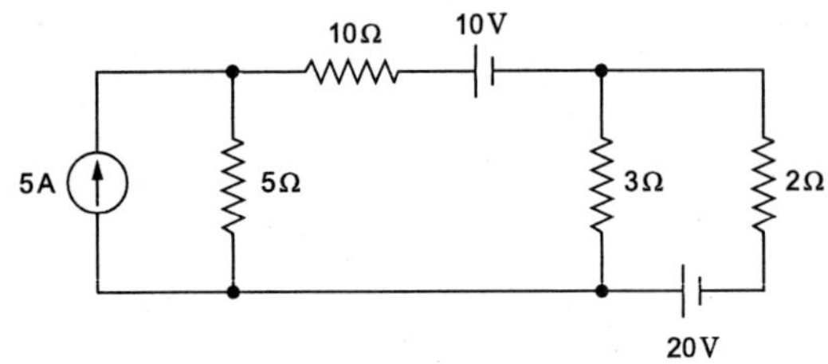

Fig. 2.23

Solution: Converting current source to equivalent voltage source, the given circuit becomes as shown in Fig. 2.24.

Apply KCL to node B we have

$$I = I_1 + I_2 \quad ...(i)$$

Applying KVL to meshes ABEFA and BCDEB respectively we have

$$(5 + 10) I + 3I_1 = 25 - 10 \quad ...(ii)$$

$$\text{and} \quad 2I_2 - 3I_1 = 20 \quad ...(iii)$$

Substituting $I = I_1 + I_2$ from Eq. (*i*) in Eq (*ii*) we have

$$15 (I_1 + I_2) + 3I_1 = 15$$

$$\text{or} \quad 18I_1 + 15I_2 = 15$$

$$\text{or} \quad 6I_1 + 5I_2 = 5 \quad ...(iv)$$

Solving Eqs. (*iii*) and (*iv*) we have

$$I_1 = \frac{-10}{3} \text{ A and } I_2 = 5\text{A}$$

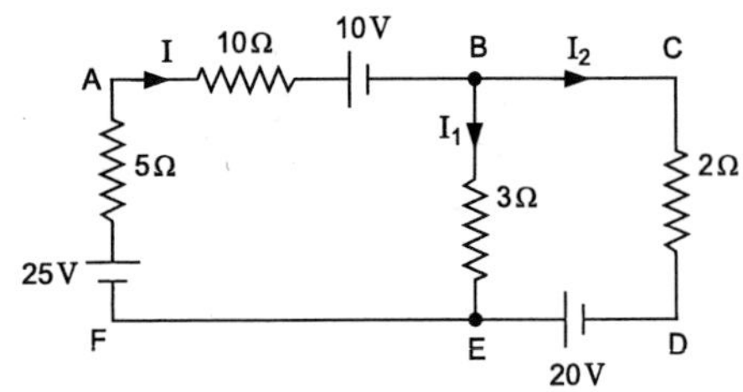

Fig. 2.24

Current through 2 Ω resistance $= I_2 = 5\text{A}$ **Ans.**

Voltage across 2Ω resistance $= I_2 \times 2 = 5 \times 2 = 10$ V **Ans.**

Example 2.6 **In Fig. 2.25 (*a*) the potential of point A = – 30 V. Using Kirchhoff's laws find (*a*) value of V and (*b*) power dissipated by 5 Ω resistance.**

Solution: The potential of point A is – 30 V and potential of point G is zero, being grounded. So the *pd* across 12 Ω resistor connected between terminals A and H is 30 V and current flowing through 12 Ω resistor is $\frac{30}{12}$ *i.e.* 2.5 A from terminal H to A.

Let the current distribution in the other branches of the circuit be, as shown in Fig. 2.25 (*b*). Applying Kirchhoff's current law (KCL) to node H we have

$$I_2 + I_3 = 2.5 \quad ...(i)$$

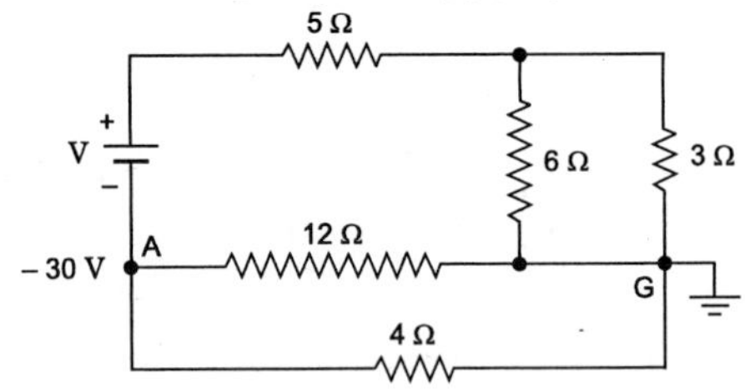

Fig. 2.25 (*a*)

Applying Kirchhoff's voltage law to meshes CHGDC and AFEGHA respectively we have

$$6 I_2 - 3 (I_1 - I_2) = 0$$

$$\text{or} \quad I_2 = \frac{I_1}{3} \quad ...(ii)$$

$$\text{and } 4 (I_1 - I_2 - I_3) - 12 \times 2.5 = 0$$

$$\text{or } I_1 - (I_2 + I_3) = 7.5$$

Substituting $I_2 + I_3 = 2.5$ from Eq. (*i*) in the above equation we have

$$I_1 - 2.5 = 7.5$$

$$\text{or} \quad I_1 = 10 \text{ A}$$

$$\text{and } I_2 = \frac{I_1}{3} = \frac{10}{3} \text{ A}$$

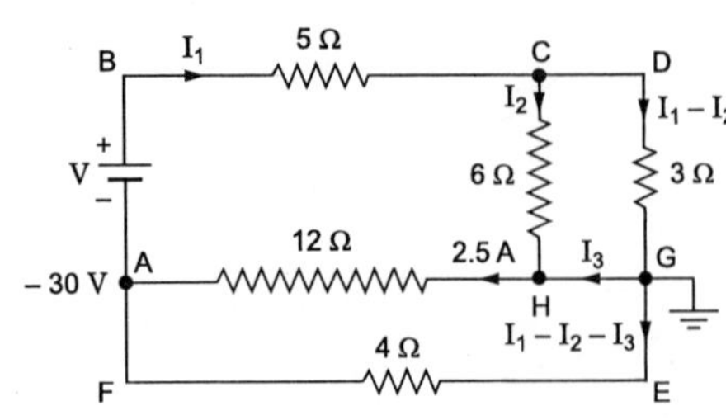

Fig. 2.25 (*b*)

Applying KVL to mesh ABCHA we have

$$5 I_1 + 6 I_2 + 12 \times 2.5 = V$$

(*a*)
$$\text{or } V = 5 I_1 + 6 I_2 + 30$$

$$= 5 \times 10 + 6 \times \frac{10}{3} + 30 = 100 \text{ V}$$ **Ans.**

(*b*) Power dissipated in 5 Ω resistor $= I_1^2 \times 5 = 10^2 \times 5 = 500$ W **Ans.**

Example 2.7 **Find the value of resistance R and the current I_R in Fig. 2.26 when the branch AD carries no current.**

Solution: Let the current through the branches AB and DB of the circuit be I_1 and I_2 amperes respectively. Applying KCL to junctions B, A and D respectively we have

Current supplied by battery or current in branch BC = $I_1 + I_2$

Current through branch CA = I_1 ∵ Current through branch AD is zero

Current through branch CD = I_2

Now applying Kirchhoff's voltage law to meshes ABDA, CADC and BCDB respectively we have

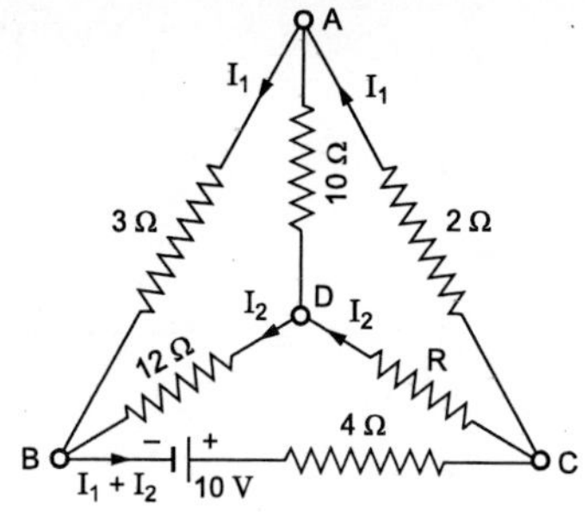

Fig. 2.26

$$3 I_1 - 12 I_2 + 0 = 0$$

$$\text{or } I_1 = 4 I_2 \quad \text{...(i)}$$

$$2 I_1 + 0 - R I_2 = 0$$

$$\text{or } R = \frac{2 I_1}{I_2} = 2 \times 4 = 8\ \Omega\ \textbf{Ans.} \qquad \because \text{ from Eq. (i) } \frac{I_1}{I_2} = 4$$

$$\text{and } 4 (I_1 + I_2) + R I_2 + 12 I_2 = 10$$

Substituting $I_1 = 4 I_2$ from Eq. (*i*) and R = 8 Ω, as determined above in the above equation we have

$$4 (4 I_2 + I_2) + 8 I_2 + 12 I_2 = 10$$

$$\text{or } I_2 = \frac{10}{40} = 0.25 \text{ A } \textbf{Ans.}$$

Example 2.8 **Applying Kirchhoff's current law, determine current I_S in the electric circuit of Fig. 2.27. Take V_0 = 16 V.** [*U.P. Technical Univ. Elec. Engineering, February 2001*]

Solution : Let the current flowing through the various branches of the circuit, be as shown in the Fig. 2.27.

Now applying Kirchhoff's current law to nodes A and B respectively we have

$$I_1 = I_2 + I_S \quad \text{...(i)}$$

$$\text{and } I_2 + I_3 = \frac{V_1}{4} \quad \text{...(ii)}$$

Also voltage of node B = V_0 = 16 V

and, therefore, $4 I_2 + V_1 = 16$ ...(*iii*)

$$I_1 = \frac{V_1}{6} \quad \text{...(iv)}$$

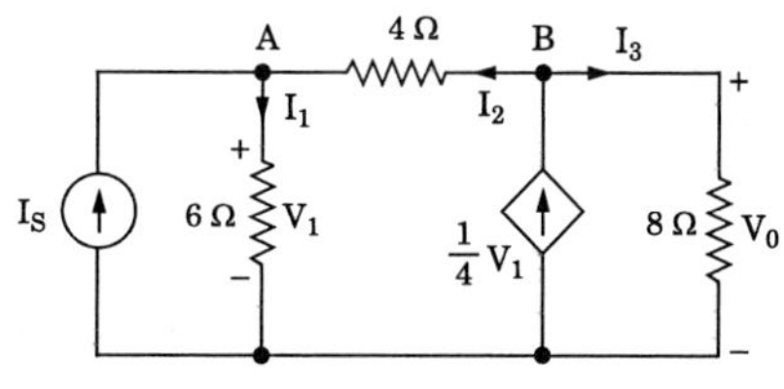

Fig. 2.27

Solving Eqs. (*i*), (*ii*), (*iii*) and (*iv*) we have

$$V_1 = 12 \text{ V}; I_1 = 2 \text{ A}; I_2 = 1 \text{ A and } I_S = I_1 - I_2 = 2 - 1 = 1 \text{ A}$$

$$\text{So } I_S = 1 \text{ A } \textbf{Ans.}$$

Example 2.9 **In Fig. 2.28 compute the voltage required between terminal *a-b* so that a voltage drop of 45 V occurs across 15 Ω resistor.** [*G.B. Technical Univ. Electrical Engineering Second Semester 2010-11*]

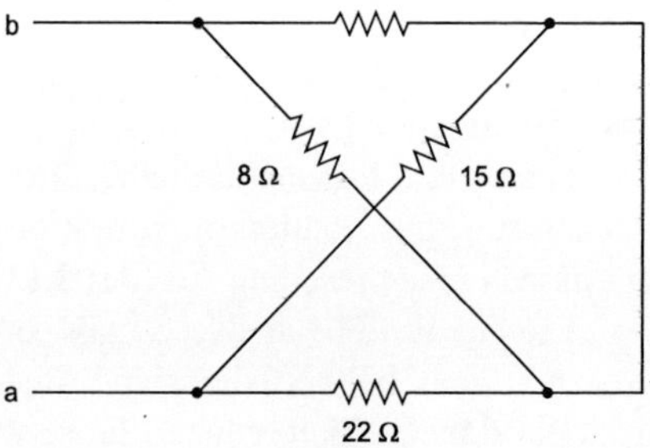

Fig. 2.28

Solution: Given circuit is redrawn as shown in Fig. 2.29. Let the currents flowing through the various branches be as shown in the circuit.

Voltage drop across 15 Ω resistor = $(I_1 - I_3) \times 15 = 45$ V

$$\text{or } I_1 - I_3 = 3 \quad \text{...(i)}$$

Applying KVL to loop ABCA, we have

$$35 I_1 - 8 I_2 = 0$$

$$\text{or } I_2 = \frac{35}{8} I_1 = 4.375 I_1 \quad \text{...(ii)}$$

Applying KVL to loop BCDB, we have

$$22(I_2 + I_3) - 15(I_1 - I_3) = 0$$

$$\text{or} \quad 22(I_2 + I_3) = 15(I_1 - I_3) = 45$$

$$\because \text{From Eq. } (i),\ 15(I_1 - I_3) = 45$$

$$\text{or} \quad I_2 + I_3 = \frac{45}{22} = 2.0454545$$

Substituting $I_2 = 4.375\, I_1$ from Eq. (*ii*) in above equation, we have

$$4.375\, I_1 + I_3 = 2.0454545 \qquad ...(iii)$$

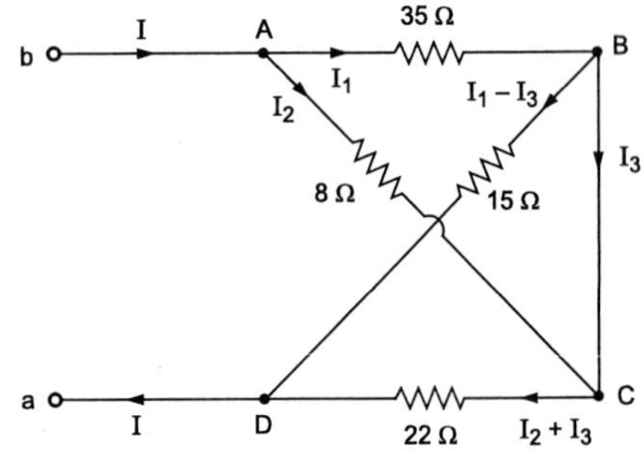

Fig. 2.29

Adding Eqs. (*i*) and (*iii*), we have

$$5.375\, I_1 = 5.0454545$$

$$\text{or} \quad I_1 = \frac{5.0454545}{5.375} = 0.9387 \text{ A}$$

Voltage required between terminal *ba* $= 35\, I_1 + 15(I_1 - I_3)$

$= 35 \times 0.9387 + 45 = 77.854$ V **Ans.**

So $V_{ab} = -V_{ba} = -77.854$ V **Ans.**

Example 2.10 **Calculate the current through the galvanometer in the following bridge [Fig. 2.30 (*a*)].**

[*U.P. Technical Univ. Electrical Engineering January 2003*]

Solution: Assume current distribution in the bridge network as shown in Fig. 2.30 (*b*).

Applying Kirchhoff's second law to meshes ABDA, BCDB and ABCA respectively we have

$$I_1 + 4\, I_3 - 2\, I_2 = 0 \qquad ...(i)$$

$$2(I_1 - I_3) - 3(I_2 + I_3) - 4\, I_3 = 0$$

$$\text{or} \quad 2\, I_1 - 3\, I_2 - 9\, I_3 = 0 \qquad ...(ii)$$

$$\text{and} \quad I_1 + 2(I_1 - I_3) = 2$$

$$\text{or} \quad 3\, I_1 - 2\, I_3 = 2 \qquad ...(iii)$$

Solving Eqs. (*i*), (*ii*) and (*iii*) we get

$$I_3 = \frac{1}{44} \text{ A};\ I_2 = \frac{17}{44} \text{ A and } I_1 = \frac{30}{44} \text{ A}$$

So current through galvanometer $= I_3 = \dfrac{1}{44}$ A **Ans.**

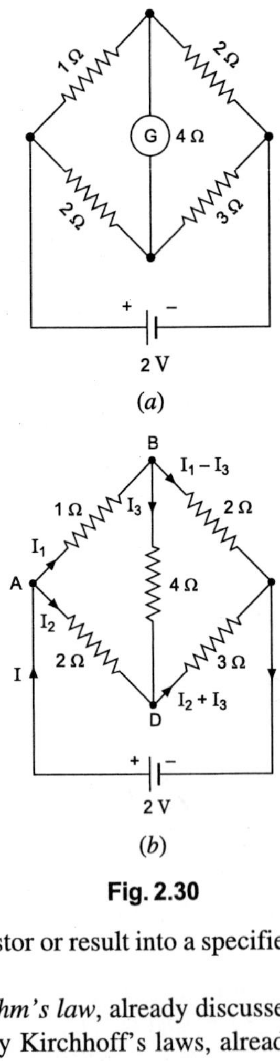

(*a*)

(*b*)

Fig. 2.30

2.12. NETWORK THEOREMS

The circuit variables, we are interested in determining, generally are (*i*) current through or (*ii*) voltage across a resistance of interest or a group of resistances. Sometimes it is required to work out the necessary source voltage or source current which will cause a specific current through or voltage across a given resistor or result into a specified power dissipation by it.

The current flowing through a circuit is governed by basic law called the *Ohm's law*, already discussed in chapter 1. Distribution of current or the voltage over a circuit is governed by Kirchhoff's laws, already discussed in the preceding Art (Art 2.11).

The circuit to be analysed may be simple or quite complex. In case of complex networks the solution procedure may be too tedious and time-consuming. Certain techniques for solution of such networks have been developed which reduces the networks to simpler form for quick solution. This may be accomplished through the use of what are called as *network theorems*. Few of these network theorems, which are relevant at this stage, will be discussed here.

2.13. SUPERPOSITION THEOREM

This theorem is applied when we are to determine the current in one particular branch of a network containing several voltage sources or current sources or both voltage sources and current sources. This scheme is to determine how much current each of the individual source contributes to the branch in question, and then add algebraically these component currents.

If there are several sources of emfs acting simultaneously in an electric circuit, then according to this theorem emf of each source acts independently of those of other sources, *i.e.* as if the other sources of emf did not exist and current in any branch or conductor of a network is equal to the algebraic sum of the currents due to each source of emf separately, all other emfs being taken equal to zero. This theorem is applicable only in linear circuits, *i.e.* circuits consisting of resistances in which Ohm's law is valid. In circuits having nonlinear resistances such as thermionic valves and metal rectifiers, this theorem is not applicable. However, superposition theorem can be applied to a circuit containing current sources and even to circuits containing both voltage sources and current sources. To remove a current source from the circuit, circuit of the source is opened leaving in place any conductance that may be in parallel with it, just as series resistance is kept in place when voltage source is removed.

Though the application of the above theorem requires a little more work than other methods such as the circulating current method but it avoids the solution of two or more simultaneous equations. After a little practice with this method, equations can be written directly from the original circuit diagram and labour in drawing extra diagrams is saved.

The superposition theorem can be stated as below :

In a linear resistive network containing two or more voltage sources, the current through any element (resistance or source) may be determined by adding together algebraically the currents produced by each source acting alone, when all other voltage sources are replaced by their internal resistances. If a voltage source has no internal resistance, the terminals to which it was connected are joined together. If there are current sources present they are removed and the network terminals to which they were connected are left open.

The procedure for applying superposition theorem is as follows:

1. Replace all but one of the sources of supply by their internal resistances. If the internal resistance of any source is very small as compared to other resistances existing in the network, the source is replaced by a short circuit. In case of a current source open the circuit leaving in place any conductance that may be in parallel with it.
2. Determine the currents in various branches using Ohm's law.
3. Repeat the process using each of the emfs turn by turn as the sole emf each time. Now the total current in any branch of the circuit is the algebraic sum of currents due to each source.

The details of the above procedure can best be understood by examining its application to the following solved examples.

Example 2.11 By means of superposition theorem find the current which flows through R_2 in the circuit of Fig. 2.31. [*G.B. Technical Univ. Electrical Engineering Odd Semester 2010-11*]

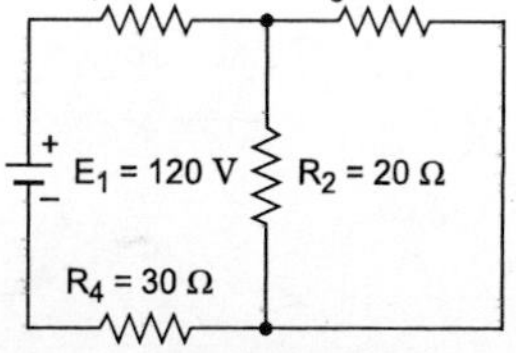

Fig. 2.31

Solution: Superposition theorem cannot be applied to given circuit because it can be applied only when the supply sources are two or more.

Equivalent resistance of the circuit, $R_{eq} = 30 + 10 + 20 \parallel 60$

$$= 40 + \frac{20 \times 60}{20 + 60} = 40 + 15 = 55\ \Omega$$

Current supplied by 120 V battery, $I = \frac{120}{55} = 2.182$ A

Current through 20 Ω resistor, $I_1 = I \times \frac{R_3}{R_2 + R_3} = 2.182 \times \frac{60}{20 + 60} = 1.636$ A by current division rule.

Example 2.12 In the circuit shown in Fig. 2.32, find the current through the 6 Ω resistor using superposition theorem. *[U.P. Technical Univ. Electrical Engineering Second Semester 2007-08]*

Solution: Replacing 60 V source by a short circuit, the circuit is reduced to a simple circuit shown in Fig. 2.33(*a*).

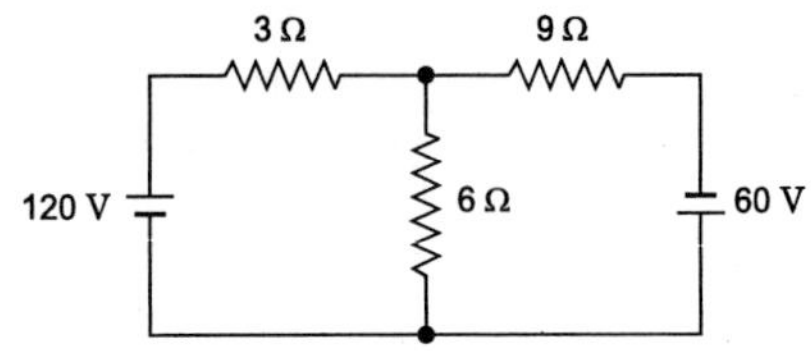

Fig. 2.32

Equivalent resistance of the circuit shown in Fig. 2.33(*a*).

$$R' = R_{AB} + R_{BC} \parallel R_{BD}$$
$$= 3\ \Omega + 6 \parallel 9 = 3 + 3.6 = 6.6\ \Omega$$

Current supplied by 120 V source

$$I_1 = \frac{120}{R'} = \frac{120}{6.6} = \frac{200}{11}\ \text{A}$$

Current in resistance, R_{BD}, $I' = I_1 \times \dfrac{R_{BC}}{R_{BC} + R_{CD}}$ By current division rule

$$= \frac{200}{11} \times \frac{9}{9+6} = \frac{120}{11}\ \text{A}$$

Similarly on replacement of 120 V source by a short circuit, the given circuit is reduced to a simple circuit shown in Fig. 2.33(*b*).

Equivalent resistance of this circuit is

$$R'' = R_{CB} + R_{BD} \parallel R_{BA} = 9 + 6 \parallel 3 = 9 + 2 = 11\ \Omega$$

Current supplied by 60 V source,

$$I_2 = \frac{60}{11} = \frac{60}{11}\ \text{A}$$

Current in resistor BD, $I'' = I_2 \times \dfrac{R_{BA}}{R_{BA} + R_{CB}} = \dfrac{60}{11} \times \dfrac{3}{3+6} = \dfrac{20}{11}\ \text{A}$

Total current, through 6 Ω resistor is equal to the algebraic sum of currents I′ and I″

$$I = I' + I'' = \frac{120}{11} + \frac{-20}{11} = \frac{100}{11}\ \text{A} \quad \textbf{Ans.}$$

∵ I″ is in opposite direction to that of I′.

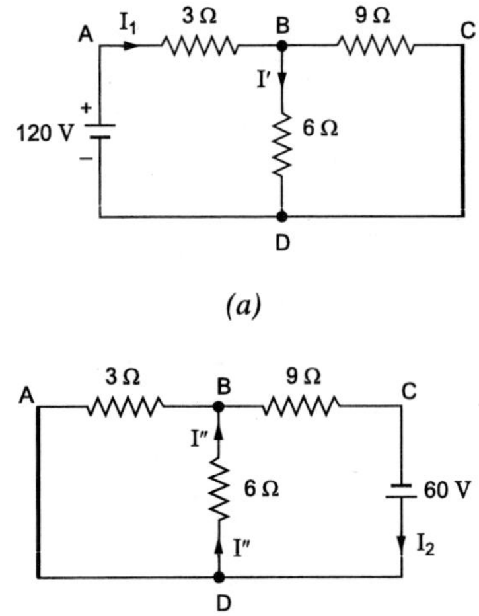

(*a*) (*b*)

Fig. 2.33

Example 2.13 Find the current flowing through 10 Ω resistance in the following circuit (Fig. 2.34) use superposition theorem. *[Electrical Engineering G.B. Technical Univ. Second Semester 2011-12; U.P. Technical Univ. Even Semester 2013-14]*

Solution: Replacing 60 V source by a short circuit, the circuit is reduced to a simple circuit shown in Fig. 2.35 (*a*).

Current flowing through 10 Ω resistance, $I_1 = 30 \times \dfrac{20}{20+10} = 20$ A upward

Replacing 30 A current source by an open circuit, the given circuit is reduced to a simple circuit shown in Fig. 2.35 (*b*).

Current flowing through 10 Ω resistance, $I_2 = \dfrac{60}{20+10} = 2$ A downward

Total current flowing through 10 Ω resistance, $I = I_1 + I_2 = 20 + (-2) = 18$ A upward **Ans.**

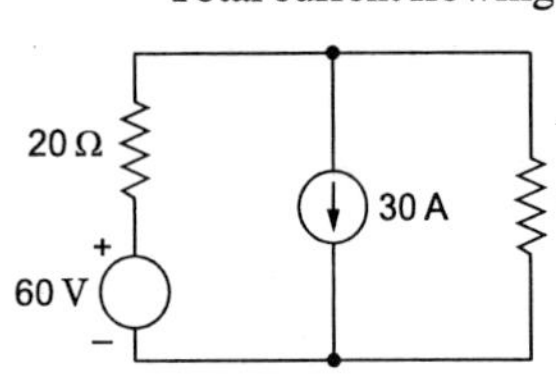

Fig. 2.34

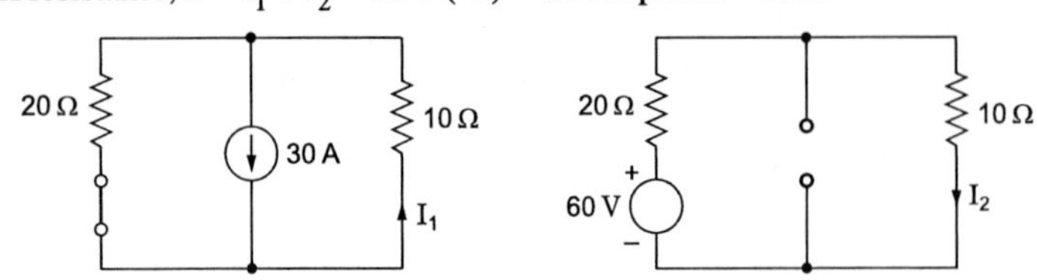

(*a*) (*b*)

Fig. 2.35

Example 2.14 **By superposition theorem, find the current in R.**

[U.P. Technical Univ. Elec. Engineering First Semester 2004-05]

Solution: Taking $E_2 = 0$, the circuit is reduced to a simple circuit shown in Fig. 2.37 (*a*) in which resistances of 0.04 Ω and 1 Ω are in parallel across points A and C and combination acts in series with resistor R_1 of 0.05 Ω. Equivalent resistance of the circuit shown in Fig. 2.37 (*a*).

$$R' = R_1 + \frac{R_2\,R}{R_2 + R} = 0.05 + \frac{0.04 \times 1}{0.04 + 1} = 0.0885\ \Omega$$

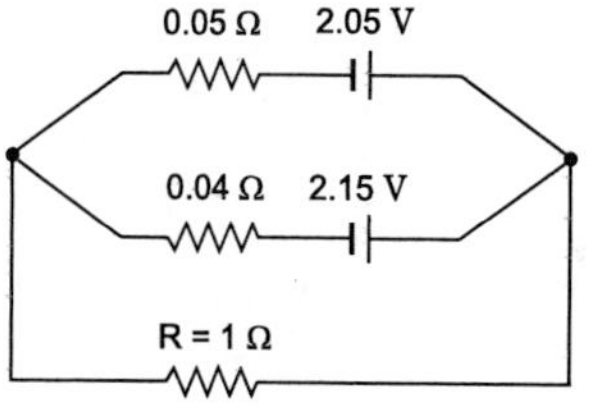

Fig. 2.36

$$\text{Current supplied by battery } E_1,\ I_1 = \frac{E_1}{R'} = \frac{2.05}{0.0885} = 23.2\text{ A}$$

$$\text{Current in 1 } \Omega \text{ resistor, } I_1' = I_1 \times \frac{R_2}{R + R_2}$$

$$= \frac{23.2 \times 0.04}{1.04} = 0.892\text{ A from C to A}$$

Taking $E_1 = 0$, the circuit is reduced to a simple circuit shown in Fig. 2.37 (*b*) in which resistors of 0.05 Ω and 1 Ω are in parallel across points A and C and the combination is in series with the resistor R_2 of 0.04 Ω. Equivalent resistance of the circuit shown in Fig. 2.37 (*b*)

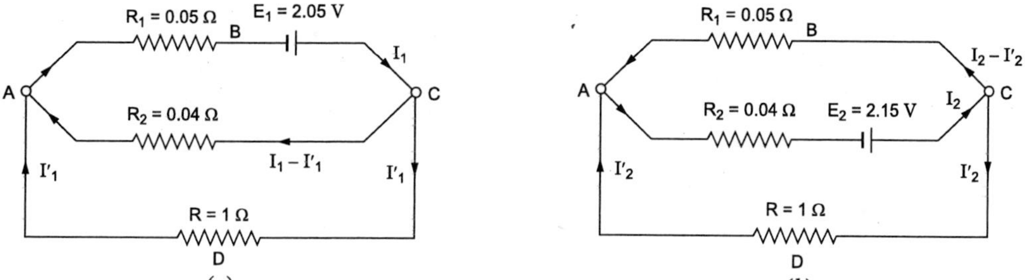

(*a*) (*b*)

Fig. 2.37

$$R'' = R_2 + \frac{R_1\,R}{R_1 + R} = 0.04 + \frac{0.05 \times 1}{0.05 + 1} = 0.0876\ \Omega$$

$$\text{Current supplied by battery } E_2,\ I_2 = \frac{E_2}{R''} = \frac{2.15}{0.0876} = 24.54\text{ A}$$

$$\text{Current in 1 } \Omega \text{ resistor, } I_2' = I_2 \times \frac{R_1}{R + R_1} = \frac{24.54 \times 0.05}{1.05} = 1.169\text{ A from C to A}$$

Total current through 1 Ω resistor when both batteries are present

$$= I_1' + I_2' = 0.892 + 1.169 = 2.06\text{ A } \textbf{Ans.}$$

Example 2.15 **Using superposition theorem find the current flowing through resistor R in Fig. 2.38 (*a*).**

[Mahamaya Technical Univ. Electrical Engineering First Semester 2011-12]

Solution: Replacing 15 V source by a short circuit, the circuit is reduced to a simple circuit as shown in Fig. 2.38(*b*). The current of 3 A is divided into resistances of 4 Ω and 4 Ω equally, as shown in Fig. 2.38(*b*). The current in resistor R will be zero, being short circuited.

Replacing 3 A current source by an open circuit, the circuit is reduced to a simple circuit as shown in Fig. 2.38(*c*). Current in resistor R will be equal to $\frac{15}{8}$ i.e., 1.875 A.

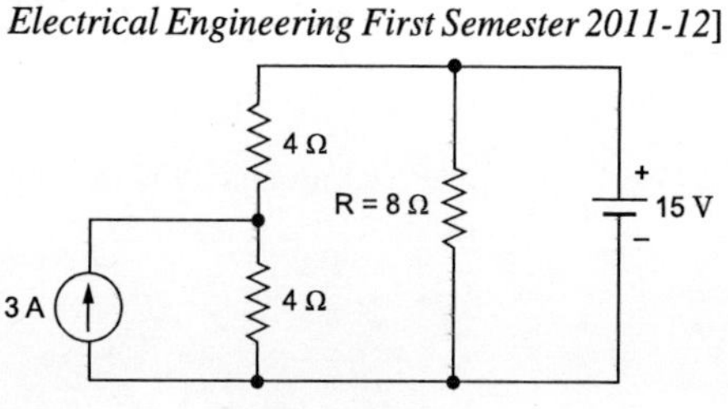

Fig. 2.38 (*a*)

Thus current through resistor R in Fig. 2.42 = 0 + 1.875 = 1.875 A **Ans.**

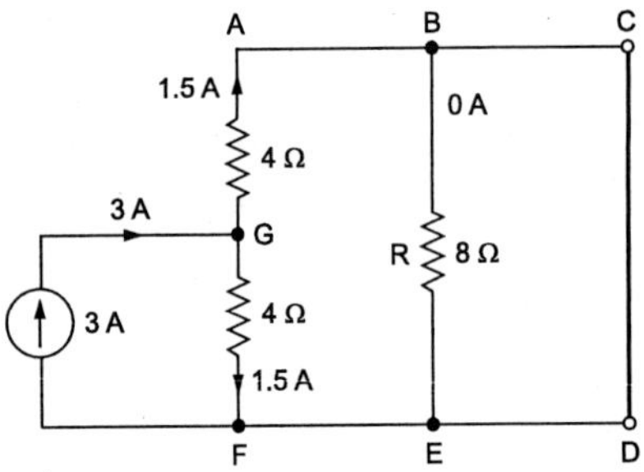

Fig. 2.38 (*b*)

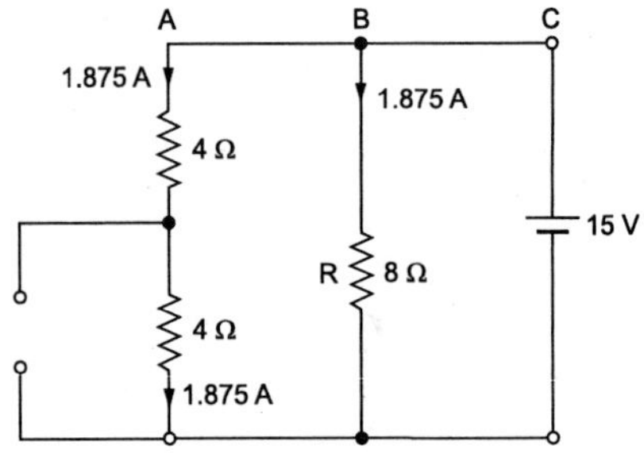

Fig. 2.38 (*c*)

Example 2.16 Using superposition theorem find the current in 20 Ω resistor of circuit shown in Fig. 2.39.

[*G.B. Technical Univ. Electrical Engineering Even Semester 2009-10*]

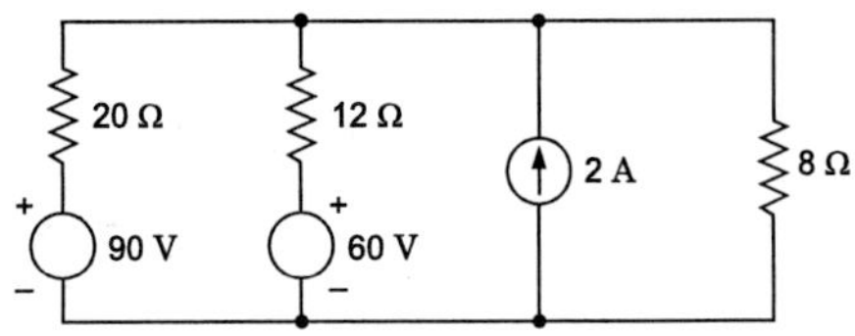

Fig. 2.39

Solution: Replacing 60 V source by a short circuit and current source by an open circuit, the circuit is reduced to a simple circuit shown in Fig. 2.40 (*a*). Equivalent resistance of the circuit shown in Fig. 2.40 (*a*).

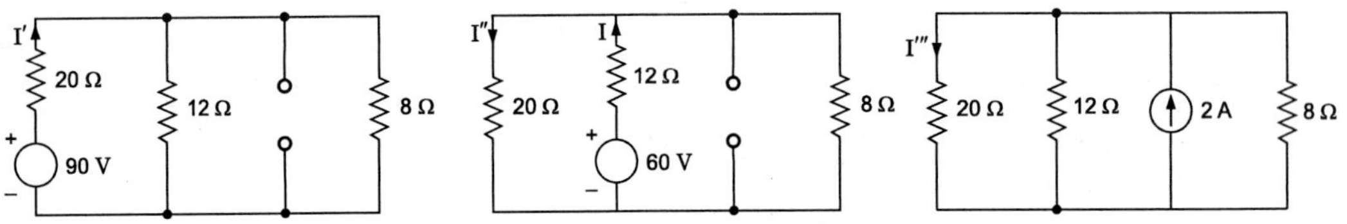

(*a*) (*b*) (*c*)

Fig. 2.40

$$R = 20 + 12 \,||\, 8 = 20 + \frac{12 \times 8}{12 + 8} = 20 + 4.8 = 24.8\,\Omega$$

Current through 20 Ω resistance, $I' =$ Current supplied by 90 V source $= \dfrac{90}{24.8} = 3.63$ A

Replacing 90 V source by a short circuit and current source by an open circuit, the circuit is reduced to a simple circuit shown in Fig. 2.40 (*b*).

Equivalent resistance of the circuit, $R = 12 + 20 \,||\, 8 = 12 + \dfrac{20 \times 8}{20 + 8} = 17.71\,\Omega.$

Current supplied by 60 V source, $I = \dfrac{60}{17.71} = 3.387$ A

Current through 20 Ω resistor, $I'' = I \times \dfrac{8}{8 + 20} = 3.387 \times \dfrac{8}{28} = 0.9677$ A   by current division rule

Replacing 90 V source and 60 V source by short circuits, the circuit is reduced to a simple circuit shown in Fig. 2.40 (*c*).

Current through 20 Ω resistor $I''' = 2 \times \dfrac{\frac{1}{20}}{\frac{1}{20} + \frac{1}{12} + \frac{1}{8}}$ By current division rule

$$= 2 \times \frac{1}{1 + \frac{20}{12} + \frac{20}{8}} = \frac{2 \times 24}{24 + 40 + 60} = \frac{48}{124} = 0.387 \text{ A}$$

Current in 20 Ω resistor, $I = I' + I'' + I''' = 3.63 + (-0.9677) + (-0.387)$ A $= 2.2753$ A **Ans.**

∵ Current I'' and I''' flow in direction opposite to that of I'.

Example 2.17 **Using superposition theorem, calculate the current in the AB in the circuit shown in Fig. 2.41.**

[*G.B. Technical Univ. Electrical Engineering Odd Semester 2012-13*]

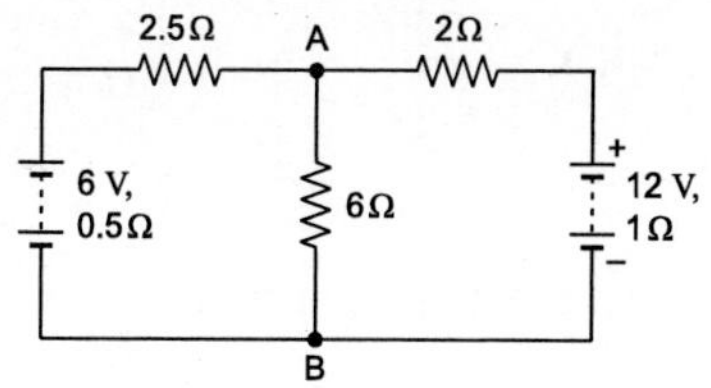

Fig. 2.41

Solution: Taking 12 V battery out of the circuit, the circuit is reduced to a simple circuit shown in Fig. 2.42(a) in which resistance of 6 Ω and (2 + 1) Ω are in parallel and the combination is in series with a resistor of 2.5 Ω and battery of 6 V and 0.5 Ω internal resistance.

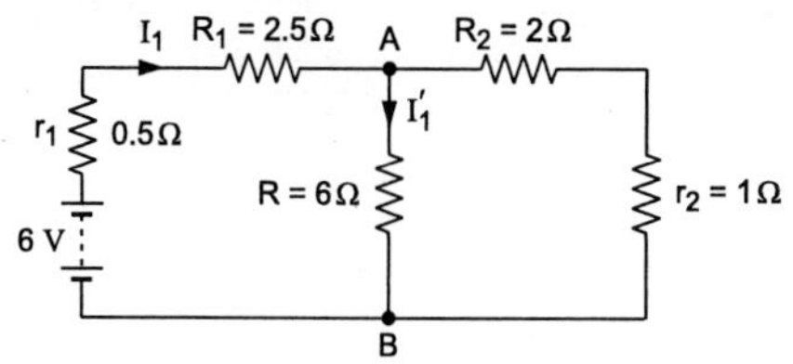

Fig. 2.42 (*a*)

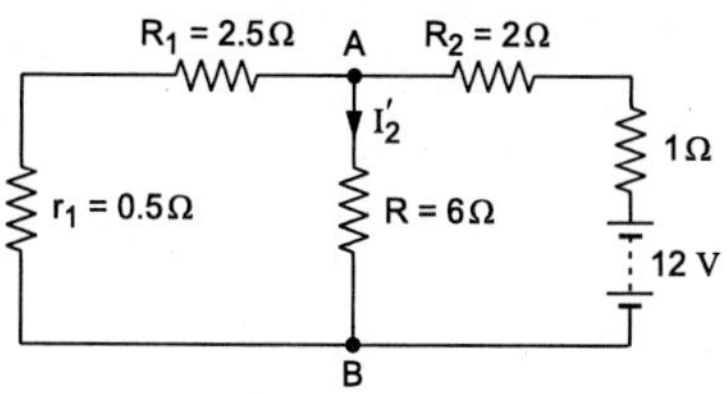

Fig. 2.42 (*b*)

Equivalent resistance of the circuit shown in Fig. 2.42 (*a*).

$$R' = (R_1 + r_1) + R \parallel (R_2 + r_2) = (2.5 + 0.5) + 6 \parallel (2 + 1) = 3 + 6 \parallel 3 = 3 + 2 = 5\ \Omega$$

Current supplied by 6 V battery

$$I_1 = \frac{E_1}{R'} = \frac{6}{5} = 1.2\text{ A}$$

$$\text{Current in branch AB, } I_1' = I_1 \times \frac{R_2 + r_2}{R_2 + r_2 + R} = 1.2 \times \frac{2+1}{2+1+6} = 0.4\text{ A from node A to node B}$$

Taking 6 V battery out of circuit, the circuit is reduced to a simple circuit shown in Fig. 2.42 (*b*) in which resistances of 6 Ω and (0.5 + 2.5) Ω are in parallel and the combination is in series with resistor of 2.5 Ω and battery of 12 V and 1 Ω internal resistance.

Equivalent resistance of the circuit shown in Fig. 2.42 (*b*).

$$R'' = (R_2 + r_2) + R \parallel (R_1 + r_1) = (1 + 2) + 6 \parallel (2.5 + 0.5) = 3 + 6 \parallel 3 = 5\ \Omega$$

$$\text{Current supplied by 12 V battery, } I_2 = \frac{E_2}{R''} = \frac{12}{5} = 2.4\text{ A}$$

$$\text{Current in branch AB, } I_2' = I_2 \times \frac{R_1 + r_1}{R_1 + r_1 + R} = 2.4 \times \frac{2.5 + 0.5}{2.5 + 0.5 + 6} = 0.8\text{ A} \quad \text{from node A to node B}$$

Total current through branch AB $= I_2' + I_2' = 0.4 + 0.8 = 1.2$ A **Ans.**

Example 2.18 **Determine current through 8 Ω resistor in the following network (Fig. 2.43) using superposition theorem:**

[*U.P. Technical Univ. Electrical Engineering, Second Semester 2002-03*]

Solution: In Fig. 2.44 (*a*), the voltage source has been replaced by a short. Using the current division rule we have

$$\text{Current through 8 }\Omega\text{ resistor, } I_1 = 2 \times \frac{2}{2+8} = 0.4\text{ A in the direction shown in Fig. 2.44 } (a)$$

In Fig. 2.44 (*b*) only voltage source has been considered.

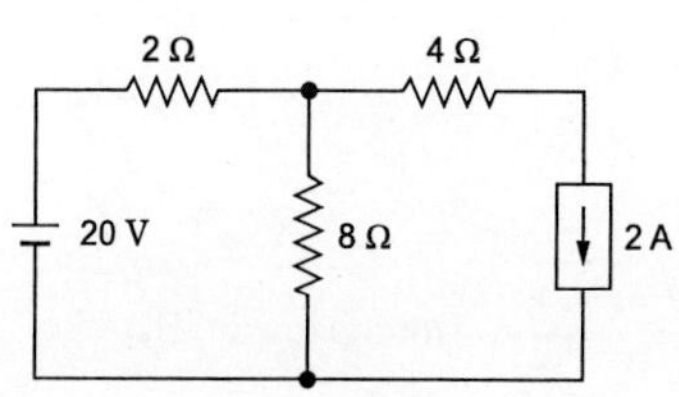

Fig. 2.43

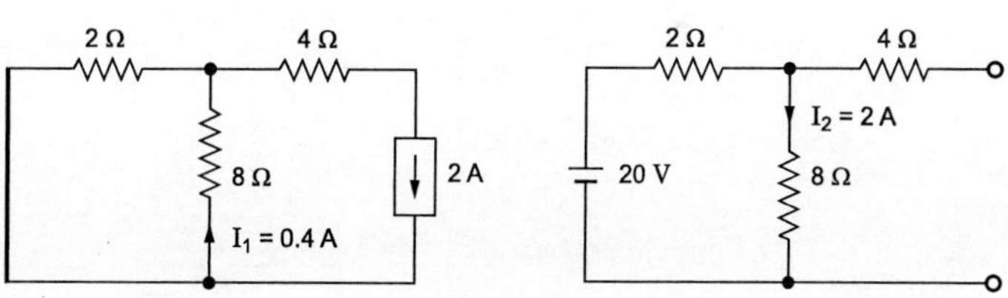

(*a*) (*b*)

Fig. 2.44

Using Ohm's law, $I_2 = \dfrac{20}{2+8} = 2$ A in the direction shown in Fig. 2.44 (*b*).

The resultant current in 8 Ω resistor, $I = I_1 + I_2 = (-0.4 + 2)$ A = 1.6 A downward **Ans.**

Example 2.19 **For the circuit of Fig. 2.45, find I using Superposition Theorem.**

[U.P. Technical Univ. Electrical Engineering, September 2001]

Solution: Replacing 64 V source by a short circuit, the circuit is reduced to a circuit shown in Fig. 2.46 (*a*).

From series-parallel circuit law equivalent resistance of the circuit shown in Fig. 2.46 (*a*) is given as

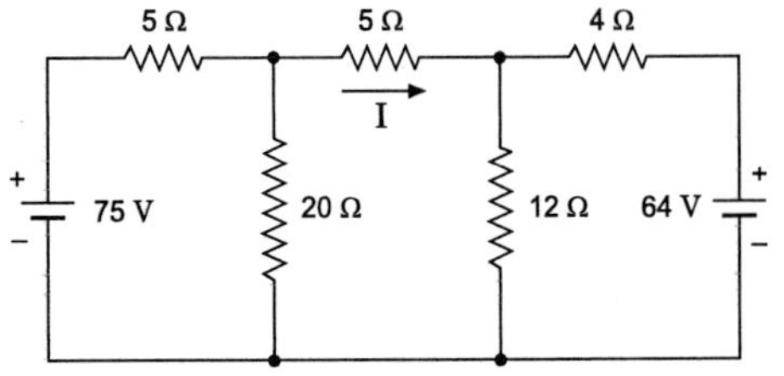

Fig. 2.45

$$R_{eq} = R_{AB} + R_{BG} \,||\, [R_{BC} + R_{CF} \,||\, R_{DE}]$$
$$= 5 + 20 \,||\, [5 + 12 \,||\, 4]$$
$$= 5 + 20 \,||\left[5 + \frac{12 \times 4}{12 + 4}\right]$$
$$= 5 + 20 \,||\, (5 + 3)$$
$$= 5 + 20 \,||\, 8 = 5 + \frac{20 \times 8}{20 + 8} = \frac{75}{7}\ \Omega$$

$$\text{Current through branch AB} = \frac{V}{R_{eq}} = \frac{75}{75/7} = 7\ \text{A}$$

$$\text{Current through branch BC, } I' = I_{AB} \times \frac{R_{BG}}{R_{BG} + R_{BF}} = 7 \times \frac{20}{20 + 8} = 5\ \text{A from B toward C}$$

Replacing 75 V source by a short circuit, the circuit is reduced to a circuit shown in Fig. 2.46 (*b*).

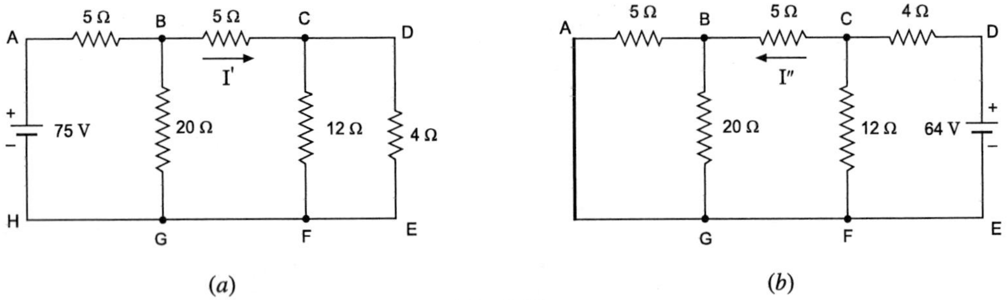

Fig. 2.46

$$\text{Equivalent resistance of the circuit, } R'_{eq} = R_{DC} + R_{CF} \,||\, [R_{CB} + R_{BG} \,||\, R_{BA}]$$
$$= 4 + 12 \,||\, [5 + 20 \,||\, 5] = 4 + 12 \,||\left[5 + \frac{20 \times 5}{20 + 5}\right]$$
$$= 4 + 12 \,||\, (5 + 4) = 4 + \frac{12 \times 9}{12 + 9} = \frac{64}{7}\ \Omega$$

$$\text{Current through branch DC, } I_{DC} = \frac{64}{64/7} = 7\ \text{A}$$

$$\text{Current in branch CB, } I'' = I_{DC} \times \frac{R_{CF}}{R_{CF} + R_{BAG}} = 7 \times \frac{12}{12 + 9} = 4\ \text{A from C toward B}$$

Hence current, I = Current in branch BC = $I' - I'' = 5 - 4 = 1$ A **Ans.**

2.14. MAXWELL CIRCULATING CURRENT THEOREM

If a network with several sources has more than two nodes the current in it may be determined by this theorem. This is one of the most universal methods for solving networks.

In a number of cases, a network may be considered as consisting of a set of adjoining loops (two in Fig. 2.3), each of which forms a polygon made up of several branches of the network (without any diagonals). Some branches of the network are common to two adjacent loops, while others form an external circuit where each branch occurs in one loop only.

This theorem involves representing a current that is assumed to circulate around a closed loop by a curved arrow and labelling the arrow with its identifying current symbol I with a subscript. By this theorem the current flowing through the branch common to two meshes will be equal to the algebraic sum of the two loop currents flowing through it. The direction of any loop current may be taken either as clockwise or counter-clockwise but for systematic solution the directions of all loop currents are assumed to be the same (say clockwise). Then Kirchhoff's second law is applied to each mesh and algebraic equations are obtained. The total number of independent equations is equal to the number of meshes (*i.e.* there are fewer equations than in a purely Kirchhoffian solution). Therefore, they can be solved as simultaneous equations to give the circulating currents and then the branch currents. Thus, this method eliminates a great deal of tedious calculation work involved in the branch current method, discussed in Art 2.11.

Application of Maxwell circulating current theorem will be more clear from the following illustrations.

Example 2.20 **Find the currents in all the resistive branches of the circuit shown in Fig. 2.47 (*a*) by KVL.**

[U.P. Technical Univ. Electrical Engineering Second Semester 2007-08]

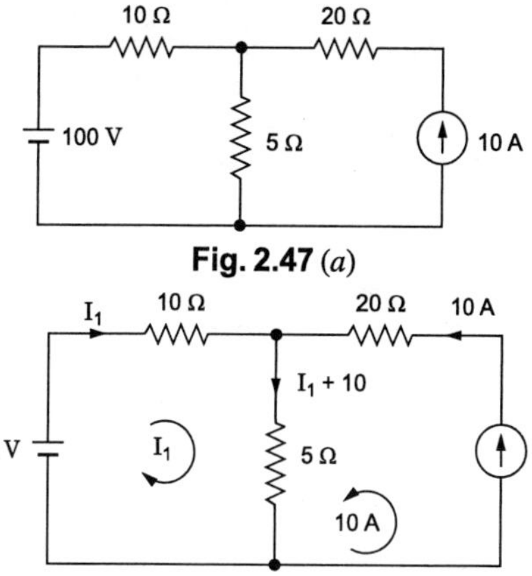

Fig. 2.47 (*a*)

Solution: Applying KVL to loop 1 we have

$$10\,I_1 + 5\,(I_1 + 10) = 100$$

$$\text{or} \quad 15\,I_1 + 50 = 100$$

$$\text{or} \quad I_1 = \frac{50}{15} = 3.33\text{ A}$$

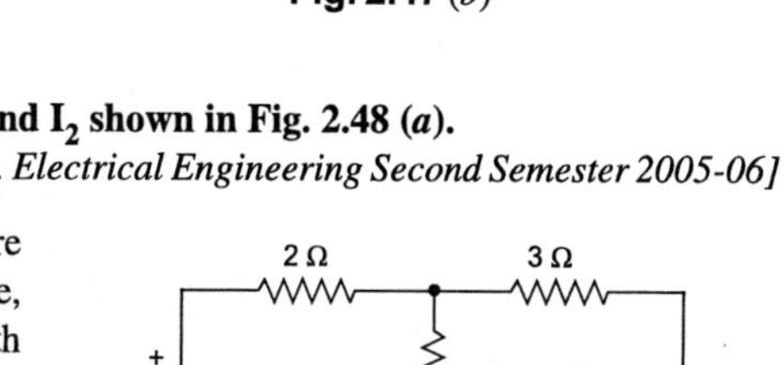

Fig. 2.47 (*b*)

Thus current in 10 Ω resistance,

$$I_1 = 3.33\text{ A} \quad \textbf{Ans.}$$

$$\text{Current in 5 Ω resistance} = I_1 + 10$$

$$= 3.33 + 10 = 13.33\text{ A} \quad \textbf{Ans.}$$

$$\text{Current in 20 Ω resistance} = 10\text{ A} \quad \textbf{Ans.}$$

Example 2.21 **Using loop current method find the current I_1 and I_2 shown in Fig. 2.48 (*a*).**

[U.P. Technical Univ. Electrical Engineering Second Semester 2005-06]

Solution: The network is redrawn, as illustrated in Fig. 2.48 (*b*). There are two independent loops. The loop currents have been taken clockwise, as marked in the given circuit. The individual branch currents along with their directions of flow are also shown in the circuit diagram. Applying Kirchhoff's voltage law to meshes I and II respectively we have

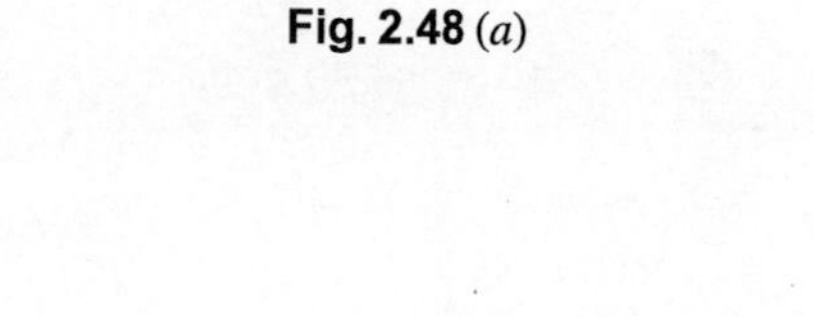

Fig. 2.48 (*a*)

$$2\,I_1 + 6\,(I_1 - I_2) + 4\,I_1 = 10 - 6$$

$$\text{or} \quad 12\,I_1 - 6\,I_2 = 4$$

$$\text{or} \quad 6\,I_1 - 3\,I_2 = 2 \qquad ...(i)$$

$$\text{and} \quad 3\,I_2 + 5\,I_2 - 6\,(I_1 - I_2) = 6 - 2$$

$$\text{or} \quad -6\,I_1 + 14\,I_2 = 4 \qquad ...(ii)$$

Adding expressions (*i*) and (*ii*) we have

$$11\,I_2 = 6$$

$$\text{or}\quad I_2 = \frac{6}{11}\ \text{A}\ \textbf{Ans.}$$

Substituting $I_2 = \frac{6}{11}$ in equation (*i*) we have

$$I_1 = \frac{2+3I_2}{6} = \frac{2+3\times(6/11)}{6} = \frac{40}{66} = \frac{20}{33}\ \text{A}\ \textbf{Ans.}$$

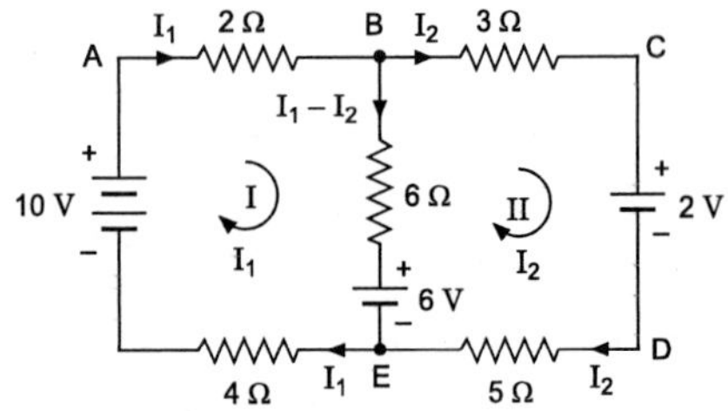

Fig. 2.48 (*b*)

Example 2.22 **Using mesh analysis, calculate the currents I_1 and I_2 in Fig. 2.49.**

[*G.B. Technical Univ. Electrical Engineering First Semester 2011-12*]

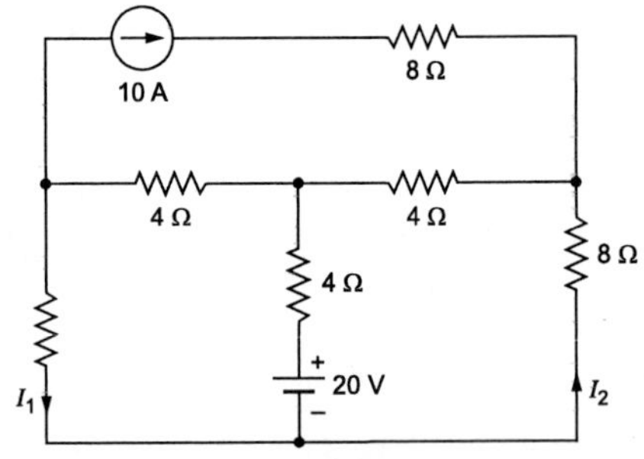

Fig. 2.49

Solution: The network shown in Fig. 2.49 is redrawn, as shown in Fig. 2.50. These are three independents loops. The loop current have been taken as shown in Fig. 2.50 satisfying the currents marked in Fig. 2.49. The individual branch currents along with their directions of flow are also shown in the circuit diagram. The resistor carrying current I_1, whose value is not given, may be assumed of 8 Ω.

Applying KVL to meshes I and II respectively, we have

$$8I_1 + 4(I_1 - 10) + 4(I_1 + I_2) = -20$$

$$\text{or } 16I_1 + 4I_2 = 20 \qquad ...(i)$$

$$\text{and } 8I_2 + 4(I_2 + 10) + 4(I_1 + I_2) = -20$$

$$\text{or } 4I_1 + 16I_2 = -60$$

$$\text{or } I_1 + 4I_2 = -15 \qquad ...(ii)$$

Subtracting Eq. (*ii*) from Eq. (*i*), we have

$$15I_1 = 35$$

$$\text{or } I_1 = \frac{7}{3}\ \text{A}\ \textbf{Ans.}$$

Substituting $I_1 = \frac{7}{3}$ A in Eq. (*ii*), we have

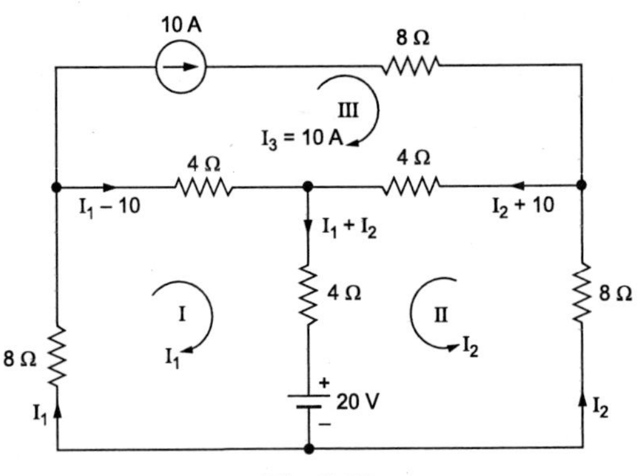

Fig. 2.50

$$I_2 = \frac{-15 - I_1}{4} = \frac{-15-\frac{7}{3}}{4} = \frac{-13}{3}\ \text{A}\ \textbf{Ans.}$$

Example 2.23 **Find mesh currents i_1 and i_2 in the electric circuit of Fig. 2.51.**

[*U.P. Technical Univ. Elec. Engg. February 2001*]

Solution: Applying Kirchhoff's voltage law to the outer mesh we have

$$i_1 \times 1 + 2 \times i_2 = 4 - 3$$

$$\text{or } i_1 + 2\,i_2 = 1 \qquad ...(i)$$

Applying Kirchhoff's voltage law to the mesh I we have

$$i_1 + (i_1 - i_2) \times 1 + 3\,i_2 = 4$$

$$\text{or } i_1 + i_2 = 2 \qquad ...(ii)$$

Solving Eqs. (*i*) and (*ii*) we have

$$i_1 = 3\ \text{A}\ \textbf{Ans.}$$

$$i_2 = -1\ \text{A}\ \textbf{Ans.}$$

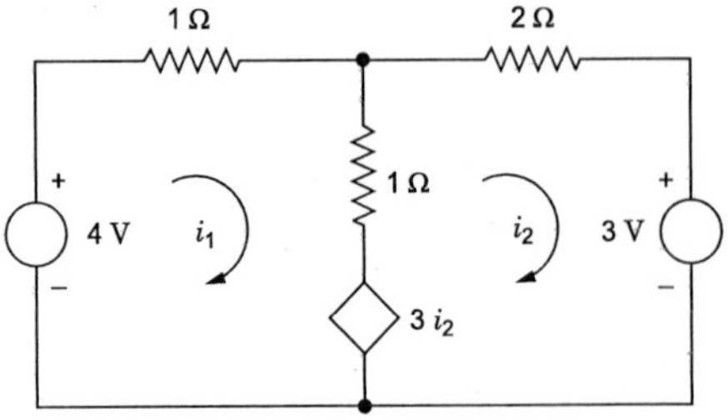

Fig. 2.51

Example 2.24 **Using mesh current method, determine current I_x in the following circuit (Fig. 2.52).**

[U.P. Technical Univ. Electrical Engineering First Semester 2005-06]

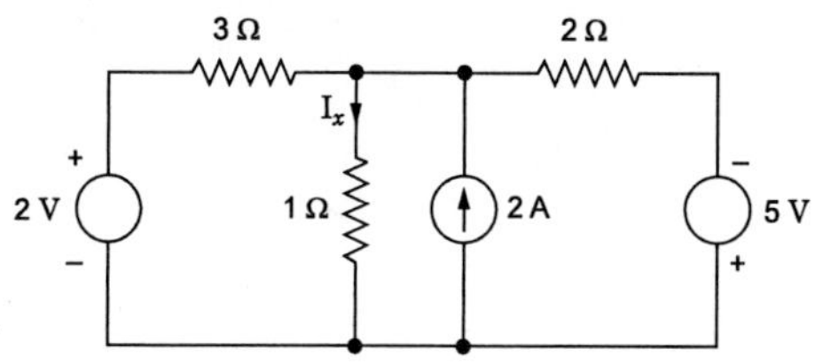

Fig. 2.52

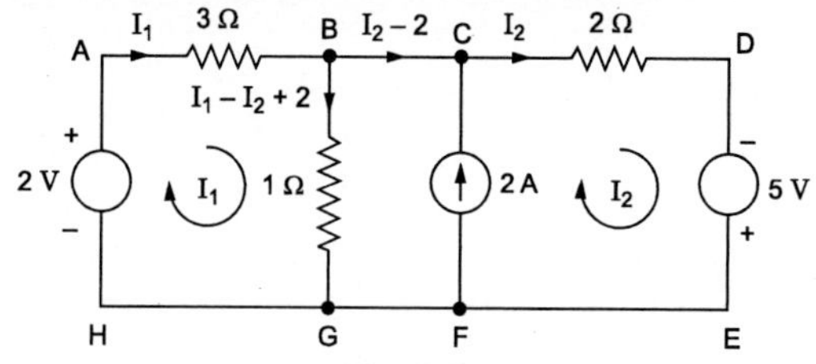

Fig. 2.53

Solution: Current distribution in different branches of the circuit satisfying Kirchhoff's first law is shown in the Fig. 2.53. Applying Kirchhoff's second law to meshes I and II respectively we have

$$3I_1 + (I_1 - I_2 + 2) = 2 \quad \text{...(i)}$$

$$\text{and } 2I_2 - (I_1 - I_2 + 2) = 5 \quad \text{...(ii)}$$

$$\text{or } 4I_1 - I_2 = 0 \text{ or } I_2 = 4I_1 \quad \text{...(iii)}$$

$$\text{and } 3I_2 - I_1 = 7 \quad \text{...(iv)}$$

Solving Eqs. (*iii*) and (*iv*) we have $I_1 = \frac{7}{11}$ A and $I_2 = \frac{28}{11}$ A

$$\text{Current } I_x = \text{Current through branch BG} = I_1 - I_2 + 2 = \frac{7}{11} - \frac{28}{11} + 2 = \frac{1}{11} \text{ A} \quad \textbf{Ans.}$$

Example 2.25 **Applying mesh analysis, obtain the current through 5 Ω resistance in the following circuit (Fig. 2.54).**

[Electrical Engineering G.B. Technical Univ. Second Semester 2011-12; U.P. Technical Univ. Even Semester 2013-14]

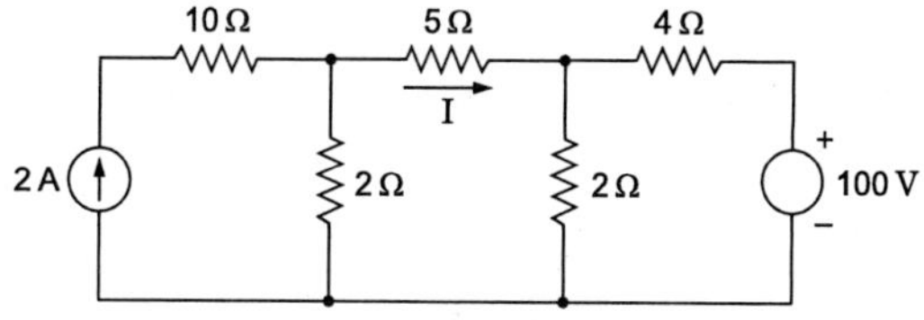

Fig. 2.54

Solution: The circuit is redrawn with branch currents marked with the flow of direction, as illustrated in Fig. 2.55.

From circuit diagram shown in Fig. 2.55, we have

$$I_1 = 2 \text{ A}$$

$$I_2 = I \text{ A}$$

Applying KVL to meshes BCFGB and CDEFC respectively we have

$$5 I_2 + 2 (I - I_3) - 2 (I_1 - I_2) = 0$$

$$\text{or } 9 I - 2 I_3 = 2 \times 2 = 4 \quad \text{...(i)}$$

$$\because I_1 = 2 \text{ A and } I_2 = I$$

$$\text{and } 4 I_3 - 2 (I_2 - I_3) = -100$$

$$\text{or } 2 I - 6 I_3 = 100 \quad \text{...(ii)}$$

Solving Eqs (*i*) and (*ii*) we have

$$I = \frac{-88}{25} = -3.52 \text{ A} \quad \textbf{Ans.}$$

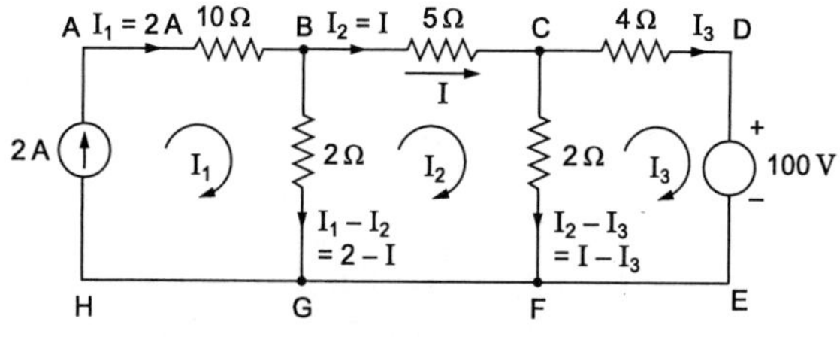

Fig. 2.55

Example 2.26 **Find voltage V_1 across 6 Ω resistance in the following circuit (Fig. 2.56) using loop analysis method.**

[Electrical Engineering G.B. Technical Univ. Odd Semester 2009-10; U.P. Technical Univ. Odd Semester 2013-14]

Solution: Current distribution in different branches of the circuit satisfying KCL is shown in Fig. 2.57.

Applying KVL to meshes I and II respectively we have

$$6 I_1 + 12 (I_1 - I_2) = 30 - 12$$

$$\text{or } 3 I_1 - 2 I_2 = 3 \quad \text{...(i)}$$

$$\text{and } 2 I_2 + 6 I_2 - 12 (I_1 - I_2) = 12$$

$$\text{or } 5 I_2 - 3 I_1 = 3 \quad \text{...(ii)}$$

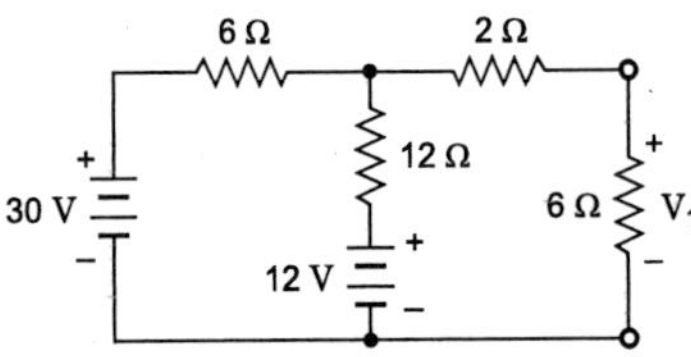

Fig. 2.56

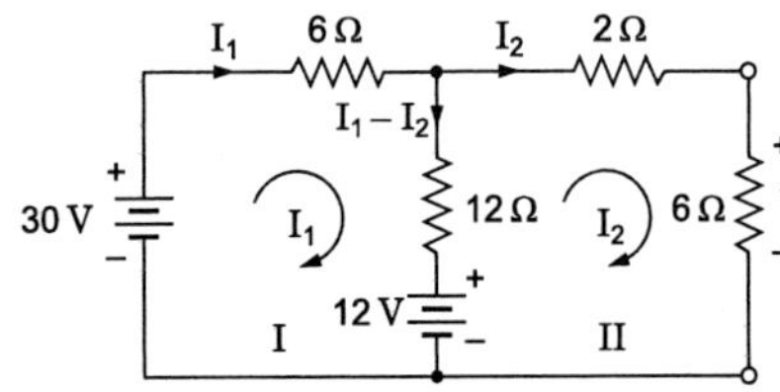

Fig. 2.57

Solving Eqs (i) and (ii) we have

$$I_2 = 2 \text{ A}$$

$$\text{and } V_1 = 6\, I_2 = 6 \times 2 = 12 \text{ V} \quad \textbf{Ans.}$$

2.15. NODE–VOLTAGE THEOREM OR NODAL ANALYSIS

The direct use of Kirchhoff's laws for determining the currents in complicated networks necessitates the simultaneous solution of a considerable number of equations, making the computations extremely time-consuming. However, there are a number of methods (such as loop-current method, node-voltage method etc.), based on the same Kirchhoff's laws, that obviate the solving of a set of equations or reduce the number of equations and, therefore, significantly cut the computation work.

For application of node voltage theorem one of the node is taken as *reference* or *zero potential* or *datum node* and the potential difference between each of the other nodes and the reference node is expressed in terms of an unknown voltage (symbolized as V_1, V_2 or V_A, V_B or V_x, V_y etc.) and at every node Kirchhoff's first (or current) law is applied assuming the possible directions of branch currents. This assumption does not change the statement of problem, since the branch currents are determined by the potential difference between respective nodes and not by absolute values of node potentials. Like Maxwell's circulating current theorem, node-voltage theorem reduces the number of equations to be solved to determine the unknown quantities. If there are *n* number of nodes, there shall be $(n - 1)$ number of nodal equations in terms of $(n - 1)$ number of unknown variables of nodal voltages. By solving these equations, nodal voltages are known to compute the branch currents.

When the number of nodes minus one is less than the number of independent meshes in the network, it is, in fact more advantageous. Moreover, it is particularly suited for networks having many parallel circuits with common ground connected node such as in electronic circuits.

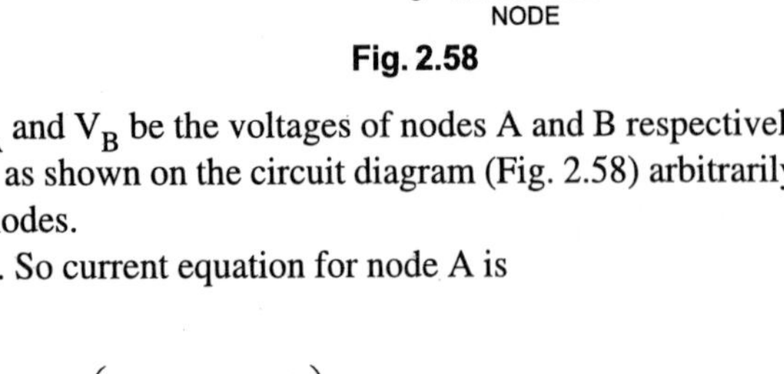

Fig. 2.58

Illustration. Consider, for example, a two node network, as illustrated in Fig. 2.58.

Node C has been taken as reference node. Let V_A and V_B be the voltages of nodes A and B respectively with respect to node C. Let the current distribution be as shown on the circuit diagram (Fig. 2.58) arbitrarily. Now let us get independent equations for these two nodes.

Node A is the junction of resistors R_1, R_2 and R_4. So current equation for node A is

$$I_1 = I_2 + I_4$$

$$\text{or} \quad \frac{E_1 - V_A}{R_1} = \frac{V_A - V_B}{R_2} + \frac{V_A}{R_4} \quad \text{or} \quad V_A\left(\frac{1}{R_1} + \frac{1}{R_2} + \frac{1}{R_4}\right) - \frac{V_B}{R_2} = \frac{E_1}{R_1} \qquad ...(2.23)$$

Node B is the junction of resistors R_2, R_3 and R_5. So current equation for node B is

$$I_5 = I_2 + I_3$$

$$\text{or} \quad \frac{V_B}{R_5} = \frac{V_A - V_B}{R_2} + \frac{E_2 - V_B}{R_3} \quad \text{or} \quad \frac{-V_A}{R_2} + V_B\left[\frac{1}{R_2} + \frac{1}{R_3} + \frac{1}{R_5}\right] = \frac{E_2}{R_3} \qquad ...(2.24)$$

The Eqs. (2.23) and (2.24) can now be solved to get the values of V_A and V_B and then the values of currents I_1, I_2, I_3, I_4 and I_5 can be computed easily.

Example 2.27 **Find the currents in all the resistive branches of the circuit shown in Fig. 2.47 (*a*) by KCL.**

[U.P. Technical Univ. Electrical Engineering Second Semester 2007-08]

Solution: Let the given circuit be redrawn as with node E as the reference node and arbitrarily assumed distribution of current as shown in Fig. 2.59.

Applying Kirchhoff's current law (KCL) to nodes C and B we have

$$I_3 = 10 \text{ A} \qquad ...(i)$$

$$\text{and} \quad I_1 + I_3 = I_2$$

$$\text{or} \quad \frac{V_A - V_B}{10} + 10 = \frac{V_B}{5}$$

$$\text{or} \quad \frac{100 - V_B}{10} + 10 = \frac{V_B}{5}$$

$$\text{or} \quad 100 - V_B + 100 = 2 V_B$$

$$\text{or} \quad 3 V_B = 200$$

$$\text{or} \quad V_B = \frac{200}{3} \text{ V}$$

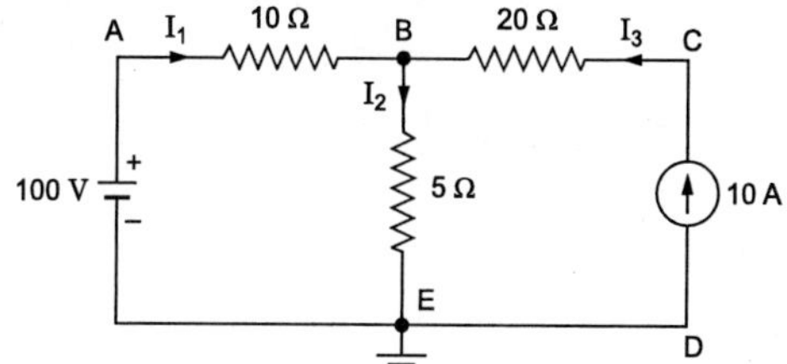

Fig. 2.59

$$\text{Thus current in 10 } \Omega \text{ resistor} = \frac{V_A - V_B}{10} = \frac{100 - \frac{200}{3}}{10} = \frac{10}{3} \text{ A} \quad \textbf{Ans.}$$

$$\text{Current in 5 } \Omega \text{ resistor} = \frac{V_B}{5} = \frac{200}{3 \times 5} = \frac{40}{3} \text{ A} \quad \textbf{Ans.}$$

$$\text{Current in 20 } \Omega \text{ resistor} = 10 \text{ A} \quad \textbf{Ans.}$$

Example 2.28 **Out of Mesh and Nodal analysis, which method will require least number of equations to solve the network in Fig. 2.60. Write the equations in matrix form following the method requiring least effort.**

[U.P. Technical Univ. Electrical Engineering First Semester 2004-05]

Solution : In the given circuit, there are three meshes and so three equations will be formed to solve the given network whereas in solving the circuit by nodal analysis only two equations will have to be solved.

Applying KCL to nodes A and B we have

$$\frac{V_A - 20}{5} + \frac{V_A - 5}{3} + \frac{V_A + 5 - V_B}{4} = 0 \qquad ...(i)$$

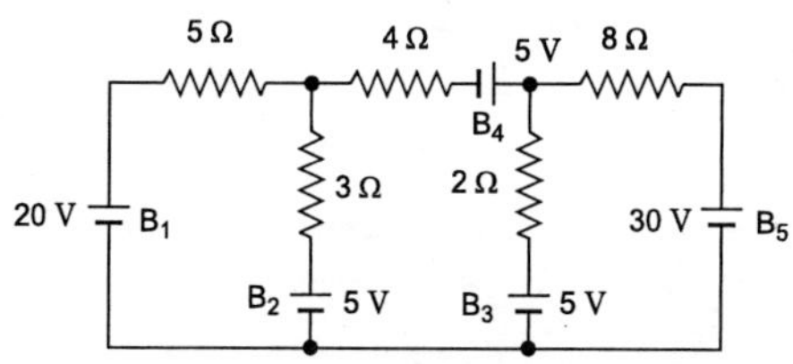

Fig. 2.60

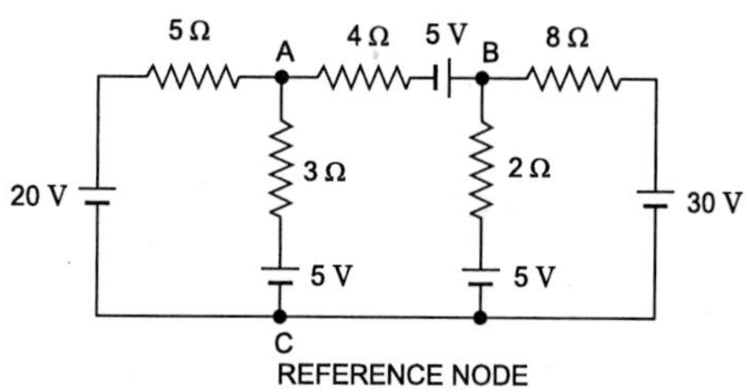

Fig. 2.61

$$\frac{V_B - 5 - V_A}{4} + \frac{V_B - 30}{8} + \frac{V_B - 5}{2} = 0 \qquad ...(ii)$$

$$\text{or } 47 V_A - 15 V_B = 265 \qquad ...(iii)$$

$$2 V_A - 7 V_B = -60 \qquad ...(iv)$$

The above equations in matrix form are written as

$$\begin{bmatrix} 47 & -15 \\ 2 & -7 \end{bmatrix} \begin{bmatrix} V_A \\ V_B \end{bmatrix} = \begin{bmatrix} 265 \\ -60 \end{bmatrix} \quad \textbf{Ans.}$$

Example 2.29 **Determine the voltage V_1 and V_2 in Fig. 2.62 (*a*).**

[*Mahamaya Technical Univ. Electrical Engineering Second Semester 2011-12*]

Solution: Let the circuit be redrawn with terminal O as the reference node and arbitrarily assumed distribution of currents, as shown in Fig. 2.62 (*b*).

Assume I = 5 A (missing data)

Applying KCL to nodes A and B respectively we have

$$I_1 + I_2 = I$$

$$\text{or} \quad \frac{V_1}{1} + \frac{V_1 - V_2}{2.5} = 5 \qquad \ldots(i)$$

$$\text{and } I_3 - I_2 = 4$$

$$\text{or} \quad \frac{V_2}{0.5} - \frac{V_1 - V_2}{2.5} = 4 \qquad \ldots(ii)$$

$$\text{or } 7\,V_1 - 2\,V_2 = 25 \qquad \ldots(iii)$$

$$\text{and } -V_1 + 6\,V_2 = 10 \qquad \ldots(iv)$$

Solving Eqs. (iii) and (iv) we have

$$V_1 = 4.25 \text{ V} \quad \textbf{Ans.}$$

$$\text{and } V_2 = 2.375 \text{ V} \quad \textbf{Ans.}$$

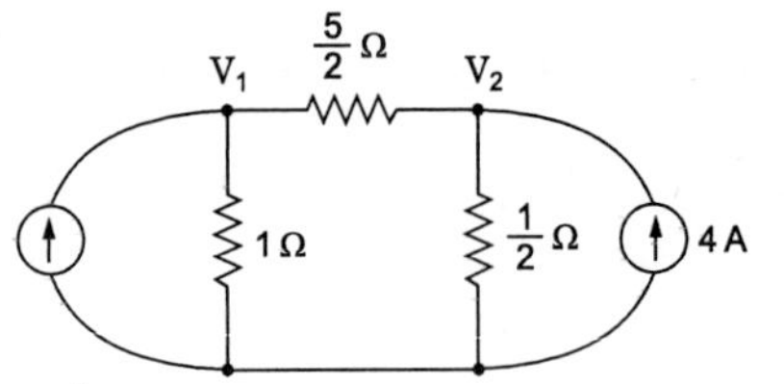

(*a*)

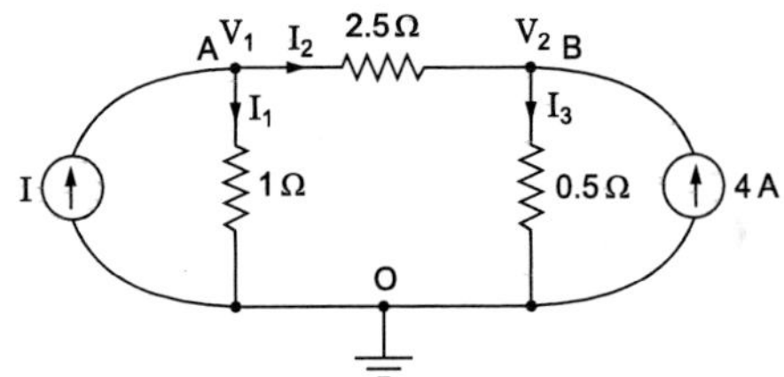

(*b*)

Fig. 2.62

Example 2.30 **Determine the currents i_1 and i_2 in the following network shown in Fig. 2.63 (*a*).**

[*Mahamaya Technical Univ. Electrical Engineering Second Semester 2010-11*]

Solution: Let the given circuit be redrawn with terminal G as the reference node and arbitrarily assumed distribution of currents as shown in Fig. 2.63(*b*).

Applying Kirchhoff's current law to node A, we have

$$i = i_1 + i_3$$

$$\text{or} \quad \frac{100 - V_A}{10} = \frac{V_A}{5} + \frac{V_A + 50 - V_B}{2}$$

$$\text{or} \quad 100 - V_A = 2\,V_A + 5\,V_A + 250 - 5\,V_B$$

$$\text{or} \quad 8\,V_A - 5\,V_B = -150 \qquad \ldots(i)$$

Applying KCL to node B, we have

$$i_3 = i_4 + i_2$$

$$\text{or} \quad \frac{V_A + 50 - V_B}{2} = \frac{V_B}{2} + \frac{V_B + 50}{4}$$

$$\text{or} \quad 2\,V_A + 100 - 2\,V_B = 2\,V_B + V_B + 50$$

$$\text{or} \quad 2\,V_A - 5\,V_B = -50 \qquad \ldots(ii)$$

Solving Eq. (*i*) and (*ii*), we have

$$V_A = \frac{-50}{3} \text{ V} \quad \text{and} \quad V_B = \frac{10}{3} \text{ V}$$

$$\text{So} \quad i_1 = \frac{V_A}{5} = \frac{-50/3}{5} = \frac{-10}{3} \text{ A or 3.333 A towards to node A} \quad \textbf{Ans.}$$

$$i_2 = \frac{V_B + 50}{4} = \frac{\frac{10}{3} + 50}{4} = \frac{40}{3} = 13.333 \text{ A with direction as shown in Fig. 2.63 (b).} \quad \textbf{Ans.}$$

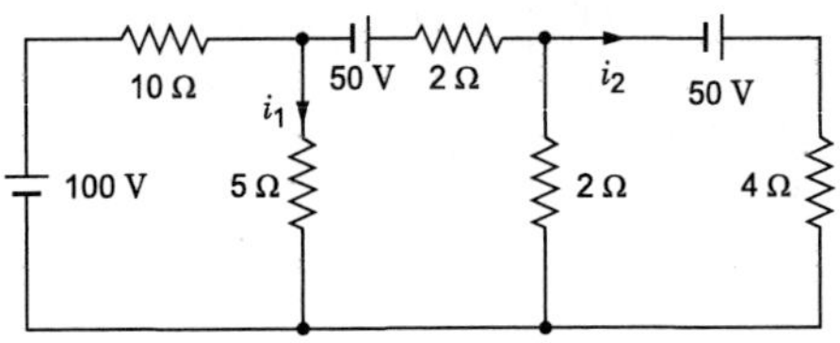

Fig. 2.63 (*a*)

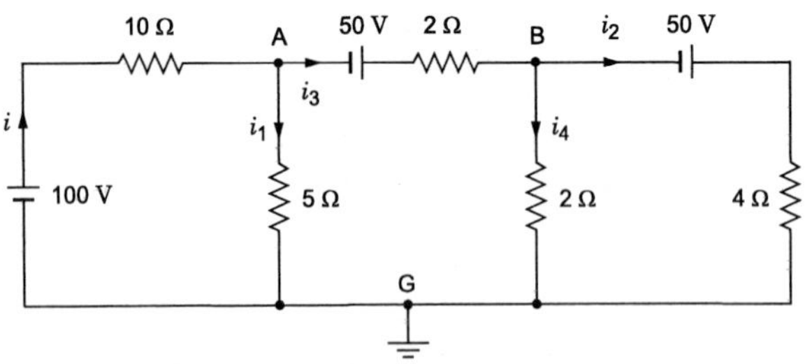

Fig. 2.63 (*b*)

Example 2.31 **Calculate currents in all the resistances of the circuit shown in Fig. 2.64 (*a*) using node analysis method.** *[U.P. Technical Univ. Electrical Engineering First Semester 2006-07]*

Solution: Let the given circuit be redrawn with terminal G as the reference node and arbitrarily assumed distribution of currents, as shown in Fig. 2.64 (*b*).

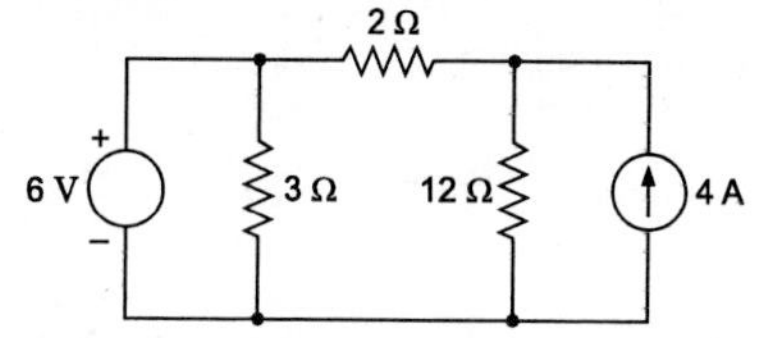

Fig. 2.64 (*a*)

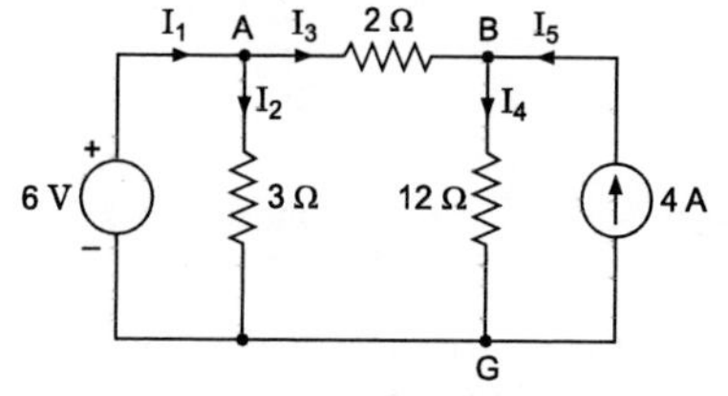

Fig. 2.64 (*b*)

Applying Kirchhoff's current law to node B we have

$$I_4 = I_3 + I_5$$

$$\text{or } \frac{V_B}{12} = \frac{V_A - V_B}{2} + 4$$

$$\text{or } \frac{V_B}{12} = \frac{6 - V_B}{2} + 4 \qquad \because V_A = 6\text{ V}$$

$$\text{or } V_B = 36 - 6V_B + 48$$

$$\text{or } 7V_B = 84$$

$$\text{or } V_B = 12\text{ V}$$

So we have

$$\text{Current in 3 Ω resistance, } I_2 = \frac{V_A}{3} = \frac{6}{3} = 2\text{ A} \quad \textbf{Ans.}$$

$$\text{Current in 2 Ω resistance, } I_3 = \frac{V_A - V_B}{2} = \frac{6-12}{2} = -3\text{ A} \quad \textbf{Ans.}$$

i.e. 3 A from node B to node A **Ans.**

$$\text{Current in 12 Ω resistance, } I_4 = \frac{V_B}{12} = \frac{12}{12} = 1\text{ A} \quad \textbf{Ans.}$$

Example 2.32 **Find current in each branch by using nodal analysis. Also calculate total power loss.** *[U.P. Technical Univ. Basic Electrical Engineering Odd Semester 2013-14]*

Solution: Taking node G as reference node and the potentials of nodes A, B and C be V_A, V_B and V_C respectively and current distribution as shown in Fig. 2.65 (*b*) arbitrarily.

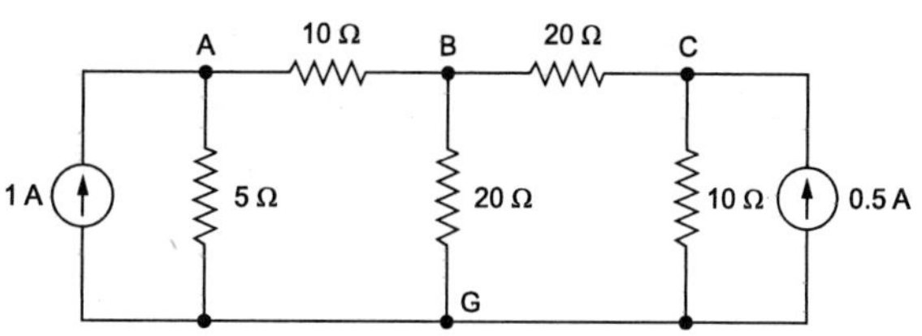

Fig. 2.65 (*a*)

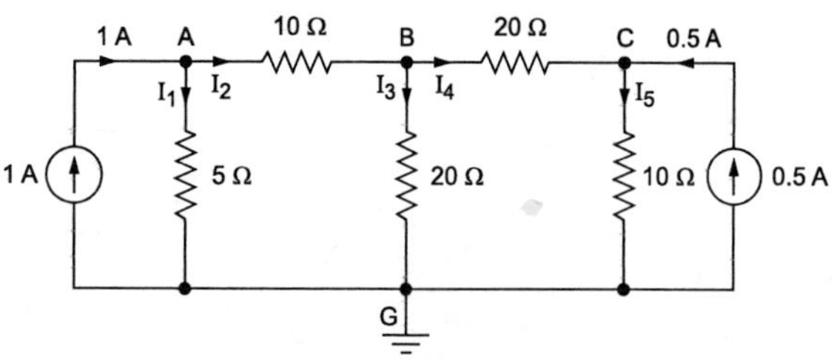

Fig. 2.65 (*b*)

Applying Kirchhoff's current law to nodes A, B and C respectively we have

$$1 = I_1 + I_2$$

$$\text{or } 1 = \frac{V_A}{5} + \frac{V_A - V_B}{10}$$

$$\text{or } 3V_A - V_B = 10 \qquad ...(i)$$

$$I_2 = I_3 + I_4$$

$$\text{or } \frac{V_A - V_B}{10} = \frac{V_B}{20} + \frac{V_B - V_C}{20}$$

$$\text{or } 2V_A - 4V_B + V_C = 0 \qquad ...(ii)$$

$$\text{and } I_4 + 0.5 = I_5$$

$$\text{or } \frac{V_B - V_C}{20} + 0.5 = \frac{V_C}{10}$$

$$\text{or } V_B - 3V_C = -10 \qquad ...(iii)$$

Solving Eqs. (*i*), (*ii*) and (*iii*) we have

$$V_A = \frac{40}{9}\text{ V},\ V_B = \frac{10}{3}\text{ V and } V_C = \frac{40}{9}\text{ V}$$

$$\text{So current through 5 Ω resistor, } I_1 = \frac{V_A}{5} = \frac{40}{9} \times \frac{1}{5} = \frac{8}{9}\text{ A} \quad \textbf{Ans.}$$

Current through 10 Ω resistor, $I_2 = \frac{V_A - V_B}{10} = \left(\frac{40}{9} - \frac{10}{3}\right) \times \frac{1}{10} = \frac{1}{9}$ A **Ans.**

Current through 20 Ω resistor, $I_3 = \frac{V_B}{20} = \frac{10}{3} \times \frac{1}{20} = \frac{1}{6}$ A

Current through 20 Ω resistor, $I_4 = \frac{V_B - V_C}{20} = \left(\frac{10}{3} - \frac{40}{9}\right) \times \frac{1}{20} = \frac{-1}{18}$ A *i.e.* $\frac{1}{18}$ A from node C to node B

Current through 10 Ω resistor, $I_5 = \frac{V_C}{10} = \frac{4}{9}$ A **Ans.**

$$\text{Total power loss} = I_1^2 \times 5 + I_2^2 \times 10 + I_3^2 \times 20 + I_4^2 \times 20 + I_5^2 \times 10$$

$$= \left(\frac{8}{9}\right)^2 \times 5 + \left(\frac{1}{9}\right)^2 \times 10 + \left(\frac{1}{6}\right)^2 \times 20 + \left(\frac{-1}{18}\right)^2 \times 20 + \left(\frac{4}{9}\right)^2 \times 10 = \frac{20}{3} \text{ watts } \textbf{Ans.}$$

Example 2.33 **Using nodal analysis, find current through 10 Ω resistance.**

[*G.B. Technical Univ. Electrical Engineering First Semester 2011-12*]

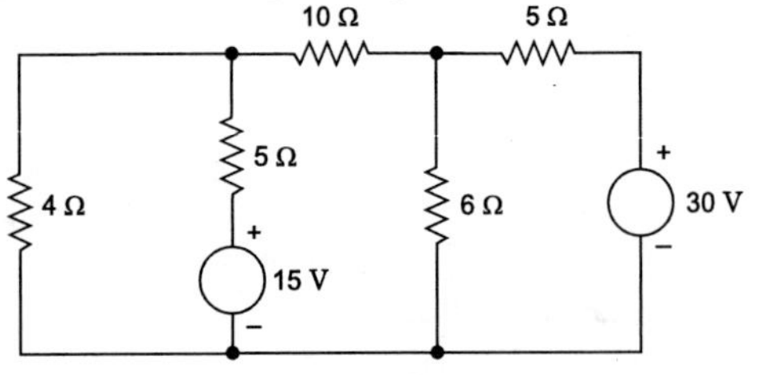

Fig. 2.66 (*a*)

Solution: Let the circuit be redrawn with its different nodes marked A and B and terminal O as the reference node and arbitrarily assumed distribution of current, as shown in Fig. 2.66 (*b*).

Applying KCL to nodes A and B respectively, we have

$$I_1 = I_2 + I_3$$

$$\text{or } \frac{15 - V_A}{5} = \frac{V_A}{4} + \frac{V_A - V_B}{10} \quad ...(i)$$

$$\text{and } I_3 + I_4 = I_5$$

$$\text{or } \frac{V_A - V_B}{10} + \frac{30 - V_B}{5} = \frac{V_B}{6} \quad ...(ii)$$

$$\text{or } 11V_A - 2V_B = 60 \quad ...(iii)$$

$$\text{and } 3V_A - 14V_B = -180 \quad ...(iv)$$

Solving Eqs. (*iii*) and (*iv*), we have

$$V_A = \frac{300}{37} \text{ V and } V_B = \frac{540}{37} \text{ V}$$

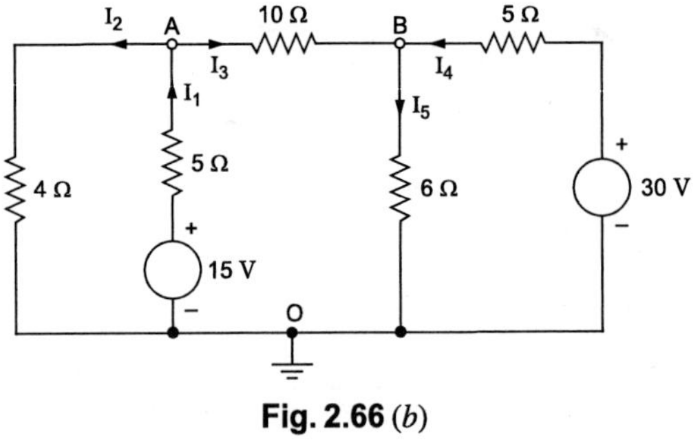

Fig. 2.66 (*b*)

$$\text{Current through 10 Ω resistance} = I_3 = \frac{V_A - V_B}{10} = \frac{\frac{300}{37} - \frac{540}{37}}{10} = \frac{-24}{37} \text{ A}$$

i.e., current through resistor of 10 Ω $= \frac{24}{37}$ A from node B to node A **Ans.**

Example 2.34 **For the circuit shown in Fig. 2.67 (*a*) find (*a*) voltage *v* (*b*) current through the 2 Ω resistor using nodal method.**

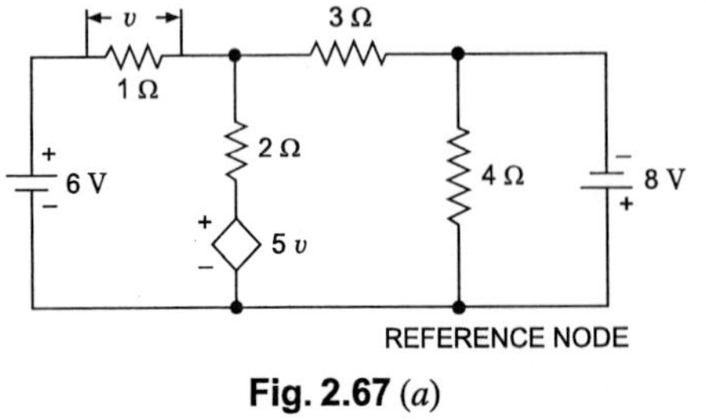

Fig. 2.67 (*a*)

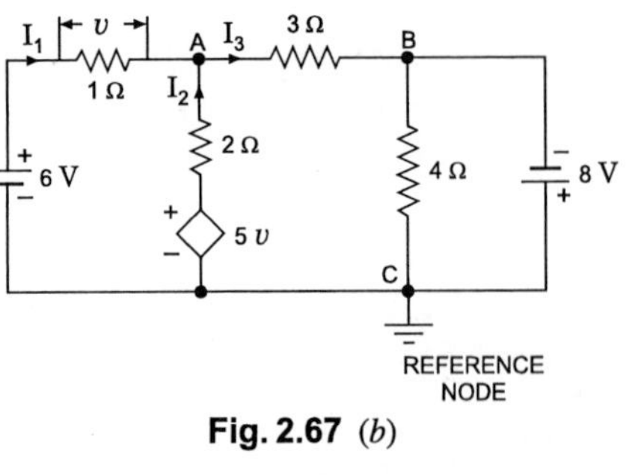

Fig. 2.67 (*b*)

Solution: Let the circuit be redrawn with node C as the reference node and arbitrarily assumed distribution of current, as shown in Fig. 2.67 (*b*). From circuit diagram shown in Fig. 2.67 (*b*) we have

$$V_B = -8\text{ V} \qquad \because V_C = 0$$

$$\text{and } v = I_1 \times 1 = I_1 = \frac{6 - V_A}{1} = (6 - V_A)$$

Applying KCL to node A we have

$$\text{or } \frac{6 - V_A}{1} + \frac{5v - V_A}{2} = \frac{V_A - V_B}{3}$$

$$\text{or } (6 - V_A) + \frac{5(6 - V_A) - V_A}{2} = \frac{V_A + 8}{3}$$

$$\text{or } 13\,V_A = 55$$

$$\text{or } V_A = \frac{55}{13}\text{ V}$$

$$(a)\text{ Voltage } v = 6 - V_A = 6 - \frac{55}{13} = \frac{23}{13}\text{ V } \textbf{Ans.}$$

$$(b)\text{ Current through 2 }\Omega\text{ resistor } = I_2 = \frac{5v - V_A}{2} = \frac{5 \times \frac{23}{13} - \frac{55}{13}}{2} = \frac{30}{13}\text{ A } \textbf{Ans.}$$

Example 2.35 Determine the currents i_x and i_y in the following network. State theorem used.

[*U.P. Technical Univ. Electrical Engineering First Semester 2008-09*]

Solution: Theorem used is Node-Voltage Theorem.

Let the given circuit be redrawn with node C as the reference node and arbitrarily assumed distribution of current as shown in Fig. 2.68 (*b*). From the circuit diagram shown in Fig. 2.68 (*b*) we have

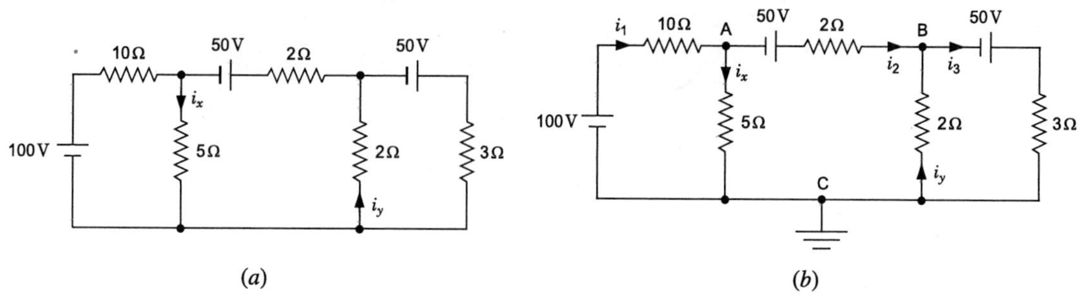

Fig. 2.68

$$i_1 = i_x + i_2 \qquad \text{by applying KCL to node A}$$

$$\text{or } \frac{100 - V_A}{10} = \frac{V_A}{5} + \frac{V_A + 50 - V_B}{2}$$

$$\text{or } 100 - V_A = 2V_A + 5V_A + 250 - 5V_B$$

$$\text{or } 8V_A - 5V_B = -150 \qquad \dots (i)$$

$$\text{and } i_2 + i_y = i_3 \qquad \text{by applying KCL to node B}$$

$$\text{or } \frac{V_A + 50 - V_B}{2} + \frac{0 - V_B}{2} = \frac{V_B + 50}{3}$$

$$\text{or } 3V_A + 150 - 3V_B - 3V_B = 2V_B + 100$$

$$\text{or } 3V_A - 8V_B = -50 \qquad \dots (ii)$$

Solving Eqs. (*i*) and (*ii*) we have

$$V_A = \frac{-2{,}850}{147}\text{ V} = -19.3878\text{V}$$

$$\text{and } V_B = \frac{-150}{147} \text{ V} = -1.02\text{V}$$

$$\text{Current } i_x = \frac{V_A}{5} = \frac{-19.3878}{5} = -3.878\text{A } \textit{i.e.} \text{ 3.878 towards node A } \textbf{Ans.}$$

$$\text{Current } i_y = \frac{0 - V_B}{2} = \frac{0-(-1.02)}{2} = 0.51\text{A } \textbf{Ans.}$$

Example 2.36 **Use nodal analysis to find the currents in various resistors of the circuit shown (Fig. 2.69).**

[U.P. Technical Univ. Electrical Engineering Second Semester 2005-06]

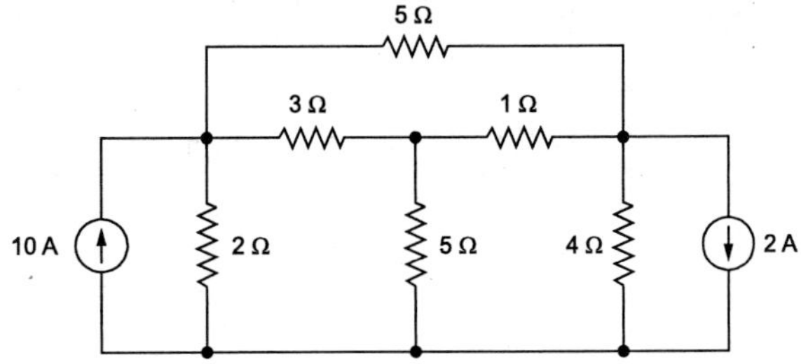

Fig. 2.69

Solution: Let the circuit be redrawn with its different nodes marked A, B, C and D, the last one being taken as reference or datum node (Fig. 2.70). Applying KCL to nodes A, B and C respectively we have

$$I_1 + I_2 + I_3 = I = 10$$

$$\text{or } \frac{V_A}{2} + \frac{V_A - V_B}{3} + \frac{V_A - V_C}{5} = 10$$

$$\text{or } 15V_A + 10V_A - 10V_B + 6V_A - 6V_C = 300$$

$$\text{or } 31V_A - 10V_B - 6V_C = 300 \quad ...(i)$$

$$I_2 = I_4 + I_5$$

$$\text{or } \frac{V_A - V_B}{3} = \frac{V_B - V_C}{1} + \frac{V_B}{5}$$

$$\text{or } 5V_A - 5V_B = 15V_B - 15V_C + 3V_B$$

$$\text{or } 5V_A - 23V_B + 15V_C = 0 \quad ...(ii)$$

$$\text{and } I_3 + I_4 - I_6 = 2$$

$$\text{or } \frac{V_A - V_C}{5} + \frac{V_B - V_C}{1} - \frac{V_C}{4} = 2$$

$$\text{or } 4V_A - 4V_C + 20V_B - 20V_C - 5V_C = 40$$

$$\text{or } 4V_A + 20V_B - 29V_C = 40 \quad ...(iii)$$

Solving Eqs. (*i*), (*ii*) and (*iii*) we have

$$V_A = \frac{6{,}572}{545} \text{ A}$$

$$V_B = \frac{556}{109} \text{V}$$

$$V_C = \frac{2{,}072}{545} \text{ V}$$

$$\text{Thus current } I_1 = \frac{V_A}{2} = \frac{6{,}572}{2 \times 545} = 6.029 \text{ A } \textbf{Ans.}$$

$$\text{Current } I_2 = \frac{V_A - V_B}{3} = \frac{6{,}572 - 2{,}780}{3 \times 545} = 2.319 \text{ A } \textbf{Ans.}$$

$$\text{Current, } I_3 = \frac{V_A - V_C}{5} = \frac{6{,}572 - 2{,}072}{5 \times 545} = 1.651 \text{ A } \textbf{Ans.}$$

$$\text{Current, } I_4 = \frac{V_B - V_C}{1} = \frac{556}{109} - \frac{2{,}072}{545} = 1.299 \text{ A } \textbf{Ans.}$$

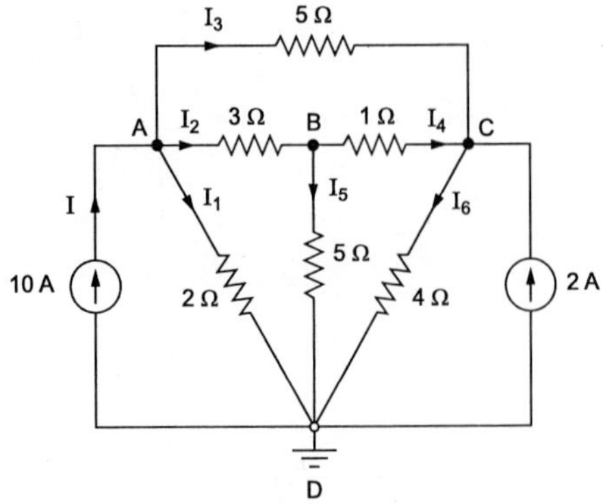

Fig. 2.70

$$\text{Current, } I_5 = \frac{V_B}{5} = \frac{556}{109 \times 5} = 1.02 \text{ A } \textbf{Ans.}$$

$$\text{Current, } I_6 = \frac{V_C}{4} = \frac{2{,}072}{4 \times 545} = 0.95 \text{ A } \textbf{Ans.}$$

Example 2.37 **Determine i_1, i_2, and i_3 using nodal analysis for a given circuit shown in Fig. 2.71.**

[U.P. Technical Univ. Electrical Engineering Odd Semester 2013-14]

Solution: Let the circuit given in Fig. 2.71 be redrawn with terminal G as the reference node as shown in Fig. 2.72. Let the current flowing through branch AB be i_x amperes flowing from node A to node B.

Applying Kirchhoff's current law to nodes A and B respectively we have

$$i_1 + i_2 = i_x$$

$$\text{or } \frac{24 - V_A}{2} + \frac{0 - V_A}{6} = i_x \quad ...(i)$$

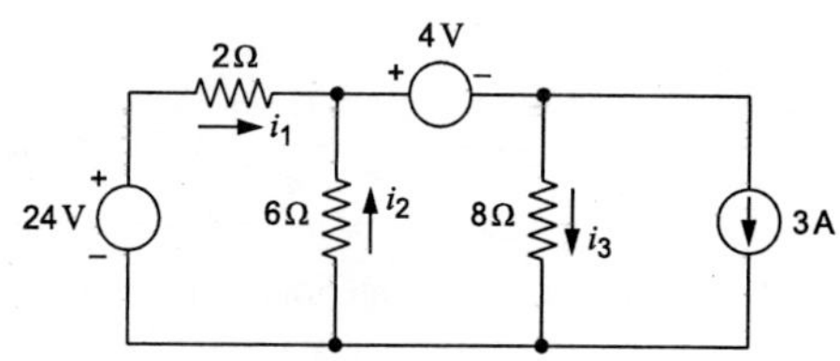

Fig. 2.71

$$\text{and } i_x - 3 = i_3$$

$$\text{or } i_3 + 3 = i_x$$

$$\text{or } \frac{V_B}{8} + 3 = i_x \quad ..(ii)$$

$$\text{Node voltage } V_B = \text{Node voltage } V_A - 4 \quad ...(iii)$$

Substituting $V_B = V_A - 4$ in Eq. *(ii)* we have

$$\frac{V_A - 4}{8} + 3 = i_x \quad ...(iv)$$

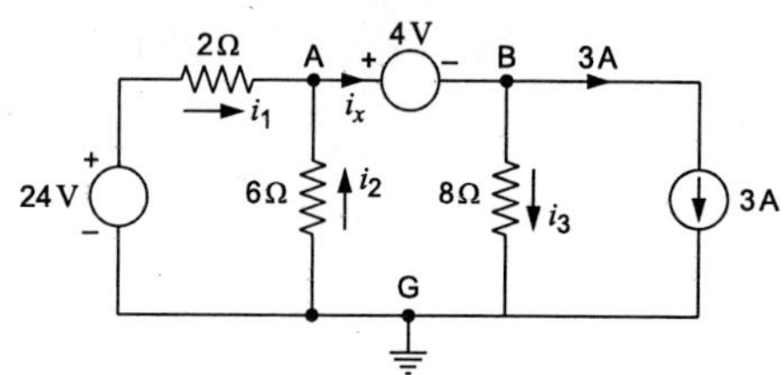

Fig. 2.72

Comparing Eqs. *(i)* and *(iv)* we have

$$\frac{24 - V_A}{2} - \frac{V_A}{6} = \frac{V_A - 4}{8} + 3$$

$$\text{or } 288 - 12V_A - 4V_A = 3V_A - 12 + 72$$

$$\text{or } 19V_A = 288 + 12 - 72 = 288$$

$$\text{or } V_A = \frac{228}{19} = 12 \text{ V}$$

$$V_B = V_A - 4 = 12 - 4 = 8 \text{ V}$$

$$\text{Current } i_1 = \frac{24 - V_A}{2} = \frac{24 - 12}{2} = 6 \text{ A} \quad \textbf{Ans.}$$

$$\text{Current } i_2 = \frac{0 - V_A}{6} = \frac{0 - 12}{6} = -2 \text{ A } \textit{i.e.} \text{ 2 A flowing away from node A} \quad \textbf{Ans.}$$

$$\text{Current } i_3 = \frac{V_B}{8} = \frac{8}{8} = 1 \text{ A} \quad \textbf{Ans.}$$

Example 2.38 **Using nodal analysis, find V_{cd} for the circuit shown below in Fig. 2.73.**

[U.P. Technical Univ. Electrical Engineering First Semester 2007-08]

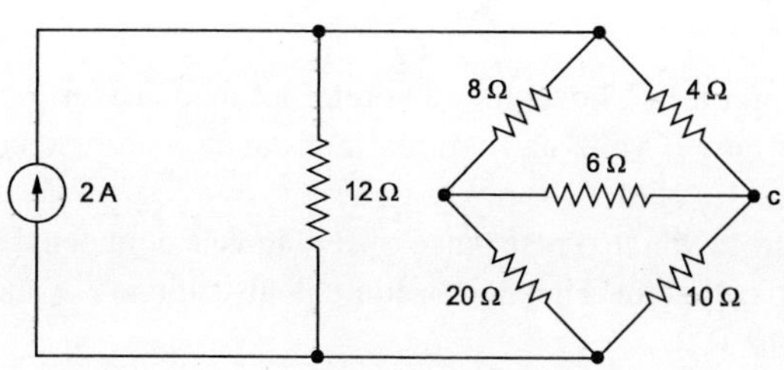

Fig. 2.73

Solution : Let the given circuit be redrawn with node *d* as the reference node and arbitrarily assumed distribution of current, as shown in Fig. 2.74.

Applying Kirchhoff's current law to nodes *a*, *b* and *c* respectively we have

$$I_1 + I_2 = 2 - I_6 \quad ...(i)$$

$$I_1 = I_3 + I_4 \quad ...(ii)$$

$$\text{and } I_2 + I_3 = I_5 \quad ...(iii)$$

$$\text{or } \frac{V_a - V_b}{8} + \frac{V_a - V_c}{4} = 2 - \frac{V_a}{12} \quad ...(iv)$$

$$\frac{V_a - V_b}{8} = \frac{V_b - V_c}{6} + \frac{V_b}{20} \quad ...(v)$$

$$\text{and } \frac{V_a - V_c}{4} + \frac{V_b - V_c}{6} = \frac{V_c}{10} \quad ...(vi)$$

$$\text{or } 11V_a - 3V_b - 6V_c = 48 \quad ...(vii)$$

$$15V_a - 41V_b + 20V_c = 0 \quad ...(viii)$$

$$15V_a + 10V_b - 31V_c = 0 \quad ...(ix)$$

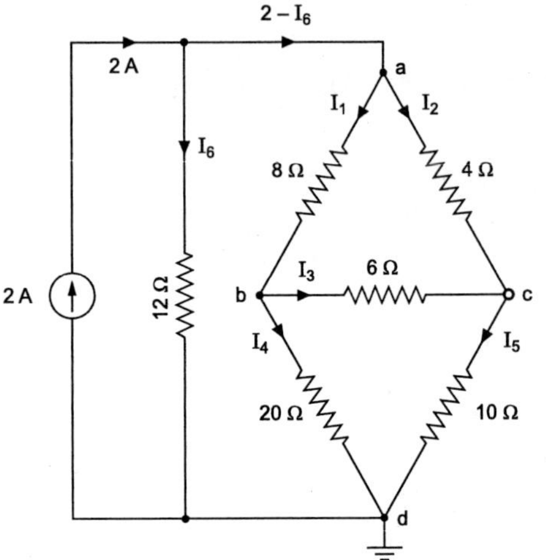

Fig. 2.74

Subtracting Eq. (*ix*) from Eq. (*viii*) we have

$$-51V_b + 51V_c = 0$$

$$\text{or } V_b = V_c$$

Substituting $V_b = V_c$ in Eq. (*viii*) or (*ix*) we have

$$V_a = \frac{21}{15}V_c = 1.4\,V_c$$

Substituting $V_a = 1.4\,V_c$ and $V_b = V_c$ in Eq. (*vii*) we have

$$15.4\,V_c - 3V_c - 6V_c = 48$$

$$\text{or } V_c = \frac{48}{6.4} = 7.5\text{ V}$$

$$\text{So } V_{cd} = 7.5\text{ V} \quad \textbf{Ans.}$$

2.16. THEVENIN'S THEOREM

This theorem provides a mathematical technique for replacing a two-terminal network by a voltage source V_T, and resistance R_T connected in series. The voltage source V_T (called the *Thevenin's equivalent voltage*) is the open-circuit voltage that appears across the load terminals when the load is removed or disconnected and resistance R_T, called the *Thevenin's equivalent resistance,* is equal to the resistance of the network looking back into the load terminals. A Thevenin's equivalent circuit is shown in Fig. 2.75. The steady-state current will be given as

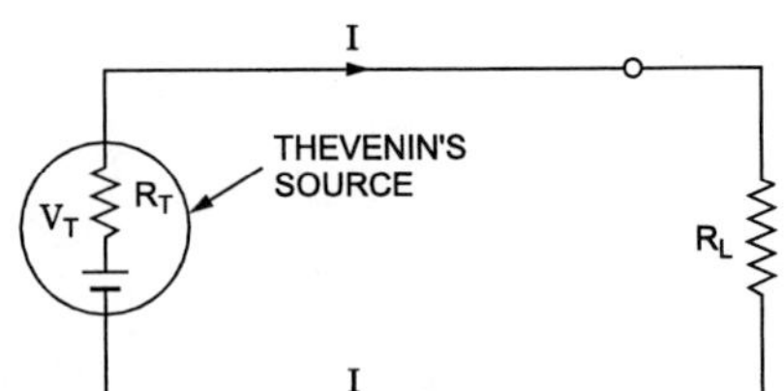

Fig. 2.75

$$I = \frac{V_T}{R_T + R_L}$$

For visualizing the application of Thevenin's theorem, let us consider a circuit shown in Fig. 2.76 (*a*) which consists of a source of emf E volts and internal resistance *r* ohms connected to an external circuit consisting of resistances R_1 and R_2 ohms in series.

So far as terminals AB across which a resistance of R_2 ohms is connected the network acts as source of open-circuit voltage V_{OC} (also called the Thevenin's equivalent voltage V_T) and internal resistance R_{in} (also called the Thevenin's resistance R_T).

For determination of open-circuit voltage V_{OC} (or V_T), disconnect the load resistance R_L from the terminals A and B to provide open circuit [Fig. 2.76 (*b*)].

Now current through resistance R_2, $I = \dfrac{E}{R_1 + R_2 + r}$

and open-circuit voltage V_{OC} or V_T = Voltage across terminals AB

$$= \text{Voltage drop across resistance } R_2 = IR_2 = \frac{ER_2}{R_1 + R_2 + r} \quad ...(2.25)$$

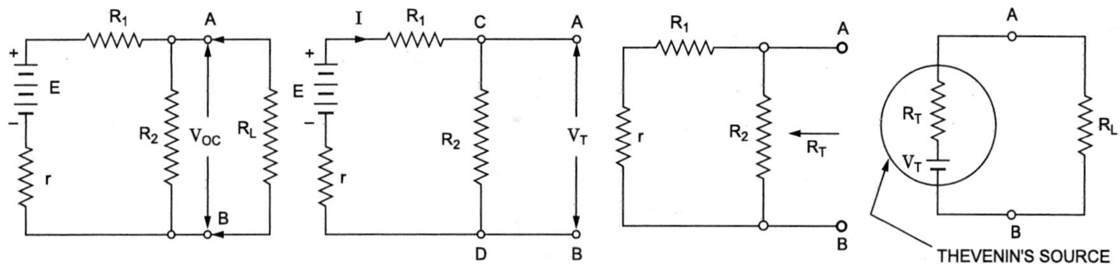

Fig. 2.76 *Illustration of Application of Thevenin's Theorem*

For determination of internal resistance R_{in} (or R_T) of the network under consideration remove the voltage source from the circuit, leaving behind only its internal resistance *r*, as illustrated in Fig. 2.76 (*c*). Now view the circuit inwards from the open terminals A and B. It is found that the circuit [Fig. 2.76 (c)] now consists of two parallel paths—one consisting of resistance R_2 only, and the other consisting of resistance R_1 and *r* in series. Thus, the equivalent resistance (R_T), as viewed from the open terminals A and B, is given as

$$R_T = \frac{R_2 (R_1 + r)}{R_2 + (R_1 + r)} \quad ...(2.26)$$

Now when load resistance R_L is connected across terminals A and B, the network behaves as a source of voltage V_T and internal resistance R_T [Fig. 2.76 (*d*)] and current flowing through the load resistance R_L is given as

$$I_L = \frac{V_T}{R_T + R_L} = \frac{ER_2/[R_1 + R_2 + r]}{\dfrac{R_2 (R_1 + r)}{R_1 + r + R_2} + R_L} = \frac{ER_2}{R_2 (R_1 + r) + R_L (R_1 + R_2 + r)} \quad ...(2.27)$$

The Thevenin's theorem can be stated as follows:

The current in any *passive circuit element* (which may be called R_L) in a network is the same as would be obtained if R_L were supplied with a voltage source V_{OC} (or V_T) in series with an equivalent resistance R_{in} (or R_T), V_{OC} being the open-circuit voltage at the terminals from which R_L has been removed and R_{in} (or R_T) being the resistance that would be measured at these terminals after all sources have been removed and each has been replaced by its internal resistance.

This theorem is advantageous when we are to determine the current in a particular element of a linear bilateral network particularly when it is desired to find the current which flows through a resistor for its different values. It makes the solution of the complicated networks (particularly electronic networks) quite simple.

Example 2.39 Using Thevenin's theorem, determine current and voltage in 2 ohm resistance in the circuit shown in Fig. 2.77. [*U.P. Technical Univ. Electrical Engineering First Semester 2006-07*]

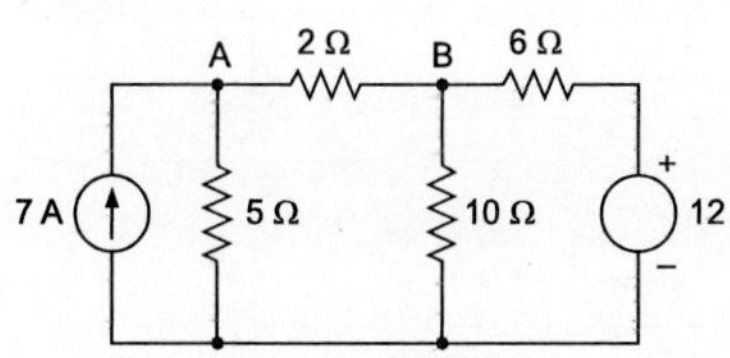

Fig. 2.77

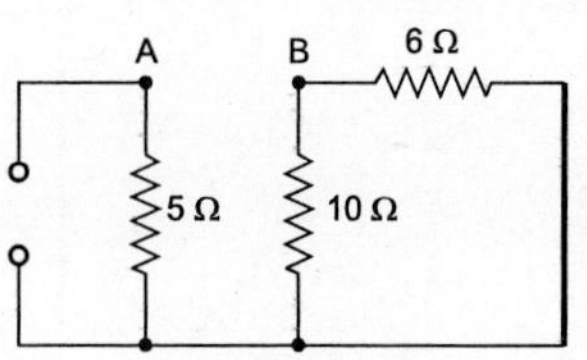

Fig. 2.78

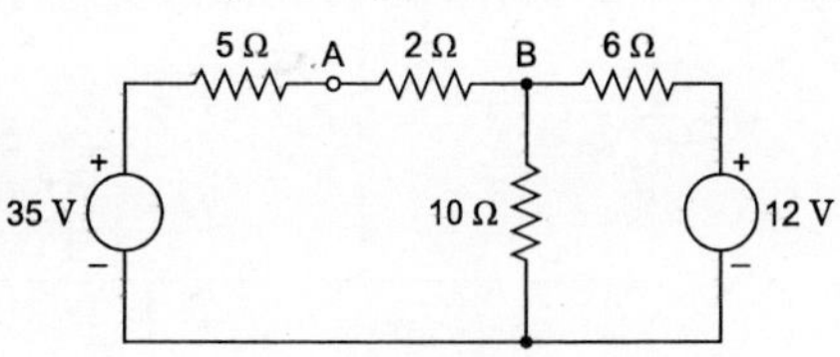

Fig. 2.79

Solution: For determination of Thevenin's equivalent resistance of the circuit w.r.t. terminals A and B, the voltage source is replaced by a short circuit and current source is replaced by an open circuit, as shown in Fig. 2.78.

$$\text{Thevenin's equivalent resistance, } R_T = 5 + 10 \parallel 6 = 5 + \frac{10 \times 6}{10 + 6} = 8.75\ \Omega$$

Converting current source into equivalent voltage source, the circuit becomes as shown in Fig. 2.79.

When terminals A and B are open,

$$\text{Thevenin's voltage, } V_T = V_A - V_B = 35 - \frac{12}{16} \times 10 = 27.5 \text{ V}$$

Current through 2 Ω resistance connected across AB,

$$I_L = \frac{V_T}{R_T + R} = \frac{27.5}{8.75 + 2} = 2.558 \text{ A } \textbf{Ans.}$$

Example 2.40 **Find the current I using Thevenin's theorem in circuit in Fig. 2.80.**

[*U.P. Technical Univ. Electrical Engineering Second Semester 2006-07*]

Solution: The equivalent resistance of the network (with voltage sources replaced by short circuits, as shown in Fig. 2.81) with reference to terminals A and B

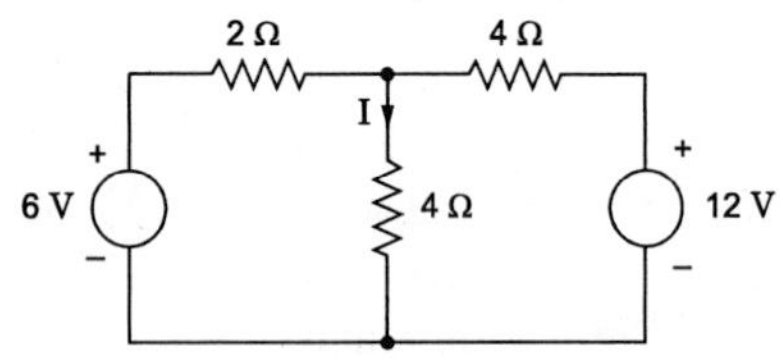

Fig. 2.80

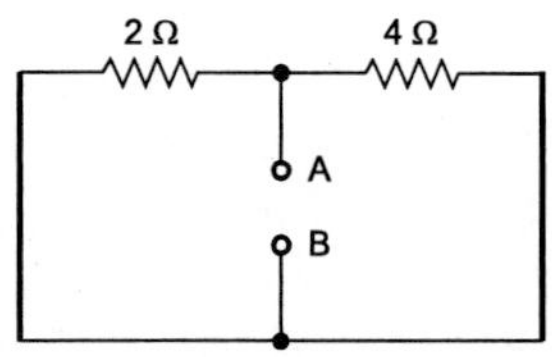

Fig. 2.81

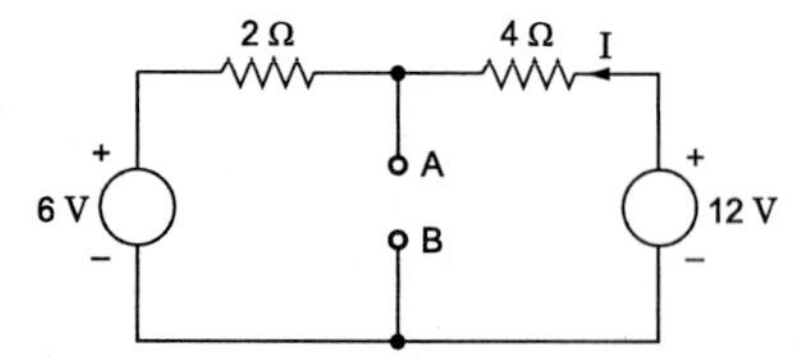

Fig. 2.82

$$R_T = 2 \parallel 4 = \frac{1}{\frac{1}{2} + \frac{1}{4}} = \frac{4}{3}\ \Omega$$

When terminals A and B are open, the current through the mesh formed by voltage sources and 2 Ω and 4 Ω resistor [Fig. 2.82]

$$I = \frac{12 - 6}{2 + 4} = 1 \text{ A}$$

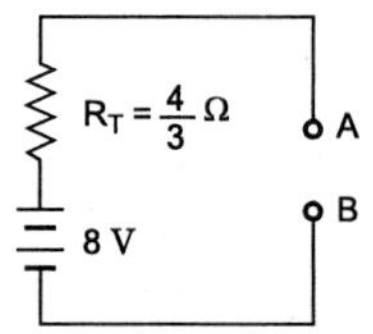

Fig. 2.83

So open-circuit voltage, $V_T = 6 + 2 \times 1$ or $12 - 4 \times 1$ *i.e.* 8 V

Thevenin's equivalent circuit is shown in Fig. 2.83.

Current through the load resistance of 4 Ω connected across terminals AB

$$I = \frac{V_T}{R_T + R_L} = \frac{8}{\frac{4}{3} + 4} = \frac{24}{16} = 1.5 \text{ A } \textbf{Ans.}$$

Example 2.41 **Replace the network of the following figure [Fig. 2.84 (*a*)] to the left of terminals *ab* by its Thevenin's equivalent circuit. Hence determine I.** [*U.P. Technical Univ. Electrical Engineering First Semester 2009-10*]

Solution: For determining Thevenin equivalent resistance of the circuit with reference to terminals *ab* the voltage sources are replaced by short circuits, as shown in Fig. 2.84 (*b*).

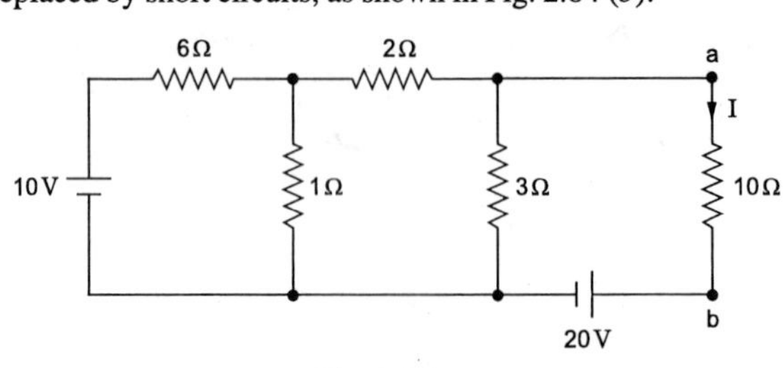

Fig. 2.84 (*a*)

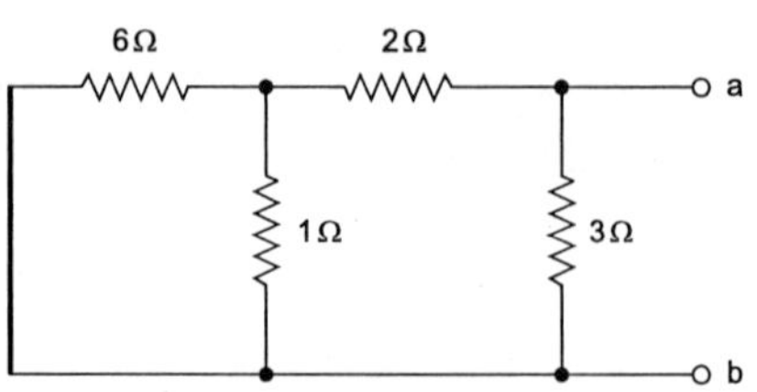

Fig. 2.84 (*b*)

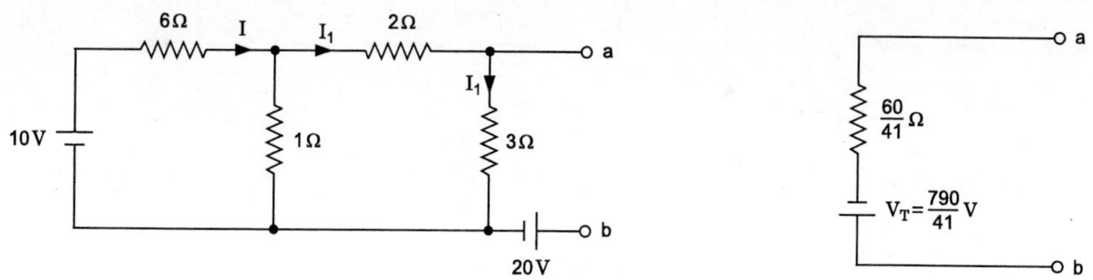

(c) (d)

Fig. 2.84

$$\text{Thevenin's resistance, } R_T = 3 \parallel (2 + 1 \parallel 6) = 3 \parallel \left(2 + \frac{1}{1 + \frac{1}{6}}\right) = 3 \parallel \left(2 + \frac{6}{7}\right) = 3 \parallel \frac{20}{7} = \frac{1}{\frac{1}{3} + \frac{7}{20}} = \frac{60}{41}\,\Omega$$

After removing 10 Ω resistance, the circuit takes the form as shown in Fig. 2.84 (*c*).

Equivalent resistance of the circuit is

$$R = 6 + 1 \parallel (2 + 3) = 6 + \frac{1}{1 + \frac{1}{5}} = 6 + \frac{5}{6} = \frac{41}{6}\,\Omega$$

$$\text{Current supplied by 10V battery, } I = \frac{V}{R} = \frac{10}{\frac{41}{6}} = \frac{60}{41}\,\text{A}$$

$$\text{Current through } 3\Omega \text{ resistance, } I_1 = I \times \frac{1}{1+2+3}$$ By current division rule

$$= \frac{60}{41} \times \frac{1}{6} = \frac{10}{41}\,\text{A}$$

$$\text{Voltage drop across } 3\,\Omega \text{ resistance} = \frac{10}{41} \times 3 = \frac{30}{41}\,\text{V}$$

$$\text{Thevenin's voltage, } V_T = V_{ba} = \left(20 - \frac{30}{41}\right)\text{V} = \frac{790}{41}\,\text{V}$$

Thevenin's equivalent circuit is shown in Fig. 2.84 (*d*).

$$\text{Current, } I_1 = \frac{V_T}{R_T + R_L} = \frac{790/41}{\frac{60}{41} + 10} = \frac{790}{470} = \frac{79}{47}\,\text{A}$$ from terminal *b* to terminal *a* **Ans.**

Example 2.42 **Use Thevenin's theorem to replace the three loop circuit of the following figure [Fig. 2.85 (*a*)] by a single loop equivalent circuit in which the identity of R_L is preserved.**

[*U.P. Technical Univ. Electrical Engineering Second Semester 2008-09*]

Solution: For determination of Thevenin's equivalent resistance of the circuit w.r.t. terminals A and B, the voltage sources are replaced by short-circuits, load resistor R_L is removed, as shown below in Fig. 2.85 (*b*).

Thevenin's equivalent resistance,

$$R_T = (30 \parallel 60)\,\Omega + (60 \parallel 10)\,\Omega$$

$$= \frac{30 \times 60}{30 + 60} + \frac{60 \times 10}{60 + 10}$$

$$= 20 + \frac{60}{7} = \frac{200}{7}\,\Omega \quad \textbf{Ans.}$$

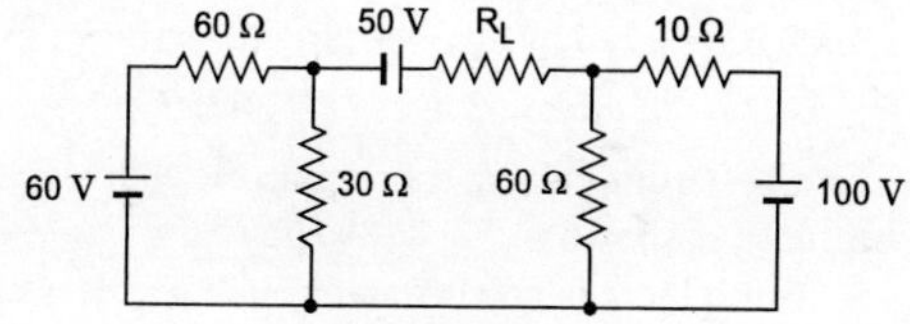

Fig. 2.85 (*a*)

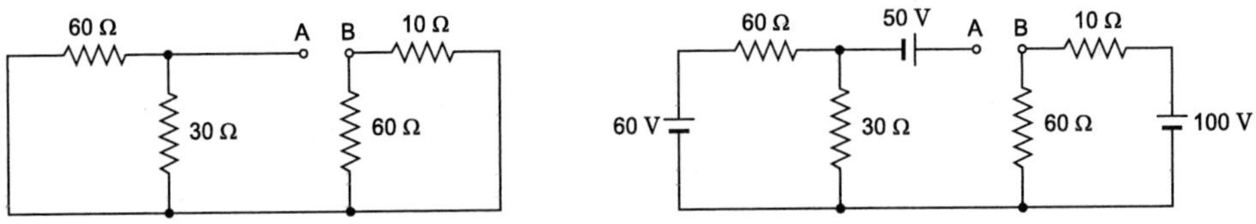

(b) (c)

Fig. 2.85

Removing R_L from the given circuit, the circuit is reduced to that shown in Fig. 2.85 (c).

Thevenin's voltage

$$V_T = V_A - V_B$$

$$= \frac{60}{60+30} \times 30 + 50 - \frac{100}{60+10} \times 60$$

$$= 70 - \frac{600}{7} = \frac{490-600}{7} = \frac{-110}{7} \text{ V}$$

i.e. $V_T = V_{BA} = \dfrac{110}{7}$ V **Ans.**

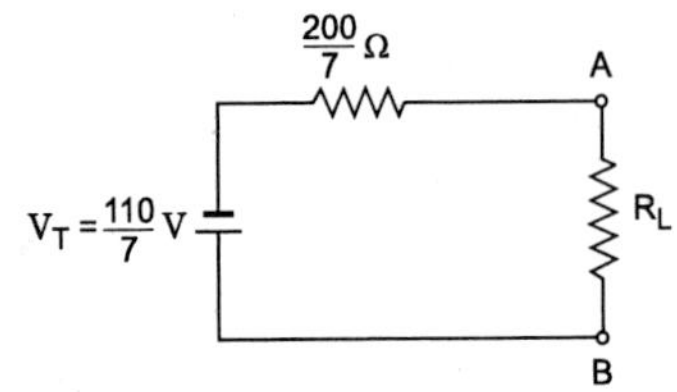

Fig. 2.85 (d)

Thevenin's equivalent circuit is shown in Fig. 2.85 (d).

Example 2.43 Find the Thevenin's equivalent of the network (Fig. 2.86).

[U.P. Technical Univ. Electrical Engineering Second Semester 2004-05]

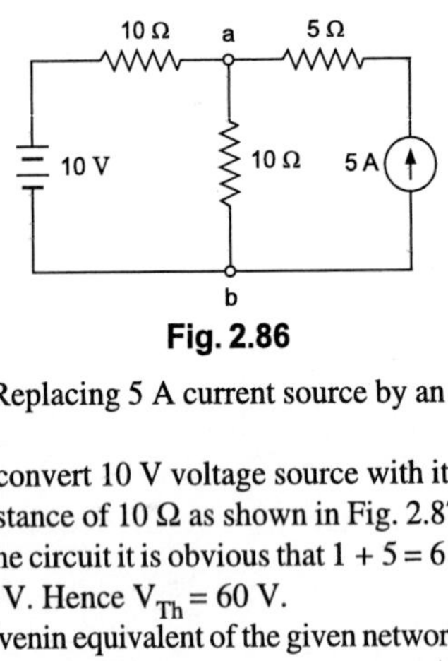

Fig. 2.86

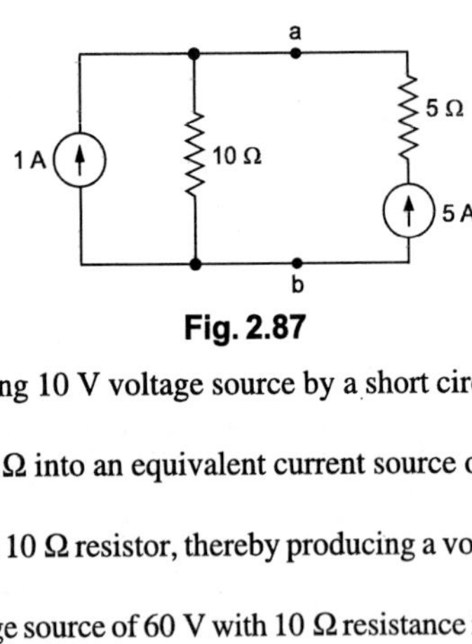

Fig. 2.87

Solution: Replacing 5 A current source by an open circuit and replacing 10 V voltage source by a short circuit we have $R_{Th} = 10\ \Omega$.

Let us convert 10 V voltage source with its series resistance of 10 Ω into an equivalent current source of 1 A with a parallel resistance of 10 Ω as shown in Fig. 2.87.

From the circuit it is obvious that 1 + 5 = 6 A current flows through 10 Ω resistor, thereby producing a voltage drop of 6 × 10 = 60 V. Hence $V_{Th} = 60$ V.

So Thevenin equivalent of the given network is equivalent to a voltage source of 60 V with 10 Ω resistance in series with it.

Example 2.44 Find Thevenin's equivalent circuit across AB shown in Fig. 2.88.

[U.P. Technical Univ. Electrical Engineering Second Semester 2005-06]

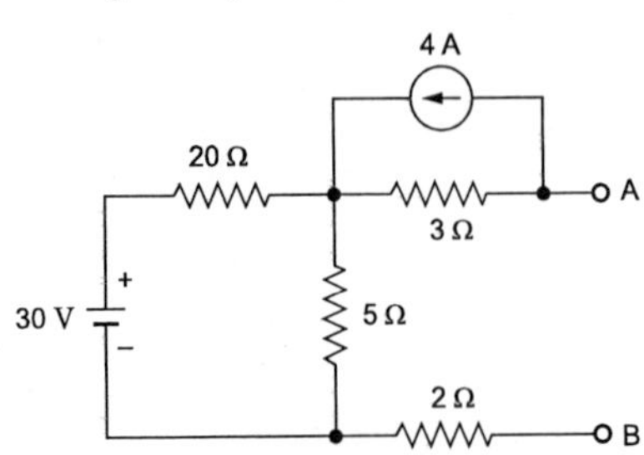

Fig. 2.88

Solution: For determining the Thevenin's equivalent resistance of the circuit w.r.t. terminals AB the voltage source is replaced by a short circuit and current source is replaced by an open circuit, as shown in Fig. 2.89 (a).

$$R_T = 3 + 2 + 5 \parallel 20 = 3 + 2 + 4 = 9\ \Omega$$

Converting current source into the equivalent voltage source the circuit becomes as shown in Fig. 2.89 (b).

Since there is no current through resistors of 3 Ω and 2 Ω, there is no voltage drop across them.

Thevenin's voltage V_{Th} equals algebraic sum of battery voltage and drop across 5 Ω resistor. Hence

$$V_{Th} = 30 \times \frac{5}{20+5} - 12 = -6 \text{ V}$$

Thevenin's equivalent circuit is shown in Fig. 2.89 (*c*).

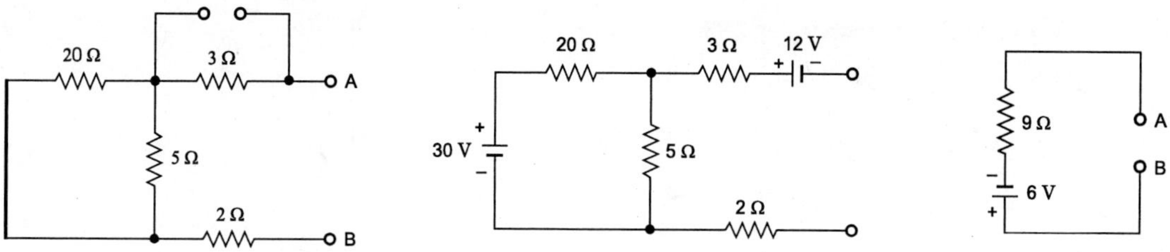

(*a*) (*b*) (*c*)

Fig. 2.89

Example 2.45 The resistance of the various arms of a Wheatstone bridge are shown in Fig. 2.90. The battery has an emf of 2 V and negligible internal resistance. Using Thevenin's theorem, determine the value and direction of the current in the galvanometer circuit BD. [*U.P. Technical Univ. Electrical Engineering July 2002*]

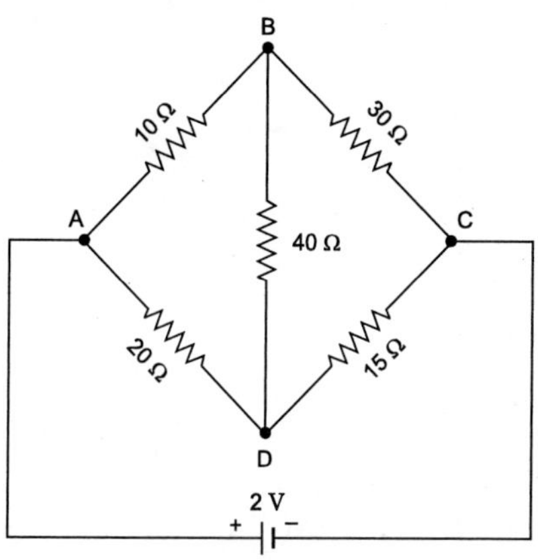

Fig. 2.90

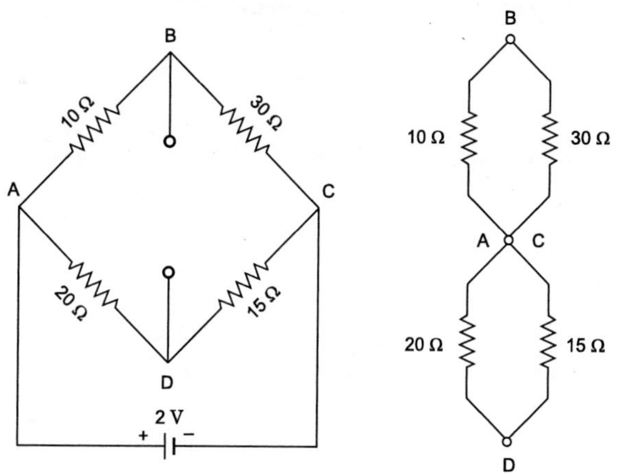

(*a*) (*b*)

Fig. 2.91

Solution: After removing the 40 Ω resistor between terminals B and D circuit takes the form shown in Fig. 2.91 (*a*).

Now in the circuit shown in Fig. 2.91 (*a*), potential of terminal B w.r.t. terminal C,

$$V_{BC} = 2 \times \frac{30}{30+10} = 1.5 \text{ V}$$

$$\text{Similarly } V_{DC} = \frac{2 \times 15}{20+15} = \frac{30}{35} \text{ V or } \frac{6}{7} \text{ V}$$

$$\text{PD between terminals B and D, } V_T = V_{BC} - V_{DC} = 1.5 - \frac{6}{7} = \frac{4.5}{7} \text{ V}$$

After removing 2 V battery from the circuit (2 V battery being replaced by a short circuit) equivalent resistance of network with reference to terminals B and D [Fig. 2.91 (*b*)]

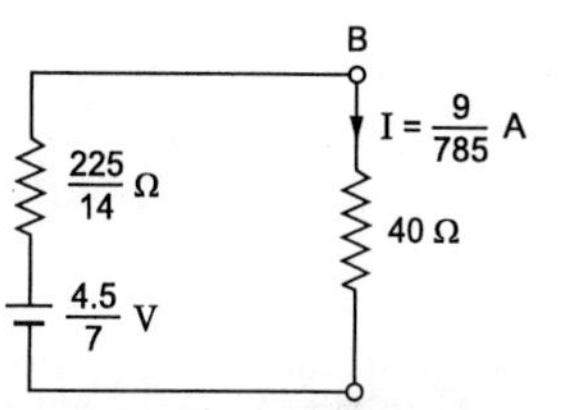

Fig. 2.92

$$R_T = 10 \,||\, 30 + 20 \,||\, 15 = \frac{10 \times 30}{10+30} + \frac{20 \times 15}{20+15} = 7.5 + \frac{60}{7} = \frac{225}{14} \ \Omega$$

Current through the galvanometer circuit

$$BD = \frac{V_T}{R_T + R_L} \ \frac{4.5/7}{\frac{225}{14} + 40} = \frac{9}{785} \text{ A from B to D } \textbf{Ans.}$$

Example 2.46 **State Thevenin's theorem and calculate current in a 1,000 Ω resistor connected between terminals A and B, as shown in Fig. 2.93.**

[*Electrical Engineering U.P. Technical Univ. June 2001; G.B. Technical Univ. Odd Semester 2012-13*]

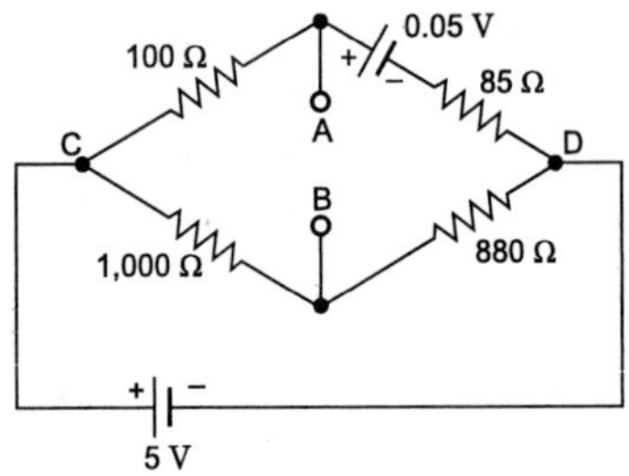

Fig. 2.93

Solution: Now in the circuit shown in Fig. 2.93,

Potential of point B w.r.t. terminal D

$$V_{BD} = \frac{5 \times 880}{1{,}000 + 880} = 2.340426 \text{ V}$$

Potential of point A w.r.t. to terminal D

$$V_{AD} = \frac{5 - 0.05}{100 + 85} \times 85 + 0.05 = 2.324324 \text{ V}$$

PD between terminals B and A,

$$V_T = 2.340426 - 2.324324 = 0.0161 \text{ V}$$

After replacing batteries by short circuits in the circuit, equivalent resistance of the network with reference to terminals A and B (Fig. 2.94),

$$R_T = 100 \,||\, 85 + 1{,}000 \,||\, 880 = \frac{1700}{17+20} + \frac{22{,}000}{22+25} = 514 \,\Omega$$

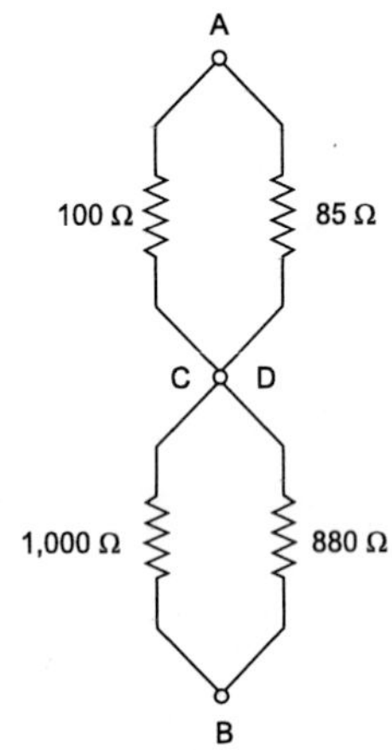

Fig. 2.94

Current in 1,000 Ω resistor connected between terminals A and B,

$$I = \frac{V_T}{R_T + R_L} = \frac{0.0161}{1{,}000 + 514}$$

$= 10.625$ μ A from terminals B to A **Ans.**

Example 2.47 **Determine current in 4 Ω resistance using Thevenin's theorem in the following circuit [Fig. 2.95 (*a*)].**

[*G.B. Technical Univ. Electrical Engineering First Semester 2009-10*]

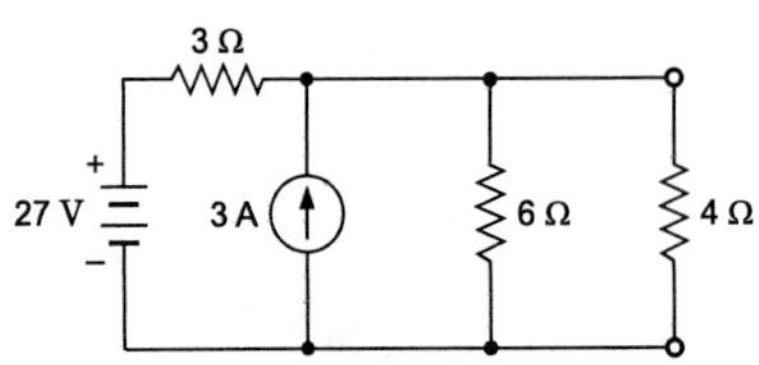

Fig. 2.95 (*a*)

Solution: For determination of Thevenin's equivalent resistance of the circuit w.r.t terminals A and B in Fig. 2.95 (*b*), the voltage source is replaced by a short circuit and current source is replaced by an open circuit, as shown in Fig. 2.95 (*b*).

$$\text{Thevenin's equivalent resistance, } R = 6 \,||\, 3 = \frac{1}{\frac{1}{6} + \frac{1}{3}} = 2\Omega$$

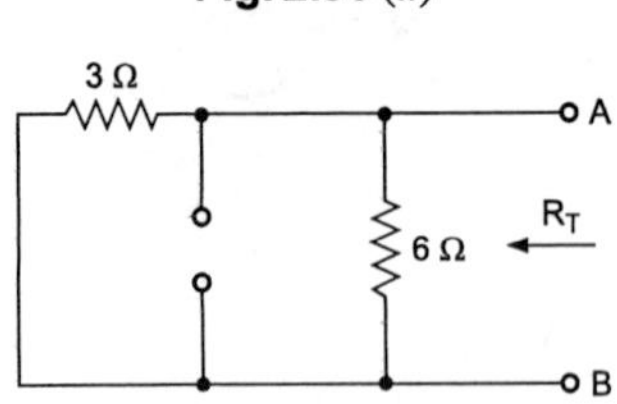

(*b*)

Removing load resistance of 4 Ω and applying kVL to the loop formed by 27 V source, 3 Ω resistor and 6 Ω resistor we have, if the current supplied by the 27 V source is I amperes as shown in Fig. 2.95 (*c*),

$$3\text{I} + 6(\text{I} + 3) = 27$$

$$\text{or I} = \frac{27 - 6 \times 3}{3 + 6} = 1 \text{ A}$$

Current flowing through 6 Ω resistor

$= 3 + \text{I} = 3 + 1 = 4$ A

Thevenin's voltage, V_T = Voltage drop across 6 Ω resistance (keeping terminals A and B open)

$= 4 \times 6 = 24$ V

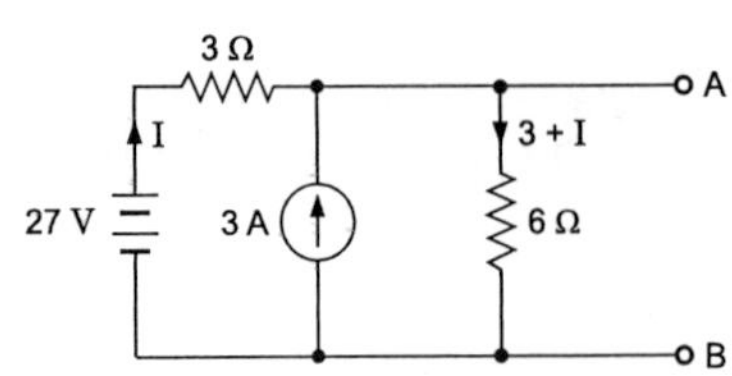

(*c*)

Fig. 2.95

Current through 4 Ω resistance,

$$I_L = \frac{V_T}{R_L + R_T} = \frac{24}{4+2} = 4\text{A} \quad \textbf{Ans.}$$

Example 2.48 **Find the magnitude and direction of the current in the 2 Ω resistor by using Thevenin's theorem for the circuit shown in Fig. 2.96.**

Solution : The circuit is redrawn as shown in Fig. 2.97 (*a*). After removing the 2 Ω resistor between terminals B and D, the circuit takes the form shown in Fig. 2.97 (*b*).

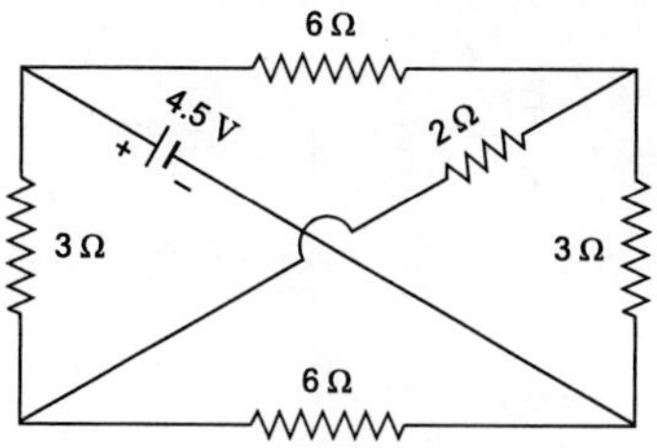

Fig. 2.96

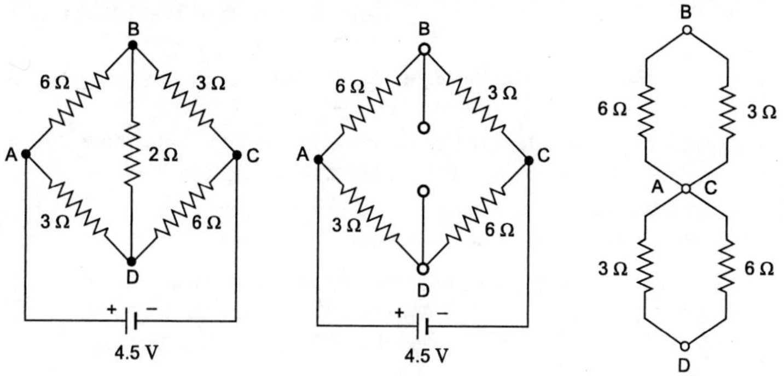

(*a*) (*b*) (*c*)

Fig. 2.97

Now in the circuit shown in Fig. 2.97 (*b*), potential of terminal B w.r.t. terminal C,

$$V_{BC} = 4.5 \times \frac{3}{6+3} = 1.5 \text{ V}$$

Similarly

$$V_{DC} = 4.5 \times \frac{6}{3+6} = 3 \text{ V}$$

PD between terminals D and B, $V_T = V_{DC} - V_{BC} = 3 - 1.5 = 1.5$ V

After removing 4.5 V battery from the circuit (4.5 V battery being replaced by a short circuit) equivalent resistance of the network with reference to terminals B and D. [Fig. 2.97 (*c*)]

$$R_T = 6||3 + 3||6 = 2 + 2 = 4 \ \Omega$$

$$\text{Current in 2 Ω resistor, } I = \frac{V_T}{R_T + R_L} = \frac{1.5}{4+2} = 0.25 \text{ A from terminal D to B}$$ **Ans.**

Example 2.49 **Find current in 6 Ω using Thevenin's theorem.**

[*U.P. Technical Univ. Basic Electrical Engineering Odd Semester 2013-14*]

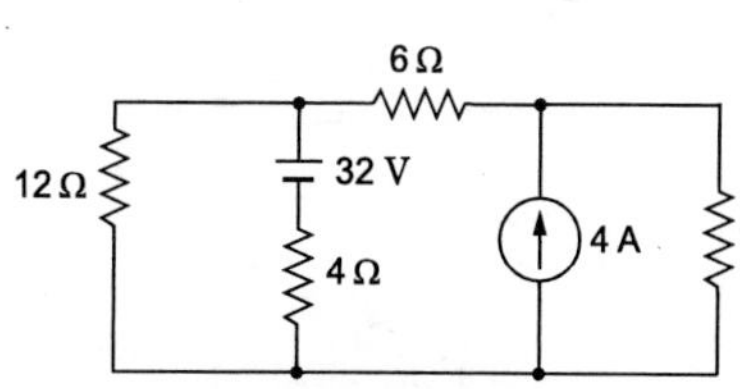

Fig. 2.98 (*a*)

Solution: The equivalent resistance of the network with voltage source replaced by a short circuit and current source by an open circuit with reference to terminals A and B [Fig. 2.98 (*b*)]

$$R_T = (12 || 4) + 9 = \frac{12 \times 4}{12 + 4} + 9 = 12 \ \Omega$$

Converting 4 A current source connected across 9 Ω resistor into equivalent voltage source of 4 × 9 i.e. 36 V in series with the resistor of 9 Ω and removing 6 Ω resistor, the given circuit is converted into a circuit shown in Fig. 2.98 (*c*).

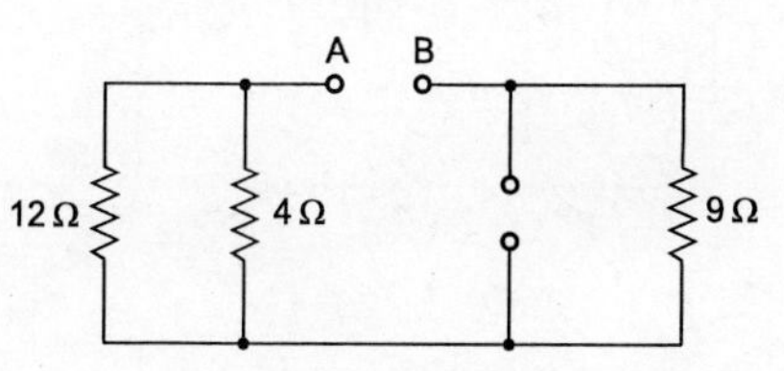

Fig. 2.98 (*b*)

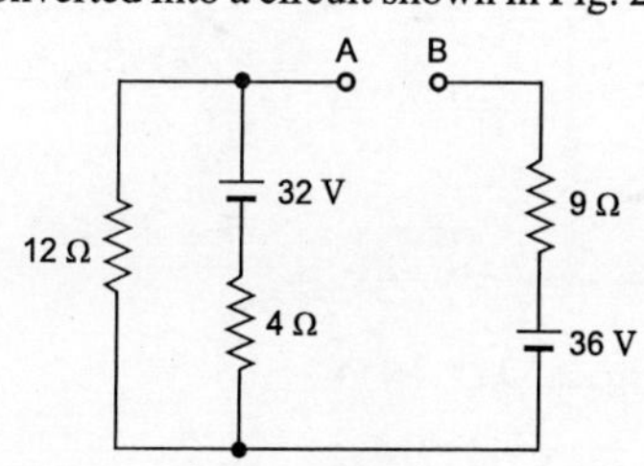

Fig. 2.98 (*c*)

Current flowing through the mesh formed by voltage source of 32 V and resistors 4 Ω and 12 Ω is given as

$$I = \frac{32}{12+4} = 2\text{ A}$$

$$\text{Voltage across terminals B and A, } V_{BA} = 36 - 12 \times 2 = 12\text{ V}$$

$$\text{Also } V_{BA} = 36 - (32 - 4 \times 2) = 12\text{ V}$$

$$\text{Current flowing through 6 Ω resistor} = \frac{V_{BA}}{R_T + R} = \frac{12}{12+6} = \frac{2}{3}\text{ A from terminal B to terminal A} \quad \textbf{Ans.}$$

Example 2.50 Consider the circuit shown in Fig. 2.99. Determine open-circuit voltage across AB terminals shown in Fig. 2.99 by Thevenin's theorem.

[U.P. Technical Univ. Electrical Engineering Odd Semester 2013-14]

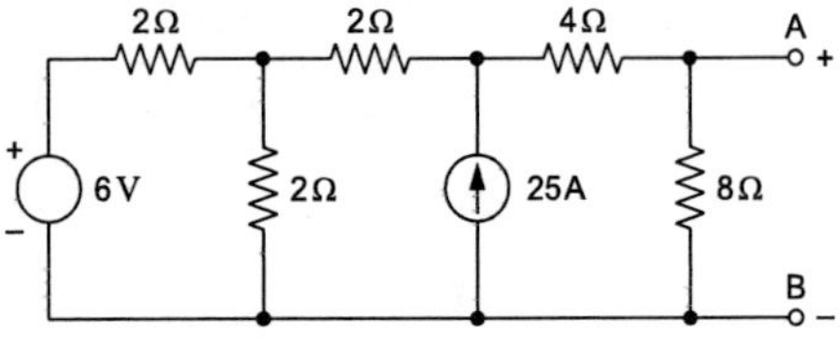

Fig. 2.99

Solution: For determining the Thevenin's equivalent resistance of the circuit with reference to terminals AB, the 6 V source is replaced by a short circuit and 25 A source is replaced by an open circuit, as shown in Fig. 2.100 (*a*).

$$\text{Thevenin's equivalent resistance, } R_T = 8 \parallel (4 + 2 + 2 \parallel 2) = 8 \parallel (4 + 2 + 1)$$

$$= 8 \parallel 7 = \frac{8 \times 7}{8+7} = \frac{56}{15}\ \Omega$$

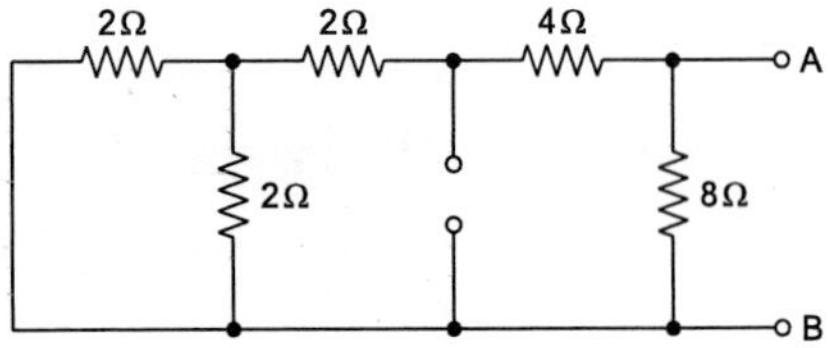

Fig. 2.100 (*a*)

For determination of short-circuit current I_{SC} *i.e.* current in zero resistance connected across terminals AB [Fig. 2.100(*b*)], Kirchhoff's voltage law is applied to loops I and II.

$$2I_1 + 2I_2 = 6$$

$$\text{or } I_1 + I_2 = 3 \qquad ...(i)$$

$$\text{and } 2(I_1 - I_2) + 4(I_1 - I_2 + 25) - 2I_2 = 0$$

$$\text{or } 6I_1 - 8I_2 = -100 \qquad ...(ii)$$

Solving Eqs. (*i*) and (*ii*) we have

$$I_2 = \frac{59}{7}\text{ A}$$

$$\text{and } I_1 = \frac{-38}{7}\text{ A}$$

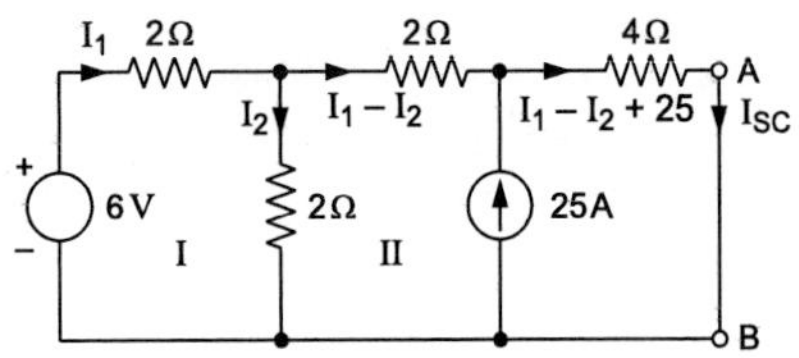

Fig. 2.100 (*b*)

Short-circuit current, I_{SC} = Current flowing through 4 Ω resistor

$$= I_1 - I_2 + 25 = \frac{-38}{7} - \frac{59}{7} + 25 = \frac{78}{7}\text{ A}$$

$$\text{Open-circuit voltage, } V_{OC} = I_{SC} \times R_T = \frac{78}{7} \times \frac{56}{15} = 41.6\text{ V} \quad \textbf{Ans.}$$

Example 2.51 Find the Thevenin's equivalent circuit for terminal pair AB of the network shown in Fig. 2.101.

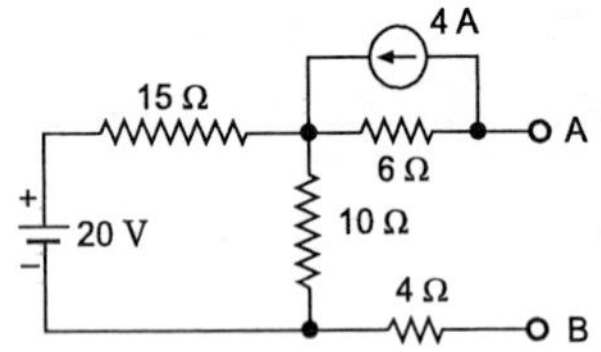

Fig. 2.101

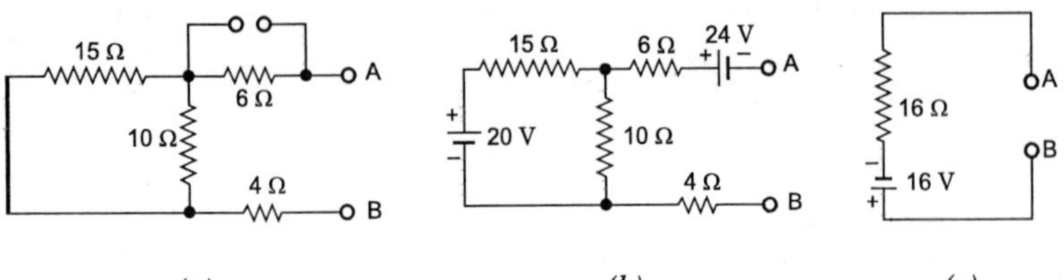

Fig. 2.102

Solution: For determining the Thevenin's equivalent resistance of the circuit with reference to terminals AB the voltage source is replaced by a short circuit and current source is replaced by an open circuit, as shown in Fig. 2.102 (*a*).

$$R_T = 6 + 4 + (15 \,||\, 10) = 16\ \Omega$$

Converting current source into the equivalent voltage source, the circuit becomes as shown in Fig. 2.102 (*b*).

Since there is no current through resistors of 4 Ω and 6 Ω (hence no voltage drop across them), Thevenin's voltage V_{Th} equals algebraic sum of battery voltage and drop across 10 Ω resistor. Hence

$$V_{Th} = 20 \times \frac{10}{10+15} - 24 = -16\ V$$

Thevenin's equivalent circuit is shown in Fig. 2.102 (*c*).

Example 2.52 **Find V_{Th} and R_{Th} for the circuit shown in Fig. 2.103.**

[U.P. Technical Univ. Electrical Engineering Odd Semester 2013-14]

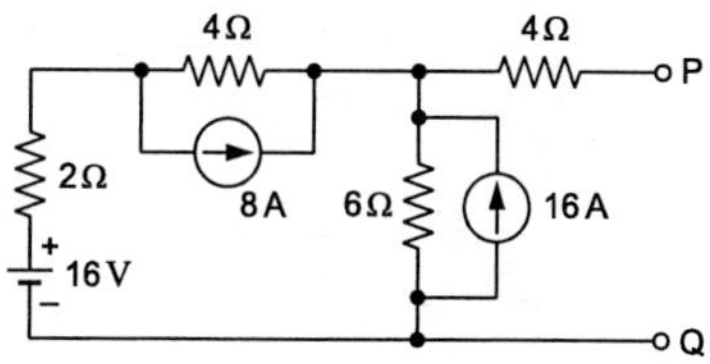

Fig. 2.103

Solution: The equivalent resistance of the network (with voltage source replaced by a short circuit and current sources replaced by open-circuits with reference to terminals P-Q [Fig. 2.104(*a*)]

$$R_1 = 4 + 6 \,||\, (4+2)$$
$$= 4 + 6 \,||\, 6 = 4 + 3 = 7\ \Omega \quad \textbf{Ans.}$$

(*a*) (*b*)

Fig. 2.104

The given circuit may be redrawn, as illustrated in Fig. 2.104 (*b*). The current sources are converted into their equivalent voltage sources, the circuit with terminals PQ kept open.

$$\text{Current flowing through closed loop, } I = \frac{16+32-96}{2+4+6} = \frac{-48}{12} = -4A$$

Thevenin's voltage V_{Th} = Voltage across branch BC = 96 + 6 I = 96 + 6 × (– 4) = 72 V **Ans.**

Example 2.53 **Determine the current in the 1 Ω resistor connected across AB of the network of Fig. 2.105 using Thevenin's theorem.**

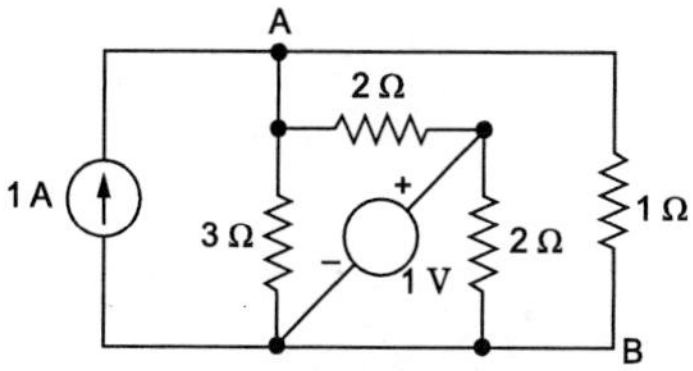

Fig. 2.105

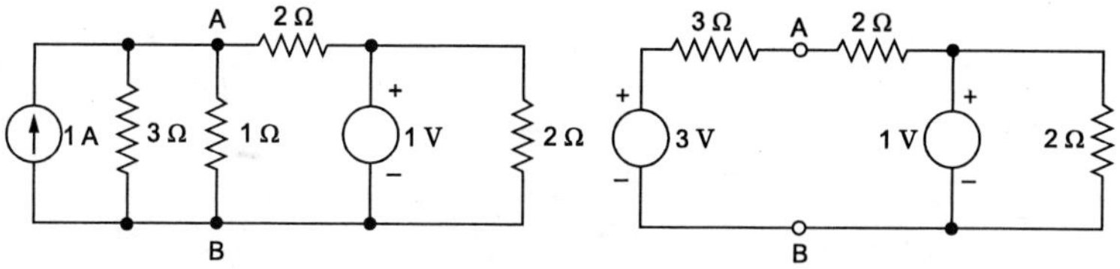

(*a*) (*b*)

Fig. 2.106

Solution: The given circuit may be redrawn, as illustrated in Fig. 2.106 (*a*). Now the current source is converted into its equivalent voltage source, the circuit, with terminals AB kept open, becomes as shown in Fig. 2.106 (*b*).

Open-circuited voltage across terminals AB,

$$V_T = 3 - \frac{3-1}{3+2} \times 3 = 1.8\ V$$

The equivalent resistance of the network (with voltage source replaced by a short circuit and current source replaced by an open circuit) with reference to terminals A-B, [Fig. 2.106 (*c*)]

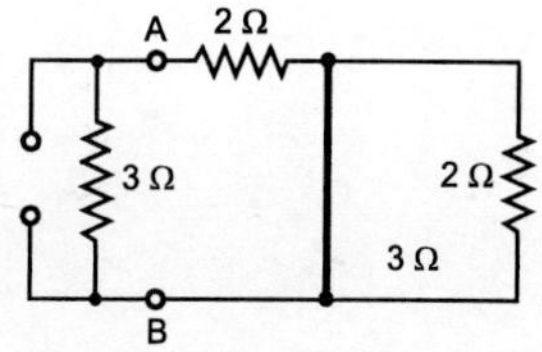

(*c*)

Fig. 2.106

$$R_T = 3 \parallel 2 = \frac{3 \times 2}{3+2} = 1.2\ \Omega$$

$$\text{Current in 1 } \Omega \text{ resistor} = \frac{V_T}{R_T + 1} = \frac{1.8}{1.2+1} = 0.82 \text{ A } \textbf{Ans.}$$

Example 2.54 **Using Thevenin's theorem obtain current in 13 Ω resistance in Fig. 2.107.**

[U.P. Technical Univ. First Semester 2007-08]

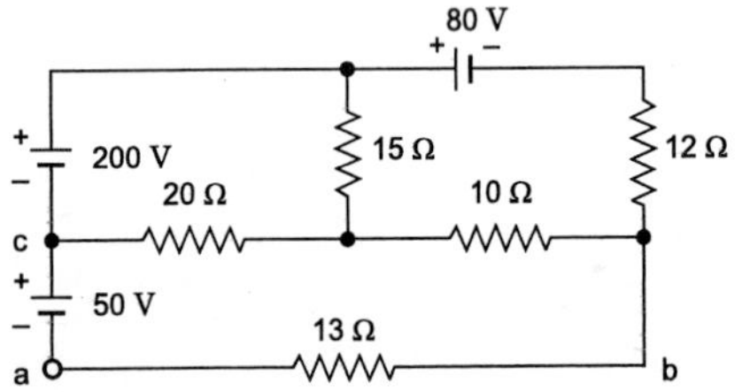

Fig. 2.107

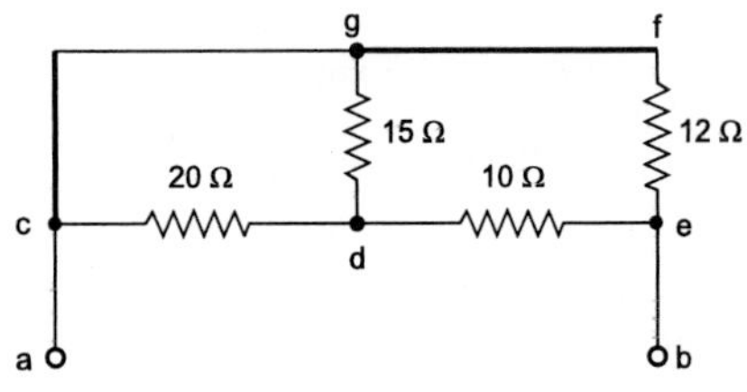

Fig. 2.108

Solution: For determining the Thevenin's equivalent resistance of the circuit w.r.t. terminals *a b*, the voltage sources are replaced by short circuits, as shown in Fig. 2.108.

Equivalent resistance of the network shown in Fig. 2.108.

$$R_T = 20 + 10 \parallel (15 + 12) = 20 + \frac{10 \times 27}{10 + 27}$$

$$= \frac{1{,}010}{37}\ \Omega = 27.3\ \Omega$$

After removing the resistance of 13 Ω, the circuit takes the form as shown in Fig. 2.109.

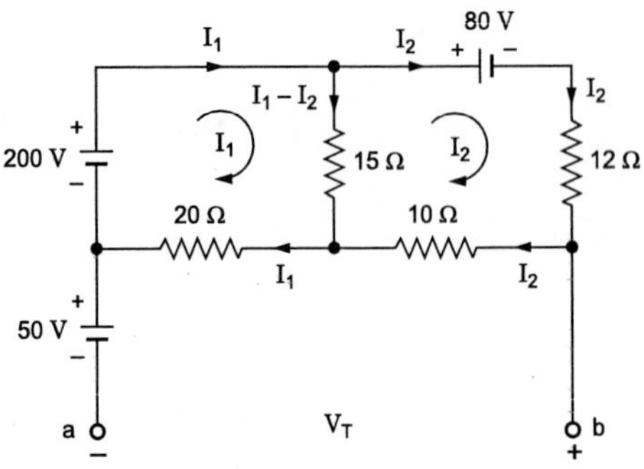

Fig. 2.109

Applying KVL to meshes I and II we have

$$20\, I_1 + 15\, (I_1 - I_2) = 200$$

$$\text{or} \quad 35\, I_1 - 15\, I_2 = 200 \qquad ...(i)$$

$$\text{and} \quad 15\, (I_1 - I_2) - 10\, I_2 - 12\, I_2 = 80$$

$$\text{or} \quad 15\, I_1 - 37\, I_2 = 80 \qquad ...(ii)$$

Multiplying Eq. (*i*) by 2 and Eq. (*ii*) by 5 we have

$$70\, I_1 - 30\, I_2 = 400 \qquad ...(iii)$$

$$75\, I_1 - 185\, I_2 = 400 \qquad ...(iv)$$

Subtracting Eq. (*iii*) from Eq. (*iv*) we have

$$5\, I_1 - 155\, I_2 = 0$$

$$\text{or} \quad I_1 = 31\, I_2$$

Substituting $I_1 = 31\, I_2$ in Eq. (*i*) we have

$$35 \times 31\, I_2 - 15\, I_2 = 200$$

$$\text{or} \quad I_2 = \frac{200}{1{,}070} \text{ A} \quad \text{and} \quad I_1 = 31\, I_2 = \frac{200 \times 31}{1{,}070} = \frac{6{,}200}{1{,}070} \text{ A}$$

$$\text{Thevenin voltage, } V_T = V_{ba} = 10\, I_2 + 20\, I_1 + 50$$

$$= 10 \times \frac{200}{1{,}070} + \frac{20 \times 6{,}200}{1{,}070} + 50 = 1.869 + 115.888 + 50 = 167.76 \text{ V}$$

$$\text{Current through 13 } \Omega \text{ resistance} = \frac{V_T}{R_T + 13} = \frac{167.76}{27.3 + 13} = 4.16 \text{ A} \quad \textbf{Ans.}$$

2.17. NORTON'S THEOREM

This theorem is in fact, an alternative to the Thevenin's theorem. Whereas by Thevenin's theorem a complex two-terminal network may be simplified for solution by reducing it into a simple circuit in which the so called open-circuit voltage and looking back resistance are connected in series with the load resistance, by Norton's theorem network is reduced into a simple circuit in which a parallel combination of constant current source and looking back resistance feeds the load resistance.

In both theorems use of resistance looking back into the network from the load terminals, with all sources removed leaving their internal resistances in the circuit is made. However, while solving circuit by Thevenin's theorem, the open-circuit voltage is determined at the load terminals with the load removed whereas in Norton's method use of a fictitious constant current source is made, the constant current delivered being equal to the current that would pass into a short circuit connected across the output terminals of the given network.

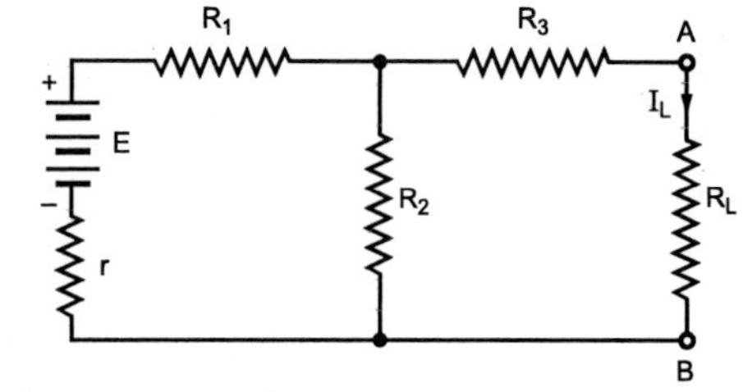

Fig. 2.110

Now for understanding this theorem let us consider a circuit shown in Fig. 2.110 in which load current I_L is to be determined.

For *determination of short-circuit current I_{SC}*, terminals A and B are short circuited by zero resistance thick wire, as illustrated in Fig. 2.111 (*a*).

$$\text{Equivalent resistance of the network, } R = (R_1 + r) + R_2 \,||\, R_3 = (R_1 + r) + \frac{R_2 R_3}{R_2 + R_3} \quad ...(2.28)$$

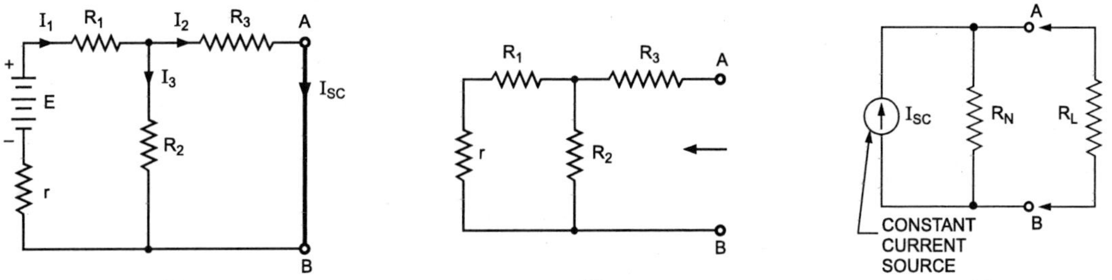

(*a*) (*b*) (c)

Fig. 2.111

$$\text{Current supplied by battery, } I_1 = \frac{E}{R} = \frac{E}{R_1 + r + \dfrac{R_2 R_3}{R_2 + R_3}} \quad ...(2.29)$$

Short-circuit current, I_{SC} = Current flowing through resistance R_3,

$$I_2 = I_1 \times \frac{R_2}{R_2 + R_3} \quad \text{By current division rule}$$

$$= \frac{E}{R_1 + r + \dfrac{R_2 R_3}{R_2 + R_3}} \times \frac{R_2}{R_2 + R_3} = \frac{E R_2}{(r + R_1)(R_2 + R_3) + R_2 R_3} \quad ...(2.30)$$

For determination of internal resistance R_{in} (or R_N) of the network under consideration, remove the load resistance R_L from terminals A and B and also remove voltage source from the circuit leaving behind only its internal resistance, as illustrated in Fig. 2.111 (*b*)

Equivalent resistance (R_N), as viewed from the open-terminals A and B is given as

$$R_N = R_3 + R_2 \,||\, (R_1 + r)$$

$$= R_3 + \frac{R_2 (R_1 + r)}{R_2 + R_1 + r} = \frac{R_2 R_3 + (R_1 + r)(R_2 + R_3)}{R_1 + R_2 + r} \quad ...(2.31)$$

Now when load resistance R_L is connected across terminals A and B, the network behaves as constant current source of current I_{SC} in parallel with a resistance R_N, as shown in Fig. 2.111 (*c*) and current flowing through the load resistance is given as

$$I_L = \frac{I_{SC} R_N}{R_N + R_L} \qquad \text{By current division rule}$$

$$= \frac{\dfrac{E R_2}{(r+R_1)(R_2+R_3)+R_2R_3} \times \dfrac{R_2R_3+(R_1+r)(R_2+R_3)}{R_1+R_2+r}}{\dfrac{R_2R_3+(R_1+r)(R_2+R_3)}{R_1+R_2+r}+R_L}$$

$$= \frac{E R_2}{R_2R_3+(R_1+r)(R_2+R_3)+R_L(R_1+R_2+r)} \qquad ...(2.32)$$

Norton's equivalent circuit is illustrated in Fig. 2.111 (*c*).

Norton's theorem can be stated as follows:

The current in any *passive circuit element* (which may be called R_L) in a network is the same as would flow in it if it were connected in parallel with R_N and the parallel pair were supplied with a constant current I_{SC}. R_N is the resistance measured "looking back" into the original circuit after R_L has been disconnected and all the sources have been replaced by their internal resistances: I_{SC} is the current which will flow in a short placed at the terminals of R_L in the original circuit.

2.18. CONVERSION OF THEVENIN'S EQUIVALENT INTO NORTON'S EQUIVALENT AND VICE VERSA

A Thevenin's equivalent can be converted into its Norton's equivalent and vice versa. A Thevenin's equivalent is depicted in Fig. 2.112 (*a*). According to statement made in Art 2.17, Norton's current source equals the current I_{SC} which flows through a short across terminals A and B.

$$\text{Hence } I_{SC} = \frac{V_{OC}}{R_{in}} \qquad ...(2.33)$$

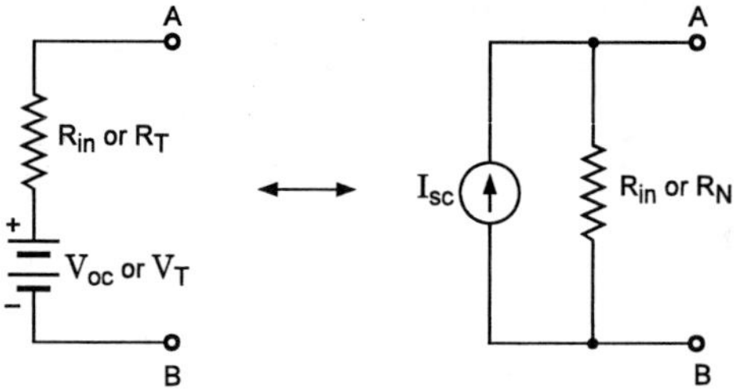

(*a*) *Thevenin's Equivalent* (*a*) *Norton's Equivalent*

Fig. 2.112 *Thevenin's Equivalent Compared With Norton's Equivalent*

Likewise a Norton's circuit can be converted into its Thevenin's equivalent. The Thevenin's equivalent source V_{OC} or V_T is the voltage on open circuit and is given as

$$V_{OC} \text{ or } V_T = I_{SC} R_{in} \qquad ...(2.34)$$

Each theorem is dual of the other.

Example 2.55 **Obtain the Norton's equivalent circuit at terminal AB of the network shown in Fig. 2.113.**

[*G.B. Technical Univ. Electrical Engineering Even Semester 2012-13; U.P. Technical Univ. Basic Electrical Engineering Second Semester 2013-14*]

Solution: Equivalent resistance of the network when viewed from terminals AB replacing 100 V and 40 V sources by short circuits [Fig. 2.114 (*a*)],

$$R_N = 2 \,||\, [5 + 4 \,||\, 2] = 2 \,||\, \left[5 + \frac{4}{3}\right]$$

$$= 2 \,||\, \frac{19}{3} = \frac{2 \times 19/3}{2 + 19/3} = \frac{38}{25}\ \Omega$$

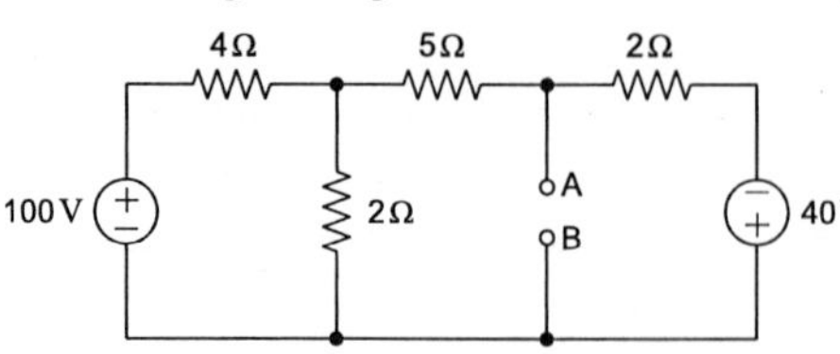

Fig. 2.113

For determination of short-circuit current I_{SC} *i.e.* current in zero resistance connected between terminals A and B, terminals A and B are short circuited as shown in Fig. 2.114 (*b*).

Current supplied by 40 V source

$$I_1 = \frac{40}{2} = 20 \text{ A}$$

Resistance of the circuit when viewed from 100 V source,

$$R = 4 + 2 \parallel 5 = 4 + \frac{2 \times 5}{2+5} = \frac{38}{7} \ \Omega$$

Current supplied by 100 V source,

$$I = \frac{100}{38/7} = \frac{350}{19} \text{ A}$$

Current flowing through 5 Ω resistance,

$$I_2 = I \times \frac{2}{2+5} = \frac{350}{19} \times \frac{2}{7} = \frac{100}{19} \text{ A}$$

Short-circuit current,

$$I_{SC} = I_1 - I_2 = 20 - \frac{100}{19} = \frac{280}{19} \text{ A}$$

Norton's equivalent circuit is shown in Fig. 2.14 (*c*).

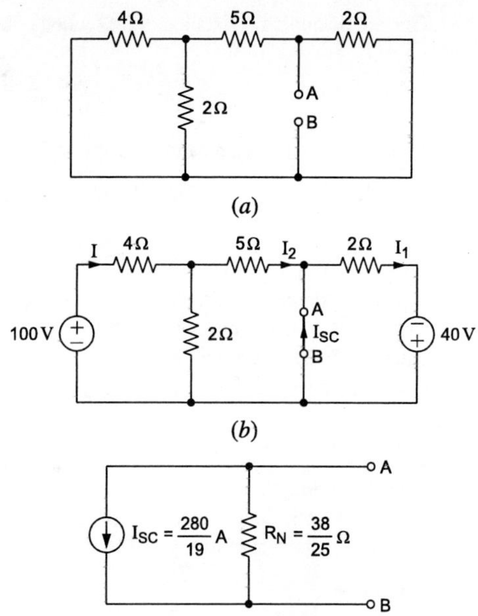

(*a*)

(*b*)

(*c*) *Norton's Equivalent Circuit*

Fig. 2.114

Example 2.56 **Determine the value of current through the 5 Ω resistance using Norton's theorem in the circuit shown in Fig. 2.115. State whether superposition theorem can be applied for the circuit with reasons.**

[*U.P. Technical Univ. Electrical Engineering Second Semester 2007-08*]

Solution: Equivalent resistance of the network when viewed from terminals *c* and *d* replacing the 100 V source by short circuit,

$$R_N = 20 + 10 \parallel (10 + 10) = 20 + \frac{10 \times 20}{10 + 20} = 26.667 \ \Omega$$

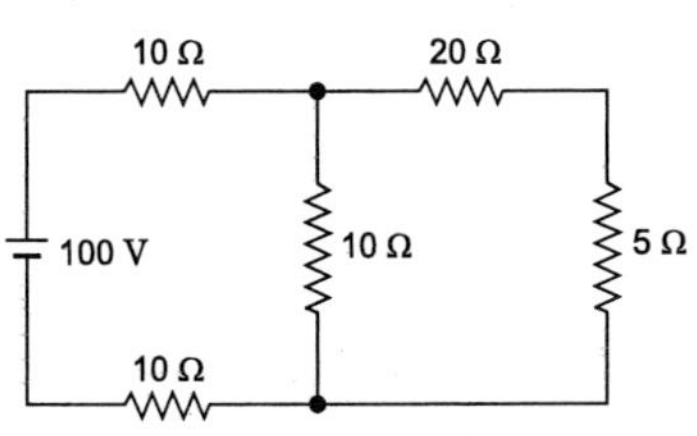

Fig. 2.115

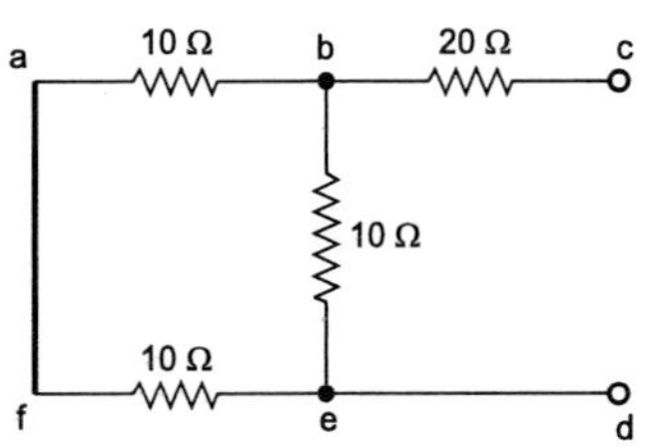

Fig. 2.116 (*a*)

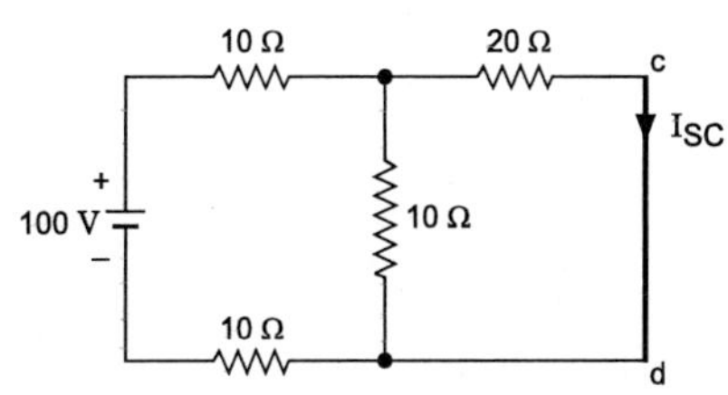

Fig. 2.116 (*b*)

For determination of short-circuit current, I_{SC} i.e. current in zero resistance connected between terminals *c* and *d*, 5 Ω resistance is replaced by zero resistance, as shown in Fig. 2.116 (*b*), current supplied by 100 V source

$$I = \frac{100}{10 + (10 \parallel 20) + 10} = \frac{100}{20 + \dfrac{10 \times 20}{10 + 20}} = \frac{100}{80/3} = 3.75 \text{ A}$$

Short-circuit current, I_{SC} = Current through 20 Ω resistance

$$= I \times \frac{10}{10 + 20} = 3.75 \times \frac{10}{30} = 1.25 \text{ A} \qquad \text{By current division rule}$$

Current through a resistance of 5 Ω when connected between terminals *c* and *d*

$$I = \frac{I_{SC}}{R_N + R_L} \times R_N = \frac{1.25 \times 26.667}{26.667 + 5} = \frac{20}{19} \text{ or } 1.053 \text{ A} \quad \textbf{Ans.}$$

Superposition theorem cannot be applied for the given circuit because the network consists only one supply source.

Example 2.57 State Norton's theorem. Find current through 15 Ω by using Norton's theorem.

[U.P. Technical Univ. Basic Electrical Engineering Odd Semester 2013-14]

Solution: Norton's equivalent resistance of the given network (i.e. equivalent resistance of the network after terminals A and B are open circuited, 30 V battery removed from the circuit leaving its internal resistance of 0.5 Ω in the circuit and removing 4 A current source leaving 0.5 Ω resistance in the circuit) as viewed from terminals A and B shown in Fig. 2.118 (*a*)

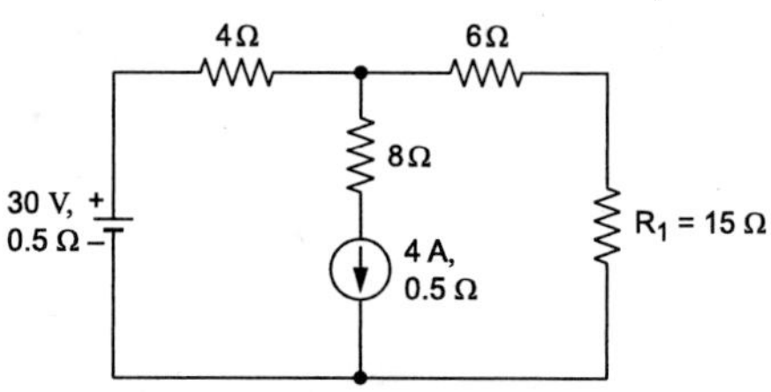

Fig. 2.117

$$R_N = 6 + 4 + 0.5 = 10.5 \ \Omega$$

For determining short-circuit current I_{SC} i.e. current in zero resistance connected across terminals A and B, resistance R_1 of 15 Ω is replaced by a zero resistance as shown in Fig. 2.118 (*b*).

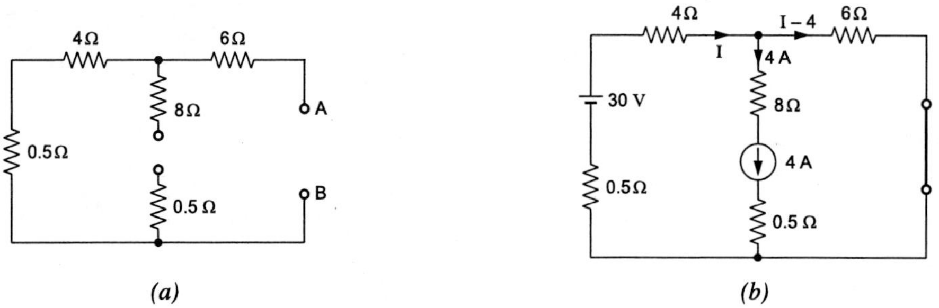

Fig. 2.118

Applying Kirchhoff's voltage law to outer loop we have

$$(0.5 + 4)\,I + 6\,(I - 4) = 30$$

$$\text{or } 10.5\,I = 54$$

$$\text{or } I = \frac{36}{7} \text{ A}$$

$$\text{Short-circuit current, } I_{SC} = I - 4 = \frac{36}{7} - 4 = \frac{8}{7} \text{ A}$$

$$\text{Current through 15 Ω resistor} = \frac{I_{SC}}{R_N + R_1} \times R_N = \frac{8/7}{10.5 + 15} \times 10.5 = \frac{8}{17} \text{ A} \quad \textbf{Ans.}$$

Example 2.58 Using Norton's theorem, find current which would flow in a 25 Ω resistor connected between points N and O in Fig. 2.119.

Solution: Equivalent resistance of network when viewed from terminals N and O, keeping all the voltage sources replaced by short circuits (Fig. 2.120)

$$R_N = 5 \ || \ 10 \ || \ 20 = \frac{1}{\frac{1}{5} + \frac{1}{10} + \frac{1}{20}} = \frac{20}{7} \ \Omega$$

Short-circuit current *i.e.* the current in zero resistance conductor connected across terminals ON (Fig. 2.121)

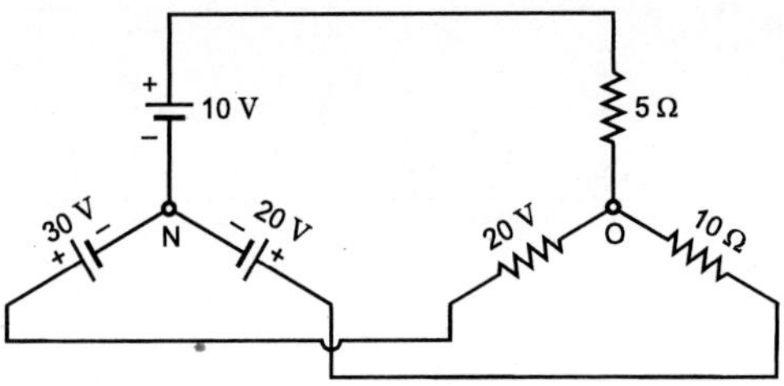

Fig. 2.119

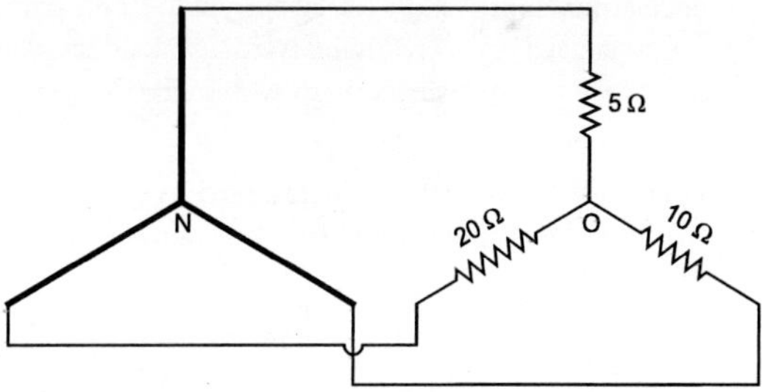

Fig. 2.120

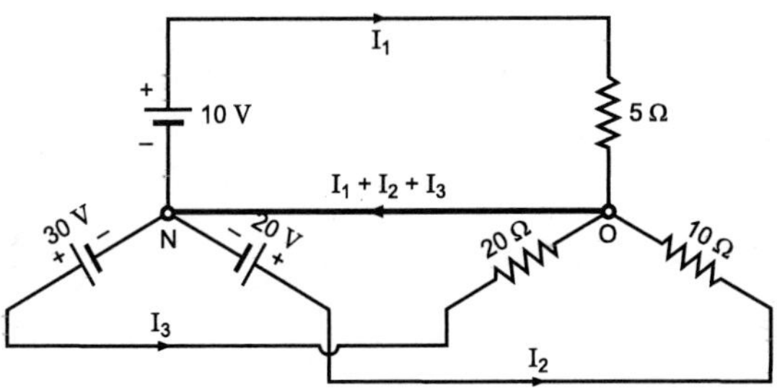

Fig. 2.121

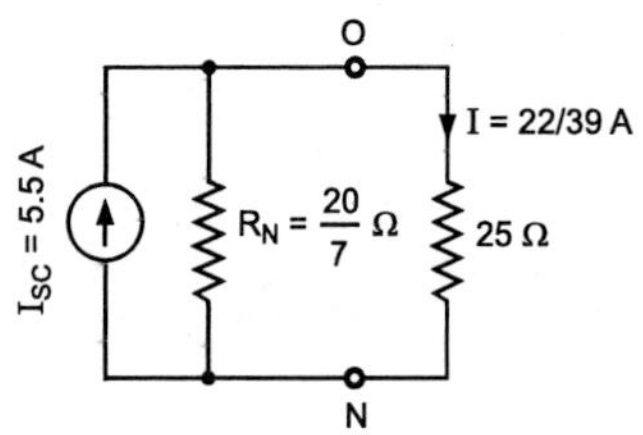

Fig. 2.122

$$I_{SC} = I_1 + I_2 + I_3 = \frac{10}{5} + \frac{20}{10} + \frac{30}{20} = 5.5 \text{ A}$$

Current through a resistance of 25 Ω connected between points O and N,

$$I = \frac{I_{SC}}{R_N + R_L} \times R_N = \frac{5.5 \times \frac{20}{7}}{\frac{20}{7} + 25} = \frac{22}{39} \text{ A Ans.}$$

Example 2.59 **Apply Norton's theorem to find the Norton's equivalent circuit as seen by R_L in the circuit shown in the following figure (Fig. 2.123).**

[U.P. Technical Univ. Electrical Engineering Second Semester 2008-09]

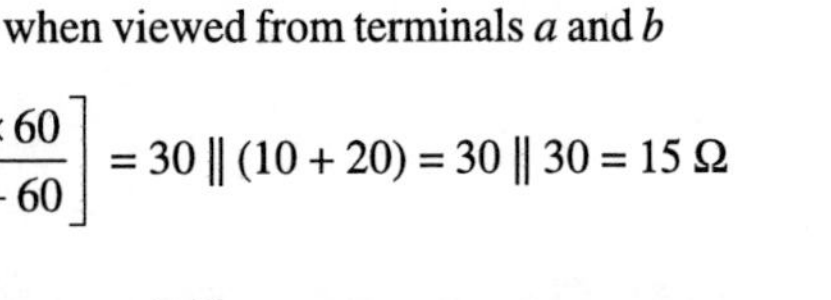

Fig. 2.123

Solution: For determination of Norton's equivalent resistance of the circuit w.r.t. terminals *a* and *b*, the voltage sources are replaced by short circuits and load resistance R_L is removed as shown in Fig. 2.124 (*a*).

Norton's resistance, R_N = Equivalent resistance of the network when viewed from terminals *a* and *b*

$$= 30 \parallel (10 + 30 \parallel 60) = 30 \parallel \left[10 + \frac{30 \times 60}{30 + 60}\right] = 30 \parallel (10 + 20) = 30 \parallel 30 = 15 \ \Omega$$

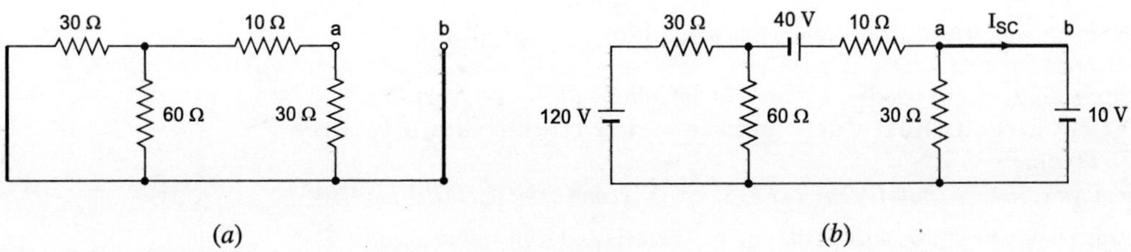

Fig. 2.124

For determination of short-circuit current I_{SC}, *i.e.* current in zero resistance connected between terminals *a* and *b*, load resistance R_L is replaced by zero ohm resistance as shown in Fig. 2.124 (*b*).

Applying superposition theorem.

Replacing 40 V and 10 V voltage sources by short circuits, the circuit is reduced to a simple circuit as shown in Fig. 2.125(*a*).

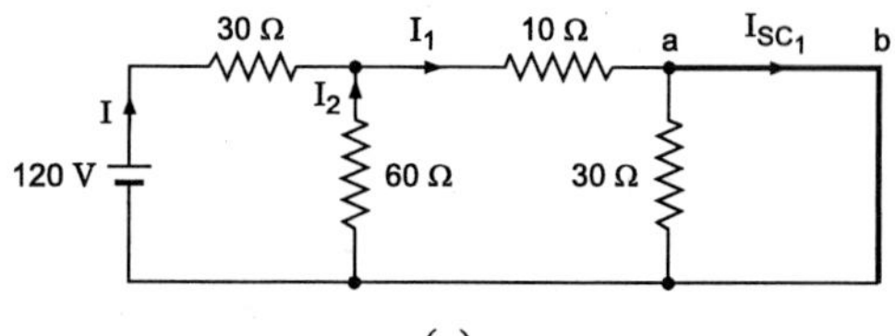

(*a*)

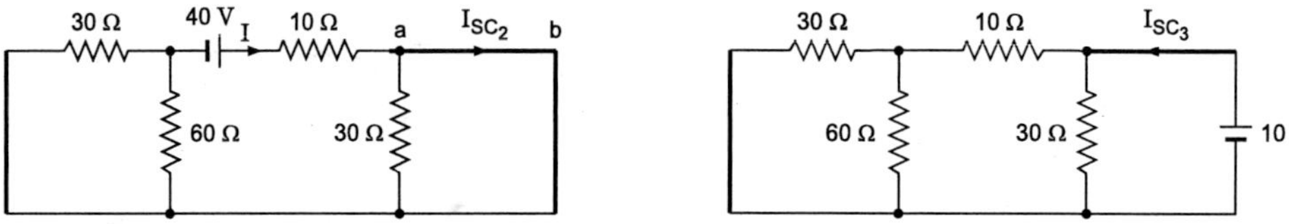

(*b*) (*c*)

Fig. 2.125

Current supplied by the 120 V

$$\text{Source, I} = \frac{120}{30+60\,||\,10} = \frac{120}{30+\dfrac{60\times 10}{70}} = \frac{120}{\dfrac{210+60}{7}} = \frac{120\times 7}{270} = \frac{28}{9}\text{ A}$$

Current through zero ohm, resistance wire connected between terminals *a* and *b*,

$$I_{SC1} = I_1 = I \times \frac{60}{60+10} = \frac{28}{9}\times\frac{60}{70} = \frac{8}{3}\text{ A} \qquad \text{By current division rule}$$

Replacing 120 V and 10 V voltage sources by short circuit, the circuit is reduced to a simple circuit, as shown in Fig. 2.125 (*b*).

Current through zero ohm resistance wire connected between terminals *a* and *b*,

$$I_{SC2} = I = \frac{40}{10+30\,||\,60} = \frac{40}{10+\dfrac{30\times 60}{30+60}} = \frac{40}{30} = \frac{4}{3}\text{ A}$$

Replacing 120 V and 40 V voltage sources by short circuit the circuit is reduced to as shown in Fig. 2.125 (*c*).

Current through zero ohm resistance connected between terminals *a* and *b*

$$I_{SC3} = \text{Current supplied by 10 V voltage source}$$

$$= \frac{10}{30\,||\,[10+(30\,||\,60)]} = \frac{10}{30\,||\left[10+\dfrac{30\times 60}{30+60}\right]} = \frac{10}{30\,||\,30} = \frac{10}{15} = \frac{2}{3}\text{ A}$$

Total current through zero ohm resistance connected between terminals *a* and *b*,

$$I_{SC} = I_{SC_1} + I_{SC_2} - I_{SC_3} = \frac{8}{3}+\frac{4}{3}-\frac{2}{3} = \frac{10}{3}\text{ A}$$

∵ The current I_{SC_3} is flowing in a direction opposite to that of currents I_{SC_1} and I_{SC_2}

Norton's equivalent circuit is shown in Fig. 2.126.

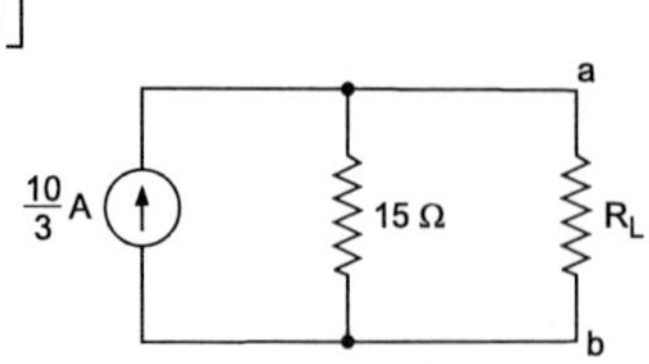

Fig. 2.126

Example 2.60 **Convert the network at terminals *ab* by its Norton's equivalent circuit. Hence determine I of the circuit shown in Fig. 2.127 (*a*).**

[*Mahamaya Technical Univ. Electrical Engineering Second Semester 2010-11*]

Solution: Equivalent resistance of the circuit when viewed from terminals *ab* with voltage sources replaced by short circuits [Fig. 2.127(*b*)],

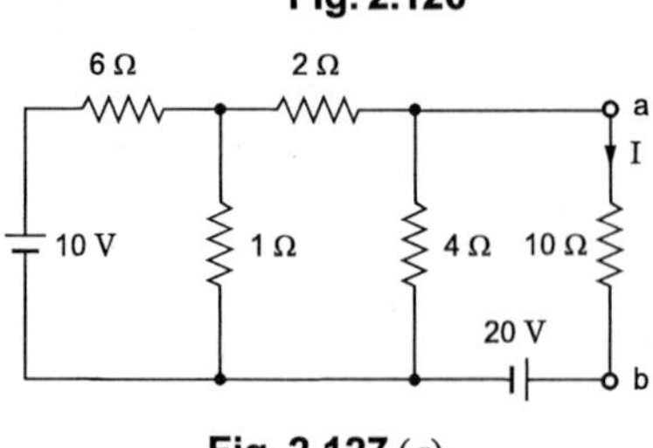

Fig. 2.127 (*a*)

$$R_N = 4 \parallel (2 + 1 \parallel 6) = 4 \parallel \left(2 + \frac{1 \times 6}{6+1}\right)$$

$$= 4 \parallel \frac{20}{7} = \frac{4 \times \frac{20}{7}}{4 + \frac{20}{7}} = \frac{80}{48} = \frac{5}{3}\ \Omega$$

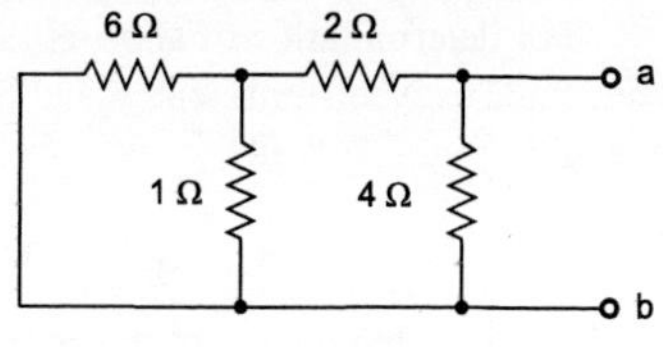

(*b*)

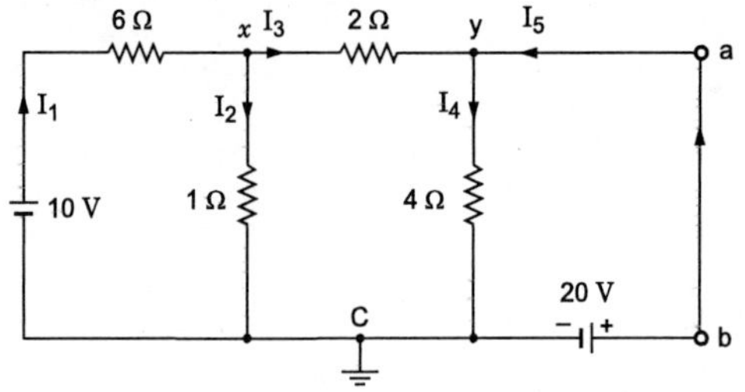

Fig. 2.127 (*c*)

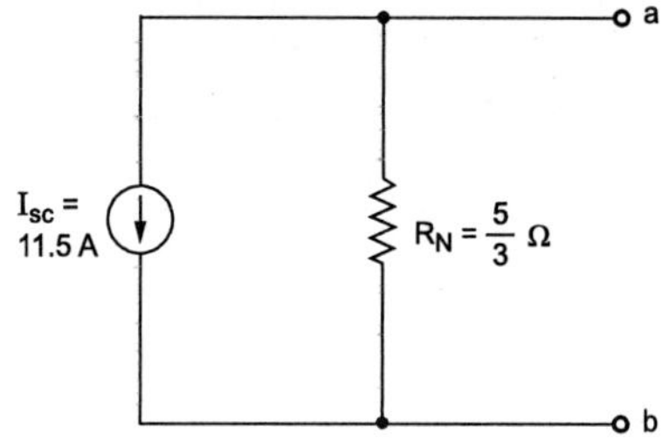

Fig. 2.127 (*d*) *Norton's Equivalent*

For determination of short-circuit current I_{SC} i.e. current in zero resistance connected between terminals *a* and *b*, R_L of 10 Ω is replaced by zero resistance as shown in Fig. 2.127 (*c*).

Let the node C be the reference node and let the arbitrary distribution of current as shown in Fig. 2.127 (*c*).

Applying KCL to node *x*, we have

$$I_1 = I_2 + I_3$$

$$\text{or} \quad \frac{10 - V_x}{6} = \frac{V_x}{1} + \frac{V_x - V_y}{2}$$

$$\text{or} \quad 10 - V_x = 6\,V_x + 3\,V_x - 3\,V_y$$

$$\text{or} \quad 10\,V_x - 3\,V_y = 10 \qquad ...(i)$$

$$\text{or} \quad V_x = \frac{10 + 3V_y}{10} = \frac{10 + 3 \times 20}{10} = 7\text{ V} \qquad \because\ V_y = 20\text{ V}$$

$$\text{So} \quad I_3 = \frac{V_x - V_y}{2} = \frac{7 - 20}{2} = -6.5\text{ A}$$

$$\text{and} \quad I_4 = \frac{V_y}{4} = \frac{20}{4} = 5\text{ A}$$

Short-circuit current, $I_{SC} = I_5 = I_4 - I_3 = 5 - (-6.5) = 11.5$ A

Norton's equivalent circuit is shown in Fig. 2.127 (*d*).

$$\text{Current through } R_L \text{ of } 10\ \Omega = I_{SC} \times \frac{R_N}{R_N + R_L} = 11.5 \times \frac{5/3}{5/3 + 10} = \frac{11.5 \times 5}{35} = 1.63\text{ A} \quad \textbf{Ans.}$$

Example 2.61 State Norton's theorem in dc circuit. Also calculate Norton's equivalent of the network shown in Fig. 2.128 (*a*) at terminal AB. Determine the current through 4 Ω resistor across AB.

[*G.B. Technical Univ. Electrical Engineering First Semester 2011-12*]

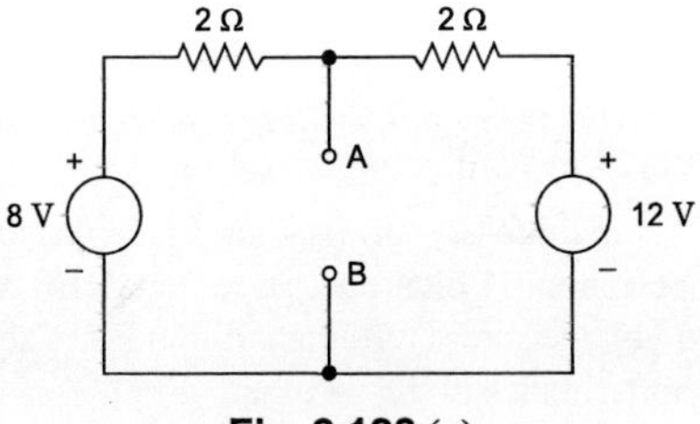

Fig. 2.128 (*a*)

Solution: Refer to Art. 2.17.

The equivalent resistance of the network, with voltage sources replaced by short circuits, as shown in Fig. 2.128 (*b*) with reference to terminals A and B.

$$R_T = 2 \parallel 2 = \frac{1}{\frac{1}{2} + \frac{1}{2}} = 1\ \Omega$$

For determination of short-circuit current I_{SC} i.e. current in zero resistance connected between terminals A and B the terminals A and B are short circuited, as shown in Fig. 2.128 (*c*).

$$I_{SC} = I_1 + I_2 = \frac{8}{2} + \frac{12}{2} = 10 \text{ A}$$

Norton's equivalent of the circuit is shown in Fig. 2.128 (*d*).

Current through 4 Ω resistor connected across AB

$$I_L = \frac{I_{SC} \times R_N}{R_N + R_L} = \frac{10 \times 1}{1 + 4} = 2 \text{ A } \textbf{Ans.}$$

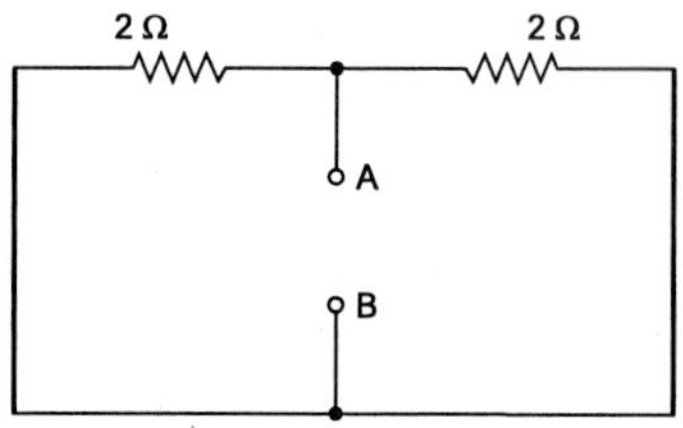

Fig. 2.128 (*b*)

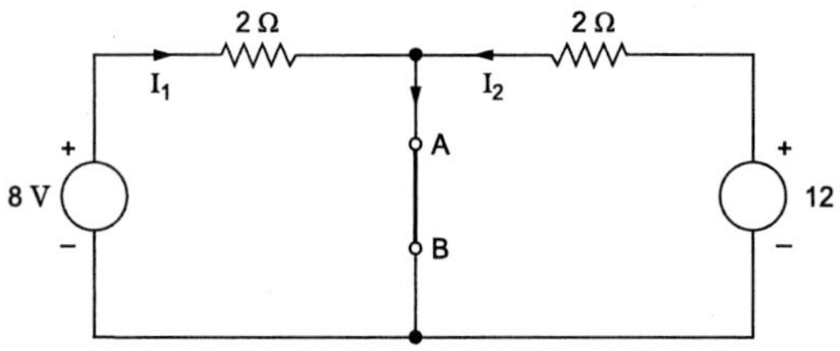

Fig. 2.128 (*c*)

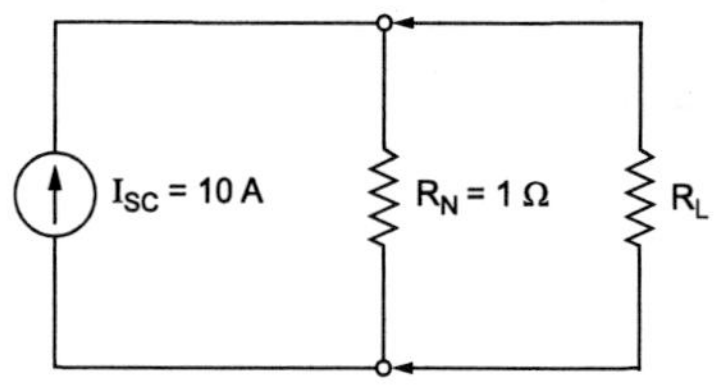

Fig. 2.128 (*d*)

Example 2.62 Draw the Norton's equivalent circuit across AB, and determine current flowing through 12 Ω resistor for the network shown in Fig. 2.129. [*U.P. Technical Univ. Electrical Engineering Second Semester 2003-04*]

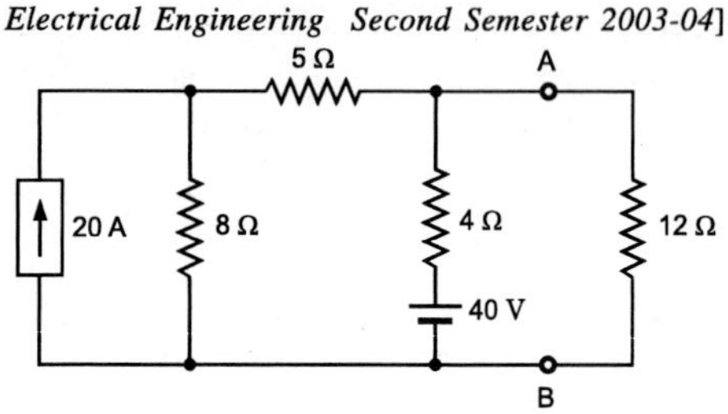

Fig. 2.129

Solution: As illustrated in Fig. 2.130 (*a*), terminals A and B have been shorted after removing 12 Ω resistor.

Now short-circuit current is determined by making use of superposition theorem.

(*i*) *When only current source is present* [Fig. 2.130 (*b*)]

In this case 40 V battery is replaced by a short circuit. The 20 A current divides at point C between parallel combination of 8 Ω and 5 Ω (no current will flow through 4 Ω resistor being short circuited at terminals A and B).

I_{SC_1} = Current through 5 Ω resistor due to current source alone

$$= 20 \times \frac{8}{8+5} = \frac{160}{13} \text{ A}$$

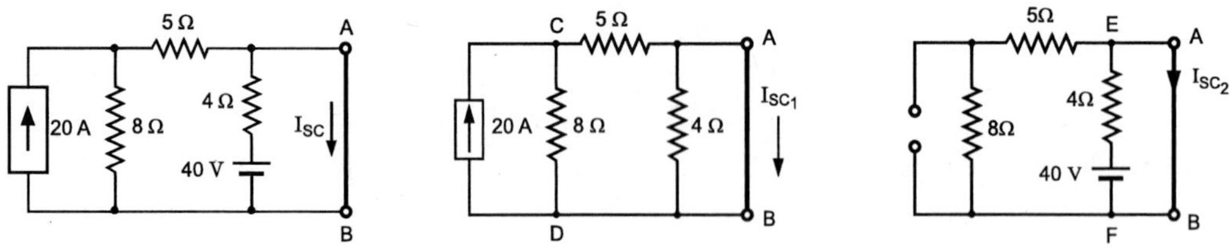

(*a*) (*b*) (*c*)

Fig. 2.130

(*ii*) *When only voltage source is present* [Fig. 2.130 (*c*)]

In this case, current source is replaced by an open circuit. Voltage across terminals EF is equal to voltage across terminals A and B *i.e.* zero. So short-circuit current,

I_{SC_2} = Current through 4 Ω resistor

$$= \frac{40}{4} = 10 \text{ A}$$

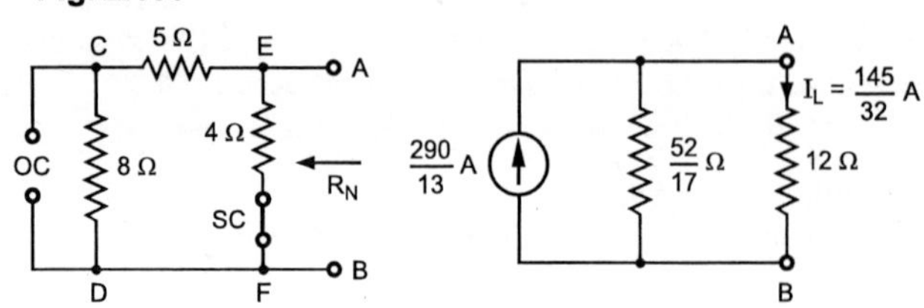

(*d*) (*e*)

Fig. 2.130

$$\text{Short-circuit current, } I_{SC} = I_{SC_1} + I_{SC_2} = \frac{160}{13} + 10 = \frac{290}{13} \text{ A}$$

As seen from Fig. 2.130 (*d*), Norton's equivalent resistance of the network,

$$R_N = 4 \parallel (5+8) = 4 \parallel 13 = \frac{1}{\frac{1}{4}+\frac{1}{13}} = \frac{52}{17}\ \Omega$$

Fig. 2.130 (*e*) shows the Norton's equivalent circuit with the load resistor of 12 Ω.

$$\text{Load current, } I_L = I_{SC} \times \frac{R_N}{R_N + R_L} = \frac{290}{13} \times \frac{52/17}{\frac{52}{17}+12} = \frac{1{,}160}{256} = 4.53125 \text{ A } \textbf{Ans.}$$

Example 2.63 **Find the Norton's equivalent circuit as seen by R_L in the following circuit:**

[U.P. Technical Univ. Electrical Engineering First Semester 2007-08]

Solution: Equivalent resistance of the circuit when viewed from terminals *ab* with voltage sources replaced by short circuits [Fig. 2.131 (*b*)].

$$R_N = 30 \parallel (10 + 30 \parallel 60) = 30 \parallel \left(10 + \frac{30 \times 60}{30+60}\right) = 30 \parallel 30 = 15\ \Omega$$

For determination of short-circuit current I_{SC} i.e. current in zero resistance connected between terminals *a* and *b*, R_L is replaced by zero resistance as shown in Fig. 2.131 (*c*).

Let the node J be the reference node and let the arbitrary distribution of current as shown in Fig. 2.131 (*c*).

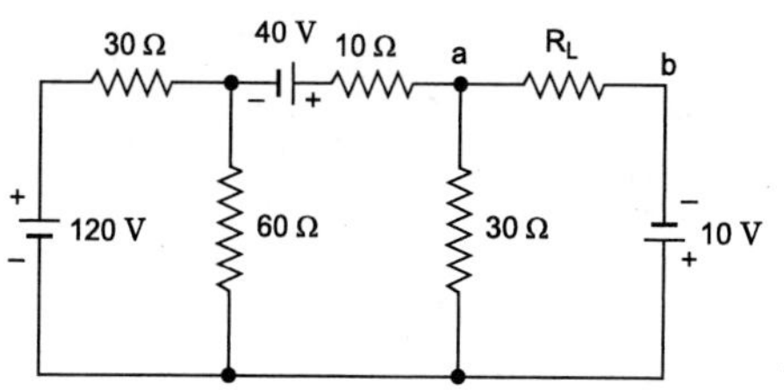

Fig. 2.131 (*a*)

Applying KCL to node *h* we have

$$I_1 = I_2 + I_3$$

$$\text{or} \quad \frac{120 - V_h}{30} = \frac{V_h}{60} + \frac{V_h + 40 - V_a}{10} \qquad \because V_a = -10 \text{ V}$$

$$\text{or} \quad 2(120 - V_h) = V_h + 6(V_h + 40 + 10)$$

$$\text{or} \quad V_h = \frac{-20}{3} \text{ V}$$

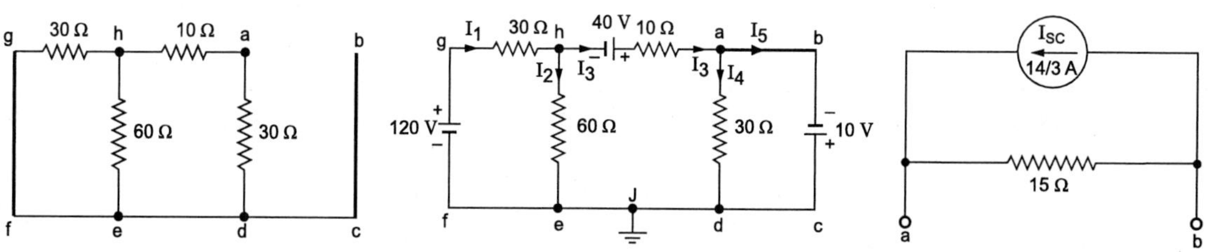

(*b*) (*c*) (*d*)

Fig. 2.131

Applying KCL to node a we have

$$I_3 = I_4 + I_5$$

$$\text{or} \quad I_5 = I_3 - I_4 = \frac{V_h + 40 - V_a}{10} - \frac{V_a}{30} = \frac{\frac{-20}{3} + 40 - (-10)}{10} - \frac{(-10)}{30} = \frac{14}{3} \text{ A}$$

$$\text{and short-circuit current } I_{SC} = I_5 = \frac{14}{3} \text{ A}$$

Thus Norton's equivalent circuit is a current source of $\frac{14}{3}$ A with a resistance of 15 Ω across it as shown in Fig. 2.131(*d*).

2.19. MAXIMUM POWER TRANSFER THEOREM

This theorem is particularly useful for analyzing communication networks because in communication engineering it is usually desirable to deliver maximum power to a load. Electronic amplifiers in radio and television receivers are so designed to do this as it does not matter that an equal amount of power is lost in the process. Obviously, it is important to make maximum use of power available for driving a loudspeaker. One can readily agree, however, that such a situation would not be tolerable in power engineering because in this situation efficiency of transmission will be very poor whereas in this case goal is high efficiency and not maximum power transfer.

Consider a generator supplying electrical power over a transmission line to a load resistor R_L, as shown in Fig. 2.132 (*a*). Let E_g be the voltage generated inside the generator, and let R_0 be the internal resistance of generator plus the resistance of the two line conductors. It is required to determine the value of R_L that will draw maximum power from the source.

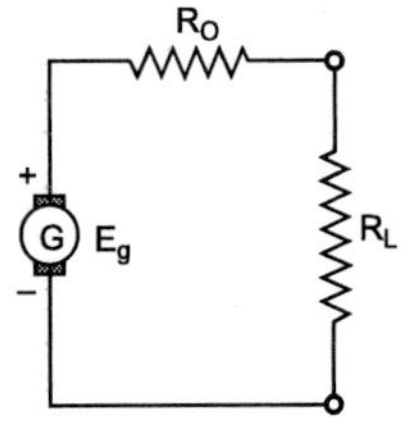

Fig. 2.132 (*a*)

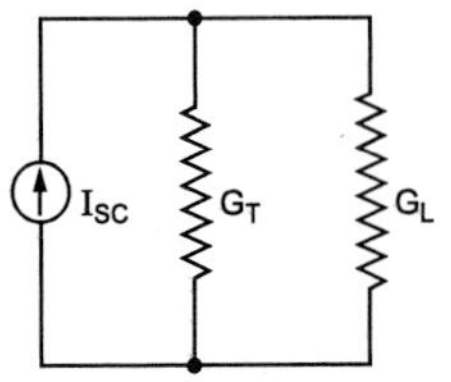

Fig. 2.132 (*b*)

The current delivered to load resistance, $I = E_g/(R_0 + R_L)$

Power delivered to load resistance, $P = I^2 R_L = \dfrac{E_g^2 R_L}{(R_0 + R_L)^2}$

To find the maximum value of power, differentiate the above expression with respect to R_L and equate to zero. Thus,

$$\frac{dP}{dR_L} = \frac{E_g^2 (R_0 + R_L)^2 - 2 R_L (R_0 + R_L) E_g^2}{(R_0 + R_L)^4} = 0$$

or $R_L = R_0$

Thus, the condition for maximum power transfer is that the load resistance R_L shall be equal to R_0.

The value of the maximum power transferred is seen to be $P_{max} = \dfrac{E_g^2}{4 R_0}$ ...(2.35)

As applied to dc networks, this theorem may be stated as follows:

A resistive load, served through a resistive network, will abstract maximum power when the load resistance value is the same as the resistance "viewed by the load as it looks back into the network". This is also called the output resistance of the network or the resistance presented to the output terminals by the network. It is the resistance we called R_T *in explaining Thevenin's theorem (Art 2.16). The load resistance* R_L *will draw maximum power when it will be equal to* R_T *i.e.* $R_L = R_T$ *and the maximum power drawn will* be equal to $\dfrac{V_{OC}^2}{4 R_L}$ where V_{OC} is the open-circuit voltage at the terminals from which R_L has been disconnected.

Similarly for a current source, as illustrated in Fig. 2.132 (*b*), the power transferred will be maximum when load conductance is equal to the source conductance *i.e.*

$$G_L = G_T$$

and the maximum power transferred is $P_{max} = \dfrac{I_{SC}^2}{4 G_L}$ ...(2.36)

Example 2.64 Find the value of load resistance R_L in the circuit of Fig. 2.133 for maximum power transfer. Determine the maximum power transferred. [*G.G.S.I.P. Delhi Univ. Electrical Science 2005-06*]

Solution: The equivalent resistance of the network, when viewed from terminals A and B after replacing the voltage source by short circuit, is given as

$$R_T = 38 + 20 \parallel 30 = 38 + \frac{60}{3+2} = 38 + 12 = 50\ \Omega$$

Maximum power will be transferred to load resistance R_L when it is equivalent to Thevenin's equivalent resistance, R_T

i.e. $R_L = R_T = 50\ \Omega$ **Ans.**

Open-circuit voltage, V_T = Voltage across 20 Ω resistor

$$= 50 \times \frac{20}{20+30} = 20\ V$$

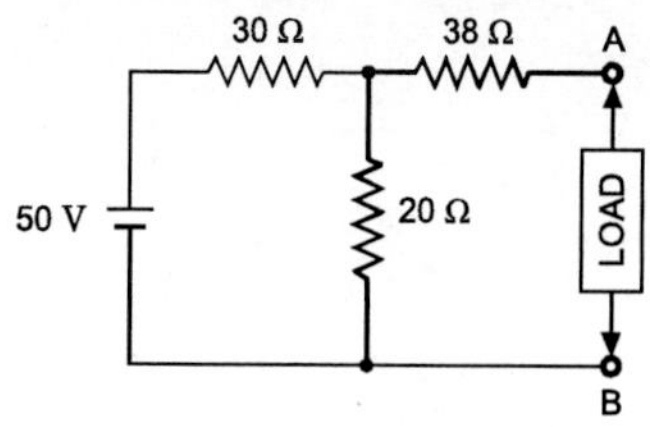

Fig. 2.133

$$\text{Maximum power transferred, } P_{max} = \frac{V_T^2}{4R_L} = \frac{20^2}{4 \times 50} = 2 \text{ watts } \textbf{Ans.}$$

Example 2.65 **Find the value of resistance 'R' to have maximum power transfer in the circuit as shown in Fig. 2.134. Also obtain the amount of maximum power.**

[U.P. Technical Univ. Electrical Engineering Second Semester 2003-04; Basic Electrical Engineering Even Semester 2013-14]

Solution: Let the current source of 2 A shunted by 15 Ω resistor be converted into its equivalent voltage source of 2 × 15 = 30 V connected in series with a resistance of 15 Ω and load resistor R be removed. The circuit is shown in Fig. 2.135 (*a*).

The equivalent resistance of the network with voltage sources replaced by short circuits when viewed from terminals A-B [Fig. 2.135 (*b*)],

$$R_T = 3 \parallel (6+15) = 3 \parallel 21 = \frac{1}{\frac{1}{3}+\frac{1}{21}} = \frac{21}{8}\ \Omega$$

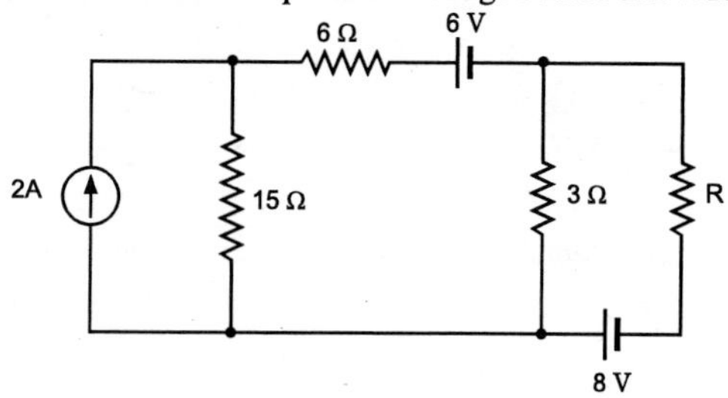

Fig. 2.134

Current flowing through a resistance of 3 Ω in the circuit shown in Fig. 2.135 (*a*).

$$I = \frac{30-6}{15+6+3} = 1\ A$$

Open-circuit voltage,

V_T = Voltage across 3 Ω resistor + 8 V

$= 1 \times 3 + 8 = 11$ V

According to maximum power transfer theorem, R will absorb maximum power when it is equal to R_T *i.e.* $\frac{21}{8}\ \Omega$ **Ans.**

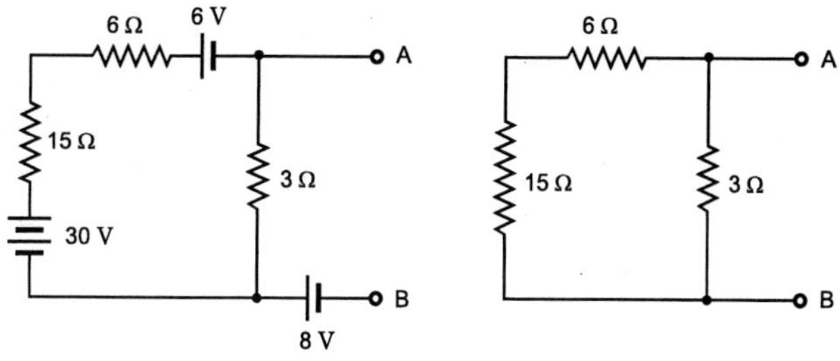

Fig. 2.135

Maximum power transferred,

$$P_{max} = \frac{V_T^2}{4\ R_L} = \frac{11^2}{4 \times \frac{21}{8}} = 11.524\ W \textbf{ Ans.}$$

Example 2.66 **For the circuit shown in Fig. 2.136, find the Thevenin's equivalent voltage source. Also give the value of load resistance to be connected across terminals A and B to get maximum power delivered.**

[U.P. Technical Univ. Electrical Engineering First Semester 2004-05]

Solution:

$$R_T = 15 + 10 \parallel 10 = 15 + 5 = 20\ \Omega$$

$$V_T = \frac{100 \times 10}{10+10} = 50\ V$$

Hence the Thevenin's equivalent voltage source is of open-circuit voltage 50 V in series with a resistance of 20 Ω **Ans.**

According to maximum power transfer theorem, load resistance will absorb maximum power when it is equal to R_T *i.e.* 20 Ω **Ans.**

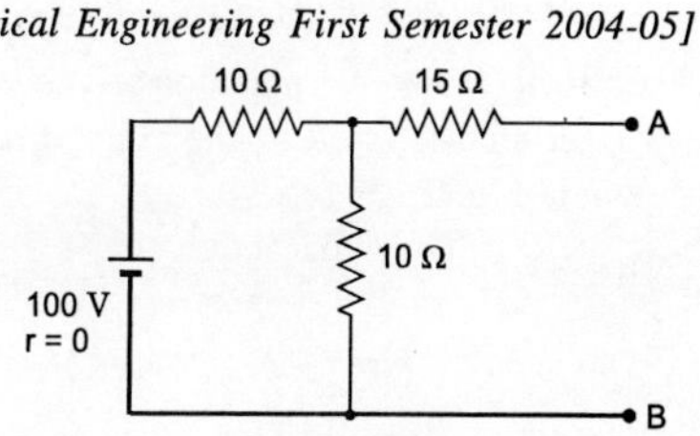

Fig. 2.136

Example 2.67 **Find the value of R in the circuit of Fig. 2.137 (*a*) such that the maximum power transfer takes place.**

[U.P. Technical Univ. Electrical Engineering September 2001]

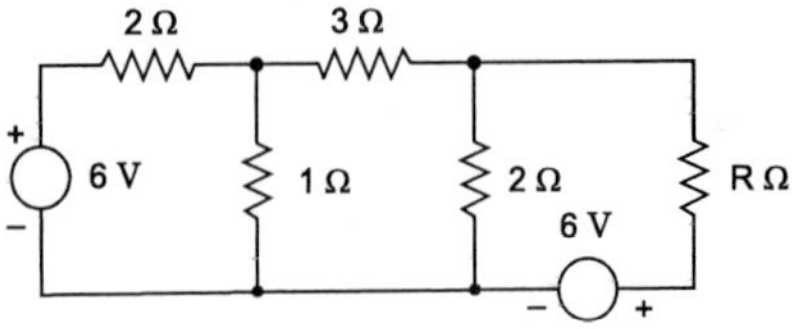

Fig. 2.137 (*a*)

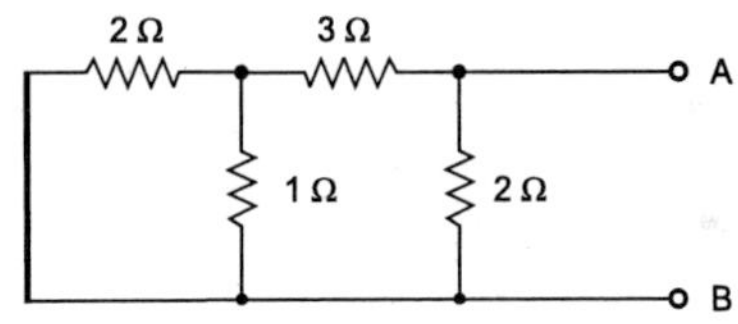

Fig. 2.137 (*b*)

Solution: The equivalent resistance of the network when viewed from terminals A and B after removing resistor R from the circuit and replacing the voltage [Fig. 2.137 (*b*)] sources by short circuits we have

$$R_T = 2 \,||\, [3 + 2 \,||\, 1]\ \Omega = 2 \,||\, \left[3 + \frac{2}{3}\right]\ \Omega = \left(2 \,||\, \frac{11}{3}\right)\ \Omega = \frac{22}{17}\ \Omega$$

Maximum power will be transferred to the resistor R when it is equal to Thevenin's equivalent resistance R_T

$$\text{i.e.}\quad R = R_T = \frac{22}{17} = 1.29\ \Omega\ \textbf{Ans.}$$

Example 2.68 **Explain maximum power transfer theorem. Using this theorem find the value of load resistance R_L for maximum power flow through it in Fig. 2.138 (*a*).**

[U.P. Technical Univ. Electrical Engineering Second Semester 2006-07]

Solution: For determining the Thevenin's equivalent resistance of the circuit with respect to terminals AB the voltage source is replaced by a short circuit and current source is replaced by an open circuit, as shown in Fig. 2.138 (*b*).

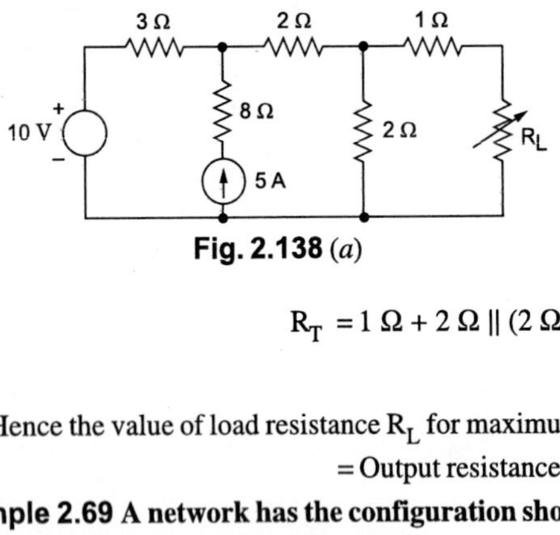

Fig. 2.138 (*a*)

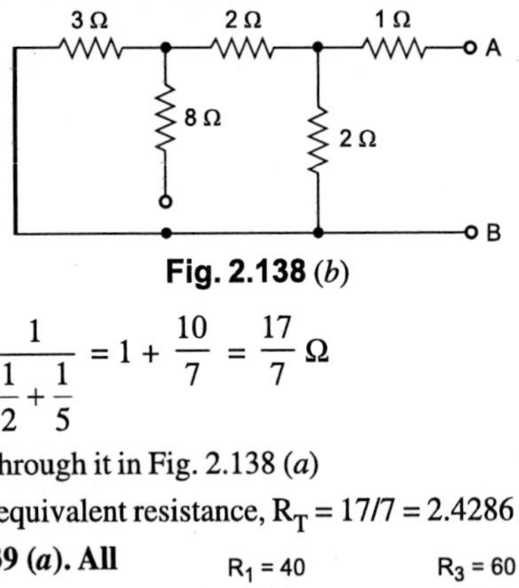

Fig. 2.138 (*b*)

$$R_T = 1\ \Omega + 2\ \Omega \,||\, (2\ \Omega + 3\ \Omega) = 1 + \frac{1}{\frac{1}{2} + \frac{1}{5}} = 1 + \frac{10}{7} = \frac{17}{7}\ \Omega$$

Hence the value of load resistance R_L for maximum power flow through it in Fig. 2.138 (*a*)

= Output resistance or Thevenin's equivalent resistance, $R_T = 17/7 = 2.4286\ \Omega$ **Ans.**

Example 2.69 **A network has the configuration shown in Fig. 2.139 (*a*). All resistance values are expressed in ohms.**

(*i*) **Find the current through R_L when it takes on values of 10, 50 and 200 Ω using Thevenin's theorem.**

(*ii*) **Determine the value of R_L corresponding to which there a maximum power is transferred to the load resistor. Compute this maximum power.**

[G.B. Technical Univ. Electrical Engineering Second Semester 2010-11]

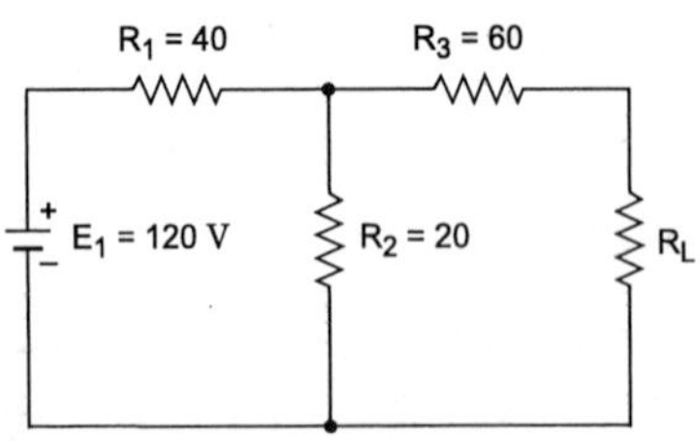

Fig. 2.139 (*a*)

Solution: (*i*) For determination of Thevenin's equivalent resistance of the circuit w.r.t. terminals A and B, the voltage source is replaced by a short circuit, as shown in Fig. 2.139 (*b*).

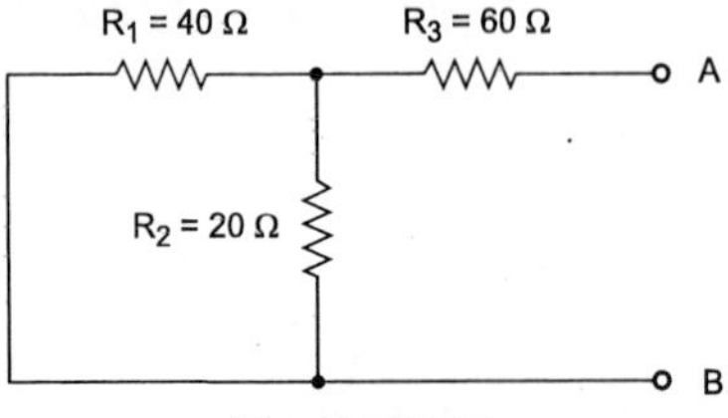

Fig. 2.139 (*b*)

Thevenin's equivalent resistance,

$$R_T = 60 + 40 \,||\, 20\ \Omega = 60 + \frac{40 \times 20}{40 + 20}$$

$$= 60 + \frac{40}{3} = \frac{220}{3}\ \Omega$$

$$\text{Thevenin's voltage, } V_T = \text{Voltage drop across } R_2 \text{ with } R_L \text{ removed}$$

$$= 120 \times \frac{20}{40+20} = 40 \text{ V}$$

$$\text{Current through } R_L \text{ will be equal to } \frac{V_T}{R_T + R_L}$$

$$\text{So current through } R_L\text{; when } R_L = 10\ \Omega,\ I_{L_1} = \frac{40}{\frac{220}{3}+10} = 0.48 \text{ A} \quad \textbf{Ans.}$$

$$\text{Current through } R_L \text{ for } R_L = 50\ \Omega,\ I_{L_2} = \frac{40}{\frac{220}{3}+50} = 0.3243 \text{ A} \quad \textbf{Ans.}$$

$$\text{Current through } R_L \text{ for } R_L = 200\ \Omega,\ I_{L_3} = \frac{40}{\frac{220}{3}+200} = 0.146 \text{ A} \quad \textbf{Ans.}$$

(*ii*) Value of R_L corresponding to transfer of maximum power to load $= R_T = \frac{220}{3}\ \Omega$ **Ans.**

$$\text{Maximum power transferred} = \frac{V_T^2}{4R_L} = \frac{40^2}{4 \times \frac{220}{3}} = 5.45 \text{ W} \quad \textbf{Ans.}$$

Example 2.70 **A circuit has an arrangement of its elements as in Fig. 2.140 (*a*). Find the Thevenin's equivalent considering R_4 as the variable resistance. Find also the value of R_4 which results a maximum power drawn from the circuit.**
[*Mahamaya Technical Univ. Electrical Engineering Second Semester 2011-12*]

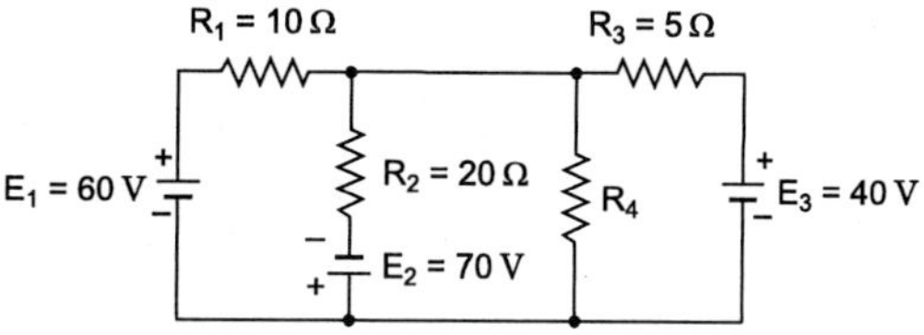

Fig. 2.140 (*a*)

Solution: For determination of the Thevenin's equivalent resistance of the circuit with respect to resistance R_4, the voltage sources are replaced by short circuits, and the circuit becomes as shown in Fig. 2.140 (*b*)

Now the circuit R_1, R_2 and R_3 are in parallel so Thevenin's equivalent resistance

$$R_T = 10 \,\|\, 20 \,\|\, 5\ \Omega = \frac{1}{\frac{1}{10}+\frac{1}{20}+\frac{1}{5}} = \frac{20}{7}\ \Omega \quad \textbf{Ans.}$$

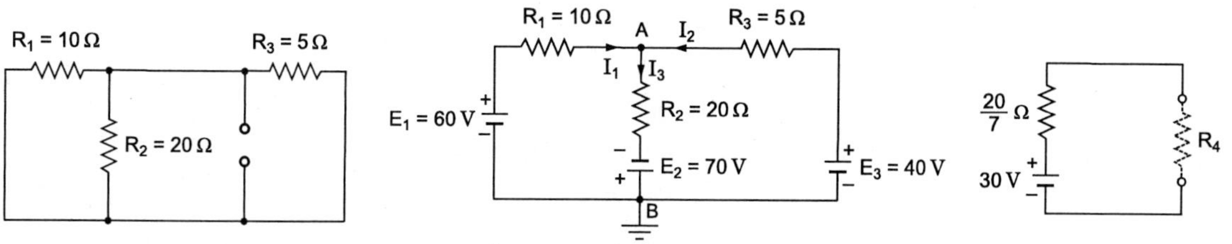

(*b*) (*c*) (*d*)

Fig. 2.140

After removing the resistor R_4, the circuit becomes as shown in Fig. 2.140 (*c*).

Let the terminal B be taken as reference node.

Applying KCL to node A we have

$$I_1 + I_2 = I_3$$

$$\frac{60 - V_A}{10} + \frac{40 - V_A}{5} = \frac{V_A + 70}{20}$$

$$\text{or } 7\,V_A = 120 + 160 - 70$$
$$\text{or } V_A = 30 \text{ V}$$
$$\text{i.e } V_{Th} = 30 \text{ V} \quad \textbf{Ans.}$$

Thevenin's equivalent circuit is shown in Fig. 2.140 (*d*).

R_4 of $\frac{20}{7}$ Ω will draw maximum power from the circuit. **Ans.**

2.20. NETWORK REDUCTION BY DELTA-STAR TRANSFORMATION OR VICE VERSA

Three resistances connected nose-to-tail as shown in Fig. 2.141 (*a*), are said to be *delta* (or Δ) or *mesh-connected* (they form a mesh). Three resistances connected together at a common point O, as shown in Fig. 2.141 (*b*), are said to *be star* (or Y)*-connected.* If the nodes (A, B, and C) to which the two sets of resistances are connected are part of a larger network, it is possible to assign values to the two sets of resistances so that they have exactly the same effect on the network. If, therefore, delta-connected resistances are part of a network it is possible to substitute them by the star-connected ones and vice versa. The obvious advantages are that a delta-star transformation eliminates a mesh which reduces by one the variables and equations necessary to solve a network by mesh analysis whereas star-delta transformation eliminates a node (node O) which reduces by one the variables and equations necessary to solve a network by node analysis.

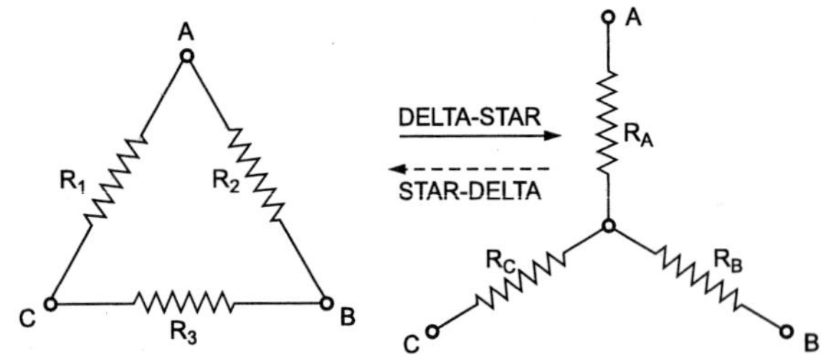

Fig. 2.141 *Delta-Star Transformation*

2.20.1. Delta-Star Transformation. *The replacement of delta or mesh by equivalent star system is known as delta-star transformation.*

The two systems will be equivalent or identical if the resistances measured between any pair of lines is same in both of the systems, when the third line is open.

Hence resistances between terminals B and C,

$$R_{BC} = R_3 \parallel (R_1 + R_2) = \frac{R_3(R_1+R_2)}{R_1+R_2+R_3} \text{ in delta system}$$

$$\text{and } R_{BC} = R_B + R_C \text{ in star system}$$

Since the two systems are identical, resistances measured between terminals B and C in both of the systems must be equal.

$$\text{So } R_B + R_C = \frac{R_3(R_1+R_2)}{R_1+R_2+R_3} \quad \text{...(2.37)}$$

Similarly resistances between terminals C and A being equal in the two systems

$$R_C + R_A = \frac{R_1(R_2+R_3)}{R_1+R_2+R_3} \quad \text{...(2.38)}$$

And resistance between terminals A and B

$$R_A + R_B = \frac{R_2(R_1+R_3)}{R_1+R_2+R_3} \quad \text{...(2.39)}$$

Adding Eqs. (2.37), (2.38) and (2.39) we have

$$2\,(R_A + R_B + R_C) = \frac{2(R_1 R_3 + R_2 R_3 + R_1 R_2)}{R_1+R_2+R_3}$$

$$\text{or } R_A + R_B + R_C = \frac{R_1 R_3 + R_2 R_3 + R_1 R_2}{R_1+R_2+R_3} \quad \text{...(2.40)}$$

Subtracting Eqs. (2.37), (2.38) and (2.39) from Eq. (2.40) we have respectively

$$R_A = \frac{R_1 R_2}{R_1 + R_2 + R_3} \quad ...(2.41)$$

$$R_B = \frac{R_2 R_3}{R_1 + R_2 + R_3} \quad ...(2.42)$$

$$R_C = \frac{R_3 R_1}{R_1 + R_2 + R_3} \quad ...(2.43)$$

These relationships may be expressed as follows:

The equivalent star resistance connected to a given terminal is equal to the product of the two delta resistances connected to the same terminal divided by the sum of the delta-connected resistances.

If the three delta-connected resistances have the same value R_D, the three resistances in the equivalent star for identical systems will be

$$R_S = \frac{R_D R_D}{R_D + R_D + R_D} = \frac{R_D}{3} \quad ...(2.44)$$

2.20.2. Star-Delta Transformation. Multiplying Eqs. (2.41) and (2.42), (2.42) and (2.43) and (2.43) and (2.41) and then adding them we get

$$R_A R_B + R_B R_C + R_C R_A = \frac{R_1 R_2^2 R_3 + R_1 R_3^2 R_2 + R_1^2 R_2 R_3}{(R_1 + R_2 + R_3)^2}$$

$$= \frac{R_1 R_2 R_3 (R_2 + R_3 + R_1)}{(R_1 + R_2 + R_3)^2} = \frac{R_1 R_2 R_3}{R_1 + R_2 + R_3} \quad ...(2.45)$$

Dividing Eq. (2.45) by Eqs. (2.41), (2.42) and (2.43) separately we have

$$R_3 = \frac{R_A R_B + R_B R_C + R_C R_A}{R_A} = R_B + R_C + \frac{R_B R_C}{R_A} \quad ...(2.46)$$

$$R_1 = \frac{R_A R_B + R_B R_C + R_C R_A}{R_B} = R_A + R_C + \frac{R_A R_C}{R_B} \quad ...(2.47)$$

and $$R_2 = \frac{R_A R_B + R_B R_C + R_C R_A}{R_C} = R_A + R_B + \frac{R_A R_B}{R_C} \quad ...(2.48)$$

The above relationship may be expressed as below:

The equivalent delta resistance between two terminals is the sum of the two star resistances connected to those terminals plus the product of the same divided by the third star resistance.

If the three star-connected resistances have the same value, say R_S, the three resistances of the equivalent delta for identical systems will be

$$R_D = R_S + R_S + \frac{R_S R_S}{R_S} = 3 R_S \quad ...(2.49)$$

The advantage of delta-star transformation may be shown by reference to network of Fig. 2.142. Fig. 2.142 (*a*) illustrates the network before conversion, where the dotted lines are drawn around the delta connection to be transformed into a star. Fig. 2.142 (*b*) illustrates the same network after transformation. The currents in the transformed form [Fig. 2.142 (*b*)] are much simpler to determine.

The advantage of star-delta transformation can be illustrated by reference to network of Fig. 2.143. Figure 2.143 (*a*) illustrates the network prior to transformation, with the dotted lines around the star to be transformed. Fig. 2.143 (*b*) illustrates the same network after transformation. The original network is now reduced to a simple series parallel connection of resistances.

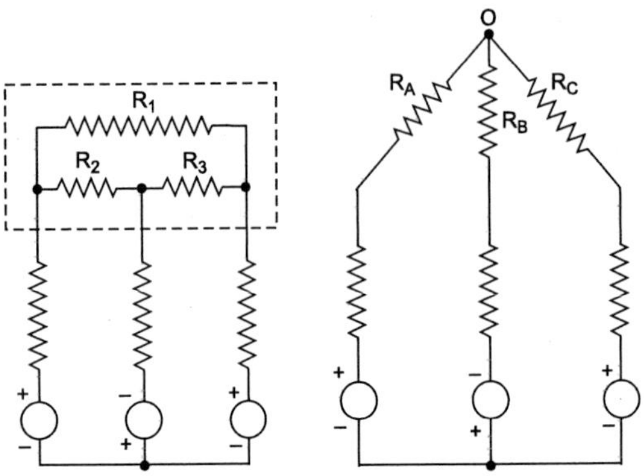

(a) (b)

Fig. 2.142

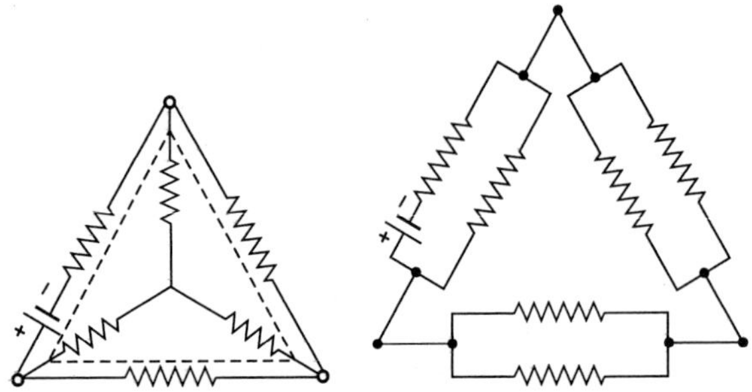
(a) (b)

Fig. 2.143

Example 2.71 **Three resistances r, $2r$, and $3r$ are connected in delta. Determine the resistances for an equivalent star connection.**

[*U.P. Technical Univ. Electrical Engineering January 2003*]

Solution: Resistances for the equivalent star-connection (Fig. 2.144) are worked out as below:

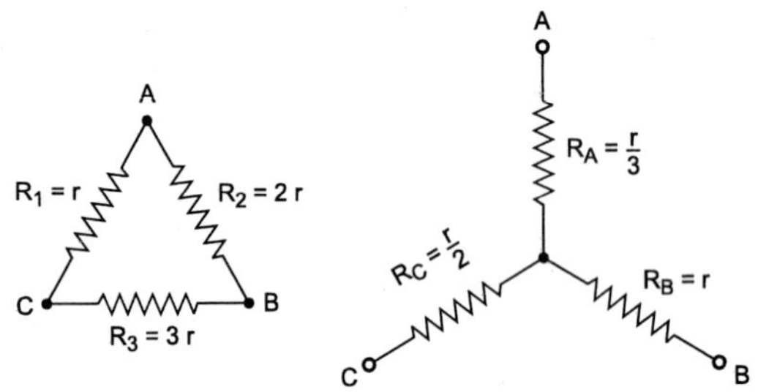

(a) Delta-Connected Resistors *(b) Equivalent Star-Connected Resistors*

Fig. 2.144

$$R_A = \frac{R_1 R_2}{R_1 + R_2 + R_3} = \frac{r \times 2r}{r + 2r + 3r} = \frac{r}{3} \quad \textbf{Ans.}$$

$$R_B = \frac{R_2 R_3}{R_1 + R_2 + R_3} = \frac{2r \times 3r}{r + 2r + 3r} = r \quad \textbf{Ans.}$$

$$R_C = \frac{R_3 R_1}{R_1 + R_2 + R_3} = \frac{3r \times r}{r + 2r + 3r} = \frac{r}{2} \quad \textbf{Ans.}$$

Example 2.72 **Find the resistance at the A–B terminals in the electric circuit of Fig. 2.145 using Δ–Y transformation.**

[U.P. Technical Univ. Elec. Engineering February 2001]

Solution: As shown in Fig. 2.146, delta ACD has been reduced to its equivalent star

$$R_A = \frac{R_{AC}\, R_{AD}}{R_{AC} + R_{CD} + R_{DA}} = \frac{30 \times 20}{30 + 50 + 20} = 6\ \Omega$$

$$\text{Similarly } R_C = \frac{30 \times 50}{100} = 15\ \Omega \quad \text{and } R_D = \frac{20 \times 50}{100} = 10\ \Omega$$

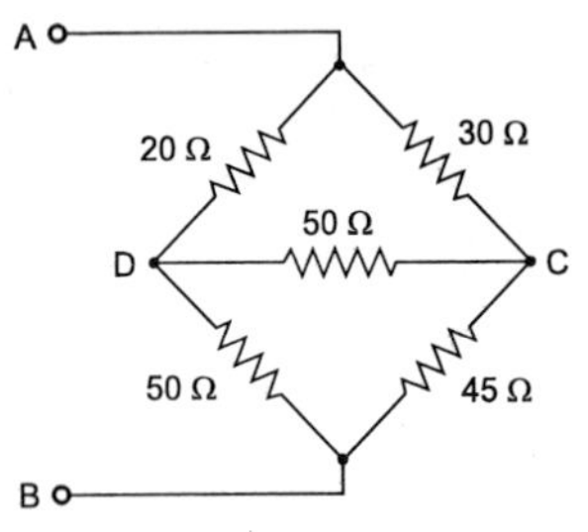

Fig. 2.145

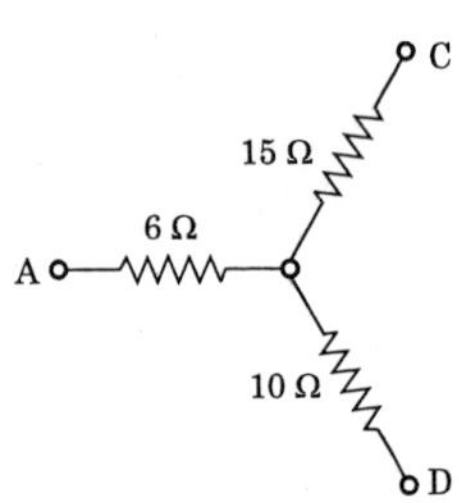

Fig. 2.146

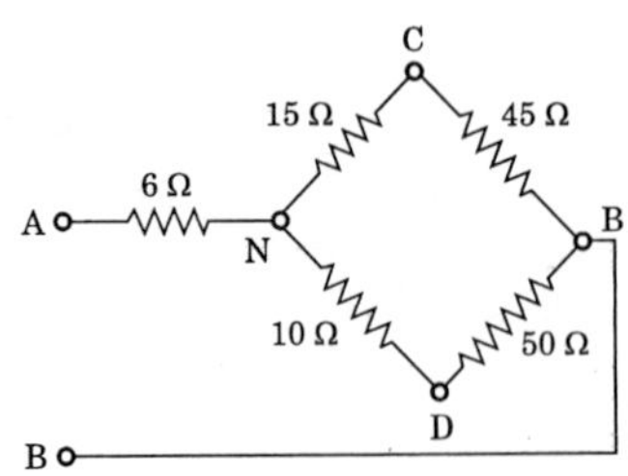

Fig. 2.147

Hence the given circuit is reduced to the circuit shown in Fig. 2.147. As seen, there are two parallel paths between points N and B, one is of resistance 15 + 45 = 60 Ω and the other is also of 10 + 50 = 60 Ω.

$$\text{Their combined resistance} = \frac{1}{\frac{1}{60}+\frac{1}{60}} = 30\ \Omega$$

So total resistance of the network between terminals A and B = 30 + 6 = 36 Ω **Ans.**

Example 2.73 **Find the current drawn I from the source voltage of 100 V in the circuit shown in Fig. 2.148.**

[U.P. Technical Univ. Elec. Engineering First Semester 2004-05]

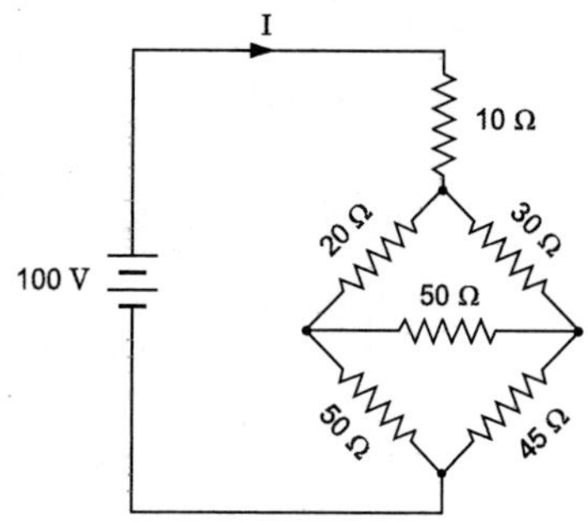

Fig. 2.148

Solution: From solution of above Example 2.72, the Wheatstone bridge is equivalent to a resistor of 36 Ω

Hence total resistance of the circuit, R = 10 + 36 = 46 Ω

$$\text{Current drawn, } I = \frac{V}{R} = \frac{100}{46} = 2.174 \text{ A } \textbf{Ans.}$$

Example 2.74 **Find the resistance between AB of the circuit shown in Fig. 2.149. Use Y-Δ (star-delta) transformation.**

[U.P. Technical Univ. Elec. Engineering Second Semester 2005-06]

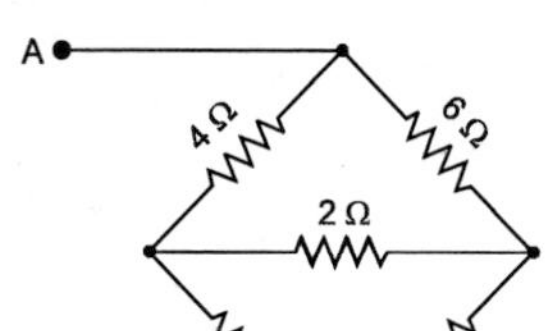

Fig. 2.149

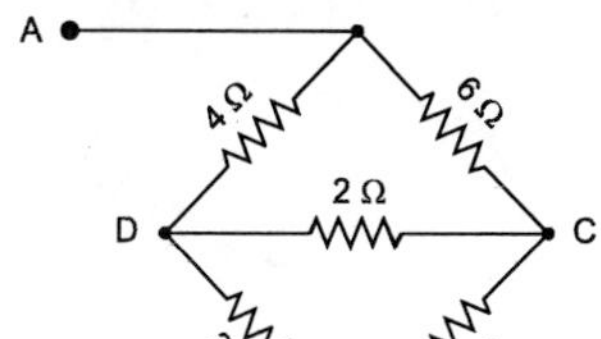

Fig. 2.150 (*a*)

Solution: As shown in Fig. 2.150 (*a*), delta ACD has been reduced to equivalent star

$$R_A = \frac{R_{AC}R_{AD}}{R_{AC}+R_{CD}+R_{DA}}$$

$$= \frac{6\times4}{6+2+4} = 2\ \Omega$$

$$\text{Similarly, } R_C = \frac{6\times2}{12} = 1\ \Omega$$

$$\text{and } R_D = \frac{4\times2}{12} = \frac{2}{3}\ \Omega$$

Hence the given circuit is reduced to the circuit shown in Fig. 2.150 (*c*). As seen, there are two parallel paths between terminals N and B', one of resistance 1 + 5 = 6 Ω and the other of resistance

$$3+\frac{2}{3} = \frac{11}{3}\ \Omega.$$

The equivalent resistance of parallel combination

$$= \frac{1}{\frac{1}{6}+\frac{3}{11}} = \frac{66}{29}\ \Omega$$

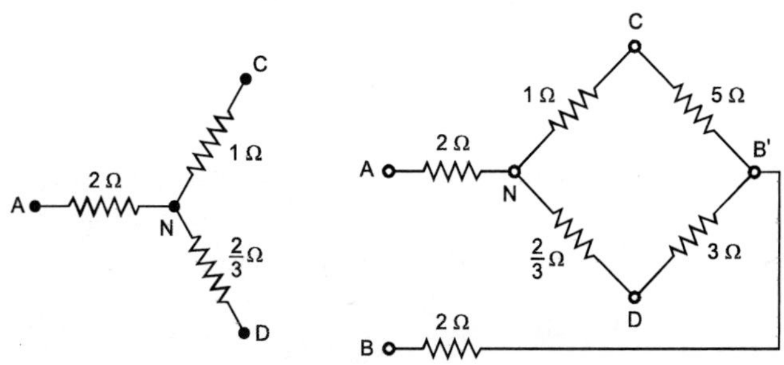

(*b*) (*c*)

Fig. 2.150

So total resistance of the network between terminals A and B

$$= \frac{66}{29}+2+2 = 6.276\ \Omega \textbf{ Ans.}$$

Example 2.75 **A star-network in which N is star point made up as follows:**

AN = 70 Ω, BN = 100 Ω, CN = 90 Ω. Find an equivalent delta network.

If the above star-delta networks are superimposed, what would be measured resistance between A and C ?

[R.G.P.V. Bhopal Basic Electrical Engineering 2001]

Solution: Resistance between terminals A and N, R_A = 70 Ω

Resistance between terminals B and N, R_B = 100 Ω

Resistance between terminals C and N, R_C = 90 Ω

Resistances for equivalent delta (Fig. 2.151) are worked out as below:

Resistance between terminals C and A,

$$R_1 = R_A + R_C + \frac{R_A R_C}{R_B}$$

$$= 70 + 90 + \frac{70 \times 90}{100} = 223\ \Omega \quad \textbf{Ans.}$$

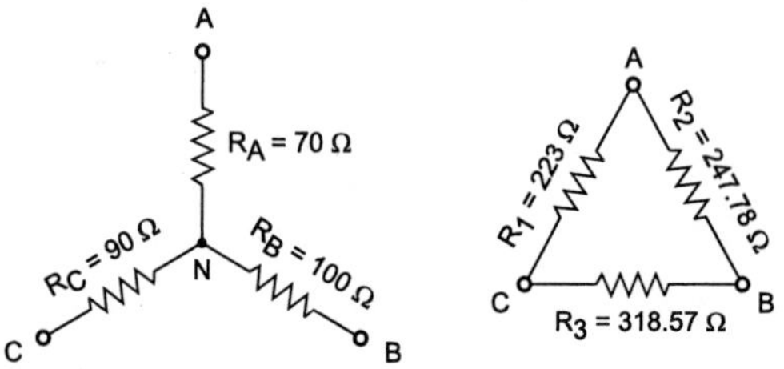

Fig. 2.151

Resistance between terminals A and B,

$$R_2 = R_A + R_B + \frac{R_A R_B}{R_C} = 70 + 100 + \frac{70 \times 100}{90} = 247.78\ \Omega \quad \textbf{Ans.}$$

Resistance between terminals B and C,

$$R_3 = R_B + R_C + \frac{R_B R_C}{R_A} = 100 + 90 + \frac{100 \times 90}{70} = 318.57\ \Omega \quad \textbf{Ans.}$$

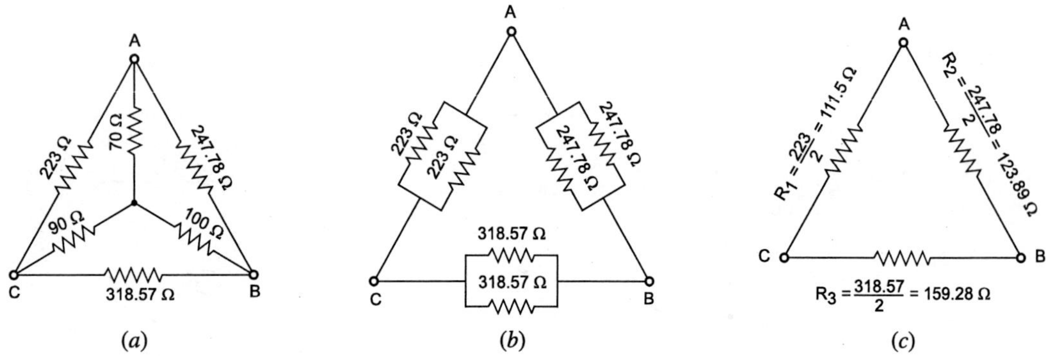

Fig. 2.152

Now, when star and delta networks are superimposed, the equivalent network becomes as shown in Figs 2.152 (*b*) and (*c*) by converting star into equivalent delta.

From network shown in Fig. 2.152 (*c*) resistance measured between terminals A and C,

$$R_{AC} = R_1 \,||\, (R_2 + R_3) = 111.5 \,||\, (123.89 + 159.285) = 111.5 \,||\, 283.175 = \frac{111.5 \times 283.175}{111.5 + 283.175} = 80\ \Omega \quad \textbf{Ans.}$$

Example 2.76 Using delta to star transformation determine the resistance between terminals *a-b* and the total power drawn from the supply in the circuit shown in Fig. 2.153.

[*U.P. Technical Univ. Electrical Engineering First Semester 2006-07*]

Solution: By transforming delta formed by resistors of 8 Ω, 7 Ω and 3 Ω into star we have

$$R_A = \frac{R_{AC} \times R_{AB}}{R_{AC} + R_{AB} + R_{BC}} = \frac{3 \times 8}{3 + 8 + 7} = \frac{4}{3}\ \Omega$$

$$\text{Similarly } R_B = \frac{8 \times 7}{18} = \frac{28}{9}\ \Omega$$

$$\text{and } R_C = \frac{3 \times 7}{18} = \frac{7}{6}\ \Omega$$

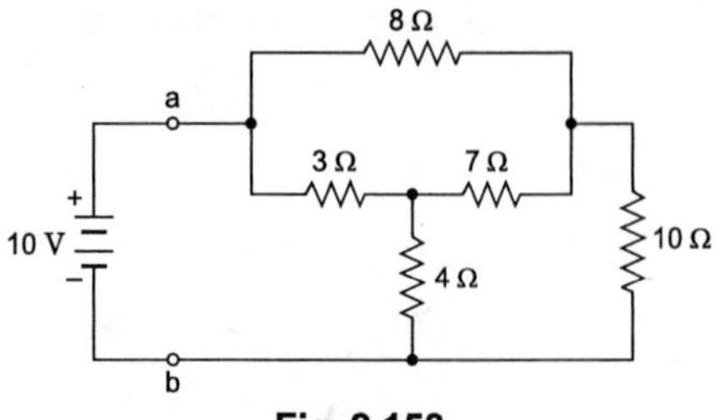

Fig. 2.153

Now the given circuit is reduced to that as shown in Fig. 2.154 (*c*).

Resistance between terminals *a* and *b*,

$$R = \frac{4}{3} + \left(\frac{7}{6} + 4\right) || \left(\frac{28}{9} + 10\right) = \frac{4}{3} + \frac{31}{7} || \frac{118}{9} = 1.3333 + 3.7062 = 5.04\ \Omega \quad \textbf{Ans.}$$

$$\text{Total power drawn, } P = \frac{V^2}{R} = \frac{10^2}{5.04} = 19.84 \text{ watts} \quad \textbf{Ans.}$$

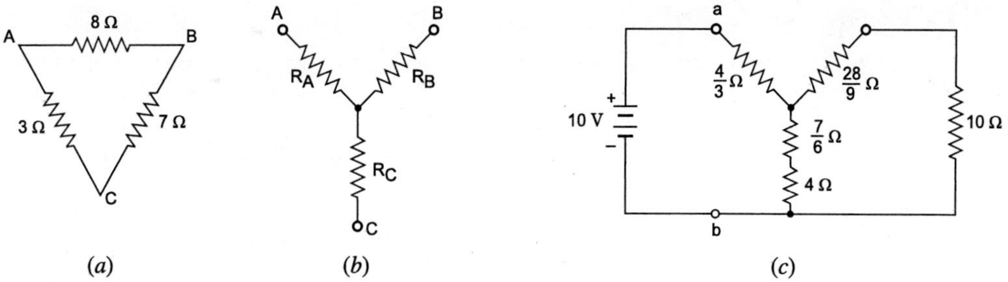

Fig. 2.154

HIGHLIGHTS

1. An *electric circuit* (or network) is an interconnection of physical electrical devices such as an *energy source* (or sources), an *energy convertor* or convertors (load or loads), and *conductors* that connect them.
2. A *junction (or node)* is a point in a network where two or more branches meet.
3. A *loop* is a closed path in a network formed by a number of connected branches. *Mesh* is a loop that contains no other loop within it.
4. Network elements may be classified into two categories viz. active elements and passive elements.

   The elements which supply energy to the network are known as *active elements*. The voltage sources like batteries, dc generators, ac generators and current sources like photoelectric cells, metadyne generators fall under the category of active elements. Most of the semiconductor devices like transistors are treated as current sources.

   The components which dissipate or store energy are known as *passive components*. Resistors, inductors and capacitors fall under the category of passive elements. The resistor is the only component which dissipates electrical energy. The inductors and capacitors are the components which store energy, the inductor stores energy by virtue of a current passing through it whereas the capacitor stores energy by virtue of potential difference across it.
5. A voltage source of voltage V_S and internal resistance R_{in} can be converted into an equivalent current source of current $I_S = V_S/R_{in}$ and a resistance R_{in} across it. Similarly a current source of output current I_S in parallel with resistance R_{in} can be converted into an equivalent voltage source of voltage $V_S = I_S R_{in}$ and a resistance R_{in} in series with it.

   A voltage source-series resistance combination is equivalent to a current source-parallel resistance combination if, and only if their respective open-circuit voltages are equal, and their respective short-circuit currents are equal.
6. According to *Kirchhoff's first law* (*or current law*), the algebraic sum of currents in two or more conductors at a point (junction) is always zero

   $$i.e.\ \Sigma I = 0$$

   While applying above law, incoming currents are taken as positive and outgoing currents as negative.

   According to *Kirchhoff's second law* (*or voltage law*), the algebraic sum of emfs acting in any closed circuit or mesh is equal to the algebraic sum of the products of currents and resistances of each part of that closed circuit or mesh *i.e.*,

   $$\Sigma IR = \Sigma \text{emf}$$
7. According to Superposition theorem if there are a number of voltage and current sources acting simultaneously in any linear bilateral network, then each source can be considered acting independently of the others.
8. In *loop method of analysis*, independent mesh currents are taken and the network is solved by framing equations according to Kirchhoff's second or voltage law (KVL).
9. In *nodal analysis* independent nodes are considered and voltages are assumed at these nodes w.r.t. one reference node, called the datum node. The equations are framed according to Kirchhoff's current law (KCL) which reveal the desired results after their solution.

10. Thevenin's theorem may be stated as follows:

The current in any passive circuit element (which may be called R_L) in a network is the same as would be obtained if R_L were supplied with a source voltage V_{OC} or V_T in series with an equivalent resistance R_{in} or R_T; V_{OC} being the open-circuit voltage at the terminals from which R_L has been removed and R_T being the resistance that would be measured at these terminals after all sources have been removed and each source has been replaced by its internal resistance.

11. *Norton's theorem* is an alternative to Thevenin's theorem. According to this theorem, any two-terminal active network, when viewed from output terminals, is equivalent to a constant current source in parallel with a resistance.

12. *Maximum power transfer theorem* states that a resistive load, served through a resistive network, will abstract maximum power when the load resistance value is the same as the resistance "viewed by the load as it looks back into the network".

13. *In delta-star transformation,* the equivalent star resistance connected to a given terminal is equal to the product of the two delta resistances connected to the same terminal divided by the sum of the delta connected resistances.

14. *In star-delta transformation,* the equivalent delta resistance between two terminals is the sum of the two star resistances connected to those terminals plus the product of the same divided by the third star resistance.

EXERCISES

1. Explain the following:
   (*i*) Active and passive elements
   (*ii*) Unilateral and bilateral elements
   (*iii*) Time variant and time invariant elements
   (*iv*) Distributed and lumped parameters. **[U.P. Technical Univ. Electrical Engineering Odd Semester 2013-14]**
2. Discuss the characteristics of the following elements:
   (*a*) Capacitor (*b*) Inductor. **[U.P. Technical Univ. Electrical Engineering Second Semester 2008-09]**
3. Discuss different types of voltage and current sources.
   [G.B. Technical Univ. Electrical Engineering First Semester 2009-10]
4. Explain voltage and current sources of a network with characteristics. Explain source transformation principle in any circuit.
   [Electrical Engineering G.B. Technical Univ. First Semester 2011-12; U.P. Technical Univ. Even Semester 2013-14]
5. Explain, what do you understand by the voltage-dependent current source and current-dependent current source.
   [U.P.S.C. I.E.S. Electrical Engineering-I, 2004]
6. Draw V-I characteristics of voltage and current sources. Explain source transformation theory in any circuit.
   [Mahamaya Technical Univ. Electrical Engineering Second Semester 2010-11]
7. State and explain Kirchhoff's laws. What are the limitations and applications of Kirchhoff's laws in circuit theory?
   [U.P. Technical Univ. Electrical Engineering Odd Semester 2013-14]
8. State and explain superposition theorem. Mention its limitations.
   [U.P. Technical Univ. Electrical Engineering First Semester 2006-07]
9. State and prove superposition theorem taking an example.
   [U.P. Technical Univ. Electrical Engineering First Semester 2007-08]
10. State and explain superposition theorem. **[G.B. Technical Univ. Electrical Engineering First Semester 2009-10]**
11. Explain the loop current method of solving a network. **[R.G.P.V. Bhopal Basic Electrical Engineering June 2007]**
12. State and explain the nodal method of analysis. **[M.D. Univ. Rohtak Electrical Technology July 2001]**
13. State Thevenin's theorem. **[U.P. Technical Univ. Electrical Engineering First Semester 2004-05]**
14. State the superposition and Norton's theorem.
   [G.B. Technical Univ. Electrical Engineering Second Semester 2009-10]

15. State the Thevenin's and Norton's theorems. **[G.B. Technical Univ. Electrical Engineering First Semester 2010-11]**

16. State and explain the Norton's theorem.

[Mahamaya Technical Univ. Electrical Engineering Second Semester 2011-12]

17. Write the statement of Norton's theorem and discuss it with the help of example.

[G.B. Technical Univ. Electrical Engineering First Semester 2010-11]

18. Explain the duality between a Thevenin's and a Norton's equivalent circuits.

[G.B. Technical Univ. Electrical Engineering Second Semester 2011-12]

19. State and prove maximum power transfer theorem.

[Electrical Engineering; U.P. Technical Univ. First Semester 2008-09; Second Semester 2013-14; Mahamaya Technical Univ. Second Semester 2010-11; G.B. Technical Univ. First Semester 2011-12]

20. State and explain maximum power transfer theorem for dc circuit.

[Mahamaya Technical Univ. Electrical Engineering First Semester 2011-12]

21. State maximum power transfer theorem. For which type of circuits is it generally applied. Derive condition for maximum power transfer to a purely resistive load. **[U.P. Technical Univ. Elec. Engineering First Semester 2005-06]**

22. State and explain maximum power transfer theorem. Mention its significance in communication circuits.

[U.P. Technical Univ. Electrical Engineering Odd Semester 2013-14]

23. Derive an expression of star to delta transformation and vice versa.

[U.P. Technical Univ. Basic Electrical Engineering Second Semester 2013-14]

24. When is the Δ-Y transformation useful in network reduction. Illustrate.

[U.P. Technical Univ. Electrical Engineering Second Semester 2008-09]

25. How a star network is converted into a delta network ? Explain with example.

[U.P. Technical Univ. Electrical Engineering Second Semester 2006-07]

26. What is the criterion for star-delta transformation? Develop the relationship for delta-to-star and star-to-delta transformation.

[G.B. Technical Univ. Electrical Engineering Even Semester 2013-14]

SHORT ANSWER TYPE QUESTIONS WITH ANSWERS

Q. 1. What is an electric network ?

Ans. An electric network is an interconnection of physical electrical devices such as an energy source (or sources), an energy convertor or convertors (load or loads), and conductors that connect them.

Q. 2. What is an energy source ?

Ans. An energy source (or source), such as primary or secondary cell, a generator, and the like, is a device that converts chemical, mechanical, thermal or some other form of energy into the electrical energy.

Q. 3. What is an energy convertor ?

Ans. An energy convertor, also called the load, such as lamp, heating appliance, or electric motor, converts electrical energy into light, heat, mechanical work etc.

Q. 4. What do you mean by 'linear bilateral circuit'? **[M.T.U. Second Semester 2011-12]**

Ans. An electric circuit, that has linear volt-ampere (V-I) characteristic and whose characteristics or properties are same in either direction (e.g., a distribution or transmission line) is called the linear bilateral circuit.

Q. 5. What do you understand by "time variant" and "time invariant" elements?

Ans. Circuit element whose characteristics or properties do not change with time is called the time invariant. Otherwise it is said to be time variant.

Q. 6. What do you understand by "lumped" and "distributed" parameter circuits?

Ans. A circuit consisting of a limited (finite) number of circuit elements is referred to as '*lumped parameter circuit*' while a circuit containing unlimited (infinite) number of circuit elements is called a *distributed parameter circuit*. Transmission line is an example of distributed parameter network.

Q. 7. What do you understand by "active" and "passive" elements? **[U.P. Technical Univ. 2013-14]**

Ans. Refer to Art. 2.3.

Q. 8. What do you understand by unilateral and bilateral elements? Give examples. **[U.P.T.U. First Semester 2008-09]**

Ans. Circuit element, whose characteristics or properties are same in either direction (e.g. a distribution or transmission line), is called the bilateral element. The distribution or transmission line can be made to perform its function equally well in either direction.

Circuit element, whose characteristics or properties change with the direction of its operation (e.g. a diode rectifier), is called the unilateral element. A diode rectifier cannot perform rectification in both directions.

Q. 9. Is 'L' a linear circuit? Explain the answer.

[G.B.T.U., Second Semester 2011-12; U.P.T.U. Second Semester 2013-14]

Ans. L is a symbol of inductance and is a linear element. Refer to Art 2.6.

Q. 10. What is meant by 'node' ?

Ans. A junction or node is a point in a network where two or more branches meet.

Q. 11. Distinguish between mesh and loop of network. **[R.G.P.V. Bhopal June 2005]**

Ans. A loop is a closed path in a network formed by a number of connected branches. Mesh is a loop that contains no other loop within it.

Q. 12. Write the properties of ideal voltage source.

[G.B.T.U., First Semester 2011-12; U.P.T.U. Second Semester 2013-14]

Ans. An ideal voltage source is a constant voltage source and is capable of supplying any current at a given voltage. The internal resistance of an ideal voltage source is zero and therefore, the terminal voltage (voltage across the load) is always equal to the source emf and is independent of the magnitude of current supplied.

There are two noteworthy points regarding ideal voltage source.

1. Ideal voltage source cannot be short circuited
2. Two ideal voltage sources of unequal output cannot be placed in parallel.

Q. 13. Enlist properties of voltage and current sources. **[M.T.U. First Semester 2011-12]**

Or

What are the properties of ideal voltage and current sources? **[U.P.T.U. Odd Semester 2013-14]**

Ans. Voltage and current sources are energy sources such as batteries, dynamos, alternators, photoelectric cells, metadyne generators etc.

The energy source (voltage or current) may be independent or dependent. A source is said to be independent when it does not depend on any other quantity in the circuit. A dependent source is one which depends on some other quantity in the circuit.

Voltage source is said to be an ideal or constant voltage source if it is capable of supplying any current at a given voltage. A voltage source having zero internal resistance is known as an ideal or constant voltage source.

Similarly current source is said to be an ideal or constant current source if it is capable of supplying a constant current to a load even if its impedance varies.

Practically, a voltage source is not different from a current source. In fact, a source can either operate as a current source or as a voltage source. It merely depends upon its operating conditions. If load impedance is very large in comparison to internal impedance of the source, it will be advantageous to treat the source as a voltage source. On the other hand, if the load impedance is very small in comparison to the internal impedance of the source, it is better to represent the source as a current source. From the circuit point of view it does not matter at all whether the source is treated as a voltage source or a current source. In fact, it is possible to convert a voltage source into a current source and vice versa.

Q. 14. What are respective values of internal resistances of ideal voltage and ideal current sources? Is it possible to convert an ideal voltage into an ideal current source or vice versa? **[M.T.U. Second Semester 2011-12]**

Ans. Internal resistances of ideal voltage and ideal current sources are zero and infinite respectively.

It is not possible to convert ideal voltage source into ideal current source and vice versa.

Q. 15. What is the utility of superposition theorem ?

Ans. This theorem is applied when we are to determine the current in one particular branch of a network containing several voltage sources or current sources or both voltage sources and current sources.

Q. 16. Name the circuit laws on which mesh-method and nodal-method are respectively based. A network is found to have a total of seven branches and four junctions (i.e. independent nodes). Which method is preferred for solving network.

[M.T.U. Second Semester 2011-12]

Ans. Mesh-method is based upon Kirchhoff's voltage law and nodal method is based upon Kirchhoff's current law.
Both methods are equal good because in both methods, four equations will be formed.

Q. 17. What is the utility of Thevenin theorem ?

Ans. Thevenin's theorem is advantageous when we are to determine the current in a particular element of a linear bilateral network particularly when it is desired to find the current which flows through a resistor for its different values. It makes the solution of the complicated networks (particularly electronic networks) quite simple.

Q. 18. What is maximum power transfer theorem ?

Ans. A resistive load, served through a resistive network, will abstract maximum power when the load resistance value is the same as the resistance "viewed by the load as it looks back into the network".

Q. 19. Give the relationship between resistances connected in delta and equivalent star systems ?

Ans. The equivalent star resistance connected to a given terminal is equal to the product of the two delta resistances connected to the same terminal divided by the sum of the delta connected resistances.

Q. 20. Give the relationship between resistances connected in star and equivalent delta systems ?

Ans. The equivalent delta resistance between two terminals is the sum of the two star resistances connected to those terminals plus the product of the same divided by the third star resistance.

Q. 21. It resistance of each branch is 3 Ω in a delta connected load, what would be the resistance of each branch in star equivalent? **[G.B.T.U. Second Semester 2011-12]**

Ans. Equivalent resistance in star-connection

= One-third resistance in delta-connection

$$= \frac{1}{3} \times 3 = 1\ \Omega \quad \textbf{Ans.}$$

Q. 22. A voltage source of 100 V has an internal impedance of 2 Ω and supplies a load having the same impedance. How much power is absorbed by the load? **[G.B. Technical Univ. Even Semester 2012-13]**

Ans. Load resistance R_L = Internal resistance of voltage source = 2 Ω

$$\text{Power absorbed by the load} = \frac{V^2}{4R_L} = \frac{100^2}{4 \times 2} = 1{,}250\ \text{W}.$$

PROBLEMS

1. Convert 4 A source with its parallel resistance of 15 Ω into its equivalent voltage source.
[**Ans.** Voltage source of 60 V in series with a resistance of 15 Ω]
2. Convert 240 V source with a series resistance of 40 Ω into its equivalent current source.
[**Ans.** Current source of 6 A is parallel with a resistor of 40 Ω]
3. Determine current flowing through 5 Ω resistor in the circuit shown in Fig. 1. Use source transformation technique.
[**Ans.** 2.667 A]

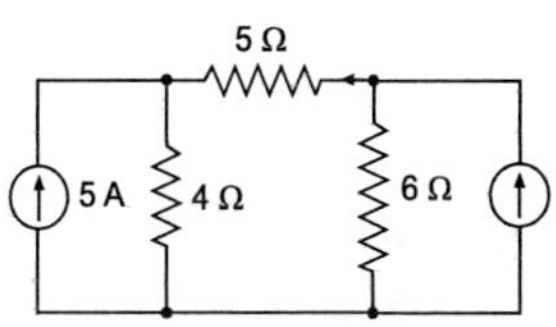

Fig. 1

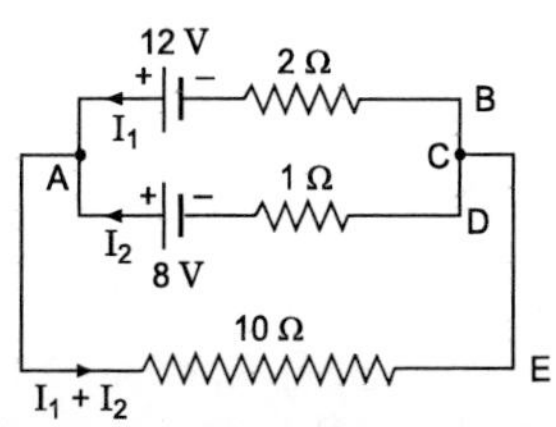

Fig. 2

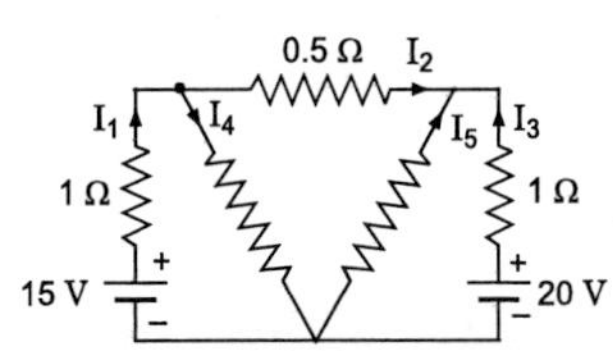

Fig. 3

4. Two batteries A and B are connected in parallel and a load of 10 Ω is connected across their terminals. A has an emf of 12 V and an internal resistance of 2 Ω and B has an emf of 8 V and an internal resistance of 1 Ω. Use Kirchhoff's laws to determine the magnitude of currents and also the directions in each of the batteries. Also determine the pd across external resistance. [**Ans.** 1.625 A, – 0.75 A and 8.75 V]
5. Find the various branch currents in the passive elements of the network, as shown in Fig. 3.
[**Ans.** I_1 = 5.75 A, I_2 = – 3.5 A; I_3 = 9 A; I_4 = 9.25 A, I_5 = 5.5 A]

6. Fig. 4 shows two batteries connected in parallel each represented by an emf along with its internal resistance. A load resistance of 6 Ω is connected across the ends of the batteries. Calculate the current through each battery and the load.
[U.P. Technical Univ. Electrical Engineering July 2002]

$$\left[\textbf{Ans. } \frac{34}{11}\text{ A}, \frac{28}{11}\text{ A}, \frac{62}{11}\text{ A}\right]$$

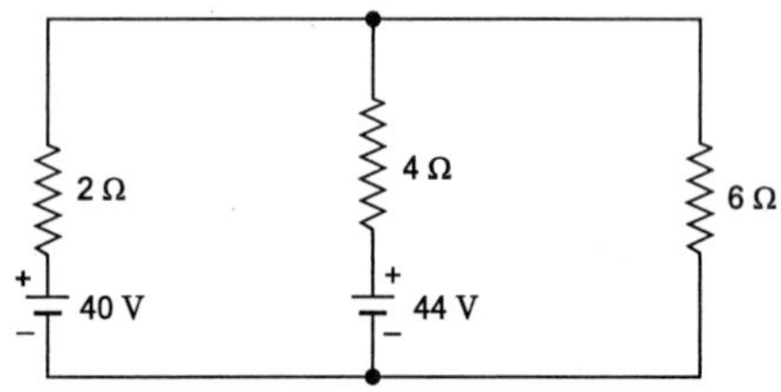

Fig. 4

7. A Wheatstone bridge consists of AB = 400 Ω, BC = 220 Ω, CD = 225 Ω and DA = 100 Ω. A 2 V cell is connected between A and C and a galvanometer of 200 Ω resistance between B and D. Estimate the current through the galvanometer by applying Kirchhoff's laws.
[**Ans.** 2/703 A from D to B]

8. Using "superposition theorem" or "Norton's theorem", find the current I in the following (Fig. 5).
[U.P. Technical Univ. Electrical Engineering January 2003]
[**Ans.** 2 A]

9. Using superposition theorem, determine currents in all the resistors of the following network (Fig. 6).
[U.P. Technical Univ. Electrical Engineering, First Semester 2005-06]
[**Ans.** 1 A, 0 A, 2 A]

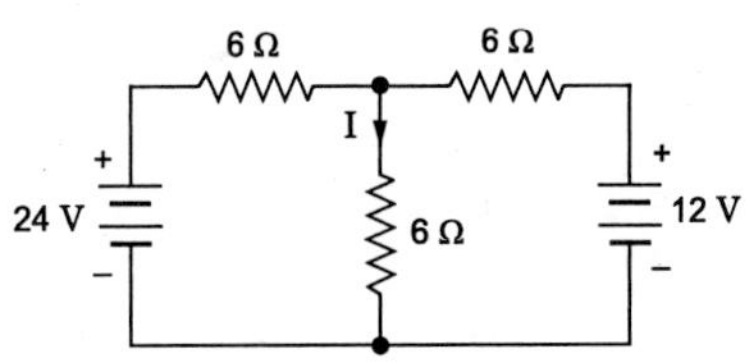

Fig. 5

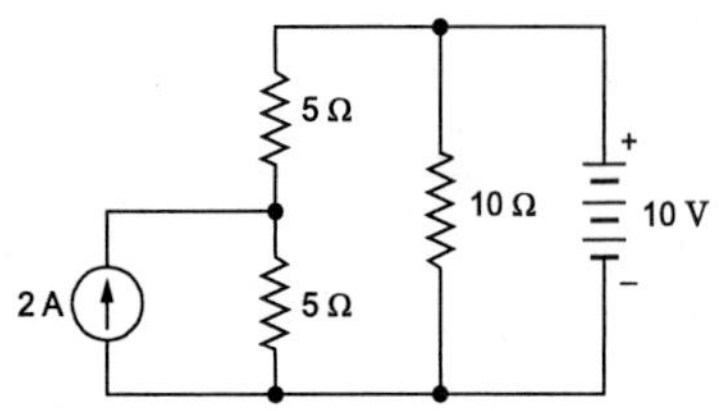

Fig. 6

10. Determine the value of current in 8 Ω resistor of the network shown in Fig. 7, using mesh current method.
[M.D. Univ. Rohtak Electrical Technology First Semester 2006-07]
[**Ans.** 0.31945 A]

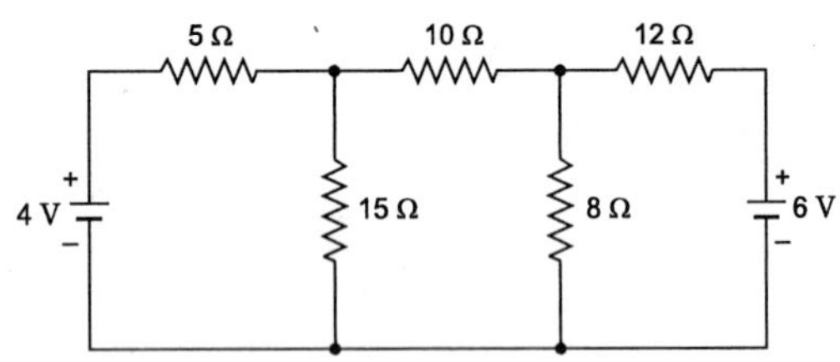

Fig. 7

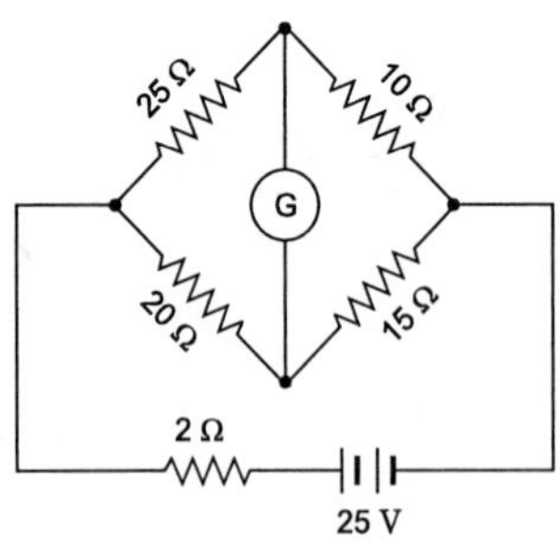

Fig. 8

11. Using mesh current analysis, find current through galvanometer G shown in Fig. 8. Internal resistance of galvanometer G is 50 Ω.
[M.D. Univ. Rohtak Electrical Technology First Semester 2006-07]
[**Ans.** 0.04875 A upward]

12. Determine the loop currents in the network (Fig. 9) by mesh analysis method.
[**Ans.** 5 A; 8 A]

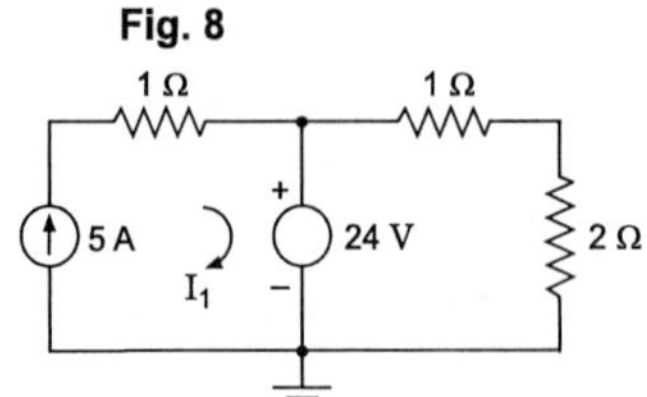

Fig. 9

13. For the circuit shown in Fig. 10, find voltages of nodes B and C, and determine current in 8 Ω resistor.

[U.P. Technical Univ. Electrical Engineering Second Semester 2003-04]

[**Ans.** 7.0133 V, 3.5376 V]

14. Using nodal analysis, determine current in each branch of the network shown in Fig. 11. Also find total power loss in the network. **[U.P. Technical Univ. Electrical Engineering February 2002]**

[**Ans.** 1 A, 0.6 A, 0.4 A, 0.6 A, 0.2 A, 0.3 A, 0.5 A, 15 W]

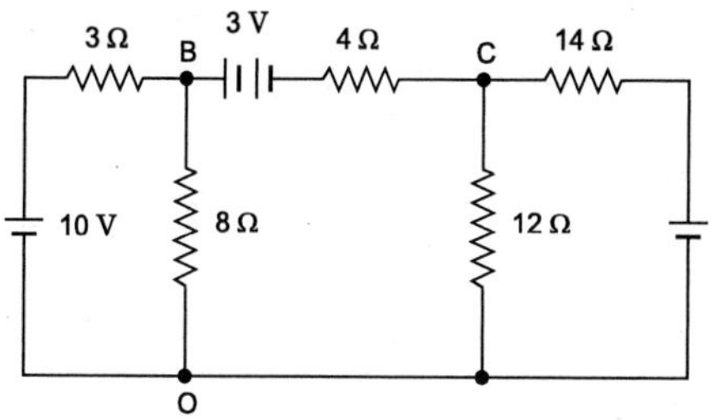

Fig. 10

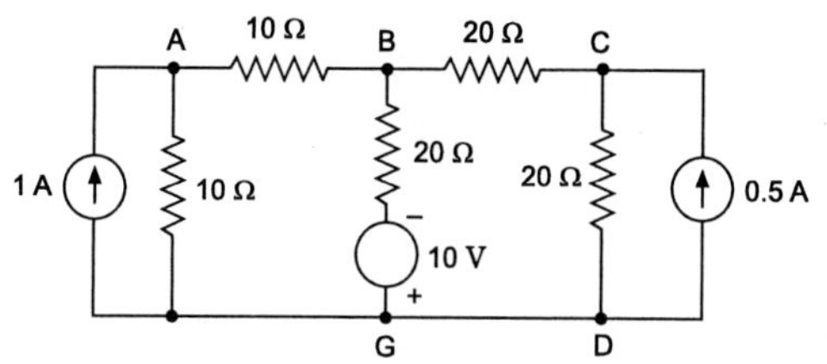

Fig. 11

15. Two batteries A and B are connected in parallel to a load of 10 ohm. Battery A has an emf of 12 V and an internal resistance of 2 ohm and battery B has an emf of 10 V and internal resistance of 1 ohm. Using nodal analysis, determine the currents supplied by each battery and load current.

[U.P. Technical Univ. Electrical Engineering First Semester 2003-04]

[**Ans.** 1 A, 0 A, 1 A]

16. Solve the circuit shown in Fig. 13 by using node-voltage method.

[**Ans.** 10 A; 10 A; 40 A; 5.45 Ω; 3.373 Ω]

17. Use the node voltage method to solve the mesh currents in the network shown in Fig. 14. **[U.P. Technical Univ. Electrical Engineering June 2001]**

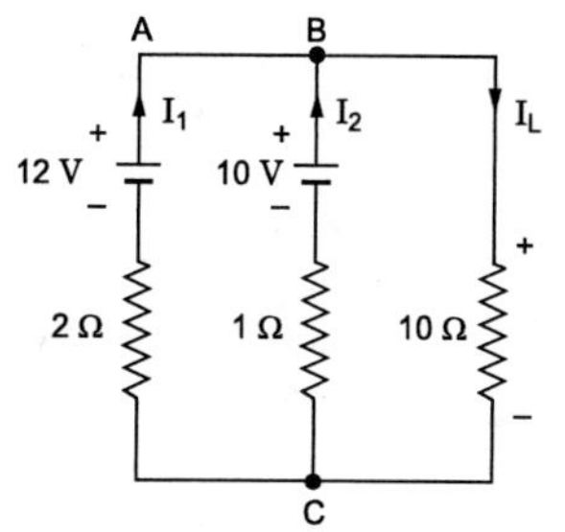

Fig. 12

[**Ans.** – 5.0373 A, – 2.0522 A and – 9.7015 A]

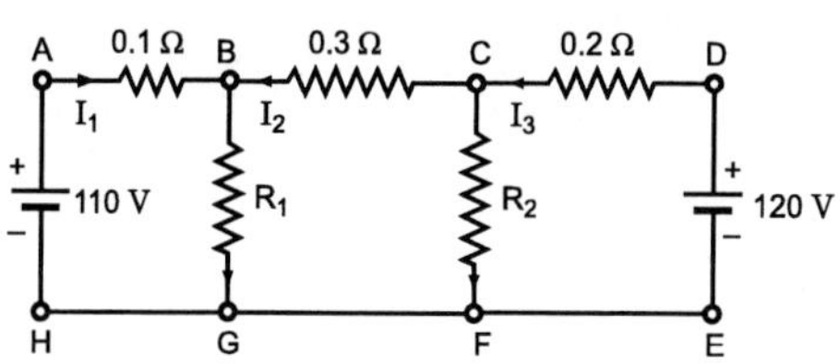

Fig. 13

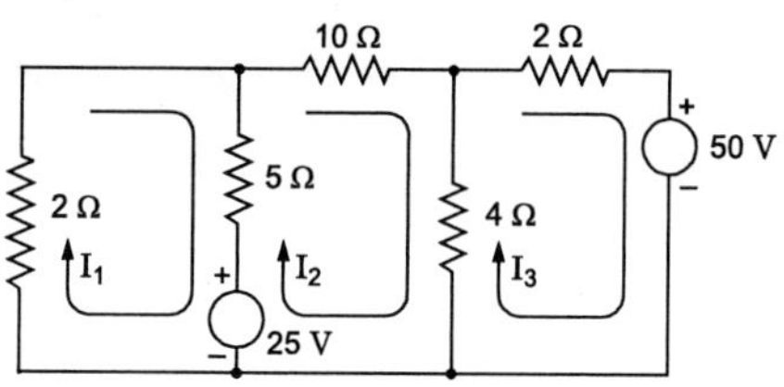

Fig. 14

18. Use Thevenin's theorem to find the current flowing through the 6 Ω resistor of the network shown in Fig. 15.

[**Ans.** 1 A from terminal A to terminal B]

19. Determine current through 6-Ω resistance connected across A-B terminals in the electric circuit of Fig. 16 using Thevenin's Theorem. **[U.P. Technical Univ. Electrical Engineering February 2001]**

[**Ans.** 1 A]

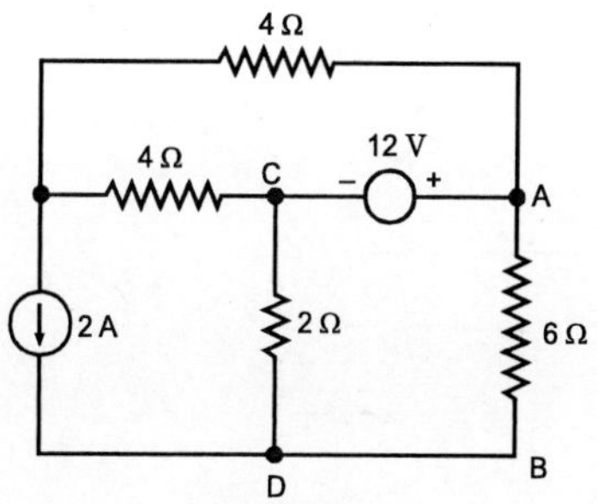

Fig. 15

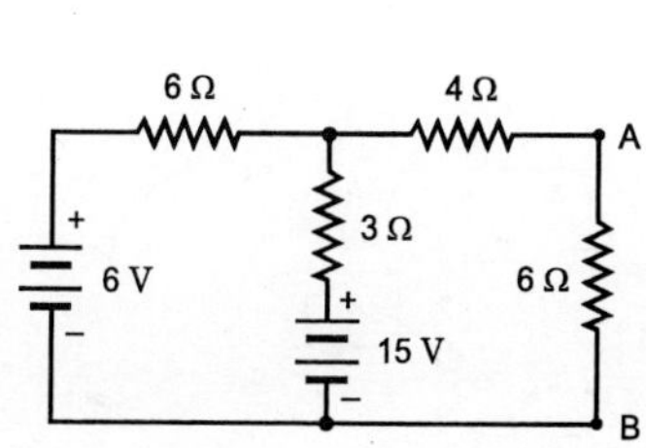

Fig. 16

20. Determine the current flowing through 5 Ω resistor in the network shown in Fig. 17 using Thevenin's theorem.

[U.P. Technical Univ. Electrical Engineering Second Semester 2002-03]

$\left[\textbf{Ans. } \frac{39}{28} \text{ A}\right]$

21. For the circuit shown in Fig. 18, calculate the current in the 6 Ω - resistance by using Norton's theorem.

[Ans. 0.5 A from B to A]

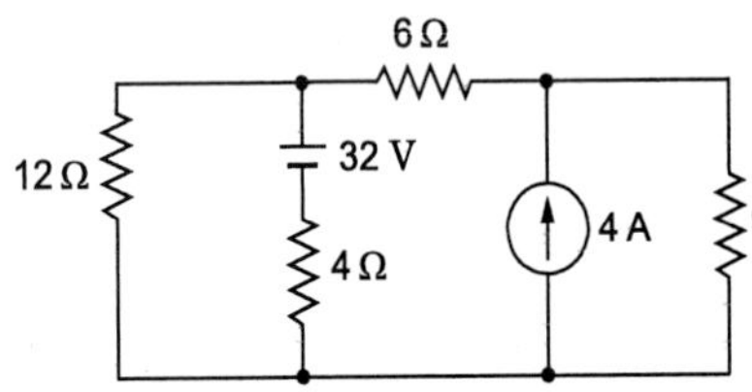

Fig. 17

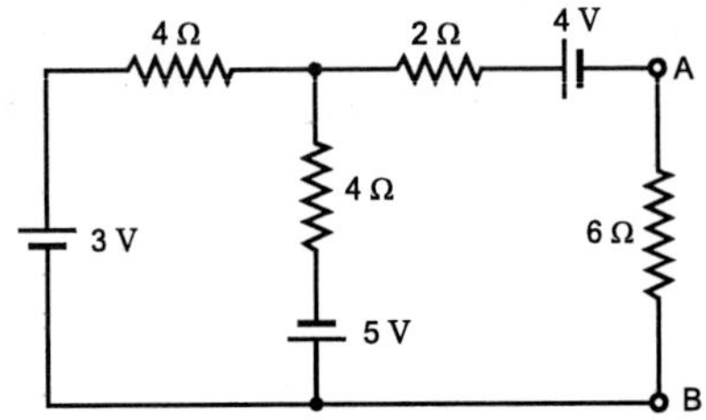

Fig. 18

22. Determine current in 10 Ω resistance using Norton's theorem in the following network (Fig. 19)

[U.P. Technical Univ. Electrical Engineering First Semester 2005-06]

$\left[\textbf{Ans. } \frac{165}{122} \text{ A}\right]$

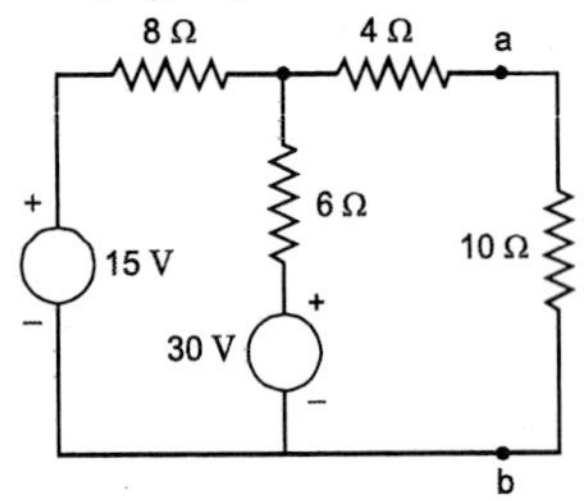

Fig. 19

23. Determine the value of R_L in the network shown in Fig. 20 to have power transferred to it maximum. Determine also the output power of the battery and the power lost in the three star-connected resistances.

[Ans. 5 Ω, 5 W, 38.33 W, 43.33 W]

24. Three resistances of 20 Ω each are connected in star. Find the equivalent delta resistance. If a source of emf of 120 V is connected across any two terminals of the equivalent delta-connected resistance find the current supplied by the source.

[Ans. 60 Ω; 3 A]

25. Three resistances of 20, 25 and 30 ohms are connected in delta. Calculate the corresponding resistances in an equivalent star connection.

[Ans. 8 Ω, 6.67 Ω and 10 Ω]

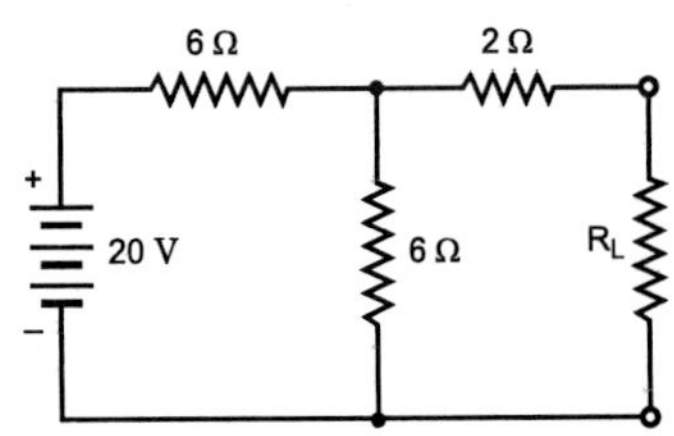

Fig. 20

26. A network of resistances is shown in Fig. 21. Compute the equivalent network resistance measured between (*i*) A and B (*ii*) B and C and (*iii*) C and A.

[Ans. $\frac{4914}{2679}$ Ω, $\frac{2160}{2679}$ Ω, $\frac{3726}{2669}$ Ω]

27. Find the resistance between terminals XY of the bridge circuit shown in Fig. 22 by using delta - star transformations.

[U.P. Technical Univ. Electrical Engineering July 2002]

[Ans. 8.234 Ω]

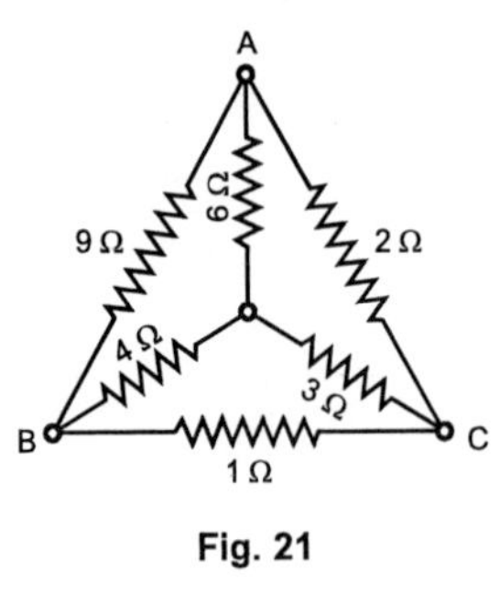

Fig. 21

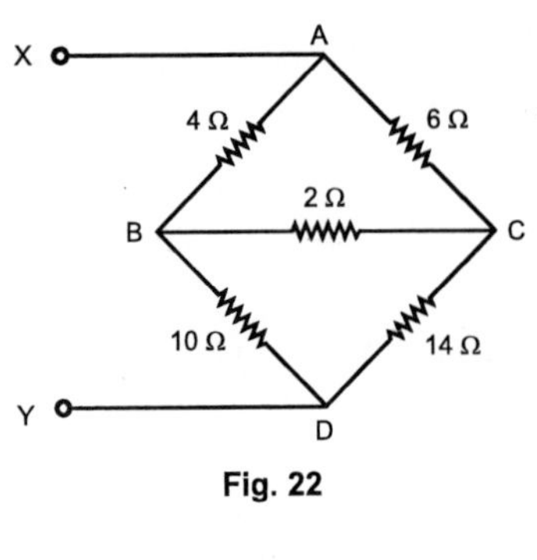

Fig. 22

UNIT-II

Steady-State Analysis of Single Phase AC Circuits

CONTENTS IN THIS UNIT

Steady-State Analysis of Single Phase AC Circuits:

AC Fundamentals: Sinusoidal, square and triangular waveforms — average and effective values, form and peak factors, Concept of phasors, Phasor representation of sinusoidally varying voltage and current.

Analysis of series, parallel and series-parallel R-L-C circuits: apparent, active and reactive powers, power factor, Causes and problems of low power factor, Power factor improvement.

Resonance in series and parallel circuits, bandwidth and quality factor (simple numerical problems)

3 CHAPTER

AC Fundamentals

3.1. INTRODUCTION

A current (or voltage) is called *alternating* if it reverses periodically in direction, and its magnitude undergoes a definite cycle of changes in definite intervals of time. Each cycle of alternating current (or voltage) consists of two half cycles, during one of which the current (or voltage) acts in one direction; while during the other in opposite direction. In more restricted sense, alternating current is a periodically varying current, the average value of which, over a period, is zero. The direct current always flows in one direction, and its magnitude remains unaltered. In order to produce an alternating current through an electric circuit, a source capable of reversing the emf periodically (ac generator) is required while for generating dc in an electric circuit, a source capable of developing a constant emf is required such as a battery or dc generator. The graphical representations of direct current and alternating current are given in Figs. 3.1(*a*) and (*b*) respectively.

At present a large percentage of the electrical energy (nearly all) being used for domestic and commercial purposes is generated as alternating current. In fact, almost the whole of the vast amount of electrical energy used throughout the world for every imaginable purpose is generated by alternating current generators. This is not due to any superiority of alternating current over direct current in the sphere of applicability to industrial and domestic use. In fact there are certain types of works for which alternating current is unsuitable and, therefore, direct current is absolutely necessary such as for electroplating, charging of storage batteries, refining of copper, refining of aluminium, electrotyping, production of industrial gases by electrolysis, municipal traction etc. In some power applications, the ac motor is unsatisfactory such as for metal rolling mills, paper making machines, high-speed gearless elevators, automatic machine tools and high-speed printing presses. Direct current required for these applications is nowadays derived from an ac supply by the use of suitable convertors or rectifiers. For lighting and heating dc and ac are equally useful. The reasons for generation of electrical energy in the form of alternating current are given on next page:

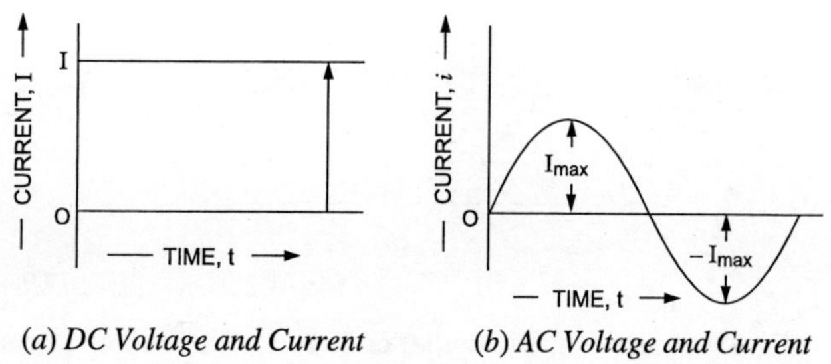

(*a*) *DC Voltage and Current* (*b*) *AC Voltage and Current*

Fig. 3.1

1. AC generators have no commutator and can, therefore, be built in very large units to run at high speeds producing high voltages (as high as 33,000 volts), so that the construction and operating cost per kW is low, whereas dc generator capacities and voltages are limited to comparatively low values.
2. Alternating current can be generated at comparatively high voltages and can be raised and lowered readily by a static machine called the *transformer*, which makes the transmission and distribution of electrical energy economical. In direct current use of transformers is not possible.
3. AC induction motor is cheaper in initial cost and in maintenance since it has got no commutator and is more efficient than dc motor for constant speed work, so it is desirable to generate power as alternating current.
4. The high transmission efficiency in ac makes the generation of electrical energy economical by generating it in large quantities in a single station and distributing over a large territory.
5. The switchgear (*e.g.* switches, circuit breakers etc.) for ac system is simpler than that required in a dc system.
6. The maintenance cost of ac equipment is less.

3.2. GENERATION OF ALTERNATING EMF

We know that an alternating emf can be generated either by rotating a coil within a stationary magnetic field, as illustrated in Fig. 3.2 (*a*) or by rotating a magnetic field within a stationary coil, as illustrated in Fig. 3.2 (*b*). The emf generated, in either case, will be of sinusoidal waveform. The magnitude of emf generated in the coil depends upon the number of turns on the coil, the strength of magnetic field and the speed at which the coil or magnetic field rotates. The former method is employed in case of small ac generators while the later one is employed for large sized ac generators.

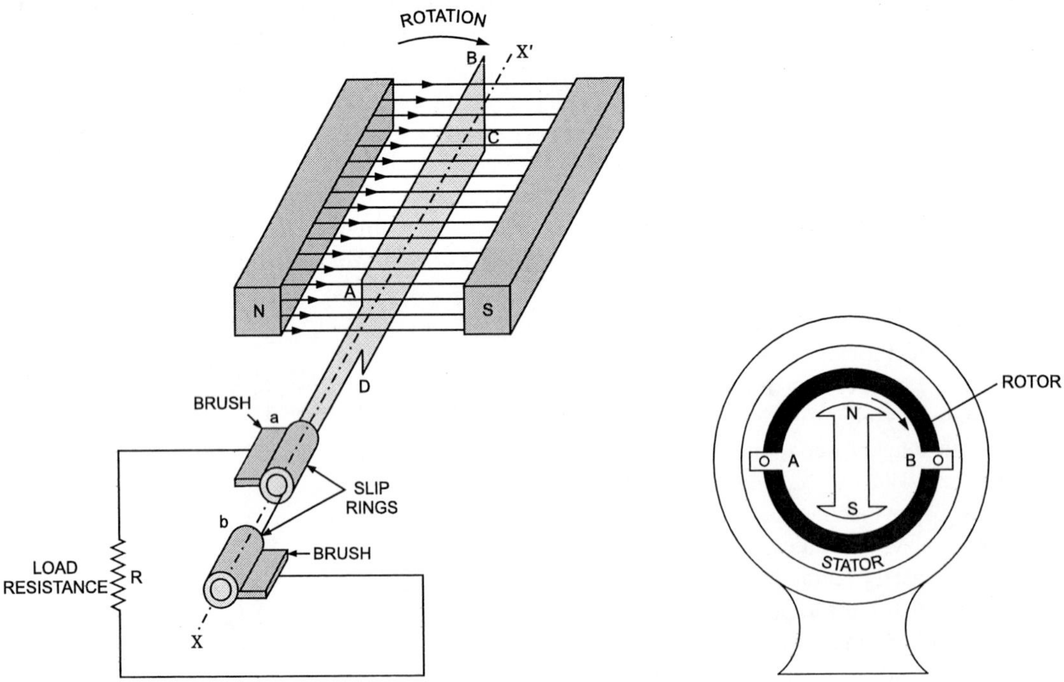

(*a*) (*b*)

Fig. 3.2 *Generation of Alternating EMF*

Now consider a rectangular coil of N turns rotating in counter-clockwise direction with angular velocity of ω radians per second in a uniform magnetic field, as illustrated in Fig. 3.3.

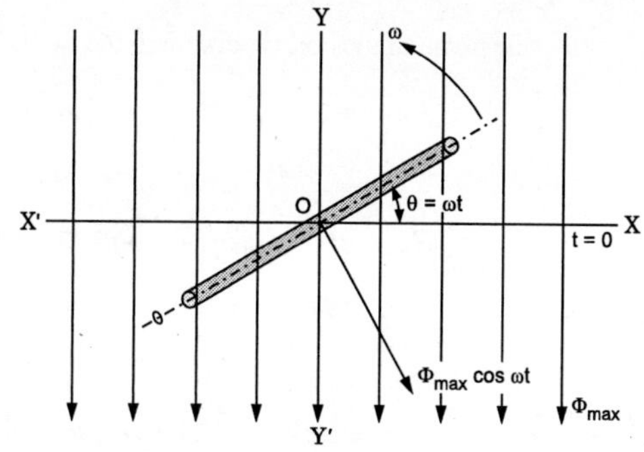

Fig. 3.3

Let the time be measured from the instant of coincidence of the plane of the coil with the X-axis. At this instant maximum flux, Φ_{max} links with the coil. Let the coil assume the position, as shown in Fig. 3.3, after moving in counter-clockwise direction for t seconds.

The angle θ through which the coil has rotated in t seconds = ωt

In this position, the component of flux along perpendicular to the plane of coil = $\Phi_{max} \cos \omega t$.

Hence flux linkages of the coil at this instant

= Number of turns on coil × linking flux

i.e; instantaneous flux linkages = $N \Phi_{max} \cos \omega t$

Since emf induced in a coil is equal to the rate of change of flux linkages with minus sign,

EMF induced at any instant,

$$e = -\frac{d}{dt}[N \Phi_{max} \cos \omega t]$$

$$= \Phi_{max} N \frac{d}{dt}[-\cos \omega t] = \Phi_{max} N \omega \sin \omega t \quad ...(3.1)$$

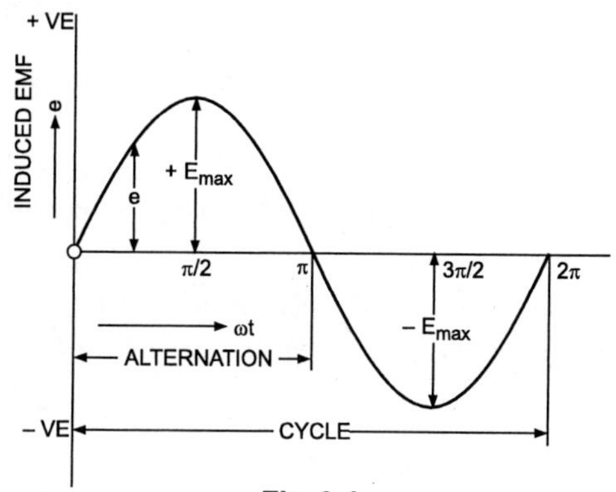

Fig. 3.4

when $\omega t = 0$, $\sin \omega t = 0$, therefore, induced emf is zero,

when $\omega t = \frac{\pi}{2}$, $\sin \frac{\pi}{2} = 1$, therefore, induced emf is maximum, which is denoted by E_{max} and is equal to $\Phi_{max} N \omega$

Substituting $\Phi_{max} N\omega = E_{max}$ in Eq. (*i*) we have

Instantaneous emf, $e = E_{max} \sin \omega t$ ...(3.2)

So the emf induced varies as the sine function of the time angle ωt, and if emf induced is plotted against time, a curve of sine waveshape is obtained as illustrated in Fig. 3.4. Such an emf is called the *sinusoidal* emf. The sine curve is completed when the coil rotates through an angle of 2π radians. The induced emf e will have maximum value, represented by E_{max}, when the coil has turned through $\frac{\pi}{2}$ radians (or 90°) in counter-clockwise direction from the reference axis (*i.e.* OX axis).

Example 3.1 **A coil having 200 turns and area of x-section 250 cm² is rotated about its axis at right angle to a uniform magnetic field of strength 0.5 T at a speed of 1,200 rpm. Determine (*i*) maximum value of emf induced (*ii*) equation for instantaneous induced emf and (*iii*) instantaneous values of induced emf when (*a*) the plane of the coil is at right angle to the field. (*b*) the plane of the coil is parallel to the field and (*c*) the plane of the coil is at an angle of 60° to the field.**

Solution: Angular velocity, $\omega = \frac{2\pi \times RPM}{60} = \frac{2\pi \times 1,200}{60} = 40\,\pi$ radians/s

Maximum flux linking with the coil, $\Phi_{max} = B_{max} \times$ area of coil $= 0.5 \times 250 \times 10^{-4} = 0.0125$ Wb

(*i*) Maximum value of induced emf, $E_{max} = \Phi_{max} N\omega = 0.0125 \times 200 \times 40\,\pi = 314$ V **Ans.**

(*ii*) Equation for instantaneous induced emf $= E_{max} \sin \omega t = 314 \sin 40\,\pi t$ **Ans.**

(*iii*) Instantaneous values of induced emf when

(*a*) the plane of the coil is at right angle to the field *i.e.* when $\theta = \omega t = 0$

$\therefore e_0 = 314 \sin 0° = 0$ **Ans.**

(*b*) the plane of the coil is parallel to the field, *i.e.* when $\theta = \omega t = \frac{\pi}{2}$ radians

$$\therefore \qquad e' = 314 \sin \frac{\pi}{2} = 314 \text{ V } \textbf{Ans.}$$

(*c*) the plane of the coil makes an angle of 60° to the field *i.e.* when $\theta = \frac{\pi}{2} - \frac{\pi}{3} = \frac{\pi}{6}$ radians

$$\therefore \qquad e'' = 314 \sin \frac{\pi}{6} = 314 \times \frac{1}{2} = 157 \text{ V } \textbf{Ans.}$$

Example 3.2 **A square coil of 10 cm side and 100 turns is rotated at a uniform speed of 1,000 revolutions per minute, about an axis at right angles to a uniform magnetic field of 0.5 Wb/m². Calculate the instantaneous value of the induced electromotive force, when the plane of the coil is (*i*) at right angle to the field (*ii*) in the plane of the field.**

Solution:

$$\text{Angular velocity, } \omega = \frac{2\pi \times \text{RPM}}{60} = \frac{2\pi \times 1{,}000}{60} = 33.33\,\pi \text{ radians/second}$$

$$\text{Maximum flux linking with the coil, } \Phi_{max} = B_{max} \times \text{area of coil} = 0.5 \times (0.1)^2 = 0.005 \text{ Wb}$$

$$\text{Number of turns on the coil, } N = 100$$

$$\text{Maximum value of emf induced, } E_{max} = \Phi_{max} N\omega = 0.005 \times 100 \times 33.33\,\pi = 52.36 \text{ V}$$

Instantaneous value of induced emf when the plane of the coil is,

(*i*) at right angle to field *i.e.* when $\theta = \omega t = 0$

$$e_0 = 52.36 \sin 0° = 0 \textbf{ Ans.}$$

(*ii*) in the plane of the field *i.e.* when $\theta = \omega t = \frac{\pi}{2}$ radians

$$e' = 52.36 \sin \frac{\pi}{2} = 52.36 \text{ V } \textbf{Ans.}$$

3.3. SINUSOIDAL QUANTITIES (EMF, VOLTAGE OR CURRENT)

It is not an accident that the bulk of electric power generated in electric power stations throughout the world and distributed to the consumers appears in the form of sinusoidal variations of voltage and current. There are many technical and economical advantages associated with the use of sinusoidal voltages and currents. For example, it will be learned that the use of sinusoidal voltages applied to appropriately designed coils results in a revolving magnetic field which has the capacity to do work. As a matter of fact it is this principle which underlies the operation of almost all the electric motors found in home appliances and about 90% of all electric motors found in commercial and industrial applications. Although other waveforms can be used in such devices, none leads to an operation which is as efficient and economical as that achieved through the use of sinusoidal quantities.

The other advantages of using sinusoidal voltages and currents are:

1. The waveform from generation to utilization remains the same if a sinusoidal waveform is generated.
2. Electromagnetic torque developed in three-phase machines (generators and motors) with balanced three-phase currents is uniform (constant), and therefore, there are no oscillations in developed torque and absence of noise in operation.
3. Nonsinusoidal voltages which contain harmonic frequencies, according to Fourier analysis, are harmful to the system on account of
   (*i*) increased losses in generators, motors, transformers, and transmission and distribution systems,
   (*ii*) more interference (noise) to nearby communication circuits,
   (*iii*) resonance may result in overvoltages or overcurrents at many pockets on the way from generating station to consumer's premises which may damage the equipment and increase losses.
   (*iv*) increased current through power factor improvement capacitors.

In practical electrical engineering it is assumed that the alternating voltages and currents are sinusoidal, though they may slightly deviate from it. The advantage of this assumption is that calculations become simple. *It may be noted that alternating voltage and current mean sinusoidal voltage and current unless stated otherwise.*

Alternating emf following sine law (*i.e.* sinusoidal emf) is illustrated in Fig. 3.4 and is expressed in the form

$$e = E_{max} \sin \omega t \qquad ...(3.3)$$

where e is the instantaneous value of alternating emf (or voltage), E_{max} is the maximum value of the alternating emf (or voltage) and ω is angular velocity of the coil.

The rotating coil moves through an angle of 2π radians in one cycle, so angular velocity $\omega = 2\pi f$ where f is the number of cycles completed per second.

Substituting $\omega = 2\pi f$ in Eq. (3.3) we have

$$e = E_{max} \sin 2\pi ft \qquad ...(3.4)$$

If the alternating emf (or voltage) given by Eq. (3.3) is applied across a load, alternating current flows through the circuit which would also vary sinusoidally *i.e.* following a sine law. The expression for alternating current is given as

$$i = I_{max} \sin \omega t = I_{max} \sin 2\pi ft \qquad ...(3.5)$$

provided the load is pure resistive.*

3.4. TERMINOLOGY

An *alternating quantity* (voltage or current) is one which changes continuously in magnitude and alternates in direction at regular intervals of time. It rises from zero to maximum positive value, falls to zero, increases to a maximum in the reverse direction and falls back to zero again, as illustrated in Fig. 3.4. The emf (or voltage) and current repeat the procedure.

The important ac terms are defined below:

3.4.1. Waveform. The shape of the curve of the voltage or current when plotted against time as abcissa (base) is called the *waveform*. The waveform of an alternating voltage varying sinusoidally is shown in Fig. 3.4. The waveform of the induced emf in an alternator differs slightly from that of sine wave but for calculation purposes it is treated as such.

3.4.2. Instantaneous Value. The value of alternating quantity (emf, voltage or current) at any particular instant is called the *instantaneous value* and is designated by a small italic letter (e for emf, v for voltage and i for current). The instantaneous values of an alternating quantity can be determined either from the curve or from an equation of the alternating quantity. For example, the instantaneous values of emf represented by the curve shown in Fig. 3.4 at 0, $\frac{\pi}{2}$, π and $\frac{3\pi}{2}$ are zero, $+ E_{max}$, zero and $- E_{max}$ respectively.

3.4.3. Alternation and Cycle. When a periodic wave, such as sinusoidal wave, goes through one complete set of positive or negative values, it completes one *alternation* and when it goes through one complete set of positive and negative values it is said to have completed one *cycle*.

Alternation and cycle can also be defined in terms of angular measure. One alternation corresponds to 180°(or π radians) while one cycle corresponds to 360° (or 2π radians).

3.4.4. Time Period and Frequency. The time taken in seconds by an alternating quantity to complete one cycle is known as *time period* or *periodic time* and is denoted by T.

---

* The load may be resistive, inductive or capacitive. It will be shown in chapter 4 that if the load is inductive or capacitive the current equation differs in time angle.

The number of cycles completed per second by an alternating quantity is known as *frequency* and is denoted by *f*. In SI system the frequency is expressed in hertz (pronounced as hurts). One hertz (or Hz) is equal to one cycle per second.

The number of cycles completed per second $= f$

$$\text{Time period, T} = \text{Time taken in completing one cycle} = 1/f$$

$$\text{or } f = 1/\text{T} \qquad ...(3.6)$$

The commercial ac power is generated at frequency of 50 Hz or 60 Hz.* The reasons of suitability of frequency of this range are:

1. The output of the equipment increases with the increase in frequency. For a given output, smaller size machines are required as compared to those for lower frequency output. Because of high power-weight ratio, relative cost of the equipment is also reduced.
2. Lower regulation, lower skin effect resulting in lower ohmic losses, lower magnetic and dielectric losses resulting in higher efficiency, lower corona loss and higher power transmission line capacity as compared to those at higher frequency.

3.4.5. Angular Velocity and Frequency. A glance at Fig. 3.4 indicates that each cycle spans 2π radians. Hence, if this quantity is divided by the time period, *angular velocity* of the sine function is obtained. It is denoted by ω and is expressed in radians per second.

$$\therefore \quad \text{Angular velocity, } \omega = \frac{\text{Angle turned}}{\text{Time taken}} = \frac{2\pi}{\text{T}} = \frac{2\pi}{1/f} = 2\pi f \text{ radians per second} \qquad ...(3.7)$$

3.4.6. Electrical Time Degrees and Mechanical Degrees. It is seen that the coil must revolve past a pair of poles in order to carry the generated emf through one complete cycle. In circuit work, one complete cycle of voltage or current is designated as 360 electrical degrees or 2π electrical radians. To correlate with this designation the arc through which a coil of dynamo must rotate in order to generate one cycle of emf is called 360 electrical degrees. In a 2-pole machine one complete revolution of coil produces one cycle of emf. But in a multipolar machine, such as four, six or eight pole machine, the emf completes one cycle or 360 electrical degrees or 2π electrical radians as soon as the coil passes a pair of poles and a mechanical degree will be equal to as many electrical degrees as there are pairs of poles in the dynamo structure. In a multipolar machine, the number of cycles completed per second by generated emf,

$$f = \text{Pair of poles} \times \text{number of revolutions made per second} = \frac{\text{P}}{2} \times \frac{\text{N}}{60} = \frac{\text{PN}}{120} \qquad ...(3.8)$$

where N is the speed of rotation of the coil in rpm.

3.4.7. Amplitude. The maximum value, positive or negative, which an alternating quantity attains during one cycle is called the *amplitude* of the alternating quantity. The amplitude of an alternating emf (or voltage) and current is designated by E_{max} (or V_{max}) and I_{max} respectively.

Example 3.3 **What is the time period of the wave produced by a 6-pole alternator which is driven at 1,000 rpm ?**

[Pb Technical Univ. Basic Electrical Engineering, June 2003]

Solution:

$$\text{Number of poles on alternator, P} = 6$$

$$\text{Speed of alternator, N} = 1{,}000 \text{ rpm}$$

$$\text{Number of cycles completed per second by generated emf, } f = \frac{PN}{120} = \frac{6 \times 1{,}000}{120} = 50$$

$$\text{Time period, T} = \frac{1}{f} = \frac{1}{50} = 0.02 \text{ second} \quad \textbf{Ans.}$$

* In India the standard frequency for power supply is 50 Hz.

3.5. DETERMINATION OF MAXIMUM VALUE AND FREQUENCY FROM EMF OR CURRENT EQUATIONS

From Eq. (3.4)

$$e = E_{max} \sin 2\pi f t$$

From above expression we observe that

(*i*) *the maximum value of an alternating emf is given by the coefficient of the sine of the time angle.*

(*ii*) *the frequency is given by coefficient of time t divided by* 2π

$$i.e.\ f = \frac{\text{Coefficient of time, } t}{2\pi} \qquad ...(3.9)$$

Similarly we can also find the maximum value and frequency of the current from the equation of instantaneous values of current.

Example 3.4 **An ac voltage of 50 Hz frequency has a peak value of 220 V. Write down the expression for the instantaneous value of this voltage.** [*Pb Technical Univ. Electrical Engineering January 2000*]

Solution: Supply frequency, $f = 50$ Hz

Peak value of ac voltage, $V_{max} = 220$ V

Expression for instantaneous value of ac voltage (assumed sinusoidal) with θ as zero when time is zero, is given as

$$v = V_{max} \sin \omega t = V_{max} \sin 2\pi ft = 220 \sin 2\pi \times 50\, t = 220 \sin 314t \text{ **Ans.**}$$

Example 3.5 **An alternating current of frequency 50 Hz has a maximum value of 100 A. Calculate (*a*) its value $\frac{1}{600}$ second after the instant the current is zero and its value decreasing there afterwards (*b*) how many seconds after the instant the current is zero (increasing there afterwards) will the current attain the value of 86.6 A ?**

[*Allahabad Univ. Elec. Technology 1991*]

Solution: The current waveform is shown in Fig. 3.5.

The equation of the alternating current (assumed sinusoidal) with respect to the origin O is

$$i = 100 \sin 100\, \pi\, t \qquad ...(i)$$

(*a*) Since the current is measured from the instant the current is zero and is decreasing there afterwards (*i.e.* from point A in Fig. 3.5), the equation for the alternating current with respect to the point A becomes

$$i = 100 \sin (100\, \pi\, t + \pi) = -100 \sin 100\, \pi\, t \qquad ...(ii)$$

Substituting $t = \frac{1}{600}$ second in above equation we get the instantaneous value of current $\frac{1}{600}$ second after the instant the current is zero and decreasing there afterward

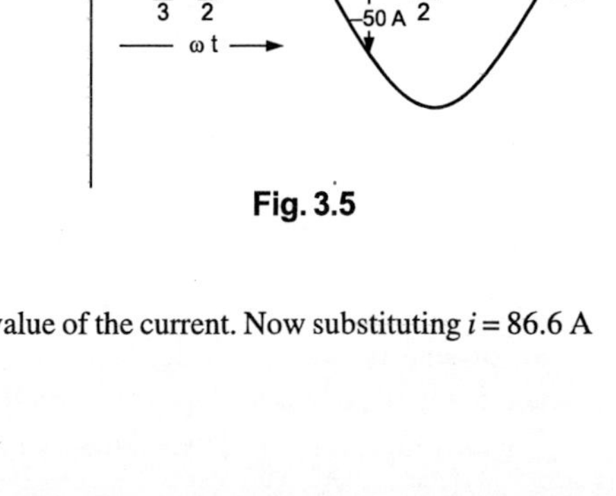

Fig. 3.5

$$\text{So } i = -100 \sin 100\, \pi \times \frac{1}{600} = -100 \sin \frac{\pi}{6} = -50 \text{ A **Ans.**}$$

(*b*) Let the current attain the value of 86.6 A, *t* seconds after the zero value of the current. Now substituting $i = 86.6$ A in Eq. (*i*) we get

$$86.6 = 100 \sin 100\, \pi\, t$$

$$\text{or } t = \frac{1}{100\, \pi} \sin^{-1} \frac{86.6}{100} = \frac{1}{100\, \pi} \times \frac{\pi}{3} = \frac{1}{300} \text{ second **Ans.**}$$

3.6. PLOTTING OF SINE WAVEFORM

Sine curve may be graphically drawn, as illustrated in Fig. 3.6. Draw a circle of radius equal to the maximum value of sinusoidal quantity. Divide the circumference of the circle drawn so into any number of equal parts, say 12, and draw a horizontal line AB (the base on which the sine wave is to be drawn) passing through the centre of the circle. Divide the line AB into the same number of equal parts *i.e.* 12 and number the points correspondingly. Draw perpendicular ordinates from each point. Project the points on the circle horizontally to meet the perpendicular ordinates having corresponding numbers. Draw smooth curve through these points. Curve so drawn will be of sine waveform.

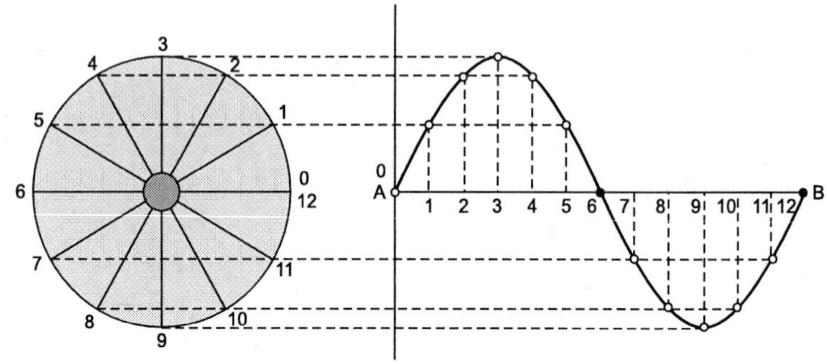

Fig. 3.6 *Simple Method of Plotting Sine Wave*

3.7. AVERAGE AND EFFECTIVE (RMS) VALUES OF ALTERNATING VOLTAGE AND CURRENT

In a dc system, the voltage and current are constant and, therefore, there is no problem in specifying their magnitude. But in case of ac system, an alternating voltage or current varies from instant to instant and so poses a problem how to specify the magnitude of an alternating voltage or current. An alternating voltage or current may possibly be expressed in terms of peak (maximum) value, average (mean) value or effective (rms) value.

In specifying an alternating voltage or current, its peak or maximum value is rarely used because it has that value only twice each cycle. Furthermore, the average or mean value cannot be used because it is positive as much as it is negative, so the average value is zero. Although the average value over half cycle might be used, it would not be as logical a choice as what we shall find *effective* (*virtual or rms*) *value* which is related to the power developed in a resistance by the alternating voltage or current.

3.7.1. Average Value of Alternating Current. The average (or mean) value of an alternating current is equal to the value of direct current which transfers across any circuit the same charge as is transferred by that alternating current during a given time.

Since in the case of a symmetrical alternating current (*i.e.* one whose two half cycles are exactly similar, whether sinusoidal or nonsinusoidal) the average or mean value over a complete cycle is zero hence for such alternating quantities average or mean value means the value determined by taking the average of instantaneous values during half cycle or one alternation only. However, for unsymmetrical alternating current, as half-wave rectified current, the average value means the value determined by taking the mean of instantaneous values over the whole cycle.

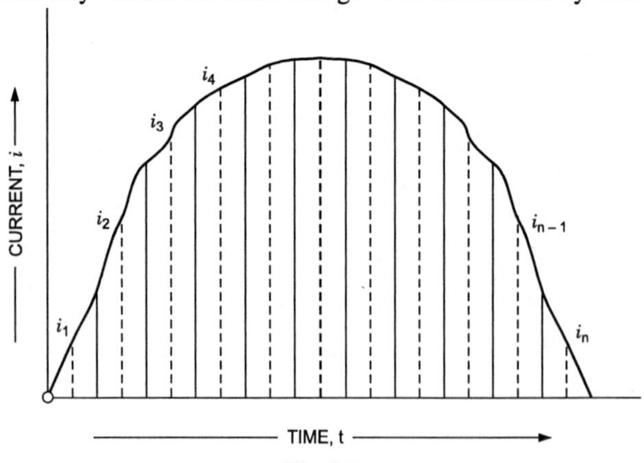

Fig. 3.7

The average value is determined by measuring the lengths of a number of equidistant ordinates and then taking their mean *i.e.* of i_1, i_2, i_3 ..... i_n etc. which are mid-ordinates.

$$\because \text{ Average value of alternating current, } I_{av} = \frac{i_1 + i_2 + i_3 + i_4 \; .... + i_n}{n}$$

$$= \frac{\text{Area of one alternation (or half cycle)}}{\text{Length of base over one alternation (or half cycle)}}$$

Using the integral calculus the average (or mean) value of a function $f(t)$ over a specific interval of time between t_1 and t_2 is given by

$$F_{av} = \frac{1}{t_2 - t_1} \int_{t_1}^{t_2} f(t)\, dt \qquad ...(3.10)$$

Any function whose cycle is repeated continuously, irrespective of its waveshape, is termed as *periodic function*, such as sinusoidal function, and its average value is given by

$$F_{av} = \frac{1}{T} \int_0^T f(t)\, dt \qquad ...(3.11)$$

where T is time period of periodic function.

In case of a symmetrical alternating current, whether sinusoidal or nonsinusoidal the average value is determined by taking average of one half cycle or one alternation only.

i.e., for symmetrical waveforms, $F_{av} = \frac{2}{T} \int_0^{T/2} f(t)\, dt \qquad ...(3.12)$

3.7.2. RMS Value or Effective Value of Alternating Current. The rms or effective value of an alternating current or voltage is given by that steady current or voltage which when flows or applied to a given resistance for a given time produces the same amount of heat as when the alternating current or voltage is flowing or applied to the same resistance for the same time.

Consider an alternating current of waveform shown in Fig. 3.8 flowing through a resistor of R ohms. Divide the base of one alternation into n equal parts and let the mid-ordinates be $i_1, i_2, i_3 ... i_n$ etc.

Heat produced during 1st interval $= i_1^2 R \times \frac{T}{n}$ joules

Heat produced during 2nd interval $= i_2^2 R \times \frac{T}{n}$ joules

Heat produced during 3rd interval $= i_3^2 R \times \frac{T}{n}$ joules

Heat produced during nth interval $= i_n^2 R \times \frac{T}{n}$ joules

Total heat produced in time T

$$= RT\left(\frac{i_1^2 + i_2^2 + i_3^2 + ... i_n^2}{n}\right) \text{ joules}$$

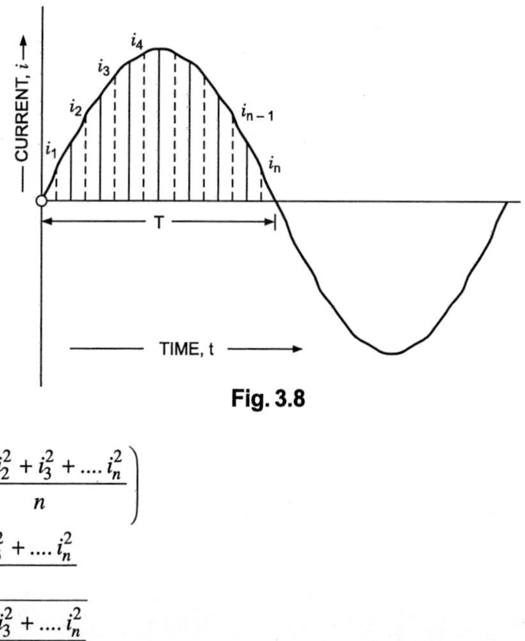

Fig. 3.8

Now if I_{eff} is the effective current, then heat produced by this current in time T = I_{eff}^2 RT joules. By definition these two expressions are equal

$$\therefore \quad I_{eff}^2 RT = RT\left(\frac{i_1^2 + i_2^2 + i_3^2 + ... i_n^2}{n}\right)$$

$$\text{or } I_{eff}^2 = \frac{i_1^2 + i_2^2 + i_3^2 + ... i_n^2}{n}$$

$$\text{or } I_{eff} = \sqrt{\frac{i_1^2 + i_2^2 + i_3^2 + ... i_n^2}{n}}$$

Hence the *effective* or *virtual value* of alternating current or voltage is equal to the square root of the mean of the squares of successive ordinates and that is why it is known as *root mean square (rms) value.*

Using the integral calculus the root mean square (rms) or effective value of an alternating quantity over a time period is given by

$$F_{rms} = \sqrt{\frac{1}{T} \int_0^T f^2(t)\, dt} \qquad (3.13)$$

3.8. AVERAGE AND EFFECTIVE (RMS) VALUES OF SINUSOIDAL CURRENT AND VOLTAGE

3.8.1. Average Value For Sinusoidal Current or Voltage. The average value of a sine wave over a complete cycle is zero. Therefore, the half cycle average value is intended.

Instantaneous value of sinusoidal current is given by

$$i = I_{max} \sin \omega t$$

Considering first half cycle i.e when ωt varies from 0 to π we get,

$$I_{av} = \frac{\text{Area of first half cycle}}{\pi}$$

$$= \frac{1}{\pi}\int_0^{\pi} i \, d(\omega t) = \frac{1}{\pi}\int_0^{\pi} I_{max} \sin \omega t \; d(\omega t)$$

$$\text{or } I_{av} = \frac{I_{max}}{\pi}\left[-\cos \omega t\right]_0^{\pi}$$

$$= \frac{2}{\pi} I_{max} = 0.637\, I_{max} \qquad ...(3.14)$$

$$\text{Similarly } E_{av} = 0.637\, E_{max}$$

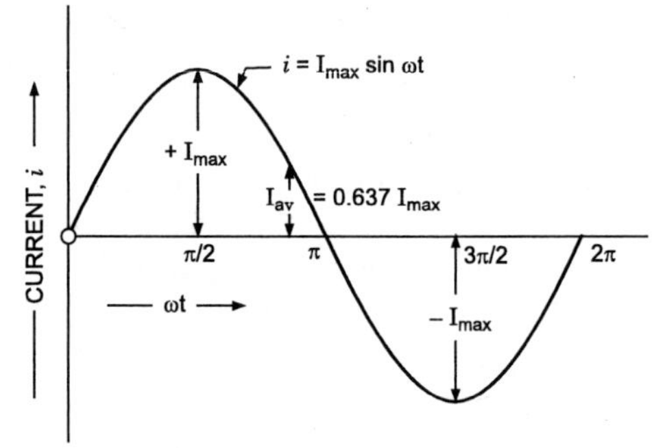

Fig. 3.9

3.8.2. Effective (RMS) Value For Sinusoidal Current or Voltage. A sinusoidal alternating current is represented by

$$i = I_{max} \sin \omega t$$

$$= \frac{\text{Area of first half cycle of } i^2}{\pi}$$

$$= \frac{1}{\pi}\int_0^{\pi} i^2 \, d(\omega t)$$

$$= \frac{1}{\pi}\int_0^{\pi} I_{max}^2 \sin^2 \omega t \, d(\omega t)$$

$$= \frac{I_{max}^2}{2\pi}\int_0^{\pi} (1 - \cos 2\omega t)\, d(\omega t)$$

$$= \frac{I_{max}^2}{2\pi}\left[\omega t - \frac{1}{2}\sin 2\omega t\right]_0^{\pi}$$

$$= \frac{I_{max}^2}{2\pi} \times \pi = \frac{I_{max}^2}{2}$$

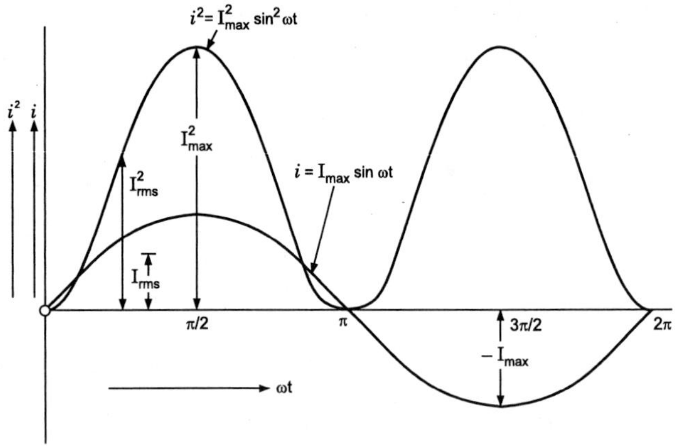

Fig. 3.10

$$\text{or } I_{rms} = \sqrt{\frac{I_{max}^2}{2}} = \frac{I_{max}}{\sqrt{2}} \qquad ...(3.15)$$

$$\text{Similarly, } E_{rms} = \frac{E_{max}}{\sqrt{2}}$$

Example 3.6 **A sinusoidally varying alternating current of frequency 60 Hz has a maximum value of 15 amperes.**

(*i*) **Write down the equation for instantaneous value,**

(*ii*) **Find the value of current after 1/200 second,**

(*iii*) **Find the time taken to reach 10 amperes for the first time, and**

(*iv*) **Find its average value.** [*U.P. Technical Univ. Electrical Engineering Second Semester 2002-03*]

Solution: Maximum value of current, $I_{max} = 15$ A

Frequency of alternating current, $f = 60$ Hz

(*i*) The equation for instantaneous value of sinusoidal alternating current with θ as zero when time is zero, is given as

$$i = I_{max} \sin 2\pi f t = 15 \sin 2\pi \times 60\, t = 15 \sin 120\pi t \text{ **Ans.**}$$

(*ii*) Substituting $t = \frac{1}{200}$ second in the expression $i = 15 \sin 120\pi t$ we have value of current after $\frac{1}{200}$ second

$$= 15 \sin 120\pi \times \frac{1}{200} = 15 \sin 0.6\pi = 15 \times 0.951 = 14.266 \text{ A } \textbf{Ans.}$$

(*iii*) Let the instantaneous value of current be 10 A for the first time *t* seconds after the instant the current is zero and becoming positive

$$\therefore \quad 10 = 15 \sin 120\pi t$$

$$\text{or} \quad t = \frac{1}{120\pi} \sin^{-1} \frac{10}{15} = \frac{1}{120\pi} \sin^{-1} 0.6667 = 0.001936 \text{ second } \textbf{Ans.}$$

(*iv*) Average value, $I_{av} = \frac{2}{\pi} I_{max} = \frac{2}{\pi} \times 15 = 9.55$ A **Ans.**

Example 3.7 **An alternating current when passed through a resistance immersed in water for 5 minutes, just raised the temperature of water to boiling point. When a direct current of 4 amperes was passed through the same resistance under identical conditions it took 8 minutes to boil the water. Find the rms value of the alternating current.**

Solution: Let the rms value of alternating current passed through the resistance be I_{rms} amperes.

Heat produced when an alternating current of I_{rms} amperes is passed through a resistance R immersed in water for 5 minutes

$$= (I_{rms})^2 \times R \times 5 \times 60 = 300\, I_{rms}^2\, R \text{ joules} \qquad ...(i)$$

Heat produced when a direct current of 4 A is passed through the same resistance R immersed in the water for 8 minutes

$$= (4)^2 \times R \times 8 \times 60 = 7{,}680 \text{ R joules} \qquad ...(ii)$$

Since heat produced in both of the cases is same, equating expressions (*i*) and (*ii*) we get

$$300\, I_{rms}^2\, R = 7{,}680\, R$$

$$\text{or} \quad I_{rms} = \sqrt{\frac{7{,}680}{300}} = \sqrt{25.6} = 5.06 \text{ A } \textbf{Ans.}$$

Example 3.8 **An alternating current is given by $i = 20 \sin 600\, t$ amperes. Find the (*i*) frequency, (*ii*) peak value of current and (iii) the time taken from $t = 0$ for the current to reach a value of 10 A.**

[*U.P. Technical Univ. Electrical Engineering First Semester 2004-05*]

Solution: (*i*) Frequency, $f = \frac{600}{2\pi} = 99.5$ Hz **Ans.**

(*ii*) Peak value of current, $I_{max} = 20$ A **Ans.**

(*iii*) Time taken from $t = 0$ for the current to reach a value of 10 A,

$$t = \frac{1}{600} \sin^{-1} \frac{10}{20} = \frac{\pi}{6 \times 600} s = \frac{\pi}{3{,}600} s = 0.873 \text{ ms } \textbf{Ans.}$$

Example 3.9 **An alternating current varying sinusoidally with a frequency of 50 Hz has an rms value of 20 A. Write down the equation for the instantaneous value and find this value (a) 0.0025 second (b) 0.0125 second after passing through a positive maximum value. At what time, measured from a positive maximum value, will the instantaneous current be 14.14 A ?**

[*Allahabad Univ. Electrical Science 1992*]

Solution : The current waveform is shown in Fig. 3.11.

Peak value of current,

$$I_{max} = \sqrt{2}\, I_{rms} = \sqrt{2} \times 20 = 28.28 \text{ A}$$

The equation of alternating current with respect to origin O is

$$i = I_{max} \sin 2\pi f t = 28.28 \sin 2\,\pi \times 50\, t$$

$$= 28.28 \sin 100\,\pi\, t$$

The above equation is valid when time is measured from the instant the current is zero and increasing in positive direction. Since the time is measured from the positive maximum value (point A in Fig. 3.11), the above equation is modified to

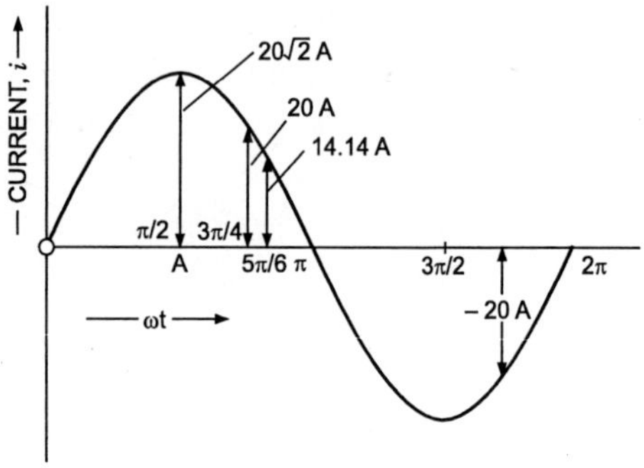

Fig. 3.11

$$i = 28.28 \sin\left(100\,\pi t + \frac{\pi}{2}\right) = 28.28 \cos 100\,\pi\, t \textbf{ Ans.}$$

(*a*) When $t = 0.0025$ s, the instantaneous value of current,

$$i_1 = 28.28 \cos 100\,\pi \times 0.0025 = 28.28 \cos \frac{\pi}{4} = 28.28 \times \frac{1}{\sqrt{2}} = 20 \text{ A} \textbf{ Ans.}$$

(*b*) When $t = 0.0125$ s, the instantaneous value of current,

$$i_2 = 28.28 \cos 100\,\pi \times 0.0125 = 28.28 \cos \frac{5\pi}{4} = 28.28 \left(\frac{-1}{\sqrt{2}}\right) = -20 \text{ A} \textbf{ Ans.}$$

(*c*) Substituting $i = 14.14$ A in expression $i = 28.28 \cos 100\,\pi\, t$ we have

$$14.14 = 28.28 \cos 100\,\pi\, t$$

$$\text{or } t = \frac{1}{100\,\pi} \cos^{-1} \frac{14.14}{28.28} = \frac{1}{100\,\pi} \cos^{-1} 0.5 = \frac{1}{100\,\pi} \times \frac{\pi}{3} = \frac{1}{300} \text{ second } \textbf{Ans.}$$

Example 3.10 An alternating voltage is given by *v* = 141.4 sin 314 *t*. Find (*i*) frequency (*ii*) rms value (*iii*) average value (*iv*) the instantaneous value of voltage when '*t*' is 3 ms (*v*) the time taken for the voltage to reach 100 V for the first time after passing through zero value. *[U.P. Technical Univ. Electrical Engineering First Semester 2006-07]*

Solution: Instantaneous value of alternating voltage is given by expression

$$v = 141.4 \sin 314\, t \qquad ...(i)$$

(*i*) The frequency is given by the coefficient of time divided by 2 π, therefore,

$$\text{frequency}, f = \frac{314}{2\pi} = 50 \text{ Hz } \textbf{Ans.}$$

Peak value of an alternating voltage is given by the coefficient of sine of time angle, therefore,

$$V_{max} = 141.4 \text{ volts}$$

(*ii*) RMS value of given alternating voltage, $V_{rms} = \frac{V_{max}}{\sqrt{2}} = \frac{141.4}{\sqrt{2}} = 100 \text{ V}$ **Ans.**

(*iii*) Average value of given alternating voltage, $V_{av} = 0.637\, V_{max} = 0.637 \times 141.4 = 90$ V **Ans.**

(*iv*) Instantaneous value of voltage at 3 ms (0.003 *s*) after the instant the voltage is zero and increasing in positive direction may be obtained by substituting $t = 0.003$ second in Eq. (*i*). So we have

$$v = 141.4 \sin 314 \times 0.003 = 141.4 \sin 0.942 = 114.4 \text{ V } \textbf{Ans.}$$

(*v*) Time taken from $t = 0$ for the voltage to reach 100 V

$$t = \frac{1}{314} \sin^{-1} \frac{100}{141.4} = \frac{\pi/4}{314} \text{ second} = \frac{1}{400}\, s = 2.5 \text{ ms } \textbf{Ans.}$$

3.9. FORM FACTOR AND PEAK FACTOR

3.9.1. Form Factor. In certain cases it is convenient to have calculations at first upon the mean value of the emf over half a period, therefore, it becomes essential to have some means of connecting this mean value with the effective or rms value. The knowledge of form factor, which is defined as the ratio of effective value to the average or mean value of periodic wave is, therefore, necessary.

Mathematically form factor is given by the relation

$$\text{Form factor} = \frac{\text{Effective value}}{\text{Average value}}$$

$$\text{Form factor for sinusoidal wave, } K_f = \frac{E_{rms}}{E_{av}} = \frac{\frac{E_{max}}{\sqrt{2}}}{\frac{E_{max}}{\pi/2}} = 1.11$$

3.9.2. Peak Factor. Knowledge of peak factor of an alternating voltage is very essential in connection with determining the dielectric strength since the dielectric stress developed in any insulating material is proportional to the maximum value of the voltage applied to it.

Peak or crest or amplitude factor of a periodic wave is defined as the ratio of maximum or peak to the effective or rms value of the wave

$$\text{i.e. Peak factor, } K_p = \frac{\text{Maximum value}}{\text{Effective value}}$$

$$\text{Peak factor for sinusoidal wave, } K_p = \frac{E_{max}}{E_{rms}} = \frac{E_{max}}{E_{max}/\sqrt{2}} = 1.414$$

Example 3.11 The equation of an alternating current $i = 42.42 \sin 628\, t$. Determine (*i*) its maximum value (*ii*) frequency (*iii*) rms value (*iv*) average value (*v*) form factor.

[U.P. Technical Univ. Electrical Engineering Second Semester 2005-06]

Solution: Instantaneous value of alternating current is given by the expression

$$i = 42.42 \sin 628\, t$$

(*i*) Maximum value of an alternating current is given by the coefficient of the sine of the time angle

$$\text{i.e. } I_{max} = 42.42 \text{ A } \textbf{Ans.}$$

(*ii*) Frequency is given by coefficient of time, *t* divided by 2π

$$\therefore \text{Frequency}, f = \frac{628}{2\pi} = 100 \text{ Hz } \textbf{Ans.}$$

$$(iii) \text{ RMS value of the alternating current, } I_{rms} = \frac{I_{max}}{\sqrt{2}} = \frac{42.42}{\sqrt{2}} = 30 \text{ A } \textbf{Ans.}$$

$$(iv) \text{ Average value of the alternating current, } I_{av} = 0.637\, I_{max} = 0.637 \times 42.42 = 27\text{A } \textbf{Ans.}$$

$$(v) \text{ Form factor } K_f = \frac{I_{rms}}{I_{av}} = \frac{30}{27} = 1.11 \textbf{ Ans.}$$

Example 3.12 An alternating voltage is $v = 100 \sin 100t$. Find (*i*) Amplitude (*ii*) Time period and frequency (*iii*) Angular velocity (*iv*) Form factor (*v*) Peak factor. *[U.P. Technical Univ. Electrical Engineering Second Semester 2007-08]*

Solution: Instantaneous value of alternating voltage is given by expression

$$v = 100 \sin 100\, t$$

(*i*) Amplitude of alternating voltage is given by the coefficient of the sine of the time angle, so

$$\text{Amplitude of given wave} = V_{max} = 100 \text{ V } \textbf{Ans.}$$

(*ii*) Frequency is given by coefficient of time, *t* divided by 2π

$$\therefore \text{Frequency}, f = \frac{100}{2\pi} = 15.9 \text{ Hz } \textbf{Ans.}$$

$$\text{Time period, T} = \frac{1}{f} = \frac{1}{15.9} = 0.063 \text{ second or 63 ms} \quad \textbf{Ans.}$$

(*iii*) Angular velocity, $\omega = 2\,\pi f = 2\,\pi \times 15.9 = 100$ radians/second **Ans.**

$$\text{RMS value, } V_{rms} = \frac{V_{max}}{\sqrt{2}} = \frac{100}{\sqrt{2}} = 70.71 \text{ V}$$

$$\text{Average value } V_{av} = 0.637\ V_{max} = 0.637 \times 100 = 63.7 \text{ V}$$

$$(iv) \text{ Form factor, } K_f = \frac{V_{rms}}{V_{av}} = \frac{70.71}{63.7} = 1.11 \quad \textbf{Ans.}$$

$$(v) \text{ Peak factor, } K_p = \frac{V_{max}}{V_{rms}} = \frac{100}{70.71} = 1.4142 \quad \textbf{Ans.}$$

Example 3.13 Find the rms value, average value and form factor of the voltage waveform shown in Fig. 3.12.

[*U.P. Technical Univ. Electrical Engineering Second Semester 2002-03*]

Solution: Average value of voltage over one cycle

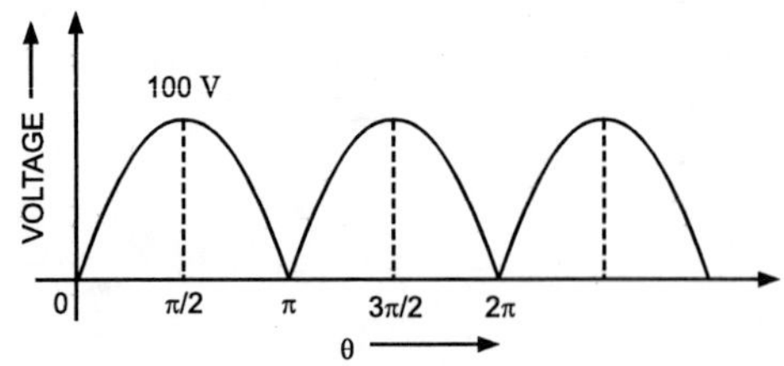

Fig. 3.12

$$= \frac{2\int_0^{\pi} V_{max} \sin\theta\, d\theta}{2\pi}$$

$$= \frac{V_{max}}{\pi}\int_0^{\pi} \sin\theta\, d\theta$$

$$= \frac{2\,V_{max}}{\pi} = \frac{2}{\pi} \times 100 = 63.66 \text{ V } \textbf{Ans.}$$

RMS value of voltage over one cycle

$$= \sqrt{\frac{2\int_0^{\pi} V^2_{max} \sin^2 d\theta}{2\pi}} = \sqrt{\frac{V^2_{max}}{2\pi}\int_0^{\pi} 2\sin^2\theta\, d\theta} = \sqrt{\frac{V^2_{max}}{2\pi}\int_0^{\pi}(1-\cos 2\theta)\, d\theta}$$

$$= \sqrt{\frac{V^2_{max}}{2\pi} \times \pi} = \frac{V_{max}}{\sqrt{2}} = \frac{100}{\sqrt{2}} = 70.71 \text{ V } \textbf{Ans.}$$

$$\text{Form factor of the waveform shown} = \frac{\text{RMS value}}{\text{Average value}} = \frac{70.7}{63.66} = 1.11 \textbf{ Ans.}$$

Example 3.14 Calculate the average and root mean square values, the form factor and peak factor of a periodic current wave having the following values for equal time intervals over half cycle, changing suddenly from one value to the next: 0, 40, 60, 80, 100, 80, 60, 40, 0.

Solution: Since the wave is symmetrical, so considering one half cycle only. Average value

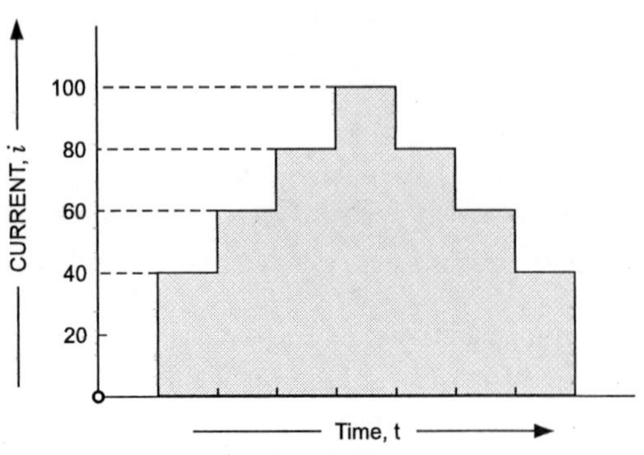

Fig. 3.13

$$I_{av} = \frac{i_1 + i_2 + i_3 + i_4 + i_5 + i_6 + i_7 + i_8}{8}$$

$$= \frac{0 + 40 + 60 + 80 + 100 + 80 + 60 + 40}{8} = 57.5 \text{ A } \textbf{Ans.}$$

RMS value, I_{rms}

$$= \sqrt{\frac{i_1^2 + i_2^2 + i_3^2 + i_4^2 + i_5^2 + i_6^2 + i_7^2 + i_8^2}{8}}$$

$$= \sqrt{\frac{0 + 1{,}600 + 3{,}600 + 6{,}400 + 10{,}000 + 6{,}400 + 3{,}600 + 1{,}600}{8}}$$

$= 64.42$ A **Ans.**

$$\text{Peak value, } I_{max} = 100 \text{ A}$$

$$\text{Form factor} = \frac{I_{rms}}{I_{av}} = \frac{64.42}{57.5} = 1.12 \textbf{ Ans.}$$

$$\text{Peak factor} = \frac{I_{max}}{I_{rms}} = \frac{100}{64.42} = 1.552 \textbf{ Ans.}$$

3.10. RMS VALUE, AVERAGE VALUE, PEAK FACTOR AND FORM FACTOR OF HALF-WAVE RECTIFIED ALTERNATING CURRENT

Half-wave rectified alternating current is one whose one-half cycle has been suppressed *i.e.* one which flows for half the time during one cycle. It is illustrated in Fig. 3.14 where suppressed half cycle is shown dotted.

As mentioned earlier, for determining rms and average values of such an alternating current summation would be carried over the period for which current actually flows *i.e.* 0 to π but would be averaged for the whole cycle *i.e.* from 0 to 2π.

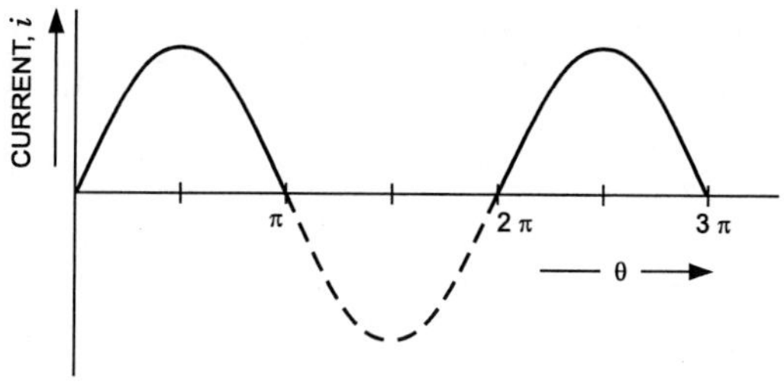

Fig. 3.14

∴ RMS value of current,

$$I_{rms} = \sqrt{\frac{1}{2\pi}\int_0^{\pi} i^2\, d\theta}$$

$$= \sqrt{\frac{1}{2\pi}\int_0^{\pi} I_{max}^2 \sin^2\theta\, d\theta}$$

$$= \frac{I_{max}}{\sqrt{2\pi}}\sqrt{\int_0^{\pi}\sin^2\theta\, d\theta}$$

$$= \frac{I_{max}}{\sqrt{4\pi}}\sqrt{\int_0^{\pi}(1-\cos 2\theta)\, d\theta} = \frac{I_{max}}{\sqrt{4\pi}}\sqrt{\left[\theta - \frac{1}{2}\sin 2\theta\right]_0^{\pi}} = \frac{I_{max}}{\sqrt{4\pi}} \times \sqrt{\pi} = \frac{I_{max}}{2}$$

Average value of current,

$$I_{av} = \frac{1}{2\pi}\int_0^{\pi} i\, d\theta = \frac{1}{2\pi}\int_0^{\pi} I_{max}\sin\theta\, d\theta = \frac{I_{max}}{2\pi}\left[-\cos\theta\right]_0^{\pi} = \frac{I_{max}}{\pi}$$

$$\text{Peak Factor} = \frac{I_{max}}{I_{rms}} = \frac{I_{max}}{I_{max}/2} = 2$$

$$\text{Form factor} = \frac{I_{rms}}{I_{av}} = \frac{I_{max}/2}{I_{max}/\pi} = \frac{\pi}{2} = 1.57$$

Example 3.15 **An alternating voltage e = 200 sin 314 t is applied to a device which offers an ohmic resistance of 20 Ω to the flow of current in one direction, while preventing the flow of current in opposite direction. Calculate rms value, average value and form factor for the current over one cycle.** [*Nagpur Univ. Elec. Engineering 1992*]

Solution: The instantaneous voltage applied to the rectifying device is given by the expression

$$e = 200 \sin 314\, t$$

Maximum value of applied voltage, E_{max} = Coefficient of the sine of time angle = 200 volts

Resistance of rectifying device, R = 20 Ω

Maximum value of half-wave rectified alternating current,

$$I_{max} = \frac{E_{max}}{R} = \frac{200}{20} = 10 \text{ amperes}$$

RMS value of half-wave rectified alternating current,

$$I_{rms} = \frac{I_{max}}{2} = \frac{10}{2} = 5 \text{ amperes } \textbf{Ans.}$$

Average value of the half-wave rectified alternating current,

$$I_{av} = \frac{I_{max}}{\pi} = \frac{10}{\pi} = 3.18 \text{ amperes } \textbf{Ans.}$$

Form factor of the half-wave rectified alternating current

$$= \frac{I_{rms}}{I_{av}} = \frac{5}{3.18} = 1.57 \textbf{ Ans.}$$

Example 13.16 A voltage $v(t) = 220\sqrt{2}$ sin 100t is applied to the circuit shown. What is the rms value of current through the resistor R of 100 Ω ? *[Pb Technical Univ. Electrical Engineering May 2002]*

Solution: The maximum value of voltage applied to the circuit

$$V_{max} = 220\sqrt{2} \text{ V}$$

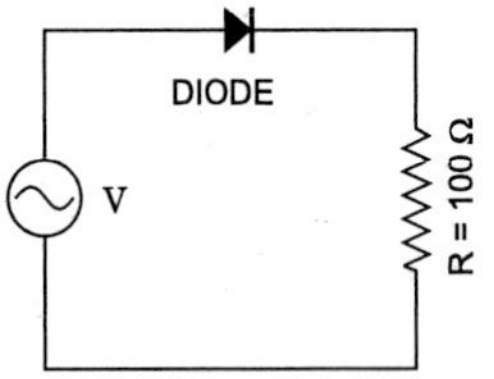

Fig. 3.15

$$\text{Maximum vlaue of circuit current, } I_{max} = \frac{V_{max}}{R} = \frac{220\sqrt{2}}{100} = 3.11 \text{ A}$$

$$\text{RMS value of circuit current, } I_{rms} = \frac{I_{max}}{2} = \frac{3.11}{2} = 1.55 \text{ A } \textbf{Ans.}$$

Example 3.17 Define average value of alternating current. Find this value for the waveform shown in Fig. 3.16.

[Pb Technical Univ. Basic Electrical Engineering December 2002]

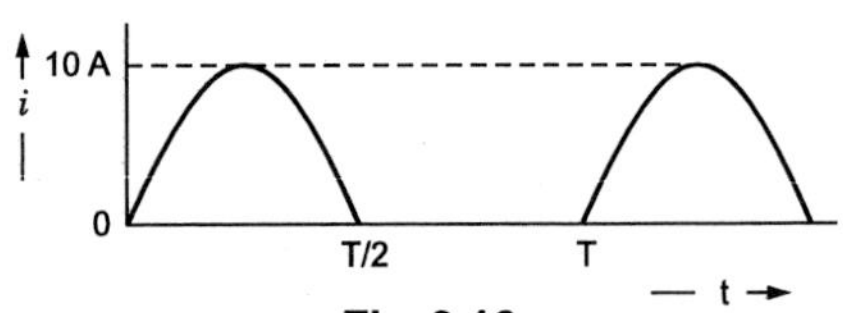

Fig. 3.16

Solution: Maximum value of alternating current, $I_{max} = 10$ A

$$\text{Average value, } I_{av} = \frac{I_{max}}{\pi} = \frac{10}{\pi} = 3.183 \text{ A } \textbf{Ans.}$$

Example 3.18 Calculate the rms value of current given by $i = 10 + 5\cos(628t + 30°)$

Solution: The waveform of the given current is shown in Fig. 3.17.

$$i = 10 + 5\cos\left(628t + \frac{\pi}{6}\right) \qquad \because 30° = \frac{\pi}{6} \text{ radians}$$

$$= 10 + 5\cos\left(\theta + \frac{\pi}{6}\right) \qquad [\text{where } \theta = 628\,t]$$

$$i^2 = \left[10 + 5\cos\left(\theta + \frac{\pi}{6}\right)\right]^2$$

$$= 100 + 100\cos\left(\theta + \frac{\pi}{6}\right) + 25\cos^2\left(\theta + \frac{\pi}{6}\right)$$

$$= 100 + 100\cos\left(\theta + \frac{\pi}{6}\right) + 12.5\left[1 + \cos 2\left(\theta + \frac{\pi}{6}\right)\right]$$

$$= 112.5 + 100\cos\left(\theta + \frac{\pi}{6}\right) + 12.5\cos 2\left(\theta + \frac{\pi}{6}\right)$$

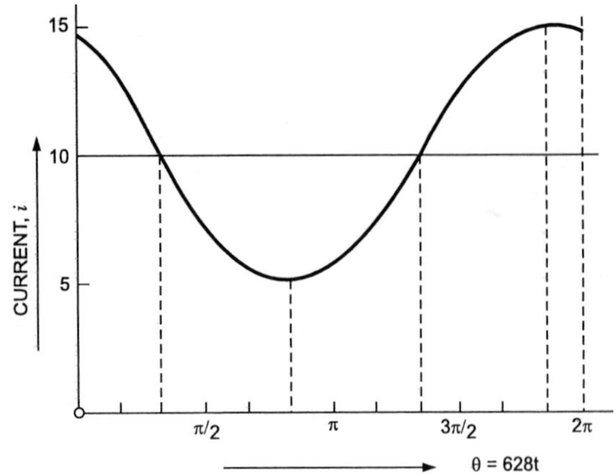

Fig. 3.17

$$i_{rms}^2 = \frac{1}{2\pi}\int_0^{2\pi} i^2\, d\theta \text{ considering complete cycle} = \frac{1}{2\pi}\int_0^{2\pi}\left[112.5 + 100\cos\left(\theta + \frac{\pi}{6}\right) + 12.5\cos 2\left(\theta + \frac{\pi}{6}\right)\right] d\theta$$

$$= \frac{1}{2\pi} \times 112.5 \times 2\pi = 112.5$$

or $I_{rms} = \sqrt{112.5} = 10.6$ A **Ans.**

Example 3.19 **Find the average and effective values of voltage for sinusoidal waveform shown in Fig. 3.18.**

[Allahabad Univ. Electrical Science-I, 1991]

Solution: The equation of the given sinusoidal waveform is $v = 100 \sin\theta$

$$V_{av} = \frac{1}{2\pi}\int_{\pi/4}^{\pi} 100\sin\theta\, d\theta = \frac{100}{2\pi}\left[-\cos\theta\right]_{\pi/4}^{\pi} = 27.17 \text{ V } \textbf{Ans.}$$

$$V_{rms}^2 = \frac{1}{2\pi}\int_{\pi/4}^{\pi} 100^2\sin^2\theta\, d\theta$$

$$= \frac{10{,}000}{4\pi}\int_{\pi/4}^{\pi}(1-\cos 2\theta)\, d\theta$$

$$= \frac{2{,}500}{\pi}\left[\theta - \frac{\sin 2\theta}{2}\right]_{\pi/4}^{\pi} = \frac{2{,}500}{\pi}\left[\pi - \frac{\pi}{4} + \frac{1}{2}\right] = 2{,}273$$

or $V_{rms} = \sqrt{2{,}273} = 47.67$ V **Ans.**

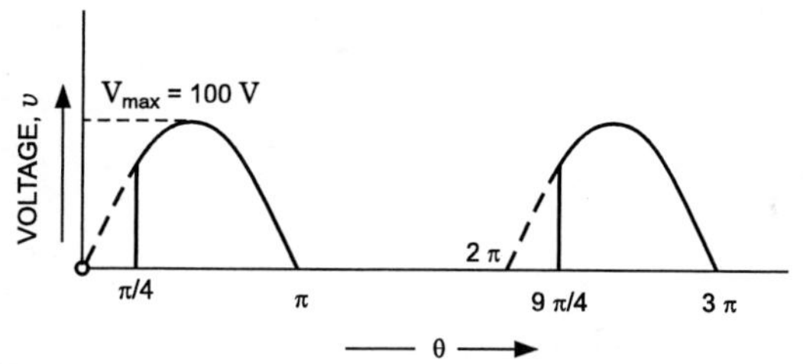

Fig. 3.18

3.11. RMS VALUE, AVERAGE VALUE, PEAK FACTOR AND FORM FACTOR OF SQUARE WAVEFORM

Square current wave is shown in Fig. 3.19. Instantaneous value of current can be expressed as

$$i = I_{max} \quad \text{for} \quad 0 < \theta < \pi$$

Average value of the square current wave is given by expression

$$I_{av} = \frac{1}{\pi}\int_0^{\pi} i\, d\theta$$

$$= \frac{1}{\pi}\int_0^{\pi} I_{max}\, d\theta$$

$$= I_{max}\frac{(\pi - 0)}{\pi} = I_{max}$$

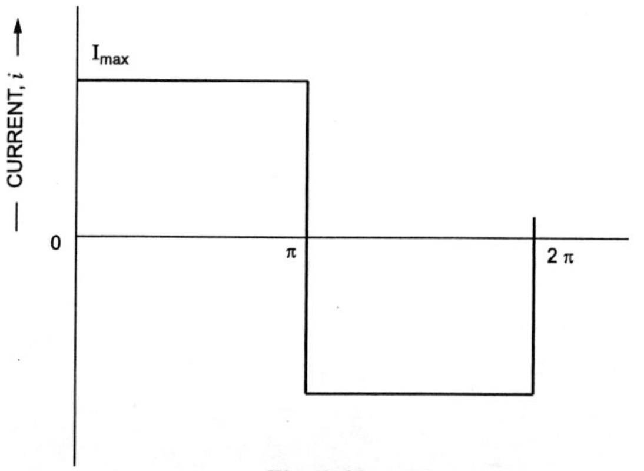

Fig. 3.19

RMS value of the square current wave is given by the expression

$$I_{rms} = \sqrt{\frac{1}{\pi}\int_0^{\pi} i^2 d\theta}$$

$$= \sqrt{\frac{1}{\pi}\int_0^{\pi} I_{max}^2\, d\theta} = \sqrt{\frac{1}{\pi} I_{max}^2 (\pi - 0)} = I_{max}$$

$$\text{Form factor, } K_f = \frac{\text{RMS value}}{\text{Average value}} = \frac{I_{max}}{I_{max}} = 1$$

$$\text{Peak factor, } K_p = \frac{\text{Peak value}}{\text{RMS value}} = \frac{I_{max}}{I_{max}} = 1$$

Example 3.20 **Find rms value, average value and form factor of the wave shown in Fig. 3.20.**

[*U.P. Technical Univ. Electrical Engineering Odd Semester 2013-14*]

Solution: The given voltage waveform may be expressed as

$$i = V_{max} \text{ for } 0 < t < 0.1$$
$$= 30 \text{ V for } 0 < t < 0.1$$

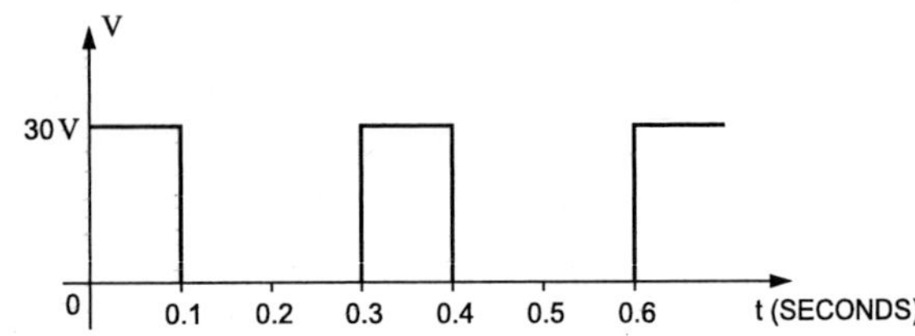

Fig. 3.20

Average value of the given voltage waveform is given as

$$V_{av} = \frac{1}{0.3}\int_0^{0.1} V_{max}\, dt$$

$$= \frac{1}{0.3}\int_0^{0.1} 30\, dt = \frac{1}{0.3}[30t]_0^{0.1} = \frac{3}{0.3} = 10 \text{ V} \quad \textbf{Ans.}$$

RMS value of the given voltage waveform is given as

$$V_{rms} = \sqrt{\frac{1}{0.3}\int_0^{0.1} V_{max}^2\, dt} = \sqrt{\frac{1}{0.3}\int_0^{0.1} 30^2\, dt} = \sqrt{\frac{1}{0.3}[900t]_0^{0.1}} = \sqrt{300} = 17.32 \text{ V} \quad \textbf{Ans.}$$

$$\text{Form factor, } k_f = \frac{V_{rms}}{V_{av}} = \frac{17.32}{10} = 1.732 \quad \textbf{Ans.}$$

3.12. RMS VALUE, AVERAGE VALUE, FORM FACTOR AND PEAK FACTOR OF A TRIANGULAR WAVEFORM

Let the maximum value of the current be I_{max} amperes.

Since $i = I_{max}$ when $\theta = \pi$. Hence, expression for the instantaneous current can be written as

$$i = \qquad \text{for } 0 < \theta < \pi$$

$$\text{RMS value of the current wave, } I_{rms} = \sqrt{\frac{1}{\pi}\int_0^{\pi} i^2\, d\theta}$$

$$= \sqrt{\frac{1}{\pi}\int_0^{\pi} \frac{I_{max}^2}{\pi^2}\theta^2\, d\theta} = \frac{I_{max}}{\sqrt{3}}$$

$$\text{Average value of the current wave, } I_{av} = \frac{1}{\pi}\int_0^{\pi} i\, d\theta$$

$$= \frac{1}{\pi}\int_0^{\pi} \frac{I_{max}}{\pi}\theta\, d\theta = \frac{I_{max}}{2}.$$

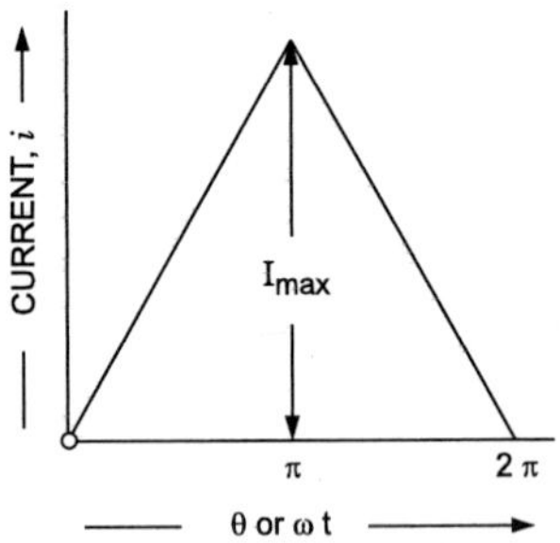

Fig. 3.21

$$\text{Form factor, } K_f = \frac{I_{rms}}{I_{av}} = \frac{I_{max}/\sqrt{3}}{I_{max}/2} = \frac{2}{\sqrt{3}} = 1.155 \quad \textbf{Ans.}$$

$$\text{Peak factor, } K_p = \frac{I_{max}}{I_{rms}} = \frac{I_{max}}{I_{max}/\sqrt{3}} = \sqrt{3} \quad \textbf{Ans.}$$

Example 3.21 **Calculate from first principles the rms value of a triangular voltage wave in which the voltage rises from 0 to V_{max} and completes the cycle by falling instantly back to zero. Give also the ratio of rms to mean value for the waveform (sawtooth wave).**

Solution: Triangular voltage wave, in which the voltage rises from 0 to V and completes the cycle by falling instantly back to zero (*i.e.* sawtooth voltage wave) is shown in Fig. 3.22.

As seen from Fig. 3.22 the instantaneous value of voltage can be expressed as follows:

$v = k\theta$ where k is the slope of the curve.

Let the maximum value of the voltage be V_{max}.

Since $v = V_{max}$ when $\theta = 2\pi$

$$\therefore k = \frac{V_{max}}{2\pi}$$

and the expression for instantaneous voltage becomes

$$v = \frac{V_{max}}{2\pi}\theta$$

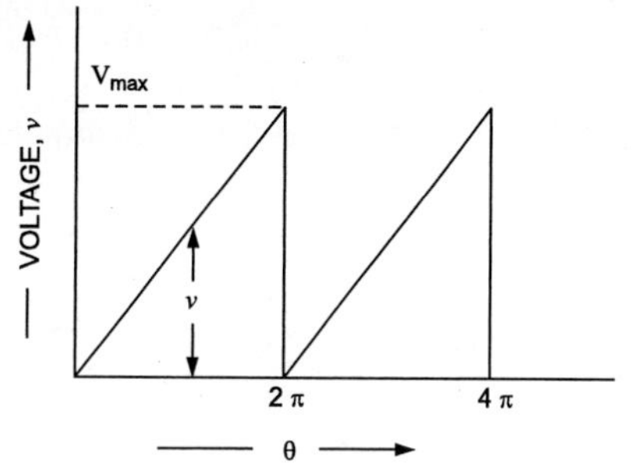

Fig. 3.22

$$\text{RMS value of the voltage wave, } V_{rms} = \sqrt{\frac{1}{2\pi}\int_0^{2\pi} v^2\, d\theta}$$

$$= \sqrt{\frac{1}{2\pi}\int_0^{2\pi}\frac{V_{max}^2}{4\pi^2}\theta^2\, d\theta} = \sqrt{\frac{V_{max}^2}{8\pi^3}\left[\frac{\theta^3}{3}\right]_0^{2\pi}} = V_{max}\sqrt{\frac{8\pi^3/3}{8\pi^3}} = \frac{V_{max}}{\sqrt{3}} \textbf{ Ans.}$$

$$\text{Mean value of the voltage wave, } V_{av} = \frac{1}{2\pi}\int_0^{2\pi} v\, d\theta = \frac{1}{2\pi}\int_0^{2\pi}\frac{V_{max}}{2\pi}\theta\, d\theta = \frac{V_{max}}{4\pi^2}\left[\frac{\theta^2}{2}\right]_0^{2\pi} = \frac{V_{max}}{2}$$

Ratio of rms value to mean value for the given waveform

$$= \frac{V_{rms}}{V_{av}} = \frac{V_{max}/\sqrt{3}}{V_{max}/2} = \frac{2}{\sqrt{3}} = 1.155 \textbf{ Ans.}$$

3.13. GRAPHICAL OR PHASOR REPRESENTATION OF ALTERNATING (SINUSOIDAL) QUANTITIES (VOLTAGES AND CURRENTS)

It has already been pointed out that it is assumed that alternating voltages and currents follow sine law and generators are designed to give emfs having sine waveform. The above said assumption makes the calculations simple. The method of representing alternating quantities by waveform or by the equations giving instantaneous values is quite cumbersome. For solution of ac problems it is advantageous to represent a sinusoidal quantity (voltage or current) by a line of definite length rotating in counter-clockwise* direction with the same angular velocity as that of the sinusoidal quantity. Such a rotating line is called the *phasor.***

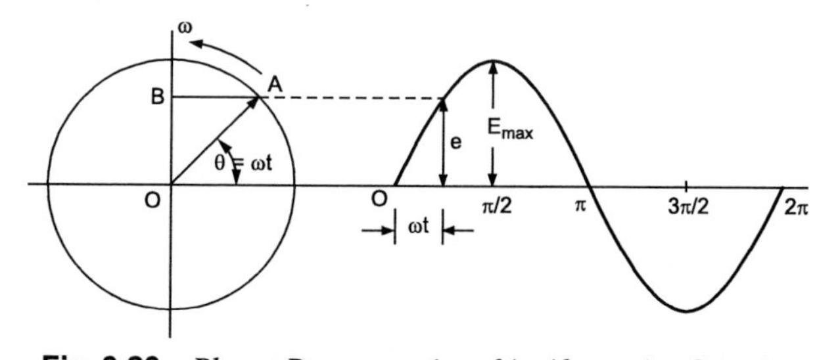

Fig. 3.23 *Phasor Representation of An Alternating Quantity*

Consider a line OA (or phasor as it is called) representing to scale the maximum value of an alternating quantity, say emf *i.e.* OA = E_{max} and rotating in counter-clockwise direction at an angular velocity

* It is a standard convention that the phasor is rotated in counter-clockwise direction — a convention that is in harmony with the general use of polar coordinates.

** Strictly speaking, sinusoidal quantities are scalar quantities varying periodically with time and, according to the definition of vector quantities, they are not true vectors. Voltage is simply energy or work per coulomb and cannot be classified as a vector. Current is also not a vector quantity because it is merely the flow of electrons through a wire. But keeping in mind that any sinusoidal quantity at a given frequency is completely specified by its amplitude and phase angle, its similarity to a vector quantity is evident, since the amplitude may be considered as the magnitude and the phase angle as the direction of a vector. To account for the difference the term phasor has been adopted, instead of the term vector, for representing graphically the magnitude and phase of a sinusoidal current or voltage.

ω radians/second about the point O, as shown in Fig. 3.23. An arrow head is put at the outer end of the phasor, partly to indicate which end is assumed to move and partly to indicate the precise length of the phasor when two or more phasors happen to coincide.

Figure 3.23 shows OA when it has rotated through an angle θ, being equal to ωt, from the position occupied when the emf was passing through its zero value. The projection of OA on Y-axis,

$$OB = OA \sin \theta = E_{max} \sin \omega t = e, \text{ the value of the emf at that instant}$$

Thus, the projection of OA on the vertical axis represents to scale the instantaneous value of emf.

It will be seen that the phasor OA rotating in counter-clockwise direction will represent a sinusoidal quantity (voltage or current) if

(*i*) its length is equal to the peak or maximum value of the sinusoidal voltage or current to a suitable scale.

(*ii*) it is in horizontal position at the instant the alternating quantity (voltage or current) is zero and increasing and

(*iii*) its angular velocity is such that it completes one revolution in the same time as taken by the alternating quantity (voltage or current) to complete one cycle.

3.14. PHASOR DIAGRAM USING RMS VALUES

Since there is definite relation between maximum value and rms value ($E_{max} = \sqrt{2}\ E_{rms}$), the length of phasor OA can be taken equal to rms value if desired. But it should be noted that in such cases, the projection of the rotating phasor on the vertical axis will not give the instantaneous value of that alternating quantity.

The phasor diagram drawn in rms values of the alternating quantities helps in understanding the behaviour of the ac machines under different loading conditions.

3.15. PHASE AND PHASE ANGLE

By phase of an alternating current is meant the fraction of the time period of that alternating current that has elapsed since the current last passed through the zero position of reference. The phase angle of any quantity means the angle the phasor representing the quantity makes with the reference line (which is taken to be at zero degrees or radians). For example the phase angle of current I_2 in Fig. 3.24 is $(-\phi)$.

3.16. PHASE DIFFERENCE

When two alternating quantities, say, two emfs, or two currents or one voltage and one current are considered simultaneously, the frequency being the same, they may not pass through a particular point at the same instant. One may pass through its maximum value at the instant when the other passes through the value other than its maximum one. These two quantities are said to have a *phase difference*. Phase difference is always given either in degrees or in radians.

The phase difference is measured by the angular distance between the points where the two curves cross the base or reference line in the same direction.

The quantity ahead in phase is said to lead the other quantity while the second quantity is said to lag behind the first one. In Fig. 3.24 (*b*) current I_1 represented by phasor OA leads the current I_2 represented by phasor OB by ϕ or current I_2 lags behind the current I_1 by ϕ. The leading current I_1 goes through its zero and maximum values first and the current I_2 goes through its zero and maximum values after time angle ϕ. The two waves representing these two currents are shown in Fig. 3.24 (*a*). If I_1 is taken as reference phasor, the two currents can be expressed as

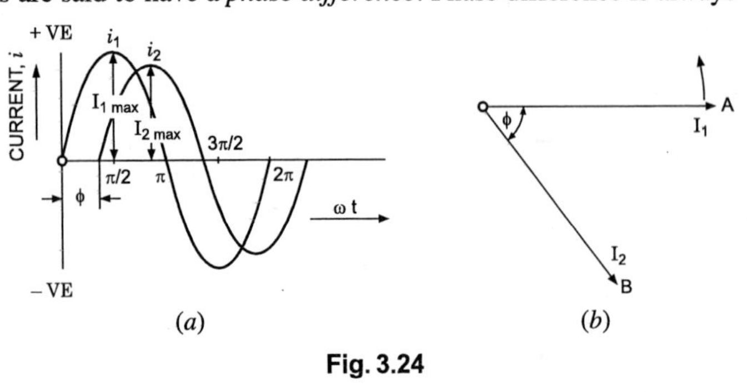

Fig. 3.24

$$i_1 = I_{1\,max} \sin \omega t$$

$$\text{and } i_2 = I_2 \sin (\omega t - \phi)$$

The two quantities are said to be in phase with each other if they pass through zero values at the same instant and rise in the same direction, as shown in Fig. 3.25. But the two quantities passing through zero values at the same instant but rising in opposite directions, as shown in Fig. 3.26 are said to be in phase opposition, *i.e.* phase difference is 180°. When the two alternating quantities have a phase difference of 90° or $\frac{\pi}{2}$ radians they are said to be in quadrature.

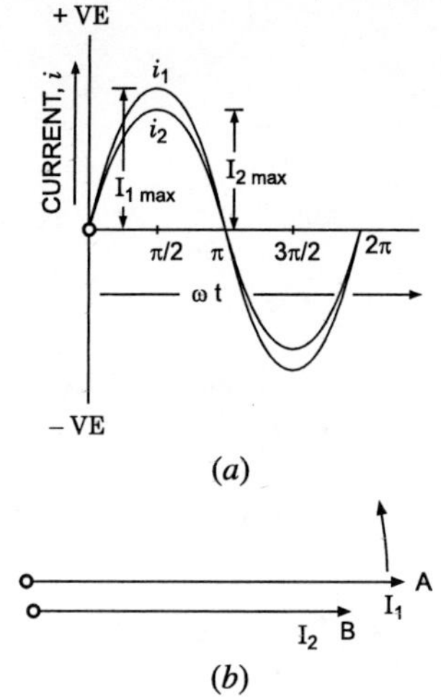

(*a*)

(*b*)

Fig. 3.25

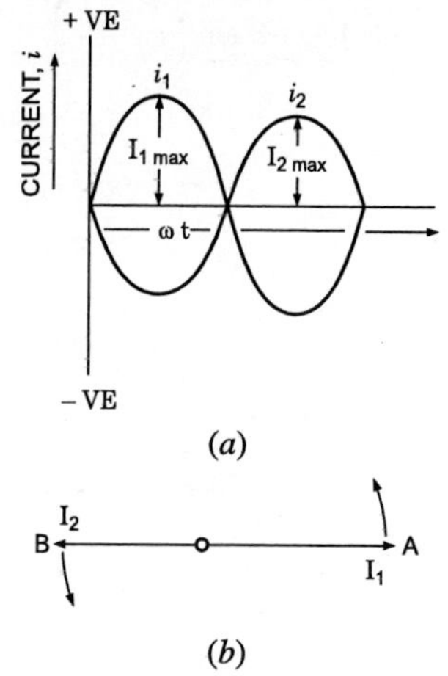

(*a*)

(*b*)

Fig. 3.26

3.17. CONVENTIONS FOR DRAWING PHASOR DIAGRAMS

As already mentioned, the alternating quantities (voltages and currents) in practice are represented by straight lines having definite direction and length. Such lines are called the *phasors* and the diagrams in which phasors represent currents, voltages and their phase difference are known as *phasor diagrams*.

Though phasor diagrams can be drawn to represent either maximum or effective values of voltages and currents but since effective values are of much more importance, phasor diagrams are mostly drawn to represent effective values.

In order to achieve consistent and accurate results it is essential to follow certain conventions. Some of the common conventions in this regard are enlisted below:

1. Counter-clockwise direction of rotation of phasors is usually taken as positive direction of rotation of phasors *i.e.* a phasor rotated in a counter-clockwise direction from a given phasor is said to lead the given phasor while a phasor rotated in clockwise direction is said to lag the given phasor.
2. For series circuits, in which the current is common to all parts of the circuit, the current phasor is usually taken as reference phasor for other phasors in the same diagram and drawn on horizontal line.
3. In parallel circuits in which the voltage is common to all branches of the circuit, the voltage phasor is usually taken as reference phasor and drawn on the horizontal line. Other phasors are referred to the common voltage phasor.
4. It is not necessary that current and voltage phasors are drawn to the same scale; in fact it is often desirable to draw the current phasor to a larger scale than the voltage phasor when the values of currents being represented are small. However, if several voltage phasors are to be used in the same phasor diagram, they should all be drawn to the same scale. Likewise all current phasors in the same diagram should be drawn to the same scale.

Example 3.22 Calculate (*a*) the maximum value and (*b*) the root mean square values of the following quantities:

(*i*) 40 sin ωt (*ii*) B sin $\left(\omega t - \frac{\pi}{2}\right)$ and (*iii*) 10 sin ωt –17.3 cos ωt.

Draw the phasors showing the phase difference with respect to A sin $\left(\omega t - \frac{\pi}{6}\right)$.

Solution: (*i*) For 40 sin ω *t*

(*a*) Maximum value = Coefficient of the sine of the time angle = 40 **Ans.**

$$(b) \text{ RMS value} = \frac{\text{Maximum value}}{\sqrt{2}} = \frac{40}{\sqrt{2}} = 28.28 \textbf{ Ans.}$$

(*ii*) For B $\sin\left(\omega t - \frac{\pi}{2}\right)$

(*a*) Maximum value = B **Ans.**

(*b*) RMS value = $\frac{B}{\sqrt{2}}$ **Ans.**

(*iii*) For 10 sin ω*t* – 17.3 cos ω*t*, which may be written as

$$20\left(\frac{1}{2}\sin\omega t - \frac{\sqrt{3}}{2}\cos\omega t\right) \quad \text{or} \quad 20\left(\cos\frac{\pi}{3}\sin\omega t - \sin\frac{\pi}{3}\cos\omega t\right)$$

or $20\sin\left(\omega t - \frac{\pi}{3}\right)$

(*a*) Maximum value = 20

(*b*) RMS value = $\frac{20}{\sqrt{2}}$ = 14.14 **Ans.**

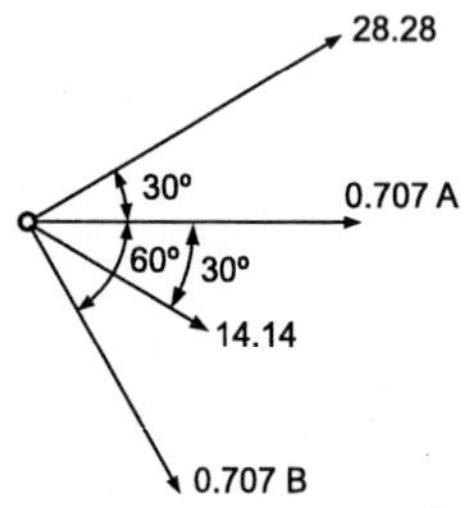

Fig. 3.27

Phasors showing the phase difference with respect to A $\sin\left(\omega t - \frac{\pi}{6}\right)$ are shown in Fig. 3.27.

3.18. ADDITION AND SUBTRACTION OF ALTERNATING QUANTITIES

In most of the ac circuits, it is necessary to consider the combined action of several emfs or voltages acting in a series circuit and action of several currents flowing through the different branches of a parallel circuit.

Let it be required to add two currents given by the equations

$$i_1 = I_{1\,max} \sin \omega t \text{ and } i_2 = I_{2\,max} \sin (\omega t - \phi)$$

The resultant sum may be expressed as

$$i_r = i_1 + i_2 = I_{1\,max} \sin \omega t + I_{2\,max} \sin (\omega t - \phi)$$

but it is too awkward and gives no idea of the peak value and phase angle of the resultant current.

The currents may be added graphically by plotting their curves in the same system of coordinates and then adding the ordinates of i_1 and i_2 point by point, according to the equation $i_r = i_1 + i_2$. Evidently this method is also too cumbersome and unwieldy to be practical. This is particularly so when situation arise where more than two sinusoidal quantities are to be added.

A simpler and more direct method consists in adding the sinusoidal quantities as phasors.

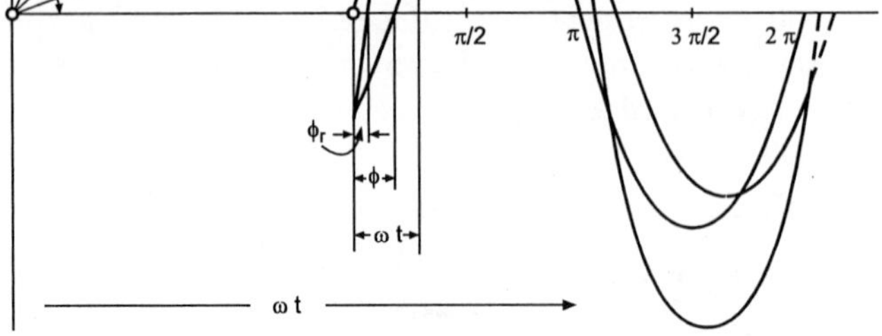

Fig. 3.28

Consider the phasors $I_{1\,max}$ and $I_{2\,max}$ that would generate the two curves i_1 and i_2 and let them be in a position, as shown in Fig. 3.28 at one particular instant of time. If we now add $I_{1\,max}$ and $I_{2\,max}$ by completing the parallelogram as shown, the diagonal $I_{r\,max}$ will, when rotated, generate a third sine curve. It remains to be shown that this third sine curve coincides with the waveform of i_r obtained by adding i_1 and i_2 point by point.

Now the vertical component of $I_{r\,max}$ is the sum of the vertical components of $I_{1\,max}$ and $I_{2\,max}$. Therefore, the waveform of i_r is the graph generated by rotating $I_{r\,max}$ in counter-clockwise direction.

It follows, therefore, that two or more alternating quantities may be added in the same way as forces are added, namely by constructing parallelograms or closed polygons and either measuring or calculating the lengths of the diagonals or closing sides and the magnitude of the phase angles.

Illustrations: The way in which the two given currents i_1 and i_2 can be added by the parallelogram rule of phasor addition is illustrated in Fig. 3.29 where the currents are shown as phasor drawn from the origin O of the system of coordinates. The resultant phasor is the diagonal of the parallelogram formed by the phasors $I_{1\,max}$ and $I_{2\,max}$.

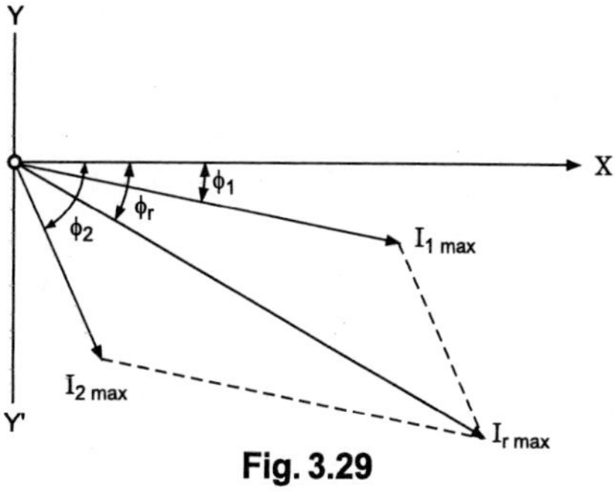

Fig. 3.29

This method is more convenient when more than two phasors are to be added, as shown in Fig. 3.30. From the end point of $I_{1\,max}$ a phasor is constructed parallel to $I_{2\,max}$ of the same magnitude and direction as the latter; then from the end point of $I_{2\,max}$ a phasor is constructed parallel to $I_{3\,max}$ and so on. Phasor $I_{r\,max}$ from the origin of the first phasor ($I_{1\,max}$) to the end point of the last phasor ($I_{5\,max}$) represents the sum of all the phasors.

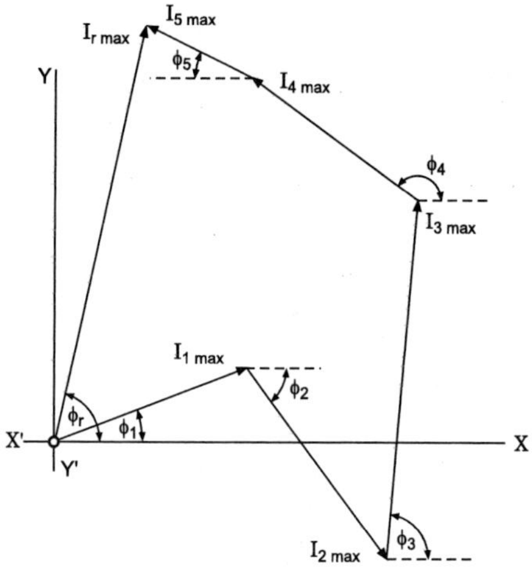

Fig. 3.30

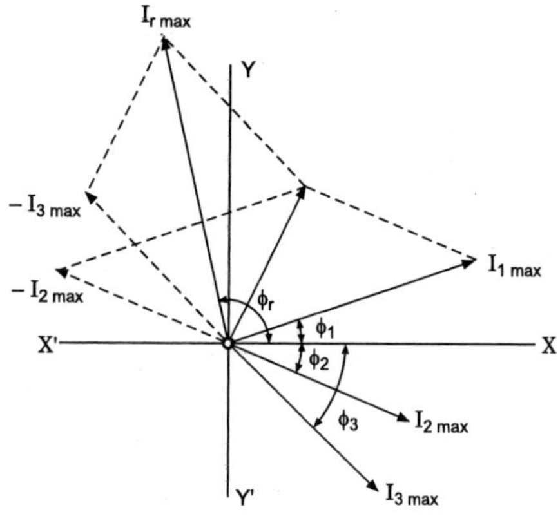

Fig. 3.31

Phasors may also be subtracted by the above method. For example if phasors $I_{2\,max}$ and $I_{3\,max}$ are to be subtracted from phasor $I_{1\,max}$ each of the two phasors be reversed in direction and then added as explained above [Fig. 3.31]. This time, the phasor drawn from the origin of the first phasor $I_{1\,max}$ to the terminal point of the last phasor $I_{3\,max}$ gives the difference of phasors $I_{2\,max}$ and $I_{3\,max}$ from phasor $I_{1\,max}$.

Example 3.23 **The two branches of parallel circuit draws currents I_1 and I_2 such that $I_1 = 10\sqrt{2}\sin\omega t$ and $I_2 = 5\sqrt{2}\sin(\omega t - 60°)$. What is the total current drawn?**

[G.B. Technical Univ. Electrical Engineering Second Semester 2010-11]

Solution:

$$i_1 = 10\sqrt{2}\sin\omega t$$

$$i_2 = 5\sqrt{2}\sin(\omega t - 60°)$$

$$\text{Resultant current } i = i_1 + i_2 = 10\sqrt{2}\sin\omega t + 5\sqrt{2}\sin(\omega t - 60°)$$

$$= 10\sqrt{2}\sin\omega t + 5\sqrt{2}\sin\omega t\cos(-60°) + 5\sqrt{2}\cos\omega t\sin(-60°)$$

$$= 10\sqrt{2}\sin\omega t + 2.5\sqrt{2}\sin\omega t - 2.5\sqrt{6}\cos\omega t = 17.678\sin\omega t - 6.124\cos\omega t$$

Multiplying and dividing RHS of the above expression by $\sqrt{(17.678)^2 + (-6.124)^2}$ i.e. 18.708 we have

$$i_r = 18.708\left(\sin \omega t \times \frac{17.678}{18.708} + \cos \omega t \times \frac{-6.124}{18.708}\right)$$

$$= 18.708 (\sin \omega t \cos \phi_r + \cos \omega t \sin \phi_r) \quad \text{where } \cos \phi_r = \frac{17.678}{18.708} \text{ and } \sin \phi_r = \frac{-6.124}{18.708}$$

$$= 18.708 \sin (\omega t + \phi_r) \quad \text{where } \phi_r = \text{Tan}^{-1}\frac{-6.124}{17.678} = -19.1°$$

So $i_r = 18.708 \sin (\omega t - 19.1°)$ **Ans.**

Analytical method. Resolving currents $I_{1\,max}$ and $I_{2\,max}$ along X-axis and Y-axis we have

Algebraic sum of X-components $= 10\sqrt{2} \cos 0° + 5\sqrt{2} \cos (-60°) = 14.1421 - 3.5355 = 17.678$

Algebraic sum of Y-components $= 10\sqrt{2} \sin 0° + 5\sqrt{2} \sin (-60°) = 0 + (-6.124) = -6.124$

Maximum value of resultant current, $I_{r\,max} = \sqrt{(\text{X-components})^2 + (\text{Y-components})^2}$

$$= \sqrt{(17.678)^2 + (-6.124)^2} = 18.708 \text{ A}$$

Phase angle, $\phi_r = \text{Tan}^{-1}\frac{\text{Y-components}}{\text{X-components}} = \text{Tan}^{-1}\frac{-6.124}{17.678} = -19.1°$

∴ Resultant current, $i_r = I_{r\,max} \sin (\omega t + \phi_r) = 18.708 \sin (\omega t - 19.1°)$ **Ans.**

Graphical Method. Take OX as reference phasor. Draw phasor OA = 4 cm in length along OX as to represent current $I_{1\,max}$ of $10\sqrt{2}$ A (to the scale 1 cm = $2.5\sqrt{2}$ A i.e. 3.5355 A) in magnitude as well as in direction. Then draw AB = 2 cm in length and making an angle of –60° with the reference phasor so as to represent $I_{2\,max}$ of $5\sqrt{2}$ A in magnitude as well as in direction. The phasor OB, obtained by joining points O and B, will represent the resultant current in magnitude as well as in direction. On measurement from Fig. 3.33, OB = 5.3 cm and angle AOB = 19.1°.

Maximum value of resultant current, $I_{r\,max} = 5.3 \times 2.5\sqrt{2} = 18.7$ A

Phase angle $\phi_r = 19.1°$

Hence the resultant current may be expressed as

$$i_r = 18.7 \sin (\omega t - 19.1°) \quad \textbf{Ans.}$$

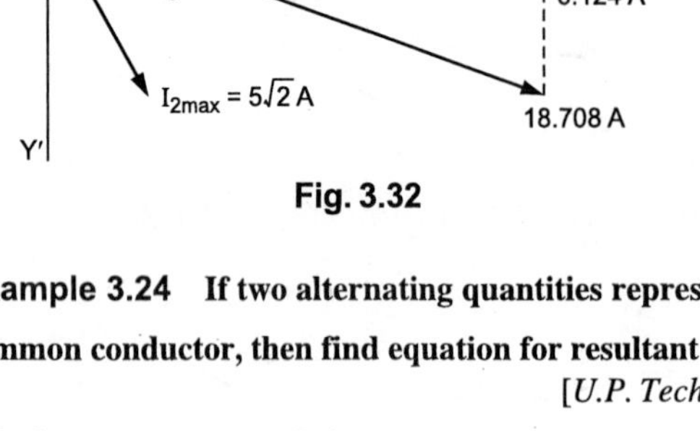

Fig. 3.32

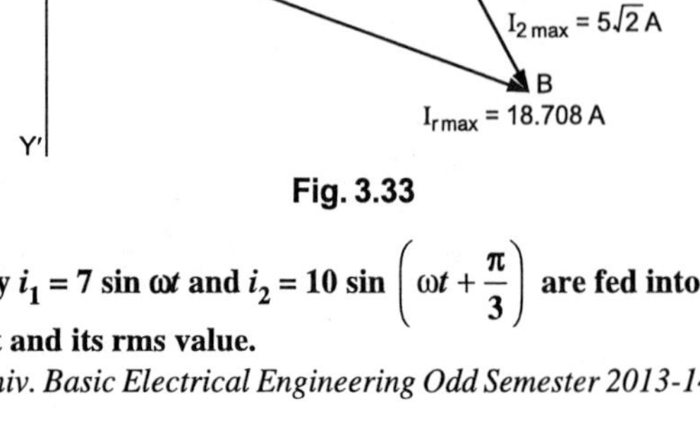

Fig. 3.33

Example 3.24 **If two alternating quantities represented by $i_1 = 7 \sin \omega t$ and $i_2 = 10 \sin \left(\omega t + \frac{\pi}{3}\right)$ are fed into a common conductor, then find equation for resultant current and its rms value.**

[U.P. Technical Univ. Basic Electrical Engineering Odd Semester 2013-14]

Solution:

$$i_1 = 7 \sin \omega t$$

$$i_2 = 10 \sin \left(\omega t + \frac{\pi}{3}\right)$$

$$\text{Resultant current, } i_r = i_1 + i_2 = 7 \sin \omega t + 10 \sin\left(\omega t + \frac{\pi}{3}\right)$$

$$= 7 \sin \omega t + 10 \sin \omega t \cos\frac{\pi}{3} + 10 \cos \omega t \sin\frac{\pi}{3} = 12 \sin \omega t + 8.66 \cos \omega t$$

Multiplying and dividing RHS of the above expression by $\sqrt{(12)^2 + (8.66)^2}$ *i.e.* 14.8, we get

$$i_r = 14.8\left(\sin \omega t \times \frac{12}{14.8} + \cos \omega t \times \frac{8.66}{14.8}\right) = 14.8\,(\sin \omega t \cos \phi_r + \cos \omega t \sin \phi_r)$$

where $\cos \phi_r = \dfrac{12}{14.8}$ and $\sin \phi_r = \dfrac{8.66}{14.8}$

or $i_r = 14.8 \sin (\omega t + \phi_r)$ where $\phi_r = \text{Tan}^{-1}\dfrac{8.66}{12} = 0.199\,\pi$ radians

or $i_r = 14.8 \sin (\omega t + 0.199\,\pi)$ **Ans.**

Peak value of resultant current, $I_{r\,max} = 14.8$ A

RMS value of resultant current, $I_{r\,rms} = \dfrac{I_{r\,max}}{\sqrt{2}} = \dfrac{14.8}{\sqrt{2}} = 10.46$ A **Ans.**

***Analytical Method*:** Resolving currents $I_{1\,max}$ and $I_{2\,max}$ along X-axis and Y-axis we get

Algebraic sum of X-components $= 7 \cos 0° + 10 \cos\dfrac{\pi}{3} = 7 + 5 = 12$ A

Algebraic sum of Y-components $= 7 \sin 0° + 10 \sin\dfrac{\pi}{3} = 8.66$ A

Maximum value of resultant current, $I_{r\,max} = \sqrt{(\text{X-components})^2 + (\text{Y-components})^2}$

$$= \sqrt{12^2 + (8.66)^2} = 14.8 \text{ A}$$

and phase angle, $\phi_r = \text{Tan}^{-1}\dfrac{\text{Y-components}}{\text{X-components}} = \text{Tan}^{-1}\dfrac{8.66}{12} = 0.199\,\pi$

∴ Resultant current, $i_r = I_{r\,max} \sin (\omega t + \phi_r) = 14.8 \sin (\omega t + 0.199\,\pi)$ **Ans.**

Graphical Method. Take OX as reference phasor. Draw phasor OA 3.5 cm in length along OX so as to represent current $I_{1\,max}$ of 7 amperes (to the scale 1 cm = 2 A) in magnitude as well as in direction. Then draw phasor AB = 5 cm in length making an angle of $\frac{\pi}{3}$ radians or 60° with the reference phasor so as to represent current $I_{2\,max}$ of 10 amperes in magnitude as well as in direction. The phasor OB, obtained by joining points O and B, will represent the resultant current in magnitude as well as in direction. On measurement from Fig. 3.35.

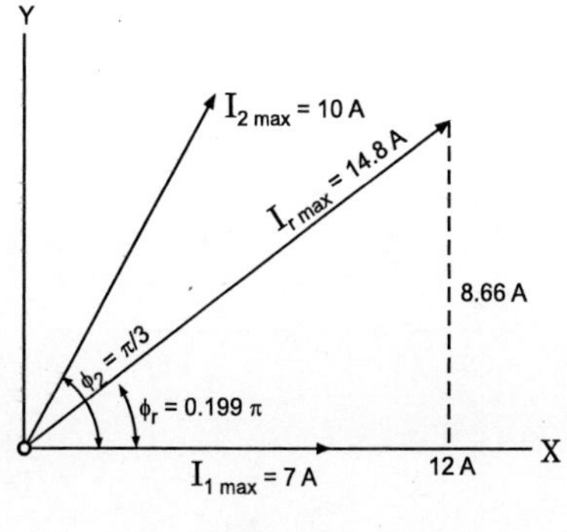

Fig. 3.34

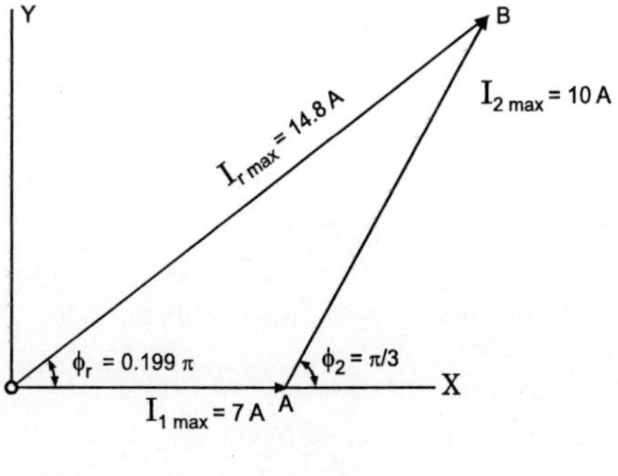

Fig. 3.35

Maximum value of resultant current, $I_{r\,max}$ = OB × 2 = 7.4 × 2 = 14.8 A

and phase angle, ϕ_r = 35.8° or 0.199 π radians

Hence the resultant current may be expressed as

$$i_r = 14.8 \sin(\omega t + 0.199\pi) \textbf{ Ans.}$$

Example 3.25 Two ac voltages are represented by —

$$v_1(t) = 30 \sin(314t + 45°)$$
$$v_2(t) = 60 \sin(314t + 60°)$$

Calculate the resultant voltage $v(t)$ and express in the form —

$$v(t) = V_m \sin(314t + \phi)$$

[U.P. Technical Univ. Electrical Engineering Second Semester 2003-04]

Solution:

$$v_1(t) = 30 \sin(314t + 45°)$$
$$v_2(t) = 60 \sin(314t + 60°)$$

Resolving voltages $V_{1\,max}$ and $V_{2\,max}$ along X-axis and Y-axis we have

$$\text{Algebraic sum of X-components} = 30\cos 45° + 60\cos 60° = 30 \times \frac{1}{\sqrt{2}} + 60 \times \frac{1}{2} = 51.21 \text{ V}$$

$$\text{Algebraic sum of Y-component} = 30\sin 45° + 60\sin 60° = 30 \times \frac{1}{\sqrt{2}} + 60 \times \frac{\sqrt{3}}{2} = 73.17 \text{ V}$$

$$\text{Maximum value of resultant voltage, } V_{r\,max} = \sqrt{\text{X-components})^2 + (\text{Y-components})^2}$$
$$= \sqrt{(51.21)^2 + (73.17)^2} = 89.31 \text{ V}$$

$$\text{Phase angle, } \phi_r = \text{Tan}^{-1}\frac{\text{Y-component}}{\text{X-component}} = \text{Tan}^{-1}\frac{73.17}{51.21} = 55°$$

So expression for resultant voltage is

$$v(t) = V_{max}\sin(\omega t + \phi_r) = 89.31 \sin(314t + 55°) \textbf{ Ans.}$$

Example 3.26 Draw a phasor diagram showing the following voltages:

$$v_1 = 100 \sin 500t$$
$$v_2 = 200 \sin(500t + 45°)$$
$$v_3 = -50 \cos 500t$$

Find rms value of the resultant voltage.

[U.P. Technical Univ. Basic Electrical Engineering Second Semester 2013-14]

Solution:

$$v_1 = 100 \sin 500t$$
$$v_2 = 200 \sin(500t + 45°)$$
$$v_3 = -50\cos 500t = -50\sin(90° - 500t)$$
$$= 50 \sin(500t - 90°)$$

Phasor diagram is shown in Fig. 3.36.

Resolving voltages V_1, V_2 and V_3 along X-axis and Y-axis we have

Algebraic sum of X-components = 100 cos 0° + 200 cos 45° + 50 cos 90°

$$= 100 + 100\sqrt{2} = 241.42 \text{ V}$$

Algebraic sum of Y-components = 100 sin 0° + 200 sin 45° + 50 sin (–90°)

$$= 0 + 141.42 - 50 = 91.42 \text{ V}$$

Maximum value of the resultant voltage,

$$V_{max} = \sqrt{(241.42)^2 + (91.42)^2} = 258.15 \text{ V}$$

RMS value of the resultant voltage,

$$V_{rms} = \frac{V_{max}}{\sqrt{2}} = \frac{258.15}{\sqrt{2}} = 182.54 \text{ V} \quad \textbf{Ans.}$$

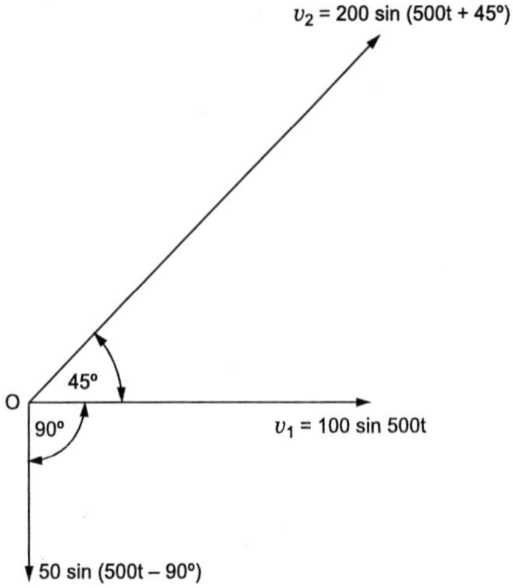

Fig. 3.36 *Phasor Diagram*

Example 3.27 **Draw a phasor diagram showing the following voltages:**

$$v_1 = 100 \sin 500\,t \qquad v_2 = 200 \sin\left(500t + \frac{\pi}{3}\right)$$

$$v_3 = -50 \cos 500\,t \qquad v_4 = 150 \sin\left(500t - \frac{\pi}{4}\right)$$

Find rms value of resultant voltage. [*U.P. Technical Univ. Elec. Engineering Second Semester 2005-06*]

Solution : $v_1 = 100 \sin 500\,t$

$$v_2 = 200 \sin\left(500t + \frac{\pi}{3}\right)$$

$$v_3 = -50 \cos 500\,t = -50 \sin\left(500t + \frac{\pi}{2}\right) = 50 \sin\left(500t - \frac{\pi}{2}\right)$$

$$v_4 = 150 \sin\left(500t - \frac{\pi}{4}\right)$$

The phasor diagram is shown in Fig. 3.37.

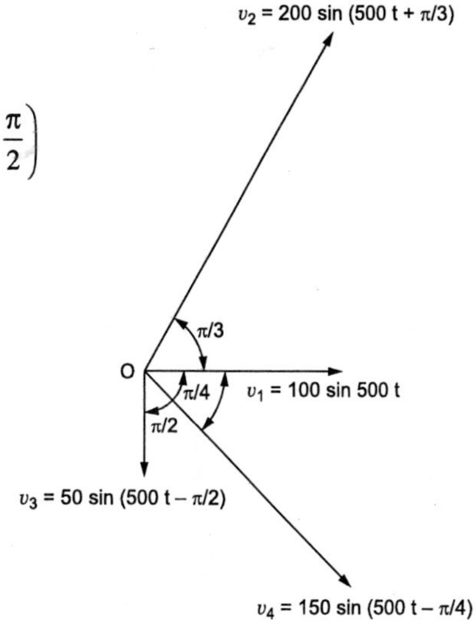

Fig. 3.37 *Phasor Diagram*

Resolving voltages V_1, V_2, V_3 and V_4 along X-axis and Y-axis we have

Algebraic sum of X-components

$$= 100 \cos 0° + 200 \cos\frac{\pi}{3} + 50 \cos\left(-\frac{\pi}{2}\right) + 150 \cos\left(-\frac{\pi}{4}\right)$$

$$= 100 + 200 \times 0.5 + 0 + \frac{150}{\sqrt{2}} = 306 \text{ V}$$

Algebraic sum of Y-components

$$= 100 \sin 0° + 200 \sin\frac{\pi}{3} + 50 \sin\left(-\frac{\pi}{2}\right) + 150 \sin\left(-\frac{\pi}{4}\right)$$

$$= 0 + 200 \times 0.866 - 50 - 106 = 17.2 \text{ V}$$

Maximum value of resultant voltage, $V_{r\,max} = \sqrt{(306)^2 + (17.2)^2} = 306.5$ V

RMS value of resultant voltage, $V_{r\,rms} = \dfrac{V_{r\,max}}{\sqrt{2}} = \dfrac{306.5}{\sqrt{2}} = 216.72$ V **Ans.**

Example 3.28 **The instantaneous values of two alternating voltages are represented by $v_1 = 60 \sin\theta$ and $v_2 = 40 \sin(\theta - \pi/3)$. Derive expressions for the instantaneous values of (*i*) the sum and (*ii*) the difference of these voltages.** [*U.P. Technical Univ. Electrical Engineering July 2002*]

Solution: (*a*) $v_1 = 60 \sin\theta$ and $v_2 = 40 \sin\left(\theta - \dfrac{\pi}{3}\right)$

(*i*) Maximum value of voltage, $V_{r\,max} = \sqrt{(\text{Sum of X-components})^2 + (\text{Sum of Y-components})^2}$

$$= \sqrt{\left[60 \cos 0° + 40 \cos\left(\frac{-\pi}{3}\right)\right]^2 + \left[60 \sin 0° + 40 \sin\left(\frac{-\pi}{3}\right)\right]^2}$$

$$= \sqrt{(60 + 20)^2 + (-34.64)^2} = \sqrt{(80)^2 + (-34.64)^2} = 87.178 \text{ V}$$

Phase angle, $\phi = \text{Tan}^{-1} \dfrac{\text{Sum of Y-components}}{\text{Sum of X-components}}$

$$= \text{Tan}^{-1} \frac{-34.64}{80} = -23.41° = -0.13\,\pi \text{ radian}$$

So expression for the instantaneous values of the sum of given voltages,

$$v_r = 87.178 \sin (\theta - 0.13\,\pi) \textbf{ Ans.}$$

(*ii*) Maximum value of voltage, $V_{rms} = \sqrt{(\text{Difference of X-components})^2 + (\text{Difference of Y-components})^2}$

$$= \sqrt{\left[60 \cos 0° - 40 \cos\left(\frac{-\pi}{3}\right)\right]^2 + \left[60 \sin 0° - 40 \sin\left(\frac{-\pi}{3}\right)\right]^2}$$

$$= \sqrt{(60-20)^2 + (34.64)^2} = \sqrt{40^2 + (34.64)^2} = 52.915 \text{ V}$$

$$\text{Phase angle, } \phi = \text{Tan}^{-1} \frac{34.64}{40} = 40.89° \text{ or } 0.227\,\pi \text{ radian}$$

So expression for the instantaneous values of the difference of given voltages,

$$v_r = 52.915 \sin (\theta + 0.227\,\pi)$$

Example 3.29 **Three voltages represented by the following equations, $e_1 = 15 \sin \omega t$, $e_2 = 5 \sin (\omega t + \pi/6)$, $e_3 = 10 \cos \omega t$ act together in an ac circuit. Represent these voltages by phasors and calculate an expression for the resultant voltage. Check the result so obtained graphically.**

[*G.B. Technical Univ. Electrical Engineering Second Semester 2009-10*]

Solution:

$$e_1 = 15 \sin \omega t$$

$$e_2 = 5 \sin\left(\omega t + \frac{\pi}{6}\right)$$

$$e_3 = 10 \cos \omega t = 10 \sin\left(\omega t + \frac{\pi}{2}\right)$$

The phasor diagram is shown in Fig. 3.38

Resolving E_1, E_2 and E_3 along X-axis and Y-axis we have

$$\text{Algebraic sum of X-components} = 15 \cos 0° + 5 \cos \frac{\pi}{6} + 10 \cos 90° = 15 + \frac{5\sqrt{3}}{2} + 0 = 19.33$$

$$\text{Algebraic sum of Y components} = 15 \sin 0° + 5 \sin \frac{\pi}{6} + 10 \sin 90° = 0 + 2.5 + 10 = 12.5$$

$$\text{Maximum value of resultant voltage, } E_{max} = \sqrt{(19.33)^2 + (12.5)^2} = 23$$

$$\text{Phase angle, } \phi = \text{Tan}^{-1} \frac{\text{Sum of Y-components}}{\text{Sum of X-components}} = \text{Tan}^{-1} \frac{12.5}{19.33} = 32.9°$$

$$\text{Expression resultant voltage, } e_r = E_{r\,max} \sin (\omega t + \phi) = 23 \sin (\omega t + 32.9°) \textbf{ Ans.}$$

$$\text{RMS value of resultant voltage, } E_{r\,rms} = \frac{E_{r\,max}}{\sqrt{2}} = \frac{23}{\sqrt{2}} = 16.26 \text{ V} \textbf{ Ans.}$$

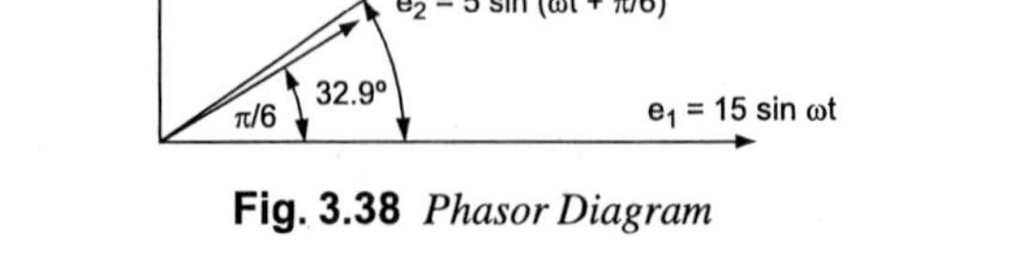

Fig. 3.38 *Phasor Diagram*

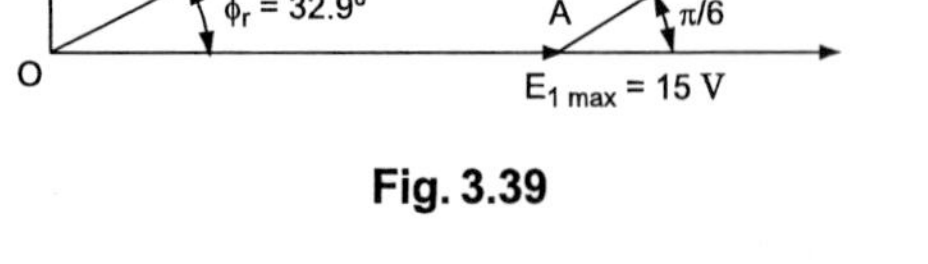

Fig. 3.39

Graphical Method: Take OX as reference phasor. Draw OA = 7.5 cm along X-axis so as to represent voltage $E_{1\,max}$ of 15V (to the scale of 1 cm = 2V) in magnitude as well as in direction. Then draw phasor AB = 2.5 cm in length making an angle of $\frac{\pi}{6}$ or 30° with the reference phasor (or X-axis) so as to represent E_{2max} of 5 V in magnitude as well as in direction. Then draw phasor BC = 5 cm in length making an angle of $\frac{\pi}{2}$ or 90° with the reference phasor to represent $E_{3\,max}$ of 10 V in magnitude as well as in direction. The phasor OC, obtained by joining points O and C, well represent the resultant voltage in magnitude as well as in direction. On measurement from Fig. 3.39.

Maximum value of resultant voltage,

$$E_{r\,max} = 2 \times 11.5 = 23 \text{ V}$$

$$\text{Phase angle } \phi_r = 32.9° \text{ or } 0.1833\,\pi \text{ radians}$$

The results are same by analysis as well as graphically.

Example 3.30 **Three voltages represented by the following equations are: $e_1 = 15 \sin \omega t$, $e_2 = 5 \sin\left(\omega t + \frac{\pi}{4}\right)$ and $e_3 = 10 \cos \omega t$ together in an ac circuit. Represent these voltages by phasor and calculate an expression for the resultant voltage. Also cross, verify it.** [*Mahamaya Technical Univ. Electrical Engineering Second Semester 2010-11*]

Solution:

$$e_1 = 15 \sin \omega t$$

$$e_2 = 5 \sin\left(\omega t + \frac{\pi}{4}\right)$$

$$e_3 = 10 \cos \omega t = 10 \sin\left(\omega t + \frac{\pi}{2}\right)$$

The phasor diagram is shown in Fig. 3.40

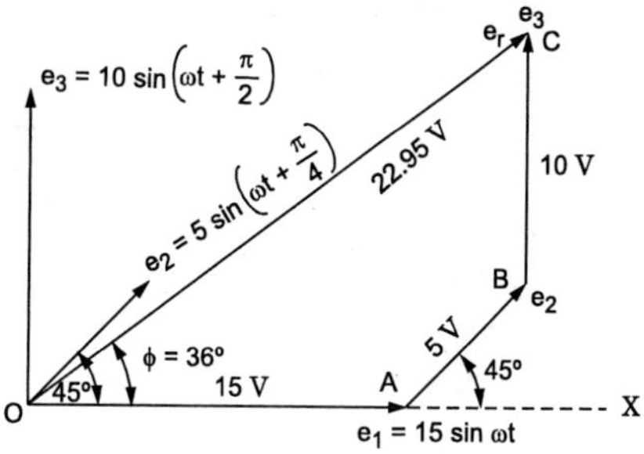

Fig. 3.40 *Phasor Diagram*

Resolving e_1, e_2 and e_3 along X-axis and Y-axis, we have

Algebraic sum of X-components $= 15 \cos 0° + 5 \cos 45° + 10 \cos 90°$

$= 15 + 2.5\sqrt{2} + 0 = 18.536$ V

Algebraic sum of Y-components $= 15 \sin 0° + 5 \sin 45° + 10 \sin 90° = 13.536$ V

Maximum value of resultant voltage, $E_{r\,max} = \sqrt{18.536^2 + 13.536^2} = 22.95$ V

Phase angle, $\phi = \text{Tan}^{-1}\frac{\text{Y-component}}{\text{X-component}} = \text{Tan}^{-1}\frac{13.536}{18.536} = 36.14°$ or $0.2\,\pi$

So expression for resultant voltage, $e_r = E_{r\,max} \sin(\omega t + \phi) = 22.95 \sin(\omega t + 0.2\pi)$ **Ans.**

Graphical method is illustrated in Fig. 3.40.

Resultant voltage is represented by phasor OC

On measurement OC represents 22.95 V at an angle of 36°

Thus the result is verified.

HIGHLIGHTS

1. An *alternating quantity* (voltage or current) is one which changes continuously in magnitude and alternates in direction at regular intervals of time.

   An alternating quantity that varies sinusoidally (*i.e.* according to the sin of angle θ) is called the *sinusoidal quantity*. Sinusoidal quantities are expressed as

   $$e = E_{max} \sin \theta = E_{max} \sin \omega t = E_{max} \sin 2\pi ft$$

   $$\text{and } i = I_{max} \sin \theta = I_{max} \sin \omega t = I_{max} \sin 2\pi ft$$

2. The shape of the curve of the voltage or current when plotted against time as abscissa (base) is called the *waveform.*
3. When a periodic wave, such as sinusoidal wave, goes through one complete set of positive and negative values it is said to have completed one *cycle*. One cycle corresponds to 360° or 2 π radians.

   Alternation is one half of cycle and corresponds to 180° or π radians.
4. The time taken in seconds by an alternating quantity to complete one cycle is known as *time period* or *periodic time* (T), while the number of cycles completed per second by an alternating quantity is known as *frequency* (f).

$$f = \frac{1}{T}$$

   In India the standard frequency for power supply is 50 Hz (cycles per second).
5. Angular velocity, $\omega = 2\pi f$ radians per second.
6. In a multipolar machine, the number of cycles completed per second by generated emf,

$$f = \text{Pairs of poles} \times \text{number of revolutions made per second} = \frac{P}{2} \times \frac{N}{60} = \frac{PN}{120}$$

7. The maximum value, positive or negative, which an alternating quantity attains during one cycle is called the *amplitude* of the alternating quantity.
8. The general expression for an alternating voltage and current are given as

$$v = V_{max} \sin(\omega t \pm \phi)\ ; \quad i = I_{max} \sin(\omega t \pm \phi)$$

   where ϕ is the phase angle in degrees or radians.
   The coefficient of the sine of the time angle gives maximum or peak value of the alternating quantity (emf, voltage or current)
   Coefficient of time t divided by 2 π gives the frequency of the periodic wave.
9. The *average (or mean) value* of an alternating current is expressed by that steady current (dc) which transfers across any circuit the same charge as transferred by that alternating current during a given time duration.

   Average or mean value of alternating current is given as

$$I_{av} = \frac{i_1 + i_2 + i_3 + \ldots\ldots + i_n}{n} = \frac{\text{Area of one alternation}}{\text{Length of base over one alternation}}$$

$$\text{For sinusoidal current } I_{av} = \frac{2}{\pi} I_{max} = 0.637\ I_{max}.$$

10. *Effective* or *virtual value* of alternating current or voltage is equal to the square root of the mean of the squares of successive ordinates and that is why it is known as *root mean square* (rms) value.

$$\textit{i.e. } I_{eff} \text{ or } I_{rms} = \sqrt{\frac{i_1^2 + i_2^2 + i_3^2 + \ldots + i_n^2}{n}}$$

   The *rms* or *effective value* of an alternating current or voltage is given by that steady current or voltage which when flows or applied to a given resistance for a given time produces the same amount of heat as when the alternating current or voltage flowing or applied to the same resistance for the same time.

$$\text{For sinusoidal current, } I_{rms} = \frac{I_{max}}{\sqrt{2}} = 0.707\ I_{max}$$

11. The ratio of rms value to the average value of a waveform is known as *form factor.* For sinusoidal wave its value is 1.11.
12. The ratio of maximum value to the rms value of a waveform is known as peak factor. For sinusoidal wave its value is $\sqrt{2}$ or 1.414.

13. The effective and average values of different waves are given below in tabular form.

Waveform	*Sinusoidal*	*Half-Wave Rectified AC*	*Square*	*Triangular*
Effective Value	$0.707\ I_{max}$	$0.5\ I_{max}$	I_{max}	$\frac{1}{\sqrt{3}} I_{max}$
Average Value	$0.637\ I_{max}$	$0.318\ I_{max}$	I_{max}	$0.5\ I_{max}$

14. By *phase* of an alternating quantity is meant the fraction of time period of that alternating quantity that has elapsed since the quantity last passed through zero position of reference.

    The angular displacement between the maximum positive values of two alternating quantities having the same frequency is called the phase difference between them.

    An alternating quantity that attains its positive maximum value prior to the other is called the *leading quantity.*

    An alternating quantity that attains its positive maximum value after the other is called the *lagging quantity.*

15. The resultant current of two currents repersented by

$$i_1 = I_{max_1} \sin \omega t \text{ and } i_2 = I_{max_2} \sin (\omega t + \theta)$$

flowing through a common conductor is given as

$$i_r = I_{r\,max} \sin (\omega t \pm \phi)$$

where
$$I_{r\,max} = \sqrt{I^2_{max_1} + I^2_{max_2} + 2 I_{max_1} I_{max_2} \cos \theta}$$

$$\text{and } \phi = \text{Tan}^{-1} \frac{I_{max_2} \sin \theta}{I_{max_1} + I_{max_2} \cos \theta}.$$

The sign of the phase angle ϕ of the resultant current will depend upon its position w.r.t. the reference phasor.

EXERCISES

1. Explain why a large percentage of electrical energy used for commercial purposes is generated as ac.
2. With a neat sketch briefly explain how an alternating voltage is produced when a coil is rotated in a magnetic field. **[V.T.U. Belgaum Karnataka Univ., Summer 2003]**
3. Derive an expression for the instantaneous value of alternating sinusoidal emf in terms of its maximum value, angular frequency and time. **[Gujarat Univ. June/July 2003]**
4. Define and explain the following:
   (*i*) Time period (*ii*) Amplitude (*iii*) Phase difference. **[Gujarat Univ. June/July 2003]**
5. Define and explain the following:
   (*i*) Frequency (*ii*) Time period (*iii*) Instantaneous value (*iv*) Average value (*v*) RMS value with respect to alternating quantities (*vi*) Amplitude. **[R.G.P.V. Bhopal Basic Electrical Engineering 2002]**
6. Define the terms: Instantaneous value, Average value, RMS value, frequency and phase. **[Pb. Technical Univ. Electrical Engineering June 2000; December 2000; May 2001]**
7. What is an rms value of an ac quantity ? Obtain expression for the rms value of a sinusoidal current in terms of its maximum value. **[V.T.U. Belgaum Univ. February 2002]**
8. Derive expressions for average value and rms value of a sinusoidally varying ac voltage. **[V.T.U. Belgaum Univ. Summer 2003]**
9. Define rms value, average value and form factor of a sinusoidal alternating voltage. **[Pb. Technical Univ. Basic Electrical and Electronics Engineering First Semester 2004-05]**
10. Find average value, rms value and form factor of half-wave rectified alternating current. **[U.P. Technical Univ. Electrical Engineering Second Semester 2006-07]**

11. Calculate the average value, effective value and form factor of the output voltage wave of a half-wave rectifier as shown in Fig. 1. **[U.P. Technical Univ. Electrical Engineering Second Semester 2003-04]**

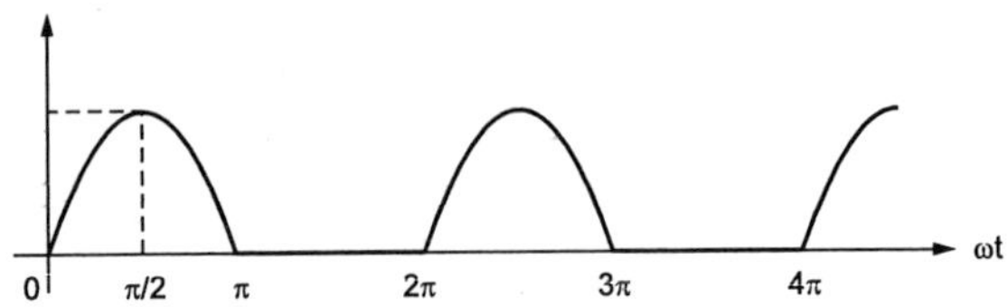

Fig. 1

12. How is the phasor of a sinusoidal quantity defined? Mention specifically the information that is conveyed by the phasor about the corresponding sinusoidal function. **[U.P. Technical Univ. Electrical Engineering First Semester 2008-09]**
13. What do you understand by phasor representation of alternating quantity ? What is an alternating quantity ? When is an alternating quantity said to be lagging or leading with respect to another quantity ?
14. Explain with diagrams what do you understand by (i) in-phase (*ii*) lagging and (*iii*) leading as applied to sinusoidal ac quantities.
15. Explain the difference between ac and dc quantities. Draw the waveform and phasor diagram of two alternating quantities having a phase difference of 90°.
16. Explain the component method for the addition and subtraction of alternating quantities.

SHORT ANSWER TYPE QUESTIONS WITH ANSWERS

Q. 1. Why sinusoidal waveshape is insisted for voltages and currents while generating, transmitting and utilising ac electric power ? **[P.T.U. January 2000]**

Ans. Sinusoidal waveshape is insisted for voltages and currents while generating, transmitting and utilising ac electrical power because it has following advantages.

(*i*) Sinusoidal voltages and currents produce minimum disturbance in the electrical circuits during operation.

(*ii*) Sinusoidal voltages and currents cause less interference (noise) to nearby communication circuits (telephone lines etc.).

(*iii*) Sinusoidal voltages and currents result in low iron as well as low copper losses in transformers, and ac rotating machines for a given output. So ac machines with sinusoidal voltages and currents operate at higher efficiency.

Q. 2. What do you understand by ω ?

Ans. Each cycle of a sinusoidal wave spans 2π radians. Hence, if this quantity is divided by the time period, angular velocity of the sinusoidal wave is obtained. It is denoted by ω and is expressed in radians per second.

Q. 3. Why the rms value of an alternating current or voltage is used to denote its amplitude ?

Ans. RMS value of an alternating current or voltage is used to denote its amplitude because it is related to the power developed in a resistance by the alternating current or voltage.

Q. 4. What is rms value of an alternating current ? **[P.T.U. December 2003]**

Ans. The effective or rms value of an alternating current is given by that steady current which when flows through a given resistance for a given time produces the same amount of heat as when the alternating current is flowing through the same resistance for the same time duration.

Q. 5. Do waveshapes other than sine wave have effective value ? Explain. **[P.T.U. December 2000]**

Ans. Yes, all the waveshapes including the sine wave have effective value since in each half cycle of the wave, work is being done. Actually the effective value of an alternating wave (sinusoidal or nonsinusoidal) is defined as below:

The rms or effective value of an alternating current or voltage is given by that steady current or voltage which when flows or applied to a given resistance for a given time produces the same amount of heat as when the alternating current or voltage is flowing or applied to the same resistance for the same time.

Q. 6. Define form factor and peak factor. **[U.P. Technical Univ. 2013-14]**

Ans. Form factor is defined as the ratio of the effective value to the average or mean value of the periodic wave.

i.e. Form factor, $K_f = \dfrac{\text{RMS value}}{\text{Average value}}$

Peak factor is defined as the ratio of peak or maximum value to the effective or rms value of the periodic wave

i.e. Peak factor, $K_p = \dfrac{\text{Peak value}}{\text{RMS value}}$.

Q. 7. Define form factor in ac circuit. **[M.T.U. First Semester 2011-12]**

Ans. Form factor is a means of relating the mean value with the effective or rms value of alternating quantity and it is useful in determination of effective or rms values of the alternating quantities whose mean or average values over half a period can be determined conveniently.

Q. 8. Determine the form factor of ac current $i = 400 \sin\left(157t + \dfrac{\pi}{6}\right)$. **[U.P. Technical Univ. Second Semester 2013-14]**

Ans. Given ac current waveform is sinusoidal, so its form factor is 1.11.

Q. 9. What is the significance of peak factor ?

Ans. Knowledge of peak factor of an alternating voltage is very essential in connection with determining the dielectric strength since the dielectric stress developed in an insulating material is proportional to the peak value of the voltage applied to it.

Q. 10. Calculate the form factor of a square wave. **[G.B. Technical Univ. Odd Semester 2012-13]**

Ans. For a square wave

$$\left.\begin{array}{r}\text{Average value, } I_{av} = I_{max}\\ \text{RMS value, } I_{rms} = I_{max}\end{array}\right\}\text{Refer to Art. 3.11}$$

$$\text{So form factor, } K_f = \frac{I_{rms}}{I_{av}} = \frac{I_{max}}{I_{max}} = 1$$

Q. 11. What is the significance of the phasor representation of an alternating quantity ?

Ans. The phasor representation of an alternating quantity enables us to understand its magnitude and position with respect to reference line.

Q. 12. What is the significance of phasor and vector in circuit analysis? **[U.P. Technical Univ. Odd Semester 2013-14]**

Ans. Refer to Answer of Q. 11.

Q. 13. What do you mean by phase and phase difference ?

Ans. The phase of an alternating quantity (voltage or current) at any instant is defined as the fractional part of a cycle through which the quantity has advanced while the phase difference may be defined as the angular displacement between the maximum positive values of the two phasors representing the two quantities having the same frequency.

Q. 14. Draw suitable wave forms to show the following: $v = V_m \sin \omega t$; $i = I_m \sin(\omega t + \theta)$. Level the quantities suitably.

[G.B.T.U. Second Semester 2011-12]

Ans. The required wave form is shown in Fig. 1.

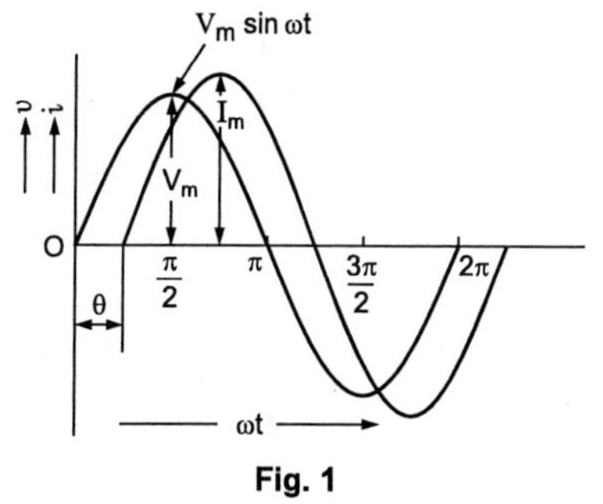

Fig. 1

PROBLEMS

1. For ac single phase 230 V, 50 Hz, assuming sinusoidal waveform find (*i*) maximum value (*ii*) rms value (*iii*) frequency (*iv*) instantaneous value voltage equation (*v*) voltage 0.003 second after starting from zero and rising in + ve direction.
   [**Ans.** (*i*) 325.3 V (*ii*) 230 V (*iii*) 50 Hz (*iv*) 325.3 sin 314 t (*v*) 263.17 V]

2. $e = 300 \sin 628\, t$
   Find maximum value, rms value, average value and frequency. Find the instantaneous value of voltage at $t = 0.001$ second after e is zero and rising in positive direction. [**Ans.** 300 V, 212.13 V, 190.98 V, 100 Hz, 176.26 V]

3. An alternating current is represented by $i = 400 \sin\left(157t + \dfrac{\pi}{6}\right)$.

Determine: (*i*) Peak value (*ii*) Average value (*iii*) RMS value (*iv*) Frequency (*v*) Time period (*vi*) Form factor (*vii*) Peak factor, and (*viii*) Phase angle of the current represented.

[**Ans.** (*i*) 400 A (*ii*) 254.65 A (*iii*) 282.84 A, (*iv*) 25 Hz (*v*) 0.04 second (*vi*) 1.11 (*vii*) 1.414 (*viii*) 30°]

4. An alternating current of frequency 50 Hz has a maximum value of 100 A. Calculate (i) its value $\frac{1}{600}$ second after the instant the current is zero (ii) in how many seconds after the zero value the current attains the value of 86.6 A ?

[**Ans.** (*i*) 50 A (*ii*) $\frac{1}{300}$ second]

5. The equation of an alternating current $i = 42.42 \sin 628\,t$. Determine (*i*) its maximum value (*ii*) frequency (*iii*) rms value (*iv*) average value (*v*) form factor. [**Ans.** (*i*) 42.42 A (*ii*) 100 Hz (*iii*) 30 A (*iv*) 27 A (*v*) 1.11]

6. Determine the rms value of a triangular wave in which the voltage rises uniformly from 0 to V volts and completes the cycle by falling instantaneously to zero. [**Ans.** $V_{max}/\sqrt{3}$]

7. A sinusoidal alternating voltage of amplitude 100 V is applied across a circuit containing a rectifying device which entirely prevents current flowing in one direction and offers a resistance of 10 Ω to the flow of current in the other direction. A hot wire ammeter is used to measure the current. Find the reading of the instrument. [**Ans.** 5 A]

8. Three circuits in parallel take currents which can be represented by

$$i_1 = 10 \sin 314\,t;\ i_2 = 7.5 \sin\left(314t - \frac{\pi}{3}\right)$$

and

$$i_3 = 12 \sin\left(314t - \frac{\pi}{4}\right)$$

Sketch a phasor diagram to represent the three currents and their resultant. Express the resultant in the same form as the three individual currents expressed above. What is the rms value and the frequency of the resultant current ?

[**Ans.** 26.35 sin (314 *t* – 0.584); 18.63 A; 50 Hz]

9. Three voltages represented by $e_1 = 20 \sin \omega t$; $e_2 = 30 \sin\left(\omega t - \frac{\pi}{4}\right)$ and $e_3 = 40 \cos\left(\omega t + \frac{\pi}{6}\right)$ act together in a circuit. Find an expression for the resultant voltage. Represent them by appropriate phasors.

[Madras Univ. Electrotechnics 1981; Nagpur Univ. Elec. Circuits 1991]

[**Ans.** 25.1 sin (ω *t* + 0.18 π)]

10. Three sinusoidal voltages acting in series are given by—

$$V_1 = 10 \sin 440\,t$$
$$V_2 = 10\sqrt{2} \sin (440\,t - 45^\circ)$$
$$V_3 = 20 \cos 440\,t$$

Determine—

(i) an expression for the resultant voltage,

(ii) the frequency and rms. value of the resultant voltage. **[U.P. Technical Univ. Elec. Engineering February 2001]**

[**Ans.** $v_r = 22.36 \sin (440\,t + 0.1476\,\pi)$; 70 Hz; 15.81 V]

❑❑❑

4 CHAPTER

Single Phase Series Circuits

4.1. INTRODUCTION

In a dc circuit the relationship between the applied voltage V and current flowing through the circuit I is a simple one and is given by the expression I = V/R but in an a c circuit this simple relationship does not hold good. Variations in current and applied voltage set up magnetic and electrostatic effects respectively and these must be taken into account with the resistance of the circuit while determining the quantitative relations between current and applied voltage. With comparatively low-voltage, heavy-current circuits magnetic effects may be very large, but electrostatic effects are usually negligible. On the other hand with high-voltage circuits electrostatic effects may be of appreciable magnitude, and magnetic effects are also present.

In this chapter we will discuss how do the magnetic effects due to variations in current and electrostatic effects due to variations in the applied voltage affect the relationship between the applied voltage and current.

4.2. PURE RESISTIVE CIRCUITS

A pure resistive or a non-inductive circuit is a circuit which has inductance so small that at normal frequency its reactance is negligible as compared to its resistance. Ordinary filament lamps, water resistances etc. are the examples of non-inductive resistances. If the circuit is non-inductive, no reactance emf *(i.e.* self- induced or back emf) is set up and whole of the applied voltage is utilised in overcoming the ohmic resistance of the circuit.

Consider an ac circuit containing a non-inductive resistance of R ohms connected across a sinusoidal voltage represented by $v = V_{max} \sin \omega t$, as shown in Fig. 4.1 (*a*).

As already said, when the current flowing through a pure resistance changes, no back emf is set up, therefore, applied voltage has to overcome the ohmic drop of i R only

$$i.e. \quad i R = v$$

$$\text{or} \quad i = \frac{v}{R} = \frac{V_{max}}{R} \sin \omega t$$

$$\text{Current will be maximum when } \omega t = \frac{\pi}{2} \text{ or } \sin \omega t = 1$$

$$\therefore \quad I_{max} = \frac{V_{max}}{R}$$

and instantaneous current may be expressed as

$$i = I_{max} \sin \omega t$$

From the expressions of instantaneous applied voltage and instantaneous current, it is evident that in a pure resistive circuit, the applied voltage and current are in phase with each other, as shown by wave and phasor diagrams in Figs. 4.1 (*b*) and (*c*) respectively.

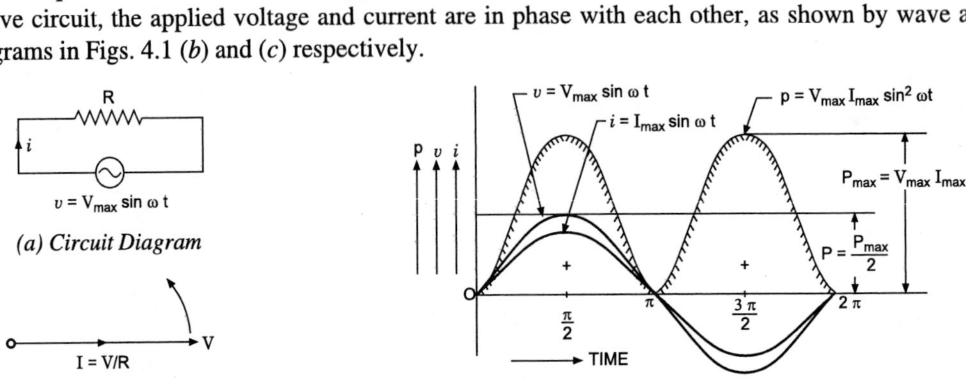

(*a*) *Circuit Diagram*

(*c*) *Phasor Diagram*

(*b*) *Wave Diagram*

Fig. 4.1 *Pure Resistive Circuit*

4.2.1. Power in Pure Resistive Circuit. The instantaneous power delivered to the circuit in question is the product of the instantaneous values of applied voltage and current.

$$i.e.\ p = v\,i = V_{max} \sin \omega t\, I_{max} \sin \omega t = V_{max}\, I_{max} \sin^2 \omega t$$

$$\text{or } p = \frac{V_{max}\, I_{max}}{2} (1 - \cos 2\, \omega t) \qquad \text{Since } \sin^2 \omega\, t = \frac{1 - \cos 2\, \omega t}{2}$$

$$= \frac{V_{max}\, I_{max}}{2} - \frac{V_{max}\, I_{max}}{2} \cos 2\, \omega t$$

$$\text{Average power, } P = \text{Average of } \frac{V_{max}\, I_{max}}{2} - \textit{average of } \frac{V_{max}\, I_{max}}{2} \cos 2\, \omega t$$

Since average of $\frac{V_{max}\, I_{max}}{2}$ cos 2 ω*t* over a complete cycle is zero,

$$P = \frac{V_{max}\, I_{max}}{2} = \frac{V_{max}}{\sqrt{2}} \cdot \frac{I_{max}}{\sqrt{2}} = VI \text{ watts}$$

where V and I are the rms values of applied voltage and current respectively.

Thus, for pure resistive circuits, the expression for power is the same as for d c circuits. From the power curve for a pure resistive circuit shown in Fig. 4.1 (*b*) it is evident that power consumed in a pure resistive circuit is not constant, it is fluctuating. However, it is always positive. This is so because the instantaneous values of voltage and current are always either both positive or negative and, therefore, the product is always positive. This means that the voltage source constantly delivers power to the circuit and the circuit consumes it.

Example 4.1 **A 250 V (rms), 50 Hz voltage is applied across a circuit consisting of a pure (non-inductive) resistance of 20 Ω. Determine (*i*) the current flowing through the circuit and (*ii*) power absorbed by the circuit. Give the expressions for the voltage and current.**

Solution: (*i*) Current flowing through the circuit, $I = \frac{V}{R} = \frac{250}{20} = 12.5$ A **Ans.**

(*ii*) Power absorbed, P = VI = 250 × 12.5 = 3,125 W or 3.125 kW **Ans.**

Peak value of applied voltage, $V_{max} = \sqrt{2}\, V_{rms} = \sqrt{2} \times 250 = 353.6$ V

Peak value of current, $I_{max} = \sqrt{2}\, I_{rms} = \sqrt{2} \times 12.5 = 17.68$ A

Expression for voltage, $v = V_{max} \sin 2\, \pi f t$

Assuming that at $t = 0$, $v = 0$ and is increasing in positive direction

$$\text{so } v = 353.6 \sin 2\pi \times 50\, t = 353.6 \sin 314\, t \textbf{ Ans.}$$

$$\text{Expression for current, } i = I_{max} \sin 2\pi f t = 17.68 \sin 314\, t \textbf{ Ans.}$$

Example 4.2 **A 100 ohm resistance is carrying a sinusoidal current given by 3 cos ω *t*.**

Determine: (*i*) instantaneous power taken by resistance, (*ii*) average power.

[*U.P. Technical Univ. Electrical Engineering First Semester 2003-04*]

Solution: (*i*) Instantaneous power taken by the resistance, $p = v\,i = i\,R \times i = 100 \times 3 \cos \omega t \times 3 \cos \omega t$

$$= 900 \cos^2 \omega t = 450\,(1 + \cos 2\,\omega t) \textbf{ Ans.}$$

$$(ii) \text{ Average power} = 450 \text{ watts } \textbf{Ans.}$$

Example 4.3 **A 100 V, 60 watt bulb is to be operated from 220 V supply. What resistance must be connected in series with the bulb to glow normally?** [*Mahamaya Technical Univ. Electrical Engineering First Semester 2011-12*]

Solution: Compared with the resistance of the filament of a lamp its inductance is negligible. As such a lamp is considered to be a non-inductive resistance.

Rated power of lamp, P = 60 W

Rated voltage of lamp, V = 100 V

$$\text{Current drawn by lamp, } I = \frac{P}{V} = \frac{60}{100} = 0.6 \text{ A}$$

Let a non-inductive resistance of R Ω be connected in series with the lamp, as shown in Fig. 4.2 (*a*).

Since in this case voltage across the lamp and that across the resistance, V_R are in phase, their arithmetic sum is equal to applied voltage i.e.,

$$V_R + 100 = 220$$

$$\text{or } V_R = 220 - 100 = 120 \text{ V}$$

$$\text{Resistance, } R = \frac{V_R}{I} = \frac{120}{0.6} = 200\ \Omega \textbf{ Ans.}$$

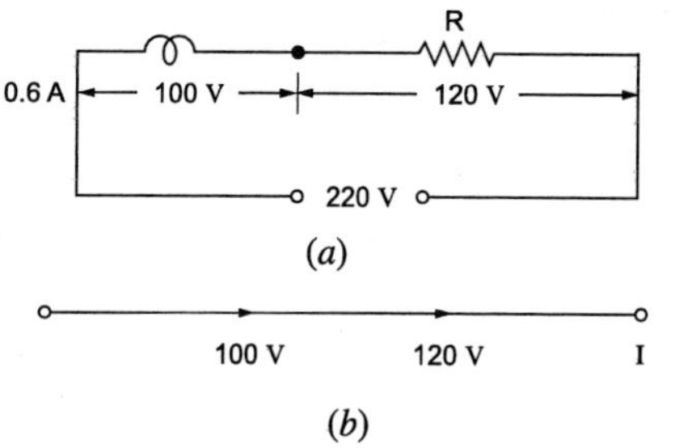

Fig. 4.2

4.3. PURE INDUCTIVE CIRCUITS

An inductive circuit is a coil with or without an iron core having negligible resistance. Practically pure inductance can never be had as the inductive coil has always small resistance. However, a coil of thick copper wire wound on a laminated iron core has negligible resistance and is known as a *choke coil.**

When an alternating voltage is applied to a pure inductive coil, an emf, known as self-induced emf, is induced in the coil which opposes the applied voltage. Since coil has no resistance, at every instant applied voltage has to overcome this self-induced emf only.

$$\text{Let the applied voltage } v = V_{max} \sin \omega t$$

$$\text{and self-inductance of coil} = L \text{ henry}$$

$$\text{Self-induced emf in the coil, } e_L = -L\frac{di}{dt}$$

Since applied voltage at every instant is equal and opposite to the self-induced emf *i.e.* $v = -e_L$

$$\therefore V_{max} \sin \omega t = -\left(-L\frac{di}{dt}\right)$$

$$\text{or } di = \frac{V_{max}}{L} \sin \omega t\, dt$$

* Inductors or choke coils are made of iron core because large valued flux densities can be produced in iron cores and so inductances of large values can be had. Air-cored inductors become too much bulky to give an inductance of a given value.

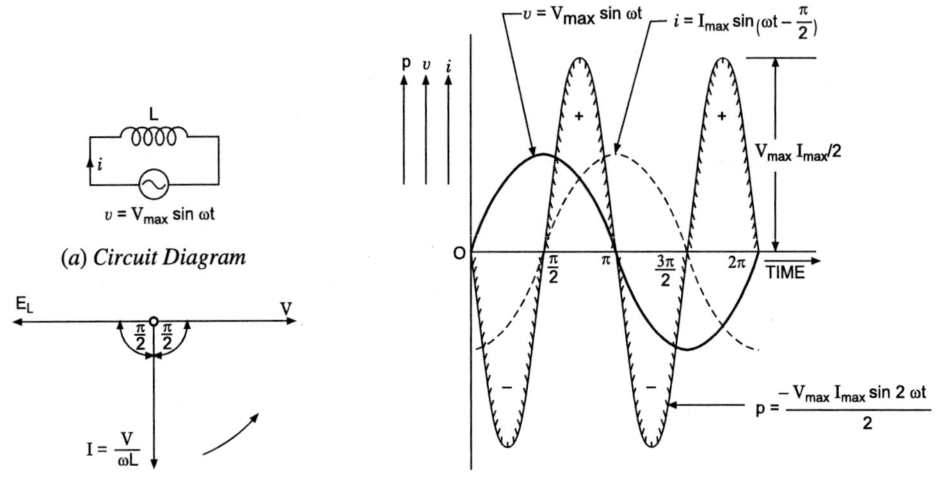

(*a*) *Circuit Diagram*

(*c*) *Phasor Diagram* (*b*) *Wave Diagram*

Fig. 4.3 *Pure Inductive Circuit*

$$\text{Integrating both sides we get } i = \frac{V_{max}}{L} \int \sin \omega t \, dt = \frac{V_{max}}{\omega L} (-\cos \omega t) + A$$

where A is a constant of integration, which is found to be zero from initial conditions

$$i.e. \quad i = \frac{-V_{max}}{\omega L} \cos \omega t = \frac{V_{max}}{\omega L} \sin \left(\omega t - \frac{\pi}{2} \right)$$

Current will be maximum when $\sin \left(\omega t - \frac{\pi}{2} \right) = 1$, hence, maximum value of current,

$$I_{max} = \frac{V_{max}}{\omega L}$$

and instantaneous current may be expressed as

$$i = I_{max} \sin \left(\omega t - \frac{\pi}{2} \right)$$

From the expressions of instantaneous applied voltage and instantaneous current flowing through a pure inductive coil it is observed that the current lags behind the applied voltage by $\frac{\pi}{2}$, as shown in Fig. 4.3 (*b*) by wave diagram and in Fig. 4.3 (*c*) by phasor diagram.

4.3.1. Inductive Reactance. ωL in the expression $I_{max} = \frac{V_{max}}{\omega L}$ is known as *inductive reactance* and is denoted by X_L *i.e.* $X_L = \omega L$

If L is in henry and ω is in radians per second then X_L will be in ohms.

4.3.2. Power in Pure Inductive Circuit.

$$\text{Instantaneous power, } p = v \times i = V_{max} \sin \omega t \, I_{max} \sin \left(\omega t - \frac{\pi}{2} \right)$$

$$\text{or } \quad p = -V_{max} I_{max} \sin \omega t \cos \omega t = -\frac{V_{max} I_{max}}{2} \sin 2 \omega t$$

The power measured by wattmeter is the average value of p which is zero since average of a sinusoidal quantity of double frequency over a complete cycle is zero. *Hence, in a pure inductive circuit power absorbed is zero.*

Physically the above fact can be explained as below:

During the second quarter of a cycle the current and the magnetic flux of the coil increases and the coil draws power from the supply source to build up the magnetic field (the power drawn is positive and the energy drawn by the coil from the supply source is represented by the area between the curve p and the time axis). The energy stored in the magnetic field during build up is given as $W_{max} = \frac{1}{2} LI^2_{max}$.

In the next quarter the current decreases. The emf of self-induction will, however, tend to oppose its decrease. The coil acts as a generator of electrical energy, returning the stored energy in the magnetic field to the supply source (now the power drawn by the coil is negative and the curve p lies below the time axis).

The chain of events repeats itself during the next half cycles. Thus, a proportion of power is continually exchanged between the field and the inductive circuit and the power consumed by a pure inductive coil is zero.

Example 4.4 **The voltage and current through a circuit element are**

$$v = 100 \sin (314\,t + 45°) \text{ volts}$$

$$i = 10 \sin (314\,t + 315°) \text{ amperes.}$$

(*i*) Identify the circuit element (*ii*) Find the value (*iii*) Obtain expression for power.

[*U.P. Technical Univ. Electrical Engineering September 2001*]

Solution: $v = 100 (\sin 314\,t + 45°)$ volts

$i = 10 \sin (314\,t + 315°) = 10 \sin (314\,t + 315° - 360°) = 10 \sin (314\,t - 45°)$ amperes

(*i*) From equations for voltage and current it is observed that the circuit current i lags behind the applied voltage v by 90°. It means the circuit element is an *inductor*. **Ans.**

(*ii*) Inductive reactance of the inductor, $X_L = \frac{V_{rms}}{I_{rms}} = \frac{100/\sqrt{2}}{10/\sqrt{2}} = 10\ \Omega$ **Ans.**

and $L = \frac{X_L}{\omega} = \frac{10}{314} = 0.0318$ H or 31.8 mH **Ans.**

(*iii*) Expression for power is given as $p = \frac{-V_{max}\, I_{max}}{2} \sin 2\,\omega t$

$$= \frac{-100 \times 10}{2} \sin (2 \times 314\,t) = -500 \sin 628\,t$$ **Ans.**

4.4. PURE CAPACITIVE CIRCUITS

When a dc voltage is impressed across the plates of a perfect condenser, it will become charged to full voltage almost instantaneously. The charging current will flow only during the period of "build up" and will cease to flow as soon as the capacitor has attained the steady voltage of the source. This implies that for a direct current, a capacitor is a break in the circuit or an infinitely high resistance.

In Fig. 4.4 a sinusoidal voltage is applied to a capacitor. During the first quarter-cycle, the applied voltage increases to the peak value, and the capacitor is charged to that value. The current is maximum in the beginning of the cycle and becomes zero at the maximum value of the applied voltage, so there is a phase difference of 90° between the applied voltage and current. During the first quarter-cycle the current flows in the normal direction through the circuit; hence the current is positive.

In the second quarter-cycle, the voltage applied across the capacitor falls, the capacitor loses its charge, and current flows through it against the applied voltage because the capacitor discharges into the circuit. Thus, the current is negative during the second quarter-cycle and attains a maximum value when the applied voltage is zero.

The third and fourth quarter-cycles repeat the events of the first and second, respectively, with the difference that the polarity of the applied voltage is reversed, and there are corresponding current changes. In other words, an alternating current flows in the circuit because of the charging and discharging of the capacitor. As illustrated in Figs. 4.4 (*b*) and (*c*) the current begins its cycle 90 degrees ahead of the voltage, so the current in a capacitor leads the applied voltage by 90 degrees – the opposite of the inductance current-voltage relationship.

Let an alternating voltage represented by $v = V_{max} \sin \omega t$ be applied across a capacitor of capacitance C farads. The expression for instantaneous charge is given as

$$q = C V_{max} \sin \omega t$$

Since the capacitor current is equal to the rate of change of charge, the capacitor current may be obtained by differentiating the above equation

$$i = \frac{dq}{dt} = \frac{d}{dt} [C V_{max} \sin \omega t] = \omega C V_{max} \cos \omega t = \frac{V_{max}}{1/\omega C} \sin \left(\omega t + \frac{\pi}{2}\right)$$

Current is maximum when $t = 0$

$$\therefore \ I_{max} = \frac{V_{max}}{1/\omega C}$$

Substituting $\frac{V_{max}}{1/\omega C} = I_{max}$ in the above equation for instantaneous current, we get

$$i = I_{max} \sin \left(\omega t + \frac{\pi}{2}\right)$$

From the equations of instantaneous applied voltage and instantaneous current flowing through capacitance, it is observed that the current leads the applied voltage by $\frac{\pi}{2}$, as shown in Figs. 4.4 (*b*) and (*c*) by wave and phasor diagrams respectively.

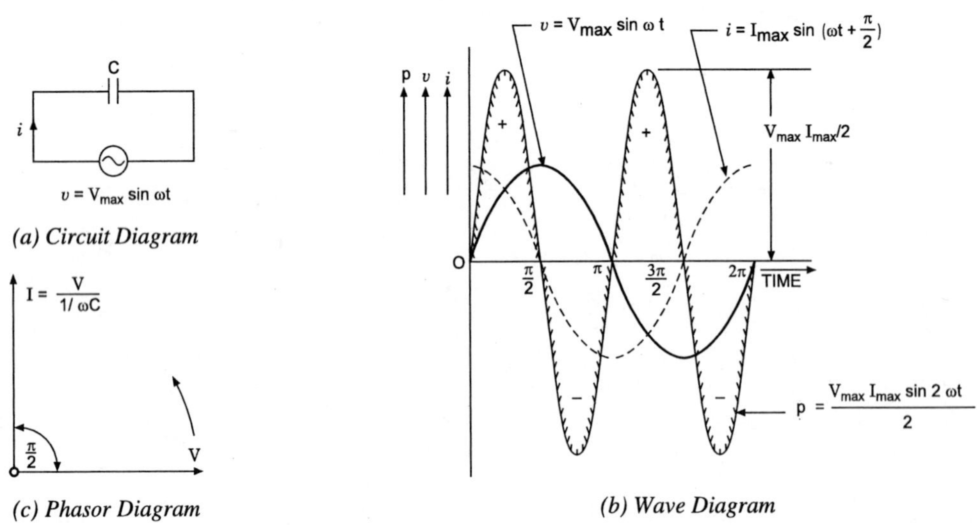

Fig. 4.4 *Pure Capacitive Circuit*

4.4.1. Capacitive Reactance. $\frac{1}{\omega C}$ in the expression $I_{max} = \frac{V_{max}}{1/\omega C}$ is known as *capacitive reactance* and is denoted by X_C *i.e.* $X_C = \frac{1}{\omega C}$

If C is in farads and ω is in radians/s, then X_C will be in ohms.

4.4.2. Power in Pure Capacitive Circuit.

Instantaneous power, $p = vi = V_{max} \sin \omega t . I_{max} \sin \left(\omega t + \frac{\pi}{2}\right) = V_{max} I_{max} \sin \omega t \cos \omega t$

$$= \frac{V_{max} I_{max}}{2} \sin 2\omega t$$

$$\text{Average power, } P = \frac{V_{max} I_{max}}{2} \times \text{average of } \sin 2\omega t \text{ over a complete cycle} = 0.$$

Hence, power absorbed in a pure capacitive circuit is zero. The same is shown graphically in Fig. 4.4 (*b*). The energy taken from the supply circuit is stored in the capacitor during the first quarter-cycle and returned during the next.

The energy stored by a capacitor at maximum voltage across its plates is given by the expression

$$W_C = \frac{1}{2} CV^2_{max}$$

This can be realized when it is recalled that no heat is produced and no work is done while current is flowing through a capacitor. As a matter of fact, in commercial capacitors, there is a slight energy loss in the dielectric in addition to a minute I^2R loss due to flow of current over the plates having definite ohmic resistance. The power curve is a sine wave of double the supply frequency. Although it raises the power factor from zero to 0.002 or even a little more, but for ordinary purposes the power factor is taken to be zero. Obviously the phase angle due to dielectric and ohmic losses decreases slightly.

Example 4.5 **A 50-Hz voltage of 230 V effective value is impressed on a capacitance of 26.5 μF. (*a*) Write the time equations for the voltage and the resulting current. Let the zero axis of the voltage wave be at *t* = 0. (*b*) Show the voltage and current on a time diagram. (*c*) Show the voltage and current on a phasor diagram. (*d*) Find the maximum energy stored in the capacitor. Find the relative heating effects of two current waves of equal peak value, the one sinusoidal and the other rectangular in waveform.** [*Allahabad Univ. Electrical Technology 1991*]

Solution: (*a*) The time equation for the applied voltage is given by

$$v(t) = V_{max} \sin (2\pi f t + \theta) = 230\sqrt{2} \sin (2\pi \times 50\,t + 0^\circ) = 325 \sin 314\,t \quad \textbf{Ans.}$$

$$\because V_{max} = 230\sqrt{2}\ \text{V}; f = 50 \text{ Hz and } \theta = 0^\circ$$

$$\text{Capacitive reactance, } X_C = \frac{1}{2\pi f C} = \frac{1}{2\pi \times 50 \times 26.5 \times 10^{-6}}\ \Omega = 120\ \Omega$$

Maximum value of current flowing through the capacitor,

$$I_{max} = \frac{V_{max}}{X_C} = \frac{325}{120} = 2.71 \text{ A}$$

Phase angle $\phi = 90^\circ$ (lead)

So the time equation for the resulting current is given by

$$i(t) = 2.71 \sin (314\,t + 90^\circ) = 2.71 \cos 314\,t \quad \textbf{Ans.}$$

(*b*) Voltage and current on a time diagram are shown in Fig. 4.4(*b*).

(*c*) Voltage and current on a phasor diagram are shown in Fig. 4.4(*c*).

(*d*) Maximum energy stored in the capacitor,

$$W_C = \frac{1}{2} C (V_{max})^2 = \frac{1}{2} \times 26.5 \times 10^{-6} \times \left(230\sqrt{2}\right)^2 = 1.4 \text{ J} \quad \textbf{Ans.}$$

For equal peak value, say I_{max},

RMS value of current in case of rectangular waveform, $I_{rms1} = I_{max}$

RMS value of current in case of sinusoidal waveform, $I_{rms2} = \dfrac{I_{max}}{\sqrt{2}}$

The ratio of heating effects of two current waves is given by

$$\frac{H_{rectangular}}{H_{sinusoidal}} = \frac{(I_{rms_1})^2 \times R}{(I_{rms_2})^2 \times R} = \frac{I_{max}^2\ R}{\left(\frac{I_{max}}{\sqrt{2}}\right)^2 R} = 2 \textbf{ Ans.}$$

4.5. RESISTANCE—INDUCTANCE (R-L) SERIES CIRCUIT

This is the case most generally met within practice, nearly all circuits contain both resistance and inductance.

Consider an ac circuit consisting of resistance of R ohms and inductance of L henrys connected in series, as shown in Fig. 4.5 (*a*).

Let the supply frequency be f and current flowing through the circuit be of I amperes (rms value).

Now voltage drop across resistance, $V_R = I R$ in phase with the current.

Voltage drop across inductance, $V_L = I X_L = I\omega L$ leading I by $\pi/2$ radians, as shown in Fig 4.5 (*c*)

The applied voltage, being equal to the phasor sum of V_R and V_L, will be given by the diagonal of the parallelogram.

$$\text{Applied voltage, } V = \sqrt{(V_R)^2 + (V_L)^2} = \sqrt{(I R)^2 + (I X_L)^2} = I\sqrt{R^2 + X_L^2} = I Z$$

where $X_L = \omega L = 2\pi f L$

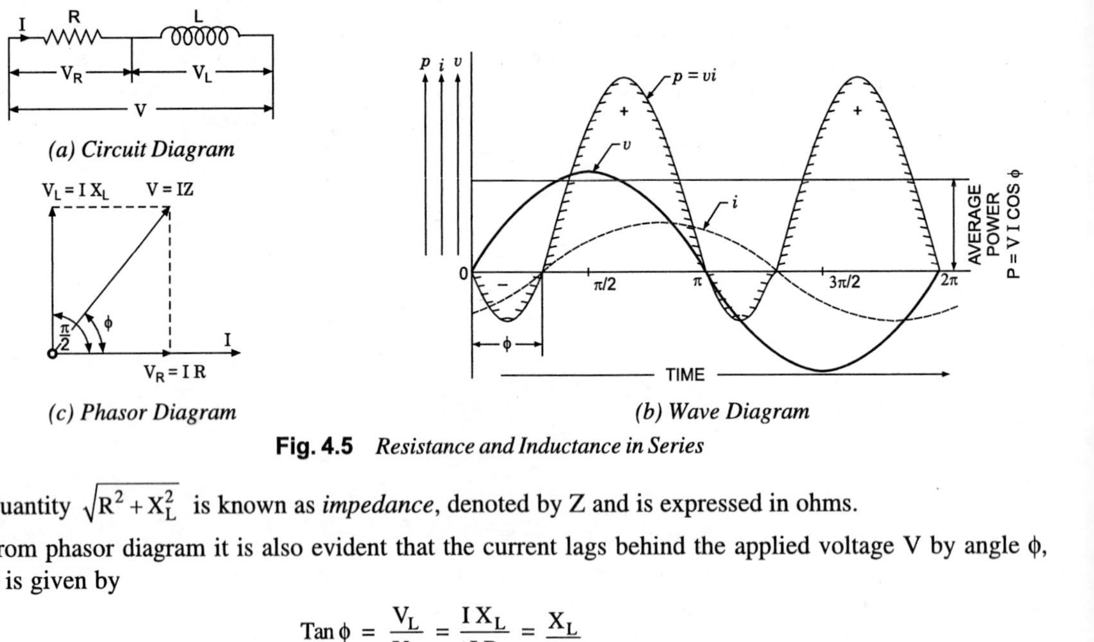

(*a*) *Circuit Diagram*

(*c*) *Phasor Diagram*

(*b*) *Wave Diagram*

Fig. 4.5 *Resistance and Inductance in Series*

Quantity $\sqrt{R^2 + X_L^2}$ is known as *impedance*, denoted by Z and is expressed in ohms.

From phasor diagram it is also evident that the current lags behind the applied voltage V by angle ϕ, which is given by

$$\text{Tan}\,\phi = \frac{V_L}{V_R} = \frac{I X_L}{I R} = \frac{X_L}{R}$$

$$\text{or } \phi = \text{Tan}^{-1}\frac{X_L}{R}$$

Since X_L and R are known, the value of phase angle ϕ can be computed. If the applied voltage

$v = V_{max} \sin \omega t$, then expression for the circuit current will be

$i = I_{max} \sin (\omega t - \phi)$

where $I_{max} = \frac{V_{max}}{Z}$ and $\phi = \text{Tan}^{-1}\frac{X_L}{R}$.

4.5.1. Impedance Triangle. If a triangle ABC is drawn so that AB = $\frac{V_R}{I}$ = R, BC = $\frac{V_L}{I}$ = X_L and AC = $\frac{V}{I}$ = Z , it is a triangle similar to the voltage triangle shown in Fig. 4.6 (*a*). Such a triangle is called an *impedance triangle,* which is most useful in letting one see at a glance how R, X, and Z are related to each other. The angle between Z and R sides of the impedance triangle is known as *phase angle of the circuit* and cos of this angle is known as *power factor of the circuit*.

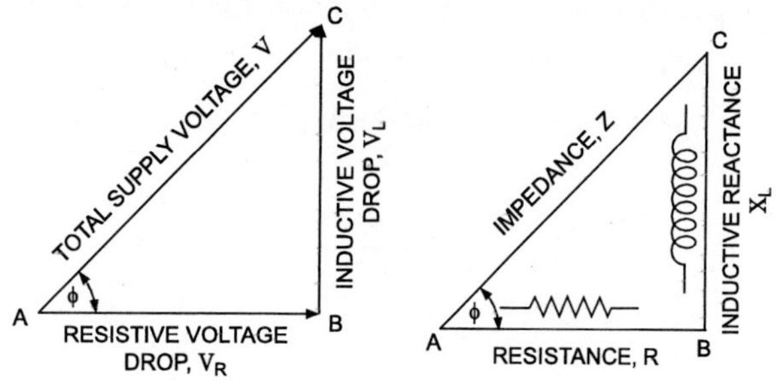

$$\text{Power factor} = \text{Cos}\ \phi = \frac{R}{Z}$$

(*a*) *Voltage Triangle* (*b*) *Impedance Triangle*

Fig. 4.6

4.5.2. Power in Resistance—Inductance (R-L) Circuit.

Instantaneous power, $p = v\,i = V_{max} \sin \omega t \times I_{max} \sin (\omega t - \phi)$

$$= \frac{1}{2} V_{max} I_{max} [2 \sin \omega t . \sin (\omega t - \phi)]$$

$$= \frac{1}{2} V_{max} I_{max} [\cos \phi - \cos (2\,\omega t - \phi)]$$

$$= \frac{1}{2} V_{max} I_{max} \cos \phi - \frac{1}{2} V_{max} I_{max} \cos (2\,\omega t - \phi)$$

Since average value of pulsating component $\frac{1}{2} V_{max} I_{max} \cos (2\,\omega t - \phi)$ over a complete cycle is zero,

average power of the circuit, P = Constant component, $\frac{1}{2} V_{max} I_{max} \cos \phi = \frac{V_{max}}{\sqrt{2}} . \frac{I_{max}}{\sqrt{2}} \cos \phi = VI \cos \phi$

where V and I are the rms values of voltage and current and ϕ is the phase angle between applied voltage V and circuit current I.

Alternatively power consumed by the circuit,

P = Power consumed by resistance $\quad\because$ Power consumed by inductance is zero

$$= I^2 R = I\,(I\,R) = \frac{V}{Z} . I\,R = VI \frac{R}{Z} = VI \cos \phi$$

Since from impedance triangle $\cos \phi = \frac{R}{Z}$

So the power in an a c circuit is given by the product of rms values of current and voltage and cosine of the phase angle between voltage and current. Cosine of the phase angle between the voltage and current, cos ϕ is known as the *power factor* of the circuit, and is equal to $\frac{R}{Z}$ which is obvious from impedance triangle [Fig. 4.6 (*b*)]

4.6. APPARENT POWER, TRUE POWER, REACTIVE POWER AND POWER FACTOR

The product of rms values of current and voltage, VI is called the *apparent power* and is measured in volt-amperes or kilovolt amperes (kVA).

The *true power* in an a c circuit is obtained by multiplying the apparent power by the power factor and is expressed in watts or kilowatts (kW).

The product of apparent power, VI and the sine of the angle between voltage and current, sin ϕ is called the *reactive power.* This is also known as *wattless power* and is expressed in reactive volt-amperes or kilovolt amperes reactive (kVAR).

i.e Apparent power, $S = VI$ volt-amperes or $\frac{VI}{1,000}$ kVA

True power, $P = VI \cos \phi$ watts or $\frac{VI \cos \phi}{1,000}$ kW

Reactive power, $Q = VI \sin \phi$ VAR or $\frac{VI \sin \phi}{1,000}$ kVAR

and $\text{kVA} = \sqrt{(\text{kW})^2 + (\text{kVAR})^2}$

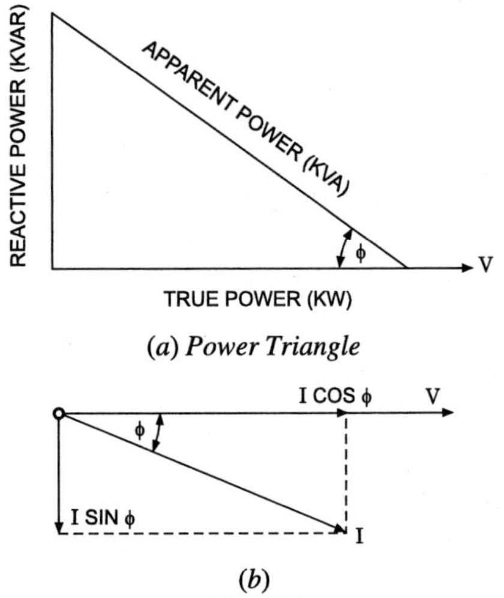

(*a*) *Power Triangle*

(*b*)

Fig. 4.7

The above relations can easily be followed by referring to the power diagram shown in Fig. 4.7 (*a*).

Power factor may be defined as

(*i*) cosine of the phase angle between voltage and current or

(*ii*) the ratio of the resistance to impedance or

(*iii*) the ratio of true power to apparent power.

The power factor can never be greater than unity. The power factor is expressed either as fraction or as a percentage. It is usual practice to attach the word 'lagging' or ' leading' with the numerical value of power factor to signify whether the current lags behind or leads the voltage.

4.6.1. Active Component of Current. The current component which is in phase with circuit voltage (*i.e.* I cos ϕ) and contributes to active or true power of the circuit is called the *active* (*wattful* or *in-phase*) *component* of current.

4.6.2. Reactive Component of Current. The current component which is in quadrature (or 90° out of phase) to circuit voltage (*i.e.* I sin ϕ) and contributes to reactive power of the circuit, is called the *reactive* (or *wattless*) *component* of current.

4.7. Q-FACTOR OF COIL

Reciprocal of power factor is known as Q-factor of the coil. It is also called the quality factor or figure of merit of a coil.

$$\text{Mathematically Q-factor} = \frac{1}{\text{Power factor}} = \frac{1}{\text{Cos}\,\phi} = \frac{Z}{R}$$

If R is very small in comparison to inductive reactance X_L, the

$$\text{Q-factor} = \frac{X_L}{R} = \frac{\omega L}{R}$$

$$\text{Also } Q = 2\pi \times \frac{\text{Maximum energy stored}}{\text{Energy dissipated per cycle}}$$

Example 4.6 **An alternating current of 1.5 A flows in a circuit when applied voltage is 300 V. The power consumed is 225 W. Find resistance and reactance of the circuit.**

[*U.P. Technical Univ. Electrical Engineering First Semester 2004-05*]

Solution :

$$\text{Circuit resistance, } R = \frac{\text{Power consumed}}{I^2} = \frac{225}{(1.5)^2} = 100\ \Omega \quad \textbf{Ans.}$$

$$\text{Circuit impedance, } Z = \frac{V}{I} = \frac{300}{1.5} = 200\ \Omega$$

$$\text{Circuit reactance, } X = \sqrt{Z^2 - R^2} = \sqrt{(200)^2 - (100)^2} = 173.2\ \Omega \ \textbf{Ans.}$$

Example 4.7 **$V = \sqrt{2} \times 200 \cos 500t$, $P_{av} = 250$ W, power factor = 0.7 lagging. Calculate the reactive power of the system.**

[G.B. Technical Univ. Electrical Engineering Even Semester 2012-13]

Solution :

$$\text{RMS value of supply voltage, } V = \frac{V_{max}}{\sqrt{2}} = \frac{\sqrt{2} \times 200}{\sqrt{2}} = 200 \text{ V}$$

$$\text{Power factor} = 0.7 \text{ lagging}$$

$$\text{Phase angle, } \phi = \text{Cos}^{-1}\, 0.7 = 45.573°$$

$$\text{Sin}\, \phi = \text{Sin}\, 45.573° = 0.714$$

$$\text{Average power, } P = 250 \text{ W}$$

$$\text{Current drawn, } I = \frac{P}{V \cos \phi} = \frac{250}{200 \times 0.7} = 1.7857 \text{ A}$$

$$\text{Reactive power of the system, } Q = VI \sin \phi = 200 \times 1.7857 \times 0.714 = 255 \text{ VAR (inductive)}$$ **Ans.**

Example 4.8 **The voltage and current through a circuit element are**

$$v = 50 \sin (314\,t + 55°) \text{ volts}$$

$$i = 10 \sin (314\,t + 325°) \text{ amperes}$$

Find the value of power drawn by the element.

[U.P. Technical Univ. Electrical Engineering Second Semester 2006-07]

Solution: Phase angle between voltage and current = 325° – 55° = 270° *i.e.* current leads the voltage by 270° or voltage leads current by 90° *i.e.* current lags behind the voltage by 90°

$$\text{Power drawn by the element} = VI \cos \phi = \frac{V_{max}}{\sqrt{2}} \times \frac{I_{max}}{\sqrt{2}} \cos \phi = \frac{50}{\sqrt{2}} \times \frac{10}{\sqrt{2}} \times \cos 90° = 0$$ **Ans.**

Example 4.9 **A resistance and inductance are connected in series across a voltage $v = 283 \sin 314t$. An expression of current is found to be $i = 4 \sin (314t - 45°)$. Find the values of resistance, inductance and power factor.**

[G.B. Technical Univ. Electrical Engineering Even Semester 2012-13; U.P. Technical Univ. Basic Electrical Engineering Second Semester 2013-14]

Solution:

$$\text{Supply frequency, } f = \frac{\text{Coefficient of time, } t}{2\pi} = \frac{314}{2\pi} = 50 \text{ Hz}$$

$$\text{RMS value of applied voltage, } V = \frac{V_{max}}{\sqrt{2}} = \frac{283}{\sqrt{2}} = 200.11$$

$$\text{RMS value of current, } I = \frac{I_{max}}{\sqrt{2}} = \frac{4}{\sqrt{2}} = 2.828 \text{ A}$$

$$\text{Phase angle, } \phi = 45° \text{ (lagging)}$$

$$\text{Impedance of the circuit, } Z = \frac{V}{I} = \frac{200.11}{2.828} = 70.75\ \Omega$$

$$\text{Resistance of the circuit, } R = Z \cos \phi = 70.75 \cos 45° = 50\ \Omega$$ **Ans.**

$$\text{Reactance of the circuit, } X_L = Z \sin \phi = 70.75 \sin 45° = 50\ \Omega$$

$$\text{Inductance, } L = \frac{X_L}{2\pi f} = \frac{50}{2\pi \times 50} = 0.159 \text{ H}$$ **Ans.**

$$\text{Power factor} = \text{Cos}\, \phi = \text{Cos}\, 45° = 0.707 \text{ (lagging)}$$ **Ans.**

Example 4.10 **If load draws a current of 10 A at 0.8 pf lagging when connected to 100 V supply, calculate the value of real, reactive and apparent powers. Also find out the resistance of the load.**

[G.B. Technical Univ. Electrical Engineering Second Semester 2011-12]

Solution:

$$\text{Supply voltage, } V = 100 \text{ V}$$

$$\text{Current, } I = 10 \text{ A}$$

$$\text{Power factor, } \cos \phi = 0.8 \text{ (lagging)}$$

Phase angle, $\phi = \text{Cos}^{-1} 0.8 = 36.87°$ (lagging)

Real power, $P = VI \cos \phi = 100 \times 10 \times 0.8 = 800$ W **Ans.**

Reactive power, $Q = VI \sin \phi = 100 \times 10 \sin(-36.87°)$

$= 100 \times 10 \times 0.6 = 600$ VAR (lagging) **Ans.**

Apparent power, $S = \sqrt{P^2 + Q^2} = \sqrt{800^2 + 600^2} = 1{,}000$ VA **Ans.**

Load impedance, $Z = \dfrac{V}{I} = \dfrac{100}{10} = 10\ \Omega$

Resistance of load, $R = Z \cos \phi = 10 \times 0.8 = 8\ \Omega$ **Ans.**

Example 4.11 **Given $v = 200 \sin 377t$ V and $i = 8 \sin(377t - 30°)$ A for an ac circuit.**

Determine : (*a*) The power factor (*b*) True power (*c*) Apparent power (*d*) Reactive power.

[*U.P. Technical Univ. Electrical Engineering First Semester 2008-09*]

Solution:

$v = 200 \sin 377t$ V

$i = 8 \sin(377t - 30°)$ A

Phase angle, $\phi = 30°$ (lagging)

RMS value of applied voltage, $V = \dfrac{V_{max}}{\sqrt{2}} = \dfrac{200}{\sqrt{2}} = 141.42$ V

RMS value of current, $I = \dfrac{I_{max}}{\sqrt{2}} = \dfrac{8}{\sqrt{2}} = 5.657$ A

(*a*) Power factor $= \text{Cos}\ \phi = \text{Cos}\ 30° = 0.866$ (lagging) **Ans.**

(*b*) True power, $P = VI \cos \phi = 141.42 \times 5.657 \times 0.866 = 692.82$ W **Ans.**

(*c*) Apparent power, $S = VI = 141.42 \times 5.657 = 800$ VA **Ans.**

(*d*) Reactive power, $Q = VI \sin \phi = 141.42 \times 5.657 \times \sin 30° = 400$ VAR (lagging) **Ans.**

Example 4.12 **For an ac circuit, the voltage and current are given by:**

$v = 200 \sin 377t$ volts, and

$i = 8 \sin(377t - 30°)$ amperes.

Determine the power factor, true power, apparent power and reactive power of the circuit. Also cross verify from power triangle.

[*Mahamaya Technical Univ. Electrical Engineering First Semester 2011-12*]

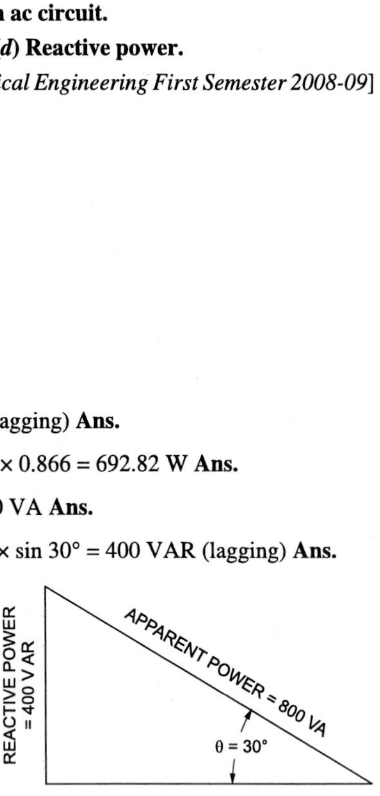

Fig. 4.8

Solution: Refer to Solution of Example 4.11.

Power triangle is shown in Fig. 4.8.

From power triangle

$$S = \sqrt{P^2 + Q^2} = \sqrt{(692.82)^2 + 400^2} = 800 \text{ VA, same value as determined in Example 4.11}$$

Phase angle, $\phi = \text{Tan}^{-1}\dfrac{Q}{P} = \text{Tan}^{-1}\dfrac{400}{692.8} = 30°$

Example 4.13 **An emf, $e = 141.4 \sin(377t + 30°)$ is impressed on the impedance coil having a resistance of 4 Ω and an inductive reactance of 1.25 Ω, measured at 25 Hz. What is the equation of the current ? Sketch the waves for i, e_R, e_L and e.**

Solution:

Applied emf, $e = 141.4 \sin(377t + 30°)$

Supply frequency, $f = \dfrac{\text{Coefficient of time, } t}{2\pi} = \dfrac{377}{2\pi} = 60$ Hz

Resistance of coil, $R = 4\ \Omega$

Inductive reactance of coil at supply frequency of 60 Hz,

$$X_L = \text{Coil inductive reactance at 25 Hz} \times \frac{60}{25} = 1.25 \times \frac{60}{25} = 3\ \Omega \quad \because X_L \alpha f$$

$$\text{Coil impedance, } Z = \sqrt{R^2 + X_L^2} = \sqrt{4^2 + 3^2} = 5\ \Omega$$

$$\text{Maximum value of coil current, } I_{max} = \frac{E_{max}}{Z} = \frac{141.4}{5} = 28.28 \text{ A}$$

$$\text{Phase angle, } \phi = \text{Tan}^{-1}\frac{X_L}{R} = \text{Tan}^{-1}\frac{3}{4} = 36.87°$$

It means that the circuit current lags behind the impressed voltage by 36.87°. Hence the circuit current is given by the equation

$$i = I_{max} \sin(377\,t + 30° - \phi) = 28.28 \sin(377\,t + 30° - 36.87°) = 28.28 \sin(377\,t - 6.87°) \text{ **Ans.**}$$

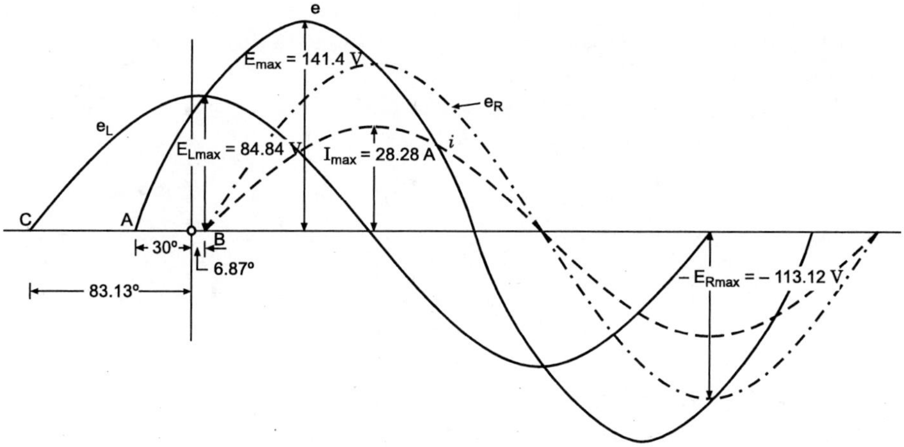

Fig. 4.9

Resistive drop e_R, being in phase with circuit current i, is given by the equation

$$e_R = i\,R = 4 \times 28.28 \sin(377\,t - 6.87°) = 113.12 \sin(377\,t - 6.87°)$$

The inductive voltage, that leads the circuit current by 90°, is given by the equation

$$e_L = i\,X_L = 3 \times 28.28 \sin(377\,t - 6.87° + 90°) = 84.84 \sin(377\,t + 83.13°)$$

The waves for i, e_R, e_L and e are shown in Fig. 4.9.

Example 4.14 **In a particular circuit a voltage of 10 V at 25 Hz produces 100 mA, while the same voltage at 75 Hz produces 60 mA. Draw the circuit diagram and insert values of the constants. At what frequency will the value of the impedance be twice that at 25 Hz ?** [*R.G.P.V. Bhopal Basic Electrical Engineering June 2003*]

Solution: Since the current at high frequency decreases with the increase in frequency, the circuit is inductive.

Let the circuit parameters be resistance of R ohms and inductance L Henry.

At 25 Hz supply frequency circuit impedance, $\mathbf{Z}_1 = R + j2\pi f L = R + j\,50\pi L$

$$\text{and current } \mathbf{I}_1 = \frac{\mathbf{V}}{\mathbf{Z}_1} = \frac{10}{\sqrt{R^2 + (50\,\pi L)^2}} = 0.1 \text{ A} \qquad ...(i)$$

At 75 Hz supply frequency circuit impedance, $\mathbf{Z}_2 = R + j150\pi L$

$$\text{and current } \mathbf{I}_2 = \frac{\mathbf{V}}{\mathbf{Z}_2} = \frac{10}{\sqrt{R^2 + (150\,\pi L)^2}} = 0.06 \text{ A} \qquad ...(ii)$$

Solving Eqs. (*i*) and (*ii*) we have R = 88.2 Ω **Ans.**

and L = 0.3 H **Ans.**

Let the impedance at a frequency of f' be double of that at 25 Hz

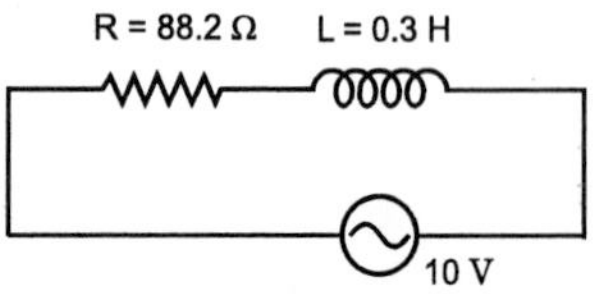

Fig. 4.10

$$\text{i.e. } \sqrt{R^2 + (2\pi f'L)^2} = 2\sqrt{R^2 + (2\pi \times 25 \times L)^2}$$

$$\text{or } R^2 + 4\pi^2 f'^2 L^2 = 4R^2 + 10{,}000\,\pi^2 L^2$$

$$\text{or } f' = \sqrt{\frac{3R^2 + 10{,}000\,\pi^2 L^2}{4\pi^2 L^2}}$$

$$= \sqrt{\frac{3 \times (88.2)^2 + 10{,}000\,\pi^2 \times (0.3)^2}{4\pi^2 (0.3)^2}} = 95 \text{ Hz } \textbf{Ans.}$$

Example 4.15 **A non-inductive resistance of 10 Ω is connected in series with an inductive coil across 200 V, 50 Hz ac supply. The current drawn by the series combination is 10 A. The resistance of the coil is 2 Ω. Determine (*i*) inductance of the coil (*ii*) power factor (*iii*) voltage across the coil.**

[U.P. Technical Univ. Electrical Engineering First Semester 2005-06]

Solution: Total resistance of the circuit, R = Non-inductive resistance + resistance of coil = 10 + 2 = 12 Ω

Voltage drop across the resistance of the whole circuit,

$$V_R = IR = 10 \times 12 = 120 \text{ V}$$

Let the voltage drop across the inductance of the coil be V_L volts.

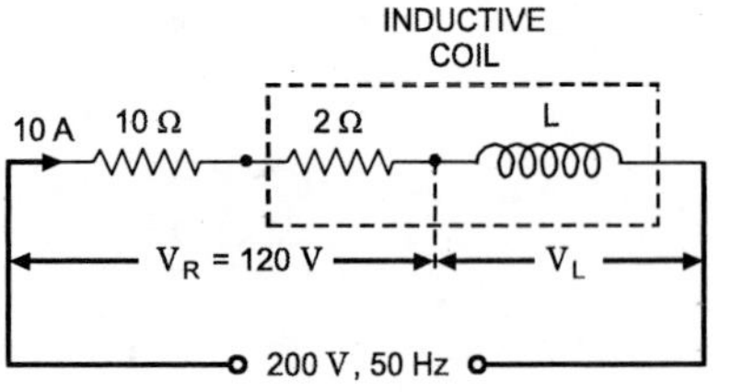

Fig. 4.11

$$\text{Now } \sqrt{V_R^2 + V_L^2} = \text{Supply voltage, } V_S$$

$$\text{or } V_L = \sqrt{V_S^2 - V_R^2}$$

$$= \sqrt{200^2 - 120^2} = 160 \text{ V}$$

$$\text{Inductive reactance of the coil, } X_L = \frac{V_L}{I} = \frac{160}{10} = 16\ \Omega$$

$$(i) \text{ Inductance of coil, } L = \frac{X_L}{2\pi f} = \frac{16}{2\pi \times 50} = 0.051 \text{ H or } 51 \text{ mH } \textbf{Ans.}$$

$$(ii) \text{ Power factor of the coil} = \text{Cos}\left(\text{Tan}^{-1}\frac{16}{2}\right) = 0.124 \text{ (lagging) } \textbf{Ans.}$$

$$\text{Power factor of the circuit} = \frac{R}{Z} = \frac{12}{\sqrt{12^2 + 16^2}} = 0.6 \text{ (lagging) } \textbf{Ans.}$$

$$(iii) \text{ Voltage across the coil, } V_C = I\sqrt{2^2 + 16^2} = 10 \times \sqrt{260} = 161.245 \text{ V } \textbf{Ans.}$$

Example 4.16 **A coil is connected in series with a non-inductive resistance of 30 Ω across 240 V, 50 Hz supply. The reading of a voltmeter across the coil is 180 V and across the resistance is 130 V. Calculate (*a*) power consumed by the coil, and (*b*) *pf* of the whole circuit.**

[Pb Technical Univ. Basic Electrical and Electronics Engineering Second Semester 2004-05]

Solution: $$\text{Circuit current, } I = \frac{\text{Voltage across non-inductive resistance}}{\text{Magnitude of non-inductive resistance}} = \frac{V_R}{R} = \frac{130}{30} = 4.333 \text{ A}$$

$$\text{Impedance of choke coil, } Z = \frac{\text{Voltage across choke coil}}{\text{Circuit current, I}} = \frac{180}{4.333} = 41.54\ \Omega$$

From Δ OAB shown in Fig. 4.13

$$OB^2 = OA^2 + AB^2 + 2\ OA \times AB \times \cos\theta$$

$$\text{or } \cos\theta = \frac{OB^2 - OA^2 - AB^2}{2 \times OA \times AB}$$

$$= \frac{240^2 - 130^2 - 180^2}{2 \times 130 \times 180} = 0.177735$$

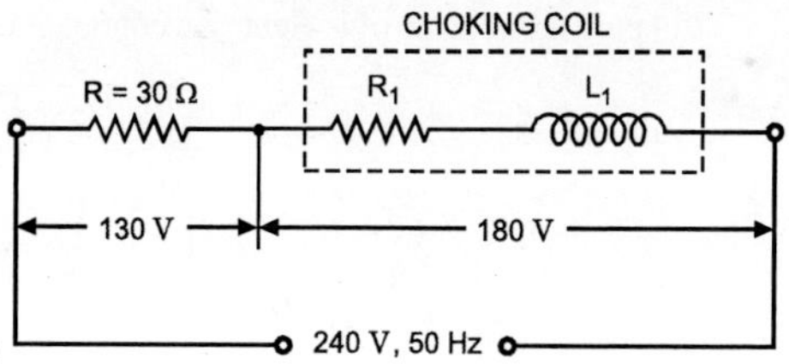

Fig. 4.12

Resistance of coil, $R_1 = Z \cos\theta = 41.54 \times 0.17735 = 7.367\ \Omega$

Reactance of coil $X_L = \sqrt{Z^2 - R_1^2} = \sqrt{(41.54)^2 - (7.367)^2}$

$= 40.88\ \Omega.$

(*a*) Power consumed by the coil

$= I^2R_1 = (4.333)^2 \times 7.367 = 138.34$ W **Ans.**

(b) Resistance of whole circuit

$= R + R_1 = 30 + 7.367 = 37.367\ \Omega$

Reactance of whole circuit = Reactance of coil = 40.88 Ω

Impedance of whole circuit = $\sqrt{(37.367)^2 + (40.88)^2} = 55.38\ \Omega$

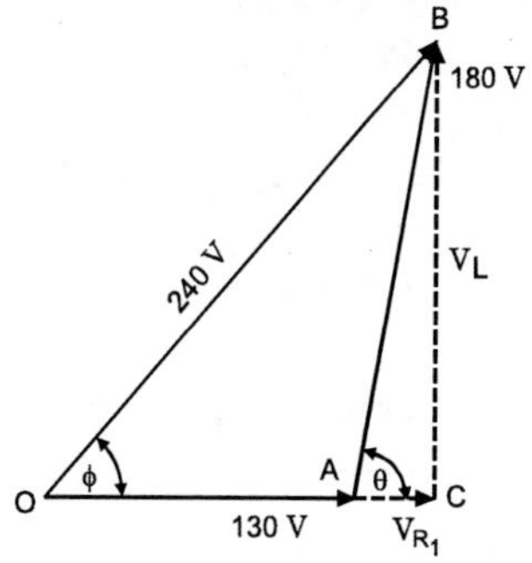

Fig. 4.13

Power factor of the whole circuit, $\cos\phi = \frac{37.367}{55.38} = 0.6747$ (lagging) **Ans.**

Example 4.17 A 500 W discharge lamp takes a current of 4 A at unity power factor. Calculate the inductance of the choke coil required to enable the lamp to work on a 250 V, 50 Hz ac mains.

[G.G.S.I.P. Delhi Univ. Electrical Science May-June 2006]

Solution: Voltage rating of lamp, $V = \frac{P}{I} = \frac{500}{4} = 125$ V

Let the inductance of choke coil, required to enable the lamp to work on 250 V, 50 Hz mains, be L Henry.

Voltage drop across choke coil (neglecting its resistance),

$$V_L = I \times X_L = I \times 2\pi f L = 4 \times 2\pi \times 50 \times L = 400\ \pi L \text{ volts}$$

Since voltage drop across the inductance leads the current by 90° or $\frac{\pi}{2}$ radians while that across the lamp will be in phase with it, so

$$\sqrt{(125)^2 + (400\ \pi L)^2} = 250$$

$$\text{or } 400\ \pi\ L = \sqrt{(250)^2 - (125)^2} = 216.506 \text{ V } \textbf{Ans.}$$

$$L = \frac{216.506}{400\pi} = 0.1723 \text{ H } \textbf{Ans.}$$

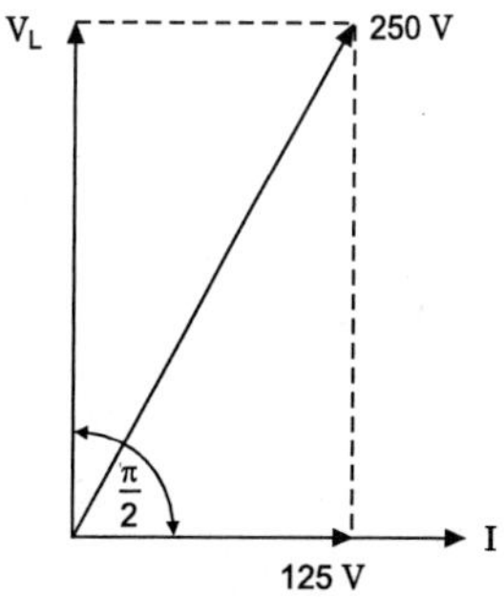

Fig. 4.14

Example 4.18 A 120 V, 100 W lamp is to be connected to a 220 V, 50 Hz ac supply. What value of pure inductance should be connected in series in order that lamp is run on the rated voltage?

[U.P. Technical Univ. Electrical Engineering First Semester 2003-04, Odd Semester 2013-14]

Solution : Assume that the lamp is pure resistance

Current drawn by the lamp when operated on rated voltage,

$$I = \frac{100}{120} = \frac{5}{6} \text{ A}$$

Let pure inductance of L Henry be connected in series with lamp, the voltage across inductance,

$$V_L = \sqrt{220^2 - 120^2} = 184.39\ \Omega$$

$$\text{and } L = \frac{V_L}{2\pi f I} = \frac{184.39}{2\pi \times 50 \times \frac{5}{6}} = 0.704 \text{ H} \quad \textbf{Ans.}$$

Example 4.19 A 120 V, 60 W lamp is to be operated on 220 V, 50 Hz supply mains. In order that lamp should operate on correct voltage, calculate value of—

(*a*) non inductive resistance, (*b*) pure inductance.

[U.P. Technical Univ. Elec. Engineering Second Semester 2005-06]

Solution: Compared with the resistance of the filament of a lamp its inductance is negligible. As such a lamp is considered to be a non-inductive resistance.

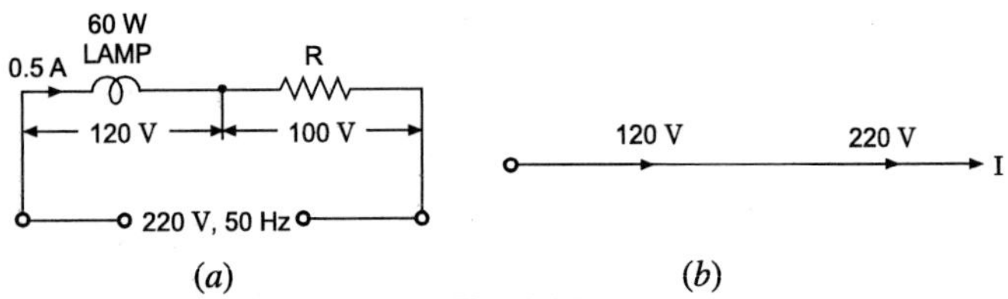

Fig. 4.15

Rated power of lamp, P = 60 W

Rated voltage of lamp, V = 120 V

Current drawn by the lamp when operated on rated voltage

$$I = \frac{P}{V} = \frac{60}{120} = 0.5 \text{ A}$$

(*a*) Let a non-inductive resistance of R ohms be connected in series with the lamp, as shown in Fig. 4.15 (*a*)

Since in this case voltage across the lamp and that across the resistance V_R, both are in phase, their arithmetic sum is equal to applied voltage *i.e.*

$$V_R + 120 = 220$$

$$\text{or } V_R = 220 - 120 = 100 \text{ V}$$

$$\text{Resistance, } R = \frac{V_R}{I} = \frac{100}{0.5} = 200\ \Omega \quad \textbf{Ans.}$$

(*b*) Let pure inductance of L henry be connected in series with the lamp, as shown in Fig. 4.16 (*a*)

Since in this case voltage drop across the inductance leads the current by 90° or $\pi/2$ radians while that across the lamp will be in phase with it, therefore, from phasor diagram shown in Fig. 4.16 (*b*) we have

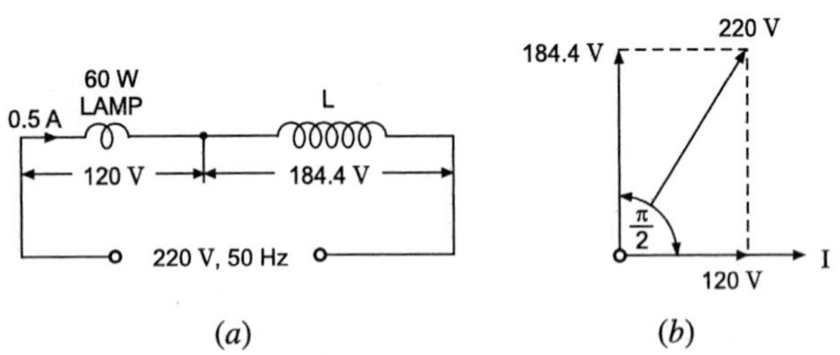

Fig. 4.16

$$120^2 + V_L^2 = 220^2$$

$$\text{or } V_L = \sqrt{220^2 - 120^2} = 184.4 \text{ volts}$$

$$\text{or } 2\pi f L I = 184.4 \text{ V}$$

$$\text{or } L = \frac{184.4}{2\pi \times 50 \times 0.5} = 1.174 \text{ H} \quad \textbf{Ans.}$$

Method (*b*) is preferable to method (*a*) because in method (*b*) there is no power loss whereas in method (a) power loss comes out to be $0.5^2 \times 200$ *i.e.* 50 watts.

Example 4.20 **Three coils are connected in series across a 100 V, 50 Hz supply. Values of their resistances are as follows:**

$$R_1 = 15\ \Omega, \quad L_1 = 0.01\ H$$
$$R_2 = 10\ \Omega, \quad L_2 = 0.03\ H$$
$$R_3 = 3\ \Omega, \quad L_3 = 0.06\ H$$

Calculate the potential drop and phase angle for each coil.

Solution: Total resistance in the circuit, $R = R_1 + R_2 + R_3 = 15 + 10 + 3 = 28\ \Omega$

Total inductance in the circuit, $L = L_1 + L_2 + L_3 = 0.01 + 0.03 + 0.06 = 0.1\ H$

$$\text{Impedance of first coil, } Z_1 = \sqrt{R_1^2 + (2\pi f L_1)^2} = \sqrt{(15)^2 + (2\pi \times 50 \times 0.01)^2} = 15.325\ \Omega$$

$$\text{Impedance of second coil, } Z_2 = \sqrt{(10)^2 + (2\pi \times 50 \times 0.03)^2} = 13.742\ \Omega$$

$$\text{Impedance of third coil, } Z_3 = \sqrt{(3)^2 + (2\pi \times 50 \times 0.06)^2} = 19.087\ \Omega$$

$$\text{Impedance of whole circuit, } Z = \sqrt{R^2 + (2\pi f L)^2} = \sqrt{(28)^2 + (2\pi \times 50 \times 0.1)^2} = 42.083\ \Omega$$

$$\text{Current through the circuit, } I = \frac{V}{Z} = \frac{100}{42.083} = 2.376\ A$$

Potential drop across first coil, $V_1 = I Z_1 = 2.376 \times 15.325 = 36.417$ V **Ans.**

Potential drop across second coil,

$V_2 = I Z_2 = 2.376 \times 13.742 = 32.65$ V **Ans.**

Potential drop across third coil,

$V_3 = I Z_3 = 2.376 \times 19.087 = 45.35$ V **Ans.**

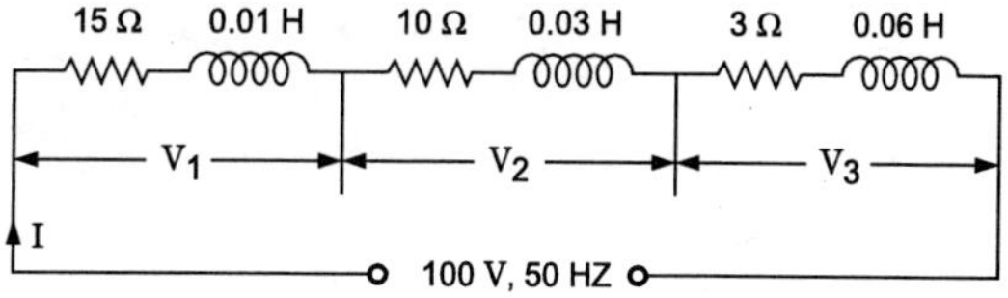

Fig. 4.17 *Circuit Diagram*

Phase angle of first coil,

$$\phi_1 = \cos^{-1}\frac{R_1}{Z_1} = \cos^{-1}\frac{15}{15.325} = 11°51' \text{ **Ans.**}$$

Phase angle of second coil,

$$\phi_2 = \cos^{-1}\frac{R_2}{Z_2} = \cos^{-1}\frac{10}{13.742} = 43°18' \text{ **Ans.**}$$

Phase angle of third coil, $\phi_3 = \cos^{-1}\frac{R_3}{Z_3} = \cos^{-1}\frac{3}{19.087} = 80°57'$ **Ans.**

4.8. POWER IN IRON-CORED CHOKING COIL

The power drawn by a choking coil is used up to supply the iron losses in the core in addition to that used up for heating the coil.

Power consumed by a choking coil,

P = Iron losses (hysteresis and eddy current loss), P_i + power loss in ohmic resistance *i.e.* $I^2 R$ (R being the dc resistance or true resistance of the coil)

$$\text{or } P = P_i + I^2 R$$

$$\text{or } \frac{P}{I^2} = \frac{P_i}{I^2} + R$$

$\frac{P}{I^2}$ is called the *effective resistance* of the choking coil and it is equal to the true resistance plus resistance representing iron loss $\frac{P_i}{I^2}$

$$\therefore R_{eff} = \frac{P}{I^2} = R + \frac{P_i}{I^2}$$

Example 4.21 **A coil takes 4 A when connected to 24 V dc supply. If this coil is connected to 40 V, 50 Hz ac supply then same amount of power is consumed. Explain why are different voltages required for same power consumption. Calculate inductance of the coil and the phase angle between voltage and current ?**

Solution: When dc voltage is applied to the coil, there is no induced emf due to self-inductance of the coil because the current is steady and frequency is zero. Hence the applied voltage is to overcome only the resistive drop I × R, where R is the resistance of the coil. But when ac voltage is applied, the coil offers resistance as well as reactance, the combination of the two *i.e.* impedance of the coil is certainly greater in magnitude as compared to resistance. Hence to cause the same current to flow through the coil greater voltage is required to overcome the impedance of the coil.

When dc voltage is applied

$$\text{Applied voltage, } V_{dc} = 24 \text{ V}$$

$$\text{Current through coil, } I_{dc} = 4 \text{ A}$$

$$\text{Resistance of coil, } R = \frac{V_{dc}}{I_{dc}} = \frac{24}{4} = 6\ \Omega$$

When ac voltage is applied

$$\text{Applied voltage, } V_{ac} = 40 \text{ V}$$

$$\text{Resistance of coil, } R = \text{The same as when dc is applied } i.e.\ 6\ \Omega$$

$$\text{Current, } I_{ac} = 4 \text{ A}$$

$$\text{Coil impedance, } Z = \frac{V_{ac}}{I_{ac}} = \frac{40}{4} = 10\ \Omega$$

$$\text{Inductive reactance of coil, } X_L = \sqrt{Z^2 - R^2} = \sqrt{(10)^2 - (6)^2} = 8\ \Omega$$

$$\text{Inductance of the coil, } L = \frac{X_L}{2\pi f} = \frac{8}{2\pi \times 50} = 0.0255 \text{ H } \textbf{Ans.}$$

$$\text{The angle between the applied pd and current, } \phi = \text{Tan}^{-1}\frac{X_L}{R} = \text{Tan}^{-1}\frac{8}{6} = 53.13° \text{ (lag) } \textbf{Ans.}$$

Example 4.22 **An iron choke coil takes 4 A when connected to a 20 V dc supply and takes 5 A from 65 V, 50 Hz ac supply.**

Determine—

(*i*) the resistance and inductance of the coil,
(*ii*) the power drawn by the coil,
(*iii*) the power factor. [*U.P. Technical Univ. Elec. Engineering, February 2001*]

Solution:

$$(i) \text{ Resistance of coil, } R = \frac{V_{dc}}{I_{dc}} = \frac{20}{4} = 5\ \Omega \textbf{ Ans.}$$

$$\text{Impedance of coil, } Z = \frac{V_{ac}}{I_{rms}} = \frac{65}{5} = 13\ \Omega \textbf{ Ans.}$$

$$\text{Inductive reactance of coil, } X_L = \sqrt{Z^2 - R^2} = \sqrt{13^2 - 5^2} = 12\ \Omega$$

$$\text{Inductance of coil, } L = \frac{X_L}{2\pi f} = \frac{12}{2\pi \times 50} = 38.2 \times 10^{-3} \text{ H or } 38.2 \text{ mH } \textbf{Ans.}$$

$$(ii) \text{ Power drawn by the coil, } P = (I_{rms})^2 \times R = 5^2 \times 5 = 125 \text{ watts } \textbf{Ans.}$$

$$(iii) \text{ Power factor, } \cos\phi = \frac{R}{Z} = \frac{5}{13} = 0.3846 \text{ (lagging) } \textbf{Ans.}$$

Example 4.23 **An iron cored coil has a dc resistance of 8 ohms. When it is connected to 230 volt, 50 Hz mains, the current taken is 3.4 A at a power factor of 0.5. Determine (*a*) the effective resistance of the coil (*b*) inductance of the coil and (*c*) the resistance which represents the effect of the iron loss.**

Solution: Total power consumed by an iron-cored choke coil,

$$P = \text{Power loss in ohmic resistance + iron loss in core.} = I^2 R + P_i$$

$$\text{or} \quad \frac{P}{I^2} = R + \frac{P_i}{I^2}$$

where $\frac{P}{I^2}$ is known as *effective resistance* of the coil

(*a*) Now since power consumed, $P = VI\cos\phi = 230 \times 3.4 \times 0.5 = 391$ watts

Current, $I = 3.4$ A

$$\therefore \text{Effective resistance, } R_e = \frac{P}{I^2} = \frac{391}{(3.4)^2} = 33.8\ \Omega \textbf{ Ans.}$$

(*b*) Impedance of choke coil, $Z = \frac{V}{I} = \frac{230}{3.4} = 67.65\ \Omega$

$$\text{Inductive reactance of coil, } X_L = \sqrt{Z^2 - R_e^2} = \sqrt{(67.65)^2 - (33.8)^2} = 58.6\ \Omega$$

$$\text{Inductance of choke coil, } L = \frac{X_L}{2\pi f} = \frac{58.6}{2\pi \times 50} = 0.1865 \text{ H } \textbf{Ans.}$$

$$(c) \text{ Since } \frac{P}{I^2} = R + \frac{P_i}{I^2}$$

or Effective resistance, R_e = True resistance R + resistance representing iron loss

$\therefore$ Resistance representing iron loss $= R_e - R = 33.8 - 8 = 25.8\ \Omega$ **Ans.**

4.9. RESISTANCE — CAPACITANCE (R-C) SERIES CIRCUIT

Consider an ac circuit consisting of resistance of R ohms and capacitance of C farads connected in series, as shown in Fig. 4.18 (*a*).

Let the supply frequency be of *f* Hz and current flowing through the circuit be of I amperes (rms value).

Voltage drop across resistance,

$V_R = IR$ in phase with the current.

Voltage drop across capacitance,

$V_C = IX_C$ lagging behind I by $\frac{\pi}{2}$ radians or 90°, as shown in Fig. 4.18 (*b*).

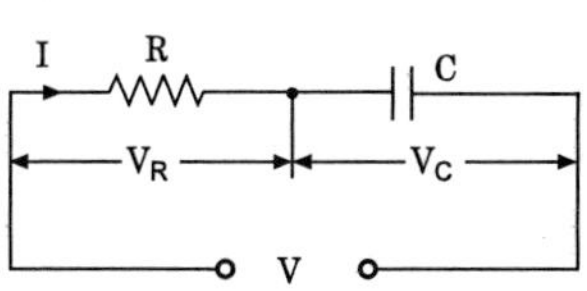

(*a*) *Circuit Diagram*

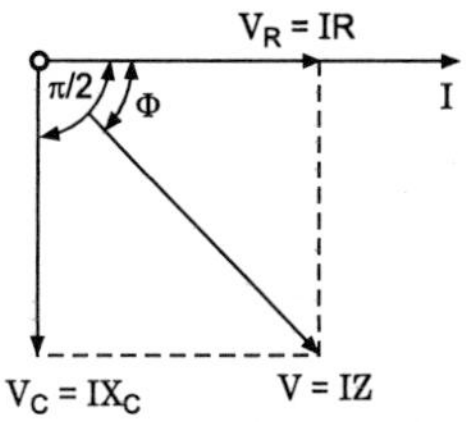

(*b*) *Phasor Diagram*

Fig. 4.18

The applied voltage, being equal to phasor sum of V_R and V_C, is given in magnitude by

$$V = \sqrt{(V_R)^2 + (V_C)^2} = \sqrt{(IR)^2 + (IX_C)^2} = I\sqrt{R^2 + X_C^2} = IZ \text{ where } Z^2 = R^2 + X_C^2$$

The applied voltage lags behind the current by an angle ϕ

where $$\tan\phi = \frac{V_C}{V_R} = \frac{IX_C}{IR} = \frac{X_C}{R} = \frac{1}{\omega RC}$$

$$\text{or } \phi = \text{Tan}^{-1} \frac{1}{R\omega C}$$

$$\text{Power factor, } \cos\phi = \frac{R}{Z}$$

If instantaneous voltage is represented by

$$v = V_{max} \sin \omega t$$

then instantaneous current will be expressed as

$$i = I_{max} \sin(\omega t + \phi)$$

and power consumed by the circuit is given by

$$P = VI \cos\phi$$

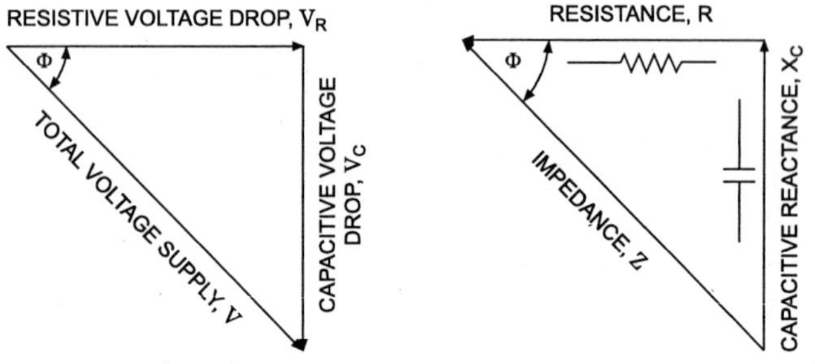

(a) Voltage Triangle *(b) Impedance Triangle*

Fig. 4.19

Voltage triangle and impedance triangle are shown in Figs. 4.19 (*a*) and 4.19 (*b*) respectively.

Example 4.24 **A circuit consists of 20 Ω resistance in series with capacitance of 200 μF, connected across 50 Hz supply. The current through the circuit is 10.8 sin 314*t* amperes. Determine the voltage across each component and across the circuit.** *[Pb. Technical Univ. Basic Electrical Engineering December 2002]*

Solution: Maximum value of circuit current, I_{max} = Coefficient of the sine of the time angle = 10.8 A

$$\text{RMS value of circuit current, } I_{rms} = \frac{I_{max}}{\sqrt{2}} = \frac{10.8}{\sqrt{2}} = 7.637 \text{ A}$$

$$\text{Voltage across 20 } \Omega \text{ resistance} = I_{rms} \times R = 7.637 \times 20 = 152.74 \text{ V } \textbf{Ans.}$$

$$\text{Voltage across 200 } \mu\text{ F capacitor} = I_{rms} X_C = \frac{7.637}{2\pi \times 50 \times 200 \times 10^{-6}} = 121.55 \text{ V } \textbf{Ans.}$$

$$\text{Voltage across the circuit, } V = \sqrt{(152.74)^2 + (121.55)^2} = 195.2 \text{ V } \textbf{Ans.}$$

Example 4.25 **The voltage applied to a circuit is v = 100 sin (ωt + 30°) and current flowing in the circuit is i = 20 sin (ωt + 60°). Determine the impedance, resistance, reactance, power and power factor of the circuit.**

[G.B. Technical Univ. Electrical Engineering First Semester 2011-12]

Solution: Voltage applied to a circuit, $v = 100 \sin(\omega t + 30°)$

Current flowing in the circuit, $i = 20 \sin(\omega t + 60°)$

Current leads the applied voltage by (60° – 30°) i.e. 30°

$$\text{Circuit impedance, } Z = \frac{V_{max}}{I_{max}} = \frac{100}{20} = 5 \ \Omega \ \textbf{Ans.}$$

$$\text{Circuit resistance, } R = Z\cos\phi = 5 \cos 30° = 4.33 \ \Omega \ \textbf{Ans.}$$

$$\text{Circuit reactance, } X = Z\sin\phi = 5 \sin 30° = 2.5 \ \Omega \text{ (capacitive) } \textbf{Ans.}$$

∵ phase angle φ is leading one

$$\text{Power factor of the circuit} = \text{Cos}\,\phi = \text{Cos } 30° = 0.866 \text{ (leading) } \textbf{Ans.}$$

$$\text{Power of the circuit, } P = V_{rms} \times I_{rms} \cos\phi = \frac{V_{max}}{\sqrt{2}} \times \frac{I_{max}}{\sqrt{2}} \times \cos 30°$$

$$= \frac{100}{\sqrt{2}} \times \frac{20}{\sqrt{2}} \times 0.866 = 866 \text{ watts } \textbf{Ans.}$$

Example 4.26 **A supply of 400 V, 50 Hz is applied to a series R-C circuit. Find the value of C if the power absorbed by the resistor be 500 W at 150 V. What is the energy stored in capacitor?**

[R.G.P.V. Bhopal Basic Electrical Engineering June 2005]

Solution: Supply voltage, V = 400 V

Power absorbed, P = 500 W

$$\text{Voltage across resistor, } V_R = 150 \text{ V}$$

$$\text{Current through R-C series circuit, } I = \frac{P}{V_R} = \frac{500}{150} = 3.333 \text{ A}$$

$$\text{Voltage drop across capacitor, } V_C = \sqrt{400^2 - 150^2} = 370.8\text{V}$$

$$\text{Capacitive reactance of the circuit, } X_C = \frac{V_C}{I} = \frac{370.8}{3.333} = 111.25\ \Omega$$

$$\text{Capacitance of circuit, } C = \frac{1}{2\pi f X_C} = \frac{1}{2\pi \times 50 \times 111.25} = 28.6\ \mu\text{F} \quad \textbf{Ans.} \qquad \because\ X_C = \frac{1}{2\pi f C}$$

$$\text{Energy stored in capacitor} = \frac{1}{2} C \times V_{C\max}^2 = \frac{1}{2} \times 28.6 \times 10^{-6} \times (370.8\sqrt{2})^2 = 3.934 \text{ J} \quad \textbf{Ans.}$$

Example 4.27 **A voltage wave $e(t) = 141.4 \sin 120t$ produces a current $i(t) = 14.14 \sin 120t + 7.07 \cos(120t + 30°)$ in a circuit. Determine (*i*) the resultant time expression of current (*ii*) the power factor and the power delivered by the source.** *[U.P. Technical Univ. Electrical Engineering February 2002]*

Solution: Current $i(t) = 14.14 \sin 120t + 7.07 \cos(120t + 30°)$

$$= 14.14 \sin 120t + 7.07 \sin(120t + 30° + 90°)$$

$$\text{Maximum value of current, } I_{r\max} = \sqrt{(\text{X-component})^2 + (\text{Y-component})^2}$$

$$= \sqrt{(14.14 \cos 0° + 7.07 \cos 120°)^2 + (14.14 \sin 0° + 7.07 \sin 120°)^2}$$

$$= \sqrt{(14.14 - 3.535)^2 + (6.1228)^2} = 12.25 \text{ A}$$

Phase angle of resultant current is given as

$$\theta = \text{Tan}^{-1} \frac{\text{Y-component}}{\text{X-component}} = \text{Tan}^{-1} \frac{6.1228}{10.605} = 30°$$

(*i*) So the resultant time expression of current is

$$i(t) = 12.25 \sin(120t + 30°) \ \textbf{Ans.}$$

Phase angle between V and I is $\phi = 30°$ (leading)

Power factor = Cos 30° = 0.866 (leading) **Ans.**

$$\text{Power delivered by source, } P = V_{rms} I_{rms} \cos\phi = \frac{V_{max}}{\sqrt{2}} \cdot \frac{I_{max}}{\sqrt{2}} \cos\phi = \frac{V_{max} I_{max}}{2} \cos\phi$$

$$= \frac{141.4 \times 12.25}{2} \cos 30° = 750 \text{ W} \ \textbf{Ans.}$$

Example 4.28 **Voltage and current for a circuit with two elements in series are expressed as follows:**

$$v(t) = 170 \sin\left(6{,}280\,t + \frac{\pi}{3}\right)$$

$$i(t) = 8.5 \sin\left(6{,}280\,t + \frac{\pi}{2}\right)$$

(*i*) Plot the two waveforms (*ii*) Determine the frequency in Hz. (*iii*) Determine the power factor stating its nature (*iv*) What are the values of the elements ? *[Nagpur UNiv., April 1996]*

Solution: Peak value of applied voltage, $V_{max} = 170$ V

Peak value of circuit current, $I_{max} = 8.5$ A

(*ii*) Frequency,

$$f = \frac{\text{Coefficient of time in expression of voltage or current}}{2\pi}$$

$$= \frac{6,280}{2 \times 3.14} = 1,000 \text{ Hz } \textbf{Ans.}$$

(*iii*) Phase angle between voltage and current,

$$\phi = \left(\frac{\pi}{2} - \frac{\pi}{3}\right) = \frac{\pi}{6} \text{ or } 30°.$$

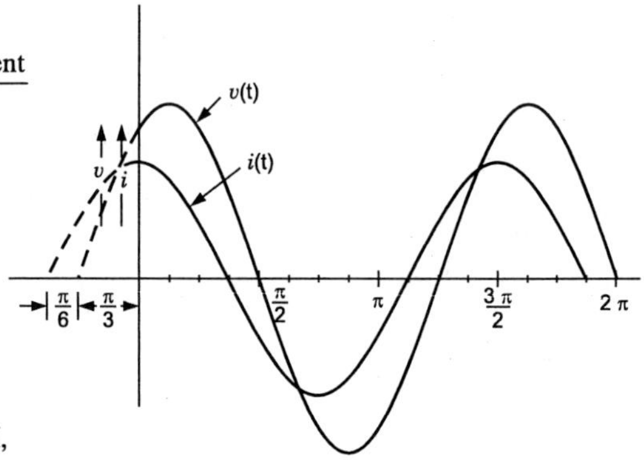

Fig. 4.20

Since current leads the voltage, phase angle between V and I,

$$\phi = 30° \text{ (leading)}$$

Power factor $= \cos\phi = \cos 30° = 0.866$ (leading) **Ans.**

(*i*) Voltage and current waveforms are shown in Fig. 4.20.

$$\text{Impedance, Z} = \frac{V_{rms}}{I_{rms}} = \frac{V_{max}/\sqrt{2}}{I_{max}/\sqrt{2}} = \frac{V_{max}}{I_{max}} = \frac{170}{8.5} = 20\ \Omega$$

$$\text{Resistance, R} = Z\cos\phi = 20 \times 0.866 = 17.32\ \Omega \quad \textbf{Ans.}$$

$$\text{Reactance, X} = Z\sin\phi = 20 \sin 30° = 10\ \Omega$$

Since the current leads the voltage, the reactive element is capacitance

$$\text{and } C = \frac{1}{2\pi f X_C} = \frac{1}{2\pi \times 1,000 \times 10} = 15.92\ \mu F \quad \textbf{Ans.}$$

Example 4.29 A metal-filament lamp, rated at 750 W, 100 V is to be connected in series with a capacitance across a 230 V, 50 Hz supply. Calculate the value of capacitance required. Draw phasor diagram.

[U.P. Technical Univ. Electrical Engineering July 2002; G.B. Technical Univ. First Semester 2011-12; U.P. Technical Univ. Basic Electrical Engineering First Semester 2013-14]

Solution : Let a pure capacitance of C farads be connected in series with the lamp, as shown in Fig. 4.21 (*a*).

Since in this case voltage drop across the capacitor lags behind the current by 90° or $\frac{\pi}{2}$ radians while that across the lamp will be in phase with it, from phasor diagram shown in Fig. 4.21 (*b*), we have

$$V_C = \sqrt{(230)^2 - (100)^2} = 207 \text{ V}$$

$$I = \frac{P}{V} = \frac{750}{100} = 7.5\text{A}$$

$$X_C = \frac{V_C}{I} = \frac{207}{7.5} = 27.6\ \Omega$$

$$\text{and } C = \frac{1}{2\pi f X_C} = \frac{1}{2\pi \times 50 \times 27.6} = 115.33\ \mu F \quad \textbf{Ans.}$$

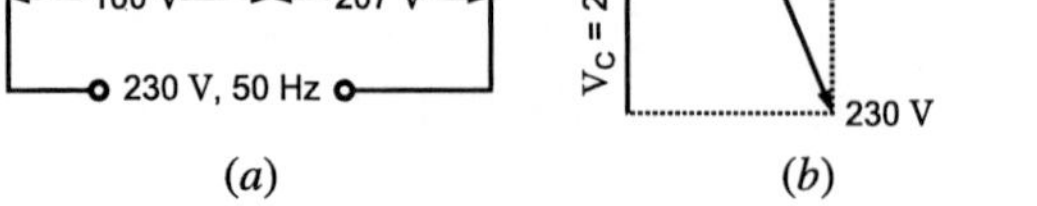

Fig. 4.21

4.10. RESISTANCE-INDUCTANCE-CAPACITANCE (R-L-C) SERIES CIRCUIT

Consider an ac circuit containing resistance of R ohms, inductance of L henries and capacitance of C farads connected in series, as shown in Fig. 4.22 (*a*).

Let the current flowing through the circuit be of I amperes and supply frequency be *f* Hz.

Voltage drop across resistance, V_R = IR in phase with I

Voltage drop across inductance, V_L = $I\omega L$ leading I by $\pi/2$ radians or 90°

Voltage drop across capacitance, $V_C = \frac{I}{\omega C}$ or IX_C lagging behind I by $\pi/2$ radians or 90°

V_L and V_C are 180° out of phase with each other (or reverse in phase), therefore, when combined by parallelogram they cancel each other. The circuit can either be effectively inductive or capacitive depending upon which voltage drop (V_L or V_C) is predominant. Let us consider the case when V_L is greater than V_C.

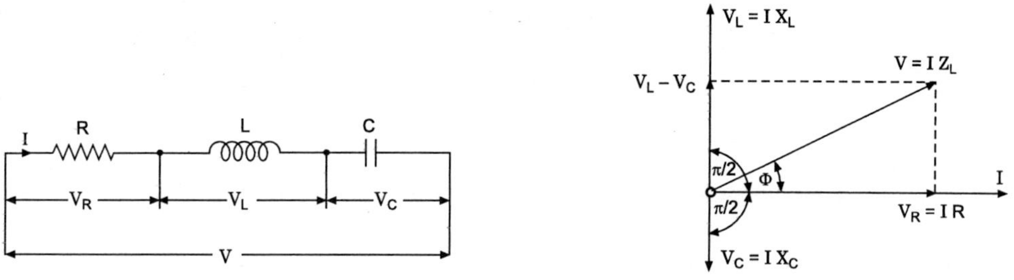

(a) Circuit Diagram *(b) Phasor Diagram*

Fig. 4.22

The applied voltage *V*, being equal to the phasor sum of V_R, V_L and V_C is given in magnitude by

$$V = \sqrt{(V_R)^2 + (V_L - V_C)^2} = \sqrt{(IR)^2 + (IX_L - IX_C)^2} = I\sqrt{(R)^2 + (X_L - X_C)^2}$$

The term $\sqrt{R^2 + (X_L - X_C)^2}$ is known as *impedance* of the circuit and is represented by Z. Its unit is ohm.

Phase angle ϕ between voltage and current is given by

$$\phi = \text{Tan}^{-1}\frac{V_L - V_C}{V_R} = \text{Tan}^{-1}\frac{IX_L - IX_C}{IR} = \text{Tan}^{-1}\frac{X_L - X_C}{R} = \text{Tan}^{-1}\frac{X}{R}$$

ϕ will be + ve *i.e.* applied voltage will lead the current if $X_L > X_C$ and ϕ will be – ve *i.e.* applied voltage will be behind the current if $X_L < X_C$.

Power factor of the circuit is given by

$$\text{Cos}\,\phi = \frac{R}{Z} = \frac{R}{\sqrt{R^2 + (X_L - X_C)^2}}$$

Power consumed in the circuit, $P = I^2R$ or $VI\cos\phi$.

4.10.1. Reactance. Inductive reactance, X_L is directly proportional to frequency being equal to ωL or $2\pi f L$ and capacitive reactance, X_C is inversely proportional to frequency being equal to $\frac{1}{\omega C}$ or $\frac{1}{2\pi f C}$.

Inductive reactance causes the current to lag behind the applied voltage, while the capacitive reactance causes the current to lead the voltage. So when inductance and capacitance are connected in series, their effects neutralize each other and their combined effect is then their difference. The combined effect of inductive reactance and capacitive reactance is called the reactance and is found by subtracting the capacitive reactance from the inductive reactance or according to equation

$$X = X_L - X_C$$

When $X_L > X_C$ *i.e.* $X_L - X_C$ is positive, the circuit is inductive and phase angle ϕ is positive.

When $X_L < X_C$ *i.e.* $X_L - X_C$ is negative, the circuit is capacitive and phase angle ϕ is negative.

When $X_L = X_C$ *i.e.* $X_L - X_C = 0$, the circuit is purely resistive and phase angle ϕ is zero.

If the expression for applied voltage is taken as

$$v = V_{max} \sin \omega t$$

then expression for the current will be

$$i = I_{max} \sin (\omega t \pm \phi) \qquad \text{where } I_{max} = \frac{V_{max}}{\text{Circuit impedance, Z}}$$

The value of ϕ will be negative when current leads the applied voltage V *i.e.* when $X_C > X_L$ and positive when current lags behind the applied voltage V *i.e.* when $X_L > X_C$.

Example 4.30 **A coil having a resistance of 6 ohm and an inductance of 0.0255 H is connected across a 230 V, 50 Hz ac supply. Calculate**

(*i*) current **(*ii*) power factor**
(*iii*) active power **(*iv*) reactive power**
(*v*) apparent power
(*vi*) It is desired to improve power factor to 0.8. What value of capacitance is to be connected in series with R and what is reduction in reactive power?

[*G.B. Technical Univ. Electrical Engineering First Semester 2011-12*]

Solution: Resistance of coil, R = 6 Ω

Inductance of coil, L = 0.0255 H

Supply voltage, V = 230

Supply frequency, f = 50 Hz

Inductive reactance of coil, $X_L = 2\pi f L = 2\pi \times 50 \times 0.0255 = 8\ \Omega$

$$\text{Impedance of coil, } Z = \sqrt{R^2 + X_L^2} = \sqrt{6^2 + 8^2} = 10\ \Omega$$

(*i*) $\text{Current, } I = \frac{V}{Z} = \frac{230}{10} = 23\ \text{A}$

(*ii*) $\text{Power factor, } \cos \phi = \frac{R}{Z} = \frac{6}{10} = 0.6 \text{ (lagging)}$ **Ans.**

(*iii*) Active power, $P = VI \cos \phi = 230 \times 23 \times 0.6 = 3{,}174$ W or 3.174 kW **Ans.**

(*iv*) Reactive power, $Q = VI \sin \phi = 230 \times 23 \times 0.8 = 4.232$ VAR or 4.232 kVAR **Ans.**

(*v*) Apparent power, $S = VI = 230 \times 230 = 5{,}290$ VA or 5.29 kVA **Ans.**

(*vi*) New power factor, $\cos \phi' = 0.8$ (lagging)

$$\text{New impedance, } Z' = \frac{R}{\text{Cos}\phi} = \frac{6}{0.8} = 7.5\ \Omega$$

$$\text{Reactance of circuit, } X_L - X_C = \sqrt{Z^2 - R^2} = \sqrt{7.5^2 - 6^2} = 4.5\ \Omega$$

$$X_C = X_L - 4.5 = (8 - 4.5)\ \Omega = 3.5\ \Omega$$

Capacitance to be connected in series with R to improve *pf* to 0.8 (lag)

$$= \frac{1}{2\pi f X_C} = \frac{1}{2\pi \times 50 \times 3.5} = 910\ \mu\text{F} \quad \textbf{Ans.}$$

$$\text{New current } I' = \frac{V}{Z'} = \frac{230}{7.5} = 30.667\ \text{A}$$

New reactive power, $Q' = VI' \sin \phi' = 230 \times 30.667 \times 0.6 = 4{,}232$ VAR $\qquad \because \text{Sin } \phi' = 0.6$

i.e. no reduction in reactive power **Ans.**

Example 4.31 **A series R-L-C circuit has 100 Ω resistor, 0.318 H inductor and C, $v = 230 \times \sqrt{2}\sin \omega t$ V, $i = 2.3\sqrt{2}\sin \omega t$ A.**

Find (*i*) C (*ii*) V_L (*iii*) power. Take ω = 314.15 *r*/second.

[U.P. Technical Univ. Electrical Engineering First Semester, 2007-08]

Solution:

$$\text{Resistance, R} = 100\ \Omega$$

$$\text{Inductive reactance, } X_L = \omega L = 314.5 \times 0.318 = 100\ \Omega \qquad \because \omega = 314.5$$

$$\text{Capacitive reactance} = X_C \text{ (say)}$$

$$\text{Now impedance, Z} = \sqrt{R^2 + (X_L - X_C)^2}$$

$$= \frac{V_{max}}{I_{max}} = \frac{230\sqrt{2}}{2.3\sqrt{2}} = 100\ \Omega$$

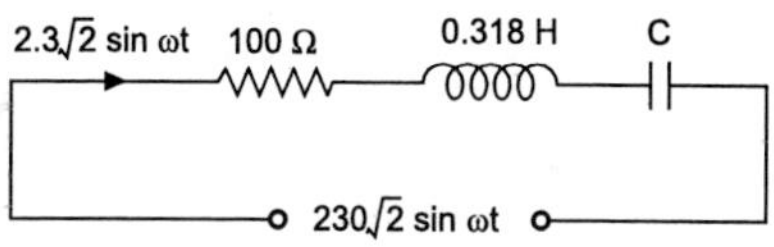

Fig. 4.23

$$\text{or } \sqrt{100^2 + (100 - X_C)^2} = 100$$

$$\text{or } X_C = 100\ \Omega$$

$$(i) \text{ Capacitance of the capacitor, C} = \frac{1}{\omega X_C} = \frac{1}{314.5 \times 100} = 31.8\ \mu\text{F } \textbf{Ans.}$$

$$(ii) \text{ Voltage across inductor, } V_L = \frac{I_{max}}{\sqrt{2}} \times X_L = \frac{2.3\sqrt{2}}{\sqrt{2}} \times 100 = 230 \text{ V } \textbf{Ans.}$$

$$(iii) \text{ Total power consumed, P} = \text{VI watts} = \frac{230\sqrt{2}}{\sqrt{2}} \times \frac{2.3\sqrt{2}}{\sqrt{2}} = 529 \text{ W } \textbf{Ans.}$$

Example 4.32 **A resistance R, an inductance L = 0.01 H and a capacitance C are connected in series. When a voltage $v = 400\cos(3{,}000\,t - 10^\circ)$ V is applied to the series combination, the current flowing is $10\sqrt{2}\cos(3{,}000\,t - 55^\circ)$ A. Find R and C.** *[Nagpur Univ. Electric Circuits 1992]*

Solution: $v = 400\cos(3{,}000\,t - 10^\circ) = 400\sin(3{,}000\,t - 10^\circ + 90^\circ) = 400\sin(3{,}000\,t + 80^\circ)$

$$i = 10\sqrt{2}\cos(3{,}000\,t - 55^\circ) = 10\sqrt{2}\sin(3{,}000\,t - 55^\circ + 90^\circ) = 10\sqrt{2}\sin(3{,}000\,t + 35^\circ)$$

$$\text{Phase angle, } \phi = 80^\circ - 35^\circ = 45^\circ$$ *i.e.* circuit is inductive and current lags behind voltage

$$\omega = 3{,}000$$

$$\text{Inductive reactance, } X_L = \omega L = 3{,}000 \times 0.01 = 30\ \Omega$$

$$\text{Circuit impedance, Z} = \frac{V_{max}}{I_{max}} = \frac{400}{10\sqrt{2}} = 20\sqrt{2} = 28.28\ \Omega$$

$$\text{Net reactance, X} = X_L - X_C = Z\sin\Phi = 28.28\sin 45^\circ = 28.28 \times \frac{1}{\sqrt{2}} = 20\ \Omega$$

$$\text{or Capacitive reactance } X_C = X_L - X = 30 - 20 = 10\ \Omega$$

$$\text{Circuit Capacitance, C} = \frac{1}{\omega X_C} = \frac{1}{3{,}000 \times 10} = 33.33\ \mu\text{F } \textbf{Ans.}$$

$$\text{Circuit resistance, R} = Z\cos\phi = 28.28\cos 45^\circ = 20\ \Omega \textbf{ Ans.}$$

Example 4.33 **A coil of 0.8 pf is connected in series with 110 μF capacitor. Supply frequency is 50 Hz. The potential difference across the coil is found to be equal to that across the capacitor. Calculate the resistance and inductance of the coil. Calculate the net power factor.** *[Nagpur University, November 1997]*

Solution: Let the resistance and inductance of the coil be of R ohms and L henrys. The circuit is given in Fig. 4.24.

Reactance of capacitor,

$$X_C = \frac{1}{\omega C} = \frac{1}{2\pi \times 50 \times 110 \times 10^{-6}} = 28.94\ \Omega$$

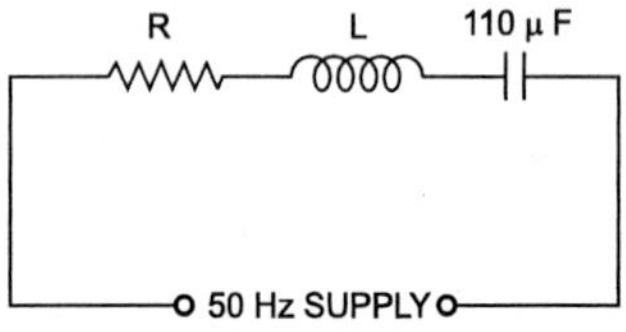

Fig. 4.24

Since voltage drop across the coil is equal to that across the capacitor which are connected in series, impedance of the coil will be equal to that of capacitor. Thus

$$\text{Coil impedance, } Z = X_C = 28.94\ \Omega$$

$$\text{Coil resistance, } R = Z \cos\phi = 28.94 \times 0.8 = 23.15\ \Omega \ \textbf{Ans.}$$

$$\text{Coil reactance, } X_L = \sqrt{Z^2 - R^2} = \sqrt{(28.94)^2 - (23.15)^2} = 17.36\ \Omega$$

$$\text{Coil inductance, } L = \frac{X_L}{2\pi f} = \frac{17.36}{2\pi \times 50}\ \text{H} = 0.0553\ \text{H or } 55.3\ \text{mH} \ \textbf{Ans.}$$

$$\text{Total impedance of circuit, } Z_T = \sqrt{R^2 + (X_L - X_C)^2} = \sqrt{(23.15)^2 + (17.36 - 28.94)^2} = 25.885\ \Omega$$

$$\text{New power factor} = \frac{R}{Z_T} = \frac{23.15}{25.885} = 0.8943\ \text{(leading)} \ \textbf{Ans.}$$

Example 4.34 Consider series R-L-C circuit as shown in Fig. 4.25(*a*). When $V_R = 8$ V, $V_L = 18$ V, $V_C = 12$ V. Find V in the circuit. [*U.P. Technical Univ. Electrical Engineering Odd Semester 2013-14*]

Solution: Phasor diagram is shown in Fig. 4.25 (*b*).

$$\text{Voltage across resistor R, } V_R = 8\ \text{V}$$

$$\text{Voltage across inductor L, } V_L = 18\ \text{V}$$

$$\text{Voltage across capacitor C, } V_C = 12\ \text{V}$$

From phasor diagram shown in Fig. 4.25 (*b*).

$$\text{Applied voltage, } V = \sqrt{V_R^2 + (V_L - V_C)^2} = \sqrt{8^2 + (18 - 12)^2} = 10\ \text{V} \ \textbf{Ans.}$$

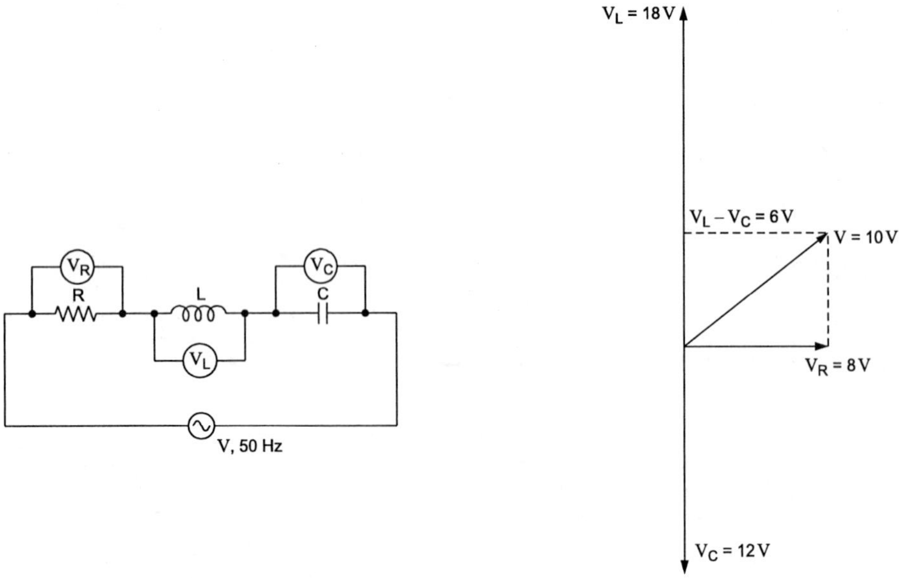

(*a*) Circuit Diagram (*b*) Phasor Diagram

Fig. 4.25

Example 4.35 Find applied voltage and power loss in a circuit shown below in Fig. 4.26. The value of C is 20 μF, current I = 0.345 A. [*U.P. Technical Univ. Electrical Engineering First Semester 2007-08*]

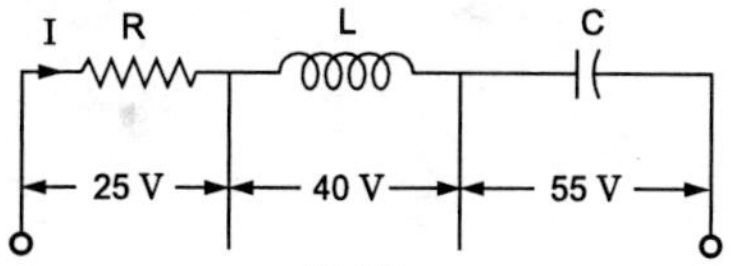

Fig. 4.26

Solution: From circuit diagram we have

$$R = \frac{\text{Voltage drop across R}}{\text{Current flowing through R}} = \frac{25 \text{ V}}{0.345} = 72.46\ \Omega$$

Power loss in the circuit = Power loss in resistor

∵ power loss consumed in L and C is zero.

$= I^2 R = (0.345)^2 \times 72.46 = 8.625$ W **Ans.**

$$\text{Applied voltage, V} = \sqrt{V_R^2 + (V_L - V_C)^2} = \sqrt{25^2 + (40 - 55)^2} = 29.155 \text{ V} \textbf{ Ans.}$$

HIGHLIGHTS

1. The path of flow of alternating current is called the ac circuit.
2. When an alternating voltage given by expression $v = V_{max} \sin \omega t$ is applied across a pure resistive circuit, the alternating current flowing through the circuit is given as
$$i = I_{max} \sin \omega t \quad \text{where } I_{max} = \frac{V_{max}}{R}$$
Power in pure resistive circuit, P $= V_{rms} \times I_{rms}$ or VI watts
Power factor of pure resistive circuit is unity.
3. When an alternating voltage given by expression $v = V_{max} \sin \omega t$ is applied across a pure inductive circuit (a coil with or without an iron core having negligible resistance), the alternating current flowing through the circuit is given as
$$i = I_{max}\left(\omega t - \frac{\pi}{2}\right)$$
where $I_{max} = \dfrac{V_{max}}{X_L}$; Inductive reactance X_L being equal to $2\pi f L$ or ωL ohms
and power absorbed by a pure inductor = 0
PF of a pure inductive circuit is zero.
4. In a pure capacitive circuit, if $v = V_{max} \sin \omega t$
$$i = I_{max} \sin\left(\omega t + \frac{\pi}{2}\right)$$
where $I_{max} = \dfrac{V_{max}}{X_C}$; the capacitive reactance X_C being equal to $\dfrac{1}{\omega C}$ or $\dfrac{1}{2\pi f C}$ ohms
Power absorbed, P = 0
PF is also zero.
5. An ac circuit can have three elements, namely R, L and C. Power is only consumed in R. The power consumed in elements L and C are zero.

   The current through pure inductor lags behind the applied voltage by $\dfrac{\pi}{2}$ radians while the current through a pure capacitor leads the applied voltage by $\dfrac{\pi}{2}$ radians.
6. In an R-L series ac circuit, if $v = V_{max} \sin \omega t$
$$i = I_{max} \sin(\omega t - \phi)$$
$$I_{max} = \frac{V_{max}}{Z};\ Z = \sqrt{R^2 + \omega^2 L^2};\ \phi = \text{Tan}^{-1}\frac{\omega L}{R};$$
$$P = VI\cos\phi = I^2 R$$

Power factor, $\cos \phi = \frac{R}{Z}$ (lagging)

7. Power factor of an circuit is given by

   (*i*) cosine of the phase angle between voltage and current *i.e.* $\cos \phi$

   or (*ii*) the ratio of resistance to impedance *i.e.* $\frac{R}{Z}$

   or (*iii*) the ratio of true power to apparent power *i.e.* $\frac{\text{True Power, P}}{\text{Apparent power, S}}$

   The *apparent power* of an ac circuit is given by the product of rms values of current and voltage *i.e.* S = V I volt-amperes.

   The *true power* in an ac circuit is obtained by multiplying the apparent power by the power factor *i.e.*

   $$P = S \cos \phi = V I \cos \phi \text{ watts.}$$

   The product of apparent power VI and sine of angle between voltage and current, $\sin \phi$, is called the reactive power, Q.

   $$\text{and } S = \sqrt{P^2 + Q^2}$$

8. Reciprocal of power factor of a coil is known as Q-factor of the coil. It is also called the quality factor or figure of merit of a coil.

   Mathematically, $\text{Q-factor} = \frac{1}{PF} = \frac{1}{\cos \phi} = \frac{Z}{R} = \frac{\omega L}{R}$ if R is very small in comparison with inductive reactance X_L.

9. In an R-C series ac circuit, if $v = V_{max} \sin \omega t$

   $$i = I_{max} \sin (\omega t + \phi)$$

   $$I_{max} = \frac{V_{max}}{Z}; \; Z = \sqrt{R^2 + \left(\frac{1}{\omega C}\right)^2} = \sqrt{R^2 + X_C^2}$$

   $$\phi = \text{Tan}^{-1} \frac{X_C}{R} = \text{Tan}^{-1} \frac{1}{\omega CR}$$

   $$P = VI \cos \phi = I^2 R$$

   Power factor, $\cos \phi = \frac{R}{Z}$ (leading)

10. In an R-L-C series ac circuit

   $$\text{Impedance, } Z = \sqrt{R^2 + (X_L - X_C)^2}$$

   Circuit behaves as an inductive circuit if $X_L > X_C$; as a capacitive if $X_L < X_C$ and as a purely resistive if $X_L = X_C$

   $$\text{Circuit current, } I = \frac{V}{Z}$$

   $$\text{Phase angle, } \phi = \text{Tan}^{-1} \frac{X_L - X_C}{R}$$

   $$\text{Power consumed, } P = V I \cos \phi = I^2 R$$

   $$\text{Power factor, } \cos \phi = \frac{R}{Z}$$

EXERCISES

1. Show that the instantaneous power consumed in a pure resistive circuit is not constant but it is fluctuating.

   [U.P. Technical Univ. Elec. Engineering February 2001]

2. In a pure inductive circuit, prove from first principles, that the current lags behind applied voltage by quarter of a cycle and also show that the average power demand is zero.

   [U.P. Technical Univ. Electrical Engineering First Semester 2004-05]

3. Prove that the average power consumption in a pure inductor is zero when ac voltage is applied. **[Gujarat Univ., June/July 2003]**
4. Explain that power consumed in a pure capacitor is zero. **[R.G.P.V. Bhopal Basic Electrical Engineering 2001]**
5. Deduce an expression for the average power in a single phase series R-L circuit and therefrom explain the term power factor. **[V.T.U. Belgaun Karnataka Univ. February 2002]**
6. Derive expressions for instantaneous power and active power in a circuit in which the voltage and current are represented by $v = V_m \sin \theta$ and $i = I_m \sin (\theta - \phi)$ respectively. **[U.P. Technical Univ. Electrical Engineering July 2002]**
7. Differentiate among real, reactive and apparent powers. **[U.P. Technical Univ. Electrical Engineering June 2001]**
8. Define : (*i*) Q factor of a coil (*ii*) Power factor (*iii*) Active, and reactive power and their significance. **[M.D. Univ. Rohtak Electrical Technology May 2002]**
9. A single phase sinusoidal ac voltage supply is fed to a series R-C circuit. Determine (*i*) the instantaneous expression of current flowing in the circuit (*ii*) the impedance (*iii*) the power factor (*iv*) the power consumed and (*v*) the reactive volt-amperes. **[U.P. Technical Univ. Electrical Engineering First Semester 2006-07]**
10. Draw wave diagram and phasor diagram to illustrate clearly the relation between *v* and *i* in the case of (*i*) R-L series circuit (*ii*) R-C series circuit (*iii*) R-L-C series circuit. **[U.P Technical Univ. Electrical Engineering June 2001]**
11. Explain the behaviour of series R-L-C circuit with sinusoidal input. **[M.D. Univ. Rohtak Electrical Technology May 2003]**
12. Derive expressions for impedance, current and power factor for an R-L-C series circuit when applied with ac voltage. Draw also the vector diagram. **[Gujarat University June/July 2003]**

SHORT ANSWER TYPE QUESTIONS WITH ANSWERS

Q. 1. What is the phase relationship between the supply voltage and current flowing through a pure resistive circuit ?

Ans. In a pure resistive circuit, both supply voltage and circuit current are in phase with each other.

Q. 2. What is the phase relationship between the supply voltage and current flowing through a pure inductive circuit ?

Ans. In a pure inductive circuit, current lags behind the supply voltage by 90° or $\frac{\pi}{2}$ radians.

Q. 3. Why are the inductors usually iron cored ?

Ans. Inductors or choke coils are made of iron core because large valued flux densities can be produced in iron cores and so inductances of large value can be had. Air-cored inductors become too much bulky to provide an inductance of a required value.

Q. 4. Does an inductance draw instantaneous power as well as average power ? **[R.G.P.V. December 2002]**

Ans. An inductance draws instantaneous power but no average power. The power drawn by inductance during one quarter cycle is released in another quarter cycle, and so average power drawn by an inductance is zero.

Q. 5. Draw the phasor diagrams for pure inductive and pure capacitive circuits. What will be the power drawn from the source? **[Mahamaya Technical Univ. Second Semester 2011-12]**

Ans. Phasor diagrams for pure inductive and capacitive circuits are shown in Figs. 4.3 (*c*) and 4.4 (*c*) respectively. Power drawn in both the circuits will be zero.

Q. 6. What do you understand by power factor of a circuit ?

Ans. Power factor may be defined as the cosine of the phase angle between voltage and current. It may also be defined as the ratio of resistance to impedance or the ratio of true power to apparent power.

Q. 7. The maximum and minimum values of power factor can be ..... . **[G.B.T.U. Odd Semester 2012-13]**

Ans. Unity and zero respectively.

Q. 8. What do you understand by active and reactive components of current in an ac circuit ?

Ans. The current component which is in phase with circuit voltage (i.e. I cos ϕ) and contributes to active or true power of the circuit is called the active (wattfull or in-phase) component of current.

The current component which is in quadrature (or 90° out of phase) to the circuit voltage (i.e. I sin ϕ) and contributes to reactive power of the circuit is called the reactive (or wattless) component of current.

Q. 9. What are active and reactive powers ? Draw the power triangle. **[P.T.U. December 2003]**

Ans. The power which is actually consumed or utilized in an ac circuit is called the true or active power of the circuit. Power is consumed only in resistance. It is given by the product of the circuit voltage V, current I and power factor cos ϕ *i.e.* P = VI cos ϕ. It is expressed in watts.

A pure inductor and a pure capacitor do not consume any power, as in a quarter cycle whatsoever power is drawn from the supply source by these components, the same is returned to the supply source in the other quarter cycle. This power which flows back and forth (i.e. in both directions in the circuit) or reacts itself is called the reactive power. This is also known as wattless power.

The reactive power of an ac circuit is given by the product of voltage V, current I and sine of the phase angle ϕ i.e. Reactive power, Q = VI sin ϕ. It is expressed in reactive volt-amperes.

The power triangle is shown in Fig 4.7(a)

Q. 10. What is the relationship between supply voltage and circuit current in a pure capactive circuit ?

Ans. In a pure capactive circuit, current leads the applied voltage by 90° or $\pi/2$ radians.

Q. 11. What is the effect of frequency on inductive reactance ?

Ans. Inductive reactance of an inductor increases proportionately with the increase in supply frequency *i.e.* $X_L \alpha f$.

Q. 12. What is the effect of frequency on capactive reactance ?

Ans. Capactive reactance of a given capacitor decreases with the increase in supply frequency. Capacitive reactance is inversely proportional to supply frequency *i.e.* $X_C \alpha \dfrac{1}{f}$.

Q. 13. Given are the source voltage, source current and the power factor of the network, write the expressions for apparent and real powers drawn from the source. **[M.T.U. Second Semester 2011-12]**

Ans. The expressions for apparent and real power are given as

$$\text{Apparent power, S} = \text{VI volt-amperes}$$
$$\text{Real power, P} = \text{S} \times \text{pf} = \text{VI} \times \text{pf} = \text{VI} \cos\phi \text{ watts}$$

where V is source voltage, I is the source current and ϕ is phase angle.

Q. 14. What are the values of power factor for (*i*) pure resistive circuit (*ii*) pure inductive circuit and (*iii*) pure capacitive circuit ?

Ans. (*i*) Unity, (*ii*) Zero (lagging) (*iii*) Zero (leading).

Q. 15. What are active and reactive powers? Why is the term reactive power not encountered when dc sources are used in an electric circuit? **[U.P.T.U. Second Semester 2008-09]**

Or

What do you understand by active and reactive powers? Why reactive power is not present in dc circuits? **[U.P. Technical Univ. Odd Semester 2013-14]**

Ans. The product of rms values of current and voltage, VI is called the *apparent power* and is measured in volt-amperes or kilovolt amperes (kVA).

The *true power* in an a c circuit is obtained by multiplying the apparent power by the power factor and is expressed in watts or kilowatts (kW).

The product of apparent power, VI and the sine of the angle between voltage and current, sin ϕ is called the *reactive power*. This is also known as *wattless power* and is expressed in reactive volt-amperes or kilovolt amperes reactive (kVA R).

Inductance does not offer any opposition to the flow of direct current while capacitance does not allow flow of direct current through it, therefore, either current is in phase with the source voltage or it does not exist. That is why, the term reactive power is not encountered with dc sources.

Q. 16. What is the relationship between apparent power, true power and reactive power of an ac circuit ?

Ans. Apparent power = $\sqrt{(\text{True power})^2 + (\text{Reactive power})^2}$

Q. 17. For a series R-L-C circuit draw the phasor diagram and show the power factor angle in that. Assume that the circuit draws inductive current. **[G.B.T.U. Second Semester 2011-12]**

Ans. Refer to Fig. 4.22 (*b*).

Q. 18. In an R-L-C circuit on what factors depends whether the phase angle between voltage V and current I is lagging or leading ?

Ans. Phase angle in an R-L-C circuit is given as

$$\phi = \text{Tan}^{-1}\frac{X}{R} = \text{Tan}^{-1}\frac{X_L - X_C}{R}$$

i.e. the phase angle between voltage V and current I will depend upon whether the inductive reactance is greater or smaller than capacitive reactance. When X_L exceeds X_C voltage V will lead current I and when X_C exceeds X_L applied voltage V will lag behind the current I.

Q. 19. Explain 'resistance', 'reactance' and 'impedance'. **[R.G.P.V. December 2002]**

Ans. Resistance may be defined as that property of a substance which opposes (or restricts) the flow of an electric current (or electrons) through it. It is expressed in ohms.

Reactance may be inductive or capacitive.

Inductive reactance is capable of storing energy by virtue of a current flowing through it. It opposes the change of current flowing through it and induces a voltage when the current flowing through it varies in magnitude and/or direction. In inductive circuit current does not change instantaneously. It is highly resistive to ac but it does not oppose dc.

Capacitive reactance is capable of storing energy by virtue of a volage existing across it. It does not oppose the flow of ac but it is highly resistive to dc. Voltage across it cannot change instantaneously.

Inductive reactance and capacitive reactance in series or in parallel form a tuned circuit. Both of the reactances are expressed in ohms.

Impedance is a property of an ac circuit due to which it opposes the flow of current through it. It may be defined as the ratio of voltage applied across the circuit and current flowing in that circuit. It is expressed in ohms. It is given as

$$\text{Impedance} = \sqrt{(\text{Resistance})^2 + (\text{Reactance})^2}$$

Q. 20. The dimension of L/CR is .... **[G.B.T.U. Odd Semester 2012-13]**

Ans. The dimension of L/CR is that of resistance R.

$\because$ L has the dimension of R/ω and C has the dimension of $\frac{1}{\omega R}$.

PROBLEMS

1. A 50 Hz voltage of 115 V (rms) is impressed on a 50 Ω resistance. Write the time equations for the voltage and resultant current. Let the zero point of the voltage wave be at $t = 0$ **[Ans.** 162.6 sin 314*t* and 3.25 sin 314*t*]
2. A 50 Hz voltage of 230 V (rms) is impressed on an inductance of 0.2 H. Give the time equations for the voltage and the resultant current. Let the zero point of the voltage wave be at $t = 0$. Determine the maximum energy stored in the inductance.

   $$\left[\textbf{Ans. } 325.3\sin 314t;\ 5.2\sin\left(314t - \frac{\pi}{2}\right);\ 2.7\text{ J}\right]$$
3. A 50 Hz voltage of 230 V rms is impressed on a capacitance of 20 μ F. Write the time equation for the voltage and the resultant current. Let the zero point of the voltage wave be at $t = 0$. Determine the energy stored in the capacitor.

   $$\left[\textbf{Ans. } 325.3\sin 314t;\ 1.45\sin\left(314t + \frac{\pi}{2}\right);\ 1.06\text{ J}\right]$$
4. A 120 V ac circuit contains 10 Ω resistance and 30 Ω reactance in series. What would be the average power in the circuit? **[Ans.** 144 watts]
5. A coil has a resistance of 10 Ω and draws a current of 5 A when connected across a 100 V, 50 Hz source. Determine the power factor of the circuit. **[Pb. Technical Univ. Electrical Engineering June 2000]**

[**Ans.** 0.5 (lagging)]

6. A coil has a resistance of 10 Ω and draws a current of 5 A when connected across a 100 V, 50 Hz source. Determine the inductance of the coil. **[Pb. Technical Univ. Electrical Engineering December 2000]**

[**Ans.** 55.13 mH]

7. A 20 mH inductor is connected in series with a 10 Ω resistor across a 220 V, 50 Hz supply. Find the current and power factor of the circuit. **[Pb. Technical Univ. Electrical Engineering December 2001]**

[**Ans.** 18.628 A; 0.8467 (lagging)]

8. The element of 500 W electric press is designed for use on 200 V, 50 Hz supply. What value of resistance is needed to be connected in series in order that press can be operated at 230 V ac supply ?

[Pb. Technical Univ. Basic Electrical and Electronics Engineering First Semester 2004-05]

[**Ans.** 12 Ω]

9. In a coil, the inductance being 50 mH: the impedance is 20 ohm. If the power factor of the circuit be 0.5 lag, find the values of angular frequency and resistance of the coil. **[R.G.P.V. Bhopal Basic Electrical Engineering June 2005]**

[**Ans.** 346.41 radians/second; 10 Ω]

10. The voltage applied to a circuit is v = 100 sin (ωt + 30°) and the current flowing in the circuit is i = 15 sin (ωt + 60°). Determine the impedance, resistance, reactance, power and the power factor of the circuit.

[U.P. Technical Univ. Electrical Engineering Second Semester 2003-04]

[**Ans.** 6.667 Ω; 5.774 Ω; 3.333 Ω (capacitive); 649.52 W; 0.866 (leading)]

11. In an ac circuit the current is given by $i = 22 \sin (314 - \pi/6)$ A. If the voltage of the circuit is the reference quantity, determine (*i*) power factor (*ii*) rms value of current, (*iii*) frequency of current.

[U.P. Technical Univ. Electrical Engineering Jaunary 2003]

[**Ans.** 0.866 (lagging); 15.56 A; 50 Hz]

12. A choke coil takes a current of 2 A lagging 60° behind the applied voltage of 220 Vat 50 Hz. Calculate the inductance and resistance of coil. **[V.T. U. Belgaum Karnataka Univ. Winter 2003]**

[**Ans.** 0.3032 H; 55 Ω]

13. The instantaneous values of the voltage across a two element series circuit and the current flowing through it are given by $v = 100 \sin\left(314t - \frac{\pi}{4}\right)$ V, $i = 20 \sin (314\,t - 90°)$A. Find the frequency and the circuit elements.

[V.T. U. Belgaum Karnataka Univ. Winter 2003]

[**Ans.** 50 Hz; 3.536 Ω, 11.26 mH]

14. A 100 V, 80 W lamp is to be operated on 230 V, 50 Hz ac. Calculate the inductance to be connected in series with the lamp. Lamp can be taken as a pure resistance. **[U.P. Technical Univ. Electrical Engineering First Semester 2004-05]**

[**Ans.** 0.824 H]

15. An iron-cored choking coil takes 5 A when connected to a 20-V dc and takes 5 A at 100 V ac and consumes 250 W. Determine (*a*) impedance (*b*) the power factor (*c*) the iron loss. [**Ans.** (*i*) 20 Ω (*ii*) 0.5 (lag) (*iii*) 150 W]

16. An iron core choke coil has a d c resistance of 10 Ω. When it is connected to 250 V, 50 Hz mains the current taken is 5.0 A at a power factor of 0.4. Calculate:-

(*i*) The resistance which represents the effect of iron loss and

(*ii*) the inductance of the coil. Give also the phasor diagram. [**Ans.** (*i*) 10 Ω (*ii*) 0.146 H]

17. A voltage $v(t) = 141.4 \sin (314\,t + 10°)$ is applied to a circuit and the steady current given by $i(t) = 14.14 \sin (314\,t - 20°)$ is found to flow through it. Determine (*i*) the pf of the circuit (*ii*) the power delivered to the circuit (*iii*) Draw the phasor diagram. **[Nagpur Univ., Summer 2000]**

[**Ans.** (*i*) 0.866 (lag) (*ii*) 866 W]

18. A voltage of 125 V at 60 Hz is applied across a non-inductive resistor connected in series with a condensor. The current is 2.2 A. The power loss in the resistor is 96.8 W and that in the condensor is negligible. Calculate the resistance and the capacitance. [**Ans.** 20 Ω ; 49.88 μ F]

19. A voltage wave e (t) = 141.4 sin 120 t produces a current i (t) = 14.14 sin 120 t + 7.07 cos (120 t + 30°) in a circuit. Determine (*i*) the resultant time expression of current (*ii*) the power factor and the power delivered by the source.

[U.P. Technical Univ. Electrical Engineering February 2002]

[**Ans.** (*i*) $i(t) = 12.25 \sin (120\,t + 30°)$; (*ii*) 0.866 (leading); 750 W]

20. A coil of resistance 8 Ω and inductance 0.12 henry is connected in series with a condenser of capacitance 140 microfarads across a 230 volts, 50 Hz supply. Determine:

(*i*) Impedance of the entire circuit,

(*ii*) Current flowing through the condenser,

(*iii*) Power factor of the circuit,

(*iv*) Voltage across the condenser. **[U.P. Technical Univ. Electrical Engineering Second Semester 2002-03]**

[**Ans.** (*i*) 16.97 Ω (*ii*) 13.55 A (*iii*) 0.4714 (lagging) (*iv*) 308.2 V]

21. A coil of resistance 12 Ω and inductance 0.05 H, a non-inductive resistance of 20 Ω resistance and a loss-free 40 μF capacitor are connected across a 240 V, 50 Hz sinusoidal supply. Calculate (i) the current and (ii) the power factor of the circuit. **[U.P. Technical Univ. Electrical Engineering September 2001]**

[**Ans.** (*i*) 3.36 A (*ii*) 0.448 (lagging)]

22. A series R-L-C circuit consisting of a resistance of 20 Ω, inductance 0.2 H and capacitive of 150 μ F is connected across a 230 V 50 Hz source. Calculate (i) the impedance (ii) the current (iii) the magnitude and nature of the power factor (iv) the frequency of supply to be adjusted to make power factor unity.

[U.P. Technical Univ. Electrical Engineering February 2002]

[**Ans.** (*i*) 46.167 Ω (*ii*) 4.98 A (*iii*) 0.433 (lagging) (*iv*) 29.06 Hz]

❑❑❑

5 CHAPTER

Complex Notations and Circuit Analysis

5.1. INTRODUCTION

We have seen that ac circuits cannot be solved by use of simple algebra since geometrical relations are to be taken into consideration. Though phasor diagram method is quite satisfactory for fairly simple circuits but becomes complicated when a circuit or network is made up of several branches. AC circuits can be conveniently solved by use of complex algebra. The results obtained by this method are of the same order of accuracy as those obtained by trigonometrical methods although calculations are usually much simpler.

The above system enables equations representing alternating voltages and currents and their phase relationships to be expressed in simple algebraic form. It is based upon the idea that a phasor can be resolved into two components along two axis at right angles to each other.

There are four forms or methods of representing the vector or phasor quantities viz., (*i*) symbolic form, (*ii*) trigonometrical form (*iii*) exponential form and (*iv*) polar form. All of these methods of expressing vector or phasor quantities enable those operations which are carried out graphically in a phasor diagram, to be performed analytically.

5.2. SYMBOLIC FORM

A phasor, as already mentioned above, can be resolved into two components along two axis at right angles to each other. It is preferable and convenient to resolve it along X-axis and Y-axis.

In symbolic form, a phasor is resolved into two components at right angles to each other and the component lying along Y axis is designated by $+j$ or $-j$ according to its direction.

In Fig. 5.1 phasors OA_1, OA_2, OA_3 and OA_4 are shown lying in first, second, third and fourth quadrants respectively. The phasor OA_1 lying in first quadrant may be resolved into two components a_1 along X-axis (sometimes called the axis of reals) and b_1 along Y-axis (sometimes called the axis of imaginaries or *j*-axis). The component along Y-axis should be designated by *j* and should be written as jb_1. Thus we represent the phasor OA_1

$$\boldsymbol{OA_1}^* = a_1 + j\,b_1$$

The phasor, when expressed in above form, is said to be expressed in *complex form*. In electrical engineering, the horizontal component (*i.e.* a_1) is called the *in-phase* or *active* component while the vertical component is called the *quadrature* or *reactive* component.

* Symbols representing phasors are printed in bold face italics, while those representing only the magnitude are printed in ordinary italics, thus:

$\boldsymbol{V}, \boldsymbol{I}, \boldsymbol{Z}$, phasors or complex numbers

V, I, Z, magnitude or module

Other methods of denoting phasors are by putting bar above, dot above or below letters in ordinary italics as $\bar{V}, \bar{I}, \bar{Z}, \dot{V}, \dot{I}, \dot{Z}$ or $\underset{\cdot}{V}, \underset{\cdot}{I}, \underset{\cdot}{Z}$.

The length of line OA_1 represents, to adopted scale, the absolute value or modulus (magnitude) of the phasor and is equal to $\sqrt{a_1^2 + b_1^2}$, and its angle with the X-axis, θ_1, called the *argument* of the phasor, is given by the expression

$$\theta_1 = \text{Tan}^{-1} \frac{b_1}{a_1}.$$

Similarly phasors OA_2, OA_3 and OA_4 can be resolved into two components along axis of reals and axis of imaginaries and can be expressed as

$$OA_2 = -a_2 + j\,b_2$$
$$OA_3 = -a_3 - j\,b_3$$
$$OA_4 = a_4 - j\,b_4$$

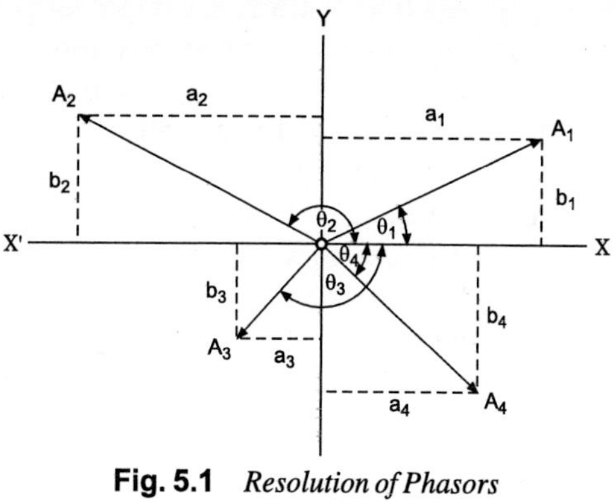

Fig. 5.1 *Resolution of Phasors*

5.2.1. Significance of Operator *j*. The letter *j* used in the above expressions is an operator, the application of which to phasor *a*, or phasor component *a*, causes its rotation through 90° in the positive (counter-clockwise) direction. It is assigned the numerical value $\sqrt{-1}$. The double operation of *j* on phasor *a* rotates it in counter-clockwise direction through 180° and hence reverses its sense because $j\,j = j^2 = \left(\sqrt{-1}\right)^2 = -1$. The triple operation of *j* on phasor *a* rotates it in counter-clockwise direction through 270° and so on.

For example, if phasor V lying along X-axis is operated with *j*, the phasor V will rotate through an angle of 90° in counter-clockwise direction and lie along Y-axis. If this phasor V is operated twice with operator *j* it will rotate through 180° in counter-clockwise direction from X-axis and will lie along X-axis in negative direction, as illustrated in Fig. 5.2. Application of the operator *j* thrice to this phasor V will rotate it through 270° in counter-clockwise direction from X-axis and now the phasor V will lie along Y-axis in negative direction. On operating the phasor V with *j* four times, it will rotate through 360° in counter-clockwise direction from the reference axis and so will come to the original position.

Hence it is seen that *j* is an operator which turns phasor *a* through an angle of 90° in counter-clockwise direction, each time that it is applied and has a numerical value of $\sqrt{-1}$ so that

$$j^2 = \left(\sqrt{-1}\right)^2 = -1;$$
$$j^3 = \left(\sqrt{-1}\right)^3 = -\sqrt{-1}$$
$$\text{and } j^4 = \left(\sqrt{-1}\right)^4 = (-1)^2 = 1$$

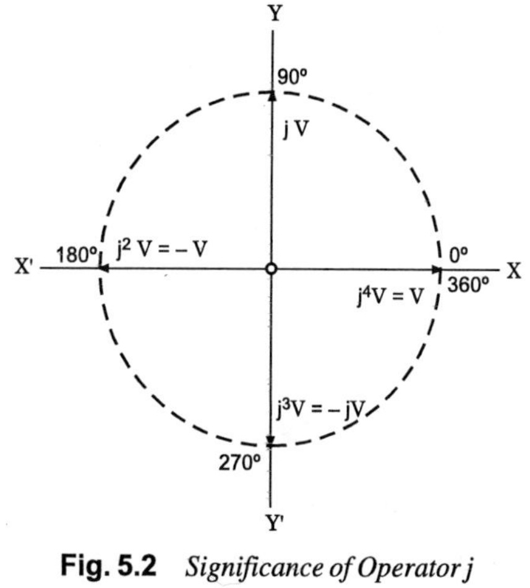

Fig. 5.2 *Significance of Operator j*

5.3. CONJUGATE COMPLEX NUMBERS

Two complex numbers are said to be conjugate when they are of same modulus and have arguments of equal magnitude and opposite sign. For example, the numbers $(a + j\,b)$ and $(a - j\,b)$ are conjugate. The sum of two conjugate numbers gives in-phase or active component and their difference gives quadrature or reactive component. The product of two conjugate numbers gives in-phase or active component only (no quadrature component) equal to the sum of the square of their magnitudes.

i.e. $(a + j\,b) + (a - j\,b) = 2a$, active component

$(a + j\,b) - (a - j\,b) = j\,2b$, reactive component

$(a + j\,b) \times (a - j\,b) = (a^2 + b^2)$, active component

5.4. TRIGONOMETRICAL FORM OF PHASOR REPRESENTATION

As may be seen from Fig. 5.3, X-component of $\boldsymbol{V} = V\cos\theta$ and Y-component of $\boldsymbol{V} = V\sin\theta$. Hence phasor V can be expressed in the trigonometrical form as

$$\boldsymbol{V} = V\cos\theta + jV\sin\theta = V(\cos\theta + j\sin\theta)$$

This is equivalent to the symbolic or rectangular form $\boldsymbol{V} = (a + jb)$ because $a = V\cos\theta$ and $b = V\sin\theta$

In general $\boldsymbol{V} = V(\cos\theta \pm j\sin\theta)$

The above expression represents a phasor of numerical value V and having phase angle of $\pm\theta$ with the reference axis, positive angle being measured in counter-clockwise direction from the reference axis and negative angle in clockwise direction from the reference axis.

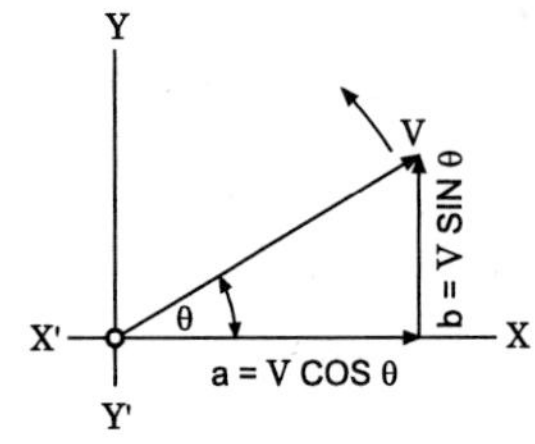

Fig. 5.3

The phasors OA_1, OA_2, OA_3 and OA_4 (Fig 5.1) may be represented thus:

$$\boldsymbol{OA_1} = OA_1(\cos\theta_1 + j\sin\theta_1)$$
$$\boldsymbol{OA_2} = OA_2(\cos\theta_2 + j\sin\theta_2)$$
$$\boldsymbol{OA_3} = OA_3(\cos\theta_3 - j\sin\theta_3)$$
$$\boldsymbol{OA_4} = OA_4(\cos\theta_4 - j\sin\theta_4)$$

5.5. POLAR FORM OF PHASOR REPRESENTATION

The phasor V (Fig. 5.3) may be alternatively expressed as $V\angle\theta$ where V is the magnitude of the phasor and θ is the angle measured in counter-clockwise direction from reference axis, called the *argument* of the phasor.

In general, the phasor V may be represented as

$$\boldsymbol{V} = V\angle\pm\theta$$

The above form of phasor representation is known as *polar form.*

The phasors OA_1, OA_2, OA_3 and OA_4 (Fig. 5.1) may be represented thus $\boldsymbol{OA_1} = OA_1\angle\theta_1$; $\boldsymbol{OA_2} = OA_2\angle\theta_2$; $\boldsymbol{OA_3} = OA_3\angle-\theta_3$; $\boldsymbol{OA_4} = OA_4\angle-\theta_4$.

5.6. EXPONENTIAL FORM OF PHASOR REPRESENTATION

By Euler's equation $e^{j\theta} = \cos\theta + j\sin\theta$

Hence phasor V may be expressed as

$$\boldsymbol{V} = Ve^{j\theta}$$ where e is the base of natural logarithm.

In general, the phasor V may be represented as

$$\boldsymbol{V} = Ve^{\pm j\theta}$$

The above form is known as *exponential form* of representing a phasor. It represents a phasor of modulus (or magnitude) V and argument $\pm\theta$.

Note: The phasor representation in either of four mathematical forms convey the same information *i.e.* magnitude (modulus) of the phasor and its phase angle (argument). So it is possible to convert one form to the other.

Example 5.1 Give the voltage expressed in the symbolic form $V = (5 + j\,12)$ volts in trigonometrical, polar and exponential forms.

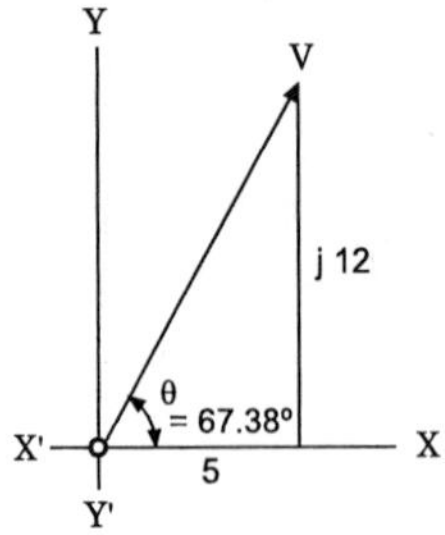

Fig. 5.4

Solution: The modulus or magnitude of the given voltage, $V = \sqrt{5^2 + 12^2} = 13$ V

The argument or phase angle of the given voltage, $\theta = \text{Tan}^{-1}\dfrac{12}{5}$

$= 67.38°$ or $0.374\,\pi$ radians.

So given voltage in

(*i*) Trigonometrical form is 13 (cos 67.38° + j sin 67.38°) volts **Ans.**

(*ii*) Polar form is 13 $\angle 67.38°$ volts **Ans.**

(*iii*) Exponential form is 13 $e^{+j\,0.374\pi}$ volts **Ans.**

5.7. ADDITION AND SUBTRACTION OF PHASORS

The symbolic or rectangular form is most suitable form for addition or subtraction of phasors. If the phasors are given in polar or exponential form, they should be first converted to rectangular form and then addition or subtraction be carried out.

5.7.1. Addition of Phasors. For the addition of phasors in the rectangular form in-phase (or active) components are added together and quadrature components are added together.

Let the two voltage phasors V_1 and V_2 represented as $\mathbf{V}_1 = a_1 + j\,b_1$, and $\mathbf{V}_2 = a_2 - j\,b_2$, be added.

Then resultant voltage will be given as

$$\mathbf{V} = \mathbf{V}_1 + \mathbf{V}_2 = (a_1 + j\,b_1) + (a_2 - j\,b_2) = (a_1 + a_2) + j\,(b_1 - b_2)$$

Magnitude of resultant voltage,

$$V = \sqrt{(a_1 + a_2)^2 + (b_1 - b_2)^2}$$

Argument or phase angle of the resultant voltage,

$$\theta = \text{Tan}^{-1}\frac{b_1 - b_2}{a_1 + a_2}$$

Addition process by graphic representation is illustrated in Fig. 5.5.

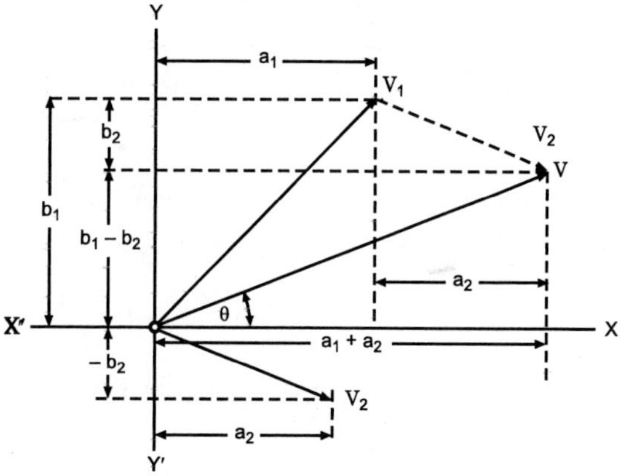

Fig. 5.5 *Addition of Phasors*

5.7.2. Subtraction of Phasors. By the rule of phasor subtraction

$$\mathbf{V} = \mathbf{V}_1 - \mathbf{V}_2 = (a_1 + j\,b_1) - (a_2 - j\,b_2) = (a_1 - a_2) + j\,(b_1 + b_2)$$

Magnitude of resultant voltage,

$$V = \sqrt{(a_1 - a_2)^2 + (b_1 + b_2)^2}$$

Argument or phase angle of resultant voltage,

$$\theta = \text{Tan}^{-1}\frac{b_1 + b_2}{a_1 - a_2}$$

Subtraction process by graphic representation is illustrated in Fig. 5.6.

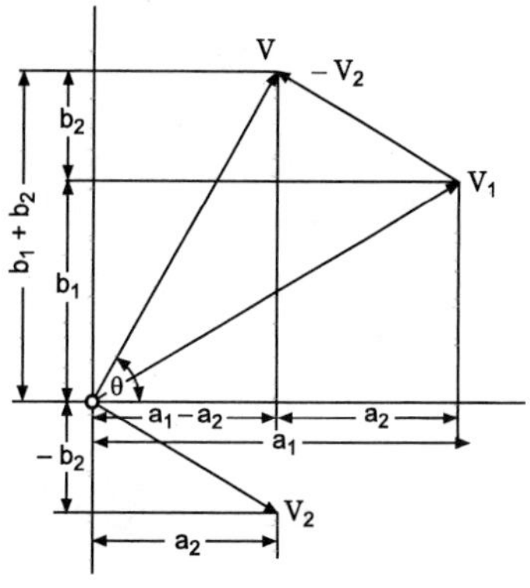

Fig. 5.6 *Subtraction of Phasors*

5.8. MULTIPLICATION AND DIVISION OF PHASORS

The $(a + j\,b)$ method of representing a phasor is particularly well suited to the process of addition and subtraction, as illustrated above, but it also lends itself readily to multiplication and division as illustrated below:

However, it is noteworthy that it is easier to multiply and divide phasors when they are expressed in polar or exponential form.

Consider two phasors given by $V_1 = (a_1 + j\,b_1)$ or $V_1 \angle \theta_1$ or $V_1 e^{j\theta_1}$

and $V_2 = (a_2 - j\,b_2)$ or $V_2 \angle -\theta_2$ or $V_2 e^{-j\theta_2}$

5.8.1. Multiplication of Phasors

(*i*) Rectangular Form.

$$V_1 \times V_2 = (a_1 + j\,b_1)(a_2 - j\,b_2) = a_1 a_2 - j\,a_1 b_2 + j\,a_2 b_1 - j^2 b_1 b_2 = (a_1 a_2 + b_1 b_2) + j\,(a_2 b_1 - a_1 b_2)$$

$$\text{Magnitude of resultant} = \sqrt{(a_1 a_2 + b_1 b_2)^2 + (a_2 b_1 - a_1 b_2)^2}$$

$$\text{Argument, } \theta = \text{Tan}^{-1} \frac{a_2 b_1 - a_1 b_2}{a_1 a_2 + b_1 b_2}$$

***(ii)* Polar and Exponential Forms.** For multiplication of phasors in polar forms, their magnitude or modulus are multiplied and arguments or phase angles are added as proved below:

$$\boldsymbol{V}_1 = \text{V}_1 \angle\theta_1 = \text{V}_1 e^{j\theta_1} = \text{V}_1 (\cos\theta_1 + j\sin\theta_1)$$

$$\boldsymbol{V}_2 = \text{V}_2 \angle-\theta_2 = \text{V}_2 e^{-j\theta_2} = \text{V}_2 (\cos\theta_2 - j\sin\theta_2)$$

$$\begin{aligned}\text{Their product} &= \boldsymbol{V}_1 \boldsymbol{V}_2 = [\text{V}_1 (\cos\theta_1 + j\sin\theta_1)]\,[\text{V}_2 (\cos\theta_2 - j\sin\theta_2)] \\ &= \text{V}_1\text{V}_2 [\cos\theta_1 \cos\theta_2 - j\cos\theta_1 \sin\theta_2 + j\sin\theta_1 \cos\theta_2 - j^2 \sin\theta_1 \sin\theta_2] \\ &= \text{V}_1 \text{V}_2 [\cos\theta_1 \cos\theta_2 + \sin\theta_1 \sin\theta_2 + j(\sin\theta_1 \cos\theta_2 - \sin\theta_2 \cos\theta_1)] \\ &= \text{V}_1 \text{V}_2 [\cos(\theta_1 - \theta_2) + j\sin(\theta_1 - \theta_2)]\end{aligned}$$

We thus see that, the product of two phasors is a phasor having modulus equal to the product of the modulus of the two phasors and the argument is equal to the algebraic sum of the arguments of the two phasors.

In polar form we have the product $\boldsymbol{V}_1\boldsymbol{V}_2 = \text{V}_1\text{V}_2 \angle\theta_1 - \theta_2$

In exponential form we have the product $\boldsymbol{V}_1\boldsymbol{V}_2 = \text{V}_1 \text{V}_2 e^{j(\theta_1 - \theta_2)}$

5.8.2. Division of Phasors.

***(i)* Rectangular Form.** $\dfrac{\boldsymbol{V}_1}{\boldsymbol{V}_2} = \dfrac{a_1 + j b_1}{a_2 - j b_2}$

Multiplying both the numerator and denominator by conjugate of $(a_2 - j b_2)$ *i.e.* by $(a_2 + j b_2)$ we have

$$\frac{\boldsymbol{V}_1}{\boldsymbol{V}_2} = \frac{a_1 + j b_1}{a_2 - j b_2} \times \frac{a_2 + j b_2}{a_2 + j b_2} = \frac{(a_1 a_2 - b_1 b_2) + j(a_1 b_2 + a_2 b_1)}{a_2^2 + b_2^2} = \frac{(a_1 a_2 - b_1 b_2)}{a_2^2 + b_2^2} + j\,\frac{a_1 b_2 + a_2 b_1}{a_2^2 + b_2^2}$$

***(ii)* Polar and Exponential Forms.**

$$\frac{\boldsymbol{V}_1}{\boldsymbol{V}_2} = \frac{a_1 + j b_1}{a_2 - j b_2} = \frac{\text{V}_1(\cos\theta_1 + j\sin\theta_1)}{\text{V}_2(\cos\theta_2 - j\sin\theta_2)} = \frac{\text{V}_1 \angle\theta_1}{\text{V}_2 \angle-\theta_2} = \frac{\text{V}_1 e^{j\theta_1}}{\text{V}_2 e^{-j\theta_2}}$$

The quotient is

$$\frac{\boldsymbol{V}_1}{\boldsymbol{V}_2} = \frac{\text{V}_1 (\cos\theta_1 + j\sin\theta_1)}{\text{V}_2 (\cos\theta_1 - j\sin\theta_2)} = \frac{\text{V}_1}{\text{V}_2}\,\frac{(\cos\theta_1 + j\sin\theta_1)}{(\cos\theta_2 - j\sin\theta_2)} \times \frac{\cos\theta_2 + j\sin\theta_2}{(\cos\theta_2 + j\sin\theta_2)}$$

$$= \frac{\text{V}_1}{\text{V}_2}\left[\frac{\cos\theta_1 \cos\theta_2 - \sin\theta_1 \sin\theta_2 + j(\sin\theta_1 \cos\theta_2 + \sin\theta_2 \cos\theta_1)}{\cos^2\theta_2 + \sin^2\theta_2}\right]$$

$$= \frac{\text{V}_1}{\text{V}_2}[\cos(\theta_1 + \theta_2) + j\sin(\theta_1 + \theta_2)]$$

We thus see that, when we divide one phasor by another phasor, the quotient is a phasor having modulus equal to the quotient of the modulus of the two phasors and an argument equal to the algebraic difference of the arguments of the two phasors.

In polar form we have

$$\frac{\boldsymbol{V}_1}{\boldsymbol{V}_2} = \frac{\text{V}_1}{\text{V}_2} \angle\theta_1 - (-\theta_2) = \frac{\text{V}_1}{\text{V}_2} \angle\theta_1 + \theta_2$$

In exponential form we have

$$\frac{\boldsymbol{V}_1}{\boldsymbol{V}_2} = \frac{\text{V}_1}{\text{V}_2}\, e^{j(\theta_1 + \theta_2)}$$

Example 5.2 **Perform the following operation and the final result may be given in the polar form :**

$$(8 + j6) \times (-10 - j7.5)$$

[Bangalore Univ. Electrical Engineering and Electronics, 1990]

Solution: $(8 + j\,6) \times (-10 - j\,7.5) = (-80 - j\,60 - j\,60 + 45) = (-35 - j\,120)$

The resultant phasor lies in third quadrant because both the components of the phasor are negative. The magnitude and phase angle are given by

$$\text{Magnitude} = \sqrt{(-35)^2 + (-120)^2} = 125$$

$$\text{Phase angle } \theta = \text{Tan}^{-1}\frac{-120}{-35} = \angle 180° + 73.74° = \angle 253.74°$$

Hence resultant may be given as $125 \angle 253.74°$ **Ans.**

Example 5.3 **Perform the following operation and express the final result in polar form :**

$$5 \angle 30° + 8 \angle -30°$$

[Bangalore Univ. Electrical Engineering and Electronics, 1989]

Solution:

$$5 \angle 30° = 5\,(\cos 30° + j \sin 30°) = (4.33 + j\,2.5)$$

$$8 \angle -30° = [8\,(\cos(-30°) + j \sin(-30°)) = (6.928 - j\,4)$$

So $5 \angle 30° + 8 \angle -30° = (4.33 + j\,2.5) + (6.928 - j\,4) = (11.258 - j\,1.5)$

The resultant phasor lies in fourth quadrant and its magnitude and phase angle are given by

$$\text{Magnitude} = \sqrt{(11.258)^2 + (-1.5)^2} = 11.358$$

$$\text{Phase angle } \theta = \text{Tan}^{-1}\frac{-1.5}{11.258} = \angle -7.59°$$

Hence resultant may be given as $11.358 \angle -7.59°$ **Ans.**

Example 5.4 **Calculate the following in polar form :**

(*i*) **Add (40 + *j* 20) to (20 + *j* 120)**
(*ii*) **Subtract (10 + *j* 30) from (20 − *j* 20)**
(*iii*) **Multiply (15 + *j* 20) with (20 + *j* 30)**
(*iv*) **Divide (6 + *j* 7) by (5 + *j* 3)**

[Gujarat Univ. June/July 2003]

Solution: (*i*) $(40 + j\,20) + (20 + j\,120) = (60 + j\,140) = \sqrt{(60^2 + 140^2)} \angle \tan^{-1} 140/60 = 152.32 \angle 66.8°$ **Ans.**

(*ii*) $(20 - j\,20) - (10 + j\,30) = (10 - j\,50) = \sqrt{10^2 + (-50)^2} \angle \tan^{-1} -50/10 = 50.99 \angle -78.69°$ **Ans.**

(*iii*) $(15 + j\,20) \times (20 + j\,30) = 300 + j\,450 + j\,400 - 600 = (-300 + j\,850) = 901.4 \angle 109.44°$ **Ans.**

Alternatively $(15 + j\,20) \times (20 + j\,30) = 25 \angle 53.13° \times 36.056 \angle 56.31° = 901.4 \angle 109.44°$ **Ans.**

$$(iv)\ \frac{6 + j\,7}{5 + j\,3} = \frac{(6 + j\,7)(5 - j\,3)}{(5 + j\,3)(5 - j\,3)} = \frac{30 - j\,18 + j\,35 + 21}{34}$$

$$= \frac{51 + j\,17}{34} = (1.5 + j\,0.5) = 1.581 \angle 18.44° \quad \textbf{Ans.}$$

$$\text{Alternatively } \frac{6 + j\,7}{5 + j\,3} = \frac{9.22 \angle 49.40°}{5.83 \angle 30.96°} = 1.581 \angle 18.44° \quad \textbf{Ans.}$$

Example 5.5 **Three phasors are A = 10 + *j* 10 ; B = 50 $\angle 120^\circ$ and C = 8 – *j* 6. Evaluate $\frac{AB}{C}$ and express it in rectangular and polar forms.**

Solution : Phasor $A = (10 + j10)$ have modulus of $\sqrt{10^2 + 10^2}$ *i.e.* $10\sqrt{2}$ and argument of $\tan^{-1}\frac{10}{10}$ *i.e* 45° or $\frac{\pi}{4}$ radians.

So phasor *A* in polar form will be $10\sqrt{2}\ \angle 45^\circ$.

Phasor *B* in polar form is $50\ \angle 120^\circ$.

Phasor $C = (8 - j\,6)$ have modulus of $\sqrt{8^2 + (-6)^2}$ *i.e.* 10 and argument of $\tan^{-1}\frac{-6}{8}$ or – 36.87°

$$\text{So } C = 10\ \angle -36.87^\circ$$

$$\text{Hence } \frac{AB}{C} = \frac{10\sqrt{2}\ \angle 45^\circ \times 50\ \angle 120^\circ}{10\ \angle -36.87^\circ} = \frac{10\sqrt{2} \times 50}{10}\ \angle 45^\circ + 120^\circ - (-36.87^\circ)$$

$$= 50\sqrt{2}\ \angle 201.87^\circ \textbf{ Ans.}$$

In rectangular form $\frac{AB}{C}$ will be $= 50\sqrt{2}\cos 201.87^\circ + j\,50\sqrt{2}\sin 201.87^\circ = -65.622 - j\,26.34$ **Ans.**

5.9. POWERS AND ROOTS OF PHASORS

Powers and roots of phasors can be determined conveniently in polar form. If the phasor is not in polar form then it will be advisable to convert it first to polar form and then carry out algebraic operations.

5.9.1. Powers of Phasors. Let us find the fourth power of the phasor $5\ \angle 20^\circ$, for this purpose, the phasor has to be multiplied by itself four times.

$$\text{So } \left(5\ \angle 20^\circ\right)^4 = 5 \times 5 \times 5 \times 5\ \angle 20^\circ + 20^\circ + 20^\circ + 20^\circ = 625\ \angle 80^\circ$$

$$\text{In general } \mathbf{A}^n = \mathrm{A}^n\ \angle n\,\theta$$

5.9.2. Roots of Phasors. Let us find the one-fourth power of the phasor $5\ \angle 20^\circ$

$$\left(5\ \angle 20^\circ\right)^{1/4} = (5)^{1/4} \angle 20/4 = 1.496\ \angle 5^\circ \textbf{ Ans.}$$

$$\text{In general } \sqrt[n]{\mathbf{A}} = \sqrt[n]{\mathrm{A}}\ \angle \theta/n.$$

5.10. PHASOR ALGEBRA APPLIED TO SINGLE PHASE SERIES CIRCUITS

A simple series circuit having non-inductive resistance R and inductive reactance X_L is shown in Fig. 5.7 (*a*). Let the current phasor be taken as reference phasor and be represented as $\boldsymbol{I} = I + j\,0$ or $I\ \angle 0^\circ$.

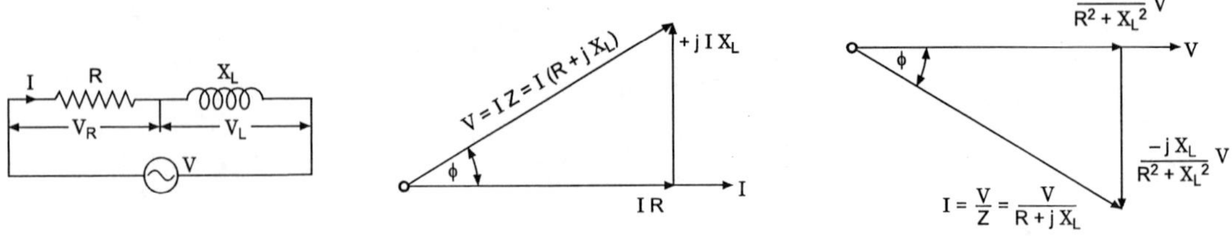

(*a*) Circuit Diagram (*b*) Phasor Diagram (*c*) Phasor Diagram

Fig. 5.7

Since voltage drop across resistance R, V_R = IR in phase with current I, it is expressed as (IR $+ j\,0$) and the voltage drop across the inductive reactance, $V_L = I\,X_L$ in quadrature with current and leading, therefore, it is expressed as ($0 + j\,IX_L$). Hence the voltage applied across the circuit can be expressed as

$$V = V_R + V_L = (IR + j\,0) + (0 + j\,IX_L) = I\,(R + j\,X_L) = I\,Z^* = I\,Z\,\angle\phi$$

Hence impedance **Z** can be written as ($R + j\,X_L$) or $Z\angle\phi$ where $\phi = \text{Tan}^{-1}\dfrac{X_L}{R}$.

Thus, impedance can be written as complex quantity though it is not a complex quantity. Since impedance causes resistance and reactance voltage drops to be at right angle to each other, it is actually a complex operator.

Now let the applied voltage be taken as reference phasor and expressed as $V(1 + j\,0)$ or $V\angle 0°$ then current be determined by using impedance as phasor

$$I = \frac{V}{Z} = \frac{V + j0}{R + j\,X_L} \times \frac{R - j\,X_L}{R - j\,X_L} = \frac{VR}{R^2 + X_L^2} - j\frac{VX_L}{R^2 + X_L^2} \text{ using complex notation}$$

$$\text{and } I = \frac{V}{Z} = \frac{V\angle 0°}{Z\angle\phi} = \frac{V}{Z}\angle -\phi \qquad \text{where } \phi = \text{Tan}^{-1}\frac{X_L}{R} \text{ using polar notation.}$$

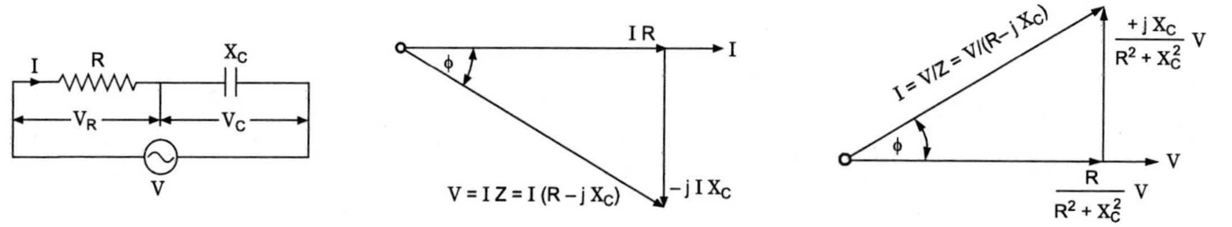

(*a*) *Circuit Diagram* (*b*) *Phasor Diagram* (*c*) *Phasor Diagram*

Fig. 5.8

In Fig. 5.8 (*a*) circuit having resistance *R* and capacitive reactance X_C connected in series are shown. Taking current *I* as reference phasor *i.e.* along *X*-axis, as shown in phasor diagram [Fig. 5.8 (*b*)]

Applied voltage,

$$V = V_R + V_C = (I\,R + j\,0) + (0 - j\,I\,X_C) = I\,(R - j\,X_C) = I\,Z = I\,Z\,\angle -\phi$$

Hence impedance Z can be written as $(R - j\,X_C)$ or $Z\angle -\phi$ where $\phi = \text{Tan}^{-1}\dfrac{X_C}{R}$

Taking applied voltage as reference phasor, as shown in Fig. 5.8 (*c*), the current flowing through the circuit is given as

$$I = \frac{V}{Z} = \frac{V + j\,0}{R - j\,X_C} \times \frac{R + j\,X_C}{R + j\,X_C} = \frac{VR}{R^2 + X_C^2} + j\frac{VX_C}{R^2 + X_C^2} \text{ using complex notation}$$

$$\text{and } I = \frac{V}{Z} = \frac{V\angle 0°}{Z\angle -\phi} = \frac{V}{Z}\angle\phi \qquad \text{where } \phi = \text{Tan}^{-1}\frac{X_C}{R} \text{ using polar notation.}$$

Thus, general expression for phasor impedance can be written as

$$Z = (R \pm j\,X) \text{ in complex notation and } \mathbf{Z} = Z\angle\pm\phi; \qquad \text{where } \phi = \text{Tan}^{-1}\frac{X}{R} \text{ in polar notation.}$$

Plus sign is used for inductive circuit as inductive reactance is considered positive and minus sign is used for capacitive circuit as capacitive reactance is considered negative.

* In the expression **V** = **I Z**, **V** and **I** differ from **Z** in that they are associated with time varying quantities, whereas **Z** is a complex number independent of time and is, therefore, not a phasor in the sense that **V** and **I** are phasors. It has, however, become the practice to refer all complex numbers used in ac circuit calculations as phasors.

Example 5.6 **State in symbolic notation, the impedance of each of the following circuits at a frequency of 50 Hz.**
(*a*) A resistance of 20 Ω in series with an inductance of 0.1 H
(*b*) A resistance of 50 Ω in series with a capacitance of 40 μ F
(c) Circuits (*a*) and (*b*) in series.

Solution: *For Circuit (a)*

$$\text{Resistance, } R_1 = 20\ \Omega$$

$$\text{Reactance, } X_1 = 2\pi f L = 2\pi \times 50 \times 0.1 = 31.4\ \Omega \text{ (being inductive)}$$

$$\therefore \text{ Impedance, } \mathbf{Z}_1 = (R_1 + jX_1) = (20 + j\,31.4)\ \Omega \textbf{ Ans.}$$

For Circuit (b)

$$\text{Resistance, } R_2 = 50\ \Omega$$

$$\text{Reactance, } X_2 = \frac{1}{2\pi f C} = \frac{1}{2\pi \times 50 \times 40 \times 10^{-6}} = -79.6\ \Omega \text{ (being capacitive)}$$

$$\therefore \text{ Impedance, } \mathbf{Z}_2 = (50 - j\,79.6)\ \Omega \textbf{ Ans.}$$

When circuits (*a*) and (*b*) are connected in series, total impedance of the circuit,

$$\mathbf{Z} = \mathbf{Z}_1 + \mathbf{Z}_2 = (20 + j\,31.4) + (50 - j\,79.6) = (70 - j\,48.2) \text{ ohms } \textbf{Ans.}$$

Example 5.7 **The voltage and current of a circuit are given as $V_1 + j\,V_2$ and $I_1 - j\,I_2$ respectively. Determine the circuit impedance in symbolic notation.**

Solution: Voltage, $V = V_1 + j\,V_2$ and Current, $I = I_1 - j\,I_2$

$$\text{Impedance, } Z = \frac{V}{I} = \frac{V_1 + j\,V_2}{I_1 - j\,I_2} = \frac{V_1 + j\,V_2}{I_1 - j\,I_2} \times \frac{I_1 + j\,I_2}{I_1 + j\,I_2} = \frac{V_1 I_1 - V_2 I_2}{I_1^2 + I_2^2} + j\,\frac{V_2 I_1 + I_2 V_1}{I_1^2 + I_2^2} \textbf{ Ans.}$$

Example 5.8 **A resistance of 4 Ω is connected in series with an inductive reactance of 5 Ω. Draw the impedance phasor for this circuit and express the impedance in the symbolic form, polar form and exponential form.**

Solution: Circuit resistance, R = 4 Ω

Circuit reactance, X = 5 Ω (inductive)

The impedance phasor of the circuit is shown in Fig. 5.9.

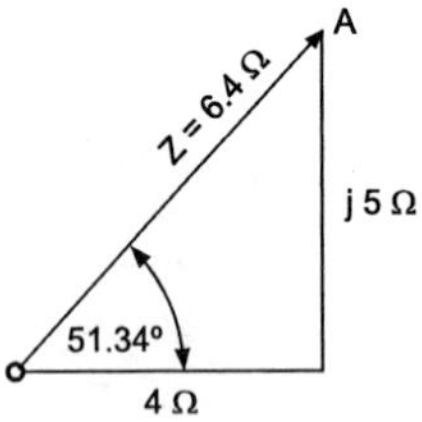

Fig. 5.9

$$\text{Magnitude of impedance, Z} = \sqrt{4^2 + 5^2} = 6.4\ \Omega$$

$$\text{Argument, } \theta = \text{Tan}^{-1}\,\frac{5}{4} = 51.34° \text{ or } 0.285\,\pi \text{ radians } \textbf{Ans.}$$

$$\therefore \text{ Impedance, } \mathbf{Z} = (4 + j\,5)\ \Omega \text{ in symbolic form } \textbf{Ans.}$$

$$\text{or } \mathbf{Z} = 6.4\ \angle 51.34°\ \Omega \text{ in polar form } \textbf{Ans.}$$

$$\text{or } \mathbf{Z} = 6.4\,e^{j\,0.285\pi} \text{ in exponential form } \textbf{Ans.}$$

Example 5.9 **$V = 200\ \angle 36.9°$ volts ; Z = 20 $\angle -53.1°$ ohms. Find *I*.**

Solution : Circuit voltage, $V = 200\ \angle 36.9°$ volts

and Circuit impedance, $\mathbf{Z} = 20\ \angle -53.1°$ ohms

$$\text{Circuit current, } I = \frac{V}{Z} = \frac{200\ \angle 36.9°}{20\ \angle -53.1°} = \frac{200}{20}\ \angle 36.9° - (-53.1°) = 10\ \angle 90° \text{ A } \textbf{Ans.}$$

Example 5.10 **A load having impedance of $1 + j1\ \Omega$ is connected to an ac voltage represented as $v = 20\sqrt{2}\cos(\omega t + 10°)$ volt.**

(*i*) **Find the current in load, expressed in the form of $i = I_m \sin(\omega t + \phi)$ A.**

(*ii*) **Find real power consumed by the load.** [*U.P. Technical Univ. Elec. Engineering Second Semester 2004-05*]

Solution:

Load impedance, $\mathbf{Z} = (1 + j1)\ \Omega = \sqrt{2}\angle 45°\ \Omega$

Maximum value of applied voltage, $V_{max} = 20\sqrt{2}$ V

RMS value of applied voltage, $V_{rms} = \dfrac{V_{max}}{\sqrt{2}} = \dfrac{20\sqrt{2}}{\sqrt{2}} = 20$ V

Phase angle of applied voltage, $\phi = 100°$ $\quad \because \text{Cos}(\omega t + 10°) = \text{Sin}(\omega t + 90° + 10°)$

So applied voltage, $\mathbf{V} = 20\angle 100°$ V

Load current, $\mathbf{I} = \dfrac{\mathbf{V}}{\mathbf{Z}} = \dfrac{20\angle 100°}{\sqrt{2}\angle 45°} = 10\sqrt{2}\angle 55°$ A

Peak value of load current, $I_{max} = \sqrt{2} \times 10\sqrt{2} = 20$ A

and equation for load current is given as $i = I_{max}\sin(\omega t + \phi) = 20\sin(\omega t + 55°)$ **Ans.**

Example 5.11 **In a given R-L circuit, R = 35 Ω and L = 0.1 H. Find (*i*) current through the circuit (*ii*) power factor if a 50 Hz frequency, voltage $V = 220\angle 30°$ is applied across the circuit.**

[*R.G.P.V. Bhopal Basic Electrical Engineering 2001*]

Solution:

Resistance of R-L circuit, $R = 35\ \Omega$

Inductive reactance of R-L circuit, $X_L = 2\pi f L = 2\pi \times 50 \times 0.1 = 31.42\ \Omega$

Magnitude of impedance, $Z = \sqrt{R^2 + X_L^2} = \sqrt{35^2 + 31.42^2} = 47.03\ \Omega$

Argument, $\theta = \text{Tan}^{-1}\dfrac{X_L}{R} = \text{Tan}^{-1}\dfrac{31.42}{35} = 41.91°$

Hence impedance, $\mathbf{Z} = (35 + j\,31.42)\ \Omega = 47.03\angle 41.9°\ \Omega$

Applied voltage, $\mathbf{V} = 220\angle 30°$ V

(*i*) Current through the circuit, $\mathbf{I} = \dfrac{\mathbf{V}}{\mathbf{Z}} = \dfrac{220\angle 30°}{47.03\angle 41.9°} = 4.678\angle -11.9°$ A

Magnitude of current, I = 4.678 A **Ans.**

(*ii*) Power factor $= \text{Cos}\,\theta = \text{Cos}\,[30° - (-11.9°)] = \text{Cos}\,41.9° = 0.7443$ (lagging) **Ans.**

Example 5.12 **A coil of resistance 8 Ω and inductance 0.1 H is connected in series with a condenser of 160 μF across 230 V, 50 Hz supply. Calculate complex impedance, current and power factor.**

[*Mahamaya Technical Univ. Electrical Engineering Second Semester 2011-12*]

Solution:

Coil resistance, $R = 8\ \Omega$

Inductive reactance of coil, $X_L = 2\pi f L = 2\pi \times 50 \times 0.1 = 31.42\ \Omega$

Capacitive reactance of condenser, $X_C = \dfrac{1}{2\pi f C} = \dfrac{1}{2\pi \times 50 \times 160 \times 10^{-6}} = 19.89\ \Omega$

Complex impedance, $\mathbf{Z} = R + j(X_L - X_C) = 8 + j(31.42 - 19.89) = (8 + j\,11.53)\ \Omega$ **Ans.**

Taking voltage as reference phasor

Supply voltage, $\mathbf{V} = (230 + j\,0)$ volts

Current, $\mathbf{I} = \dfrac{\mathbf{V}}{Z} = \dfrac{(230 + j\,0)}{8 + j\,11.53} = (9.343 - j\,13.5)\ \text{A} = 16.42\angle -55.31°$ A **Ans.**

Power factor, $\cos\phi = \text{Cos}(-55.31°) = 0.569$ (lagging) **Ans.**

Example 5.13 A coil of resistance 10 Ω and inductance 0.1 H is connected in series with a condenser of capacitance 150 μ F across a 200 V, 50 Hz supply. Determine the following:

(*i*) Impedance (*ii*) Current (*iii*) Power factor (*iv*) Voltage across the coil; and (*v*) Voltage across the condenser.

Solution: Resistance of the circuit, $R = 10\ \Omega$

Inductive reactance of the circuit, $X_L = 2\pi f L = 2\pi \times 50 \times 0.1 = 31.42\ \Omega$

$$\text{Capacitive reactance of the circuit, } X_C = \frac{1}{2\pi f C} = \frac{1}{2\pi \times 50 \times 150 \times 10^{-6}} = 21.22\ \Omega$$

$$\text{Magnitude of impedance, } Z = \sqrt{R^2 + (X_L - X_C)^2} = \sqrt{(10^2) + (31.42 - 21.22)^2} = 14.28\ \Omega$$

$$\text{Phase angle, } \phi = \text{Tan}^{-1}\frac{X_L - X_C}{R} = \text{Tan}^{-1}\frac{31.42 - 21.22}{10} = 45.57^\circ$$

(*i*) Hence impedance, $\mathbf{Z} = (10 + j\ 10.2)$ or $14.28\ \angle 45.57^\circ\ \Omega$ **Ans.**

$$(ii)\ \text{Current, } \mathbf{I} = \frac{\mathbf{V}}{\mathbf{Z}} = \frac{200\ \angle 0^\circ}{14.28\ \angle 45.57^\circ} = 14\ \angle -45.57^\circ\ \text{A}\ \textbf{Ans.}$$

(*iii*) Power factor $= \text{Cos}\ \phi = \text{Cos}\ 45.57^\circ = 0.7$ (lagging) **Ans.**

$$(iv)\ \text{Voltage across the coil} = \mathbf{I}\ (R + jX_L) = 14\ \angle -45.57^\circ \times (10 + j\ 31.42)$$

$$= 14\ \angle -45.57^\circ \times 32.97\ \angle 72.34^\circ$$

$$= 14 \times 32.97\ \angle -45.57^\circ + 72.34^\circ$$

$$= 461.58\ \angle 26.77^\circ \text{ volts } \textbf{Ans.}$$

$$(v)\ \text{Voltage across the capacitor} = IX_C = 14\ \angle -45.57^\circ \times 21.22\ \angle -90^\circ$$

$$= 297.08\ \angle -135.57^\circ \text{ volts } \textbf{Ans.}$$

Example 5.14 In a circuit, the applied voltage is found to lag the current by 30°.

(*a*) Is the power factor lagging or leading ? (*b*) What is the value of power factor (*c*) Is the circuit inductive or capacitive ?

In the diagram of Fig. 5.10, the voltage drop across Z_1 is $(10 + j0)$ volts. Find out (*i*) the current in the circuit (*ii*) the voltage drops across Z_2 and Z_3 (*iii*) the voltage of the generator.

[*Bombay Univ. Electrical Engineering-I, 1991*]

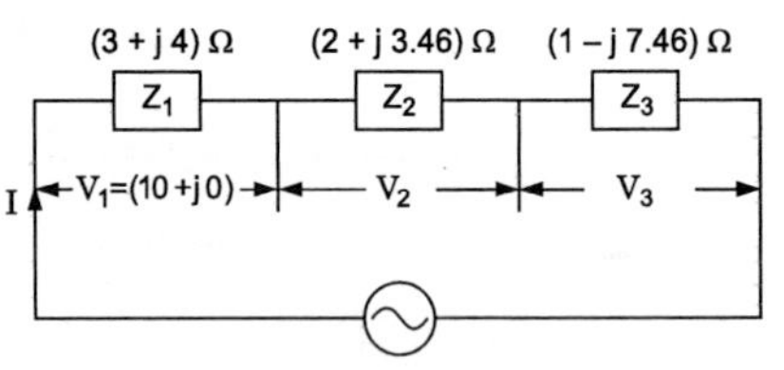

Fig. 5.10

Solution: (*a*) Since the applied voltage lags behind the current, current leads the applied voltage. It means that power factor is **leading. Ans.**

(*b*) Power factor = Cos φ = Cos 30° = 0.866 (leading) **Ans.**

(*c*) Since the current leads the applied voltage, the circuit is *capacitive.* **Ans.**

(*i*) Circuit current, **I** = Current through impedance $\mathbf{Z_1}$

$$= \frac{\text{Voltage drop across } \mathbf{Z_1}, \mathbf{V_1}}{\text{Impedance } \mathbf{Z_1}} = \frac{10 + j\,0}{3 + j\,4} = \frac{10\ \angle 0^\circ}{5\ \angle 53.15^\circ} = 2\ \angle -53.13^\circ\ \text{A}$$

$$= (1.2 - j1.6)\ \text{A}\ \textbf{Ans.}$$

$$(ii)\ \text{Voltage drop across } \mathbf{Z_2}, \mathbf{V_2} = \mathbf{I} \times \mathbf{Z_2} = 2\ \angle -53.13^\circ \times (2 + j\,3.46) = 2\ \angle -53.13^\circ \times 3.996\ \angle 59.97^\circ$$

$$= 7.993\ \angle 6.84^\circ = (7.936 + j\,0.952)\ \text{V}\ \textbf{Ans.}$$

$$\text{Voltage drop across } \mathbf{Z_3}, \mathbf{V_3} = \mathbf{I} \times \mathbf{Z_3} = 2\ \angle -53.13^\circ \times (1 - j\,7.46) = 2\ \angle -53.13^\circ \times 7.527\ \angle -82.365^\circ$$

$$= 15.05\ \angle -135.495^\circ\ \text{V}$$

$$= (-10.74 - j\,10.55)\ \text{V}\ \textbf{Ans.}$$

(*iii*) Generator voltage, $\mathbf{V} = \mathbf{V}_1 + \mathbf{V}_2 + \mathbf{V}_3 = (10 + j\,0) + (7.936 + j\,0.952) + (-10.74 - j\,10.55)$

$= (7.196 - j\,9.598) = 12 \angle -53.13°$ V **Ans.**

Example 5.15 In the circuit of Fig. 5.11, applied voltage is given by (0 + *j*10) and the current is (0.8 + *j* 0.6) A. Determine the values of R and X and also indicate if X is inductive or capacitive.

[*Nagpur Univ. Electrical Technology 1991*]

Solution:

Applied voltage, $V = (0 + j\,10) = 10 \angle 90°$ volts

Circuit current, $I = (0.8 + j\,0.6) = 1 \angle 36.9°$ A

Current lags behind applied voltage by $\phi = (90° - 36.9°)$ *i.e.* 53.1°

Hence the circuit shown is inductive.

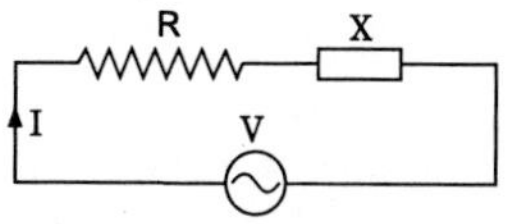

Fig. 5.11

$$\text{Circuit impedance, } Z = \frac{V}{I} = \frac{10 \angle 90°}{1 \angle 36.9°} = 10 \angle 53.1° = (6 + j\,8)\ \Omega$$

Hence R = 6 Ω and X = 8 Ω (inductive)

5.11. POWER DETERMINATION

If the voltage and current in a circuit are expressed in complex form, power can be determined as follows:

Let the current be expressed as

$I = I_1 - j\,I_2$ and voltage as $V = V_1 + j\,V_2$.

From Fig. 5.12 it is evident that the real component of voltage, V_1 and real component of current I_1 are in phase. Since components of voltage and current may be treated as if they were acting alone, the power contributed by V_1 and I_1, $P_1 = V_1\,I_1$.

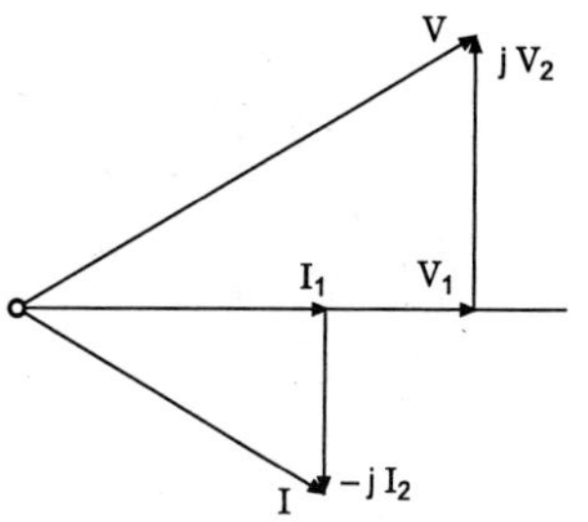

Fig. 5.12

The coefficient of V_2 is $+j$ and coefficient I_2 is $-j$, these two components of voltage and current are opposite in phase and therefore, their product gives negative power.

So power $P = -V_2\,I_2$

Voltage component V_1 and current component I_2 are in quadrature so power contributed by V_1 and I_2 is zero, similarly voltage component V_2 and current component I_1 are in quadrature and, therefore, contribute no power.

∴ Total power in the circuit, $P = P_1 + P_2 = V_1\,I_1 - V_2\,I_2$

The total power is the algebraic sum of the product of the real quantities V_1 and I_1 and the product of imaginary quantities V_2 and I_2. The sign being determined in the ordinary algebraic method. The operator j must not be included in the multiplication.

Example 5.16 Determine the power consumed in a circuit carrying a current of (3 − *j* 4) A and connected across (80 + *j* 40) volt supply mains.

Solution: Supply voltage, $V = (80 + j\,40)$ volts

Circuit current, $I = (3 - j\,4)$ amperes

Power consumed, $P = V_1\,I_1 + V_2\,I_2 = 80 \times 3 + 40\,(-4) = 80$ watts **Ans.**

5.12. CONJUGATE METHOD OF POWER DETERMINATION

In any circuit let the applied voltage and current flowing be $V = V(\cos\alpha + j\sin\alpha)$ and $I = \mathrm{I}(\cos\beta + j\sin\beta)$ respectively. $(\alpha - \beta)$ is the phase difference between V and I so power factor of the circuit is $\cos(\alpha - \beta)$.

Multiplying V and I we have

$$VI = \mathrm{V}(\cos\alpha + j\sin\alpha) \times \mathrm{I}(\cos\beta + j\sin\beta) = \mathrm{VI}[\cos(\alpha + \beta) + j\sin(\alpha + \beta)].$$

The real part of the above expression is $VI\cos(\alpha + \beta)$ and evidently it cannot be true power of the circuit, because power factor of the circuit is $\cos(\alpha - \beta)$. Similarly $VI\sin(\alpha + \beta)$ is not correct expression of the reactive volt-amperes. Hence to determine the power by j method, it is necessary to change the sign of the imaginary part of either V or I. Thus if we change the sign of imaginary part of V, leaving I as before, we have

$$V_C = V(\cos\alpha - j\sin\alpha)$$

Complex quantity so obtained is known as *conjugate* of the original complex quantity.

Now $V_C I = V(\cos\alpha - j\sin\alpha) \times I(\cos\beta + j\sin\beta) = VI[\cos(\alpha-\beta) - j\sin(\alpha-\beta)]$.

The real part of the above expression $VI\cos(\alpha-\beta)$ gives the true power and imaginary part $VI\sin(\alpha-\beta)$ gives the reactive power. It should, however, be noted that the real power obtained by the conjugate method is the same regardless of whether V or I is reversed but sign of reactive volt-amperes will depend upon it. When the conjugate of voltage is used, capacitive VARs have a positive sign and inductive VARs a negative sign. If the conjugate of the current rather than of the voltage is used the capacitive VARs have a negative sign and inductive VARs a positive sign.

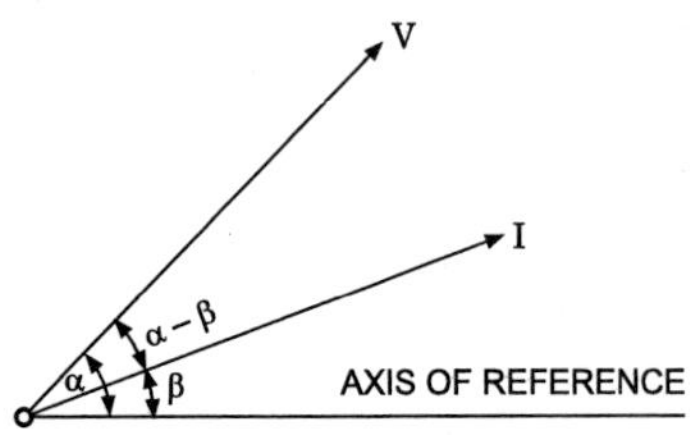

Fig. 5.13

The product of V and I_C is called the *complex power* where I_C is the conjugate of I.

Example 5.17 **In an alternating circuit, the impressed voltage is given by V = (100 – *j* 50) volts, and the current in the circuit is I = (3 – *j* 4) A. Determine the real and reactive power in the circuit.**

[Calcutta Univ. Electrical Engineering 1991]

Solution: Impressed voltage, $V = (100 - j\,50)$

Conjugate of current, $I_C = (3 + j\,4)$ amperes

$VI_C = (100 - j\,50)(3 + j\,4) = 500 + j\,250$

True power = Real part of $VI_C = 500$ watts **Ans.**

Reactive power = j part of $VI_C = 250$ VAR (inductive) **Ans.**

Example 5.18 **The current in a circuit is given by (5 + *j* 10) A when the applied voltage is (225 + *j* 150) *V*. Determine (*i*) the complex expression for the impedance stating whether it is inductive or capacitive (*ii*) power (*iii*) phase angle between current and voltage. Draw the complete phasor diagram.**

Solution: Circuit voltage, $V = (225 + j\,150)$ volts or $270.42 \angle 33.7^\circ$ V

Circuit current, $I = (5 + j\,10)$ amperes or $11.18 \angle 63.4^\circ$ A

$$\text{Circuit impedance, } Z = \frac{V}{I} = \frac{270.42}{11.18} \angle 33.7^\circ - 63.4^\circ = 24.2 \angle -29.7^\circ\ \Omega$$

or $Z = 24.2\,[\cos(-29.7^\circ) + j\sin(-29.7^\circ)] = (21 - j\,12)\ \Omega$ **Ans.**

(*i*) Since the j part of the impedance expression is negative hence the impedance is capacitive. **Ans.**

(*ii*) Conjugate current,

$I_C = (5 - j\,10)$

$VI_C = (225 + j\,150)(5 - j\,10) = (2{,}625 - j\,1{,}500)$

Power of the circuit,

P = Real part of $VI_C = 2{,}625$ W **Ans.**

Power of the circuit can also be determined as follows :

$P = VI\cos\phi = 270.42 \times 11.18 \cos(63.4^\circ - 33.7^\circ)$

$= 270.42 \times 11.18 \cos 29.7^\circ = 2{,}625$ W **Ans.**

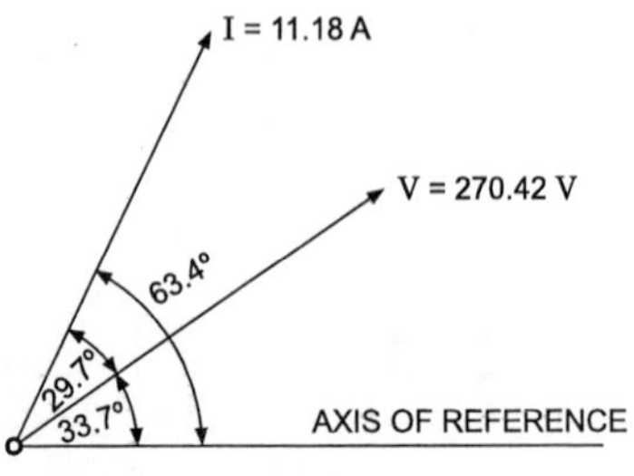

Fig. 5.14

(*iii*) Phase angle between current and voltage,

ϕ = Argument of current – argument of voltage = 63.4° – 33.7° = 29.7° (leading) **Ans.**

Phasor diagram is shown in Fig. 5.14.

HIGHLIGHTS

1. j is an operator which indicates an operation of rotating the phasor through 90° in counter-clockwise direction. It has numerical value of $(\sqrt{-1})$ so that

$$j^2 = (\sqrt{-1})^2 = -1; \quad j^3 = (\sqrt{-1})^3 = -\sqrt{-1} = -j \text{ and } j^4 = (\sqrt{-1})^4 = 1$$

2. Impedance can be expressed as

$$\mathbf{Z} = (R \pm jX) \text{ in rectangular form}$$

$$\mathbf{Z} = Z \angle \pm\theta \text{ in polar form} \qquad \text{where } \theta = \text{Tan}^{-1}\frac{X}{R} \text{ and } Z = \sqrt{R^2 + X^2}$$

$$\mathbf{Z} = Z(\cos\theta \pm j\sin\theta) \text{ in trigonometrical form}$$

and
$$\mathbf{Z} = Z\, e^{\pm j\theta} \text{ in exponential form}$$

3. In symbolic method or j-method of solution of ac circuits, the quantities are represented in complex, triangular or polar form and then the solution is obtained.

If applied voltage is represented as

$$\mathbf{V} = V \pm j\,0 \text{ or } V\angle 0°$$

and I is a current lagging behind the voltage by angle ϕ, then it is represented as

$$\mathbf{I} = (a - j\,b) \text{ or } I(\cos\phi - j\sin\phi) \text{ or } I\angle -\phi$$

For addition and subtraction, complex form is used while for multiplication and division, polar form is used.

4. Apparent power in complex form is given as

$$S = \text{Phasor voltage} \times \text{conjugate of phasor current} = \mathbf{V}\,\mathbf{I_C}$$

EXERCISES

1. What is the use of operator j ? Explain briefly. **[G.G.S.I.P. Delhi Univ. Electrical Science May 2005]**
2. Explain the method of representing a phasor by symbolic or j notation.
3. For two phasors $A = a_1 + jb_1$ and $B = a_2 - jb$, obtain their multiplication and division using polar form of representation. **[G.B. Technical Univ. First Semester 2009-10]**
4. Explain how alternating quantities may be represented by complex numbers ?
5. Is impedance a phasor quantity, if not then how it is expressed in phasor or complex form ?
6. Explain the conjugate method of determining power in ac circuits.
7. Write short note on j operator. **[R.G.P.V. Bhopal December 2002]**

SHORT ANSWER TYPE QUESTIONS WITH ANSWERS

Q. 1. What are the values of j^2 and j^4 ?

Ans. $j^2 = -1$ and $j^4 = +1$

Q. 2. What is meant by complex power ?

Ans. The product $S = V.I_C$ is called the complex power where I_C is the conjugate of I.

Q. 3. Why is impedance represented by a complex number ? **[R.G.P.V. Bhopal December 2002]**

Ans. In an expression **V** = **IZ**, the impedance **Z** is not associated with time varying quantities and is not a phasor in the sense that **V** and **I** are phasors. However, it is represented as a complex number. It is represented as

$\mathbf{Z} = R \pm jX$ +ve sign for inductive reactance and – ve sign for capacitive reactance.

Q. 4. How is complex impedance dependent on frequency ?

[R.G.P.V. Bhopal December 2002]

Ans. Complex impedance has two components viz. real or horizontal component and imaginary or vertical component. The reactance is represented by imaginary component. The inductive and capacitive reactances are given as

Inductive reactance, $X_L = \omega L = 2\pi f L$ where *f* is supply frequency in Hz.

Capacitive reactance, $X_C = \dfrac{1}{\omega C} = \dfrac{1}{2\pi f C}$ where *f* is supply frequency in Hz.

Since imaginary component of impedance depends upon the supply frequency so the impedance.

PROBLEMS

1. Give the current expressed in the polar form $\mathbf{I} = 10 \angle 53.1°$ amperes in symbolic, trigonometrical and exponential forms. **[Ans.** $(6 + j\,8)$ A; $10(\cos 53.1° + j \sin 53.1°)$ A; $10\, e^{j\,0.295\,\pi}$ A]
2. Divide phasor $(5 + j\,6)$ by phasor $(4 - j\,3)$. **[Ans.** $0.08 + j1.56$]
3. Add phasors $30 \angle 45°$ and $7.5 \angle -30°$. **[Ans.** $27.7 + j\,17.46$]
4. Determine $\sqrt[5]{243 \angle 120°}$. **[Ans.** $3 \angle 24°$]
5. Determine the current flowing through a circuit having impedance of $(5 + j\,12)\ \Omega$ connected across 230 V, 50 Hz supply. **[Ans.** $17.69 \angle -67.38°$ A]
6. A resistance of 50 Ω, inductance of 100 mH and a capacitor of 100 μ F are connected in series across 200 V, 50 Hz supply. Determine the impedance, current, phase angle, *p f*, active and reactive powers.

   [Ans. $50 \angle -0.5°\ \Omega$; $4 \angle 0.5°$ A ; 0.5° (lead), 0.99997 (lead); 799.97 W; 7 VARs (capacitive)]
7. A coil of resistance 10 Ω and inductance 0.1 H is connected in series with a capacitor of 150 μ F across 200 V, 50 Hz, supply. Determine the impedance, circuit current, active and reactive powers.

   [Ans. $14.28 \angle 45.57°\ \Omega$; $14 \angle -45.57°$ A; 1,960 W; 1,960 VARs (inductive)]
8. In an ac circuit the impressed voltage is $(100 + j\,50)$ V and the current flowing through the circuit is $(4 - j\,3)$ A. Determine the real and reactive power in the circuit. **[Ans.** 250 W; 500 VARs (inductive)]

❑❑❑

6

CHAPTER

Single Phase Parallel and Series-Parallel Circuits

6.1. PARALLEL CIRCUITS

In practice we come across the parallel circuits more often because multiple system of transmission and distribution is extensively used nowadays.

Practically all lighting and power circuits are constant voltage circuits with the loads connected in parallel. As in parallel dc circuits, the voltage is the same across each branch of a parallel ac circuit. The total line current supplied to the circuit is equal to the phasor sum of the branch currents. The total line current supplied to a parallel circuit may be determined by determining the currents in each branch and adding them, taking into account their phase relations, or by determining an equivalent impedance and dividing the applied voltage by the equivalent impedance.

Parallel ac circuits can be solved by the following methods

1. Phasor method
2. Admittance method
3. Symbolic method or *j*-method

The application of a particular method will depend upon the conditions of the problem. However, in general, that method should be used which gives quick results.

6.2. PHASOR METHOD

Consider a parallel circuit consisting of three branches of impedances Z_1 (R_1, L_1 C_1), Z_2 (R_2, L_2, C_2) and Z_3 (R_3, L_3, C_3) respectively and connected in parallel across an alternating voltage of *V* volts (rms), as shown in Fig. 6.1 (*a*).

Since all the three branches are connected in parallel, the voltage across each branch is the same and equal to supply voltage V, but currents through them will be different.

Branch I

$$\text{Impedance, } Z_1 = \sqrt{(R_1)^2 + \left(X_{C_1} - X_{L_1}\right)^2}$$

$$\text{Current, } I_1 = \frac{V}{Z_1}$$

$$\text{Phase angle, } \phi_1 = \text{Tan}^{-1} \frac{X_{C_1} - X_{L_1}}{R_1}$$

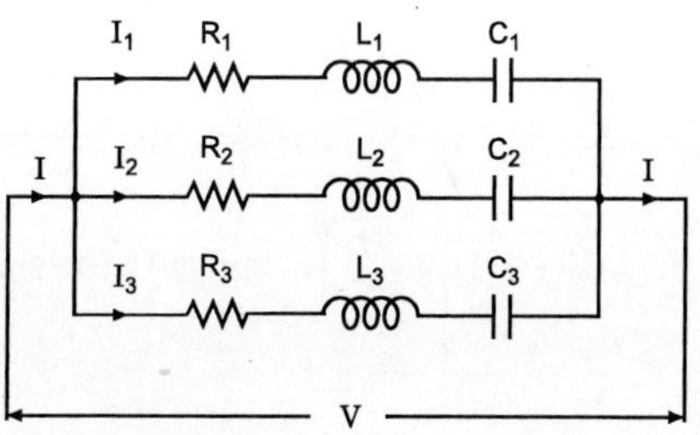

Fig. 6.1 (*a*) *Circuit Diagram*

Current I_1 will lead or lag behind the applied voltage if ϕ_1 is + ve or – ve respectively.

Similarly in Branch 2

$$\text{Impedance, } Z_2 = \sqrt{(R_2)^2 + (X_{C_2} - X_{L_2})^2}$$

$$\text{Current, } I_2 = \frac{V}{Z_2}$$

$$\text{Phase angle, } \phi_2 = \text{Tan}^{-1}\frac{X_{C_2} - X_{L_2}}{R_2}$$

and in Branch 3

$$\text{Impedance, } Z_3 = \sqrt{(R_3)^2 + (X_{C_3} - X_{L_3})^2}$$

$$\text{Current, } I_3 = \frac{V}{Z_3}$$

$$\text{Phase angle, } \phi_3 = \text{Tan}^{-1}\frac{X_{C_3} - X_{L_3}}{R_3}$$

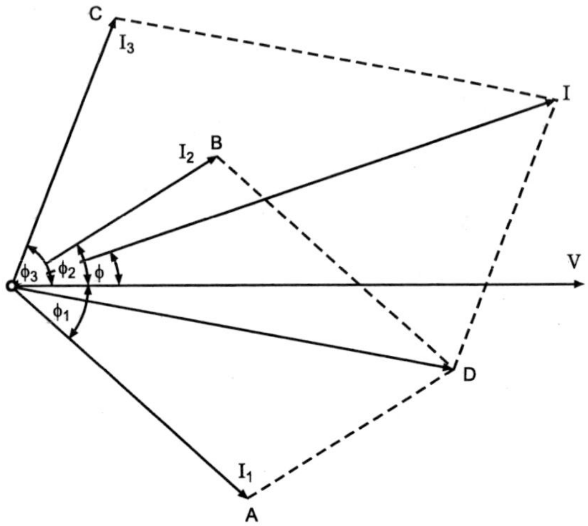

(*b*) *Phasor Diagram*

Fig. 6.1 *Single Phase Parallel Circuit*

Resultant current, I, which is phasor sum of currents I_1, I_2 and I_3, can be determined either by using parallelogram law of phasors, as shown in Fig. 6.1 (*b*) or by resolving branch currents I_1, I_2, and I_3 along X-axis and Y-axis and then determining the resultant of these components analytically.

Component of resultant current I along X-axis

= Sum of components of branch currents I_1, I_2 and I_3 along X-axis

$$\text{or } I\cos\phi = I_1\cos\phi_1 + I_2\cos\phi_2 + I_3\cos\phi_3$$

Similarly component of resultant current I along Y-axis

= Sum of components of branch currents I_1, I_2 and I_3 along Y-axis

$$\text{or } I\sin\phi = I_1\sin\phi_1 + I_2\sin\phi_2 + I_3\sin\phi_3$$

$$\text{Resultant current, } I = \sqrt{(I\cos\phi)^2 + (I\sin\phi)^2}$$

$$= \sqrt{(I_1\cos\phi_1 + I_2\cos\phi_2 + I_3\cos\phi_3)^2 + (I_1\sin\phi_1 + I_2\sin\phi_2 + I_3\sin\phi_3)^2}$$

$$\text{and Tan}\,\phi = \frac{I\sin\phi}{I\cos\phi}$$

$$\text{or } \phi = \text{Tan}^{-1}\frac{I_1\sin\phi_1 + I_2\sin\phi_2 + I_3\sin\phi_3}{I_1\cos\phi_1 + I_2\cos\phi_2 + I_3\cos\phi_3}$$

If ϕ is + ve, current I will lead the applied voltage V and if ϕ is – ve, current I will lag behind the applied voltage.

Example 6.1 **A voltage source of $e(t) = 141 \sin 377t$ is applied to two parallel branches. The time expression for the current in the first branch is $i_1(t) = 7.07\left(\omega t - \frac{\pi}{3}\right)$. In the second branch it is $i_2(t) = 10\sin\left(\omega t + \frac{\pi}{6}\right)$. Compute the total power supplied by the source.**

[*U.P. Technical Univ. Electrical Engineering Second Semester 2008-09*]

Solution: RMS value of applied voltage, $V_{rms} = \frac{V_{max}}{\sqrt{2}} = \frac{141}{\sqrt{2}} = 99.7\text{ V}$

$$\text{RMS value of current in first branch, } I_{rms1} = \frac{7.07}{\sqrt{2}} = 5 \text{ A}$$

$$\text{Phase angle of current } I_1, \phi_1 = -\frac{\pi}{3} = -60° = 60° \text{ (lagging)}$$

$$\text{RMS value of current in second branch, } I_{rms2} = \frac{10}{\sqrt{2}} = 7.071 \text{ A}$$

$$\text{Phase angle of current } I_2, \phi_2 = \frac{\pi}{6} = 30° \text{ (leading)}$$

$$\text{Total power supplied by source, } P = V_{rms} \times I_{rms} \times \cos\phi_1 + V_{rms} \times I_{rms2} \times \cos\phi_2$$
$$= 99.7 \times 5 \cos 60° + 99.7 \times 7.071 \times \cos 30°$$
$$= 249.25 + 610.55 = 859.8 \text{ W} \quad \textbf{Ans.}$$

Example 6.2 **The following circuit (Fig. 6.2) shows a parallel R-L arrangement connected across 200 volts, 50 Hz ac supply. Calculate:**

(*i*) the current drawn from the supply,
(*ii*) apparent power,
(*iii*) real power, and
(*iv*) reactive power.

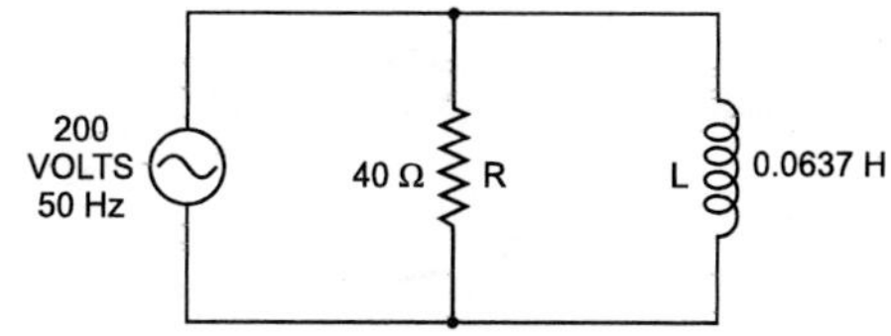

Fig. 6.2

[*U.P. Technical Univ. Electrical Engineering Second Semester 2002-03*]

Solution: Resistance of resistive branch, R = 40 Ω

$$\text{Inductive reactance of inductive branch, } X_L = 2\pi f L = 2\pi \times 50 \times 0.0637 = 20\ \Omega$$

$$\text{Current drawn by resistive branch, } I_R = \frac{V}{R} = \frac{200}{40} = 5 \text{ A}$$

$$\text{Current drawn by inductive branch, } I_L = \frac{V}{X_L} = \frac{200}{20} = 10 \text{ A}$$

$$(i) \text{ Current drawn from the supply, } I = \sqrt{I_R^2 + I_L^2} = \sqrt{5^2 + 10^2} = 11.18 \text{ A} \quad \textbf{Ans.}$$

$$(ii) \text{ Apparent power, } S = V \times I = 200 \times 11.18 = 2{,}236 \text{ VA or } 2.236 \text{ kVA} \quad \textbf{Ans.}$$

$$(iii) \text{ Real power, } P = V \times I_R = 200 \times 5 = 1{,}000 \text{ W or } 1.0 \text{ kW} \quad \textbf{Ans.}$$

$$(iv) \text{ Reactive power, } Q = V \times I_L = 200 \times 10 = 2{,}000 \text{ VAR or } 2.0 \text{ kVAR} \quad \textbf{Ans.}$$

Example 6.3 **A voltage $v(t)$ = 400 sin 314 is applied to the circuit shown (Fig. 6.3). Find the currents and their phase angles w.r.t. the voltages for the three branches.**

C = 50 μF ; R_1 = 100 Ω ; R_2 = 50 Ω, L = 0.1 H

[*Pb. Technical Univ. Electrical Engineering May 2002*]

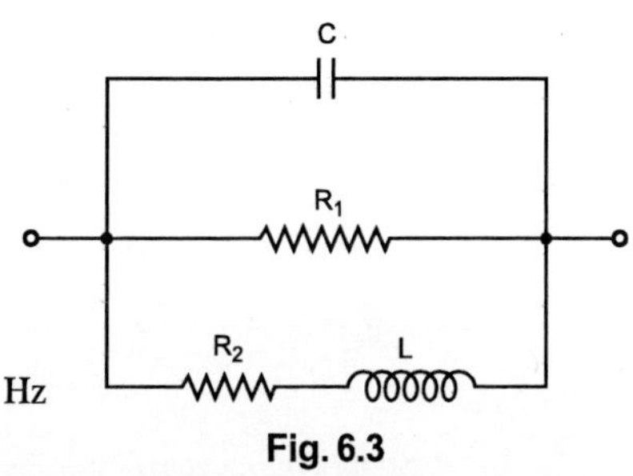

Fig. 6.3

Solution: Maximum value of applied voltage,

V_{max} = Coefficient of the sine of time angle = 400 V

$$\text{RMS value of applied voltage, } V_{rms} = \frac{V_{max}}{\sqrt{2}} = \frac{400}{\sqrt{2}} = 282.84 \text{ V}$$

$$\text{Supply frequency, } f = \frac{\text{Coefficient of time } t}{2\pi} = \frac{314}{2\pi} = 50 \text{ Hz}$$

$$\text{Current in capacitive branch, } I_1 = \frac{V}{X_C} = 2\pi f C V = 2\pi \times 50 \times 50 \times 10^{-6} \times 282.84 \text{ A}$$

$$= 4.44 \text{ A leading applied voltage by } \frac{\pi}{2} \text{ radians } \textbf{Ans.}$$

i.e. Phase angle = 90° (leading)

Current in resistive branch, $I_2 = \frac{V}{R_1} = \frac{282.84}{100} = 2.83$ A in phase with applied voltage **Ans.**

i.e. phase angle = 0°

Current in R-L branch, $I_3 = \frac{V}{Z} = \frac{V}{\sqrt{R_2^2 + (2\pi f L)^2}} = \frac{282.84}{\sqrt{50^2 + (2\pi \times 50 \times 0.1)^2}} = \frac{282.84}{59.05} = 4.79$ A **Ans.**

Phase difference between voltage and current I_3,

$$\phi_3 = \text{Tan}^{-1} \frac{2\pi f L}{R_2} = \text{Tan}^{-1} \frac{2\pi \times 50 \times 0.1}{50} = 32.14° \text{ (lag) } \textbf{Ans.}$$

Example 6.4 **A 159.23 μF capacitor in parallel with a resistance R draws a current of 25 A from 300 V, 50 Hz mains. Using phasor diagrams find the frequency *f* at which this combination draws the same current from a 360 V mains.** [*GATE 1994*]

Solution: The circuit and phasor diagrams are shown in Figs. 6.4 (*a*), (*b*) and (*c*) respectively.

When supply voltage V = 300 volts and supply frequency *f* = 50 Hz

Capacitive reactance,

$$X_C = \frac{1}{2\pi f C} = \frac{1}{2 \times \pi \times 50 \times 159.23 \times 10^{-6}} = 20\ \Omega$$

Supply current, I = 25 A

$$\text{Capacitive current, } I_C = \frac{V}{X_C} = \frac{300}{20}$$

$$= 15 \text{ A, leading supply voltage by } 90°$$

From phasor diagram, shown in Fig. 6.4 (*b*), drawn for supply voltage of 300 V

$$\text{Resistive current, } I_R = \sqrt{I^2 - I_C^2} = \sqrt{25^2 - 15^2} = 20 \text{ A}$$

$$\text{Circuit resistance, } R = \frac{V}{I_R} = \frac{300}{20} = 15\ \Omega$$

When the supply voltage is 360 V and the supply frequency is *f'* (say)

Supply current, I' = I = 25 A, as before

$$\text{Resistive current, } I'_R = \frac{V'}{R} = \frac{360}{15} = 24 \text{ A}$$

From phasor diagram, shown in Fig. 6.4 (*c*), drawn for supply voltage of 360 V

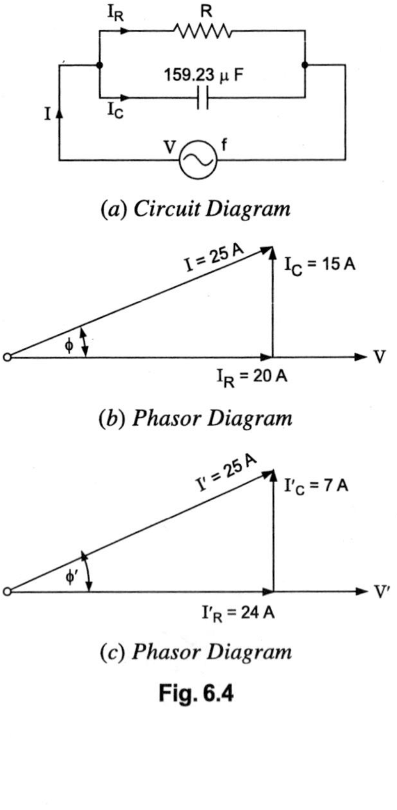

(*a*) Circuit Diagram

(*b*) Phasor Diagram

(*c*) Phasor Diagram

Fig. 6.4

$$I'_C = \sqrt{I'^2 - I_R'^2} = \sqrt{25^2 - 24^2} = 7 \text{ A}$$

$$\text{Capacitive reactance, } X_C = \frac{V'}{I'_C} = \frac{360}{7} = 51.43\ \Omega$$

$$\text{Supply frequency, } f' = \frac{1}{2\pi C X_C} = \frac{1}{2\pi \times 159.23 \times 10^{-6} \times 51.43} = 19.4 \text{ Hz } \textbf{Ans.}$$

Example 6.5 **An inductive coil of resistance 15 Ω and inductive reactance 42 Ω is connected in parallel with a capacitive reactance of 47.6 Ω. The combination is energized from a 200 V, 33.5 Hz ac supply. Find the total current drawn by the circuit and its power factor. Draw to the scale the phasor diagram of the circuit.**

[*Bombay University, 2000*]

Solution: Impedance of coil, $Z = \sqrt{R^2 + X_L^2} = \sqrt{15^2 + 42^2} = 44.6\ \Omega$

Current in inductive coil, $I_1 = \dfrac{V}{Z} = \dfrac{200}{44.6}\ \text{A} = 4.48\ \text{A}$

Phase angle, $\phi_1 = \text{Tan}^{-1}\dfrac{X_L}{R} = \text{Tan}^{-1}\dfrac{42}{15} = 70.35°$ (lagging)

Current through capacitance, $I_2 = \dfrac{V}{X_C} = \dfrac{200}{47.6} = 4.2\ \text{A}$

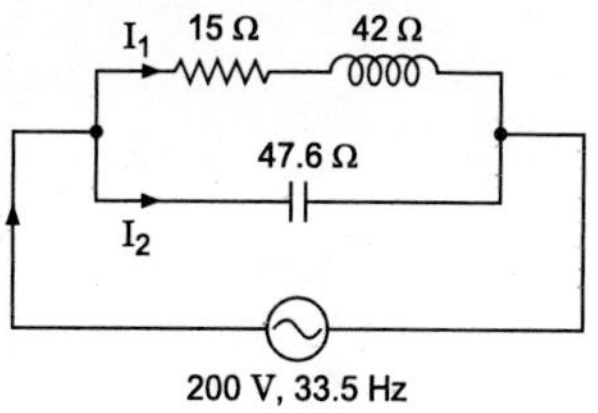

(a) Circuit Diagram

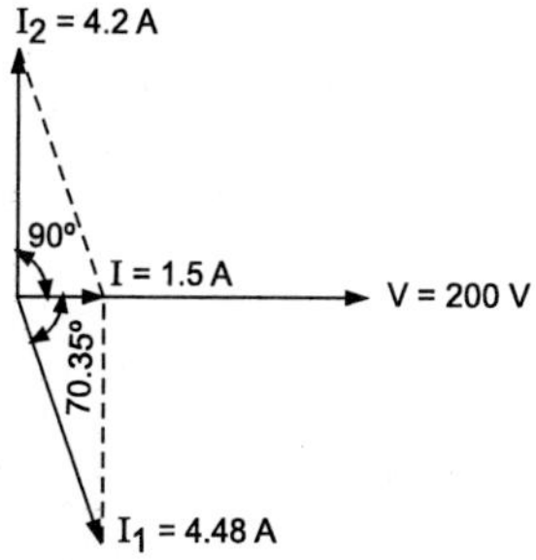

(b) Phasor Diagram

Fig. 6.5

Phasor diagram is given in Fig. 6.5(*b*).

Resolving current in in-phase and quadrature components we have

$$I\cos\phi = I_1\cos 70.35° + I_2\cos 90°$$
$$= 4.48 \times 0.3363 + 4.2 \times 0 = 1.5\ \text{A}$$
$$I\sin\phi = I_1\sin\phi_1 + I_2\sin\phi_2$$
$$= 4.48\sin(-70.35°) + 4.2\sin 90° = -4.2 + 4.2 = 0$$

Total current drawn by the circuit, $I = \sqrt{(I\cos\phi)^2 + I(\sin\phi)^2}$

$$= \sqrt{(1.5)^2 + 0^2} = 1.5\ \text{A}\ \textbf{Ans.}$$

Power factor of the circuit, $\cos\phi = \dfrac{I\cos\phi}{I} = \dfrac{1.5}{1.5} = 1.0$ **Ans.**

Example 6.6 **Two circuits having the same numerical ohmic impedance are joined in parallel. The power factor of one circuit is 0.8 (lag) and that of the other is 0.6 (lag). What is the power factor of the combination?**

Solution: As the impedances connected in parallel are of the same numerical value, therefore, the currents flowing through them will also be of the same numerical value, say, I amperes. However, their phase angles will be different as given below:

Phase angle of circuit I, $\phi_1 = \text{Cos}^{-1} 0.8 = 36.87°$ (lag)

Phase angle of circuit II, $\phi_2 = \text{Cos}^{-1} 0.6 = 53.13°$ (lag)

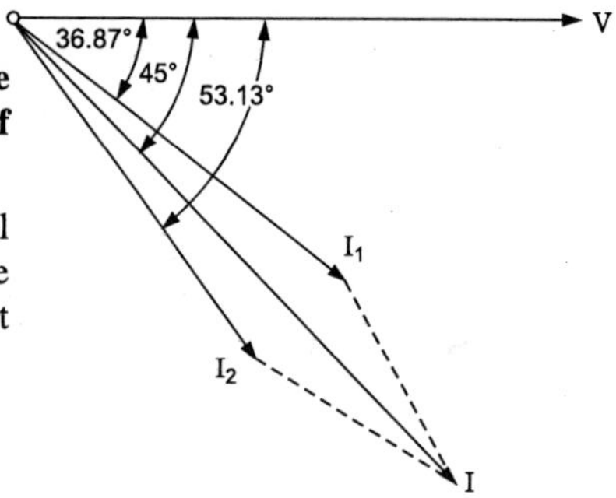

Fig. 6.6

Resolving currents in active and reactive components we have

Active component of resultant current, $I_r\cos\phi_r = I\cos\phi_1 + I\cos\phi_2 = I\cos(-36.87°) + I\cos(-53.13°)$

$$= I \times 0.8 + I \times 0.6 = 1.4\,I \quad ...(i)$$

Reactive component of resultant current, $I_r\sin\phi_r = I\sin\phi_1 + I\sin\phi_2 = I\sin(-36.87°) + I\sin(-53.13°)$

$$= I \times (-0.6) + I \times (-0.8) = -1.41 \quad ...(ii)$$

Dividing Eq. (*ii*) by Eq. (*i*) we have $\text{Tan}\,\phi_r = \dfrac{-1.4I}{1.4I} = -1.0$

or $\phi_r = \text{Tan}^{-1}(-1) = -45°$

Power factor of the combination $= \text{Cos}\,\phi_r = \text{Cos}(-45°) = 0.707$ (lag)

Phasor diagram is shown in Fig. 6.6.

Example 6.7 **Find the active and reactive components of current taken by a series circuit consisting of a coil of inductance 0.1 H, resistance 8 Ω and a capacitor of 120 μ F connected to a 240 V, 50 Hz supply mains. Find the value of the capacitor that has to be connected in parallel with the above series circuit so that the pf of the entire circuit is unity.**

Solution: Inductive reactance, $X_L = 2\pi fL = 2\pi \times 50 \times 0.1 = 31.416\ \Omega$

Capacitive reactance, $X_C = \dfrac{1}{2\pi fC} = \dfrac{1}{2\pi \times 50 \times 120 \times 10^{-6}} = 26.526\ \Omega$

$$\text{Circuit impedance, } Z = \sqrt{R^2 + (X_L - X_C)^2} = \sqrt{8^2 + (31.416 - 26.526)^2} = 9.376\ \Omega$$

$$\text{Circuit current, } I = \frac{V}{Z} = \frac{240}{9.376} = 25.6\ \text{A}$$

$$\text{Phase angle, } \phi = \text{Tan}^{-1}\frac{X_L - X_C}{R} = \text{Tan}^{-1}\frac{31.416 - 26.526}{8} = 31.435°$$

Active component of current $= I \cos \phi = 25.6 \cos 31.435° = 21.84$ A **Ans.**

Reactive component of current $= I \sin \phi = 25.6 \sin 31.435° = 13.35$ A (lagging) **Ans.**

The power factor of the whole circuit will be unity if the current drawn by the capacitor, connected across the series circuit, equals the reactive (lagging) component of the series circuit.

i.e. when $2\pi f C V = I \sin \phi$

$$\text{or } C = \frac{I \sin \phi}{2\pi f V} = \frac{13.35}{2\pi \times 50 \times 240} = 177\ \mu\text{F} \quad \textbf{Ans.}$$

6.3. ADMITTANCE METHOD

In the circuit shown in Fig. 6.1 (*a*), the branch currents

$$\mathbf{I}_1 = \frac{\mathbf{V}}{\mathbf{Z}_1} = \frac{\mathbf{V}}{\sqrt{R_1^2 + \left(X_{L_1} - X_{C_1}\right)^2}}$$

$$\mathbf{I}_2 = \frac{\mathbf{V}}{\mathbf{Z}_2} = \frac{\mathbf{V}}{\sqrt{R_2^2 + \left(X_{L_2} - X_{C_2}\right)^2}}$$

$$\mathbf{I}_3 = \frac{\mathbf{V}}{\mathbf{Z}_3} = \frac{\mathbf{V}}{\sqrt{R_3^2 + \left(X_{L_3} - X_{C_3}\right)^2}}$$

If **Z** is equivalent impedance of the circuit then total current flowing through the circuit,

$$\mathbf{I} = \frac{\mathbf{V}}{\mathbf{Z}}$$

$$\text{Since } \mathbf{I} = \mathbf{I}_1 + \mathbf{I}_2 + \mathbf{I}_3 \quad \text{(phasor sum)}$$

$$\therefore \frac{\mathbf{V}}{\mathbf{Z}} = \frac{\mathbf{V}}{\mathbf{Z}_1} + \frac{\mathbf{V}}{\mathbf{Z}_2} + \frac{\mathbf{V}}{\mathbf{Z}_3} \quad \text{(phasor sum)}$$

$$\text{or } \frac{1}{\mathbf{Z}} = \frac{1}{\mathbf{Z}_1} + \frac{1}{\mathbf{Z}_2} + \frac{1}{\mathbf{Z}_3} \quad \text{(phasor sum)}$$

The reciprocal of impedance is known as admittance and is represented by English capital letter **Y** and expressed in siemens (S)

$$\therefore \mathbf{Y} = \frac{1}{\mathbf{Z}};\ \mathbf{Y}_1 = \frac{1}{\mathbf{Z}_1};\ \mathbf{Y}_2 = \frac{1}{\mathbf{Z}_2} \text{ and } \mathbf{Y}_3 = \frac{1}{\mathbf{Z}_3}$$

$$\text{and } \mathbf{Y} = \mathbf{Y}_1 + \mathbf{Y}_2 + \mathbf{Y}_3 \quad \text{(phasor sum)}$$

$$\text{Resultant current, } \mathbf{I} = \mathbf{V} \times \frac{1}{\mathbf{Z}} = \mathbf{V} \times \mathbf{Y}$$

Just as impedance has two components, resistance R and reactance X similarly the admittance has two components known as *conductance*, G and *susceptance*, B respectively. The unit of conductance and susceptance is siemens (S).

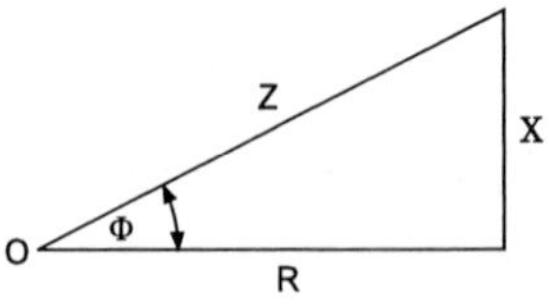

(*a*) *Impedance Triangle*

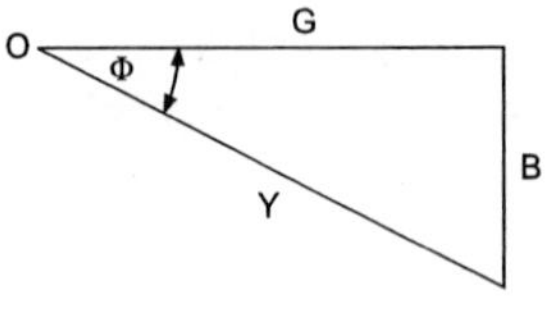

(*b*) *Admittance Triangle*

Fig. 6.7

From admittance and impedance triangles (Fig. 6.7)

$$\text{Conductance, } G = Y\cos\phi = \frac{1}{Z}\cdot\frac{R}{Z} = \frac{R}{Z^2} = \frac{R}{R^2+X^2}$$

$$\text{Susceptance, } B = Y\sin\phi = \frac{1}{Z}\cdot\frac{X}{Z} = \frac{X}{Z^2} = \frac{X}{R^2+X^2}$$

$$\text{Admittance, } Y = \sqrt{(Y\cos\phi)^2 + (Y\sin\phi)^2} = \sqrt{G^2+B^2}$$

$$\text{Phase angle, } \phi = \text{Tan}^{-1}\frac{B}{G}$$

6.3.1 Application of Admittance Method in Solution of Single Phase Parallel Circuits

Determine conductance and susceptance of individual branches from the relation

$$G = \frac{R}{Z^2} \text{ and } B = \frac{X}{Z^2}$$

taking B as + ve if X is capacitive and as –ve if X is inductive. Let the conductances of the three branches of circuit shown in Fig. 6.1 (*a*) be G_1, G_2 and G_3 respectively and susceptances be B_1, B_2 and B_3 respectively. Find the equivalent conductance and susceptance of the circuit which will be equal to algebraic sum of conductances and susceptances of individual branches respectively.

For the circuit shown in Fig. 6.1 (*a*)

$$\text{Total conductance, } G = G_1 + G_2 + G_3$$

$$\text{and total susceptance, } B = B_1 + B_2 + B_3$$

$$\text{Total admittance of the circuit, } Y = \sqrt{G^2+B^2}$$

The current in the circuit may be determined from the relation

$$I = V \times Y \text{ having phase angle } \phi = \text{Tan}^{-1}\frac{B}{G}$$

Note: Capacitive susceptance will be taken as + ve and inductive susceptance as – ve though capacitive reactance is taken as – ve and inductive reactance as + ve. If ϕ is + ve, current will lead and if ϕ is – ve, current will lag behind the applied voltage.

Example 6.8 The active and lagging reactive components of current taken by an ac circuit from a 250 V supply are 50 A and 25 A respectively. Calculate the conductance, susceptance, admittance and power factor of the circuit.

Solution: Active component of current, $I\cos\phi = 50$ A

Lagging reactive component of current, $I\sin\phi = 25$ A

$$\text{Circuit current, } I = \sqrt{(I\cos\phi)^2 + (I\sin\phi)^2} = \sqrt{50^2+25^2} = 55.9 \text{ A}$$

$$\text{Admittance of the circuit, } Y = \frac{I}{V} = \frac{55.9}{250} = 0.224 \text{ S } \textbf{Ans.}$$

$$\text{Power factor of the circuit, } \cos\phi = \frac{I\cos\phi}{I} = \frac{50}{55.9} = 0.8944 \text{ (lagging) } \textbf{Ans.}$$

$$\text{Conductance of the circuit, } G = Y\cos\phi = 0.224 \times 0.8944 = 0.2 \text{ S } \textbf{Ans.}$$

$$\text{Susceptance of the circuit, } B = \sqrt{Y^2 - G^2} = \sqrt{(0.224)^2 - 0.2^2} = 0.1 \text{ S (inductive) } \textbf{Ans.}$$

Example 6.9 A series ac circuit has a resistance of 15 Ω and inductive reactance of 10 Ω. Calculate the value of a capacitor which is connected across this series combination so that system has unity power factor. The frequency of ac supply is 50 Hz. [*U.P. Technical Univ. Electrical Engineering Second Semester 2005-06*]

Solution: Series circuit resistance, R = 15 Ω

Series circuit reactance, X = 10 Ω

$$\text{Series circuit conductance, } G = \frac{R}{R^2 + X^2} = \frac{15}{(15)^2 + (10)^2} = \frac{15}{325} \text{ S}$$

$$\text{Series circuit susceptance, } B = \frac{X}{R^2 + X^2} = \frac{10}{15^2 + 10^2} = \frac{10}{325} \text{ S (inductive)}$$

The power factor of the whole circuit will be unity only if capacitive susceptance of the capacitor to be connected across the series combination is equal to the equivalent inductive susceptance of the series combination *i.e.* equal to $\frac{10}{325}$ S.

Reactance of the capacitance connected across the series combination,

$$X_C = \frac{1}{B_C} = \frac{1}{10/325} = 32.5\ \Omega$$

$$\text{Capacitance of the required capacitor, } C = \frac{1}{2\pi f X_C} = \frac{1}{2\pi \times 50 \times 32.5} = 98\ \mu\text{F} \textbf{ Ans.}$$

Example 6.10 **A resistor of 12 Ω and an inductance of 0.025 H are connected in series across a 50 Hz supply. What values of resistance and inductance when connected in parallel will have the same resultant impedance and pf. Find the current in each case when the supply voltage is 230 V.** [*Nagpur University, November 1996*]

Solution: Series circuit resistance, $R_S = 12\ \Omega$

Series circuit reactance, $X_S = 2\pi f L = 2\pi \times 50 \times 0.025 = 7.854\ \Omega$

$$\text{Series circuit conductance, } G_S = \frac{R_S}{R_S^2 + X_S^2} = \frac{12}{12^2 + 7.854^2} = 0.05834 \text{ S}$$

$$\text{Series circuit susceptance, } B_S = \frac{X_S}{R_S^2 + X_S^2} = \frac{7.854}{12^2 + 7.854^2} = 0.03818 \text{ S (inductive)}$$

The circuits shown in Figs. 6.8 (*a*) and (*b*) will be equivalent (the resultant impedance and pf will be same) if

Conductance of parallel circuit,

G_p = Conductance of series circuit, G_S

i.e. $\frac{1}{R_p} = 0.05834$ S

or $R = 17.14\ \Omega$ **Ans.**

Susceptance of parallel circuit,

B_p = Susceptance of series circuit, B_S

i.e. $\frac{1}{X_p} = 0.03818$ S or $X_p = 26.192\ \Omega$

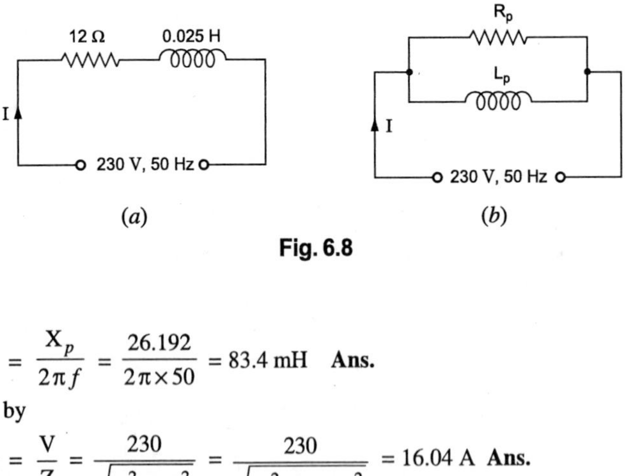

Fig. 6.8

$$\text{Inductance of equivalent parallel circuit, } L = \frac{X_p}{2\pi f} = \frac{26.192}{2\pi \times 50} = 83.4 \text{ mH} \quad \textbf{Ans.}$$

Current will be same in both cases and is given by

$$I = \frac{V}{Z} = \frac{230}{\sqrt{R_S^2 + X_S^2}} = \frac{230}{\sqrt{12^2 + 7.854^2}} = 16.04 \text{ A} \textbf{ Ans.}$$

$$\text{Phase angle, } \phi = \text{Tan}^{-1}\frac{X_S}{R} = \text{Tan}^{-1}\frac{7.854}{12} = 33.2° \text{ (lagging)} \quad \textbf{Ans.}$$

Example 6.11 **A variable capacitor is connected in series with a circuit consisting of a non-inductive resistance of 50 Ω in parallel with a coil across a 200 V, 50 Hz supply. The coil has an inductance of 0.2 H and negligible resistance.**

Calculate (*a*) The capacitance of the capacitor when the pf of the whole circuit is unity. (*b*) The corresponding pd across the capacitor.

Solution: $\text{Conductance of parallel circuit, } G = \frac{1}{R} = \frac{1}{50} = 0.02 \text{ S}$

$$\text{Susceptance of parallel circuit, } B = \frac{1}{X_L} = \frac{1}{2\pi f L} = \frac{1}{2\pi \times 50 \times 0.2} = 0.0159 \text{ S (inductive)}$$

$$\text{Equivalent resistance of parallel circuit, } R_{eq} = \frac{G}{G^2 + B^2} = \frac{0.02}{(0.02)^2 + (0.0159)^2} = 30.64\ \Omega$$

$$\text{Equivalent reactance of parallel circuit, } X_{eq} = \frac{B}{G^2 + B^2} = \frac{0.0159}{(0.02)^2 + (0.0159)^2} = 24.36\ \Omega \text{ (inductive)}$$

(*a*) The power factor of the whole circuit will be unity only if capacitive reactance is equal in magnitude to the equivalent inductive reactance of the parallel circuit.

i.e. $\frac{1}{\omega C} = 24.36\ \Omega$

$$\text{or } C = \frac{1}{\omega \times 24.36} = \frac{1}{2\pi \times 50 \times 24.36} = 130.7\ \mu\text{F } \textbf{Ans.}$$

$$I = \frac{V}{R_{eq}} = \frac{200}{30.64} \text{ A}$$

PD across the capacitor

$$= IX_C = \frac{200}{30.64} \times 24.36 = 159 \text{ volts } \textbf{Ans.}$$

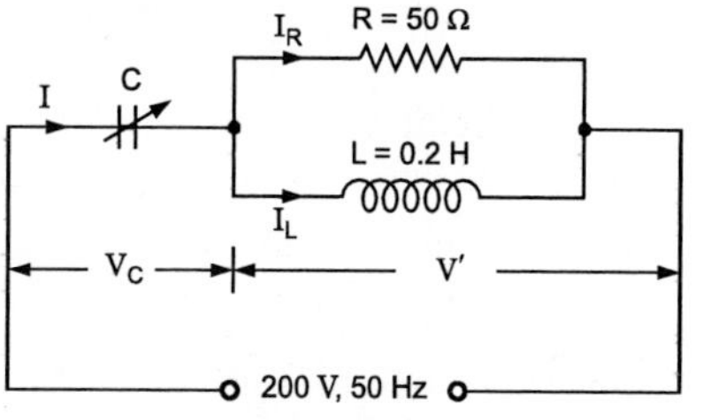

Fig. 6.9

Example 6.12 **For the circuit shown in Fig. 6.10, calculate I_1, I_2 and I_3. The values marked on the inductance and capacitance give their reactances.**

Solution:

$$\text{Impedance, } Z_2 = \sqrt{R_2^2 + X_2^2} = \sqrt{4^2 + 2^2} = \sqrt{20}\ \Omega$$

$$\text{Conductance, } G_2 = \frac{R_2}{Z_2^2} = \frac{4}{20} = 0.2 \text{ S}$$

$$\text{Susceptance, } B_2 = \frac{X_2}{Z_2^2} = \frac{2}{20} = 0.1 \text{ S (inductive)}$$

$$\text{Impedance, } Z_3 = \sqrt{R_3^2 + X_3^2} = \sqrt{1^2 + 5^2} = \sqrt{26}\ \Omega$$

$$\text{Conductance, } G_3 = \frac{R_3}{Z_3^2} = \frac{1}{26} = 0.0384 \text{ S}$$

$$\text{Susceptance, } B_3 = \frac{X_3}{Z_3^2} = \frac{5}{26} = 0.1923 \text{ S (capacitive)}$$

Total conductance of parallel combination, $G_P = G_2 + G_3 = 0.2 + 0.0384 = 0.2384$ S

Total susceptance of parallel combination, $B_P = B_2 + B_3 = (-0.1) + 0.1923 = 0.0923$ S (capacitive)

$$\text{Equivalent resistance of parallel combination, } R_P = \frac{G_P}{Y_P^2} = \frac{G_P}{G_P^2 + B_P^2} = \frac{0.2384}{(0.2384)^2 + (0.0923)^2} = 3.65\ \Omega$$

$$\text{Equivalent reactance of parallel combination, } X_P = \frac{B_P}{Y_P^2} = \frac{0.0923}{(0.2384)^2 + (0.0923)^2} = 1.41\ \Omega \text{ (capacitive)}$$

Equivalent resistance of whole circuit,

$$R = R_1 + R_{BC} = R_1 + R_P = 2 + 3.65 = 5.65\ \Omega$$

Equivalent reactance of whole circuit,

$$X = X_1 + X_P = 3 + (-1.41) = 1.59\ \Omega$$

Equivalent impedance of whole circuit,

$$Z = \sqrt{R^2 + X^2} = \sqrt{(5.65)^2 + (1.59)^2} = 5.87\ \Omega$$

$$\text{Current, } I_1 = \frac{V}{Z} = \frac{10}{5.87} = 1.7 \text{ A } \textbf{Ans.}$$

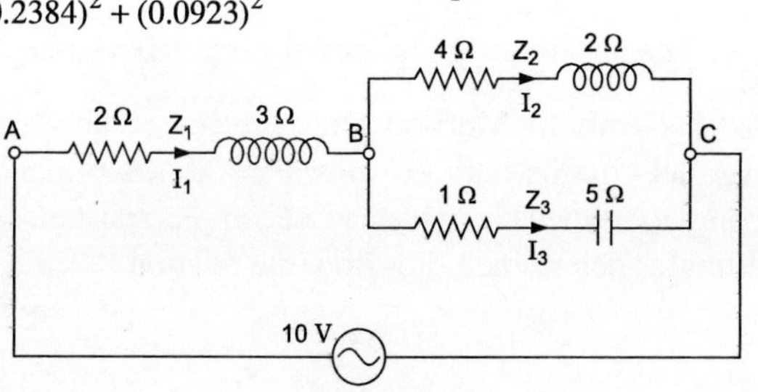

Fig. 6.10

Voltage drop across parallel circuit, $V_P = I \times Z_P = 1.7 \times \sqrt{(3.65)^2 + (1.41)^2} = 6.652$ V

Current through impedance Z_2, $I_2 = \frac{V_P}{Z_2} = \frac{6.652}{\sqrt{20}} = 1.49$ A **Ans.**

Current through impedance Z_3, $I_3 = \frac{V_P}{Z_3} = \frac{6.652}{\sqrt{26}} = 1.3$ A **Ans.**

6.4. SYMBOLIC OR j-METHOD

When two or more circuits are connected in parallel, the current in each branch may be found in complex form and then total current in the circuit may be determined by addition of complex numbers, as detailed in Art. 5.7.

In the circuit shown in Fig. 6.11.

Current in branch 1, $\mathbf{I}_1 = \frac{\mathbf{V}}{R_1 + jX_{L_1}} = \frac{\mathbf{V}}{R_1 + jX_{L_1}} \times \frac{R_1 - jX_{L_1}}{R_1 - jX_{L_1}} = \mathbf{V}\left(\frac{R_1}{R_1^2 + X_{L_1}^2} - j\frac{X_{L_1}}{R_1^2 + X_{L_1}^2}\right)$

Similarly current in branch 2,

$$\mathbf{I}_2 = \mathbf{V}\left(\frac{R_2}{R_2^2 + X_{L_2}^2} - j\frac{X_{L_2}}{R_2^2 + X_{L_2}^2}\right)$$

and current in branch 3, $\mathbf{I}_3 = \mathbf{V}\left(\frac{R_3}{R_3^2 + X_C^2} + j\frac{X_C}{R_3^2 + X_C^2}\right)$

Total circuit current, $\mathbf{I} = \mathbf{I}_1 + \mathbf{I}_2 + \mathbf{I}_3$

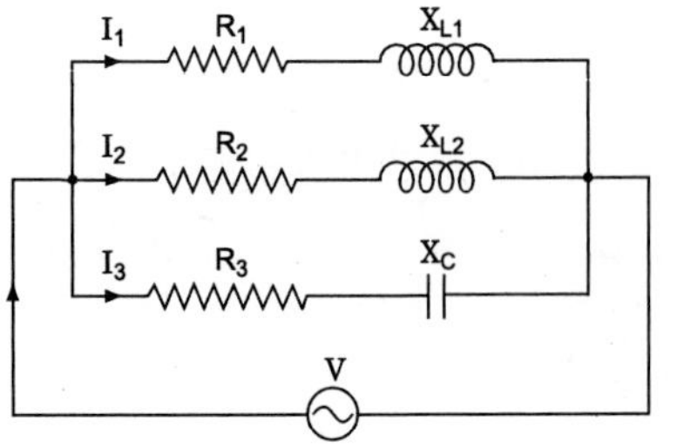

Fig. 6.11

6.4.1. Equivalent Parallel Impedance. The ac parallel circuit, may be solved in the same way as dc parallel circuit except that complex impedances rather than simple resistances are involved. With several impedances connected in parallel the equivalent impedance is expressed as

$$\frac{1}{\mathbf{Z}} = \frac{1}{\mathbf{Z}_1} + \frac{1}{\mathbf{Z}_2} + \frac{1}{\mathbf{Z}_3} + \ldots\ldots$$

Since admittance is reciprocal of impedance, therefore,

$$\mathbf{Y} = \frac{1}{\mathbf{Z}} = \mathbf{Y}_1 + \mathbf{Y}_2 + \mathbf{Y}_3 + \ldots\ldots$$

After finding equivalent impedance or admittance, current in the circuit may be determined from the relation

$$\mathbf{I} = \frac{\mathbf{V}}{\mathbf{Z}} = \mathbf{V} \times \mathbf{Y}$$

where $\mathbf{Y} = \frac{1}{\mathbf{Z}} = \frac{1}{(R \pm jX)} = \frac{R \mp jX}{(R \pm jX)(R \mp jX)} = \frac{R}{R^2 + X^2} \mp j\frac{X}{R^2 + X^2} = G \mp B$

The negative sign is used for inductive susceptance and positive sign for capacitive susceptance.

6.4.2. Symbolic Method For Solution of Single Phase Series-Parallel Circuits. Single phase ac series-parallel circuits may be solved in the same manner as dc series-parallel circuits are solved except that complex impedances instead of simple resistances are involved. The equivalent impedance of a parallel circuit is determined first from the relation

$$\frac{1}{\mathbf{Z}_{eq}} = \frac{1}{\mathbf{Z}_1} + \frac{1}{\mathbf{Z}_2} + \frac{1}{\mathbf{Z}_3} + \ldots$$

then it is added (in complex form) to the impedance of the circuit connected in series with the combined parallel circuit and thus total impedance of the whole circuit is determined. The problems can be solved as discussed before.

Example 6.13 Consider an electric circuit shown in Fig. 6.12. Determine (*i*) the current and power consumed in each branch, (*ii*) the supply current and power factor. [*U.P. Technical Univ. Electrical Engineering, February, 2001*]

Solution: Current in 10 Ω branch, $\mathbf{I}_1 = \frac{\mathbf{V}}{\mathrm{R}} = \frac{100\angle 45^\circ}{10\angle 0^\circ} = 10\angle 45^\circ$ A **Ans.**

Current in inductive branch, $\mathbf{I}_2 = \frac{\mathbf{V}}{\mathbf{Z}_2} = \frac{100\angle 45^\circ}{5 + j5\sqrt{3}} = \frac{100\angle 45^\circ}{10\angle 60^\circ} = 10\angle -15^\circ$ A **Ans.**

Current in capacitive branch, $\mathbf{I}_3 = \frac{\mathbf{V}_3}{\mathbf{Z}_3} = \frac{100\angle 45^\circ}{5 - j5\sqrt{3}} = \frac{100\angle 45^\circ}{10\angle -60^\circ} = 10\angle 105^\circ$ A **Ans.**

Power consumed in resistive branch, $\mathrm{P}_1 = \mathrm{I}_1^2\,\mathrm{R} = 10^2 \times 10$

$= 1{,}000$ W **Ans.**

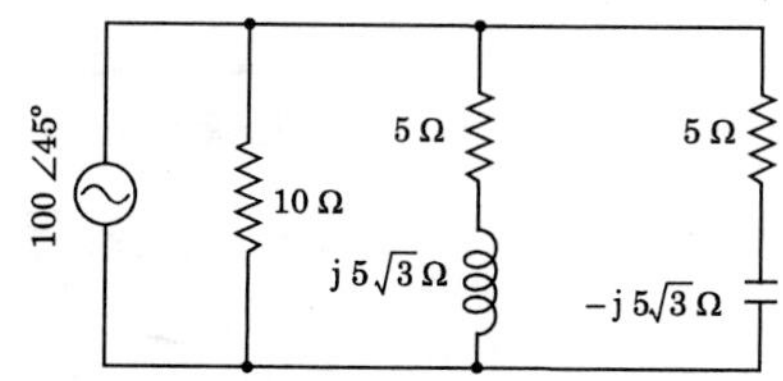

Fig. 6.12

Power consumed in inductive branch, $\mathrm{P}_2 = \mathrm{I}_2^2\,\mathrm{R}_2 = 10^2 \times 5$

$= 500$ W **Ans.**

Power consumed in capacitive branch, $\mathrm{P}_3 = \mathrm{I}_3^2\,\mathrm{R}_3 = 10^2 \times 5$

$= 500$ W **Ans.**

(*ii*) Supply current, $\mathbf{I} = \mathbf{I}_1 + \mathbf{I}_2 + \mathbf{I}_3$

$= 10\angle 45^\circ + 10\angle -15^\circ + 10\angle 105^\circ$

$= (7.071 + j\,7.071) + (9.659 - j\,2.588) + (-2.588 + j\,9.659)$

$= 14.142 + j\,14.142 = 20\angle 45^\circ$ amperes **Ans.**

Power factor, $\cos\phi = \mathrm{Cos}\,(45^\circ - 45^\circ) = 1.0$ **Ans.**

Example 6.14 Two impedances $\mathbf{Z}_1 = (10 + j5)$ and $\mathbf{Z}_2 = (8 + j6)$ are connected in parallel across a voltage of $\mathbf{V} = 200 + j0$. Calculate the circuit current, power factor and reactive power.

[*U.P. Technical Univ. Elec. Engineering First Semester 2004-05*]

Solution : Equivalent impedance of parallel combination,

$$\mathbf{Z} = \frac{\mathbf{Z}_1\,\mathbf{Z}_2}{\mathbf{Z}_1 + \mathbf{Z}_2} = \frac{(10 + j5)\,(8 + j6)}{(10 + j5) + (8 + j6)} = \frac{(10 + j5)(8 + j6)}{(18 + j11)}$$

$$= \frac{11.18\angle 26.56^\circ \times 10\angle 36.87^\circ}{21.095\angle 31.43^\circ} = 5.3\angle 32^\circ\ \Omega$$

Circuit current, $\mathbf{I} = \frac{\mathbf{V}}{\mathbf{Z}} = \frac{200\angle 0^\circ}{5.3\angle 32^\circ} = 37.736\angle -32^\circ$ A

The circuit current, I = 37.736 A **Ans.**

Phase angle, $\phi = -32^\circ$

Power factor $= \mathrm{Cos}\,\phi = \mathrm{Cos}(-32^\circ) = 0.848$ (lagging) **Ans.**

Reactive power, $\mathrm{Q} = \mathrm{VI}\sin\phi = 200 \times 37.736 \times \sin(-32^\circ) = -4{,}000$ VAR or 4 kVAR (lagging) **Ans.**

Example 6.15 **Obtain the power factor of a two branch parallel circuit where the first branch has $\mathbf{Z_1} = (2 + j\,4)$ and second $\mathbf{Z_2} = (6 + j\,0)$. To what value must the 6 Ω resistor be changed to result in the overall power factor 0.9 lagging ?** [*Electrical Engineering U.P. Technical Univ. June 2001; G.B. Technical Univ. Even Semester 2012-13*]

Solution:

$$\text{Admittance of branch I, } \mathbf{Y_1} = \frac{1}{\mathbf{Z_1}} = \frac{1}{2 + j4} = \frac{2 - j4}{2^2 + 4^2} = (0.1 - j\,0.2)\text{ S}$$

$$\text{Admittance of branch II, } \mathbf{Y_2} = \frac{1}{6 + j0} = (0.167 + j\,0)\text{ S}$$

$$\text{Equivalent admittance of the circuit, } \mathbf{Y} = \mathbf{Y_1} + \mathbf{Y_2} = (0.1 - j\,0.2) + (0.167 + j\,0) = (0.267 - j\,0.2)\text{ S}$$

$$\text{The magnitude of equivalent admittance, } \mathbf{Y} = \sqrt{(0.267)^2 + (0.2)^2} = 0.333\text{ S}$$

$$\text{Power factor of the circuit, } \cos\phi = \frac{0.267}{0.333} = 0.8\text{ (lagging)} \quad \textbf{Ans.}$$

Let the resistance of the resistor of Z_2 be changed to R ohms so as to make the power factor of whole circuit 0.9 (lagging)

$$\text{Now total conductance of the circuit, G} = \left(\frac{1}{R} + 0.1\right)\text{ S}$$

$$\text{Total susceptance of the circuit, } B' = B = -0.2\text{ S}$$

$$\text{Phase angle of the circuit, } \phi' = \cos^{-1} 0.9 = -25.84\text{, being lagging}$$

$$\text{Phase angle of the circuit is given as } \phi' = \tan^{-1}\frac{B'}{G'}$$

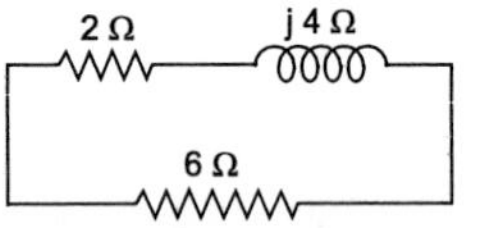

Fig. 6.13

$$\text{or } \tan^{-1}\frac{-0.2}{\frac{1}{R} + 0.1} = -25.84^\circ \quad \text{or} \quad R = 3.2\ \Omega \quad \textbf{Ans.}$$

Example 6.16 **An inductive coil of $(6 + j\,8)$ Ω impedance is connected to 100 V, 50 Hz ac supply. It is desired to improve power factor of supply current to 0.8 lagging by connecting a capacitor (*i*) is in series of the coil (*ii*) in parallel of the coil. Determine the value of the capacitance in each case.**

[*U.P. Technical Univ. Electrical Engineering February 2002*]

Solution: Let a capacitor of capacitance C be connected in series with the coil so as to raise the pf of supply current to 0.8 lagging, as shown in Fig. 6.14.

$$\text{Resistance of the circuit, R} = 6\ \Omega$$

$$\text{Inductive reactance of the series circuit, X} = 8 - \frac{1}{2\pi f C}$$

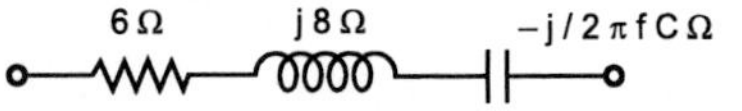

Fig. 6.14

$$\text{Improved pf of the circuit, } \cos\phi = 0.8\text{ lagging}$$

$$\tan\phi = \tan\cos^{-1} 0.8 = 0.75$$

$$\text{Since } \frac{X}{R} = \tan\phi$$

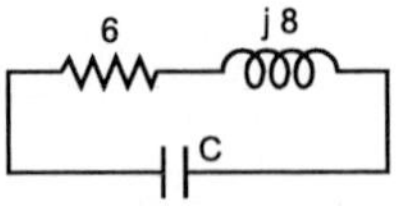

Fig. 6.15

$$\text{So } \frac{8 - \frac{1}{2\pi f C}}{6} = 0.75$$

$$\text{or } \frac{1}{2\pi f C} = 8 - 6 \times 0.75 = 3.5\ \Omega$$

$$\text{or } C = \frac{1}{2\pi f \times 3.5} = \frac{1}{2\pi \times 50 \times 3.5} = 909.4568\ \mu\text{F} \quad \textbf{Ans.}$$

(*ii*) Let a capacitor of capacitance C be connected across the coil so as to raise the pf of supply current to 0.8 lagging, as shown in Fig. 6.15.

$$\text{Conductance of coil, G} = \frac{R}{R^2 + X^2} = \frac{6}{6^2 + 8^2} = 0.06\text{ S}$$

$$\text{Susceptance of coil, } B_L = \frac{X}{R^2 + X^2} = \frac{8}{6^2 + 8^2} = 0.08 \text{ S (inductive)}$$

$$\text{Capacitive susceptance of capacitor C, } B_C = 2\pi f C$$

$$\text{Since } \frac{B}{G} = \text{Tan}\,\phi$$

$$\text{or } \frac{B_L - B_C}{G} = 0.75$$

$$\text{or } B_L - B_C = 0.75 \times G = 0.75 \times 0.06 = 0.045 \text{ S}$$

$$\text{or } B_C = B_L - 0.045 = 0.08 - 0.045 = 0.035 \text{ S}$$

$$\text{or } 2\pi f C = 0.035$$

$$\text{or } C = \frac{0.035}{2\pi \times 50} = 111.4\ \mu\text{F} \textbf{ Ans.}$$

Example 6.17 **Two impedances $Z_1 = (10 + j15)$ ohms and $Z_2 = (6 - j8)$ ohms are connected in parallel. The total current supplied is 15 A. What is the power taken by each impedance?**

[G.B. Technical Univ. Electrical Engineering Second Semester 2009-10]

Solution: Equivalent impedance of parallel combination,

$$\mathbf{Z} = \frac{\mathbf{Z}_1 \mathbf{Z}_2}{\mathbf{Z}_1 + \mathbf{Z}_2} = \frac{(10 + j15)(6 - j8)}{(10 + j15) + (6 - j8)} = \frac{(10 + j15)(6 - j8)}{(16 + j7)}$$

$$\text{Total current supplied, I} = 15 \text{ A}$$

$$\text{Taking supply current I as reference phasor, } \mathbf{I} = (15 + j0) \text{ A}$$

$$\text{Current through impedance } Z_1, \mathbf{I}_1 = \frac{\mathbf{IZ}}{\mathbf{Z}_1} = \frac{15(10 + j15)(6 - j8)}{(16 + j7)(10 + j15)} = \frac{15(6 - j8)}{(16 + j7)} = \frac{15 \times 10 \angle -53.13^\circ}{17.464 \angle -23.63^\circ}$$

$$= 8.589 \angle -76.76^\circ = (1.967 - j\,8.361) \text{ A}$$

$$\text{Current through impedance } Z_2, \mathbf{I}_2 = \frac{\mathbf{I} \times \mathbf{Z}}{\mathbf{Z}_2} = \frac{15(10 + j15)(6 - j8)}{(16 + j7)(6 - j8)} = \frac{15(10 + j15)}{(16 + j7)} = \frac{15 \times 18.028 \angle 56.31^\circ}{17.464 \angle 23.63^\circ}$$

$$= 15.484 \angle 32.68^\circ = (13.033 + j\,8.361) \text{ A}$$

$$\text{Power absorbed in branch 1, } P_1 = I_1^2 R_1 = (8.589)^2 \times 10 = 738 \text{ W} \textbf{ Ans.}$$

$$\text{Power absorbed in branch 2, } P_2 = I_2^2 R_2 = (15.484)^2 \times 6 = 1{,}438 \text{ W} \textbf{ Ans.}$$

Example 6.18 **Two impedances $Z_1 = 5 + j\,10\ \Omega$ and $Z_2 = 10 - j\,15\ \Omega$ are connected in parallel. If total current is 20 A then find: (*a*) current taken by each branch, (*b*) power factor (*c*) power consumed in each branch.**

[U.P. Technical Univ. Basic Electrical Engineering Odd Semester 2013-14]

Solution: Equivalent impedance of parallel combination,

$$\mathbf{Z} = \frac{\mathbf{Z}_1 \mathbf{Z}_2}{\mathbf{Z}_1 + \mathbf{Z}_2} = \frac{(5 + j10)(10 - j15)}{(5 + j10) + (10 - j15)}$$

$$= \frac{200 + j25}{(15 - j5)} = \frac{201.556 \angle 7.1250^\circ}{15.811 \angle -18.435^\circ}$$

$$= 12.748 \angle 7.125^\circ - (-18.435^\circ) = 12.748 \angle 25.56^\circ$$

$$\text{Total current supplied, I} = 20 \text{ A}$$

$$\text{Taking supply current as reference phasor, } \mathbf{I} = (20 + j0) \text{ A}$$

$$(a) \text{ Current through impedance } Z_1, \mathbf{I}_1 = \frac{\mathbf{IZ}}{\mathbf{Z}_1} = \frac{(20 + j0)(5 + j10)(10 - j15)}{(5 + j10) + (10 - j15)} \times \frac{1}{(5 + j10)}$$

$$= \frac{20(10-j15)}{(15-j5)} = \frac{20\times 18.0278 \angle -56.31^\circ}{15.811 \angle -18.435^\circ} = 22.804 \angle -37.945^\circ \text{ A}$$

$$= (18.0 - j\,14^\circ) \text{ A } \textbf{Ans.}$$

Current through impedance Z_2, $\mathbf{I_2} = \dfrac{\mathbf{IZ}}{\mathbf{Z_2}} = \dfrac{(20+j0)(5+j10)(10-j15)}{(15-j5)(10-j15)} \times \dfrac{1}{10-j15}$

$$= \frac{20(5+j10)}{15-j5} = \frac{20\times 11.18 \angle 63.435^\circ}{15.811 \angle -18.435^\circ}$$

$$= 14.142 \angle 81.87^\circ \text{ A} = (2 + j\,14.0) \text{ A } \textbf{Ans.}$$

(*b*) Power factor = Cos 25.56° = 0.9 (lagging) **Ans.**

(*c*) Power consumed in branch 1 = $I_1^2R_1 = (22.804)^2 \times 5 = 2{,}600$ W **Ans.**

Power consumed in branch 2 = $I_2^2R_2 = (14.142)^2 \times 10 = 2{,}000$ W **Ans.**

Example 6.19 The parallel circuit shown in Fig. 6.16 is connected across a single phase 100 V, 50 Hz ac supply. Calculate (*i*) the branch currents (*ii*) the total current (*iii*) the supply power factor (*iv*) the active and reactive power supplied by the supply. *[U.P. Technical Univ. Electrical Engineering First Semester 2006-07]*

Solution: Admittance of branch 1,

$$\mathbf{Y}_1 = \frac{1}{\mathbf{Z}_1} = \frac{1}{(8+j\,6)} = (0.08 - j\,0.06) \text{ S}$$

Admittance of branch 2,

$$\mathbf{Y}_2 = \frac{1}{\mathbf{Z}_2} = \frac{1}{(6-j\,8)} = (0.06 + j\,0.08) \text{ S}$$

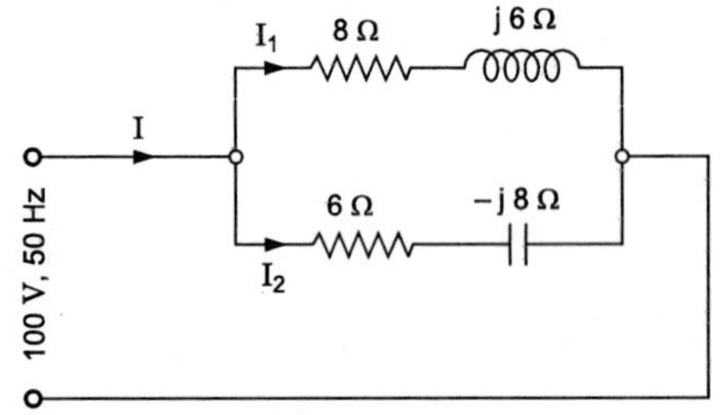

Fig. 6.16

Total admittance of the circuit,

$$\mathbf{Y} = \mathbf{Y}_1 + \mathbf{Y}_2 = (0.08 - j\,0.06) + (0.06 + j\,0.08) = (0.14 + j\,0.02) \text{ S}$$

(*i*) Branch currents, $\mathbf{I}_1 = \mathbf{V} \times \mathbf{Y}_1 = (100 + j\,0)(0.08 - j\,0.06) = (8 - j\,6)$ A or $10 \angle -36.87^\circ$ A **Ans.**

$\mathbf{I}_2 = \mathbf{V} \times \mathbf{Y}_2 = (100 + j\,0)(0.06 + j\,0.08) = (6 + j\,8)$ A or $10 \angle 53.13^\circ$ A **Ans.**

(*ii*) Total current, $\mathbf{I} = \mathbf{V} \times \mathbf{Y} = (100 + j\,0)(0.14 + j\,0.02) = (14 + j\,2)$ A or $14.42 \angle 8.13^\circ$ A **Ans.**

(*iii*) Supply power factor = Cos ϕ = Cos 8.13° = 0.99 (leading) **Ans.**

(*iv*) Active power supplied, P = V I cos ϕ = 100 × 14.142 × 0.99 = 1400 W **Ans.**

Reactive power supplied, Q = V I sin ϕ = 100 × 14.142 sin 8.13° = 200 VAR (capacitive) **Ans.**

Example 6.20 A 15 mH inductor is in series with a parallel combination of an 80 Ω resistor and 20 μ F capacitor. If the angular frequency of the applied voltage is ω = 1,000 radians/second, find the admittance of the network. *[Osmania Univ. Basic Circuit Analysis Jan/Feb. 1992]*

Solution: Inductive reactance, $X_L = \omega L = 1{,}000 \times 15 \times 10^{-3} = 15\ \Omega$

Capacitive reactance, $X_C = \dfrac{1}{\omega C} = \dfrac{1}{1{,}000 \times 20 \times 10^{-6}} = 50\ \Omega$

$R = 80\ \Omega$

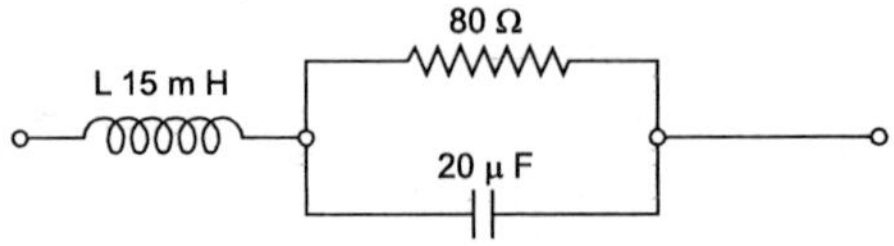

Fig. 6.17

Admittance of parallel circuit, $\mathbf{Y}_P = \dfrac{1}{80} + j\dfrac{1}{50}$

$= (0.0125 + j\,0.02)$ S

Impedance of parallel circuit, $\mathbf{Z}_P = \dfrac{1}{\mathbf{Y}_P} = \dfrac{1}{(0.0125 + j\,0.02)} = (22.48 - j\,35.96)\ \Omega$

Total impedance of the circuit, $\mathbf{Z} = (0 + j\,X_L) + \mathbf{Z}_P = (0 + j\,15) + (22.48 - j\,35.96) = (22.48 - j\,20.96)\ \Omega$

Admittance of the circuit, $\mathbf{Y} = \dfrac{1}{\mathbf{Z}} = \dfrac{1}{22.48 - j20.96} = (0.238 + j\,0.022)$ siemens **Ans.**

Example 6.21 **In the circuit shown in Fig. 6.18 (*a*), the reactance of the capacitor C_1 is 4 Ω, the reactance of capacitor C_2 is 8 Ω and the reactance of L is 8 Ω. A sinusoidal voltage of 120 V is applied to the circuit. Find (*i*) current in each branch (*ii*) power loss in the circuit.** [*U.P. Technical Univ. Electrical Engineering February 2002*]

Solution: The circuit given in Fig. 6.18 (*a*) is redrawn with given values marked on it and is shown in Fig. 6.18 (*b*).

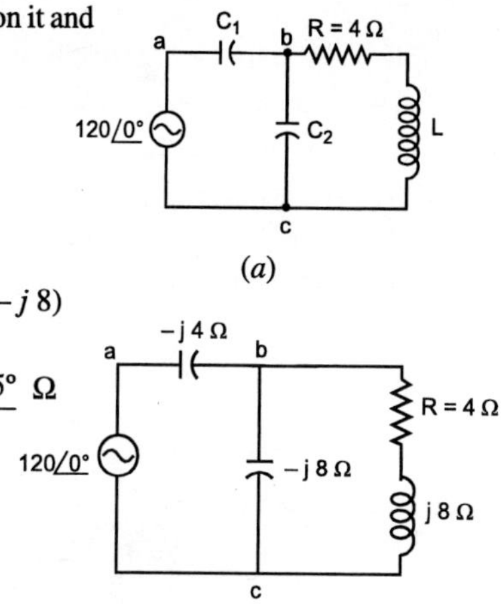

(*a*)

(*b*)

Fig. 6.18

$$\text{Impedance of parallel branch, } \mathbf{Z}_{bc} = \frac{1}{\dfrac{1}{-j8}+\dfrac{1}{(4+j8)}} = \frac{-j\,8\times(4+j\,8)}{-j\,8+4+j\,8}$$

$$= \frac{64-j\,32}{4} = (16-j\,8)$$

$$= 17.89\,\angle{-26.565°}\ \Omega$$

$$\text{Total impedance of circuit, } \mathbf{Z} = \mathbf{Z}_{ab}+\mathbf{Z}_{bc} = (-j\,4)+(16-j\,8)$$

$$= (16-j\,12) = 20\,\angle{-36.87°}\ \Omega$$

$$\text{Circuit current, } \mathbf{I} = \frac{\mathbf{V}}{\mathbf{Z}} = \frac{120\,\angle 0°}{20\,\angle{-36.87°}} = 6\,\angle 36.87°\text{ A}$$

$$= (4.8+j\,3.6)\text{ A}$$

$$\text{Current in branch } ab,\ \mathbf{I}_{ab} = \mathbf{I} = (4.8+j\,3.6)\text{ A} \quad \textbf{Ans.}$$

$$\text{Voltage, } \mathbf{V}_{bc} = \mathbf{I}\times\mathbf{Z}_{bc} = 6\,\angle 36.87° \times 17.89\,\angle{-26.565°}\text{ V}$$

$$= 107.33\,\angle 10.305°\text{ V} = (105.6+j\,19.2)\text{ V}$$

$$\text{Current in 8 } \Omega \text{ capacitor} = \frac{\mathbf{V}_{bc}}{-j\,8} = \frac{107.33\,\angle 10.305°}{8\,\angle{-90°}} = 13.416\,\angle 100.305°\text{ A} = (-2.4+j\,13.2)\text{ A} \quad \textbf{Ans.}$$

Current in 4 Ω resistor = Current in 8 Ω inductor

$$= \frac{\mathbf{V}_{bc}}{(4+j\,8)} = \frac{107.33\,\angle 10.305°}{8.944\,\angle 63.435°}\text{ A} = 12\,\angle{-53.13°}\text{ A} = (7.2-j\,9.6)\text{ A} \quad \textbf{Ans.}$$

$$\text{Power loss in the circuit, P} = \text{V}\times\text{I}\times\cos\phi = 120\times 6\times\cos 36.87° = 576\text{ W} \quad \textbf{Ans.}$$

Example 6.22 **Determine the power factor for the circuit given in Fig. 6.19.**
[*G.B. Technical Univ. Electrical Engineering Second Semester 2010-11*]

Solution: Equivalent impedance of the circuit.

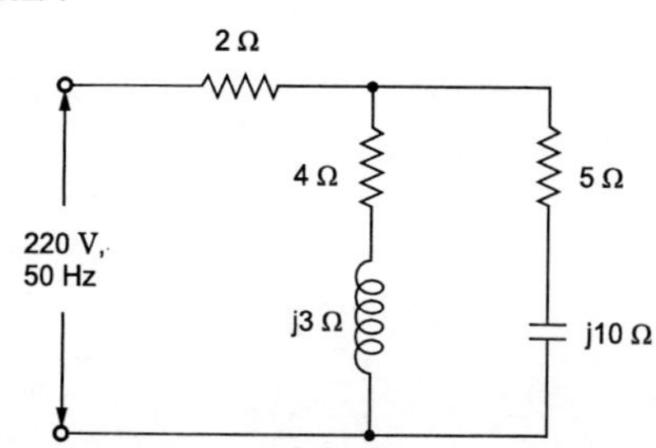

Fig. 6.19

$$Z = (2+j0)+\frac{1}{\dfrac{1}{4+j3}+\dfrac{1}{5-j10}}$$

$$Z = (2+j0)+\frac{(4+j3)(5-j10)}{(4+j3)+(5-j10)}$$

$$= (2+j0)+\frac{20-j25+30}{(9-j7)}$$

$$= (2+j0)+\frac{(50-j25)}{(9-j7)}\times\frac{(9+j7)}{(9+j7)} = (2+j0)+\frac{450+175-j125}{130}$$

$$= (2+j0)+\frac{625-j\,125}{130} = \frac{885-j\,125}{130}\ \Omega$$

$$\text{Phase angle of the given circuit, } \phi = \text{Tan}^{-1}\frac{-125/130}{885/130} = -8.039°$$

Power factor of the given circuit, $\cos\phi$ = Cos (–8.039°) = 0.99 (lagging) **Ans.**

Example 6.23 **Figure 6.20 (*a*) shows a series-parallel circuit. Find:**

(*i*) **Admittance of each parallel branch.**

(*ii*) **Total circuit impedance.**

(*iii*) **Supply current and power factor.**

(*iv*) **Total power supplied by the source.**

[U.P. Technical Univ. Elec. Engineering First Semester 2005-06]

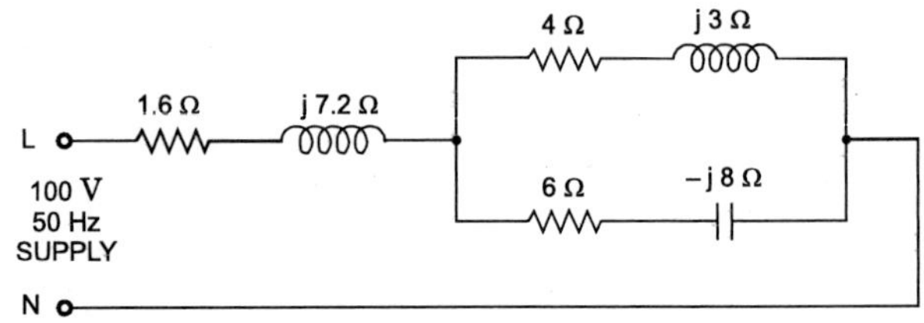

Fig. 6.20 (*a*)

Solution: Admittance of parallel branches,

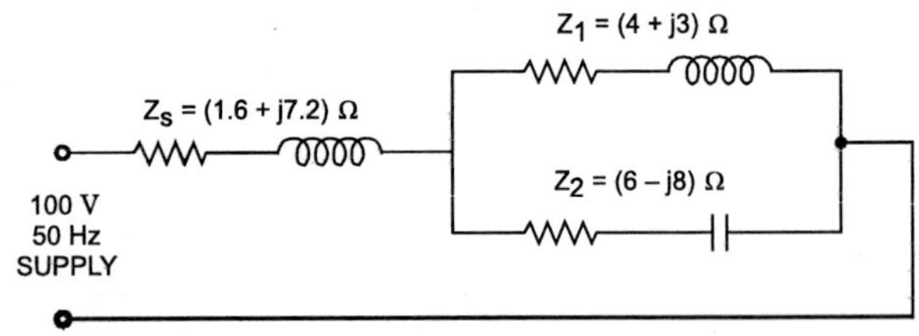

Fig. 6.20 (*b*)

$$\mathbf{Y}_1 = \frac{1}{\mathbf{Z}_1} = \frac{1}{4+j3} = \frac{4-j3}{4^2+3^2} = (0.16 - j\,0.12)\text{S} = 0.2\angle{-36.87°}\ \text{S} \quad \textbf{Ans.}$$

$$\mathbf{Y}_2 = \frac{1}{\mathbf{Z}_2} = \frac{1}{6-j8} = \frac{6+j8}{6^2+8^2} = (0.06 + j0.08)\text{S or } 0.1\angle 53.13°\ \text{S} \quad \textbf{Ans.}$$

Admittance of parallel branches, $\mathbf{Y}_p = \mathbf{Y}_1 + \mathbf{Y}_2 = (0.16 - j\,0.12) + (0.06 + j\,0.08) = (0.22 - j\,0.04)$ S

$$\text{Impedance of the parallel circuit, } \mathbf{Z}_p = \frac{1}{\mathbf{Y}_p} = \frac{1}{0.22 - j0.04} = \frac{0.22 + j0.04}{(0.22)^2 + (0.04)^2} = (4.4 + j0.8)\ \Omega$$

(*ii*) Total circuit impedance, $Z = \mathbf{Z}_s + \mathbf{Z}_p = (1.6 + j7.2) + (4.4 + j0.8) = (6 + j8)\ \Omega = 10\angle 53.13°\ \Omega$ **Ans.**

$$(iii)\ \text{Supply current, } I = \frac{V}{Z} = \frac{100}{10} = 10\ \text{A} \quad \textbf{Ans.}$$

Power factor, $\cos\phi$ = Cos 53.13° = 0.6 (lagging) **Ans.**

(*iv*) Total power supplied by the source, $P = VI\cos\phi = 100 \times 10 \times 0.6 = 600$ W **Ans.**

Example 6.24 **Determine the following in the circuit shown in Fig. 6.21.**

(*i*) **The current phasors I, I_1 and I_2.**

(*ii*) **Active power dissipated in the three resistive branches.**

(*iii*) **Power factor of the circuit.** *[U.P. Technical Univ. Electrical Engineering Second Semester 2007-08]*

Solution: Impedance of branch I of parallel circuit,

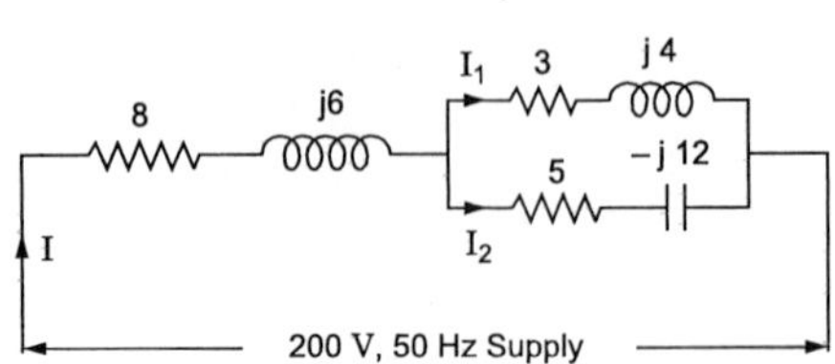

Fig. 6.21

$$\mathbf{Z}_1 = R_1 + jX = (3 + j\,4)\ \Omega = 5\angle 53.13°\ \Omega$$

Impedance of branch II of parallel circuit, $\mathbf{Z}_2 = (5 - j\,12)\ \Omega = 13\angle{-67.38°}\ \Omega$

Impedance of series branch, $\mathbf{Z}_s = (8 + j\,6)\ \Omega = 10\angle 36.87°\ \Omega$

Equivalent impedance of parallel circuit, $\mathbf{Z}_p = \dfrac{\mathbf{Z}_1 \mathbf{Z}_2}{\mathbf{Z}_1 + \mathbf{Z}_2} = \dfrac{5\angle 53.13^\circ \times 13\angle -67.38^\circ}{(3 + j\,4) + (5 - j\,12)}$

$$= \frac{65\angle -14.25^\circ}{(8 - j\,8)} = \frac{65\angle -14.25^\circ}{11.314\angle -45^\circ} = 5.745\angle 30.75^\circ\ \Omega$$

$$= (4.937 + j\,2.937)\ \Omega$$

Total impedance of the circuit, $\mathbf{Z} = \mathbf{Z}_s + \mathbf{Z}_p = (8 + j\,6) + (4.937 + j\,2.937)\ \Omega$

$$= (12.937 + j\,8.937)\ \Omega = 15.7237\angle 34.64^\circ\ \Omega$$

Let the supply voltage be the reference phasor i.e. $\mathbf{V} = 200\angle 0^\circ$ V

(*i*) Circuit current, $\mathbf{I} = \dfrac{\mathbf{V}}{\mathbf{Z}} = \dfrac{200\angle 0^\circ}{15.7237\angle 34.64^\circ} = 12.72\angle -34.64^\circ$ A **Ans.**

(*iii*) Power factor of the circuit, Cos ϕ = Cos (– 34.64°) = 0.8228 (lagging) **Ans.**

$$\text{Current } \mathbf{I}_1 = \mathbf{I}\frac{\mathbf{Z}_2}{\mathbf{Z}_1 + \mathbf{Z}_2} = \frac{12.72\angle -34.64^\circ \times 13\angle -67.38^\circ}{(3 + j\,4) + (5 - j\,12)}$$

$$= \frac{165.36\angle -102.02^\circ}{11.314\angle -45^\circ}$$

$$= 14.616\angle -57.02^\circ \text{ A} \quad \textbf{Ans.}$$

$$\text{Current } \mathbf{I}_2 = \mathbf{I}\frac{\mathbf{Z}_1}{\mathbf{Z}_1 + \mathbf{Z}_2} = \frac{12.72\angle -34.64 \times 5\angle 53.13^\circ}{11.314\angle -45^\circ}$$

$$= 5.62\angle 63.49^\circ \text{ A} \quad \textbf{Ans.}$$

(*ii*) Active power dissipated in

$$\text{Series branch} = I^2R_s = 12.72^2 \times 8 = 1{,}294.4 \text{ W}$$

$$\text{Parallel branch 1} = I_1^2R_1 = 14.616^2 \times 3 = 641 \text{ W}$$

$$\text{Parallel branch 2} = I_2^2R_2 = 5.62^2 \times 5 = 158 \text{ W}$$

$$\text{Total power loss, P} = 1{,}294.4 + 641 + 158 = 2{,}093.4 \text{ W}$$

(*iii*) Power factor of the circuit, cos $\phi = \dfrac{P}{VI} = \dfrac{2{,}093.4}{200 \times 12.72} = 0.8228$ (lagging) **Ans.**

6.5. POWER FACTOR IMPROVEMENT

The power factor for lighting and heating loads supplied from 3-phase power supply ranges from 0.95 to unity but the power factor for motor loads ranges from 0.5 to 0.9. Power factor for single phase motors may be as low as 0.4 (lagging) and for electric welding may be even 0.2 or 0.3 (lagging).

6.5.1. Disadvantages of Low Power Factor. The current for a given load supplied at a constant voltage will be higher at lower power factor and low at higher power factor. For example if load P is supplied at terminal voltage V and at power factor cos ϕ by a 1-phase system then load current is given by

$$I = \frac{P}{V\cos\phi}$$

If P and V are constant, the load current I is inversely proportional to power factor, cos ϕ *i.e.* lower the power factor, higher the current and vice versa. The higher current due to poor power factor affects the system and results in the following disadvantages:

(*i*) Rating of generators and transformers are proportional to their output current hence inversely proportional to power factor, therefore, large generators and transformers are required to deliver same load but at a low power factor.

(*ii*) The x-sectional area of the bus-bar and the contact surface of switchgear is required to be enlarged for the same power to be delivered but at a low power factor.

(*iii*) For the same power to be transmitted but at low power factor, the transmission line or the distributor or cable will have to carry more current. Thus more conductor material is required for transmission lines, distributors and cables to deliver the same load but at a low power factor.

(*iv*) Copper losses are proportional to the square of the current hence inversely proportional to the square of the power factor *i.e.* more copper losses incur at low power factor, which results in poor efficiency.

(*v*) Low lagging power factor results in large voltage drop in generators, transformers, transmission lines and distributors which results in poor regulation. Hence extra regulating equipment is required to keep the voltage drop within permissible limits.

Thus we see that the low power factor leads to a high capital cost for the alternators, switchgears, transformers, transmission lines, distributors and cables etc.

6.5.2. Causes of Low Power Factor.

(*i*) All ac motors (except overexcited synchronous motors and certain types of commutator motors) and transformers operate at lagging power factor. The power factor decreases with the decrease in load. For example an induction motor has a reasonable higher power factor of 0.85 at full load, 0.8 at 75% of full-load, 0.7 at half-full load, 0.5 at 25% of full-load and as low as 0.1 on no-load.

(*ii*) Arc lamps operate at low power factor (lagging) due to typical characteristic of the arc.

(*iii*) Industrial heating furnaces such as arc and induction furnaces operate on very low lagging power factor.

(*iv*) Due to increase in supply main voltage, which usually occurs during low load periods, such as lunch hours, night hours etc. the magnetizing current of inductive reactances increases and the power factor of the electrical plant as a whole comes down.

The power factor can be improved by (i) using induction motors with phase advancers (ii) connecting the static capacitors in parallel with the equipment operating at lagging power factor such as induction motors, fluorescent tubes etc or (iii) by using synchronous condensers* like shunt capacitors connected across the supply.

The capacitors draw current leading the supply voltage V by $\pi/2$ and neutralize the quadrature or wattless component of current drawn by the equipment across which they are connected.

Let the current drawn by an inductive load be of I amperes lagging behind the applied voltage V by an angle ϕ_1 and pf be improved to $\cos \phi_2$ (lagging) by connecting a capacitor across the inductive load, as illustrated in Fig. 6.22 (*a*).

Active component of current drawn by load $= I \cos \phi_1$ represented by phasor OB in phasor diagram.

Reactive (lagging) component of current drawn by load $= I \sin \phi_1 = I \cos \phi_1 \times \tan \phi_1$, represented by phasor AB in phasor diagram.

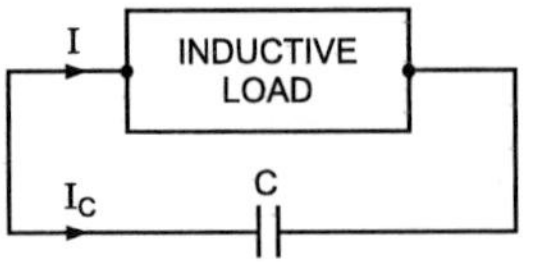

(*a*) Circuit Diagram

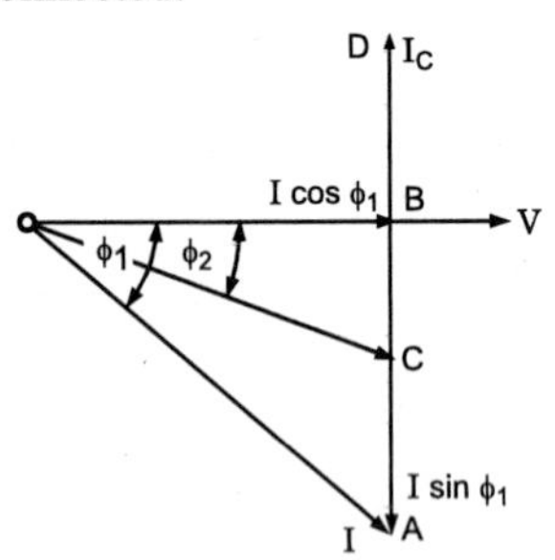

(*b*) Phasor Diagram

Fig. 6.22

Let the current drawn by capacitor C be of I_C amperes leading the applied voltage V by $\pi/2$, as represented by phasor BD in Fig. 6.22 (*b*).

Lagging component of current drawn from supply mains, as represented by phasor BC in phasor diagram $= OB \tan \phi_2 = I \cos \phi_1 \tan \phi_2$.

* An overexcited synchronous motor operating on no load is called the *synchronous condenser* or *synchronous phase advancer* and behaves like a capacitor.

The current drawn by capacitor C, I_C should neutralize the reactive component (lagging), represented by phasor AC in phasor diagram i.e. I_C should be equal to current represented by phasor AB less current represented by phasor BC. So

$$I_C = \text{Current represented by phasor AB} - \text{current represented by phasor BC}$$
$$= I \cos \phi_1 \tan \phi_1 - I \cos \phi_1 \tan \phi_2 = I \cos \phi_1 (\tan \phi_1 - \tan \phi_2)$$

The value of capacitance required, $C = \dfrac{I_C}{2\pi f V}$ where V is supply voltage and f is the supply frequency.

6.5.4. Advantages of Power Factor Improvement. If the power factor of the supply or power station is raised to unity, the current for the same amount of power to be supplied is reduced to minimum. This results in reduction of transmission line copper losses and reduction of voltage drop in transmission line and in alternator windings as copper losses are directly proportional to the square of supply current and voltage drop is directly proportional to the supply current.

The advantages of good power factor are that (*i*) load output of a given plant is better utilized (*ii*) line losses are reduced to minimum, therefore, efficiency of the plant is improved (*iii*) voltage drop or voltage regulation is considerably reduced.

Example 6.25 A single phase load of 5 kW operates at a power factor of 0.6 lagging. It is proposed to improve the power factor to 0.95 lagging by connecting a capacitor across a load. Calculate the kVAR rating of the capacitor.

Solution: Load input, $P = 5$ kW

Initial power factor, $\cos \phi_1 = 0.6$; $\tan \phi_1 = 1.333$

Power factor after improvement,

$\cos \phi_2 = 0.95$;

$\tan \phi_2 = 0.328684$

Leading kVAR to be supplied by the capacitor to be connected across the load for power factor improvement

$$= kVAR_1 - kVAR_2 = P(\tan \phi_1 - \tan \phi_2) = 5(1.333 - 0.328684)$$
$$= 5.023 \text{ kVAR } \textbf{Ans.}$$

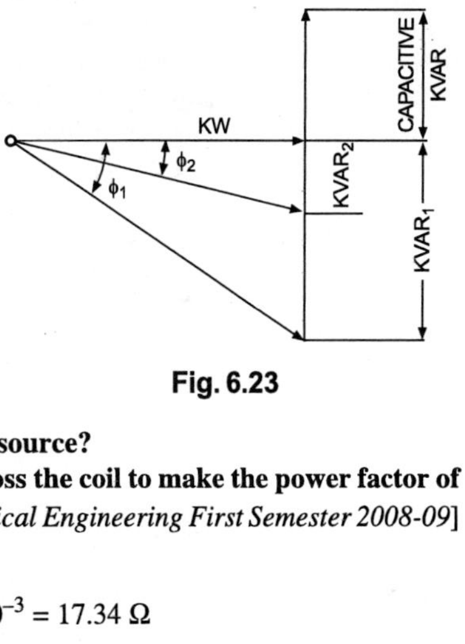

Fig. 6.23

Example 6.26 A 46 mH inductive coil has a resistance of 10 Ω.

(*a*) How much current will it draw if connected across a 100 V, 60 Hz, source?

(*b*) Determine the value of the capacitance that must be connected across the coil to make the power factor of the overall circuit unity. [*U.P. Technical Univ. Electrical Engineering First Semester 2008-09*]

Solution: (*a*) Resistance of coil, $R = 10\ \Omega$

Inductive reactance of coil, $X_L = 2\pi f L = 2\pi \times 60 \times 46 \times 10^{-3} = 17.34\ \Omega$

Impedance of the circuit, $Z = \sqrt{R^2 + X_L^2} = \sqrt{10^2 + 17.34^2} = 20\ \Omega$

Current drawn, $I = \dfrac{V}{Z} = \dfrac{100}{20} = 5\text{A}$ **Ans.**

(*b*) Phase angle, $\phi = \tan^{-1} \dfrac{X_L}{R} = \tan^{-1} \dfrac{17.34}{10} = 60°$ (lagging)

Reactive component of current $= I \sin \phi = 5 \sin 60° = 4.33$ A

The power factor of the overall circuit will be unity if the current drawn by the capacitor, connected across the coil, is equal to the reactive (lagging) component of current drawn by coil

i.e. when $2\pi f C V = I \sin \phi$

$$\text{or } C = \frac{4.33}{2\pi \times 60 \times 100} = 114.86\ \mu\text{F } \textbf{Ans.}$$

Example 6.27 **A 46 mH inductive coil has a resistance of 10 Ω. How much current will it draw, if connected across 100 V, 50 Hz source? Also determine the value of capacitance that must be connected across the coil to make the power factor of the circuit be unity.** [*Mahamaya Technical Univ. Electrical Engineering Second Semester 2010-11*]

Solution:

Resistance of the coil, $R = 10\ \Omega$

Inductive reactance of the coil, $X_L = 2\pi f L = 2\pi \times 50 \times 46 \times 10^{-3} = 14.45\ \Omega$

Impedance of the coil, $Z = \sqrt{R^2 + X_L^2} = \sqrt{10^2 + 14.45^2} = 17.574\ \Omega$

Current drawn by the coil, $I_1 = \dfrac{V}{Z} = \dfrac{100}{17.574} = 5.69$ A

Phase angle $\phi = \text{Tan}^{-1} \dfrac{X_L}{R} = \text{Tan}^{-1} \dfrac{14.45}{10} = 55.32°$ (lagging)

So current drawn from inductive coil, $\mathbf{I_1} = 5.69 \angle -55.32°$ A **Ans.**

Reactive component of current drawn by the coil $= I_1 \sin \phi = 5.69 \sin 55.32° = 4.679$ A

Power factor of the circuit shown in Fig. 6.24 will be unity when the current drawn by the capacitive will neutralize the reactive component of current drawn by the coil i.e.,

$$I_2 = 2\pi f CV = I_1 \sin \phi$$

or $2\pi \times 50 \times C \times 100 = 4.679$

or $C = \dfrac{4.679}{2\pi \times 50 \times 100} = 148.94\ \mu F$ **Ans.**

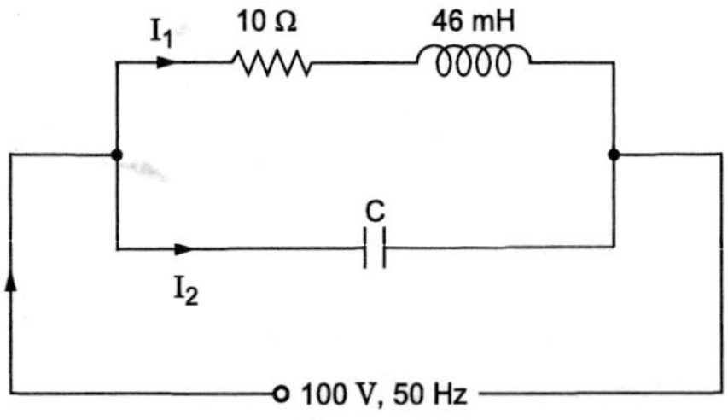

Fig. 6.24

Example 6.28 **An industrial consumer is operating 3-phase, 1.0 kW induction motor at a lagging pf of 0.8 and a source voltage of 400 V_{rms}. He wishes to raise the pf to 0.95 lagging by connecting a circuit element in parallel to the load (motor). Indicate the type of circuit element (inductive or capacitive) and find its value if the operating frequency is 50 Hz. Give your comments on the new line and motor currents.**

Solution: Motor input, $P = 1\text{ kW} = 1{,}000$ W

Power factor, $\cos \phi_1 = 0.8$ (lagging),

$\phi_1 = \text{Cos}^{-1}\ 0.8 = 36.87°$ (lagging)

Improved pf, $\cos \phi_2 = 0.95$ (lagging),

$\phi_2 = \text{Cos}^{-1}\ 0.95 = 18.195°$ (lagging)

Reactive power to be supplied by the pf improving device

$$= P(\tan \phi_1 - \tan \phi_2) = 1{,}000\ (\tan 36.87° - \tan 18.195°)$$

$$= 1{,}000\ (0.75 - 0.3287) = 421.3 \text{ VAR}$$

Since leading VARs are to be supplied by the power factor improving device, it must be capacitive.

If C is the capacitance of each capacitor then leading VARs supplied will be equal to $3 \times 2\pi f V_P C \times V_P = 6\pi f C V_P^2 = 6\pi f C V_L^2$ (assuming delta-connection)

or $C = \dfrac{421.3}{6\pi f V_L^2} = \dfrac{421.3}{6\pi \times 50 \times (400)^2} = 2.794\ \mu F$ **Ans.**

HIGHLIGHTS

AC Parallel Circuits. An ac circuit having two or more branches, each branch having different components in series, is known as an ac parallel circuit.

AC parallel circuits can be solved by the following methods:

1. Phasor method 2. Admittance method 3. Symbolic or *j* method.

In *phasor method*, the magnitude and direction of currents flowing through different branches are determined and represented as phasors. The phasor sum of the branches currents gives the resultant current. In *admittance method*, the component values are determined as per admittance triangle which contains conductance and susceptance.

$$\text{Conductance, G} = \text{Y}\cos\phi = \frac{\text{R}}{\text{Z}^2}$$

$$\text{Susceptance, B} = \text{Y}\sin\phi = \frac{\text{X}}{\text{Z}^2}$$

$$\text{Admittance, Y} = \sqrt{\text{G}^2 + \text{B}^2}, \quad \text{where G} = \text{G}_1 + \text{G}_2 + ... + \text{G}_n \text{ and B} = \text{B}_1 + \text{B}_2 + ... + \text{B}_n$$

$$\text{Current, I} = \text{V} \times \text{Y}$$

EXERCISES

1. Define the terms: Impedance, Admittance, Power factor, Active power and Reactive power.
[Pb. Technical Univ. Elec. Engineering June 2000 ; December 2000; May 2001]
2. Derive the sinusoidal response of parallel R-C circuit.
[G.B. Technical Univ. Electrical Engineering First Semester 2010-11]
3. Write the expression of currents in different branches and the total current of a circuit consisting of an inductance, a capacitance and a non-inductive resistance connected in parallel. Draw also the phasor diagram.
4. Derive the expression of impedance, admittance, conductance and susceptance of R-L series circuit.
[R.G.P.V. Bhopal Basic Electrical Engineering Jan-Feb 2006]
5. What is admittance triangle ? Explain.
6. What are the causes and disadvantages of low power factor?
[U.P. Technical Univ. Basic Electrical Engineering Odd Semester 2013-14]
7. Explain the main causes of large kVAR in a system.
[U.P. Technical Univ. Basic Electrical Engineering Second Semester 2013-14]
8. Why power factor improvement is important? What are causes and problems of low power factor? How power factor can be improved. **[G.B. Technical Univ. Electrical Engineering Odd Semester 2012-13]**
9. What is meant by power factor? What is its significance? How will you obtain power factor from kVA triangle?
[G.B. Technical Univ. Electrical Engineering First Semester 2009-10]
10. Define power factor. Also discuss the reasons for low power factor and ways to improve it.
[G.B. Technical Univ. Electrical Engineering Second Semester 2010-11]
11. What are the causes of low power factor in supply system ? Discuss its harming effects. How can the power factor be improved ? **[U.P. Technical Univ. Electrical Engineering January 2003]**
12. Write two problems of low power factor in a system.
[U.P. Technical Univ. Basic Electrical Engineering Second Semester 2013-14]

PROBLEMS

1. A circuit containing a resistance of 40 Ω in parallel with an inductive reactance of 30 Ω is connected across 240 V ac mains. Calculate (*i*) the current in each branch and the total current drawn from the mains (*ii*) the power factor of each branch and power factor of the whole circuit (*iii*) the power drawn by each branch and the total power drawn from the mains. [**Ans.** (*i*) 6A, 8A, 10A, (*ii*)1.0, zero, 0.6 (lag) (*iii*)1,440 watts, zero 1,440 watts]
2. An inductance of 0.6 H, a resistor of 100 Ω and a 30 μF capacitor are connected in parallel across a 230 V, 50 Hz supply. Calculate the line current, circuit phase angle, power dissipated, total circuit impedance and parameters of equivalent series circuit. [**Ans.** 2.49 A, 22.39° (lead), 529 W, 92.53 Ω, 80.51°, 35.25°]
3. A resistance of 100 Ω, an inductance of 0.1H and a capacitance of 150 μF are connected in parallel across 200 V, 50 Hz ac supply. Determine the branch currents, total current and the total power taken from the supply.
[**Ans.** 2A, 6.366 A, 9.42A, 3.655 A; 400 watts]

4. A series combination of R and C is in parallel with a 20 Ω resistor across 50 Hz source results in a total current of 7 A, a current through the 20 Ω resistor of 5 A and a current in the R-C branch of 3 A. Determine R and C
[**Ans. R** = 16.667Ω; X_C = 28.8675 Ω]

5. An inductor of 0.5 H inductance and 90 Ω resistance is connected in parallel with a 2 μF capacitor. A voltage of 230 V at 50 Hz is maintained across the circuit. Determine the total power taken form the source [**Ans.**146 watts]

6. Two parallel circuits comprise respectively of (*i*) a coil of resistance 20 Ω and inductance 0.07 H and (*ii*) a capacitor or capacitance 60 μF in series with resistance of 50 Ω. Calculate the current in each branch and the current taken from the supply and power factor when connected to 220 V, 50 Hz supply. [**Ans.** 7.4 A, 3A, 7.8 A, 0.907 (lag)]

7. Two circuits, the impedances of which are given by $\mathbf{Z}_1 = (15 + j\,12\ \Omega)$ and $\mathbf{Z}_2 = (8 - j\,5)\ \Omega$ are connected in parallel. If the potential difference across one of the impedances is (250 + *j*) V, calculate (*i*) total current and branch currents (*ii*) total power and power consumed in each branch and (*iii*) overall power factor and power factor of each branch.

[**Ans.** (*i*) 33.15 $\angle 10.36^\circ$ A; 13 $\angle -38.6^\circ$ A; 26.5 $\angle 32^\circ$ A (*ii*) 8.153 kW; 2.535 kW, 5.618 kW, (*iii*) 0.984 lead; 0.78 lag; 0.848 lead]

8. Three impedances $\mathbf{Z}_1 = (8 + j\,6)\ \Omega$, $\mathbf{Z}_2 = (2 - j\,1.5)\ \Omega$ and $\mathbf{Z}_3 = 2\ \Omega$ are connected in parallel across a 50 Hz supply. If the current through $\mathbf{Z}_1$ is (3 + *j* 4) A, calculate the current through the other impedances and also the power absorbed by this parallel circuit. **[V.T.U. Belgaum Karnataka Univ. Winter 2004]**
[**Ans.** (– 12 + *j* 16) A, (0 + *j* 25) A, 2.25 kW]

9. For the circuit shown in Fig. 1, find (*i*) total impedance (*ii*) total current (*iii*) total power absorbed and power factor. Draw a phasor diagram. [**Ans.** (*i*) 14.12 $\angle 29.2^\circ$ Ω (*ii*) 7.08 $\angle -29.2^\circ$ (*iii*) 618.2 W (*iv*) 0.873 (lag)]

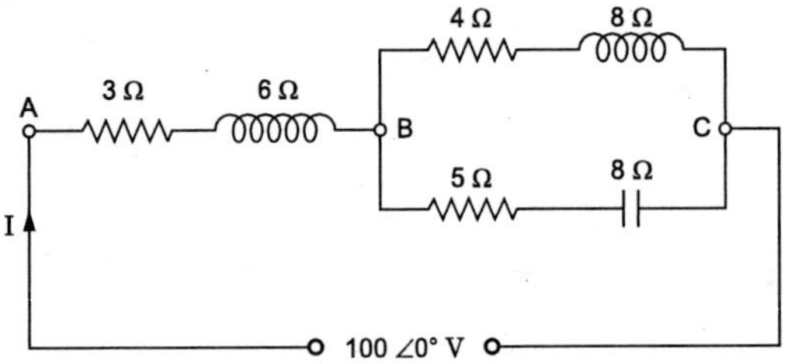

Fig. 1

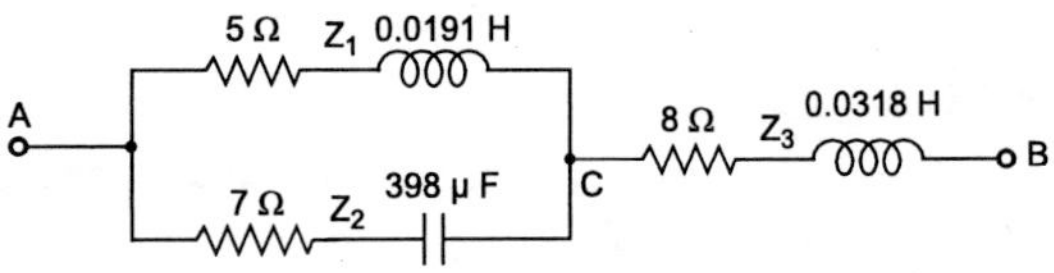

Fig. 2

10. In the circuit shown in Fig. 2, determine the voltage at 50 Hz to be applied across AB in order that a current of 10 A flows in the capacitor. **[U.P. Technical Univ. Electrical Engineering June 2001]**
[**Ans.** 288.62 $\angle 22.15^\circ$ V]

❑❑❑

7 CHAPTER

Resonance in R-L-C Circuits

7.1. INTRODUCTION

Resonance is identified with engineering situations which involve energy-storing elements subjected to a forcing function of varying frequency. Specifically, *resonance* is the term employed for describing the steady-state operation of a circuit or system at that frequency for which the resultant response is in time phase with the source function despite the presence of energy-storing elements. Resonance cannot take place when only one type of energy-storing element is present, *e.g.* capacitance or spring. There must exist two types of independent energy-storing elements capable of interchanging energy between them—for example, inductance and capacitance or mass and spring. Thus resonance is a phenomenon found in any system involving two independent energy-storing elements, be they electrical, mechanical, pneumatic, hydraulic or whatever.

If we have an ac circuit having a resistance R, an inductance L, and a capacitance C, connected in series (Fig. 7.1) and apply a small voltage V from a source that can keep the magnitude of V constant but can vary its frequency, we find that the magnitude of current drawn from the source of supply varies with the variation in frequency of supply source. There will be a value of frequency at which the current is maximum. *Electrical resonance* is said to exist when this condition is reached.

In this chapter we will discuss this phenomenon (more precisely called the *series* or *voltage resonance*) and also the situation when a constant voltage of variable frequency is applied to a *parallel circuit*. The simplest parallel circuit encountered in practice is a coil having resistance R and inductance L connected in parallel with a capacitor C. The resonant condition in this case is called *parallel resonance,* and sometimes *anti-resonance*. The latter name is prompted due to the fact that at resonance the input current to the parallel circuit is *minimum*.

Under the conditions of resonance, such a network becomes purely resistive in its effects, and the voltage and the current in the network are in phase. For this to occur, the inductive reactance and the capacitive reactance should be made equal.

7.2. SERIES OR VOLTAGE RESONANCE

Consider an ac circuit containing a resistance R, an inductance L and a capacitance C connected in series, as shown in Fig. 7.1.

Impedance of the circuit, $Z = \sqrt{R^2 + (X_L - X_C)^2} = \sqrt{R^2 + \left(\omega L - \dfrac{1}{\omega C}\right)^2}$

If for some frequency of applied voltage, $X_L = X_C$ in magnitude then

(*i*) Net reactance is zero *i.e.* $X = 0$

(*ii*) Impedance of the circuit, Z = R

(*iii*) The current flowing through the circuit is maximum and in phase with the applied voltage. The magnitude of the current will be equal to V/R

(*iv*) The voltage drop across the inductance is equal to the voltage drop across capacitance and is maximum

(*v*) The power factor is unity and

(*vi*) The power expended = VI watts.

When this condition exists, the circuit is said to be in resonance and the frequency at which it occurs is known as *resonant frequency*.

If resonant frequency is denoted by f_r, then

$$X_L = \omega L = 2\pi f_r L$$

$$\text{and } X_C = \frac{1}{2\pi f_r C}$$

$$\text{Since for resonance } X_L = X_C$$

$$\therefore\ 2\pi f_r L = \frac{1}{2\pi f_r C}$$

$$\text{or } f_r = \frac{1}{2\pi\sqrt{LC}}$$

$$\text{or } \omega_r = \sqrt{1/LC} \quad ...(7.1)$$

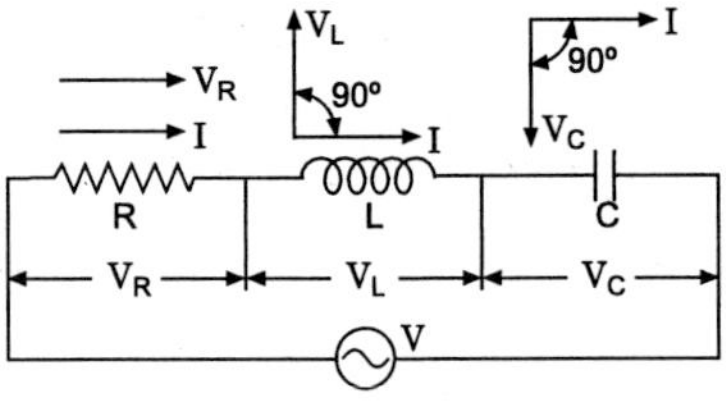

Fig. 7.1

From the above expression it is obvious that the value of resonance frequency depends on the parameters of the two energy-storing elements.

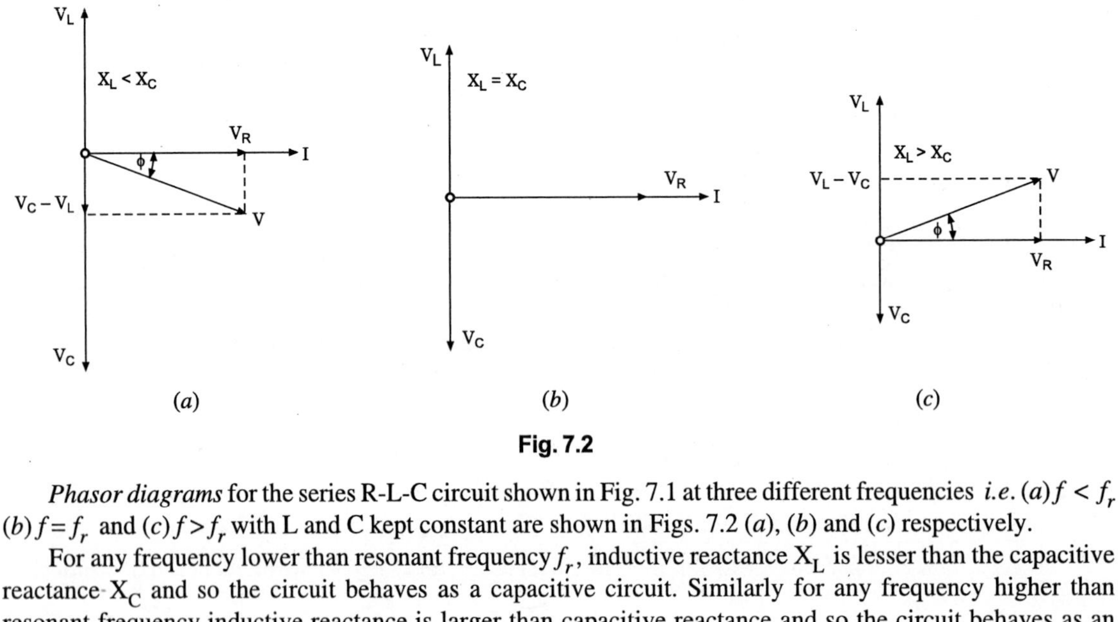

Fig. 7.2

Phasor diagrams for the series R-L-C circuit shown in Fig. 7.1 at three different frequencies *i.e.* (*a*) $f < f_r$ (*b*) $f = f_r$ and (*c*) $f > f_r$ with L and C kept constant are shown in Figs. 7.2 (*a*), (*b*) and (*c*) respectively.

For any frequency lower than resonant frequency f_r, inductive reactance X_L is lesser than the capacitive reactance X_C and so the circuit behaves as a capacitive circuit. Similarly for any frequency higher than resonant frequency inductive reactance is larger than capacitive reactance and so the circuit behaves as an inductive circuit. When the frequency of applied voltage is equal to the resonant frequency, the inductive reactance is equal to the capacitive reactance, the voltage drop across inductor is equal to voltage drop across capacitor in magnitude, but opposite in phase and, therefore, the circuit current I is in phase with the applied voltage *i.e.* the circuit behaves as a resistive circuit.

When the circuit is in resonance, the current is too large and will produce large voltage drop across inductance and capacitance, which will be equal in magnitude but opposite in phase and each may be several times greater than the applied voltage. If resistance R would have not been present in the circuit, such a circuit, would act like a *short circuit* to currents of frequency to which it resonates.

Since in this resonance the voltage is maximum, it is called the *voltage resonance*. The series resonance is also called an *acceptor circuit* because such a circuit accepts currents at one particular frequency but rejects currents of other frequencies. Such circuits are used in radio receivers.

7.2.1. Graphical Representation of Resonance in an R-L-C Series Circuit. The circuit can be made resonant in two different ways namely (*i*) by varying L and C parameters (either one or both) at a constant supply frequency or (*ii*) by varying the supply frequency *f* with parameters L and C constant. In our study of series resonance phenomenon we shall keep the applied voltage to the circuit and parameters L and C constant and allow the frequency to vary.

The circuit *resistance* R is independent of supply frequency and, therefore, remains constant. It has been represented by a straight line, parallel to the X-axis (or frequency axis) in Figure (Fig. 7.3). *Inductive reactance* X_L, being equal to ωL, increases in direct proportion to the supply frequency and is represented by a straight line passing through the origin (as X_L is regarded positive, so it lies in the first quadrant). *Capacitive reactance*, being equal to 1/ωC decreases inversely with the increase in frequency and is represented by a rectangular hyperbola lying in the fourth quadrant, below the frequency axis (capacitive reactance is considered negative).

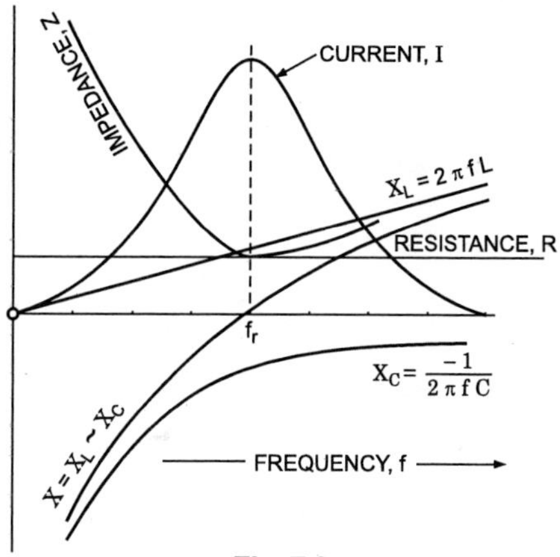

Fig. 7.3

The *net reactance* is the difference of inductive reactance X_L and capacitive reactance X_C and the curve drawn between the net reactance ($X_L \sim X_C$) and frequency will be a hyperbola (not rectangular) as shown in Fig. 7.3. The frequency at which the reactance curve crosses the frequency axis is called the *resonant frequency*, f_r (or f_o).

The impedance of the circuit, Z, being equal to $\sqrt{R^2 + (X_L \sim X_C)^2}$ is minimum at resonant frequency f_r.

At frequencies lower than resonant frequency f_r the impedance Z is large and capacitive as $X_C > X_L$ and the power factor is leading and at frequencies higher than resonant frequency f_r, the impedance Z is again large but inductive as $X_L > X_C$ and the power factor is lagging. The power factor has the maximum value of unity at resonant frequency.

7.2.2. Resonance Curve. Current varies inversely with the variation in impedance and, therefore, it is maximum at resonant frequency when impedance is minimum and decreases with the variation in frequency on both sides of the resonant frequency (as impedance Z is large), as illustrated in Fig. 7.4. The curve drawn between circuit current and the frequency of the applied voltage is called the *resonance curve* and its shape depends upon the value of circuit resistance R, as shown in the figure For smaller values of R, the resonance curve is sharply peaked, but for larger values of R, the curve is flat (Fig. 7.4).

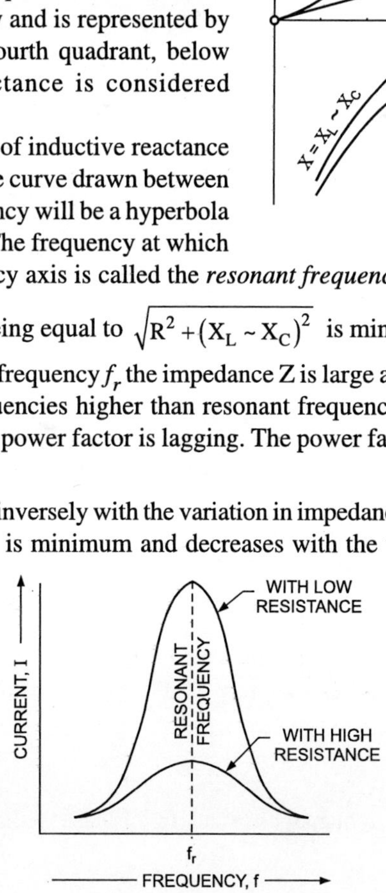

Fig. 7.4 *Resonance Curve*

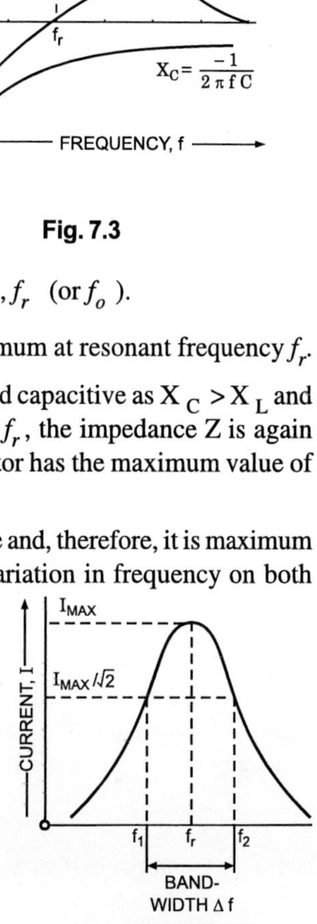

Fig. 7.5

7.2.3. Selectivity and Bandwidth. We have seen that for low resistance circuit the resonance curve is sharply peaked and such a circuit is said to be *sharply resonant* or *highly selective*. On the other hand, high resistance circuit have flat resonance curve and is said to have poor selectivity. Selectivity of different resonant circuits are compared in terms of their bandwidths, which are given by the bands of frequencies which lie between two points on either side of the resonant frequency where current is $1/\sqrt{2}$ times of maximum current I_{max}

$$\text{Bandwidth, } \Delta f = f_2 - f_1 \quad \text{...(7.2)}$$

The actual power input at frequencies f_1 and f_2

$$P = I^2 R = \left(\frac{I_{max}}{\sqrt{2}}\right)^2 R = \frac{I_{max}^2 R}{2} = \frac{1}{2} P_{max} = \frac{1}{2} \times \text{power input at resonance} \quad \text{...(7.3)}$$

That is why the frequencies f_1 and f_2 at the limits of the bandwidth are called the *half- power points* on the frequency scale, and the corresponding value of the bandwidth is termed as *half-power bandwidth* (B_{hp}) or -3 dB bandwidth.

The noteworthy points regarding half-power points are

1. The current is $\frac{I_{max}}{\sqrt{2}}$;
2. Impedance is $\sqrt{2}$ R or $\sqrt{2}\, Z_{min}$;
3. $P_1 = P_2 = \frac{P_{max}}{2}$;
4. The circuit phase angle $\phi = \pm 45°$ or $\frac{\pi}{4}$ radian;
5. Tangent of the circuit phase angle at the off-resonance frequencies f_1 and f_2, Q = Tan θ = Tan 45° =1

7.2.4. Determination of Bandwidth and Edge Frequencies. The impedance of the tuned circuit must be $\sqrt{2}$ times its impedance at resonance so that the current is $\frac{I_{max}}{\sqrt{2}}$. But the impedance at resonance, Z = R so at half-power points the impedance is $\sqrt{2}$ R. Since $Z = \sqrt{R^2 + X^2}$, therefore, $\sqrt{2}\,R = \sqrt{R^2 + X^2}$ or X = R. This information can now be used to obtain an expression for the bandwidth in terms of the series circuit parameters. Calling the reactance at the lower half-power frequency X_1, we have

$$X_1 = \omega_1 L - \frac{1}{\omega_1 C} = -R$$

The minus sign appears on the right hand side of the equation because below resonance the capacitive reactance exceeds the inductive reactance. The above equation may also be rewritten as

$$\omega_1^2 + \frac{R}{L}\omega_1 - \frac{1}{LC} = 0$$

$$\text{or } \omega_1 = \frac{-R}{2L} \pm \sqrt{\left(\frac{R}{2L}\right)^2 + \frac{1}{LC}} = -\alpha \pm \sqrt{\alpha^2 + \omega_r^2} \qquad \text{where } \alpha = \frac{R}{2L} \because \text{ From Eq. (7.1) } \omega_r = \sqrt{\frac{1}{LC}}$$

Although above equation provides two values of ω_r but only one of these is physically realizable. The negative frequency is meaningless and so may be discarded and so the equation for the lower half-power frequency is

$$\omega_1 = -\alpha + \sqrt{\alpha^2 + \omega_r^2} \quad \text{...(7.4)}$$

Similarly the upper half-power frequency is found. In this case we have

$$X_2 = \omega_2 L - \frac{1}{\omega_2 C} = +R$$

$$\text{or } \omega_2^2 - \frac{R}{L}\omega_2 - \frac{1}{LC} = 0$$

$$\text{or } \omega_2 = \alpha + \sqrt{\alpha^2 + \omega_r^2} \quad ...(7.5)$$

So the equation for the bandwidth becomes

$$\omega_{b\omega} = \omega_2 - \omega_1 = \alpha + \sqrt{\alpha^2 + \omega_r^2} - \left[-\alpha + \sqrt{\alpha^2 + \omega_2^2}\right] = 2\alpha = \frac{R}{L}$$

$$\text{and bandwidth, } \Delta f = f_2 - f_1 = \frac{\omega_2 - \omega_1}{2\pi} = \frac{R}{2\pi L}\text{ Hz} \quad ...(7.6)$$

$$\text{Lower half-power frequency, } f_1 = f_r - \frac{\Delta f}{2} = f_r - \frac{R}{4\pi L}\text{ Hz} \quad ...(7.7)$$

$$\text{Upper half-power frequency, } f_2 = f_r + \frac{\Delta f}{2} = f_r + \frac{R}{4\pi L}\text{ Hz} \quad ...(7.8)$$

Equation (7.6) reveals that the bandwidth of the series R–L–C circuit depends solely upon the R/L ratio. Note that it is not the individual values of R or L but rather their ratio that is important. Note too that the bandwidth is independent of the circuit capacitance C.

If Eqs. (7.4) and (7.5) are multiplied we have

$$\omega_1 \omega_2 = \left(-\alpha + \sqrt{\alpha^2 + \omega_r^2}\right)\left(\alpha + \sqrt{\alpha^2 + \omega_r^2}\right)$$

$$= \alpha^2 + \omega_r^2 - \alpha^2 = \omega_r^2$$

$$\text{or } \frac{\omega_1}{2\pi} \cdot \frac{\omega_2}{2\pi} = \frac{\omega_r^2}{(2\pi)^2}$$

$$\text{or } f_1 f_2 = f_r^2 \quad ...(7.9)$$

i.e. the resonant frequency f_r or f_o is the root of the product of the half-power frequencies of a series resonant circuit.

7.2.5. Quality Factor of a Series Resonant Circuit. The Q-factor of an R-L-C series circuit can be defined in any of the following ways.

It may be given as the voltage magnification that the circuit produces at resonance.

We have seen that at resonance, the circuit current is maximum and is equal to V/R

$$\text{or supply voltage, } V = I_{max} R$$

$$\text{and Voltage across inductance or capacitance} = I_{max} X_L = I_{max} X_C$$

$$\therefore \text{Voltage magnification} = \frac{I_{max} X_L}{I_{max} R} = \frac{\text{Reactive power}}{\text{Active power}} = \frac{X_L}{R} = \frac{\omega_r L}{R} = \frac{\text{Reactance}}{\text{Resistance}}$$

$$= \frac{I_{max} X_C}{I_{max} R} = \frac{\text{Reactive power}}{\text{Active power}} = \frac{X_C}{R} = \frac{\text{Reactance}}{\text{Resistance}} = \frac{1}{\omega_r RC}$$

$$\text{or Q-factor at resonance, } Q_r = \frac{\omega_r L}{R} = \frac{2\pi f_r L}{R} = \frac{2\pi L}{R} \times \frac{1}{2\pi\sqrt{LC}} = \frac{1}{R}\sqrt{\frac{L}{C}} \quad ...(7.10) \quad \because f_r = \frac{1}{2\pi\sqrt{LC}}$$

$$\text{Also } Q_r = \frac{2\pi f_r L}{R} = \text{Tan}\,\phi \quad ...(7.11)$$ where ϕ is the power factor angle of the coil.

The Q-factor may also be defined as

$$\text{Q-factor} = 2\pi \times \frac{\text{Maximum stored energy}}{\text{Energy dissipated per cycle}}$$

$$= 2\pi \times \frac{\frac{1}{2} L (I_{max})^2}{\left(\frac{I_{max}}{\sqrt{2}}\right)^2 R T_r} \qquad \because \text{At half power points the current} = \frac{I_{max}}{\sqrt{2}}$$

$$= \frac{(I_{max})^2 \times 2\pi f_r L}{I_{max}^2 R} = \frac{2\pi f_r L}{R} = \frac{\omega_r L}{R} = \frac{1}{\omega_r RC} \qquad \because T_r = \frac{1}{f_r}$$

i.e., Q_r = Energy stored/energy lost in the circuit.

In case of series resonance, higher value of Q-factor means not only higher voltage magnification, but also a higher selectivity of the tuning coil and that is why it is necessary that the coil be of high inductance and low resistance. In fact, Q-factor of a series resonant circuit may be defined as the ratio of resonant frequency to bandwidth

$$i.e. \quad Q_r = \frac{f_r}{\Delta f} = \frac{f_r}{f_2 - f_1} = \frac{\frac{1}{2\pi\sqrt{LC}}}{R/2\pi L} = \frac{1}{R}\sqrt{\frac{L}{C}}$$

as before [Refer to Eq. (7.10)]

7.2.6. Determination of Frequency For Circuit Current Half as Much as at Resonance.

$$\text{At resonance, } I_{max} = \frac{V}{R}$$

At any other frequency above or below the resonant frequency, the current is given by

$$I = \frac{V}{Z} = \frac{V}{\sqrt{R^2 + \left(\omega L \sim \frac{1}{\omega C}\right)^2}}$$

This current will lag behind or lead the applied voltage by a certain angle ϕ (lag behind the applied voltage in case frequency exceeds the resonant one and lead the applied voltage when the frequency is lower than resonant frequency. Thus

$$\frac{I_{max}}{I} = \frac{V}{R} \times \frac{\sqrt{R^2 + \left(\omega L \sim \frac{1}{\omega C}\right)^2}}{V} = \sqrt{1 + \frac{1}{R^2}\left(\omega L \sim \frac{1}{\omega C}\right)^2}$$

Substituting $I = \frac{1}{2} I_{max}$ we have

$$2 = \sqrt{1 + \frac{1}{R^2}\left(\omega L \sim \frac{1}{\omega C}\right)^2}$$

$$\text{or} \quad \omega L \sim \frac{1}{\omega C} = \sqrt{3} R$$

Substituting $C = \frac{1}{\omega_r^2 L}$ from Eq. (7.1) in above equation we have

$$\omega L \sim \frac{\omega_r^2 L}{\omega} = \sqrt{3} R$$

$$\text{or} \quad (\omega^2 \sim \omega_r^2) = \frac{\sqrt{3} R}{L} \omega \qquad ...(7.12)$$

From above Eq. (7.12) the values of the frequency at which the circuit current would be half as much as at resonance can be determined.

Example 7.1 **A coil of resistance 40 Ω and inductance 0.75 H forms part of a series circuit for which resonant frequency is 55 Hz. If the supply is 250 V, 50 Hz, find (*i*) line current (*ii*) power factor (*iii*) power consumed and (*iv*) voltage across the coil.** *[U.P. Technical Univ. Electrical Engineering June 2001]*

Solution: Resonant frequency, $f_r = 55$ Hz

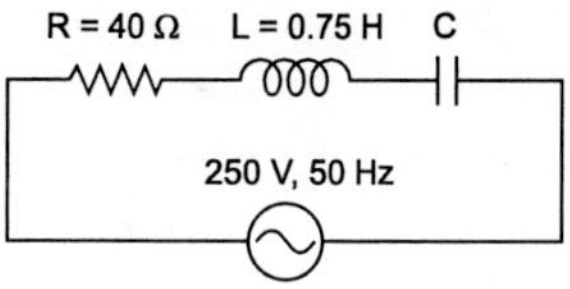

Fig. 7.6

At resonant frequency

Inductive reactance = Capacitive reactance

$$\text{or } 2\pi f_r L = \frac{1}{2\pi f_r C}$$

$$\text{or } C = \frac{1}{(2\pi f_r)^2 L} = \frac{1}{(2\pi \times 55)^2 \times 0.75} = 11.165\ \mu F$$

AT 250 V, 50 Hz

$$\text{Reactance of circuit, } X = \omega L - \frac{1}{\omega C} = 2\pi \times 50 \times 0.75 - \frac{1}{2\pi \times 50 \times 11.165 \times 10^{-6}}$$

$$= 235.62 - 285.1 = -49.48\ \Omega$$

$$\text{Impedance of the circuit, } Z = \sqrt{R^2 + X^2} = \sqrt{40^2 + (-49.48)^2} = 63.63\ \Omega$$

(*i*) Line current, $I = \frac{V}{Z} = \frac{250}{63.63} = 3.93$ A **Ans.**

(*ii*) Power factor, $\cos\phi = \frac{R}{Z} = \frac{40}{63.63} = 0.629$ (leading) ∵ because the circuit is capacitive, X being negative

(*iii*) Power consumed, $P = I^2 R = (3.93)^2 \times 40 = 618$ W **Ans.**

(*iv*) Voltage across the coil, $V_{R-L} = I Z_{R-L} = 3.93\sqrt{40^2 + (2\pi \times 50 \times 0.75)^2} = 939.2$ V **Ans.**

Example 7.2 **A coil of resistance 40 Ω and inductance 0.75 H are in a series circuit. The resonant frequency is 55 Hz. If supply is 250 V, 50 Hz find (i) line current (ii) power factor (iii) power consumed.**

[U.P. Technical Univ. Electrical Engineering Second Semester 2006-07]

Solution: Refer to Solution of Example 7.1.

Example 7.3 **Voltages across resistance, inductance and capacitance connected in series are 3 V, 4 V and 5 V respectively. If supply voltage has 50 Hz frequency what is the magnitude of supply voltage. Find the resonant frequency of this series RLC circuit.** *[U.P. Technical Univ. Electrical Engineering Second Semester 2004-05]*

Solution: Voltage across resistance, $V_R = 3$ V

Voltage across inductance, $V_L = 4$ V

Voltage across capacitance, $V_C = 5$ V

$$\text{Supply voltage, } V = \sqrt{V_R^2 + (V_C - V_L)^2} = \sqrt{3^2 + (5-4)^2} = \sqrt{10} = 3.162 \text{ V } \textbf{Ans.}$$

Voltage drop across an inductor is given as $2\pi f L I$ where f is supply frequency, L is the inductance of the inductor in Henries and I is the current flowing through it in amperes.

$$\text{So } 2\pi f L I = 4 \quad ...(i)$$

Voltage drop across a capacitor is given as $I/2\pi f C$ where f is supply frequency, C is the capacitance of capacitor in farads and I is the current flowing through it in amperes.

$$\text{So } \frac{I}{2\pi f C} = 5 \quad ...(ii)$$

Dividing Eq. (*i*) by Eq. (*ii*) we have

$$2\pi f L I \times \frac{2\pi f C}{I} = \frac{4}{5}$$

$$\text{or } 4\pi^2 f^2 LC = 0.8$$

$$\text{or } LC = \frac{0.8}{4\pi^2 \times 50^2} = 8.1051 \times 10^{-6} \qquad \because f = 50 \text{ Hz}$$

$$\text{and resonant frequency is given as } f_r = \frac{\omega_r}{2\pi} = \frac{1}{2\pi}\sqrt{\frac{1}{LC}} = \frac{1}{2\pi}\sqrt{\frac{1}{8.1051 \times 10^{-6}}} = 55.9 \text{ Hz } \textbf{Ans.}$$

Example 7.4 **A series R-L-C circuit consisting of a resistance of 20 Ω, inductance 0.2 H and a capacitance of 150 μF is connected across a 230 V, 50 Hz source. Calculate: (*i*) the impedance (*ii*) the current (*iii*) power factor (*v*) the frequency of supply to be adjusted to make power factor unity.**

[G.B. Technical Univ. Electrical Engineering Odd Semester 2012-13]

Solution: Resistance of the circuit, = 20 Ω

$$\text{Inductance reactance of the circuit, } X_L = 2\pi f L = 2\pi \times 50 \times 0.2 = 62.83\ \Omega$$

$$\text{Capacitive reactance of the circuit, } X_C = \frac{1}{2\pi f C} = \frac{1}{2\pi \times 50 \times 150 \times 10^{-6}} = 21.22$$

$$\text{Net reactance of the circuit, } X = X_L - X_C = (62.83 - 21.22) = 41.61\ \Omega \text{ (inductive)}$$

$$(i) \text{ Net impedance of the circuit, } Z = \sqrt{R^2 + X^2} = \sqrt{20^2 + 41.61^2} = 46.167\ \Omega \quad \textbf{Ans.}$$

$$(ii) \text{ The current, } I = \frac{V}{Z} = \frac{230}{46.167} = 4.98 \text{ A} \quad \textbf{Ans.}$$

$$(iii) \text{ Power factor, } \cos\phi = \frac{R}{Z} = \frac{20}{46.167} = 0.4332 \text{ (lagging)} \qquad \because \text{ the circuit is inductive}$$

(*iv*) Power factor will be unity when $X_L = X_C$ *i.e.* at resonance frequency which is given as

$$f_r = \frac{1}{2\pi\sqrt{LC}} = \frac{1}{2\pi\sqrt{0.2 \times 150 \times 10^{-6}}} = 29.06 \text{ Hz} \quad \textbf{Ans.}$$

Example 7.5 **Voltage across R, L, C connected in series are 5, 8 and 10 V respectively. Calculate the value of supply voltage at 50 Hz. Also find the frequency at which this circuit would resonate.**

[Electrical Engineering G.B. Technical Univ. Second Semester 2011-12; U.P. Technical Univ. Second Semester 2013-14]

Solution: Voltage across R, $V_R = 5$ V

Voltage across L, $V_L = 8$ V

Voltage across C, $V_C = 10$ V

$$\text{Supply voltage, } V = \sqrt{V_R^2 + (V_L - V_C)^2} = \sqrt{5^2 + (8-10)^2} = \sqrt{29} = 5.39 \text{ V} \quad \textbf{Ans.}$$

$$\text{Inductance, } L = \frac{V_L}{2\pi f I} = \frac{8}{2\pi \times 50 \times I} = \frac{8}{100\,\pi I} \qquad \because V_L = 2\pi f L I$$

$$\text{Capacitance, } C = \frac{I}{2\pi f V_C} = \frac{I}{2\pi \times 50 \times 10} = \frac{I}{1{,}000\,\pi} \qquad \because V_C = \frac{I}{2\pi f C}$$

$$\text{Resonant frequency, } f = \frac{1}{2\pi}\sqrt{\frac{1}{LC}} = \frac{1}{2\pi}\sqrt{\frac{100\,\pi I}{8} \times \frac{1{,}000\,\pi}{I}} = 55.9 \text{ Hz } \textbf{Ans.}$$

Example 7.6 **A 10 mH coil is connected in series with a loss free capacitor to a variable frequency source of 20 V. The current in the circuit has maximum value of 0.2 A at a frequency of 100 kHz. Calculate: (*i*) the value of capacitance (*ii*) the Q-factor of the coil (*iii*) the half-power frequencies**

[U.P. Technical Univ. Electrical Engineering First Semester 2006-07]

Solution: Supply voltage, V = 20 V

Supply frequency, f = 100 kHz = 10^5 Hz

Maximum current in series R-L-C circuit will flow under resonant condition and under resonant condition,

$$\text{Current, } I_{max} = \frac{V}{R},$$

$$\text{so circuit resistance, } R = \frac{V}{I_{max}} = \frac{20}{0.2} = 100\ \Omega$$

$$\text{Resonant frequency, } f_r = \text{Supply frequency, } f = 10^5 \text{ Hz}$$

$$\text{and } \omega L = \frac{1}{\omega C},$$

$$\text{so } C = \frac{1}{\omega^2 L} = \frac{1}{(2\pi \times 10^5)^2 \times 10 \times 10^{-3}} = 253.3 \text{ pF } \textbf{Ans.}$$

$$(ii) \text{ Q-factor of the coil} = \frac{\omega_r L}{R} = \frac{2\pi \times 10^5 \times 10 \times 10^{-3}}{100} = 62.83 \quad \textbf{Ans.}$$

$$(iii) \text{ Lower half-power frequency, } f_1 = f_r - \frac{R}{4\pi L} = 10^5 - \frac{100}{4\pi \times 0.01} \text{ Hz} = 99.204 \text{ kHz } \textbf{Ans.}$$

$$\text{Upper half-power frequency, } f_2 = f_r + \frac{R}{4\pi L} = 10^5 + \frac{100}{4\pi \times 0.01} \text{ Hz} = 100.796 \text{ kHz} \quad \textbf{Ans.}$$

Example 7.7 **A series circuit has R = 10 Ω, L = 0.05 H, C = 10 μF. Calculate Q-factor of the circuit.**

[U.P. Technical Univ. Electrical Engineering Odd Semester 2013-14]

Solution: Q-factor of the circuit at resonance $= \frac{1}{R}\sqrt{\frac{L}{C}} = \frac{1}{10}\sqrt{\frac{0.05}{10 \times 10^{-6}}} = 7.07$ **Ans.**

Example 7.8 **A series R-L-C circuit has R = 10 Ω, L = 0.1 H and C = 8 μF. Determine**

(i) resonant frequency

(ii) Q-factor of the circuit at resonance

(iii) the half power frequencies.

[U.P. Technical Univ. Elec. Engineering First Semester 2005-06]

Solution:

$$(i) \text{ Resonant frequency, } f_r = \frac{1}{2\pi}\sqrt{\frac{1}{LC}} \text{ Hz} = \frac{1}{2\pi}\sqrt{\frac{1}{0.1 \times 8 \times 10^{-6}}} = 177.94 \text{ Hz } \textbf{Ans.}$$

$$(ii) \text{ Q-factor of circuit at resonance} = \frac{1}{R}\sqrt{\frac{L}{C}} = \frac{1}{10}\sqrt{\frac{0.1}{8 \times 10^{-6}}} = 11.18 \text{ } \textbf{Ans.}$$

$$(iii) \text{ Half-power frequencies } f = f_r - \frac{R}{4\pi L} = 177.94 - \frac{10}{4\pi \times 0.1} = 170 \text{ Hz } \textbf{Ans.}$$

$$f_2 = f_r + \frac{R}{4\pi L} = 177.94 + \frac{10}{4\pi \times 0.1} = 185.9 \text{ Hz } \textbf{Ans.}$$

Example 7.9 **If the bandwidth of a resonant circuit is 10 kHz and the lower half power frequency is 120 kHz, find out the value of the upper half power frequency and the quality factor of the circuit.**

[U.P. Technical Univ. Electrical Engineering Second Semester 2003-04]

Solution: Bandwidth of the resonant circuit, $\Delta f = 10$ kHz

Lower half power frequency, $f_1 = 120$ kHz

$$\text{Since lower half-power frequency, } f_1 = f_r - \frac{\Delta f}{2}$$

$$\text{Resonant frequency, } f_r = f_1 + \frac{\Delta f}{2} = 120 + \frac{10}{2} = 125 \text{ kHz}$$

$$\text{Upper half-power frequency, } f_2 = f_r + \frac{\Delta f}{2} = 125 + \frac{10}{2} = 130 \text{ kHz } \textbf{Ans.}$$

$$\text{Quality factor of the resonant circuit, } Q_r = \frac{f_r}{\Delta f} = \frac{125}{10} = 12.5 \text{ } \textbf{Ans.}$$

Example 7.10 **A 20 Ω resistor is connected in series with an inductor and a capacitor, across a variable frequency 25 V supply. When the frequency is 400 Hz, the current is at its maximum value of 0.5 A and the potential difference across the capacitor is 150 V. Calculate the resistance and inductance of the inductor.**

[U.P. Technical Univ. Electrical Engineering Junuary 2003]

Solution: Supply voltage, V(rms) = 25 V

Frequency, f = 400 Hz

Maximum current in a series R-L-C circuit will flow under resonant condition and current under resonant condition is given as

$$I_{max} = \frac{V}{R}$$

$$\text{So circuit resistance, } R = \frac{V}{I_{max}} = \frac{25}{0.5} = 50\ \Omega \ \textbf{Ans.}$$

$$\text{Voltage across the capacitor, } V_C = \frac{I_{max}}{2\pi f C}$$

$$\text{So circuit capacitance, } C = \frac{I_{max}}{2\pi f V_C} = \frac{0.5}{2\pi \times 400 \times 150} = 1.3263\ \mu F$$

$$\text{Since under resonant condition, } X_L = X_C$$

$$\text{or } 2\pi f L = \frac{1}{2\pi f C}$$

$$\therefore \text{ Inductance of inductor, } L = \frac{1}{4\pi^2 f^2 C} = \frac{1}{4\pi^2 \times (400)^2 \times 1.3263 \times 10^{-6}} = 0.119\ \text{H}\ \textbf{Ans.}$$

Example 7.11 **Determine the frequency at which the voltage V_{out} is zero in Fig. 7.7 shown below.**

[U.P. Technical Univ. Tutorial Question Bank]

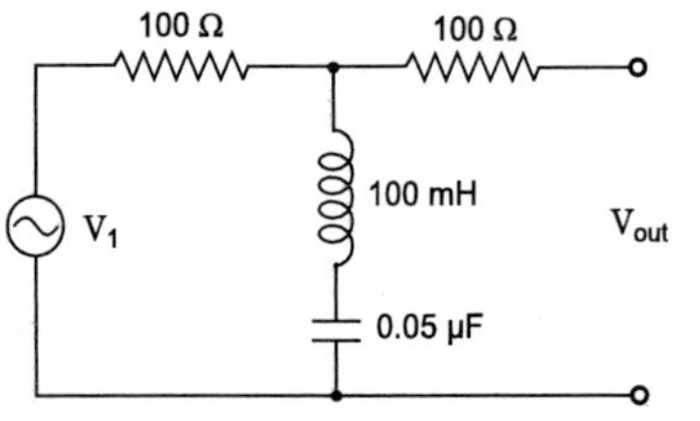

Fig. 7.7

Solution: The voltage V_{out} will be zero in the circuit when voltage drop across 100 mH inductor becomes equal to voltage drop across 0.05 μF capacitor *i.e.* when the frequency becomes equal to resonant frequency given as

$$f_r = \frac{1}{2\pi}\sqrt{\frac{1}{LC}} = \frac{1}{2\pi}\sqrt{\frac{1}{100 \times 10^{-3} \times 0.05 \times 10^{-6}}}$$

$$= 2{,}250\ \text{Hz}\ \textbf{Ans.}$$

Example 7.12 **A voltage of $v(t) = 10 \sin \omega t$ is applied to a series R-L-C circuit. At resonant frequency of the circuit the maximum voltage across the capacitor is found to be 500 V. Moreover the bandwidth is known to be 400 rad/sec, and the impedance at resonance is 100 Ω.**

Attempt any two parts of the following:

(a) Find the resonant frequency.

(b) Compute the upper and lower limits of bandwidth.

(c) Determine the values of L and C for this circuit.

[Mahamaya Technical Univ. Electrical Engineering Second Semester 2011-12]

Solution: Resistance, R = Impedance at resonance = 100 Ω

$$\text{Bandwidth, } \omega_{bW} = \frac{R}{L}$$

$$\text{So } 400 = \frac{100}{L}$$

$$\text{or } L = \frac{100}{400} = 0.25\ \text{H} \quad \textbf{Ans.}$$

Maximum value of supply voltage, $V_{max} = 10$ V

Peak value of current at resonance, $I_{max} = \dfrac{V_{maximum} \text{ value of supply voltage}}{R} = \dfrac{10}{100} = 0.1$ A

At resonance

Peak value of voltage across inductance = Peak value of voltage across capacitor = 500 V

$$\text{So } 2\pi f L I_{max} = 500$$

$$\text{or } f = \frac{500}{2\pi L I_{max}} = \frac{500}{2\pi \times 0.25 \times 0.1} = 3{,}183 \text{ Hz}$$

(*a*) So resonant frequency, $f_r = 3{,}183$ Hz **Ans.**

(*b*) Lower half-power frequency, $f_1 = f_r - \dfrac{\omega_{bW}}{4\pi} = 3{,}183 - \dfrac{400}{4\pi} = 3{,}151.17$ Hz

Upper half power frequency, $f_2 = f_r + \dfrac{\omega_{bm}}{4\pi} = 3{,}183 + \dfrac{400}{4\pi} = 3{,}214.83$ Hz **Ans.**

(*c*) Capacitance, $C = \dfrac{1}{\omega_r^2 L} = \dfrac{1}{(2\pi \times 3{,}183)^2 \times 0.25} = 0.01\ \mu\text{F}$ **Ans.**

Example 7.13 **A constant voltage at a frequency of 1 MHz is applied to an inductor in series with a variable capacitor. When the capacitor is set 500 pF, the current has its maximum value while it is reduced to one-half when the capacitance is 600 pF. Find (*i*) resistance (*ii*) the inductance, and (*iii*) the Q-factor of the inductor.**

Solution: Since circuit current has maximum value at a resonant frequency, resonant frequency,

$$f_r = 1 \text{ MHz} = 1 \times 10^6 \text{ Hz for } C = 500 \times 10^{-12} \text{ F}$$

$$\text{Inductance, } L = \frac{1}{(2\pi f)^2 C} = \frac{1}{(2\pi \times 1 \times 10^6)^2 \times 500 \times 10^{-12}} = 50.66\ \mu\text{H} \textbf{ Ans.}$$

Impedance of circuit with C = 500 *p*F, $Z_1 = R$ $\quad\because$ At resonance Z = R

Impedance of circuit with C = 600 pF,

$$Z_2 = \sqrt{R^2 + \left(\omega L - \frac{1}{\omega C}\right)^2} = 2R \text{ since current is halved}$$

$$\text{or } \left(\omega L - \frac{1}{\omega C}\right)^2 = 3R^2$$

$$\text{or } R = \frac{1}{\sqrt{3}}\left(\omega L - \frac{1}{\omega C}\right) = \frac{1}{\sqrt{3}}\left(2\pi \times 10^6 \times 50.66 \times 10^{-6} - \frac{1}{2\pi \times 10^6 \times 600 \times 10^{-12}}\right)$$

$$= 30.62\ \Omega \textbf{ Ans.}$$

$$\text{Q-factor of inductor} = \frac{\omega_r L}{R} = \frac{2\pi \times 1 \times 10^6 \times 50.66 \times 10^{-6}}{30.62} = 10.4 \textbf{ Ans.}$$

7.3. PARALLEL OR CURRENT RESONANCE

When an inductive reactance and a capacitive reactance are connected in parallel, as shown in Fig. 7.8, conditions may reach under which *current resonance* (also known as *parallel or anti-resonance*) will take place.

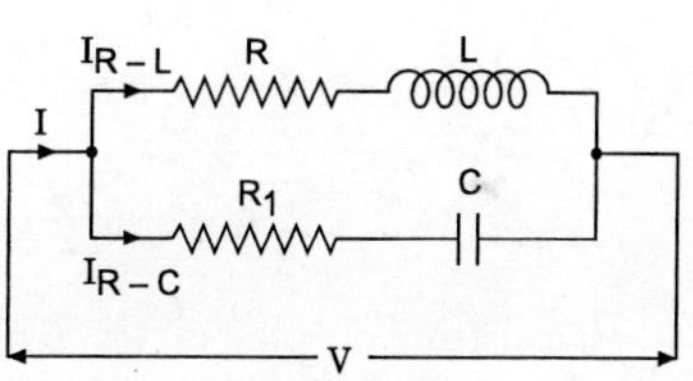

Fig. 7.8

Consider a practical case of a coil in parallel with a condenser, as shown in Fig. 7.8. Let the coil be of resistance R ohms and inductance L henrys and the condenser of resistance R_1 ohms and capacitance C farads.

Such a circuit is said to be in electrical resonance when the reactive (or wattless) component of line current becomes zero. The frequency at which this happens is known as *resonant frequency.*

Circuit will be in electrical resonance if

reactive component of R-L branch current, $I_{R-L} \sin \phi_{R-L}$

$$= \text{Reactive component of R-C branch current, } I_{R-C} \sin \phi_{R-C}$$

Now since $I_{R-L} = \dfrac{V}{\sqrt{R^2 + (\omega_r L)^2}}$

and $\sin \phi_{R-L} = \dfrac{X_L}{Z_{R-L}} = \dfrac{\omega_r L}{\sqrt{R^2 + (\omega_r L)^2}}$

$$I_{R-C} = \frac{V}{Z_{R-C}} = \frac{V}{\sqrt{R_1^2 + \left(\frac{1}{\omega_r C}\right)^2}}$$

and $\sin \phi_{R-C} = \dfrac{X_C}{Z_{R-C}} = \dfrac{1/\omega_r C}{\sqrt{R_1^2 + \left(\frac{1}{\omega_r C}\right)^2}}$

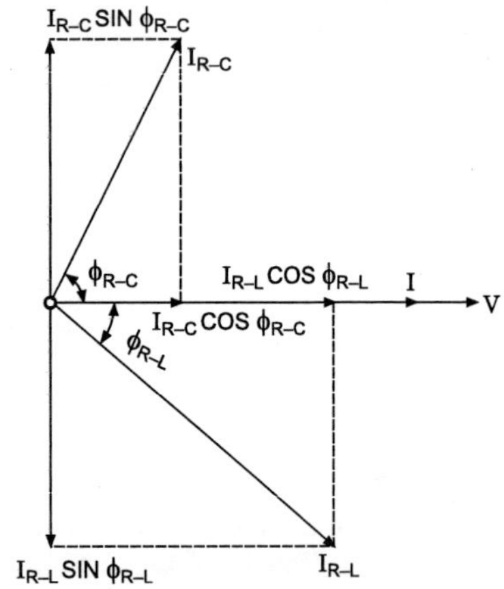

Fig. 7.9

$$\therefore \quad \frac{V}{\sqrt{R^2 + (\omega_r L)^2}} \times \frac{\omega_r L}{\sqrt{R^2 + (\omega_r L)^2}} = \frac{V}{\sqrt{R_1^2 + \left(\frac{1}{\omega_r C}\right)^2}} \times \frac{1/\omega_r C}{\sqrt{R_1^2 + \left(\frac{1}{\omega_r C}\right)^2}}$$

$$\text{or} \quad \frac{\omega_r L}{R^2 + (\omega_r L)^2} = \frac{1/\omega_r C}{R_1^2 + (1/\omega_r C)^2}$$

$$\text{or} \quad \frac{\omega_r L}{R^2 + \omega_r^2 L^2} = \frac{\omega_r C}{\omega_r^2 R_1^2 C^2 + 1}$$

$$\text{or} \quad L\left(\omega_r^2 R_1^2 C^2 + 1\right) = C\left(R^2 + \omega_r^2 L^2\right)$$

$$\text{or} \quad \omega_r^2 LC\left(R_1^2 C - L\right) = CR^2 - L$$

$$\text{or} \quad \omega_r = \frac{1}{\sqrt{LC}}\sqrt{\frac{CR^2 - L}{CR_1^2 - L}}$$

$$\text{or Resonant frequency, } f_r = \frac{1}{2\pi}\omega_r = \frac{1}{2\pi\sqrt{LC}}\sqrt{\frac{CR^2 - L}{CR_1^2 - L}} \qquad ...(7.13)$$

If resistance of capacitor is negligible *i.e.* $R_1 = 0$, as is usually the case,

$$\text{Resonant frequency, } f_r = \frac{1}{2\pi\sqrt{LC}}\sqrt{1 - \frac{CR^2}{L}} = \frac{1}{2\pi}\sqrt{\frac{1}{LC} - \frac{R^2}{L^2}} \qquad ...(7.14)$$

If resistance of coil R is zero

$$\text{Resonant frequency, } f_r = \frac{1}{2\pi}\sqrt{\frac{1}{LC}} \qquad ...(7.15)$$

Current at Resonance.

Assuming $R_1 = 0$, as usually, at resonant frequency

$$I_{R-C} = \frac{V}{Z} = \frac{V}{\sqrt{\left(R_1^2\right)^2 + \frac{1}{(2\pi f_r C)^2}}} = 2\pi f_r C V \qquad ...(7.16)$$

$$\phi_{R-C} = \cos^{-1}\frac{R}{Z} = 90° \text{ or } \frac{\pi}{2} \text{ radians}$$

Active component of R-C branch current, $I_{R-C} \cos\phi = 0$

Reactive component of R-C branch current, $I_{R-C} \sin\phi = 2\pi f_r CV$

$$\text{R-L branch current, } I_{R\text{-}L} = \frac{V}{Z} = \frac{V}{\sqrt{R^2 + (2\pi f_r L)^2}}$$

$$\phi_{R-L} = \cos^{-1}\frac{R}{Z}$$

$$\text{Active component of R-L branch current, } = I_{R-L}\cos\phi_{R-L} = \frac{V}{Z}.\frac{R}{Z} = \frac{VR}{Z^2}$$

$$\text{Reactive component of R-L branch current } = I_{R-L}\sin\phi_{R-L} = \frac{V}{Z}.\frac{2\pi f_r L}{Z}$$

$$\text{At resonance, } I_{R-L}\sin\phi_{R-L} = I_{R-C}\sin\phi_{R-C}$$

$$\text{or } \frac{V}{Z}.\frac{2\pi f_r L}{Z} = 2\pi f_r CV$$

$$\text{or } \frac{L}{Z^2} = C$$

$$\text{or } Z = \sqrt{\frac{L}{C}} \qquad ...(7.17)$$

$$\text{Line current, } I = I_{R-L}\cos\phi_{R-L} + I_{R-C}\cos\phi_{R-C}$$

$$= \frac{V}{Z}.\frac{R}{Z} + 0 = \frac{VR}{Z^2} = VR \times \frac{C}{L} = \frac{V}{L/CR} \qquad ...(7.18)$$

The denominator $\frac{L}{CR}$ is known as *effective or equivalent* or *dynamic impedance* of the parallel circuit at resonance.

7.3.1. Important Points About Current or Parallel Resonance

1. Net susceptance is zero *i.e.* $\frac{1}{X_C} = \frac{X_L}{Z^2}$

   or $\omega_r C = \frac{\omega_r L}{Z^2}$

   or $Z = \sqrt{\frac{L}{C}}$
2. The admittance is equal to conductance.
3. Reactive or wattless component of line current is zero, hence circuit power factor is unity.
4. Impedance is purely resistive, maximum in magnitude and is equal to L/CR.
5. Line current is minimum and is equal to $\frac{V}{L/CR}$ in magnitude and is in phase with the applied voltage.
6. Frequency is equal to $\frac{1}{2\pi}\sqrt{\frac{1}{LC} - \frac{R^2}{L^2}}$ Hz.

Note: *Parallel resonant circuit is sometimes called the rejector circuit because at resonant frequency the line current is minimum or it almost rejects it.*

Since in parallel resonant circuits circulating current between the branches is many times the line current, such type of resonance is sometimes known as current resonance.

An inductive coil of inductance L connected in parallel with a capacitance C is called the tank circuit.

7.3.2. Graphical Representation of Current or Parallel Resonance. Now we will discuss the effect of variation in frequency on the susceptances of the two parallel branches. The variations are illustrated in Fig. 7.10.

Inductive susceptance, being equal to $\frac{1}{\omega L}$ or $\frac{1}{2\pi f L}$, decreases inversely with the increase in frequency and is represented by a rectangular hyperbola lying in the fourth quadrant, below the frequency axis (inductive susceptance being considered negative). *Capacitive susceptance,* being equal to ωC or $2\pi f C$, increases in direct proportion to the supply frequency and is represented by a straight line passing through origin. As capacitive susceptance is considered positive, so it lies in first quadrant. The net susceptance B is the difference of capacitive susceptance and inductive susceptance and the curve drawn between net susceptance and frequency of the applied voltage is a hyperbola (not rectangular) as illustrated in Fig. 7.10.

The frequency at which the net susceptance curve crosses the frequency axis is called the *resonant frequency*. At this point impedance is maximum or admittance is minimum and is equal to G, consequently line current is minimum. Obviously at frequency lower than resonant frequency the inductive susceptance is more than capacitive susceptance, therefore, the circuit is inductive and the line current lags behind the applied voltage. But for frequencies exceeding resonant frequency capacitive susceptance predominates, therefore, the circuit is capacitive and the line current leads the applied voltage.

If the resistance is relatively low the current will fall considerably at resonant frequency and if the resistance is high, decrease in current will be less pronounced, as illustrated in Fig. 7.11.

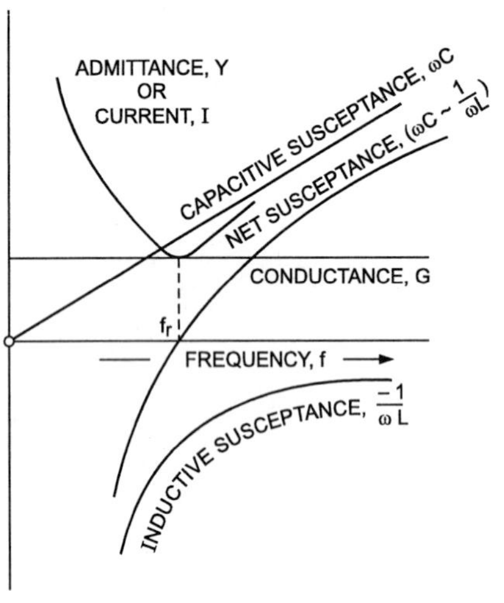

Fig. 7.10

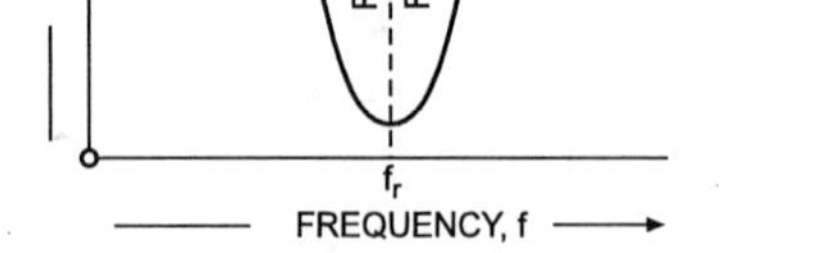

Fig. 7.11

7.3.3. Bandwidth in Case of a Parallel Resonant Circuit. The bandwidth in case of a parallel circuit is defined in the same way as in case of a series circuit. In this case also there are upper and lower half-power points where power absorbed is half of that at resonant frequency.

At bandwidth frequencies, the net susceptance B is equal to the conductance G. So at frequency f_1 the net susceptance $B_{L_1} - B_{C_1} = G$ and at frequency f_2, $B_{C_2} - B_{L_2} = G$. Thus admittance $Y = \sqrt{G^2 + B^2} = \sqrt{2}\,G$ and phase angle $\phi = \tan^{-1}\frac{B}{G} = \tan^{-1} 1 = 45°$ or $\frac{\pi}{4}$ radian.

7.3.4. Q-Factor or Current Magnification Factor. Q-factor of a parallel circuit is defined as the ratio of the circulating current to the line current or as the current magnification.

$$\therefore \text{ Q-factor} = \frac{\text{Circulating current between L and C}}{\text{Line current}}$$

Since from Eq. (7.16), the circulating current, $I_C = 2\pi f_r C V$

and from Eq. (7.18), the line current, $I = \dfrac{V}{L/CR} = VCR/L$

$$\therefore \text{ Q-factor} = \frac{I_C}{I} = \frac{2\pi f_r CV}{VCR/L} = \frac{2\pi f_r L}{R}$$

$$\text{but } f_r = \frac{1}{2\pi}\sqrt{\frac{1}{LC}} \text{ neglecting R}$$

$$\text{So Q-factor} = 2\pi \times \frac{1}{2\pi}\sqrt{\frac{1}{LC}} \times \frac{L}{R}$$

$$= \frac{1}{R}\sqrt{\frac{L}{C}} \text{, It is the same as for series resonance.}$$

It should be noted that in series circuits, Q-factor provides the voltage magnification, whereas in parallel circuits, it provides current magnification.

Example 7.14 **Find the resonant frequency for the circuit shown in Fig. 7.12.**

[Pb. Technical Univ. Basic Electronics and Elec. Engineering Second Semester 2004-05]

Solution: Resonant frequency of the given circuit

$$f_r = \frac{1}{2\pi\sqrt{LC}}\sqrt{\frac{CR^2 - L}{CR_1^2 - L}}$$

$$= \frac{1}{2\pi\sqrt{10\times10^{-3}\times40\times10^{-6}}} \times \sqrt{\frac{40\times10^{-6}\times5^2 - 10\times10^{-3}}{40\times10^{-6}\times2^2 - 10\times10^{-3}}}$$

= 263 Hz **Ans.**

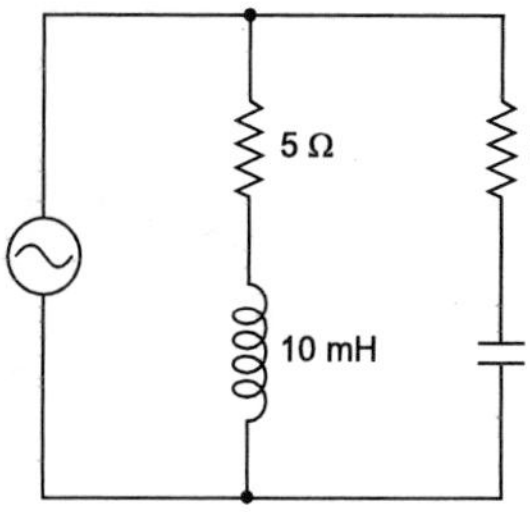

Fig. 7.12

Example 7.15 **For the circuit shown below, determine**

(*i*) **Resonant frequency.**
(*ii*) **Total impedance of the circuit at resonance.**
(*iii*) **Bandwidth**
(*iv*) **Quality factor.** *[U.P. Technical Univ. Elec. Engineering First Semester 2005-06]*

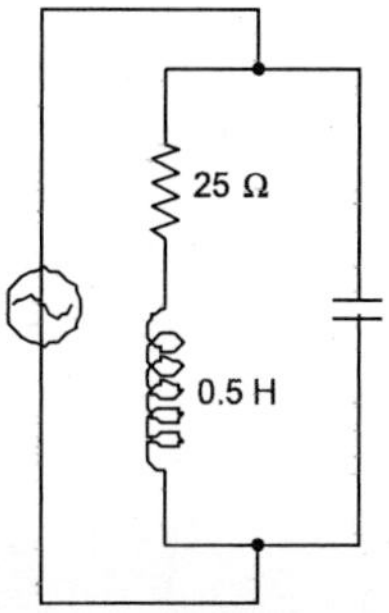

Fig. 7.13

Solution: R = 25 Ω, L = 0.5 H and C = 5 × 10⁻⁶ F

(*i*) Resonant frequency, $f_r = \dfrac{1}{2\pi}\sqrt{\dfrac{1}{LC} - \dfrac{R^2}{L^2}}$ Hz

$$= \frac{1}{2\pi}\sqrt{\frac{1}{0.5\times5\times10^{-6}} - \frac{25^2}{(0.5)^2}} \text{ Hz} = 100.34 \text{ Hz } \textbf{Ans.}$$

(*ii*) Impedance of the circuit at resonance, $Z = \dfrac{L}{CR} = \dfrac{0.5}{5\times10^{-6}\times25} = 4{,}000\ \Omega$ **Ans.**

(*iii*) Bandwidth $= \dfrac{R}{2\pi L} = \dfrac{25}{2\pi\times0.5} = 7.958$ Hz **Ans.**

(*iv*) Quality factor, $Q = \dfrac{1}{R}\sqrt{\dfrac{L}{C}} = \dfrac{1}{25}\sqrt{\dfrac{0.5}{5\times10^{-6}}} = 12.65$ **Ans.**

Example 7.16 **In the circuit shown in Fig. 7.14, find the value of R such that the impedance of the whole circuit should be independent of the frequency of the supply. If voltage = 200 V, L = 0.16 H and C = 100 μ F, calculate the power loss in the circuit.**

[*U.P. Technical Univ. Electrical Engineereing February 2002*]

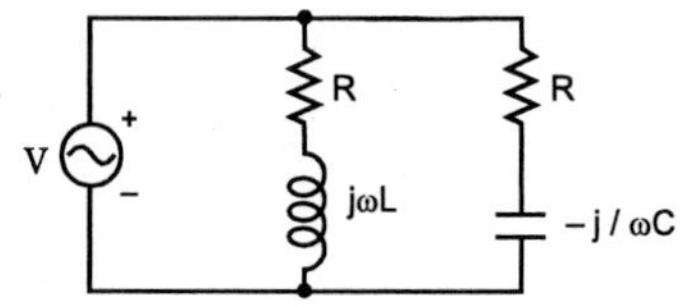

Fig. 7.14

Solution: Impedance of inductive branch, $Z_L = (R + j\,\omega\,L)$

$$\text{Impedance of capacitive branch, } Z_C = \left(R - \frac{j}{\omega C}\right)$$

Impedance of the whole circuit shown in Fig. 7.14,

$$\mathbf{Z} = \frac{\mathbf{Z_L\,Z_C}}{\mathbf{Z_L + Z_C}} = \frac{(R + j\omega L)\left(R - \dfrac{j}{\omega C}\right)}{(R + j\omega L) + \left(R - \dfrac{j}{\omega C}\right)} = \frac{R^2 + \dfrac{L}{C} + jR\left(\omega L - \dfrac{1}{\omega\,C}\right)}{2R + j\left(\omega L - \dfrac{1}{\omega C}\right)}$$

We note that *j* term in the numerator of the above expression is R times the denominator so if the real term *i.e.* $\left(R^2 + \dfrac{L}{C}\right)$ is also R times of the real term in denominator *i.e.* R × 2 R, then the term consisting of ω will disappear. Thus the impedance of the circuit will be independent

$$\text{if } R^2 + \frac{L}{C} = R \times 2\,R \quad \text{or } R^2 = \frac{L}{C} \quad \text{or} \quad R = \sqrt{\frac{L}{C}} \quad \textbf{Ans.}$$

Substituting L = 0.16 H and C = 100 × 10^{-6} F in above expression we have

$$R = \sqrt{\frac{0.16}{100 \times 10^{-6}}} = 40\ \Omega\ \textbf{Ans.}$$

$$\text{Power loss in the circuit, } P = \frac{V^2}{R} = \frac{(200)^2}{40} = 1{,}000 \text{ W or } 1 \text{ kW } \textbf{Ans.}$$

Example 7.17 **What are the features of resonance in parallel circuits ? Calculate the value of C which results in resonance for the circuit shown in Fig. 7.15 when frequency is 1,000 Hz and find Q-factor for each branch.**

[*U.P. Technical Univ. Electrical Engineering June 2001*]

$$\textbf{Solution: } \text{Admittance of inductive branch, } \mathbf{Y}_{R-L} = \frac{1}{4 + j\,8} = (0.05 - j\,0.1) \text{ Siemen}$$

$$\text{Admittance of capacitive branch, } \mathbf{Y}_{R-C} = \frac{1}{5 - jX_C} = \left(\frac{5}{25 + X_C^2} + j\frac{X_C}{25 + X_C^2}\right) \text{ Siemen}$$

For resonance, *j* parts of admittances must be equal in magnitude i.e.

$$\frac{X_C}{25 + X_C^2} = 0.1$$

$$\text{or} \quad 0.1\,X_C^2 - X_C + 2.5 = 0$$

$$\text{or} \quad X_C = 5\ \Omega$$

$$\text{and } C = \frac{1}{2\,\pi f X_C} = \frac{1}{2\pi \times 1{,}000 \times 5}$$

$$= 31.83\ \mu\text{F} \quad \textbf{Ans.}$$

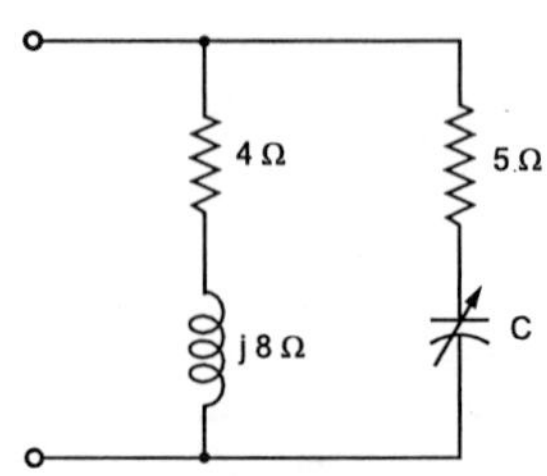

Fig. 7.15

$$\text{Q factor for R-L branch} = \frac{\omega L}{R} = \frac{8}{4} = 2 \ \textbf{Ans.}$$

$$\text{Q factor for R-C branch} = \frac{1}{\omega R\,C} = \frac{1}{\omega C\,R} = \frac{5}{5} = 1.0 \ \textbf{Ans.} \qquad \left[\because \frac{1}{\omega C} = X_C = 5\,\Omega\right]$$

Example 7.18 **Calculate the resonance frequency of the circuit shown in Fig. 7.16.**

[*Electrical Engineering G.B. Technical Univ. First Semester 2011-12; U.P. Technical Univ. Second Semester 2013-14*]

Solution: The impedance of the given circuit is given by

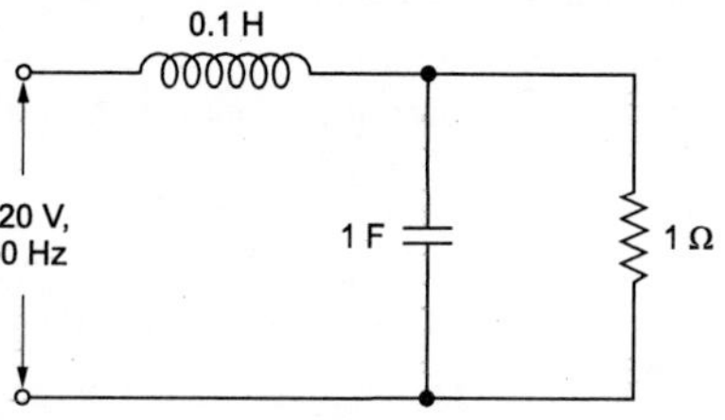

Fig. 7.16

$$Z = j\omega L + \frac{1}{\frac{1}{R} + j\omega C} = j\omega L + \frac{R}{1 + j\omega RC}$$

$$= j\omega L + \frac{R(1 - j\omega RC)}{1 + \omega^2 R^2 C^2}$$

$$= \frac{j\omega L + j\omega^3 LR^2C^2 + R - j\omega R^2 C}{1 + \omega^2 R^2 C^2}$$

For resonance, sum of *j* parts must be equal to zero

$$\text{So } \omega_r L + \omega_r^3 L R^2C^2 - \omega_r R^2 C = 0$$

$$\text{or } \omega_r (L + \omega_r^2 L R^2C^2 - R^2C) = 0$$

$$\text{or } L + \omega_r^2 LR^2C^2 - R^2C = 0$$

$$\text{or } \omega_r = \sqrt{\frac{R^2C - L}{LR^2C^2}} = \sqrt{\frac{1}{LC} - \frac{1}{R^2C^2}}$$

Substituting L = 0.1 H, C = 1 F and R = 1 Ω, we have

$$\omega_r = \sqrt{\frac{1}{0.1 \times 1} - \frac{1}{1^2 \times 1^2}} = \sqrt{10 - 1} = 3$$

$$\text{and resonance frequency, } f_r = \frac{\omega_r}{2\pi} = \frac{3}{2\pi} = 0.4775 \text{ Hz } \textbf{Ans.}$$

7.4. COMPARISON BETWEEN VOLTAGE RESONANCE AND CURRENT RESONANCE

The comparative results of voltage resonance and current resonance studied so far are summarized below in the tabular form.

S.No.	*Particulars*	*Voltage Resonance*	*Current Resonance*
1.	Impedance under *Resonance* conditions	Z = R, the minimum	Z = L/CR, the maximum
2.	Magnitude of current under resonant condition	I = V/R, the maximum	I = V/(L/CR), the minimum
3.	Power factor at resonance	Unity	Unity
4.	Resonant frequency	$f_r = \frac{1}{2\pi\sqrt{LC}}$	$f_r = \frac{1}{2\pi}\sqrt{\frac{1}{LC} - \frac{R^2}{L^2}}$
5.	Magnification occurs of	voltage	current
6.	Magnification	ω L/R	ω L/R
7.	Bandwidth	R/2πL Hz	R/2πL Hz

HIGHLIGHTS

1. An R-L-C series circuit is said to be in electrical resonance when

$$X_L = X_C \textit{ i.e. } \text{the net reactance } X = 0$$

Under resonant conditions

(i) Circuit impedance, Z = R

(ii) The current flowing through the circuit is maximum and is in phase with the applied voltage and will be equal to V/R

(iii) The voltage drop across inductance is equal to the voltage drop across capacitance and is maximum.

(iv) The power factor is unity.

(v) The power expended, P = VI watts and

(vi) The resonant frequency, $f_r = \dfrac{1}{2\pi\sqrt{LC}}$ Hz

Q-factor of a series resonant circuit is given as

$$\text{Q-factor} = \frac{\omega_r L}{R} = \frac{2\pi f_r L}{R} = \frac{1}{R}\sqrt{\frac{L}{C}}$$

$$\text{Bandwidth, BW} = \frac{R}{2\pi L} = \frac{f_r}{\text{Q-factor}}$$

2. A parallel ac circuit is said to be in resonance when it draws no reactive current.

Under resonant condition of parallel ac circuit

$$\text{Impedance, Z} = \frac{L}{CR}, \text{ the maximum}$$

$$\text{Current, I} = \frac{V}{L/CR}, \text{ the minimum}$$

$$\text{Power factor} = \text{Unity}$$

$$\text{Resonant frequency}, f_r = \frac{1}{2\pi}\sqrt{\frac{1}{LC} - \frac{R^2}{L^2}} \simeq \frac{1}{2\pi\sqrt{LC}} \quad \text{if R is negligibly small}$$

$$\text{Q-factor} = \frac{1}{R}\sqrt{L/C}$$

$$\text{Bandwidth} = \frac{R}{2\pi L} \text{ Hz}$$

EXERCISES

1. Derive the response of R-L-C series circuit to sinusoidal input. Also derive the condition for resonance.

[G.B. Technical Univ. Electrical Engineering First Semester 2010-11]

2. Derive the condition for resonance in series R-L-C circuit. What are the different applications of resonance?

[U.P. Technical Univ. Basic Electrical Engineering Second Semester 2013-14]

3. Explain the concept of bandwidth and quality factor for a series R-L-C circuit. Derive their expressions.

[U.P. Technical Univ. Electrical Engineering Second Semester 2007-08]

4. Derive the quality factor Q of the series R-L-C circuit at resonance. Define the bandwidth for the same. **[G.B. Technical Univ. Electrical Engineering First Semester 2011-12]**

5. Derive bandwidth for series resonance. **[U.P. Technical Univ. Basic Electrical Engineering Odd Semester 2013-14]**

6. Explain the following terms:

Bandwidth and quality factors. **[G.B. Technical Univ. Electrical Engineering First Semester 2010-11]**

7. Deduce the formula for half-power frequencies for a series R-L-C circuit under resonance. Why are they called "half-power frequencies"? **[G.B. Technical Univ. Electrical Engineering Second Semester 2011-12]**

8. Define series resonance and resonant frequency. Why in series circuit voltage across L and C is very high? Draw resonance curve. **[U.P. Technical Univ. Electrical Engineering Odd Semester, 2013-14]**

9. Explain series resonance in R-L-C circuit. What are bandwidth and quality factor of the circuit ?

[U.P. Technical Univ. Electrical Engineering Second Semester, 2006-07]

10. Explain the selectivity of a series RLC resonance circuit. Prove that selectivity is reciprocal of quality of the circuit.

[U.P. Technical Univ. Electrical Engineering Second Semester, 2004-05]

11. Show that for an RLC series circuit the resonant frequency $\omega_0 = \sqrt{\omega_1 \omega_2}$, where ω_1 and ω_2 are the lower and upper half power frequencies respectively. **[U.P. Technical Univ. Electrical Engineering Second Semester, 2004-05]**

12. Derive an expression for parallel resonance and mention its salient features.

[U.P. Technical Univ. Electrical Engineering First Semester 2006-07]

13. Define resonance in parallel RLC circuit. Draw resonance curve.

[U.P. Technical Univ. Electrical Engineering Odd Semester 2013-14]

14. Explain parallel resonance. Why is parallel resonance called the current resonance? Show the graphical representation of current in parallel resonance. **[U.P. Technical Univ. Electrical Engineering Second Semester, 2005-06]**

15. Define parallel resonance. Calculate at resonance the resultant current and quality factor in terms of the parameters of a circuit. **[U.P. Technical Univ. Electrical Engineering First Semester, 2007-08]**

16. Derive an expression for the resonance frequency of the parallel resonance circuit, one branch consisting of inductor L, resistance R and other branch contain capacitor C. Also draw the phasor diagram for the same.

[Mahamaya Technical Univ. Electrical Engineering Second Semester 2010-11]

17. Explain parallel resonance and draw graphs of Z, X_L and X_C against frequency.

[G.B. Technical Univ. Electrical Engineering First Semester 2009-10]

18. Show that the condition for resonance in a parallel R-L-C circuit is same as that in a series R-L-C circuit. State the application of series as well as parallel resonance.

[G.B. Technical Univ. Electrical Engineering Second Semester 2009-10]

19. Derive the quality factor Q_p of the parallel RLC circuit at resonance. Define bandwidth for the same.

[U.P. Technical Univ. Electrical Engineering First Semester 2008-09]

20. Derive the expression for Q-factor in the R-L-C parallel circuit.

[G.B. Technical Univ. Electrical Engineering Second Semester 2010-11]

21. What do you understand by "series resonance" and "parallel resonance"? What are the applications of tank circuits?

[U.P. Technical Univ. Electrical Engineering Odd Semester 2013-14]

22. Explain the following:

(*i*) Series resonance circuits.
(*ii*) Parallel resonance circuits. **[U.P. Technical Univ. Electrical Engineering Odd Semester 2013-14]**

SHORT ANSWER TYPE QUESTIONS WITH ANSWERS

Q. 1. What is meant by resonant frequency ?

Ans. The frequency of supply at which two reactances (inductive reactance and capacitive reactance) are equal is called resonant frequency.

Q. 2. What is series resonance ?

Ans. When in an R-L-C series circuit the inductive reactance equals capacitive reactance, the circuit is said to be in resonance, called the series resonance.

Q. 3. Why is the series resonance called the voltage resonance?

Ans. Since in series resonance the voltage across inductance and capacitance is maximum, it is called the voltage resonance.

Q. 4. Why is the series R-L-C circuit also called the acceptor circuit ?

Ans. Series R-L-C circuit accepts currents at one particular frequency but rejects currents of other frequencies and therefore, series R-L-C circuit is called the acceptor circuit.

Q. 5. Write an expression of resonance frequency for high Q-series R-L-C circuit. **[G.B.T.U. First Semester 2011-12]**

Ans. The expression of resonance frequency for high Q-series RLC circuit is

$$\text{Resonant frequency, } f_r = \frac{1}{2\pi\sqrt{LC}} \text{ Hz}$$

where L is the circuit inductance in henries and C is the circuit capacitance in farads.

Q. 6. In a series RLC circuit, the voltage across L and C at resonance may exceed even the supply voltage. Why ?

[U.P. Technical Univ. First Semester 2005-06]

Ans. When the series circuit is in resonance, the voltage drops across inductance and capacitance will be equal in magnitude but opposite in phase and so nullify each other and supply voltage will be equal to voltage drop across resistance. So the current will be too large and the voltage drops may exceed the supply voltage in case the inductive reactance (or capacitive reactance) exceed the resistance of the circuit.

Q. 7. What do you understand by the term bandwidth of a resonant circuit ?

Ans. The frequency band over which the magnitude of current is equal to $1/\sqrt{2}$ times the peak current is referred to as bandwidth of the circuit.

Q. 8. Define bandwidth of a resonant circuit. Give the relationship of quality factor in terms of bandwidth and resonant frequency. **[M.T.U. Second Semester 2011-12]**

Ans. The difference of upper and lower half-power frequencies is called the bandwidth.

The relationship between Q_r, Δf and f_r is given as

$$Q_r = \frac{f_r}{\Delta f}$$

Q. 9. What is selectivity? **[U.P. Technical Univ. Second Semester 2013-14]**

Ans. We have seen that for low resistance circuit the resonance curve is sharply peaked and such a circuit is said to be *sharply resonant* or *highly selective*. On the other hand, high resistance circuit have flat resonance curve and is said to have poor selectivity. Selectivity of different resonant circuits are compared in terms of their bandwidths, which are given by the bands of frequencies which lie between two points on either side of the resonant frequency where current is $1/\sqrt{2}$ times of maximum current I_{max}.

Q. 10. A series R-L-C circuit has R = 10 Ω, L = 0.1 H, C = 8 μF. Calculate half-power frequencies for the circuit.

[U.P. Technical Univ. Odd Semester 2013-14]

Ans. 170 Hz, 185.9 Hz ... Refer to Solution of Example 7.8.

Q. 11. Give the relationship between quality factor, resonant frequency and bandwidth for a series R-L-C circuit.

[G.B.T.U. Second Semester 2011-12]

Ans. The relationship between Q_r, Δf and f_r is given as

$$Q_r = \frac{f_r}{\Delta f}$$

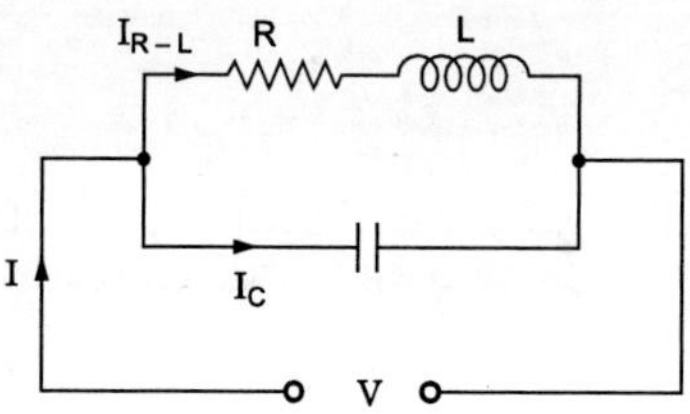

Fig. 1

Q. 12. What is a resonance in a parallel circuit ?

Ans. A parallel electrical circuit (Fig. 1) is said to be in electrical resonance when the reactive or wattless component of line current is zero i.e. when reactive component of R-L branch current is equal to current drawn by capacitance.

Q. 13. Why is parallel resonance called the current resonance ?

Ans. Since in parallel resonant circuit, circulating current between the branches is many times the line current, such type of resonance is called the current resonance.

Q. 14. Why a series resonant circuit is called as an acceptor circuit and a parallel resonant circuit as a rejector circuit ?

[U.P. Technical Univ. First Semester 2000-01]

Ans. The series resonant circuit is called an acceptor circuit because such a circuit accepts currents at one particular frequency but rejects currents of other frequencies. Parallel resonant circuit is called the rejector circuit because at resonant frequency the line current is minimum or it almost rejects it.

Q. 15. What is meant by Q-factor of series resonant circuit ?

Ans. The Q-factor of a series resonant circuit may be given as the voltage magnification that the circuit produces at resonance.

$$\text{Q-factor at resonance, } Q_r = \frac{\omega_r L}{R} = \frac{2\pi f_r L}{R} = \frac{1}{R}\sqrt{\frac{L}{C}}$$

Q. 16. What is Q-factor of a parallel resonant circuit ?

Ans. Q-factor of a parallel resonant circuit is defined as the ratio of the circulating current to the line current at resonance or as the current magnification. Q-factors for series resonance and parallel resonance are the same.

Q. 17. What is the power factor of an R-L-C circuit under resonant condition ?

Ans. Unity.

Q. 18. At resonance the current is maximum in a series circuit and minimum in parallel circuit. Why ?

[U.P. Technical Univ. First Semester 2005-06]

Ans. At resonance in series circuit the inductive reactance is equal to capacitive reactance in magnitude but opposite in phase and so nullify each other and thus impedance is minimum and equal to the resistance of the circuit and, therefore, current is maximum.

But at resonance in parallel circuit, net susceptance is zero, the admittance is, therefore, minimum and is equal to conductance of the circuit and the current, being the product of supply voltage and circuit admittance is minimum.

Q. 19. The shape of resonance curve depends on Q of the coil. Why ? **[U.P. Technical Univ. First Semester 2005-06]**

Ans. Bandwidth of the series RLC circuit, being equal to f_r/Q_r, depends upon the Q of the coil i.e. with high-Q coil the resonance curve will be of smaller bandwidth (i.e. sharply peaked) while with small-Q coil the resonance curve will be of larger bandwidth (flat). Thus the shape of the resonance curve depends upon the Q of the coil.

Q. 20. What will be the current at resonance in series and parallel R-L-C circuit?

[G.B. Technical Univ. Even Semester 2012-13]

Ans. Current at resonance in series is given as

$I = \frac{V}{R}$ where V is supply voltage in volts and R is circuit resonance in Ω.

Current at resonance in parallel is given as

$I = \frac{V}{L/CR}$ where V is supply voltage in volts, R is circuit resistance in ohms, L is circuit inductance in henries and C is capacitance in Farads.

PROBLEMS

1. A series R-L-C circuit has R = 10 Ω, L = 0.1 H and C = 8 μ F. Determine (*i*) resonant frequency (*ii*) Q-factor of the circuit at resonance and (*iii*) the half-power frequencies.

[U.P. Technical Univ. Electrical Engineering September 2001]

[**Ans.** (*i*) 177.94 Hz (*ii*) 11.18 (*iii*) 170 Hz, 185.8 Hz]

2. A 20 Ω resistor is connected in series with an inductor and a capacitor, across a variable frequency 25 V supply. When the frequency is 400 Hz, the current is at its maximum value of 0.5 A and the potential difference across the capacitor is 150 V. Calculate the resistance and inductance of the inductor.

[U.P. Technical Univ. Electrical Engineering Junuary 2003]

[**Ans.** 50 Ω; 0.119 Ω]

3. An inductive coil of resistance 10 Ω and inductance 0.1 H is connected in parallel with a 150 μF capacitor to a variable frequency, 200 V supply. Find the resonant frequency at which the total current taken from the supply is in phase with the supply voltage. Also find the value of this current. **[G.G.S.I.P. Univ. Delhi Electrical Science June 2006]**

[**Ans.** 37.886 Hz; 3 A]

4. A series RLC circuit with R = 250 Ω, and L = 0.6 H results in a leading phase angle of 60° at a frequency of 40 Hz. At what frequency will the circuit resonate ? **[Pb. Technical Univ. Basic Electrical Engineering December 2006]**

[**Ans.** 78.70 Hz]

5. An R-L-C series circuit consists of a resistance of 1,000 Ω, an inductance of 100 mH and a capacitance of 10 μμF. If the voltage of 100 V is applied across the combination, find (*i*) the resonance frequency (*ii*) Q-factor of the circuit and (*iii*) the half-power points. [**Ans.** (*i*) $10^6/2\pi$ Hz; (*ii*) 100; (*iii*) 158.4 kHz and 160 kHz]

6. An inductive coil when connected across 500 V, 50 Hz supply draws 1.0 A at 0.8 p f. What capacitance is connected in parallel with the coil to make the power factor of the combination unity ? What will be the current (*i*) drawn from the supply; and (*ii*) in each branch with this new capacitor. [**Ans.** 3.82 μ F; (*i*) 0.8 A (*ii*) 1.0 A, 0.6 A]

7. An inductive circuit of resistance 2 Ω and inductance 0.01 H is connected to 250-V, 50 Hz supply. What capacitance placed in parallel will produce resonance ? Find the total current taken from the supply and the current in the branch circuits.

[**Ans.** 721 μ F; 36.05 A; 125 A; 67.13 A $\angle -57.52°$ A; 56.63 $\angle 90°$ A

8. In the circuit given below, if the value of $R = \sqrt{\frac{L}{C}}$, then prove that the impedance of the entire circuit is equal to R only and is independent of the frequency of supply. Find the value of impedance for L = 0.02 H and C = 100 μ F. [**Ans.** 14.142 Ω]

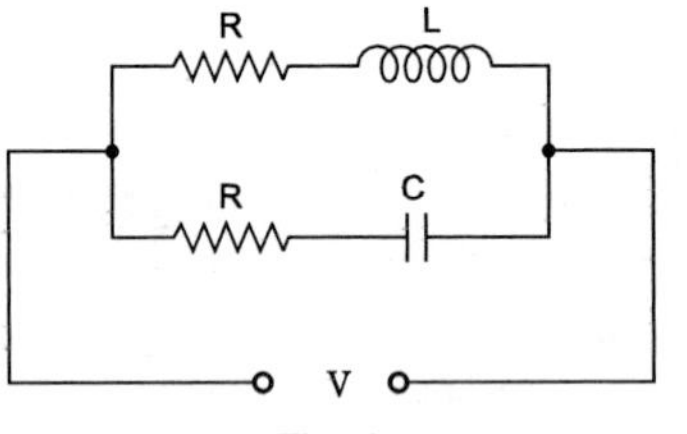

Fig. 1

UNIT-III

Three-Phase AC Circuits and Measuring Instruments

CONTENTS IN THIS UNIT

Three-Phase AC Circuits:

Three-phase system – Its necessity and advantages, Star and delta connections, Balanced supply and balanced load, Line and phase voltage/current relations, Three-phase power and its measurements (Simple numerical problems).

Measuring Instruments:

Types of instruments, Construction and working principles of PMMC and moving iron type voltmeters and ammeters, Single-phase dynamometer wattmeter and induction type energy meter, Use of shunts and multipliers (Simple numerical problems on energy meter, shunts and multipliers).

8 CHAPTER

Three-Phase AC Circuits

8.1. INTRODUCTION

The type of alternating currents and voltages discussed so far in the book are termed as *single-phase currents and voltages* as they consist of a single alternating current and voltage waves. Single-phase systems involving single-phase currents and voltages are quite satisfactory for domestic applications. Even the motors employed in domestic applications are mostly single phase, for example, motors for mixers, coolers, fans, air-conditioners, refrigerators. However, the single-phase system has its own limitations and, therefore, has been replaced by polyphase system. For supplying power to electric furnaces 2-phase system is generally employed. Six-phase system is usually employed in connection with converting machinery and apparatus. For general supply three-phase system is universally used. For generation, transmission and distribution of electric power 3-phase system has been universally adopted. The two-phase supply and six-phase supply are obtained from 3-phase supply.

Polyphase system means the system which consists of numerous (poly means numerous or multiple) windings or circuits (phase* means winding or a circuit).

A polyphase system is essentially a combination of several single-phase voltages having same magnitude and frequency but displaced from one another by equal angle (electrical), which depends upon the number of phases and can be determined from the following relation

$$\text{Electrical displacement} = \frac{\text{360 electrical degrees}}{\text{Number of phases}} \quad ...(8.1)$$

The above relation does not hold good for two-phase windings, which are displaced by 90 electrical degrees apart.

A supply system is said to be symmetrical when several voltages of the same frequency have equal magnitude and are displaced from one another by equal time angle. 3-phase, 3-wire or 4-wire supply system will be symmetrical when the line to line voltages are equal in magnitude and displaced in phase by 120 electrical degrees with respect to each other. Further in a four-wire system, the voltage with respect to neutral of the three phase wires should be equal to one another in magnitude and displaced in phase by 120° with respect to each other.

A 3-phase supply will be unbalanced when either of the three-phase voltages are unequal in magnitude or the phase angle between these phases is not equal to 120°.

A load circuit is said to be *balanced* when the loads (impedances) connected in various phases are same in magnitude as well as in phase. Any three-phase load in which the impedances in one or more phases differ from the impedances of other phases is called an *unbalanced three-phase load.*

* Each emf of a polyphase system is referred to as a *phase.*

In case one phase out of three phases of a 3-phase supply connected to the 3-phase load is not available, then such a condition is called the *single phasing*.

8.1.1. Merits and Demerits of Polyphase System Over Single-Phase System. The advantages of a polyphase system over single-phase system are enumerated below:

(*i*) In a single-phase circuit the power delivered is pulsating. Even when current and voltage are in phase, the power is zero twice in each cycle, and when the current leads or lags behind the voltage, the power is negative twice and zero four times during each cycle. This is not objectionable for lighting and small motors but with large motors it causes excessive vibrations. In polyphase system the total power delivered is constant if loads are balanced though the power of any one phase or circuit may be negative. So polyphase system is highly desirable particularly for power loads.

(*ii*) The rating of a given machine increases with the increase in number of phases. For example output of a 3-phase motor is 1.5 times the output of a single phase motor of same size.

(*iii*) Single-phase induction motors have no starting torque and so it is necessary to provide these motors with an auxiliary means of starting , but in case of three-phase motors except synchronous motors, there is no need of providing an auxiliary means for starting.

(*iv*) Power factor of a single-phase motor is lower than that of a polyphase motor of the same rating (output and speed). The efficiency of a polyphase motor is also higher than that of a single-phase motor.

(*v*) Three-phase system requires 3/4th weight of copper of that required by single-phase system to transmit the same amount of power at a given voltage and over a given distance.

(*vi*) Rotating magnetic field can be set up by passing polyphase currents through stationary coils.

(*vii*) Polyphase system is more capable and reliable than single-phase system .

(*viii*) Parallel operation of polyphase alternators is simple as compared to that of single-phase alternators because of pulsating reaction in single-phase alternators.

However, three-phase operation is not as practical for domestic applications where motors are usually smaller than 1 kW and where lighting circuits supply most of the load.

8.1.2. Commonly Used Polyphase System. Although there are several polyphase systems such as two-phase, three-phase but 3-phase system is invariably adopted because of its inherent advantages over all other polyphase systems .

The demand for two-phase system has almost disappeared because that does not have any advantage that is not equalled or surpassed by three-phase system in generation, transmission or utilization. Three-phase system is universally employed for generation, transmission and distribution of electric power. Two-phase supply and six-phase supply, when required, is obtained from 3-phase supply.

Systems with number of phases, more than three, increase complexity and cost of transmission and utilisation hardware and becomes uneconomical.

A knowledge of 3-phase power systems is, therefore, essential for understanding of power technology. Fortunately, the basic circuit technique used in solving single-phase circuits are directly applicable to 3-phase circuits because three phases are identical and one phase represents the behaviour of all the above. In this chapter we will discuss only 3-phase systems.

8.2. GENERATION OF EMFs

8.2.1. Generation of Single-Phase EMF. As already discussed when a coil rotates in a uniform magnetic field, an alternating emf is induced in the coil. At the instant the coil is at right angles to the direction of field, as in position shown in Fig. 8.1 (*a*) the linking flux in the coil is maximum, but the rate of change of flux is zero, therefore induced emf is zero. But when the plane of the coil is in the plane of the field as in position CD shown in Fig. 8.1 (*a*) the linking flux is zero but the rate of change of flux is maximum so emf induced in the coil at this instant is maximum. The emf induced in the coil in various *positions of the coil with respect to the field is shown in Fig. 8.1 (b).*

8.2.2. Generation of Two-Phase EMF. When two coils fastened rigidly together 90 degrees (electrical) apart from each other rotate in a uniform magnetic field, the emf induced in each coil will have phase difference of 90° or $\pi/2$ radians.

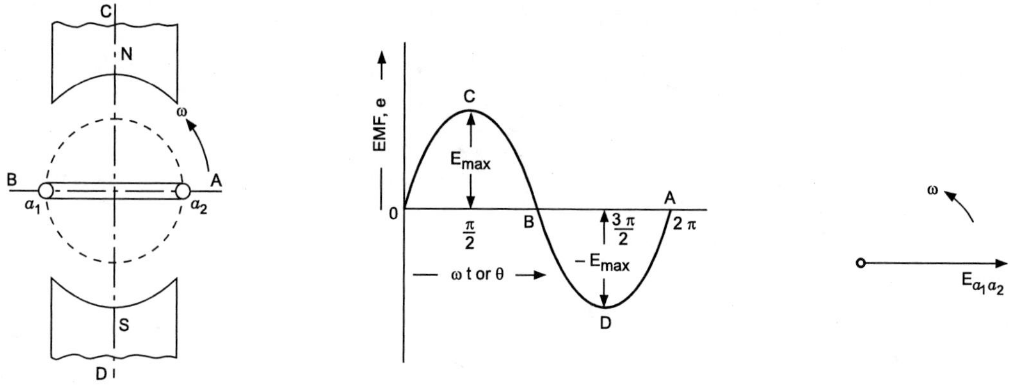

(*a*) (*b*) (*c*)

Fig. 8.1 *Generation of Single-Phase EMF*

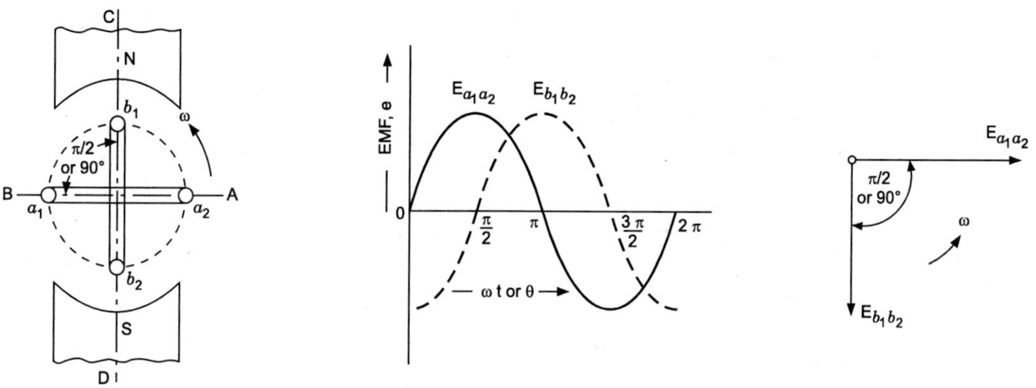

(*a*) (*b*) (*c*)

Fig. 8.2 *Generation of Two-Phase EMF*

Consider two identical coils a_1 a_2 and b_1 b_2 fastened together at right angles to each other, rotating in counter-clockwise direction in a bipolar field, as shown in Fig. 8.2 (*a*). The terminal a_1 of coil a_1 a_2 leads the terminal b_1 of coil b_1 b_2 by 90° or $\pi/2$ radians. At the instant, the coil a_1 a_2 is in position AB, the coil b_1 b_2 is in position CD and, therefore, at this instant the induced emfs in coil a_1 a_2 and b_1 b_2 are zero and maximum (in negative direction)respectively. When the coils rotate through $\pi/2$ radians , the coil a_1 a_2 and b_1 b_2 come in position CD and AB respectively, therefore, emfs induced in them will be maximum (in positive direction) and zero respectively.

Since each coil has an equal number of turns and is wound with the wire of the same type and same x-section, the induced emfs in both of the coils are, therefore, of equal magnitude. The induced emf in each coil is also of same frequency and same waveform (sinusoidal in this case) but is displaced from each other by $\pi/2$ radians or 90°, as illustrated in Fig. 8.2(*b*) by waveforms. The emf induced in coil a_1a_2 leads the emf induced in coil b_1b_2 by $\pi/2$ radians or 90°. If instantaneous induced emf in coil a_1a_2 is given as

$e_{a_1a_2} = E_{max} \sin \omega t$ then instantaneous induced emf in coil b_1b_2 will be given as

$$e_{b_1b_2} = E_{max} \sin\left(\omega t - \frac{\pi}{2}\right) \text{ or } E_{max} \sin(\omega t - 90°) \quad ...(8.2)$$

In symbolic notation, the real axis is usually made to coincide with the voltage or emf of coil a_1a_2 (say phase A). Then the effective values of the system are represented by the complex numbers as

$$\mathbf{E}_{a_1a_2} = \mathrm{E_P}$$

$$\mathbf{E}_{b_1b_2} = \mathrm{E_P} \angle -90^\circ \text{ or } \mathrm{E_P}\, e^{-j\,90^\circ} \text{ or } -j\,\mathrm{E_P} \qquad \text{...(8.3)}$$

8.2.3. Generation of 3-Phase EMF. When three coils fastened rigidly together and 120° (electrical) apart rotate about the same axis in a uniform magnetic field, the induced emf in each of them will have a phase difference of 120° or $\frac{2}{3}\pi$ radians.

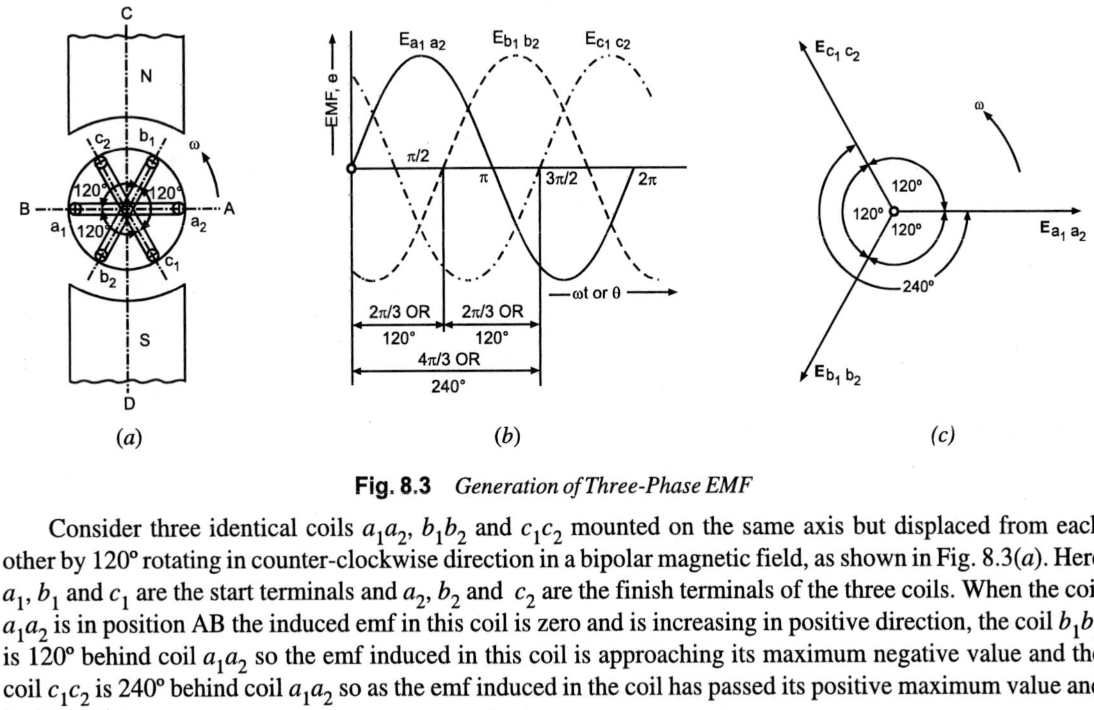

Fig. 8.3 *Generation of Three-Phase EMF*

Consider three identical coils a_1a_2, b_1b_2 and c_1c_2 mounted on the same axis but displaced from each other by 120° rotating in counter-clockwise direction in a bipolar magnetic field, as shown in Fig. 8.3(*a*). Here a_1, b_1 and c_1 are the start terminals and a_2, b_2 and c_2 are the finish terminals of the three coils. When the coil a_1a_2 is in position AB the induced emf in this coil is zero and is increasing in positive direction, the coil b_1b_2 is 120° behind coil a_1a_2 so the emf induced in this coil is approaching its maximum negative value and the coil c_1c_2 is 240° behind coil a_1a_2 so as the emf induced in the coil has passed its positive maximum value and is decreasing.

Since each coil being identical has an equal number of turns and is wound with the wire of the same type and same x-section, the induced emfs in each of the coils are of same magnitude. The induced emf in each coil is also of same frequency and same waveform (sinusoidal in this case)but displaced from each other by $\frac{2\pi}{3}$ radians or 120°, as illustrated in Fig. 8.3 (*b*) by waveforms .

Accordingly, the instantaneous values of the emfs induced in coils a_1a_2, b_1b_2 and c_1c_2 may be given as

$$\left.\begin{aligned} e_{a_1a_2} &= \mathrm{E_{max}} \sin \omega t \\ e_{b_1b_2} &= \mathrm{E_{max}} \sin\left(\omega t - \frac{2\pi}{3}\right) \text{ or } \mathrm{E_{max}} \sin(\omega t - 120^\circ) \\ e_{c_1c_2} &= \mathrm{E_{max}} \sin\left(\omega t - \frac{4\pi}{3}\right) \text{ or } \mathrm{E_{max}} \sin(\omega t - 240^\circ) \\ &\qquad \text{ or } \mathrm{E_{max}} \sin(\omega t + 120^\circ) \end{aligned}\right\} \qquad \text{...(8.4)}$$

if $t = 0$ corresponds to the instant when the voltage or emf of coil a_1a_2 passes through zero and increases in positive direction.

In symbolic notation, the real axis is usually made to coincide with the emf (or voltage) of coil $a_1\ a_2$ (say phase A). Then the effective values of the symmetrical system are given in complex numbers as

$$\left.\begin{aligned} E_{a_1a_2} &= E_P \\ E_{b_1b_2} &= E_P \angle -120^\circ \text{ or } E_P\, e^{-j\,120^\circ} \\ &\quad \text{or } E_P(-0.5 - j\,0.866) \\ E_{c_1c_2} &= E_P \angle +120^\circ \text{ or } E_P\, e^{+j\,120^\circ} \\ &\quad \text{or } E_P(-0.5 + j\,0.866) \end{aligned}\right\} \qquad ...(8.5)$$

It is easy to prove that, at any instant, the algebraic sum of the three emfs is zero.

This result can be established analytically with the aid of the relations in Eq. (8.4)

$$\begin{aligned} \text{Resultant} &= e_{a_1a_2} + e_{b_1b_2} + e_{c_1c_2} \\ &= E_m\ [\sin \omega t + \sin(\omega t - 120^\circ) + \sin(\omega t - 240^\circ)] \\ &= E_m\ [\sin \omega t - 2\sin(\omega t - 180^\circ)\cos 60^\circ] \\ &= E_m\ [\sin \omega t - 2\sin \omega t \cos 60^\circ] = 0 \end{aligned}$$

It is easier to establish the result by phasor diagram shown in Fig. 8.4.

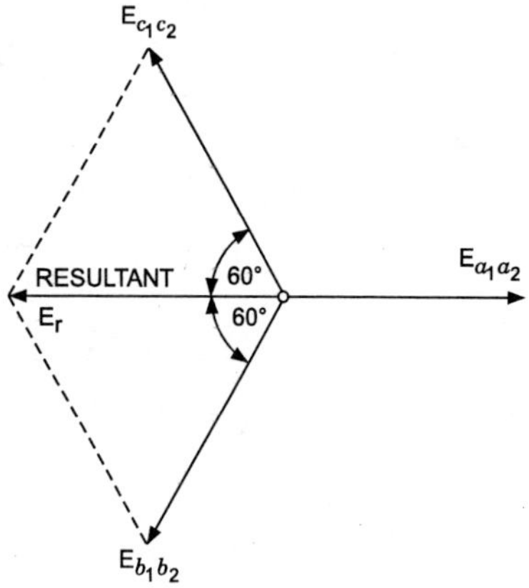

Fig. 8.4

Indeed, combining the phasors $E_{b_1b_2}$ and $E_{c_1c_2}$ we find that, due to symmetry, the phasor $\mathbf{E}_{b_1b_2} + \mathbf{E}_{c_1c_2}$ is directed horizontally to the left, *i.e.* it is opposite to phasor $E_{a_1a_2}$.

$$\textit{i.e. } \mathbf{E}_{a_1a_2} + \mathbf{E}_{b_1b_2} + \mathbf{E}_{c_1c_2} = \mathbf{E}_{a_1a_2} - (\mathbf{E}_{a_1a_2}) = 0$$

The result can be established by using complex notation–relations (8.5)

$$\mathbf{E}_{a_1a_2} + \mathbf{E}_{b_1b_2} + \mathbf{E}_{c_1c_2} = E_P + E_P(-0.5 - j\,0.866) + E_P(-0.5 + j\,0.866) = 0$$

The result is also obvious from wave diagram shown in Fig. 8.3 (*b*).

8.3. DOUBLE SUBSCRIPT NOTATION

The solution of problems involving circuits and systems containing a number of voltages and currents is simplified and less susceptible to error if the voltage and current phasors are designated by some systematic notation. The double subscript notation is a very useful concept from this point of view. In this notation two letters are placed at the foot of the symbol for voltage or current, the order in which the subscripts are written, indicates the direction in which the voltage acts or current flows. For example, if a voltage in a circuit acts in such a direction as to cause a current to flow from A to B, the positive direction of voltage is from A to B, and the voltage may be represented by V_{AB} or E_{AB}, the order of the subscripts denoting that the voltage or emf is acting from A to B. If the voltage is indicated by V_{BA} or E_{BA} it means that point B is positive w.r.t. point A (during its positive half cycle) and the voltage causes the current to flow from B to A *i.e.* V_{BA} or E_{BA} indicates that voltage or emf acts in a direction opposite to that in which V_{AB} or E_{AB} acts.

$$\text{So } \mathbf{V}_{BA} = -\mathbf{V}_{AB} \qquad ...(8.6)$$

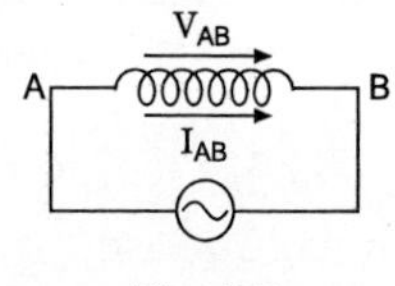

Fig. 8.5

Similarly I_{AB} indicates that current is flowing from A to B but I_{BA} indicates that current is flowing in direction from B to A

$$\textit{i.e. } \mathbf{I}_{BA} = -\mathbf{I}_{AB}$$

8.4. PHASE SEQUENCE

Phase sequence is the order or sequence in which the currents or voltages in different phases attain their maximum values one after the other.

In Fig. 8.3 (*a*) three coils a_1a_2, b_1b_2 and c_1c_2 rotating in counter-clockwise direction are shown. Since the coil b_1b_2 is 120° behind coil a_1a_2 and coil c_1c_2 is 240° behind coil a_1a_2, therefore, first coil a_1a_2 attains maximum or peak value of induced emf, the coil b_1b_1 attains maximum or peak value of induced emf, when the coils rotate further 120° (electrical) and coil c_1c_2 attains peak value of induced emf when the coils rotate through 240° (electrical). Since the induced emfs in three coils a_1a_2, b_1b_2 and c_1c_2 attain maximum values in order *a, b, c*, phase sequence is *a b* c.

If the direction of rotation of the coils is reversed *i.e.* clockwise, the three-phase emfs would attain their maximum values in order *a, c, b* and, therefore, phase sequence would be *acb*. Since the coils can be rotated in clockwise or in counter-clockwise direction, there can be only two possible phase sequences.

The knowledge of phase sequence of a 3-ϕ supply is essential for making connections for alternators and transformers to operate them in parallel. Reversal of phase sequence of a 3-phase alternator which is to be paralleled with another similar alternator may cause extensive damage to both the machines. The direction of rotation of an induction motor depends on the phase sequence. If the phase sequence is reversed by interchanging any two terminals of the 3-ϕ supply, the motor would rotate in the opposite direction.

The phase sequence of the voltage applied to the load, in general, is determined by the order in which three phase lines are connected. In the case of 3-phase unbalanced loads, the effect of reversal of phase sequence, in general, causes a completely different set of values of the currents. So while working on such systems it is essential to specify the phase sequence clearly to avoid unnecessary confusion.

8.5. INTERCONNECTION OF THREE PHASES

Each coil of three phases has two terminals (one 'start' and another 'finish') and if individual phase is connected to a separate load circuit, as illustrated in Fig. 8.6, we get a non-interlinked three-phase, six-wire system. Then, by Ohm's law, the current in individual phase will be

$$I_P = E_P / Z_P \qquad ...(8.7)$$

where I_P is the phase current, E_P is the phase emf or voltage.

In such a system, each circuit will require two conductors, therefore, 6 conductors in all.

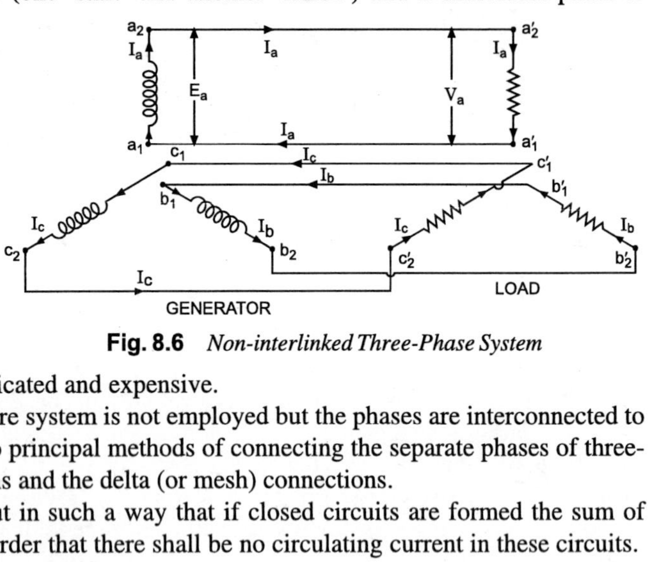

Fig. 8.6 *Non-interlinked Three-Phase System*

Hence such a system will be very complicated and expensive.

In practice non-interlinked 3-phase, 6-wire system is not employed but the phases are interconnected to provide 3 or 4-wire, 3-phase system. The two principal methods of connecting the separate phases of three-phase system are the star (or wye) connections and the delta (or mesh) connections.

The interconnections must be carried out in such a way that if closed circuits are formed the sum of instantaneous emfs in them must be zero in order that there shall be no circulating current in these circuits.

8.6. STAR OR WYE (Y)-CONNECTED SYSTEM

This system is obtained by joining together similar ends, either the start or the finish, the other ends are joined to the line wire, as shown in Fig. 8.7 (*a*). The common point N at which similar (start or finish) ends are connected is called the *neutral or star point*. Ordinarily only three wires are carried to the external circuit giving 3-phase 3-wire star-connected system but sometimes a fourth wire, known as *neutral wire* is carried to the neutral point of the external load circuit giving 3-phase, 4-wire star-connected system. The voltage between any line and the neutral point *i.e.* voltage across the phase winding is called the *phase voltage*; while the voltage between any two outers is called the *line voltage*. The neutral point is usually earth connected.

In Fig. 8.7 (*a*) positive directions of emfs are taken from star point outwards. The arrowheads on emfs and currents indicate the positive direction and not their directions at any instant. The three phases be either numbered as 1, 2, 3, or a, b, c or R, Y, B. In Fig. 8.7 (*a*) the three phases are numbered as usually done; R,Y and B indicate the three natural colours red, yellow and blue respectively.

It is worthwhile to state here that there are only two possible phase sequences, namely RYB and RBY. By convention sequence RYB is taken as positive and RBY as negative.

In Fig. 8.7 (*b*) the emfs induced in 3 phases are shown by phasors. In star connection, there are two windings between each pair of outers and due to joining of similar ends together the emf induced in them are in opposition.

Hence the potential difference between the two outers, known as *line voltage,* is the phasor difference of phase emfs of the two phases concerned.

For example, the potential difference between outers R and Y or line voltage E_{RY} is the phasor difference of phase emfs E_R and E_Y or phasor sum of phase emfs E_R and $(-E_Y)$.

i.e. $\mathbf{E}_{RY} = \mathbf{E}_R - \mathbf{E}_Y$ (phasor difference)

or $\mathbf{E}_{RY} = \mathbf{E}_R + (-\mathbf{E}_Y)$ (phasor sum)

Since phase angle between phasors E_R and $(-E_Y)$ is 60°,

From phasor diagram shown in Fig. 8.7 (*b*)

$$E_{RY} = \sqrt{E_R^2 + E_Y^2 + 2\,E_R\,E_Y \cos 60°}$$

Let $E_R = E_Y = E_B = E_P$ (phase voltage) (Assuming balanced system) then line voltage

$$E_{RY} = \sqrt{E_P^2 + E_P^2 + 2E_P\,E_P \times (1/2)} = \sqrt{3}\;E_P$$

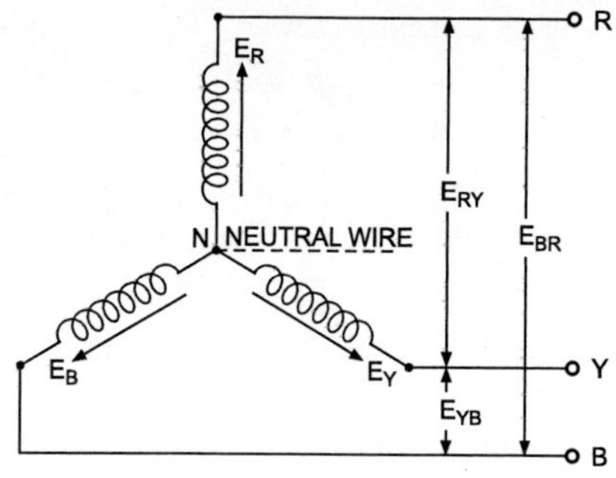

(*a*) *Connection Diagram*

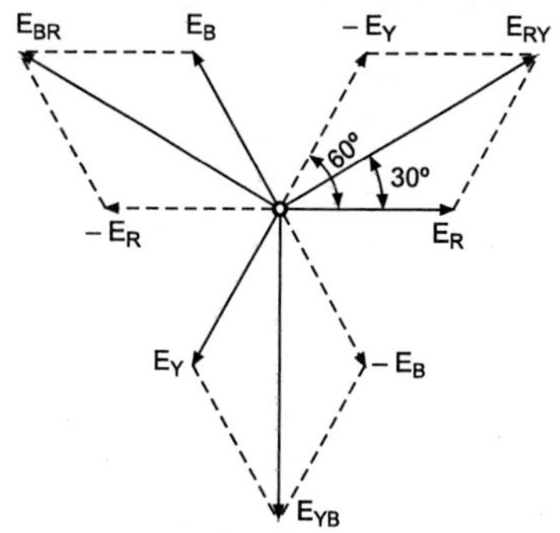

(*b*) *Phasor Diagram of Line and Phase Voltages*

Fig. 8.7 *Three-Phase Star-Connected System*

Similarly potential difference between outers Y and B

$$\text{or line voltage } \mathbf{E}_{YB} = \mathbf{E}_Y - \mathbf{E}_B = \sqrt{3}\;E_P \text{ in magnitude}$$

and potential difference between outers B and R

$$\text{or line voltage, } \mathbf{E}_{BR} = \mathbf{E}_B - \mathbf{E}_R = \sqrt{3}\;E_P \text{ in magnitude}$$

In balanced star (Y) system E_{RY}, E_{YB} and E_{BR} are equal in magnitude and are known as *line voltages*. Let each of them be equal to E_L

$$\therefore\; E_L = \sqrt{3}\;E_P \text{ (in balanced star-connected system)} \qquad ...(8.8)$$

Since in star-connected system each line conductor is connected to separate phase, current flowing through the lines and phases are same

$$\textit{i.e.} \text{ Line current, } I_L = \text{Phase current, } I_P \qquad ...(8.9)$$

8.6.1. 3-Phase Power. Phase voltages and currents in a balanced three-phase circuit (star or delta) may be written in the instantaneous form as below :

$$e_R = \sqrt{2}\;E_P \sin \omega t \qquad ...(8.10\;a)$$

$$e_Y = \sqrt{2}\;E_P \sin (\omega t - 120°) \qquad ...(8.10\;b)$$

$$e_B = \sqrt{2}\;E_P \sin (\omega t + 120°) \qquad ...(8.10\;c)$$

$$\text{and } i_R = \sqrt{2}\;I_P \sin (\omega t - \phi) \qquad ...(8.11\;a)$$

$$i_Y = \sqrt{2}\;I_P \sin (\omega t - \phi - 120°) \qquad ...(8.11\;b)$$

$$i_B = \sqrt{2}\;I_P \sin (\omega t - \phi + 120°) \qquad ...(8.11\;c)$$

where ϕ is the phase angle between phase voltage and phase current.

Total instantaneous power,

$$p = e_R i_R + e_Y i_Y + e_B i_B$$
$$= E_P I_P [2 \sin \omega t \sin (\omega t - \phi)] + E_P I_P [2 \sin (\omega t - 120^\circ) \sin (\omega t - \phi - 120^\circ)$$
$$+ E_P I_P [2 \sin (\omega t + 120^\circ) \sin (\omega t - \phi + 120^\circ]$$
$$= E_P I_P [\cos \phi - \cos (2 \omega t - \phi) + \cos \phi - \cos (2 \omega t - \phi - 240^\circ) + \cos \phi - \cos (2 \omega t - \phi + 240^\circ)]$$

or Average power, $P = 3 E_P I_P \cos \phi$ ...(8.12)

∵ the sum of the three second harmonic oscillating terms which have a progressive phase difference of 120° is zero.

Thus the instantaneous power in a 3-phase balanced system is constant and equal to three times the average power per phase. This is in contrast to power in a single-phase system which has a second harmonic oscillating component. Constancy of power in a balanced 3-phase system affords the unique advantage of uniform torque in 3-phase electric machines.

Substituting $E_P = \frac{E_L}{\sqrt{3}}$ and $I_P = I_L$ for a 3-phase star-connected balanced system in Eq. (8.12) we have

Total circuit power, $P = 3 \frac{E_L}{\sqrt{3}} \times I_L \cos \phi = \sqrt{3} E_L I_L \cos \phi$ ...(8.13)

i.e. Total power, P = $\sqrt{3}$ × line voltage × line current × power factor.

Total apparent power, $S = 3 \times$ apparent power per phase $= 3 E_P I_P = 3 \frac{E_L}{\sqrt{3}} I_L = \sqrt{3} E_L I_L$ ...(8.14)

Total reactive power, $Q = 3 E_P I_P \sin \phi = 3 \frac{E_L}{\sqrt{3}} I_L \sin \phi = \sqrt{3} E_L I_L \sin \phi = \frac{P}{\cos \phi} \sin \phi = P \tan \phi$ ...(8.15)

It should be noted that in a balanced star-connected system

(*i*) Line voltages are 120° apart.

(*ii*) Line voltages are 30° ahead of the respective phase voltages.

(*iii*) Line voltages are $\sqrt{3}$ times of phase voltages.

(*iv*) Line currents are equal to phase currents.

(*v*) The angle between line currents and the corresponding line voltages is (30° ± ϕ); + ve for lagging currents and –ve for leading currents.

(*vi*) True power output = $\sqrt{3} E_L I_L \cos \phi$, where ϕ is the angle between the respective phase current and phase voltage (not between line current and line voltage).

(*vii*) Apparent power = $\sqrt{3} E_L I_L$

(*viii*) In balanced system, the potential of neutral or star point is zero because potential at neutral or star point,

$$E_{NR} + E_{NY} + E_{NB} = 0$$

Example 8.1 **Phase voltages of a star-connected alternator are $E_R = 231 \angle 0°$ V; $E_Y = 231 \angle -120°$ V; and $E_B = 231 \angle +120°$ V. What is the phase sequence of the system ? Compute the line voltages E_{RY} and E_{YB}.**

Solution: The phase voltages of the star- connected alternator are given as $E_R = 231 \angle 0°$; $E_Y = 231 \angle -120°$ V and

$$E_B = 231 \angle +120° \text{ V or } 231 \angle -240° \text{ V}$$

From the above expressions of phase voltages in polar form, it is obvious that E_R is the reference voltage, E_Y lags behind it by 120° and E_B lags behind E_R by 240° and E_Y by 120°. So phase sequence is RYB.

Since phase voltages are of equal magnitude and are displaced from one another by equal time angle, it is a symmetrical 3-phase voltage system and so the line voltage

$$E_{RY} = E_{YB} = \sqrt{3} \times 231 = 400 \text{ V } \textbf{Ans.}$$

Example 8.2 **If the phase voltage of a 3-phase star-connected system is 200 V, what will be the line voltages (a) when the phases are correctly connected (b) when connections to one of phases are reversed ?**

[U.P. Technical Univ. Electrical Engineering First Semester 2000-01; First Semester 2003-04]

Solution: (*i*) When the phases are correctly connected *i.e.* similar ends are connected together, as shown in Fig. 8.7 (*a*) all line voltages will be equal in magnitude and will be given as

Each line voltage, $E_L = \sqrt{3}\ E_P = \sqrt{3} \times 200 = 346.41$ V **Ans.**

(*ii*) When the connections to one of the phases are reversed, as shown in Fig. 8.8 (*a*), voltage in phase R is in reverse direction. The voltage phasor diagram for such a case is shown in Fig. 8.8 (*b*).

It should be noted that E_R has been drawn in reverse direction, so the angles between the three phase voltages are of 60°, as shown in Fig. 8.8(*b*)

So $\mathbf{E_{RY}} = \mathbf{E_R} - \mathbf{E_Y}$

$= 2\ E_P \cos 60° = E_P = 200$ V **Ans.**

$\mathbf{E_{YB}} = \mathbf{E_Y} - \mathbf{E_B}$

$= 2E_P \times \cos 30°$

$= 2 \times 200 \times \dfrac{\sqrt{3}}{2} = 346.41$ V **Ans.**

and $\mathbf{E_{BR}} = \mathbf{E_B} - \mathbf{E_R} = 2\ E_P \cos 60° = 200$ V **Ans.**

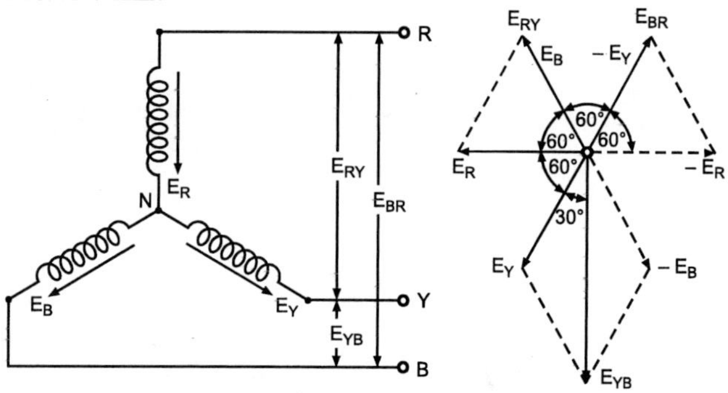

(*a*) *Connection Diagram*

(*b*) *Phasor Diagram of Line and Phase Voltages*

Fig. 8.8

Example 8.3 **A balanced star-connected load of (8 + *j*6) Ω per phase is connected to a 3-phase 400 V supply. Find the line current, power factor, power and three-phase VAs.**

[G.B. Technical Univ. Electrical Engineering Even Semester 2012-13]

Solution:

$$\text{Phase voltage, } V_P = \frac{\text{Line voltage, } V_L}{\sqrt{3}} = \frac{400}{\sqrt{3}} = 231 \text{ V}$$

$$\text{Impedance per phase, } Z_P = \sqrt{R_P^2 + X_P^2} = \sqrt{8^2 + 6^2} = 10\ \Omega$$

$$\text{Phase current, } I_P = \frac{V_P}{Z_P} = \frac{231}{10} = 23.1 \text{ A}$$

$$\text{Line current, } I_L = I_P = 23.1 \text{ A } \textbf{Ans.}$$

$$\text{Power factor, } \cos\phi = \frac{R_P}{Z_P} = \frac{8}{10} = 0.8 \text{ (lagging) } \textbf{Ans.}$$

$$\text{Total power, } P = \sqrt{3}\ V_L I_L \cos\phi = \sqrt{3} \times 400 \times 23.1 \times 0.8 = 12{,}800 \text{ W } \textbf{Ans.}$$

$$\text{Total volt-amperes, } S = \sqrt{3}\ V_L I_L = \sqrt{3} \times 400 \times 23.1 = 16{,}000 \text{ VA } \textbf{Ans.}$$

$$\text{Also total power, } P = 3\ I_P^2 R_P = 3 \times (23.1)^2 \times 8 = 12{,}800 \text{ W } \textbf{Ans.}$$

Example 8.4 **A 3-phase voltage source has a phase voltage of 120 V and supplies star-connected load having impedance 36 + *j*48Ω per phase. Calculate :**

(*a*) **The line voltage.**

(*b*) **The line current.**

(*c*) **The power factor.**

(*d*) **The total 3-phase power supplied to the load.**

[U.P. Technical Univ. Electrical Engineering First Semester 2008-09]

Solution:

$$\text{Phase voltage, } V_P = 120\text{V}$$

$$\text{Load impedance, } Z = 36 + j48 = 60\ \angle 53.13°\ \Omega$$

$$(a) \text{ Line voltage, } V_L = \sqrt{3}\ V_P = \sqrt{3} \times 120 = 207.85\text{V } \textbf{Ans.}$$

$$(b) \text{ Line current, } I_L = \text{Phase current, } I_P = \frac{V_P}{Z} = \frac{120}{60} = 2\text{A } \textbf{Ans.}$$

$$(c) \text{ Power factor} = \cos\phi = \cos 53.13° = 0.6 \text{ (lagging) } \textbf{Ans.}$$

$$(d) \text{ Power} = \sqrt{3}\ V_L I_L \cos\phi = \sqrt{3} \times 207.85 \times 2 \times \cos 53.13° = 432 \text{ watts}$$

Power can also be determined from the relation

$$P = 3\ I^2R = 3 \times 2^2 \times 36 = 432 \text{ watts } \textbf{Ans.}$$

Example 8.5 **A 3-phase, 400 V supply is connected to a 3-phase star connected balanced load. The line current is 20 A and the power consumed by the load is 12 kW. Calculate the impedance of the load, phase current and power factor.** *[U.P. Technical Univ. Electrical Engineering Second Semester 2006-07; Odd Semester 2013-14]*

Solution:

$$\text{Load supplied, } P = 12 \text{ kW} = 12{,}000 \text{ W}$$

$$\text{Line voltage, } V_L = 400 \text{ V}$$

$$\text{Line current, } I_L = 20 \text{ A}$$

$$\text{Phase voltage, } V_P = \frac{V_L}{\sqrt{3}} = \frac{400}{\sqrt{3}} = 230.94 \text{ V}$$

$$\text{Phase current, } I_P = I_L = 20 \text{ A } \textbf{Ans.}$$

$$\text{Load impedance, } Z_L = \frac{V_P}{I_P} = \frac{230.94}{20} = 11.547\ \Omega \textbf{ Ans.}$$

$$\text{Load power factor, } \cos\phi = \frac{P}{\sqrt{3}\ V_L I_L} = \frac{12{,}000}{\sqrt{3} \times 400 \times 20} = 0.866 \textbf{ Ans.}$$

Example 8.6 **A balanced 3-phase star-connected load of 18 kW taking a leading current of 60 amperes when connected across a 3-phase 440 V, 50 Hz supply. Find the values and nature of the load.**

[Electrical Engineering U.P. Technical Univ. Second Semester 2007-08; G.B. Technical Univ. Odd Semester 2012-13]

Solution:

$$\text{Load supplied, } P = 18 \text{ kW} = 18{,}000 \text{ W}$$

$$\text{Line voltage, } V_L = 440 \text{ V}$$

$$\text{Line current, } I_L = 60 \text{ A}$$

$$\text{Phase voltage, } V_P = \frac{V_L}{\sqrt{3}} = \frac{440}{\sqrt{3}} = 254 \text{ V}$$

$$\text{Phase current, } I_P = I_L = 60 \text{ A}$$

$$\text{Load impedance, } Z = \frac{V_P}{I_P} = \frac{254}{60} = 4.234\ \Omega$$

$$\text{Load power factor, } \cos\phi = \frac{P}{\sqrt{3}\ V_L I_L} = \frac{18{,}000}{\sqrt{3} \times 440 \times 60} = 0.39365 \text{ (leading) } \textbf{Ans.} \qquad \because \text{ Current is leading}$$

$$\text{Resistance of load, } R = Z\cos\phi = 4.234 \times 0.39365 = 1.6667\ \Omega \textbf{ Ans.}$$

$$\text{Reactance of load, } X = Z\sin\phi = 4.234\sqrt{1 - 0.39365^2} = 4.234 \times 0.9193 = 3.89242\ \Omega \text{ (capacitive) } \textbf{Ans.}$$

$\because$ Current is leading

Example 8.7 **A 3-ϕ, 3-wire Y-connected system has 150 V between two phases. Each phase has $Z = 5\angle{-30°}\ \Omega$. Find: (i) Current in each phase (ii) total power (iii) Draw phasor diagram.**

[U.P. Technical Univ. Electrical Engineering First Semester 2007-08]

Solution:

$$\text{Line voltage, } V_L = \text{Phase-to-phase voltage} = 150 \text{ V}$$

Phase voltage, $V_P = \frac{V_L}{\sqrt{3}} = \frac{150}{\sqrt{3}} = 86.6$ V

(*i*) Current in each phase, $I_P = \frac{V_P}{Z_P} = \frac{86.6}{5} = 17.32$ A **Ans.**

Power factor of the circuit, $\cos\phi$ = Cosine of phase angle

$= \cos 30° = 0.866$ (leading)

(*ii*) Total power, $P = 3\, V_P\, I_P \cos\phi$

$= 3 \times 86.6 \times 17.32 \times 0.866$

$= 3{,}897$ W or 3.897 kW **Ans.**

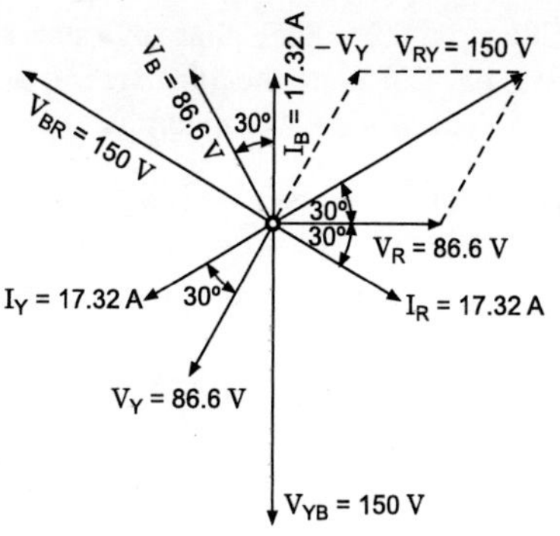

Fig. 8.9

Phasor diagram is shown in Fig. 8.9.

Example 8.8 A balanced star-connected load of (8 + *j*6) Ω per phase is connected to a 3-phase, 230 V, 50 Hz supply. Find the line current, pf, power, volt-amperes and reactive power. Draw the phasor diagram for the above circuit. [*Pb. Technical Univ. Electrical Engineering, June 1999*]

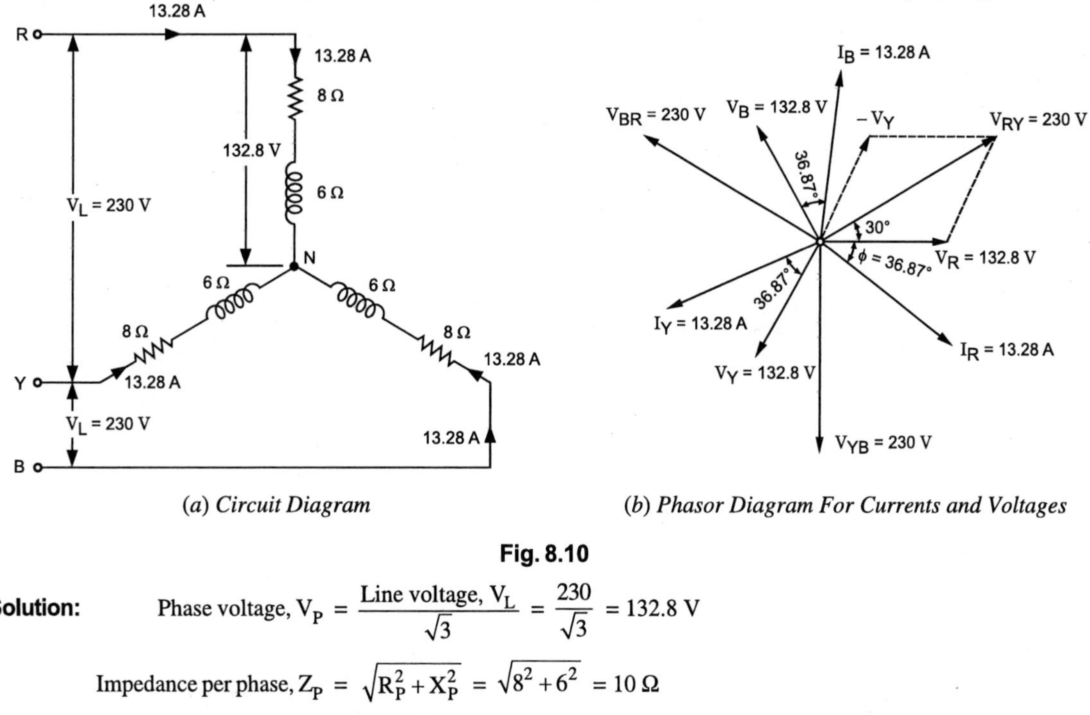

(*a*) *Circuit Diagram* (*b*) *Phasor Diagram For Currents and Voltages*

Fig. 8.10

Solution: Phase voltage, $V_P = \frac{\text{Line voltage, } V_L}{\sqrt{3}} = \frac{230}{\sqrt{3}} = 132.8$ V

Impedance per phase, $Z_P = \sqrt{R_P^2 + X_P^2} = \sqrt{8^2 + 6^2} = 10\ \Omega$

Phase current, $I_P = \frac{V_P}{Z_P} = \frac{132.8}{10} = 13.28$ A

Line current, $I_L = I_P = 13.28$ A **Ans.**

Power factor, $\cos\phi = \frac{R_P}{Z_P} = \frac{8}{10} = 0.8$ (lagging) **Ans.**

Power, $P = \sqrt{3}\ V_L\, I_L \cos\phi = \sqrt{3} \times 230 \times 13.28 \times 0.8 = 4{,}232$ watts or 4.232 kW **Ans.**

Total volt-amperes, $S = \sqrt{3}\ V_L\, I_L = \sqrt{3} \times 230 \times 13.28 = 5{,}290$ VA **Ans.**

Reactive power, $Q = \sqrt{3}\, V_L\, I_L \sin\phi = \sqrt{3} \times 230 \times 13.28 \sin \cos^{-1}(0.8)$

$= \sqrt{3} \times 230 \times 13.28 \times 0.6 = 3{,}174$ VAR or 3.174 kVAR **Ans.**

Phasor diagram for currents and voltages is shown in Fig. 8.10 (*b*).

Example 8.9 **Each phase of a star-connected load consists of a resistance of 100 Ω in parallel with a capacitance 31.8 μF. Calculate the line current, power absorbed, total kVA and power factor of the load when connected to a 416 V, 3ϕ, 4 wire 50 Hz supply.** *[Pb. Technical Univ. Basic Electrical Engineering December 2003]*

Solution:

$$\text{Phase voltage, } V_P = \frac{V_L}{\sqrt{3}} = \frac{416}{\sqrt{3}} = 240.2 \text{ V}$$

Taking phase voltage as reference phasor, $\mathbf{V_P} = 240.2 \angle 0° \text{ V}$

$$\text{Admittance per phase, } \mathbf{Y_P} = \frac{1}{R} + j\omega C$$

$$= \frac{1}{100} + j\,2\pi \times 50 \times 31.8 \times 10^{-6}$$

$$= (0.01 + j0.01) \text{ or } 0.01414 \angle 45° \text{ S}$$

$$\text{Line current, } \mathbf{I_L} = \mathbf{I}$$

$$= \mathbf{V_P}\,\mathbf{Y_P} = 240.2 \angle 0° \times 0.01414 \angle 45°$$

$$= 3.397 \angle 45° \text{ A}$$

So line current, I_L = 3.397 A **Ans.**

Power factor, cos ϕ = Cos 45° = 0.707 (leading) **Ans.**

$$\text{Power absorbed, } P = \sqrt{3}\,V_L I_L \cos\phi = \sqrt{3} \times 416 \times 3.397 \times 0.707$$

= 1,730 watts or 1.73 kW **Ans.**

$$\text{Total kVA, } S = \frac{\sqrt{3} V_L I_L}{1{,}000} = \frac{\sqrt{3} \times 416 \times 3.397}{1{,}000} = 2.45 \text{ Ans.}$$

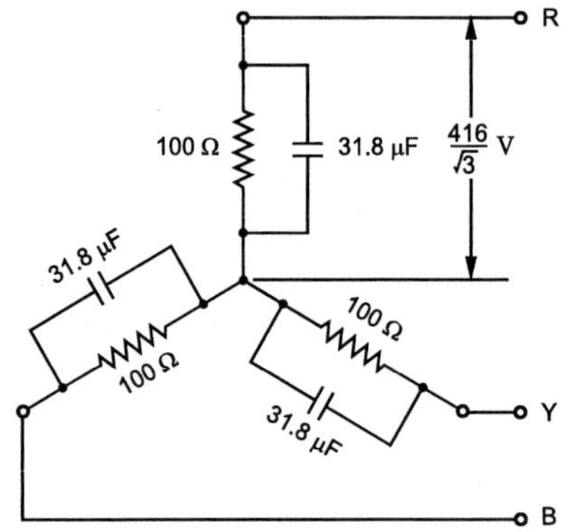

Fig. 8.11

Example 8.10 **Three-phase star-connected load when supplied from a 400 V, 50 Hz source takes a line current of 10 A at 66.86° w.r.t. to its line voltage. Calculate : (*i*) impedance parameters (*ii*) pf and active power consumed. Draw the phasor diagram.** *[Nagpur University, April 1998]*

Solution: As mentioned in Art 8.6, line voltages are 30° ahead of respective phase voltage *i.e.* phase voltages lag behind respective line voltages by 30°. It is also illustrated in phasor diagram shown in Fig. 8.12. Thus phase angle between phase current and its phase voltage is (66.86° – 30°) = 36.86° (lagging). This is because a current differing in phase with respect to its line voltage by 66.86° can only be lagging.

$$(i) \text{ Impedance } Z = \frac{V_P}{I_P} = \frac{V_L/\sqrt{3}}{I_L} = \frac{400}{\sqrt{3} \times 10} = 23.1\ \Omega \text{ Ans.}$$

Resistance R = Z cos ϕ = 23.1 cos 36.86° = 18.48 Ω **Ans.**

Reactance X_L = Z sin ϕ = 23.1 sin 36.86° = 13.86 Ω **Ans.**

(*ii*) Power factor = Cos ϕ = Cos 36.86° = 0.8 (lagging) **Ans.**

Total active power consumed

$$= \sqrt{3}\,V_L I_L \cos\phi = \sqrt{3} \times 400 \times 10 \times 0.8$$

= 5,544 W or 5.544 kW **Ans.**

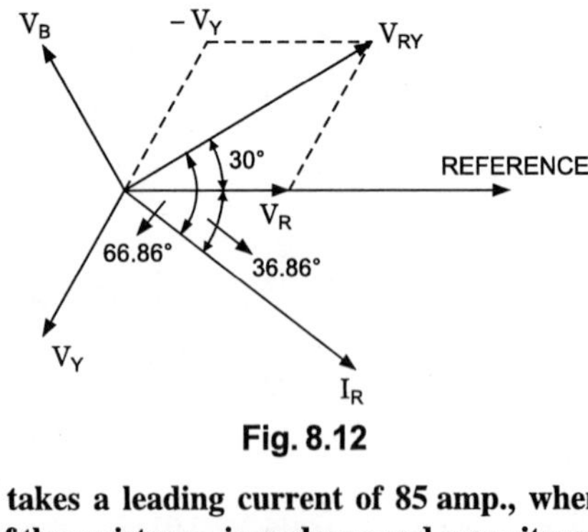

Fig. 8.12

Example 8.11 **A balanced 3-phase star-connected load of 120 kW takes a leading current of 85 amp., when connected across a 3-phase 1,100 volts, 50 Hz supply. Obtain the values of the resistance, impedance and capacitance of the load per phase and also calculate the power factor of the load.**

[U.P. Technical Univ. Electrical Engineering Second Semester 2003-04]

Solution:

Supply voltage, V_L = 1,100 V

Supply current, I_L = 85 A

Power supplied, P = 120 kW or 1,20,000 W

$$\text{Power factor of the load, } \cos\phi = \frac{P}{\sqrt{3}\,V_L I_L} = \frac{1{,}20{,}000}{\sqrt{3} \times 1{,}100 \times 85} = 0.741 \text{ (leading) Ans.}$$

$$\text{Impedance of the load per phase, } Z = \frac{V_P}{I_P} = \frac{V_L/\sqrt{3}}{I_L} = \frac{1{,}100/\sqrt{3}}{85} = 7.472\ \Omega \textbf{ Ans.}$$

$$\text{Resistance of the load per phase, } R = Z\cos\phi = 7.472 \times 0.741 = 5.536\ \Omega \textbf{ Ans.}$$

$$\text{Reactance of the load per phase, } X_C = \sqrt{Z^2 - R^2} = \sqrt{(7.472)^2 - (5.536)^2} = 5.02\ \Omega$$

$$\text{Capacitance of the load per phase, } C = \frac{1}{2\pi f X_C} = \frac{1}{2\pi \times 50 \times 5.02} = 634.3\ \mu\text{F} \textbf{ Ans.}$$

Example 8.12 **A 3-phase balanced load draws 10 kW power from a 400 V, 3-phase, 50 Hz, 4-wire supply at 0.8 lagging power factor.**

(*i*) Determine the line current.
(*ii*) What would be the line current if power factor were raised to unity, power remaining the same?
(*iii*) How can the power factor be improved to unity?

[*R.G.P.V. Bhopal Basic Electrical Engineering December 2004*]

Solution:

Power, P = 10 kW
Line voltage, V_L = 400 V
Power factor, $\cos\phi$ = 0.8 (lagging)

$$(i)\text{ Line current, } I_L = \frac{P}{\sqrt{3}\, V_L \cos\phi} = \frac{10 \times 1{,}000}{\sqrt{3} \times 400 \times 0.8} = 18.04 \text{ A } \textbf{Ans.}$$

Line current when power factor is raised to unity, power remaining the same

$$I_L' = \frac{10 \times 1{,}000}{\sqrt{3} \times 400 \times 1} = 14.43 \text{ A } \textbf{Ans.}$$

Power factor can be improved to unity by connecting capacitors of suitable capacity across the load.

Example 8.13 **A three-phase, three-wire RYB system, with effective line voltage of 150 V has a balanced Y-connected load of Z = 5 $\angle -30°$ Ω in each phase. Obtain the currents, power supplied to load and draw voltage-current phasor diagram.**

[*U.P. Technical Univ. Elec. Engineering June 2001*]

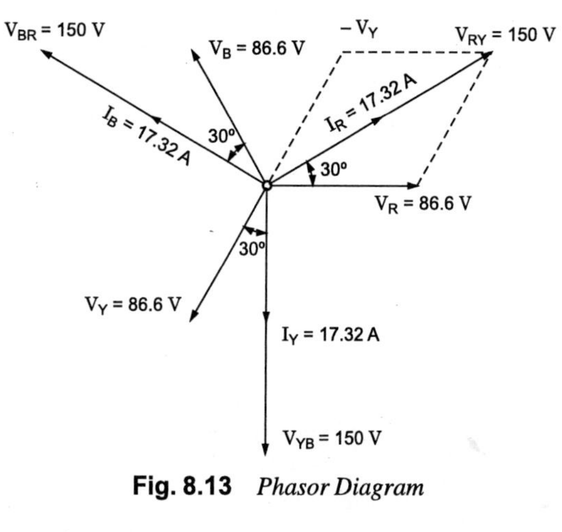

Fig. 8.13 *Phasor Diagram*

Solution: Phase voltage, $V_P = \dfrac{\text{Line voltage, } V_L}{\sqrt{3}}$

$$= \frac{150}{\sqrt{3}} = 86.6 \text{ V}$$

Taking phase voltage, V_P as reference phasor

$$\mathbf{V_P} = 86.6 \angle 0° \text{ V}$$

$$\text{Impedance per phase, } \mathbf{Z_P} = 5 \angle -30°\ \Omega$$

$$\text{Line current, } \mathbf{I_L} = \text{Phase current, } \mathbf{I_P} = \frac{\mathbf{V_P}}{\mathbf{Z_P}} = \frac{86.6 \angle 0°}{5 \angle -30°} \text{ A} = 17.32 \angle 30° \text{ A } \textbf{Ans.}$$

$$\text{Power supplied to load, } P = 3V_P I_P \cos\phi = 3 \times 86.6 \times 17.32 \cos 30° = 3{,}897 \text{ W } \textbf{Ans.}$$

Phasor diagram is shown in Fig. 8.13.

8.7. MESH OR DELTA (Δ)-CONNECTED SYSTEM

When the starting end of one coil is connected to the finishing end of another coil, as shown in Fig. 8.14 (*a*), the *delta or mesh connection is obtained.* The direction of emfs in the coils have been taken as positive from start end to finish end in Fig. 8.14 (*a*).

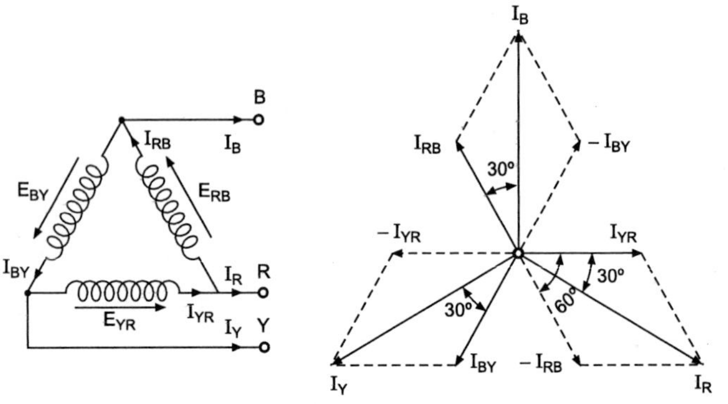

(*a*) *Connection Diagram* (*b*) *Phasor Diagram*

Fig. 8.14 *Three-Phase Delta-Connected System*

From Fig. 8.14 (*b*) it is obvious that line current is phasor difference of phase currents of two phases concerned. For example the line current in red outer I_R will be equal to the phasor difference of phase currents I_{YR} and I_{RB}. The current phasors are shown in Fig. 8.14 (*b*).

From Figs. 8.14 (*a*) and 8.14 (*b*) we have

$$\text{Line current, } \mathbf{I}_R = \mathbf{I}_{YR} - \mathbf{I}_{RB} = \mathbf{I}_{YR} + (-\mathbf{I}_{RB})$$

Since phase angle between phase current phasors $\mathbf{I}_{YR}$ and $-\mathbf{I}_{RB}$ is 60°,

$$I_R = \sqrt{I^2_{YR} + I^2_{RB} + 2\, I_{YR}\, I_{RB} \cos 60^\circ}$$

Assuming balanced load

The phase current in each winding is equal and let each be equal to I_P

$$\therefore \text{ Line current } I_R = \sqrt{I^2_P + I^2_P + 2\, I_P\, I_P \times \frac{1}{2}} = \sqrt{3}\, I_P \quad \text{...(8.16)}$$

$$\text{Similarly line current } \mathbf{I}_Y = \mathbf{I}_{BY} - \mathbf{I}_{YR} = \sqrt{3}\, I_P \text{ in magnitude}$$

$$\text{and line current } \mathbf{I}_B = \mathbf{I}_{RB} - \mathbf{I}_{BY} = \sqrt{3}\, I_P \text{ in magnitude}$$

Since in delta-connected system, only one phase is included between any pair of line outers, potential difference between the line outers, called the *line voltage*, is equal to phase voltage.

$$\textit{i.e.} \text{ Line voltage, } E_L = \text{Phase voltage, } E_P \quad \text{...(8.17)}$$

$$\text{Power output per phase} = E_P\, I_P \cos\phi$$ where $\cos\phi$ is the power factor of the load

$$\text{Total power output} = 3\, E_P\, I_P \cos\phi = 3\, E_L \frac{I_L}{\sqrt{3}} \cos\phi = \sqrt{3}\, E_L\, I_L \cos\phi \quad \text{...(8.18)}$$

i.e. Total power output $= \sqrt{3} \times$ line voltage $\times$ line current $\times$ power factor

Apparent power of 3-phase delta-connected system $= 3 \times$ apparent power per phase

$$= 3\, E_P\, I_P = 3\, E_L \frac{I_L}{\sqrt{3}} = \sqrt{3}\, E_L\, I_L \quad \text{...(8.19)}$$

$$\text{Total reactive power, } Q = 3\, E_P\, I_P \sin\phi = \sqrt{3}\, E_L\, I_L \sin\phi = P \tan\phi \quad \text{...(8.20)}$$

It should be noted that in balanced delta-connected system

(*i*) Line currents are 120° apart.

(*ii*) Line currents are 30° behind the respective phase currents.

(*iii*) Line currents are $\sqrt{3}$ times phase currents.

(*iv*) The angle between line currents and corresponding line voltage is (30° ± ϕ) as in star system.

(*v*) Line voltages are equal to phase voltages.

(*vi*) Power = $\sqrt{3}\, E_L\, I_L \cos\phi$, where ϕ is the phase angle between respective phase current and phase voltage.

(*vii*) Apparent power $= \sqrt{3}\, E_L\, I_L$

(*viii*) In balanced system the resultant emf in the closed circuit will be zero.

$$\textit{i.e.}\ \mathbf{E}_{RY} + \mathbf{E}_{YB} + \mathbf{E}_{BR} = 0$$

Hence there will be no circulating current in the mesh if no load is connected to the lines.

Example 8.14 **Three identical coils connected in delta across 400 V, 50-Hz, 3-phase ac supply, take a line current of 17.32 A at power factor of 0.8 lagging. Calculate (*i*) the phase current (*ii*) the resistance and inductance of each coil (*iii*) the power drawn by each coil.** [*U.P. Technical Univ. Electrical Engineering February 2002*]

Solution: Phase voltage, $V_P = V_L = 400$ V

(*i*) Phase current, $I_P = \dfrac{I_L}{\sqrt{3}} = \dfrac{17.32}{\sqrt{3}} = 10$ A **Ans.**

Impedance of each coil, $Z_P = \dfrac{V_P}{I_P} = \dfrac{400}{10} = 40\ \Omega$

(*ii*) Resistance of each coil, $R_P = Z_P \cos\phi = 40 \times 0.8 = 32\ \Omega$ **Ans.**

Inductive reactance of each coil, $X_P = \sqrt{Z_P^2 - R_P^2} = \sqrt{40^2 - 32^2} = 24\ \Omega$

Inductance of each coil, $L = \dfrac{X_P}{2\pi f} = \dfrac{24}{2\pi \times 50} = 0.0764$ H or 76.4 m H **Ans.**

(*iii*) Power drawn by each coil, $P = I_P^2 R_P = 10^2 \times 32 = 3{,}200$ W **Ans.**

Example 8.15 **A balanced Δ-connected load of impedance 16 + *j* 12 Ω/phase is connected to a 3-phase 400 V supply. Find the phase current, line current, power factor, power, reactive VA and total VA.**
[*U.P. Technical Univ. Electrical Engineering Second Semester 2008-09*]

Solution: Phase voltage, $\mathbf{V_P} = \mathbf{V_L} = 400 \angle 0°$ V   Taking phase voltage as reference phasor

Phase impedance, $\mathbf{Z_P} = (16 + j\,12)\ \Omega = 20 \angle 36.87°\ \Omega$

Phase current, $\mathbf{I_P} = \dfrac{\mathbf{V_P}}{\mathbf{Z_P}} = \dfrac{400 \angle 0°}{20 \angle 36.87°} = 20 \angle -36.87°$ A

Phase current, $I_P = 20$ A **Ans.**

Line current, $I_L = 20\sqrt{3} = 34.64$ A **Ans.**

Power factor, pf = Cos (– 36.87°) = 0.8 (lagging) **Ans.**

Power, $P = \sqrt{3}\, V_L I_L \cos\phi = \sqrt{3} \times 400 \times 34.64 \times 0.8 = 19{,}200$ W

Reactive volt-amperes, $Q = \sqrt{3}\, V_L I_L \sin\phi = \sqrt{3} \times 400 \times 34.64 \times 0.6 = 14{,}400$ VA (lagging) **Ans.**

Total volt-amperes, $S = \sqrt{3}\, V_L I_L = \sqrt{3} \times 400 \times 34.64 = 24{,}000$ VA **Ans.**

Example 8.16 **A balanced delta connected load of (12 + *j*9) Ω/phase is connected to 3-phase 400 V supply. Find:**

(*i*) **Line current**
(*ii*) **Power factor**
(*iii*) **Power drawn**
(*iv*) **Reactive Volt-Amperes**
(*v*) **Total Volt-Amperes.** [*Electrical Engineering Mahamaya Technical Univ. Second Semester 2010-11; U.P. Technical Univ. Second Semester 2013-14*]

Solution: Circuit resistance, $R = 12\ \Omega$

Circuit inductive reactance, $X_L = 9\ \Omega$

Circuit impedance, $Z = \sqrt{R^2 + X_L^2} = \sqrt{12^2 + 9^2} = 15\ \Omega$

Phase angle of the circuit, $\phi = \text{Tan}^{-1}\dfrac{X_L}{R} = \text{Tan}^{-1}\dfrac{9}{12} = 36.87°$ (lagging)

Phase current, $I_P = \dfrac{V_P}{Z} = \dfrac{V_L}{Z} = \dfrac{400}{15} = 26.667$ A $\because$ in Δ system, $V_P = V_L$

(*i*) Line current, $I_L = \sqrt{3}\, I_P = \sqrt{3} \times 26.667 = 46.2$ A **Ans.**

(*ii*) Power factor, $\cos\phi = \text{Cos } 36.87° = 0.8$ (lagging) **Ans.**

(*iii*) Power drawn, $P = 3\,I_P^2\,R = 3 \times 26.667^2 \times 12 = 25{,}600$ W or 25.6 kW **Ans.**

Also, $P = \sqrt{3}\ VI\cos\phi = \sqrt{3} \times 400 \times 46.2 \times 0.8 = 25{,}600$ W Same as above

(*iv*) Reactive volt amperes, $Q = \sqrt{3}\ VI\sin\phi = \sqrt{3} \times 400 \times 46.2 \times 0.6 = 19{,}200$ VAR or 19.2 kVAR **Ans.**

(*v*) Total volt-amperes, $S = \sqrt{3} \times VI = \sqrt{3} \times 400 \times 46.2 = 32{,}000$ VA or 32 kVA **Ans.**

Example 8.17 **A balanced delta connected load of (12 + *j*9) Ω/phase is connected to 3-phase 400 V supply. Calculate line current, power factor and power drawn by it.**

[*G.B. Technical Univ. Electrical Engineering First Semester 2011-12*]

Solution: Refer to Solution of Example 8.16.

Example 8.18 **Each phase of a delta-connected load has a resistance of 25 Ω, an inductance of 0.15 H and a capacitance of 120 μF in series. The load is connected across a 400 V, 50 Hz, 3-phase supply. Determine the line current, active power and reactive volt-amperes.** [*U.P. Technical Univ. Electrical Engineering July 2002*]

Solution: Resistance per phase, $R_P = 25\ \Omega$

Inductive reactance per phase, $X_{LP} = 2\pi f L = 2\pi \times 50 \times 0.15 = 47.124\ \Omega$

Capacitive reactance per phase, $X_{CP} = \dfrac{1}{2\pi f C} = \dfrac{1}{2\pi \times 50 \times 120 \times 10^{-6}} = 26.526\ \Omega$

Net reactance per phase, $X_p = X_{LP} - X_{CP} = 47.124 - 26.526 = 20.6\ \Omega$

Impedance per phase $= Z_P = \sqrt{R_P^2 + X_P^2} = \sqrt{25^2 + 20.6^2} = 32.4\ \Omega$

Phase angle, $\phi = \text{Tan}^{-1}\dfrac{X_p}{R_p} = \text{Tan}^{-1}\dfrac{20.6}{25} = 39.5°$ (lag)

Phase current, $I_p = \dfrac{V_P}{Z_P} = \dfrac{V_L}{Z_P} = \dfrac{400}{32.4} = 12.35$ A

Line current, $I_L = \sqrt{3}\ I_p = \sqrt{3} \times 12.35 = 21.38$ A **Ans.**

Active power, $P = \sqrt{3}\ V_L I_L \cos\phi = \sqrt{3} \times 400 \times 21.38 \times \cos 39.5°$
$= 11{,}430$ W or 11.43 kW **Ans.**

Reactive volt-amperes, $Q = \sqrt{3}\ V_L I_L \sin\phi = \sqrt{3} \times 400 \times 21.38 \times \sin 39.5°$
$= 9{,}420$ VAR or 9.42 kVAR **Ans.**

Example 8.19 **A 220-V 3-ϕ voltage is applied to a balanced delta-connected 3-ϕ load of phase impedance (15 + *j* 20) Ω.**

(*a*) **Find the phasor current in each line.**

(*b*) **What is the power consumed per phase ?**

(*c*) **What is the phasor sum of the three line currents ? Why does it have this value ?**

[*Banaras Hindu Univ. Elec. Circuits & Instruments 1985*]

Solution: Phase voltage, $V_P = V_L = 220$ V

Phase impedance, $Z_P = \sqrt{R_P^2 + X_P^2} = \sqrt{15^2 + 20^2} = 25\ \Omega$

Phasor current, $I_P = \dfrac{V_P}{Z_P} = \dfrac{220}{25} = 8.8$ A

Line current, $I_L = \sqrt{3}\ I_P = \sqrt{3} \times 8.8 = 15.24$ A

Phase angle, $\phi = \text{Cos}^{-1}\dfrac{R_P}{Z_P} = \text{Cos}^{-1}\dfrac{15}{25} = \text{Cos}^{-1} 0.6 = 53.13°$ (lag)

(*a*) Since line currents are 30° behind their respective phase currents, line currents are expressed as, taking phase voltage V_{RY} as reference phasor,

$$\mathbf{I}_R = 15.24 \angle -83.13^\circ \text{ A}, \mathbf{I}_Y = 15.24 \angle -203.13^\circ \text{ A}$$

$$\text{or } \mathbf{I}_Y = 15.24 \angle 156.87^\circ \text{ and } \mathbf{I}_B = 15.24 \angle 36.87^\circ \text{ A } \textbf{Ans.}$$

(*b*) Power consumed per phase $= V_P I_P \cos\phi = 220 \times 8.8 \times 0.6 = 1{,}161.6$ watts **Ans.**

(*c*) Phasor sum of the three line currents $= \mathbf{I}_R + \mathbf{I}_Y + \mathbf{I}_B = 15.24 \angle -83.13^\circ + 15.24 \angle 156.87^\circ + 15.24 \angle 36.87^\circ$

$$= (1.823 - j\,15.131) + (-14.015 + j\,5.987) + (12.192 + j\,9.144)$$

$$= 0 \textbf{ Ans.}$$

As expected, phasor sum of three line currents drawn by a balanced load is zero because they are equal in magnitude and have a phase difference of 120° amongst themselves.

8.8. COMPARISON BETWEEN STAR AND DELTA SYSTEMS

1. In star-connected system similar ends are joined together whereas in delta-connected system dissimilar ends are joined together.
2. In a star-connected system phase voltage is equal to $\frac{1}{\sqrt{3}}$ times the line voltage, but in delta-connected system phase voltage is equal to line voltage, therefore, star-connected alternator needs less number of turns per phase and less insulation than a delta-connected alternator for the same line voltage. These are the reasons that 3-phase alternators are usually star-connected.
3. In star-connected system, phase current is equal to line current *i.e.* $I_P = I_L$ but in delta-connected system, the phase current is $\frac{1}{\sqrt{3}}$ times of line current *i.e.* $I_P = \frac{I_L}{\sqrt{3}}$
4. It is possible to carry neutral wire to the load in star-connected system only *i.e.* only star-connected system can give 3-phase 4-wire arrangement and hence only star-connected system can be used for lighting as well as power loads *i.e.* for an unbalanced load. Lighting loads are connected between neutral and on any one of the outers and power circuits can be connected to the lines.
5. Neutral point of a star-connected system can be connected to earth, so relays and protective devices can be provided in star-connected system for protection of the system against ground faults and it is usually done in practice.
6. Delta connections are mostly used in transformers for operating small low-voltage 3-phase motors and are best suited for rotary convertors. Most of the three-phase induction motors are delta-connected, 3-phase loads are also usually delta-connected as it provides flexibility for addition or removal of a load on a single phase in case of an unbalanced load.

8.9. CONVERSION OF BALANCED LOAD SYSTEM FROM STAR TO DELTA AND VICE VERSA

Any balanced star-connected system can be completely replaced by an equivalent delta-connected system or vice versa because of their relationships between phase and line voltages and currents. For example, a balanced star-connected load having an impedance of magnitude Z with a power factor $\cos\phi$ (or $Z\angle\phi$) in each phase can be replaced by an equivalent delta-connected load having an impedance of magnitude 3Z and power factor $\cos\phi$ (*i.e.* $3Z\angle\phi$) in each phase.

This may be established as follows:

For a balanced star-connected load, let the line voltage be V_L, line current be I_L and impedance per phase be Z

$$\text{Phase voltage, } V_P = \frac{V_L}{\sqrt{3}}$$

$$\text{Phase current, } I_P = I_L$$

$$\text{So impedance per phase, } Z_S = \frac{V_P}{I_P} = \frac{V_L/\sqrt{3}}{I_L} = \frac{1}{\sqrt{3}}\frac{V_L}{I_L} \quad ...(8.21)$$

Now in equivalent delta-connected system for the same line values of voltage and current as in case of star-connected system

$$\text{Phase voltage, } V_P = V_L$$

$$\text{Phase current, } I_P = \frac{I_L}{\sqrt{3}}$$

$$\text{So impedance per phase, } Z_D = \frac{V_P}{I_P} = \frac{V_L}{I_L/\sqrt{3}} = \sqrt{3}\frac{V_L}{I_L} \quad ...(8.22)$$

Comparing Eqs. (8.21) and (8.22) we have

$$Z_S = \frac{1}{3}Z_D \text{ or } Z_D = 3\,Z_S \quad ...(8.23)$$

8.10. BALANCED PARALLEL LOADS

A combination of balanced 3-ϕ loads connected in parallel may be solved by any one of the following methods.

1. All the given loads may be converted into either equivalent Y or Δ-connected loads and then combined together according to the law governing parallel circuits.
2. Alternative method is of working out volt-amperes. The real power and reactive power of various loads may be added arithmetically and algebraically respectively in order to give total volt-amperes according to expression

$$VA = \sqrt{W^2 + VAR^2}$$

$$\text{or } S = \sqrt{P^2 + Q^2}$$

where P is the power is watts (or kW), Q is the reactive power in reactive volt-amperes (or kVAR) and S is the volt-amperes (or kVA)

Example 8.20 A 3-phase star-connected 1,000 V alternator supplies power to 500 kW delta-connected induction motor. If the motor power factor is 0.8 lagging and its efficiency 0.9, find the current in each alternator and motor phase.

Solution:

$$\text{Motor input, } P = \frac{\text{Motor output}}{\text{Motor efficiency}} = \frac{500}{0.9} = 555.55 \text{ kW}$$

Motor power factor, $\cos\phi = 0.8$ (lagging)

$$\text{Line current, } I_L = \frac{P \text{ in watts}}{\sqrt{3}V_L \cos\phi} = \frac{555.55 \times 1{,}000}{\sqrt{3} \times 1{,}000 \times 0.8} = 401 \text{ A}$$

Current in star-connected alternator phase, $I_P = I_L = 401$ A **Ans.**

Current in delta-connected motor phase, $I_P = \frac{I_L}{\sqrt{3}} = \frac{401}{\sqrt{3}} = 231.5$ A **Ans.**

Example 8.21 Three similar resistors connected in star draw a line current of 5 A from a 400 V, 3-phase mains. To what value should the line voltage be changed to obtain the same line current with the resistors connected in delta ?

Solution: **When star-connected**

$$\text{Phase voltage, } V_P = \frac{V_L}{\sqrt{3}} = \frac{400}{\sqrt{3}} = 231 \text{ V}$$

$$\text{Phase current, } I_P = I_L = 5 \text{ A}$$

$$\text{Phase resistors, } R_P = \frac{V_P}{I_P} = \frac{231}{5} = 46.2\ \Omega$$

When delta-connected

$$\text{Line current, } I_L = 5 \text{ A, the same as in star-connection}$$

$$\text{Phase current, } I_P = \frac{I_L}{\sqrt{3}} = \frac{5}{\sqrt{3}} = 2.887 \text{ A}$$

$$\text{Phase voltage, } V_P = I_P R_P = 2.887 \times 46.2 = 133.33 \text{ V}$$

$$\text{Line voltage, } V_L = V_P = 133.33 \text{ V,}$$ one-third of line voltage for star-connection **Ans.**

Example 8.22 **A 3-phase delta-connected load, each phase of which has an inductive reactance of 40 Ω and a resistance of 25 Ω, is fed from the secondary of a 3-phase star-connected transformer, which has phase voltage of 240 V. Draw the circuit diagram of the system and calculate:**

(*a*) The current in each phase of the load;
(*b*) The voltage across each phase of the load;
(*c*) The current in the transformer secondary windings;
(*d*) The total power taken from the supply and its power factor.

Solution:

$$\text{Resistance, } R_P = 25 \ \Omega$$

$$\text{Reactance, } X_P = 40 \ \Omega$$

$$\text{Impedance, } Z_P = \sqrt{R_P^2 + X_P^2} = \sqrt{(25)^2 + (40)^2} = 47.2 \ \Omega$$

Phase voltage across transformer secondary = 240 V

Line voltage across delta-connected load = Line voltage across transformer secondary = $\sqrt{3} \times 240 = 415.7$ V

(*b*) Voltage across each phase of the load,

$$V_P = V_L = 415.7 \text{ V } \textbf{Ans.}$$

(*a*) Current in each phase of the load,

$$I_P = \frac{V_P}{Z_P} = \frac{415.7}{47.2} = 8.8 \text{ A } \textbf{Ans.}$$

(*c*) Current in the transformer secondary

= Line current of the load

$= \sqrt{3} \times 8.8 = 15.3$ A **Ans.**

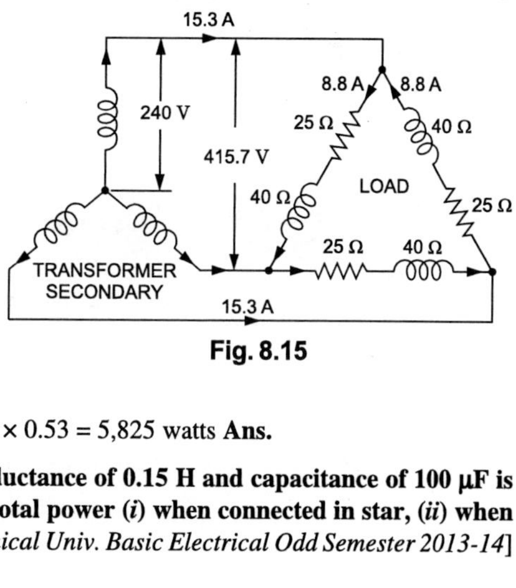

Fig. 8.15

(*d*) Power factor, $\cos \phi = \frac{R_p}{Z_p} = \frac{25}{47.2} = 0.53$ (lag) **Ans.**

Total power drawn, $P = \sqrt{3}\, V_L I_L \cos \Phi = \sqrt{3} \times 415.7 \times 15.3 \times 0.53 = 5{,}825$ watts **Ans.**

Example 8.23 **A 3-phase load consisting of resistance 25 Ω, inductance of 0.15 H and capacitance of 100 μF is connected to 400 V, 50 Hz. Calculate line current, power factor, total power (*i*) when connected in star, (*ii*) when connected in delta.** [*U.P. Technical Univ. Basic Electrical Odd Semester 2013-14*]

Solution:

$$\text{Resistance per phase, } R_p = 25 \ \Omega$$

$$\text{Inductive reactance per phase, } X_{Lp} = 2\pi f L = 2\pi \times 50 \times 0.15 = 47.124 \ \Omega$$

$$\text{Capacitive reactance per phase, } X_{Cp} = \frac{1}{2\pi f C} = \frac{1}{2\pi \times 50 \times 100 \times 10^{-6}} = 31.83 \ \Omega$$

$$\text{Net reactance per phase, } X_p = X_{Lp} - X_{Cp} = 47.124 - 31.83 = 15.294 \ \Omega$$

$$\text{Impedance per phase, } Z_p = \sqrt{R_p^2 + X_p^2} = \sqrt{25^2 + 15.294^2} = 29.3 \ \Omega$$

$$\text{Power factor, } \cos \phi = \frac{R_p}{Z_p} = \frac{25}{29.3} = 0.853 \text{ (lagging) } \textbf{Ans.}$$

(*i*) **When connected in star**

$$\text{Line current, } I_L = \text{Phase current, } I_P = \frac{V_P}{Z_P} = \frac{400/\sqrt{3}}{29.3} = 7.88 \text{ A} \quad \textbf{Ans.}$$

$$\text{Total power, } P = \sqrt{3}\, V_L I_L \cos\phi = \sqrt{3} \times 400 \times 7.88 \times 0.853 = 4{,}658 \text{ W or } 4.658 \text{ kW} \quad \textbf{Ans.}$$

(*ii*) **When connected in delta**

$$\text{Line current, } I_L = \sqrt{3} \times I_P = \sqrt{3} \times \frac{V_P}{Z_P} = \sqrt{3} \times \frac{V_L}{Z_P} = \frac{\sqrt{3} \times 400}{29.3} = 23.64 \text{ A} \quad \textbf{Ans.}$$

$$\text{Total power, } P = \sqrt{3}\, V_L I_L \cos\phi = \sqrt{3} \times 400 \times 23.64 \times 0.853 = 13{,}974 \text{ W or } 13.974 \text{ kW} \quad \textbf{Ans.}$$

Note: *For same load in delta-connection line current and power drawn are three times when connected in star.*

Example 8.24 **A 400 V, 3-phase source supplies power to both delta and a wye-connected loads connected in parallel. All the phase impedances are identical and equal to (5 + *j* 8.66) Ω. Compute total line current and power drawn from source.** [*Mahamaya Technical Univ. Electrical Engineering Second Semester 2011-12*]

Solution:

$$\text{Phase impedances, } Z_P = \sqrt{5^2 + 8.66^2} = 10\ \Omega$$

$$\text{Power factor, } \cos\phi = \frac{R}{Z} = \frac{5}{10} = 0.5 \text{ (lagging)}$$

Star-connected Load

$$\text{Phase voltage, } V_P = \frac{V_L}{\sqrt{3}} = \frac{400}{\sqrt{3}} = 230.94 \text{ V}$$

$$\text{Line current, } I_L = I_P = \frac{V_P}{Z} = \frac{230.94}{10} = 23.094 \text{ A}$$

$$\text{Power drawn, } P_1 = \sqrt{3}\, V_L I_L \cos\phi = \sqrt{3} \times 400 \times 23.094 \times 0.5 = 8{,}000 \text{ W} \quad \textbf{Ans.}$$

Delta-connected Load

$$\text{Phase voltage, } V_P = V_L = 400 \text{ V}$$

$$\text{Line current, } I_L = \sqrt{3}\, I_P = \sqrt{3} \times \frac{V_P}{Z} = \sqrt{3} \times \frac{400}{10} = 69.282 \text{ A}$$

$$\text{Power drawn, } P_2 = \sqrt{3}\, V_2 I_L \cos\phi = \sqrt{3} \times 400 \times 69.282 \times 0.5 = 24{,}000 \text{ W}$$

$$\text{Total power drawn from source, } P = P_1 + P_2 = 8{,}000 + 24{,}000 = 32{,}000 \text{ W or } 32 \text{ kW} \quad \textbf{Ans.}$$

$$\text{Total line current, } I_L = \frac{P}{\sqrt{3}\, V_L \cos\phi} = \frac{32{,}000}{\sqrt{3} \times 400 \times 0.5} = 92.376 \quad \textbf{Ans.}$$

Example 8.25 **A 3- wire, 3-phase supply feeds a load consisting of three equal resistors. By how much is the load reduced if one of the resistor be removed ? (*i*) When the load is star-connected (*ii*) When the load is delta-connected.** [*R.G.P.V. Bhopal Basic Electrical Engineering June 2002*]

Solution: Let the line voltage of 3-phase supply be V_L volts and resistance of each phase be of R ohms.

(*i*) **In Star**

$$\text{Power consumed, } P = 3 \times \frac{V_P^2}{R} = 3 \times \left(\frac{V_L}{\sqrt{3}}\right)^2 \times \frac{1}{R} = \frac{V_L^2}{R}$$

After disconnection or removal of one of the resistances, the circuit is no longer 3-phase but converted into single phase circuit, having two resistors, each of R ohms connected in series across supply of V_L volts, as shown in Fig. 8.16(*b*).

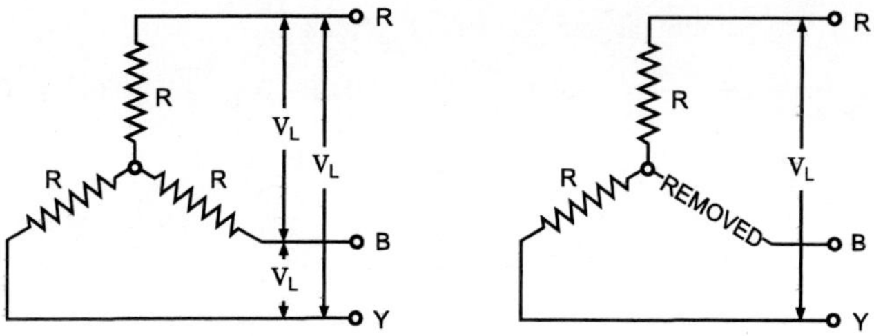

(*a*) (*b*)

Fig. 8.16

$$\text{Hence total power consumed } = 2\,\frac{\left(\frac{V_L}{2}\right)^2}{R} = \frac{V_L^2}{2R}$$

$$\therefore \text{ Reduction in load } = \frac{\frac{V_L^2}{R}-\frac{V_L^2}{2R}}{\frac{V_L^2}{R}} \times 100 = 50\%\ \textbf{Ans.}$$

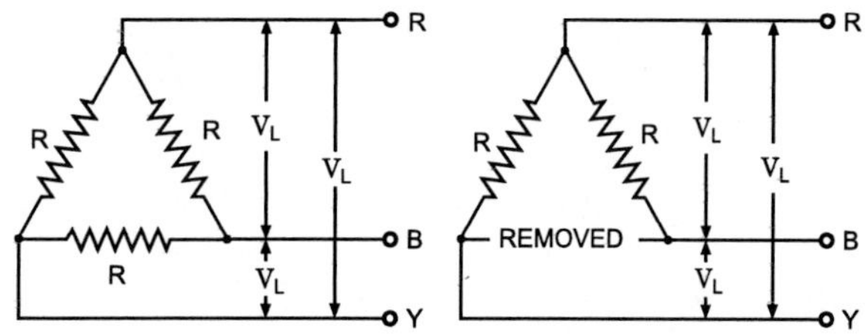

(*a*) (*b*)

Fig. 8.17

(*ii*) **In Delta**

$$\text{Power consumed } = 3 \times \frac{V_P^2}{R} = 3 \times \frac{V_L^2}{R} = 3\,\frac{V_L^2}{R}$$

After removal of one of the resistance, each of the remaining two resistances is connected across supply of V_L volts as shown in Fig. 8.17(*b*).

$$\therefore \text{ Power consumed } = 2 \times \frac{V_L^2}{R} = \frac{2V_L^2}{R}$$

$$\text{So reduction in load } = \frac{\frac{3V_L^2}{R}-\frac{2V_L^2}{R}}{\frac{3V_L^2}{R}} \times 100 = 33.33\%\ \textbf{Ans.}$$

Example 8.26 **Three identical resistors of 20 Ω each are connected in star to a 415 V, 50 Hz, three-phase supply. Calculate (*i*) the total power consumed, (*ii*) the total power consumed, if they are connected in delta (*iii*) the power consumed, if one of the resistors is opened.** [*U.P. Technical Univ. Electrical Engineering January 2003*]

Solution: Total power consumed from 415 V, 50 Hz 3-phase supply, when three 20 Ω resistors are connected in

$$(i)\ \text{star, } P_S = \frac{3\times\left(V_L/\sqrt{3}\right)^2}{R} = \frac{V_L^2}{R} = \frac{415^2}{20} = 8{,}611.25 \text{ W or } 8.61125 \text{ kW } \textbf{Ans.}$$

$$(ii)\ \text{delta, } P_D = 3 \times \frac{V_L^2}{R} = 3\times\frac{415^2}{20} = 25{,}833.75 \text{ W or } 25.83375 \text{ kW } \textbf{Ans.}$$

(*iii*) In the event of one of the three resistors getting open,

$$(a)\ \text{Power drawn in star-connections } = \frac{2\times(V_L/2)^2}{R} = \frac{V_L^2}{2R} = \frac{415^2}{2\times 20} = 4{,}305.625 \text{ W or } 4.306 \text{ kW } \textbf{Ans.}$$

$$(b)\ \text{Power drawn in delta-connections } = \frac{2\times V_L^2}{R} = \frac{2\times 415^2}{20} = 17{,}222.5 \text{ W or } 17.2225 \text{ kW } \textbf{Ans.}$$

Example 8.27 **Three identical impedances of 30 $\angle 30°$ ohms are connected in delta to a 3-phase, 3-wire, 208 V abc system by conductors which have impedances of $(0.8 + j\,0.6)$ ohm. Find the magnitude of the voltage at the load end.** [*Pb Univ. Electrical Engineering May 1990*]

Solution: The equivalent star load of the delta-connected load is given as

$$Z_S = \frac{Z_D}{3} \text{ i.e. } \frac{30\angle 30°}{3} \text{ or } 10\angle 30°\ \Omega \qquad \text{(Refer to Art. to 8.9)}$$

From circuit diagram shown in Fig. 8.18.

Impedance per phase, $\mathbf{Z}_{an} = (0.8 + j\,0.6) + 10\ \angle 30^\circ = (0.8 + j\,0.6) + (8.66 + j\,5)$

$= (9.46 + j\,5.6)$ or $10.993\ \angle 30.62^\circ\ \Omega$

Phase voltage, $\mathbf{V}_{an} = \dfrac{208}{\sqrt{3}}\ \angle 0^\circ = 120.1\ \angle 0^\circ$ V

taking $\mathbf{V}_{an}$ as reference phasor

So phase current, $\mathbf{I}_{an} = \dfrac{\mathbf{V}_{an}}{\mathbf{Z}_{an}} = \dfrac{120.1\ \angle 0^\circ}{10.993\ \angle 30.62^\circ}$

$= 10.925\ \angle -30.62^\circ$ A

Voltage across load, $V_{a'n} = \mathbf{I}_{an} \times \mathbf{Z}_L$

$= 10.925\ \angle -30.62^\circ \times 10\ \angle 30^\circ$

$= 109.25\ \angle 0.62^\circ$ V

Line voltage at the load $= \sqrt{3}\ V_{a'n} = \sqrt{3} \times 109.25 = 189.2$ V **Ans.**

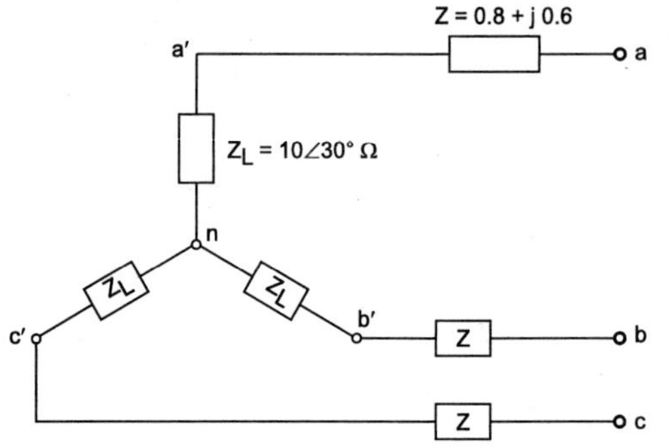

Fig. 8.18

8.11. MEASUREMENT OF POWER IN THREE-PHASE CIRCUITS

Wattmeter consists of a low resistance coil, called the *current coil* (*CC*) which is inserted in series with line carrying current to be measured and a high resistance coil, called the *pressure coil* (*PC*), which is connected in series with a high resistance between the two points across which potential difference is to be measured, as shown is Fig. 8.19. The wattmeter gives reading which is proportional to the product of the current flowing through its current coil, potential difference across its pressure coil and cosine of the phase angle between voltage and current.

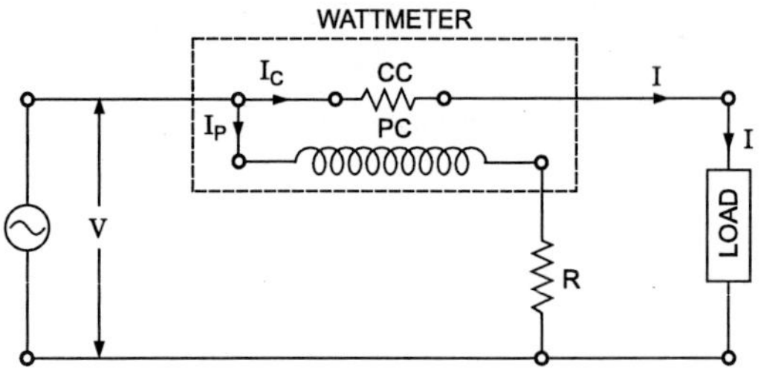

Fig. 8.19 *Wattmeter Connections*

8.11.1. Three-Wattmeter Method. This method is employed for measurement of power in 3-phase, 4-wire circuits, the connections being as shown in Fig. 8.20 (*a*). As the neutral wire is common to the three phases,

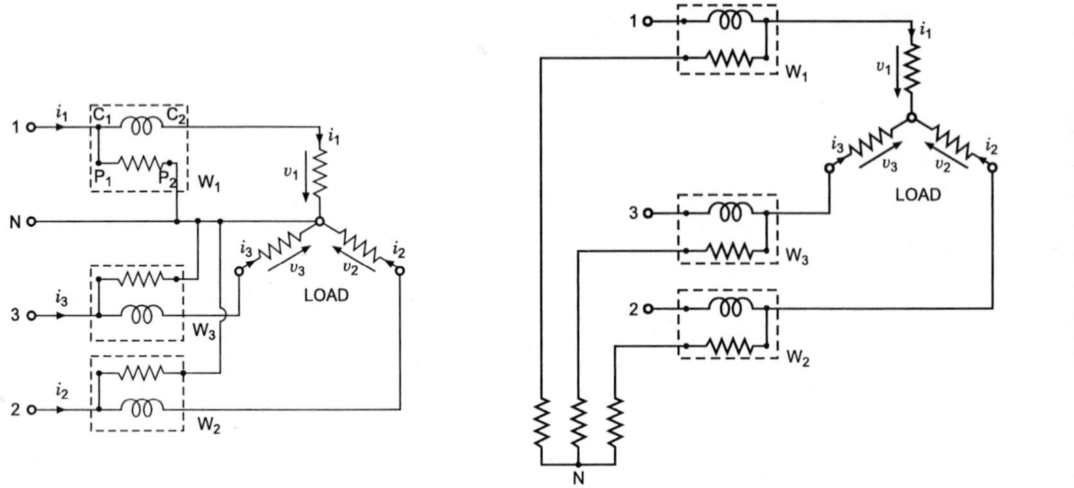

(*a*) (*b*)

Fig. 8.20

each wattmeter reads power in its own phase, and the total power of the load circuit is given by the sum of the readings of the three wattmeters

i.e. Total power of load circuit, $P = W_1 + W_2 + W_3$

The above method is only useful for measuring power in 3-phase, 4-wire load circuits. In case of 3-phase, 3-wire star-connected circuits difficulty is experienced in getting neutral. In special cases when it is necessary to employ this method for measurement of power in 3-phase, 3-wire circuits an "artificial star" can be formed by connecting three equal high resistances in star to the three line conductors, as shown in Fig. 8.20 (*b*). (In case of low-voltage circuits three potential coils may be connected to form the common star).

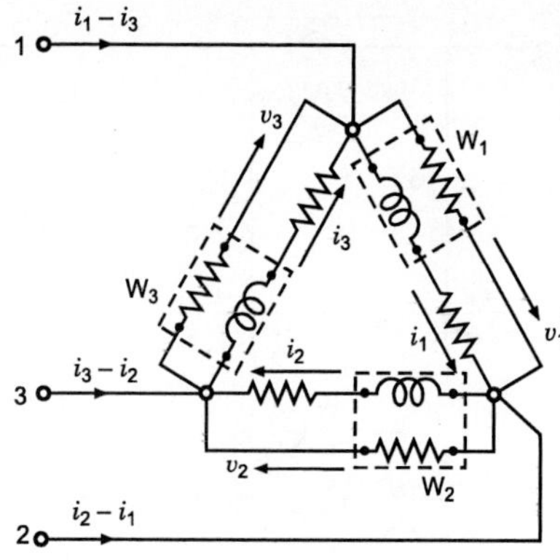

(*c*)

Fig. 8.20 *Three-Wattmeter Method of Measuring 3-Phase Power*

In case of delta-connected circuits the difficulty in adopting the above mentioned method for measurement of power is due to the fact that the phase coils are required to be broken for inserting current coils of wattmeters, as shown in Fig. 8.20 (*c*).

8.11.2. One-Wattmeter Method. In a balanced 3-wire, 3-phase load circuit the power in each phase is equal and, therefore, total power of the circuit can be determined by multiplying the power measured in any one phase by three.

Figure 8.21 (*a*) shows connections for measurement of power in one phase of the 3-phase, 3-wire star-connected balanced load circuit. Total power of the load circuit will be 3 times the reading of wattmeter.

Figure 8.21 (*b*) shows connections for measurement of power in 3-phase, 3-wire delta-connected balanced load circuits by one-wattmeter method. In this method of measurement of power, resistance coils of resistance equal to that of pressure coil is connected in each of remaining two phases, as shown in Fig. 8.21 (*b*). The pressure coil and resistance coils form a balanced star-connection.

It should be noted that the one-wattmeter method has the disadvantage that even a slight degree of unbalance in the loading produce a large error in the measurement.

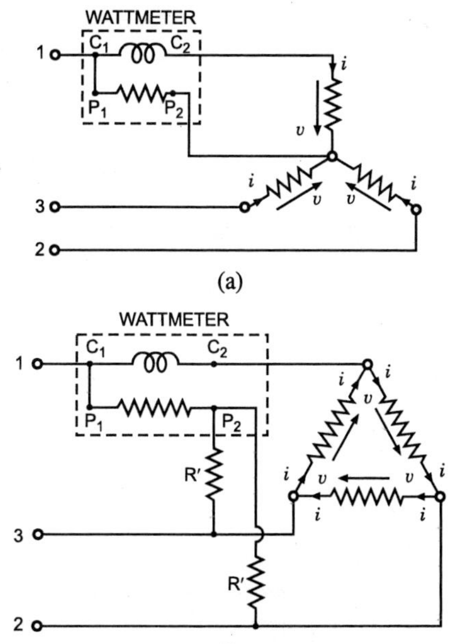

(*b*)

Fig. 8.21 *One-Wattmeter Method of Measuring Power in 3-Phase 3-Wire Balanced Load Circuits*

8.11.3. Two-Wattmeter Method. This is the generally used method for measurement of power in 3-phase, 3-wire load circuits. The current coils of two wattmeters are inserted in any two lines and pressure coil is connected from its own current coil to the line without a current coil. The connections are shown in Fig. 8.22.

Let v_1, v_2, v_3 and i_1, i_2, i_3 be the voltages and currents of the three loads connected across three different phases at any particular instant. These being instantaneous values, the power at the instant under consideration is equal to the sum of their products regardless of power factor *i.e.* instantaneous power,

$$p = v_1 i_1 + v_2 i_2 + v_3 i_3 \text{ watts} \quad ...(8.24)$$

(*a*) **Star-Connected System.** Since all the three phases meet at a star point so according to Kirchhoff's first law, the algebraic sum of three instantaneous currents is zero.

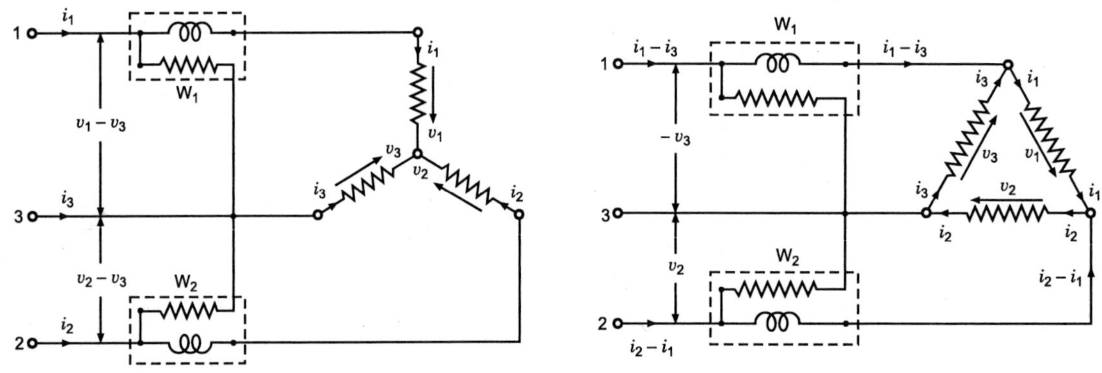

Fig. 8.22 *Two-Wattmeter Method of Measuring 3-Phase 3-Wire Power*

$$i.e.\ i_1 + i_2 + i_3 = 0 \text{ or } i_3 = -(i_1 + i_2)$$

Substituting $i_3 = -(i_1 + i_2)$ in Eq. (8.24) we get

$$\text{Instantaneous power, } p = v_1 i_1 + v_2 i_2 - v_3 (i_1 + i_2) = i_1 (v_1 - v_3) + i_2 (v_2 - v_3)$$

Since i_1 is the instantaneous current flowing through the current coil and $(v_1 - v_3)$ is the instantaneous potential difference across pressure coil of wattmeter W_1, $(v_1 - v_3)\, i_1 = w_1$, the instantaneous power measured by wattmeter W_1.

Similarly i_2 is the instantaneous current flowing through the current coil and $(v_2 - v_3)$ is the instantaneous potential difference across pressure coil of wattmeter W_2, therefore, $(v_2 - v_3)\, i_2 = w_2$ the instantaneous power measured by wattmeter W_2.

$$\text{Hence total instantaneous power, } p = w_1 + w_2$$

$$\text{or Total average power, } P = W_1 + W_2 \qquad ...(8.25)$$

Hence the algebraic sum of two wattmeter readings gives the total power in the 3-phase, 3-wire star-connected load circuits whether the load is balanced or unbalanced.

(*b*) **Delta-Connected System.** In delta-connected system the three phases form a closed loop and according to Kirchhoff's second law.

$$v_1 + v_2 + v_3 = 0 \text{ or } v_1 = -(v_2 + v_3)$$

$$\text{Instantaneous power, } p = v_1 i_1 + v_2 i_2 + v_3 i_3 = -(v_2 + v_3)\, i_1 + v_2 i_2 + v_3 i_3 = -v_3 (i_1 - i_3) + v_2 (i_2 - i_1) \qquad ...(8.26)$$

Since $-v_3$ is the instantaneous pd across pressure coil and $(i_1 - i_3)$ is the instantaneous current flowing through current coil of wattmeter W_1, wattmeter W_1 reads average of $-v_3 (i_1 - i_3)$ and similarly wattmeter W_2 reads average of $v_2 (i_2 - i_1)$

$$\text{Hence total power, } P = W_1 + W_2 \qquad ...(8.27)$$

Hence the algebraic sum of two wattmeter readings gives the total power of the circuit irrespective of the fact that the circuit is balanced or unbalanced and star-connected or delta-connected.

Determination of Power Factor From Wattmeter Readings. If load is balanced then pf of the load can also be determined from the wattmeter readings.

The phasor diagram for a balanced star-connected inductive load is shown in Fig. 8.23. Let V_1, V_2, and V_3 be the rms values of phase voltages and I_1, I_2 and I_3 be the rms values of phase currents.

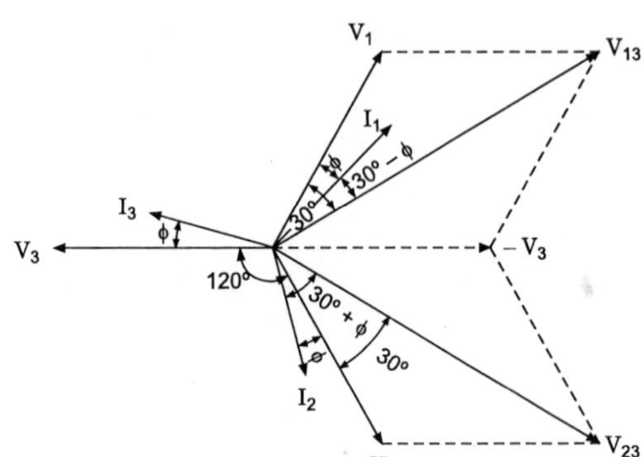

Fig. 8.23 *Phasor Diagram For Balanced Star-connected Inductive Load*

Since load is balanced,

(*i*) Phase voltages V_1, V_2 and V_3 will be equal (say, equal to V_P)

(*ii*) Phase currents I_1, I_2 and I_3 will be equal (say, equal to I_P or I_L)

(*iii*) Phase angles between respective phase voltages and phase currents will be equal, say ϕ

The current in current coil of wattmeter $W_1 = I_1 = I_L$ lagging behind V_1 by ϕ.

The pd across pressure coil of wattmeter $W_1 = V_{13} = \sqrt{3}\ V_P = V_L$ lagging behind V_1 by 30°.

Therefore, phase angle between voltage across potential coil and current through current coil of wattmeter W_1 is $(30° - \phi)$ and reading of wattmeter

$$W_1 = V_L I_L \cos(30° - \phi) \quad \text{...(8.28)}$$

The current in current coil of wattmeter $W_2 = I_2 = I_L$ lagging behind V_2 by ϕ.

The pd across pressure coil of wattmeter $W_2 = V_{23} = \sqrt{3}\ V_P = V_L$ leading V_2 by 30°.

Therefore, phase angle between pd across potential coil and current through current coil of wattmeter W_2 is $(30° + \phi)$.

Hence reading of wattmeter $W_2 = V_L I_L \cos(30° + \phi)$ ...(8.29)

$$\text{The sum of two wattmeter readings } W_1 + W_2 = V_L I_L \cos(30° - \phi) + V_L I_L \cos(30° + \phi)$$

$$= V_L I_L \times 2\cos 30° \cos\phi = \sqrt{3}\ V_L I_L \cos\phi$$

$$= \text{True power of load} \quad \text{...(8.30)}$$

$$\text{and } W_1 - W_2 = V_L I_L \cos(30° - \phi) - V_L I_L \cos(30° + \phi)$$

$$= V_L I_L \times 2\sin 30° \sin\phi = V_L I_L \sin\phi \quad \text{...(8.31)}$$

Dividing Eq. (8.31) by Eq. (8.30) we get

$$\frac{W_1 - W_2}{W_1 + W_2} = \frac{\text{Tan}\,\phi}{\sqrt{3}}$$

$$\text{or } \phi = \text{Tan}^{-1}\left(\frac{W_1 - W_2}{W_1 + W_2}\right) \times \sqrt{3}$$

$$\text{and pf of load circuit, } \cos\phi = \text{Cos}\tan^{-1}\frac{\sqrt{3}(W_1 - W_2)}{W_1 + W_2} \quad \text{...(8.32)}$$

Hence phase angle ϕ and pf, $\cos\phi$ can be determined from readings of two wattmeters.

Alternative Method of Determination of Power Factor From Wattmeter Readings. This is convenient method for determining the power factor from wattmeter readings for a balanced load. In this method of determination of power factor from the wattmeter readings, for various values of phase angles the ratio of two wattmeter readings, $\frac{\text{Cos } 30°+\phi}{\text{Cos } 30°-\phi}\left(i.e.\ \frac{\text{smaller reading}}{\text{larger reading}}\right)$ and the corresponding power factor values are determined and plotted as shown in Fig. 8.24. Now by using the watt-ratio curve (Fig. 8.24) the power factor is read directly, the ratio, $\frac{W_2}{W_1}$ being known.

It is seen that, when $\frac{W_2}{W_1} = 1.0$, the power factor is unity; when $\frac{W_2}{W_1} = 0$, the power factor is 0.5; when $\frac{W_2}{W_1}$ is negative i.e. when it becomes necessary to reverse W_2, the power factor is less than 0.5.

Variation in Wattmeter Readings. Wattmeter readings for inductive loads

$$W_1 = V_L I_L \cos(30° - \phi)$$

$$W_2 = V_L I_L \cos(30° + \phi)$$

(*i*) When pf of load is unity *i.e.* $\phi = 0$ then

$$W_1 = V_L I_L \cos 30° = \frac{\sqrt{3}}{2} V_L I_L = \text{Half of total power}$$

$$W_2 = V_L I_L \cos 30° = \frac{\sqrt{3}}{2} V_L I_L = \text{Half of total power}$$

Therefore, readings of both wattmeters are same, positive and equal to half the total power of the circuit.

(*ii*) When pf of load is 0.5 *i.e.* $\phi = 60°$ then

$$W_1 = V_L I_L \cos (30° - 60°) = \frac{\sqrt{3}}{2} V_L I_L = \text{Total power}$$

$$W_2 = V_L I_L \cos (30° + 60°) = 0$$

Hence reading of wattmeter W_1 will give total power of the load.

When the pf of load is zero *i.e.* $\phi = 90°$ then

$$W_1 = V_L I_L \cos (30° - 90°) = \frac{1}{2} V_L I_L$$

$$W_2 = V_L I_L \cos (30° + 90°) = -\frac{1}{2} V_L I_L$$

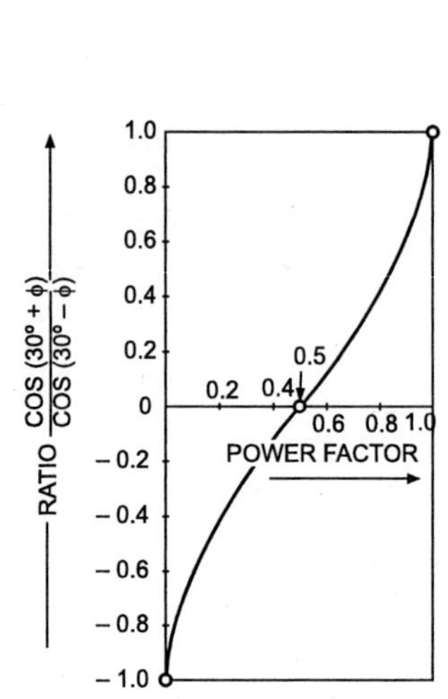

Fig. 8.24 *The Watt-ratio Curve*

Hence the readings of two wattmeters are equal but of opposite sign.

So wattmeter W_2 gives –ve reading when phase angle ϕ varies from 60° to 90°. For obtaining the reading of wattmeter W_2 either the connection of current coil or pressure coil should be changed and readings obtained after the reversal of connections should be subtracted from the other wattmeter reading in order to get the total power.

Wattmeter Readings For Capacitive Loads. In the above discussion inductive load has been considered. If the circuit is capacitive the expression for readings of two meters are obtained by substituting in ϕ by $(-\phi)$ in the above expressions for wattmeter readings.

$$W_1 = V_L I_L \cos [30° - (-\phi)] = V_L I_L \cos (30° + \phi)$$

$$W_2 = V_L I_L \cos [30° + (-\phi)] = V_L I_L \cos (30° - \phi)$$

i.e. readings of wattmeters are interchanged.

Example 8.28 **Explain two wattmeter method to measure three phase power. If in a 2 wattmeter method, readings of two wattmeters are 1,200 W and 300 W, find the power factor of the load.**

[*U.P. Technical Univ. Electrical Engineering Odd Semester 2013-14*]

Solution: $W_1 = 1{,}200$ watts and $W_2 = 300$ watts

$$\text{Power factor of load, } \cos\phi = \text{Cos} \tan^{-1} \frac{W_1 - W_2}{W_1 + W_2}\sqrt{3} = \text{Cos} \tan^{-1} \frac{1{,}200 - 300}{1{,}200 + 300}\sqrt{3} = 0.6934 \text{ (lag) } \textbf{Ans.}$$

Example 8.29 **Two wattmeters connected to measure the total power in a 3-phase balanced circuit. One measures 4,800 W, while the other reads backwards. On reversing the latter it is found to read 400 W. What is the total power and power factor ?**

[*R.G.P.V. Bohpal Basic Electrical Engineering 2001*]

Solution: Wattmeter reading, $W_1 = 4{,}800$ W

Wattmeter reading $W_2 = -400$ W

Total power of the circuit, $P = W_1 + W_2 = 4{,}800 + (-400) = 4{,}400$ W **Ans.**

$$\text{Power factor of the circuit} = \text{Cos} \tan^{-1} \frac{W_1 - W_2}{W_1 + W_2}\sqrt{3} = \text{Cos} \tan^{-1} \frac{4{,}800 - (-400)}{4{,}800 + (-400)}\sqrt{3} = \text{Cos} \tan^{-1} 2.047$$

$$= 0.438 \text{ (lagging) } \textbf{Ans.}$$

Example 8.30 **In a 2-wattmeter method power measured was 30 kW at 0.7 pf lagging. Find the reading of each wattmeter.** [*U.P. Technical Univ. Electrical Engnieering Second Semester 2006-07; Mahamaya Technical Univ. First Semester 2011-12*]

Solution: Total load supplied, P = 30 k W

Power factor, $\cos\phi$ = 0.7 (lagging)

Phase angle, $\phi = \text{Cos}^{-1}\, 0.7 = 45.57°$ (lagging)

$$\text{So}\;\; V_L\, I_L = \frac{\text{P in kW} \times 1{,}000}{\sqrt{3}\cos\phi} = \frac{30 \times 1{,}000}{\sqrt{3} \times 0.7} = 24{,}743.6 \text{ V A}$$

Reading of wattmeter $W_1 = V_L\, I_L \cos(30° - \phi) = 24{,}743.6 \cos(30° - 45.57°) = 23{,}835$ W or 23.835 k W **Ans.**

Reading of wattmeter, $W_2 = V_L\, I_L \cos(30° + \phi) = 24{,}743.6 \cos(30° + 45.57°) = 6{,}165$ W or 6.165 kW **Ans.**

Example 8.31 **A 3-phase balanced load connected across a 3-ϕ, 400 V ac supply draws a line current of 10 A. Two wattmeters are used to measure input power. The ratio of two wattmeter readings is 2 : 1. Find the readings of the two wattmeters.** [*U.P. Technical Univ. Electrical Engineering February 2002; Odd Semester 2013-14*]

Solution: Let the ratio of wattmeter readings be *r i.e.* $\frac{W_2}{W_1} = r$

$$\text{and } \tan\phi = \sqrt{3}\left(\frac{W_1 - W_2}{W_1 + W_2}\right) = \sqrt{3}\left(\frac{1 - \frac{W_2}{W_1}}{1 + \frac{W_2}{W_1}}\right) = \sqrt{3}\left(\frac{1-r}{1+r}\right)$$

$$\text{and power factor, } \cos\phi = \frac{1}{\sec\phi} = \frac{1}{\sqrt{1+\tan^2\phi}} = \frac{1}{\sqrt{1+3\left(\frac{1-r}{1+r}\right)^2}}$$

Substituting $r = \frac{W_2}{W_1} = \frac{1}{2} = 0.5$ in above expression we get

$$\text{Power factor of the load, } \cos\phi = \frac{1}{\sqrt{1+3\left(\frac{1-0.5}{1+0.5}\right)^2}} = 0.866$$

or phase angle, $\phi = \text{Cos}^{-1}\, 0.866 = 30°$

Wattmeter reading, $W_1 = V_L I_L \cos(30° - 30°) = 400 \times 10 \cos 0° = 4{,}000$ W **Ans.**

Wattmeter reading, $W_2 = V_L\, I_L \cos(30° + 30°) = 400 \times 10 \times \cos 60° = 2{,}000$ W **Ans.**

Example 8.32 **A balanced star-connected inductive load is connected to a 400 V, 50 Hz ac supply. Two wattmeters used to measure supply power indicate 8,000 W and 4,000 W respectively. Determine:**

(*i*) line current **(*ii*) impedance of each phase** **(*iii*) resistance and inductance of each phase.**

[*G.B. Technical Univ. Electrical Engineering First Semester 2009-10*]

Solution: Wattmeter reading W_1 = 8,000 W

Wattmeter reading W_2 = 4,000 W

Supply voltage, V_L = 400 V

Supply frequency, f = 50 Hz

$$\text{Power factor of circuit, } \cos\phi = \text{Cos}\tan^{-1}\frac{W_1 - W_2}{W_1 + W_2} \times \sqrt{3} = \text{Cos}\tan^{-1}\frac{8{,}000 - 4{,}000}{8{,}000 + 4{,}000} \times \sqrt{3} = \text{Cos } 30° = 0.866$$

Total, $P = W_1 + W_2 = 8{,}000 + 4{,}000 = 12{,}000$ W

$$(i)\quad \text{Line current, } I_L = \frac{P}{\sqrt{3}\, V_L \cos\phi} = \frac{12{,}000}{\sqrt{3} \times 400 \times 0.866} = 20 \text{ A} \quad \textbf{Ans.}$$

(*ii*) Impedance in each phase, $Z = \frac{V_p}{I_p} = \frac{400/\sqrt{3}}{20} = 11.547\ \Omega$ **Ans.**

(*iii*) Resistance in each phase, $R = Z \cos \phi = 11.547 \times 0.866 = 10\ \Omega$ **Ans.**

Inductive reactance in each phase, $X_L = \sqrt{Z^2 - R^2} = \sqrt{11.547^2 - 10^2} = 5.7735\ \Omega$

Inductance in each phase, $L = \frac{X_L}{2\pi f} = \frac{5.7735}{2\pi \times 50} = 18.38\ \text{mH}$ **Ans.**

Example 8.33 **A star-connected three-phase load has a resistance of 8 Ω and an inductive reactance of 6 Ω in each phase. It is fed from a 400 V, 3-phase balanced supply. Determine the line current, power factor, active and reactive powers. Draw phasor diagram showing phase and line voltages and currents. If power measurement is made using two-wattmeter method, what will be the readings of both wattmeters ?**

[*U.P. Technical Univ. Electrical Engineering First Semester 2005-06*]

Solution:

Phase voltage, $V_P = \frac{\text{Line voltage, } V_L}{\sqrt{3}} = \frac{400}{\sqrt{3}} = 231\ \text{V}$

Impedance per phase, $Z_P = \sqrt{R_P^2 + X_P^2} = \sqrt{8^2 + 6^2} = 10\ \Omega$

Line current, I_L = Phase current, $I_P = \frac{V_P}{Z_P} = \frac{231}{10} = 23.1\ \text{A}$ **Ans.**

Power factor, $\cos \phi = \frac{V_P}{Z_P} = \frac{8}{10}$

= 0.8 (lagging) **Ans.**

Phase angle, $\phi = \text{Cos}^{-1}\ 0.8$

= 36.87° (lagging)

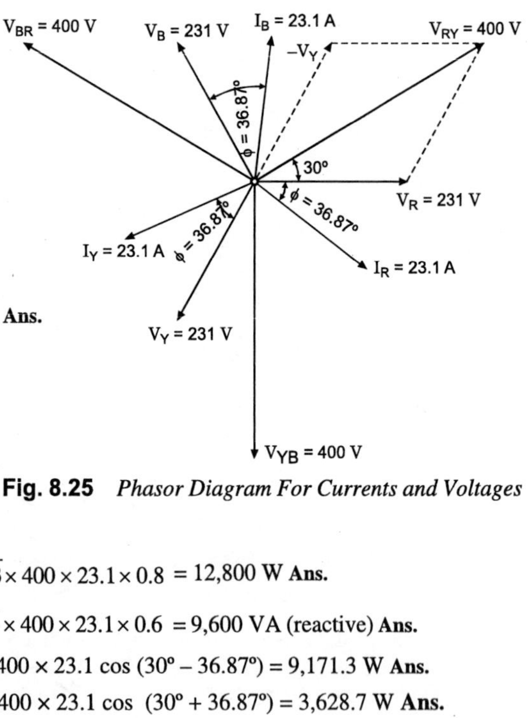

Fig. 8.25 *Phasor Diagram For Currents and Voltages*

Active power, $P = \sqrt{3} V_L I_L \cos\phi = \sqrt{3} \times 400 \times 23.1 \times 0.8 = 12{,}800$ W **Ans.**

Reactive power, $Q = \sqrt{3} V_L I_L \sin\phi = \sqrt{3} \times 400 \times 23.1 \times 0.6 = 9{,}600$ VA (reactive) **Ans.**

Wattmeter readings, $W_1 = V_L I_L \cos (30° - \phi) = 400 \times 23.1 \cos (30° - 36.87°) = 9{,}171.3$ W **Ans.**

$W_2 = V_L I_L \cos (30° + \phi) = 400 \times 23.1 \cos (30° + 36.87°) = 3{,}628.7$ W **Ans.**

Phasor diagram for currents and voltages is shown in Fig. 8.25.

Example 8.34 **A star-connected system has impedance as Z = 8 + *j*6 in each phase. It is fed from 400 V 3-phase ac supply. Find active and reactive powers. If power measurement is done by two wattmeter method then calculate readings of each wattmeter. Also draw connection diagram for this power measurement.**

[*U.P. Technical Univ. Basic Electrical Engineering Odd Semester 2013-14*]

Solution: Refer to Solution of Example 8.33.

Example 8.35 **The power input to a synchronous motor is measured by two wattmeters both of which indicate 50 kW. If the pf of the motor be changed to 0.866 leading determine the readings of the wattmeters, the total input power remaining the same. Draw the phasor diagram for the second condition of the load.**

[*Nagpur Univ. Elec. Technology 1992*]

Solution: Let the readings of wattmeters in second conditions be W_1 and W_2

$$\text{Now } W_1 + W_2 = \text{Total power of the load} = 100 \text{ kW} \quad ...(i)$$

as before change of pf of operation

$$\text{Phase angle, } \phi = \text{Cos}^{-1} \text{ pf} = \text{Cos}^{-1} 0.866 = 30° \text{ (lead)}$$

$$\text{Since } \phi = \text{Tan}^{-1}\left(-\sqrt{3}\,\frac{W_1 - W_2}{W_1 + W_2}\right)$$

$$\text{So } \tan\phi = -\sqrt{3}\,\frac{W_1 - W_2}{W_1 + W_2}$$

$$\text{or } W_1 - W_2 = (W_1 + W_2) \times \frac{\tan 30°}{-\sqrt{3}}$$

$$= 100 \times \frac{1/\sqrt{3}}{-\sqrt{3}} = -33.34 \text{ kW} \quad ...(ii)$$

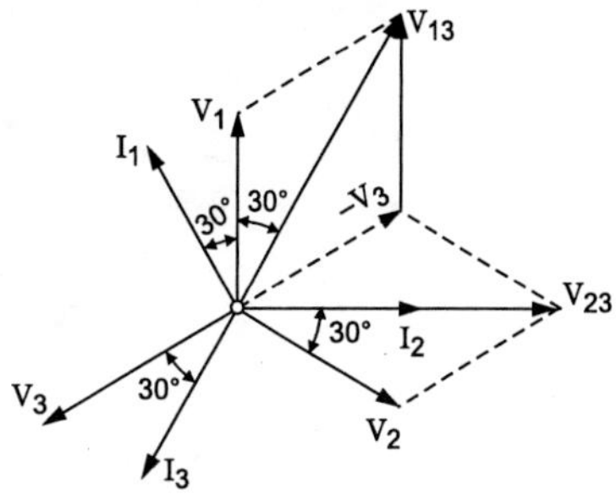

Fig. 8.26 *Phasor Diagram*

From Eqs. (*i*) and (*ii*) we get

W_1 = 33.33 kW **Ans.** and W_2 = 66.67 kW **Ans.**

Phasor diagram is shown in Fig. 8.26.

HIGHLIGHTS

1. *Polyphase system* means the system which consists of numerous windings or circuits.
2. A supply system is said to be *symmetrical* when different voltages of the same frequency have equal magnitude and are displaced from one another by equal time angle. 3-phase, 3-wire supply system will be symmetrical when line-to-line voltages are equal in magnitude and displaced by 120 electrical degrees w.r.t. each other.
3. A 3-phase supply will be *unbalanced* when either of three phase voltages are unequal in magnitude or the phase angles between these phases is not equal to 120°.
4. A load circuit is said to be balanced when the loads (impedances) connected in different phases are same in magnitude as well as in phase.
5. For a balanced system, the phasor sum of the three phase voltages is always zero.
6. In case one phase out of three phases of a 3-phase supply connected to the three-phase load is not available, then such a condition is called the single phasing.
7. Commonly used polyphase system is 3-phase system.
8. *Phase sequence* is the order in which the currents or voltages in different phases attain their maximum values.
9. The three-phase system is connected in two ways namely star connection and delta connection.
10. Star-connected system is obtained by joining together similar ends, either the start or the finish, the other ends are joined to the line wires. The common point at which similar ends are connected is called the neutral or star point.
11. In a balanced star-connected system
    (*i*) Line voltages are 120° apart.
    (*ii*) Line voltages are 30° ahead of the respective phase voltages.
    (*iii*) Line voltages are $\sqrt{3}$ times of phase voltages.
    (*iv*) Line currents are equal to phase currents.
    (*v*) The angle between line currents and the corresponding line voltages is $(30° \pm \phi)$; + ve for lagging currents and –ve for leading currents.
    (*vi*) True power output = $\sqrt{3}\, E_L I_L \cos\phi$, where ϕ is the angle between the respective phase current and phase voltage (not between line current and line voltage).
    (*vii*) Apparent power = $\sqrt{3}\, E_L I_L$

(*viii*) In balanced system, the potential of neutral or star point is zero because potential at neutral or star point,

$$E_{NR} + E_{NY} + E_{NB} = 0$$

12. When the starting end of one coil is joined to the finishing end of another coil, the delta or mesh connection is obtained. The direction of emf in the coils are taken as positive from start to finished ends.

13. In balanced delta connected system

(*i*) Line currents are 120° apart.

(*ii*) Line currents are 30° behind the respective phase currents.

(*iii*) Line currents are $\sqrt{3}$ times phase currents.

(*iv*) The angle between line currents and corresponding line voltage is (30° ± φ) as in star system.

(*v*) Line voltages are equal to phase voltages.

(*vi*) Power = $\sqrt{3}\ E_L\ I_L \cos\phi$, where φ is the phase angle between respective phase current and phase voltage.

(*vii*) Apparent power = $\sqrt{3}\ E_L\ I_L$

(*viii*) In balanced system the resultant emf in the closed circuit will be zero.

$$i.e.\ E_{RY} + E_{YB} + E_{BR} = 0$$

Hence there will be no circulating current in the mesh if no load is connected to the lines.

14. Two-wattmeter method is generally used to measure 3-phase power. In this method, the current coils of the two wattmeters are connected in any two lines and potential coil is connected from its own current coil to the line without a current coil. The sum of the two wattmeter readings gives the total power of the circuit.

This method of power measurement is applicable to the 3-phase, 3-wire circuit irrespective of the fact that the circuit is balanced or unbalanced, and star-connected or delta-connected.

If load is balanced, then its power factor can also be calculated from the readings of the two wattmeters, which are given as

$$W_1 = V_L I_L \cos(30° - \phi)$$

$$\text{and } W_2 = V_L I_L \cos(30° + \phi) \text{ in case of inductive loads}$$

$$\text{and power factor } \cos\phi = \text{Cos}\left[\tan^{-1}\frac{\sqrt{3}\,(W_1 - W_2)}{W_1 + W_2}\right]$$

$$W_1 = V_L I_L \cos(30° + \phi)$$

$$\text{and } W_2 = V_L I_L \cos(30° - \phi) \text{ in case of capacitive circuits.}$$

EXERCISES

1. What do you understand by the following terms.

(*i*) Three phase balanced supply.

(*ii*) Three phase balanced load.

(*iii*) Three phase unbalanced load.

(*iv*) Advantages of 3-phase over single phase. **[M.D. Univ. Rohtak Electrical Technology May 2002]**

2. What is a 3-phase system? Give its necessity and advantages. What is meaning of phase sequence and how can it be changed? **[G.B. Technical Univ. Electrical Engineering First Semester 2009-10]**

3. Show that power in 3-phase, balanced system is constant at every instant and is given by 3 $V_p I_p \cos\phi$, where V_p, I_p and φ have usual meanings. **[G.B. Technical Univ. Electrical Engineering First Semester 2011-12; U.P. Technical Univ. Basic Electrical Engineering Second Semester 2013-14]**

4. What is the necessity and advantage of 3-phase system? Derive $V_L = \sqrt{3}\ V_{ph}$ for star-connected system.

[U.P. Technical Univ. Electrical Engineering First Semester 2008-09]

5. Show that in a 3-phase star-connected system, the line voltage is $\sqrt{3}$ times of the phase voltage.
[Electrical Engineering G.B. Technical Univ. First Semester 2011-12; U.P. Technical Univ. Second Semester 2013-14]

6. With the help of phasor diagram or otherwise, explain how the power drawn from 3-phase source to a balanced 3-phase network is constant at all instants. **[Mahamaya Technical Univ. Electrical Engineering Second Semester 2011-12]**

7. Derive the relationship between line current, phase current, line voltage and phase voltage in a 3-phase star-connected and delta-connected circuits. **[U.P. Technical Univ. Electrical Engineering Second Semester 2008-09]**

8. Derive the relationship between the line and phase voltages of an alternator.
[U.P. Technical Univ. Electrical Engineering First Semester 2006-07]

9. For star-connected system in a 3-phase circuit prove that $V_L = \sqrt{3}\ V_{ph}$ and $I_L = I_{ph}$.
[U.P. Technical Univ. Electrical Engineering Second Semester 2006-07]

10. Derive the relationship between line and phase voltages and currents for a delta-connected 3-phase balanced system.
[U.P. Technical Univ. Electrical Engineering First Semester 2004-05]

11. What are the necessity and advantages of 3-phase system? Derive $I_L = \sqrt{3}\ I_{ph}$ for delta-connected system.
[Mahamaya Technical Univ. Electrical Engineering Second Semester 2011-12]

12. Prove that in a 3-phase balanced delta connected system $I_L = \sqrt{3}\ I_{ph}$.
[U.P. Technical Univ. Electrical Engineering Odd Semester 2013-14]

13. Derive the relationship between line and phase current for a delta- and star-connected 3-phase balanced system.
[Mahamaya Technical Univ. Electrical Engineering First Semester 2011-12]

14. Draw and explain the neat circuit diagrams of 2-wattmeter method and 3-wattmeter method for measurement of power and power factor in 3-ϕ balanced circuit.
[U.P. Technical Univ. Electrical Engineering Odd Semester 2013-14]

15. Explain the two-wattmeter method to determine the power in three-phase system.
[G.B. Technical Univ. Electrical Engineering Second Semester 2010-11]

16. Explain two-wattmeter method to determine power in 3-phase system.
[U.P. Technical Univ. Electrical Engineering First Semester 2008-09]

17. Explain the two-wattmeter method for determination of power and power factor of three-phase load with suitable diagram. **[Mahamaya Technical Univ. Electrical Engineering Second Semester 2010-11]**

18. Power measurement by two wattmeters explain the significance of (*i*) equal wattmeters readings (*ii*) zero reading on one wattmeters using suitable phasor diagram.

    Show that the power intake by a three-phase circuit can be measured by two wattmeters connected properly in a circuit. **[G.B. Technical Univ. Electrical Engineering Second Semester 2009-10]**

19. With the help of connection diagram and the related phasor diagram show that $W_1 + W_2$ gives the total 3-phase power in a two-wattmeter method of power measurement.
[G.B. Technical Univ. Electrical Engineering Second Semester 2011-12]

20. Derive the expression for the power delivered and power factor for the balanced 3-phase network. If there is an indication that one of the wattmeter readings is negative, then what corrective action is to be taken to read the power correctly and what is the indication of pf of network at this situation.
[Mahamaya Technical Univ. Electrical Engineering Second Semester 2011-12]

21. Explain two-wattmeter method to measure 3-ϕ power in Y-connected load. Derive the phase angle in terms of wattmeter readings. Draw phasor diagram also.
[U.P. Technical Univ. Electrical Engineering First Semester 2007-08]

22. Derive and explain the two-wattmeter method of measurement of three-phase power for a balanced star-connected load. How is the three phase power determined ? Discuss the variations in readings for different power factors of loads from unity to zero. **[U.P. Technical Univ. Electrical Engineering Second Semester 2007-08]**

23. Explain the methods to measure power in 3-phase circuits.

[U.P. Technical Univ. Electrical Engineering Second Semerter 2006-07]

24. Prove that the power in a three-phase balanced circuit can be deduced from the readings of two wattmeters with the help of suitable vector diagrams. Discuss the nature of power factor when, (*i*) the two wattmeters read equal and positive values, (*ii*) the two wattmeters read equal and opposite values, (*iii*) the one wattmeter reads zero value.

[U.P. Technical Univ. Elec. Engineering February 2001]

SHORT ANSWER TYPE QUESTIONS WITH ANSWERS

Q. 1. What do you mean by three-phase balanced load ?

Ans. Any three-phase load will be balanced when the loads (impedances) connected in three phases are same in magnitude as well as in phase.

Q. 2. What is three-phase unbalanced load ?

Ans. Any three-phase load will be unbalanced if impedances in one or more phases differ from the impedance (s) of the remaining phase (s).

Q. 3. Why is an unbalanced load not normally used on a 3-phase 3-wire system ? Any line/phase voltages equal in such situations ? **[P.T.U. 1999]**

Ans. Normally, unbalanced load is not employed on 3-phase, 3-wire ungrounded star-connected system because unbalanced loading may cause different voltage drops in lines. Consequently the three phase voltages are different (or unbalanced) in magnitude as well as in phase and it is quite possible that one phase voltage may exceed the line voltage. Such a condition is undesirable, since some loads may operate inefficiently due to lowering of the voltage and the other equipment may get damaged due to overvoltage.

Q. 4. What is phase sequence in a 3-phase system ? How is given phase sequence reversed ? **[P.T.U. 1999]**

Ans. Phase sequence is the order or sequence in which the currents or voltages in different phases attain their maximum values one after the other.

The given phase sequence can be reversed by interchanging any two terminals of the supply.

Q. 5. Why interconnection of 3-phase system is necessary ?

Ans. Non-interlinked three-phase system will require 6 wires while interconnected 3-phase system will require 3 or 4 wires. So non-interlinked system will be very complicated and expensive.

Q. 6. Differentiate between star and delta connections. **[P.T.U. June 2003]**

Ans. In star-connected system similar (either start or finish) terminals of the three phases are connected together to provide star or neutral point [refer to Fig. 8.7 (*a*)] while in delta-connected system start terminal of one phase is connected to the finish terminal of the second phase and start terminal of second phase is connected to finish terminal of the third phase and so on to provide a closed circuit in 3-phase system [refer to Fig. 8.14 (*a*)].

Q. 7. Discuss, why the stator windings of alternator are generally star-connected ? **[P.T.U. December 2000]**

Ans. In a star-connected system phase voltage is equal to $1/\sqrt{3}$ times the line voltage, but in a delta-connected system phase voltage is equal to line voltage, therefore, star-connected alternator needs less number of turns per phase and less insulation than a delta-connected alternator for the same line voltage.

Neutral point of a star-connected system can be connected to earth, so relays and protective devices can be provided in star-connected system for protection of the system against ground faults and it is usually done in practice. These are the reasons that 3-phase alternators are usually star-connected.

Q. 8. Why does star-connected alternator need less number of turns per phase as compared to that required by delta-connected one to provide the same line voltage ?

Ans. In a star-connected system phase voltage is equal to $1/\sqrt{3}$ times the line voltage, but in a delta-connected system phase voltage is equal to line voltage, therefore, star-connected alternator needs less number of turns per phase than a delta-connected alternator for the same line voltage.

Q. 9. Why delta-connected system cannot be employed for supplying lighting as well as power loads ?

Ans. Delta-connected system cannot provide 3-phase 4-wire system (i.e. neutral wire cannot be carried) which is essential for supplying lighting as well as power loads.

Q. 10. Draw the circuit diagram for 3-phase power measurement by two wattmeters. Under what operating conditions, one of the wattmeters will show a negative reading ? **[P.T.U. December 2001]**

Ans. The circuit diagrams for 3-phase power measurement by two wattmeters are shown in Figs 8.22 (*a*) and 8.22 (*b*). One of the wattmeters will show a negative reading when the power factor of the load circuit is below 0.5.

Q. 11. Why two-wattmeter method for power measurement is universal one? **[G.B.T.U. First Semester 2011-12]**

Or

Two-wattmeter method for three phase power measurement is universal one. Why? **[M.T.U. First Semester 2011-12]**

Ans. Two-wattmeter method can be used for measurement of power in 3-phase 3-wire circuit irrespective of the fact that the load is star-connected or delta-connected, balanced or unbalanced. As such two-wattmeter method for power measurement is universal one.

Q. 12. While measuring power in a balanced three-phase circuit, one of the wattmeters keeps showing zero as the load is varied. How do you account for it ? **[P.T.U. May 2002]**

Ans. One of the two wattmeters, in a two-wattmeter method of measuring power in a balanced 3-phase circuit, will keep showing zero as the load is varied if and only if the power factor of the load remains constant at 0.5. The reading of other wattmeter will vary with the variation in load.

PROBLEMS

1. Three similar coils each of resistance 20 Ω and an inductance of 0.5 H are connected in star across a three-phase supply source of 400 V, 50 Hz. Calculate the line current and power absorbed by the circuit.
[Pb. Technical Univ. Basic Electrical Engineering, December 2002]
[**Ans.** 1.458 A; 127.6 W]

2. A star-connected load has impedance of $(6 + j\,8)$ Ω in each phase and is connected across a balanced 400 V, 3-phase supply. Obtain the line currents, total real power and reactive power consumed by the load.
[R.G.P.V. Bhopal Basic Electrical Engineering December 2003]
[**Ans.** 23.1 A; 9.6 kW, 12.8 kVAR]

3. A balanced star-connected load of $(8 + j\,6)$ Ω is connected to a 3-phase 230 V supply. Find the line current, power factor, power, reactive volt-amperes and total volt-amperes.
[V.T.U. Belgaum Karnataka Univ., July/August 2003]
[**Ans.** 13.28 A, 0.8 (lagging), 4.232 kW, 3.174 kVAR, 5.29 kVA]

4. A balanced three-phase, star-connected load of 100 kW takes a leading current of 80 A, when connected across a 3-phase 1100 V, 50 Hz supply. Find the circuit constants of the load per phase. **[Nagpur University April 1996]**
[**Ans.** 7.94 Ω; 5.21 Ω; 530.8 μF]

5. A 3-phase 230 V system supplies a total load of 2,000 W at a line current of 6 A when three identical impedances are in star-connection across the line terminals of the system : Determine the resistive and reactive components of each impedance. **[Nagpur University Winter 2002]**
[**Ans.** 18.52 Ω; 12.12 Ω]

6. A 3-phase star-connected load when supplied from 440 V, 50 Hz source takes a line current of 12 A lagging w.r.t. line voltage by 70°. Calculate (*i*) impedance parameters (*ii*) power factor and its nature (*iii*) Draw phasor diagram indicating all voltages and currents. **[Nagpur University Summer 2004]**
[**Ans.** (*i*) 16.217 Ω; 13.6 Ω; (*ii*) 0.766 (lagging)]

7. A balanced Δ–connected load of $(8 + j6)$ ohm per phase is connected to 3-phase 230 V supply. Find the line current.
[U.P. Technical Univ. Electrical Engineering First Semester 2004-05]
[**Ans.** 39.84 A]

8. Define phase sequence in a 3-phase system.
Three 100 Ω resistors are connected in (*a*) star and (*b*) delta across 440 V, 50 Hz, 3-phase lines : Calculate the line currents and phase currents and power taken from supply mains in each case.
[Pb. Technical Univ. Basic Electrical and Electronics Engineering Second Semester 2004-05]
[**Ans.** (*a*) 2.54 A, 2.54 A, 1,936 W (*b*) 4.4 A, 7.62 A, 5808 W]

9. Three similar coils each having a resistance of 8 ohm and an inductance of 0.0191 H in series in each phase is connected across a 400 V, three phase, 50 Hz supply. Calculate the line current, power input, kVA and kVAR taken by the load. **[U.P. Technical Univ. Electrical Engineering First Semester 2006-07]**

[**Ans.** 23.0 A, 12.8 kW, 16, 9.6 (lagging), 69.0 A, 38.4 kW, 48, 28.8 (lagging)]

10. In the two-wattmeter method of power measurement in a three-phase, the readings of the wattmeter are 1,000 W and 550 W. What is the power factor of the load ? **[U.P. Technical Univ. Electrical Engineering September 2001]**

[**Ans.** 0.893 (lagging)]

11. For a certain load, one of the wattmeter reads 20 kW and the other 5 kW after the voltage of this wattmeter has been reversed. Calculate the power and power factor of the load. **[U.P. Technical Univ. Electrical Engineering July 2002]**

[**Ans.** 15 kW, 0.3273 (lagging)]

12. The power readings of two wattmeters are + 15 kW and – 4 kW for a 3-phase balanced load. If the supply voltage is balanced 440 V, find the true power drawn by the load, the power factor and line current. **[R.G.P.V. Bhopal Basic Electrical Engineering June 2005]**

[**Ans.** 11 kW, 0.317 (lagging), 45.53 A]

13. Phase voltage and current of a star-connected inductive load is 150 V and 25 A. Power factor of the load is 0.707 (lag). Assuming that the system is 3-wire and power is measured using two wattmeters , find the readings of wattmeters. **[Nagpur Univ. Elec. Instruments & Measurements 1993]**

[**Ans.** 6,247 W, 1,681 W]

14. Each of the two wattmeters connected to measure the input to a three-phase circuit reads 20 kW. What does each instrument reads, when the load pf is 0.866 lagging with the total three phase power remaining unaltered in the altered condition ? **[V.T.U. Belgaum Karnataka Univ. January/February 2003]**

[**Ans.** 26.667 kW; 13.333 kW]

15. Three identical coils, each having a reactance of 20 Ω and resistance of 20 Ω are connected in (*a*) star (*b*) delta across a 440 V, 3-phase line. Calculate for each method of connection the line current and readings on each of two wattmeters connected to measure the power. **[Allahabad University 1992]**

[**Ans.** (*a*) 8.98 A; 3,817 W; 1,023 W (*b*) 26.94 A, 11,451; 3,069 W]

16. Three inductive coils each having a resistance of 20 Ω and reactance of 15 Ω are connected in star across a 3-phase, 400 V, 50 Hz supply. Calculate the readings of two wattmeters connected in the circuit to measure the power input. **[R.G.P.V. Bhopal Basic Electrical Engineering February 2005]**

[**Ans.** 3.6685 kW, 1.4515 kW]

17. Three equal impedances, each consisting of R and L in series are connected in star and are supplied from a 400 volts, 50 Hz, 3-phase, 3-wire balanced supply system. The power input to the load is measured by 2-wattmeter method and the two wattmeters read 3 kW and 1 kW. Determine the values of R and L connected in each phase. **[U.P. Technical Univ. Electrical Engineering Second Semester 2002-03]**

[**Ans.** 22.856 Ω; 63 mH]

18. Power in a balanced 3-phase system is measured by the two wattmeter method and it is found that the ratio of the two readings is 2 to 1. What is the power factor of the system ? **[Allahabad Univ. Electrical Science-I 1991]**

[**Ans.** 0.866 (lag)]

19. The power input to a 2,000 V, 50 Hz, 3-phase motor running on full load at an efficiency of 90% is measured by two wattmeters which indicate 300 kW and 100 kW respectively. Calculate (*i*) input (*ii*) power factor (*iii*) line current and (*iv*) hp output. [**Ans.** (*i*) 400 kW (*ii*) 0.756 (lag) (*iii*) 152 A (*iv*) 490 (metric)]

20. A 3-phase motor draws a line current of 50 A from 220 V source while starting. The pf is 0.4. Find the readings of the two wattmeters connected to measure power. **[Pb. Univ. Elec. Measurements-I, June 1993]**

[**Ans.** 8.85 kW ; (–) 1.23 kW]

9

CHAPTER

Measuring Instruments

9.1. INTRODUCTION

In electrical engineering, we generally come across current, voltage, power, energy, flux, frequency etc., which are not visible to eyes. The instruments, which are used to measure these quantities, are known as *electrical measuring instruments*. The instruments measuring current, voltage, power, energy and frequency are known as *ammeter*, *voltmeter*, *wattmeter*, *energy meter* (*watthour meter or kWh meter*) and *frequency meter* respectively.

9.2. CLASSIFICATION OF MEASURING INSTRUMENTS

The various electrical measuring instruments may, in a very broad sense, be divided into two classes namely, (*i*) absolute and (*ii*) secondary instruments.

9.2.1. Absolute Instruments. The instruments of this type give the quantity to be measured in terms of instrument constant and its deflection. Such instruments do not require any comparison with any other standard instrument. The example of this type of instrument is *tangent galvanometer,* which gives the value of current to be measured in terms of tangent of the angle of deflection produced, the horizontal component of the earth's magnetic field, the radius and the number of turns of wire used. Rayleigh current balance and absolute electrometer are other examples of absolute instruments.

Such instruments are seldom used except in standard laboratories and in similar institutions as standardizing instruments.

9.2.2. Secondary Instruments. The deflection of such instruments gives the magnitude of electrical quantity to be measured directly. These instruments are required to be calibrated by comparison with either an absolute instrument or with secondary instrument, which has already been calibrated before putting into use. These instruments are generally used in practice.

Absolute instruments are mostly used in standard laboratories and similar institutions and secondary instruments are widely used in practice so we will concentrate only on secondary instruments in the book.

Electrical measuring instruments may also be classified according to the kind of quantity being measured, kind of current for which they are designed, the principle of operation of the moving system (Art. 9.4), their accuracy class, protection against the influence of external fields, service conditions, stability against mechanical effects, field of application, method of installation and mounting, or shape and size of the instrument cases and the degree of enclosure they provide.

According to the type of current that can be measured electrical measuring instruments may be classified as instruments for dc, ac or dc and ac measurements.

9.3. TYPES OF SECONDARY INSTRUMENTS

Secondary instruments are further classified into three groups namely (*i*) indicating instruments (*ii*) recording instruments and (*iii*) integrating instruments.

9.3.1. Indicating instruments are those which indicate the magnitude of an electrical quantity at the time when it is being measured. The indications are given by a pointer moving over a graduated dial. Ordinary ammeters, voltmeters, wattmeters, frequency meters, power factor metres etc., fall into this class.

9.3.2. Recording instruments are those which keep a continuous record of the variations of the magnitude of an electrical quantity to be observed over a definite period of time. In such instruments the moving system carries an inked pen which touches lightly a sheet of paper wrapped over a drum moving with uniform slow motion in a direction perpendicular to that of the deflection of the pointer. Thus a curve is traced which shows the variations in the magnitude of the electrical quantity under observation over a definite period of time. Such instruments are generally used in powerhouses where the current, voltage and power etc., are to be maintained within certain specified values.

9.3.3. Integrating instruments are those which measure the total amount of either quantity of electricity (ampere-hours) or electrical energy supplied over a period of time. The summation, given by such an instrument is the product of time and an electrical quantity under measurement. The ampere-hour meters and energy meters fall into this class.

9.4. PRINCIPLE OF OPERATION OF ELECTRICAL MEASURING INSTRUMENTS

The principle of working of all electrical measuring instruments depend upon the various effects of electric current or voltage. The effects utilised in the manufacturing of electrical instruments are magnetic, heating, chemical, electrostatic and electromagnetic induction effects.

The classification of the instruments according to the effects utilized in their operation is given below.

Effects	*Instrument utilizing the effect*	*Suitability for type of measurements*
Magnetic effect	Ammeters, voltmeters, wattmeters integrating metres and other electrical instruments.	Current, voltage, power and energy on both ac and dc systems
Thermal effect	Ammeters and voltmeters.	Current and voltage for both dc and ac systems.
Chemical effect	Integrating meters.	For measurement of dc ampere-hours.
Electrostatic effect	Voltmeters only.	Voltage only, on both ac and dc systems.
Electromagnetic induction effect	Voltmeters, ammeters, wattmeters and energy meters.	For measurement of voltage, current, power and energy in ac system only.

9.5. ESSENTIALS OF INDICATING INSTRUMENTS

Indicating instruments consist, essentially of a pointer moving over a calibrated scale and attached to the moving system pivoted on jewelled bearings. For satisfactory working of indicating instruments the torques required are:

1. Deflecting torque 2. Controlling torque and 3. Damping torque.

1. *Deflecting Torque.* The deflecting torque is produced by making use of one of the magnetic, heating, chemical, electrostatic and electromagnetic induction effects of current or voltage and causes the moving system of the instrument to move from its zero position when the instrument is connected in an electrical circuit to measure the electrical quantity. The method of producing this torque depends upon the type of the instrument and will be discussed individually in succeeding Arts.

2. *Controlling Torque.* The magnitude of the movement of the moving system would be somewhat indefinite under the influence of deflecting torque unless some controlling torque existed. This torque opposes the deflecting torque and increases with the increase in deflection of the moving system, thus limits the movement and ensures that the magnitude of the deflection is always the same for a given value of quantity to be measured. Under the influence of the controlling torque the pointer (or moving system) will return to its zero position on removing the source producing the deflecting torque. Without controlling system, the pointer would swing over its maximum deflected position irrespective of magnitude of current and moreover, once deflected it would not return to its zero position on removing the source producing the deflecting torque. The controlling torque in indicating instruments is created either by a spring or by gravity as given below:

(*a*) *Spring Control.* The spring control is illustrated in Fig. 9.1. The phosphor bronze spiral hair springs A and B coiled in opposite directions and acting one against the other are used in spring control. One end of each spring is attached to the spindle. The outer end of spring A is attached to a lever at point C pivoted at P while that of B is fixed, as shown in Fig. 9.1 (*b*). Under the influence of deflecting torque when the pointer moves, one of the springs unwinds itself while the other gets twisted. The twist, in fact produces controlling torque which is directly proportional to the angle of the deflection of the moving system. When deflecting torque T_d and controlling torque T_c are equal, the pointer comes to rest in its final deflected position.

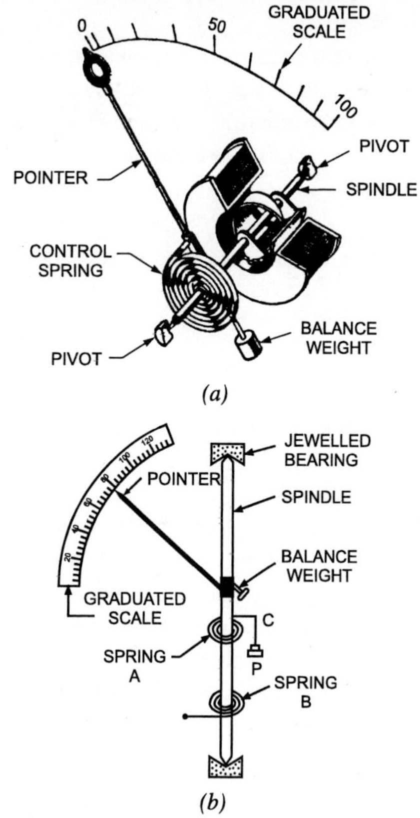

Fig. 9.1 *Spring Control*

Since controlling torque T_c is directly proportional to the angle of twist θ and deflecting torque T_d is proportional to the current flowing through it, as in case of a permanent magnet moving coil instruments, and in final deflected position of the moving system $T_c = T_d$, deflection $\theta \propto I$, the current flowing through the instrument. Since in spring controlled instruments the deflection is directly proportional to the current, these instruments have uniform scales.

(*b*) *Gravity Control.* Temperature coefficient of stiffness of the spring results in a temperature coefficient in the indication of the instrument; and inelastic yield in the spring results in displacement in the zero position of the moving system. So it is advantageous to substitute gravity control for spring control in electrical measuring instruments. Gravity control is free from the effects mentioned above.

In gravity-controlled instruments, a small weight is attached to the moving system in such a way that it produces a controlling torque, when the moving system is in deflected position. The controlling torque can be varied quite

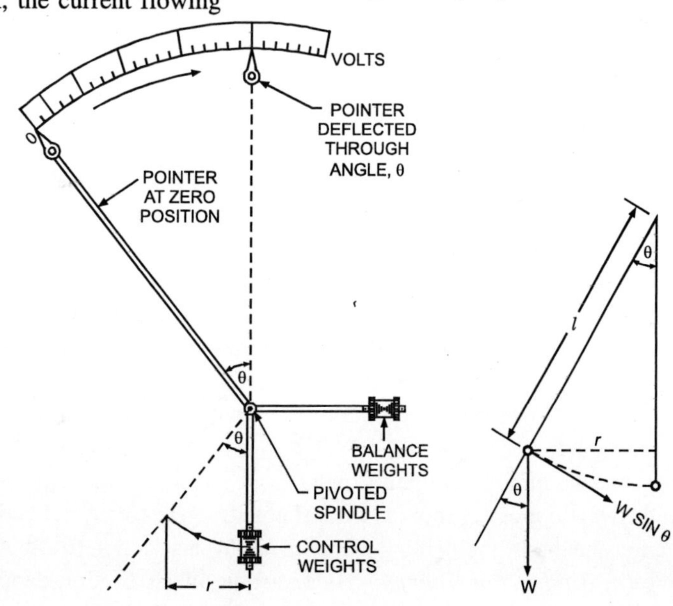

Fig. 9.2 *Gravity Control*

easily by adjusting the position of the controlling weight upon the arm. The usual arrangement is shown in Fig. 9.2.

In zero position of the pointer the control weight is vertical. When the pointer is deflected through an angle θ from its zero position, the control weight will be in a position shown dotted in Fig. 9.2. In deflected position, the controlling torque T_c will be $Wl \sin\theta$ where W is the control weight, l is distance from the axis of rotation and θ is deflection.

$$\text{or } T_c \propto \sin\theta$$

$$\text{If } T_d \propto I$$

Then at final deflected position $T_d = T_c$

$$\text{or } I \propto \sin\theta$$

$$\text{or } I = k\sin\theta \text{ (say), then}$$

$$\theta = \text{Sin}^{-1}\frac{I}{k} \qquad ...(9.1)$$

Hence, in gravity-controlled instruments scales are not uniform but are crowded in the beginning. This is a disadvantage when the pointer lies at the lower scale values, when it will not be possible to read the instrument scale accurately on account of it being cramped.

Gravity-controlled instruments must be used in vertical position in order that the control may operate.

Gravity control is cheap, unaffected by change in temperature and is free from fatigue or deterioration with time but it gives a cramped scale and the instrument has to be kept in a vertical position.

Such instruments usually have a bubble level mounted in the case to indicate the reference plane in which they have been calibrated. By returning the instrument to the reference plane the control forces resulting from any residual unbalance of the moving system act in precisely the same way that they did at the time of calibration.

3. *Damping Torque.* Damping force or torque is also necessary to avoid oscillations of the moving system about its final deflected position owing to the inertia of the moving parts and to bring the moving system to rest in its final deflected position quickly.

In the absence of damping, the moving system of an instrument would oscillate about the position at which the deflecting and restoring torques are equal. The function of damping is to absorb energy from the oscillating system and to bring it to rest promptly in its equilibrium position so that its indication may be observed.

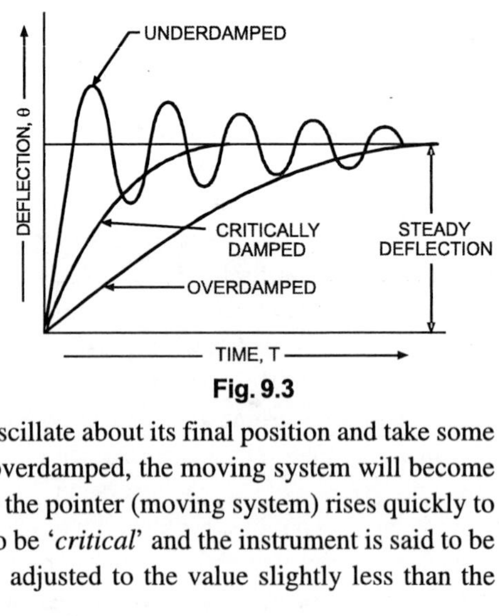

Fig. 9.3

If the instrument is underdamped the moving system will oscillate about its final position and take some time to come to rest in its steady position. If the instrument is overdamped, the moving system will become slow and lethargic and when the degree of damping is such that the pointer (moving system) rises quickly to its deflected position without oscillations; the damping is said to be '*critical*' and the instrument is said to be "*dead beat*". In practice to obtain best results the damping is adjusted to the value slightly less than the critical value.

The damping torque must operate only while the moving system of the instrument is actually moving and always oppose its motion. It must not affect the steady deflection produced by the deflecting torque. The various methods of obtaining damping are (*a*) air friction (*b*) fluid friction and (*c*) eddy current.

(*a*) *Air Friction Damping.* In first type of air friction damping system, a light aluminium piston is attached to the moving system, as shown in Fig. 9.4 (*a*). The piston moves with a very small clearance in fixed air chamber closed at one end. Cross section of the chamber may be either circular or rectangular. If the piston is moving rapidly into the chamber, the air inside the chamber is compressed and thus the pressure inside the chamber opposes the motion of the piston *i.e.* of the moving system to which piston is attached. If the piston is moving outside, pressure inside the chamber falls and, therefore, the outside pressure becomes greater than that inside the chamber and thus the motion of the piston or of the moving system is again opposed. In this system care must be taken to ensure that the arm carrying piston is not bent and the piston does not touch the sides of chamber during its movement. When once bent it is often difficult to straighten the piston arm so that it does not touch the sides of the chamber at any point during deflection. The solid friction occurring between piston and sides of air chamber may result in serious error in the deflection.

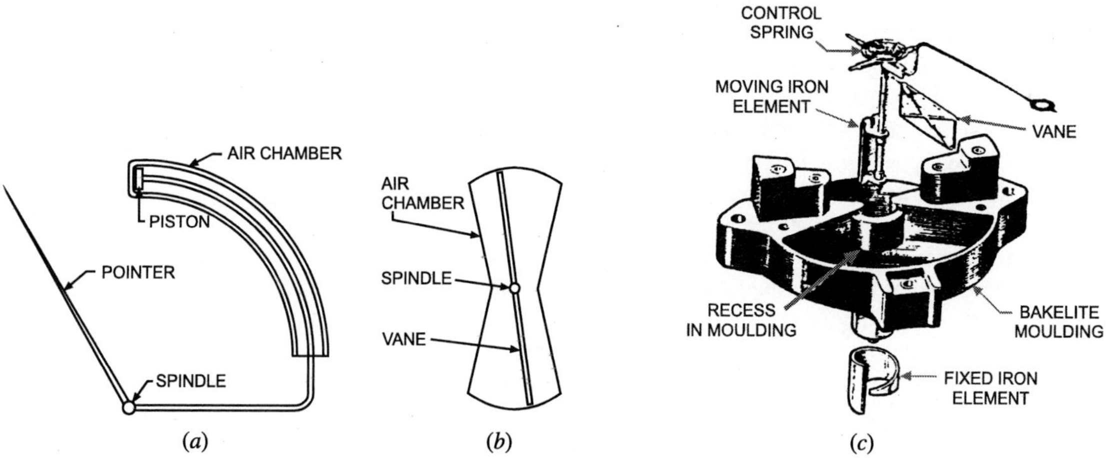

Fig. 9.4 *Air Friction Damping*

In second type of air friction damping system, a vane mounted on the spindle of the moving system is utilized. The vane is of thin aluminium sheet and moves in a closed sector shaped box, as shown in Fig. 9.4 (*b*).

The modern form of air-friction damping brake is shown in the exploded view of Fig. 9.4 (*c*). An aluminium vane is enclosed in the quadrant shaped recess cast in a bakelite moulding, or die casting. The air brake chamber is completed by a cover plate, not shown in the figure.

Air friction damping provides a very simple and cheap method and has the advantage, that it does not need the use of permanent magnet whose introduction may lead to distortion of operating field. Therefore, this type of damping is used in moving iron and dynamometer type instruments where the working magnetic field is weak and is likely to get distorted with the introduction of permanent magnet.

(*b*) *Fluid Friction Damping.* In this method of damping light vanes or disc are attached to the spindle of the moving system and they move in a damping oil, as shown in Figs. 9.5 (*a*) and (*b*). The damping oil employed must be good insulator, non-evaporating, non-corrosive upon the metal of disc or vane and of viscosity not subject to change with the change of temperature.

There are two methods of fluid friction damping as shown in Fig. 9.5.

In first method, light vanes are attached to the spindle of the moving system, as shown in Fig. 9.5 (*a*). The vanes are dipped into a pot of damping oil and are completely submerged by the oil. The motion of the moving system is always opposed by the friction of damping oil on the vanes. The damping force thus created always increases with the increase in velocity of vanes. There is no damping force when the vanes are stationary.

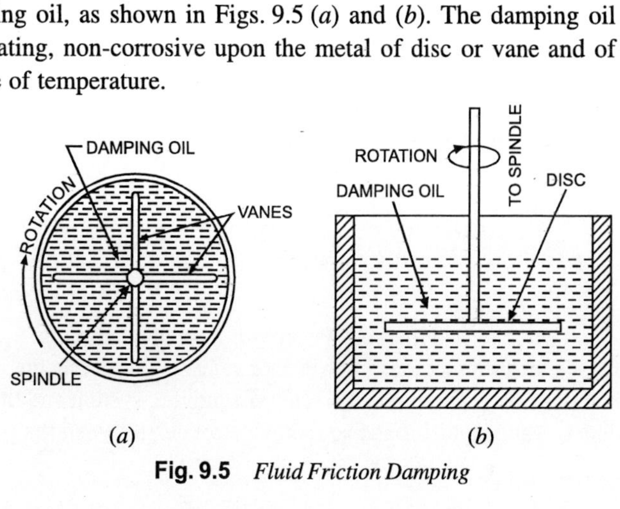

Fig. 9.5 *Fluid Friction Damping*

In the second method of fluid friction damping, a disc attached to the spindle of moving system is immersed in damping oil, as shown in Fig. 9.5 (*b*). The frictional drag developed during the motion of the disc always opposes the motion of the moving system and becomes zero when the disc is stationary. The suspension system of the disc should be cylindrical and of small diameter so that the surface tension effect is negligible. The former method gives increased damping.

Fluid friction damping has the advantage that the oil required for damping can be used for insulation purposes in some types of instruments which are submerged in oil. A vane moving in oil instead of air does not require the same small clearances to provide effective damping and, therefore, this method is quite suitable for instruments having movements suspended rather than pivoted such as electrostatic type instruments. It has another advantage that due to the upthrust of oil, the load on bearings or suspension is reduced thus reducing frictional errors.

It has several drawbacks such as creeping of oil, the necessity of using the instruments always in vertical position. Hence the use of this type of damping is restricted to laboratory type electrostatic instruments.

(*c*) *Eddy Current Damping.* In electrical measuring instruments, the use of eddy currents has been made as an electromagnetic braker or damper. The eddy current damping is the most efficient and is based on the principle that whenever a sheet of conducting but non-magnetic material like copper or aluminium moves in a magnetic field so as to cut through lines of force, eddy currents are set up in the sheet. Due to these eddy currents, a force opposing the motion of the sheet is experienced between them and the magnetic field. This force is proportional to eddy currents and the strength of the magnetic field. The eddy currents are proportional to the velocity of the moving system. Hence if the strength of the magnetic field is constant, the damping force is proportional to the velocity of the moving system and is zero when the moving system is at rest.

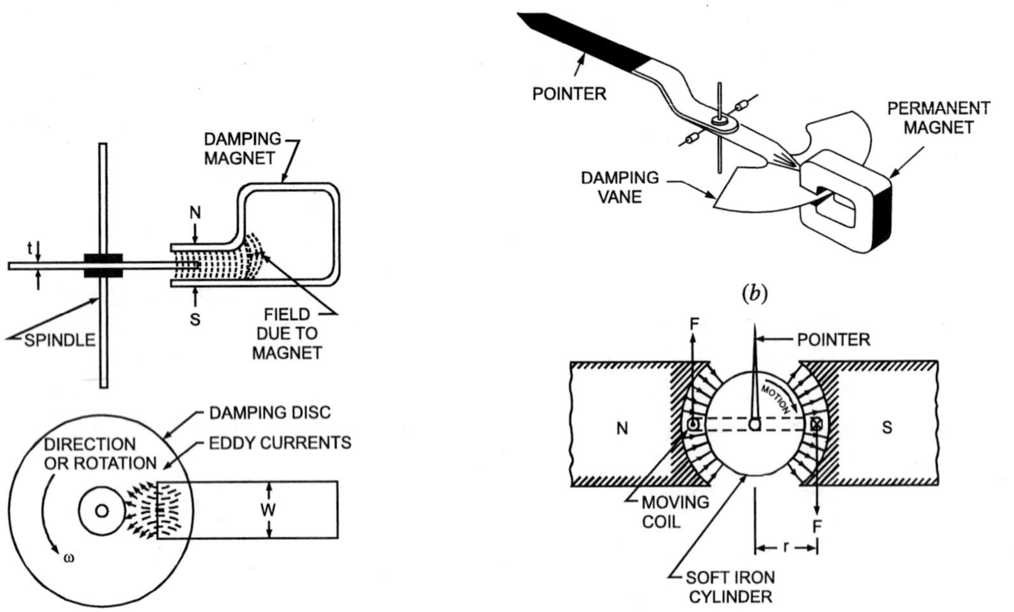

(*a*) (*c*) *Eddy Current Damping Torque With a Metal Former*

Fig. 9.6 *Eddy Current Damping*

There are two methods of providing such a damping.

In the first method, a thin disc or vane of conducting but non-magnetic material like copper or aluminium is mounted on the spindle carrying the moving system and the pointer, as shown in Figs. 9.6 (*a*) and (*b*). The disc or vane is positioned so that edge rotates between the poles of a permanent magnet.

The magnitude of damping torque may be varied by adjusting the radial position of the poles. Damping torque decreases with the movement of the magnet towards the edge of the disc and becomes zero when the centres of the poles are at the edge of the disc. The above phenomenon is utilized in the induction type instruments by placing a damping magnet (permanent type) at the opposite side of the disc from the electro-magnet, so that the disc serves for damping as well as for operating purposes.

In the second method of providing eddy current damping, as used in permanent magnet moving coil instruments, the coil is wound on a light metal former in which eddy currents are produced due to movement of the coil in the field of a permanent magnet, as shown in Fig. 9.6 (*c*). Damping can be varied by varying the thickness of the metal former.

Eddy current damping is most efficient form of damping. It can be very conveniently used in instruments where a metallic disc or former and a permanent magnet already form part of the operating system. That is why, this method is used in hot-wire, moving coil and induction type instruments. This method cannot be employed in the instruments in which a permanent magnet required for causing eddy currents will distort the existing magnetic field as in moving iron or dynamometer type instruments.

9.6. AMMETERS AND VOLTMETERS

Ammeters and voltmeters, except electrostatic voltmeters, operate on the same principle. The ammeter carries the current to be measured or a definite fraction of it and this current or its definite fraction produces the deflecting torque whereas voltmeter carries the current proportional to the voltage to be measured which produces the deflecting torque.

Ammeter, which is connected in series with the circuit carrying the current under measurement, must be of *very low resistance* so that the voltage drop across the ammeter and power absorbed from the circuit are as low as possible.

Voltmeter, which is connected in parallel with the circuit across which the voltage is to be measured, must be of very *high resistance* so that the current flowing through the voltmeter and the power absorbed from the circuit are minimum possible.

Apparently it is, therefore, impracticable in most of the cases, and dangerous in some, for an instrument manufactured as a voltmeter to be employed as an ammeter, or alternatively to use an ammeter as a voltmeter. The low resistance winding of an ammeter will suffer serious damage if connected directly across a high voltage source (*i.e.* more than a fraction of volt) and the high resistance winding of the voltmeter will prevent all but the smallest current flowing and, therefore, would normally be quite unsuitable to be used as an ammeter, except of course in the special case where it is required to read only very small currents.

However, an ammeter of low range may be used as a voltmeter by connecting a high resistance in series with it provided the current through the series combination is within the range of the ammeter when connected across the voltage to be measured.

9.7. TYPES OF INSTRUMENTS

Some of the instruments are used for dc measurements only, some are used for ac measurements only and some are used for dc as well as ac measurements.

1. *DC Instruments.* The instruments, whose deflections are proportional to the current or voltage under measurement are used for dc measurements only. If such an instrument is connected in an ac circuit, the pointer will deflect up-scale for one half cycle of the input waveform and down-scale for the next half cycle. At lower frequencies of 50 Hz, the pointer will not be able to follow the variations in direction and will quiver slightly around the zero mark, seeking the average value of ac *i.e.*, zero. The permanent magnet moving coil (PMMC) instruments fall under the category of dc instruments.

2. *AC Instruments.* The instruments utilising the electromagnetic induced currents for their operation are used for ac measurements only. These instruments cannot be used for dc measurements because the electromagnetic induced currents are not generally available in dc circuits. The induction instruments fall under the category of ac instruments.

3. *DC/AC Instruments.* The instruments having deflection proportional to the square of the current or voltage under measurement can be used for dc as well as ac measurements. Moving iron (attraction as well as repulsion types), dynamometer type moving coil, hot-wire, electrostatic instruments fall under this category.

The various types of indicating instruments used for measurement of current, voltage, power, and energy along with suitability for the type of measurements, type of control, type of damping, special features and applications are given below in tabular form.

S.No.	*Type of Instrument*	*Suitablity For Type of Measurements*	*Type of Control*	*Type of Damping*	*Remarks*
1.	Moving iron	Direct or alternating current measurements (current and voltage)	Spring or gravity control	Air friction	It is cheaper to manufacture and mostly used as an indicating instrument. It is very accurate on ac as well as dc, if properly designed.
2.	Moving coil				
	(*i*) Permanent magnet type	DC measurements (Current and voltage) only	Spring	Eddy current	It is most accurate type for dc measurements and most widely used for measurements of dc voltage, current and resistance.
	(*ii*) Dynamometer type	DC or AC measurements (current, voltage and power)	Do	Air friction	The instruments of this type are mainly used as wattmeter. Also used as standard meter for calibration and as transfer instruments.
3.	Thermal				
	(*i*) Hot wire	DC or AC measurements (current, voltage and power)	Spring	Eddy current	These instruments have same calibration for both dc and ac. These are free from errors due to frequency, waveform and external fields when used on ac, therefore, these are particularly used for ac measurements. Since its disadvantages outweigh its advantages hence it is now little used.
	(*ii*) Thermocouple	DC or AC measurements (current and voltage)	Do	Do	These instruments are free from errors due to frequency, waveform and external fields when used on ac and are used for measurement of current and voltage at power frequencies up to 100 MHz.
4.	Electrostatic	DC or AC (voltage only)	Gravity or spring	Air friction	These instruments have very low power consumption and can be made to cover a wide range of voltage. Usually range is above 500 volts.
5.	Induction	AC measurements (current, voltage, power and energy)	Spring	Eddy current	Ammeters and voltmeters of this type are expensive and not of high degree of accuracy. These instruments are mainly used for measurements of power and energy in ac circuits.
6.	Rectifier	DC or AC (current and voltage)	Do	Do	These instruments are nothing but permanent magnet moving coil instruments used in conjunction with rectifying devices for ac measurements (current and voltage) from about 20 Hz to 20 kHz.

9.8. ERRORS COMMON TO ALL TYPES OF INSTRUMENTS

The errors common to various types of instruments are on account of friction, change in temperature, mechanical unbalance and variation of the electrical and mechanical properties of the materials employed with time.

1. *Temperature Error.* This type of error, apart from the change in room temperature, is mainly due to heating of the working coils and other resistance coils connected internally in the instrument casing by the operating current. As a result there is a change in resistance of the working coils and other resistance coils. Such an error is not serious in case of an ammeter but it is very important in case of a voltmeter which is of high resistance and where constancy of resistance is very important. For reducing such an error the power lost in a voltmeter should be kept very small and arrangement for ventilation and cooling should be provided. Sometimes the error is minimized by winding the series resistance coil of material of very small temperature coefficient of resistance and having the working coil, which is wound with copper wire, of resistance very small in comparison with that of series coil. Consequently the overall percentage change of the two together connected in series will be much reduced below that of the working coil alone. Series resistance and shunts housed internally in instrument cases should be well separated from working elements and well ventilated.

Other errors resulting from heating may be caused by expansion of the control spring or of other parts of the instruments but such errors are usually negligibly small.

2. *Friction Error.* The friction of the pivot in the jewel produces a frictional torque which opposes the deflection of the moving element and as a result the pointer comes to rest a short distance from its correct position and so there is an error in the instrument readings. This error is small for a bearing in good condition, but may become very large in case of badly worn or damaged pivot. Friction error is more serious for sensitive instruments designed for low operating torque. Such errors may be minimized by adopting a moving system of a light construction and large deflecting torque *i.e.*, large torque-weight ratio and mounting the moving system on a vertical, rather than a horizontal, spindle for preference, between jewelled bearings. The user can minimize frictional effects by tapping gently on the case with his finger tip before taking an observation. Vigorous tapping will defeat the purpose and may even injure the bearing.

3. *Observational Error.* Such errors are due to misreading of the scale, parallax in readings and errors of estimation. Precision of readings becomes difficult if the pointer has a broad tip. Interpolation errors may enter, particularly if the scale readings are too large or too small.

Parallax error is caused by the observer not having his line of sight on the pointer exactly at right angle to the plane of the scale. Such an error is eliminated by providing a mirror beneath the scale and a knife-edged pointer.

Lack of balance in the moving system and variations in the strength of permanent magnets (if used) are other possible sources of error which are common to several types of instruments.

9.9. MOVING IRON INSTRUMENTS

These instruments are widely used in laboratories and switchboards at commercial frequencies because these are cheaper in cost, robust in construction and can be manufactured with required accuracy.

There are two general types of such instruments *i.e.* (*i*) the *attraction type* and (*ii*) the *repulsion type*. The attraction type instruments operate on the principle of attraction of a single piece of soft iron into a magnetic field and repulsion type instruments operate on the principle of repulsion of two adjacent iron pieces magnetised by the same magnetic field. Repulsion type instruments are more sensitive as in these instruments large operating torque is developed by having two iron elements positioned close together inside the field coil where the magnetizing effect is maximum. In both types of these instruments the current under measurement (or a definite fraction of it or proportional to the voltage under measurement), is passed through a coil of wire. This current carrying coil sets up the necessary field. Depending on the magnitude of the current to be measured, the coil may be of a few turns of very heavy conductor or of many turns of fine wire. The instrument to be used as an ammeter is provided with a coil of few turns of thick wire in order to have low resistance and carry large current and that to be used as a voltmeter is provided with a coil of large number of turns of fine wire in order to have high resistance and draw as small current as possible.

1. *Attraction Type Moving Iron Instruments.* The earliest and simplest form of attraction moving iron instrument uses a solenoid and moving oval shaped soft iron pivoted eccentrically as shown in Fig. 9.7. To this iron a pointer is attached so that it may deflect along with the moving iron over a graduated scale. The iron is made of sheet metal specially shaped to give a scale as nearly uniform as possible. The moving iron is drawn into the field of solenoid when current flows through it. The movement of the iron is always from weaker magnetic field outside the coil into the stronger magnetic field inside the coil regardless the direction of flow of current.

When the current to be measured (or a definite fraction of the current to be measured or proportional to the voltage to be measured) is passed through the solenoid, a magnetic field is set up inside the solenoid, which in turn magnetises the iron. Thus the iron is attracted into the coil causing the spindle and the pointer to rotate. Such instruments normally have spring control and pneumatic damping. In Fig. 9.7 gravity control attraction type instrument, which have been now superceeded by spring-controlled instruments owing to their merits over gravity-controlled instruments, is shown.

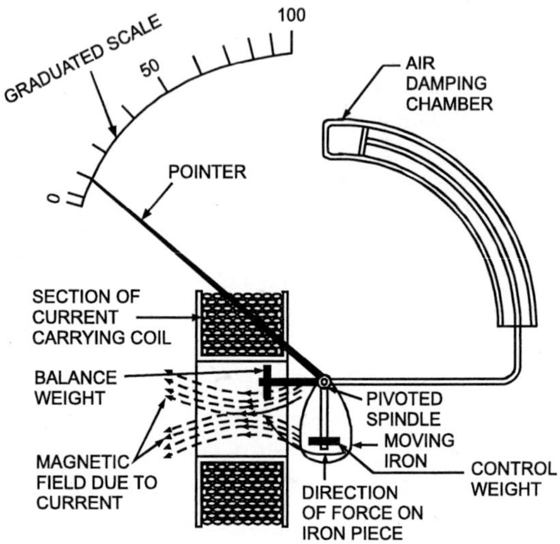

Fig. 9.7 *Moving Iron Attraction Type Instrument*

Such an instrument has a scale cramped at the lower end and greatly expanded at the upper end, as the operating torque is quite low when the moving iron is just entering the solenoid and increases rapidly as iron is drawn further in. However, it does permit the development of large working torque and is still used in some instruments requiring high operating torque.

2. ***Repulsion Type Moving Iron Instruments.*** Repulsion type moving iron instrument, in its simplest form, is shown in Fig. 9.8. In this type of instrument there are two irons. A curved 'iron' of soft iron or of mumetal is fixed to the inside of the bobbin former and another curved 'iron' is mounted on a spindle which passes axially through the solenoid. The two irons lie in the magnetic field produced by a solenoid. When there is no current in the solenoid the two irons, moving one and fixed one, are almost touching each other and the pointer rests on zero position. When the current to be measured (or a definite fraction of it or proportional to the voltage to be measured) is passed through the solenoid, which is wound with insulated copper wire on a cylindrical non-magnetic former (a few turns of thick wire in case of an ammeter and a large number of turns of fine wire in case of a voltmeter), a magnetic field is set up inside the solenoid and the two irons are magnetized in the same direction. This sets up a repulsive force so moving iron is repelled by fixed iron, thereby results in the motion of the moving iron carrying the pointer. The pointer comes to rest in a deflected position when equilibrium is attained between the repulsion force of the working elements and the controlling force. Control of the movement is either by hair springs, when the instrument is to be employed in any position, or by gravity, when the instrument is to be operated only vertically. Damping used is air friction (pneumatic) one. Eddy current damping is not possible because of presence of permanent magnet needed for such purpose would affect the deflection i.e. reading of the instrument. The repulsion of irons is proportional to the square of current, and so the scale is uneven, being crowded at the lower values and wide spread at the higher values. The iron may be so shaped and arranged that the scale is spread out to some extent at the lower values of current.

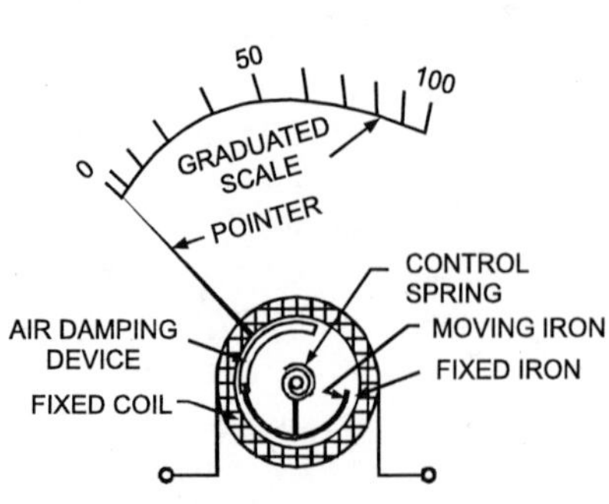

Fig. 9.8 *Repulsion Type Moving Iron Instrument*

Deflecting Torque. Let L be the self-inductance corresponding to a total angular deflection of θ radians and change in inductance be dL corresponding to small change in deflection angle $d\,\theta$ due to small change in current.

The change in the energy of the magnetic field, $d\,\mathrm{E} = \frac{1}{2}\mathrm{I}^2\,d\,\mathrm{L}$

Also if T_d is the deflecting torque, the work done during the movement of the iron,

$$dw = T_d \, d\theta$$

Since change in energy, dE = Work done, dw,

$$T_d \, d\theta = \frac{1}{2} I^2 \, dL$$

$$\text{or } T_d = \frac{1}{2} I^2 \frac{dL}{d\theta} \text{ N-m if I is in amperes, L in henrys and } \theta \text{ in radians.}$$

Thus the torque is proportional to the square of the instrument current and to the rate of change of inductance with deflection.

An attraction type instrument will usually have a lower inductance than the corresponding repulsion type instrument, and voltmeters will, therefore, be accurate over a wider range of frequency and there is a greater possibility of using shunts with ammeters. On the other hand, repulsion type instruments are more suitable for economical production in manufacture and a nearly uniform scale is more easily obtained; they are, therefore, much more common than the attraction type.

Ranges.

Ammeters — from about 20 mA to 750 A maximum without current transformer.

Voltmeters — from about 0-1 V to 750 V maximum without potential transformer.

Errors in Moving Iron Instruments. The errors usually occurring in moving iron instruments are (*i*) friction (*ii*) temperature (*iii*) stray magnetic field (*iv*) hysteresis (*v*) frequency and (*vi*) waveform. The first four errors affect the instruments on ac and dc while last two affect only on ac. Friction and temperature errors have already been dealt earlier. *Stray magnetic field errors* may be serious if not guarded against because of the weakness of the operating magnetic field in case of moving iron instruments. Such errors are minimized by shielding the instrument with the help of steel case or sheet steel lining to the case. *Hysteresis error* is serious type of error in moving iron instruments specially when used on dc circuits. This error may be reduced by making the iron parts short so that they demagnetize themselves or by choosing low value of flux density in the iron or by using mumetal or permalloy having negligible hysteresis. Change in frequency may cause errors due to change of reactance of the operating coil and also to the changes in magnitude of eddy currents set up in the metal parts of the instrument near the operating portion. The change in instrument coil reactance owing to change in frequency causes serious error in case of voltmeters (not in case of ammeters). The eddy current error affects both ammeters and voltmeters equally. The frequency error may be compensated for by connecting a suitable capacitor C in parallel with the swamping resistance R (C being equal to L/R^2 if frequency is not too high).

Moving iron instruments may be seriously affected by waveform both on account of the change in the form of flux waveform and in case of voltmeters, the effect of harmonics upon the inductance.

Merits and Demerits of Moving Iron Instruments.

Merits. (*i*) These instruments can be used both in dc as well as in ac circuits. This is due to the fact that the deflecting torque depends on the square of the current.

(*ii*) These instruments are robust, owing to the simple construction of the moving parts and the fact that there are no current leads to these parts.

(*iii*) The stationary parts of the instruments are also simple.

(*iv*) The simplicity of construction, as described in (*ii*) and (*iii*) results in low cost.

(*v*) These instruments possess high operating torque.

(*vi*) These instruments can withstand overload momentarily.

(*vii*) These instruments are capable of giving an accuracy within the limits of both precision and industrial grades. Instruments which are required to have industrial grade accuracy on both ac and dc require very considerable care in design.

The moving iron instruments, owing to the above advantages, have become most widely used instruments in ac circuits, in addition, they are largely used on dc circuits in cases where the superior accuracy of moving coil instruments is not an essential requirement.

Demerits. (*i*) Scales of these instruments are not uniform.

(*ii*) Power consumption is higher for low voltage range.

(*iii*) The stiffness of the spring decreases with the increase in temperature.

(*iv*) The errors are caused due to hysteresis in the iron of the operating system and due to stray magnetic field.

(*v*) Change in frequency and in waveform causes serious errors in ac measurements.

(*vi*) There is a difference between dc and ac calibrations because of inductance effect of the instruments and eddy currents on ac. The instruments should be calibrated for frequencies at which these are to be used.

9.10. MOVING COIL INSTRUMENTS

There are two types of moving coil instruments namely the *permanent magnet type* and *dynamometer type.*

Permanent magnet moving coil (PMMC) *instrument* is the most accurate and useful for dc measurements. In fact, were it not for the need of other instruments for ac measurements, it may be doubted whether any other type of indicating instruments would have survived except where cheapness is main requirement, in which case the simple moving iron instrument holds its place. *Dynamometer type moving coil instruments* can be used both on dc as well on ac. Their calibration is the same for both dc as well as ac and so these instruments are very useful as transfer instruments.

9.10.1. Permanent Magnet Moving Coil (PMMC) Instruments. A permanent magnet type moving coil instrument is shown in Fig. 9.9. It consists of permanent powerful magnet with soft iron pole pieces. A cylindrical iron core is mounted between the two poles of the magnet giving very narrow air gap in which the sides of a pivoted light rectangular coil lies. The rectangular coil is wound of many turns from fine wire on light aluminium or copper former and acts as moving element. The purpose of using core is to make the field uniform and to reduce the reluctance of the magnetic circuit. A low reluctance helps to retain permanence of the magnet for a longer period.

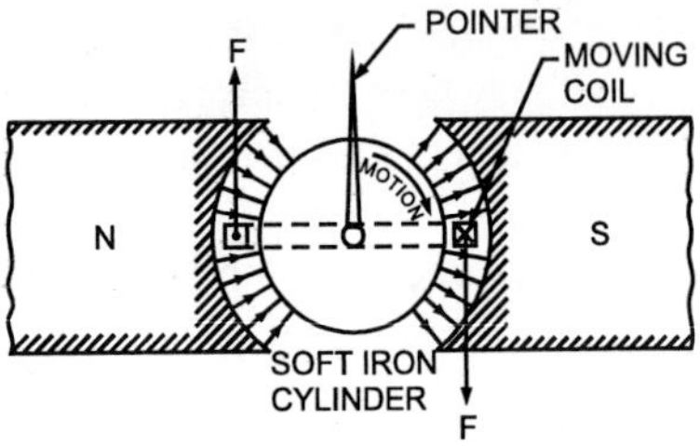

(*a*)

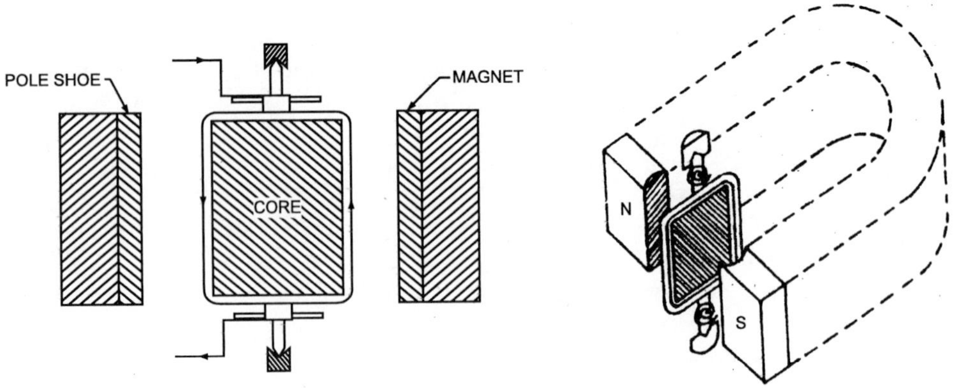

(*b*) (*c*)

Fig. 9.9 *Moving Coil (Permanent Magnet Type) Instrument*

The current is led into and out of the coil by means of phosphor bronze hair springs provided at both ends. The springs also provide the controlling torque. The springs are spiralled in opposite directions to neutralise the effect of change in temperature.

Torque Equation. When the current to be measured (or a definite fraction of it or proportional to the voltage to be measured) is passed through the coil, say in the direction shown in Fig. 9.9, a deflecting torque is produced on account of reaction of the permanent magnetic field with the coil magnetic field. The direction of deflecting torque can be determined by applying Fleming's left hand rule.

If i is the current in amperes flowing through the coil of turns N and length l *metres and* B *is the flux density in teslas in the air gap*

Then deflecting force, $F = BilN$ newtons

If r is the distance in metres between the centre of the coil and the force F

Then deflecting torque, $T_d = F \times r = BilNr$ N-m

From the above expression it is obvious that if flux density B in the air gap is constant, then

Deflecting torque, $T_d \; \alpha \; i$

Since such instruments are spring controlled,

Controlling torque, $T_c \; \alpha$ deflection θ

Since in steady deflection position, $T_c = T_d$

$\therefore \theta \; \alpha \; i$

Since θ is directly proportional to current, the scale markings of the basic dc PMMC instrument are usually linearly spaced. If low frequency ac is applied to the moving coil, the pointer will deflect up-scale for one half cycle of the input waveform and down-scale for the next half cycle. At power frequencies of 50 Hz and above, the pointer will not be able to follow the rapid variations in direction and will quiver slightly around the zero mark, seeking the average value of ac i.e. zero. That is why, PMMC instrument is unsuitable for ac measurements, unless the current is rectified before applying it to the moving coil.

Ranges. (*i*) *DC Ammeters* : 0 – 5 μA and up to 20 mA with internal shunts and 0 – 5,000 A with external shunts.

(*ii*) *DC Voltmeters* : 0 – 100 mV without series resistance and up to 20 kV or 30 kV with external series resistance.

Voltmeter Sensitivity and Loading Effect. We know that PMMC movement requires series resistors of different ranges. For example a movement requiring 5 mA for full-scale deflection can be converted into a voltmeter of range 250 V, 25 V and 2.5 V by connecting series resistances of different values so as to make total resistance of 50,000 Ω, 5,000 Ω and 500 Ω respectively. In either case if we divide the number of ohms by the scale range *i.e.* 50,000/250 or 5,000/25 or 500/2.5 we have the same quotient in ohms per volt *i.e.* 200 Ω/V. This is called the sensitivity of the voltmeter. The current drawn at full scale is the reciprocal of this number *i.e.* 0.005 A.

Merits And Demerits of PMMC Instruments.

Merits. (*i*) Uniform scale.

(*ii*) Low power consumption (say from 25 microwatt to 200 microwatt) because of small driving power.

(*iii*) No hysteresis loss as the former is of copper or aluminium.

(*iv*) Very effective and reliable eddy current damping.

(*v*) High torque-weight ratio resulting in high accuracy (2-5% of full-scale reading).

(*vi*) Use of single instrument for measurement of currents and voltages by employing shunts and multipliers of different resistances.

(*vii*) No effect of stray magnetic fields because of use of intensive polarised or unidirectional field.

(*viii*) The instrument using core magnet is very suitable in aircraft and aerospace applications, where a multiplicity of instruments are mounted in close proximity to each other. This is because of self-shielding property of core magnets.

Demerits. (*i*) These instruments cannot be used for ac measurements.

(*ii*) These are costlier in comparison to moving iron instruments because of delicate construction and the necessary accurate machining and assembly of various parts.

(*iii*) Friction and temperature might introduce errors, as in case of other instruments, which can be reduced as already discussed.

(*iv*) Ageing of control springs and of the permanent magnets might cause errors which can be considerably obviated by careful choice of material and pre-ageing during manufacture.

9.10.2. Dynamometer Type Instruments. Dynamometer type instruments are very similar to PMMC instruments except that the permanent magnetic field is replaced by a coil (usually two fixed air-cored coils are used) which carry the current to be measured (or a definite fraction of it or proportional to the voltage to be measured). The coils are usually air-cored to avoid hysteresis, eddy currents and other errors when the instrument is used on ac. The fixed coil is divided into two halves, connected in series with the moving coil; and placed together and parallel to each so as to provide a fairly uniform field within the range of the movement of the moving coil. However, the space between the two halves of the fixed coil must be sufficient enough to allow the movement of the moving coil shaft.

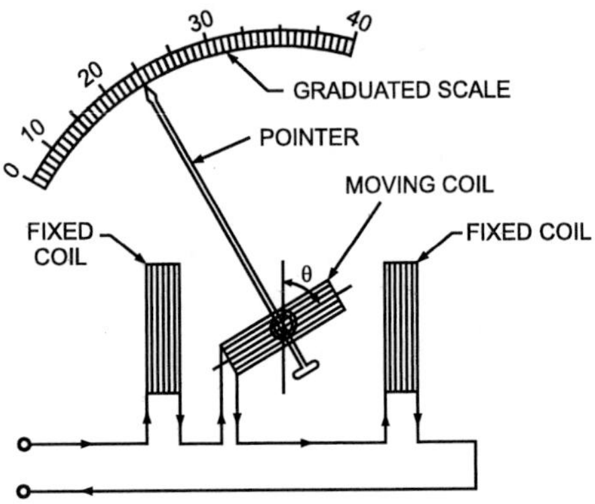

Fig. 9.10 *Dynamometer Type Instrument*

This instrument, like other instruments already discussed, develops deflecting torque by the interaction of magnetic fields, one field due to current in a moving coil and the other due to current in the fixed coil. The important difference is that the field due to fixed coil is not constant, but varies with the magnitude of current flowing through the fixed coil. Hence the deflecting torque of this instrument is determined not only by the moving coil current but also by the fixed coil current.

This instrument, with its fixed and moving coils connected in series, is readily adapted for voltage measurement, provided the proper series resistor is used, as shown in Fig. 9.11 (*d*).

This instrument is seldom used as an ammeter because the lead-in spirals to the moving coil can carry limited current, frequency variations influence the inductance of the coils and introduce error and the resistance of the moving and fixed coils in series may cause an undesirably high voltage drop across the shunt.

Since the deflecting torque is determined by variations in either the fixed or the moving coil currents, the dynamometer instrument is a versatile measuring device for several other applications such as for measurement of power, reactive volt-amperes and, with some modifications, for measuring power factor and frequency in ac circuits.

Though these instruments can be used both, for dc and ac measurements, but most of its practical uses are in ac as these instruments have higher cost, higher power consumption, lower torque-weight ratio, non-uniform scale and other several drawbacks in comparison to PMMC instruments.

The dynamometer type ammeters and voltmeters are usually portable and laboratory standard instruments of the highest attainable precision, but owing to the fact that they are costly to build in consequence of the care to be taken in their manufacture and calibration and their operating power required are usually considerably higher than those of moving iron instruments they are not usually used as ammeters and voltmeters for general laboratory and shop measurements at power frequency i.e. where a sturdier and cheaper instruments are adequate.

Such instruments may also serve as *transfer instruments** because they have precision grade accuracy (approximately 5% of full-scale deflection) both for dc and ac measurements and can be calibrated on dc and then used directly on ac establishing a direct means of equating ac and dc measurements of voltage and current.

A dynamometer type moving coil instrument is shown in Fig. 9.10.

To use the instrument as an ammeter for measurement of small currents, the moving coil and fixed coils are connected in series, as shown in Fig. 9.11 (*a*).

* A transfer instrument is that which may be calibrated with dc source and then used on ac without any modification.

In this case current in fixed coils = Current in moving coil.

When the instrument is to be used for measurement of higher currents, the fixed coils are wound with thick wire. The moving coil cannot be wound of thick wire and if it is wound it will be very heavy coil and, therefore, a strong magnetic field will be required to move it. The difficulty is overcome by connecting the moving coil to the terminals of a shunt, as shown in Fig. 9.11 (*b*)

Thus the shunt resistance limits the current in the moving coil to only a small fraction of the current being measured.

For measurements of higher currents by dynamometer type instrument, the instrument may be used by connecting the moving coil across a shunt and the fixed coils, as shown in Fig. 9.11 (*c*)

To use the instrument as a voltmeter, the coils are wound with high resistance fine wire and fixed and moving coils are connected in series along with a high non-inductive resistance *R*, as shown in Fig. 9.11 (*d*). In this case same current flows through both fixed and moving coils.

The torque of this instrument is dependent upon the strength of the magnetic field created by fixed coils and the moving coil which are proportional to current. Hence in ammeters the torque is proportional to the square of the current under measurement and in voltmeters the torque is proportional to the square of the voltage under measurement.

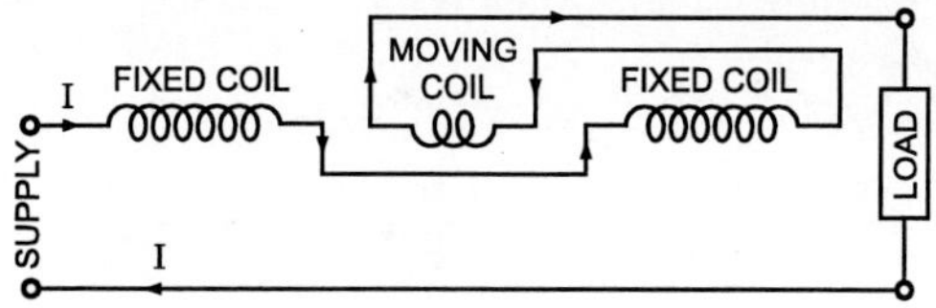

(a) Arrangement of Coils of Dynamometer Instrument For Measurement of Small Currents

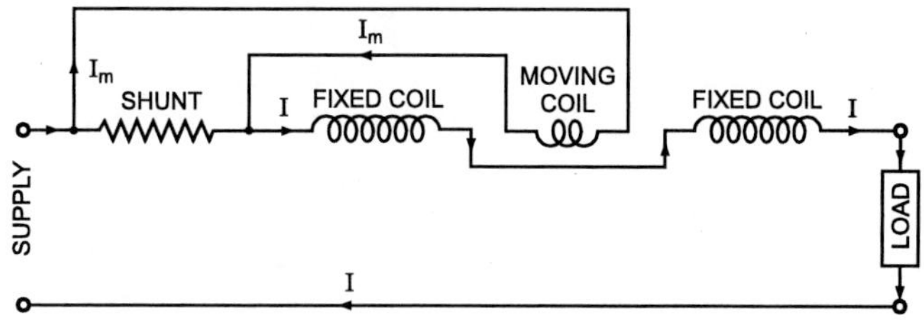

(b) Arrangement of Coils of Dynamometer Instrument For Measurement of Higher Currents

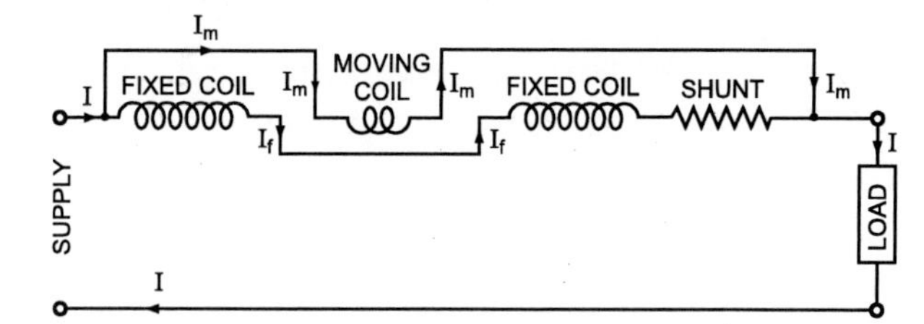

(c) Arrangement of Coils of Dynamometer Instrument For Measurement of Higher Currents

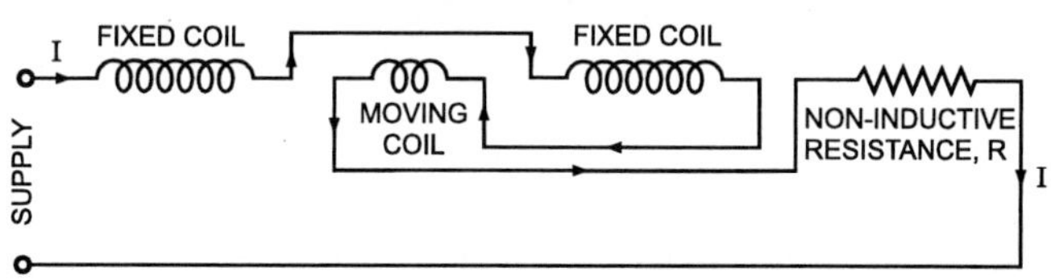

(d) Arrangement of Coils of Dynamometer Instrument For Measurement of Voltage

Fig. 9.11

Two hair springs are used to create controlling torque in the instrument which also serve the purpose of leads to the moving coil. Air friction damping is usually provided in such instruments.

Deflecting Torque. If i_f and i_m are the currents flowing through the fixed and moving coils respectively then

$$\text{Deflecting torque, } T_d \propto i_f i_m$$

In spring-controlled instruments controlling torque,

$$T_c \propto \theta, \text{ the deflection}$$

$$\text{But in steady deflected position, } T_c = T_d$$

$$\therefore \theta \propto i_f i_m$$

$$\text{or } \theta \propto i^2 \qquad \text{since } i_f \propto i \text{ and } i_m \propto i$$

The instrument, therefore, has a "square-law" response and it indicates in terms of effective (or rms) current or voltage.

Ranges. (1) Ammeters : 0 – 0.01 A up to 0 – 0.05 A with fixed and moving coils connected in series [Fig. 9.11 (a)] and up to 0 – 20 A with moving coil shunted [Fig. 9.11 (b)] or parallel connections [Fig. 9.11 (*c*)].

(2) Voltmeters: Up to 0 – 750 V

Merits and Demerits of Dynamometer Type Instruments.

Merits. (*i*) As the instrument has square-law response, so can be used on both dc as well as on ac.

(*ii*) These instruments are free from hysteresis and eddy current errors because of absence of iron in the operating parts of the instrument.

(*iii*) Ammeters up to 10 A and voltmeters up to 600 V can be constructed with precision grade accuracy.

(*iv*) Dynamometer type voltmeters are very useful for accurate measurement of rms values of voltages irrespective of waveforms.

(*v*) Because of precision grade accuracy and same calibration for dc and ac measurements these instruments are used as transfer and calibration instruments.

Demerits. (*i*) The scale is not uniform as the instrument has square-law response.

(*ii*) These instruments have small torque-weight ratio, as already explained, so the friction error is considerable.

(*iii*) Owing to heavy moving system, friction losses in these instruments are somewhat more than those in other instruments.

(*iv*) As a result of measures taken to reduce the frictional error; their cost is more in comparison to moving iron and PMMC instruments and they are more sensitive to overloads and mechanical impacts and are to be handled with care.

(*v*) Adequate screening of the movements against stray magnetic fields is essential.

(*vi*) The sensitivity of the instrument is typically very low due to poor deflecting torque. The sensitivity of dynamometer type voltmeter is 10 to 30 Ω/V in comparison to the sensitivity of 20 k Ω/V in case of D' Arsonval movement.

(*vii*) Power consumption of these instruments is comparatively high because of their construction.

Errors in Dynamometer Type Instruments. The errors that usually occur in dynamometer type instruments are (i) frictional error (ii) temperature error (iii) error due to stray magnetic fields and (iv) frequency error.

(*i*) *Frictional Error.* Since the coils are air-cored, therefore, magnetic field produced is of small strength (near about 0.005 T – 0.006 T) thereby requiring large number of ampere-turns to create necessary deflecting torque. This results in heavy moving system and, therefore, small torque-weight ratio. Thus the frictional losses in dynamometer type instruments are somewhat larger than those in other types. However, frictional error is minimised by a reasonable reduction in weight of the moving element, proper selection of materials for the bearings and spindle and adequate polishing of these parts. These measures, however, increase the cost of instrument and make it more sensitive to overloads and mechanical impacts.

(*ii*) *Temperature Error.* Since the operation of dynamometer type instruments require considerable power, self heating in these instruments is appreciable and the error due to self heating may be as much as 1 % of full-scale deflection.

(*iii*) *Error Due To Stray Magnetic Fields.* Since the operating magnetic field produced by the fixed coil in these instruments is somewhat weaker in comparison to that in instruments of other types, the operation of these instruments is more sensitive to the stray magnetic field. This effect may be reduced by proper shielding—surrounding the complete assembly by a laminated steel shield (to reduce eddy currents induced in the shield) and surroundings this shield by a drawn steel cylinder at a clearance of a few mm.

In case of instruments provided with eddy current damping (usually not) the moving coil will have also to be shielded from the field of permanent magnet used for eddy current damping.

(*iv*) *Frequency Error*. As already explained in case of moving iron instruments, the change in frequency causes errors (*i*) due to change in reactance of the operating coils and (*ii*) due to change in the magnitude of eddy currents set up in the metal parts of the instrument near to the operating portion.

(*a*) *The reactance error* may be reduced by making the series resistance very high or alternatively by connecting a capacitor of proper value across the series or swamping resistance, as shown in Fig. 9.12. The impedance of the instrument will be independent of frequency if $C = L/R^2$ as in case of moving iron instruments.

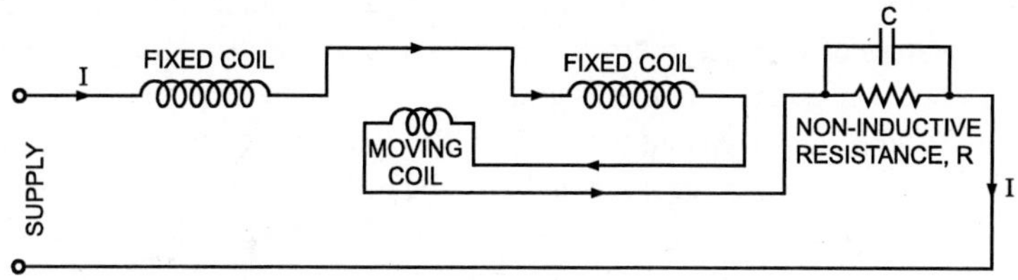

Fig. 9.12 *Compensation For Reactance Error in Dynamometer Type Voltmeter*

(*b*) *Eddy Current Error*. The eddy currents are induced in the metal parts of the instrument and develop a torque because of coupling between the moving coil and the neighbouring metal parts. This error, however, can be reduced by keeping the metal in the supports of the coil and other structural parts to the minimum possible and selecting the metal, to be used, of high resistivity.

9.10.3. Comparison Between Moving Iron and Moving Coil Instruments

1. *Construction*. Moving iron instruments are simple and robust in construction and are therefore, cheaper in cost. Moving coil instruments are costlier in comparison to moving iron instruments because of delicate construction and the necessary accurate machining and assembly of various parts.
2. *Damping torque*. In moving iron instruments, air friction damping is provided which is not effective while in moving coil instruments very effective and reliable eddy current damping is provided.
3. *Power consumption*. Moving coil instruments consume less power as compared to moving iron instruments because of small driving power.
4. *Scale*. Moving coil instruments have uniform scale while moving iron instruments have non-uniform scale (crowded at the beginning).
5. Sensitivity. Moving coil instruments are more sensitive as compared to moving iron instruments because of high torque-weight ratio.
6. *Accuracy and reliability*. As compared to moving iron instruments, moving coil instruments are more accurate and reliable because they have no hysteresis loss and they are not affected by stray magnetic fields.
7. *Applications*. Moving iron instruments can be used both in dc as well as in ac circuits but moving coil instruments cannot be used in ac circuits.

9.10.4. Comparison Between Moving Iron and Dynamometer Type Instruments

1. *Torque-Weight Ratio*. Dynamometer type instruments have small torque-weight ratio.
2. *Friction Error*. Dynamometer type instruments have considerable friction error.
3. *Friction Loss*. Owing to heavy moving system, dynamometer type instruments have more friction losses.
4. *Cost and Sensitivity To Overloads*. As a result of measures taken to reduce the frictional error, the cost of dynamometer type instruments is more in comparison to moving iron. Dynamometer type instruments are more sensitive to overloads and mechanical impacts and are to be handled with care.
5. *Sensitivity*. The sensitivity of dynamometer instruments is typically very poor due to poor deflecting torque.
6. *Power Consumption*. Dynamometer instruments have comparatively higher power consumption.
7. *Effect of Stray Magnetic Field*. There is no effect of stray magnetic field on moving iron instruments because of use of intensive polarised or unidirectional field while dynamometer type instruments are more sensitive to the stray magnetic field and need adequate screening.

8. *Hysteresis and Eddy Current Errors.* Dynamometer type instruments are free from hysteresis and eddy current errors because of absence of iron in the operating parts while moving iron instruments have these errors.
9. *Effect of Waveform.* Dynamometer type voltmeters are very useful for accurate measurement of rms values of voltages irrespective of waveforms while change in frequency and in waveform causes serious errors in ac measurements with moving iron instruments.
10. *Calibration.* Dynamometer type instruments have same calibration for dc and ac measurements and have precision grade accuracy and are, therefore, widely used as transfer and calibration instruments while moving iron instruments have a difference between dc and ac calibrations.

Dynamometer type instruments are superior because of its inherent advantages given above in paras (8), (9) and (10).

9.11. EXTENSION OF INSTRUMENT RANGE

The range of an electrical measurement is limited by the current, which can be carried by the coil of the instrument safely. For example, the moving coil and the spiral springs used as coil connections can be designed for a maximum current of only about 50 m A and a potential drop of about 50 millivolts. Hence for measurement of large currents or voltage or power, some means for increasing the range of the instrument is to be adopted.

There are four common devices employed for extending the range of instruments; these are shunts, multipliers, current transformers and potential transformers. When instruments are supplied with such external devices, the instrument is calibrated and its scale is correspondingly marked, over the range of the associated shunt, multiplier or transformer.

9.11.1. *Ammeter Shunts.* An ammeter shunt is merely a low resistance that is placed in parallel with the coil circuit of the instrument in order to measure fairly large currents. The greater part of the current in main circuit is then diverted around the coil through the shunt.

The connection diagram for a shunt and milliammeter for measuring large currents is shown in Fig. 9.13. The shunt has four terminals, two of large current carrying capacity known as *current terminals* for inserting it in series with the main circuit and the other two of smaller size, known as *potential terminals,* for connecting the ammeter across it.

Theory. Let I be the current of the circuit to be measured, R_s be the resistance of the shunt and R_m be the resistance of the ammeter plus its leads to the potential terminals.

The main circuit current will be split in two parts, I_m, the current flowing through the ammeter and I_s, the current through the shunt.

$$\text{Thus } I = I_m + I_s \qquad ...(i)$$

Since the voltage drop across the shunt and the instrument is same,

$$I_m R_m = I_s R_s$$

$$\text{or } R_s = \frac{I_m R_m}{I_s} = \frac{I_m R_m}{I - I_m}$$

$$= \frac{R_m}{I/I_m - 1} \qquad ...(ii)$$

Since from expression (*i*) $I_s = I - I_m$

The ratio of the total current to the instrument current (*i.e.* I/I_m) is known as the 'multiplying power' of the shunt. Denoting the ratio by N, we have shunt resistance,

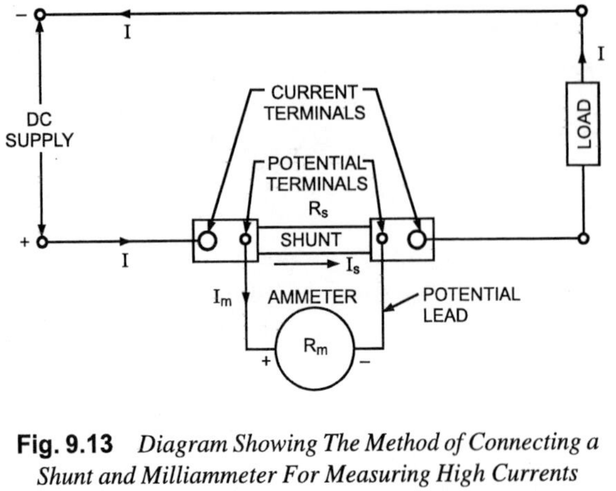

Fig. 9.13 *Diagram Showing The Method of Connecting a Shunt and Milliammeter For Measuring High Currents*

$$R_s = \frac{R_m}{N-1} \quad ...(iii)$$

Construction of Shunts. The shunts used for extension of range of instruments must have temperature coefficient of resistance very low and as far as possible equal to that of the instrument coil so that multiplying power of the shunt is independent of temperature. The resistance of the shunt should also not vary with the time of usage. Other requirements are that the thermoelectric emf of the shunt should be very small and it must be capable of carrying required current without appreciable heating.

The material employed for shunts is manganin, which has a low temperature coefficient, in the form of thin strips, the ends of which are soldered to two large copper blocks. Each copper block carry two terminals — one current terminal and other potential terminal. The strips, which form the shunt are spaced from each other to promote a good circulation of air and thus efficient cooling.

Precautions. The resistance of the leads, as already mentioned, used with the instruments forms the part of the resistance of the instrument circuit and, therefore, must be considered while determining the required shunt resistance. A special pair of leads is usually supplied with the instrument by the manufacturer intended to be used. These leads should always be used in order to obtain the stated accuracy as the instrument is calibrated with these leads in use.

Advantages. (*i*) The same instrument can be used for measurement of currents of different ranges by employing different shunts.

(*ii*) The shunt enables the ammeter to be mounted remote from the main circuit, which results in reduction of the possible effect of stray field and have the advantage of remote indication.

9.11.2. *Voltmeter Multipliers.* Voltmeter multiplier is a high non-inductive resistance connected in series with the voltmeter coil and is used for increasing the range of the voltmeter. A voltmeter connected in series with the high non-inductive resistance (voltmeter multiplier) for measurement of high voltage is shown in Fig. 9.14.

Theory. Let R_v be the resistance of voltmeter, I_m be the full-scale deflection current, V be the voltage of the circuit to be measured and R be the external series resistance.

Now since voltage across supply leads = Voltage drop across the voltmeter + voltage drop across external resistance

$$\therefore V = I_m R_v + I_m R$$

$$\text{or } R = \frac{V}{I_m} - R_v$$

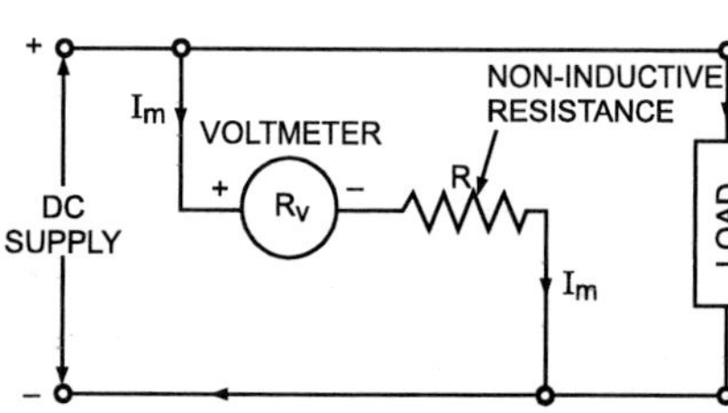

Fig. 9.14

The main requirements of voltage multipliers to be used for dc measurements are that (*i*) resistance should not change with time of usage (*ii*) their temperature coefficient of resistance must be very low so that change in temperature may not affect their resistance values and (*iii*) should have ample provision for cooling to dissipate the heat developed, as an appreciable amount of power is developed in such resistors.

The series resistance is usually of manganin, either in the form of a bobbin or in the form of strip wound on a narrow sheet of bakelite or other insulating material. It may be inside the instrument case or particularly for higher voltages, in a separate ventilated case.

The question of temperature compensation does not normally arise in the case of a voltmeter, as any slight change in the coil resistance is negligible in comparison with the total resistance, the major portion of which consists of manganin wire. The series resistance, in fact acts as a completely effective swamp resistance.

Example 9.1 An ammeter has an internal resistance of 2 Ω and it is designed to measure 1 A current. What would you do to make this meter read 15 A current?

[*G.B. Technical Univ. Electrical Engineering Second Semester 2011-12*]

Solution: Full-scale deflection current, $I_m = 1$ A

Instrument internal resistance, $R_m = 2\ \Omega$

Total current to be measured, $I = 15$ A

Shunt resistance required to be connected across ammeter to measure current up to 15 A

$$R_s = \frac{R_m}{\frac{I}{I_m} - 1} = \frac{2}{\frac{15}{1} - 1} = \frac{1}{7}\ \Omega \text{ **Ans.**}$$

Example 9.2 **A moving coil instrument having a resistance of 50 Ω has a full-scale deflection current of 1 mA. Calculate:**

(*i*) **Shunt resistance to convert the instrument into an ammeter of 2 A range.**

(*ii*) **Net resistance of the meter.** *[G.B. Technical Univ. Electrical Engineering, Odd Semester 2012-13]*

Solution: Instrument resistance, $R_m = 50\ \Omega$

Full-scale deflection current, $I_m = 1$ mA = 0.001 A

Total current to be measured, $I = 2$ A

$$\text{Multiplying power of shunt, } N = \frac{I}{I_m} = \frac{2}{0.001} = 2{,}000$$

(*i*) Shunt resistance required for full-scale deflection at 2 A,

$$R_s = \frac{R_m}{N-1} = \frac{50}{2{,}000-1} = 0.0250125\ \Omega$$

(*ii*) Net resistance of the meter is equivalent to parallel combination of instrument resistance R_m and shunt resistance R_s *i.e.*

$$R = R_m \parallel R_{sh} = 50 \parallel 0.0250125 = \frac{50 \times 0.0250125}{50 + 0.0250125} = 0.025\ \Omega \quad \textbf{Ans.}$$

Example 9.3 **A moving coil instrument having internal resistance of 50 Ω indicate full-scale deflection with a current of 10 mA. How can it be made to work as (*a*) voltmeter to read 100 V (*b*) ammeter to read 1 A on full scale?**

[U.P. Technical Univ. Electrical Engineering Odd Semester 2013-14]

Solution: Instrument resistance, $R_m = 50\ \Omega$

Full-scale deflection current, $I_m = 10$ mA = 0.01 A

(*a*) Series resistance required for full-scale deflection with 100 V,

$$R = \frac{V}{I_m} - R_m = \frac{100}{0.01} - 50 = 9{,}950\ \Omega \quad \text{or} \quad 9.95\ \text{k}\Omega \quad \textbf{Ans.}$$

$$(b) \text{ Multiplying power of shunt, } N = \frac{1}{0.01} = 100$$

Shunt resistance required for full-scale deflection at 1 A,

$$R_s = \frac{R_m}{N-1} = \frac{50}{100-1} = 0.505\ \Omega \quad \textbf{Ans.}$$

Example 9.4 **A moving coil instrument has a resistance of 2 Ω and it reads up to 250 V when a resistance of 5,000 Ω is connected in series with it. Find the current range of the instrument when it is used as an ammeter with the coil connected across a shunt resistance of 2 milli ohms.**

[U.P. Technical Univ. Electrical Engineering First Semester 2001-02 and 2004-05; Basic Electrical Engineering First Semester 2013-14]

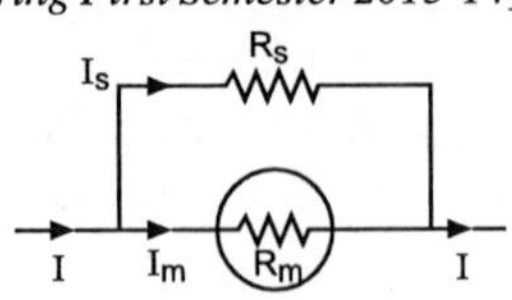

Fig. 9.15

Solution: Instrument coil resistance, $R_m = 2\ \Omega$

Current flowing through the instrument for full-scale deflection,

$$I_m = \frac{\text{Full-scale reading in volts}}{R_m + \text{Series resistance}}$$

$$= \frac{250}{2 + 5{,}000} = 0.04998 \text{ A or } 49.98 \text{ m A}$$

$$\text{Shunt resistance, } R_S = 2 \times 10^{-3}\ \Omega$$

$$\text{Current through shunt resistance, } I_s = \frac{I_m R_m}{R_s} = \frac{49.98 \times 10^{-3} \times 2}{2 \times 10^{-3}} = 49.98 \text{ A}$$

Current range of instrument = Full-scale deflection current = $I_m + I_s = 0.04998 + 49.98 = 50$ A **Ans.**

Example 9.5 **A moving coil meter has full-scale deflection current of 15 mA and a resistance of 5 Ω. Find the resistance which should be suitably connected with instrument to make the meter usable as an ammeter of 10 A range and the same meter as a voltmeter of 100 V range. Indicate their suitable connection.**

[*Mahamaya Technical Univ. Electrical Engineering Second Semester 2011-12*]

Solution: Full-scale deflection current, $I_m = 15 \text{ mA} = 15 \times 10^{-3}$ A

Instrument resistance, $R_m = 5\ \Omega$

Current to be measured (maximum), I = 10 A

$$\text{Multiplying power of shunt, } N = \frac{I}{I_m} = \frac{10}{15 \times 10^{-3}} = 666.667$$

Resistance to be connected in parallel (Fig. 9.13) to enable the instrument to read up to 10 A

$$R_s = \frac{R_m}{N-1} = \frac{5}{666.667 - 1} = \frac{5}{665.667}\ \Omega = 0.0075113\ \Omega \textbf{ Ans.}$$

Voltage to be measured (maximum range), V = 100 V

Series resistance to be connected in series (Fig. 9.14) to enable the instrument to read up to 100 V,

$$R = \frac{V}{I_m} - R_m = \frac{100}{15 \times 10^{-3}} - 5 = 6{,}661.6667\ \Omega \textbf{ Ans.}$$

Example 9.6 **A moving coil instrument has a resistance of 5 ohm and gives a full-scale deflection of 100 mV. Show how the instrument may be used to measure:**

(*i*) voltages upto 50 V

(*ii*) currents upto 10 A.

[*U.P. Technical Univ. Electrical Engineering First Semester 2006-07*]

Solution: Full-scale deflection current, $I_m = \frac{100 \text{ mV}}{5\Omega} = 20$ mA

(*i*) The instrument can be used to measure voltages up to 50 V by connecting resistance R in series with the instrument coil and the required series resistance is given by

$$R = \frac{V}{I_m} - R_m = \frac{50}{20 \times 10^{-3}} - 5 = 2{,}495\ \Omega \textbf{ Ans.}$$

(*ii*) The instrument can be used to measure currents upto 10 A by connecting a resistance R_S across the instrument coil and the required instrument shunt resistance is given by

$$R = \frac{R_m}{\frac{I}{I_m} - 1} = \frac{5}{\frac{10}{20 \times 10^{-3}} - 1} = \frac{5}{499}\ \Omega \textbf{ Ans.}$$

9.12. DYNAMOMETER TYPE WATTMETER

This instrument is similar in design and principle to the dynamometer type ammeter and voltmeter already described in Art 9.10.2. When the instrument of this type is used as a wattmeter, the fixed coil, which is divided into two equal portions in order to provide uniform field, is employed as current coil and the moving coil is used as a pressure coil i.e. the fixed coil carries the current flowing through the circuit and the moving coil carries the current proportional to the voltage across the circuit. A high non-inductive resistance is connected in series with the moving coil in order to limit the current. The magnetic fields of the fixed and moving coils react on one another causing the moving coil to turn about its axis. The movement is controlled by hair springs which also lead the current into and out of the moving element. Damping is provided by light aluminium vanes moving in an air dash pot. The pointer is fixed to the moving-coil spindle and moves over a suitably calibrated scale.

Theory. Let v be the supply voltage, i the load current and R the total resistance of the moving-coil circuit.

$$\text{Current through fixed coil, } i_f = i$$

$$\text{Current through moving coils, } i_m = \frac{v}{R}$$

$$\text{Deflecting torque, } T_d \; \alpha \; i_f i_m \; \alpha \; \frac{iv}{R}$$

For a dc circuit the deflecting torque is thus proportional to the power and for any circuit with fluctuating torque, the instantaneous torque is proportional to the instantaneous power. In the latter case due to inertia of moving parts the deflection will be proportional to the average torque *i.e.* the deflection will be proportional to the average power. For sinusoidal alternating quantities the average power is VI cos ϕ where V and I are rms values of voltage and current and ϕ is the phase angle between voltage V and current I.

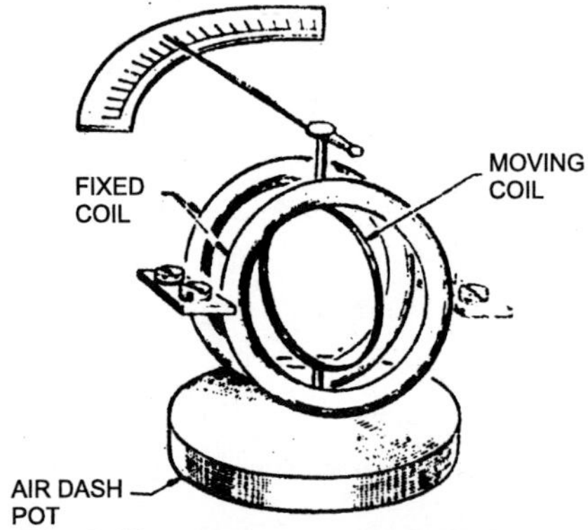

Fig. 9.16 *Dynamometer Type Wattmeter*

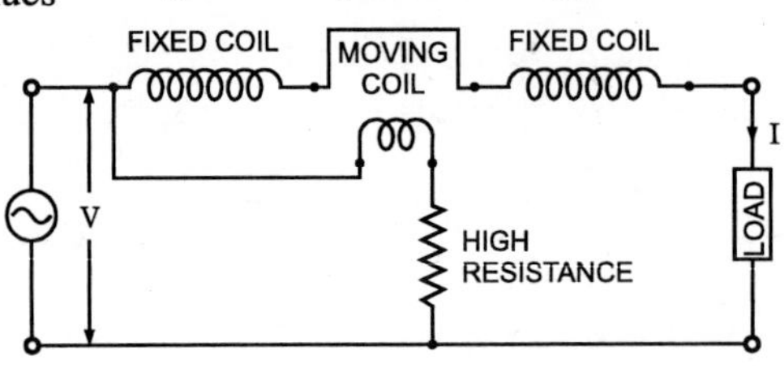

Fig. 9.17 *Connection Diagram of Wattmeter For Measurement of Power in AC circuits*

Thus an electrodynamic instrument, when connected, as shown in Fig. 9.17, indicates the power, irrespective of the fact it is connected in an ac or a dc circuit.

Ranges. (*i*) Current circuit 0.25 to 100 A without employing CTs.

(*ii*) Potential circuit 0.5 to 750 V without employing PTs.

Advantages. (*i*) Since deflecting torque is proportional to true power in both the cases *i.e.* dc and ac and the instrument is spring controlled, the scale of the instrument is uniform.

(*ii*) high degree of accuracy can be obtained by careful design, hence these are used as standard for calibration purposes.

Disadvantages. (*i*) At low power factor the error due to the inductance of pressure coil is very serious unless special features are incorporated to reduce its effect.

(*ii*) The reading of the instrument may be affected by stray field acting on the moving coil. To reduce it, magnetic shielding is provided by enclosing the instrument in an iron case.

9.13. ENERGY METERS

The essential difference between an energy meter and a wattmeter is that the former is fitted with some type of registration mechanism whereby all the instantaneous readings of power are summed over a definite period of time whereas the latter indicates the value at particular instant when it is read.

Energy meters are generally of the following three types *viz* (*i*) electrolytic meters (*ii*) motor meters and (*iii*) clock meters.

Out of above three types of energy meters, the motor meters are the most commonly used meters, where as the clockmeters, because of their complexity, have practically disappeared from use, although they have some application as standards because of their inherent high accuracy. Electrolytic meters are still used to some extent.

9.13.1. Motor Meters. These meters can be used on dc as well as on ac circuits. In principle the motor meter is a small motor of dc or ac type whose instantaneous speed of rotation is proportional to the circuit current in case of an ampere-hour meter and to the power of the circuit in case of a watt-hour meter. The number of revolutions made by the revolving portion (rotor) in any given time is proportional, in case of A- H meter, to the quantity of electricity supplied during that time; and in case of energy meter to the energy supplied. A train of wheels connected to the spindle (rotor) count the number of revolutions made and thus register the AH or kWh directly.

These meters essentially consist of the following parts:

1. *An operating system,* which develops a torque, called the *driving torque*, that causes rotation of the moving system continuously. The driving torque may be compared with the deflecting torque in an indicating instrument.

2. *A braking device* is used to produce a braking torque. It is due to this torque that the moving system of the instrument (armature) rotates at a particular speed for a particular rate of energy consumption at any time. The braking torque is usually produced by a permanent magnet, called the *brake magnet*, that induces current in some part of the moving system which in turn produces the braking torque. Therefore, the braking torque is proportional to the induced currents whereas induced currents are proportional to the speed of the moving system and hence the braking torque is proportional to the speed of the moving system (disc). The moving system attains a steady speed when the braking torque is equal to the driving torque.

3. A device for registering the number of revolutions made by rotating element. This is obtained by having a worm cut on the spindle of the instrument. The worm engages with a pinion and thus drives the train of wheels and registers A-H or W-H directly.

Mechanical Details of Construction. The construction of the meter must be such that the functioning of the meter is not readily interfered with by tilting or any external means. For safety the meters are enclosed in an insulating or metal casing provided with glass windows for the purpose of reading the meter readings. House service meters are so designed that both the terminal block and the remainder of the meter can be sealed off separately in order to prevent access to both the leads to the meter and to the internal working parts. Thus complete protection is provided aganist electric shock by accidental contact with the meter.

Errors Common To All Motor Meters. There are two principal errors common to all motor meters namely friction errors and braking errors.

1. ***Friction Errors.*** The friction error, which is due to friction at the pivots and bearings, plays a considerably more important part than the corresponding error in most indicating instruments, since it continuously operates and affects the speed of the rotating element (disc) for any given value of current or power. The friction torque, which exists when the disc is just starting to rotate, may prevent it from starting at all, if load is small and will cause its registration to be low on small loads. This part of the friction torque may be assumed to remain constant when the moving system of the meter is in rotation and may be compensated for by providing a small constant driving torque on the moving system independent of load. When the moving system of the meter is rotating normally, a frictional torque proportional to the speed of rotating element also exists, but this torque is not of considerable importance since it merely adds a retarding torque and helps the braking magnet action.

In some meters, such as those of the mercury motor type, a frictional torque which is proportional to the square of the speed also exists and has to be compensated for. In order to reduce the frictional torque to the minimum, the weight of the rotating system should be made as small as possible, the jewels and pivots should be kept well in order and spring-loaded bearing should be employed as their use reduces the friction and increases the life of the jewels.

2. ***Braking Errors.*** Variation in braking action affects the speed of the rotating element of the meter, for a given driving torque, and so affects the number of revolutions made in a given time.

The rotating element of the meter attains steady speed when the braking torque, which is proportional to speed, is equal to the driving torque. The braking torque is also proportional to the strength of the brake magnet.

The braking torque, $T_b \propto \Phi iR$

where Φ is the flux of the brake magnet, i is the eddy current induced in the rotating element (disc) due to rotation in the field of the brake magnet and R is the effective radius of the disc from its axis.

Since the induced emf, $e \propto \Phi n$ where n is the speed of rotation,

Induced current, $i = \frac{e}{r}$ where r is the resistance of the path of induced current i.

or $i \propto \frac{\Phi n}{r}$ and braking torque $T_b \propto \Phi^2 \frac{NR}{r}$

Since at steady speed (say of N) the braking torque is equal to the driving torque T_d,

$$T_d \propto \Phi^2 \frac{NR}{r}$$

$$\text{or} \quad N \propto T_d \frac{r}{\Phi^2 R}$$

From the above expression of steady speed it is obvious that the steady speed attained for a constant driving torque T_d is directly proportional to the resistance of eddy current path and inversely proportional to the square of the flux of the brake magnet. Hence it is necessary that the strength of the brake magnet should remain constant during the use of the meter. Careful design and treatment of brake magnet during its manufacture are necessary in order to ensure this constancy.

With the increase in temperature, the resistance of eddy current path r will increase and, therefore, the braking torque will decrease and cause an error in the instrument registration. Though it is somewhat difficult to compensate completely the reduction in braking torque due to increase in temperature but in some meter driving torque also decreases with the increase in temperature and thus automatically compensates partially.

It is necessary that the resistance of the path of induced current, r should be low, the field produced by the permanent magnet should be strong and the effective radius of the disc large so that the steady speed of the meter be low. The radius of the disc cannot be made unduly large, otherwise the size and weight of the disc will increase and so result in increase in friction torque. An aluminium disc is usually preferred to a copper disc in order to have resistance per unit weight smaller. Strong magnets having large pole shoes and smallest possible air gap from mechanical consideration are used.

The braking torque can be adjusted by adjusting the effective radius R.

There are three types of motor meters, which are common in use, *viz* mercury motor meter, commutator motor meter and induction motor meter. The first two types are used for dc circuits.

9.14. INDUCTION TYPE ENERGY METER

The most commonly used meter in ac circuits for the measurement of energy is the induction type watt-hour meter. The principle of operation of induction type energy meter is exactly the same as that of induction type wattmeter. The construction is also same except that the spring control and pointer of wattmeter are replaced by a brake magnet, the disc is allowed to rotate continuously instead of rotating through only a fraction of revolution as in the wattmeter and a train of wheels driven by a worm on the spindle of the meter is provided. *Single-phase induction type energy meter* will be discussed here in detail.

Single-Phase Induction Type Energy Meter

Construction. The construction of a typical single phase induction type energy meter is shown in Fig. 9.18, the brake magnet and recording wheel train being omitted for sake of clarity.

Induction type energy meter (single phase) essentially consists of (*a*) driving system (*b*) moving system (*c*) braking system and (*d*) registering system.

The *driving system* of an energy meter essentially consists of two electromagnets, called the *shunt magnet* and *series magnet.* The core of these electromagnets is made up of silicon steel laminations and the laminations are rivetted together to form a rigid mechanical structure in which definite alignment of parts maintains the correct magnetic flux distribution for consistent performance. The *shunt magnet* consists of a number of M-shaped iron laminations assembled together to form a core. A coil having large number of turns of fine wire is fitted on the middle limb of the shunt magnet. The coil is known as *pressure coil* (PC) and is connected across the supply mains. The potential coil is highly inductive as it has very large number of turns and the reluctance of the magnetic circuit is very small. The *series electromagnet* consists of a number of U-shaped iron laminations assembled together to form a core. Each of the two limbs is wound with a few turns of heavy gauge wire. This wound coil is known as *current coil* (CC) and is connected in one of the lines and in series with load to be metered. The series magnet is energized and set up a magnetic field cutting through the rotating disc, when load current flows through the current coil (CC). Copper shading bands are provided on central limb of shunt magnet so that phase displacement of 90° between the magnetic field set up by shunt magnet and applied voltage is achieved by their adjustment. The copper shading bands are also called the *power factor compensator or compensating loop.*

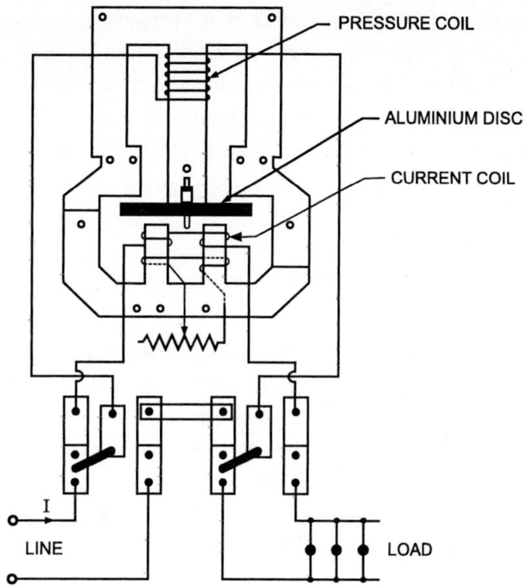

Fig. 9.18 *Induction Type Single-Phase Energy Meter*

The *moving system* essentially consists of a rotating aluminium disc mounted on a vertical spindle and supported on a sapphire cup contained in a bottom bearing screw. The bottom pivot, which is usually removable, is of hardened steel, and the end, which is hemispherical in shape, rests in the sapphire cup. The top pivot merely serves to maintain the spindle in a vertical position under working conditions and does not support any weight or exert appreciable thrust in any direction. A pinion engages the shaft with counting or registering mechanism.

The magnetic field produced by shunt electromagnet is pulsating in character and cuts through the rotating disc and induces eddy currents there in, but normally does not in itself produce any driving force. The reaction between these two magnetic fields and eddy currents set up a driving torque in the disc.

The *braking system* essentially consists of a permanent magnet, called *the brake magnet.* The brake magnet consists of C-shaped piece of alloy steel bent round to form a complete magnetic

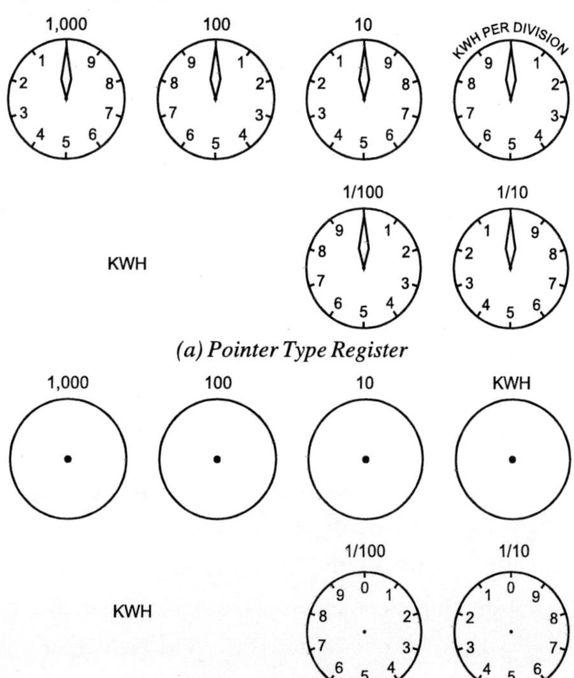

(a) Pointer Type Register

(b) Cyclometer Register

Fig. 9.19

circuit, with the exception of a narrow gap between the poles. This magnet is mounted so that the disc revolves in the air gap between the polar extremities. The movement of rotating disc through the magnetic field crossing the air gap sets up eddy currents in the disc which react with the field and exerts a braking effect. By changing the position of the brake magnet or by diverting some of the flux there from, the speed of the rotating disc may be adjusted.

The *registering or counting system* essentially consists of a train of gears, driven by either a worm or pinion gear on the disc shaft, which turns pointers that indicate on dials the number of times the disc has turned. The energy meter thus determines and adds together or integrates all the instantaneous power values so that total energy used over a period of time is known.

The right-hand dial pointer in the register indicates one kWh after the disc has made the required number of revolutions. The gearing is so arranged that each division on the right-hand dial represents one kWh, on the second dial from the right each division represents 10 kWh; on the third dial from the right each division represents 100 kWh, while on the extreme left hand dial each division represents 1,000 kWh. In some of the meters cyclo-meter register is used in place of pointer type register shown in Fig. 9.19 (*b*).

Theory of Operation. A simplified functional diagram of the driving system of the induction type energy meter is shown in Fig. 9.20. The corresponding phasor diagram is shown in Fig. 9.21.

Let the load current lag the circuit voltage V by an angle ϕ, as shown in Fig. 9.21. The flux Φ_{sh} of the shunt magnet is made to lag behind the circuit voltage, V by 90° by adjusting the position of shading bands.

Assuming that hysteresis and saturation effects in the iron are negligible the series magnet flux Φ_{se} is taken proportional to and in phase with load current I.

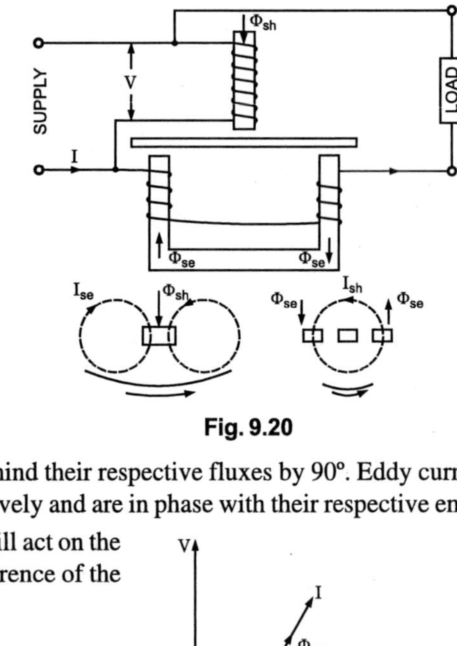

Fig. 9.20

EMFs E_{sh} and E_{se} are induced in disc by shunt magnet flux Φ_{sh} and series magnet flux Φ_{se} respectively and lags behind their respective fluxes by 90°. Eddy current I_{sh} and I_{se} are set up by the induced emfs, E_{sh} and E_{se} respectively and are in phase with their respective emfs.

Now two opposite torques given by $\Phi_{sh} I_{se}$ and $\Phi_{se} I_{sh}$ will act on the disc. The instantaneous values of resultant torque is the difference of the two *i.e.*,

Since phase angle between Φ_{sh} and I_{se} is ϕ

and phase angle between Φ_{se} and I_{sh} is $(180° - \phi)$

$$\therefore \text{ Average torque } \propto [\Phi_{sh} I_{se} \cos\phi - \Phi_{se} I_{sh} \cos(180° - \phi)]$$

$$\propto [\Phi_{sh} I_{se} \cos\phi + \Phi_{se} I_{sh} \cos\phi]$$

$$\propto (\Phi_{sh} I_{se} + \Phi_{se} I_{sh}) \cos\phi$$

$$\text{Since } \Phi_{sh} \propto V;\ \Phi_{se} \propto I;\ I_{se} \propto I \text{ and } I_{sh} \propto V$$

$$\therefore \Phi_{sh} I_{se} \propto VI = K_1 VI$$

$$\text{and } \Phi_{se} I_{sh} \propto VI = K_2 VI$$

Substituting above values in the expression of average torque we get

$$T_d \propto (K_1 VI + K_2 VI) \cos\phi$$

$$\text{or } T_d \propto VI \cos\phi$$

$$\text{or } T_d \propto \text{True power of circuit.}$$

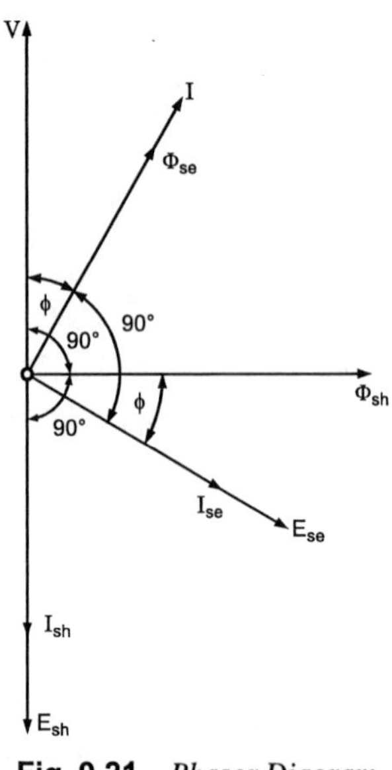

Fig. 9.21 *Phasor Diagram*

The braking torque is due to eddy currents induced in the aluminium disc. Since the magnitude of eddy currents is proportional to the speed of the disc, the braking torque T_b will also be proportional to the disc speed, N *i.e*, $T_b \propto N$.

For steady speed of rotation, $T_d = T_b$, so power $\propto$ N

Hence in a given period of time, the total number of revolutions, $\int_0^t N\,dt$ is proportional to $\int_0^t P\,dt$ *i.e.* electrical energy consumed.

The counting mechanism is so arranged that the meter indicates kWh (units) directly and not the revolutions.

Meter Constant. We have seen that N $\propto$ energy

or N = K × energy where K is a constant, called the *meter constant*

$$\text{So meter constant } K = \frac{N}{\text{Energy}} = \frac{\text{Number of revolutions}}{\text{kWh}}$$

Hence the number of revolutions made by the aluminium disc for 1 kWh of energy consumption is called the *meter constant.*

Errors and Adjustments.

1. *Phase and Speed Errors.* It is necessary that the energy meter should give correct reading on all power factors, which is only possible when the field set up by the shunt magnet lags behind the applied voltage exactly by 90°. Ordinarily the flux due to shunt magnet does not lag behind the applied voltage by 90° because of winding resistance and iron losses. The flux due to shunt magnet is made to lag behind the applied voltage by 90° with the help of copper shading band provided on the central limb. An error due to incorrect adjustment of shading band will be evident when the meter is tested on a load of power factor less than unity.

An error on the fast side under these conditions can be eliminated by bringing the shading band nearer to the disc and vice versa.

An error in the speed of the meter when tested on non-inductive load can be eliminated by adjustment of the position of the brake magnet. Movement of the brake magnet in the direction of the spindle will reduce the braking torque and vice versa.

2. *Friction Compensation and Creeping Error.* Two shading bands embrace the flux contained in the two outer limbs of the shunt electromagnet,and thus eddy currents are induced in them which cause a phase displacement between the enclosed flux and the main gap flux. As a result, a small driving torque is exerted on the disc, this torque being adjusted, by variation of the position of these bands, to compensate for friction torque in the instrument.

In some energy meters, the disc continues rotating when the potential coils are excited but with no load current flowing. This defect is known as *creeping* and is prevented by cutting two holes or slots in the disc on opposite sides of the spindle. The disc tends to remain stationary when one of the holes comes under one of the poles of the shunt magnet.

In some cases, a small piece of iron wire is attached to the edge of the disc. The force of attraction of the brake magnet upon this wire is sufficient to prevent continuous rotation of the disc under no-load conditions.

3. *Temperature and Frequency Errors.* The errors due to variation in temperature are very small, since the various effects due to change in temperature tend to neutralize each other on unity power factor if not on low pf *(lagging)*. Since the meters are used normally at fixed frequency, they can be adjusted to have a minimum error at declared supply frequency which is normally 50 Hz.

Example 9.7 **A correctly adjusted single phase 230 V induction watt-hour meter has a meter constant of 800 revolutions / kWh. Determine the speed of the disc for a current of 5 A at a lagging pf of 0.9.**

Solution: Supply voltage, V = 230 V

Load current, I = 5 A

$$\text{Power factor, } \cos\phi = 0.9$$

$$\text{Energy supplied per minute} = \frac{VI\cos\phi \times \text{time}}{1{,}000} = \frac{230 \times 5 \times 0.9}{1{,}000} \times \frac{1}{60} = 0.01725 \text{ kWh}$$

$$\text{Speed of the disc} = \text{Energy supplied per minute} \times \text{meter constant in rev/kWh}$$

$$= 0.01725 \times 800 = 13.8 \text{ rev/minute}$$ **Ans.**

Example 9.8 **An energy meter revolves 10 revolutions of disc for one unit of energy. Find the number of revolutions made by it during an hour when connected across load which takes 20A at 210V and 0.8 power factor leading. If energy meter revolves 350 revolutions, find the percentage error.**

[*U.P. Technical Univ. Electrical Engineering Second Semester 2004-05*]

Solution:

$$\text{Energy consumed in one hour} = \frac{V \times I \times \cos\phi}{1{,}000} = \frac{210 \times 20 \times 0.8}{1{,}000} = 3.360 \text{ kWh}$$

Number of revolutions the meter should make if it is correct

$$= 3.360 \times \text{registration constant in revolutions per kWh}$$

$$= 3.360 \times 100^* = 336$$ **Ans.**

$$\text{Number of revolutions actually made} = 350$$

$$\text{Percentage error} = \frac{350 - 336}{336} \times 100 = 0.04166\% \text{ fast.}$$

* It seems that there is a printing mistake in question paper, otherwise the percentage error will be too large.

Example 9.9 **A 40 amperes, 230 volts energy meter on full-load test makes 60 revolutions in 46 seconds. If the normal disc speed is 500 revolutions per kWh, find the percentage error with proper sign by assuming the load to be purely resistive.**

[*U.P. Technical Univ. Electrical Engineering Second Semester 2003-04*]

Solution:

$$\text{Energy consumed in 46 seconds} = \frac{VI\cos\phi}{1{,}000} \times t \text{ kWh} = \frac{230 \times 40 \times 1}{1{,}000} \times \frac{46}{60 \times 60} = 0.117555 \text{ kWh}$$

∵ for purely resistive load the power factor is unity.

$$\text{Energy consumption registered by the meter} = \frac{\text{Number of revolutions made}}{\text{Meter constant}} = \frac{60}{500} = 0.12 \text{ kWh}$$

$$\text{Percentage error} = \frac{\text{Actual registration} - \text{true energy consumption}}{\text{True energy consumption}} \times 100$$

$$= \frac{0.12 - 0.117555}{0.117555} \times 100 = +2.08\ \% \text{ i.e. } 2.08\ \% \text{ fast}$$ **Ans.**

Example 9.10 **A single-phase energy meter has a constant of 1,200 revolutions/kWh. When a load of 200 W is connected, the disc rotates at 4.2 revolutions/minute. If the load is on for 10 hours, how many units are recorded as an error ? Also find percentage error.** [*U.P. Technical Univ. Electrical Engineering Second Semester 2008-09*]

Solution:

$$\text{Actual energy consumed in 10 hours} = \text{Load in kW} \times \text{time in hours} = \frac{200}{1{,}000} \times 10 = 2 \text{ kWh}$$

$$\text{Revolutions made by disc in 10 hours} = \text{Speed of disc in rpm} \times \text{time in minutes} = 4.2 \times 10 \times 60 = 2{,}520$$

$$\text{Energy consumption recorded by meter} = \frac{\text{Revolutions made by the disc}}{\text{Meter constant}} = \frac{2{,}520}{1{,}200} = 2.1 \text{ kWh}$$

$$\text{Recording error} = \text{Energy recorded by meter} - \text{actual energy consumption}$$

$$= 2.1 - 2 = 0.1 \text{ kWh (excess)}$$ **Ans.**

$$\text{Percentage error} = \frac{2.1 - 2}{2} \times 100 = 5\%$$ **Ans.**

Example 9.11 **A 20 A, 220 V single-phase energy meter is tested. The number of revolutions per kWh is 240. When the load is 3,000 W, the disc makes 12 revolutions in 62 seconds. Find the percentage error.**

[*Mahamaya Technical Univ. Electrical Engineering Second Semester 2011-12*]

Solution: Actual energy consumed in 62 seconds = Load in kW × time in hours

$$= \frac{3{,}000}{1{,}000} \times \frac{62}{60 \times 60} = \frac{31}{600} \text{ kWh}$$

Revolutions made by disc in 62 seconds = 12

$$\text{Energy consumption recorded by meter} = \frac{\text{Revolution made by disc}}{\text{Meter constant}} = \frac{12}{240} = \frac{1}{20} \text{ kWh}$$

$$\text{Percentage error} = \frac{\text{Recording error}}{\text{Actual energy consumption}} \times 100$$

$$= \frac{\frac{1}{20} - \frac{31}{600}}{\frac{31}{600}} \times 100 = -3.226\% \text{ i.e. } 3.226\% \text{ (slow) } \textbf{Ans.}$$

Example 9.12 **A 50 A, 230 V energy meter on full-load test makes 61 revolutions in 37 seconds. If the normal disc speed is 500 revolutions per kWh, find the percentage error.**

[*Electrical Engineering U.P. Technical Univ. Second Semester 2005-06; G.B. Technical Univ. Even Semester 2012-13*]

Solution: Energy consumed in 37 seconds $= \frac{VI \cos \phi}{1{,}000} \times t$ kWh

$$= \frac{230 \times 50 \times 1}{1{,}000} \times \frac{37}{60 \times 60} \quad \text{assuming pf to be unity}$$

$$= 0.11819 \text{ kWh}$$

$$\text{Energy consumption registered by the meter} = \frac{\text{Number of revolutions made}}{\text{Meter constant}} = \frac{61}{500} = 0.122 \text{ kWh}$$

$$\text{Percentage error} = \frac{\text{Actual registration} - \text{true energy consumption}}{\text{True energy consumption}} \times 100$$

$$= \frac{0.122 - 0.11819}{0.11819} \times 100 = 3.22\% \text{ (fast). } \textbf{Ans.}$$

HIGHLIGHTS

1. **Moving iron instruments** are widely used in laboratories and switchboards at commercial frequencies because these are cheaper in cost, robust in construction and can be manufactured with required accuracy.
2. There are two general types of moving iron instruments *i.e.* (*i*) the *attraction type* and (*ii*) the *repulsion type*. The attraction type instruments operate on the principle of attraction of a single piece of soft iron into a magnetic field and repulsion type instruments operate on the principle of repulsion of two adjacent iron pieces magnetised by the same magnetic field.

   An **attraction type instrument** will usually have a lower inductance than the corresponding repulsion type instrument, and voltmeters will, therefore, be accurate over a wider range of frequency and there is a greater possibility of using shunts with ammeters. On the other hand, repulsion type instruments are more suitable for economical production in manufacture and a nearly uniform scale is more easily obtained; they are, therefore, much more common than the attraction type.
3. In a **PMMC instrument** the deflecting torque T_d is given by

   $T_d = BilNr$ N-m $= BiNa$ where $a = lr$, the area of the coil in m^2

4. **Dynamometer type instruments** are very similar to PMMC instruments except that the permanent magnetic field is replaced by a coil (usually two fixed air-cored coils are used) which carry the current to be measured (or a definite fraction of it or proportional to the voltage to be measured).

   Though these instruments can be used both, for dc and ac measurements but most of its practical uses are in ac as these instruments have higher cost, higher power consumption, lower torque-weight ratio, non-uniform scale and other several drawbacks in comparison to PMMC instruments.

   Such instruments may also serve as *transfer instruments* because they have precision grade accuracy (approximately 5% of full-scale deflection) both for dc and ac measurements and can be calibrated on dc and then used directly on ac establishing a direct means of equating ac and dc measurements of voltage and current.
5. The range of a PMMC ammeter can be extended by using a shunt. The value of resistance of shunt is given by

$$R_S = \frac{R_m}{\frac{I}{I_m} - 1} = \frac{R_m}{N-1}$$

   where R_m is the resistance of ammeter, I_m is the current flowing through the meter and I is the main circuit current. The ratio of the total current to the instrument current (i.e. I/I_m) is known as the *multiplying power* of shunt.
6. The range of a PMMC voltmeter can be extended by using a voltmeter multiplier (non-inductive series resistance). The value of voltmeter multiplier is given as

$$R = \frac{V}{I_m} - R_V$$

   where V is the voltage to be measured, I_m is the full-scale deflection current and R_V is the voltmeter resistance.
7. A **wattmeter** is essentially an inherent combination of an ammeter and a voltmeter and, therefore, consists of two coils, known as *current coil* and *pressure coil.* The operation torque is produced due to interaction of fluxes on account of currents in current and pressure coils.

   Dynamometer wattmeter is similar in design and principle to the dynamometer ammeter and voltmeter.

   The dynamometer instrument can be used on both dc and ac systems.
8. The essential difference between an energy meter and a wattmeter is that the former is fitted with some type of registration mechanism whereby all the instantaneous readings of power are summed over a definite period of time whereas the latter indicates the value at particular instant when it is read.

   Meter constant of energy meter is given as

$$\text{Meter constant, } K = \frac{\text{Number of revolutions}}{\text{kWh}}$$

   Percentage error in energy meter is given as

$$\text{Percentage error} = \frac{\text{Energy recorded by the meter} - \text{actual energy consumption}}{\text{Actual energy consumption}} \times 100$$

EXERCISES

1. Classify instruments and clearly differentiate between absolute and secondary instruments.
   [U.P. Technical Univ. Tutorial Question Bank]
2. Differentiate between recording and integrating type of instruments. Give two examples of each of these.
   [U.P. Technical Univ. Tutorial Question Bank]
3. What are the different torques required in an indicating type instruments? Draw and explain the working of attraction type moving iron instrument. **[G.B. Technical Univ. Electrical Engineering First Semester 2010-11]**
4. Explain the essential torques of an indicating instrument in detail.
   [G.B. Technical Univ. Electrical Engineering Even Semester 2012-13]

5. Describe the following in case of measuring instruments.

   (a) Deflecting torque (b) Controlling torque (c) Damping torque.

   [U.P. Technical Univ. Electrical Engineering First Semester 2003-04; Basic Electrical Engineering Second Semester 2013-14]

6. Describe the operating principle of measuring instruments. What are the various torques acting on the moving mechanism of the instruments ? How are these obtained and what are their roles in the operation of the instruments ?

   [U.P. Technical Univ. Electrical Engineering September 2001]

7. Discuss common features of indicating instruments. **[U.P. Technical Univ. Electrical Engineering February 2002]**

8. Discuss the principle, construction and operation of moving iron type measuring instruments.

   [U.P. Technical Univ. Electrical Engineering First Semester 2008-09]

9. Discuss construction and principle of operation of moving iron type measuring instruments. Also enlist advantages and disadvantages of it. **[Mahamaya Technical Univ. Electrical Engineering Second Semester 2010-11]**

10. Describe the construction and principle of operation of attraction type moving iron instrument.

    [Electrical Engineering G.B. Technical Univ. First Semester 2011-12; U.P. Technical Univ. Second Semester 2013-14]

11. Explain the principle and construction of attraction type moving iron instruments.

    [U.P. Technical Univ. Basic Electrical Engineering Second Semester 2013-14]

12. Explain the principle of operation of one type of moving iron instrument, showing how it is suitable for ac and dc measurement. **[U.P.Technical Univ. Electrical Engineering First Semester 2004-05]**

13. Explain the working of a moving iron type instrument. Derive the expression for torque produced.

    [U.P.Technical Univ. Electrical Engineering Second Semester 2001-02, 2004-05]

14. Explain the principle of operation of a repulsion type moving-iron instrument. Discuss its merits and demerits.

    [U.P. Technical Univ. Elec. Engineering February 2001]

15. Explain the construction and working principle of a moving iron attraction type instrument. Why is its scale non-uniform? **[U.P. Technical Univ. Electrical Engineering Second Semester 2007-08]**

16. Explain why moving iron type of instrument is suitable both on DC and AC. Also differentiate between moving iron type instrument and moving coil permanent magnet instrument.

    [G.B. Technical Univ. Electrical Engineering Second Semester 2009-10]

17. Discuss the construction and working principle of PMMC type measuring instruments.

    [G.B. Technical Univ. Electrical Engineering Second Semester 2010-11]

18. Explain the PMMC instrument. Also derive the expression for deflecting torque.

    [G.B. Technical Univ. Electrical Engineering First Semester 2010-11]

19. Explain the construction and working principle of a PMMC type instrument. What will it give as output if a half-wave rectified ac having peak value of 100 is given as input ?

    [U.P. Technical Univ. Electrical Engineering Second Semester 2007-08]

20. Describe the working principle of PMMC instruments. Why is the scale linear?

    [G.B. Technical Univ. Electrical Engineering Even Semester 2012-13]

21. Discuss the constructional details and working principle of PMMC type of voltmeters. What are the advantages and disadvantages of PMMC type of voltmeter? Explain why PMMC type of instruments belong to linear scale instrument.

    [U.P. Technical Univ. Electrical Engineering Odd Semester 2013-14]

22. Compare merits and demerits of moving coil and moving iron instruments.

    [U.P. Technical Univ. Electrical Engineering February 2002]

23. Derive the expression for torque produced in a moving coil type of instrument and explain, briefly, its working.

    [U.P. Technical Univ. Electrical Engineering Second Semester 2003-04]

24. Describe a PMMC instrument in detail. Also discuss its advantages and disadvantages.

    [U.P. Technical Univ. Electrical Engineering September 2001]

25. Enlist the advantages and disadvantages of permanent magnet moving coil (PMMC) type instruments.

    [U.P. Technical Univ. Electrical Engineering Second Semester 2002-03]

26. Describe with the aid of a labelled diagram, the constructional details of a moving coil instrument and explain how 'control' and 'damping' forces are obtained. **[U.P. Technical Univ. Electrical Engineering July 2002]**

27. Explain construction and principle of operation of dynamometer instrument. Why is this instrument suitable for both dc and ac measurements? **[U.P. Technical Univ. Electrical Engineering First Semester 2006-07]**

28. Explain the construction and working of electrodynamic type instrument with diagram. **[U.P. Technical Univ. Basic Electrical Engineering Odd Semester 2013-14]**

29. Compare merits and demerits of moving iron instruments and dynamometer type instruments. Which one is superior and why? **[U.P.Technical Univ. Elec. Engineering First Semester 2005-06]**

30. Describe deflecting torque, controlling torque and damping torque in measuring instruments. How range is extended in moving coil ammeter and voltmeter? **[Mahamaya Technical Univ. Electrical Engineering First Semester 1011-12]**

31. Discuss the following:
Use of shunt and multipliers in measuring instruments.
[G.B. Technical Univ. Electrical Engineering Second Semester 2010-11]

32. Give the construction and working of dynamometer wattmeter.
[G.B. Technical Univ. Electrical Engineering Second Semester 2009-10]

33. Describe the construction and working principle of a dynamometer type wattmeter and show how its deflecting force is proportional to the average value of power. **[U.P. Technical Univ. Electrical Engineering July 2002]**

34. With appropriate mathematical derivations, explain the principle of electrodynamic wattmeter. **[U.P. Technical Univ. Electrical Engineering January 2003]**

35. What is accuracy and resolution of an instrument? Briefly explain the induction type energy meter. **[G.B. Technical Univ. Electrical Engineering First Semester 2010-11]**

36. Explain construction, working and torque in energy meter with neat sketch. **[U.P. Technical Univ. Basic Electrical Engineering Odd Semester 2013-14]**

37. Explain construction and working principle of a single-phase induction type energy meter. How is energy measured? **[G.B. Technical Univ. Electrical Engineering First Semester 2009-10]**

38. Explain working of single-phase induction type energy meter with neat diagram. **[U.P. Technical Univ. Electrical Engineering Second Semester 2006-07]**

39. Describe the construction of a single-phase induction type energy meter. Show that the number of revolutions made by the disc is proportional to the energy supplied. **[U.P. Technical Univ. Elec. Engineering February 2001]**

40. Describe the working principle and construction of single phase energy meter and its applications. **[U.P. Technical Univ. Electrical Engineering Odd Semester 2013-14]**

41. Explain how the following adjustments are made in a single phase induction type energy meter ?
(*i*) Adjustment for friction compensation (*ii*) overload. **[U.P. Technical Univ. Tutorial Question Bank]**

42. Write short notes on the following :
(*i*) Moving iron instrument.
(*ii*) Permanent magnet moving coil (PMMC) instrument.
(*iii*) Extension of instrument range.
(*iv*) Dynamometer wattmeter.
(*v*) Induction type single phase energy meter.

SHORT ANSWER TYPE QUESTIONS WITH ANSWERS

Q. 1. What are the essentials of indicating instruments ?

Ans. Indicating instruments consist, essentially of a pointer moving over a calibrated scale and attached to the moving system pivoted on jewelled bearings. For satisfactory working of indicating instruments the torques required are: (*i*) deflecting torque, (*ii*) controlling torque and (*iii*) damping torque.

Q. 2. Why is controlling mechanism provided in indicating instruments ?

Ans. The controlling torque performs two functions. Firstly, it balances the deflecting torque and hence, insures that the magnitude of the deflection is always the same for a particular value of the measurand and secondly, it brings back the moving element to its zero position on removal of the source producing the deflecting torque.

Q. 3. How gravity control is inferior to spring control ?

Ans. Gravity-controlled instrument's scales are not uniform but are crowded in the beginning. This is drawback when the pointer lies at the lower scale values. Secondly these are to be used in vertical position in order that the control may operate.

Q. 4. Why is damping mechanism provided in electrical measuring instruments ?

Ans. Damping mechanism is provided in order to avoid oscillations of the moving element about its final deflected position owing to inertia of the moving parts and to bring the moving element to rest in its deflected position quickly.

Q. 5. When the instrument is said to be dead beat ?

Ans. When the degree of damping is such that the pointer (moving element) rises quickly to its deflected position without oscillations, the damping is said to be 'critical' and the instrument is said to be dead beat.

Q. 6. What are the various techniques by which damping torque is produced in electrical instruments ?

Ans. The various techniques of obtaining damping are (*i*) air friction, (*ii*) fluid friction and (*iii*) eddy current.

Q. 7. Which technique of obtaining damping is most efficient ?

Ans. Eddy current damping is most efficient.

Q. 8. Give the application of air friction damping.

Ans. Air friction damping is used in moving iron and dynamometer type instruments where the working magnetic field is weak and is likely to get distorted with the introduction of permanent magnet.

Q. 9. How does eddy current damping torque vary with the variation in the velocity of the moving element ?

Ans. The damping torque varies linearly with the variation in velocity of the moving element provided that the strength of the magnetic field is constant.

Q. 10. How does eddy current damping torque vary with the radial position of the poles relative to the axis of the disc ?

Ans. Damping torque decreases with the movement of the magnet towards the edge of the disc and becomes zero when the centres of the poles are at the edge of the disc.

Q. 11. What is difference between an ammeter and a voltmeter ?

Ans. Ammeter is a low resistance indicating instrument while the voltmeter is high resistance one.

Q. 12. Why an ammeter should be of very low resistance ?

Ans. *Ammeter,* which is connected in series with the circuit carrying the current under measurement, must be of *very low resistance* so that the voltage drop across the ammeter and power absorbed from the circuit are as low as possible.

Q. 13. Why a voltmeter should be of very high resistance ?

Ans. *Voltmeter,* which is connected in parallel with the circuit across which the voltage is to be measured, must be of very *high resistance* so that the current flowing through the voltmeter and the power absorbed from the circuit are minimum possible.

Q. 14. How an ammeter can be changed into a voltmeter ?

Ans. An ammeter of low range can be converted into a voltmeter by connecting a high resistance in series with it provided the current through the series combination is within the range of the ammeter when connected across the voltage under measurement.

Q. 15. What happens when an ammeter is connected across the circuit ?

Ans. If an ammeter is connected in parallel to the circuit like a voltmeter, a very high current will flow through it which will produce such an excessive heat that the insulation of the wire carrying the current will be destroyed. The wire may itself melt away. Thus the instrument will get damaged.

Q. 16. What happens when a voltmeter is connected in series with the circuit ?

Ans. If a voltmeter is connected in series with the circuit, the circuit resistance will become too large and consequently a very small current will flow through it. The instrument will, however, read almost the same emf acting in the circuit.

Q. 17. Enlist various types of moving iron instruments. **[G.B.T.U. Odd Semester 2012-13; Even Semester 2012-13]**

Ans. There are two general types of moving iron instruments *i.e.* (*i*) the attraction type and (*ii*) the repulsion type.

Q. 18. Why are the moving iron indicating instruments popular for the ordinary measurement work ?

Ans. Moving iron indicating instruments are popular for the ordinary measurement work because these are cheaper in cost, robust in construction, and can be manufactured with required accuracy.

Q. 19. Can the same moving iron instrument be used for both dc and ac measurements ?

Ans. Yes, the same moving iron instrument can be used for both dc and ac measurements because it has a sqaure-law response.

Q. 20. Why is the graduation of scale of moving iron instrument not uniform throughout ?

Ans. Since the moving iron instrument has a square-law response, the graduation of scale is not uniform throughout.

Q. 21. Why is eddy current damping not possible in moving iron instruments ?

Ans. Eddy current damping is not possible in moving iron instruments because of presence of permanent magnet required for such purpose would affect the deflection i.e. the reading of the instrument.

Q. 22. Why are irons made of mumetal employed in moving iron instruments ?

Ans. With the use of mumetal irons, the effects of saturation are avoided in the instruments and adequate torque can be obtained with low values of flux density since it has high permeability.

Q. 23. Why are repulsion type moving iron instruments more commonly used than the attraction type ones ?

Ans. Repulsion type moving iron instruments are more suitable for commercial production in manufacture and a nearly uniform scale is more easily obtained, they are, therefore, much more common than the attraction type.

Q. 24. What are the major differences between attraction and repulsion type of moving iron instruments?

Ans. An attraction instrument will usually have a lower inductance than the corresponding repulsion instrument, and voltmeters will, therefore, be accurate over a wider range of frequency and there is a greater possibility of using shunts with ammeters. On the other hand, repulsion instruments are more suitable for economical production in manufacture, and a nearly uniform scale is more easily obtained; they are, therefore, much more common than the attraction type.

Q. 25. Among moving-coil and moving-iron instruments, which one is suitable for both ac and dc measurements? Why?

[M.T.U. Second Semester 2011-12]

Ans. Among moving-coil and moving iron instruments, moving-iron instrument is suitable for both dc and ac measurements. This is because deflecting torque depends on the square of current or voltage under measurement.

Q. 26. Why permanent magnet moving coil instruments cannot be used for ac measurement of high frequency ?

Ans. At high frequencies, the pointer of a PMMC instrument will not be able to follow the rapid variations in direction and will quiver slightly around the zero mark, seeking the average value of ac i.e. zero. That is why, PMMC instruments are unsuitable for ac measurements unless the current is rectified before applying it to the PMMC instrument.

Q. 27. How is controlling torque provided in PMMC instruments ?

Ans. In PMMC instruments, controlling torque is provided by two phosphor-bronze hair springs, either helical or spiral, coiled in opposite directions, normally equal in strengths.

Q. 28. What type of damping is provided in PMMC instruments ?

Ans. Usually eddy current damping is used in PMMC instruments.

Q. 29. What is the major difference between PMMC type and dynamometer type instruments?

[G.B.T.U. Second Semester 2011-12]

Ans. Dynamometer type instruments are very similar to PMMC instruments except that the permanent magnetic field is replaced by a coil (usually two fixed air-cored coils are used) which carry the current to be measured (or a definite fraction of it or proportional to the voltage to be measured). The coils are usually air-cored to avoid hysteresis, eddy currents and other errors when the instrument is used on ac. The fixed coil is divided into two halves, connected in series with the moving coil; and placed together and parallel to each so as to provide a fairly uniform field within the range of the movement of the moving coil. However, the space between the two halves of the fixed coil must be sufficient enough to allow the movement of the moving coil shaft.

Q. 30. Why are dynamometer type instruments suitable both for dc and ac measurements?

Ans. The dynamometer type instruments are suitable both for dc and ac measurements because they have square-law response.

Q. 31. Why dynamometer type instruments are most suitable as transfer instruments?

Ans. A transfer instrument is one that may be calibrated with a dc source and then used without modification to measure ac. This requires the transfer instrument to have same accuracy for both dc and ac measurements. As the dynamometer type instruments have same accuracy on dc and ac both, they are most suitable as transfer instruments.

Q. 32. Why dynamometer type instruments are not usually used for dc measurements?

Ans. Dynamometer type instruments are usually not used for dc measurements as these instruments have higher cost, higher power consumption, low torque-weight ratio, non-uniform scale and other several drawbacks in comparison to PMMC instruments.

Q. 33. How is range extended in moving coil ammeter and voltmeter? **[U.P. Technical Univ. Second Semester 2013-14]**

Ans. The range of a PMMC ammeter can be extended by using a shunt. The value of resistance of shunt is given by

$$R_s = \frac{R_m}{\dfrac{I}{I_m} - 1} = \frac{R_m}{N - 1}$$

where R_m is the resistance of ammeter, I_m is the current flowing through the meter and I is the main circuit current. The ratio of the total current to the instrument current (i.e. I/I_m) is known as the *multiplying power* of shunt.

The range of a PMMC voltmeter can be extended by using a voltmeter multiplier (non-inductive series resistance). The value of voltmeter multiplier is given as

$$R = \frac{V}{I_m} - R_V$$

where V is the voltage to be measured, I_m is the full-scale deflection current and R_V is the voltmeter resistance.

Q. 34. What do you understand by ammeter shunt ?

Ans. An ammeter shunt is merely a low resistance that is placed across the coil circuit of the instrument in order to measure fairly large currents (Fig. 9.13)

Q. 35. What do you understand by voltmeter multiplier ?

Ans. Voltmeter multiplier is a high non-inductive resistance connected in series with the voltmeter coil and is used for increasing the range of a voltmeter (Fig. 9.14).

Q. 36. What is meant by instrument 'wattmeter' ?

Ans. A wattmeter is essentially an inherent combination of an ammeter and a voltmeter and, therefore, consists of two coils known as current coil (CC) and potential coil (PC).

Q. 37. Explain why you cannot measure power in ac circuits by using an ammeter and voltmeter.

Ans. Power in ac circuit is given by $P = VI \cos \phi$ where V and I are rms values of voltage and current respectively and ϕ is the phase angle between V and I. The fact that the power factor ($\cos \phi$) is involved in the expression for power, we cannot measure power in ac circuits by using an ammeter and a voltmeter.

Q. 38. State the nature of scale distribution for a dynamometer instrument used as wattmeter.

Ans. Almost linear scale in the usual range of the instrument.

Q. 39. Why an ordinary electrodynamometer wattmeter is not suitable for measurement of power in low power-factor circuits ?

Ans. Ordinary electrodynamometer wattmeter is not suitable for measurement of power in low-power factor circuits owing to (*i*) small deflecting torque on the moving system even when the current and pressure coils are fully excited and (*ii*) introduction of large error due to inductance of pressure coil at low power factor.

Q. 40. Name the meter used for measurement of electrical energy of consumer. **[M.T.U. First Semester 2011-12]**

Ans. Energy meter (kWh meter) is used for measurement of electrical energy of consumer.

Q. 41. What is the difference between an energy meter and a wattmeter ?

Ans. The essential difference between an energy meter and a wattmeter is that the former is fitted with some type of registration mechanism whereby all the instantaneous readings of power are summed over a definite period of time whereas the latter indicates the value of power at particular instant when it is read.

Q. 42. Why the rotating system of the energy meter is made as small as possible ? **[Aligarh Univ. 1986]**

Ans. The rotating system of the energy meter is made as small as possible so that frictional torque is reduced to the minimum.

Q. 43. Why use of aluminium disc is preferred over copper disc ?

Ans. An aluminium disc is usually preferred to a copper disc in order to have resistance per unit weight smaller.

Q. 44. What is meant by meter constant of an energy meter ?

Ans. The number of revolutions made by the aluminium disc for 1 kWh of energy consumption is called the meter constant of an energy meter.

Q. 45. Explain the term creep in energy meter. **[U.P. Technical Univ. Second Semester 2013-14]**

Ans. In some energy meters, the disc continues rotating when the potential coils are excited but with no load current flow. This defect is known as creeping.

Q. 46. How creeping can be avoided in energy meters ?

Ans. Creeping error can be avoided by either cutting two holes (or slots) in the disc on the opposite sides of the spindle or attaching a small piece of iron wire to the edge of the disc.

Q. 47. What is the effect of bringing the shading band nearer to the disc on the speed of the disc?

Ans. The speed of the disc will decrease.

Q. 48. What is self braking torque? Why is it important on heavy loads ?

Ans. The disc rotates continuously in the series magnetic field under load conditions and so there is a dynamically induced emf in the disc which induces eddy currents. Due to interaction of eddy currents with the field of series magnet a braking torque proportional to the square of the energizing current is developed. The braking torque so developed is called the self braking torque.

The self braking torque assumes increasing importance with the increase in load current and thus at high values of load current the registration tends to be low.

Q. 49. Why the speed of the rotating disc in energy meter is kept as small as possible ?

Ans. The full-load speed of the rotating disc in energy meter is kept as small as possible so that the self braking action is reduced to minimum.

PROBLEMS

1. If the moving coil consists of 150 turns wound on a square former which has a length of 4 cm and the flux density in the air gap is 0.06 T, calculate the turning moment acting on the coil when it is carrying a current of 12 m A.

[**Ans.** 172.8×10^{-6} Nm]

2. A moving coil instrument gives a full-scale deflection of 10 m A when the potential difference across its terminals is 100 mV. Calculate the series resistance to measure 1,000 V on full scale. [**Ans.** 99,990 Ω]

3. A moving-coil milliammeter having a resistance of 8 Ω gives full-scale deflection when a current of 5 mA is passed through it. Explain how this instrument can be used for measurement of (*i*) current up to 2 A (*ii*) voltages up to 8 V.

[G.B. Technical Univ. Electrical Engineering Second Semester 2009-10]

[**Ans.** (*i*) 20.5 mΩ (*ii*) 1.592 kΩ]

4. A moving coil ammeter has a resistance of 0.01 Ω and full-scale deflection current of 0.25 A. How this meter can be made to read (*i*) voltage up to 250 V (*ii*) current up to 20 A ? [**Ans.** (*i*) 999.99 Ω (*ii*) 1.266×10^{-4} Ω]

5. A moving coil milliammeter gives full-scale deflection with 15 m A and has a resistance of 5 Ω. Calculate the resistance to be connected in (*a*) parallel to enable the instrument to read up to 1 A, (*b*) series to enable it to read upto 10 V.

[U.P. Technical Univ. Electrical Engineering July 2002]

[**Ans.** (*a*) 0.07614 Ω (*b*) 661.6667 Ω]

6. An ammeter gives full-scale deflection on 100 m A and has a resistance of 10 Ω. Show how it can be used to measure (*i*) current up to 5 A and (*ii*) voltage up to 100 V. Calculate the values of additional resistances also.

[**Ans.** (*i*) 0.20408 Ω (*ii*) 990 Ω]

7. A single-phase energy meter makes 500 revolutions per kWh. It is found on testing as making 40 revolutions in 58.1 seconds at full load of 5 kW. Determine the percentage error. [**Ans.** – 0.855%]

8. A single-phase energy meter has a constant speed of 1,300 revolutions/kWh. The disc revolves at 3.5 rpm when a load of 150 W is connected to it. If the load is on for 11 hours, how many units are recorded as error ? What is the percentage error ? **[U.P. Technical Univ. Electrical Engineering September 2001]**

[**Ans.** 0.127 kWh (excess); 7.69% (fast)]

❑❑❑

UNIT-IV

Introduction To Earthing and Electrical Safety, Magnetic Circuit and Single-Phase Transformers

CONTENTS IN THIS UNIT

Introduction To Earthing and Electrical Safety: Need of earthing of equipment and devices, Important electrical safety issues.

Magnetic Circuit: Magnetic circuit concept, Analogy between electric and magnetic circuits, B-H curve, Hysteresis and eddy current losses, Mutual coupling and dot convention, Magnetic circuit calculations.

Single-Phase Transformer: Principle of operation, Construction, EMF equation, Equivalent circuit, Power losses, Efficiency (simple numerical problems), Introduction to autotransformer.

10 CHAPTER

Introduction To Earthing and Electrical Safety

10.1. GENERAL

There is no danger in operating or servicing electrical equipment, provided attending personnel observe all rules and regulations of operation and safety.

Protective facilities may be classified into basic and auxiliary. Basic facilities are those whose insulation will stand the operating voltage of the plant/equipment under service and allow contact with live (or energized) parts. Those for use at any voltage include insulating rods for routine switchings, measurements, applications of grounds, etc., and also fuse pullers. Those for low voltages additionally include rubber gloves and gauntlets, and tools with insulated handles.

Auxiliary facilities are those which by themselves protect from any electric shock and only serve as adjuncts to basic facilities, although they are capable of protecting the attending personnel from touch voltage, step voltage and burns due to an electric arc. Auxiliary devices for work at high voltages are rubber gloves and gauntlets, rubber boots, rubber mats, and insulating supports. In all work on high-voltage equipment, use must be made of basic facilities in conjunction with auxiliary facilities. All facilities, both in use and in storage, must be registered and regularly checked for their condition.

Electrical equipment is classed as low voltage (*lv*) if the voltage between any line and earth does not exceed 250 V and high-voltage (*hv*) if the voltage between any line and earth exceeds 250 V.

In *lv* equipment, all current carrying parts must be protected to avoid accidental contact. In damp locations and also when working on solidly grounded structures, like boilers, bridges, etc., step-down transformers (for 12 V and 24 V) should be used.

All metal parts which are located closed to live parts and may come in contact with a latter must be grounded.

Installation and maintenance work is performed only on de-energized equipment. If the equipment cannot be de-energized for one reason or another, work on live equipment must be done in conformity with the relevant safety rules, using protecting facilities (insulating supports, rubber boots, "live" tools with insulated handles, temporary grounds and short-circuits, insulating sticks, and protective goggles).

Wherever required, caution signs must be posted (skull crossbones) and the inscription "Death on Contact" or "High Voltage–Dangerous to Life", or other similar inscription. Caution signs must be sufficiently large, with inscriptions in large clear letters made with indelible paint. The smallest size for caution signs is 20 cm × 10 cm.

In work on *hv* equipment, the precautions to be observed are:

(*i*) Work must be done by a crew of not less than two, so that one of the attendants can help in case of an accident.

(*ii*) The attendants must be well insulated from earth:

(*iii*) While working, the men must not touch persons not standing on insulating supports and must not come in contact with metal parts like machines, piping etc.

(*iv*) The workers must inspect all protective facilities before beginning any job.

Maintenance work on *hv* equipment (substations, transformer rooms and parts, cables, etc.) is begun only after verbal or written permission has been taken and the maintenance workers have informed that the equipment to be serviced has been de-energized.

Before commencement of maintenance work, a check should be made with appropriate testing equipment to see that the equipment to be serviced is dead. Then collecting buses and transformer cables must be discharged, tested for a short-circuit, short-circuited and reliably grounded.

Also prior to work, ladders, scaffoldings, cradles and ropes must be checked for possible defects.

10.2. EARTHING

The term '*earthing*' means connecting of the non-current carrying parts of the electrical equipment or the neutral point of the supply system to the general mass of earth in such a manner that at all times an immediate discharge of electrical energy takes place without danger. Earthing is achieved by electrically connecting the respective parts in the installations to some system of electrical conductors or electrodes placed in intimate contact with the soil some distance below the ground level. This contacting assembly is called the *earthing*. The metallic conductors connecting the parts of the installations with the earthing are called *electrical connections*. When speaking of the earthing and the earthing connections as a whole, it is customary to use the term the *earthing system.*

The earthing is provided with the following objectives:

1. For the safety of the personnel from the electric shock—insuring that non-current carrying parts, such as equipment frames are always safely at ground potential even though insulation fails.
2. For the safety of the equipment and personnel against lightning and voltage surges providing the discharge path for surge arresters, gaps and other similar devices.
3. For providing the ground connections for grounded neutral systems.
4. For providing a means of positively discharging and de-energizing feeders or equipment before proceeding with maintenance on them.

Power system earthing is highly important. A *substantial and adequate ground* that will not burn off or permit dangerous rise in voltage under abnormal conditions is essential. Extensive damage and dangerous conditions have arisen when inadequate grounds have been provided. Multiple grounds and multiple connections to them are usually desirable to ensure ground protection, even though one ground or connection opens owing to burn-off or other condition such as high resistance. The station earthing system should have low resistance.

The earthing can be divided into *neutral* and *equipment* earthing. Neutral earthing (or grounding) deals with earthing of the system neutral to ensure system security and protection. Neutral earthing is also called the *system earthing*. Equipment earthing or safety earthing deals with earthing of non-current carrying parts of the equipment to ensure safety of personnel and protection against lightning.

Equipment earthing also helps in the earth fault protection. The earth fault current from the equipment flows through the earthing system to the earth and is sensed by protection system and circuit breakers are opened. The faulty equipment is then repaired and recommissioned. The earthed parts remain at approximately earth potential even during flow of fault current. In *neutral earthing* (or *neutral grounding*), the neutral points (or star points) of star-connected 3-phase windings of power transformers, generators, motors, earthing transformers are connected to low resistance earth. Neutral earthing has numerous advantages.

System earthing and equipment earthing have common earth mat* from which flat iron risers are taken out to connect all the non-current carrying metal parts of the equipment. At the time of fault the earth mat conductors rise to a voltage, which is equal to the resistance of the earth mat multiplied by ground fault current. This difference of potential results in voltage gradients**. The earth mat of the substation is so designed that the *step potential* (potential difference between two feet of a person at the time of fault) and *touch potential* (potential difference to which body is subjected to when a person comes in contact with an earthed body) are within safe limits. The average resistance of human body is 1,000 Ω and a current of 9 to 25 mA may be painful for the body for long duration. As fault is cleared very fast, *i.e.,* within 2 to 4 cycles (40 to 80 milliseconds) therefore 9 mA current is considered safe.

In electrical installations the following components must be earthed :

(*i*) The frames, tanks and enclosures of electric machines, transformers, and apparatus, lighting fittings, and other items of equipment;

(*ii*) the operating mechanism of the switchgear;

(*iii*) the framework of the switchboards, control boards, individual panel boards, cubicles;

(*iv*) the structural steelwork of indoor and outdoor substations, metal cable jointing boxes, the metal sheaths of the cable, the rigid metal conduit runs, and similar metalwork.

10.3. DEFINITIONS

1. Earth Electrode. Any wire, rod, pipe, plate or an array of conductors, embedded in ground horizontally or vertically is known as the *earth electrode.* In distribution systems the earth electrode may consist of a rod, about 1 metre in length and driven vertically into the ground. In generating stations and substations, grounding mats are used rather than individual rods.

2. Resistance-To-Earth or Earth Resistance. The resistance offered by the earth electrode to the flow of current into the ground is known as *resistance-to-earth.* The term resistance-to-earth (or earth resistance) does not imply the contact resistance between the earthing electrode and the soil (which is insignificant), but mainly the resistance of soil between the electrode and the point of zero potential. Numerically it is equal to the ratio of the potential of earth electrode with respect to remote point, to the current dissipated by it.

Mathematically, earth resistance is given as $\frac{V}{I}$ where V is the measured voltage between the electrode and the voltage spike and I is the injected current during the earth resistance measurement through the electrode.

As a general rule the lower the value of earth resistance better it is but even then the following values of earth resistance (maximum permissible) values will give satisfactory results.

Large power stations	— 0.5 Ω
Major power stations	— 1.0 Ω
Small substations	— 2.0 Ω
In all other cases	— 8.0 Ω

3. Earthing Lead. The wire which connects earth wire or any other apparatus to be earthed to the earth electrode is known as the earthing lead.

4. Earth Current. The current dissipated by the earth electrode into the ground is called the *earthing current.*

* A number of rods when joined together through copper conductors constitute an earthing mat or earthing grid.

** The resistance of an earthing electrode is not concentrated at one point but is distributed over the soil around the electrode.

5. Step Potential. Step potential is the potential difference between the feet of a person standing on the floor of the substation, with 0.5 m spacing between two feet (one step), during the flow of earth fault current through the grounding system.

6. Touch Potential. Touch potential is a potential difference between the fingers of raised hand touching the faulted structure and the feet of the person standing on substation floor. The person should not get a shock even if the grounded structure is carrying fault current, *i.e.,* the touch potential should be very small.

7. Mesh Potential. It is the maximum touch potential within a mesh of the grid.

8. Transferred Potential. It is a special case of touch potential where a potential is transferred into or out of the substation.

10.4. EFFECTS OF ELECTRIC CURRENTS ON HUMAN BODY

Bruner (1967) states that the threshold of perception of electric shock is about 1 mA. At this level a tingling sensation is felt by the subject when there is a contact with an electrified object through intact skin. With the increase in magnitude of ac, the sensation of tingling gives way to contraction of muscles. The muscular contractions increase as the current is increased and finally a value of current is reached at which the subject cannot release his grip on the current carrying conductor. The maximum current at which the subject is still capable of releasing a conductor by using muscles directly stimulated by the current is called "*let go current*". The value of this current is significant because an individual can withstand, without serious after-effects, repeated exposures to his 'let go current' for at least the time required for him to release the conductor. Also, currents slightly in excess of 'let go current' would not permit the individual to release his grip from the conductor supplying current.

Based on the experiments conducted on males and females, it is generally accepted that the safe 'let go current' could be taken approximately 9 mA and 6 mA for men and women respectively.

At current levels higher than the 'let go current' the subject loses ability to control his own muscle actions and he is unable to release his grip on the electrical conductor. Such currents are very painful and hard to bear. This type of accident is called 'hold-on-type' accident, and is caused by currents in the range 20-100 mA. These currents may also cause physical injury due to powerful contraction of the skeletal muscles. However, the heart and respiratory function usually continue because of uniform spread of current through the trunk of the body.

If current contacts intact skin and passes through the trunk, at about 100 mA and above, there is a likelihood of pulling the heart into ventricular fibrillation. In this condition, the rhythmic action of the heart ceases, pumping action stops and the pulse disappears. Ventricular fibrillation is a serious cardiac emergency because once it starts, it practically never stops spontaneously. It proves *fatal* unless corrected within minutes, since the brain begins to die 2 to 4 minutes after it is robbed of its supply of oxygenated blood.

At very high currents of the order of 6 A and above, there is a danger of temporary respiratory paralysis and also of serious burns. However, if the shock duration is of only a very few seconds, there is a possibility of heart reverting to the normal rhythmic action.

The threshold of perception depends largely on the current density in the body tissues. It may vary widely depending upon the size of the current contact. At very small point contact, it is probable that even 0.3 mA current may be felt whereas a current in excess of perhaps 1 mA may not produce sensation if the contacts are somewhat larger. Similarly, depending on the size of contact, the threshold of pain may also be considerably above 1 mA, probably 10 mA if the contacts are large enough.

Besides the magnitude of current, the current duration and the relationship of current flow resistance are also important. Duration of less than 10 ms typically does not produce fibrillation whereas duration of 0.1 s or longer does. It has been found experimentally that the safe value of current in amperes (rms) which a human body can tolerate is given as

$$I = \frac{0.165}{\sqrt{t}} \text{ for } t < 3 \text{ s and } I = 9 \text{ mA for } t > 3 \text{ s.}$$

where t is the time duration in seconds of the flow of current.

The tolerable currents mentioned above are for power-frequency currents. It has been found that human body can tolerate about 5 times higher direct current. At high frequencies (3-10 kHz) still higher currents are tolerable.

10.5. EARTHING CONDUCTOR

Earthing conductor is of high conductivity copper and is of either stranded, flat strips, or circular or rectangular bar. It is protected against mechanical injury. Bare conductor is protected against corrosion. Galvanised solid iron or steel wire or rod or any other suitable approved material can be used provided the conductivity is not less than the copper earthing conductor. The cross-sectional area of copper earthing conductor is never smaller than half that of the largest current carrying conductor subject to an upper limit of 65 mm^2. However, for equipment of capacity larger than 750 kVA, the size is determined as in Appendix A of IS 1886-1961 [Code of Practice for Installation and Maintenance of Transformers].

Earthing conductors are so placed and connected that they may not get accidentally damaged or cut. They are fixed over their entire length by clamps, clips, saddles, staples, clouts etc. which in no way damage the conductor. Aerial earthing conductor is supported on suitable insulators and is clearly identified. It is protected against mechanical damage, when buried. Joints in main earthing conductors are made by soldering, brazing or welding for conductors of size up to 7/1.60 mm, for larger main earthing conductors mechanical clamping is used. Joints in other earthing conductors are made by soldering or by mechanical coupling.

10.6. METHODS OF EARTHING

The various methods of earthing are:

1. Strip or Wire Earthing. In this system of earthing, strip electrodes of cross section not less than 25 mm × 1.6 mm if of copper and 25 mm × 4 mm if of galvanised iron or steel are buried in horizontal trenches of minimum depth 0.5 metre. If round conductors are used, their cross-sectional area shall not be smaller than 3.0 mm^2 if of copper and 6 mm^2 if of galvanised iron or steel. The length of buried conductor shall be sufficient to give the required earth resistance. It shall, however, be not less than 15 metres. The electrodes shall be as widely distributed as possible, preferably in a single straight or circular trench or in a number of trenches radiating from a point. If conditions require use of more than one strip, they shall be laid either in parallel trenches or in radial trenches.

This type of earthing is used at places which have rocky soil earth bed because at such places excavation work of plate earthing is difficult.

2. Rod Earthing. In this system of earthing, 12.5 mm diameter solid rods of copper or 16 mm diameter solid rods of galvanised iron or steel or hollow section 25 mm GI pipes of length not less than 2.5 metres are driven vertically into the earth either manually or by pneumatic hammer. In order to increase the embedded length of electrodes under the ground, which is sometimes necessary to reduce the earth resistance to desired value, more than one rod sections are hammered one above the other.

This system of earthing is suitable for areas which are sandy in character. This system of earthing is very cheap as no excavation work is involved.

3. Pipe Earthing. This is the most common and best system of earthing as compared to other systems suitable for the same earth and moisture conditions.

In this method of earthing, a galvanised steel and perforated pipe of approved length and diameter is placed upright in a permanently wet soil, as shown in Fig. 10.1.

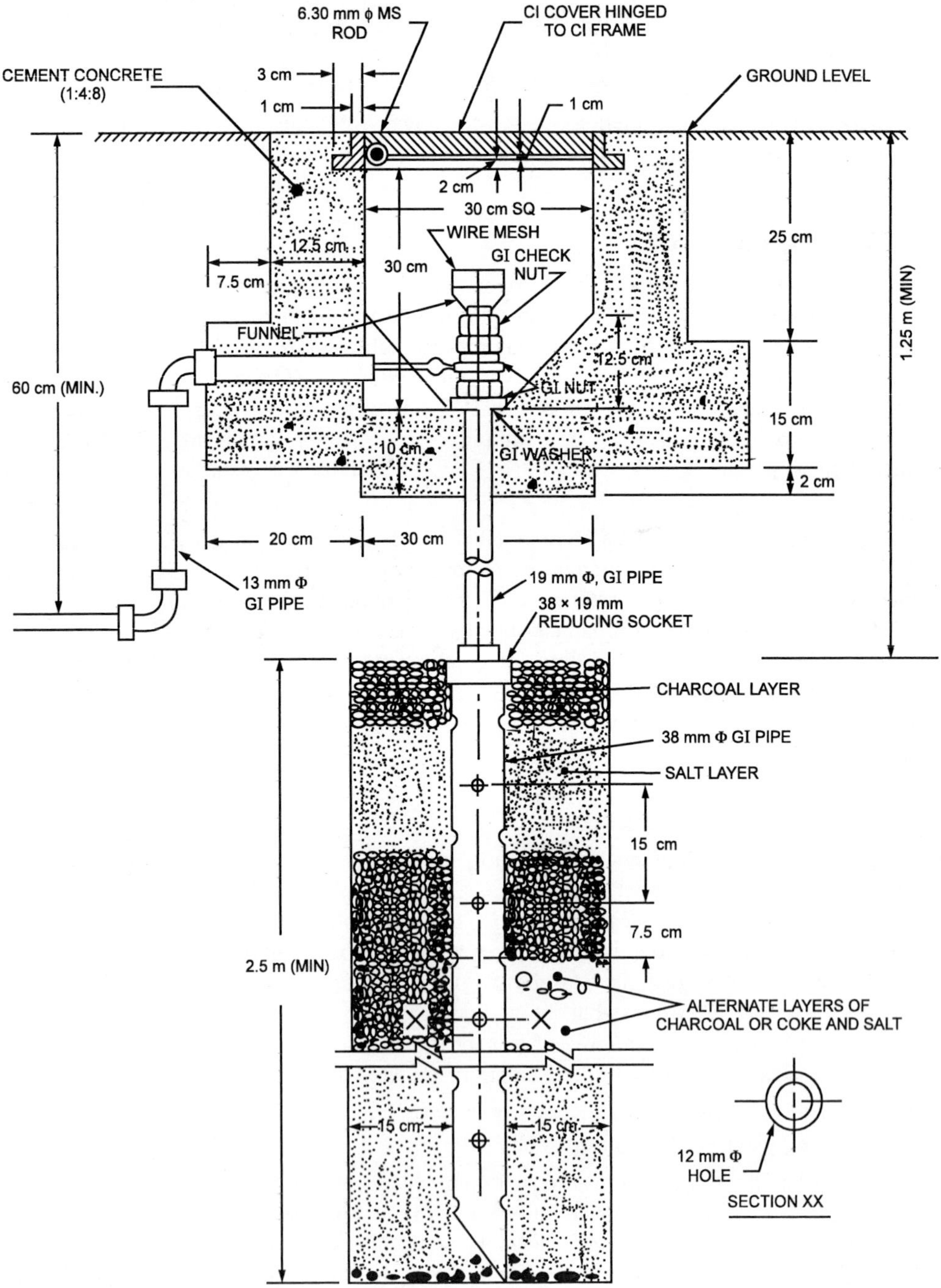

Note-Three or four buckets of water to be poured into sump every few days to keep the soil surrounding the earth pipe permanently moist.

Fig. 10.1 *A Typical Illustration of Pipe Earth Electrode*

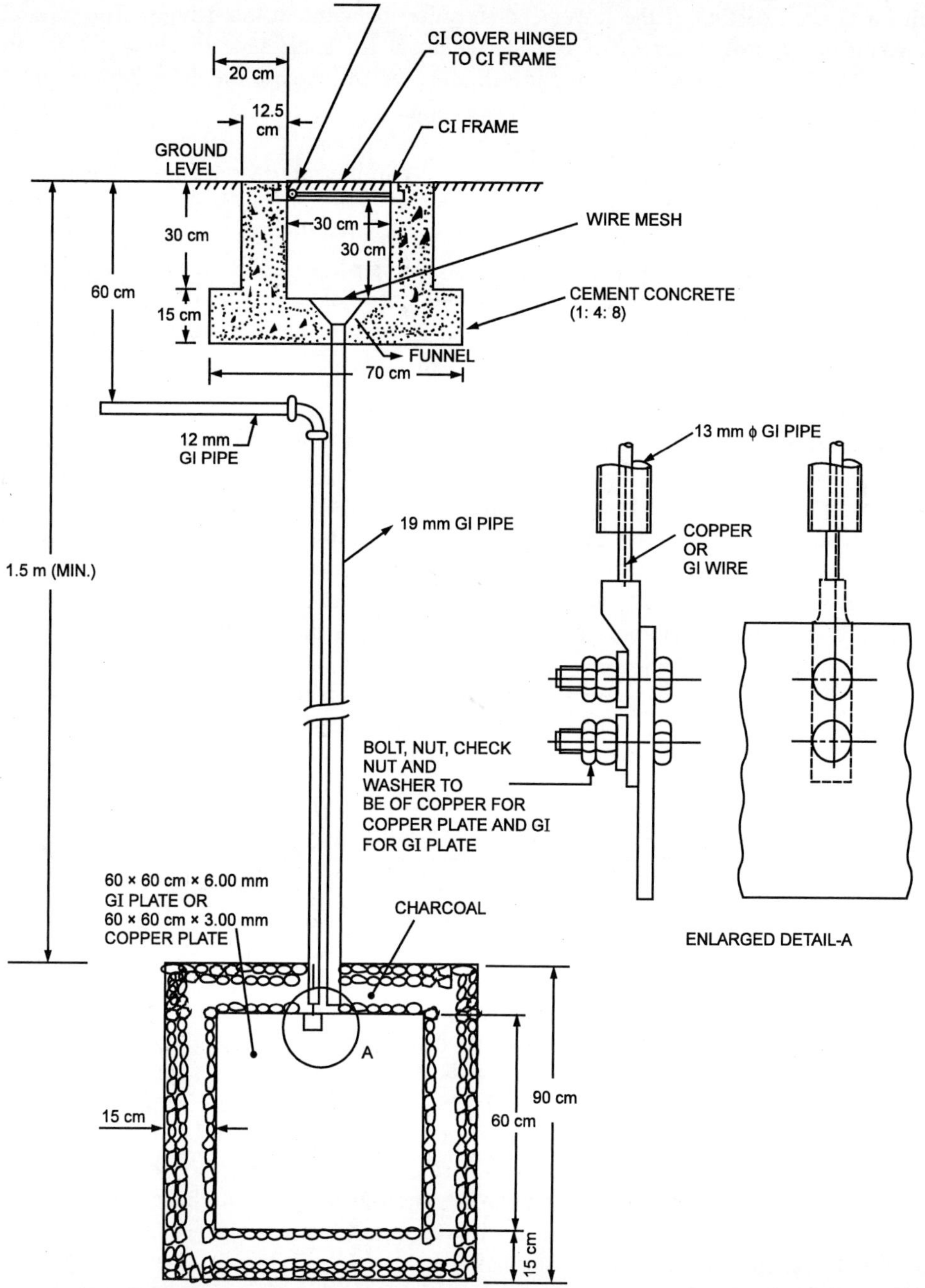

Note-Three or four buckets of water to be poured into sump every few days to keep the soil surrounding the earth pipe permanently moist.

Fig. 10.2 *A Typical Illustration of Plate Earth Electrode*

The size of the pipe depends upon the current to be carried and type of soil. Usually the pipe used for this purpose is of diameter 40 mm and 2.5 metres in length for ordinary soil or of greater length in case of dry and rocky soil. The depth at which the pipe must be buried depends upon the moisture of the ground. The pipe is placed at a depth of 3.75 metres (minimum). The pipe is provided with a tapered casting at the lower end in order to facilitate the driving. The pipe at the bottom is surrounded by broken pieces of coke or charcoal for a distance of about 15 cm around the pipe. Generally alternate layers of coke and salt are used to increase the effective area of the earth and to decrease the earth resistance respectively. Another pipe of 19 mm diameter and minimum length 1.25 metres is connected at the top of GI pipe through reducing socket.

In our country in summer season the moisture in the soil decreases which causes increase in earth resistance. So a cement concrete work, as shown in Fig. 10.1, is done in order to keep the water arrangement accessible, and in summer to have an effective earth, 3 or 4 buckets of water are put through the funnel connected to 19 mm diameter pipe, which is further connected to GI pipe.

The earth wire (either GI wire or GI strip of sufficient cross section to carry faulty current safely) is carried in a GI pipe of diameter 12 mm at a depth of about 60 cm from the ground.

Care should be taken that earth wire is well protected from mechanical injury, when it is carried over from one machine to another.

4. Plate Earthing. This is another common system of earthing. In plate earthing an earthing plate either of copper of dimensions 60 cm × 60 cm × 3 mm or of galvanised iron of dimensions 60 cm × 60 cm × 6 mm is buried into the ground with its face vertical at a depth of not less than 3 metres from ground level. The earth late is embedded in alternate layers of coke and salt for a minimum thickness of 15 cm. The earth wire (GI wire for GI plate earthing and copper wire for copper plate earthing) is securely bolted to an earth plate with the help of a bolt, nut and washer made of material of that of earth plate (made of copper in case of copper plate earthing and of galvanised iron in case of GI plate earthing).

A small masonry brick wall enclosure with a cast iron cover on top or an RCC pipe round the earth plate is provided to facilitate its identification and for carrying out periodical inspection and tests.

The copper plate and copper wire are usually not employed for grounding purposes because of their higher cost.

10.7. PROTECTIVE DEVICES FOR WIRING INSTALLATIONS AND MOTORS

The most frequent causes for abnormal conditions are overloads and short circuits, therefore, protection is to be provided to wiring installations and motors against overloading and short circuiting. The devices used are fuses, relays, miniature circuit breakers (MCB), earth leakage circuit breakers.

Fuses are used for low and high voltage circuits of moderate to high capacity, where frequent operations are not expected e.g. for the protection of distribution transformers, small and medium size motors, lighting circuits, branch circuits of distribution lines, and industrial plants and commercial buildings up to 600 A at 250 V, 500 A at 600 V and in high voltages up to 100 A at 132,000 V. The protection by fuses is advantageous from the point of view of cost but it does not permit supervision and are to be replaced in the field. Moreover, when a fuse blows, it takes quite some time to renew it and restore the supply. Circuit breakers are much more convenient and quick and are very useful where slight overloads occur often or where heavy currents have to be broken.

Circuit breakers are mechanical devices designed to close or open contact members, thus closing or opening an electric circuit under normal and abnormal conditions. Automatic circuit breakers, which are usually employed for the protection of electric circuits, are equipped with a trip coil connected to a relay or other means designed to open the circuit breaker automatically under abnormal conditions, such as overcurrent. The automatic circuit breakers perform the following duties :

(*i*) It carries the full-load current continuously without overheating or damage.
(*ii*) It opens and closes the circuit on load.
(*iii*) It makes and breaks the normal operating current.
(*iv*) It makes and breaks the short-circuit currents of magnitude up to which it is designed for.

A circuit breaker consists of fixed and moving contacts which are touching each other under normal conditions i.e. when circuit breaker is closed. Whenever a fault occurs, the trip coils get energized, the moving contacts are pulled by some mechanism and therefore the circuit breaker is opened and the circuit is broken.

A more economical device used for protection of a motor is the externally mounted **thermal overload relay.** The construction of a thermal overload relay, used in magnetic starters, is shown diagrammatically in Fig. 10.3. The key element of a thermal overload relay is bimetallic strip and placed next to the bimetallic strip is a heater coil. When the coil gets sufficiently heated due to excessive current, the bimetallic strip flexes upwards at its outer end and releases a latch. A spring rotates the latch about its axis to actuate the tripping member which opens the normally closed contacts. The relay is manually reset after tripping by means of a reset push button.

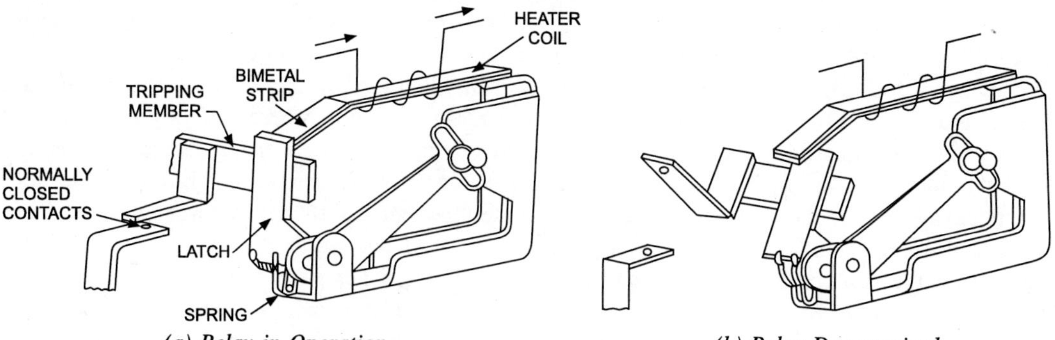

(a) Relay in Operation *(b) Relay De-energized*

Fig. 10.3 *Thermal Overload Relay*

Thermal overload relays provide protection for the motor only in the case of an overload and do not provide protection against short circuits. This is because a short circuit is required to be cleared instantaneously, while the heater coil takes some time to heat the metal strip to the set temperature.

Miniature circuit breaker (MCB) is a device that provides definite protection to the wiring installations and sophisticated equipment against overcurrents and short-circuit faults. The outer and

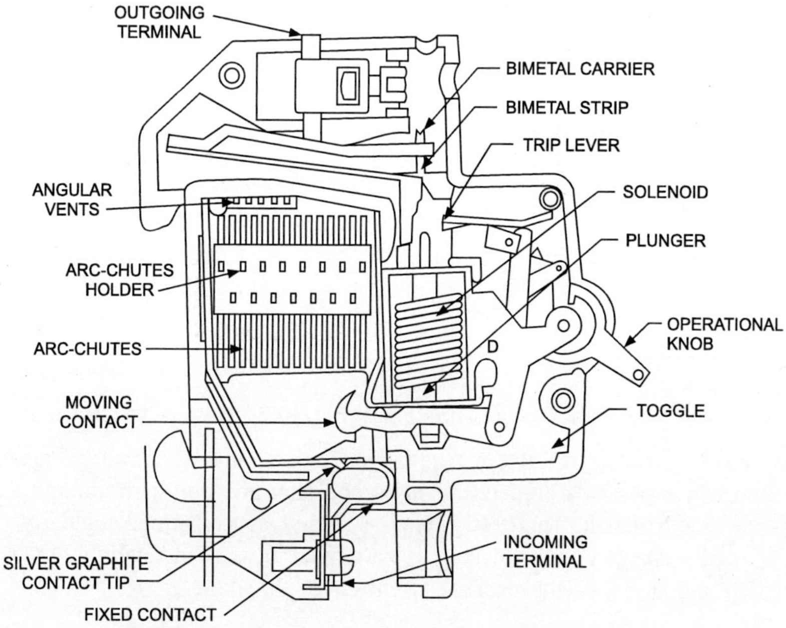

Fig. 10.4 *Miniature Circuit Breaker (Courtesy Havells)*

interior views of an MCB are shown in Fig. 10.4. Thermal operation (overload protection) is achieved with a bimetallic strip, which deflects when heated by any overcurrents flowing through it. In doing so, releases the latch mechanism and causes the contacts to open. Inverse time-current characteristics result, *i.e.*, greater the overload or excessive current, shorter the time required to operate the MCB. On the occurrence of a short circuit, the rising current energizes the solenoid, operating the plunger to strike the trip lever causing immediate release of the latch mechanism. Rapidity of the magnetic solenoid operation causes instantaneous opening of contacts.

Miniature circuit breakers are available with different current ratings of 0.5, 1, 2, 2.5, 3, 4, 5, 6, 7.5, 10, 16, 20, 25, 32, 35, 40, 63, 100, 125, 160 A and voltage ratings of 240/415 V ac and up to 220 V dc. Operating time is very short (less than 5 ms). So they are very suitable for the protection of important and sophisticated equipment, such as air-conditioners, refrigerators, computers etc.

Earth-leakage circuit breaker (ELCB) is a device that provides protection against earth leakage. These are of two types viz, the current operated type and the voltage operated type.

Current operated earth-leakage circuit breaker is used when the product of the operating current in amperes and the earth-loop impedance in ohms does not exceed 40. Where such a circuit breaker is used, the consumer's earthing terminal is connected to a suitable earth electrode. A current-operated earth leakage applied to a 3-ϕ, 3-wire circuit is shown in Fig. 10.5. In normal conditions when there is no earth leakage the algebraic sum of the currents in the three coils of the current transformers (CTs) is zero, and no current flows through the trip coil. In case of any earth leakage, the currents are unbalanced and the trip coil is energized and thus the circuit breaker is tripped.

Voltage-operated earth leakage circuit breaker is suitable for use when the earth-loop impedance exceeds the values applicable to fuses or excess-current circuit breaker or to current-operated earth-leakage circuit breaker. Such an earth-leakage trip in a 2-wire circuit is shown in Fig. 10.6. When the voltage between the earth continuity conductor (ECC) and earth electrode rises to a sufficient value, the trip coil will carry the required current to trip the circuit breaker. With such a circuit breaker the earthing lead between the trip-coil and the earth electrode must be insulated; in addition, the earth electrode must be placed outside the resistance area of any other parallel earths which may exist.

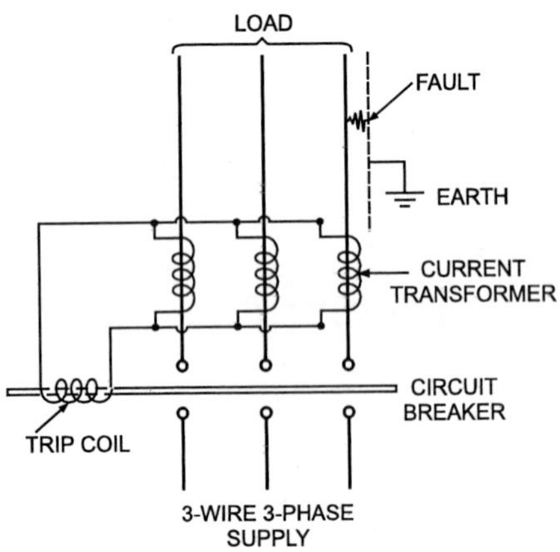

Fig. 10.5 *Current-Operated Earth-Leakage Trip*

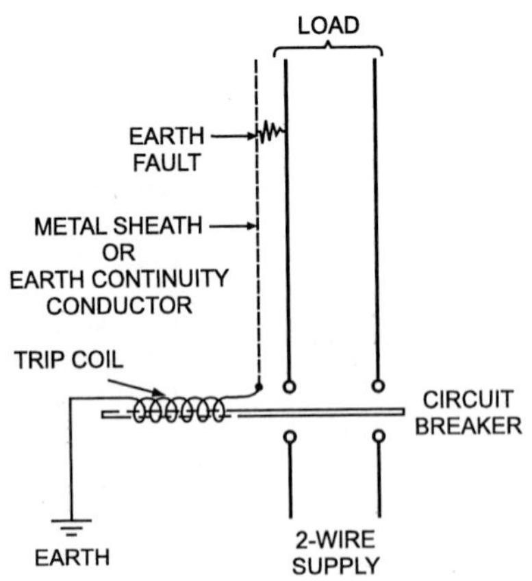

Fig. 10.6 *Voltage-Operated Earth-Leakage Trip*

In both the above types of ELCB the tripping operation may be tested by means of a finger-operated test button which passes a predetermined current from the line wire through a high resistance to trip the coil and thus to earth. This test operation should be performed regularly.

Both types of earth-leakage circuit breakers are arranged to work manually and may take the place of the linked switch and fuses, or the excess-current circuit breaker.

10.8. FUSES

Fuse is perhaps the simplest and cheapest device used for interrupting an electrical circuit under short circuit, or excessive overload, current magnitudes. As such, it is used for overload and/or short-circuit protection in high voltage (up to 66 kV) and low voltage (up to 400 V) installations/circuits. In high voltage circuits their use is confined to those applications where their performance characteristics are particularly suitable for current interruption.

The action of a fuse is based upon the heating effect of the electric current. In normal operating conditions, when the current flowing through the circuit is within safe limits, the heat developed in the fuse element carrying this current is readily dissipated into the surrounding air, and therefore, fuse element remains at a temperature below its melting point. However, when some fault, such as short circuit occurs or when load connected in a circuit exceeds its capacity, the current exceeds the limiting value, the heat generated due to this excessive current cannot be dissipated fast enough and the fusible element gets heated, melts and breaks the circuit. It thus protects a machine or apparatus or an installation from damage due to excessive current.

The time for blowing out of fuse depends upon the magnitude of the excessive current. Larger the current, the more rapidly the fuse will blow i.e., the fuse has inverse time-current characteristic, as shown in Fig. 10.7. Such a characteristic is desirable for protective gear.

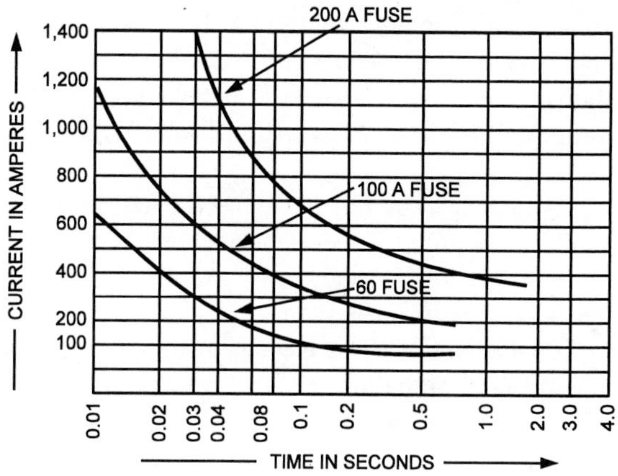

Fig. 10.7 *Time-Current Characteristic*

Essentially, a fuse consists of a fusible element in the form of a metal conductor of specially selected small cross-sectional area, a case or cartridge to hold the fusible element, and in some cases, provided with a means to aid arc extinction. The part which actually melts and opens the circuit is known as the *fuse element.* It forms a series part of the circuit to be protected against short circuit or excessive overloads.

Fuses have following advantages and disadvantages.

Advantages. (*i*) It is the cheapest form of protection available.

(*ii*) It needs no maintenance.

(*iii*) Its operation is inherently completely automatic unlike a circuit breaker which requires an elaborate equipment for automatic action.

(*iv*) It interrupts enormous short circuit currents without noise, flame, gas or smoke.

(*v*) The minimum time of operation can be made much smaller than that with the circuit breakers.

(*vi*) The smaller sizes of fuse element impose a current limiting effect under short circuit conditions.

(*vii*) Its inverse time-current characteristic enables its use for overload protection.

Disadvantages. (*i*) Considerable time is lost in rewiring or replacing a fuse after operation.

(*ii*) On heavy short circuits, discrimination between fuses in series cannot be obtained unless there is considerable differences in the relative sizes of the fuses concerned.

(*iii*) The current-time characteristic of a fuse cannot always be correlated with that of the protected device.

The function of fuse wire is (*i*) to carry the normal working current safely without heating and (*ii*) to break the circuit when the current exceeds the limiting current.

Distribution circuits are protected from ground and short-circuit currents by fuses or circuit breakers so arranged as to disconnect the faulted equipment promptly from its source of supply. Fuses are used almost exclusively for the protection of cables in low-voltage light and power circuits and for transformers of rating not exceeding 200 kVA, in primary distribution systems. Circuit breakers are employed for larger amounts of power and in cases where the operation of the overload device is so frequent as to make the use of fuses impractical.

Nowadays several types of fuses are available which find extensive use in low and moderate voltage applications where frequent operations are not expected or where the use of circuit breakers is uneconomical. Most important among these is the *motor protective fuse* which may be chosen to have a suitable time-delay to permit starting inrush current to flow and also carrying of moderate motor overloads without blowing, and yet be capable of blowing early enough to avoid excessive winding temperature due to overload and at the same time blow very quickly in case of a short circuit.

Modern HRC cartridge fuses provide reliable discrimination and accurate characteristics. In some respects HRC fuses are superior to circuit breakers.

10.8.1. Fuse Element Materials. The materials used for fuse elements must be of low melting point, low ohmic loss, high conductivity (or low resistivity), low cost and free from deterioration. Experience has shown that the most generally suitable material for the fuse element is a low melting point material such as tin, lead or zinc. A low melting point is, however, available with a high specific resistance metal (refer to Table 10.1).

A high specific resistance metal entails, for a given rating of fuse element, a greater mass metal than would be required if the elements were made from high melting point metal of low specific resistance. Thus the use of low melting point metals introduces the problems of handling large mass of vaporised metal set free on fusion.

Oxidation of the fuse element tends to generate an outer layer which may be strong enough to support the metallic core of the fuse element when in a molten state. The effect is more pronounced with certain high melting point metals such as copper and iron than it is with low melting point metals such as tin or lead.

The above discussion reveals that no material possesses all the desirable characteristics. For instance lead has low melting point but it has high specific resistance and it is liable to oxidation. Similarly, copper has high conductivity and low cost but oxidises rapidly. Thus a compromise is required to be made in the selection of material for a fuse element.

The materials commonly used for fuse elements are tin, lead, silver, copper, zinc, aluminium and alloy of lead and tin. An alloy of lead and tin (lead 37% and tin 63%) is used for small current rating fuses (say not beyond 15 A). For current exceeding 15 A this alloy is not used as the diameter of the wire will be larger and after fusing the metal released will be excessive. Beyond 15 A rating circuits copper wire fuses are employed. Copper wire suffers from the drawback that it operates at a rather high temperature if a reasonably low fusing factor is desired. There is thus a tendency for the wire to overheat with the result that its cross-sectional area and fusing currents are gradually reduced and premature melting of the wire may occur. It is usual practice to tin the copper for its protection against oxidation. Zinc (in strip form only) is good if a fuse with considerable time-delay is required i.e., one which does not melt very quickly with a small overload.

The present trend is to use silver as fuse element material despite its higher cost owing to its following advantages.

(*i*) It does not get oxidised and its oxide is unstable.

(*ii*) It does not deteriorate when used in dry air and it remains bright but when the air is moist and contains hydrogen sulphide, the silver surface is attacked and a layer of silver sulphide is formed at the top which shields the metal from further attack.

(*iii*) Owing to its very small coefficient of expansion no critical fatigue occurs and thus continuous full normal current capacity rating is assured for a long time.

(*iv*) The conductivity of silver does not deteriorate with oxidation.

(*v*) Owing to its high conductivity the mass of molten metal to be handled is minimum and thus operating speed is fast.

(*vi*) Vaporization temperature of silver is very much below the one at which the vapour will readily ionize. This means that on the formation of an arc through the vaporized portion of the element, the arc path has comparatively high resistance and the short-circuit current is quickly reduced.

(*vii*) Owing to its low specific heat silver fusible elements can be raised from normal operating temperature to vaporization very much quicker than other fusible elements. Also as the temperature of the silver reaches the melting point its resistance increases abruptly and thus the transition from melting to vaporization becomes nearly instantaneous. As a result, operation becomes very much faster at higher currents.

Either copper or lead-tin alloy is mostly used as an ordinary fuse wire.

TABLE 10.1

Metal	*Melting Point in °C*	*Specific Resistance in μΩ-mm*	*Value of Fuse Constant k For d in m*
Silver	980	16	—
Tin	240	112	12.8
Zinc	419	60	—
Lead	328	210	10.8
Copper	1,090	17	80
Aluminium	665	28	59

10.8.2. Necessity of Fuse in an Electric Circuit. If no fuse or other similar device is provided in the circuit then a dangerous situation would be created on developing of faults such as overload, short circuit or earth faults.

In case of overload, short circuit and heavy earth faults a heavy current will continue to flow through the consuming apparatus, current carrying cables or wires and other current carrying equipment. Due to continuous flow of heavy current through the cables or wires, apparatus etc., these will get heated up and so get damaged. The fire may also break out.

In case of earth leakage fault, (i.e. on the body of the electrical apparatus becoming alive), the body of the electrical apparatus will continue to be alive and at much higher potential above that of the earth. In such circumstances any person coming in contact with the metal body of the apparatus is liable to get an electric shock, even if it is earthed.

The main function of a fuse is to blow out under a fault and isolate the faulty section from the live side. If the fuse is provided on neutral wire, in place of live wire, then in abnormal conditions though the fuse will blow out but the lamp or other apparatus still remains connected to the live wire and in case of leakage some trouble will arise and cause a considerable damage. In case the earth fault takes on the neutral wire between lamp and fuse provided in it, the fuse will blow out because the neutral wire is slightly at a higher potential with respect to earth and so the fault current flows through the neutral wire and fuse melts itself. The current will flow through the live wire, lamp, neutral wire and earth fault, even after the fuse has blown out and this may cause serious damage to the wiring, the apparatus connected or building itself.

If fuses of same capacity are provided on the phase wire and neutral, then in case of short-circuit fault, one of them will blow out first. If the fuse on neutral wire blows out first, the fuse in phase line remains intact and faulty apparatus still remains connected to the live. If some person comes in contact with the faulty apparatus, he is liable to get electric shock. In case the installation is connected to 3-phase 4-wire supply system, and fuses are provided on both live and neutral wire and fuse on neutral wire blows out then voltage of each phase to neutral will become considerably different, which is not desirable. *Hence the fuse is provided only in phase or live pole, never on neutral pole.*

10.8.3. Types of Fuses.

1. ***Supply Main Fuse.*** This fuse is provided by the supplier and is fixed just after the service meter and sealed by him. The seal can be broken only by authorised person of the supply authority in case of blowing out of fuse for the purpose of replacement. The rating of supply main fuse will be as per load current of the consumer.
2. ***Consumer Main Fuse.*** This is another fuse of rating slightly less than that of supply main fuse provided by the supplier and placed after the consumer's main switch. The rating of consumer fuse is kept slightly lower than that of supply main fuse so that in case of exceeding of current than normal current (that may occur due to overload, short circuit or earth fault) consumer fuse, which can be replaced by him, may blow and supply main fuse may remain intact.
3. ***Sub-Circuit Fuses.*** The total wiring system is divided into a number of sub-circuits or branch circuits. A separate fuse is provided for each branch circuit and is known as *sub-circuit or branch circuit fuse.*
4. ***Point Fuses.*** In good quality indoor wiring of buildings, every light and plug point is provided with its individual fuse known as *point fuse.*

10.9. FUSE UNITS

A fuse unit essentially consists of the metal fuse element or link, a set of contacts between which it is fixed and a body to support and isolate them. Many types of fuses also have some means for extinguishing the arc which appears when the fuse element melts.

The various types of fuse units, most commonly available are :

1. Round type fuse unit.
2. Kit-kat type fuse unit.
3. Cartridge type fuse unit
4. HRC (High rupturing capacity) fuse units and
5. Semiconductor fuse units

1. Round Type Fuse Unit. This type of fuse unit consists of a porcelain or bakelite box and two separated wire terminals for holding the fuse wire between them. This type of fuse is not in common use on account of its following disadvantages :

1. One of the terminals remain always energised and, therefore, for replacement of fuse either the worker will have to touch the live mains or open the main switch.
2. Appreciable arcing takes place at the instant of blowing off fuse and thus damage the terminals. After two or three arcing the fuse unit becomes unusable.

2. Rewirable or Kit-Kat Type Fuses. The most commonly used fuse in 'house wiring' and small current circuits is the semi-enclosed or rewirable fuse (also sometimes known as kit-kat type fuse). It consists of a *porcelain base* carrying the fixed contacts to which the incoming and outgoing live or phase wires are connected and a *porcelain fuse carrier* holding the fuse element, consisting of one or more strands of fuse wire, stretched between its terminals. The fuse carrier is a separate part and can be taken out or inserted in the base without risk, even without opening the main switch. If fuse holder or carrier gets damaged during use, it may be replaced without replacing the complete unit.

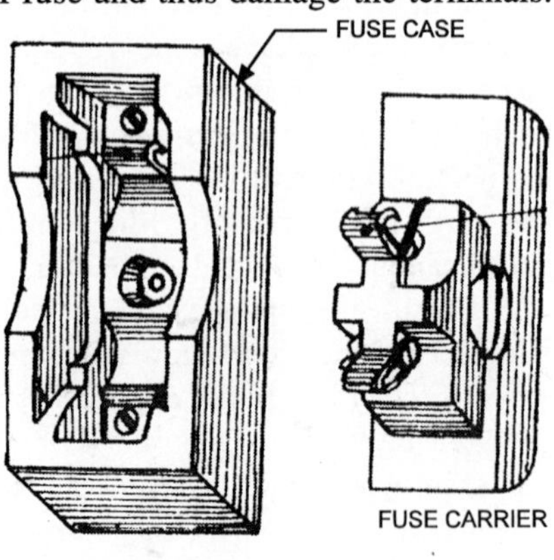

Fig. 10.8 *Rewirable or Kit-Kat Type Fuse*

The fuse wire may be of lead, tinned copper, aluminium or an alloy of tin-lead. The actual fusing current will be about twice the rated current. When two or more fuse wires are used, the wires should be kept apart and a derating factor of 0.7 to 0.8 should be used to arrive at the total fuse rating.

The specifications for rewirable fuses are covered by IS : 2086-1963. Standard ratings are 6, 16, 32, 63 and 100 A. A fuse wire of any rating not exceeding the rating of the fuse may be used in it i.e., a 80 A fuse wire can be used in a 100 A fuse, but not in the 63 A fuse.

On occurrence of fault, the fuse element blows off and the circuit is interrupted. The fuse carrier is pulled out, the blown out fuse element is replaced by new one and the supply is restored by re-inserting the fuse carrier in the base.

Though such fuses have the advantages of easy removal or replacement without any danger of coming into contact with a live part and negligible replacement cost but they suffer from the following disadvantages:

(*a*) ***Unreliable Operation.*** The operation of the rewirable fuses is unreliable because of the following factors :

(*i*) There is a possibility of renewal by the fuse wire of wrong size.

(*ii*) The fuse wire deteriorates, over a period, due to oxidation through the continuous heating up of the element. After a relatively short period the current causes the metal to deteriorate and the fuse operates at a lower current than originally rated.

(*iii*) A fuse with normal rated capacity of say 60 A would fuse at nearly 120 amperes. Should that amount of current damage the apparatus in the circuit, obviously the size of the fuse wire must be smaller but, as its rated capacity will also be lower, the same dependence could not be placed upon it.

(*iv*) The protective capacity is uncertain i.e., no dependence can be placed upon the wire to interrupt the circuit when a given current flows. For instance although theoretically a No.17 SWG wire should interrupt a circuit when a current of 135 A is flowing, the circuit may be interrupted when a lower current flows or in other circumstances when only a much higher current flows.

(*v*) Accurate calibration of the fuse wire is impossible, as a longer fuse operates earlier than one of shorter length.

Single phasing of three-phase induction motors is a common occurrence where rewirable type fuses are employed in motor circuits.

(*b*) ***Lack of Discrimination.*** Due to unreliable operation, discrimination cannot be always ensured. However, it will be possible to achieve some measure of discrimination by using a fuse which has twice the rating of the next fuse ahead of it in series. For example, a fuse of 80 A will discriminate with a fuse of 40 A, but not with a fuse of 55 A.

(*c*) ***Small Time Lag.*** Due to the small time lag, such fuses can blow with large transient currents which are encountered during the starting of motors and switching-on operation of transformers, capacitors, fluorescent lights etc., unless fuses of sufficiently high rating are used.

(*d*) ***Low Rupturing Capacity.*** Rewirable fuses have limited breaking or rupturing capacity. For example, according to IS : 2086-1963 the rewirable fuse of 16 A normal current have a breaking current of 2 kA and those up to 200 A normal current have a breaking current of 4 kA.

(*e*) ***No Current-Limiting Feature.***

(*f*) ***Slow Speed of Operation.*** No special means are employed to extinguish the arc that blows after the fuse melts. Thus arcing time is more in such fuses.

3. Cartridge Type Fuse. This is a totally enclosed type fuse unit. It essentially consists of an insulating container of bulb or tube shape and sealed at its ends with metallic cap known as *cartridge* enclosing the fuse element and filled up with powder or granular material known as filler. There are various types of materials used as filler like sand, calcium carbonate, quartz etc. There is sometimes a blow out device in the side of the tube to indicate when the fuse is blown. On overloads or short circuits, the fusible element is heated to a high temperature causing it to vaporize. The powder in the fuse cartridge cools and condenses the vapour and quenches the arc thereby interrupting the flow of current.

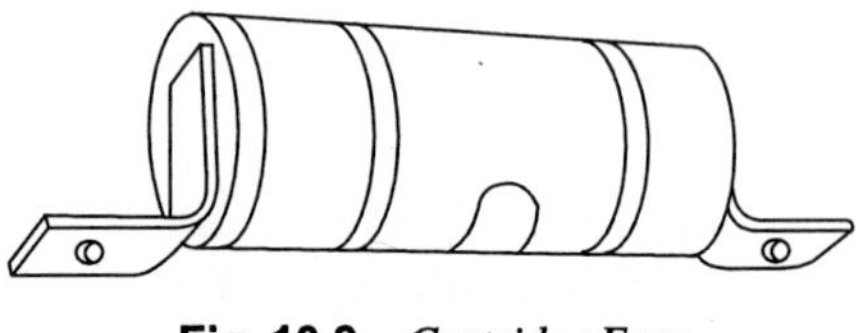

Fig. 10.9 *Cartridge Fuse*

Since it is totally enclosed it will not be possible to rewire and, therefore, the whole unit will have to be replaced, once it blows out. It provides complete security against fire risk, as it is a totally enclosed unit. The filling powder provides good insulating path and helps to extinguish the arc at the time of blowing up of fuse. This type of fuse is available up to 660 V and the current rating up to 800 A.

4. High Rupturing Capacity (HRC) Fuses. With a very heavy generating capacities of the modern power stations, extremely heavy currents would flow into the fault and the fuse clearing the fault would be required to withstand extremely high stresses in this process. A rewirable fuse may not be useful in this case and, therefore, high rupturing capacity fuses commonly known as HRC fuses, designed and developed after intensive research for use in medium and high voltage installations, are used for such duties. Their rupturing capacity is as high as 500 MVA up to 66 kV and above. The main advantages of HRC fuses are :

(*i*) They are cheaper as compared with other types of circuit interrupters of same breaking capacity.
(*ii*) No maintenance is required.
(*iii*) The operation is quick and sure.
(*iv*) They do not deteriorate with time.
(*v*) They have inverse time-current characteristic.
(*vi*) They are capable of clearing high as well as low currents.
(*vii*) They are quite reliable and can be selected for proper discrimination.

HRC fuses suffer from the following disadvantages :

(*i*) They are required to be replaced after each operation.
(*ii*) Interlocking is not possible.
(*iii*) They lack relays in complete discrimination.

Since the advantages outweigh the disadvantages to a very large extent, HRC fuses are extensively used.

(*a*) *Cartridge Type HRC Fuse.* The high rupturing capacity cartridge fuse is most popular. In its simplest form an HRC cartridge fuse consists of a heat resisting ceramic body having metal end-caps to which are welded fusible silver (or bimetallic) current-carrying elements. The complete space within the body surrounding the elements is filled with a powder which acts as an arc extinguishing agent.

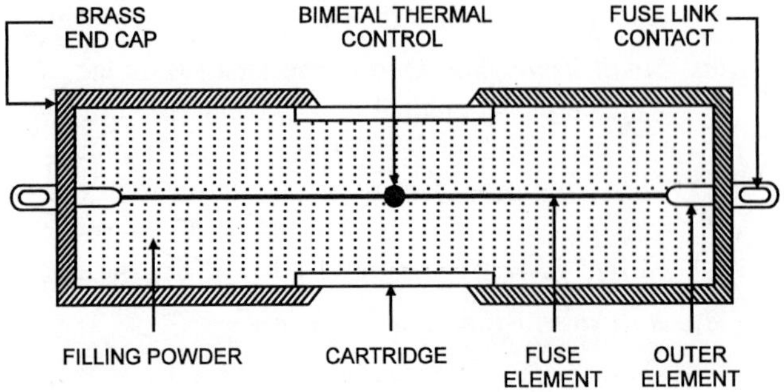

Fig. 10.10 *HRC Cartridge Fuse*

The process of fusing comprises the following operations :

1. Pre-arcing operation i.e. melting of silver elements.
2. Arcing operation i.e. vaporisation of the elements.

3. Fusion of silver vapours and the filling powder, and
4. Extinction of arc under fusion process.

On the occurrence of a fault, short-circuit current flows through the fuse element and element is thus heated up to its melting point. When the melting process is completed, an arc is formed. The chemical reaction between silver vapour and filling powder tends to establish high resistance. The high resistance acts as an insulator because the fault current decreases along with the high pressure created within the fuse by the fault (excessive) current. Thereafter a transient voltage is created at the instant of fault current interruption on account of sudden release of energy.

The physical phenomena associated with the process of fusing include a sudden rise of temperature and the generation of a high internal pressure.

(*b*) *Tetra Chloride Type HRC Fuse.* It essentially consists of a glass tube filled with carbon tetrachloride solution and sealed at both ends with brass caps. Inside the tube a high resistance fuse wire is sealed at one end of the glass tube and the other end of the fuse wire is held by a strong phosphor bronze spiral spring fixed to the other end of the glass tube. A flexible copper wire is also similarly held between the spring and the other end of cap in parallel with the high resistance wire. On short circuit or overload, the high resistance fuse wire melts and spring pulls back the flexible copper wire. The arc produced is extinguished by carbon tetrachloride vapour.

5. *Semiconductor Fuses.* These are very fast acting fuses for protection of thyristor and other electronic circuits.

10.10. PROTECTION OF BUILDING AGAINST LIGHTNING

A lightning protection system is designed to protect a structure from damage due to lightning strikes by intercepting such strikes and safely passing their extremely high voltage currents to "ground". A lightning protection system includes a network of air terminals, bonding conductors, and ground electrodes designed to provide a low impedance path to ground for potential strikes.

Because of the high energy and current levels associated with lightning (currents can be in excess of 150,000 amps), and the very rapid rise time of a lightning strike, no lightning protection system can guarantee absolute safety from lightning. Lightning current will divide to follow every conductive path to ground, and even the divided current can cause damage. Secondary "side-flashes" can be enough to ignite a fire, blow apart brick, stone, or concrete, or injure occupants within structure or building. However, the benefits of basic lightning protection systems have been evident for well over a century.

The parts of a lightning protection system are air terminals (lightning rods or strike termination devices), bonding conductors, ground terminals (ground or "earthing" rods, plates, or mesh), and all of the connectors and supports to complete the system. The air terminals are typically arranged at or along the upper points of a roof structure, and are electrically bonded together by bonding conductors (called "down conductors" or "downleads"), which are connected by the most direct route to one or more grounding or earthing terminals. Connections to the earth electrodes must not only have low resistance, but must have low self-inductance.

A lightning rod or lightning conductor is a metal rod or metallic object mounted on top of a building, electrically bonded using a wire or electrical conductor to interface with ground or "earth" through an electrode, engineered to protect the building in the event of lightning strike. If lightning targets the building it will preferentially strike the rod and be conducted to ground through the wire, instead of passing through the building, where it could start a fire or cause electrocution. Lightning rods come in many different forms, including hollow, solid, pointed, rounded, flat strips of copper or aluminium. The main attribute of all lightning rods is that they are conductive.

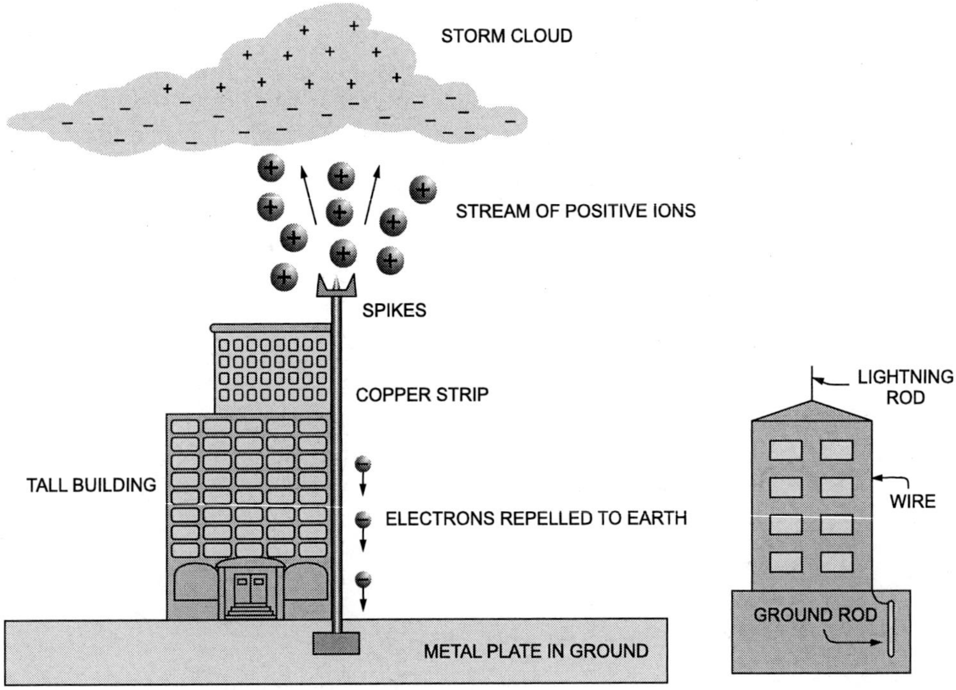

Fig. 10.11

10.10.1. Lightning Conductors and Grounding Precautions. Ideally, the underground part of the assembly should reside in an area of high ground conductivity. If the underground cable is able to resist corrosion well, it can be covered in salt to improve its electrical connection with the ground. While the electrical resistance of the lightning conductor between the air terminal and the Earth is concerning, the inductive reactance of the conductor could be more important. For this reason, the down conductor route is kept short, and any curves have a large radius. If these measures are not taken, lightning current may arc over an obstruction, resistive or reactive, that it encounters in the conductor. At the very least, the arc current will damage the lightning conductor and can easily find another conductive path, such as building wiring or plumbing, and cause fires or other disasters.

HIGHLIGHTS

1. The term '*earthing*' means connecting of the non-current carrying parts of the electrical equipment or the neutral point of the supply system to the general mass of earth in such a manner that at all times an immediate discharge of electrical energy takes place without danger. Earthing is achieved by electrically connecting the respective parts in the installations to some system of electrical conductors or electrodes placed in intimate contact with the soil some distance below the ground level. This contacting assembly is called the *earthing*. The metallic conductors connecting the parts of the installations with the earthing are called *electrical connections*. When speaking of the earthing and the earthing connections as a whole, it is customary to use the term the *earthing system*.
2. The resistance offered by the earth electrode to the flow of current into the ground is known as *resistance-to-earth*. The term resistance-to-earth (or earth resistance) does not imply the contact resistance between the earthing electrode and the soil (which is insignificant), but mainly the resistance of soil between the electrode

and the point of zero potential. Numerically it is equal to the ratio of the potential of earth electrode with respect to remote point, to the current dissipated by it.

3. Currents in the range of 20–100 mA are very painful and hard to bear. These currents may also cause physical injury due to powerful contraction of the skeletal muscles. If current contacts intact skin and passes through the trunk, at about 100 mA and above, there is a likelihood of pulling the heart into ventricular fibrillation. It proves *fatal* unless corrected within minutes, since the brain begins to die 2 to 4 minutes after it is robbed of its supply of oxygenated blood.

At very high currents of the order of 6 A and above, there is a danger of temporary respiratory paralysis and also of serious burns. However, if the shock duration is of only a very few seconds, there is a possibility of heart reverting to the normal rhythmic action.

4. Various methods of earthing are: (*i*) strip or wire earthing (*ii*) rod earthing and (*iii*) pipe earthing. *Strip* or *wire earthing* is used at places which have rocky soil earth bed because at such places excavation work required for plate earthing is difficult. *Rod earthing* is suitable for areas which are sandy in character. This system of earthing is very cheap as no excavation work is involved. *Pipe earthing* is the most common and best system of earthing as compared to other systems suitable for the same earth and moisture conditions. *Plate earthing* is another common system of earthing.

5. The most frequent causes for abnormal conditions are overloads and short circuits, therefore, protection is to be provided to wiring installations and motors against overloading and short circuiting. The devices used are fuses, relays, miniature circuit breakers (MCB), earth leakage circuit breakers.

Circuit breakers are mechanical devices designed to close or open contact members, thus closing or opening an electric circuit under normal and abnormal conditions. Automatic circuit breakers, which are usually employed for the protection of electric circuits, are equipped with a trip coil connected to a relay or other means designed to open the circuit breaker automatically under abnormal conditions, such as overcurrent.

Miniature circuit breaker (MCB) is a device that provides definite protection to the wiring installations and sophisticated equipment against overcurrents and short-circuit faults.

Earth-leakage circuit breaker (ELCB) is a device that provides protection against earth leakage. These are of two types viz, the current operated type and the voltage operated type.

6. Fuse is perhaps the simplest and cheapest device used for interrupting an electrical circuit under short circuit, or excessive overload, current magnitudes. As such, it is used for overload and/or short-circuit protection in high voltage (up to 66 kV) and low voltage (up to 400 V) installations/circuits. In high voltage circuits their use is confined to those applications where their performance characteristics are particularly suitable for current interruption.

If no fuse or other similar device is provided in the circuit then a dangerous situation would be created on developing of faults such as overload, short circuit or earth faults.

EXERCISES

1. Explain the need of earthing electrical devices. What are important safety issues?
   [U.P. Technical Univ. Basic Electrical Engineering Odd Semester 2013-14]
2. What is the purpose of earthing ? Distinguish between system earthing and equipment earthing. Define the terms earthing electrode and earth resistance.
3. Define the following terms :
   (*i*) Earth electrode (*ii*) Earth resistance (*iii*) Earthing lead (*iv*) Step potential and (*v*) Touch Potential.
4. Describe the effects of electric currents on human body.
5. Why earthing is needed in an electrical system. Enlist the various types of earthing systems.
   [U.P. Technical Univ. Basic Electrical Engineering Second Semester 2013-14]
6. What do you understand by earthing ? Why is it necessary to earth electrical installation ? Draw out a neat sketch for earthing by GI pipe.
7. Discuss methods of earthing electrical installations.

8. Enlist the various protective devices used in wiring installations and motors. Describe any one of these in detail.
9. What is fuse ? What is its purpose ? Discuss.
10. What is a fuse wire and what materials can be used for making it ? What will happen if no fuse is provided in the electric circuit ?
11. What is the function of a fuse ? Why is it connected in the phase wire ?
12. According to IE rules, a cutout (fuse) shall not be placed in the earthed neutral conductor of two-wire system. Discuss the reason to justify this rule.
13. What are the different types of fuse used ? Why is a fuse inserted in the phase wire, not in the neutral wire ? Why are HRC fuses used ?
14. Why is it advisable to use HRC fuse (cartridge fuse) for motor protection instead of rewirable fuse ? Why is a fuse inserted in the live wire and not in the neutral wire ?
15. Write short notes on the following:
    (*i*) Earthing and its objective.
    (*ii*) Effects of electric currents on human body.
    (*iii*) Plate earthing.
    (*iv*) Protection of buildings against lightning.

SHORT ANSWER TYPE QUESTIONS WITH ANSWERS

Q.1. What is difference between system earthing and equipment earthing ?

Ans. System earthing deals with earthing of the system neutral to ensure system security and protection while equipment earthing deals with earthing of non-current carrying parts of the equipment to ensure safety of personnel and protection against lightning.

Q.2. What is average resistance of a human body ?

Ans. About 1,000 Ω.

Q.3. What is the current that is considered safe for human body ?

Ans. About 9 mA.

Q.4. What is meant by an earth electrode ?

Ans. Any wire, rod, pipe, plate and an array of conductors embedded in ground horizontally or vertically is called the earth electrode.

Q.5. Give the maximum permissible values of ground resistance for (*a*) large power stations (*b*) small substations.

Ans. (*a*) 0.5 Ω; (*b*) 2.0 Ω.

Q.6. What is meant by 'let go current' ?

Ans. The maximum current at which the subject is capable of releasing an electrical conductor by using muscles directly stimulated by the current is called 'let go current'.

Q.7. What is meant by earth continuity conductor ?

Ans. The conductor, by means of which the metal body of an equipment or an appliance is connected to the earth, is known as earth continuity conductor (ECC).

Q.8. What is an earthing lead ?

Ans. The wire which connects overhead earth wire or any other apparatus to be earthed to the earth electrode is known as earthing lead.

Q.9. Why charcoal and salt is used in earthing ?

Ans. Generally alternate layers of charcoal and salt are used to increase the effective area of the earth and to reduce the earth resistance respectively.

Q.10. Give the applications of rod earthing.

Ans. Rod earthing is suitable for areas which are sandy in character.

Q.11. What is meant by a fuse ?

Ans. Fuse is a device used in an electrical circuit for protecting electrical equipment against overloads and/or short circuits.

Q.12. On what principle does the fuse operate ?

Ans. The action of a fuse is based upon the heating effect of the electric current. During overloads or short circuits the fuse gets heated up to melting point due to excessive current and breaks the circuit.

Q.13. What kind of protection is offered by a fuse ?

Ans. The fuse provides protection against short circuits and overloads.

Q.14. What function does the fuse perform ?

Ans. The function of a fuse wire is (i) to carry the normal working current safely without heating and (ii) to break the circuit when current exceeds the limiting current.

❑❑❑

11

CHAPTER

Magnetic Circuits

11.1. INTRODUCTION

Practically all electric power machinery (*e.g.* transformers, generators, motors) depend for their operation upon the magnetism produced by the magnetic circuits. The closed path followed by magnetic flux is called a *magnetic circuit* (Fig. 11.1) just as the closed path followed by current is called an electric circuit. There is a lot of similarity between the magnetic and electric circuits, and will be shown later on during discussion.

A magnetic circuit consists of a structure composed for the most part of high permeability magnetic material. The presence of high permeability material causes the magnetic flux to be confined to the paths defined by the structure, much as currents are confined to the conductors of an electric circuit. A simple example of a magnetic circuit is shown in Fig. 11.1. The core is assumed to be composed of magnetic material whose permeability is much greater than that of the surrounding air. The core is of uniform cross section and is excited by a winding having N turns and carrying a current of I amperes. This winding develops a magnetic field in the core, as illustrated in the figure. The magnetic field can be visualized in terms of flux lines, which form closed loops interlinking with the winding. According to basic law of magnetic field, called the *Ampere's circuital law* (sometimes referred to as *Ampere's work law*) the line integral of **H** around a closed path is equal to the net current enclosed by that path *i.e.*

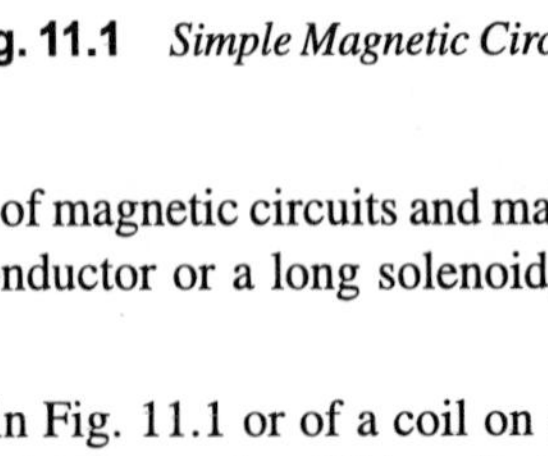

Fig. 11.1 *Simple Magnetic Circuit*

$$\oint \mathbf{H}\, dl = \sum \mathrm{I} \qquad ...(11.1)$$

The above law is very comprehensive and provides a basis for calculation of magnetic circuits and makes it possible in some cases (such as in the cases of a long current carrying conductor or a long solenoid) to determine readily the strength of the magnetic field.

11.1.1. Magnetic Circuit Concepts. The flux producing ability of the coil in Fig. 11.1 or of a coil on any other magnetic circuit is proportional to the number of turns N and the current I. The product NI is called the *magnetomotive force* (mmf) and determines the amount of flux developed in the magnetic circuit.

MMF of the magnetic circuit is the magnetic potential difference that tends to force flux around the magnetic circuit and is analogous to the electromotive force (emf) in an electric circuit.

$$\text{MMF} = \text{NI ampere-turns} \qquad ...(11.2)$$

The resulting flux Φ in the magnetic circuit, besides being dependent on the mmf, is also a function of the opposition of the iron to carrying flux. This opposition is called the *reluctance* S of the magnetic circuit.

As in the case with resistance in the electric circuit, reluctance S of the magnetic circuit is directly proportional to length l, inversely proportional to cross-sectional area a, and dependent on the nature of material of the magnetic circuit. The reluctance of the magnetic circuit is given as

$$S = \frac{l}{\mu a} \text{ ampere-turns/weber} \qquad ...(11.3)$$

when the flux Φ is constant over the length and uniform over the area. The quantity μ expresses the property of the magnetic material called its *permeability*. Permeability is a measure of the receptiveness of the material of having magnetic flux developed in it. For free space, the permeability μ_0 equals $4\pi \times 10^{-7}$ H/m in the SI system. The relative permeability of magnetic material μ_r may range up to thousands.

Total flux developed in the circuit is given as

$$\Phi = \frac{\text{MMF}}{S} \text{ webers} \qquad ...(11.4)$$

The above Eq. (11.4) is sometimes referred to as *Ohm's law* for the magnetic circuit. It serves to emphasize the mathematical analogy between the magnetic circuit and the electric circuit. Analogous quantities in the two circuits are listed in Art 11.1.3.

Magnetic circuits differ from electric circuits in one important respect. The reluctance of the magnetic circuit containing iron or ferromagnetic material depends upon the flux carried by it. With the increase in flux, a large change in mmf is required to develop the same change in flux.

11.1.2. Determination of Ampere-Turns. In any magnetic circuit, flux created is given as

$$\Phi = \frac{\text{MMF}}{\text{Reluctance}} = \frac{\text{AT}}{l/\mu a} = \frac{\text{AT}}{l/\mu_0 \mu_r a}$$

$$\text{or AT required} = \frac{\Phi}{a} \times \frac{1}{\mu_0 \mu_r} \times l = \frac{B}{\mu_0 \mu_r} \times l = H \times l \text{ ampere-turns} \qquad ...(11.5)$$

Hence for determination of AT for a magnetic circuit

(*i*) First find field strength H in each part of the magnetic circuit, (*ii*) find the length of various parts of magnetic circuit, (*iii*) find the number of ampere-turns required for the various parts of magnetic circuit from the relation AT = Hl where l is the length of the part in metres and lastly, (*iv*) find total number of ampere-turns for the whole series magnetic circuit by adding ampere-turns determined for various paths in magnetic circuit.

This will be more clear to the students from examples given in this chapter.

As both permeability and reluctance may change from one operating condition to another. Hence direct numerical application of the reluctance concept and Eqs. (11.4) and (11.5) is rare. For quantitative analysis, graphical methods are generally used because they can easily be adapted to the nonlinearities involved (Art 11.2).

11.1.3. Comparison Between Magnetic and Electric Circuits

	Magnetic Circuit	**Electric Circuit**
	SIMILARITIES	
Circuit	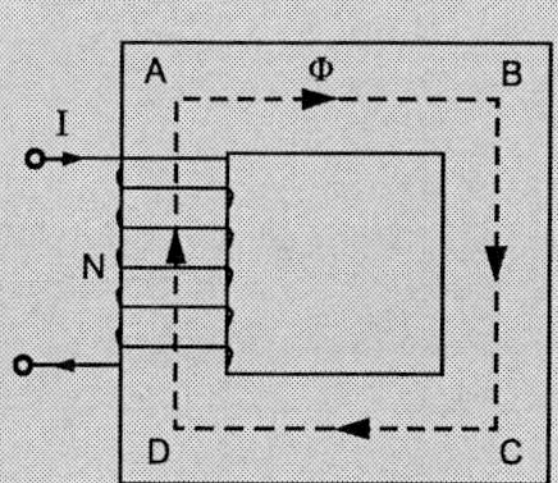	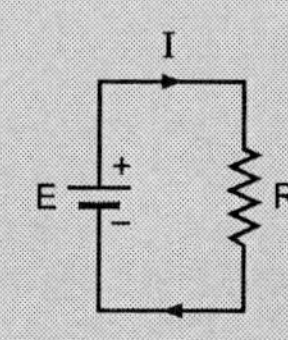

	Magnetic circuit	Electric circuit
1. Definition	The closed path followed by magnetic flux is called a magnetic circuit	The closed path followed by an electric current is called an *electric circuit*
2. Driving force	**MMF** is required to establish flux Φ in the magnetic circuit and is measured in ampere-turns (AT) or amperes	**EMF** is required to cause flow of current in an electric circuit and is measured in volts
3. Response	Flux, $\Phi = \frac{\text{Driving force}}{\text{Magnetic reluctance}} = \frac{AT}{S}$ webers	Current, $I = \frac{\text{Driving force}}{\text{Electric resistance}} = \frac{E}{R}$ amperes
4. Impedance	Reluctance, $S = \frac{l}{\mu_0 \mu_r a} = \frac{l}{\mu a}$ AT/Wb	Resistance, $R = \rho \frac{l}{a}$ ohms
(a) For series circuits	$S = S_1 + S_2 + S_3 + \ldots$	$R = R_1 + R_2 + R_3 + \ldots$
(b) For parallel circuits	$S = \frac{1}{\frac{1}{S_1} + \frac{1}{S_2} + \frac{1}{S_3} + \ldots}$	$R = \frac{1}{\frac{1}{R_1} + \frac{1}{R_2} + \frac{1}{R_3} + \ldots}$
5. Admittance	Permeance $= \frac{1}{\text{Reluctance}}$ in Wb/AT	Conductance $= \frac{1}{\text{Resistance}}$ in siemens
6. Proportionality constant	Reluctivity $= \frac{1}{\text{Permeability}} = \frac{1}{\mu}$ in m/H	Resistivity, $\rho = \frac{1}{\text{Conductivity}}$ in ohm-metre
7. Density	Flux density, $B = \mu \times$ magnetic field intensity $= \mu H$ is tesla or Wb/m^2	Current density, $J = \frac{\text{Electric field intensity}}{\rho}$ in A/m^2
8. Field intensity	Magnetic field intensity, $H = \frac{MMF}{l} = \frac{NI}{l}$ in AT/m	Electric field intensity $= \frac{E}{l}$ in volts/metre
9. Drop	MMF drop $= \Phi \times$ reluctance $= \Phi S$	Voltage drop $= I \times$ Resistance $= IR$

DISSIMILARITIES	
1. Flux does not actually flow in a magnetic circuit.	Current does flow in an electric circuit.
2. Permeability does not largely vary from material to material and there is hardly any material which can act as an insulator to the magnetic flux.	Conductivity varies largely from material to material so much so that some materials are insulators to electric current and some are very good conductors.
3. For a particular temperature the permeability depends up on the flux density (or total flux).	For a particular temperature, conductivity (or resistivity) is constant and independent of current strength.
4. Flux can pass through air.	Current would not flow through the air until an arc is struck.
5. Residual flux persists after removal of mmf.	The current is reduced to zero after removal of source of emf.
6. There is no waste of energy due to reluctance in the magnetic circuit and, therefore, energy is required only to create the magnetic flux but not to maintain it.	In an electric circuit, resistance causes heat to be generated resulting in waste of energy and, therefore, energy is required as long as the current flows.

11.1.4. Magnetic Circuits With Air Gaps. Energy conversion devices which incorporate a moving element have necessarily air gaps in their magnetic circuits. Air gaps are also provided in the magnetic circuits to avoid saturation. A magnetic circuit with an air gap is shown in Fig. 11.2. An air gap is nothing else but a volume of air between two magnetic surfaces. The length of air gap l_g equals the distance between the two magnetic surfaces. The area of x-section of any one of the surfaces gives the air gap area, a_g. When the air gap length l_g is much smaller than the dimensions of the adjacent core faces, the magnetic flux Φ is constrained essentially to reside in the core and the air gap and is continuous through out the magnetic circuit. Thus the configuration shown in Fig. 11.2 can be analysed as a magnetic circuit with two series components, a magnetic or iron core of permeability μ and mean length l_i and an air gap of permeability μ_0 and length l_g. Since the permeability of air is constant, the air gap is a linear part of the magnetic circuit and the flux density in the air gap is proportional to the mmf across the air gap. The necessary mmf is calculated separately for the air gap and the iron portions and then added to determine the total mmf.

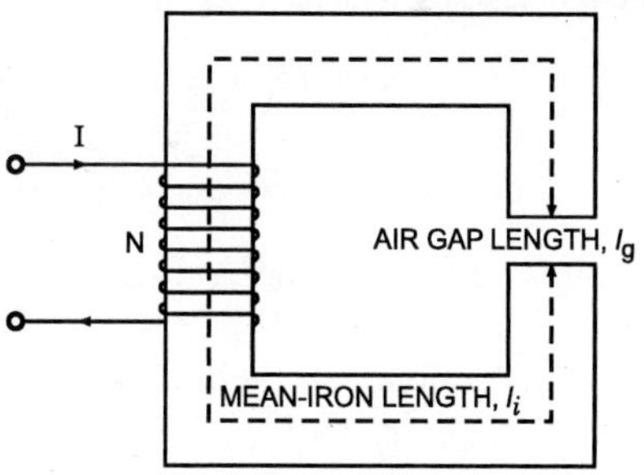

Fig. 11.2 *Magnetic Circuit With Air Gap*

11.1.5. Composite Magnetic Circuits. Consider a circular ring made from different materials of lengths l_1, l_2 and l_3, cross-sectional areas a_1, a_2 and a_3 and relative permeability μ_{r1}, μ_{r2} and μ_{r3} respectively with a cut of length l_g known as air gap. The total reluctance is the arithmetic sum of individual reluctances as they are joined in series.

∴ Total reluctance,

$$S = \frac{l_1}{\mu_0 \mu_{r_1} a_1} + \frac{l_2}{\mu_0 \mu_{r_2} a_2} + \frac{l_3}{\mu_0 \mu_{r_3} a_3} + \frac{l_g}{\mu_0 a_g}$$

$$\text{Total mmf} = \Phi \times S$$

$$= \Phi \left[\frac{l_1}{\mu_0 \mu_{r_1} a_1} + \frac{l_2}{\mu_0 \mu_{r_2} a_2} + \frac{l_3}{\mu_0 \mu_{r_3} a_3} + \frac{l_g}{\mu_0 a_g} \right]$$

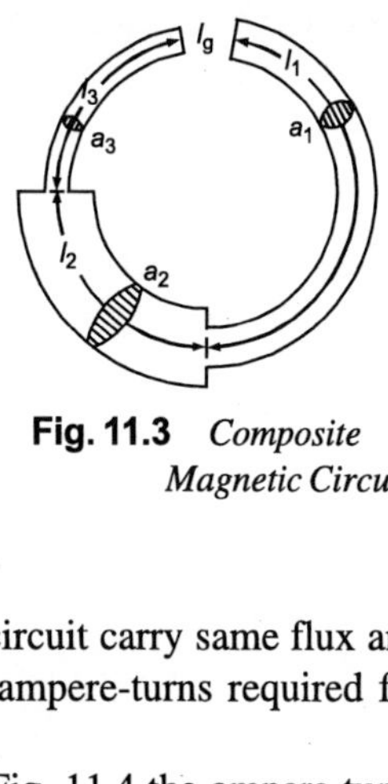

Fig. 11.3 *Composite Magnetic Circuit*

or Total ampere-turns required

$$= H_1 l_1 + H_2 l_2 + H_3 l_3 + H_g l_g \qquad ...(11.6)$$

= Sum of ampere-turns required for individual parts of the magnetic circuit.

11.1.6. Parallel Magnetic Circuits. In series circuits, all parts of the magnetic circuit carry same flux and total ampere-turns required to create a given flux is the arithmetic sum of the ampere-turns required for individual parts of the circuit.

But if the various paths of the magnetic circuit are in parallel, as shown in Fig. 11.4 the ampere-turns required for the combination is equal to the ampere-turns required to create the given flux in one path.

For example for the circuit shown in Fig. 11.4 paths ABCD and AFED are in parallel, so ampere-turns required to create flux Φ_1 in path ABCD is equal to ampere-turns required to create flux Φ_2 in path AFED and also equal to the ampere-turns required for both of the paths.

Hence total ampere-turns required for magnetic circuit shown in Fig. 11.4.

= AT for path DA + AT for path ABCD

= AT for path DA + AT for path AFED

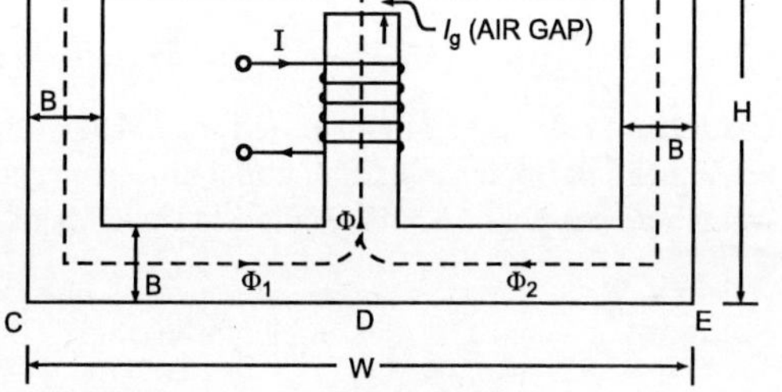

Fig. 11.4

11.1.7. Magnetic Leakage and Fringing. Leakage flux is the flux, which follows a leaking path, as shown in Fig. 11.5. Flux in the air gap is known as *useful flux* which is utilized for various useful purposes. For the purpose of calculations the iron is supposed to carry whole of the flux throughout its entire length.

The ratio of total flux (flux in the iron path) to the useful flux (flux in the air) is known as *leakage factor*.

$$i.e.\ \text{Leakage factor} = \frac{\text{Total flux}}{\text{Useful flux}}$$

$$\text{or Leakage factor} = \frac{\text{Flux in the iron path}}{\text{Flux in the air gap}}$$

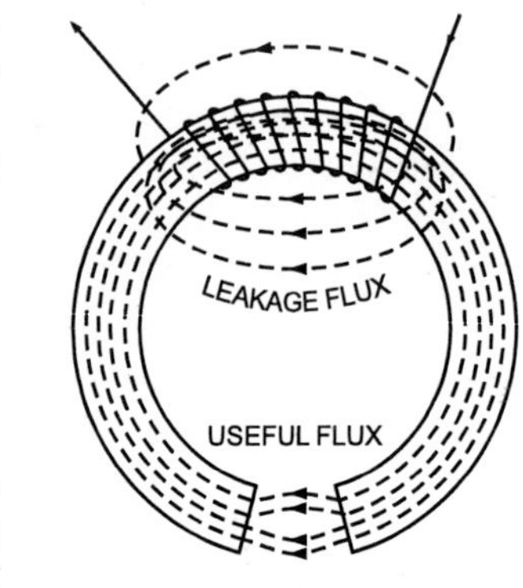

Fig. 11.5

It is also seen from Fig. 11.5 that the useful flux passing across the gap tends to bulge outwards, thereby increasing the effective area of gap and reducing the flux density in the gap. This effect is referred to as *fringing*; and the longer the air gap, the greater is the fringing.

In electrical machines (such as in generators and motors), magnetic leakage is undesirable as it causes increase in weight (not decrease in their power efficiency) and cost of manufacture. Though magnetic leakage cannot be avoided completely but can be reduced to the minimum by placing the magnetizing or exciting coils as close as possible to the air gap or to the points in the magnetic circuit where the flux is to be utilized for useful purposes.

The value of leakage factor for modern electric machinery is about 1.2.

Example 11.1 **An electromagnet has an gap of 4 mm and flux density in the gap is 1.3 Wb/m². Determine the ampere-turns for the gap.** [*U.P. Technical Univ. Electrical Engineering Second Semester 2006-07; Mahamaya Technical Univ. First Semester 2011-12*]

Solution:

$$\text{Magnetising force, } H_g = \frac{B}{\mu_0 \mu_r} = \frac{1.3}{4\pi \times 10^{-7} \times 1} = 1.035 \times 10^6 \text{ AT/m} \quad \because \mu_r \text{ for air} = 1$$

$$\text{Ampere-turns required for the gap} = H_g\, l_g = 1.035 \times 10^6 \times 4 \times 10^{-3} = 4{,}140 \textbf{ Ans.}$$

Example 11.2 **A rectangular shaped core is made of mild steel plate 15 mm × 20 mm cross section. The mean length of the magnetic path is 18 cm. The exciting coil has 300 turns and current 0.7 A. Calculate: (i) Magnetising force (ii) Flux density (iii) Reluctance (iv) Flux of magnetic circuit. Assume relative permeability of mild steel as 940.** [*RGPV Basic Electrical Engineering Jan./Feb; 2006*]

Solution:

$$\text{Mean length of magnetic path, } l = 18 \text{ cm} = 0.18 \text{ m}$$

$$\text{Cross-sectional area of mild steel plate, } a = 15 \text{ mm} \times 20 \text{ mm} = 300 \text{ mm}^2 = 3 \times 10^{-4} \text{ m}^2$$

$$(i) \text{ Magnetising force, } H = \frac{NI}{l} = \frac{300 \times 0.7}{0.18} = 1{,}166.6667 \text{ AT/m } \textbf{Ans.}$$

$$(ii) \text{ Flux density, } B = \mu_0 \mu_r H = 4\pi \times 10^{-7} \times 940 \times 1{,}166.6667 = 1.378 \text{ T } \textbf{Ans.}$$

$$(iii) \text{ Reluctance, } S = \frac{l}{\mu_0 \mu_r a} = \frac{0.18}{4\pi \times 10^{-7} \times 940 \times 3 \times 10^{-4}} = 50.794 \times 10^4 \text{ AT/Wb } \textbf{Ans.}$$

$$(iv) \text{ Flux, } \Phi = B \times a = 1.378 \times 3 \times 10^{-4} = 4.134 \times 10^{-4} \text{ Wb } \textbf{Ans.}$$

Example 11.3 **A coil of insulated wire of 500 turns and of resistance 4 Ω is closely wound on iron ring. The ring has a mean diameter of 0.25 m and a uniform cross-sectional area of 700 mm². Calculate the total flux in the ring when a dc supply of 6 V is applied to the ends of the winding. Assume a relative permeability of 550.** [*U.P. Technical Univ. Electrical Engineering July 2002*]

Solution:

$$\text{Length of iron ring, } l = \pi D = \pi \times 0.25 = 0.25\,\pi \text{ m}$$

$$\text{Area of cross section of iron path, } a = 700 \text{ mm}^2 = 0.7 \times 10^{-3} \text{ m}^2$$

$$\text{Current flowing through the winding, I} = \frac{\text{Voltage applied across coil}}{\text{Resistance of coil}} = \frac{6}{4} = 1.5 \text{ A}$$

$$\text{Total flux in the ring, } \Phi = \frac{\text{N I} \times \mu_r \mu_0 a}{l} = \frac{500 \times 1.5 \times 550 \times 4\pi \times 10^{-7} \times 0.7 \times 10^{-3}}{0.25\,\pi}$$

$$= 0.462 \text{ mWb } \textbf{Ans.}$$

Example 11.4 **A wrought iron bar 30 cm long and 2 cm in diameter is bent into a circular shape as given in Fig. 11.6. It is then wound with 500 turns of wire. Calculate the current required to produce a flux of 0.5 mWb in magnetic circuit with an air gap of 1 mm; μ_r (iron) = 4,000 (assume constant).**

[*U.P. Technical Univ. Electrical Engineering Second Semester 2004-05*]

Solution:

Flux to be produced, $\Phi = 0.5 \text{ mWb} = 0.5 \times 10^{-3}$ Wb

Area of cross section of core, $a = \pi \times 10^{-4}$ m^2

$$\text{Flux density required, B} = \frac{\Phi}{a} = \frac{0.5 \times 10^{-3}}{\pi \times 10^{-4}} = \frac{5}{\pi} = 1.5915 \text{ T}$$

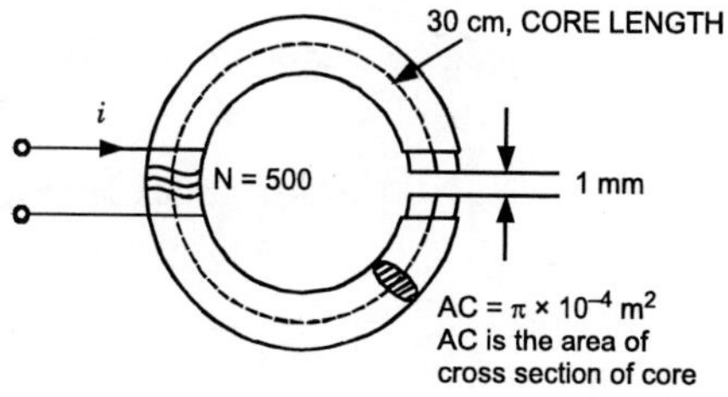

Fig. 11.6

Air-gap length, $l_g = 1 \text{ mm} = 0.001$ m

Core length, $l_i = 30 \text{ cm} = 0.3$ m

$$\text{Total ampere-turns required, AT} = \frac{\text{B}}{\mu_0} l_g + \frac{\text{B}}{\mu_0 \mu_r} l_i = \frac{1.5915}{4\pi \times 10^{-7}} \times 0.001 + \frac{1.5915}{4\pi \times 10^{-7} \times 4{,}000} \times 0.3$$

$$= 1{,}266 + 95 = 1{,}361$$

$$\text{Current required, I} = \frac{\text{AT}}{\text{N}} = \frac{1{,}361}{500} = 2.72 \text{ A } \textbf{Ans.}$$

Example 11.5 **A circular ring 20 cm in diameter has an air gap 1 mm wide cut in it. The area of cross section of the ring is 3.6 cm^2. Calculate the value of the direct current needed in a coil of 1000 turns uniformly wound round the ring to create a flux of 0.5 mWb in the air gap. Neglect fringing and assume relative permeability for the iron as 650.**

[*U.P. Technical Univ. Electrical Engineering First Semester 2006-07*]

Solution:

Flux to be produced, $\Phi = 0.5 \text{ mWb} = 0.5 \times 10^{-3}$ Wb

Area of x-section, $a = 3.6 \text{ cm}^2 = 3.6 \times 10^{-4}$ m^2

$$\text{Flux density required, B} = \frac{\Phi}{a} = \frac{0.5 \times 10^{-3}}{3.6 \times 10^{-4}} = 1.389 \text{ T}$$

Air-gap length, $l_g = 1 \text{ mm} = 1 \times 10^{-3}$ m

Length of iron path, $l_i = \pi d - l_g = \pi \times 0.20 - 1 \times 10^{-3} = 0.62732$ m

$$\text{Total ampere-turns required, AT} = \frac{\text{B}}{\mu_0} l_g + \frac{\text{B}}{\mu_0 \mu_r} l_i = \frac{1.389}{4\pi \times 10^{-7}} \times 1 \times 10^{-3} + \frac{1.389}{4\pi \times 10^{-7} \times 650} \times 0.62732$$

$$= 1{,}105 + 1{,}067 = 2{,}172$$

$$\text{Current required, I} = \frac{\text{AT}}{\text{N}} = \frac{2{,}172}{1{,}000} = 2.172 \text{ A } \textbf{Ans.}$$

Example 11.6 **A ring of ferromagnetic material has a rectangular cross section. The inner diameter is 7.4 in., the outer diameter is 9 in., and the thickness is 0.8 in. There is a coil of 600 turns wound on the ring. When the coil carries a current of 2.5 A, the flux produced in the ring is 1.2 × 10^{-3} Wb. Find: (i) Magnetic field intensity (ii) Reluctance (iii) Permeability.**

[U.P. Technical Univ. Electrical Engineering Even Semester 2007-08; Odd Semester 2013-14]

Solution: Number of turns in the coil, N = 600

Current carried by coil, I = 2.5 A

Flux produced, Φ = 1.2 × 10^{-3} Wb

Mean length of magnetic flux path is that of a circle midway between the inside and outside diameter i.e.

$$l = \frac{\pi(7.4+9)}{2} = 25.76 \text{ in} = 25.76 \times 0.0254 = 0.6543 \text{ m}$$

$$\text{Cross-sectional area of the toroidal core, } a = \frac{9-7.4}{2} \times 0.8 = 0.64 \text{ in}^2 = 4.13 \times 10^{-4} \text{ m}^2$$

$$(i) \text{ Magnetic flux density, B} = \frac{\Phi}{a} = \frac{1.2\times10^{-3}}{4.13\times10^{-4}} = 2.9 \text{ Wb/m}^2 \textbf{ Ans.}$$

$$\text{(ii) Reluctance, S} = \frac{\text{MMF}}{\Phi} = \frac{\text{NI}}{\Phi} = \frac{600\times2.5}{1.2\times10^{-3}} = 1{,}250 \times 10^3 \text{ AT/Wb } \textbf{Ans.}$$

$$\text{(iii) Permeability, } \mu = \frac{l}{Sa} = \frac{0.6543}{1{,}250\times10^3\times4.13\times10^{-4}} = 1.267 \times 10^{-3} \text{ H/m } \textbf{Ans.}$$

$$\text{Relative permeability, } \mu_r = \frac{\mu}{\mu_0} = \frac{1.267\times10^{-3}}{4\pi\times10^{-7}} = 1{,}008 \textbf{ Ans.}$$

Example 11.7 **A cast steel electromagnet has an air gap length of 3 mm and an iron path of length 40 cm. Find the number of ampere-turns necessary to produce a flux density of 0.7 Wb/m^2 in the gap. Neglect leakage and fringing. Assume ampere-turns required for air gap to be 70% of the total ampere-turns.**

[R.G.P.V. Basic Electrical Engineering 2001]

Solution: Air-gap length, l_g = 3 mm = 3 × 10^{-3} m

Flux density required in the air gap, B_g = 0.7 Wb/m^2

$$\text{Magnetising force, } H_g = \frac{B}{\mu_0\mu_r} = \frac{0.7}{4\pi\times10^{-7}\times1} = 5.57 \times 10^5 \text{ AT/m} \qquad \because \mu_r \text{ for air} = 1$$

Ampere-turns required for the air gap, $AT_g = H_g l_g = 5.57 \times 10^5 \times 3 \times 10^{-3} = 1{,}671$

Since ampere-turns required for the air gap are 70% of the total ampere-turns

i.e. AT_g = 0.7 AT (total ampere-turns required)

$$\text{So total ampere-turns required, AT} = \frac{AT_g}{0.7} = \frac{1{,}671}{0.7} = 2{,}387 \textbf{ Ans.}$$

Example 11.8 **A steel ring of 25 cm diameter and of circular section 3 cm in diameter has an air gap of 1.5 mm length. It is wound uniformly with 750 turns of wire carrying a current of 2.1 A. Calculate (i) magnetomotive force (ii) flux density in air gap (iii) magnetic flux (iv) relative permeability of steel ring. Assume that iron path takes about 35% of the total magnetomotive force.** *[R.G.P.V. Basic Electrical Engineering February 2005]*

Solution: Area of x-section, $a = \frac{\pi}{4}(d)^2 = \frac{\pi}{4}(3\times10^{-2})^2 = 0.0007 \text{ m}^2$

(*i*) Magnetomotive force, AT = N × I = 750 × 2.1 = 1,575 **Ans.**

Ampere-turns provided for iron path, AT_i = 35% of total AT = 0.35 × 1,575 = 551

Ampere-turns provided for air gap, AT_g, = AT – AT_i = 1,575 – 551 = 1,024

$$(ii) \text{ Flux density in the air gap, } B_g = \frac{AT_g \times \mu_0}{l_g} = \frac{1{,}024 \times 4\pi \times 10^{-7}}{1.5 \times 10^{-3}} = 0.858 \text{ T } \textbf{Ans.} \qquad \because AT_g = \frac{B_g}{\mu_0} l_g$$

$$(iii) \text{ Magnetic flux, } \Phi = B_g \times a = 0.858 \times 0.0007 = 0.6 \text{ mWb } \textbf{Ans.}$$

Assuming negligible leakage we have

$$\text{Flux density in iron path, } B_i = B_g = 0.858 \text{ T}$$

$$\text{Length of iron path, } l_i = \pi \times D - l_g = \pi \times 0.25 - 0.0015 = 0.784 \text{ m}$$

$$(iv) \text{ Relative permeability of steel, } \mu_r = \frac{B_i l_i}{\mu_0 AT_i} = \frac{0.858 \times 0.784}{4\pi \times 10^{-7} \times 551} = 970 \textbf{ Ans.} \qquad \because AT_i = \frac{B_i}{\mu_0 \mu_r} l_i$$

Example 11.9 **An iron ring of mean length 50 cm and relative permeability 300 has an air gap of 1 mm. If the ring is provided with a winding of 200 turns and a current of 1 A is allowed to flow through, find the flux density across the air gap.** [*U.P. Technical Univ. Elec. Engineering June 2001*]

Solution: Let the flux density through the iron as well as air gap be B Wb/m^2

$$\text{Total ampere-turns provided, AT} = N \times I = 200 \times 1 = 200$$

$$\text{Total ampere-turns required} = \text{AT required for air gap} + \text{AT required for iron ring}$$

$$\text{or Total ampere-turns provided} = \frac{B}{\mu_0} l_g + \frac{B}{\mu_0 \mu_r} l_i$$

$$\text{or } 200 = \frac{B}{\mu_0}\left(l_g + \frac{l_i}{\mu_r}\right) = \frac{B}{\mu_0}\left(0.001 + \frac{0.5}{300}\right)$$

$$\text{or Flux density, } B = \frac{200 \times \mu_0}{0.001 + \frac{0.5}{300}} = \frac{200 \times 4\pi \times 10^{-7}}{0.001 + \frac{0.5}{300}} = 0.0942 \text{ T } \textbf{Ans.}$$

Example 11.10 **An iron ring is made up of three parts:**
$l_1 = 10$ cm, $A_1 = 8$ cm^2, $l_2 = 8$ cm, $A_2 = 3$ cm^2, $l_3 = 6$ cm, $A_3 = 2.5$ cm^2.
It is wound with a coil of 250 turns. Calculate current required to produce flux of 0.4 mWb.
$\mu_1 = 2{,}670$, $\mu_2 = 1{,}050$, $\mu_3 = 600$. [*U.P. Technical Univ. Electrical Engineering Second Semester 2007-08*]

Solution: Total reluctance,

$$S = \frac{l_1}{\mu_0 \mu_{r_1} a_1} + \frac{l_2}{\mu_0 \mu_{r_2} a_2} + \frac{l_3}{\mu_0 \mu_{r_3} a_3}$$

$$= \frac{0.1}{4\pi \times 10^{-7} \times 2{,}670 \times 8 \times 10^{-4}} + \frac{0.08}{4\pi \times 10^{-7} \times 1{,}050 \times 3 \times 10^{-4}} + \frac{0.06}{4\pi \times 10^{-7} \times 600 \times 2.5 \times 10^{-4}}$$

$$= 0.37255 \times 10^5 + 2.021 \times 10^5 + 3.183 \times 10^5 = 5.57655 \times 10^5 \text{ AT/Wb}$$

$$\text{MMF} = \text{Flux } \Phi \times \text{Reluctance S} = 0.4 \times 10^{-3} \times 5.57655 \times 10^5 = 223 \text{ ampere-turns}$$

$$\text{Current required, } I = \frac{\text{MMF}}{N} = \frac{223}{250} = 0.892 \text{ A } \textbf{Ans.}$$

Example 11.11 **A rectangular magnetic core shown in Fig 11.7 has square cross section of area 16 cm^2. An air gap of 2 mm is cut across one of its limbs. Find the exciting current needed in the coil having 1,000 turns wound on the core to create an air-gap flux of 4mWb. The relative permeability of the core is 2,000.**
[*U.P. Technical Univ. Electrical Engineering February 2002*]

Solution: Flux to be created, $\Phi = 4 \text{ mWb} = 0.004 \text{ Wb}$

Area of x-section, $a = 16\ \text{cm}^2$

$= 16 \times 10^{-4}\ \text{m}^2$

Flux density required, $B = \dfrac{\Phi}{a}$

$= \dfrac{4 \times 10^{-3}}{16 \times 10^{-4}}$

$= 2.5\ \text{T}$

Length of air gap $l_g = 2\ \text{mm}$

$= 2 \times 10^{-3}\ \text{m}$

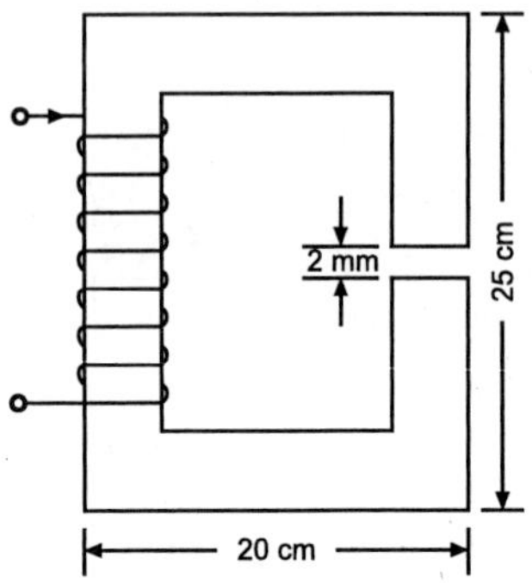

Fig. 11.7

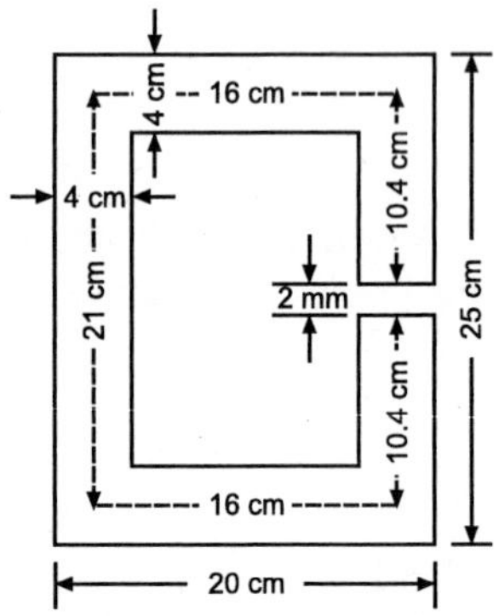

Fig. 11.8

Each side of cross section $= \sqrt{16} = 4\ \text{cm}$

Length of iron path, $l_i = \left(25 - 2 \times \dfrac{4}{2} + 20 - 2 \times \dfrac{4}{2}\right) \times 2 - 0.2 = 73.8\ \text{cm} = 0.738\ \text{m}$

Total ampere-turns required, $AT = \dfrac{B}{\mu_0} l_g + \dfrac{B}{\mu_0 \mu_r} l_i = \dfrac{2.5 \times 2 \times 10^{-3}}{4\pi \times 10^{-7}} + \dfrac{2.5 \times 0.738}{2{,}000 \times 4\pi \times 10^{-7}} = 3{,}979 + 734 = 4{,}713$

Number of turns on the coil = 1,000

Exciting current required, $I = \dfrac{AT}{N} = \dfrac{4{,}713}{1{,}000} = 4.713\ \text{A}$ **Ans.**

Example 11.12 **The magnetic circuit frame shown in Fig. 11.9 is built of iron of square cross section of 3 cm. Each air gap is 2 mm wide and each of the coil is wound with 1,000 turns. The relative permeability of part A and B may be taken as 1,000 and 1,200 respectively. Calculate (a) reluctance of part A (b) reluctance of two air gaps (c) total reluctance of complete magnetic path (d) total mmf (e) total flux and (f) flux density.**

Solution: Cross-sectional area,

$a = (0.03)^2 = 0.0009\ \text{m}^2$

Mean length flux path in part A,

$l_A = 20 - (1.5 + 1.5) = 17\ \text{cm or } 0.17\ \text{m}$

Mean length flux path in part B,

$l_B = (20 - 1.5 - 1.5) + (10 - 1.5) + (10 - 1.5)$

$= 34\ \text{cm or } 0.34\ \text{m}$

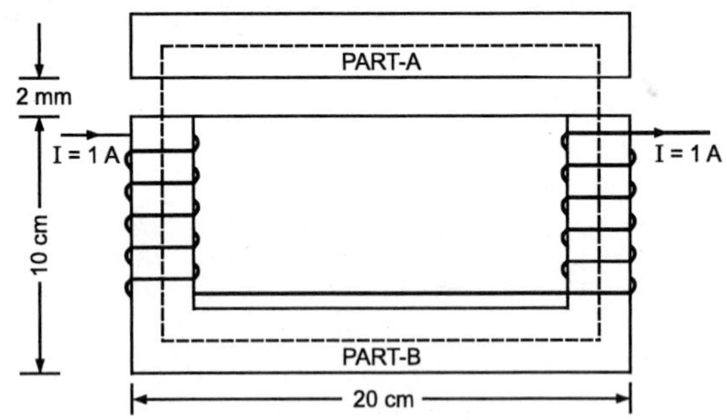

Fig. 11.9

Relative permeability of part A, $\mu_{rA} = 1{,}000$

Relative permeability of part B, $\mu_{rB} = 1{,}200$

Length of two air-gaps, $l_g = 2 \times 2 = 4\ \text{mm} = 0.004\ \text{m}$

(*a*) Reluctance of part A, $S_A = \dfrac{0.17}{4\pi \times 10^{-7} \times 1{,}000 \times 0.0009} = 1{,}50{,}313\ \text{AT/Wb}$ **Ans.**

Reluctance of part B, $S_B = \dfrac{l_B}{\mu_0 \mu_{rB} \times a} = \dfrac{0.34}{4\pi \times 10^{-7} \times 1{,}200 \times 0.0009} = 2{,}50{,}521\ \text{AT/Wb}$

(*b*) Reluctance of two air-gaps, $S_g = \dfrac{l_g}{\mu_0 a} = \dfrac{0.004}{4\pi \times 10^{-7} \times 0.0009} = 35{,}36{,}776\ \text{AT/Wb}$ **Ans.**

$$(c)\text{ Total reluctance, } S = S_A + S_B + S_g = 1{,}50{,}313 + 2{,}50{,}521 + 35{,}36{,}776$$
$$= 39{,}37{,}610 \text{ AT/Wb } \textbf{Ans.}$$

$$(d)\text{ Total mmf } = 2 \times 1{,}000 \times 1 = 2{,}000 \text{ AT } \textbf{Ans.}$$

because mmf is produced by two coils on part B each having 1,000 turns and carrying a current of 1 A

$$(e)\text{ Total flux } = \frac{\text{MMF}}{\text{Reluctance}} = \frac{2{,}000}{39{,}37{,}610} = 5.08 \times 10^{-4} \text{ Wb.}$$

$$(f)\text{ Flux density, } B = \frac{\Phi}{a} = \frac{5.08 \times 10^{-4}}{0.0009} = 0.564 \text{ Wb/m}^2 \text{ (or Tesla) } \textbf{Ans.}$$

Example 11.13 A cast steel magnetic structure made of a bar of section 2 cm × 2 cm is shown in Fig. 11.10. Determine the current that the 500 turn magnetizing coil on the left limb should carry so that a flux of 2 m Wb is produced in the right limb. Take μ_r = 600 and neglect leakage.

Solution: Flux created by magnetizing coil, say Φ, is divided at junction point into two paths depending upon the reluctances of portions B and C respectively.

$$\text{Reluctance of portion B, } S_B = \frac{l_B}{\mu_0\, \mu_r \times a} = \frac{0.15}{\mu_0\, \mu_r\, a}$$

$$\text{Reluctance of portion C, } S_C = \frac{l_C}{\mu_0\, \mu_r\, a} = \frac{0.25}{\mu_0\, \mu_r\, a}$$

$$\text{Flux in portion C, } \Phi_C = 2\text{mWb} = 0.002 \text{ Wb (given)}$$

$$\text{Flux in portion B, } \Phi_B = \Phi_C \times \frac{S_C}{S_B} = 2 \times 10^{-3} \times \frac{0.25/\mu_0\mu_r a}{0.15/\mu_0\mu_r a} = 3.33 \times 10^{-3} \text{ Wb}$$

$$\text{Total flux in portion A, } \Phi = \Phi_B + \Phi_C$$
$$= 3.33 \times 10^{-3} + 2 \times 10^{-3}$$
$$= 5.33 \times 10^{-3} \text{ Wb}$$

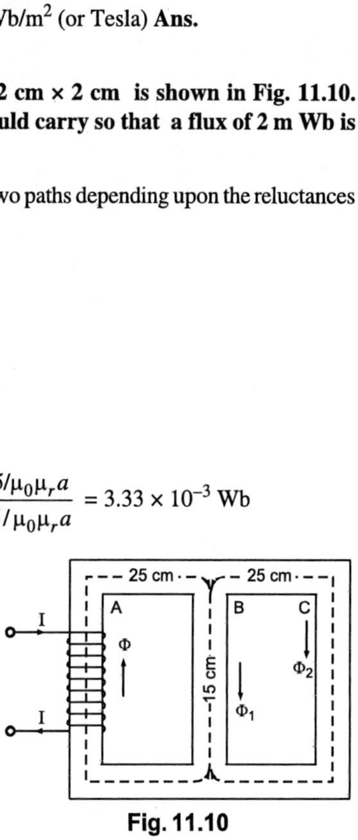

Fig. 11.10

$$\text{Flux density in portion A, } B_A = \frac{\Phi}{a}$$
$$= \frac{5.33 \times 10^{-3}}{4 \times 10^{-4}}$$
$$= 13.33 \text{ Wb/m}^2$$

$$\text{AT required for portion A, } AT_A = \frac{B_A \times l_A}{\mu_0\, \mu_r} = \frac{13.33 \times 0.25}{4\pi \times 10^{-7} \times 600} = 4{,}416$$

$$\text{Flux density in path B, } B_B = \frac{3.33 \times 10^{-3}}{4 \times 10^{-4}} = 8.33 \text{ Wb/m}^2$$

$$\text{AT required for portion B, } AT_B = \frac{B_B\, l_B}{\mu_0\, \mu_r} = \frac{8.33 \times 0.15}{4\pi \times 10^{-7} \times 600} = 1{,}658$$

$$\text{Total ampere-turns required, AT } = AT_A + AT_B = 4{,}416 + 1{,}658 = 6{,}074 \textbf{ Ans.}$$

11.2. MAGNETISATION OR B-H CURVES

The curve drawn giving relationship between induction density B and magnetising force H is known as *B-H curve* or *magnetising curve.* Fig. 11.11 shows the general shape of B-H curve of a magnetic material. In general it has four distinct regions *oa*, *ab*, *bc* and the region beyond *c*. During the region *oa* the increase in flux density is very small, in region *ab* the flux density B increases almost linearly with the magnetising force H, in region *bc* the increase in flux density is again small and in region beyond point *c*, the flux density B is almost constant. The flat part of the B-H curve corresponds to *magnetic saturation* of the material. The nonlinearity of the curve indicates that the relative permeability $\mu_r \left(\dfrac{B}{\mu_0 H}\right)$ of a magnetic material is not* constant but depends very largely upon the flux density B.

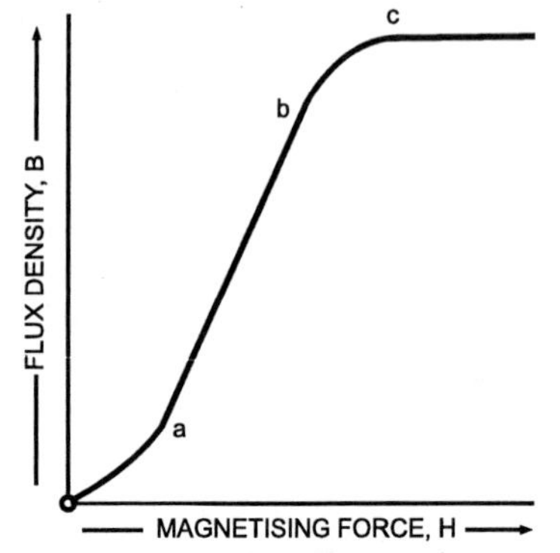

Fig. 11.11

The B-H curve depends on the material only and not upon the dimensions of a specific piece. Magnetising or B-H curves for silicon alloy (electrical sheet) steel, cast steel and cast iron etc., are shown in Fig. 11.12. In making magnetic calculations, values of H and μ_r corresponding to working flux density B are to be taken. The use of B-H curves permits the calculations of magnetic circuits with a fair degree of accuracy.

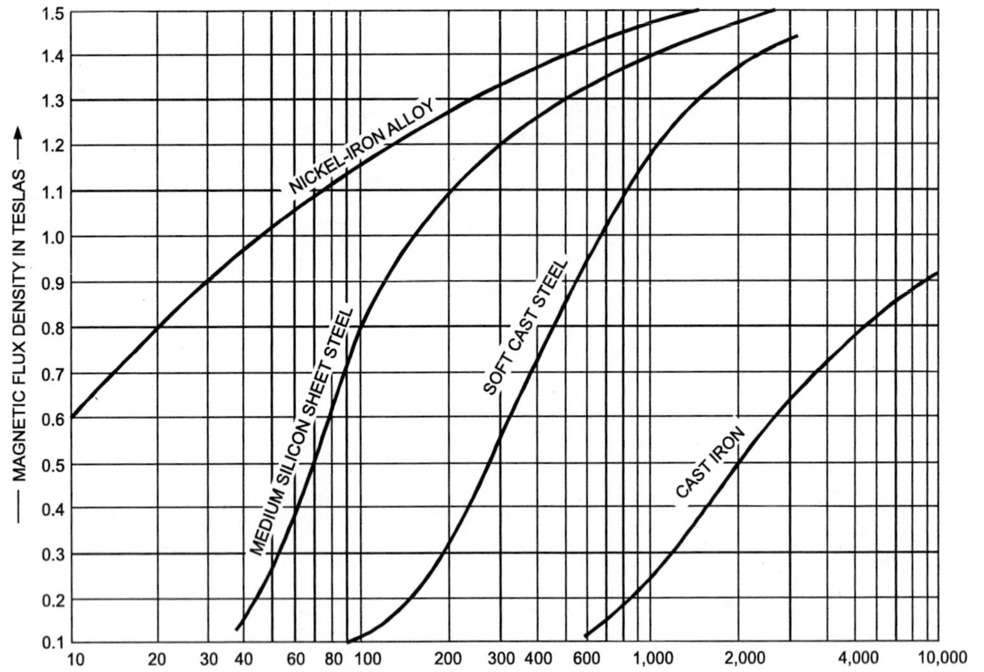

Fig. 11.12 *Magnetization Curves of Typical Ferromagnetic Materials*

11.3. MAGNETIC HYSTERESIS

If a magnetic substance is magnetised in a strong magnetic field, it retains a considerable portion of magnetism after the magnetic force has been withdrawn. The phenomenon of lagging of magnetisation or induction flux density behind the magnetising force is known as *magnetic hysteresis.*

* Relative permeability μ_r for all non-magnetic materials (*e.g.* brass, air etc.) is constant; being equal to 1. Thus for non-magnetic materials $B = \mu_0 H$. Since μ_0 is constant, B α H or B-H curve for a non-magnetic material is a straight line passing through origin.

Let a core of specimen of iron be wound with a number of turns of a wire and current be passed through the solenoid. A magnetic field of intensity H proportional to the current flowing through the solenoid is produced. Let magnetising force H is increased from zero to a certain maximum value and then gradually reduced to zero. If the values of flux density B in the core corresponding to various values of magnetising force H are determined and B-H curves are drawn for increasing and decreasing values of magnetising force H then it will be observed that B-H curve obtained for decreasing values of H lies above that obtained for increasing values of H.

While decreasing the magnetising force H, when H is brought to zero the induction density B, is represented by OC and is called as *residual magnetism.* The power of retaining the residual magnetism is called the *retentivity* of the material.

Now if the direction of flow of current is reversed, the magnetising force H is reversed. Let the current be increased in the negative direction until the induction density B becomes zero. At this instant *i.e.* when B = 0, the demagnetising force H = OD, which is required to neutralize the residual magnetism, and is known as *coercive force.* If the demagnetising force H is further increased to the previous maximum value and again gradually decreased to zero, reversed and further increased in original or positive direction to the maximum value, a closed loop ACDEFGA is obtained which is usually known as *hysteresis loop* or *magnetic cycle.*

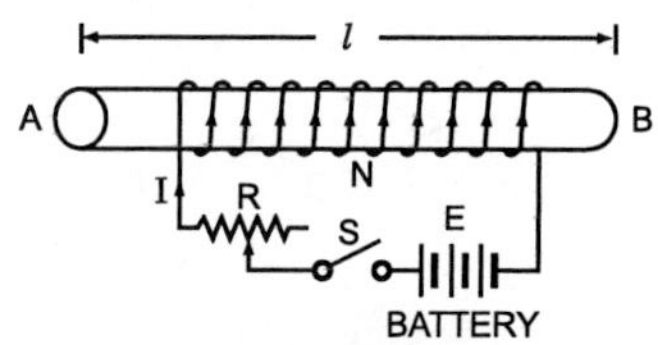

Fig. 11.13

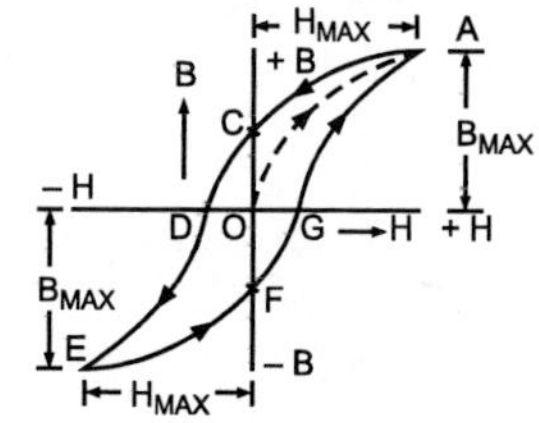

Fig. 11.14

It is to be noted from the hysteresis loop that B lags behind H. The two never attain zero value simultaneously.

Hysteresis is especially pronounced in materials of high residual magnetism, such as hard steel. In most cases, hysteresis is a liability as it causes dissipation of heat, waste of energy, and humming due to change in polarity and rotation of element magnets in the material.

Hysteresis loops for hard steel, wrought iron and cast steel and for alloyed sheet steel are shown in Fig. 11.15.

Loop (*a*) is for hard steel. Due to its high retentivity power and large coercive force, this material is well suited for permanent magnets. Since the area of hysteresis loop for hard steel is large, hard steel is not suitable for rapid reversals of magnetisation. Certain alloys of steel, aluminium and nickel known as *alnico alloys* are extremely suitable for permanent magnets,

Loop (*b*) is for wrought iron and cast steel which rises steeply. Hence these materials have high magnetic permeability and good retentivity, therefore, these materials are suitable for cores of electromagnets.

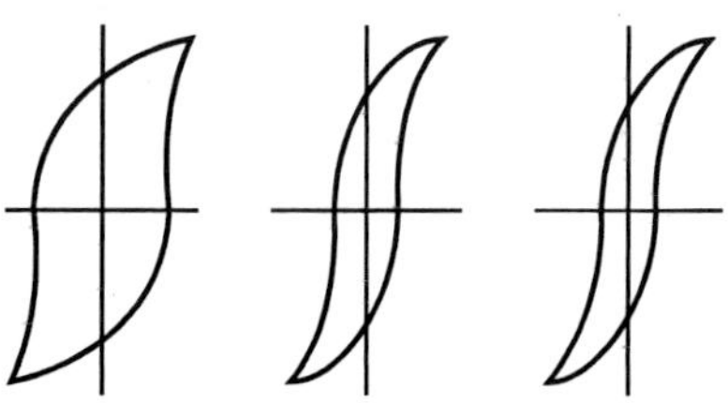

(*a*) (*b*) (*c*)

Fig. 11.15 *Hysteresis Loops*

Loop (*c*) is for iron, low-carbon steel, silicon alloys, permalloy or Mumetal sheets. Since the permeability of these materials is very high and hysteresis losses are very low, these materials are most suitable for transformer cores and armatures, which are subjected to rapid reversals of magnetisation. Silicon alloys and permalloy (78.5% Ni; 21% iron with small quantities of copper, molybdenum, chromium, cobalt and manganese etc.) are better for use as compared to iron and low-carbon steel.

11.3.1. Hysteresis Loss. When magnetic circuits are subjected to time-varying flux densities, there are two causes of power loss in the form of heat in the iron core. These losses are significant in determining the heating, rating and efficiency of rotating machines, transformers, and ac-operated magnetic devices. The first loss is associated with the phenomenon of hysteresis, explained above. It has been found experimentally that the variation of hysteresis loss with variations in the number of reversals per second (*f*) and maximum flux density (B_{max}) may be expressed with sufficient accuracy by the equation

$$\text{Hysteresis loss, } P_h = \eta\, (B_{max})^{1.6} f\, v \text{ joules per second or watts} \quad ...(11.7)$$

where η = Steinmetz hysteresis coefficient, v is volume of core in m^3, B_{max} is the peak magnitude of the flux density attained in a magnetic cycle and *f* is the frequency of magnetic cycles per second.

11.4. MAGNETICALLY INDUCED EMFS (OR VOLTAGES)

A very important effect of a magnetic field on an electric circuit is that when the flux linking the circuit changes, an emf is induced. Electromagnetic induction of emf (or voltage) is basic to the operation of transformers, generators (ac or dc) and motors (ac or dc). The effect is described by Faraday's law, which states that *the magnitude of emf (or voltage) is directly proportional to the rate of change of flux linkage or to the product of number of turns and rate of change of flux linking the coil.*

$$i.e. \text{ induced emf, } e = N \frac{d\Phi}{dt} \quad ...(11.8)$$

where $N \frac{d\Phi}{dt}$ is the product of number of turns and rate of change of linking flux and is termed as rate of change of flux linkage.

The direction of induced emf is governed by Lenz's law which states that *the direction of induced emf or voltage is such that the current produced by it sets up a magnetic field opposing the flux change.*

A minus (–) sign is required to be placed before the right hand side quantity of Eq. (11.8) just to indicate the phenomenon explained by Lenz's law. Thus the expression for induced emf becomes as

$$e = -N \frac{d\Phi}{dt} \text{ volts} \quad ...(11.9)$$

where Φ is in webers and time *t* is in seconds.

The induced emf or voltage appears in any circuit which may be linked by the changing flux, including the circuit giving rise to flux. Thus, current in a circuit produces a magnetic field linking the circuit and permeating the medium around it. Growth, decay or any other change of current with time causes corresponding changes of the magnetic flux and induces an emf in the circuit.

Example 11.14 **A coil has 400 turns. Find the induced voltage in it if the flux changes from 0.2 m Wb to 1 m Wb in 0.2 second.**

Solution:

$$\text{Change of flux in 0.2 } s,\ d\Phi = 0.001 - 0.0002 = 0.0008 \text{ Wb}$$

$$\text{Rate of change of flux, } \frac{d\Phi}{dt} = \frac{0.0008}{0.2} = 0.004 \text{ Wb/s}$$

$$\text{Induced voltage, } e = N\frac{d\Phi}{dt} = 400 \times 0.004 = 1.6 \text{ V } \textbf{Ans.}$$

11.5. DYNAMICALLY INDUCED EMF

We have learnt that when the flux linking with the coil or circuit changes, an emf is induced in the coil or circuit.

EMF can be induced by changing the flux linking in two ways :

(*i*) By increasing or decreasing the magnitude of the current producing the linking flux. In this case there is no motion of the conductor or of coil relative to the field and, therefore, emf induced in this way is known as *statically induced emf.*

(*ii*) By moving a conductor in a uniform magnetic field and emf produced in this way is known as *dynamically induced emf.*

Consider a conductor of length *l* metres placed in a uniform magnetic field of density B Wb/m^2, as shown in Fig. 11.16 (*a*).

Let this conductor be moved with velocity *v* m/s in the direction of the field, as shown in Fig. 11.16 (*b*). In this case no flux is cut by the conductor, therefore, no emf is induced in it.

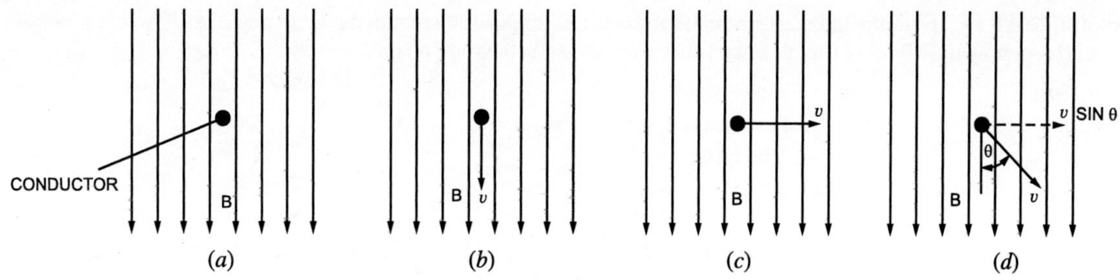

Fig. 11.16

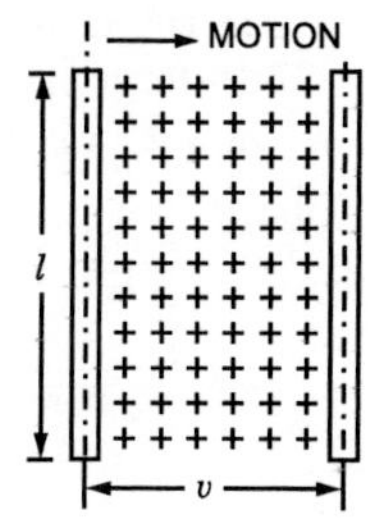

Fig. 11.17

Now if this conductor is moved with velocity v m/s in a direction perpendicular to its own length and perpendicular to the direction of the magnetic field, as shown in Fig. 11.16 (*c*) flux is cut by the conductor, therefore, an emf is induced in the conductor.

$$\text{Area swept per second by the conductor} = l \times v \text{ m}^2/\text{s}$$

$$\text{Flux cut per second} = \text{Flux density} \times \text{area swept per second} = \text{B}lv$$

$$\text{Rate of change of flux, } \frac{d\Phi}{dt} = \text{Flux cut per second} = \text{B}lv \text{ Wb/s}$$

$$\text{Induced emf, } e = \frac{d\Phi}{dt} = \text{B}lv \text{ volts}$$

If the conductor is moved with velocity v metres per second in a direction perpendicular to its own length and at an angle θ to the direction of magnetic field, as shown in Fig. 11.16 (*d*).

The magnitude of emf induced is proportional to the component of the velocity in a direction perpendicular to the direction of the magnetic field and induced emf is given by

$$e = \text{B}lv \sin\theta \text{ volts} \qquad ...(11.10)$$

The direction of this induced emf is given by Fleming's right hand rule.

If the thumb, forefinger and middle finger of right hand are held mutually perpendicular to each other, forefinger pointing into the direction of the field and thumb in the direction of motion of conductor then the middle finger will point in the direction of the induced emf as shown in Fig. 11.18 (*a*).

Fig. 11.18 (*b*) illustrates another way of determination of induced emf, known as *right hand flat palm* rule. This law states that if right hand is so placed in a magnetic field along the conductor that the magnetic lines of force emerging from the north pole enter the palm and the thumb points in the direction of motion of conductor, the other four fingers will give the direction of induced emf or current.

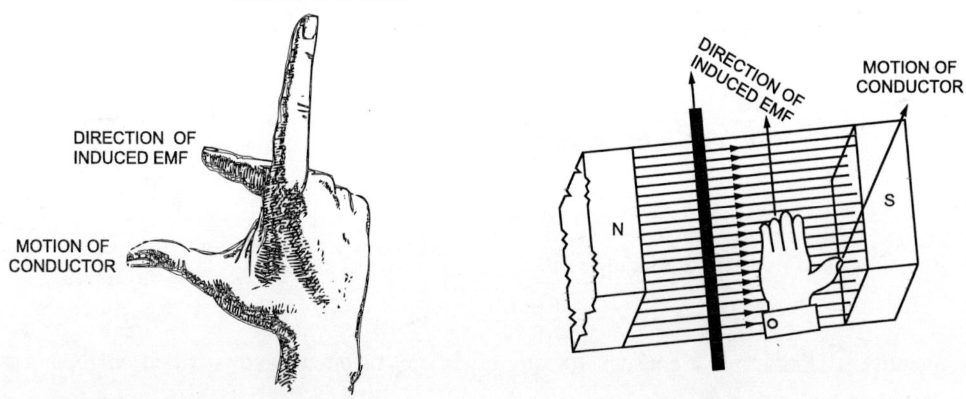

(*a*) *Fleming's Right Hand Rule* (*b*) *Right Hand Flat Palm Rule*

Fig. 11.18 *Determination of Direction of Induced EMF*

Example 11.15 **Find the induced emf in a conductor of length 150 cm moving at an angle of 30° to the direction of uniform magnetic field of flux density 1.2 Wb/m² with a velocity of 60 m/s.**

Solution:

Flux density B = 1.2 T

Length of conductor, l = 150 cm = 1.5 m

Velocity of conductor, v = 60 m/s

Angle of movement of conductor from the direction of magnetic field,

θ = 30°

EMF induced, $e = Blv \sin\theta = 1.2 \times 1.5 \times 60 \times \sin 30° = 108 \times 0.5 = 54$ V **Ans.**

Example 11.16 **A conductor of active length 30 cm carries a current of 100 A and lies at right angles to a magnetic field of strength 0.4 Wb/m². Calculate the force in newtons exerted on it.**

If the force causes the conductor to move at a velocity of 10 m/s, calculate (*i*) the emf induced in it and (*ii*) the power in watts developed by it.

Solution:

Length of conductor, l = 30 cm = 0.3 m

Current flowing through conductor, I = 100 A

Field strength, B = 0.4 Wb/m²

Force exerted on the conductor, F = B I l = 0.4 × 100 × 0.3 = 12 N **Ans.**

Conductor velocity, v = 10 m/s

Induced emf, e = B l v = 0.4 × 0.3 × 10 = 1.2 V **Ans.**

Power developed, P = F × v = 12 × 10 = 120 watts **Ans.**

11.6. STATICALLY INDUCED EMF

Statically induced emf may be (*a*) self-induced emf or (*b*) mutually induced emf.

11.6.1. Self-Induced EMF. When the current flowing through the coil is changed, the flux linking with its own winding changes and due to the change in linking flux with the coil, an emf, known as *self-induced emf, is induced.*

Since according to Lenz's law, an induced emf acts to oppose the change that produces it, a self-induced emf is always in such a direction as to oppose the change of current in the coil or circuit in which it is induced. This property of the coil or circuit due to which it opposes any change of the current in the coil or circuit, is known as *self-inductance*.

Consider a solenoid of N turns, length l metres, area of x-section a square metres and of relative permeability μ_r. When the solenoid carries a current of i amperes, a magnetic field of flux $\dfrac{Ni}{l/\mu_0 \mu_r a}$ webers is set up around the solenoid and links with it.

If the current flowing through the solenoid is changed, the flux produced by it will change and, therefore, an emf will be induced.

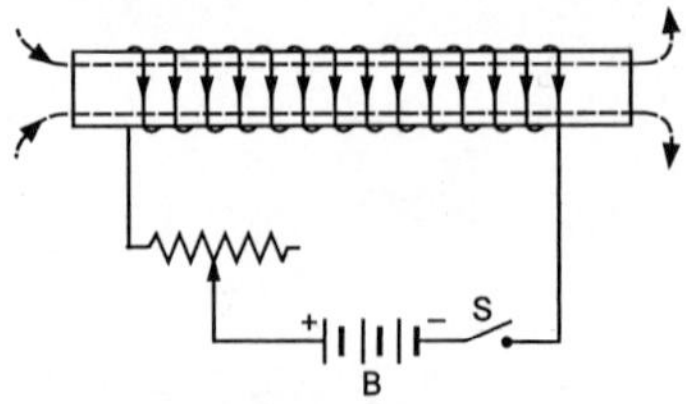

Fig. 11.19

$$\text{Self induced emf, } e = -N\frac{d\Phi}{dt} = -N\frac{d}{dt}\left[\frac{Ni}{l/\mu_r \mu_0 a}\right]$$

$$= -N\frac{N}{l/\mu_r \mu_0 a}\frac{di}{dt}$$

$$= -\frac{N^2 \mu_r \mu_0 a}{l}\cdot\frac{di}{dt} \qquad ...(11.11)$$

The quantity $\dfrac{N^2 \mu_r \mu_0 a}{l}$ is a constant for any given coil or circuit and is called *coefficient of self-inductance.* It is represented by symbol L and is measured in henries.

Hence self-induced emf, $e = -L\dfrac{di}{dt}$ where $L = \dfrac{N^2 \mu_r \mu_0 a}{l}$ henrys

11.6.2. Coefficient of Self-Induction. The coefficient of self-induction (L) can be determined from any one of the following three relations.

First Method. In case the dimensions of the solenoid are given, the coefficient of self-induction may be determined from the relation

$$L = \frac{N^2 a \mu_r \mu_0}{l} \text{ henrys} \quad ...(11.12) \quad \text{[Refer to Art. 11.6.1]}$$

Second Method. In case the magnitude of induced emf in a coil for a given rate of change of current in the coil is known, self-inductance of the coil may be determined from the following relation.

$$e = L\frac{di}{dt} \quad ...\text{[Refer to Art. 11.6.1]}$$

$$\text{or } L = \frac{e}{di/dt} \quad ...(11.13)$$

Third Method. In case the number of turns of the coil and flux produced per ampere of current in the coil is known, the self-inductance of the coil may be determined from the following relation

$$L = \frac{N\Phi}{i} \quad ...(11.14)$$

The above relation can be derived as follows:

Magnetic flux produced in a coil of N turns, length l metres, area of x-section a metres^2 and relative permeability μ_r when carrying a current of i amperes is given by

$$\Phi = \frac{Ni}{l/\mu_r \mu_0 a} = \frac{N}{l/\mu_r \mu_0 a} i$$

and self-inductance of the coil $L = \dfrac{N^2 a \mu_r \mu_0}{l} = \dfrac{N}{i}\dfrac{N a \mu_r \mu_0}{l} i = \dfrac{N\Phi}{i}$

From the above relation it is obvious that the self-inductance of a coil or circuit is equal to weber-turns per ampere in the coil or circuit.

In the above relation if N Φ = 1 Wb-turn and i = 1 A then L = 1 H.

Hence, a coil is said to have a self-inductance of one henry if a current of 1 A, when flowing through it, produces flux linkage of 1 wb-turn in it.

11.6.3. Mutually Induced EMF. Consider two coils A and B placed closed together so that the flux created by one coil completely links with the other coil. Let coil A have a battery and switch S and coil B be connected to the galvanometer G.

When switch S is opened, no current flows through coil A, so no flux is created in coil A, *i.e.* no flux links with coil B, therefore, no emf is induced across coil B, the fact is indicated by galvanometer zero deflection. Now when the switch S is closed current in coil A starts rising from zero value to a finite value, the flux is produced during this period and increases with the increase in current of coil A, therefore, flux linking with the coil B increases and an emf, known as *mutually induced emf,* is produced in coil B, the fact is indicated by galvanometer deflection. As soon as the current in coil A reaches its finite value, the flux produced or flux linking with coil B becomes constant, so no emf is induced in coil B, and galvanometer pointer returns back to zero position. Now if the switch S is opened, current will start decreasing, resulting in decrease in flux linking with coil B, an emf will be again induced but in direction opposite to previous one, this fact will be shown by the galvanometer deflection in opposite direction.

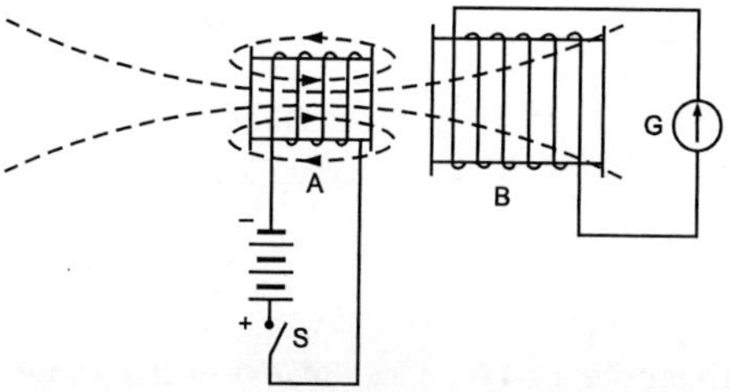

Fig. 11.20

Hence, whenever the current in coil A changes, the flux linking with coil B changes and an emf, known as mutually induced emf, is induced in coil B.

Consider coil A of turns N_1 wound on a core of length l metres, area of cross section a square metres and relative permeability μ_r. When the current of i_1 amperes flows through it, a flux of $\dfrac{N_1 i_1}{l/\mu_0 \mu_r a}$ webers is set up around the coil. Let whole of the flux produced due to flow of current in coil A be linked with the coil B having N_2 turns and placed near by coil A.

$$\text{Mutually induced emf, } e_m = -\text{Rate of change of flux linkage of coil B}$$

$$= -N_2 \times \text{rate of change of flux in coil A}$$

$$= -N_2 \frac{d}{dt}\left[\frac{N_1 i_1}{l/\mu_0 \mu_r a}\right] = -\frac{N_1 N_2 a \mu_0 \mu_r}{l} \cdot \frac{di_1}{dt} \quad \text{...(11.15)}$$

The quantity $\dfrac{N_1 N_2 a \mu_0 \mu_r}{l}$ is called the *coefficient of mutual induction* of coil B with respect to coil A. It is represented by symbol M and is measured in henrys.

Hence mutually inducd emf,

$$e_m = -M \frac{di_1}{dt} \qquad \text{where } M = \frac{N_1 N_2 a \mu_0 \mu_r}{l} \text{ henrys}$$

11.6.4. Coefficient of Mutual Induction. Mutual inductance may be defined as the ability of one coil or circuit to induce an emf in a nearby coil by induction when the current flowing in the first coil is changed. The action is also reciprocal i.e. the change in current flowing through second coil will also induce an emf in the first coil. The ability of reciprocal induction is measured in terms of the coefficient of mutual induction M.

The coefficient of mutual induction (M) can be determined from any one of the following three relations.

First Method. In case the dimensions of the coils are given, the coefficient of mutual induction may be determined from the relation

$$M = \frac{N_1 N_2 a \mu_0 \mu_r}{l} \text{ henrys} \quad \text{...(11.16) [Refer to Art. 11.6.3]}$$

Second Method. In case the magnitude of induced emf in the second coil for a given rate of change of current in the first coil is known, mutual inductance between the coil may be determined from the following relation

$$e_m = M \frac{di_1}{dt} \quad \text{[Refer to Art. 11.6.3]}$$

$$\text{or } M = \frac{e_m}{di_1/dt} \quad \text{...(11.17)}$$

Third Method. In case the number of turns of the coil and flux linking with this coil per ampere of current in another coil is known, the mutual inductance of the coil may be determined from the following relation

$$M = N_2 \frac{\Phi_2}{i_1} \text{ Henry} \quad \text{...(11.18)}$$

Example 11.17 A coil of 300 turns, wound on a core of non-magnetic material, has an inductance of 10 mH. Calculate (*i*) the flux produced by a current of 5 A and (*ii*) the average value of emf induced when a current of 5 A is reversed in 8 milliseconds.

Solution:

$$(i) \text{ Flux produced, } \Phi = \frac{iL}{N} = \frac{5 \times 0.01}{300} = 0.1667 \text{ mWb } \textbf{Ans.} \qquad \because L = \frac{N\Phi}{i}$$

$$\text{Rate of change of current, } \frac{di}{dt} = \frac{5-(-5)}{0.008} = 1{,}250 \text{ A/s}$$

$$\text{Induced emf, } e = L\frac{di}{dt} = 0.01 \times 1{,}250 = 12.5 \text{ V } \textbf{Ans.}$$

Example 11.18 **A coil of 1,000 turns is wound on a laminated core of steel having a cross section of 5 cm². The core has an air gap of 2 mm cut at right angle. What value of current is required to have an air gap flux density of 0.5 T? Permeability of steel may be taken as infinity. Determine the coil inductance.**

[U.P. Technical Univ. Electrical Engineering First Semester 2003-04]

Solution:

$$\text{Total ampere-turns required, AT} = \frac{B}{\mu_0 \mu_r} l_i + \frac{B}{\mu_0} l_g = 0 + \frac{0.5}{4\pi \times 10^{-7}} \times 2 \times 10^{-3} = 796 \qquad \because \mu_r = \infty$$

$$\text{Current required, } i = \frac{AT}{N} = \frac{796}{1,000} = 0.796 \text{ A } \textbf{Ans.}$$

$$\text{Inductance of coil, L} = \frac{N\phi}{i} = \frac{1,000 \times 0.5 \times 5 \times 10^{-4}}{0.796} = 0.314 \text{ H } \textbf{Ans.}$$

Example 11.19 **A toroid has a core of square x-section 2,500 mm² in area and a mean diameter of 250 mm. The core material has a relative permeability 1,000. Calculate the number of turns wound on the core to obtain an inductance of 1 H.**

Solution:

Length of core, $l = \pi \times D = \pi \times 0.25 = 0.25\,\pi$ metres

Area of x-section of core, $a = 2,500 \text{ mm}^2 = 0.0025 \text{ m}^2$

Relative permeability of core material, $\mu_r = 1,000$

$\mu_0 = 4\pi \times 10^{-7}$ H/m

Self inductance of coil, L = 1 H

$$\because L = \frac{N^2 a \mu_0 \mu_r}{l},$$

$$\text{Number of turns on the coil, N} = \sqrt{\frac{L \times l}{\mu_0 \mu_r a}} = \sqrt{\frac{1 \times 0.25\pi}{4\pi \times 10^{-7} \times 1,000 \times 0.0025}} = 500 \textbf{ Ans.}$$

Example 11.20 **The core of a magnetic circuit is of mean length 40 cm and uniform cross-sectional area 4 cm². The relative permeability of the core material is 1,000. An air gap of 1 mm is cut in the core, and 1,000 turns are wound on the core. Determine the inductance of the coil if fringing is negligible.**

[U.P. Technical Univ. Electrical Engineering First Semester 2008-09]

Solution:

Mean length of core, $l_i = 40 - 0.1 = 39.9 \text{ cm} = 0.399$ m

x-sectional area of core, $a = 4 \text{ cm}^2 = 0.0004 \text{ m}^2$

Relative permeability of core material, $\mu_r = 1,000$

Air-gap length, $l_g = 1 \text{ mm} = 0.001$ m

Number of turns wound on the core, N = 1,000

Assuming flux density in the core to be B Wb/m²

$$\text{Total ampere-turns required, AT} = \frac{B}{\mu_o \mu_r} l_i + \frac{B}{\mu_o} l_g$$

$$= \frac{B}{4\pi \times 10^{-7} \times 1,000} \times 0.399 + \frac{B}{4\pi \times 10^{-7}} \times 0.001$$

$$= 317.5 \text{ B} + 795.8 \text{ B} = 1,113 \text{ B}$$

$$\text{Current required, I} = \frac{AT}{N} = \frac{1,113}{1,000} B = 1.113 \text{ B}$$

$$\text{Flux, } \Phi = B \times a = B \times 0.0004 = 0.0004 \text{ B}$$

$$\text{Inductance of coil, L} = \frac{N\Phi}{I} = \frac{1,000 \times 0.0004 \text{ B}}{1.113 \text{ B}} = 0.359 \text{ H } \textbf{Ans.}$$

Example 11.21 **Find the inductance of a coil in which a current of 0.2 A increasing at the rate of 0.4 A per second represents a power flow of 0.4 watt.**

Solution:

$$\text{Power flow, } p = 0.4 \text{ watt}$$

$$\text{Current, } i = 0.2 \text{ A}$$

$$\text{Induced emf, } e = \frac{p}{i} = \frac{0.4}{0.2} = 2 \text{ V} \qquad \because p = ei$$

$$\text{Rate of increase of current, } \frac{di}{dt} = 0.4 \text{ A/s}$$

$$\text{Inductance of coil, L} = \frac{e}{di/dt} = \frac{2}{0.4} = 5 \text{ H } \textbf{Ans.}$$

Example 11.22 **Two coils having 100 and 50 turns respectively are wound on a core with $\mu = 4{,}000\ \mu_0$. Effective core length = 60 cm and core area = 9 cm^2. Find the mutual inductance between the coils.**

[U.P. Technical Univ. Electrical Engineering Second Semester 2004-05]

Solution:

$$\text{Number of turns, } N_1 = 100 \text{ and } N_2 = 50$$

$$\text{Core area, } a = 9 \text{ cm}^2 = 9 \times 10^{-4} \text{ m}^2$$

$$\text{Core length, } l = 60 \text{ cm} = 0.6 \text{ m}$$

$$\mu = 4{,}000\ \mu_0 = 4{,}000 \times 4\pi \times 10^{-7} = 5.0265 \times 10^{-3} \text{ H/m}$$

$$\text{Mutual inductance between the coils, M} = \frac{N_1 N_2 a \mu}{l} = \frac{100 \times 50 \times 9 \times 10^{-4} \times 5.0265 \times 10^{-3}}{0.6} = 37.7 \text{ mH } \textbf{Ans.}$$

11.7. DOT CONVENTION

Three coils A, B and C wound on a common magnetic circuit are shown in Fig. 11.21. Coil A has N_A turns with terminals A_1 and A_2. The sense of winding is shown in the figure. Similarly the coil B has N_B turns with terminals B_1 and B_2 and coil C has N_C turns with terminals C_1 and C_2. The terminals are marked, arbitrarily on the coils. Let us consider that a current I_A enters terminal A_1. Looking from top, the current encircles the core counter-clockwise. According to corkscrew rule the flux is directed upwards through the coil. The flux through the core is directed clockwise through the core. A dot is placed quite arbitrarily at terminal A_1. This can be done for any coil. Once the dot is placed on one coil, dots on the terminals of other remaining coils cannot be placed arbitrarily now. Dots on other coils are now automatically decided according to the sense of the winding. Now we are to determine the position of dots on the terminals of the remaining windings corresponding to the dot placed at one of the terminals of winding A, which is at A_1. Dot on the other winding is placed on such a terminal that a current entering through the dotted terminal magnetises the core in the same direction as the flux created by current entering the dot on the first coil A. Thus, if the dot were placed at B_2, a current entering this terminal would produce a flux in opposition to the flux produced by current I_A. Thus, B_2 is not the dotted terminal corresponding to the dot at terminal A_1. While, if the current enters terminal B_1, the magnetization of the core is also clockwise. Thus, terminal B_1 is the dotted terminal on coil B. Similarly C_1 is the dotted terminal on coil C. Thus, A_1, B_1 and C_1 are dotted terminals on the three coils which are mutually coupled. Now we can make a statement that "currents entering through the dotted terminals of all the coupled coils develop the flux (apply mmf) so that the fluxes due to each coil are additive."

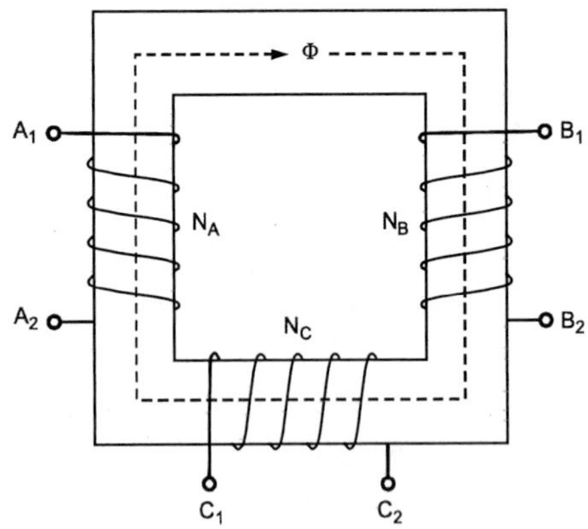

Fig. 11.21

This also means, that due to rate of change of the common flux emfs are induced in all the coils which have similar polarity at all the dotted terminals.

11.8. MUTUAL COUPLING IN A SIMPLE MAGNETIC CIRCUIT

The self-inductance of a circuit is intimately associated with the magnetic field linking the circuit. The self-inductance emf may be thought of as the emf induced in the circuit by a magnetic field produced by the circuit current.

Since a magnetic field exists in the region around the current that develops it, there is also may a possibility that an emf be induced in the other circuits linked by the field. Two circuits linked by the same magnetic field are said to be *coupled* to each other. The circuit element used to represent *magnetic coupling* is shown in Fig. 11.22 (*a*) and is called *mutual inductance.* It is represented by symbol M and is measured in henrys, already stated in Art 11.6.3. The volt-ampere relationship is one which gives the induced emf in one circuit by a current in another and is given as

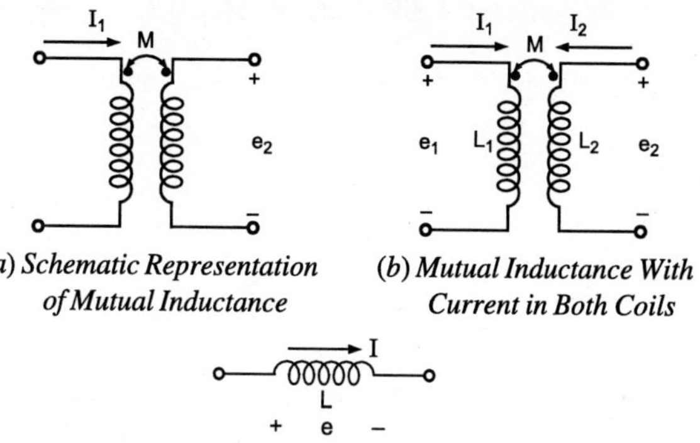

(*a*) *Schematic Representation of Mutual Inductance*

(*b*) *Mutual Inductance With Current in Both Coils*

(*c*) *Schematic Representation of Self Inductance*

Fig. 11.22

$$e_2 = M\frac{di_1}{dt}$$

A similar equation can, of course, be written giving an emf e_1 induced by a current i_2. The two dots, called *polarity markings,* in Fig. 11.22 (*a*) are used for indicating the direction of the magnetic coupling between the two coils : note from the figure and from equation given above that by matching the dots, the directions of currents and voltage drops are made to correspond with those of Fig. 11.22 (*b*) and Eq. (11.16) for mutual inductance.

If currents are present in both coupled circuits, emfs of self-inductance and mutual inductance are induced in each circuit. The self-induced emf, have the directions shown in Fig. 11.22 (*c*) the mutually induced emfs, follow the pattern of Fig. 11.22 (*a*). Fig. 11.22 (*b*) shows two such elements. Each coil is characterized by its own self-inductance; the combination has a mutual inductance indicating the coupling between the coils. The volt-ampere relationships for this case are given below :

$$e_1 = L_1\frac{di_1}{dt} + M\frac{di_2}{dt} \qquad ...(11.19)$$

$$\text{and } e_2 = L_2\frac{di_2}{dt} + M\frac{di_1}{dt} \qquad ...(11.20)$$

Coupling between two closed circuits permits the transfer of energy between the circuits through the medium of the mutual magnetic field. This phenomenon is the basis on which transformers operate.

11.9. COEFFICIENT OF COUPLING

When two coils are placed near each other, all the flux produced by one coil does not link with the other coil, as assumed in Art. 11.6.3, only a certain portion (say, K) of flux produced by one coil links with other coil, K being less than unity. K is called the *coefficient of coupling.*

Flux created in coil A due to current of i_1 amperes,

$$\Phi_1 = \frac{N_1 i_1}{l/\mu_0 \mu_r a} \text{ Wb}$$

$$\text{Flux linking with coil B, } \Phi_2 = K\,\Phi_1 = \frac{K N_1 i_1}{l/\mu_0 \mu_r a} \text{ Wb}$$

$$\text{Coefficient of mutual inductance, } M = \frac{N_2 \Phi_2}{i_1} = \frac{K N_1 N_2}{l/\mu_0 \mu_r a} \qquad ...(i)$$

$$\text{Coefficient of self inductance of coil A, } L_1 = \frac{N_1 \Phi_1}{i_1} = \frac{N_1^2}{l/\mu_0 \mu_r a} \quad ...(ii)$$

$$\text{Coefficient of self inductance of coil B, } L_2 = \frac{N_2 \Phi_2}{i_2} = \frac{N_2^2}{l/\mu_0 \mu_r a} \quad ...(iii)$$

Multiplying Eqs. (*ii*) and (*iii*) and taking square root of both sides we get

$$\sqrt{L_1 L_2} = \frac{N_1 N_2}{l/\mu_0 \mu_r a} \quad ...(iv)$$

Comparing Eqs. (*i*) and (*iv*) we get $M = K \sqrt{L_1 L_2}$

$$\text{or } K = \frac{M}{\sqrt{L_1 L_2}} \quad ...(11.21)$$

When the coils are tightly coupled *i.e.* when the flux due to one coil links with the other coil completely, the coefficient of coupling, K is unity and the coefficient of mutual inductance M is given as

$$M = \sqrt{L_1 L_2} \quad ...(11.22)$$

When the flux due to one coil does not link with the other coil at all, the value of coefficient of coupling, K is zero.

Example 11.23 **Two coils A of 1,200 turns and B of 800 turns lie near each other so that 60 per cent of the flux produced in one links the other. It is found that a current of 5 A in A produces a flux of 0.25 mWb while the same current in B produces a flux of 0.15 mWb. Determine the mutual inductance and coefficient of coupling between the coils.**

Solution:

Number of turns on coil A, $N_1 = 1{,}200$

Number of turns on coil B, $N_2 = 800$

$$\text{Flux per ampere of current in coil A, } \frac{\Phi_1}{i_1} = \frac{0.25 \times 10^{-3}}{5} = 0.05 \times 10^{-3} \text{ Wb}$$

$$\text{Self inductance of coil A, } L_1 = N_1 \frac{\Phi_1}{i_1} = 1{,}200 \times 0.05 \times 10^{-3} = 0.06 \text{ H}$$

$$\text{Flux per ampere of current in coil B, } \frac{\Phi_2}{i_2} = \frac{0.15 \times 10^{-3}}{5} = 0.03 \times 10^{-3} \text{ Wb}$$

$$\text{Self inductance of coil B, } L_2 = N_2 \frac{\Phi_2}{i_2} = 800 \times 0.03 \times 10^{-3} = 0.024 \text{ H}$$

$$\text{Flux linking coil B per ampere of current in coil A, } \frac{\Phi_2}{i_1} = \frac{0.25 \times 10^{-3}}{5} \times 0.6 = 0.03 \times 10^{-3} \text{ Wb}$$

$$\text{Mutual inductance between the coils, } M = N_2 \frac{\Phi_2}{i_1} = 800 \times 0.03 \times 10^{-3} = 0.024 \text{ H } \textbf{Ans.}$$

$$\text{Coefficient of coupling, } K = \frac{M}{\sqrt{L_1 L_2}} = \frac{0.024}{\sqrt{0.06 \times 0.024}} = 0.6325 \textbf{ Ans.}$$

Example 11.24 **Define coefficient of coupling and calculate it for two coils having self inductances of 60 mH and 80 mH. The mutual inductance between them is 40 mH**

Solution:

Self inductance, $L_1 = 60 \text{ mH} = 0.06 \text{ H}$

Self inductance, $L_2 = 80 \text{ m H} = 0.08 \text{ H}$

Mutual inductance, $M = 40 \text{ mH} = 0.04 \text{ H}$

$$\text{Coefficient of coupling, } K = \frac{M}{\sqrt{L_1 L_2}} = \frac{0.04}{\sqrt{0.06 \times 0.08}} = 0.577 \textbf{ Ans.}$$

11.10. EDDY CURRENTS AND EDDY CURRENT LOSSES

We have seen that whenever the flux linking with a closed electric circuit changes, an emf is induced in the circuit and a current flows, the value of which depends on the emf around the circuit and the resistance of the circuit. It is not necessary that the circuit be a wire and that the flux passes entirely through it. If a solid block of metal is traversed by a varying flux, metallic circuits in the block itself, which are linked by the flux, will carry current. If the magnetic circuit is made up of iron and if the flux in the circuit is variable, currents will be induced by induction in the iron circuit itself. All such currents are known as *eddy currents.*

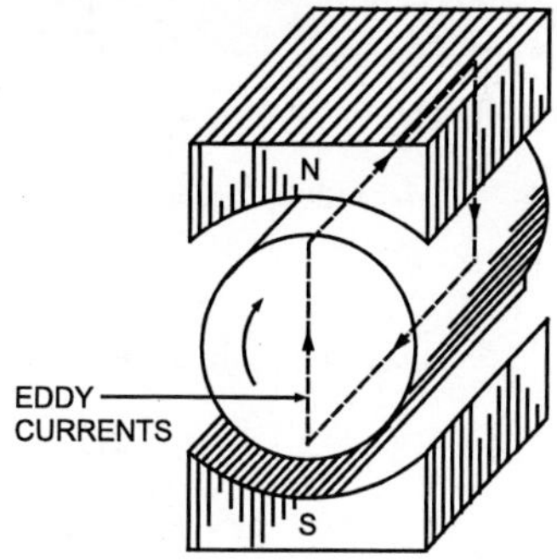

(a) Eddy Currents Induced in The Solid Armature of an Electric Machine

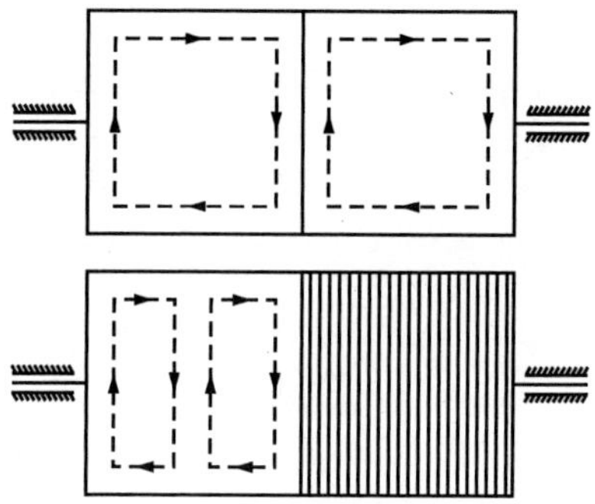

(b) Effect of a Split of Magnetic Path on Eddy Currents (Direction of Eddy Currents is Shown By Arrows)

Fig. 11.23

Eddy currents result in a loss of power, with consequent heating of the material. The magnitude of power loss due to eddy currents is often a matter of considerable importance in electrical engineering. Electrical machinery usually involves varying fluxes. In dc as well as ac machines, we have an armature built of iron, which is either revolved in a permanent magnetic field or through which a revolving magnetic field passes. In ac engineering the flux linking the core of transformer also varies. In either case there will be losses in the iron due to hysteresis as well as due to eddy currents.

Eddy currents always tend to flow in planes perpendicular to the magnetic flux, as they are induced due to variation of this flux through a circuit. If the material of the magnetic circuit is subdivided into thin sheets *i.e.*, if the material is laminated, and these sheets are placed parallel to the flux and insulated from one another, the eddy current loss will be reduced. When the core body is one continuous solid iron piece, as shown in Fig. 11.23 (*a*), cross section of armature is large, therefore, resistance is very small and consequently magnitude of eddy currents is large resulting in large power loss. If the same core is splitted up into thin insulated sheets, as shown in Fig. 11.23 (*b*), the path of eddy currents is splitted up into several paths of high resistance because cross section of such paths is very less. Hence magnitude of eddy currents and eddy current loss is considerably reduced. That is why armature of electrical machines (generators and motors) are built up of sheet-steel laminations (usually between 0.45 mm and 0.50 mm thick) assembled on a shaft so that the armature is subdivided into insulated sheets which are parallel to the working flux of the machine. The core of transformer is also built up of thin laminations due to the same reasons.

11.10.1. Eddy Current Loss. Eddy current loss in addition to hysteresis loss occurs in a magnetic material when it is subjected to a changing magnetic field. Like hysteresis loss, the eddy current loss results in the rise of temperature of the material. The power loss due to eddy currents is proportional to i^2r where i is the magnitude of eddy currents and r is the resistance of eddy current path, but it is difficult to determine it because the magnitude of eddy currents and resistance of eddy current path cannot be determined directly. Experiments have shown that the power loss due to eddy currents is proportional to the square of the flux density (B^2) and the square of number of complete magnetization cycles per second (f^2) and the eddy current loss is given by the equation

$$P_e = K_e (B_{max})^2 f^2 t^2 v \text{ watts} \quad ...(11.23)$$

where K_e is the *eddy current coefficient* and depends upon the type of magnetic material used, f is the number of complete magnetization cycles per second, B_{max} is the maximum flux density in tesla (Wb/m^2), t is the thickness of laminations in metres and v is the volume of material in m^3.

Example 11.25 **The eddy current loss in a magnetic material subjected to maximum flux density of 1.2 Wb/m², alternating sinusoidally at a frequency of 50 Hz is 1,600 watts. Determine the eddy current loss when the maximum flux density is 1.5 W/b m² and the frequency is 60 Hz.**

Solution:

$$\text{Eddy current loss, } P_{e_1} = 1{,}600 \text{ watts}$$

$$\text{Maximum flux density, } B_{max\,1} = 1.2 \text{ T}$$

$$\text{Frequency, } f_1 = 50 \text{ Hz}$$

$$\text{Maximum flux density, } B_{max\,2} = 1.5 \text{ T}$$

$$\text{Frequency, } f_2 = 60 \text{ Hz}$$

$$\text{Since eddy current loss, } P_e \propto B_{max}^2 \times f^2$$

$$\therefore\ P_{e_2} = P_{e_1} \times \left(\frac{B_{max2}}{B_{max1}}\right)^2 \times \left(\frac{f_2}{f_1}\right)^2 = 1{,}600 \times \left(\frac{1.5}{1.2}\right)^2 \times \left(\frac{60}{50}\right)^2 = 3{,}600 \text{ watts } \textbf{Ans.}$$

Example 11.26 **If in a transformer core, the hysteresis and eddy current losses are 80 W and 50 W at normal voltage and frequency, then calculate the losses when voltage and frequency are increased by 20%.**

[U.P. Technical Univ. Elec. Engineering First Semester 2004-05; Basic Electrical Engineering Odd Semester 2013-14]

Solution: When voltage and frequency are increased in the same ratio, flux density remains unchanged. Thus

$$\text{Hysteresis loss, } P_h' = P_h \times \frac{f_2}{f_1} = 80 \times 1.2 = 96 \text{ W } \textbf{Ans.}$$

$$\text{Eddy current loss, } P_e' = P_e \times \left(\frac{f_2}{f_1}\right)^2 = 50 \times (1.2)^2 = 72 \text{ W } \textbf{Ans.}$$

Example 11.27 **The total core loss (hysteresis plus eddy current) for specimen of magnetic sheet steel is found to be 1,800 W at 60 Hz. If the flux density is kept constant and frequency of supply is increased 50%, the total core loss is found to be 3,000 W. Compute the separate hysteresis and eddy current losses at both frequencies.**

[U.P. Technical Univ. Electrical Engineering Second Semester 2008-09]

Solution: Since hysteresis loss, $P_h \alpha f$ and eddy current loss αf^2, core loss can be represented as

$$P_i = P_h + P_e = Af + Bf^2 \text{ where A and B are constants}$$

Substituting P_i = 1,800 W for f_1 = 60 Hz and P_i = 3,000 W for $f_2 = 1.5 f_1 = 1.5 \times 60 = 90$ Hz in above equation we have

$$60\text{ A} + (60)^2 \text{ B} = 1{,}800$$

$$\text{and } 90\text{ A} + (90^2)\text{ B} = 3{,}000$$

$$\text{or A} + 60\text{ B} = 30 \quad ...(i)$$

$$\text{and A} + 90\text{ B} = \frac{100}{3} = 33.333 \quad ...(ii)$$

Solving Eqs. (*i*) and (*ii*) we have

$$\text{A} = 23.334 \text{ and B} = 0.1111$$

Hysteresis loss at 60 Hz = 60 A = 60 × 23.334 = 1,400 W **Ans.**

Hysteresis loss at 90 Hz = 90 A = 90 × 23.334 = 2,100 W **Ans.**

Eddy current loss at 60 Hz = 60^2 B = 3,600 × 0.1111 = 400 W **Ans.**

Eddy current loss at 90 Hz = 90^2 B = 8,100 × 0.1111 = 900 W **Ans.**

HIGHLIGHTS

1. The closed path followed by magnetic flux is called a *magnetic circuit.* There is a lot of similarity between the magnetic and electric circuits, as mentioned in Art 11.1.3.
2. **MMF** of the magnetic circuit is a magnetic potential that tends to force flux around the magnetic circuit and is analogous to the emf in an electric circuit.

$$\text{MMF} = \text{NI ampere-turns}$$

3. **Reluctance** is the name given to that property of material which opposes the creation of magnetic flux in it. It is measured in ampere-turns/Wb or amperes/Wb. It may be given as

$$S = \frac{l}{\mu_o \mu_r a} \text{ AT/Wb}$$

where l is the length of magnetic path, a is x-sectional area, μ_o is the permeability of the evacuated space or air and is equal to $4\pi \times 10^{-7}$ H/m, and μ_r is the relative permeability of the medium with respect to evacuated space or air.

4. Flux produced Φ, mmf and reluctance are related as follows

$$\Phi = \frac{\text{MMF}}{\text{Reluctance}} = \frac{\text{NI}}{l} \times \mu_o \mu_r a$$

5. Permeance of a magnetic circuit is the reciprocal of reluctance and is given as

$$\text{Permeance} = \frac{1}{\text{Reluctance}} = \frac{\mu_o \mu_r a}{l}$$

6. For a series magnetic circuit

$$\text{Total ampere-turns required} = H_1 l_1 + H_2 l_2 + H_3 l_3 + \ldots$$

where H_1 is the magnetizing force required to create flux through part l_1 of the magnetic circuit and so on.

7. For parallel magnetic circuit, AT required = AT required for either parallel magnetic path.
8. Leakage factor is the ratio of total flux (flux in the iron path) and useful flux (flux in the air gap).
9. The curve drawn giving relationship between induction density B and magnetizing force H is known as B-H curve or magnetizing curve.
10. The phenomenon of lagging of magnetization or induction flux density B behind the magnetizing force H is known as magnetic hysteresis.
11. The energy loss associated with the phenomenon of hysteresis is called the hysteresis loss and is given as

$$\text{Hysteresis loss, } P_n = \eta (B_{max})^{1.6} f \text{ v joules/second or watts}$$

where η = Steinmetz hysteresis coefficient, v is the volume of core in m^3, B_{max} is the peak magnitude of flux density and f is the frequency of magnetic cycles per second.

Hysteresis loss is also given as

$$P_h = xy \times (\text{area of loop in mm}^2) \times f \text{ v joules}$$

$$\text{where } x \text{ and } y = \text{Scale factors of H and B}$$

12. When the flux linking the circuit changes, an emf is induced, and its magnitude is given as

$$e = N\frac{d\Phi}{dt} = \frac{N(\Phi_2 - \Phi_1)}{dt}$$

13. The magnitude of dynamically induced emf in a conductor moving at right angles to a uniform magnetic field is given as $e = Blv$ volts

If the conductor moves at an angle θ to the magnetic field, then

$$e = Blv \sin \theta \text{ volts}$$

The direction of dynamically induced emf is given by Fleming's right hand rule. According to Fleming's right hand rule:

If the thumb, forefinger and middle finger of right hand are held mutually perpendicular to each other, forefinger pointing into the direction of the field and thumb in the direction of motion of conductor then the middle finger will point in the direction of the induced emf as shown in Fig 1.

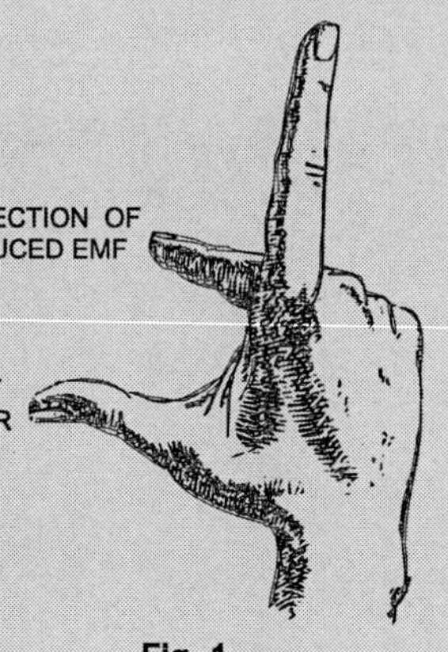

Fig. 1

14. When the current flowing through the coil is changed, the flux linking with its own winding changes and due to the change in linking flux with the coil, an emf, known as *self-induced emf, is induced.* The magnitude of self-induced emf, e in a coil of N turns is given by

$$e = N\frac{d\Phi}{dt} = L\frac{di}{dt}$$

where L is known as *self-inductance* of the coil and is given as

$$L = \frac{e}{di/dt} = \frac{N\Phi}{i} = \frac{N^2\mu_o\mu_r a}{l} = \frac{N^2}{\text{Reluctance, S}}$$

15. Whenever the current in coil A (placed near another coil B) changes, the flux linking with coil B changes and an emf, known as mutually induced emf, is induced in coil B.

The magnitude of mutually induced emf is given as

$$e_m = M\frac{di_1}{dt} \text{ volts}$$

where $\frac{di_1}{dt}$ is the rate of change of current in coil A and M is mutual inductance of the two coils in henry and is given as

$$M = \frac{e_m}{di_1/dt} = \frac{N_2\Phi_2}{i_1} = \frac{N_1N_2 a\mu_o\mu_r}{l} = \frac{N_1N_2}{\text{Reluctance, S}}$$

16. The coefficient of coupling between two coils is given as

$$K = \frac{M}{\sqrt{L_1L_2}}$$

where L_1 and L_2 are the individual inductances of the coils and M is the mutual inductance between the coils.

17. If the magnetic circuit is made up of iron and if the flux in the circuit is variable, currents, known as *eddy currents,* will be induced in the iron circuit itself.

Power loss due to eddy currents is given as

$$P_e = K_e(B_{max})^2 f^2 t^2 \text{ v watts}$$

where K_e is the eddy current coefficient, f is the number of complete magnetization cycles per second, t is the thickness of laminations, v is the volume of material and B_{max} is the maximum value of flux density.

EXERCISES

1. Define terms: mmf, flux and reluctance and give their relationship.
[Mahamaya Technical Univ. Electrical Engineering Second Semester 2011-12]
2. Define following terms as applied to magnetic circuit:
(*i*) MMF (*ii*) Flux density (*iii*) Reluctance, (*iv*) Permeability.
[U.P. Technical Univ. Basic Electrical Engineering Second Semester 2013-14]
3. Define mmf, reluctance, flux in magnetic circuit. **[U.P. Technical Univ. Electrical Engineering Odd Semester 2013-14]**
4. Describe the analogies between electric and magnetic circuits in brief.
[U.P. Technical Univ. Basic Electrical Engineering Odd Semester 2013-14]
5. What are similarities and dissimilarities between electric and magnetic circuits? Enlist any five applications of magnetic circuits in engineering fields. **[U.P. Technical Univ. Electrical Engineering Odd Semester 2013-14]**
6. Describe the analogies that can be made between electric and magnetic circuit regarding the following items: driving force, field intensity, impedance drops, equivalent circuits.
[U.P. Technical Univ. Electrical Engineering First Semester 2008-09]
7. Give analogous characteristics of electrical and magnetic circuits.
[U.P. Technical Univ. Electrical Engineering First Semester 2007-08]
8. Draw and explain B-H curve. What is meant by saturation, coercive force, residual magnetism? Show them in diagram.
[G.B. Technical Univ. Electrical Engineering Second Semester 2011-12]
9. How is B-H curve of a ferromagnetic material different from that of a non-magnetic material ? Name all the salient regions of B-H curve of magnetic material. **[U.P. Technical Univ. Electrical Engineering First Semester 2003-04]**
10. Draw and explain hysteresis loop. What is its significance?
[G.B. Technical Univ. Electrical Engineering Second Semester 2009-10]
11. State and explain Faraday's laws of electromagnetic induction, Lenz's law, Fleming's right hand rule and Fleming's left hand rule. **[V.T.U. Belgaum Karnataka Univ. Summer 2003]**
12. State the following :
(*i*) Magnetic flux and its properties (*ii*) Flux density
(*iii*) Fleming's right hand rule (*iv*) Fleming's left hand rule
(*v*) Lenz's law **[G.B. Technical Univ. Electrical Engineering Second Semester 2009-10]**
13. What do you understand by self-inductance? Derive its expression.
[Mahamaya Technical Univ. Electrical Engineering First Semester 2011-12]
14. Explain the concept of mutual inductance. Define coefficient of coupling and derive the expression between self-inductances of two coils, mutual inductance between them and the coefficient of coupling.
15. Define mutual inductance (M) and show that
$$K = \frac{M}{\sqrt{L_1 L_2}}, K \le 1$$
where L_1 and L_2 are inductances of coil 1 and coil 2 respectively and K is coupling coefficient.
[Pb Technical Univ. Basic Electrical and Electronics Engineering First Semester 2005-06]
16. Differentiate between :
(*i*) Statically and Dynamically induced emfs (*ii*) Electric and Magnetic circuits.
[U.P. Technical Univ. Electrical Engineering Second Semester 2003-04]
17. What is B-H curve? Also explain the hysteresis and eddy current loss.
[G.B. Technical Univ. Electrical Engineering First Semester 2010-11]
18. Explain eddy current loss and its remedy in magnetic circuit.
[U.P. Technical Univ. Basic Electrical Engineering Odd Semester 2013-14]

SHORT ANSWER TYPE QUESTIONS WITH ANSWERS

Q. 1. What is magnetomotive force and write its unit. **[U.P. Technical Univ. Odd Semester 2013-14]**

Ans. The flux producing ability of the coil in Fig. 11.1 or of a coil on any other magnetic circuit is proportional to the number of turns N and the current I. The product NI is called the *magnetomotive force* (mmf) and determines the amount of flux developed in the magnetic circuit.

MMF of the magnetic circuit is the magnetic potential difference that tends to force flux around the magnetic circuit and is analogous to the electromotive force (emf) in an electric circuit.

$$\text{MMF} = \text{NI ampere-turns}$$

Q. 2. On what factors does the magnetic reluctance depend ?

Ans. The reluctance of the magnetic circuit is directly proportional to length l, inversely proportional to cross-sectional area and dependent on the nature of material of the magnetic circuit.

Q. 3. What is meant by leakage and fringing ? **[R.G.P.V. December 2001]**

Ans. Leakage flux is the flux, which follows a leaking path, as shown in Fig. 11.5.

Useful flux passing across the gap tends to bulge outwards, as shown in Fig. 11.5., thereby increasing the effective area of gap and reducing the flux density in the gap. This effect is referred to as fringing; and the longer the air gap, the greater is the fringing.

Q. 4. What is meant by leakage factor ?

Ans. The ratio of total flux (flux in the iron path) to the useful flux (flux in the air gap) is known as leakage factor.

Q. 5. What is magnetization curve ?

Ans. The curve drawn between induction density B and the magnetizing force H is known as B-H or magnetization curve.

Q. 6. Show the residual magnetism in a B-H curve.

[G.B.T.U. Second Semester 2011-12; U.P.T.U. Second Semester 2013-14]

Ans. OC in Fig. 11.14 page 329.

Q. 7. What information is obtained from the loop? **[U.P. Technical Univ. First Semester 2005-06]**

Ans. The area enclosed by the hysteresis loop gives the following information regarding hysteresis loss occurring in the sample under testing. Hysteresis loss/m^3/cycle = Area of hysteresis loss in B-H scales.

Q. 8. Draw hysteresis loops for (*i*) permanent magnets (*ii*) transformer and (*iii*) Ferrite.

[U.P. Technical Univ. First Semester 2005-06]

Ans. Hysteresis loops for permanent magnets, transformer core and ferrite are shown below in Figs. 1(*a*), (*b*), (*c*) respectively.

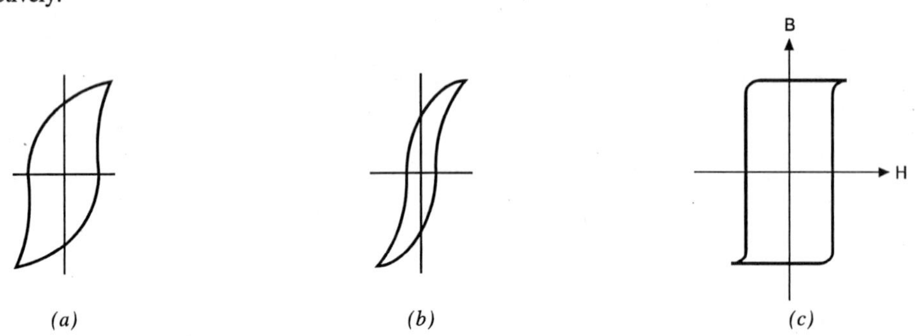

Fig. 1

Q. 9. What is hysteresis loss ? On what factors does it depend ?

Ans. The energy expended in taking a specimen through a magnetic cycle is known as hysteresis loss.

Hysteresis loss per cubic metre per cycle of magnetization of a magnetic material depends upon (*i*) the maximum value of flux density and (*ii*) the magnetic quality of the material.

Q. 10. What is indicated by the negative sign before N $\frac{d\Phi}{dt}$ in expression $e = -N \frac{d\Phi}{dt}$?

Ans. The negative sign before N $\frac{d\Phi}{dt}$ in expression $e = -N \frac{d\Phi}{dt}$ indicates the phenomenon explained by Lenz's law.

Q. 11. What is Lenz's law ?

Ans. The direction of emf induced in a conductor or coil is governed by Lenz's law which states that the direction of induced emf or voltage is such that the current produced by it sets up a magnetic field opposing the flux change.

Q. 12. What is the difference between a statically induced emf and a dynamically induced emf ? **[P.T.U. January 2000]**

Ans. When flux linking with any coil or circuit changes, an emf is induced in the coil or circuit, called the *electromagnetically induced emf.*

EMF induced in a coil or circuit by increasing or decreasing magnitude of the linking flux (no relative motion between coil or circuit and magnetic field) is known as statically *induced emf.*

The emf induced in a conductor or coil due to relative motion between it and the magnetic field, that produces flux, is known as dynamically induced emf.

Q. 13. What is the application of Fleming's right hand rule ?

Ans. Fleming's right hand rule is used in determination of direction of emf induced in a conductor moving in a uniform magnetic field.

Q. 14. What is Fleming's right hand rule ?

Ans. According to Fleming's right hand rule if the thumb, forefinger and middle finger of the right hand are held mutually perpendicular to each other, forefinger pointing into the direction of the field and thumb in the direction of motion of conductor then middle finger will point in the direction of the induced emf, as shown in Fig. 11.18 (*a*).

Q. 15. A conductor of effective length $l(m)$ moves in the direction of a uniform magnetic field B (Tesla) at a constant speed v (ms^{-1}). What is the magnitude of the emf induced in the conductor ?

Ans. No., as no flux is cut by the conductor.

Q. 16. Define coefficient of coupling. **[U.P. Technical Univ. Odd Semester 2013-14]**

Ans. When the two coils are placed near each other, all the flux produced by one coil does not link with the other coil, only a certain portion (say, K) of flux produced by one coil link with the other coil, K being less than unity. K is called the coefficient of coupling.

Q. 17. Write an expression of hysteresis loss in a transformer. **[G.B.T.U. First Semester 2011-12]**

Ans. The hysteresis loss in a transformer is given by an expression

$$P_h = \eta'(B_{max})^x f v \text{ joules per second or watts.}$$

where f is supply frequency in Hz, v is the volume of transformer core in m^3, η' is the hysteresis coefficient, B_{max} is the peak value of flux density in the core and x lies between 1.5 and 2.5 depending upon the material and is often taken as 1.6.

Q. 18. What are the undesirable effects of eddy current and hysteresis loss ? **[R.G.P.V. Bhopal June 2002]**

Ans. Eddy current and hysteresis loss appears in the form of heat and so raises the temperature of that portion of magnetic circuit, in which they occur.

Q. 19. How can eddy current and hysteresis losses be minimised ?

Ans. Hysteresis loss can be minimised by choosing a core material with low hysteresis coefficient such as low carbon steel, silicon alloys, alloy steel.

Eddy current loss can be minimised by using thin laminations.

Q. 20. Mention some applications of eddy currents. **[R.G.P.V. Bhopal June 2002]**

Ans. Use of eddy currents is made in (*i*) eddy current heating (*ii*) eddy current braking and (*iii*) to develop damping torque in measuring instruments.

PROBLEMS

1. A magnetic circuit has a uniform cross-sectional area of 5 cm^2 and a length of 25 cm. A coil of 120 turns is wound uniformly over the magnetic circuit. When the current in the coil is 1.5 A, the total flux is 0.3 mWb; when the current in the coil is 5 A, the total flux is 0.6 mWb. For each value of current, calculate (*i*) the magnetising force (*ii*) the relative permeability of iron 1,500. **[Ans.** 2,546 AT]

2. Calculate the relative permeability of an iron when the exciting current taken by the 600 turn coil is 1.2 A, and the total flux produced is 1 mWb. The mean circumference of the ring is 0.5 m and the area of cross-section is 10 cm^2. **[Ans.** 552]

3. A cast iron ring with a mean diameter of 10 cm and a cross-sectional area of 3 cm^2 has a radial gap of 0.15 cm. Calculate the current required through the winding of 1,000 turns wound uniformly over the ring to produce a flux of 2×10^{-4} Wb in the air gap. Take relative permeability of cast iron as 250. **[Ans.** 1.462 A]

4. A cast steel ring has a circular x-section 3 cm diameter and a mean circumference of 80 cm. The ring is uniformly wound with a coil of 600 turns.
   (*i*) Estimate the current required to produce a flux of 0.5 mWb in the ring.
   (*ii*) If a saw cut of 2 mm wide is made in the ring, find approximately the flux produced by the current found in (*i*).
   Take ampere-turns per metre at the above flux density 670. **[Ans.** (*i*) 0.893A (*ii*) 0.1614 m Wb]

5. An iron ring has a mean circumferential length of 60 cm with an air gap of 1 mm and a uniform winding of 300 turns. When a current of 1 A flows through the coil, find the flux density. The relative permeability of iron is 300. Assume $\mu = 4\pi \times 10^{-7}$ H/m. **[Ans.** 0.1256 T]

6. A magnetic ring has a mean circumference of 1.5 m and is of 0.01 m^2 in cross-section and is wound with 175 turns. A saw cut of 4 mm width is made in the ring. Calculate the magnetising current required to produce a flux of 0.8 m Wb in the air-gap. Assume permeability of iron as 400 and leakage factor of 1.25. **[Ans.** 3.16 A]

7. An iron ring of mean circumference 80 cm is made from round iron of area 8 cm^2. It has a saw cut of 2 mm wide and is wound with 500 turns. Find the current required to produce a flux of 0.8 m Wb across the air-gap. Assume a relative permeability for iron as 625 and leakage factor of 1.25. **[Ans.** 5.73 A]

8. The magnetic circuit shown in Fig. 1 is built-up of iron of square cross section, 2 cm wide. Each air gap is 2 mm wide, μ for part A may be taken as 900 and for B as 1250. Each of the exciting coils is wound with 1200 turns and the exciting coil is carrying current 0.8 ampere. Find:
   (*i*) Reluctance of A
   (*ii*) Reluctance of B
   (*iii*) Reluctance of two air gaps
   (*iv*) Total reluctance
   (*v*) Total flux
   (*vi*) Flux density (neglecting leakage). **[R.G.P.V. Bhopal Basic Electrical Engineering June, 2007]**

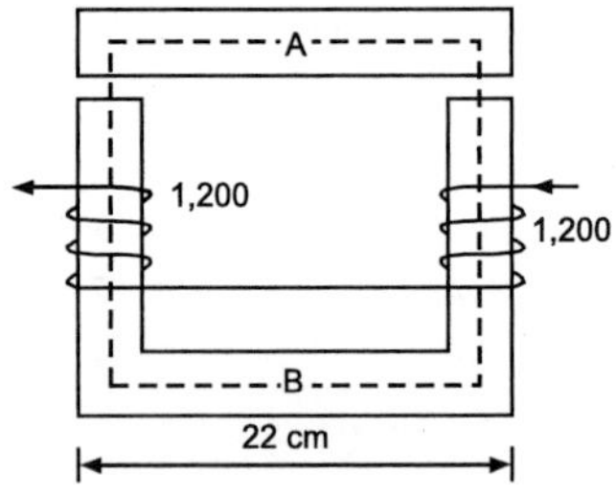

Fig. 1

[Ans. (*i*) 442,100 AT/Wb (*ii*) 636,620 AT/Wb (*iii*) 7,957,750 AT/Wb (*iv*) 9,036,470 AT/Wb (*v*) 0.2125 mWb (*vi*) 0.53 Wb/m^2 or Tesla]

9. A 0.5 m long wire moves at right angles to its length at 40 m/s in uniform magnetic field of 1 Wb/m^2. Calculate the emf induced in the conductor when the direction of motion is (*i*) perpendicular to field, (*ii*) inclined at 30° to the direction of field. **[Ans.** (*i*) 20 V (*ii*) 10 V]

10. A coil of 150 turns is linked with a flux of 0.001 Wb when carrying a current of 10 A. If the current is uniformly reversed in 0.1 s, calculate the induced emf. **[Ans.** 3 V]

11. The resistance and inductance of a coil are 3 Ω and 0.1 mH respectively. What potential difference exists at the terminals of the solenoid at the instant when the current is 1 A but is increasing at the rate of 10,000 A per second ? **[Ans.** 4 V]

12. A coil of 300 turns, wound on a core of non-magnetic material has an inductance of 10 mH. Calculate (*i*) the flux produced by a current of 5 A and (*ii*) the average value of emf induced when a current of 5 A is reversed in 8 milli-seconds. **[Ans.** (*i*)0.1667mWb (*ii*)12.5 V]

13. A circuit has 1,000 turns enclosing a magnetic circuit 20 cm^2 in section. With 4 A, the flux density is 1.0 Wb/m^2 and with 9 A it is 1.4 Wb/m^2. Find the mean value of the inductance between these current limits and the induced emf if the current falls from 9 A to 4 A in 0.05 s. **[Ans.** 0.16 H H,16 V]

14. Two coils having 30 and 600 turns respectively are wound side by side on a closed iron circuit of area of cross-section 100 cm^2 and mean length 200 cm. Estimate the mutual inductance between the coils if the relative permeability of the iron is 2,000. If a current of zero ampere grows to 20 A in a time of 0.02 second in the first coil, find the emf induced in the second coil. **[Ans.** 226 mH; 226 V]

15. Two coils A and B of 600 and 100 turns respectively are wound uniformly around a wooden ring of mean circumference 80 cm. The x-sectional area of the ring is 4 cm^2. Calculate (*a*) self inductance of each coil (*b*) mutual inductance between the coils (*c*) emf induced in the coil B when a current of 2 A in coil A is reversed in 0.01 second.

[Ans. (*a*) 226.2 μ H ; 6.283 μ H; (b) 37.7 μ H; (c) 15.08 mV]

16. Two coils having 50 and 500 turns respectively are wound side by side on a closed iron circuit of section 50 cm^2 and mean length 120 cm. Estimate the mutual inductance between the coils if the permeability of iron is 1,000. Also find the self-inductance of each coil and the coefficient of coupling. **[Ans.** 0.13 H; 0.013 H; 1.3 H; 1.0]

❑❑❑

12 CHAPTER

Single-Phase Transformers

12.1. INTRODUCTION

Electrical energy is generated at places where it is easier to get water head, oil or coal for hydroelectric, diesel or thermal power stations respectively. Then energy is to be transmitted at considerable distances for use in villages, towns and cities located at distant places. As transmission of electrical energy at high voltages is economical, some means are required for stepping up the voltage at generating stations and stepping down the same at the places where it is to be used. Electric machine used for this purpose is "*transformer*". In our country the electrical energy is usually generated at 6.6, 11 or 33 kV, stepped up to 132, 220, 400, or 765 kV with the help of step-up transformers for transmission and then stepped down to 66 kV or 33 kV at grid substations for feeding various substations, which further step down the voltage to 11 kV for feeding distributing transformers stepping down the voltage further to 433/250 volts for the consumer uses.

Transformers are extensively used like the *instrument transformers* for metering purposes. One-to-one ratio transformers (*i.e.* the transformers having equal primary and secondary voltages) are used to electrically isolate the two parts of an electrical circuit. In high-voltage laboratories the transformers are used to provide very high voltages for testing purposes, termed as *testing transformers*. In electric communication circuits transformers are employed for a variety of purposes e.g., as an impedance transformation device to allow maximum transfer of power from the input circuit to the output device. In radio and television circuits input transformers, interstage transformers and output transformers are widely employed. Transformers are also employed in telephone circuits, instrumentation circuits and control circuits. Indeed the transformer is a device which plays a vital and essential role in many facets of electrical engineering.

Transformer is an ac machine that (*i*) transfers electrical energy from one electric circuit to another (*ii*) does so without a change of frequency (*iii*) does so by the principle of electromagnetic induction and (*iv*) has electric circuits that are linked by a common magnetic circuit. The energy transfer usually takes place with a change of voltage, although this is not always necessary. When the transformer raises the voltage *i.e.* when the output voltage of a transformer is higher than its input voltage, it is called the *step-up transformer* and when it lowers the voltage it is called the *step-down transformer*.

Since its basic construction requires no moving parts, so it is often called the '*static transformer*' and it is very rugged machine requiring the minimum amount of repair and maintenance. Owing to the lack of rotating parts there are no friction or windage losses. Further, the other losses are relatively low, so that the efficiency of a transformer is high. Typical transformer efficiencies at full load lie between 96% and 97% and with extremely large capacity transformers the efficiencies are as high as 99 %. The cost per kVA output of transformers is quite low as compared with other electrical machines. As there are no teeth, slots or rotating parts, and the windings can be immersed in oil, it is not difficult to insulate transformers for very high voltages.

12.2. OPERATING PRINCIPLE

An elementary transformer consists of a soft iron or silicon steel core and two windings placed on it. The windings are insulated from both the core and each other. The core is built up of thin soft iron or silicon steel laminations to provide a path of low reluctance to the magnetic flux. The winding connected to the supply main is called the *primary* and the winding connected to the load circuit is called the *secondary*. The winding connected to higher voltage circuit is called the *high-voltage (hv) winding* while that connected to the lower voltage circuit is called the *low-voltage (lv) winding*. In case of a step-up transformer, low-voltage winding is the primary and high-voltage winding is the secondary while in case of a step-down transformer the high-voltage winding is the primary and low-voltage winding is the secondary.

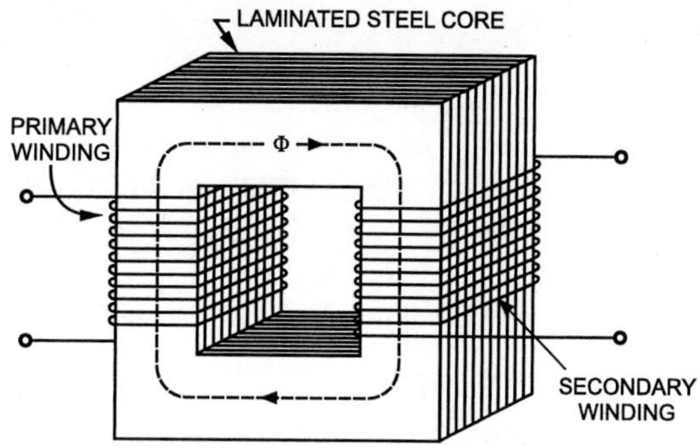

Fig. 12.1 *Simple Transformer*

The action of a transformer is based on the principle that energy may be efficiently transferred by induction from one set of coils to another by means of a varying magnetic flux, provided that both the sets of coils are on a common magnetic circuit. In a transformer, the coils and magnetic circuit are all stationary with respect to one another. The emfs are induced by the variation in the magnitude of flux with time, as illustrated in Fig. 12.2.

Although in the actual construction the two windings are usually wound one over the other, for the sake of simplicity, the figures for analysing transformer theory show the windings on opposite sides of the core as in Fig. 12.1.

When the primary winding is connected to an ac supply mains, a current flows through it. Since this winding links with an iron core, so current flowing through this winding produces an alternating flux Φ in the core. Since this flux is alternating and links with the secondary winding also, so induces an emf in the secondary winding. The frequency of induced emf in secondary winding is the same as that of the flux or that of the supply voltage. The induced emf in the secondary winding enables it to deliver current to an external load connected across it. Thus the energy is transformed from primary winding to the secondary winding by means of electromagnetic induction without any change in frequency. The flux Φ of the iron core links not only with the secondary winding but also with the primary winding, so produces self-induced emf in the primary winding. This induced emf in the primary winding opposes the applied voltage and, therefore, sometimes it is known as *back emf* of the primary. In fact the induced emf in the primary winding limits the primary current in much the same way as the back emf in a dc motor limits the armature current.

12.3. TRANSFORMER ON DC

A transformer cannot operate on dc supply and never be connected to a dc source. If a rated dc voltage is applied to the primary of a transformer, the flux produced in the transformer core will not vary but remain constant in magnitude and, therefore, no emf will be induced in the secondary winding except at the moment of switching *on*. Thus the transformer is not capable of raising or lowering the dc voltage. Also there will be no self-induced emf in the primary winding, which is only possible with varying flux linkage, to oppose the applied voltage and since the resistance of primary winding is quite low, a heavy current will flow through the primary winding which may result in the burning out of the primary winding. This is reason that *dc is never applied to a transformer*.

12.4. IDEAL TRANSFORMER

For a better understanding and an easier explanation of a practical transformer, certain idealizing assumptions are made which are close approximations for a practical transformer. A transformer having these ideal properties is hypothetical (has no real existence) and is referred to as the *ideal transformer*. It possesses certain essential features of a real transformer but some details of minor significance are ignored which will be introduced step by step while analysing a transformer. The idealizing assumptions made are as follows:

(*i*) *No winding resistance i.e.* the primary and secondary windings have zero resistance. It means that there is no ohmic power loss and no resistive voltage drop in an ideal transformer.

(*ii*) *No magnetic leakage i.e.* there is no leakage flux and all the flux set up is confined to the core and links both the windings.

(*iii*) *No iron loss i.e.* hysteresis and eddy current losses in transformer core are zero.

(*iv*) *Zero magnetizing current i.e.* the core has infinite permeability and zero reluctance so that zero magnetizing current is required for establishing the requisite amount of flux in the core.

From the above discussion an ideal transformer is supposed to consists of two pure inductive coils wound on a loss-free core.

12.5. EMF EQUATION

When an alternating (sinusoidal) voltage is applied to the primary winding of a transformer, an alternating (sinusoidal) flux, as shown is Fig. 12.2., is set up in the iron core which links both the windings (primary and secondary windings).

Let Φ_{max} = Maximum value of flux in webers

and f = Supply frequency in hertz.

As illustrated in Fig. 12.2, the magnetic flux increases from zero to its maximum value Φ_{max} in one-fourth of a cycle *i.e.* in $\frac{1}{4f}$ second.

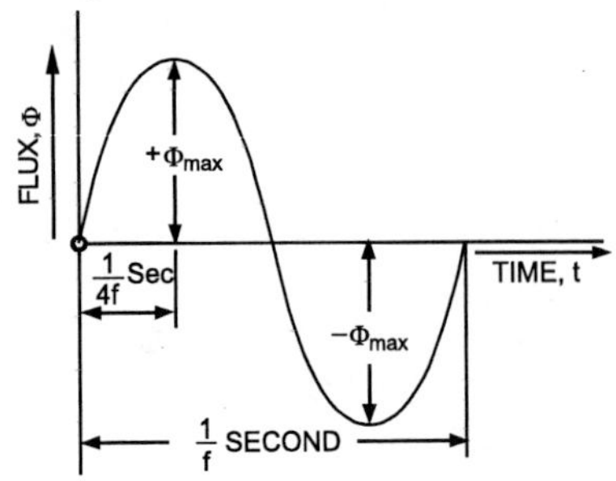

Fig. 12.2 *Sinusoidal Variation of Flux With Time*

So average rate of change of flux, $\frac{d\Phi}{dt} = \frac{\Phi_{max}}{1/4f} = 4f\Phi_{max}$

Since average emf induced per turn in volts is equal to the average rate of change of flux,

average emf induced per turn = $4f\Phi_{max}$ volts

Since flux Φ varies sinusoidally, emf induced will be sinusoidal and form factor for sinusoidal wave is 1.11 *i.e.* the rms or effective value is 1.11 times the average value.

$\therefore$ RMS value of emf induced per turn = $1.11 \times 4f\Phi_{max}$ volts ...(12.1)

If the number of turns on primary and secondary windings are N_1 and N_2 respectively, then

RMS value of emf induced in primary, E_1 = EMF induced per turn × number of primary turns

$= 4.44f\Phi_{max} \times N_1 = 4.44f\,N_1\,\Phi_{max}$ volts ...(12.2)

Similarly rms value of emf induced in secondary,

$E_2 = 4.44f\Phi_{max} \times N_2$ volts ...(12.3)

The above relations for emf induced in primary and secondary windings can be derived alternatively as below

The instantaneous value of sinusoidally varying flux may be given as

$$\Phi = \Phi_{max} \sin \omega t$$

$\therefore$ Instantaneous value of emf induced per turn $= \frac{-d\Phi}{dt}$ volts $= -\omega\,\Phi_{max} \cos \omega t = \omega\Phi_{max} \sin\left(\omega t - \frac{\pi}{2}\right)$ volts

It is clear from the above equation that the maximum value of emf induced per turn

$= \omega\,\Phi_{max} = 2\pi f\Phi_{max}$ volts $\quad \because \omega = 2\pi f$

and rms value of emf induced per turn $= \frac{1}{\sqrt{2}} \times 2\pi f\Phi_{max} = 4.44f\Phi_{max}$ volts

Hence rms value of emf induced in primary, $E_1 = 4.44 f N_1 \Phi_{max}$ volts

and rms value of emf induced in secondary, $E_2 = 4.44 f N_2 \Phi_{max}$ volts

In an ideal transformer the voltage drops in primary and secondary windings are negligible, so

EMF induced in primary winding, E = Applied voltage to primary, V_1

and terminal voltage, V_2 = EMF induced in secondary, E_2

Note: If B_{max} is the maximum allowable flux density in Wb/m^2 (or T) and a is the area of x-section of iron core in square metres, then in Eqs. (12.1), (12.2) and (12.3), Φ_{max} is given as

$$\Phi_{max} = B_{max}\, a \text{ webers}$$

12.6. VOLTAGE AND CURRENT TRANSFORMATION RATIOS

Referring to Eq. (12.1), it is clear that the *volts per turn* is exactly the same for both the primary and secondary windings *i.e.* in any transformer, the secondary and primary induced emfs are related to each other by the ratio of the number of secondary and primary turns. Thus,

$$\frac{V_2}{V_1} = \frac{E_2}{E_1} = \frac{N_2}{N_1} = K \qquad \text{...(12.4)}$$

The same relationship can be derived by dividing Eq. (12.3) by Eq. (12.2).

The constant K in Eq. (12.4) is called the *voltage transformation ratio.*

For step-up transformer, $V_2 > V_1$ or voltage transformation ratio, $K > 1$.

For step-down transformer, $V_2 < V_1$ or voltage transformation ratio, $K < 1$.

In an ideal transformer, the losses are negligible, so the volt-ampere input to the primary and volt-ampere output from secondary can be approximately equated *i.e.*

$$\text{Output VA} = \text{Input VA}$$

$$\text{or } V_2 I_2 = V_1 I_1$$

$$\text{or } \frac{I_2}{I_1} = \frac{V_1}{V_2} = \frac{E_1}{E_2} = \frac{N_1}{N_2} = \frac{1}{K} \qquad \text{...(12.5)}$$

i.e. Primary and secondary currents are inversely proportional to their respective turns.

Note: For a transformer the true ratio of transformation (or turn ratio) is constant, while the voltage ratio $\left(\frac{V_2}{V_1}\right)$ varies about 1 to 8 per cent depending upon the load and its power factor.

Example 12.1 **It is desired to have a 4.13 mWb maximum core flux in a transformer at 110 V and 50 Hz. Determine the required number of turns in the primary.**

[Pb. Technical Univ. Electrical Engineering December, 2000]

Solution: EMF induced in primary, $E_1 = 110$ V

Supply frequency, $f = 50$ Hz

Maximum core flux, $\Phi_{max} = 4.13 \text{ mWb} = 4.13 \times 10^{-3}$ Wb

Required number of turns on primary, $N_1 = \dfrac{E_1}{4.44 f \Phi_{max}}$ ...Refer to Eq. (12.2)

$$= \frac{110}{4.44 \times 50 \times 4.13 \times 10^{-3}} = 120 \textbf{ Ans.}$$

Example 12.2 **The emf per turn of a single phase 10 kVA, 2200 / 220 V, 50 Hz transformer is 10 V. Calculate (i) the number of primary and secondary turns, (ii) the net cross-sectional area of core for a maximum flux density of 1.5 T.**

Solution: EMF per turn = 10 V

Primary induced emf, $E_1 = V_1 = 2{,}200$ V

$$\text{Secondary induced emf, } E_2 = V_2 = 220 \text{ V}$$
$$\text{Supply frequency, } f = 50 \text{ Hz}$$
$$\text{Maximum flux density, } B_{max} = 1.5 \text{ T}$$

$$(i) \text{ Number of primary turns, } N_1 = \frac{E_1}{\text{EMF per turn}} = \frac{2{,}200}{10} = 220 \textbf{ Ans.}$$

$$\text{Number of secondary turns, } N_2 = \frac{E_2}{\text{EMF per turn}} = \frac{220}{10} = 22 \textbf{ Ans.}$$

$$\text{Maximum value of flux, } \Phi_{max} = \frac{\text{EMF per turn}}{4.44 f} = \frac{10}{4.44 \times 50} = 0.045 \text{ Wb} \quad \text{...refer to Eq. (12.1)}$$

$$(ii) \text{ Net cross-sectional area of core, } a = \frac{\Phi_{max}}{B_{max}} = \frac{0.045}{1.5} = 0.03 \text{ m}^2 \textbf{ Ans.}$$

Example 12.3 **A single-phase transformer has 350 primary and 1,050 secondary turns. The net cross-sectional area of the core is 55 cm². If the primary winding be connected to a 400 V, 50 Hz single phase supply, calculate (*i*) maximum value of the flux density in the core and (*ii*) the voltage induced in the secondary winding.**

[*R.G.P.V. Basic Electrical Engineering 2001; G.G.S.I.P. Delhi Univ; Electrical Science May-June 2006*]

Solution: Net cross-sectional area of core, $a = 55 \text{ cm}^2 = 0.0055 \text{ m}^2$

$$\text{Maximum value of flux, } \Phi_{max} = \frac{E_1}{4.44 f N_1} = \frac{400}{4.44 \times 50 \times 350} = 5.148 \times 10^{-3} \text{ Wb} \text{ ... refer to Eq. (12.2)}$$

$$(i) \text{ Peak value of flux density in the core, } B_{max} = \frac{\Phi_{max}}{a} = \frac{5.148 \times 10^{-3}}{0.0055} = 0.936 \text{ T} \quad \textbf{Ans.}$$

$$(ii) \text{ Voltage induced in the secondary winding, } E_2 = E_1 \times \frac{N_2}{N_1} = 400 \times \frac{1{,}050}{350} = 1{,}200 \text{ V} \quad \textbf{Ans.} \quad \text{...refer to Eq. (12.4)}$$

Example 12.4 **A 25 kVA, single-phase transformer has 250 turns on the primary and 40 turns on the secondary winding. The primary is connected to 1,500 V, 50-Hz mains calculate (*i*) primary and secondary currents on full load (*ii*) secondary emf (*iii*) maximum flux in the core.**

Solution:

$$\text{Supply voltage, } V_1 = 1{,}500 \text{ V}$$
$$\text{Primary induced emf, } E_1 \simeq V_1 = 1{,}500 \text{ V}$$

$$(ii) \text{ Secondary emf, } E_2 = \frac{E_1 \times N_2}{N_1} = \frac{1{,}500 \times 40}{250} = 240 \text{ V} \quad \textbf{Ans.}$$

(*i*) Approximate value of primary current on full load,

$$I_1 = \frac{\text{Rated kVA} \times 1{,}000}{V_1} = \frac{25 \times 1{,}000}{1{,}500} = 16.667 \text{ A} \quad \textbf{Ans.}$$

Approximate value of secondary current on full load,

$$I_2 = \frac{\text{Rated kVA} \times 1{,}000}{V_2 (\text{or } E_2)} = \frac{25 \times 1{,}000}{240} = 104.167 \text{ A} \quad \textbf{Ans.} \quad \because \; E_2 \simeq V_2$$

$$(ii) \text{ Maximum value of flux in the core, } \Phi_{max} = \frac{E_1}{4.44 f N_1} = \frac{1{,}500}{4.44 \times 50 \times 250} = 0.027 \text{ Wb or } 27 \text{ mWb} \quad \textbf{Ans.}$$

Example 12.5 **A 200/50 V, 50 Hz single-phase transformer is connected to a 200 V, 50 Hz supply with secondary winding open. Primary winding has 400 turns.**

(i) What is the value of maximum flux through the core, if the primary winding has 400 turns ?

(ii) What is the peak value of flux if the primary voltage is 200V, 25 Hz ?

(iii) What happens to no-load current ? [*R.G.P.V. Basic Electrical Engineering December 2002*]

Solution: (*i*) Maximum value of flux, $\Phi_{max} = \frac{E_1}{4.44 f N_1} = \frac{V}{4.44 f N_1} = \frac{200}{4.44 \times 50 \times 400}$

$= 2.25 \times 10^{-3}$ Wb or 2.25 m Wb **Ans.**

(*ii*) When primary voltage is 200 V and supply frequency is 25 Hz

$$\Phi'_{max} = \frac{200}{4.44 \times 25 \times 400} = 4.5 \text{ m Wb} \quad \textbf{Ans.}$$

(*iii*) The no-load primary current I_0, called the exciting current, is very small in comparison to the full-load primary current (2–5% of full-load primary current). This current is made up of a relatively larger quadrature or magnetising component I_m and a comparatively small in-phase or energy component I_e.

When supply frequency is reduced, keeping supply voltage constant, maximum value of flux is increased in inverse ratio of supply frequencies, and therefore, the magnetising component of no-load current and thus the no-load current.

Example 12.6 **A single-phase, 50 Hz, core type transformer has square cores of 20 cm side, permissible maximum flux density is 1 Wb/m^2. Calculate the number of turns per limb on the high- and low-voltage sides for a 3,000/220 V ratio.**

Solution: Core area $= 20 \times 20 = 400 \text{ cm}^2 = 0.04 \text{ m}^2$

Permissible flux density, $B_{max} = 1.0 \text{ Wb/m}^2$

Maximum value of permissible flux, $\Phi_{max} = B_{max} \times a = 1 \times 0.04 = 0.04$ Wb

Number of turns on low-voltage (secondary) winding,

$$N_2 = \frac{E_2}{4.44 f \Phi_{max}} = \frac{220}{4.44 \times 50 \times 0.04} = 24.77; \ 26 \text{ (say)}$$

Number of turns on high-voltage (primary) winding,

$$N_1 = \frac{E_1}{E_2} \times N_2 = \frac{3{,}000}{220} \times 26 = 354$$

$$\text{Number of } hv \text{ turns on each limb} = \frac{354}{2} = 177 \quad \textbf{Ans.}$$

$$\text{Number of } lv \text{ turns on each limb} = \frac{26}{2} = 13 \quad \textbf{Ans.}$$

Note: LV turns are calculated first, rounded one figure off to the next higher even number in order that maximum flux density does not exceed the permissible maximum flux density. Then *hv* turns are calculated with corrected number of turns by transformation ratio. Further there are two limbs and each limb accommodates half of the *lv* and half of the *hv* turns from the view point of reducing leakage reactance.

Example 12.7 **A single-phase transformer has a core whose cross–sectional area is 150 cm^2; operates at a maximum flux density of 1.1 Wb/m^2 from a 50 Hz supply. If the secondary winding has 66 turns, determine the output in KVA when connected to a load of 4 Ω impedance. Neglect any voltage drop in the transformer.**

Solution: Peak value of flux density, $B_{max} = 1.1 \text{ Wb/m}^2$

Cross-sectional area of core, $a = 150 \text{ cm}^2 = 0.015 \text{ m}^2$

Maximum value of flux in the core, $\Phi_{max} = B_{max} \times a = 1.1 \times 0.015 = 0.0165$ Wb

Number of turns on secondary, $N_2 = 66$

Supply frequency, $f = 50$ Hz

RMS value of induced emf in the secondary winding,

$$E_2 = 4.44 f \Phi_{max} N_2 = 4.44 \times 50 \times 0.0165 \times 66 = 240 \text{ volts}$$

Secondary voltage, $V_2 = E_2 = 240$ V neglecting voltage drop in secondary winding

$$\text{Secondary current, } I_2 = \frac{V_2}{\text{Load impedance}} = \frac{240}{4} = 60 \text{ A}$$

$$\text{Output in kVA} = \frac{V_2 I_2}{1{,}000} = \frac{240 \times 60}{1{,}000} = 14.4 \ \textbf{Ans.}$$

Example 12.8 **A single-phase transformer is rated 25 kVA, 600/200 V, 50 Hz. Calculate the impedance of load in ohms to fully load the transformer when connected to (a) 600 V side (b) 240 V side.**

[*Mahamaya Technical Univ. Electrical Engineering Second Semester 2011-12*]

Solution: (*a*) **When load is connected to 600 V side**

$$\text{Load current, } I_1 = \frac{\text{Rated kVA} \times 1{,}000}{V_1} = \frac{25 \times 1{,}000}{600} = 41.6667 \text{ A}$$

$$\text{Load impedance, } Z_{L_1} = \frac{V_1}{I_1} = \frac{600}{41.6667} = 14.4 \ \Omega \ \textbf{Ans.}$$

(*b*) **When load is connected to 200 V side**

$$\text{Load current, } I_2 = \frac{25 \times 1{,}000}{200} = 125 \text{ A}$$

$$\text{Load impedance, } Z_{L_2} = \frac{200}{125} = 1.6 \ \Omega \ \textbf{Ans.}$$

12.7. EFFECTS OF VOLTAGE AND FREQUENCY VARIATIONS

Power transformers are not ordinarily subjected to frequency variations and usually are subject to only modest voltage variations, but it is interesting to consider the effects thereof.

Variations in voltage and/or frequency affects the iron losses (hysteresis and eddy current losses) in a transformer. As long as the flux variations are sinusoidal with respect to line, hysteresis loss (P_h), and eddy current loss (P_e) varies according to the following relations

$$P_h \ \alpha \ f \, (\Phi_{max})^x$$

where x lies between 1.5 and 2.5 depending on the grade of iron used in transformer core

$$\text{and } P_e \ \alpha \ f^2 \, (\Phi_{max})^2$$

If the transformer is operated with the frequency and voltage changed in the same proportion, the flux density will remain unchanged as obvious from Eq. (12.2) and apparently the no-load current will also remain unaffected.

The transformer can be operated safely at frequency less than rated one with correspondingly reduced voltage. In this case iron losses will be reduced. But if the transformer is operated with increased voltage and frequency in the same proportion, the core losses may increase to an intolerable level. Increase in frequency with constant supply voltage will cause reduction in hysteresis loss and leave the eddy current losses unaffected. Some increase in voltage could, therefore, be tolerated at higher frequencies, but exactly how much depends on the relative magnitude of the hysteresis and eddy current losses and the grade of iron used in the transformer core.

12.8. TRANSFORMER ON NO LOAD

When the primary of a transformer is connected to the source of ac supply and the secondary is open, the transformer is said to be at no load (there is no load on secondary).

Consider an ideal transformer whose secondary side is open and the primary winding is connected to a sinusoidal alternating voltage V_1. The alternating voltage applied to the primary winding will cause flow of alternating current in the primary winding. Since the primary coil is

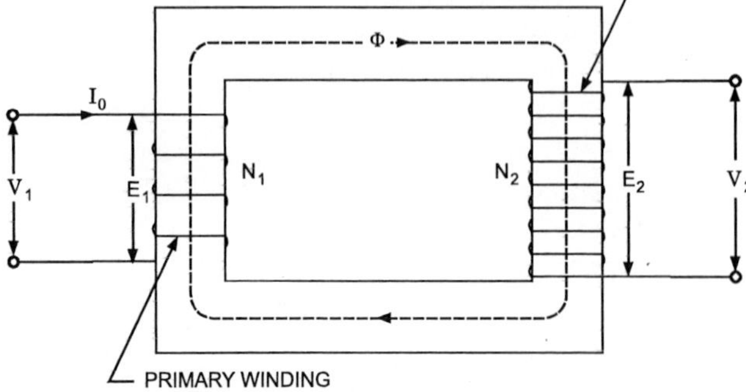

Fig. 12.3 *Transformer on No Load*

pure inductive and there is no output (secondary being open), the primary draws the magnetising current I_m only. The function of this current is merely to magnetize the core. If the transformer is truely ideal, the magnitude of I_m should be zero by virtue of assumption (*iv*) made in Art 12.4. Since the reluctance of the magnetic circuit is never zero, I_m has definite magnitude. The magnetising current, I_m is small in magnitude and lags behind supply voltage V_1 by 90 °. This magnetising current I_m produces an alternating flux Φ which is, at all times, proportional to the current (assuming permeability of the magnetic circuit to be constant) and, hence, in phase with it.

Let the instantaneous linking flux be given as

$$\Phi = \Phi_{max} \sin \omega t \quad ..(12.6)$$

The varying flux is linked with both of the windings (primary and secondary) and so induces emfs in the primary and secondary windings. The instantaneous values of induced emfs in the primary and secondary windings will be

$$e_1 = -N_1 \frac{d\Phi}{dt} = -N_1 \frac{d}{dt}(\Phi_{max} \sin \omega t) = -N_1 \omega \Phi_{max} \cos \omega t = N_1 \omega \Phi_{max} \sin\left(\omega t - \frac{\pi}{2}\right) \quad ...(12.7)$$

Similarly, $e_2 = N_2 \omega \Phi_{max} \sin\left(\omega t - \frac{\pi}{2}\right)$ ...(12.8)

Since, primary winding has no ohmic resistance, (as assumed), applied voltage to primary winding is to only oppose the induced emf in the primary winding, hence instantaneous applied voltage to primary will be given by

$$v_1 = -e_1 = -N_1 \omega \Phi_{max} \sin\left(\omega t - \frac{\pi}{2}\right) \quad ...(12.9)$$

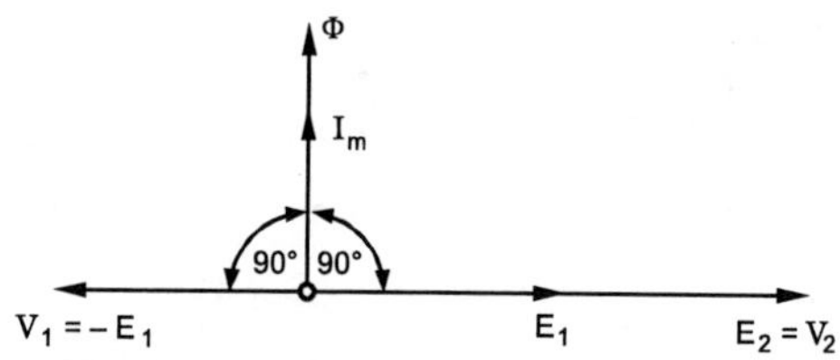

Fig. 12.4 *No-Load Phasor Diagram For an Ideal Transformer*

Comparing Eqs. (12.6), (12.7), (12.8), and (12.9) we conclude that

(*i*) Induced emfs in primary and secondary windings, E_1 and E_2 lag behind the main flux Φ by $\frac{\pi}{2}$, so these emfs (E_1 and E_2) are in phase with each other, as shown in Fig. 12.4 vectorially.

(*ii*) Applied voltage to the primary winding leads the main flux by $\frac{\pi}{2}$ and is in phase opposition to the induced emf in the primary winding, as shown in Fig. 12.4 vectorially.

(*iii*) Secondary voltage $V_2 = E_2$ as there is no voltage drop in secondary.

The instantaneous value of applied voltage, induced emfs, flux and magnetising current, in case of an ideal transformer, are illustrated by sinusoidal waves in Fig. 12.5.

However, when a varying flux is set up in magnetic material, there will be power loss, called the *iron or core loss.* So the input current to the primary under no-load condition has also to supply the hysteresis and eddy current losses (iron losses) occuring in the core in addition to small amount of copper loss occuring in primary winding (no copper loss occurs in secondary winding on open circuit or on no load). Hence, the no-load primary current I_0 does not lag behind applied voltage V_1 by 90 ° but lags behind V_1 by angle $\phi_0 < 90$ °.

Input power on no load, $P_0 = V_1 I_0 \cos \phi_0$ where $\cos \phi_0$ is the primary power factor under no-load condition.

As seen from phasor diagram shown in Fig. 12.6, input current to the primary I_0, called the *exciting current,* has two components (*i*) in-phase, active or energy component, I_e used to meet the iron loss in addition to small amount of copper loss occurring in the primary winding and (*ii*) quadrature component or wattless component called the magnetizing component, I_m used to create the alternating flux in the core.

$$\text{Thus, } I_e = I_0 \cos \phi_0$$

$$\text{and } I_m = I_0 \sin \phi_0$$

$$\text{and } \sqrt{I_e^2 + I_m^2} = I_0$$

$$\text{Angle of lag, } \phi_0 = \text{Tan}^{-1} \frac{I_m}{I_e}$$

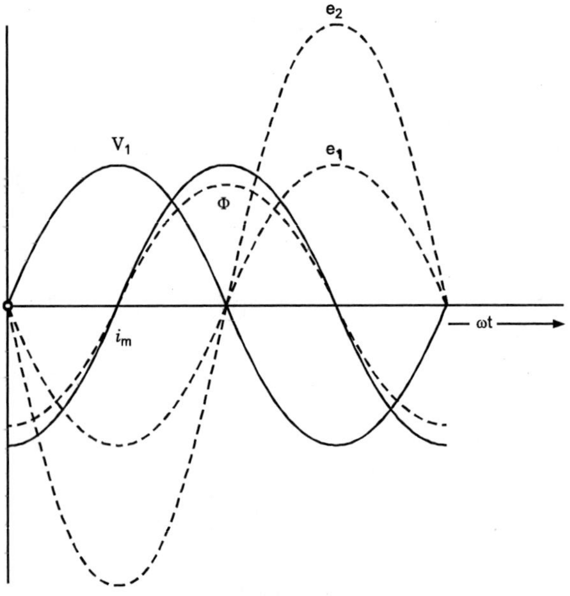

Fig. 12.5

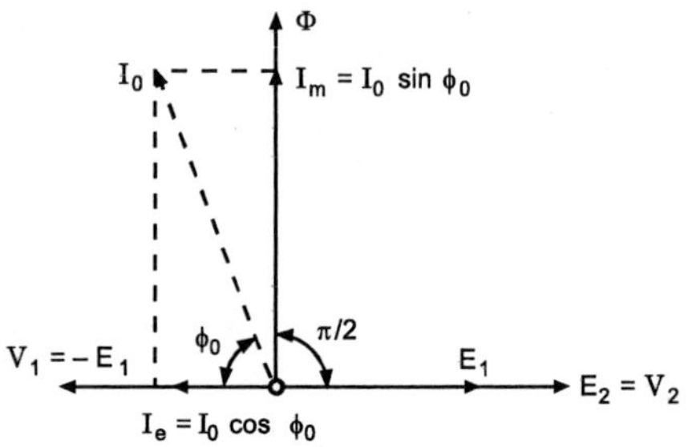

Fig. 12.6

The equivalent circuit of a transformer on no load is illustrated in Fig. 12.7, wherein two components of no-load current; I_e and I_m are represented by currents drawn by a non-inductive resistance R_0 and a pure inductive reactance X_0 respectively. Both these currents are drawn at induced emf $E_1 = V_1$ for resistance-less, no-leakage primary coil; even otherwise $E_1 = V_1$

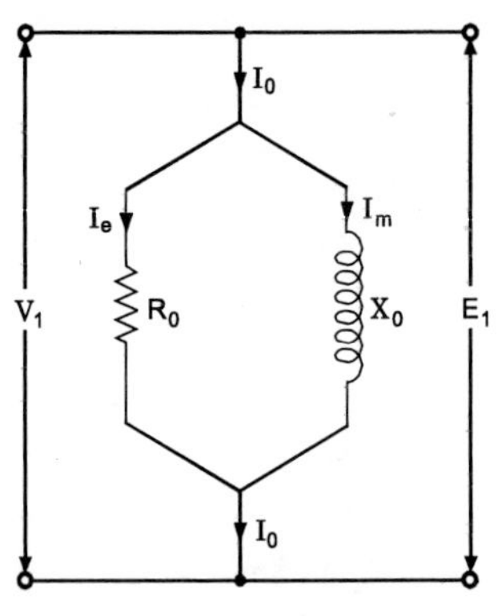

Fig. 12.7

The noteworthy points are given below:

1. The no-load primary current I_0, called the *exciting current*, is very small in comparison to the full-load primary current. It ranges from 2 to 5 per cent of full–load primary current.
2. The exciting or no-load current I_0 is made up of a relatively large quadrature or magnetizing component I_m and a comparatively small in-phase or energy component I_e, so the power factor of a transformer on no load is very small (usually varies between 0.1 and 0.2 lag). The phase angle between I_0 and V_1 is about 78 ° to 87 °.
3. No-load primary copper loss *i.e* $I_0^2 R_1$ is very small and may be neglected. Thus the no-load primary input power is practically equal to the iron loss occuring in the core of the transformer.
4. Phase angle ϕ_0 is also known as the *hysteresis angle of advance* because it is principally the core loss which is responsible for shift in the current phasor.
5. Since the permeability of the core material varies with the instantaneous value of the exciting current, the waveform of the exciting or magnetizing current is not truely sinusoidal. As such, it should not be represented by a phasor because only sinusoidally varying quantities are represented by rotating phasors. But, in practice, it makes no appreciable difference.

Example 12.9 **A voltage $v = 200 \sin 314\, t$ is applied to the transformer winding in a no-load test. The resulting current is found to be**

$$i = 3 \sin (314\, t - 60°)$$

Determine the core loss and the parameters of no-load approximate equivalent circuit.

[*U.P. Technical Univ. Electrical Machines 2002-03; Mahamaya Technical Univ. Electrical Engineering First Semester 2011-12*]

Solution: No-load current, $I_0 = \frac{3}{\sqrt{2}} = 2.12$ A

Core loss = Input at no load = $V I_0 \cos \phi_0 = \frac{200}{\sqrt{2}} \times \frac{3}{\sqrt{2}} \cos 60° = 150$ W **Ans.**

Energy component of no-load current, $I_e = I_0 \cos \phi_0 = \frac{3}{\sqrt{2}} \cos 60° = 1.06$ A

Magnetising component of no-load current, $I_m = \sqrt{I_0^2 - I_e^2} = \sqrt{(2.12)^2 - (1.06)^2} = 1.836$ A

No-load resistance, $R_0 = \frac{V}{I_e} = \frac{200/\sqrt{2}}{1.06} = 133.42\ \Omega$ **Ans.**

No-load reactance, $X_0 = \frac{V}{I_m} = \frac{200/\sqrt{2}}{1.836} = 77\ \Omega$ **Ans.**

Example 12.10 **A 50 Hz transformer takes 75 W input at 1.5 A and 120 V. If the primary winding resistance is 0.4 Ω, find the no-load power factor.** *[Pb. Technical Univ. Electrical Engineering June 2000; May 2001]*

Solution: No-load power input to transformer, $P_0 = 75$ W

No-load primary current, $I_0 = 1.5$ A

Primary voltage, $V_1 = 120$ V

No-load power factor, $\cos \phi_0 = \frac{P_0}{V_1 \times I_0} = \frac{75}{120 \times 1.5} = 0.417$ (lag) **Ans.**

Example 12.11 **A 100 kVA, 2400/240 V, 50 Hz single-phase transformer has an exciting current of 0.64 A and a core loss of 700 W, when its high voltage side is energised at rated voltage and frequency. Calculate the components of no-load current and no-load branch parameters.**

[Pb. Technical Univ. Basic Electrical Engineering December 2002]

Solution: Primary voltage, $V_1 = 2{,}400$ V

Exciting or no-load current, $I_0 = 0.64$ A

Iron loss = 700 W

Iron-loss current, $I_e = I_0 \cos \phi_0 = \frac{\text{Iron loss}}{V_1} = \frac{700}{2{,}400} = 0.292$ A **Ans.**

Magnetising component of no-load current, $I_m = \sqrt{I_0^2 - I_e^2} = \sqrt{(0.64)^2 - (0.292)^2} = 0.57$ A **Ans.**

No-load branch parameters, $R_0 = \frac{V_1}{I_e} = \frac{2{,}400}{0.292} = 8{,}220\ \Omega$ or $8.22\ \text{k}\Omega$ **Ans.**

$X_0 = \frac{V_1}{I_m} = \frac{2{,}400}{0.57} = 4{,}210\ \Omega$ or $4.21\ \text{k}\Omega$ **Ans.**

Example 12.12 **The no-load current of a transformer is 4.0 A at 0.25 pf when supplied at 250 V, 50 Hz. The number of turns on the primary winding is 200. Calculate**

(*i*) flux in the core (*ii*) the core loss, and (*iii*) magnetising current.

Solution: (*i*) Maximum value of flux in the core,

$$\Phi_{max} = \frac{E_1}{4.44 f N_1} = \frac{250}{4.44 \times 50 \times 200} = 0.00563 \text{ Wb or } 5.63 \text{ mWb} \quad \textbf{Ans.}$$

(*ii*) Core loss = Input on no load = $V_1 I_0 \cos \phi_0 = 250 \times 4 \times 0.25 = 250$ W **Ans.**

(*iii*) Magnetising current, $I_m = I_0 \sin \phi_0 = I_0 \sqrt{1 - (\cos \phi_0)^2} = 4\sqrt{1 - (0.25)^2} = 3.873$ A **Ans.**

12.9. TRANSFORMER ON LOAD

When the secondary circuit of a transformer is completed through an impedance or load, the transformer is said to be loaded and current flows through the secondary and the load. The magnitude and phase of secondary current I_2 with respect to secondary terminal voltage V_2 will depend upon the characteristic of load *i.e.* current I_2 will be in phase, lag behind and lead the terminal voltage V_2 respectively when the load is non-inductive, inductive and capacitive.

When the transformer is on no load, as shown in Fig. 12.3, it draws no-load current I_0 from the supply mains. The no-load current I_0 sets up an mmf N_1I_0 which produces flux Φ in the core. When an impedance is connected across the secondary terminals, as shown in Fig. 12.8, current I_2 flows through the secondary winding. The secondary current I_2 sets up its own mmf and hence creates a secondary flux Φ_2. The secondary flux Φ_2 opposes the main flux Φ set up by the exciting current I_0 according to Lenz's law. The opposing secondary flux Φ_2 weakens the main flux Φ momentarily, so primary back emf E_1 tends to be reduced. So difference of applied voltage V_1 and back emf E_1 increases, therefore, more current is drawn from the source of supply flowing through the primary winding until the original value of flux Φ is obtained. It again causes increase in back emf E_1 and it adjusts itself as such that there is a balance between applied voltage V_1 and back emf E_1. Let the additional primary current be I_1'. The current I'_1 is in phase opposition with secondary current I_2 and is called the *counterbalancing current.* The additional current I'_1 sets up an mmf $N_1 I'_1$ producing flux Φ_1' in the same direction as that of main flux Φ and cancels the flux Φ_2 produced by secondary mmf $N_2 I_2$ being equal in magnitude.

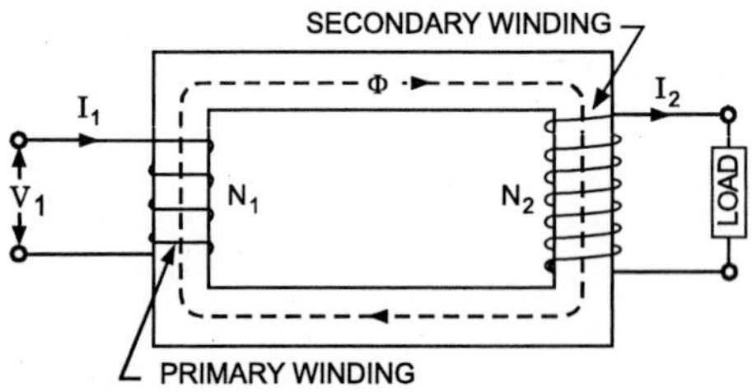

Fig. 12.8 *An Ideal Transformer On Load*

$$\text{So } N_1 I_1' = N_2 I_2$$

$$\text{or } I_1' = \frac{N_2}{N_1} I_2$$

The total primary current I_1 is, therefore, phasor sum of primary counterbalancing current I_1' and no-load current I_0, which will be approximately equal to I'_1 as I_0 is usually very small in comparison to I_1'.

$$\therefore I_1 = I_1' = \frac{N_2}{N_1} I_2$$

$$\text{or } \frac{I_1}{I_2} = \frac{N_2}{N_1} = K \qquad \text{(transformation ratio)}$$

Hence primary and secondary currents are inversely proportional to their respective turns.

Since the secondary flux Φ_2 produced by secondary mmf $N_2 I_2$ is neutralized by the flux Φ'_1 produced by mmf $N_1 I_1'$ set up by counterbalancing primary current I'_1, so the flux in the transformer core remains almost constant from no load to full load.

The phasor diagrams for transformer on non-inductive, inductive and capacitive loads are shown in Figs. 12.9 (*a*), (*b*) and (*c*) respectively.

Since the voltage drops in both of the windings of the transformer are assumed to be negligible, therefore

$$\mathbf{V_2 = E_2 \text{ and } V_1 = -E_1}$$

The secondary current I_2 is in phase, lags behind and leads the secondary terminal voltage V_2 by an angle ϕ_2 for pure resistive, inductive and capacitive loads respectively.

The induced primary current I_1', also known as *counterbalancing current,* is always in opposition to secondary current I_2 and since no-load current I_0 is very small, the total primary current I_1 is almost opposite in phase to I_2 and K times the secondary current I_2, where K is transformation ratio.

Note. In phasor diagrams shown in Figs. 12.9 (*a*), 12.9 (*b*) and 12.9 (*c*) no-load current has been drawn on exaggerated scale for sake of clarity.

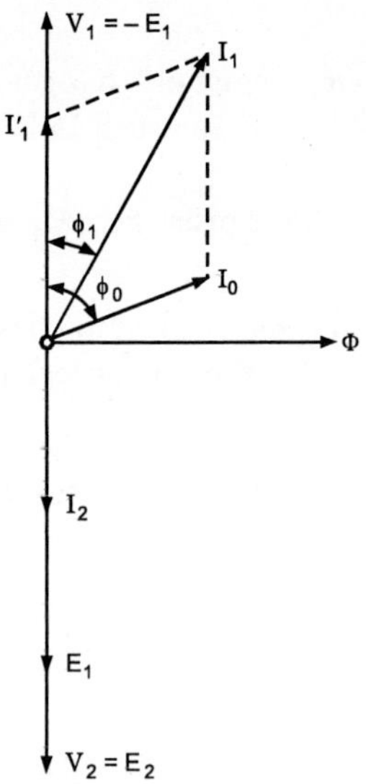

(a) For Pure Resistive Load

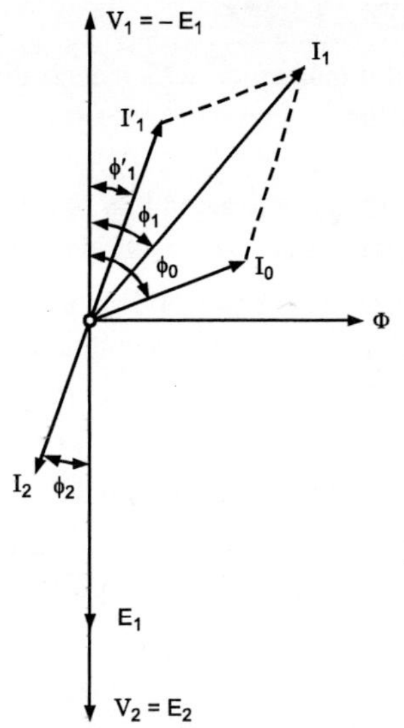

(b) For Inductive Load

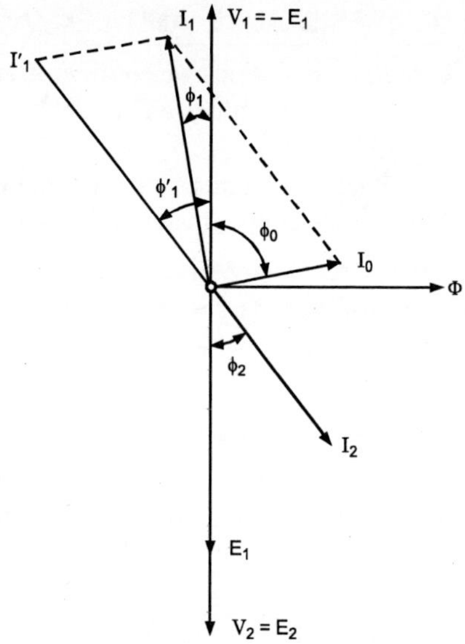

(c) For Capacitive Load

Fig. 12.9 *Phasor Diagram For an Ideal Transformer On Load*

Example 12.13 **A single-phase 440/110 V transformer takes a no-load current of 4 A at 0.2 power factor. If the secondary supplies a current of 100 A at a power factor of 0.8 lagging determine (i) the current taken by the primary winding (ii) the magnetising reactance and resistance representing core losses.**

[U.P. Technical Univ. Electrical Engineering February 2002]

Solution:

$$\text{Primary emf, } E_1 = 440 \text{ V}$$

$$\text{Secondary emf, } E_2 = 110 \text{ V}$$

$$\text{No-load power factor, } \cos\phi_0 = 0.2 \text{ (lag)}$$

$$\text{Sin } \phi_0 = \sqrt{1-0.2^2} = 0.98$$

$$\text{Secondary current, } I_2 = 100 \text{ A}$$

$$\text{Load power factor, } \cos\phi_2 = 0.8 \text{ (lag)}$$

$$\text{Transformation ratio, } K = \frac{V_2}{V_1} = \frac{110}{440} = \frac{1}{4}$$

$$\text{Primary counterbalancing current, } I'_1 = K\,I_2 = \frac{1}{4} \times 100 = 25 \text{ A}$$

lagging behind the supply voltage V_1 by an angle ϕ'_1 where $\phi'_1 = \phi_2 = \text{Cos}^{-1}\,0.8 = 36.87°$

No-load current, I_0 = 4 A lagging behind the supply voltage V_1 by an angle ϕ_0 $= \text{Cos}^{-1}\,0.2 = 78.46°$

$$(i) \text{ Primary current, } \mathbf{I_1} = I'_1 + I_0 = (I'_1 \cos\phi'_1 + I_0\cos\phi_0) + j\,(I'_1\sin\phi'_1 + I_0\sin\phi_0)$$

$$= (25 \times 0.8 + 4 \times 0.2) + j\,(25 \times 0.6 + 4 \times 0.98)$$

$$= (20.8 + j\,18.92) \text{ A} = 28.12 \angle 42.29° \text{ A } \textbf{Ans.}$$

$$\text{Energy component of no-load current, } I_e = I_0\cos\phi_0 = 4 \times 0.2 = 0.8 \text{ A}$$

$$\text{Magnetising component of no-load current, } I_m = I_0\sin\phi_0 = 4 \times 0.98 = 3.92 \text{ A}$$

$$(ii) \text{ Resistance representing core losses, } R_0 = \frac{V_1}{I_e} = \frac{440}{0.8} = 550\ \Omega \ \textbf{Ans.}$$

$$\text{Magnetising reactance, } X_0 = \frac{V_1}{I_m} = \frac{440}{3.92} = 112.25\ \Omega \ \textbf{Ans.}$$

12.10. RESISTANCE AND LEAKAGE REACTANCE

In preceding discussions we considered an ideal transformer, which according to our assumptions, has got no resistance in the windings and no leakage flux but in actual practice it is impossible to obtain such an ideal transformer.

In actual transformer both the windings, primary and secondary windings have finite resistances R_1 and R_2 which cause copper losses and voltage drop in them. The result is that:

(*i*) The secondary terminal voltage V_2 is less than the secondary induced emf E_2 and is equal to phasor difference of secondary induced emf E_2 and voltage drop in the secondary winding $I_2 R_2$, if magnetic leakage is negligible, *i.e.*

$$\mathbf{V_2 = E_2 - I_2 R_2}$$

where I_2 is the secondary current and R_2 is the secondary winding resistance.

(*ii*) Similarly the counter emf of primary, $-E_1$ is equal to phasor difference of voltage applied to the primary winding V_1 and voltage drop in the primary winding $I_1 R_1$, provided magnetic leakage is negligible, i.e.,

$$\mathbf{-E_1 = V_1 - I_1 R_1}$$

where I_1 is the primary current and R_1 is the primary resistance.

It was previously assumed that the entire flux Φ, developed by the primary winding, links with and cut every turn of both the primary and secondary windings. In practice, it is impossible to realize this condition. However, part of the flux set up by the primary winding links only the primary turns, as illustrated in Fig. 12.10 by flux Φ_{L1}. Also, some of the flux set up by the secondary winding links only the secondary turns, as illustrated in Fig. 12.10 by Φ_{L2}. These two fluxes Φ_{L1} and Φ_{L2} are known as leakage flux *i.e. that flux which leaks out of the core and does not link both windings.* The flux which does pass completely through the core and links both the windings is known as the *mutual flux* and is illustrated as Φ.

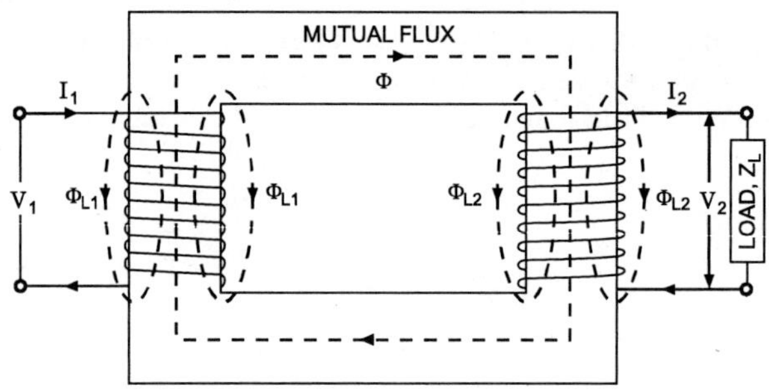

Fig. 12.10 *Magnetic Fluxes in a Transformer*

The primary leakage flux Φ_{L1} linking with the primary winding is produced by primary ampere-turns only, therefore, it is proportional to primary current, number of primary turns being fixed. On no load primary current is so small that leakage flux Φ_{L1} produced by it can be neglected but on load primary current increases resulting in increase in ampere-turns and hence leakage flux Φ_{L1} increases. The primary leakage flux Φ_{L1} is in phase with I_1 and produces self-induced emf E_{L1} given by $E_{L1} = 2\pi f L_1 I_1$ in primary winding, where L_1 is the self-inductance of the primary winding produced by primary leakage flux Φ_{L1}.

The self-induced emf E_{L1}, due to primary leakage flux, in the primary winding must lag leakage flux Φ_{L1} and primary current I by 90°. The emf necessary to balance this counter emf is opposite and equal to it and, therefore, leads the primary current I_1 by 90°. As this emf, induced by the primary leakage flux, is proportional to the current and lags it by 90°, it is nothing more than a reactance voltage and is denoted by $-I_1 X_1$. The component of line voltage that balances this emf is $+I_1 X_1$. The effect of the primary leakage flux, therefore, is to induce an emf that opposes the flow of current to the transformer. The reactance of the primary winding, X_1 can be obtained by dividing self-induced emf E_{L1} by the primary current I_1 *i.e.*

$$X_1 = \frac{E_{L1}}{I_1} = \frac{2\pi f L_1 I_1}{I_1} = 2\pi f L_1$$

Similarly secondary leakage flux Φ_{L2} is set up by secondary ampere-turns and is proportional to secondary current I_2. On no load there is no current in secondary winding and, therefore, no leakage flux exists across the secondary winding on no load. On load leakage flux Φ_{L2} in phase with secondary current

I_2 and produces self-induced emf $E_{L2} = 2\,\pi f L_2 I_2$ in the secondary winding where L_2 is self-inductance of secondary winding due to leakage flux Φ_{L2}. This is also a reactance voltage, and the component that balances it leads the secondary current by 90 °. The secondary reactance X_2 opposes the current flowing out of the transformer and can be obtained by dividing self induced emf in secondary winding, E_{L2} by the secondary current I_2 *i.e.*

$$X_2 = \frac{E_{L2}}{I_2} = \frac{2\pi f L_2 I_2}{I_2} = 2\pi f L_2$$

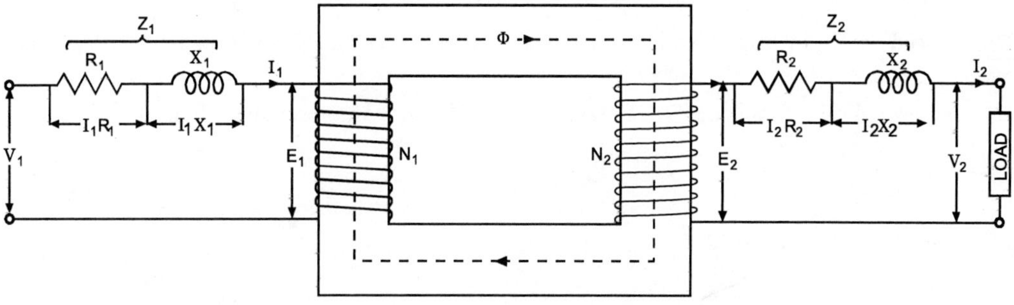

Fig. 12.11 *An Equivalent Diagram of Actual Transformer*

The effect of magnetic leakage is, thus to produce in their respective windings emfs of self-inductance which are proportional to the current, and are, therefore, equivalent in effect to the addition of an inductive coil in series with each winding, the reactance of which is called the leakage reactance.

A transformer with magnetic leakage and winding resistance is equivalent to an ideal transformer (having no resistance and leakage reactance) having inductive and resistive coils connected in series with each winding as shown in Fig. 12.11.

Few important points to be kept in mind are given below:

1. The leakage flux links one or the other winding but not both, hence it in no way contributes to the transfer of energy from the primary winding to the secondary winding.
2. The applied voltage to the primary winding, V_1 will have to meet the reactive drop $I_1 X_1$ in addition to $I_1 R_1$. Similarly induced emf in the secondary winding E_2 will have to meet the resistive and reactive drops $I_2 R_2$ and $I_2 X_2$ respectively.
3. Transformation ratio is reduced* due to resistance and magnetic leakage, since these reduce the secondary terminal voltage V_2 on load for a given primary applied voltage V_1.
4. The main useful flux Φ decreases slightly with the increase in load but the leakage fluxes are practically proportional to the currents in the respective winding.
5. In actual transformers, the primary and secondary windings are not placed on separate legs, as shown in Fig. 12.1, as due to their being widely separate, large primary and secondary leakage fluxes would result. The leakage fluxes are reduced to a minimum by sectionalizing and interleaving the primary and secondary windings.

12.11. PHASOR DIAGRAM OF ACTUAL TRANSFORMER ON LOAD

Consider a transformer shown in Fig. 12.11 having primary and secondary windings of resistances R_1 and R_2 and reactances X_1 and X_2 respectively. The impedance of primary winding is given by $\mathbf{Z}_1 = R_1 + j\,X_1$ and impedance of secondary winding is given by $\mathbf{Z}_2 = R_2 + j\,X_2$.

The phasor diagrams of above transformer on (*i*) pure resistive, (*ii*) resistive-inductive, and (*iii*) resistive-capacitive loads are shown in Fig. 12.12 (*a*), (*b*) and (*c*) respectively.

* Assuming lagging power factor

Draw *OA* representing secondary terminal voltage V_2 and OI_2 representing secondary current I_2 in phase as well as magnitude. Since voltage drops due to secondary winding resistance and reactance are $I_2 R_2$ in phase with current I_2 and $I_2 X_2$ leading current I_2 by $\pi/2$respectively, so draw AB parallel to OI_2 and equal to $I_2 R_2$ in magnitude representing resistive drop in secondary winding and draw BC perpendicular to AB and equal to $I_2 X_2$ in magnitude representing reactive drop of secondary winding. Since phasor sum of terminal voltage V_2, secondary resistive drop $I_2 R_2$ and secondary reactive drop $I_2 X_2$ is equal to induced emf E_2 in secondary winding, phasor OC represents secondary induced emf E_2. Hence we have

$$\mathbf{E}_2 = \mathbf{V}_2 + \mathbf{I}_2 (R_2 + j X_2) = \mathbf{V}_2 + \mathbf{I}_2 \mathbf{Z}_2 \quad \text{...(12.10)}$$

The induced emf E_1 in primary winding is in phase with E_2 and equal to $\frac{N_1}{N_2} E_2$ in magnitude, so take OD $= \frac{N_1}{N_2}$ OC representing E_1. Produce DO to D′ taking OD′ = OD hence representing $(-E_1)$.

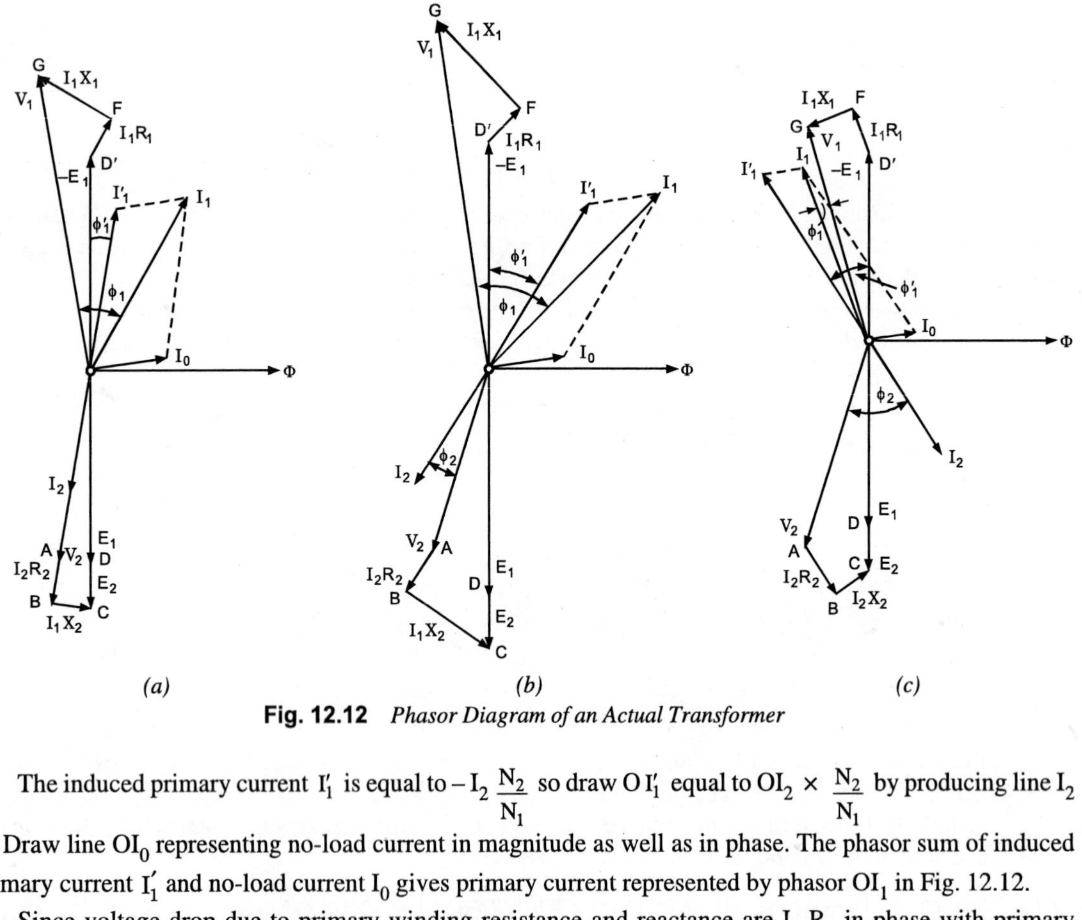

Fig. 12.12 *Phasor Diagram of an Actual Transformer*

The induced primary current I_1' is equal to $-I_2 \frac{N_2}{N_1}$ so draw OI_1' equal to $OI_2 \times \frac{N_2}{N_1}$ by producing line I_2 O. Draw line OI_0 representing no-load current in magnitude as well as in phase. The phasor sum of induced primary current I_1' and no-load current I_0 gives primary current represented by phasor OI_1 in Fig. 12.12.

Since voltage drop due to primary winding resistance and reactance are $I_1 R_1$ in phase with primary current I_1 and $I_1 X_1$ leading current I_1 by $\frac{\pi}{2}$ respectively, so draw D′F parallel to OI_1 and equal to $I_1 R_1$ in magnitude representing resistive drop in primary winding and draw FG perpendicular to D′F equal to $I_1 X_1$ in magnitude representing reactive drop in primary winding. As the phasor sum of $(-E_1)$, primary resistive drop and primary reactive drop gives the applied voltage V_1 to primary winding, hence phasor OG represents the applied voltage V_1 in magnitude as well as in phase.

$$i.e. \quad \mathbf{V}_1 = -\mathbf{E}_1 + \mathbf{I}_1 (R_1 + j X_1) = -\mathbf{E}_1 + \mathbf{I}_1 \mathbf{Z}_1 \quad \text{...(12.11)}$$

The phase angle ϕ_1 between V_1 and I_1 gives the power factor angle of the transformer.

Since no-load current I_0, resistive drops $I_1 R_1$ and $I_2 R_2$ and reactive drops, $I_1 X_1$ and $I_2 X_2$ are very small, so neglecting these we have $\phi_2 = \phi_1' = \phi_1 = \phi$, the phase angle of the load. In Fig. no-load current, resistive drops and reactive drops are shown, for clarity, on exaggerated scales.

From phasor diagrams we have

(*a*) *For pure resistive load [phasor diagram 12.12 (a)]*

$$E_2 = \sqrt{(V_2 + I_2 R_2)^2 + (I_2 X_2)^2} \approx V_2 + I_2 R_2 \quad \text{...(12.12)}$$

$$E_1 = \frac{E_2}{K}$$

$$\text{and } V_1 = \sqrt{(E_1 + I_1 R_1)^2 + (I_1 X_1)^2} \approx E_1 + I_1 R_1 \quad \text{...(12.13)}$$

(*b*) *For resistive-inductive load* [*phasor diagram* 12.12 (*b*)]

$$E_2 = \sqrt{(V_2 + I_2 R_2 \cos\phi + I_2 X_2 \sin\phi)^2 + (I_2 X_2 \cos\phi - I_2 R_2 \sin\phi)^2}$$

$$\approx V_2 + I_2 R_2 \cos\phi + I_2 X_2 \sin\phi \quad \text{...(12.14)}$$

$$E_1 = \frac{E_2}{K} \text{ and}$$

$$V_1 = \sqrt{(E_1 + I_1 R_1 \cos\phi + I_1 X_1 \sin\phi)^2 + (I_1 X_1 \cos\phi - I_1 R_1 \sin\phi)^2}$$

$$\approx E_1 + I_1 R_1 \cos\phi + I_1 X_1 \sin\phi \quad \text{...(12.15)}$$

(*c*) *For resistive-capacitive load [phasor diagam* 12.15 (*c*)]

$$E_2 = \sqrt{(V_2 + I_2 R_2 \cos\phi - I_2 X_2 \sin\phi)^2 + (I_2 X_2 \cos\phi + I_2 R_2 \sin\phi)^2}$$

$$\approx V_2 + I_2 R_2 \cos\phi - I_2 X_2 \sin\phi \quad \text{...(12.16)}$$

$$E_1 = \frac{E_2}{K} \text{ and}$$

$$V_1 = \sqrt{(E_1 + I_1 R_1 \cos\phi - I_1 X_1 \sin\phi)^2 + (I_1 X_1 \cos\phi + I_1 R_1 \sin\phi)^2}$$

$$= E_1 + I_1 R_1 \cos\phi - I_1 X_1 \sin\phi \quad \text{...(12.17)}$$

Example 12.14 **A 230/460 V transformer has a primary resistance of 0.2 Ω and a reactance of 0.5 Ω and the corresponding values for the secondary are 0.75 Ω and 1.8 Ω respectively. Find the secondary terminal voltage when supplying (*a*) 10 A at 0.8 pf lagging (*b*) 10 A at pf 0.8 leading.**

Solution: Transformation ratio, $K = \frac{V_2}{V_1} = \frac{460}{230} = 2$

(*i*) **When Load is 10 A at 0.8 pf lagging** i.e. when

$I_2 = 10$ A and $\cos\phi = 0.8$ lagging.

Primary current $I_1 = K I_2 = 2 \times 10 = 20$ A

From Eq. (12.15)

Primary induced emf, $E_1 = V_1 - (I_1 R_1 \cos\phi + I_1 X_1 \sin\phi)$

$= 230 - (20 \times 0.2 \times 0.8 + 20 \times 0.5 \times 0.6) = 220.8$ V

Secondary induced emf, $E_2 = KE_1 = 2 \times 220.8 = 441.6$ V

From Eq. (12.14)

Secondary terminal voltage, $V_2 = E_2 - (I_2 R_2 \cos\phi + I_2 X_2 \sin\phi)$

$= 441.6 - (10 \times 0.75 \times 0.8 + 10 \times 1.8 \times 0.6)$

$= 441.6 - (6 + 10.8) = 424.8$ V **Ans**

(*ii*) When load is 10 A at 0.8 pf leading

From Eq. (12.17)

$$\text{Primary induced emf, } E_1 = V_1 - (I_1 R_1 \cos\phi - I_1 X_1 \sin\phi)$$
$$= 230 - (20 \times 0.2 \times 0.8 - 20 \times 0.5 \times 0.6) = 232.8 \text{ V}$$

and from Eq. (12.16)

$$\text{Secondary terminal voltage, } V_2 = KE_1 - (I_2 R_2 \cos\phi - I_2 X_2 \sin\phi)$$
$$= 2 \times 232.8 - (10 \times 0.75 \times 0.8 - 10 \times 1.8 \times 0.6)$$
$$= 465.6 - (6 - 10.8) = 470.4 \text{ V}$$ **Ans.**

12.12. EQUIVALENT RESISTANCE AND REACTANCE

The two independent circuits of a transformer can be resolved into an equivalent circuit to make the calculations simple.

Let resistances and reactances of primary and secondary windings be R_1 and R_2 and X_1 and X_2 ohms respectively and let transformation ratio be K.

Resistive drop in secondary winding $= I_2 R_2$

Reactive drop in secondary winding $= I_2 X_2$

Resistive drop in primary winding $= I_1 R_1$

Reactive drop in primary winding $= I_1 X_1$

Referred To Secondary Side. Since transformation ratio is K, primary resistive and reactive drops as referred to secondary will be K times, *i.e.* $K I_1 R_1$ and $K I_1 X_1$ respectively. If I_1 is substituted equal to KI_2 then we have primary resistive and reactive drops referred to secondary equal to $K^2 I_2 R_1$ and $K^2 I_2 X_1$ respectively.

Total resistive drop in a transformer

$$= K^2 I_2 R_1 + I_2 R_2$$
$$= I_2 (K^2 R_1 + R_2) = I_2 R_{02}$$

Total reactive drop in a transformer

$$= K^2 I_2 X_1 + I_2 X_2$$
$$= I_2 (K^2 X_1 + X_2) = I_2 X_{02}$$

The term $(K^2 R_1 + R_2)$ and $(K^2 X_1 + X_2)$ represent the equivalent resistance and reactance respectively of the transformer referred to secondary and let these be represented by R_{02} and X_{02} respectively. Equivalent circuit referred to secondary has been shown in Fig. 12.13 (*a*).

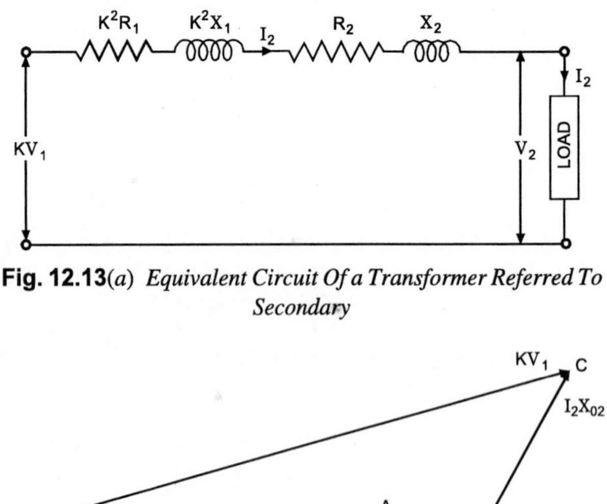

Fig. 12.13(*a*) *Equivalent Circuit Of a Transformer Referred To Secondary*

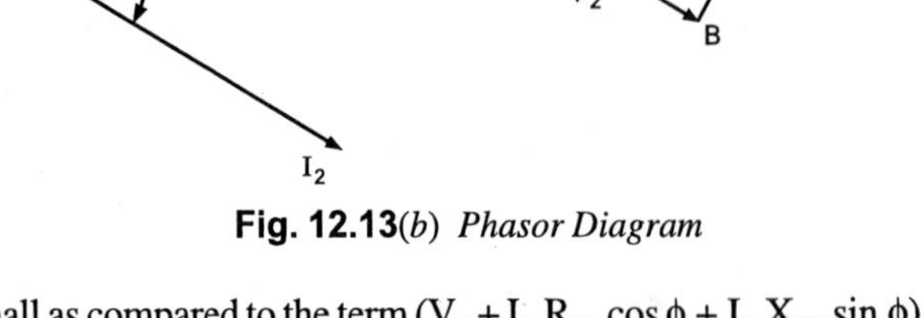

Fig. 12.13(*b*) *Phasor Diagram*

From phasor diagram [Fig. 12.13 (*b*)]

$$KV_1 = \sqrt{(V_2 + I_2 R_{02} \cos\phi + I_2 X_{02} \sin\phi)^2 + (I_2 X_{02} \cos\phi - I_2 R_{02} \sin\phi)^2}$$

where V_2 is secondary terminal voltage, I_2 is secondary current lagging behind the terminal voltage V_2 by ϕ.

Since term $(I_2 X_{02} \cos\phi - I_2 R_{02} \sin\phi)$ is very small as compared to the term $(V_2 + I_2 R_{02} \cos\phi + I_2 X_{02} \sin\phi)$, neglecting the former we have

$$K V_1 = V_2 + I_2 R_{02} \cos \phi + I_2 X_{02} \sin \phi$$

$$\text{or } V_2 = KV_1 - I_2 R_{02} \cos \phi - I_2 X_{02} \sin \phi \qquad ...(12.18)$$

where V_1 is applied voltage to primary winding.

If load is pure resistive, $\phi = 0$ and $V_2 = KV_1 - I_2 R_{02}$ ...(12.19)

If load is capacitive then ϕ should be taken as –ve hence we have

$$V_2 = KV_1 - I_2 R_{02} \cos \phi + I_2 X_{02} \sin \phi \qquad ...(12.20)$$

Referred To Primary Side. Secondary resistive drop referred to primary

$$= \frac{I_2 R_2}{K} = \frac{I_1 R_2}{K^2} \qquad \text{since } I_2 = \frac{I_1}{K}$$

Secondary reactive drop referred to primary $= \dfrac{I_2 X_2}{K} = \dfrac{I_1 X_2}{K^2}$

Total resistive drop in the transformer referred to primary

$$= I_1 R_1 + \frac{I_1 R_2}{K^2} = I_1 \left(R_1 + \frac{R_2}{K^2}\right) = I_1 R_{01}$$

Total reactive drop in the transformer referred to primary

$$= I_1 X_1 + \frac{I_1 X_2}{K^2} = I_1 \left(X_1 + \frac{X_2}{K^2}\right) = I_1 X_{01}$$

The terms $\left(R_1 + \dfrac{R_2}{K^2}\right)$ and $\left(X_1 + \dfrac{X_2}{K^2}\right)$ represent the total resistance and reactance of the transformer referred to primary respectively. Let these be represented by R_{01} and X_{01} respectively.

Equivalent circuit referred to primary is shown in Fig. 12.14 (*a*)

From phasor diagram shown in Fig. 12.14 (*b*) we have

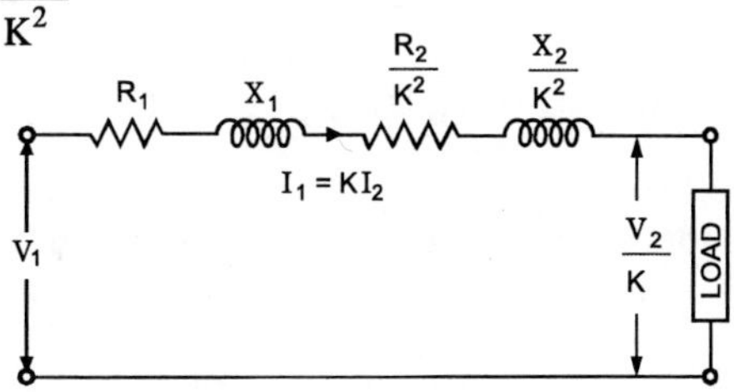

(a) Equivalent Circuit of a Transformer Referred To Primary

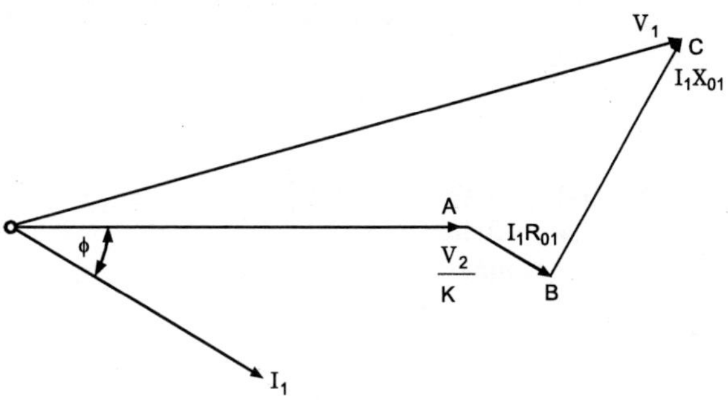

(b) Phasor Diagram

Fig. 12.14

$$V_1 = \sqrt{\left(\frac{V_2}{K} + I_1 R_{01} \cos \phi + I_1 X_{01} \sin \phi\right)^2 + (I_1 X_{01} \cos \phi - I_1 R_{01} \sin \phi)^2}$$

Since term $(I_1 X_{01} \cos \phi - I_1 R_{01} \sin \phi)$ is very small as compared to the term $\left(\dfrac{V_2}{K} + I_1 R_{01} \cos \phi + I_1 X_{01} \sin \phi\right)$, neglecting the former we have

$$V_1 = \frac{V_2}{K} + I_1 R_{01} \cos \phi + I_1 X_{01} \sin \phi \qquad ...(12.21)$$

If load is pure resistive *i.e.* $\phi = 0$ then

$$V_1 = \frac{V_2}{K} + I_1 R_{01} \qquad ...(12.22)$$

If load is capacitive then ϕ should be taken as –ve so we have

$$V_1 = \frac{V_2}{K} + I_1 R_{01} \cos \phi - I_1 X_{01} \sin \phi \qquad ...(12.23)$$

Example 12.15 **A 25 kVA, 2,200/220 V, 50-Hz, 1-phase transformer has following parameters:**

$$R_1 = 1.75\ \Omega,\ R_2 = 0.0045\ \Omega,\ X_1 = 2.6\ \Omega,\ X_2 = 0.0075\ \Omega.$$

Calculate: (*i*) Equivalent resistance referred to primary and secondary. (*ii*) Equivalent reactance referred to primary and secondary. [*U.P. Technical Univ. Electrical Engineering Odd Semester 2013-14*]

Solution: Transformation ratio, $K = \frac{V_2}{V_1} = \frac{220}{2{,}200} = 0.1$

(*i*) Equivalent resistance referred to primary, $R_{01} = R_1 + \frac{R_2}{K^2} = 1.75 + \frac{0.0045}{(0.1)^2} = 2.2\ \Omega$ **Ans.**

Equivalent resistance referred to secondary, $R_{02} = K^2R_1 + R_2 = (0.1)^2 \times 1.75 + 0.0045 = 0.022\ \Omega$ **Ans.**

(*ii*) Equivalent reactance referred to primary, $X_{01} = X_1 + \frac{X_2}{K^2} = 2.6 + \frac{0.0075}{(0.1)^2} = 3.35\ \Omega$ **Ans.**

Equivalent reactance referred to secondary, $X_{02} = K^2X_1 + X_2 = (0.1)^2 \times 2.6 + 0.0075 = 0.0335\ \Omega$ **Ans.**

Example 12.16 **A 5 kVA, 240/120 V, single-phase transformer supplies rated current to a load at 120 V. Determine the magnitude of load impedance as seen from input terminals of transformer. The load impedance is (5 + *j*15) Ω.** [*Pb. Technical Univ. Basic Electrical Engineering December 2002*]

Solution: Transformation ratio, $K = \frac{V_2}{V_1} = \frac{120}{240} = 1/2$

Load impedance referred to secondary, $Z_L = (5 + j15)\ \Omega$

Load impedance referred to primary, $Z_L = \frac{Z_L}{K^2} = \frac{5 + j15}{(½)^2} = (20 + j\,60)\ \Omega$ **Ans.**

Example 12.17 **A 10 kVA, 1-phase transformer with 2,000/400 V at no load, has resistance and leakage reactance of primary winding 5.5 Ω and 12 Ω respectively, the corresponding values of secondary winding being 0.2 Ω and 0.45 Ω. Determine the value of secondary voltage at full load, 0.8 pf lagging, when the primary applied voltage is 2,000 V.** [*U.P. Technical Univ. Electrical Engineering July 2002*]

Solution: Transformation ratio, $K = \frac{400}{2{,}000} = 0.2$

Equivalent resistance referred to secondary, $R_{02} = K^2 R_1 + R_2 = (0.2)^2 \times 5.5 + 0.2 = 0.42\ \Omega$

Equivalent reactance referred to secondary, $X_{02} = K^2 X_1 + X_2 = (0.2)^2 \times 12 + 0.45 = 0.93\ \Omega$

Secondary current, $I_2 = \frac{\text{kVA} \times 1{,}000}{V_2} = \frac{10 \times 1{,}000}{400} = 25\ \text{A}$

Power factor, $\cos\phi_2 = 0.8$ (lagging)

$\text{Sin}\,\phi_2 = \text{Sin}\cos^{-1}(0.8) = 0.6$

Secondary terminal voltage, $V_2 = kV_1 - I_2 R_{02} \cos\phi_2 - I_2 X_{02} \sin\phi_2$

$= 0.2 \times 2{,}000 - 25 \times 0.42 \times 0.8 - 25 \times 0.93 \times 0.6$

$= 400 - 8.4 - 13.95 = 377.65$ V **Ans.**

Example 12.18 **A transformer has a primary winding of 600 turns and a secondary winding of 150 turns. When the load current on the secondary is 60 A at 0.8 power factor lagging, the primary current is 20 A at 0.707 power factor lagging. Determine the no-load current of the transformer and its phase with respect to the voltage.** [*G.B. Technical Univ. Electrical Engineering Second Semester 2009-10*]

Solution: Turn ratio $K = \frac{N_2}{N_1} = \frac{150}{600} = \frac{1}{4}$

$$\text{Secondary current, } \mathbf{I}_2 = 60 \angle \cos^{-1} 0.8 = 60 \angle -36.87° \text{ A}$$

$$\text{Load current referred to primary, } \mathbf{I'}_1 = \text{KI}_2 = \frac{1}{4} \times 60 \angle -36.87° = 15 \angle -36.87° \text{ A}$$

$$\text{Primary current, } \mathbf{I}_1 = 20 \angle \cos^{-1} 0.707 = 20 \angle -45° \text{ A}$$

$$\text{No-load current, } \mathbf{I}_0 = \mathbf{I}_1 - \mathbf{I}'_1 \qquad \because \text{I}_1 = \text{I}_0 + \text{I}'_1$$

$$= 20 \angle -45° - 15 \angle -36.87°$$

$$= (14.14 - j\,14.14) - (12 - j\,9) = (2.2 - j\,5.14) \text{ A} = 5.59 \angle -66.83° \text{ A} \quad \textbf{Ans.}$$

Example 12.19 **A 100 kVA, 1100/220 V, 60 Hz transformer has a high-voltage winding resistance of 0.1 Ω and a leakage reactance of 0.3 Ω. The low-voltage winding resistance is 0.004 Ω and the leakage reactance is 0.012 Ω. The source is applied to high-voltage side:**

(*i*) **Find the equivalent winding resistance and reactance referred to high-voltage side and low-voltage side.**

(*ii*) **Find the equivalent resistance and reactance drops in volts and in per cent of the rated winding voltages expressed in terms of the primary quantities.**

(*iii*) **Calculate leakage impedances of the transformer referred to the primary and secondary sides.**

[*U.P. Technical Univ. Electrical Engineering Second Semester 2008-09*]

Solution: Transformation ratio, $K = \frac{V_2}{V_1} = \frac{220}{1{,}100} = \frac{1}{5}$

(*i*) Equivalent resistance referred to primary (*hv*),

$$R_{01} = R_1 + \frac{R_2}{K^2} = 0.1 + \frac{0.004}{(1/5)^2} = 0.1 + 0.004 \times 25 = 0.1 + 0.1 = 0.2\ \Omega \quad \textbf{Ans.}$$

Equivalent reactance referred to primary (*hv*)

$$X_{01} = X_1 + \frac{X_2}{K^2} = 0.3 + \frac{0.012}{(1/5)^2} = 0.3 + 0.012 \times 25 = 0.3 + 0.3 = 0.6\ \Omega \quad \textbf{Ans.}$$

Equivalent resistance referred to secondary (*lv*)

$$R_{02} = K^2 R_1 + R_2 = \left(\frac{1}{5}\right)^2 \times 0.1 + 0.004 = \frac{0.1}{25} + 0.004 = 0.004 + 0.004 = 0.008\ \Omega \quad \textbf{Ans.}$$

Equivalent reactance referred to secondary (*lv*)

$$X_{02} = K^2 X_1 + R_2 = \left(\frac{1}{5}\right)^2 \times 0.3 + 0.012 = \frac{0.3}{25} + 0.012 = 0.012 + 0.012 = 0.024\ \Omega \quad \textbf{Ans.}$$

$$(ii)\ \text{Full-load primary current, I} = \frac{\text{Output in kVA} \times 1{,}000}{\text{Primary voltage}} = \frac{100 \times 1{,}000}{1{,}100} = \frac{1{,}000}{11} \text{ A}$$

$$\text{Primary rated voltage, } V_1 = 1{,}100 \text{ V}$$

$$\text{Equivalent resistance drop referred to primary} = I_1 R_{01} = \frac{1{,}000}{11} \times 0.2 = \frac{200}{11} = 18.18 \text{ V} \quad \textbf{Ans.}$$

$$\text{Equivalent reactance drop referred to primary} = I_1 X_{01} = \frac{1{,}000}{11} \times 0.6 = \frac{600}{11} = 54.54 \text{ V} \quad \textbf{Ans.}$$

$$\text{Percentage resistance drop} = \frac{I_1 R_{01}}{V_1} = \frac{18.18}{1{,}100} \times 100 = 1.653\% \quad \textbf{Ans.}$$

$$\text{Percentage reactance drop} = \frac{I_1 X_{01}}{V_1} = \frac{54.54}{1{,}100} \times 100 = 4.96\% \quad \textbf{Ans.}$$

(*iii*) Leakage impedance of transformer referred to primary (*hv*)

$$Z_{01} = \sqrt{(R_{01})^2 + (X_{01})^2} = \sqrt{0.2^2 + 0.6^2} = 0.6325\ \Omega \quad \textbf{Ans.}$$

Leakage impedance of transformer referred to secondary (*lv*)

$$Z_{02} = \sqrt{(R_{02})^2 + (X_{02})^2} = \sqrt{(0.008)^2 + (0.024)^2} = 0.0253\ \Omega \quad \textbf{Ans.}$$

12.13. EQUIVALENT CIRCUIT OF A TRANSFORMER

The equivalent circuit of any device can be quite helpful in predetermination of the behaviour of the device under various conditions of operation and it can be drawn if the equations describing its behaviour are known. If any electrical device is to be analysed and investigated further for suitable modifications, its appropriate equivalent circuit is necessary. The equivalent circuit for electromagnetic devices consists of a combination of resistances, inductances, capacitances, voltage etc. Such an equivalent circuit (or circuit model) can, therefore, be analysed and studied easily by the direct application of electric circuit theory.

As stated above, equivalent circuit is simply a circuit representation of the equa-tions describing the perfor-mance of the device. Eqs. (12.10) and (12.11) describe the behaviour of the trans-former under load and are helpful in arriving at the transformer equivalent circuit.

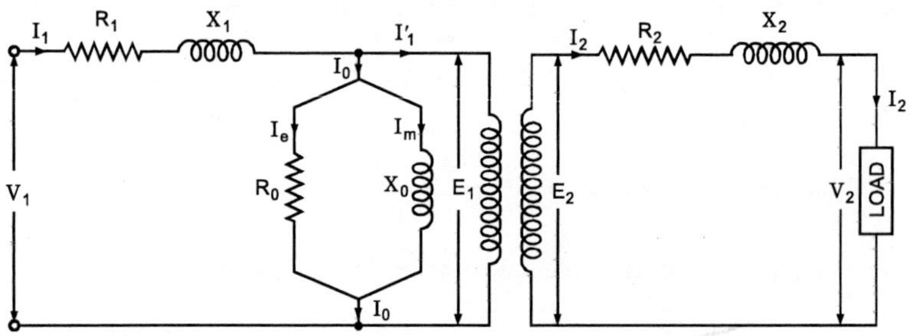

Fig. 12.15 *Equivalent Circuit of a Transformer*

Equivalent circuit of a transformer having transformation ratio $K = E_2/E_1$ is shown in Fig. 12.15.

The induced emf in primary winding E_1 is primary applied voltage V_1 less primary voltage drop. This voltage causes iron loss current I_e and magnetising current I_m and we can, therefore, represent these two components of no-load current by the current drawn by a non-inductive resistance R_0 and pure reactance X_0 having the voltage E_1 or (V_1—primary voltage drop) applied across them, as shown in Fig. 12.15.

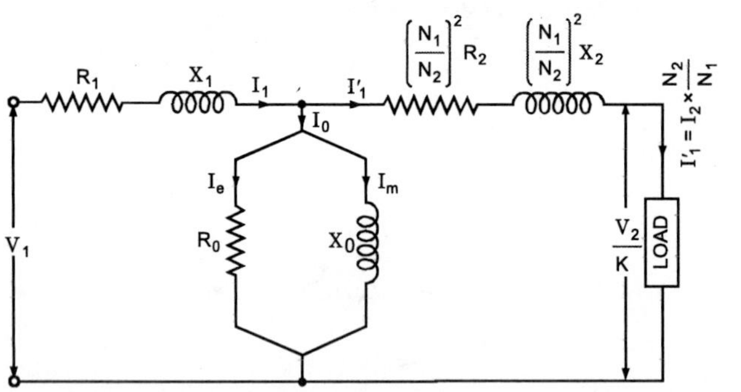

Fig. 12.16 *Equivalent Circuit of a Transformer With All Secondary Impedances Transferred To Primary Side*

$$\text{Secondary current, } I_2 = \frac{I'_1}{K} = \frac{I_1 - I_0}{K}$$

Terminal voltage V_2 across load is induced emf E_2 in secondary winding less voltage drop in secondary winding.

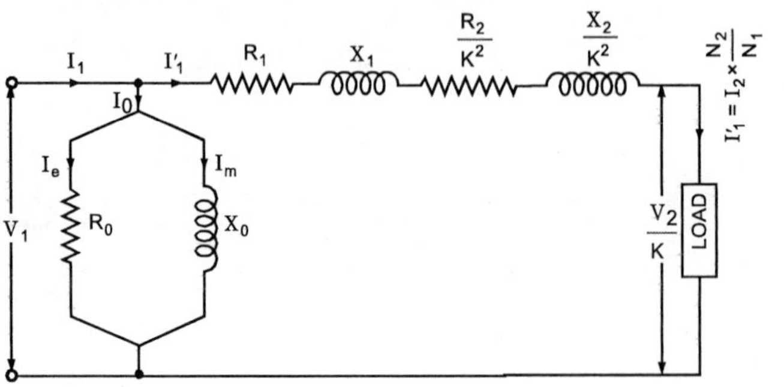

Fig. 12.17 *Approximate Equivalent Circuit of a Transformer Referred To Primary Side*

The equivalent circuit can be simplified by transferring the voltage, current and impedance to the primary side. After transferring the secondary voltage, current and impedance to primary side equivalent circuit is reduced to that shown in Fig. 12.16.

The equivalent circuit diagram can further be simplified by transferring the resistance R_0 and reactance X_0 towards left end, as shown in Fig. 12.17. The error introduced by doing so is very small and can be neglected.

No-load current I_0 is hardly 3 to 5 per cent of the full-load rated current, the parallel branch consisting of resistance R_0 and reactance X_0 can be omitted without introducing any appreciable error in the behaviour of the transformer under loaded condition. Such a circuit is shown in Fig. 12.14 (*a*). The equivalent circuit referred to secondary side (neglecting no-load current I_0) is illustrated in Fig. 12.13 (*a*).

Example 12.20 **The ohmic values of the circuit parameters of a transformer, having a turns ratio of 5, are $R_1 = 0.5\ \Omega$, $R_2 = 0.021\Omega$, $X_1 = 3.2\ \Omega$, $X_2 = 0.12\ \Omega$, $R_c = 350\ \Omega$, referred to the primary and $X_m = 98\ \Omega$ referred to primary. Draw the approximate equivalent circuit of the transformer referred to secondary. Show the numerical values of the circuit parameters.** [*U.P. Technical Univ. Electrical Engineering First Semester 2008-09*]

Solution:

$R_1 = 0.5\ \Omega$

$R_2 = 0.021\ \Omega$

$X_1 = 3.2\ \Omega$

$X_2 = 0.12\ \Omega$

$R_c = 350\ \Omega$ referred to primary

$X_m = 98\ \Omega$ referred to primary

Turn ratio,

$$\frac{N_1}{N_2} = 5$$

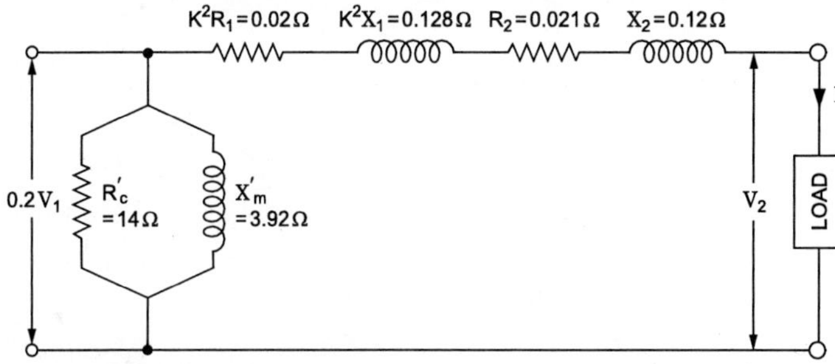

Fig. 12.18 *Phasor Diagram Referred To Secondary*

$$\text{Transformation ratio, K} = \frac{N_2}{N_1} = \frac{1}{5} = 0.2$$

$$\text{Resistance of primary winding referred to secondary} = K^2\ R_1 = 0.2^2 \times 0.5 = 0.02\ \Omega$$

$$\text{Reactance of primary winding referred to secondary} = K^2\ X_1 = 0.2^2 \times 3.2 = 0.128\ \Omega$$

$$R_c \text{ referred to secondary} = K^2\ R_c = 0.2^2 \times 350 = 14\Omega$$

$$X_m \text{ referred to secondary} = 0.2^2 \times 98 = 3.92\ \Omega$$

Approximate equivalent circuit of transformer referred to secondary is shown in Fig. 12.18.

Example 12.21 **A 25 kVA, 2200/220 V, 50 Hz single-phase transformer has the following parameters.**

$r_1 = 1.2\ \Omega$; $x_1 = 6.0\ \Omega$; $r_c = 1{,}000\ \Omega$; $x_m = 300\ \Omega$ hv side

$r_2 = 0.03\ \Omega$; $x_2 = 0.06\ \Omega$ (*lv* side)

Draw the equivalent circuit referred to lv side for a 0.8 pf lagging load being supplied at 220 V.

[*U.P. Technical Univ. Electrical Engineering June 2001*]

Solution:

$$\text{Rated secondary current, } I_2 = \frac{\text{Rated kVA} \times 1{,}000}{\text{Rated secondary voltage}} = \frac{25 \times 1{,}000}{220} = 113.64\ \text{A}$$

$$\text{Secondary resistance, } r_2 = 0.03\ \Omega$$

$$\text{Secondary leakage reactance, } x_2 = 0.06\ \Omega$$

$$\text{Power factor } \cos\phi = 0.8 \text{ ; } \sin\phi = 0.6$$

$$\text{Secondary induced emf, } E_2 = \sqrt{(V_2 + I_2 R_2 \cos\phi + I_2 X_2 \sin\phi)^2 + (I_2 X_2 \cos\phi - I_2 R_2 \sin\phi)^2}$$

$$= \sqrt{\begin{array}{l}(220 + 113.64 \times 0.03 \times 0.8 + 113.64 \times 0.06 \times 0.6)^2 \\ \quad + (113.64 \times 0.06 \times 0.8 - 113.64 \times 0.03 \times 0.6)^2\end{array}}$$

$$= \sqrt{(226.82)^2 + (3.41)^2} = 226.85\ \text{V}$$

$$\mathbf{E}_2 = (226.82 + j\,3.41)\text{ V} = 226.85\ \angle 0.86^\circ\text{ V}$$

$$\mathbf{I}_2 = 113.64\,(0.8 - j\,0.6)\text{ A} = (90.91 - j\,68.2)\text{ A}$$

$$\text{Transformation ratio, K} = \frac{E_2}{E_1} = \frac{220}{2{,}200} = 0.1$$

$$\text{Shunt branch resistance referred to } lv \text{ side } r_c' = r_c \times K^2 = 1{,}000 \times (0.1)^2 = 10\ \Omega$$

$$\text{Shunt branch reactance referred to } lv \text{ side } x_m' = x_m \times K^2 = 300 \times (0.1)^2 = 3\ \Omega$$

$$\text{Shunt admittance referred to } lv \text{ side, } Y_0 = \frac{1}{r_c} - \frac{j}{x_m} = \frac{1}{10} - \frac{j}{3} = (0.1 - j\,0.333)\,\mho = 0.348\ \angle -73.3^\circ\ \mho$$

$$\text{Exciting current, } I_0 = Y_0\,E_2 = 0.348\ \angle -73.3^\circ \times 226.85\ \angle 0.86^\circ$$

$$= 78.94\ \angle -72.44^\circ = (23.82 - j\,75.26)\text{ A}$$

$$\text{Equivalent secondary current, } \mathbf{I}_2' = \mathbf{I}_2 + \mathbf{I}_0 = (90.91 - j\,68.2) + (23.82 - j\,75.26)$$

$$= (114.73 - j\,143.46) = 183.7\ \angle -51.35^\circ\text{ A}$$

$$\mathbf{KV}_1 = \mathbf{E}_2 + K^2\,\mathbf{I}_2'\,(R_1 + j\,X_1)$$

$$= (226.82 + j\,3.41) + (114.73 - j\,143.46)\,(0.1)^2\,(1.2 + j\,6) = 237\text{ V}$$

Equivalent circuit is shown in Fig. 12.19.

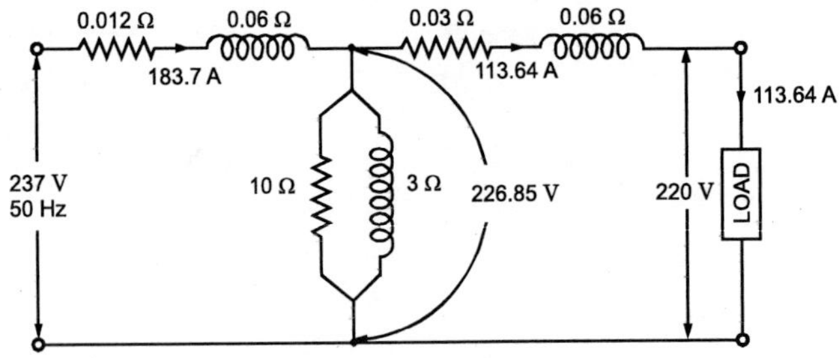

Fig. 12.19

12.14. VOLTAGE REGULATION

The way in which the secondary terminal voltage varies with the load depends upon load current, the internal impedance and the load power factor. The change in secondary terminal voltage from no load to full load, with primary voltage and frequency held constant, is termed the *inherent regulation*.

The figure of merit which determines the voltage characteristic of a transformer is termed as *voltage regulation*. It is usually expressed as a percentage of the rated terminal (secondary) voltage. It is defined as the change in magnitude of terminal (secondary) voltage when full load (rated load) of specified power factor supplied at rated voltage is thrown off (reduced to no-load) with primary voltage and frequency held constant, as percentage of the rated terminal voltage.

If V_2 is the secondary (terminal) voltage at full load at specified power factor and E_2 is the secondary (terminal) voltage at no load, then percentage regulation is given as

$$\text{Voltage regulation} = \frac{E_2 - V_2}{\text{Secondary rated voltage}} \times 100$$

As per IS, the secondary rated voltage of a transformer is equal to the secondary voltage at no load i.e. E_2

$$\text{So percentage regulation} = \frac{E_2 - V_2}{E_2} \times 100 = \frac{\text{Voltage drop in transformer at full load}}{\text{No-load rated secondary voltage, } E_2} \times 100$$

$$= \frac{I_2\,R_{02}\cos\phi + I_2\,X_{02}\sin\phi}{E_2} \times 100 \qquad ...(12.24)$$

When the power factor is leading, the percentage regulation is given as

$$\text{Percentage regulation} = \frac{I_2 R_{02} \cos\phi - I_2 X_{02} \sin\phi}{E_2} \times 100 \quad ...(12.25)$$

The percentage regulation can be given as

$$\% \text{ regulation} = \frac{I_2 R_{02} \cos\phi \pm I_2 X_{02} \sin\phi}{E_2} 100$$

+ ve sign for lagging power factor

and – ve sign for leading power factor

$$= \left(\frac{I_2 R_{02}}{E_2} \cos\phi \pm \frac{I_2 X_{02}}{E_2} \sin\phi\right) \times 100$$

$$= (\text{PU resistance} \times \cos\phi \pm \text{PU reactance} \sin\phi) \times 100 \quad ...(12.26)$$

Voltage regulation of a transformer, on an average, is about 4 per cent.

From the consumer's view point, voltage regulation due to variations in load is undesirable and should be kept smallest possible.

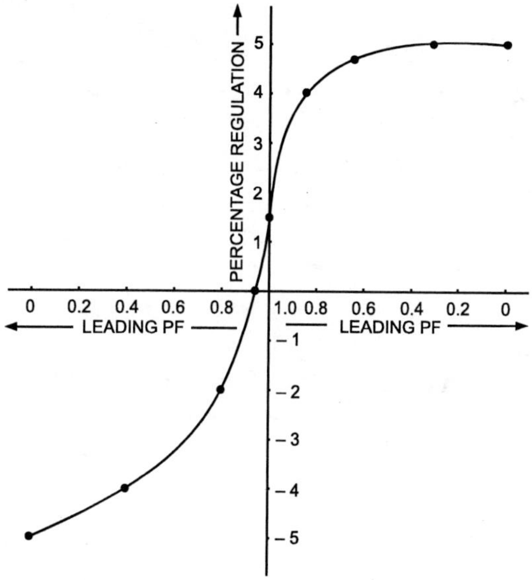

Percentage Regulation Vs Power Factor
(R = 0.02 PU, X = 0.06 PU)
Fig. 12.20

12.14.1 Condition For Zero Regulation.

$$\text{Regulation} = \frac{I_2 R_{02} \cos\phi + I_2 X_{02} \sin\phi}{E_2}$$

Regulation will be zero if the numerator will be equal to zero

$$\text{or } I_2 R_{02} \cos\phi + I_2 X_{02} \sin\phi = 0$$

$$\text{or } \tan\phi = \frac{-R_{02}}{X_{02}} \quad ...(12.27)$$

The –ve sign indicates that zero regulation occurs at a leading power factor.

Example 12.22 **Calculate the regulation of a transformer in which ohmic drop is 1% and reactance drop 5 % of the voltage at the following power factors (*i*) 0.8 lagging (*ii*) 0.8 leading.**

Solution:

$$\text{Per unit resistance} = \text{Per unit resistance drop} = \frac{1}{100} = 0.01$$

$$\text{Per unit reactance} = \text{Per unit reactance drop} = \frac{5}{100} = 0.05$$

(*i*) When power factor, $\cos\phi = 0.8$ lagging; $\sin\phi = 0.6$

Percentage voltage regulation $= (\text{PU resistance} \times \cos\phi + \text{PU reactance} \times \sin\phi) \times 100$

$= (0.01 \times 0.8 + 0.05 \times 0.6) \times 100 = 3.8\%$ **Ans.**

(*ii*) When power factor, $\cos\phi = 0.8$ (leading); $\sin\phi = -0.6$

and Percentage voltage regulation $= [0.01 \times 0.8 + 0.05 \times (-0.6)] \times 100 = -2.2\%$ **Ans.**

Example 12.23 **Define voltage regulation of a transformer. A transformer has copper loss of 1.5 % and reactance drop of 3.5 % when tested at full load. Calculate its full-load regulation at :**

(i) 0.8 pf lagging, **(ii) 0.8 pf leading.**

Solution:

$$\text{Per unit resistance} = \frac{\text{Percentage copper loss}}{100} = \frac{1.5}{100} = 0.015$$

$$\text{Per unit reactance} = \text{Per unit reactance drop} = \frac{3.5}{100} = 0.035$$

(*i*) When power factor, $\cos \phi$ = 0.8 lagging; $\sin \phi = 0.6$

Percentage voltage regulation = $(0.015 \times 0.8 + 0.035 \times 0.6) \times 100 = 3.3\%$ **Ans.**

(*ii*) When power factor, $\cos \phi$ = 0.8 leading; $\sin \phi = -0.6$

Percentage voltage regulation = $[0.015 \times 0.8 + 0.035\,(-0.6)] \times 100 = -0.9\%$ **Ans.**

12.15. TRANSFORMER LOSSES

The transformer is a static machine and, therefore, there are no friction or windage losses. The various power losses occuring in a transformer are enumerated below:

1. Iron or Core Losses. Iron loss is caused by the alternating flux in the core and consists of hysteresis and eddy current losses.

(*a*) Hysteresis Loss. The core of a transformer is subjected to an alternating magnetizing force and for each cycle of emf a hysteresis loop is traced out. The hysteresis loss per second is given by the equation

Hysteresis loss, $P_h = \eta'(B_{max})^x f\, v$ joules per second or watts ...(12.28) [Refer to Eq (11.7)]

where f is the supply frequency in Hz, v is the volume of core in cubic metres, η' is the hysteresis coefficient, B_{max} is peak value of flux density in the core and x lies between 1.5 and 2.5 depending upon the material and is often taken as 1.6.

(*b*) Eddy Current Loss. If the magnetic circuit is made up of iron and if the flux in the circuit is variable, currents will be induced by induction in the iron circuit itself. All such currents are known as *eddy currents.*

Eddy currents result in a loss of power, with consequent heating of the material.

The eddy current loss is given by equation

$P_e = K_e\,(B_{max})^2 f^2 t^2$ v watts or joules per second ...(12.29) [Refer to Eq (11.23)]

From the above expression for eddy current loss in a thin sheet it is obvious that eddy current loss varies (*i*) as the square of maximum flux density (*ii*) as the square of the frequency and (*iii*) as the square of thickness of laminations.

The hysteresis and eddy current losses depend upon the maximum flux density in the core and supply frequency. Since it has been determined that the mutual flux varies somewhat with the load (its variation being 1 to 3 % from no load to full load), the core losses will vary somewhat with the load and its power factor. It may be emphasized here that *core losses are assumed to remain constant from no load to full load,* the variations in losses from no load to full load being very small and negligible.

These losses are determined from the open-circuit test (refer to Art 12.16.3)

The input to the transformer with rated voltage applied to the primary and secondary open circuited is equal to the core loss.

These losses are minimized by using steel of high silicon content for the core and by using very thin laminations (0.3 mm to 0.5 mm) insulated from each other either by insulating varnish or by layer of papers.

2. Copper or Ohmic Losses. These losses occur due to ohmic resistance of the transformer windings. If I_1 and I_2 are the primary and secondary currents respectively and R_1 and R_2 are the respective resistances of primary and secondary windings then copper losses occuring in primary and secondary windings will be $I_1^2 R_1$ and $I_2^2 R_2$ respectively. So total copper losses will be $(I_1^2 R_1 + I_2^2 R_2)$. These losses vary as the square of the load current or kVA. For example if the copper losses at full load are P_c then copper losses at one-half or one-third of full load will be $\left(\frac{1}{2}\right)^2 P_c$ or $\left(\frac{1}{3}\right)^2 P_c$ i.e. $\frac{P_c}{4}$ or $\frac{P_c}{9}$ respectively.

Copper losses are determined on the basis of constant equivalent resistance R_{eq} determined from the short-circuit test (refer to Art 12.16.4) and then corrected to 75 °C (since the standard operating temperature of electrical machines is taken 75 °C).

12.16. TESTING OF TRANSFORMERS

The main difficulties encountered in testing of large power transformers by direct loading are (*i*) wastage of large amount of energy and (*ii*) a stupendous (impossible for large transformers) task of arranging a load large enough for direct loading. The performance characteristics of a transformer can be conveniently computed from the knowledge of its equivalent circuit parameters which, in turn, may be determined by conducting simple tests called the *open-circuit or no-load test* and *short-circuit or impedance test* involving very little power cosumption (power needed to supply the losses incurred).

Other tests to be conducted on transformers are *polarity test* and *voltage ratio test.*

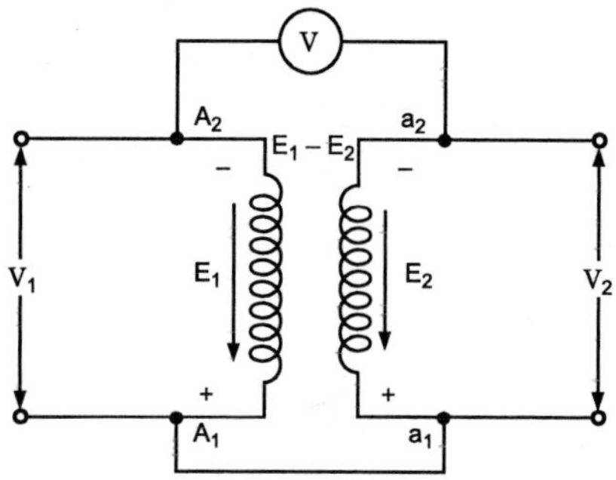

Fig. 12.21 (*a*) *Polarity Test*

12.16.1 Polarity Test. On the primary side of a two winding transformer, one terminal is positive with respect to the other one at any instant. At the same instant, one terminal of the secondary winding is positive with respect to the other one. *Polarity test* is performed to determine the terminals having the same instantaneous polarity (assuming that the terminals are not marked). The relative polarities of the primary and secondary terminals at any instant must be known for connecting windings of the same transformer in parallel, or series, or for interconnecting two or more transformers in parallel, or for connecting single phase transformers for polyphase transformation of voltages.

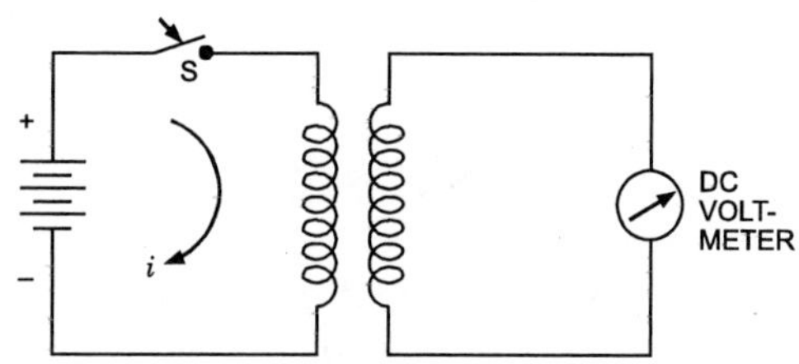

Fig. 12.21 (*b*)

For determination of relative polarity of the two windings of a transformer the two windings are connected in series across a voltmeter, while one of the windings (preferably hv winding) is excited from a suitable ac voltage source, as illustrated in Fig. 12.21 (*a*). If the polarities of the windings are as marked on the diagram i.e. when the transformer has a subtractive polarity, the voltmeter will read the difference of E_1 and E_2. If the voltmeter reads $E_1 + E_2$, the polarity markings of one of the windings must be interchanged.

In subtractive polarity, the voltage between A_1 and a_1 (or A_2 and a_2) is reduced. The leads connected to these terminals and the two windings are, therefore, not subjected to high voltage stress. On the other hand in additive polarity the two windings and leads connected to A_1, A_2, a_1 and a_2 are subjected to high voltage stresses. This is the reason that subtractive polarity is preferred over additive polarity.

The above test may not be convenient for determination of relative polarities of the terminals of the two windings of a transformer in the field. Polarity test in the field can be conveniently carried out by using a dc battery, a switch and a dc (permanent mangnet type) voltmeter. The circuit diagram is shown in Fig. 12.21 (*b*). As the switch on the primary side is closed, the primary current increases and so do the flux linkages of both the windings, inducing emfs in them. The positive polarity of this induced emf in the primary is at the end to which the battery is connected (according to Lenz's law). The end of secondary which simultaneously acquires positive polarity, as determined by the dc voltmeter is the similar polarity end. The reverse happens on opening of the switch *i.e.* the similar polarity end is that end which acquires negative potential.

12.16.2. Voltage Ratio Test. The true ratio is based on turn ratio, as explained in Art 12.6. If the secondary and primary voltages are measured on no load, their ratio is very nearly equal to the true value. Measurement of primary and secondary currents in short-circuit test also gives fairly accurate results $\left(\text{Voltage ratio } \frac{V_2}{V_1} = \frac{I_1}{I_2}\right)$, especially if the transformer has little leakage flux and low core reluctance.

12.16.3. Open-Circuit Test (or No-Load Test). The purpose of this test is to determine the core (or iron or excitation) loss, P_i and no-load current I_0 and thereby the shunt branch parameters R_0 and X_0 of the equivalent circuit.

In this test, one of the windings (usually high voltage winding) is kept open circuited and the rated voltage at rated frequency is applied to the other winding, as shown Fig. 12.22. No doubt, the core loss will be the same whether the measurements are made on lv winding or hv winding so long as the rated voltage of that winding is applied to it but in case the measurements are made on hv winding, the voltage required to be applied would often be inconveniently large while the current I_0 would be inconveniently small.

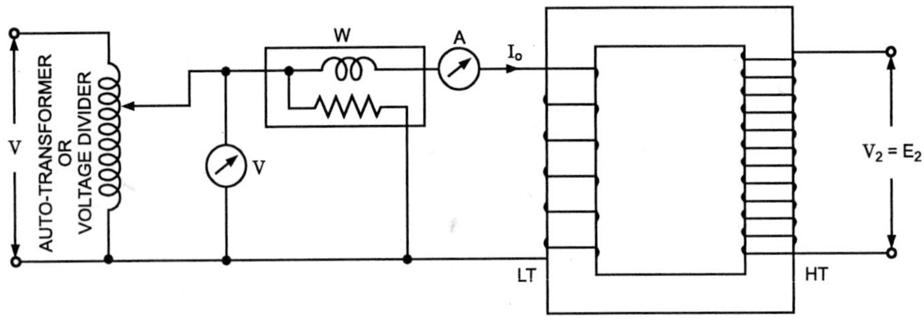

Fig. 12.22 *Circuit Diagram For Open-Circuit Test*

Either an auto-transformer or a voltage divider (VD) is used for varying the voltage applied to the low-voltage winding. Ammeter A and wattmeter W are connected to measure no-load current I_0 and input power W_0. Voltmeter V is connected to measure the applied voltage.

Since no current flows in the open-circuited secondary, the current in the primary will be merely that necessary to magnetize the core at normal voltage. Moreover, this magnetising current is a very small fraction of the full-load current (usually 3 to 10 % of full-load current) and may be neglected as far as the copper loss is concerned consequently, the test gives core loss alone practically.

With normal voltage applied to the primary, normal flux will be set up in the core and, therefore, normal iron (or core) loss will occur which are recorded by a wattmeter W.

The open-circuit test gives enough data to compute the equivalent circuit constants R_0, X_0, no–load power factor cos ϕ_0, no-load current I_0 and no-load power loss (iron loss) of a transformer.

$$\text{Iron loss, } P_i = \text{Input power on no load} = W_0 \text{ watts (say)}$$

$$\text{No-load current} = I_0 \text{ amperes}$$

$$\text{Applied voltage to primary} = V_1 \text{ volts}$$

$$\text{Angle of lag, } \phi_0 = \text{Cos}^{-1} \frac{W_0}{V_1 I_0} \qquad \text{...(12.30)}$$

$$\text{No-load current energy component, } I_e = I_0 \cos \phi_0 = \frac{W_0}{V_1} \qquad \text{...(12.31)}$$

$$\text{No-load current magnetizing component, } I_m = \sqrt{I_0^2 - I_e^2} \qquad \text{...(12.32)}$$

$$\text{Equivalent circuit parameter, } R_0 = \frac{V_1}{I_e} = \frac{V_1^2}{W_0} \qquad \text{...(12.33)}$$

$$\text{Equivalent circuit parameter, } X_0 = \frac{V_1}{I_m} = \frac{V_1}{\sqrt{I_0^2 - I_e^2}} \qquad \text{...(12.34)}$$

Note. 1. Since no-load current I_0 is very small, pressure coils of wattmeter and the voltmeter should be connected such that the currents drawn by them do not flow through the current coils of the wattmeter and ammeter.

2. Since power factor at no load is quite low (in the range of 0.1–0.2 lag), a low power factor wattmeter should be used to ensure accurate measurements.
3. The error due to power loss in ammeter can be eliminated by short circuiting the ammeter while reading wattmeter.
4. Sometimes a high resistance voltmeter is connected across the secondary to indicate the emf induced in the secondary (hv winding). This helps in determination of transformation ratio K.
5. It must, however, be remembered that in making this test, hv side is hot and, therefore, its terminals must be properly insulated.

12.16.4. Short-Circuit Test (or Impedance Test). The purpose of this test is to determine full-load copper loss and equivalent resistance and equivalent reactance referred to metering side.

In this test, the terminals of secondary winding (usually of low-voltage winding*) are short circuited by a thick wire or strip or through an ammeter (which may serve the additional purpose of indicating secondary rated load current) and variable low voltage is applied to the primary through an auto-transformer or potential divider, as shown in Fig. 12.23. The transformer now becomes equivalent to a coil having an impedance equal to impedance of both the windings.

The applied voltage, V_s to the primary is gradually increased till the ammeter A indicates the full-load (rated) current of the metering side. Since applied voltage is very low (5–8% of the rated voltage), flux linking with the core is very small and, therefore, iron losses are so small that these can be neglected. Thus the power input (reading of wattmeter W) gives total copper loss at rated load, output being nil. Let the readings of voltmeter, ammeter and wattmeter be V_s, I_s and W_s respectively.

$$\text{Full-load copper loss, } P_c = I_s^2 R_{eq} = W_s \quad ...(12.35)$$

$$\text{Equivalent resistance, } R_{eq} = \frac{W_s}{I_s^2} \quad ...(12.36)$$

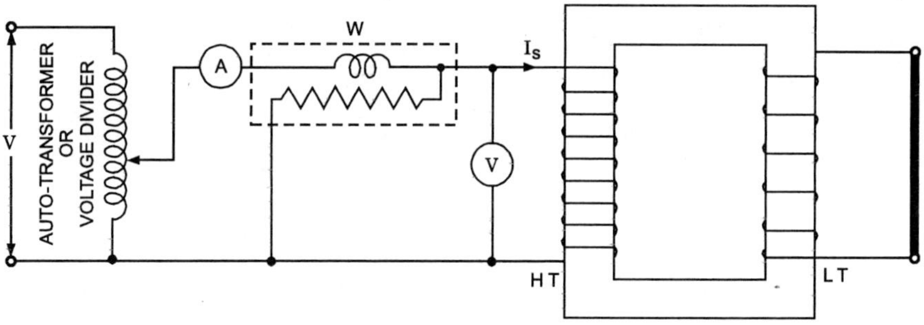

Fig. 12.23 *Circuit Diagram For Short-Circuit Test*

* Voltage required for the short-circuit test is about 5 per cent of the rated value. For a 200 kVA, 2200/220 V transformer, test on high voltage side would need the voltage (to be applied) of $2{,}200 \times \frac{5}{100}$ i.e. 110 V (which is standard voltage for instrument coils) and a current of $\frac{200 \times 1{,}000}{2{,}200} = 91\text{A}$.

If the test is conducted on low voltage side of the above transformer, the voltage needed would be $220 \times \frac{5}{100} = 11$ V (very small) and the current would be $\frac{200 \times 1{,}000}{220} = 910$ A (very high). At this low voltage, high precision would not be readily obtainable with ordinary instruments.

Thus we see that the if the measurements were made on low voltage side, the voltage needed would be inconveniently low, while the current would often be inconveniently large.

$$\text{Equivalent impedance, } Z_{eq} = \frac{V_s}{I_s} \quad ...(12.37)$$

$$\text{Equivalent reactance, } X_{eq} = \sqrt{Z_{eq}^2 - R_{eq}^2} \quad ...(12.38)$$

The above values are referred to the metering side (high voltage side in above case). If desired, the values could be easily determined referred to the other side, as discussed in Art 12.12.

Example 12.24 **Single-phase transformer gave following test results: Rating 100 kVA, 11 kV/220 V, 50 Hz**

OC test : 220 V, 45 A, 2 kW

SC test : 500 V, 9.09 A, 3 kW

Determine equivalent circuit parameters referred to *lv* side.

[*U.P. Technical Univ. Electrical Engineering Second Semester 2005-06*]

Solution: Assuming low-voltage winding as primary

$$\text{Energy component of no-load current, } I_e = \frac{W_0}{V} = \frac{2 \times 1{,}000}{220} = 9.091 \text{ A}$$

$$\text{Magnetising component of no-load current, } I_m = \sqrt{I_0^2 - I_e^2} = \sqrt{45^2 - 9.091^2} = 44.07 \text{ A}$$

$$\text{Equivalent circuit parameter, } R_0 = \frac{V_1}{I_e} = \frac{220}{9.091} = 24.2\ \Omega \textbf{ Ans.}$$

$$\text{Equivalent circuit parameter, } X_0 = \frac{V_1}{I_m} = \frac{220}{44.07} = 4.992\ \Omega \textbf{ Ans.}$$

$$\text{Equivalent impedance referred to hv side, } Z_{02} = \frac{V_s}{I_s} = \frac{500}{9.09} = 55\ \Omega$$

$$\text{Equivalent resistance referred to hv side, } R_{02} = \frac{W_s}{I_s^2} = \frac{3 \times 1{,}000}{(9.09)^2} = 36.3\ \Omega$$

$$\text{Equivalent reactance referred to hv side, } X_{02} = \sqrt{(Z_{02})^2 - (R_{02})^2} = \sqrt{55^2 - 36.3^2} = 41.31\ \Omega$$

$$\text{Transformation ratio, } K = \frac{11{,}000}{220} = 50$$

$$\text{Equivalent resistance referred to lv side} = \frac{R_{02}}{K^2} = \frac{36.3}{50^2} = 0.0145\ \Omega \textbf{ Ans.}$$

$$\text{Equivalent reactance referred to lv side} = \frac{X_{02}}{K^2} = \frac{41.31}{50^2} = 0.0165\ \Omega \textbf{ Ans.}$$

Example 12.25 **A 200/100 V, 50 Hz transformer has an impedance of (0.3 + j 0.8) ohm in the 200 V winding and impedance of (0.1 + j 0.25) ohm in the 100 V winding. What are the currents on the high and low voltage sides, if a short circuit occurs on the 100 V side with 200 V applied to the hv side.**

[*U.P. Technical Univ. Electrical Engineering September 2001*]

Solution:

$$\text{Transformation ratio, } K = \frac{V_2}{V_1} = \frac{100}{200} = 0.5$$

$$\text{Equivalent impedance referred to lv side, } Z_{02} = K^2 \mathbf{Z}_1 + \mathbf{Z}_2 = 0.5^2(0.3 + j\,0.8) + (0.1 + j\,0.25)$$

$$= (0.175 + j\,0.45) = 0.483 \angle 68.8^\circ\ \Omega$$

$$\text{Short-circuit current on 100 V side, } I_2 = \frac{V_2}{Z_{02}} = \frac{100}{0.483} = 207 \text{ A } \textbf{Ans.}$$

$$\text{Short-circuit current on 200 V side, } I_1 = K\,I_2 = 0.5 \times 207 = 103.5 \text{ A } \textbf{Ans.}$$

Example 12.26 **A 20 kVA, 2000 V/200 V, single-phase, 50 Hz transformer has a primary resistance of 1.5 ohm and reactance of 2 ohm. The secondary resistance and reactance are 0.015 ohm and 0.02 ohm respectively. The no load current of transformer is 1 A at 0.2 power factor. Determine:**

(*i*) equivalent resistance, reactance and impedance referred to primary

(*ii*) supply current **(*iii*) total copper loss.**

Draw approximate equivalent circuit.

[*G.B. Technical Univ. Electrical Engineering First Semester 2009-10*]

Solution: Transformation ratio, $K = \frac{200}{2,000} = 0.1$

(*i*) Equivalent resistance referred to primary, $R_{01} = R_1 + \frac{R_2}{K^2} = 1.5 + \frac{0.015}{(0.1)^2} = 3\Omega$ **Ans.**

Equivalent reactance referred to primary, $X_{01} = X_1 + \frac{X_2}{K^2} = 2 + \frac{0.02}{(0.1)^2} = 4\Omega$ **Ans.**

Equivalent impedance referred to primary, $Z_{01} = \sqrt{R_{01}^2 + X_d^2} = \sqrt{3^2 + 4^2} = 5\Omega$ **Ans.**

(*ii*) Full-load primary current, $I_1 = \frac{kVA \times 1,000}{V} = \frac{20 \times 1,000}{2,000} = 10\text{ A}$ **Ans.**

(*iii*) Total copper loss in transformer $= I_1^2 R_{01} = 10^2 \times 3 = 300\text{ W}$ **Ans.**

No-load primary current, $I_0 = 1\text{ A}$

No-load power factor, $\cos\phi_0 = 0.2$ (lagging)

Energy component of no-load current, $I_e = I_0 \cos\phi_0 = 1 \times 0.2 = 0.2\text{ A}$

Magnetising component of no-load current, $I_m = \sqrt{I_0^2 - I_e^2} = \sqrt{1^2 - 0.2^2} = 0.98\text{ A}$

No-load branch parameters, $R_0 = \frac{V_1}{I_e} = \frac{2,000}{0.2} = 10,000\ \Omega$ or $10\text{ k}\Omega$

$$X_0 = \frac{V_1}{I_m} = \frac{2,000}{0.98} = 2,000\ \Omega \text{ or } 2\text{ k}\Omega$$

Approximate equivalent circuit is shown in Fig. 12.24.

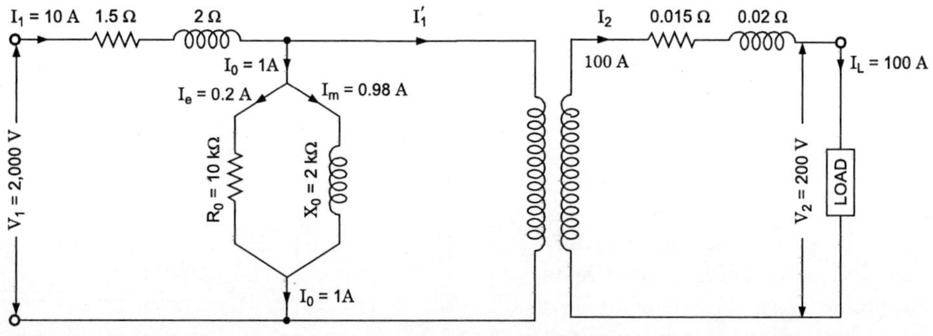

Fig. 12.24

12.17. TRANSFORMER EFFICIENCY

The rated capacity of a transformer is defined as the product of rated voltage and full-load (rated) current on the output side. The power output depends upon the power factor of the load.

The efficiency (η) of a transformer, like that of any other apparatus, is defined as the ratio of useful power output to the input power, the two being measured in same units (either in watts or killowatts).

$$\textit{i.e.}\ \text{Transformer efficiency, } \eta = \frac{\text{Output}}{\text{Input}} = \frac{\text{Output}}{\text{Output + losses}} = \frac{\text{Output}}{\text{Output + iron loss + copper loss}}$$

$$= 1 - \frac{\text{iron loss + copper loss}}{\text{Output + iron loss + copper loss}}$$

$$\text{Now power output} = V_2 I_2 \cos \phi$$

where V_2 is the secondary terminal voltage on load, I_2 is the secondary current at load and $\cos \phi$ is the power factor of the load.

$$\text{Iron loss, } P_i = \text{Hysteresis loss + eddy current loss}$$

$$\text{Copper loss} = I_1^2 R_1 + I_2^2 R_2 = I_1^2 R_{01} = I_2^2 R_{02}$$

12.17.1. Determination of Transformer Efficiency. The ordinary transformer has a very high efficiency (in the range of 96—99%). Hence the transformer efficiency cannot be determined with high precision by direct measurement of output and input, since the losses are of the order of only 1–4 %. The difference between the readings of output and input instruments is then so small that an instrument error as low as 0.5 % would cause an error of the order of 15 % in the losses. Further, it is inconvenient and costly to have the necessary loading devices of the correct current and voltage ratings and power factor to load the transformer. There is also a wastage of large amount of power (equal to that of power output + losses) and no information is available from such a test about the proportion of copper and iron losses.

The best and accurate method of determining of efficiency of a transformer would be to compute losses from open-circuit and short-circuit tests and determine efficiency as follows :

$$\text{Iron loss, } P_i = W_0 \text{ or } P_0, \text{ determined from open-circuit test}$$

$$\text{Copper loss at full load, } P_c = W_s \text{ or } P_s, \text{ determined from short-circuit test}$$

$$\text{Copper loss at a load } x \text{ times full load} = I_2^2 R_{02} = x^2 P_c$$

where x is the ratio of load current I_2 to full-load secondary current.

$$\text{And transformer efficiency, } \eta = \frac{V_2 I_2 \cos \phi}{V_2 I_2 \cos \phi + P_i + x^2 P_c} \qquad ...(12.39)$$

In Eq. (12.39) the effect of instrument readings is confined to losses only so that overall efficiency obtained from it is far more accurate than that obtained by direct loading. Another great advantage of this method is that it is not necessary for the transformer to be loaded to its full-load rating during testing, and the kW rating of the test plant need be equal to the value of the individual transformer losses.

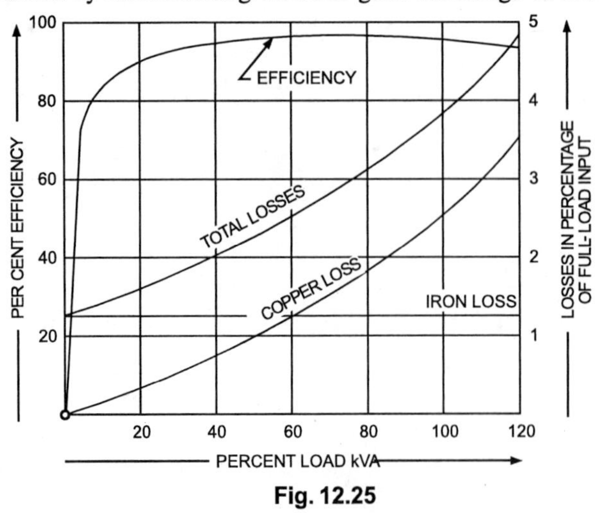

Fig. 12.25

12.17.2. Efficiency Versus Load. It has been pointed out that with constant voltage the mutual flux of the transformer is practically constant from no load to full load (maximum variation is from 1 to 3 %). The core or iron loss is, therefore, considered constant regardless of load. Copper loss varies as the square of the load current or kVA output. The variations in copper loss with the increase in load current (or kVA) is shown in Fig. 12.25. The efficiency vs load curve as deduced from these is also shown in the Fig. From the efficiency-load curve shown in Fig. 12.25 it is obvious that the efficiency is very high even at light load, as low as 10 % of rated load. The efficiency is practically constant from about 20% rated load to about 20% overload. At light loads the efficiency is poor because of constant iron loss whereas at high loads the efficiency falls off due to increase in copper loss as the square of load. From Fig. 12.25 it is also obvious that the transformer efficiency is maximum at the point of intersection of copper loss and iron loss curves *i.e.* when copper loss equals iron loss.

12.17.3. Condition For Maximum Efficiency.

From Eq. (12.39)

$$\text{Transformer efficiency, } \eta = \frac{V_2 I_2 \cos\phi}{V_2 I_2 \cos\phi + P_i + x^2 P_c} = \frac{x\,P}{x\,P + P_i + x^2 P_c} \quad ...(12.40)$$

where P is equal to the full load (rated output in volt-amperers or kVA) × load power factor (cos ϕ), P_i is the total iron loss, P_c is the full-load copper loss and x is the fraction of full-load kVA at which efficiency is maximum.

Differentiating both sides of Eq. (12.40) we get

$$\frac{d\eta}{dx} = \frac{\left(x\,P + x^2 P_c + P_i\right)P - x\,P\left(P + 2x\,P_c\right)}{\left(x\,P + P_i + x^2 P_c\right)^2} = \frac{P\left(P_i - x^2 P_c\right)}{\left(x\,P + x^2 P_c + P_i\right)^2}$$

$$\text{Efficiency } \eta \text{ will be maximum if } \frac{d\eta}{dx} = 0$$

$$\text{or } P_i - x^2 P_c = 0$$

$$\text{or } x^2 P_c = P_i \quad ...(12.41)$$

Hence the efficiency will be maximum when variable loss (copper loss) is equal to constant loss (iron loss).

Output kVA corresponding to maximum efficiency = x × Rated

$$\text{kVA} = \text{Rated kVA} \times \sqrt{\frac{P_i}{P_c}} \quad ...(12.42)$$

Since from Eq. (12.41) $x = \sqrt{\frac{P_i}{P_c}}$

Output current I_2 corresponding to maximum efficiency is determined as below:

Copper loss at given load at which efficiency is maximum

$$= I_2^2 R_{02}$$

and since for maximum efficiency to occur it is necessary that copper loss equals P_i

$$\text{So } I_2^2 R_{02} = P_i$$

So current corresponding to maximum efficiency,

$$I_2 = \sqrt{\frac{P_i}{R_{02}}} \quad ...(12.43)$$

Power transformers employed for bulk power transmission are operated continuously near about full load and are, therefore, designed to have maximum efficiency at full load. On the other hand, the distribution transformers, which supply load varying over the day through a wide range, are designed to have maximum efficiency at about three-fourths the full load.

It is important to appreciate that, since copper loss depends on current and the iron loss depends on voltage, the total loss in the transformer depends on the volt-ampere product, and not on the phase angle between voltage and current, *i.e.* is independent of the load power factor. The transformers are, therefore, rated in kilovolt-amperes (kVA) and not in kilowatts.

12.17.4. Efficiency Versus Power Factor. Transformer efficiency is given as

$$\eta = \frac{\text{Output}}{\text{Output+losses}}$$

$$= 1 - \frac{\text{losses}}{\text{output + losses}}$$

$$= 1 - \frac{\text{losses}}{V_2 I_2 \cos\phi + \text{losses}}$$

$$= 1 - \frac{\text{losses}/V_2 I_2}{\cos\phi + \text{losses}/V_2 I_2}$$

Substituting $\frac{\text{losses}}{V_2 I_2} = x$ in above equation we get

$$\eta = 1 - \frac{x}{\cos\phi + x} = 1 - \frac{x/\cos\phi}{1 + x/\cos\phi} \quad ...(12.44)$$

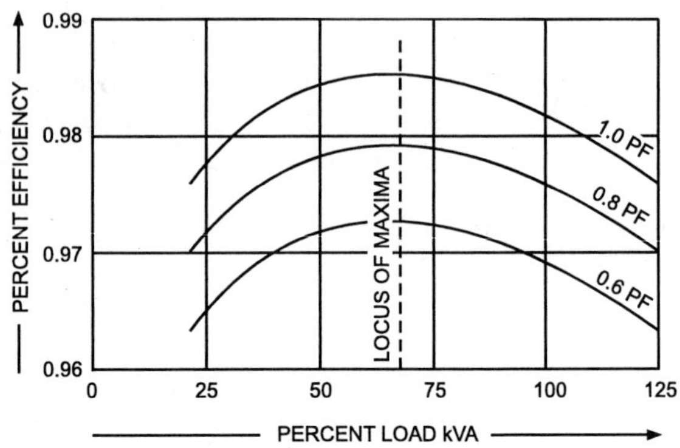

Fig. 12.26

The variations of efficiency with power factor at different loadings for a typical transformer are illustrated in Fig. 12.26.

Example 12.27 **A transformer rated 200 V/50 V, 10 kVA has a core loss of 100 watts. What is the maximum efficiency of the transformer at 0.8 lagging power factor? Assume full-load copper loss as 200 watts. At what load this maximum efficiency is obtained ?**

Solution:

Iron loss, $P_i = 100 \text{ W} = 0.1 \text{ kW}$

Full-load copper loss, $P_c = 200 \text{ W} = 0.2 \text{ kW}$

The load corresponding to maximum efficiency $= \text{Rated kVA} \times \sqrt{\frac{P_i}{P_c}} = 10 \times \sqrt{\frac{0.1}{0.2}} = 7.07 \text{ kVA}$ **Ans.**

Maximum efficiency at 0.8 lagging power factor $= \frac{7.07 \times 0.8}{7.07 \times 0.8 + 0.1 + 0.1} \times 100 = 96.585\%$ **Ans.**

Maximum efficiency when pf is improved to unity $= \frac{7.07 \times 1.0}{7.07 \times 1.0 + 0.1 + 0.1} \times 100 = 97.249\%$ **Ans.**

Example 12.28 **The maximum efficiency of a 100 kVA transformer is 98.40% and operates at 90% full load unity power factor. Calculate the efficiency of a transformer at unity power factor at full load.**

[G.B. Technical Univ. Electrical Engineering First Semester 2010-11]

Solution:

Output of transformer = 100 kVA

Load power factor, $\cos\phi = 1.0$

Power output at 90% of full load

$$P_{out} = 0.9 \times 100 \times 1.0 = 90 \text{ kW}$$

Power input at 90% of full load, $P_{in} = \frac{P_{out}}{\eta} = \frac{90}{0.984} = 91.463 \text{ kW}$

Total losses $= P_{in} - P_{out} = 91.463 - 90 = 1.463 \text{ kW}$

Since maximum efficiency occurs when iron loss equals copper loss, so

$$P_i = 0.9^2 P_c = \frac{P_{in} - P_{out}}{2} = \frac{1.463}{2} = 0.7315 \text{ kW}$$

So full-load copper loss, $P_c = \frac{0.7315}{0.9^2} = 0.903 \text{ kW}$

Power output at full load and unity power factor $= 100 \times 1.0 = 100 \text{ kW}$

$$\text{Total losses at full load, } P_i + P_c = 0.7315 + 0.903 = 1.6346 \text{ kW}$$

$$\text{Transformer efficiency at unity pf and at full load, } \eta = \frac{\text{Power output}}{\text{Power output + total losses}} \times 100$$

$$= \frac{100}{100 + 1.6346} \times 100 = 98.39\% \textbf{ Ans.}$$

Example 12.29 **A 200 kVA transformer has an efficiency of 98 % at full load. If the maximum efficiency occurs at three quarters of full load, calculate the efficiency at half load. Assume negligible magnetising current and pf of 0.8 at all loads.**

Solution:

$$\text{Full-load output at 0.8 pf} = 200 \times 0.8 = 160 \text{ kW}$$

$$\text{Full-load efficiency, } \eta = 98\ \% \text{ or } 0.98$$

$$\text{Full-load input} = \frac{\text{Output}}{\eta} = \frac{160}{0.98} = 163.265 \text{ kW}$$

$$\text{Total losses on full load} = P_i + P_c = \text{Input} - \text{output} = 163.265 - 160 = 3.265 \text{ kW} \quad ...(i)$$

Since maximum efficiency occurs at three quarters of full load, so

$$\left(\frac{3}{4}\right)^2 P_c = P_i$$

$$\text{or } \frac{9}{16} P_c = P_i \quad ...(ii)$$

Solving Eqs. (*i*) and (*ii*) we get

$$\text{Full-load copper loss, } P_c = 2.09 \text{ kW}$$

$$\text{and iron loss, } P_i = 1.175 \text{ kW}$$

At half -load and 0.8 pf

$$\text{Output} = \frac{1}{2} \times 200 \times 0.8 = 80 \text{ kW}$$

$$\text{Total losses} = P_i + x^2 P_c = 1.175 + \left(\frac{1}{2}\right)^2 \times 2.09 = 1.6975 \text{ kW}$$

$$\text{Efficiency, } \eta = \frac{\text{Output}}{\text{Output+total losses}} \times 100 = \frac{80}{80 + 1.6975} \times 100 = 97.92\ \% \textbf{ Ans.}$$

Example 12.30 **The primary and secondary windings of a 500 kVA, 1-ph transformer have resistances of 0.4 Ω and 0.0015 Ω respectively. The primary and secondary voltages are 6,000 V and 400 V respectively and the iron loss is 3.2 kW. Calculate the efficiency on (*i*) full load and (*ii*) half load, assuming the pf of the load to be 0.8.**

Solution:

$$\text{Rated output, S} = 500 \text{ kVA}$$

$$\text{Secondary voltage, } V_2 = 400 \text{ V}$$

$$\text{Full-load (or rated) secondary current, } I_2 = \frac{S \times 1{,}000}{V_2} = \frac{500 \times 1{,}000}{400} = 1{,}250 \text{ A}$$

$$\text{Total resistance referred to secondary, } R_{02} = K^2 R_1 + R_2 = \left(\frac{400}{6{,}000}\right)^2 \times 0.4 + 0.0015 = 0.003278\ \Omega$$

$$\text{Full-load copper loss, } P_c = I_2^2 R_{02} = (1{,}250)^2 \times 0.003278 = 5{,}122 \text{ W or } 5.122 \text{ kW}$$

$$\text{Iron loss, } P_i = 3.2 \text{ kW}$$

$$(i) \text{ Full-load output at 0.8 pf} = S \times \cos\phi = 500 \times 0.8 = 400 \text{ kW}$$

$$\text{Transformer efficiency at full-load and 0.8 pf, } \eta = \frac{\text{Output}}{\text{Output} + P_i + P_c} \times 100$$

$$= \frac{400}{400 + 3.2 + 5.122} \times 100 = 97.962\% \textbf{ Ans.}$$

$$(ii) \text{ Output at half load and 0.8 pf} = \frac{1}{2} S \times \cos\phi = \frac{1}{2} \times 500 \times 0.8 = 200 \text{ kW}$$

$$\text{Transformer efficiency at half load and 0.8 pf} = \frac{\text{Output}}{\text{Output} + P_i + \left(\frac{1}{2}\right)^2 P_c} \times 100$$

$$= \frac{200}{200 + 3.2 + \frac{1}{4} \times 5.122} \times 100 = 97.81\ \%\ \textbf{Ans.}$$

Example 12.31 **In a 25 kVA, 2,000/200 V transformer, the iron and copper losses are 350 watts and 400 watts respectively. Calculate the efficiency at full load and 0.8 power factor lagging. Determine the maximum efficiency and the corresponding load.** [*U.P. Technical Univ. Elec. Engineering February 2001*]

Solution:

Full-load copper loss, P_c = 400 watts or 0.4 kW

Iron loss, P_i = 350 watts or 0.35 kW

Output at full-load and 0.8 *pf*, P_{out} = 25 × 0.8 = 20 kW

Input, P_{in} = P_{out} + iron loss + full-load copper loss

= 20 + 0.35 + 0.4 = 20.75 kW

$$\text{So, efficiency at full-load and 0.8 } pf = \frac{P_{out}}{P_{in}} \times 100 = \frac{20}{20.75} \times 100 = 96.39\%$$

$$\text{The load corresponding to maximum efficiency} = \text{Rated kVA} \times \sqrt{\frac{P_i}{P_e}} = 25\sqrt{\frac{350}{400}} = 23.38 \text{ kVA } \textbf{Ans.}$$

Maximum efficiency, assuming power factor unity

$$= \frac{23.38 \times 1.0}{23.38 + 0.35 + 0.35} \times 100 = 97.1\%\ \textbf{Ans.}$$

Example 12.32 **A 250/500 volts, single-phase transformer gave the following test results :**

Short-circuit test (LV short-circuited): 20 V, 12 A, 100 Watts

Open-circuit test (Low voltage side): 250 V, 1 A, 80 Watts

Determine the efficiency of the transformer when the output is 12 Amp, 500 Volts at 0.85 power factor lagging. [*U.P. Technical Univ. Electrical Engineering Second Semester 2003-04*]

Solution:

Output current, I = 12 A

Output voltage, V = 500 V

Power factor, cos ϕ = 0.85 (lagging)

Output power, P_{out} = V I cos ϕ = 500 × 12 × 0.85 = 5,100 W

Iron loss, P_i = Power input with primary winding connected to its rated voltage = 80 W

Copper loss, P_c = 100 W with 12 A in *hv* winding

So copper loss, P_c = 100 W

$$\text{Transformer efficiency, } \eta = \frac{\text{Output}}{\text{Output} + P_i + P_c} \times 100 = \frac{5{,}100}{5{,}100 + 80 + 100} \times 100 = 96.59\ \%\ \textbf{Ans.}$$

Example 12.33 **The following test results were obtained on a 20 kVA, 2200/220 V transformer.**

OC Test (LV): 220 V, 1.1 A, 125 W

SC Test (HV): 52.7 V, 8.4 A, 287 W

The transformer is loaded at 0.8 pf on secondary side with a voltage of 220 V. Determine the efficiency at 80% of full load, the maximum efficiency and load at which it occurs.

[*Mahamaya Technical Univ. Electrical Engineering Second Semester 2011-12*]

Solution:

Iron loss, P_i = Power input with its *lv* winding connected to its rated voltage

= 125 W

$$\text{Copper loss, } P_c = 287 \text{ W with 8.4 A in } hv \text{ winding}$$

$$\text{Full-load primary current, } I_{1f} = \frac{20 \times 1{,}000}{2{,}200} = 9.09 \text{ A}$$

$$\text{Full-load copper loss, } P_c = \text{Power loss with short-circuit current of 8.4 A} \times \left(\frac{I_{1f}}{I_{sc}}\right)^2$$

$$= 287 \times \left(\frac{9.09}{8.4}\right)^2 = 336 \text{ W}$$

$$\text{Output at 80\% of full load} = 0.8 \times \text{rated kVA} \times \text{pf} = 0.8 \times 20 \times 0.8 = 12.8 \text{ kW}$$

$$\text{Copper loss at 80\% of full load} = (0.8)^2 \times 336 = 215 \text{ W}$$

$$\text{Input at 80\% of full load} = 12.8 + 0.125 + 0.215 = 13.14 \text{ kW}$$

$$\text{Efficiency, } \eta = \frac{\text{Output}}{\text{Input}} \times 100 = \frac{12.8}{13.14} \times 100 = 97.41\% \textbf{ Ans.}$$

Let maximum efficiency occur at x times full load, then

$$x = \sqrt{\frac{P_i}{\text{full-load copper loss}}} = \sqrt{\frac{125}{336}} = 0.61$$

$$\text{Load at which maximum efficiency occurs} = x\,P = 0.61 \times 20 = 12.2 \text{ kVA}$$

$$\text{Output at which maximum efficiency occurs} = 12.2 \times 0.8 = 9.76 \text{ kW}$$

$$\text{Maximum efficiency } \eta_{max} = \frac{9.76}{9.76 + 0.125 + 0.125} \times 100 = 97.5\% \textbf{ Ans.}$$

Example 12.34 **A 40 kVA transformer has a core loss of 400 watts and full-load copper loss of 800 watts. If the power factor of the load is 0.9 lagging, calculate :**

(*i*) the full-load efficiency, and

(*ii*) percentage of the full-load at which the maximum efficiency occurs.

[*U.P. Technical Univ. Electrical Engineering Second Semester 2002-03*]

Solution:

$$\text{Full-load copper loss, } P_c = 800 \text{ W or } 0.8 \text{ kW}$$

$$\text{Core loss, } P_i = 400 \text{ W or } 0.4 \text{ kW}$$

Output at full-load and at 0.9 pf (lagging)

$$P_{out} = \text{Rated kVA} \times pf = 40 \times 0.9 = 36 \text{ kW}$$

$$P_{in} = P_{out} + P_i + P_c = 36 + 0.4 + 0.8 = 37.2 \text{ kW}$$

(*i*) So efficiency at full load and 0.9 pf (lagging)

$$\eta = \frac{P_{out}}{P_{in}} \times 100 = \frac{36}{37.2} \times 100 = 96.774\% \textbf{ Ans.}$$

(*ii*) The percentage of full load corresponding to maximum efficiency

$$= \sqrt{\frac{P_i}{P_c}} \times 100 = \sqrt{\frac{400}{800}} \times 100 = 70.71\% \textbf{ Ans.}$$

Example 12.35 **In a 50 kVA, 1-phase 3300/230 V, transformer, the iron losses and full-load copper losses were found to be as 500 W and 650 W respectively. Calculate efficiency at half load and 0.6 power factor. Determine maximum efficiency and corresponding load.** [*U.P. Technical Univ. Electrical Engineering First Semester 2004-05*]

Solution:

$$\text{Iron loss, } P_i = 500 \text{ W}$$

$$\text{Copper loss at full load, } P_c = 650 \text{ W}$$

$$\text{Output at half full load} = \frac{1}{2} \times \text{rated kVA} \times \text{pf} = \frac{1}{2} \times 50 \times 0.6 = 15 \text{ kW} = 15{,}000 \text{ W}$$

$$\text{Input at half full load} = P_{out} + P_i + \left(\frac{1}{2}\right)^2 P_c = 15{,}000 + 500 + \frac{650}{4} = 15{,}662.5 \text{ W}$$

$$\text{Efficiency at half full load} = \frac{P_{out}}{P_{in}} \times 100 = \frac{15{,}000}{15{,}662.5} \times 100 = 95.77\% \textbf{ Ans.}$$

$$\text{Load corresponding to maximum efficiency} = \text{Rated kVA} \sqrt{\frac{P_i}{P_c}} = 50 \times \sqrt{\frac{500}{650}} = 43.853 \text{ kVA}$$

$$\text{Maximum efficiency, } \eta_{max} = \frac{43{,}853 \times 1.0}{43{,}853 + 500 + 500} \times 100 = 97.77\% \textbf{ Ans.} \text{ assuming unity pf}$$

Example 12.36 **In a 25 kVA, 2000 V/200 V transformer the iron and copper losses are 200 W and 400 W respectively. Calculate the efficiency at half load and 0.8 power factor lagging. Determine also the maximum efficiency and the corresponding load.** [*U.P. Technical Univ. Electrical Engineering First Semester 2006-07*]

Solution:

$$\text{Full-load copper loss, } P_c = 400 \text{ W} = 0.4 \text{ kW}$$

$$\text{Iron loss, } P_i = 200 \text{ W} = 0.2 \text{ kW}$$

$$\text{Output at half-full load and 0.8 pf} = \frac{1}{2} \times 25 \times 0.8 = 10 \text{ kW}$$

$$\text{Copper losses at half-full load} = \left(\frac{1}{2}\right)^2 \times 0.4 \text{ kW} = 0.1 \text{ kW}$$

$$\text{So efficiency at half-full load and 0.8 pf, } \eta = \frac{P_{out}}{P_{in}} \times 100 = \frac{10}{10 + 0.1 + 0.2} \times 100 = 97.087\% \textbf{ Ans.}$$

$$\text{Load corresponding to maximum efficiency,} = \text{Rated kVA} \times \sqrt{\frac{P_i}{P_c}} = 25 \times \sqrt{\frac{0.2}{0.4}} = 17.68 \text{ kVA} \textbf{ Ans.}$$

$$\text{Maximum efficiency, assuming power factor unity} = \frac{17.68 \times 1.0}{17.68 + 0.2 + 0.2} \times 100 = 97.788\% \textbf{ Ans.}$$

Example 12.37 **The maximum efficiency of a 100 kVA, 1100/440 V 50 Hz transformer is 96%. This occurs at 75% of full load at 0.8 pf lagging. Find the efficiency of transformer at $\frac{3}{4}$ FL at 0.6 pf leading.**

[*G.B. Technical Univ. Electrical Engineering Second Semester 2011-12*]

Solution: Power output corresponding to maximum efficiency.

$$P_{out} = 100 \times 0.75 \times 0.8 = 60 \text{ kW}$$

$$\text{Since efficiency, } \eta = \frac{\text{Output}}{\text{Output} + P_i + P_c}$$

$$\text{or } 0.96 = \frac{60}{60 + P_i + P_c}$$

$$\text{or } P_i + P_c = \frac{60}{0.96} - 60 = 2.5 \text{ kW}$$

$$\text{Iron loss, } P_i = \frac{2.5}{2} = 1.25 \text{ kW} \quad \because \text{Corresponding to maximum efficiency } P_c = P_i$$

$$\text{At } \frac{3}{4} \text{th full-load copper loss} = 1.25 \text{ kW as determined above}$$

$$\text{Iron loss} = 1.25 \text{ kW}$$

$$\text{Power output at } \frac{3}{4} \text{th full load and 0.6 pf leading} = \frac{3}{4} \times 100 \times 0.6 = 45 \text{ kW}$$

$$\text{Efficiency, } \eta = \frac{\text{Output}}{\text{Output} + P_i + P_c} \times 100 = \frac{45}{45 + 1.25 + 1.25} \times 100 = 94.74\% \textbf{ Ans.}$$

Example 12.38 **The efficiency of a 400 kVA, single phase transformer is 98.77% at full load 0.8 power factor and 99.13% at half full load unity power factor. Find:**

(i) Iron losses at full and half full-loads **(ii) Cu losses at full and half full loads.**

[*U.P. Technical Univ. Electrical Engineering Second Semester 2007-08*]

Solution: Let the iron loss and full-load copper losses be P_i and P_c respectively.

At full-load and 0.8 power factor

$$\text{Output power, P} = 400 \times 0.8 = 320 \text{ kW}$$

$$\text{Transformer efficiency, } \eta = 98.77\% \text{ or } 0.9877$$

$$\text{Transformer input power} = \frac{P}{\eta} = \frac{320}{0.9877} = 323.985 \text{ kW}$$

$$\text{Transformer losses, } (P_i + P_c) = \text{Input} - \text{output} = 323.985 - 320 = 3.985 \text{ kW} \quad ...(ii)$$

At half-full load and unity power factor

$$\text{Output power, P}' = \frac{1}{2} \times 400 \times 1 = 200 \text{ kW}$$

$$\text{Transformer efficiency, } \eta' = 99.13\% \text{ or } 0.9913$$

$$\text{Transformer input power} = \frac{P}{\eta} = \frac{200}{0.9913} = 201.755 \text{ kW}$$

$$\text{Total losses} = P_i + \frac{1}{2}^2 P_c = \text{Input} - \text{output} = 201.755 - 200 = 1.755 \text{ kW}$$

$$\text{or } P_i + \frac{1}{4} P_c = 1.755 \text{ kW} \quad ...(i)$$

Solving Eqs. (i) and (ii) we have

(i) Iron loss at full-load = Iron loss at half load = P_i = 1.0117 kW **Ans.**

(ii) Copper loss at full-load, Pc = 2.9733 kW **Ans.**

$$\text{Copper loss at half-full load} = \left(\frac{1}{2}\right)^2 \times 2.9733 = 0.743325 \text{ kW} \quad \textbf{Ans.}$$

Example 12.39 **The efficiency of a 400 kVA, single phase transformer is 98.77% at full load 0.8 pf and 99.13% at half load unity pf. Find iron and copper losses at half load.**

[*U.P. Technical Univ. Basic Electrical Engineering First Semester 2013-14*]

Solution: Refer to Solution of Example 12.38.

12.18. ALL-DAY EFFICIENCY (OR ENERGY EFFICIENCY)

The transformer efficiency discussed so far is the ordinary, also called *commercial efficiency,* which is defined as the ratio of power output to power input.

There are certain types of transformers whose performance cannot be judged by ordinary or commercial efficiency. For instance, distribution transformers are energized for 24 hours, but they deliver very light loads for major portion of the day. Thus iron or core loss occur for the whole day but the copper loss occurs only when the transformer is loaded. The performance of such a transformer must be judged by its *all-day efficiency,* also called the *energy efficiency* or *operational efficiency* which is computed on the basis of energy consumed during the whole day (24 hours).

The all-day efficiency is defined as the ratio of energy (kWh) output over 24 hours to the energy input over the same period.

$$i.e. \text{ All-day efficiency} = \frac{\text{Output in kWh}}{\text{Input in kWh}} \quad ...(12.45)$$

Since the distribution transformer does not supply the rated load for the whole day so the all-day efficiency of such a transformer will be less than ordinary or commercial efficiency.

For determination of all-day efficiency of a transformer, it is necessary, of course, to know how the load varies from hour to hour during the day.

Higher energy efficiencies are achieved by designing distribution transformers to yield maximum commercial efficiency at less than full load (usually 50–75 per cent of full load). This is achieved by restricting the core flux density to the lower values by using a relatively larger *x*-section. Thus the ratio of iron loss to copper loss is reduced.

Example 12.40 **A 100 kVA transformer has its maximum efficiency of 98% at full load and unity *pf*. During the day it is loaded as follows:**

12 hrs — 20 kW at a pf of 0.5 lag
6 hrs — 50 kW at a pf of 0.9 lag
6 hrs — 75 kW at a pf of 0.8 lag

Calculate the "all day efficiency" of the transformer.

Solution: For maximum efficiency

$$\text{Output} = 100 \times 1.0 = 100 \text{ kW}$$

$$\text{Input} = \frac{\text{Output}}{\eta} = \frac{100}{0.98} = 102.04 \text{ kW}$$

$$\text{Total losses} = \text{Input} - \text{output} = 102.04 - 100 = 2.04 \text{ kW}$$

$$\text{Full-load copper loss, } P_c = \text{Iron loss, } P_i = \frac{\text{Total losses}}{2} = \frac{2.04}{2} = 1.02 \text{ kW}$$

$$\text{All day output} = 20 \times 12 + 50 \times 6 + 75 \times 6 = 990 \text{ kWh}$$

$$\text{Iron loss for 24 hours} = 24 \times 1.02 = 24.48 \text{ kWh}$$

$$\text{Copper loss for 24 hours} = \left(\frac{20/0.5}{100}\right)^2 \times 1.02 \times 12 + \left(\frac{50/0.9}{100}\right)^2 \times 1.02 \times 6 + \left(\frac{75/0.8}{100}\right)^2 \times 1.02 \times 6$$

$$= 1.9584 + 1.8889 + 5.3789 = 9.226 \text{ kWh}$$

$$\text{All-day efficiency, } \eta = \frac{\text{Output}}{\text{Input}} \times 100 = \frac{990}{990 + 24.48 + 9.226} \times 100 = 96.7\% \quad \textbf{Ans.}$$

Example 12.41 **The daily variation of load on a 100 kVA transformer is as follows:**

8.00 AM to 1.00 PM — 65 kW, 35 kVAR
1.00 PM to 6.00 PM — 80 kW, 50 kVAR
6.00 PM to 1.00 AM — 30 kW, 30 kVAR
1.00 AM to 8.00 AM — No - load

This transformer has no-load core loss of 270 watts and a full–load ohmic loss of 1,200 watts. Determine the all-day efficiency of the transformer.

Solution:

$$\text{All-day output} = 65 \times 5 + 80 \times 5 + 30 \times 7 = 935 \text{ kWh}$$

$$\text{Iron loss for 24 hours} = \frac{270 \times 24}{1{,}000} = 6.48 \text{ kWh}$$

$$\text{Load in kVA from 8.00 AM to 1.00 PM} = \sqrt{\text{kW}^2 + \text{kVAR}^2} = \sqrt{65^2 + 35^2} = 73.824 \text{ kVA}$$

$$\text{Load in kVA from 1.00 PM to 6.00 PM} = \sqrt{80^2 + 50^2} = 94.34 \text{ kVA}$$

$$\text{Load in kVA from 6.00 PM to 1.00 AM} = \sqrt{30^2 + 30^2} = 42.426 \text{ kVA}$$

$$\text{Copper losses for 24 hours} = \left(\frac{73.824}{100}\right)^2 \times 1{,}200 \times 5 + \left(\frac{94.34}{100}\right)^2 \times 1{,}200 \times 5 + \left(\frac{42.426}{100}\right)^2 \times 1{,}200 \times 7$$

$$= 10{,}122 \text{ Wh or } 10.122 \text{ kWh}$$

$$\text{All-day efficiency, } \eta = \frac{935}{935 + 6.48 + 10.122} \times 100 = 98.26\ \% \ \textbf{Ans.}$$

12.19. TRANSFORMER CONSTRUCTION

The transformer is very simple in construction and consists of *magnetic circuit* linking with two windings known as *primary* and *secondary windings*. Besides magnetic circuit and windings it consists of a suitable container for the assembled core and windings, such as a *tank*, a suitable medium for insulating the core and windings from its container such as *transformer oil,* suitable *bushings* (either of porcelain, oil filled or condenser type) for insulating and bringing out terminals of the windings from the container, *temperature gauge* for measurement of temperature of hot oil or hottest spot temperature, and *oil gauge* to indicate the oil level inside the tank. Some transformers are also provided with *conservator tank* in order to slow down deterioration of oil and keep the main tank full of oil, emergency vent to relieve the pressure inside the tank in case the pressure inside the transformer rises to a danger point and *gas operated relay* (*Buchholz relay*) in order to give alarm to indicate the presence of gas in case of some minor fault and take out the transformer out of circuit in case of serious fault.

Magnetic circuit consists of an iron core. Since core is magnetic link between the two systems in a transformer and it itself contains a lot a energy, it is not by any means the passive component it would first appear.

Ideally the steel, of which the transformer core is made, is simple in chemical composition, the hot rolled steel has 4-5% silicon and the cold rolled has 3%. Nowadays cold rolled steel is used almost exclusively inspite of its high price, because flux density 30% higher can be used. This is because of high permeability and lower losses at given flux density. Since flux is an alternating one, in order to reduce eddy current losses the core is laminated. The thickness of laminations or stampings varies from 0.35 mm to 0.5 mm. The laminations are insulated from each other by coating them with a thin coat of varnish. The iron core is rectangular in cross section for small size transformers and circular for large size ones to reduce the quantity of copper required.

As already mentioned, the electric circuit consists essentially of two-windings, each being split up into two equal number of coils and the coils of two windings are arranged on each limb of iron core to reduce the magnetic leakage. In order to minimise the amount of high voltage insulation, low voltage coils are placed adjacent to the iron.

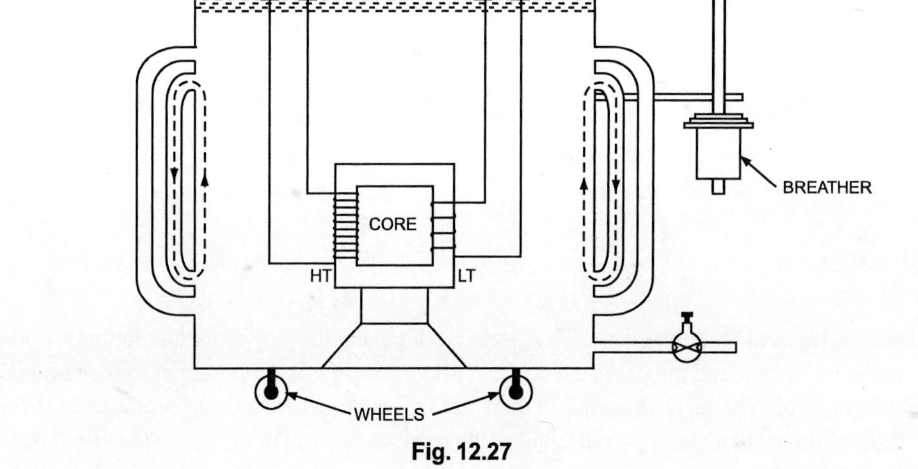

Fig. 12.27

12.20. AUTO-TRANSFORMER

The operating principle and general construction of an auto-transformer is same as that of conventional two winding transformer. The auto-transformer differs from a conventional two-winding transformer in the way in which the primary and secondary are interrelated. In the conventional transformer, the primary and secondary windings are completely insulated from each other but are magnetically linked by a common core. In the auto-transformer, the two windings primary and secondary are connected electrically as well as magnetically, in fact, a part of the single continuous winding is common to both primary and secondary.

The auto-transformers are of two types in construction. In one type of auto-transformers, there is a continuous winding with taps brought out at convenient points determined by the desired secondary voltages, and in the other type of auto-transformers there are two or more distinct coils which are electrically connected to form a continuous winding. In either case, the same laws governing conventional two winding transformers apply equally well to auto-transformers.

In the auto-transformer shown in Fig. 12.28 (*a*), the primary winding is *AB* consisting of whole turns between terminals *A* and *B*, let it be N_1. Voltage V_1 is applied across primary terminals *A* and *B*, as shown in Fig. 12.28 (*a*). To obtain secondary winding, the winding *AB* is tapped at *C* and, therefore, secondary winding consists of turns between terminals *B* and *C*, let it be N_2 and secondary voltage is available across terminals *B* and *C*.

As in ordinary transformer, transformation ratio is equal to turn-ratio

$$i.e. \quad \frac{V_2}{V_1} = \frac{E_{BC}}{E_{AB}} = \frac{N_2}{N_1} = \frac{I_1}{I_2} = K \text{ (Transformation ratio)}$$

The current flowing in section AC of winding is primary current I_1 flowing downwards, current supplied to load is I_2 and current in section BC will be $(I_2 - I_1)$ flowing upward, as shown in Fig. 12.28 (*a*).

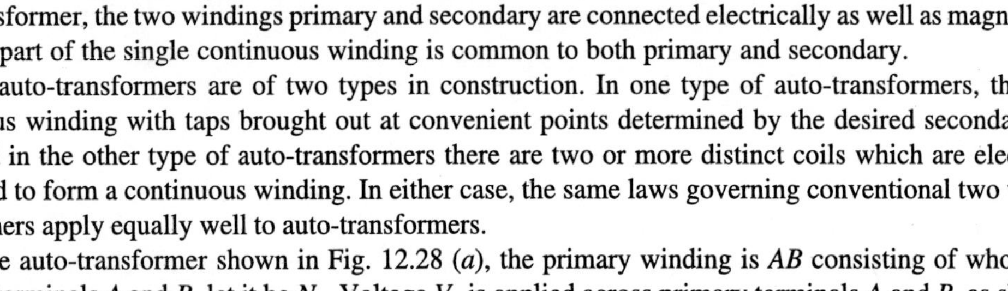

(a) *Auto-Transformer* (*b*) *Conventional Two-Winding Transformer*

Fig. 12.28

$$\text{Power delivered to load} = V_2 I_2$$

$$\text{Power in winding, AC} = E_{AC} \times I_1 = (V_1 - V_2) \times I_1$$

$$\text{Power transformed} = \text{Power in winding BC} = V_2 (I_2 - I_1) = V_2 I_2 \left(1 - \frac{I_1}{I_2}\right) = V_2 I_2 (1 - K)$$

Ratio of power transformed to the total power delivered

$$= \frac{V_2 I_2 (1-K)}{V_2 I_2} = 1 - K$$

$$\text{Power conducted directly} = \text{Power delivered to load} - \text{power transferred by transformer action}$$

$$= V_2 I_2 - V_2 I_2 (1 - K) = K V_2 I_2 = K \times \text{power output}$$

Conductor Material Requirements. The x-section of conductor (copper or aluminium) is proportional to the current to be carried and length of the conductor in a winding is proportional to the number of turns. Hence the weight of the conductor material in the winding being proportional to product of cross-sectional area and length of the conductor, is proportional to the product of number of turns and current to be carried.

In an ordinary transformer, the total weight of conductor material required is proportional to

$$(N_1 I_1 + N_2 I_2) \ \alpha \ 2 N_1 I_1 \text{ because } N_2 I_2 = N_1 I_1$$

In an auto-transformer [Fig. 12.28 (*a*)],

The top section AC has $(N_1 - N_2)$ turns and carries current I_1 and bottom section BC has turns N_2 and carries current $I_2 - I_1$.

So weight of conductor material in an auto-transformer having the same input $(V_1 I_1)$, the same output $(V_2 I_2)$ and the same voltage ratio $\left(\frac{V_2}{V_1}\right)$ as that of the two-winding transformer is proportional to $[(N_1 - N_2) I_1 + N_2 (I_2 - I_1)]$ or $[(N_1 - N_2) I_1 + N_2 I_2 - N_2 I_1]$ or $[(N_1 - N_2) I_1 + N_1 I_1 - N_2 I_1]$ or $2 (N_1 - N_2) I_1$

So $$\frac{\text{Weight of conductor in an auto-transformer}}{\text{Weight of conductor in a two winding transformer}} = \frac{2(N_1 - N_2) I_1}{2 N_1 I_1} = 1 - \frac{N_2}{N_1} = 1 - K \qquad ...(12.46)$$

Hence saving in conductor material required for winding affected by using an auto-transformer

$$= K \times \text{weight of conductor in a two winding transformer} \qquad ...(12.47)$$

From Eq. (12.47) it is obvious that nearer the ratio of transformation is to unity, the greater is the economy in conductor material requirement. If $V_2 = V_1$ i.e. when K is equal to unity, the auto-transformer, according to Eq. (12.46), needs no conductor material, for actually no transformer is required. If voltage is to be transformed from very high voltage to very low voltage i.e. when K is to be very small, there will be a little saving in the conductor material and in such cases it is preferable to use an ordinary two-winding transformer.

Advantages. 1. Continuously varying voltage can be obtained.

2. An auto-transformer requires less copper and is more efficient than a two-winding transformer of a similar rating.

Practical Applications. (*i*) The auto-transformer is used as *balance coil* to give neutral in a 3-wire ac distribution system.

(*ii*) Auto-transformers with a number of tappings are used for starting induction motors and synchronous motors. When auto- transformers are used for this purpose, these are known as *auto-starters.*

(*iii*) A continuously variable auto-transformer finds useful application in electrical testing laboratory.

(*iv*) Auto-transformers have the biggest sphere of usefulness as regulating transformers.

(*v*) Auto-transformers are also used as boosters to raise the voltage in ac feeders.

(*vi*) As furnace transformers for getting convenient supply to suit the furnace winding from 240 V ac supply.

Since in distribution transformers, the transformation ratio is low and auto-transformers loose much of their advantage when the ratio of transformation is low, it is not advantageous to use auto-transformer as distribution transformer. There is another disadvantage also, that the primary and secondary are conductively connected.

Example 12.42 **An auto-transformer having 1,250 turns is connected across a 250 V ac supply. What secondary voltage will be obtained if a tap is taken at 800 th turn ?**

Solution: Secondary voltage, $V_2 = V_1 \times \frac{N_2}{N_1} = 250 \times \frac{800}{1{,}250} = 160$ V **Ans.**

Example 12.43 **An auto-transformer supplies a load of 5 kW at 125 V and at unity power factor. If the primary voltage is 250 V, determine**

(a) transformation ratio (b) secondary current (c) primary current (d) number of turns across secondary if total number of turns is 250. (e) power transformed and (f) power conducted directly from the supply mains to load.

Solution: (*a*) Transformation ratio, $K = \frac{V_2}{V_1} = \frac{125}{250} = \frac{1}{2}$ **Ans.**

$$(b) \text{ Secondary current, } I_2 = \frac{\text{kW} \times 1{,}000}{V_2 \cos\phi} = \frac{5 \times 1{,}000}{125 \times 1} = 40 \text{ A } \textbf{Ans.}$$

$$(c) \text{ Primary current, } I_1 = K\,I_2 = \frac{1}{2} \times 40 = 20 \text{ A } \textbf{Ans.}$$

$$(d) \text{ Turns across secondary, } N_2 = K\,N_1 = \frac{1}{2} \times 250 = 125 \textbf{ Ans.}$$

$$(e) \text{ Power transformed} = \text{Load} \times (I - K) = 5\left(1 - \frac{1}{2}\right) = 2.5 \text{ kW } \textbf{Ans.}$$

(*f*) Power conducted directly from supply mains = 5 – 2.5 = 2.5 kW **Ans.**

Example 12.44 The primary and secondary voltages of an auto-transformer are 500 V and 400 V respectively. Show with the aid of a diagram the current distribution in the windings when the secondary current is 100 A and calculate the economy of copper in this particular case (in percentage).

Solution:

$$\text{Transformation voltage, } K = \frac{V_2}{V_1} = \frac{400}{500} = 0.8$$

Secondary current, $I_2 = 100$ A

Primary current, $I_1 = K\,I_2 = 0.8 \times 100 = 80$ A

The current distribution is shown in Fig. 12.29

Economy in copper = K = 0.8 or 80% **Ans.**

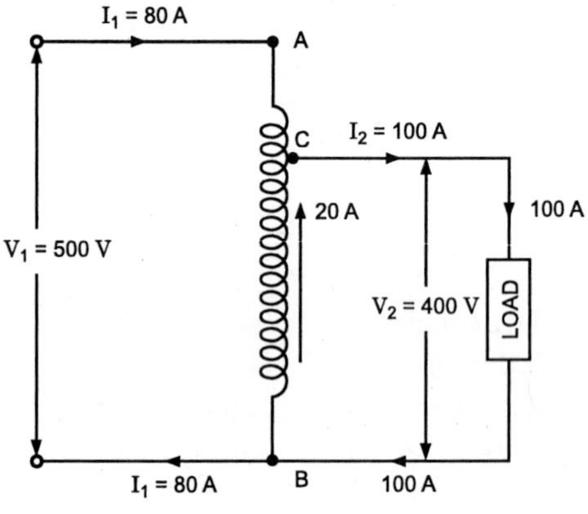

Fig. 12.29

Example 12.45 Determine the core area, the number of turns and the position of tapping point for a 400 kVA, 50 Hz, 0.6 kV/5 kV single-phase auto-transformer.

Assume emf per turn is 10 V and maximum flux density at 1.2 T.

[*R.G.P.V. Basic Electrical Engineering June 2003*]

Solution:

Primary voltage, V_1 = 0.6 kV = 600 V

Secondary voltage, V_2 = 5 kV = 5,000 V

$$\text{Number of turns on secondary, } N_2 = \frac{V_2}{\text{EMF per turn}} = \frac{5{,}000}{10} = 500 \textbf{ Ans.}$$

$$\text{Number of turns on primary, } N_1 = \frac{V_1}{\text{EMF per turn}} = \frac{600}{10} = 60$$

$$\text{Position of tapping} = \frac{60}{500} \times 100 = 12\% \text{ from the bottom of the hv winding.} \quad \textbf{Ans.}$$

$$\text{Core flux, } \Phi_{max} = \frac{\text{EMF per turn}}{4.44\,f} = \frac{10}{4.44 \times 50} = 0.045 \text{ Wb}$$

$$\text{Core area, } a = \frac{\Phi_{max}}{B_{max}} = \frac{0.045}{1.2} = 0.0375 \text{ m}^2 = 375 \text{ cm}^2 \quad \textbf{Ans.}$$

Example 12.46 An auto-transformer is used to stepdown voltage level from 230 V to 200 V while the load is 20 kW at upf. Neglecting losses and magnetising current find the current in different sections of the winding.

[*R.G.P.V. Basic Electrical Engineering June 2003*]

Solution: Transformation ratio, $K = \frac{V_2}{V_1} = \frac{200}{230} = \frac{20}{23}$

Secondary current, $I_2 = \frac{\text{Load in kW} \times 1{,}000}{V_2 \times pf}$

$= \frac{20 \times 1{,}000}{200 \times 1} = 100$ A

Primary current, $I_1 = KI_2 = \frac{20}{23} \times 100 = 86.96$ A

Current distribution is shown in Fig. 12.30.

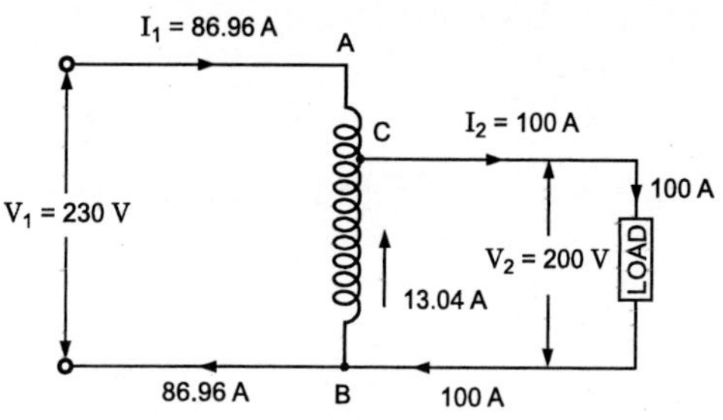

Fig. 12.30

HIGHLIGHTS

1. *Transformer* is an ac machine that (*i*) transfers electrical energy from one electric circuit to another (*ii*) does so without a change of frequency (*iii*) does so by the principle of electromagnetic induction and (*iv*) has electric circuits that are linked by a common magnetic circuit. The energy transfer usually takes place with a change of voltage, although this is not always necessary. It is a static machine and has very high efficiency (in the range of 96-99%).

   A transformer is never operated on dc. If dc supply is connected to the transformer primary or secondary, it may burn out since there is no self-induced emf and heavy current will flow through it.
2. In an *ideal transformer,* winding resistance, magnetic leakage, and iron losses are neglected and core is assumed to have infinite permeability and zero reluctance.
3. The emfs induced in the transformer windings are given as

$$E_1 = 4.44 f N_1 \Phi_{max} \text{ volts}$$
$$\text{and } E_2 = 4.44 f N_2 \Phi_{max} \text{ volts}$$

4. Voltage transformation ratio,

$$K = \frac{E_2}{E_1} = \frac{N_2}{N_1} = \frac{V_2}{V_1} = \frac{I_1}{I_2}$$

5. If the transformer is operated with the frequency and voltage changed in the same proportion, the flux density will remain unchanged and so the no-load current.
6. The current drawn by transformer on no load is called the no-load current or exciting current. It has two components (*i*) in-phase, active or *energy component* I_e ($I_0 \cos \phi_0$) used to meet the iron loss and small amount of copper loss occurring in primary winding and (*ii*) quadrature or wattless component, called the *magnetizing component,* I_m ($I_0 \sin \phi_0$) used to create alternating flux in the core. No-load phase angle, $\phi_0 = \text{Tan}^{-1} \frac{I_m}{I_e}$.
7. When a transformer is loaded, the secondary ampere-turns set up a magnetic field that opposes the main field. To neutralize this effect, transformer draws extra primary current from the mains. Flux remains almost constant from no load to full load.

$$\text{The primary current, } \mathbf{I}_1 = \mathbf{I'}_1 + \mathbf{I}_0 = K\,\mathbf{I}_2 + \mathbf{I}_0$$

8. When a practical transformer is loaded

$$\text{Induced emf in secondary winding, } \mathbf{E}_2 = \mathbf{V}_2 + \mathbf{I}_2 (R_2 + j X_2) = \mathbf{V}_2 + \mathbf{I}_2 \mathbf{Z}_2$$
$$\text{Induced emf in primary, } \mathbf{E}_1 = \mathbf{E}_2/K$$
$$\text{and applied voltage to primary winding, } \mathbf{V}_1 = -\mathbf{E}_1 + \mathbf{I}_1 (R_1 + j X_1) = -\mathbf{E}_1 + \mathbf{I}_1 \mathbf{Z}_1$$

9. The total resistance and reactance of the transformer referred to primary are given as

$$R_{01} = R_1 + \frac{R_2}{K^2}, \quad X_{01} = X_1 + \frac{X_2}{K^2}$$

and

$$Z_{01} = \sqrt{R_{01}^2 + X_{01}^2}$$

10. The total resistance and reactance of the transformer referred to secondary are given as

$$R_{02} = (K^2 R_1 + R_2),\ X_{02} = (K^2 X_1 + X_2)$$

and $$Z_{02} = \sqrt{R_{02}^2 + X_{02}^2}$$

11. Approximate voltage drop in the transformer referred to secondary = $I_2 (R_{02} \cos \phi \pm X_{02} \sin \phi)$
+ ve sign for lagging power factor and – ve sign for leading power factor.

12. The condition for zero regulation is

$$\text{Load phase angle, } \phi = \text{Tan}^{-1}\left(\frac{-R_{02}}{X_{02}}\right)$$

– ve sign indicates that zero regulation occurs at a leading power factor.

13. Percentage voltage regulation = $\dfrac{E_2 - V_2}{E_2} \times 100 = \dfrac{I_2 (R_{02} \cos \phi \pm X_{02} \sin \phi)}{\text{No-load secondary voltage, } E_2} \times 100$

14. Transformer losses are of two types namely iron loss and copper loss. Iron losses are almost constant from no load to full load while copper losses vary as the square of load in kVA.

15. Transformer efficiency is given as

$$\eta = \frac{V_2 I_2 \cos \phi}{V_2 I_2 \cos \phi + P_i + x^2 P_c}$$

Transformer efficiency will be maximum when copper losses equal iron or core losses.

Output kVA corresponding to maximum efficiency = Rated kVA $\sqrt{\dfrac{\text{Iron losses}}{\text{full-load copper losses}}}$

16. *Open-circuit lest* is performed to determine the core or iron losses, P_i, and no-load current I_0 and thereby the shunt branch parameters R_0 and X_0 of the equivalent circuit.

17. *Short-circuit test* is performed to determine the copper losses at rated load and equivalent resistance and reactance referred to metering side.

18. The performance of distribution transformers is judged by its all-day efficiency, also called the *energy efficiency* or *operational efficiency* which is computed on the basis of energy consumed during the whole day (24 hours) and is given as

$$\text{Transformer all-day efficiency, } \eta = \frac{\text{kWh output in 24 hours}}{\text{kWh input in 24 hours}} \times 100$$

In an auto-transformer:

Ratio of power transformed to total power delivered = (1 – K)

Ratio of power conducted directly to power output = K where K is transformation ratio

$$\frac{\text{Volt-ampere rating as an auto-transformer}}{\text{Volt-ampere rating as two winding transformer}} = \frac{V_1 I_1}{(V_1 - V_2) I_1} = \frac{V_2 I_2}{V_2 (I_2 - I_1)} = \frac{V_2 I_2}{V_2 (I_2 - K I_2)} = \frac{1}{1-K} \quad ...(i)$$

$$\frac{\text{Weight of conductor in an auto-transformer}}{\text{Weight of conductor in a two winding transformer}} = \frac{2 (N_1 - N_2) I_1}{2 N_1 I_1} = 1 - \frac{N_2}{N_1} = 1 - K \quad ...(ii)$$

Hence saving in conductor material required for winding affected by using an auto-transformer

= K × weight of conductor in a two winding transformer ...(*iii*)

EXERCISES

1. Explain basic principle of operation of a single-phase transformer. Where are they used ?
[U.P.Technical Univ. Elec. Engineering Second Semester 2005-06]

2. What do you understand by the term "ideal transformer"? Why the efficiency of transformer is high as compared to other machines? **[G.B. Technical Univ. Electrical Engineering Even Semester 2012-13]**

3. Derive emf equation of a single-phase transformer and obtain relation for secondary to primary winding voltages.
[G.B. Technical Univ. Electrical Engineering First Semester 2009-10]

4. Draw and explain phasor diagram of single-phase transformer on load with lagging power factor.
[U.P.Technical Univ. Elec. Engineering First Semester 2005-06]

5. Draw the phasor diagram of a single-phase transformer for leading power factor load.
[U.P. Technical Univ. Electrical Engineering Second Semester 2007-08]

6. Draw and explain the no-load and full-load phasor diagrams for a single-phase transformer.
[Electrical Engineering G.B. Technical Univ. Second Semester 2009-10; U.P. Technical Univ. Second Semester 2013-14]

7. Derive an emf expression of power transformer. Also draw an equivalent circuit of it.
[G.B. Technical Univ. Electrical Engineering First Semester 2011-12]

8. Derive an emf/voltage expression of power transformer. Also draw an equivalent circuit of it.
[Mahamaya Technical Univ. Electrical Engineering Second Semester 2010-11]

9. Develop the equivalent circuit of a single phase transformer on no load and on load conditions.
[U.P. Technical Univ. Electrical Engineering Second Semester 2009-10]

10. Draw exact equivalent circuit and corresponding phasor diagram of a single-phase transformer on load and explain them. Why no-load current is kept small and how it is reduced?
[U.P. Technical Univ. Electrical Engineering First Semester 2006-07]

11. Explain following for single phase transformer:
    (*i*) Phasor diagram for inductive load
    (*ii*) Equivalent circuit.
    (*iii*) Voltage regulation. **[U.P. Technical Univ. Electrical Engineering Second Semester 2007-08]**

12. (*i*) What is meant by leakage flux and leakage reactance in reference to a transformer ?
    (*ii*) Define voltage regulation of a transformer and deduce expression for lagging power factor load.
[U.P. Technical Univ. Electrical Engineering Second Semester 2003-04]

13. Explain the working of a transformer with the derivation of the emf equation for a transformer. Also discuss the losses in the transformer. **[G.B. Technical Univ. Electrical Engineering Second Semester 2010-11]**

14. What are different losses in transformer? Explain.
[U.P. Technical Univ. Electrical Engineering Odd Semester 2013-14]

15. Explain why the hysteresis and eddy current losses occur in the transformer. How does change in frequency affect the operation of given transformer? **[G.B. Technical Univ. Electrical Engineering Second Semester 2009-10]**

16. Explain the following:
    Efficiency of transformer. **[G.B. Technical Univ. Electrical Engineering First Semester 2010-11]**

17. Write down the expression of efficiency for a transformer. How is it affected by change in power factor?
[G.B. Technical Univ. Electrical Engineering First Semester 2011-12]

18. List various losses occuring in a transformer and mention the condition for maximum efficiency.
[U.P. Technical Univ. Electrical Engineering First Semester 2006-07]

19. Explain the OC and SC tests performed on a single-phase transformer. Determine the equivalent circuit based on the above tests. Derive the condition for maximum efficiency of a transformer.
[U.P. Technical Univ. Electrical Engineering First Semester 2007-08]

20. Prove that for a single phase transformer $(\text{kVA})_{max} = (\text{kVA})_{rated} \times \sqrt{\dfrac{P_{core}}{P_{ohmic\ (rated)}}}$.
[G.B. Technical Univ. Electrical Engineering Odd Semester 2012-13; Even Semester 2012-13]

21. Discuss in brief about auto-transformer. **[Electrical Engineering G.B. Technical Univ. Second Semester 2011-12; U.P. Technical Univ. Second Semester 2013-14]**

22. Describe the working of an auto-transformer. **[G.B. Technical Univ. Electrical Engineering First Semester 2010-11]**

23. Explain single-phase auto-transformer and give its application.
[G.B. Technical Univ. Electrical Engineering First Semester 2009-10]

24. Explain working principle of autotransformer. What are its advantages, disadvantages and applications?
[U.P. Technical Univ. Electrical Engineering Odd Semester 2013-14]
25. What are the performance parameters of 1-ϕ transformer? Differentiate between two winding transformers and autotransformers? Mention any two applications of shell type single-phase transformer.
[U.P. Technical Univ. Electrical Engineering Odd Semester 2013-14]
26. Explain working of single phase auto-transformer and mention its applications.
Show that there will be saving of copper in auto-transformer in comparison to same rating of two winding transformer.

SHORT ANSWER TYPE QUESTIONS WITH ANSWERS

Q. 1. What is a function of transformer ?

Ans. The basic function of a transformer is to transform energy from one voltage level to another (may be from low voltage to high voltage or vice versa).

Q. 2. What do you understand by step-up and step-down transformers ?

Ans. When transformer raises the voltage i.e. when the output voltage is higher than its input voltage, it is called the step-up transformer. When the transformer lowers the voltage, it is called the step-down transformer.

Q. 3. Why does a transformer has iron core ? **[P.T.U. December 2000]**

Ans. The use of iron or steel for transformer core ensures a high permeability of the magnetic circuit and because of high permeability the magnitude of exciting current necessary to create the required flux in the core is small. The presence of steel core also causes 100% of magnetic flux created by primary to be linked with secondary.

Q. 4. Why is the efficiency of a transformer maximum among electrical equipments ? Explain. **[P.T.U. 1999]**

Ans. Owing to lack of rotating parts in a transformer, there are no friction and windage losses. The other losses such as iron losses are also comparatively small because of better magnetic material and less handling of the material. So the efficiency of transformer is maximum among electrical equipments.

Q. 5. What is normally the efficiency of a transformer ?

Ans. Typical transformer efficiencies at full load lie between 96 % and 97 % and with extremely large transformers they may be as high as 99 %.

Q. 6. What name is given to the coils through which current flows from the source ?

Ans. Primary winding.

Q. 7. What name is given to the coils across which load is connected ?

Ans. Secondary winding.

Q. 8. Why the transformer is also called the static transformer ?

Ans. Since basic construction of transformer does not require any moving part, it is also called the static transformer.

Q. 9. What is one-to-one transformer ? What is its application ? **[R.G.P.V. Bhopal June 2002]**

Ans. The transformer having equal turns in primary and secondary windings is called the one-to-one transformer. Such transformers are used for isolating the load from supply.

Q. 10. What happens if dc supply is given to a transformer? **[G.B. Technical Univ. Odd Semester 2012-13]**

Ans. If a rated dc voltage is applied to the primary of the transformer, the flux produced in the transformer core will not vary but remain constant in magnitude and, therefore, there will be no self-induced emf in the primary winding (which is only possible with varying flux linkage) to oppose the applied voltage and since the resistance of the primary winding is quite low, a heavy current will flow through the primary winding that may cause burning out of the primary winding.

Q. 11. What do you understand by the term "ideal transformer"? **[U.P. Technical Univ. Second Semester 2013-14]**

Ans. The properties of an ideal transformer are (*i*) no winding resistance (*ii*) no magnetic leakage (*iii*) no iron loss and (*iv*) zero magnetizing current i.e. an ideal transformer is assumed to consist of two pure inductive coils wound on a loss-free core.

Q. 12. What is the difference between an ideal and practical transformer ? **[P.T.U. January 2000]**

Ans. An *ideal transformer* is an imaginary transformer that has no winding resistance, no magnetic leakage, no iron loss and zero reluctance while a *practical transformer* is an actual transformer which operates at a particular frequency and has winding resistance, magnetic leakage, iron loss, copper loss, and magnetic reluctance.

Q. 13. The primary of a single-phase transformer having a turn ratio 1:2 is connected to a 10 V dc source while its secondary is connected to a 10 Ω resistor. What is the value of the steady-state current in the resistor ?

Ans. Zero.

Q. 14. Is there a definite relationship between the number of turns and voltages in transformers ?

Ans. Yes; the voltage varies in exact proportion to the number of turns connected in series in each winding.

Q. 15. What is transformer and transformation ratio? **[M.D. Univ., July, 2001]**

Ans. The transformer is an ac machine that (*i*) transfers electrical energy from one electric circuit to another (*ii*) does so without a change of frequency (*iii*) does so by the principle of electromagnetic induction and (*iv*) has electric circuits that are linked by a common magnetic circuit. Voltage transformation ratio is defined as the ratio of secondary induced emf to primary induced emf.

Q. 16. Does the flux linking both the windings of a transformer remain same ?

Ans. Yes; the flux linking the primary and secondary windings is the same.

Q. 17. Does the transformer draw any current when its secondary is open ?

Ans. Yes; no-load primary current.

Q. 18. What is meant by magnetising current ?

Ans. Magnetising current is quadrature or wattless component of exciting current of transformer and is used to create alternating flux in the core.

Q. 19. What do you know about the no-load current of a transformer ?

Ans. The current drawn by primary winding of a transformer on no load is called the no-load or exciting current. It is very small in magnitude in comparison to the full-load primary current (2 to 5 per cent of full-load current). It has two components—energy component (comparatively very small) to meet the iron loss in addition to small amount of copper loss occuring in the primary winding and magnetising component relatively large to create the alternating flux in the core.

Q. 20. Does the magnetising current of a transformer lie in phase with the applied voltage ?

Ans. No; the magnetising current of a transformer lags behind the applied voltage by 90°.

Q. 21. Ordinarily, what is the phase relationship between the primary and secondary voltages of a transformer ?

Ans. 180 ° out of phase.

Q. 22. Explain how in a transformer, as the secondary current increases, primary current also increases. **[P.T.U. May 2002]**

Ans. When the secondary current of a transformer increases, the secondary ampere-turns (N_2I_2) creates more secondary flux. This increased secondary flux opposes the main flux Φ set up by the exciting current I_0. Thus main flux Φ is weakened momentarily by the opposing increased secondary flux Φ_2 and, therefore, difference of applied voltage V_1 and back emf E_1 increases, consequently, more current is drawn from the source of supply flowing through the primary winding until the original value of flux Φ restores. The extra current drawn from the supply source increases the primary ampere-turns (N_1I_1') and compensates the extra secondary ampere-turns i.e. $N_1I_1' = N_2I_2$

Thus in a transformer when the secondary current increases, primary current also increases.

Q. 23. Why the main flux in a transformer remains practically constant from no-load to full-load ?

Ans. When the transformer is on no load, it draws no-load or exciting current I_0 from the supply mains. The no-load current I_0 sets up mmf N_1I_0 which produces flux Φ in the core. When an impedance is connected across the secondary terminals, current I_2 flows through the secondary. The secondary current I_2 sets up its own mmf and hence creates a secondary flux Φ_2 opposing the main flux Φ. In other words, the secondary ampere-turns weaken the main flux in the core. Momentarily, self-induced emf E_1 decreases due to which primary draws extra current from the supply source. This extra current drawn from the supply source increases the primary ampere-turns (N_1I_1') and compensates the secondary ampere-turns N_2I_2.

Since the secondary flux Φ_2 produced by secondary mmf N_2I_2 is neutralized by the flux Φ_1' produced by mmf N_1I' set up by counterbalancing primary current I_1', the flux in the transformer core remains constant from no-load to full load.

Q. 24. How does the leakage flux occur in a transformer ? **[P.T.U. December 2000]**

Ans. In actual practice the total flux created does not link the primary and secondary windings but is divided into three parts namely the main flux Φ linking both the primary and secondary windings, primary leakage flux Φ_{L_1}, linking with primary winding only and secondary leakage flux Φ_{L_2}, linking with secondary winding only.

The flux set up by the primary or secondary which is set up in the core linking with its own turns and not linking with the other is known as *leakage* flux.

Q. 25. What is leakage reactance of transformers ? **[P.T.U. December 2003]**

Ans. Leakage fluxes of primary and secondary produce self inductances L_1 given as $\frac{N_1\Phi_{L1}}{I_1}$ and L_2 given as $\frac{N_2\Phi_{L2}}{I_2}$ in the respective windings i.e. primary and secondary. These inductances, in turn, produce inductive reactances X_1 given as $2\pi f L_1$ and X_2 given as $2\pi f L_2$ which cause voltage drop in respective windings lagging behind the currents I_1 and I_2 respectively by 90°. The reactances produced by leakage fluxes are called the leakage reactances.

Q. 26. What is the effect of leakage flux in a transformer ? What is meant by magnetising current ? **[P.T.U. June 2000]**

Ans. The effect of leakage flux is to develop into their respective windings emfs of self-inductance which are proportional to the current, and are, therefore, equivalent in effect to the addition of an inductive coil in series with each winding. The component of no-load current used to create the flux in the transformer core is known as magnetising current.

Q. 27. Explain with reasons what happens when the primary terminals of a single phase transformer are connected to the rated ac supply while secondary terminals are short circuited.

Ans. The transformer with secondary short circuited becomes equivalent to a coil having an impedance equal to impedance of both windings, which is quite small. When the primary terminals of the transformer are connected to the rated ac supply, heavy current will flow through primary and secondary windings that may cause burning out of these windings.

Q. 28. Define voltage regulation of a transformer.
[UP Technical Univ. Second Semester 2003-04; G.G.S.I.P. Univ. Delhi May 2005]

Ans. Voltage regulation of a transformer is defined as the change in magnitude of terminal or secondary voltage, when full load (rated load) of specified power factor supplied at rated voltage is thrown off (reduced to no load) with primary voltage and frequency held constant, as percentage of the rated terminal voltage.

Q. 29. Is voltage regulation better or worse at 0.6 *pf* lagging than at 0.8 *pf* lagging for the same kVA output explain.
[P.T.U. May 2001]

Ans. The voltage regulation will be the maximum and worst when power factor becomes equal to $\cos \tan^{-1} \frac{X_{02}}{R_{02}}$ (lagging).

Since equivalent reactance of transformer is usually more than double its equivalent resistance, the voltage regulation will increase (i.e. become more and more worse) till power factor falls even below $\cos \tan^{-1} 2$ i.e. 0.45 (lagging), the voltage regulation will worsen at 0.6 pf lagging than at 0.8 pf lagging.

Q. 30. Is the regulation at rated load of a transformer same at 0.8 pf lag and 0.8 pf lead ?

Ans. No; the regulation obtained is different at leading and lagging power factors.

Q. 31. On what factors does the hysteresis loss depend ? Why can this loss be assumed to be constant ?
[P.T.U. December 2000]

Ans. Hysteresis loss depends upon the type of material, volume of the material, frequency of reversal of magnetic field and maximum value of flux density in the core of the material i.e.

$P_h = \eta\,(B_{max})^{1.6} f\,v$ watts

Since from no load to full load the flux linking with the core and supply frequency remains constant, hysteresis loss is assumed to remain constant.

Q. 32. Write an expression of hysteresis loss in a transformer. Why it decreases at higher frequencies with constant V?
[U.P. Technical Univ. Odd Semester, 2013-14]

Ans. The core of a transformer is subjected to an alternating magnetizing force and for each cycle of emf a hysteresis loop is traced out. The hysteresis loss per second is given by the equation

$$\text{Hysteresis loss, } P_h = \eta' (B_{max})^x f\,v \text{ joules per second or watts}$$

where f is the supply frequency in Hz, v is the volume of core in cubic metres, η' is the hysteresis coefficient, B_{max} is peak value of flux density in the core and x lies between 1.5 and 2.5 depending upon the material and is often taken as 1.6.

Hysteresis loss is proportional to $(B_{max})^x f$ and for a transformer with constant V, the product of maximum flux density B_{max} and frequency f is constant [Refer to Eq. (12.1)]. So at higher frequencies product $(B_{max})^{1.6} f$ and therefore hysteresis loss decreases, x being more than unity.

Q. 33. On what factors does the eddy current loss depend ? Why this loss can be assumed to be constant ?

[P.T.U. June 2000]

Ans. The eddy current loss depends on the maximum value of flux density in the core and supply frequency for a given core (i.e nature of magnetic material, volume of the core, and thickness of laminations remaining unchanged). Since from no load to full load the flux linking with the core and supply frequency remain constant, the eddy current loss remains constant.

Q. 34. Classify the losses in power transformer.

[G.B.T.U. First Semester 2011-12; U.P. Technical Univ. Second Semester 2013-14]

Ans. Transformer losses may be classified into two categories namely constant losses and variable losses.

Iron losses are independent of load, and therefore, are called the constant losses. These losses consist of hysteresis and eddy current losses and occur in the core and yoke of the transformer due to alternating flux.

Copper losses vary as the square of load current or kVA, and therefore, they are called variable losses. These losses occur in primary and secondary windings of the transformer due to their resistances when the transformer is loaded.

Q. 35. Name the constant losses taking place in a transformer. **[G.B.T.U. Second Semester 2011-12]**

Ans. Iron losses are constant losses that take place in a transformer. Eddy current loss and hysteresis loss constitute the iron losses.

Q. 36. Does eddy current loss of a transformer depend on the applied voltage ?

Ans. Yes, eddy current loss depends upon the applied voltage. If the transformer is operated with increased voltage but at the same frequency the eddy current loss increases due to increase in value of peak value of flux density and if applied voltage is increased with increase in frequency, the proportion of increase being the same in both, the eddy current loss increases due to increase in supply frequency.

Q. 37. If the supply frequency is doubled, which loss component in transformer will double ?

Ans. Due to doubling of supply frequency, the hysteresis loss component will be double.

Q. 38. What is the purpose of the markings on transformer leads ?

Ans. They are there for standardization, so that the transformer polarities may be recognizable for any type of use.

Q. 39. How can eddy current loss be minimised ?

Ans. By using laminated iron core.

Q. 40. What is meant by equivalent circuit of a transformer?

Ans. An electrical circuit that can represent behaviour of the transformer to an approximation is called its equivalent circuit.

Q. 41. Draw approximate equivalent circuit of transformer referred to primary side.

[U.P. Technical Univ. Odd Semester 2013-14]

Ans. Refer to Fig. 12.17 page 370.

Q. 42. What are the performance parameters of single-phase transformers? **[U.P. Technical Univ. 2013-14]**

Ans. The performance parameters of single phase transformers are the shunt branch parameters R_0 and X_0 and equivalent resistance and reactance R_{eq} and X_{eq} of transformer primary and secondary windings.

Q. 43. How may the iron loss be reduced to a minimum in a transformer ?

Ans. The iron loss may be minimised by using steel of high silicon content for the transformer core and by using very thin laminations (0.35 to 0.5 mm thick) insulated from each other by insulating varnish or by layers of papers.

Q. 44. Why are iron or core losses assumed to remain constant from no load to full load in a power transformer ?

Ans. Since iron or core losses depend on the flux density and supply frequency for a given transformer core and flux density and supply frequency remain the same from no load to full load, iron or core losses are assumed to remain constant from no load to full load.

Q. 45. Why is the iron loss in a transformer substantially independent of the load current ? Explain.

[U.P. Technical Univ. Electrical Machines 2001-02]

Ans. Iron or core loss is caused by the alternating flux in the core and consists of hyteresis and eddy current losses which depend upon the maximum flux density in the core and supply frequency. Since it has been determined that the mutual flux varies somewhat with the load (its variation being 1 to 3 per cent from noload to fullload), the core loss will vary somewhat with the load and it's power factor. It may be emphasized here that core loss is assumed to remain constant from no load to full load, the variations in losses from no load to full load being very small and negligible.

Q. 46. Write the condition for efficiency in transformer to be maximum. **[M.T.U. First Semester 2011-12]**

Ans. Efficiency of transformer will be maximum when variable loss (copper loss) is equal to constant loss (iron loss)

$$i.e., x^2 P_c = P_i$$

where P_i is the iron loss, P_c is the copper loss on full load and x is the fraction of full-load kVA at which efficiency is maximum.

Q. 47. Is the efficiency of a transfromer same at the same load at 0.8 *pf* lag and 0.8 *pf* lead ?

Ans. Yes, the efficiency of a transformer remains the same for a given load at a given power factor irrespective of the fact whether the *pf* is lagging or leading.

Q. 48. What is meant by all-day efficiency of a power transformer and why is it lower than commercial efficiency ?

Ans. The all-day efficiency, also known as energy efficiency or operational efficiency, is defined as the ratio of energy (kWh) output over 24 hours to the energy input over the same period

$$\text{i.e.} \quad \text{All-day efficiency} = \frac{\text{Output in kWh}}{\text{Input in kWh}}$$

There are certain types of transformers, such as distribution transformers, which remain energized for 24 hours but they supply very light loads for major portion of the day. Thus iron or core loss occurs for the whole day but the copper loss occurs only when the transformer is loaded and so the all-day efficiency of transformer is lower than its commercial efficiency.

Q. 49. Why is laminated core used in the transformer? **[M.D. Univ., May 2003]**

Ans. Laminated core is used in the transformer to reduce the eddy current loss.

Q. 50. Why is an iron-silicon alloy core used in a transformer? **[P.T.U. December, 2005]**

Ans. Iron-silicon alloy core is used in a transformer to ensure a high permeability of the magnetic circuit and keep the iron loss the minimum possible.

Q. 51. Why silicon content in electrical sheet steel is limited to 4.5 to 5 % ?

Ans. Silicon content exceeding 5 % makes the sheet steel brittle and so causes problem in punchings.

Q. 52. Where may auto-transformers be used?

Ans. Auto-transformers are used when transformation ratio is nearly equal to unity and where there is no objection to direct electrical connection between primary and secondary.

Q. 53. What is typical use of an auto-transformer? **[G.B.T.U. Odd Semester 2012-13]**

Ans. Typical uses of an auto-transformer are as (*i*) auto-starters (*ii*) regulating transformer (*iii*) furnace transformer and (*iv*) continuous variable transformer in electrical testing laboratory.

Q. 54. Mention application of shell type transformer.

Ans. The shell type construction is commonly used for small transformers where a square or rectangular cross section is suitable for economic consideration. Potential transformers (PTs) are always of shell type.

Q. 55. The full-load copper loss and iron loss of a transformer are 6,400 W and 5,000 watts respectively. The above losses at half load are .... **[G.B. Technical Univ. Even Semester 2012-13]**

Ans. Copper loss and iron loss at half load are $\left(\frac{1}{2}\right)^2 \times 6{,}400$ *i.e.* 1,600 W and 5,000 W respectively.

Q. 56. A 2-winding transformer is used to step-down the voltage from 220 V to 110 V. If an auto-transformer is used for the same purpose, what would be the ratio of copper weights in them?

Ans. 0.5.

PROBLEMS

1. The maximum flux density in the core of 250/3,000 V, 50 Hz single-phasetransformer is 1.2 Wb/m^2. If the emf per turn is 8 V, determine (*i*) primary and secondary turns (*ii*) area of the core. **[Ans** (*i*) 32; 384 (*ii*) 0.03 m^2]
2. A single-phase transformer has 500 turns in the primary and 1,200 turns in the secondary. The cross-sectional area of the core is 80 cm^2. If the primary winding is connected to a 50 Hz supply at 500 V, calculate peak flux density, and voltage induced in the secondary. **[Ans.** 0.563 T; 1200 V]
3. A 200 kVA, 6600/400 V, 50 Hz single phase transformer has 80 turns on the secondary. Calculate (*i*) the approximate values of the primary and secondary currents (*ii*) the approximate number of primary turns and (*iii*) the maximum value of flux. **[Ans.** (*i*) 30.3 A, 500 A; (*ii*) 1,320 (*iii*) 22.5 mWb]

4. The no-load current of a transformer is 5 A at 0.25 pf when supplied at 235 V, 50 Hz. The number of turns on the primary winding is 200. Calculate (*i*) the maximum value of flux in the core, (*ii*) the core loss (*iii*) the magnetizing component. **[Ans.** (*i*) 5.293 mWb (*ii*) 293.75 W (*iii*) 4.84 A]

5. A single-phase transformer with a ratio of 440/110 V takes a no–load current of 5 A at 0.2 power factor lagging. If the secondary supplies a current of 120 A at a pf of 0.8 lagging, estimate the current taken by the primary.
[Ans. (*i*) 33.9 A, 42.49° (lag)]

6. A 30 kVA, 2000/200 V, single-phase, 50 Hz transformer has a primary resistance of 3.5 Ω and reactance of 4.5 Ω. The secondary resistance and reactance are 0.015 Ω and 0.02 Ω respectively. Find (*i*) equivalent resistance, reactance and impedance referred to the primary side (*ii*) total copper loss in the transformer.
[U.P. Technical Univ. Electrical Engineering First Semester 2005-06]
[Ans. (*i*) 5 Ω, 6.5 Ω, 8.2 Ω (*ii*) 1,125 watts]

7. A 10 kVA, 1-phase transformer with 2,000/400 V at no load, has resistance and leakage reactance of primary winding 5.5 Ω and 12 Ω respectively, the corresponding values of secondary winding being 0.5 Ω and 0.45 Ω. Determine the value of secondary voltage at full load, 0.8 pf lagging, when the primary applied voltage is 2,000 V.
[U.P. Technical Univ. Electrical Engineering July 2002]
[Ans. 377.65 V]

8. A 5 kVA, 2,200/220 V single phase transformer has the following parameters.
HV side : $r_1 = 3.4\ \Omega$, $x_1 = 7.2\ \Omega$
LV side : $r_2 = 0.028\ \Omega$, $x_2 = 0.060\ \Omega$
The transformer is made to deliver rated current at 0.8 lagging power factor to a load connected to LV side. If the load voltage is 220 V, calculate the terminal voltage on the HV side. Also find the voltage regulation at this load.
[R.G.P.V. Bhopal Basic Electrical Engineering December 2003]
[Ans. 2,229.28 V; 1.33%]

9. If in a transformer core, the hysteresis and eddy current losses are 80 W and 50 W at normal voltage and frequency, Calculate the losses when voltage and frequency are increased by 20%.
[U.P. Technical Univ. Elec. Engineering First Semester 2004-05]
[Ans. 96 W, 72 W]

10. A transformer is connected to 2200 V, 40 Hz supply. The core loss is 800 watts out of which 600 watts are due to hysteresis and remaining eddy current losses. Determine the core loss if the supply voltage and frequency are 3300 V and 60 Hz respectively. **[Ans.** 1,350 W]

11. The following data were obtained on a 50 kVA, 2400/120 V transformer.
Open-circuit test, instruments on *lv* side:
Wattmeter reading = 396 W; Ammeter reading = 9.65 A;
Voltmeter reading = 120 V
Short-circuit test, instruments on *hv* side:
Wattmeter reading = 810 W; Ammeter reading = 20.8 A;
Voltmeter reading = 92 V
Find the equivalent circuit parameters referred to high voltage side.
[Pb. Technical Univ. Electrical Engineering May 2002]
[Ans. 14.5 kΩ, 5.3 kΩ, 1.872 Ω, 4 Ω]

12. The parameters of the equivalent circuit of a 200 kVA, 2500/250 V single phase transformer are as follows:
Primary resistance = 0.2 Ω ;
Secondary resistance = 2 m Ω
Primary leakage reactance = 0.45 Ω ;
Secondary leakage reactance = 4.5 m Ω
Core-loss resistance = 10 k Ω ;
Magnetizing reactance = 1.55 k Ω
Using the circuit referred to primary, determine (*a*) voltage regulation (*b*) efficiency of the transformer at rated load with 0.8 lagging power factor. **[Pb. Technical Univ. Electrical Engineering May 2001]**
[Ans. (*a*) 2.686% (*b*) 98.048%]

13. In a 25 kVA, 2000/200 V transformer, the constant and variable losses are 350 W and 400 W respectively. Calculate the efficiency on unity power factor at (*i*) full load and (*ii*) half-full load.

[Pb. Technical Univ. Electrical Engineering 1999]

[**Ans.** (*i*) 97.09% (*ii*) 96.53%]

14. The no-load transformation ratio in a 200 kVA, 50 Hz single phase transformer is 4,000/250 V. For a maximum core flux of 0.06 Wb, find

(*i*) the efficiency at half the rated kVA and at unity power factor,

(*ii*) the efficiency at full load, 0.8 power factor lagging and

(*iii*) kVA load for maximum efficiency, given that the full-load copper losses are 1.8 kW and core losses are 1.5 kW.

[U.P.Technical Univ. Electrical Machines 2001-2002]

[**Ans.** (*i*) 98.087% (*ii*) 97.98% (*iii*) 182.574 kVA]

15. The maximum efficiency of a 100 kVA, single phase transformer is 98 % and occurs at 80 % of full load. If the leakage impedance of the transformer is 5 %, find the voltage regulation at rated load of 0.8 pf lagging. [**Ans.** 3.74 %]

16. A 600 kVA, single-phase transformer when working at unity power factor, has an efficiency of 92 % at full-load and also at half-load. Determine its efficiency when it operates at unity pf and 60 % of full-load.

[U.P. Technical Univ. Electrical Machines 2003-04]

[**Ans.** 93.53%]

17. A transformer rated 200 V/50 V, 10 kVA has a core loss of 100 watts. What is the maximum efficiency of the transformer at 0.8 lagging power factor ? Assume full-load copper loss as 200 watts. At what load this maximum efficiency is obtained ? How and why would your answer be different if the power factor were unity ?

[R.G.P.V. Bhopal Basic Electrical Engineering December 2004]

[**Ans.** 7.07 kVA, 96.585%, 97.249%]

18. Find all-day efficiency of a transformer having maximum efficiency of 98 % at 15 kVA at unity power factor and loaded as follows:

12 hours — 2 kW at 0.5 pf lag:

6 hours — 12 kW at 0.8 pf lag

6 hours — at no-load. [**Ans.** 95.31%]

19. A transformer has a primary voltage rating of 11,500 V and secondary voltage rating of 2,300 V. Two windings are connected in series and the primary is connected to a supply of 11,500 V, to act as a step-up auto-transformer. Determine the output voltage of the transformer. [**Ans.** 13,800 V]

20. A 11,500/2300 V transformer is rated at 100 kVA as a two-winding transformer. If the two windings are connected in series to form an auto-transformer, what will be the possible voltage ratio? [**Ans.** 13,800/2300 V, 13,800/11,500 V]

❑❑❑

UNIT-V

Electrical Machines

CONTENTS IN THIS UNIT

Electrical Machines:

Principles of Electromechanical energy conversion

DC Machines: Types, EMF equation of generator and torque equation of motor, Characteristics and applications of dc motors (simple numerical problems).

Three-Phase Induction Motor:

Types, Principle of operation, Slip-torque characteristics, Applications (Numerical problems related to slip only).

Single-Phase Induction Motor: Principle of operation and introduction to methods of starting, Applications.

Three-Phase Synchronous Machines: Principle of operation of alternator and synchronous motor and their applications.

13

CHAPTER

DC Machines-I (Generators)

13.1. PRINCIPLES OF ELECTROMECHANICAL ENERGY CONVERSION

The role of electricity in modern technology is that of an extremely versatile intermediary. Although energy is seldom directly available in electrical form, and ultimately it is seldom required in electrical form, yet conversion of other forms of energies into electrical energy is a common practice. The chief advantages of this conversion is that energy in electrical form can be transmitted, controlled and utilized with relative simplicity, reliability and efficiency. Energy conversion devices are required first to convert energy in non-electrical form to energy in electrical form and then to convert electrical energy into the desired useful form, such as light, heat, sound or mechanical energy. Thus energy conversion devices are needed at both ends of the electrical system. One typical example is generation of electrical energy (conversion of non-electrical form energy into electrical form) at nuclear power station and then transmission and distribution over lines and finally conversion into mechanical energy by an electric motor for final use. Another example is the conversion of sound energy into electrical energy at the talker's end, its transmission in electrical form over lines and then its final conversion to sound waves at the listener's end.

We are concerned here with the **electromechanical energy** conversion process which takes place through the medium of the electric or magnetic field of the conversion device. Although the various conversion devices operate on similar principles, their structures depend upon their function. Devices for measurement and control are frequently referred to as *transducers*; they usually operate under linear input-output conditions and with relatively small signals. The many examples include torque motors, microphones, pick-ups, and loudspeakers. A second category of devices encompass force-producing devices such as solenoids, electromagnets and relays. The third category is of continuous-energy-conversion equipment and includes generators and motors.

A *dynamo* is a machine that converts either mechanical energy into electrical energy or electrical energy into mechanical energy. When a dynamo is driven mechanically by a prime mover such as by a diesel engine, steam engine, steam turbine or water turbine and supplies electrical energy, it is called a *generator*. When the dynamo draws electrical energy from the supply mains and drives mechanical devices such as line shafts and machine tools it is called a *motor*. In other words it can be stated that the generator converts mechanical power ωT into electrical power EI and motor converts electrical power EI into mechanical power ωT.

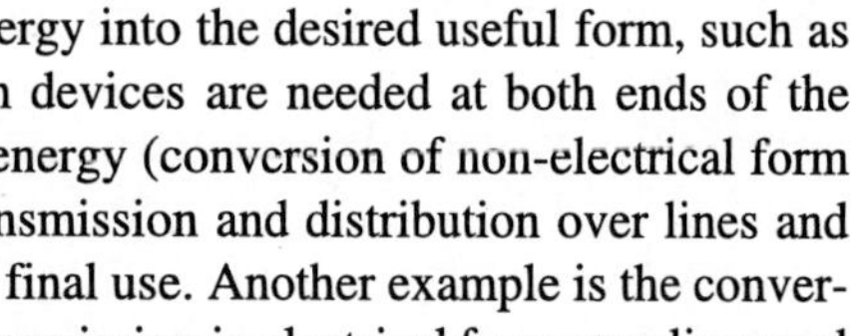

Fig. 13.1 *Block Representation of Electromechanical Energy Conversion*

The motoring and generating action is basically a reversible process *i.e.* the same machine is capable of operating either as a generator or as a motor depending upon whether the supplied power is mechanical or electrical. This is illustrated in Fig. 13.1. However, in practice, devices may be designed and constructed to suit one particular mode of conversion or the other.

An *electromechanical energy conversion device* is a link between an electrical and mechanical system and *electromechanical energy conversion* needs the presence of natural phenomena which interrelate electric and magnetic fields on one hand and mechanical force and motion on the other.

The most important phenomena which make this conversion possible are:

1. Whenever a current carrying conductor is placed in a magnetic field in such a way that a component of the length of the conductor is perpendicular to the field, a force will exist between the conductor and the field.

2. Whenever a piece of ferromagnetic material is placed in a magnetic field, a force is experienced upon the material, tending to bring it into the position of the densest portion of the field. The force tends to align the material specimen so that the reluctance of the magnetic path passing through the material will be at minimum.

3. A force of attraction is exerted upon the charged plates of a capacitor.

4. Whenever a piece of dielectric material is located in an electrostatic field, a force is experienced upon the material tending to bring it into the densest portion of the field. In case the specimen of material is of unsymmetrical configuration the force tends to align its longer axis with the electrostatic field.

All of these phenomena involve a force relation that exists between either a magnetic field or an electrostatic field and a material structure (matter). Thus, if relative motion between the field and the material structure is associated with this force, energy conversion must occur between the field and the mechanical system associated with the structure. All of these phenomena, therefore, provide a means for electromechanical energy conversion i.e. interchange of energy between a magnetic field (or an electrostatic field), and a mechanical system. The medium for the energy transfer is either a magnetic field or an electrostatic field. If the motion is in a direction opposite to that of the force resulting from the electromagnetic or electrostatic relations the motion is developed by the mechanical system, and mechanical energy is converted into energy associated with the electrical system. On the other hand, if the motion is in the direction of the force resulting from the electromagnetic or electrostatic relations, the motion is caused by the associated electrical system, and electrical energy is converted into mechanical energy.

For instance in phenomenon 1, when the motion is in the direction of the electromagnetic force, electrical energy from the electric source which causes the current to flow in the conductor is transferred through the associated magnetic field into mechanical energy. While in case the motion is opposite to the direction of the electromagnetic force, mechanical energy is transferred through the associated magnetic field into electrical energy of the system connected to the current carrying conductor.

In this text we are primarily interested in *electromechanical energy conversion by means of continually rotating machines*. The energy conversion in most of these machines with their present-day construction is produced through the presence of both phenomenon 1 and phenomenon 2. In both of these phenomena, the magnetic field is the medium for the energy transfer. Also, in all of our present-day continually rotating electrical machines, the energy of the magnetic field is directly associated with the energy of an electrical system external to the machine. Thus, the energy transfer involved always is between a mechanical system and an electrical system through the medium of a magnetic field.

Since, for all practical purposes, the mass of the materials used in the construction of an electrical machine remains constant under the conditions of operation, the principle of conservation of energy can be applied in the analysis of the energy conversion. The input energy must, therefore, be equal to the summation of the useful output energy, the energy converted into heat, and the change in the energy stored in the magnetic field.

In this energy balance equation, the useful output energy and the energy converted into heat will always be positive but the change in the energy stored in the magnetic field may be positive or negative depending upon whether the stored energy is increasing or decreasing. Some energy is always converted into heat (*i*) through the electrical (I^2R) losses in the electrical system of the machine (*ii*) through the hysteresis and eddy currents in the portions of the magnetic field which are located in magnetic materials and (*iii*) through friction losses in the mechanical system of the machine.

Thus energy balance equation may be written as

Total energy input = Energy output + increase in energy stored in the magnetic field + energy converted into heat.

$$\text{or } \int dW_{input} = \int dW_{output} + \int dW_{magnetic\ field} + \int dW_{heat} \quad ...(13.1)$$

Throughout certain lapses of a time, when a machine is started or stopped or when the load is changed, the rate of conversion of energy into heat and the energy stored in the magnetic field will vary. Such periods of operation are known as transient periods. When a machine operates at a constant speed and temperature and with constant rate of energy output, the energy stored in the magnetic field and the rate of conversion of energy into heat remain unchanged. This normal operating condition is called the *steady-state operation*. For steady-state operation, the energy and power equations become

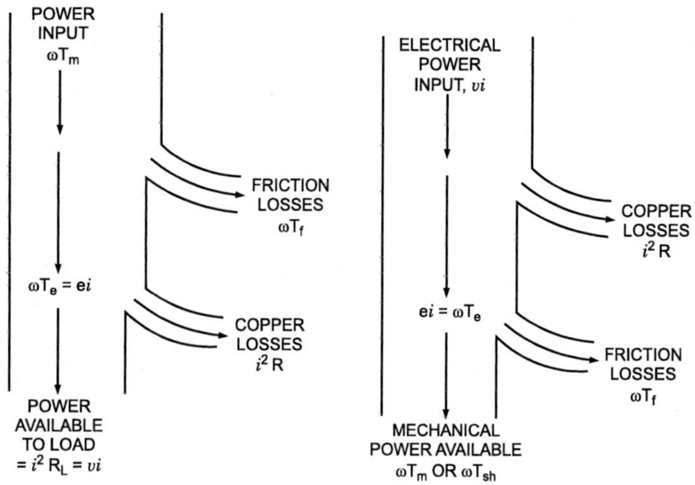

(*a*) *For Generator Operation* (*b*) *For Motor Operation*

Fig. 13.2 *Power Flow Diagrams*

$$W_{input} = W_{output} + W_{heat} \quad ...(13.2)$$

$$P_{input} = P_{output} + P_{heat} \quad ...(13.3)$$

The power flow diagrams for generator and motor are given in Figs. 13.2(*a*) and 13.2(*b*) respectively.

The flow of energy in electromechanical conversion in either direction along with irrecoverable energy losses is shown in Fig. 13.3 (*a*) and 13.3 (*b*).

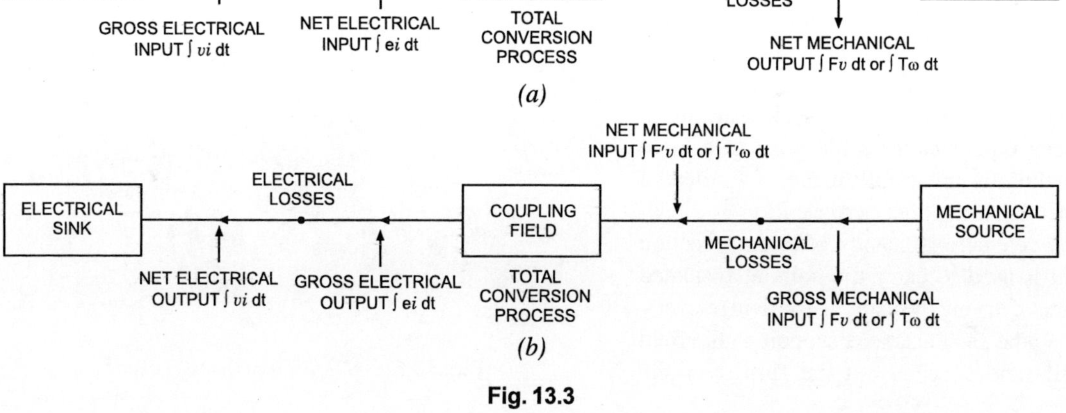

Fig. 13.3

13.2. DC MACHINES

DC machine is actually an *alternating current* machine, but furnished with a special device, called the *commutator,* which under certain conditions converts ac into dc and vice versa.

The term "generator" denotes that it generates electrical energy but actually it does not. It simply converts mechanical energy supplied to it into electrical energy. A dc generator may be compared with a water force pump to make operation of a dc generator more clear.

As the force pump does not produce water but causes a mechanical pressure which forces the existing water into an elevated reservoir against the back pressure due to its weight, in the same way the generator does not produce electricity but creates potential difference, which causes the electric current to flow from low pressure terminal to high pressure terminal in the machine and from high pressure terminal to low pressure terminal in the external circuit against the resistance of the circuit.

The generator operates on the principle of the production of dynamically induced emf *i.e.* whenever flux is cut by the conductor, dynamically induced emf is produced in it according to the laws of electromagnetic induction, which will cause a flow of current in the conductor if the circuit is closed.

For production of dynamically induced emf, three things are necessary, a magnetic field, a conductor and motion of the conductor with respect to the field. In dc generators the field is produced by the field magnets which are stationary. Permanent magnets are used for very small capacity machines and electromagnets are used for large machines to create magnetic flux. The conductors are situated on the periphery of the armature being rotated by the prime mover.

13.3. ESSENTIAL PARTS OF A DC MACHINE

DC machine (whether a generator or motor) with four poles is shown in Fig. 13.4. In construction, dc machine consists of four parts mainly 1. Field magnets 2. Armature 3. Commutator 4. Brush and brush gear. Disassembled dc machine is shown in Fig. 13.5.

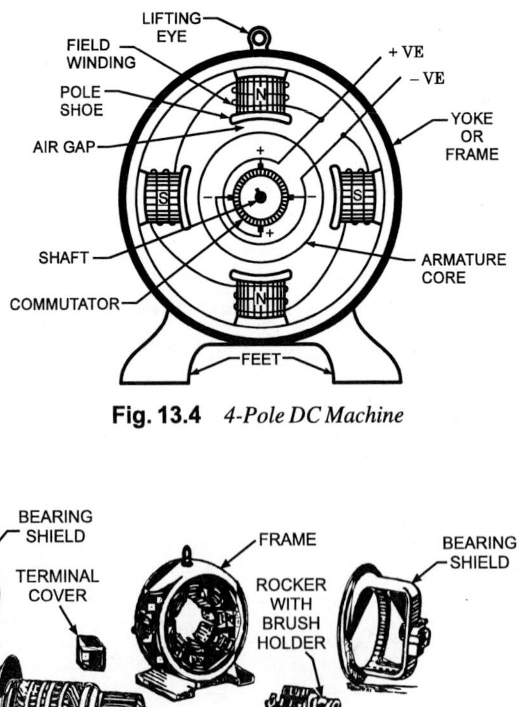

Fig. 13.4 *4-Pole DC Machine*

1. Field System. The object of the field system is to create a uniform magnetic field, within which the armature rotates.

Electromagnets are preferred in comparison with permanent magnets on account of its greater magnetic effect and its field strength regulation, which can be achieved by controlling the magnetising current.

Field magnet consists of four parts given below :

(*i*) Yoke or Frame (*ii*) Pole cores (*iii*) Pole shoes and (*iv*) Magnetising coils.

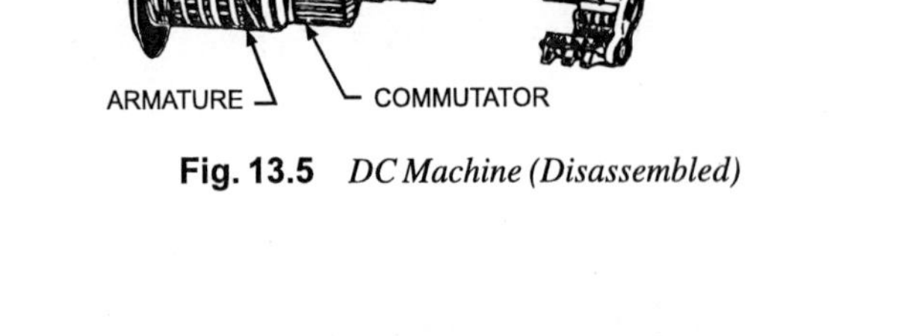

Fig. 13.5 *DC Machine (Disassembled)*

Cylindrical yoke is usually used which acts as a frame of the machine and carries the magnetic flux produced by the poles. In small machines, cast iron yokes are used, because of cheapness but yoke of a large machine is invariably made of fabricated steel due to its high permeability.

Pole core is usually of circular section and is used to carry the coils of insulated wires carrying the exciting (or field) current. The pole shoe acts as a support to the field coils and spreads out the flux over the armature periphery more uniformly and also being of larger cross section reduces the reluctance of the magnetic path.

2. Armature. It is a rotating part of a dc machine and is built up in a cylindrical or drum shape. The purpose of armature is to rotate the conductors in the uniform magnetic field. It consists of coils of insulated wires wound around an iron and so arranged that electric currents are induced in these wires when the armature is rotated in a magnetic field. In addition, its most important function is to provide a path of very low reluctance to the magnetic flux. The armature core is made from high permeability silicon steel stampings, each stamping, being separated from its neighbouring one by thin paper or thin coating of varnish as insulation.

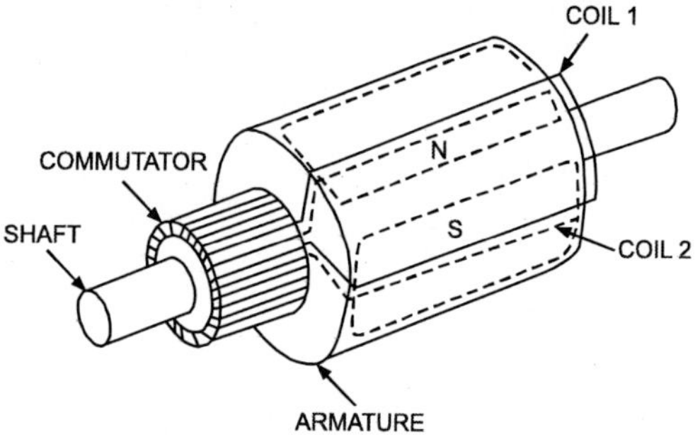

Fig. 13.6 *Longitudinal View of Armature*

A small air gap exists between the pole pieces and the armature so that there will be no rubbing in the machine. However this gap is kept as small as possible, since larger the air gap greater is the mmf required to create the required flux.

3. Commutator. The commutator is a form of rotating switch placed between the armature and the external circuit and so arranged that it will reverse the connections to the external circuit at the instant of each reversal of current in the armature coils.

It is very important part of a dc machine and serves the following purposes:

1. It provides the electrical connections between the rotating armature coils and the stationary external circuit.
2. As the armature rotates, it performs a switching action reversing the electrical connections between the external circuit and each armature coil in turn so that the armature coil voltages add together and result in a dc output voltage.
3. It also keeps the rotor or armature mmf stationary in space.

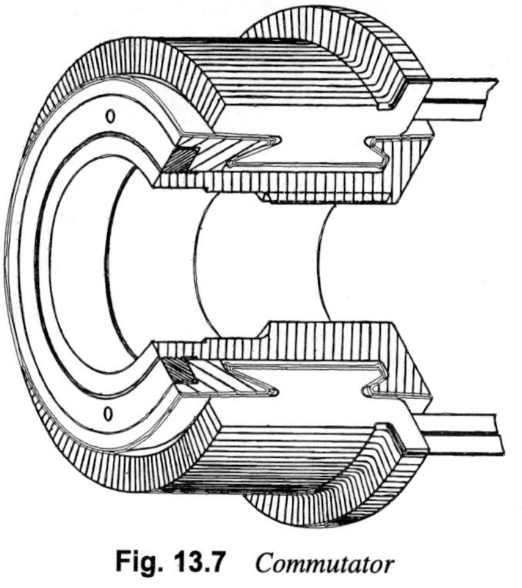
Fig. 13.7 *Commutator*

4. Brushes. The function of brushes is to collect current from the commutator and supply it to the external load circuit (the armature of the machine being connected to the external load circuit via the commutator and brushes). The brushes are rectangular in shape and rest on the commutator.

5. Armature Winding. The insulated wires housed in armature slots are suitably connected. This is called the *armature winding*. Armature winding plays vital role in a dc machine. It is a place where conversion of power takes place *i.e.* conversion of mechanical power into electrical one in case of a generator and conversion of electrical power into mechanical one in case of motor. The armature windings most commonly used in drum type armature are of two types namely *lap winding* and *wave winding*.

In *lap winding* finish end of one coil is connected to a commutator segment and to the start end of the adjacent coil under the same pole and similarly all coils are connected. The winding is known as *lap winding* because the sides of successive coils overlap each other. Single turn lap winding is shown in Fig. 13.8 (a) and (b). In lap winding there are as many parallel paths or circuits through the winding as there are field poles on the machine. Lap winding because of larger number of parallel paths and lesser number of conductors per path is suitable for large current and low voltage machines.

Wave winding is also sometimes known as *series winding*. In wave winding finish end of one coil is connected to the start of another coil as shown in Fig. 13.9. Thus in wave winding, the winding progresses,

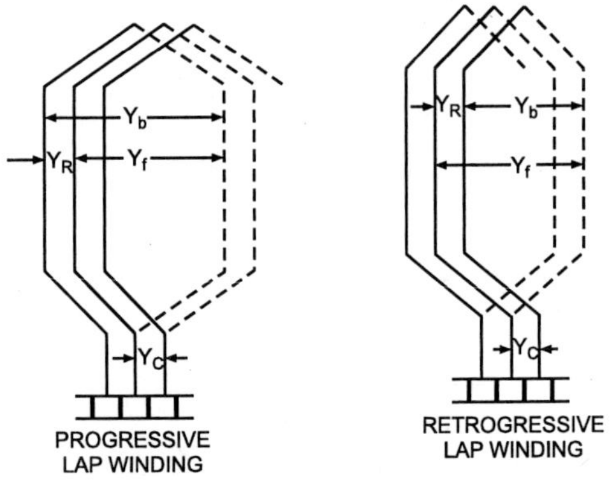

(a) (b)

Fig. 13.8

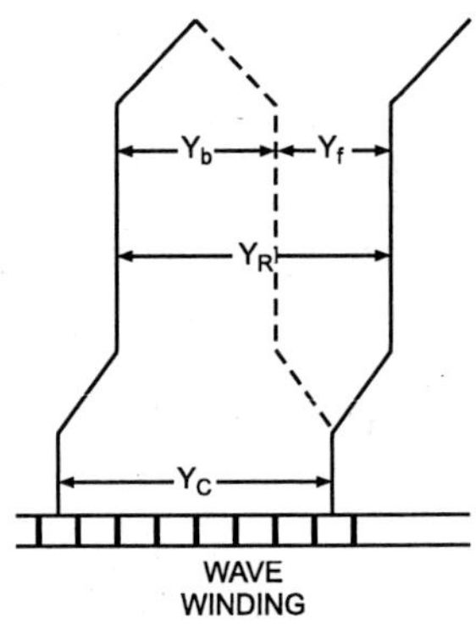

Fig. 13.9

passing every N pole and S pole till it returns to the coil side from where it was started. As the winding is wavy, the winding is, therefore, called *wave winding*. The wave winding gives always two parallel paths irrespective of number of poles and, therefore, for a given number of poles and armature conductors, it gives more emf than the lap winding. Hence it is used for high voltage and low current machines.

13.4. SIMPLE LOOP GENERATOR

A single turn rectangular copper conductor loop ABCD rotating in clockwise direction about its own axis XX' in the uniform magnetic field is shown in Fig. 13.10. As the coil is rotated in the magnetic field by some mechanical means (not shown in the fig.) the flux linking with the loop changes continually, therefore an emf is induced in it. The magnitude of emf induced at any instant is proportional to the rate of change of linking flux at that instant and its direction is given by Fleming's right hand rule.

While rotating at the instant the loop of wire ABCD assumes the position shown in Fig. 13.10 (*a*) *i.e.* when the loop ABCD is just parallel to the faces of field magnets N and S, the flux linking with the loop is maximum but the rate of change of linking flux is zero, as at this instant no flux is cut by the coil sides AB and CD which are just moving parallel to them. Hence induced emf is zero when the loop is in this position. This position of the loop is known as *neutral position* and let it be taken starting position and angle of rotation be measured from this position.

As the coil is turned in clockwise direction at a constant speed, the coil sides begin to cut across the field, slowly at first but at gradually increasing rate. Thus the magnitude of emf induced gradually increases as the loop moves and becomes maximum when the loop assumes the position shown in Fig. 13.10 (*b*). During this period the loop rotates through $\frac{\pi}{2}$ radians or 90° and induced emf in the loop increases following the sine law, as shown in Fig. 13.11. The direction of emf induced in the coil, as given by Fleming's right hand rule, is from B to A and from D to C. In the position of the loop shown in Fig. 13.10 (*b*) the coil sides are moving at right angles to the field and are, therefore, cutting across the field at maximum rate, consequently, the emf induced at this instant is of maximum value. In the next quarter of the revolution of the loop *i.e.* between $\frac{\pi}{2}$ radians and π radians (or 90° to 180°), the rate at which the conductors cut across the magnetic field gradually decreases, causing the magnitude of induced emf to fall gradually with the angular movement of the loop and becomes zero at the instant the loop becomes again parallel to the faces of the field magnets but with the sides AB and CD's position interchanged with respect to zero position [Fig. 13.10 (*c*)].

In the third quarter of the revolution of loop *i.e.* between π and $\frac{3\pi}{2}$ radians (or 180° to 270°) the rate at which the conductors cut across the magnetic field hence induced emf gradually increases as the loop moves and becomes maximum at the instant, the loop assumes the position shown in Fig. 13.10 (*d*). However, the direction of emf in the loop is now from A to B and from C to D *i.e.* opposite to that in the first two quarters.

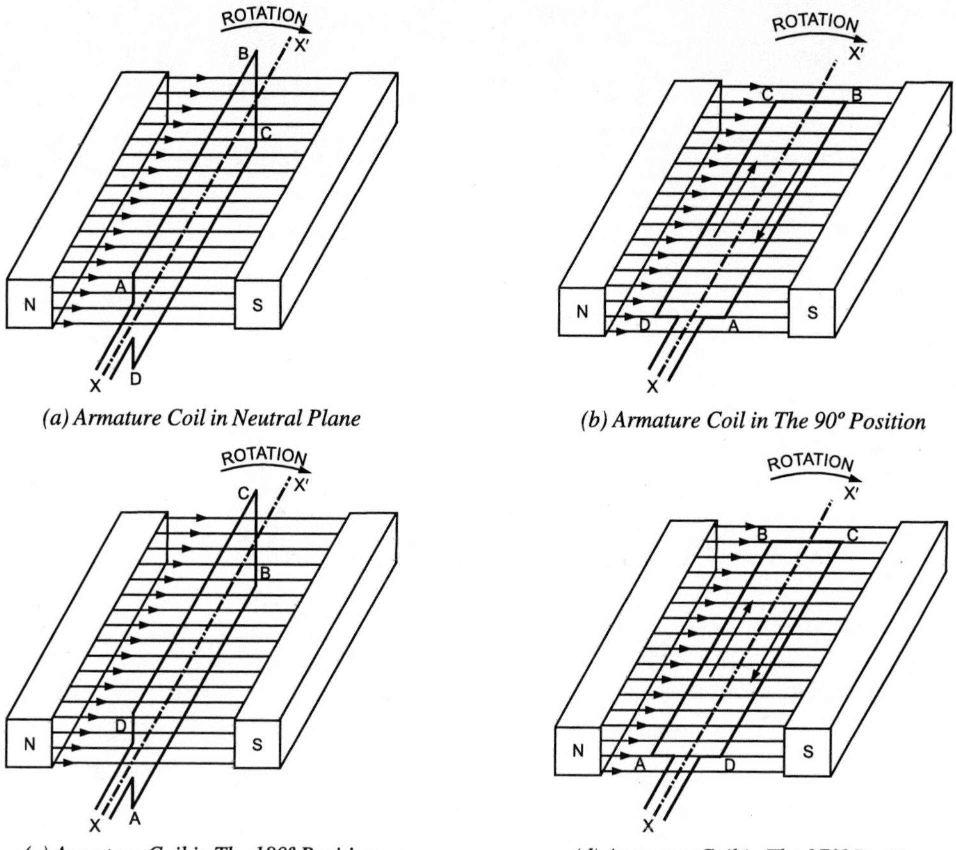

(a) Armature Coil in Neutral Plane

(b) Armature Coil in The 90° Position

(c) Armature Coil in The 180° Position

(d) Armature Coil in The 270° Position

Fig. 13.10

In the fourth quarter of the revolution of the loop *i.e.* between $\frac{3\pi}{2}$ and 2π radians (or 270° to 360°), the induced emf decreases as the coil moves and becomes zero when it completes 2π radians or 360° from the starting instant. At this instant the loop comes to its absolute original position and hence the loop is said to have completed one cycle. The cycle then repeats for each revolution of the armature.

The emf thus generated in the loop is of the form shown in Fig. 13.11 from which it is obvious that emf induced in armature conductors is of pulsating nature. Such an emf is known as an *alternating emf.*

The current induced in the coil is collected and conveyed to the external load circuit by connecting the coil terminals to two continuous and insulated rings, known as *slip rings or collector rings*, mounted on the generator shaft and making the two stationary brushes pressing against the slip rings, one brush bearing on each ring, as shown in Fig. 13.12. When the coil is rotated, the generated alternating emf causes a current to flow first in one direction and then in the other through the coil and external circuit. Such a current is called an *alternating current.*

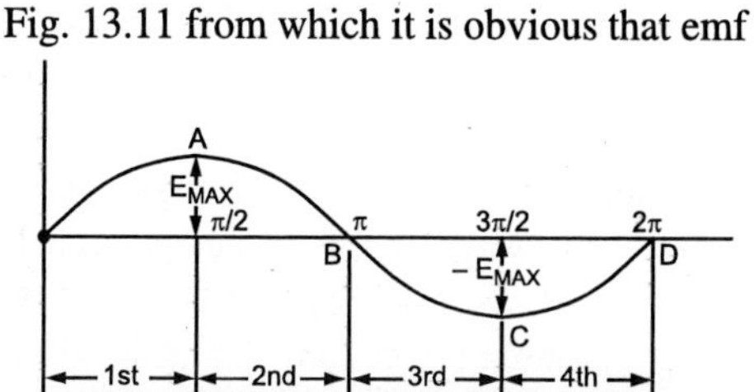

Fig. 13.11 *Curve Showing Variation of Induced EMF in a Coil Rotated in a Uniform Magnetic Field*

To obtain the unidirectional or direct current in the external circuit, the collecting arrangement is modified as shown in Fig. 13.13. In this arrangement, the slip rings are replaced by split ring made of a conducting material and splitted into two halves separated from each other by insulation and brushes are placed diametrically opposite instead of being side by side.

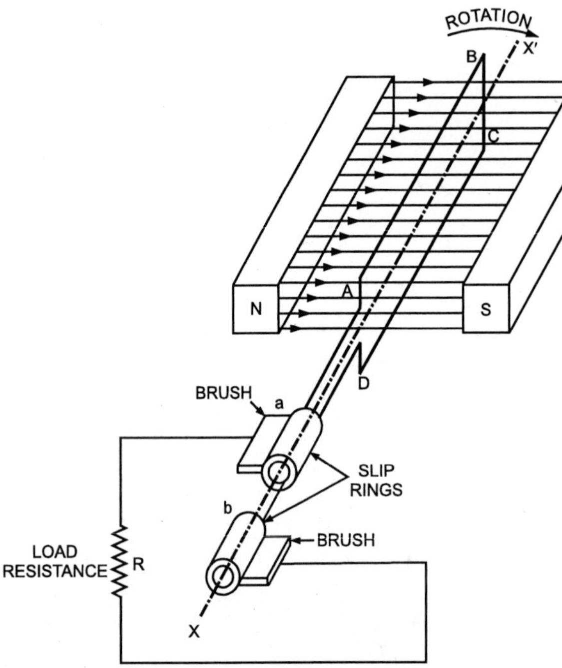

Fig. 13.12 *Elementary Generator With Brushes and Collector Rings*

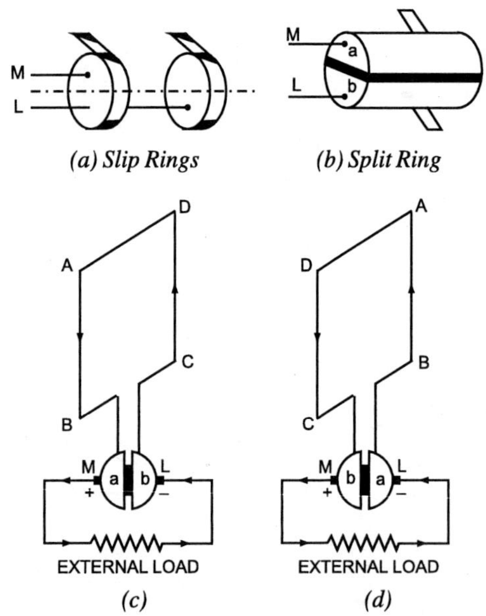

Fig. 13.13

It will be observed that in the first half revolution current flows along ABMLCDA *i.e.* the brush M in contact with segment 'a' acts as the + ve pole of the supply and brush L in contact with segment 'b' acts as the – ve pole.

In the next half revolution, the direction of induced currents in the coil is reversed, but at the same time the positions of segments 'a' and 'b' are also reversed, with the result that brushes M and L again come in contact with + ve segment 'b' and – ve segment 'a' respectively. Thus the direction of current in the external load circuit remains the same.

The position of the brushes shall be so arranged that the change of segments from one segment to the other takes place when the plane of the rotating coil is perpendicular to the plane of field, since in this position the induced emf or induced current in the coil will be zero.

Thus the emf or current induced in the armature conductors of a dc generator is alternating which is rectified by the split ring known as *commutator*.

Although after rectification the current through the load circuit is always in the same direction, but it is not a steady current since the emf generated in the armature coil and applied to the brushes, varies from zero to maximum and back to zero twice each revolution. The variation in brush voltage is shown in Fig. 13.14.

A pulsating direct current such as given by a single coil generator is not suitable for most commercial purposes. However, by employing a large number of coils and commutator segments, with the coils evenly distributed around the surface of the armature the brush voltage may be made practically constant.

The voltage generated by a single turn armature coil is small. For this reason, the coils employed in commercial generators consist of several turns in series thereby increasing the magnitude of generated emf in direct proportion to the number of the turns in the coil.

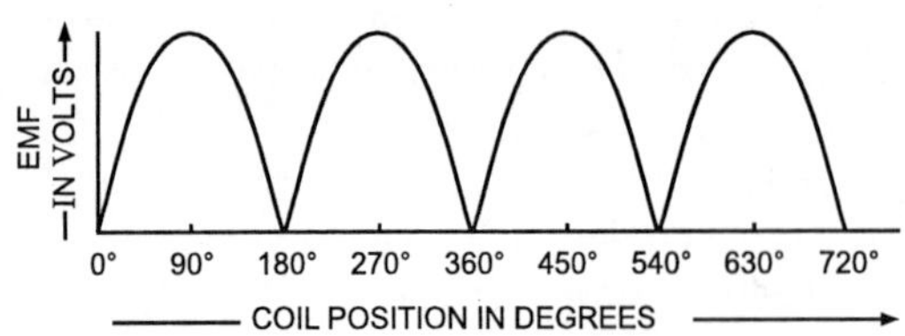

Fig. 13.14 *EMF Curve of a Simple DC Generator*

13.5. EMF EQUATION

Let Φ be the flux per pole in webers, Z the total number of armature conductors or coil sides on the armature, P the number of poles, A the number of parallel paths in the armature and N be the rotational speed of armature in revolutions per minute (rpm).

As will be recalled, the induced emf is proportional to the time rate of change of the magnetic flux *i.e.*

$$e = \frac{-d\Phi}{dt}$$

During one revolution of armature in a P-pole generator each armature conductor cuts the magnetic flux P times, so flux cut by one conductor in one revolution = P Φ webers.

Since the number of revolutions made by the armature per minute is N, number of revolutions made per second is $\frac{N}{60}$ and, therefore, flux cut by each conductor per second

= Flux cut by one conductor per revolution × number of revolutions of armature/second

$$= \Phi P \times \frac{N}{60} \text{ webers}$$

Consequently the average emf induced in one conductor will be

$$e = \Phi P \frac{N}{60} \text{ volts}$$

The number of conductors in series between a + ve brush and – ve brush is equal to the total number of conductors divided by the number of parallel paths i.e. number of armature conductors per parallel path = $\frac{Z}{A^*}$

∴ The total emf generated between the terminals,

E = Average emf induced in one conductor × number of conductors in each circuit or parallel path

$$= \Phi P \frac{N}{60} \times \frac{Z}{A} \text{ volts} = \Phi Z \frac{N}{60} \times \frac{P}{A} \text{ volts} \qquad ...(13.4)$$

For a given machine the number of poles (P) and number of armature conductors per parallel path ($\frac{Z}{A}$) are constant, therefore,

$$\text{Generated emf, } E = K \Phi N \qquad ...(13.5) \text{ where } K = \frac{PZ}{60\text{ A}}$$

$$\text{or } E \propto \Phi N$$

$$\text{or } E \propto \Phi\, \omega \text{ where } \omega = \frac{2\pi N}{60}\text{, the angular velocity in radians / second}$$

Thus we see that the induced emf is directly proportional to flux per pole Φ and speed N. Moreover, the polarity of the induced emf depends upon the direction of the magnetic field and the direction of rotation. If either of the two is reversed, the polarity of the induced emf *i.e* brushes is reversed, but when both are reversed, the polarity remains unchanged.

The induced emf is fundamental phenomenon to all dc machines whether they are operating as generators or motors. However, when the machine is operating as a generator, this induced emf is called the *generated emf,* E_g whereas in case of a machine operating as a motor it is called the *counter* or *back emf,* E_b.

Example 13.1 **A dynamo has a rated armature current at 250 A. What is the current per path of the armature if the armature winding is simplex wave wound or simplex lap wound ? The machine has 12 poles.**

Solution:

Rated armature current, I_a = 250 A

Number of poles, P = 12

* A = 2 in case of wave winding and = P in case of lap winding

With simplex wave winding

Number of parallel paths, A = 2

$$\therefore \text{Current per path, } I_c = \frac{I_a}{A} = \frac{250}{2} = 125 \text{ A } \textbf{Ans.}$$

With simplex lap winding

Number of parallel paths, A = P = 12

$$\text{Current per path, } I_c = \frac{I_a}{A} = \frac{250}{12} = 20.833 \text{ A } \textbf{Ans.}$$

Example 13.2 **Derive emf equation of a dc generator. What will be change in emf induced if flux is reduced by 20% and the speed is increased by 20%.** *[U.P. Technical Univ. Electrical Engineering Second Semester 2005-06]*

Solution: Since induced emf, $E \propto \Phi N$

$$\text{or } E_2 = E_1 \times \frac{\Phi_2}{\Phi_1} \times \frac{N_2}{N_1} = E_1 \times 0.8 \times 1.2 = 0.96\ E_1.$$

$$\text{So change in emf} = \frac{E_1 - E_2}{E_1} \times 100 = \frac{E_1 - 0.96E_1}{E_1} \times 100 = 4\% \text{ decrease } \textbf{Ans.}$$

Example 13.3 **A 6 pole lap wound dc generator has 720 conductors; a flux of 80 mweber/pole is driven at 1,000 rpm. Find the generated emf.** *[G.B. Technical Univ. Electrical Engineering Second Semester 2010-11]*

Solution:

$$\text{Generated emf, } E = \frac{\Phi Z N}{60} \times \frac{P}{A} = \frac{80 \times 10^{-3} \times 720 \times 1{,}000}{60} \times \frac{6}{6} = 960 \text{ V} \quad \textbf{Ans.}$$

Example 13.4 **A dc generator has an armature emf of 100 V when the useful flux per pole is 20 mWb, and the speed is 800 rpm. Calculate the generated emf (i) with the same flux and a speed of 1,000 rpm, (ii) with a flux per pole of 24 mWb and a speed of 900 rpm.** *[U.P. Technical Univ. Electrical Engineering January 2003]*

Solution: Generated emf, $E_1 = 100$ V

Flux per pole, $\Phi_1 = 20$ m Wb $= 20 \times 10^{-3}$ Wb

Speed, $N_1 = 800$ rpm

Since induced emf is proportional to the product of flux per pole Φ and armature speed N for a given machine i.e.

$$E \propto \Phi\ N$$

$$\text{or } \frac{E_2}{E_1} = \frac{N_2}{N_1} \times \frac{\Phi_2}{\Phi_1}$$

$$(i) \text{ or } E_2 = \frac{N_2}{N_1} \times \frac{\Phi_2}{\Phi_1} \times E_1 = \frac{1{,}000}{800} \times 1 \times 100 = 125 \text{ V } \textbf{Ans.}$$

$\because N_2 = 1{,}000$ rpm and $\Phi_2 = \Phi_1$

$$(ii) \text{ and } E_3 = \frac{N_3}{N_1} \times \frac{\Phi_3}{\Phi_1} \times E_1 = \frac{900}{800} \times \frac{24 \times 10^{-3}}{20 \times 10^{-3}} \times 100 = 135 \text{ V } \textbf{Ans.}$$

$\because N_3 = 900$ rpm and $\Phi_3 = 24 \times 10^{-3}$ Wb

Example 13.5 **A 4-pole lap wound armature has 144 slots with two coil sides per slot, each coil having two turns. If the flux per pole is 20 mWb and armature rotates at 720 rpm, what is the induced voltage ?** *[U.P. Technical Univ. Electrical Engineering Second Semester 2008-09]*

Solution: Flux per pole, $\Phi = 20$ mWb $= 20 \times 10^{-3}$ Wb

Number of armature conductors, Z = Number of slots × number of coil sides per slot × number of turns in each coil

$= 144 \times 2 \times 2 = 576$

Number of parallel paths, A = P = 4

$$\text{Induced voltage, } E = \Phi \times Z \times \frac{N}{60} \times \frac{P}{A} = 20 \times 10^{-3} \times 576 \times \frac{720}{60} \times \frac{4}{4} = 138.24 \text{ V } \textbf{Ans.}$$

Example 13.6 **An 8-pole dc generator has 500 armature conductors and a useful flux of 0.05 Wb. What will be the emf generated, if it is lap-connected and runs at 1,200 rpm ? What must be the speed at which it is to be driven to produce the same emf, if it is wave wound.** *[U.P. Technical Univ Elec. Engineering February 2001]*

Solution: EMF generated when the generator is lap-connected,

$$E_g = \frac{\Phi ZN}{60} \times \frac{P}{A} = \frac{0.05 \times 500 \times 1{,}200}{60} \times \frac{8}{8} = 500 \text{ V } \textbf{Ans.}$$

$\because$ in lap-connected armature, number of parallel paths, A = P = 8

Speed of the generator, when wave-connected, to generate an emf of 500 V

$$\text{So } N' = \frac{E_g \times 60}{\Phi \times Z} \times \frac{A}{P} = \frac{500 \times 60}{0.05 \times 500} \times \frac{2}{8} = 300 \text{ rpm } \textbf{Ans.}$$

$\because$ in wave-connected armature A = 2.

Example 13.7 **The armature of a four-pole dc machine has 100 turns and runs at 600 rpm. The EMF generated in open circuit is 220 V. Find the useful flux per pole when armature is**
(*i*) lap connected (*ii*) wave connected.

[G.B. Technical Univ. Electrical Engineering First Semester 2010-11]

Solution:

EMF generated in open circuit, E = 220 V
Speed, N = 600 rpm
Number of poles, P = 4
Number of armature conductors, Z = 2 × number of turns = 2 × 100 = 200

(*i*) When armature is lap connected

Number of parallel paths, A = P = 4

$$\text{Flux per pole, } \Phi = E \times \frac{A}{P} \times \frac{60}{N} \times \frac{1}{Z} \quad \because \quad E = \frac{\Phi Z N}{60} \times \frac{P}{A}$$

$$= 220 \times \frac{4}{4} \times \frac{60}{600} \times \frac{1}{200} = 0.11 \text{ Wb} \quad \textbf{Ans.}$$

(*ii*) When armature is wave connected,

Number of parallel paths, A′ = 2

$$\text{and Flux per pole, } \Phi' = \frac{220 \times 2}{4} \times \frac{60}{600} \times \frac{1}{200} = 0.055 \text{ Wb} \quad \textbf{Ans.}$$

13.6. TYPES OF DC GENERATORS

The mmf necessary to establish the flux in the magnetic circuit of a dc generator can be obtained by means of (*i*) a permanent magnet (*ii*) field coils excited by some external source and (*iii*) field coils excited by the generator itself.

When a permanent magnet is used for establishing the flux in the magnetic circuit, the generator is known as a *permanent magnet generator*. It consists of an armature and one or several permanent magnets encircling the armature. The field developed by the poles of such machines remains fairly constant. Although these machines are very compact but in view of the low power thus generated, permanent magnet generators have not found industrial applications. Such generators are employed only in small sizes like dynamos in motor cycles etc.

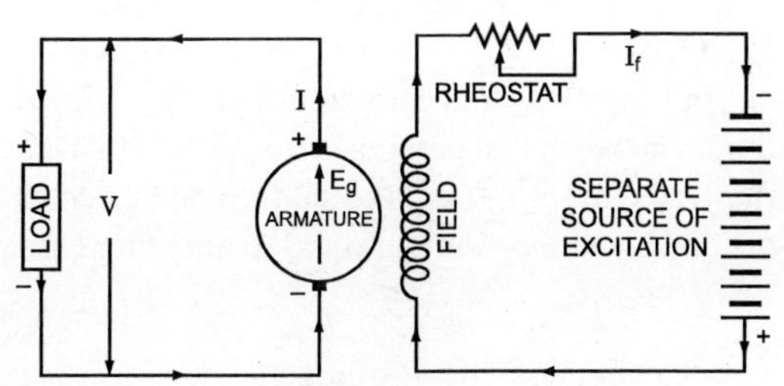

Fig. 13.15

13.6.1. Separately-Excited DC Generators. Since the operation of a dc machine depends upon a fixed polarity of the poles which does not vary with time, the field coils need energization from a dc source.

A dc generator whose field winding is excited from an independent external dc source, such as a battery, the generator is called a *separately-excited generator*. The circuit diagram is illustrated in Fig. 13.15. In this case current flowing through the armature, I_a and load, I_L is the same and the terminal voltage (voltage across the load), V is equal to the generated emf, E_g less voltage drop in armature, $I_a R_a$ *i.e.*

$$I_a = I_L = I(\text{say}) \quad ...(13.6)$$

$$V = E_g - I R_a \quad \text{where } R_a \text{ is the armature resistance} \quad ...(13.7)$$

$$\text{Power developed, } P_g = E_g I \quad ...(13.8)$$

$$\text{Power delivered to external Load, } P_L = VI \quad ...(13.9)$$

13.6.2. Self-Excited DC Generators. A dc generator whose field winding is excited by the current supplied by the generator itself, is called a *self-excited generator.* In such machines the field coils are interconnected with the armature winding. The field coils may be connected either in *series* with the armature, in parallel with the armature or *partly in series and partly in parallel* with the armature. Accordingly the self-excited generators may be classified as (*i*) *series wound generators* (*ii*) *shunt wound generators* and (*iii*) *compound wound generators.*

Due to residual magnetism, some flux is always present in the poles of such machines. When the armature is rotated, a small voltage is induced in the armature winding owing to residual flux. This induced voltage causes a small current to flow in the field coils and thus increase in flux per pole. The increase in flux causes increase in induced voltage which further increases the field current and so flux per pole. These events take place rapidly and the generator builds up to the rated voltage.

1. Series Wound Generators. In a series wound generator, there is only one field winding (one field coil on the core of each salient pole) which is connected in series with armature winding so that whole current flows through the field winding as well as load. The connection diagram of a series wound generator is given in Fig. 13.16. Since the series field winding carries the full-load current, series field winding is designed with a fewer turns of thick wire or strips. The resistance of series field winding R_{se} is, therefore, very low, say of the order of 0.5 Ω. The important relations are given below :

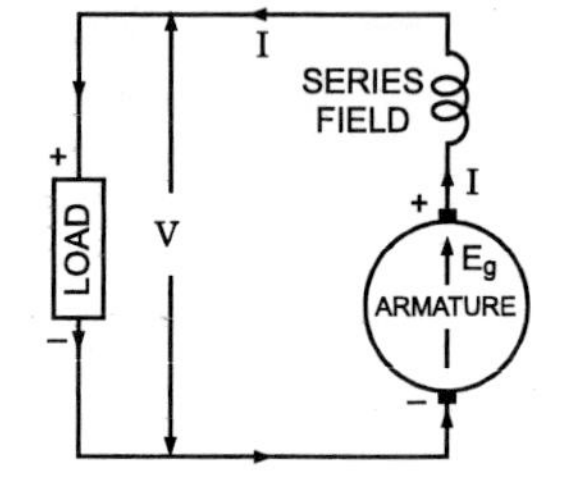

Fig. 13.16

$$I_a = I_{se} = I_L = I\,(\text{say}) \quad ...(13.10)$$

$$V = E_g - I\,(R_a + R_{se}) \quad ...(13.11)$$

$$\text{Power developed, } P_g = E_g I \quad ...(13.12)$$

$$\text{Power delivered, } P_L = VI \quad ...(13.13)$$

2. Shunt Wound Generators. In a shunt wound generator also there is only one field winding, as in case of series wound generator, but in this case it is connected across the armature circuit forming a parallel or shunt circuit. The voltage across the field winding is, therefore, the same as the terminal voltage of the generator. The connection diagram is illustrated in Fig. 13.17.

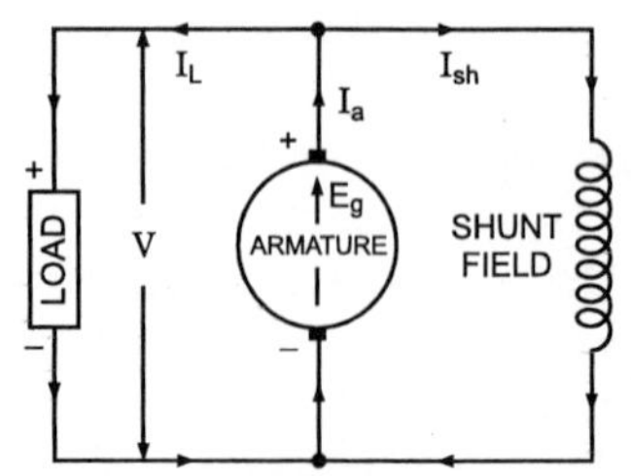

Fig. 13.17

In operation, total armature current I_a flowing from the + ve brush divides between two parallel paths; the external circuit and the field circuit. The load current I_L and the shunt field current I_{sh} close the circuit at the –ve brush and their sum is equal to armature current *i.e.*

$$I_a = I_L + I_{sh}$$

The effective power of a generator is proportional to the current delivered to the external load circuit, therefore, it is necessary to keep shunt field current as small as possible. For shunt wound generators it is kept about 2.5 per cent of the rated armature current. Shunt field coils are, therefore, designed to produce the desired mmf by means of a relatively large number of turns of fine wire. The resistance of the shunt field winding R_{sh} will naturally be high, say of the order of 100 Ω. The important relations are given below:

$$\text{Shunt field current, } I_{sh} = \frac{V}{R_{sh}} \quad ...(13.14)$$

$$\text{Armature current } I_a = I_L + I_{sh} \quad ...(13.15)$$

$$\text{Terminal voltage, } V = E_g - I_a R_a \quad ...(13.16)$$

$$\text{Power developed, } P_g = E_g I_a \quad ...(13.17)$$

$$\text{Power delivered, } P_L = V I_L \quad ...(13.18)$$

3. Compound Wound Generators. In compound wound generators, there are two field windings (two separate coils on the core of each pole). One of them (having many turns of fine wire) is connected across the armature and the other (having few turns of thick wire) is connected in series with the armature winding. Compound generators may be connected either *short shunt*, with the shunt field winding in parallel with the armature alone, as illustrated in Fig. 13.18 (a) or long shunt, with the shunt field winding in parallel with both the armature and series field winding, as shown in Fig. 13.18 (*b*).

In a compound wound generator, the major portion of the excitation is usually supplied by the shunt field. The shunt field is slightly weaker and the series field is considerably weaker than those of the corresponding machine in which the entire excitation is produced by single shunt or by a single series winding.

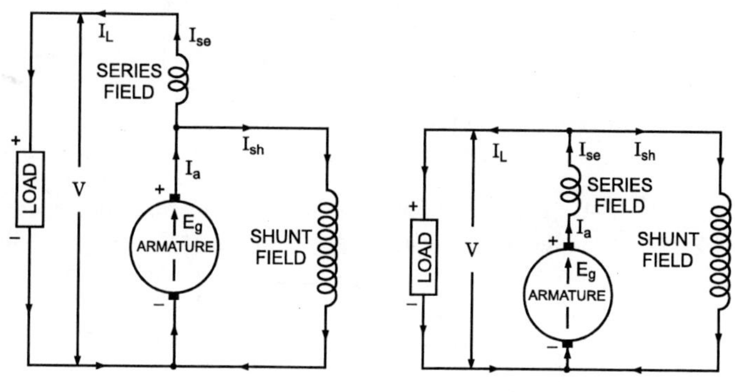

(*a*) *Short Shunt Compound Wound Generator* (*b*) *Long Shunt Compound Wound Generator*

Fig. 13.18

Compound generators are also of the two types, known as *cumulative compound wound* and *differential compound wound* generators. In cumulative compound wound generators, series field assists the shunt field, as shown in Fig. 13.19 (*a*) whereas in differential compound wound generators, series field opposes the shunt field, as shown in Fig. 13.19 (*b*).

The important relations for short shunt and long shunt compound wound generators are given below:

(a) Short Shunt Compound Wound Generators

$$\text{Series field current, } I_{se} = I_L \quad ...(13.19)$$

$$\text{Shunt field current, } I_{sh} = \frac{V + I_{se} R_{se}}{R_{sh}} \quad ...(13.20)$$

$$\text{Armature current, } I_a = I_L + I_{sh} \quad ...(13.21)$$

$$\text{Terminal voltage, } V = E_g - I_a R_a - I_{se} R_{se} \quad ...(13.22)$$

$$\text{Power developed, } P_g = E_g I_a \quad ...(13.23)$$

$$\text{Power delivered, } P_L = V I_L \quad ...(13.24)$$

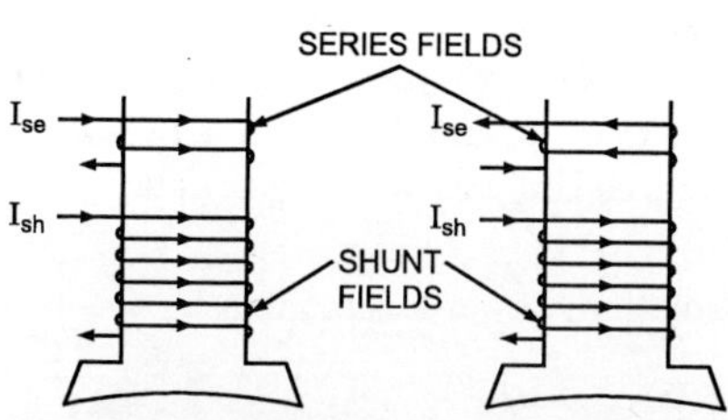

(*a*) *Cumulative Compounding* (*b*) *Differential Compounding*

Fig. 13.19

(b) Long Shunt Compound Wound Generators

$$\text{Shunt field current, } I_{sh} = \frac{V}{R_{sh}} \quad ...(13.25)$$

$$\text{Armature current, } I_a = \text{Series field current, } I_{se} \quad ...(13.26)$$
$$= I_L + I_{sh} \quad ...(13.27)$$
$$\text{Terminal voltage, } V = E_g - I_a R_a - I_{se} R_{se} = E_g - I_a (R_a + R_{se}) \quad ...(13.28)$$
$$\text{Power developed, } P_g = E_g I_a \quad ...(13.29)$$
$$\text{Power delivered, } P_L = V I_L \quad ...(13.30)$$

Example 13.8 Calculate the voltage induced in the armature winding of a 4-pole, lap wound dc machine having 728 active conductors and running at 1,800 rpm. The flux per pole is 30 mWb. If the armature is designed to carry a maximum line current of 100 A, what is the maximum electromagnetic power developed by the armature?

[U.P. Technical Univ. Electrical Engineering First Semester 2009-10]

Solution: Flux per pole, Φ = 30 mWb = 0.03 Wb

Number of armature conductors, Z = 728

Speed of rotation, N = 1,800 rpm

Number of poles, P = 4

Number of parallel paths, A = P = 4 $\quad \because$ the armature winding is lap wound

$$\text{Voltage induced, } E = \frac{\Phi ZN}{60} \times \frac{P}{A} = \frac{0.03 \times 728 \times 1{,}800}{60} \times \frac{4}{4} = 655.2 \text{ V } \textbf{Ans.}$$

Current carrying capacity of armature winding, I = 100 A

Electromagnetic power developed by armature = EI = 655.2 × 100 = 65,520 W or 65.52 kW **Ans.**

Example 13.9 A 20 kW, 200 V shunt generator has an armature resistance of 0.05 Ω and a shunt field resistance of 200 Ω. Calculate the power developed in the armature when it delivers rated output. Give sketch.

[U.P. Technical Univ. Electrical Engineering Second Semester 2006-07]

Solution:

$$\text{Load current, } I_L = \frac{P}{V} = \frac{20{,}000}{200} = 100 \text{ A}$$

Shunt field resistance, R_{sh} = 200 Ω

$$\text{Shunt field current, } I_{sh} = \frac{V}{R_{sh}} = \frac{200}{200} = 1 \text{ A}$$

Armature current, $I_a = I_L + I_{sh} = 100 + 1 = 101$ A

Generated emf, $E_g = V + I_a R_a$

= 200 + 101 × 0.05

= 205.05 V

Power developed, $P_g = E_g \times I_a$ = 205.05 × 101 = 20,710 W or 20.71 kW **Ans.**

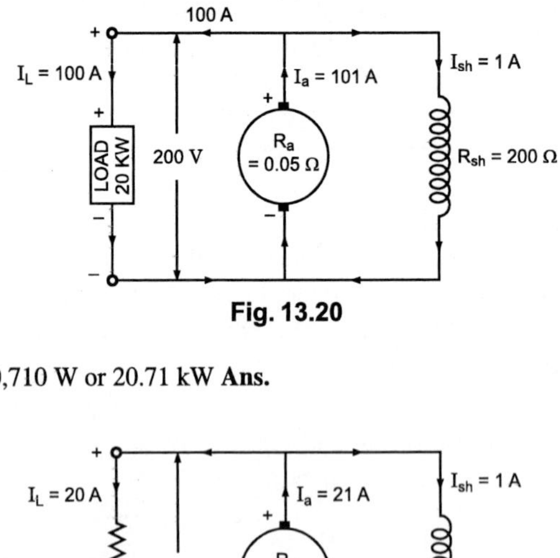

Fig. 13.20

Example 13.10 An 8-pole dc shunt generator has 778 wave-connected armature conductors running at 500 rpm, supplies a load of 12.5 Ω resistance at a terminal voltage of 250 V. The armature resistance is 0.24 Ω and the field resistance is 250 Ω. Find out the armature current, the induced emf and the flux per pole.

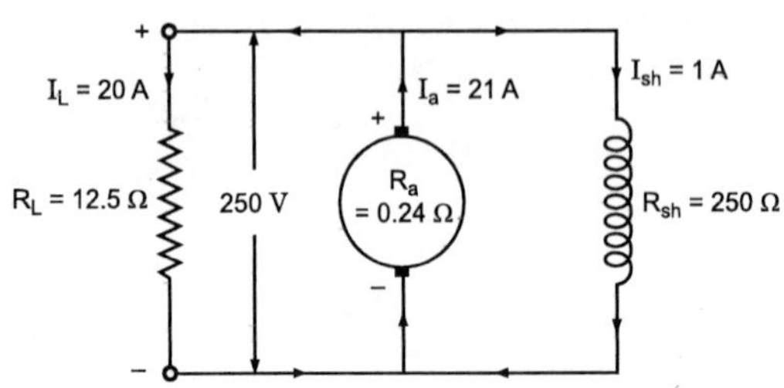

Fig. 13.21

Solution:

$$\text{Load current, } I_L = \frac{V}{R_L} = \frac{250}{12.5} = 20 \text{ A}$$

$$\text{Shunt field current, } I_{sh} = \frac{V}{R_{sh}} = \frac{250}{250} = 1 \text{ A}$$

Armature current, $I_a = I_L + I_{sh} = 20 + 1 = 21$ A **Ans.**

Induced emf, $E_g = V + I_a R_a = 250 + 21 \times 0.24 = 255.04$ V **Ans.**

$$\text{Flux per pole, } \Phi = \frac{E_g \times 60}{ZN} \times \frac{A}{P} \qquad \because E_g = \Phi Z \frac{N}{60} \times \frac{P}{A} \text{ volts}$$

$$= \frac{255.04 \times 60}{778 \times 500} \times \frac{2}{8} = 9.834 \times 10^{-3} \text{ Wb or } 9.834 \text{ m Wb } \textbf{Ans.}$$

Example 13.11 **A 4-pole dc shunt generator with lap-connected armature has field and armature resistances of 80 Ω and 0.1 Ω respectively. It supplies power to 50 lamps rated for 100 volts, 60 watts each. Calculate the total armature current and the generated emf by allowing a contact drop of 1 volt per brush.**

[U.P. Technical Univ. Electrical Engineering Second Semester 2003-04]

Solution: Total lamp load, P_L = Number of lamps × wattage of each lamp = 50 × 60 = 3,000 W

$$\text{Terminal voltage, } V = 100 \text{ V}$$

$$\text{Load current, } I_L = \frac{P_L}{V} = \frac{3{,}000}{100} = 30 \text{ A}$$

$$\text{Shunt field current, } I_{sh} = \frac{V}{R_{sh}} = \frac{100}{80} = 1.25 \text{ A}$$

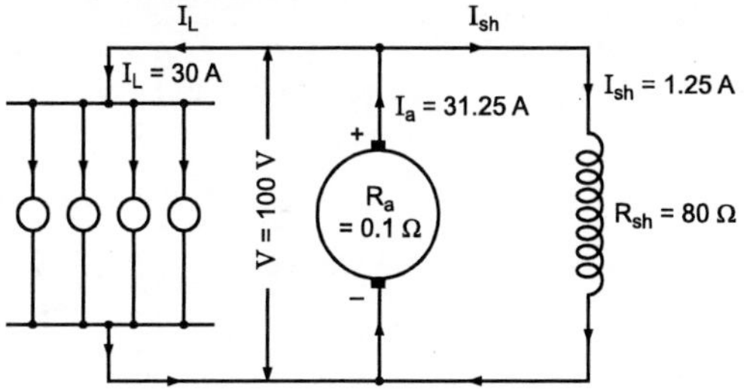

Fig. 13.22

$$\text{Total armature current, } I_a = I_L + I_{sh} = 30 + 1.25 = 31.25 \text{ A } \textbf{Ans.}$$

$$\text{Generated emf, } E_g = V + I_a R_a + \text{brush drop} = 100 + 31.25 \times 0.1 + 2 \times 1 = 105.125 \text{ V } \textbf{Ans.}$$

Example 13.12 **A 4-pole dc shunt generator with wave wound armature has 40 slots each having 12 conductors. Armature resistance is 1 ohm and shunt field resistance is 200 ohm. The flux per pole is 25 mWb. If a load of 50 ohm is connected across the armature terminals, calculate the voltage across the load when the generator is driven at 1,000 rpm. What will be the load voltage if the generator is lap wound?**

[U.P. Technical Univ. Electrical Engineering First Semester 2006-07]

Solution: *With wave-connected armature*

$$\text{Generated emf, } E_g = \frac{\Phi ZN}{60} \times \frac{P}{A} = \frac{25 \times 10^{-3} \times 40 \times 12 \times 1{,}000}{60} \times \frac{4}{2} = 400 \text{ V}$$

Let the terminal voltage be V volts

$$\text{Shunt field current, } I_{sh} = \frac{V}{R_{sh}} = \frac{V}{200} \text{ A}$$

$$\text{Load current, } I_L = \frac{V}{R_L} = \frac{V}{50} \text{ A}$$

$$\text{Armature current, } I_a = I_L + I_{sh} = \frac{V}{50} + \frac{V}{200} = \frac{V}{40} \text{ A}$$

Since generated emf, $E_g = V + I_a R_a$ neglecting brush drop,

$$400 = V + \frac{V}{40} \times 1 = \frac{41}{40} V$$

$$\text{or Terminal voltage, } V = \frac{400}{41/40} = 390.24 \text{ V } \textbf{Ans.}$$

With lap-connected armature

$$\text{Generated emf, } E_g = \frac{25 \times 10^{-3} \times 40 \times 12 \times 1{,}000}{60} \times \frac{4}{4} = 200 \text{ V}$$

$$\text{Terminal voltage, } V = \frac{200}{41/40} = 195.12 \text{ V } \textbf{Ans.}$$

Example 13.13 **A 4-pole dc generator with wave-connected armature has 41 slots, and 12 conductors/slot. Armature resistance $R_a = 0.5\ \Omega$. Shunt resistance is $R_{sh} = 200\ \Omega$. Flux per pole is 125 mWb. Speed N = 1000 rpm. Calculate voltage across 10 Ω load resistance across the armature terminal.**

[U.P. Technical Univ. Electrical Engineering First Semester 2007-08]

Solution:

$$\text{Generated emf, } E_g = \frac{\Phi ZN}{60} \times \frac{P}{A} = \frac{125 \times 10^{-3} \times 41 \times 12 \times 1{,}000}{60} \times \frac{4}{2} = 2{,}050 \text{ V}$$

Let the voltage across 10 Ω load resistance be V volts

$$\text{Load current, } I_L = \frac{V}{10} \text{ A}$$

$$\text{Shunt field current, } I_{sh} = \frac{V}{R_{sh}} = \frac{V}{200} \text{ A}$$

$$\text{Armature current, } I_a = I_L + I_{sh} = \frac{V}{10} + \frac{V}{200} = \frac{21\,V}{200} \text{ A}$$

$$\text{Since generated emf, } E_g = V + I_a R_a \quad \text{neglecting voltage drop}$$

$$2{,}050 = V + \frac{21\,V}{200} \times 0.5 = \frac{421}{400} V$$

$$\text{or } V = \frac{2{,}050 \times 400}{421} = 1{,}947.7 \text{ V } \textbf{Ans.}$$

Example 13.14 **A 4-pole dc generator with wave-connected armature has 41 slots and 12 conductors per slot. Armature resistance and shunt field resistance are 0.5 Ω and 200 Ω. Flux per pole is 125 mWb. Speed N = 1,000 rpm. Calculate the voltage drop across terminals. The load resistance is 10 Ω.**

[U.P. Technical Univ. Basic Electrical Engineering Odd Semester 2013-14]

Solution: Refer to Solution of Example 13.13.

Example 13.15 **A 4-pole, long shunt, lap wound generator supplies 25 kW at a terminal voltage of 500 V. The armature resistance is 0.03 Ω, series field resistance is 0.04 Ω and shunt field resistance is 200 Ω. The brush drop may be taken as 1 V. Determine the emf generated.**

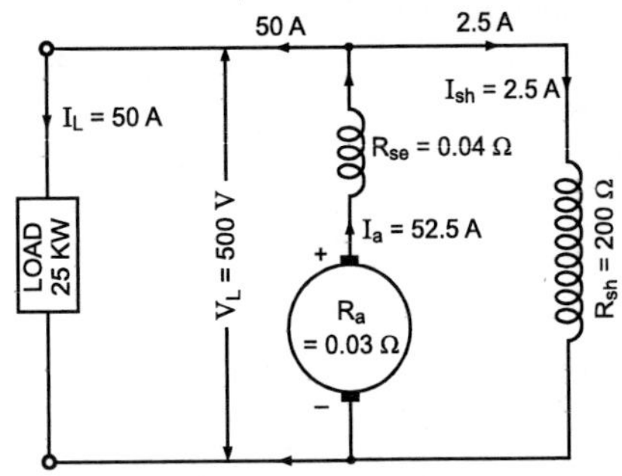

Fig. 13.23

Solution:

$$\text{Load, } P_L = 25 \times 1{,}000 = 25{,}000 \text{ W}$$

$$\text{Load voltage, } V_L = 500 \text{ V}$$

$$\text{Load current, } I_L = \frac{P_L}{V_L} = \frac{25{,}000}{500} = 50 \text{ A}$$

$$\text{Shunt field current, } I_{sh} = \frac{V_L}{R_{sh}} = \frac{500}{200} = 2.5 \text{ A}$$

$$\text{Armature current, } I_a = I_L + I_{sh} = 50 + 2.5 = 52.5 \text{ A}$$

$$\text{Generated emf, } E_g = V + I_a R_a + I_a R_{se} + \text{brush drop}$$

$$= 500 + 52.5 \times 0.03 + 52.5 \times 0.04 + 1$$

$$= 504.675 \text{ V } \textbf{Ans.}$$

13.7 MAGNETISATION CURVE

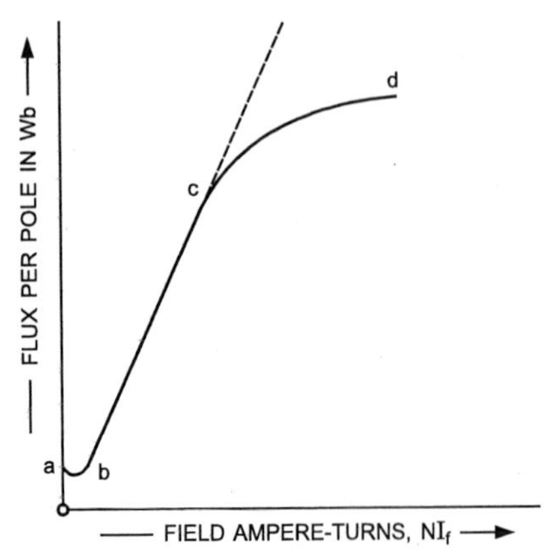

Fig. 13.24

According to molecular theory of magnetism the molecules of unmagnetised piece of iron are not arranged in any definite order. When a current is passed through a coil placed around the iron, in order to magnetise it, the molecules become arranged in a definite order. Up to certain stage relatively few ampere-turns of applied mmf are sufficient to arrange the greater part of the molecules in a definite order. In this stage of magnetisation, the amount of flux established in the iron increases almost directly with the increase in ampere-turns applied. However, above this point, which is called the *saturation point,* it becomes increasingly difficult to magnetise the iron further since the unmagnetised molecules become fewer and fewer. Above the saturation point, when much larger increase in ampere-turns are required for corresponding increase in flux in the iron, the iron is said to be saturated. The fact that iron has become saturated does not mean that a further increase in magnetism is impossible; the increase in magnetism merely requires very much larger increase in applied ampere-turns than before the iron become saturated. The curve giving relation between flux per pole and field ampere-turns per pole is known as *magnetisation or saturation curve.*

Saturation curve for a dc generator does not start from zero ordinarily but at some value slightly higher than zero due to *residual magnetism.* As discussed above for lower values of field ampere-turns, the flux is approximately proportional to the field ampere-turns and, therefore, magnetisation curve is a straight line. At point 'c' the field core begins to become saturated and the curve falls away from the straight line due to varying permeability of iron.

EMF generated in a generator is given by an expression

$$E_g = \Phi Z \frac{N}{60} \times \frac{P}{A} \text{ volts} = \left(\frac{Z}{60} \times \frac{P}{A}\right) \Phi N \text{ volts}$$

The quantity in the bracket is constant for a given machine, let it be K then $E_g = K\Phi N$

i.e. generated emf in a generator is directly proportional to the product of flux per pole and speed. If the generator runs at constant speed, then generated emf is directly proportional to the flux per pole.

Since the generated emf is proportional to flux per pole if speed is constant and field ampere-turns are proportional to field current I_f, if a curve is drawn between generated emf on no load and field current, when the machine is running at a constant speed, then curve so obtained will be similar to saturation curve. Such curves are known as *magnetic characteristics or open-circuit characteristics* (OCC). Typical OCC is shown in Fig. 13.25.

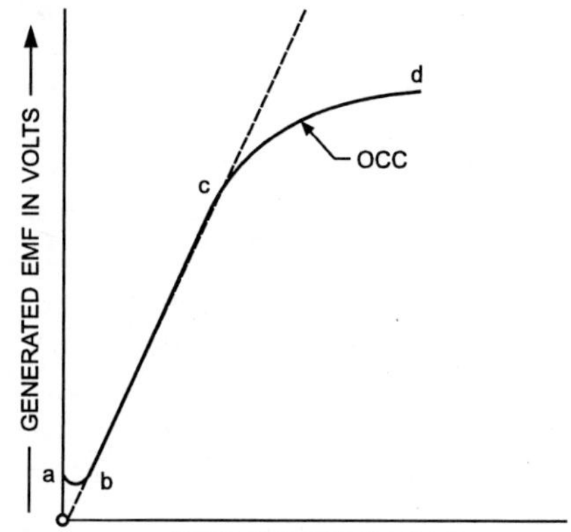

Fig. 13.25 *Magnetic or Open-Circuit Characteristic*

The curve starts from point '*a*' instead of 'O' when the field current is zero which is due to residual magnetism. The curve from point b to point c in Fig. 13.25 is practically a straight line and is called the *straight line portion* of the magnetization curve. Point *c*, at which saturation of the magnetic circuit begins is known as the *knee* of the curve. The portion of the curve from point *c* to point d is described as the curve above the knee.

13.8 EXTERNAL OR LOAD CHARACTERISTICS OF DC GENERATORS

External characteristic curve gives the relation between terminal voltage V and the load current I under the given conditions of speed and excitation. The terminal voltage V is always less than generated emf E on account of voltage drop in armature. This curve, therefore, lies below the internal characteristic. This characteristic is of utmost importance for determining the suitability of a generator for a particular service.

(i) Separately-Excited DC Generators. Since exciting current is independent of load or armature current, if a curve is drawn between flux per pole and load current, keeping field current constant and

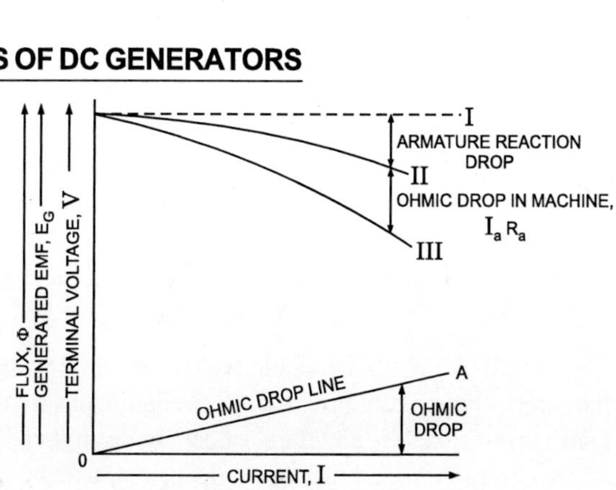

Fig. 13.26 *Characteristics of a Separately-Excited DC Generator*

neglecting armature reaction, a straight line parallel to X-axis is obtained, as indicated by curve I in Fig. 13.26. But due to armature reaction, the curve of actual flux is slightly drooping as represented by curve II in Fig. 13.26.

The second curve also gives to some other scale the emf generated in the armature and, therefore, second curve represents the internal or total characteristic* of the separately-excited dc generator. The terminal voltage on load is the emf generated less voltage drop in the armature i.e. $V = E - I_a R_a$, so if a line OA representing the armature resistance is drawn and its ordinates are deducted from curve II, curve III giving relation between terminal voltage and load current, known as *external characteristic* is obtained. From curve III, it is obvious that with the increase in load current, terminal voltage decreases slightly. This decrease can easily be neutralized by increasing field current slightly and, therefore, constant terminal voltage can be maintained.

(ii) Series Wound DC Generators. In a series wound dc generator, the armature winding, field winding and external load circuit, all are connected in series with each other, as shown in Fig. 13.16 and, therefore, the same current flows through all parts of the circuit.

At no load (the external circuit is open) no current flows through the field winding. On closing of external or load circuit, both the armature winding and the field winding carry the load current. With the increase in load current, the terminal voltage increases (the solid line in Fig. 13.27) until the magnetic circuit becomes saturated. With the further increase in load, the terminal voltage falls. This is due to increased armature reaction and larger voltage drop across the armature and field windings (the dotted line in the figure). In the case of a short circuit in the armature circuit, a fairly heavy current will be established while the terminal voltage will be zero. Because of the sharp increase in generator terminal voltage with increasing load, series wound generators are rarely used.

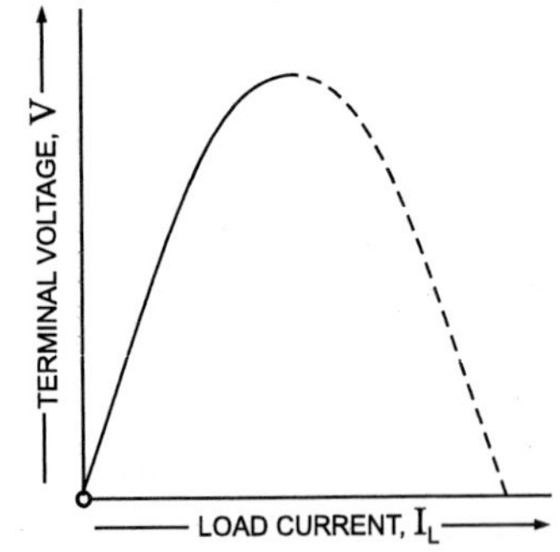

Fig. 13.27 *Load Characteristic of Series Wound DC Generator*

(iii) Shunt Wound DC Generators. The terminal voltage on no load is governed by the shunt field current (*i.e.* shunt field resistance), and let it be represented by OA in Fig. 13.28.

As the generator is loaded by decreasing the resistance of the external load circuit, the terminal voltage falls. This is due to following three causes.

1. The voltage drop across the armature winding and the contact resistance of the brushes increases with the increase in armature current.
2. As the load is increased, the current in armature conductors increases, so the effect of armature reaction increases and, therefore, field is weakened. Due to weakening of field the induced emf is reduced and, therefore, terminal voltage falls.
3. The decrease in terminal voltage as a result of the first two factors causes decrease in field current. This will in turn cause the emf and, therefore terminal voltage of the generator to drop too.

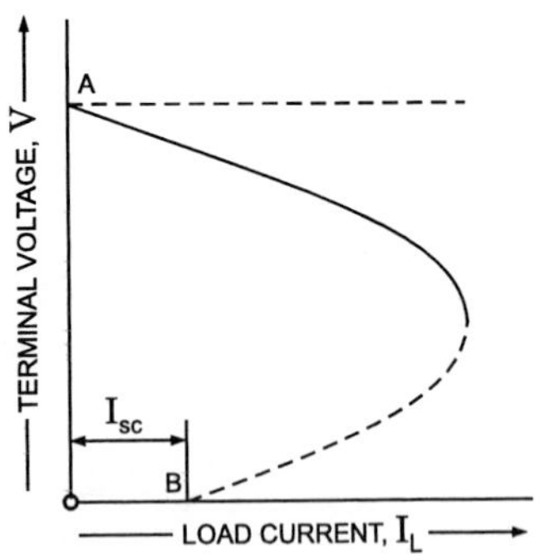

Fig. 13.28 *Load Characteristic of Shunt Wound DC Generator*

Thus the effects are cumulative and terminal voltage is reduced to a much greater extent. In the beginning the effect of decrease in resistance predominates over the effect of decreased terminal voltage but when the load current reaches a certain value (much higher than the full-load current) the demagnetising effect of armature reaction and voltage drop in the armature become so important that a further decrease in load resistance causes a decrease in current rather than increase and so the characteristic turns back.

* Curve giving relation between the emf actually generated in the armature and the armature current.

When the terminals are short circuited, there is no terminal voltage and thus the shunt winding becomes inactive but a small current is established due to a small voltage induced by the residual magnetism. Thus external characteristic meets the current axis at point B. *The shunt generators are, therefore, self-protective against accidental short circuit.*

(iv) Characteristics of Compound Wound DC Generators. As already mentioned in Art. 13.6, compound wound dc generators are of two types viz. *cumulative compound wound* and *differential compound wound* dc generators.

The cumulative compound wound generator combines the characteristics of the shunt and series generators. At no load the shunt field winding provides all the field flux since there is no current in the series field winding. With the increase in load on the generator, the series field adds an increasing amount of field flux to that of the shunt field.

If the series excitation is such that the terminal voltage of the generator on full load is the same as that on no load, the generator is said to be level or flat compounded*. If the series excitation becomes more prominent than that of the shunt field, the terminal voltage rises with the increase in load and the generator is said to be *over compounded.* Similarly when the shunt field excitation plays the prominent part and full-load terminal voltage is less than no-load terminal voltage, the generator is said to be *under compounded.*

The external characteristics of over, under and flat (or level) compound dc generators, are illustrated in Fig. 13.29.

In a **differential compound wound generator**, the induced voltage on no load would be the same as for the shunt generator because series field does not create any flux. When the generator is loaded, load current (or armature current) flows through the series field windings creating magnetic flux that opposes the flux created by shunt field. This results in decrease of induced emf and, therefore, in terminal voltage. In such generators the terminal voltage drops very rapidly with the increase in load current and, therefore it is interesting to note that a short circuit cannot cause any damage to it.

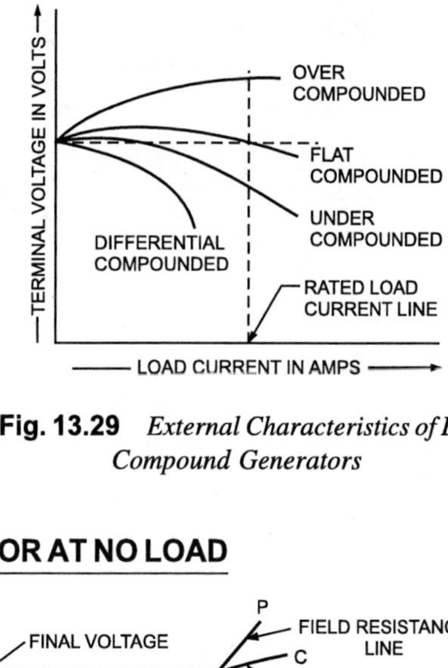

Fig. 13.29 *External Characteristics of DC Compound Generators*

13.9 BUILDING UP OF VOLTAGE OF A SHUNT GENERATOR AT NO LOAD

The curve plotted between the generated emf and shunt field current will be similar to that shown in Fig. 13.30. The generator excites itself due to residual magnetism and develops the voltage, as described below :

Line OP represents the shunt field resistance. When the generator is started a small emf (represented by OA in the Fig. 13.30) is induced due to residual magnetism. This induced emf causes a flow of current *oa'* in the field circuit. This is obtained by drawing a horizontal line from point A meeting the field resistance line at point '*a*' and then from point '*a*' drawing perpendicular line meeting current *axis at a'. When field current is oa' the generated emf is a' a" which produces a field current ob'* which in turn produces a high voltage *b' b"*. Thus it is observed that effect is cumulative and values of induced emf and field current increase until these reach point D, the point of intersection of the shunt field resistance line and magnetic characteristic obtained. In Fig. 13.30 instantaneous increases in induced emf and field current are

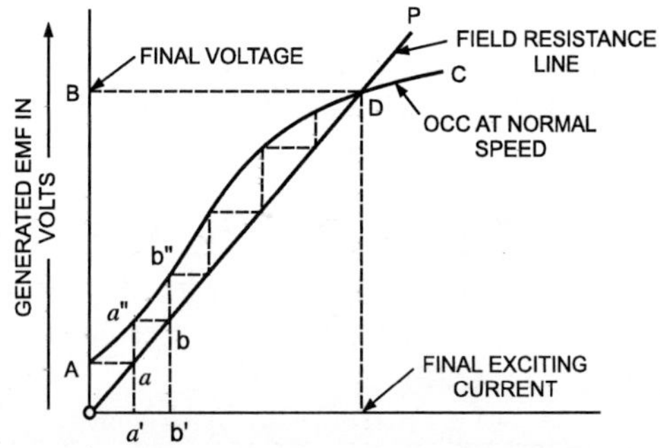

Fig. 13.30 *Building-up of Voltage of a Shunt Generator At No Load*

* Even in case of a flat-compounded generator, the terminal voltage is not constant for all loads from no load to full load. As seen from Fig. 13.29, at half-full load, the terminal votlage is more than the no-load (or full-load) terminal voltage.

shown, but in practice flux increases gradually. OB is the maximum value of emf, that can be generated and generator cannot generate more than this voltage.

13.10 CONDITIONS FOR SELF EXCITATION AND CAUSES OF FAILURE TO BUILD UP VOLTAGE

The conditions required to be fulfilled before a series or a shunt generator excites itself are given below :

1. There must be some residual magnetism in the field system.
2. The residual magnetism must be in proper direction. The field coils should be connected with the armature in such a way that current flowing through them should increase emf induced by the residual magnetism.
3. For a series wound generator, the resistance of the external circuit should be less than the critical resistance.*
4. For a shunt wound generator, the resistance in the field circuit must be less than critical resistance** for field circuit and resistance in the load circuit must be greater than critical resistance*** for the load circuit.

If a dc generator fails to build up its voltage, there may be one or more of the following reasons :

1. The residual magnetism may be zero.
2. The residual magnetism may have been reversed in polarity.
3. The terminals of the field winding may have been reversed.
4. In case of series wound generators the resistance in the circuit may be greater than the critical resistance, which may be due to :
   (*i*) open circuit;
   (*ii*) high resistance of load circuit;
   (*iii*) faulty contact between brushes and commutator; and
   (*iv*) commutator's surface dirty or greasy.
5. In case of shunt wound generator :
   (*a*) the resistance of the shunt field circuit may be greater than the critical resistance;
   (*b*) the resistance in the load circuit may be lower than the critical resistance;
   (*c*) the speed of rotation may not be equal to rated one.

Remedy. In case the generator is started up for the first time. It may be that no voltage will be built up either because the poles have no residual magnetism or the poles have retained some residual magnetism but the field winding connections are reversed so that the magnetism developed by the field winding on start has destroyed the residual magnetism and the machine cannot "build up". In both the cases, the field coils must be connected to a dc source (a storage battery) for a short while to magnetise the poles.

13.11 APPLICATIONS OF DC GENERATORS

The choice of a dc generator for a particular service depends upon its performance characteristics. Applications of various types of dc generators are given below.

---

* Critical resistance for a series wound generator is the maximum load resistance with which the generator will be able to excite. If the resistance in the external load circuit is greater than this, the generator will not excite.

** Critical field circuit resistance for a shunt wound generator is the maximum field resistance with which the generator will be able to excite. If the resistance in the field circuit exceeds this value, the generator will not excite.

*** Critical load resistance for a shunt wound generator is the minimum load resistance with which the generator will be able to excite. If the resistance of the external load circuit falls below this value, the generator will fail to build-up its voltage.

1. Separately-Excited DC Generators. These generators require separate source of supply and are, therefore, usually more expensive than self-excited generators. On this account these are generally used where self-excited generators would be relatively unsatisfactory.

2. Series Wound DC Generators. These generators have very few applications. They cannot be employed for power supply on account of their rising voltage characteristics. They can be employed as a constant current source by making use of the drooping portion of the voltage-current characteristic.

Their best application is for supplying field current for regenerative braking in dc locomotives. They are also used in series arc lighting. Another important application of these generators is as series boosters.

3. Shunt Wound DC Generators. Such generators with field regulators are used for light and power supply purposes. These are also used for charging of batteries because they can be made to give constant output voltage.

4. Cumulative Compound Wound DC Generators. Over-compounded type dc generators are generally used for lighting and power services because overcompounding compensates voltage drop in the distribution lines and thus the voltage at the consumer terminals remains more or less constant.

5. Differential Compound Wound DC Generators. It is a constant current generator and finds an useful application as an arc welding generator where the generator is practically short circuited every time the electrode touches the metal plates to be welded.

Note: *With the advent of various power electronic devices, there is no doubt that the importance of dc generators has gone down. At present time bulk of power is generated in the form of alternating current because of technical and economical reasons. Hence the use of dc generators is very limited. They find application mainly in supplying excitation to the small and medium size alternators. For industrial applications of direct current like electrolytic processes, welding processes and variable speed motor drives, the present trend is to generate ac and then convert it into dc by converters. For conversion of ac into dc the dc generator, as a part of motor (ac)-generator (dc) set, has to compete the SCR rectifiers and various other controlled power devices which usually are cheaper in cost, compact in size, relatively noise-free in operation and require minimum maintenance, though they suffer from the disadvantages of having poor power factor, harmonic generation, etc. However, some of the important applications of dc generators include dynamometers, welding, cross-field generators, for closed-loop control systems, tachogenerators (not covered in this book) etc. Separately excited dc generators still find use for a wide output-voltage control such as in Ward-Leonard system of speed control.*

HIGHLIGHTS

1. An *electromechanical energy conversion device* is a link between an electrical and mechanical system and *electromechanical energy conversion* needs the presence of natural phenomena which interrelate electric and magnetic fields on one hand and mechanical force and motion on the other.
2. A *dynamo* is a machine that converts either mechanical energy into electrical energy or electrical energy into mechanical energy. When a dynamo is driven mechanically by a prime mover such as by a diesel engine, steam engine, steam turbine or water turbine and supplies electrical energy, it is called a *generator*. When the dynamo draws electrical energy from the supply mains and drives mechanical devices such as line shafts and machine tools it is called a *motor*. In other words it can be stated that the generator converts mechanical power ωT into electrical power EI and motor coverts electrical power EI into mechanical power ωT.
3. The motoring and generating action is basically a reversible process *i.e.* the same machine is capable of operating either as a generator or as a motor depending upon whether the supplied power is mechanical or electrical.
4. The energy balance equation is given by:
$$\int dW_{input} = \int dW_{output} + \int dW_{magnetic\ field} + \int dW_{heat}$$
5. For steady-state operation, the energy and power equations are:
$$W_{input} = W_{output} + W_{heat}$$
$$P_{input} = P_{output} + P_{heat}$$

6. *DC machine* is commutator type machine. It is actually an ac machine, but furnished with a special device, called the commutator, which under certain conditions converts ac into dc and vice versa.
7. An *electrical generator* is a machine which converts mechanical energy (or power) into electrical energy (or power). Its operates on the principle of electromagnetic induction.
8. In construction a dc machine consists of four parts mainly 1. Field magnets 2. Armature 3. Commutator and 4. Brushes and brush gear.

   Field system is stationary part and it is to create a uniform magnetic field, within which the armature rotates.

   Armature is a rotating part, where mechanical energy is converted into electrical energy (in case of generator) or, conversely, electrical energy into mechanical energy (in case of motor).

   Brushes are meant for collecting current from the commutator and supply it to the external load circuit.
9. EMF equation for a generator is given as

$$E_g = \frac{\Phi Z N}{60} \times \frac{P}{A} \text{ volts;} \qquad A = Z \text{ in wave wound machines}$$

$$A = P \text{ in lap wound machines}$$

10. According to method of excitation, dc generators are classified into separately-excited generators and self-excited generators.

    Self-excited generators are further classified in accordance how the field winding is connected w.r.t. armature. Accordingly such generators are (*i*) series wound type (*ii*) shunt wound type and (*iii*) compound wound type. Compound wound generators may be long shunt type or short shunt type.

    Compound wound dc generators can also be classified into cumulative compound wound and differential compound wound.
11. Among dc generators, separately-excited and cumulatively compounded self-excited machines are the most common. Separately-excited generators have the advantage of permitting a wide range of output voltages, whereas self-excited machines may produce unstable voltages in the lower ranges. Cumulatively compounded dc generators may produce a substantially flat voltage characteristic or one which rises with load, whereas shunt or separately-excited generators produce a drooping voltage characteristic unless regulating means are used.
12. (*a*) The best application of series wound dc generators is for supplying field current for regenerative braking in dc locomotives. They are also used as series boosters and in arc lighting.

    (*b*) Cumulative compound wound dc generators are generally used for lighting and power services.

    (*c*) Differential compound generator finds an useful application as an arc welding generator.

EXERCISES

1. Explain principle of operation of electromechanical energy conversion.
   [U.P.Technical Univ. Elec. Engineering Second Semester 2005-06]
2. Explain principle of electromechanical energy conversion in motor and generator.
   [U.P.Technical Univ. Elec. Engineering First Semester 2005-06]
3. Explain the principle of energy conversion. Draw general representation of electromechanical conversion device.
   [U.P. Technical Univ. Elec. Engineering February, 2001]
4. Enumerate all the parts of a dc machine. State the material and the function of each part.
   [U.P. Technical Univ. Electrical Engineering First Semester 2003-04]
5. Explain why a commutator and brush arrangement is necessary for the operation of a dc mahine.
   [U.P. Technical Univ. Electrical Engineering Second Semester 2003-04]
6. Derive an equation for emf in a dc machine.
   [U.P. Technical Univ. Electrical Engineering First Semester 2003-04]
7. Derive emf equation of a dc generator. **[U.P.Technical Univ. Elec. Engineering Second Semester 2005-06]**
8. Develop the expression of emf generated for a dc machine and give factors on which the amount of so generated emf depends in a machine. **[Mahamaya Technical Univ. Electrical Engineering Second Semester 2011-12]**

9. Derive the emf equation of a multiple (more than two) pole dc generator.
Draw a neat labelled diagram to show the essential constructional features of a dc machine.
[U.P. Technical Univ. Electrical Engineering June 2001]

10. Derive the emf equation for a dc machine. **[U.P. Technical Univ. Electrical Engineering First Semester 2007-08]**

11. Derive the emf equation of a dc generator. What are lap and wave windings ? Which among them is used for high current and which for high voltage dc generator ? **[U.P. Technical Univ. Electrical Engineering Second Semester 2007-08]**

12. Describe the difference between the separately-excited shunt generator and self-excited one. Explain the process of voltage build up in a self-excited shunt generator.
[U.P. Technical Univ. Electrical Engineering Second Semester 2008-09]

13. Explain the generator action of a dc machine. Also describe the open-circuit characteristics of dc generator.
[Mahamaya Technical Univ. Electrical Engineering First Semester 2011-12]

14. Explain the process of "building up" of voltage in a shunt generator. Under what conditions may it fail to build up the voltage ? **[U.P. Technical Univ. Electromechanical Energy Conversion-I, 2002-03]**

SHORT ANSWER TYPE QUESTIONS WITH ANSWERS

Q. 1. On what principle does the generator operate ?
Ans. The generator operates on the principle of production of dynamically induced emf.

Q. 2. Why it is necessary to use a large number of coils and commutator segments with the coils evenly distributed around the surface of the armature of a dc machine ?
Ans. It is done so as to have practically constant unidirectional voltage at the brushes.

Q. 3 Why do the coils used in commercial generators consist of several turns in series ?
Ans. It is done so as to increase the magnitude of generated emf, being in direct proportion to the number of turns in the coil.

Q. 4. Why are electromagnets preferred over permanent magnets for use in large dc machines ?
Ans. On account of their greater magnetic effect and field strength regulation.

Q. 5. Why are cast iron yokes used in small dc machines ?
Ans. On account of low cost.

Q. 6. Why are fabricated steel yokes used in large machines ?
Ans. On account of high permeability thereby resulting in reduced weight.

Q. 7. What for field coils are provided in a dc machine ?
Ans. Field coils are provided to give, under various conditions of operation, the number of ampere-turns of excitation required to give proper flux through the armature to induce the emf.

Q. 8. Why is the armature core of a dc machine laminated ?
Ans. To reduce the eddy current loss.

Q. 9. What is the function of armature in a dc generator ?
Ans. The function of armature is to rotate the conductors in a uniform magnetic field and provide a path of very low reluctance to the magnetic flux.

Q. 10. Why is the air gap between the pole pieces and the armature kept very small ?
Ans. To reduce the reluctance of the magnetic path.

Q. 11. What is the best-suited material for commutator segments ?
Ans. Silvered copper (extruded copper) containing about 0.05% silver.

Q. 12. What for brushes are employed in dc machines ?
Ans. The brushes are employed to collect current from the commutator and supply it to the external load circuit.

Q. 13. Why are the graphite or carbon brushes preferred over copper brushes for use in dc machines ?
Ans. Due to self-lubricating property.

Q. 14. Write the emf equation of a dc generator. **[U.P. Technical Univ. Second Semester 2013-14]**
Ans. The emf equation of a dc generator is given as

$$E = \Phi Z \frac{N}{60} \times \frac{P}{A} \text{ volts}$$

Q. 15. In how many ways mmf required to establish the flux in the magnetic circuit of a dc generator can be had ?

Ans. The mmf necessary to establish the flux in the magnetic circuit of a dc generator can be obtained by means of (*i*) permanent magnets (*ii*) field coils excited by some external source and (*iii*) field coils excited by the generator itself.

Q. 16. What do you understand by self-excitation mode of dc machine? Name two dc machines working in this mode. **[P.T.U. June 2003]**

Ans. The dc machine in which field winding(s) is/are self excited by the current supplied by the machine itself is called the self-excitation mode of the machine. In such machines the field coils are interconnected with the armature winding.

The self-excited machines may be (*i*) dc series machine (*ii*) dc shunt machine (iii) compound wound.

Q. 17. Why are permanent magnet generators rarely used in industry ?

Ans. Because of very small output.

Q. 18. What is the difference between the excitation system of a dc shunt generator and that of a dc series generator ?

Ans. In a series wound generator, there is only one field winding which is connected in series with the armature winding so that whole current flows through the field winding as well as load. In a shunt wound generator also there is only one field winding, as in case of a series wound generator, but in this case it is connected across the armature winding forming a parallel or shunt circuit.

Q. 19. Why is it necessary to wind field coils of a medium or large size dc series generators with a small number of turns of thick wire ?

Ans. Since the series field winding has to carry the full-load current, series field winding is wound with a fewer turns of thick wire or strips.

Q. 20. Why is it necessary to wind field coils of a dc shunt generator with a large number of turns of fine wire.

Ans. The field coils of a dc shunt generator are wound with a large number of turns of fine wire so as to have high resistance in order to keep shunt field current as low as possible as well as to provide required mmf.

Q. 21. What is the volt drop allowed for all brushes of each polarity for carbon or graphite brushes ?

Ans. 1 V per brush.

Q. 22. What type of current flows in stator winding and rotor winding of a dc shunt generator ?

Ans. In stator : dc

In rotor : ac.

Q. 23. Why does saturation curve starts from some value higher than zero ?

Ans. Due to residual magnetism.

Q. 24. In a dc machine flux per pole decreases with increase of load. Justify the statement.

Ans. In a dc machine, flux per pole decreases with the increase of load due to demagnetization effect of armature reaction.

Q. 25. In what respect the separately-excited dc generator is superior to self-excited dc generator?

Ans. The separately-excited dc generator has the advantage of stable operation with any field excitation over self-excited generators.

Q. 26. A dc shunt genertor is developing rated terminal voltage at some speed. For this generator, answer the following and give a brief explanation in support of your answers:

(*i*) If only the direction of rotation is reversed, will the generator build up ?

(*ii*) If the direction of rotation and also the residual magnetism are reversed, will the generator build up ?

Ans. (*i*) The generator will not build up. With the direction of rotation reversed, the polarity of the generated emf will be reversed. The direction of field current will also reverse and hence the residual magnetism present in the poles will be wiped off and the generator will fail to build up.

(*ii*) When the direction of rotation and the residual magnetism both are reversed, then the small emf produced initially will strengthen the field further and hence the generator will build up voltage.

Q. 27. The terminal voltage of a dc generator at full load is more than its open-circuit voltage. What is the mode of excitation of the generator ?

Ans. Cumulative compound.

Q. 28. A dc generator can run to generate the same no load voltage with different modes of excitation. For which mode of excitation the short-circuit current will be minimum?

Ans. Differential compound wound generator.

Q. 29. Why is a dc series generator not suitable for general electric supply ?

Ans. Because of their rising V-I characteristics.

Q. 30. Which type of dc generator is used for increasing the voltage across the feeder carrying current supplied by some other source ?

Ans. Series wound dc generator, called the series booster.

Q. 31. Why a dc shunt generator is used for battery charging ?

Ans. Because a dc shunt generator can be made to provide constant output voltage.

Q. 32. What type of dc generator is suitable for electric welding ?

Ans. Differentially compounded dc generator.

Q. 33. Why is only a differentially compounded dc generator suitable for arc welding ?

Ans. Because in a differentially compounded dc generator, the terminal voltage drops very rapidly with the increase in load current and, therefore, short-circuit cannot cause any damage to it.

PROBLEMS

1. A 6-pole, lap wound armature has 840 conductors and flux per pole of 0.018 Wb. The generator is run at 1,200 rpm. Calculate the emf generated. **[Ans.** 302.4 V]
2. Calculate the emf generated by a 4-pole wave wound armature with 45 slots, with 18 conductors per slot when driven at 1,000 rpm. The flux per pole is 0.02 Wb. **[Ans.** 540 V]
3. A lap-connected 8-pole generator has 500 armature conductors and useful flux of 0.07 Wb. Determine the induced emf when it runs at 1,000 rpm. What must be the speed at which it is to be driven to produce the same emf if it is wave wound. **[U.P. Technical Univ. Tutorial Question Bank]** **[Ans.** 583.33 V; 250 rpm]
4. A lap wound dc generator having 8-poles develops an emf of 500 V at 400 rpm. The armature has 144 slots and each slot contains 6 conductors. Calculate the flux per pole. **[Ans.** 0.0868 Wb]
5. Determine the power output of a dc armature having 1,152 lap-connected conductors carrying 150 A and rotating at 300 rpm in a 12-pole field. The flux per pole is 60 m Wb. **[Ans.** 51.84 kW]
6. A 440 V dc shunt generator has an armature resistance of 0.25 Ω and the resistance of the shunt field is 220 Ω. While delivering a load current of 50 A, it has a terminal voltage of 440 V. Determine the generated emf. **[Ans.** 453 V]
7. A 20 kW, 220 V dc shunt generator has an armature resistance of 0.07 Ω and a shunt field resistance of 200 Ω. Determine the power developed in armature when it delivers rated output. **[Ans.** 20.835 kW]
8. A shunt generator has an induced voltage on open-circuit of 127 V. When the machine is loaded, the terminal voltage is 120 V. Find the load current if the field resistance is 15 Ω and the armature resistance is 0.02 Ω. Ignore armature reaction. **[Ans.** 342 A]
9. A dc shunt generator is supplying load connected to a bus-bar voltage of 220 V. It has an armature resistance of 0.025 Ω and field resistance of 110 Ω. Calculate the value of load current and load power when it generates an emf of 230 V. Neglect the effect of armature reaction. Draw circuit diagram. **[Ans.** 398 A ; 87.56 kW]

14 CHAPTER

DC Machines-II (Motors)

14.1. INTRODUCTION

An electric motor is a machine which converts electrical energy into mechanical energy where as a generator is that machine which converts mechanical energy into electrical one. As regards fundamental principles the dc motors are identical with the dc generators which have the same type of excitation *i.e.*, a machine that runs as a motor will also operate satisfactorily as a generator.

The only difference lies, however, in the mode of construction, which is due to the fact that the frame of the generator can as a rule be open but those of motors should be either partly or totally enclosed. A generator is usually placed in a suitable position and mechanical protection for the coils and armature may be reduced to minimum. Also the generator is handled by technical persons. Hence there is no risk in having the frame of the generator open, which facilitates cooling, inspection and repair. On the other hand, motors have to work in conditions of dampness, dirt, inflammable gases, chemical fumes and liability to mechanical damage and, therefore, protection must be adequate and motor frames are made flame-proof, partly enclosed or totally enclosed according to the requirements of service.

14.2. WORKING PRINCIPLE OF DC MOTOR

The principle upon which a dc motor works is very simple. If a current carrying conductor is placed in a magnetic field, mechanical force is experienced on the conductor, the direction of which is given by Fleming's left hand rule (also called *motor rule*) and hence the conductor moves in the direction of force. The magnitude of the mechanical force experienced on the conductor is given by

$$F = BI_c l_c \text{ newtons}$$

where B is the field strength in teslas (Wb/m^2), I_c is the current flowing through the conductor in amperes and l_c is the length of conductor in metres.

When the motor is connected to the dc supply mains, a direct current passes through the brushes and commutator to the armature winding; while it passes through the commutator it is converted into ac so that the group of conductors under successive field poles carry currents in the opposite directions, as shown in Fig. 14.1. Also the direction of current in the individual conductors reverses as they pass away from the influence of one pole to that of the next.

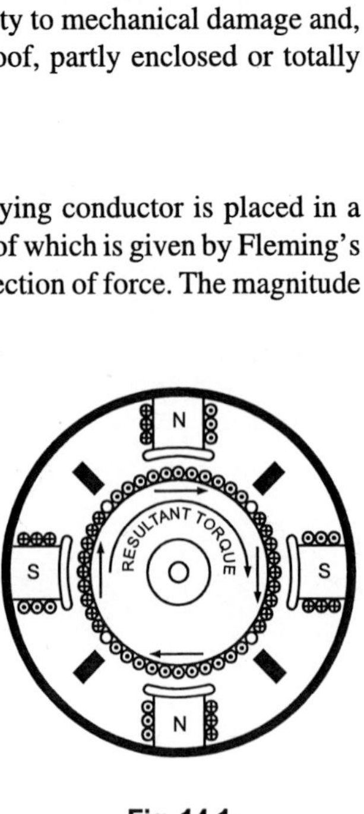

Fig. 14.1

In Fig. 14.1, a 4-pole dc motor is shown when the field and armature circuits are connected across dc supply mains. Let the current in armature conductors be outwards under the N-poles (shown by dots) and inwards under S-poles (shown by crosses). By applying Fleming's left hand rule, the direction of force on each conductor can be determined, which has been illustrated in Fig. 14.1. From Fig. 14.1. it is observed that each conductor experiences a force which tends to rotate the motor armature in clockwise direction. These forces collectively produce a driving torque.

14.2.1. Commutator Action in a DC Motor. In the case of a dc motor, it is necessary that the current through the coils of the armature winding be reversed as a particular coil leaves one pole (say, the north pole), crosses the neutral line and comes under the influence of next pole which is of opposite polarity (*i.e.* the south pole). The operation of the commutator, that serves the above purpose, is given below:

Consider a single turn coil, whose leads are soldered to commutator segments *a* and *b*, each carrying a brush, as illustrated in Fig. 14.2. The positive side of the supply line is connected to left hand brush and negative side to the right hand brush. In position I the line current arrives at the commutator segment a, flows through the bottom side 1 of the coil away from the reader (as shown by cross in the circle) and then through the upper side 2 of the coil towards the reader (as shown by dot in the circle), reaches the commutator segment *b* and flows again into the line through the brush. The coil will tend to rotate in clockwise direction, as determined by Fleming's left hand rule.

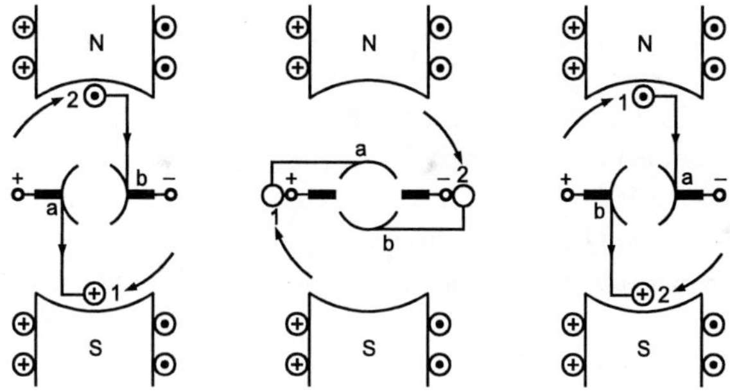

Fig. 14.2 *Commutation in a DC Motor*

In position II the coil is on the magnetic neutral line; there in no contact between the commutator segments and brushes, and there is no flow of current through the coil. The coil crosses the neutral line by inertia. In case of a multi-turn coil, the remaining turns of the coil will supply the necessary torque.

In position III, the two sides of the coil, 1 and 2, have changed poles, and the current through them has reversed. The commutator segments, however, have also changed contact with the brushes. Thus the coil will continue to rotate in the same direction as before, i.e. clockwise.

14.3. IMPORTANCE OF BACK EMF

As already explained, when the motor armature continues to rotate due to motor action, the armature conductors cut the magnetic flux and, therefore, emfs are induced in them. The direction of this induced emf, known as *back emf,* is such that it opposes the applied voltage.

Since the back emf is induced due to the generator action, the magnitude of it is, therefore, given by the same expression as that for the generated emf in a generator

$$i.e. \text{ Back emf, } E_b = \frac{\Phi ZN}{60} \times \frac{P}{A} \text{ volts} \qquad ...(14.1)$$

the symbols having their usual significance.

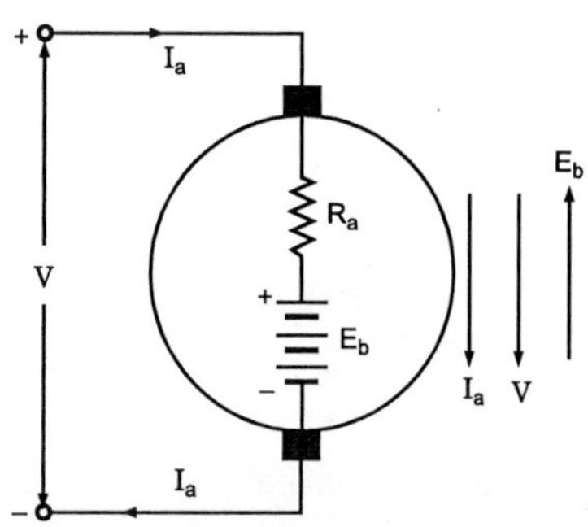

Fig. 14.3 *Equivalent Circuit of a Motor Armature*

The equivalent circuit of a motor is shown in Fig. 14.3. The armature circuit is equivalent to a source of emf, E_b in series with a resistance, R_a put across a dc supply mains of V volts. It is evident from Fig. 14.3 that the applied voltage V must be large enough to balance both the voltage drop in armature resistance and the back emf at all times *i.e.*

$$V = E_b + I_a R_a \qquad ....(14.2)$$

where V is the applied voltage across the armature, E_b is the induced emf in the armature by generator action, I_a is the armature current and R_a is the armature resistance.

The Eq. (14.2) may be rewritten as $I_a = \frac{V - E_b}{R_a}$ to give armature current in terms of applied voltage V, induced emf E_b and armature resistance R_a.

As obvious from Eqs (14.1) and (14.2) the induced emf in the armature of a motor, E_b depends among other factors upon the armature speed and armature current depends upon the back emf E_b for a constant applied voltage and armature resistance. If the armature speed is high, back emf E_b will be large and, therefore, armature current small. If the speed of the armature is low, then back emf E_b will be less and armature current I_a more resulting in development of large torque.

The presence of back emf makes the dc motor a *self-regulating* machine i.e. it makes the dc motor to draw as much armature current as is just sufficient to develop the required load torque. This is explained below:

When the motor is operating on no load, small torque is required to overcome the friction and windage losses, therefore, back emf is nearly equal to the applied voltage and armature current is small. When the motor is loaded, the driving torque of the motor is not sufficient to counter the increased retarding torque due to load and the effect is to cause the armature to slow down. With the decrease in the speed of armature back emf falls. The reduced back emf allows a larger current to flow through the armature. The increase in armature current results in higher electromagnetic driving torque. The motor continues to slow down till the electromagnetic torque developed matches the load torque and the steady-state conditions are attained. The reverse phenomenon occurs when mechanical load on the motor falls.

When the load on the motor falls, the electromagnetic torque developed is momentarily in excess of the load requirement and, therefore, the motor armature accelerates. With the increase in armature speed, back emf increases causing armature current to decrease. The decrease in armature current causes decrease in electromagnetic torque and the steady-state conditions are attained when the electromagnetic torque developed matches the load torque.

Thus it is evident that back emf E_b acts like a governor i.e. it makes a motor self-regulating so that it draws as much current as just required.

14.4. POWER RELATIONSHIP IN DC MOTOR

The Eq. (14.2) is called the *voltage equation* for a motor.

Multiplying each term of the voltage expression by I_a we get

$$V I_a = E_b I_a + I_a^2 R_a \qquad ...(14.3)$$

The above equation is known as *power equation* of the dc motor.

The term VI_a represents the power supplied to the motor armature and the term $I_a^2R_a$ represents the power lost in the armature and, therefore, the term E_bI_a must represent the power developed by the motor armature causing rotation of the armature. The power developed E_bI_a is not all available at the shaft since some of it is used to overcome the mechanical or rotational losses of the motor.

The power lost in the armature, $I_a^2R_a$ is a small portion (about 5%) of power input to the armature and the remaining portion E_bI_a is converted into mechanical power within the armature.

As mentioned above, the mechanical power developed by the motor is given by the expression

$$P_m = E_b I_a = V I_a - I_a^2 R_a \qquad ...(14.4)$$

14.5. TYPES OF DC MOTORS

Similar to dc generators, the dc motors can also be classified as (1) permanent magnet, (2) separately excited, (3) series wound (4) shunt wound and (5) compound wound dc motors.

14.5.1. Permanent Magnet Motor. It consists of an armature and one or several permanent magnets encircling the armature. Field coils are usually not required. However, some of these motors do have coils wound on the poles. If they exist, these coils are intended only for recharging the magnets in the event that they lose their strength. Schematically, this motor appears as only an armature.

14.5.2. Separately-Excited DC Motors. These motors have field coils similar to those of a shunt wound machine, but the armature and field coils are fed from different supply sources, as illustrated in Fig. 14.4 and may have different voltage ratings.

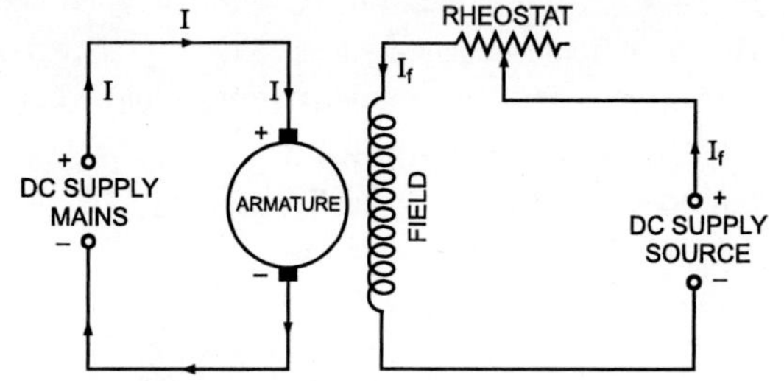

Fig. 14.4 *Separately-Excited DC Motor*

In a separately-excited dc motor,

$$\text{Armature current, } I_a = \text{Line current, } I_L = I \text{ (say)} \quad ...(14.5)$$

$$\text{Back emf developed, } E_b = V - I R_a \quad ...(14.6)$$

$$\text{Power drawn from supply mains, } P = V I \quad \text{where V is supply voltage.} \quad ...(14.7)$$

$$\text{Mechanical power developed, } P_m = \text{Power input to armature} - \text{power lost in armature}$$

$$= VI - I^2 R_a = I (V - I R_a) = E_b I \quad ...(14.8)$$

14.5.3. Series Wound DC Motors. As the name implies, the field coils, consisting of few turns of thick wire, are connected in series with the armature, as illustrated in Fig. 14.5. The cross-sectional area of the wire used for field coils has to be fairly large to carry the armature current, but owing to the higher current, the number of turns of wire in them need not be large. In a dc series motor

$$\text{Armature current, } I_a = \text{Series field current, } I_{se}$$

$$= \text{Line current, } I_L = I \text{ (say)} \quad ...(14.9)$$

$$\text{Back emf developed, } E_b = V - I (R_a + R_{se}) \quad ...(14.10)$$

$$\text{Power drawn from supply mains} = VI \quad ...(14.11)$$

Mechanical power developed, P_m

$$= \text{Power input} - \text{losses in armature and field}$$

$$= V I - I^2 (R_a + R_{se})$$

$$= I [V - I (R_a + R_{se})] = E_b I \quad ...(14.12)$$

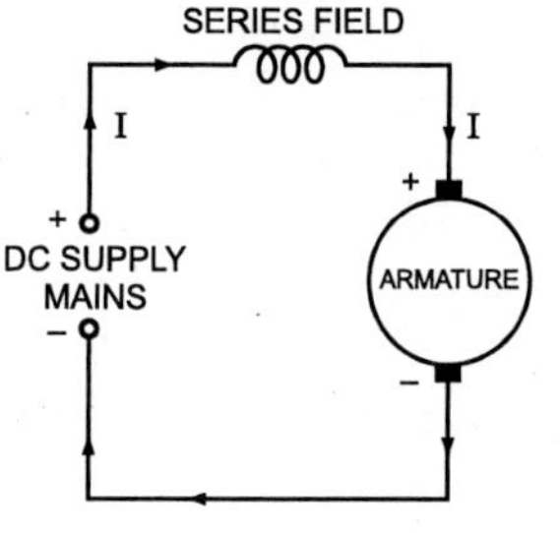

Fig. 14.5 *Series Wound DC Motor*

14.5.4. Shunt Wound DC Motors. The word "shunt" means "parallel". These motors are so named because they basically operate with the field coils connected in parallel with the armature. The field winding consists of a large number of turns of comparatively fine wire so as to provide large resistance. The field current is much less than the armature current, sometimes as low as 5%.

The connection diagram is shown in Fig. 14.6. The current supplied to the motor is divided into two paths, one through the field winding and second through the armature *i.e.*,

$$\text{Input line current, } I_L = I_a + I_{sh} \quad ....(14.13)$$

where I_a is the armature current and I_{sh} is the shunt field current and is given by the expression

$$I_{sh} = \frac{V}{R_{sh}} \quad ...(14.14)$$

where V is the supply voltage and R_{sh} is the shunt field resistance.

$$\text{Back emf developed, } E_b = V - I_a R_a \quad ...(14.15)$$

$$\text{Power drawn from supply mains, } P = V I_L \quad ...(14.16)$$

$$\text{Mechanical power developed, } P_m = \text{Power input} - \text{losses in armature and shunt field}$$

$$= V I_L - V I_{sh} - I_a^2 R_a$$

$$= V (I_L - I_{sh}) - I_a^2 R_a = V I_a - I_a^2 R_a = I_a (V - I_a R_a) = E_b I_a \quad ...(14.17)$$

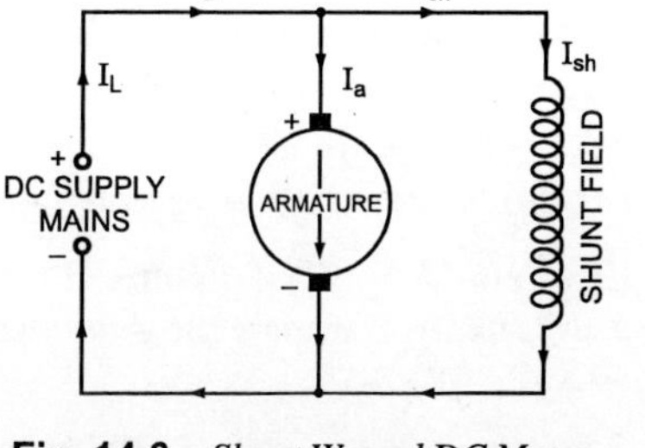

Fig. 14.6 *Shunt Wound DC Motor*

14.5.5. Compound Wound DC Motors. A compound wound dc motor has both shunt and series field coils. The shunt field is normally the stronger of the two (*i.e.*, has more ampere-turns). Compound wound motors are of two types namely cumulative compound wound and differential compound wound motors.

Cumulative compound wound motor is one in which the field windings are connected in such a way that the direction of flow of current is same in both of the field windings, as illustrated in Fig. 14.7. In the motor of this type the flux due to series field winding strengthens the field due to the shunt field winding.

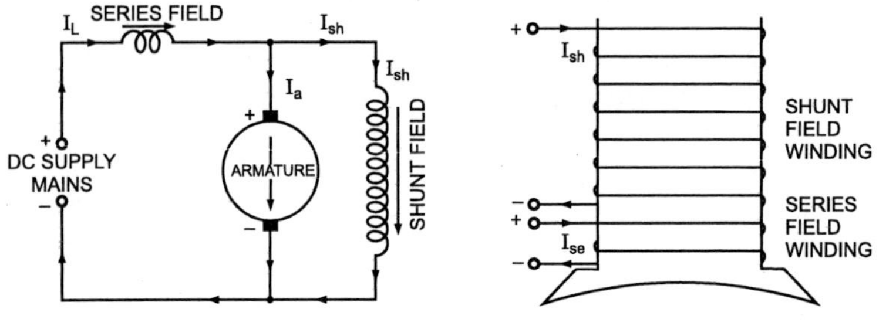

(a) Circuit Diagram *(b) Winding Connection Diagram*

Fig. 14.7 *Cumulative Compound Wound DC Motor*

Differential compound wound motor is one in which the field windings are connected in such a way that the direction of flow of current is opposite to each other in the two field windings, as illustrated in Fig. 14.8. In this type of motor the flux due to series field winding weakens the field due to shunt field winding.

Compound wound dc motors, like compound wound dc generators, may be either long shunt connected or short shunt connected. In long shunt connected (or long shunt) compound wound motors, series field and armature are connected in series with each other and in parallel with the shunt field. In short shunt connected (or short shunt) compound wound motors, the armature and shunt field are in parallel with each other and the pair is in series with the series field. Long shunt connection sometimes results in simpler wiring. Changing from long to short shunt, or vice versa, has little effect on motor performance.

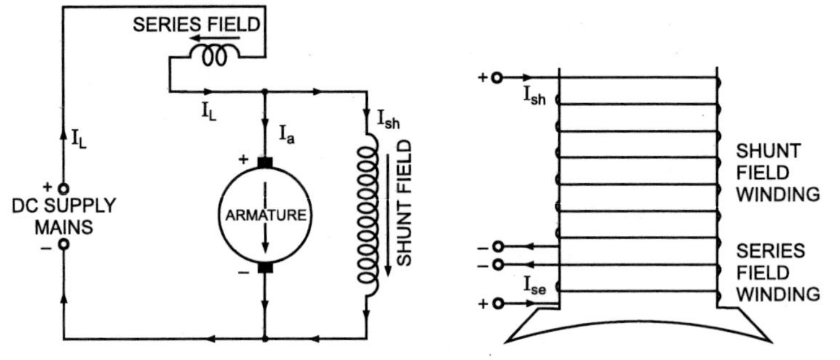

(a) Circuit Diagram *(b) Winding Connection Diagram*

Fig. 14.8 *Differential Compound Wound DC Motor*

14.6. DIRECTION OF ROTATION OF DC MOTORS

It is evident from Fig. 14.1 that if the armature current were reversed by reversing the armature leads, but leaving the field polarity the same, torque would be developed in a counter-clockwise direction. Likewise, if the field polarity were reversed leaving the armature current as shown, torque would be developed in a counter-clockwise direction. However, if both, the direction of the armature current and field polarity were reversed, torque would be developed in a clockwise direction as before. *Hence the direction of rotation of a motor can be reversed by reversing the current through either the armature winding or the field coils. If the current through both is reversed, the motor will continue to rotate in the same direction as before.*

Example 14.1 **The armature of a dc separately-excited machine has a resistance of 0.1 Ω and is connected to a 230 V supply. Calculate the generated emf when it is running (*i*) as a generator giving 80 A (*ii*) as a motor taking 60 A.** *[U.P. Technical Univ. Elec. Engineering First Semester 2004-05]*

Solution: *As a generator:* Generated emf, $E_G = V + I_a R_a = 230 + 80 \times 0.1 = 238$ V **Ans.**

As a motor: Generated emf, $E_G = E_b = V - I_a R_a = 230 - 60 \times 0.1 = 224$ V **Ans.**

Example 14.2 **A 230 V dc series motor is taking 50 A. Resistances of armature and series field winding are 0.2 Ω and 0.1 Ω respectively. Calculate (*i*) brush voltage (*ii*) back emf.**

[*U.P. Technical Univ. Basic Electrical Engineering Second Semester 2013-14*]

Solution:

$$\text{Supply voltage, } V = 230 \text{ V}$$
$$\text{Line current, } I = 50 \text{ A}$$
$$\text{Armature resistance, } R_a = 0.2\ \Omega$$
$$\text{Series field winding resistance, } R_{se} = 0.1\ \Omega$$
$$(i) \text{ Brush voltage } = V - I \times R_{se} = 230 - 50 \times 0.1 = 225 \text{ V} \quad \textbf{Ans.}$$
$$(ii) \text{ Back emf } = \text{Brush voltage} - I \times R_a = 225 - 50 \times 0.2 = 215 \text{ V} \quad \textbf{Ans}$$

Example 14.3 **A 25 kW, 250 V dc shunt generator has armature and field resistances of 0.06 Ω and 100 Ω respectively. Determine the total armature power developed when working (i) as generator delivering 25 kW output and (ii) as a motor taking 25 kW input.**

Solution:

$$\text{Shunt field current, } I_{sh} = \frac{V}{R_{sh}} = \frac{250}{100} = 2.5 \text{ A}$$

(*i*) *As generator*

$$\text{Load current, } I_L = \frac{\text{Output in kW} \times 1{,}000}{V} = \frac{25 \times 1{,}000}{250} = 100 \text{ A}$$
$$\text{Armature current, } I_a = I_L + I_{sh} = 100 + 2.5 = 102.5 \text{ A}$$
$$\text{Generated emf, } E_g = V + I_a R_a = 250 + 102.5 \times 0.06 = 256.15 \text{ V}$$
$$\text{Power developed, } P_g = \frac{E_g\, I_a}{1{,}000} = \frac{256.15 \times 102.5}{1{,}000} = 26.26 \text{ kW } \textbf{Ans.}$$

(*ii*) *As motor*

$$\text{Line current, } I_L = \frac{\text{Input in kW} \times 1{,}000}{V} = \frac{25 \times 1{,}000}{250} = 100 \text{ A}$$
$$\text{Armature current, } I_a = I_L - I_{sh} = 100 - 2.5 = 97.5 \text{ A}$$
$$\text{Back emf, } E_b = V - I_a R_a = 250 - 97.5 \times 0.06 = 244.15 \text{ V}$$
$$\text{Power developed, } P_m = \frac{E_b\, I_a}{1{,}000} = \frac{244.15 \times 97.5}{1{,}000} = 23.8 \text{ kW } \textbf{Ans.}$$

14.7. SPEED EQUATION

As mentioned in Art. 14.3, the expressions for back emf developed in the armature of a dc motor are given as

$$E_b = \frac{\Phi ZN}{60} \times \frac{P}{A} \text{ volts} \qquad \text{...(14.18)}$$

$$\text{and } E_b = V - I_a R_a \text{ volts} \qquad \text{...(14.19)}$$

Comparing expressions (14.18) and (14.19) we get

$$\frac{\Phi ZN}{60} \times \frac{P}{A} = V - I_a R_a$$

$$\text{or } N = \frac{V - I_a R_a}{\Phi Z} \times \frac{60\, A}{P}$$

$$\text{or } N = K\, \frac{V - I_a R_a}{\Phi} = K\, \frac{E_b}{\Phi}$$ since Z, A and P are constant for a particular machine.

For a dc motor, if initial values of speed, armature current, back emf and flux per pole are N_1, I_{a1}, E_{b1} and Φ_1 respectively and corresponding final values are N_2, I_{a2}, E_{b2} and Φ_2 respectively, then

$$N_1 \propto \frac{E_{b1}}{\Phi_1} \quad \text{where } E_{b1} = V - I_{a1} R_a$$

$$N_2 \propto \frac{E_{b2}}{\Phi_2} \quad \text{where } E_{b2} = V - I_{a2} R_a$$

$$\text{and } \frac{N_2}{N_1} = \frac{E_{b2}/\Phi_2}{E_{b1}/\Phi_1} = \frac{E_{b2}}{E_{b1}} \times \frac{\Phi_1}{\Phi_2} \qquad ...(14.20)$$

For a **dc shunt motor**, flux practically remains constant (*i.e.* $\Phi_2 = \Phi_1$) and

$$\frac{N_2}{N_1} = \frac{E_{b2}}{E_{b1}} = \frac{V - I_{a2} R_a}{V - I_{a1} R_a} \qquad ...(14.21)$$

In above expression since applied voltage V is constant and the voltage drop in armature ($I_a R_a$) is negligible in comparison to supply voltage V, speed of a dc shunt motor remains almost constant.

For a **dc series motor**, prior to saturation,

$$\Phi \propto I_{se} \propto I_a$$

$$\text{or } \frac{\Phi_1}{\Phi_2} = \frac{I_{a1}}{I_{a2}}$$

$$\text{and } \frac{N_2}{N_1} = \frac{E_{b2}}{E_{b1}} \times \frac{I_{a1}}{I_{a2}} \qquad ...(14.22)$$

For a **dc series motor**, after saturation

Flux Φ is independent of field current I_{se} or armature current I_a

and speed $N \propto E_b$

$$\text{or } \frac{N_2}{N_1} = \frac{E_{b2}}{E_{b1}} \qquad ...(14.23)$$

14.8. SPEED REGULATION

The speed regulation of a dc motor is defined as the change in speed when the load on the motor is reduced from full load to zero and is expressed in percentage of rated full-load speed.

$$\therefore \text{ Percentage speed regulation } = \frac{N_0 - N_f}{N_f} \times 100 \qquad ...(14.24)$$

Example 14.4 A 4-pole, 500 V dc shunt motor has 720 wave-connected conductors on its armature. The full-load armature current is 60 A and the flux per pole is 0.03 Wb. The armature resistance is 0.2 Ω and contact drop is 1 V per brush. Calculate the full-load speed of the motor. *[U.P. Technical Univ. Electrical Engineering June 2001]*

Solution: Back emf on full load, $E_{bf} = V - I_{af}R_a -$ brush contact drop $= 500 - 60 \times 0.2 - 2 \times 1 = 486$ V

Full-load speed of the motor when the armature conductors are wave- connected,

$$N_f = \frac{E_b}{\Phi} \times \frac{60}{Z} \times \frac{A}{P} = \frac{486}{0.03} \times \frac{60}{720} \times \frac{2}{4} = 675 \text{ rpm } \textbf{Ans.}$$

Example 14.5 A 4-pole dc shunt motor working on 220 V dc supply takes a line current of 3 A at no load while running at 1,500 rpm. Determine the speed when the motor takes a line current of 50 A. Assume armature and field resistances as 0.2 ohm and 400 ohm respectively. *[U.P. Technical Univ. Electrical Engineering First Semester 2003-04]*

Solution: Shunt field current, $I_{sh} = \frac{V}{R_{sh}} = \frac{220}{400} = 0.55$ A

No-load armature current, $I_{a0} = I_{L0} - I_{sh} = 3 - 0.55 = 2.45$ A

No-load back emf, $E_{b0} = V - I_{a0} R_x = 220 - 2.45 \times 0.2 = 219.51$ V

When loaded: Armature current, $I_{af} = I_{Lf} - I_{sh} = 50 - 0.55 = 49.45$ A

Back emf, $E_{bf} = V - I_{af} R_a = 220 - 49.45 \times 0.2 = 210.11$ V

$$\text{Speed, } N_f = N_0 \times \frac{E_{bf}}{E_{b0}} = 1{,}500 \times \frac{210.11}{219.51} = 1{,}436 \text{ rpm } \textbf{Ans.}$$

$\because$ flux is assumed to remain constant.

Example 14.6 **A 140 V DC shunt motor has an armature resistance of 0.2 ohm and a field resistance 70 ohm. The full-load line current is 40 A and the full-load speed is 1,800 rpm. If the brush contact drop is 3 V, find the speed of the motor at half load.** *[G.B. Technical Univ. Electrical Engineering Second Semester 2009-10]*

Solution: Full-load line current, $I_{Lf} = 40$ A

$$\text{Shunt field current, } I_{sh} = \frac{V}{R_{sh}} = \frac{140}{70} = 2 \text{ A}$$

$$\text{Full-load armature current, } I_{af} = I_{Lf} - I_{sh} = 40 - 2 = 38 \text{ A}$$

$$\text{Full-load back emf, } E_{bf} = V - I_{af} R_a - \text{brush contact drop} = 140 - 38 \times 0.2 - 3 = 129.4 \text{ V}$$

$$\text{Line current at half load, } I_{Lh} = \frac{1}{2} \times 40 = 20 \text{ A}$$

$$\text{Armature current at half load, } I_{ah} = I_{Lh} - I_{sh} = 20 - 2 = 18 \text{ A}$$

$$\text{Back emf of half load, } E_{bh} = 140 - 18 \times 0.2 - 3 = 133.4 \text{ V}$$

$$\text{Speed of motor at half load } N_h = N_f \times \frac{E_{bh}}{E_{bf}} = 1{,}800 \times \frac{133.4}{129.4} = 1{,}856 \text{ rpm} \quad \textbf{Ans.}$$

Example 14.7 **A dc shunt motor runs at 600 rpm taking 60 A from a 230 V supply. Armature resistance is 0.2 ohm and field resistance is 115 ohms. Find the speed when the current through the armature is 30 A.**
[Electrical Engineering U.P. Technical Univ. Second Semester 2007-08; G.B. Technical Univ. Odd Semester 2012-13]

Solution:

$$\text{Shunt field current, } I_{sh} = \frac{V}{R_{sh}} = \frac{230}{115} = 2 \text{ A}$$

When line current $I_{L_1} = 60$ A

$$\text{Armature current, } I_{a_1} = I_{L_1} - I_{sh} = 60 - 2 = 58 \text{ A}$$

$$\text{Back emf, } E_{b_1} = V - I_{a_1} R_a = 230 - 58 \times 0.2 = 218.4 \text{ V}$$

$$\text{Speed, } N_1 = 600 \text{ rpm}$$

When armature current $I_{a_2} = 30$ A

$$\text{Back emf, } E_{b_2} = V - I_{a_2} R_a = 230 - 30 \times 0.2 = 224 \text{ V}$$

$$\text{Speed, } N_2 = N_1 \times \frac{E_{b_2}}{E_{b_1}} = \frac{600 \times 224}{218.4} = 616.4 \text{ rpm} \quad \textbf{Ans.}$$

Example 14.8 **A 25 hp, 250 V dc series motor has armature resistance of 0.1 Ω and field resistance of 0.05 Ω and brush contact drop 3 V. When the line current is 80 A, the speed is 600 rpm. Find the speed when the line current is 100 A.**

Solution: Back emf when line current is 80 A,

$$E_{b1} = V - I(R_a + R_{se}) - \text{brush contact drop} = 250 - 80\,(0.1 + 0.05) - 3 = 235 \text{ V}$$

Back emf when line current is 100 A,

$$E_{b2} = 250 - 100\,(0.1 + 0.05) - 3 = 232 \text{ V}$$

Since $\Phi \propto I$ up to saturation point

$$\frac{\Phi_2}{\Phi_1} = \frac{I_2}{I_1} = \frac{100}{80} = 1.25$$

$$\text{Speed when line current is 100 A, } N_2 = N_1 \times \frac{E_{b2}}{E_{b1}} \times \frac{\Phi_1}{\Phi_2} = 600 \times \frac{232}{235} \times \frac{1}{1.25} = 473.9 \text{ rpm} \quad \textbf{Ans.}$$

Example 14.9 **A dc shunt generator running at 1,200 rpm supplies a load of 60 kW at 250 V. Find the speed at which it runs as a shunt motor when taking 60 kW from 250 V supply. Take armature resistance as 0.1 Ω and field winding resistance as 50 Ω. Neglect brush drop.**

[U.P. Technical Univ. Electrical Engineering First Semester 2005-06]

Solution:

$$\text{Shunt field current, } I_{sh} = \frac{\text{Terminal voltage, V}}{\text{Shunt field resistance, } R_{sh}} = \frac{250}{50} = 5 \text{ A}$$

As generator :

$$\text{Load current, } I_L = \frac{\text{Output in kW} \times 1{,}000}{V} = \frac{60 \times 1{,}000}{250} = 240 \text{ A}$$

$$\text{Armature current, } I_{ag} = I_L + I_{sh} = 240 + 5 = 245 \text{ A}$$

$$\text{Generated emf, } E_g = V + I_{ag} R_a = 250 + 245 \times 0.1 = 274.5 \text{ V}$$

$$\text{Speed, } N_g = 1{,}200 \text{ rpm}$$

As motor :

$$\text{Line current, } I_L = \frac{60 \times 1{,}000}{250} = 240 \text{ A}$$

$$\text{Armature current, } I_{am} = I_L - I_{sh} = 240 - 5 = 235 \text{ A}$$

$$\text{Back emf developed, } E_b = V - I_{am} R_a = 250 - 235 \times 0.1 = 226.5 \text{ V}$$

$$\text{Speed of motor, } N_m = \frac{E_b}{E_g} \times N_g = \frac{226.5}{274.5} \times 1{,}200 = 990 \text{ rpm } \textbf{Ans.}$$

Example 14.10 **A dc shunt machine connected to 230 volts supply has resistance of armature as 0.115 Ω and of field winding as 115 Ω. Find the ratio of the speed as a generator to the speed as a motor with the line current in each case being 100 amperes.** *[U.P. Technical Univ. Electrical Engineering Second Semester 2002-03]*

Solution:

$$\text{Shunt field current, } I_{sh} = \frac{\text{Terminal voltage, V}}{\text{Shunt field resistance, } R_{sh}} = \frac{230}{115} = 2 \text{ A}$$

$$\text{Line current, } I_L = 100 \text{ A}$$

As Generator :

$$\text{Armature current, } I_a = I_L + I_{sh} = 100 + 2 = 102 \text{ A}$$

$$\text{Generated emf, } E_g = V + I_a R_a = 230 + 102 \times 0.115 = 241.73 \text{ V}$$

As Motor :

$$\text{Armature current, } I_a = I_L - I_{sh} = 100 - 2 = 98 \text{ A}$$

$$\text{Back emf developed, } E_b = V - I_a R_a = 230 - 98 \times 0.115 = 218.73 \text{ V}$$

Ratio of speed as a generator to the speed as a motor.

$$\frac{N_g}{N_m} = \frac{E_g}{E_b} = \frac{241.73}{218.73} = 1.1052 \textbf{ Ans.}$$

$\because E \,\alpha\, \phi N \,\alpha\, N$ when ϕ or field current is constant.

i.e., $N_g : N_m :: 1.1052 : 1$ **Ans.**

Example 14.11 **A dc shunt generator delivers 50 kW at 250 V and 400 rpm. The armature and field resistances are 0.02 Ω and 50 Ω respectively. Calculate the speed of the machine running as shunt motor and taking 50 kW input at 250 V. Allow brush contact drop of 1 V per brush.**

[*Mahamaya Technical Univ. Electrical Engineering Second Semester 2011-12*]

Solution:

$$\text{Shunt field current, } I_{sh} = \frac{V}{R_{sh}} = \frac{250}{50} = 5 \text{ A}$$

(*i*) *As Generator*

$$\text{Load current, } I_{Lg} = \frac{\text{Output in kW} \times 1{,}000}{V} = \frac{50 \times 1{,}000}{250} = 200 \text{ A}$$

$$\text{Armature current, } I_{ag} = I_{Lg} + I_{sh} = 200 + 5 = 205 \text{ A}$$

$$\text{Generated emf, } E_g = V + I_{ag} R_a + \text{brush contact drop} = 250 + 205 \times 0.02 + 1 \times 2 = 256.1 \text{ V}$$

$$\text{Speed, } N_g = 400 \text{ rpm}$$

As Motor

$$\text{Load current, } I_{Lm} = \frac{50 \times 1{,}000}{250} = 200 \text{ A}$$

$$\text{Armature current, } I_{am} = I_{Lm} - I_{sh} = 200 - 5 = 195 \text{ A}$$

$$\text{Back emf developed, } E_b = V - I_{am} R_a - \text{brush contact drop} = 250 - 195 \times 0.02 - 1 \times 2 = 244.1 \text{ V}$$

$$\text{Speed of motor, } N_m = N_g \times \frac{E_b}{E_g} = \frac{400 \times 244.1}{256.1} = 381.3 \text{ rpm } \textbf{Ans.}$$

Example 14.12 **A belt driven 100 kW shunt generator running at 300 rpm on 220 V bus-bars continues to run as a motor when the belt breaks, then taking 10 kW. What will be its speed ? Given armature resistance = 0.025 Ω, field resistance = 60 Ω and contact drop under each brush = 1 V. Ignore armature reaction.**

Solution: **As Generator**

$$\text{Load current, } I_L = \frac{\text{Output in kW} \times 1{,}000}{\text{Bus-bar voltage}} = \frac{100 \times 1{,}000}{220} = 454.55 \text{ A}$$

$$\text{Shunt field current, } I_{sh} = \frac{\text{Bus-bar voltage}}{R_{sh}} = \frac{220}{60} = 3.67 \text{ A}$$

$$\text{Armature current, } I_a = I_L + I_{sh} = 454.55 + 3.67 = 458.22 \text{ A}$$

$$\text{Generated emf, } E_g = V + I_a R_a + \text{brush contact drop} = 220 + 458.22 \times 0.025 + 2 \times 1 = 233.46 \text{ V}$$

$$\text{Speed, } N_g = 300 \text{ rpm}$$

As motor:

$$\text{Line current, } I_L = \frac{\text{Output in kW} \times 1{,}000}{\text{Bus-bar voltage}} = \frac{10 \times 1{,}000}{220} = 45.46 \text{ A}$$

$$\text{Shunt field current, } I_{sh} = \frac{V}{R_{sh}} = \frac{220}{60} = 3.67 \text{ A}$$

$$\text{Armature current, } I_a = I_L - I_{sh} = 45.46 - 3.67 = 41.79 \text{ A}$$

$$\text{Back emf, } E_b = V - I_a R_a - \text{brush contact drop} = 220 - 41.79 \times 0.025 - 2 \times 1 = 216.955 \text{ V}$$

Since shunt field current is same and armature reaction is negligible, flux is same in both cases and speed of the machine as a motor,

$$N_m = N_g \times \frac{E_b}{E_g} = 300 \times \frac{216.955}{233.46} = 278.8 \text{ rpm} \quad \textbf{Ans.}$$

14.9. TORQUE

The measure of causing the rotation of a wheel or the turning or twisting moment of a force about the axis is called the torque.

Torque is measured by the product of force and the radius at which this force acts.

Consider a wheel of radius *r* metres acted by a circumferential force F newtons, as shown in Fig 14.9. Let the force F cause the wheel to rotate at *n* rps.

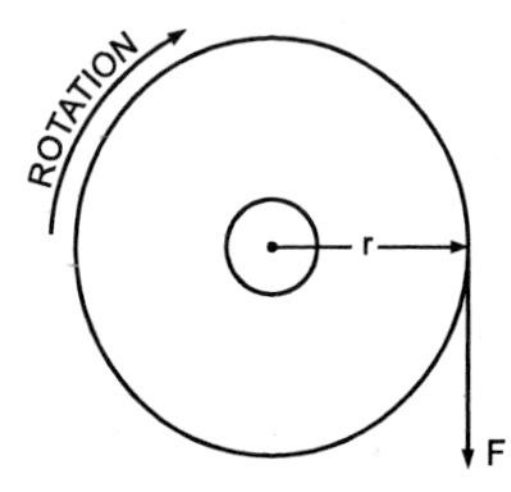

Fig. 14.9

$$\text{Torque, } T = F \times r \text{ newton-metres}$$

$$\text{Work done per revolution} = F \times \text{distance moved} = F \times 2\pi r \text{ joules}$$

$$\text{Work done per second} = F \times 2\pi r \times n = F \times r \times 2\pi n \text{ joules/second or watts}$$

Since $F \times r$ = Torque, T and $2\pi n$

= ω, angular velocity in radians per second

$$\text{So work done per second} = T\omega \text{ joules per second or watts} \qquad \text{...(14.25)}$$

$$\text{So power developed, } P = T\omega = T \times \frac{2\pi N}{60} = 0.105 \, N T \text{ watts} \qquad \text{...(14.26)}$$

14.10. ARMATURE TORQUE

Let T_e be the electromagnetic torque developed in newton-metres by the motor running at *n* rps.

$$\text{Power developed} = \text{Work done per second} = T_e \omega = T_e \times 2\pi n \text{ watts} \qquad \text{...(14.27)}$$

Electrical equivalent of mechanical power developed by the armature, as mentioned in Art 14.4, also

$$= E_b I_a \text{ watts} \qquad \text{...(14.28)}$$

Comparing expressions (14.27) and (14.28) we have

$$T_e \times 2\pi n = E_b I_a$$

$$\text{or } T_e = \frac{E_b I_a}{2\pi n} = 0.159 \frac{E_b I_a}{n} \text{ N-m} \qquad \text{....(14.29)}$$

$$\text{Also } T_e = \frac{E_b I_a}{2\pi \frac{N}{60}} = 9.55 \frac{E_b I_a}{N} \text{ N-m}$$ ... (14.30) where N is speed in rpm

Substituting $E_b = \Phi Z \frac{N}{60} \times \frac{P}{A}$ in Eq. (14.30) we have

$$T_e = 9.55 \times \Phi \times Z \times \frac{N}{60} \times \frac{P}{A} \times \frac{I_a}{N} = 0.159\, \Phi Z P \frac{I_a}{A} \text{ newton-metres} \quad ...(14.31)$$

ALTERNATIVE PROOF. Quantitatively, the force on a straight section of conductor length l metres perpendicular to a magnetic field of density B webers per square metre and carrying I_c amperes is given as

$$F_c = B l I_c \text{ newtons} \quad \text{[Refer to Eq. (1.34)]}$$

If D is the diameter of the armature, the force F_c acts at a radial distance of D/2 metres and the associated torque per conductor is given as

$$T_c = \frac{B l I_c D}{2}$$

$$\text{The average flux density, } B_{av} = \frac{\Phi P}{\pi D l}$$

$$\text{So torque developed per conductor, } T_c = \frac{\Phi P}{\pi D l} \times \frac{l I_c D}{2} = \frac{1}{2\pi} \Phi P I_c$$

The torque associated with the entire winding is the summation of the torques for the individual conductors or coil sides. Torque developed by such electromagnetic action is called the electromagnetic torque and is given as

$$T = \frac{1}{2\pi} \Phi Z P I_c$$

where Z is the number of armature conductors on the armature surface.

$$= \frac{1}{2\pi} \Phi Z P \frac{I_a}{A} = 0.159\, \Phi Z P \frac{I_a}{A} \quad \because \text{ Current per conductor, } I_c = \frac{I_a}{A}$$

Since Z, P and A are constant for a particular machine,

$$T = K \Phi I_a \text{ where } K = \frac{1}{2\pi} \frac{PZ}{A}$$

$$\text{or } T \propto \Phi I_a$$

Thus it may be concluded that

(*i*) the electromagnetic torque developed by the armature is proportional to the product of flux per pole and armature current.

(*ii*) the direction of electromagnetic torque developed by armature depends upon the direction of flux or magnetic field and the direction of flow of current in armature conductors. If either of the two is reversed the direction of torque developed will be reversed and, therefore, the direction of rotation. When both (the direction of field as well as that of armature current) are reversed the direction of torque (or rotation) will not change. This has been explained in Art 14.6 also.

In case of a series wound motor flux Φ is directly proportional to armature current I_a (before saturation) because in a series wound motor field winding and armature winding currents are same, therefore,

$$T_e \propto I_a^2$$

In case of permanent magnet motors, separately-excited motors and shunt wound motors the field strength i.e. Φ is practically constant and, therefore,

$$T_e \propto I_a$$

On the basis of the same killowatt output and speed, a dc series motor develops the highest starting torque and the dc shunt motor the least, while the cumulative compound wound dc motor falls somewhere between the first two.

14.11. SHAFT TORQUE

The armature torque is the gross torque, which is developed by the armature. Whole of this torque is not available at the pulley, since certain percentage of torque developed by the armature is lost to overcome the iron and friction losses. *Net torque i.e. gross torque less torque lost in iron and friction losses is known as shaft torque.*

If T_a is the torque developed by armature in N-m, T_f is the torque lost in iron and friction losses, and T_{sh} is the shaft torque or useful torque, then

$$T_a = \frac{E_b\, I_a}{\frac{2\pi N}{60}} = 9.55\,\frac{E_b\, I_a}{N} \text{ newton metres} \qquad \text{where N is speed of armature in rpm}$$

$$T_f = \frac{\text{Iron and frictional losses in watts}}{2\pi N/60} = \frac{9.55}{N}\text{ (iron and frictional losses in watts)}$$

$$\text{Shaft torque, } T_{sh} = T_a - T_f = \frac{E_b\, I_a - \text{iron and frictional losses}}{2\pi\, N/60} \text{ newton–metres}$$

$$T_{sh} \text{ also } = \frac{\text{Output in watts}}{2\pi\, N/60} = \frac{9.55 \times \text{output in watts}}{N} \text{ newton-metres} \qquad ...(14.32)$$

14.11.1. Brake Horse Power (BHP). In case of electric motors (ac or dc), the mechanical power available at the shaft in horse power is known as *brake horse power* (bhp). If T_{sh} is the shaft torque in newton metres and N is speed in rpm then

$$\text{Output in bhp} = \frac{2\pi\, N\, T_{sh}}{60 \times 735.5} \qquad ...(14.33)$$

Example 14.13 **The armature of a 200 V dc shunt motor is 0.12 Ω. It runs at 600 rpm at constant torque load and draws a current of 21 A. Calculate its new speed if the field current is reduced to 10%.**

[*G.B. Technical Univ. Electrical Engineering Second Semester 2011-12*]

Solution:

Supply voltage, V = 200 V

Line current, I_{L1} = 21 A

Armature current, $I_{a1} = I_{L1} = 21$ A — neglecting shunt field current

Speed, N_1 = 600 rpm

Back emf, $E_{b1} = V - I_{a1}\, R_a = 200 - 21 \times 0.12 = 197.48$ V

Since torque is same

$$\text{or } T_2 = T_1$$

$$I_{a2}\Phi_2 = I_{a1}\Phi_1$$

$$\text{or } I_{a_2} = I_{a1} \times \frac{\Phi_1}{\Phi_2} = I_{a_1} \times \frac{I_{sh_1}}{I_{sh_2}} \qquad \text{Assuming } \Phi \propto I$$

$$= 21 \times \frac{1}{0.9} = 23.333 \text{ A} \qquad \because I_{sh2} = 0.9\, I_{sh1}$$

$$\text{Back emf } E_{b2} = V - I_{a2}\, R_{a2} = 200 - 23.333 \times 0.12 = 197.2 \text{ V}$$

$$\text{Speed, } N_2 = \frac{E_{b2}}{E_{b1}} \times \frac{\Phi_1}{\Phi_2} \times N_1 = \frac{197.2}{197.48} \times \frac{1}{0.9} \times 600 = 665.7 \text{ rpm } \textbf{Ans.} \qquad \because E_b \propto \Phi N$$

Example 14.14 **A dc shunt motor develops an open-circuit emf of 250 volt at 1,500 rpm. Find its developed torque for an armature current of 20 amperes.** [*G.B. Technical Univ. Electrical Engineering First Semester 2011-12*]

Solution:

Open-circuit emf, E_0 = 250 V

Armature current, I_a = 20 A

Assuming back emf equal to open-circuit emf, though it may be slightly less due to armature reaction if field current remains unchanged.

$$\text{Torque developed, } T_e = \frac{9.55\, E_b I_a}{N} = \frac{9.55 \times 250 \times 20}{1{,}500} = 31.833 \text{ N-m } \textbf{Ans.}$$

Example 14.15 Calculate the armature torque developed by a 440 V dc shunt motor having an armature resistance of 0.25 Ω and running at 750 rpm when taking an armature current of 60 A.

Solution:

$$\text{Back emf developed, } E_b = V - I_a R_a = 440 - 60 \times 0.25 = 425 \text{ V}$$

$$\text{Torque developed, } T_e = \frac{9.55 \times E_b I_a}{N} = \frac{9.55 \times 425 \times 60}{750} = 324.7 \text{ Nm } \textbf{Ans.}$$

Example 14.16 The armature of a 6-pole lap wound dc shunt motor takes 400 A at a speed of 350 rpm. The flux per pole is 80 m Wb, the number of turns is 600, and 3% of the torque is lost in windage, friction and iron losses. Calculate the brake horse power.

Solution: Total or gross torque developed by armature,

$$T_a = 0.159 \times \Phi \times Z \times P \times \frac{I_a}{A} = 0.159 \times 80 \times 10^{-3} \times (600 \times 2) \times 6 \times \frac{400}{6} = 6{,}105.5 \text{ Nm}$$

$\because Z = 2 \times$ number of turns

Shaft torque, T_{sh} = Total torque developed – torque lost in windage, friction and iron losses

$$= T_a - T_f = T_a\left(1 - \frac{T_f}{T_a}\right) = 6{,}105.5\,(1 - 0.03) = 5{,}922.34 \text{ Nm}$$

$$\text{Brake horse power} = T_{sh} \times \frac{2\pi \times N}{60} \times \frac{1}{735.5} = 5{,}922.34 \times \frac{2\pi \times 350}{60} \times \frac{1}{735.5} = 295.13 \textbf{ Ans.}$$

Example 14.17 A 250 V dc shunt motor having an armature resistance of 0.25 Ω carries an armature current of 50 A and runs at 750 rpm. If the flux is reduced by 10 %, find the speed. Assume that the torque remains the same.

[U.P. Technical Univ. Electrical Engineering January 2003]

Solution:

Supply voltage, $V = 250$ V

Armature current, $I_{a1} = 50$ A

Back emf, $E_{b1} = V - I_{a1} R_a = 250 - 50 \times 0.25 = 237.5$ V

Flux, $\Phi_2 = 0.9\, \Phi_1$ $\quad\because$ Flux has been reduced by 10%

Since torque developed remains unchanged,

$$I_{a2}\, \Phi_2 = I_{a1}\, \Phi_1 \quad \text{as } T \propto \Phi\, I_a$$

$$\text{or Armature current, } I_{a2} = I_{a1} \times \frac{\Phi_1}{\Phi_2} = 50 \times \frac{1}{0.9} = 55.55 \text{ A}$$

$$\text{Back emf, } E_2 = V - I_{a2} R_a = 250 - 55.55 \times 0.25 = 236.11 \text{ V}$$

$$\text{Speed, } N_2 = \frac{E_{b2}}{E_{b1}} \times \frac{\Phi_1}{\Phi_2} \times N_1 = \frac{236.11}{237.5} \times \frac{1}{0.9} \times 750 = 828.5 \text{ rpm } \textbf{Ans.} \quad \because E_b \propto \Phi N$$

Example 14.18 A dc shunt generator delivers 50 kW at 250 V when running at 500 rpm. The armature and field resistances are 0.05 ohm and 125 ohm respectively. Calculate the speed of the same machine and developed torque when running as a shunt motor and taking 50 kW at 250 V. Allow 1 volt per brush for contact drop.

[G.B. Technical Univ. Electrical Engineering First Semester 2009-10]

Solution: As Generator

Generator output, $P_G = 50$ kW

Supply voltage, $V_L = 250$ V

Speed, $N_G = 500$ rpm

Armature resistance, $R_a = 0.05\ \Omega$

Shunt field resistance, $R_{sh} = 125\ \Omega$

$$\text{Line current, } I_L = \frac{P_G \times 1000}{V_L} = \frac{50 \times 1,000}{250} = 200 \text{ A}$$

$$\text{Shunt field current, } I_{sh} = \frac{V_L}{R_{sh}} = \frac{250}{125} = 2 \text{ A}$$

$$\text{Armature current, } I_{ag} = I_L + I_{sh} = 200 + 2 = 202 \text{ A}$$

$$\text{Generated emf. } E_g = V_L + I_{ag} R_a + 2 \times \text{volt per brush} = 250 + 202 \times 0.05 + 2 \times 1 = 262.1 \text{ V}$$

As Motor

$$\text{Load current, } I_L = \frac{50 \times 1,000}{250} = 200 \text{ A}$$

$$\text{Armature current, } I_{am} = I_L - I_{sh} = 200 - 2 = 198 \text{ A}$$

$$\text{Back emf developed, } E_b = V - I_{am} R_a - \text{brush contact drop} = 250 - 198 \times 0.05 - 2 \times 1 = 238.1 \text{ V}$$

$$\text{Speed of motor, } N_m = N_g \times \frac{E_b}{E_g} = \frac{500 \times 238.1}{262.1} = 454 \text{ rpm} \quad \textbf{Ans.}$$

$$\text{Motor torque, } T = \frac{E_b I_a}{2\pi N/60} = \frac{238.1 \times 198}{2\pi \times 454} \times 60 = 991.6 \text{ Nm} \quad \textbf{Ans.}$$

14.12. OPERATING CHARACTERISTICS OF DC MOTORS

The performance and, therefore, suitability of a dc motor is determined from its characteristics, known as *performance characteristics*. The important characteristics of dc motors are:

1. Torque-Armature Current Characteristic. This characteristic curve gives relation between torque developed in the armature, T and armature current I_a. This is also known as *electrical characteristic*.

2. Speed-Armature Current Characteristic. This characteristic curve gives relation between speed N and armature current I_a. This is also known *as speed characteristic*.

3. Speed-Torque Characteristic. This characteristic gives relation between speed N and torque developed in armature, T. This is also known as *mechanical characteristic*. This curve may be derived from the two characteristic curves mentioned above.

The important relations to be kept in mind while discussing motor characteristics are :

(*i*) $I_a = \frac{V - E_b}{R_a}$ (*ii*) $N \propto \frac{E_b}{\Phi}$ and (*iii*) $T \propto \Phi I_a$.

14.13. OPERATING CHARACTERISTICS OF DC SERIES MOTORS

1. Speed-Current Characteristic. In case of a dc series motor, the mmf due to the exciting coils increases in direct proportion to the line or armature current, so (neglecting armature reaction effects) the value of flux varies with the load current according to the ordinary magnetization curve. Owing to armature reaction, the actual curve representing the useful flux falls below the open-circuit magnetization curve, as illustrated in Fig. 14.10. With larger currents the magnetic circuit gets saturated and flux Φ tends to approach a constant value.

From the speed equation, it is obvious that speed is proportional to back emf E_b and inversely proportional to flux per pole Φ. With the increase in armature current voltage drop in armature circuit and series field [I (R_a + R_{se})] increases and, therefore, back emf E_b decreases, as shown in Fig. 14.10. However, under normal conditions I ($R_a + R_{se}$) drop is quite small and may be neglected. Thus if the applied voltage remains constant, speed N is inversely proportional to flux Φ. If a curve is drawn between speed and input (or line) current I, it will be a rectangular hyperbola before magnetic saturation as up to saturation point the magnetisation curve is a straight line. In this region, the speed decreases abruptly with the increase in input current. After magnetic saturation, the flux Φ tends to become constant and speed-current characteristic becomes a straight line and speed decreases slightly due to voltage drop in armature and series field, as shown in Fig. 14.10. The speed becomes zero when the input current is the normal short-circuit current of the motor *i.e.* equal to applied voltage divided by the motor resistance ($R_a + R_{se}$). Normally this current is many many times full-load value.

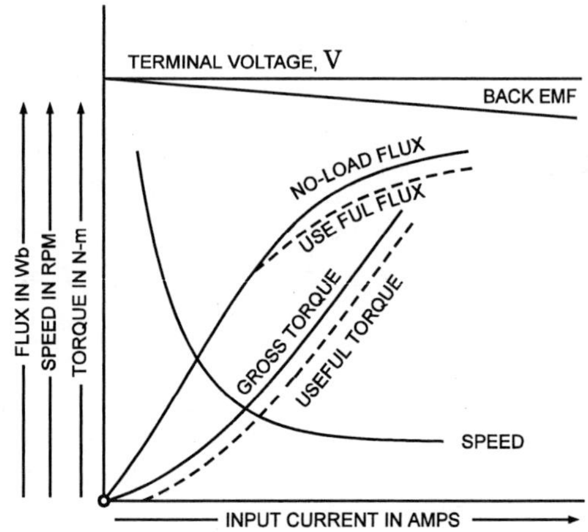

Fig. 14.10

From speed-current characteristic curve it is obvious that the series motor is a variable speed motor *i.e.* speed varies with the variation in line current. With the decrease in load on a dc series motor, the speed increases and may become dangerously high at very light loads.

Since on no load the speed is dangerously high, the machine may get damaged due to heavy centrifugal forces set up in the rotating parts. *This is the reason that series motors are never started on no load, which is explained below.*

When the motor is connected across the supply mains without load, it draws a small current from supply mains flowing through the series field and armature, the speed tends to increase so that back emf may approach the applied voltage in magnitude. The increase in back emf weakens the armature current and hence the field current. This causes again increase in speed so in back emf. Thus the field continues to weaken and speed continues to increase until the armature gets damaged.

To start a dc series motor, mechanical load is first put and then the motor is started.

Since on no load the series motor attains dangerously high speed, which causes heavy centrifugal force resulting in the damage of machine, series motors are not suitable for the services:

(*i*) where the load may be entirely removed and

(*ii*) for driving by means of belts because mishap to the belt would cause the motor run on no load.

These motors are suitable for gear drive, because gears provide some load on account of the frictional resistances of the gear teeth in case of sudden release of load.

However, very small series motors may be used with belts, since in case of mishap to belt, the comparatively large frictional resistance would represent an appreciable load on it.

The minimum load on a dc series motor should be great enough (not below 15% of full load) to keep the speed of the motor within limits. In case the speed becomes dangerously high the motor must be disconnected from the supply main.

2. Torque-Current Characteristic. From the expression of mechanical torque T it is obvious that torque is directly proportional to the product of flux per pole Φ and armature current I_a. Up to saturation point flux is proportional to field current and hence to the armature current, because $I_a = I_f$. Therefore, on light load mechanical torque T is proportional to the square of the armature current *i.e.*, $T \propto I_a^2$ and hence curve drawn between torque and armature current up to saturation point is a parabola, as shown in Fig. 14.10.

After saturation point flux Φ is almost independent of excitation current and so the torque is proportional to the armature current *i.e.* $T \propto I_a$. Hence the characteristic becomes a straight line. The useful (or shaft) torque is, of course, less than the total (or gross) torque developed. This is due to torque lost in iron and friction and windage losses.

From the torque-armature current curve it is evident that so long as the field of the motor is not saturated, the series motor exerts a torque proportional to the square of current i.e. starting torque is very high. Hence series motors are used where large starting torque is required for accelerating heavy masses quickly such as in hoists, electric railways, trolleys and electric vehicles.

From torque-current characteristic it is evident that series motor develops large starting torque to accelerate the heavy masses and from speed-current characteristic it is also evident that speed falls as the load increases, so series motor is automatically relieved from heavy excessive load, therefore, series motor is best suited for electric traction work.

3. Speed-Torque Characteristic. The speed-torque characteristic can be drawn with the help of above two characteristics as illustrated in Fig. 14.11. Speed-torque characteristic, is also known as mechanical characteristic. Speed sharply falls with the increase in torque for smaller values of load. But at higher loads, the speed drops linearly but slowly with increasing torque.

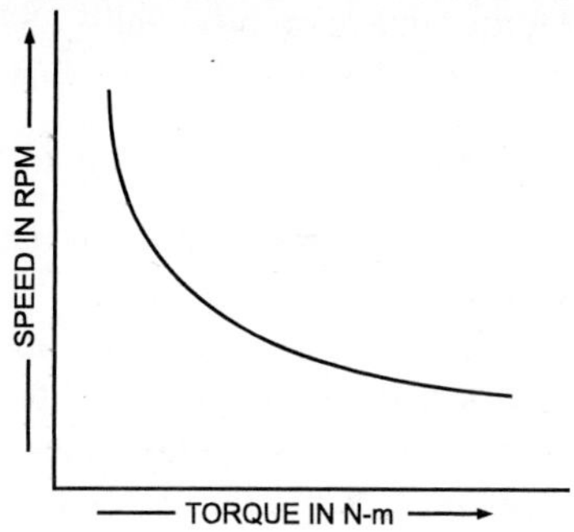

Fig. 14.11 *Speed-Torque Characteristic of DC Series Motor*

Hence series motors are best suited for services where the motor is directly coupled to the load such as fans whose speed falls with the increase in load torque.

14.14. OPERATING CHARACTERISTICS OF DC SHUNT MOTORS

1. Speed - Armature Current Characteristic. If applied voltage V is kept constant, the field current will remain constant, hence flux will have maximum value on no load but will slightly decrease due to armature reaction as the load increases but for most purposes the flux is considered to be constant, neglecting armature reaction effect.

From speed equation, speed N is directly proportional to back emf E_b or $(V - I_a R_a)$ and inversely proportional to the flux Φ. Since flux is considered to be constant as mentioned above, with the increase in armature current the speed slightly falls due to increase in voltage drop in armature and the speed-armature current curve coincides with the back emf-armature current curve. Since voltage drop in armature at full load is very small as compared to applied voltage, drop in speed from no load to full load is very small. If demagnetising effect of armature reaction is considered, the drop in speed due to voltage drop in armature is compensated for up to some extent due to decrease in flux with the increase in armature current and speed-armature current characteristic is less drooping, as shown dotted in Fig. 14.12 or even be rising if demagnetisation is high. Thus the drop in speed from no load to full load is very small and for all practical purposes the dc shunt motor is taken as a constant speed motor.

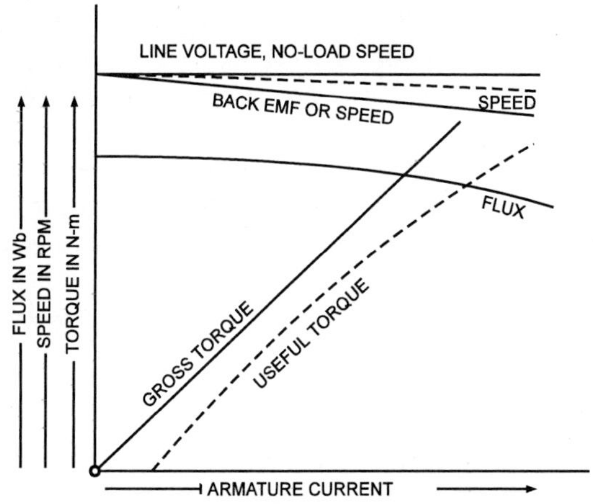

Fig. 14.12

There is a slight variation in speed of the shunt motor from no load to full load and this slight variation in speed can be made up by inserting resistance in the shunt field and so reducing the flux. Therefore, shunt motors can be used for the loads which are totally and suddenly thrown off without resulting in excessive speed. Shunt motors being constant speed motors are best suited for driving of line shafts, machine lathes, milling machines, conveyors, fans and for all purposes where constant speed is required. It is not suitable for use with flywheel or with fluctuating loads or for parallel operation due to its constant speed characteristic. It is also useful where a moderate degree of speed control is required.

2. Torque-Armature Current Characteristic. From the expression for the torque of a dc motor, torque is directly proportional to the product of flux per pole Φ and armature current I_a. Since in case of dc shunt motor the flux per pole Φ is considered to be constant, torque increases with the increase in load current following linear law, *i.e.* torque-armature current characteristic is a straight line passing through origin O

(Fig. 14.12). But the weakening of field due to armature reaction causes the torque line to droop slightly, and the iron and friction losses cause it to be slightly lower than the line representing the electromagnetic developed torque corresponding to $T = 0.159\ \Phi\ Z\ P\ \frac{I_a}{A}$ Nm.

3. Speed-Torque Characteristic. This characteristic curve can be drawn from the above two characteristics and is shown in Fig. 14.13.

This type of motor is used in applications requiring medium starting torque such as centrifugal pumps, blowers, fans, conveyors, boring mills, shapers, woodworking machines, spinning and weaving machines, printing presses, machine tools etc.

DC shunt motors should never be started on heavy loads because such loads need heavy starting current.

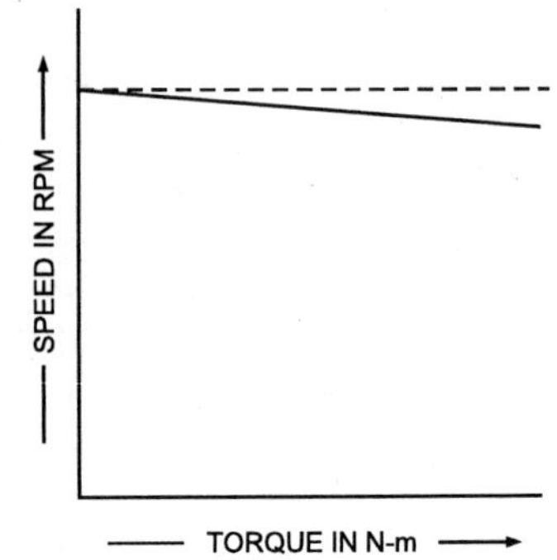

Fig. 14.13

14.15. OPERATING CHARACTERISTICS OF SEPARATELY-EXCITED DC MOTORS

The torque-armature current, speed-armature current and speed-torque characteristics of separately-excited dc motors are similar to those of dc shunt motors.

14.16. OPERATING CHARACTERISTICS OF COMPOUND WOUND DC MOTORS

As explained in Art 14.5, the compound wound dc motors are of two types viz cumulative compound wound motors and differential compound wound motors.

14.16.1. Operating Characteristics of Cumulative Compound Wound DC Motors. The characteristics of cumulative compound wound motor are combination of shunt and series characteristics and lie between those of shunt and series motors.

As the load is increased, the flux due to series field winding increases and causes the torque greater than it would have been with shunt field winding alone for a given machine and for a given current. The increase in flux due to series field winding on account of increase in load causes the speed to fall more rapidly than it would have done in shunt motor. The characteristics are shown in Fig. 14.14.

Like the series motor, the cumulative compound wound motor has excellent characteristics for starting heavy loads and yet there is no danger of "running away" at light loads when used without a speed controller.

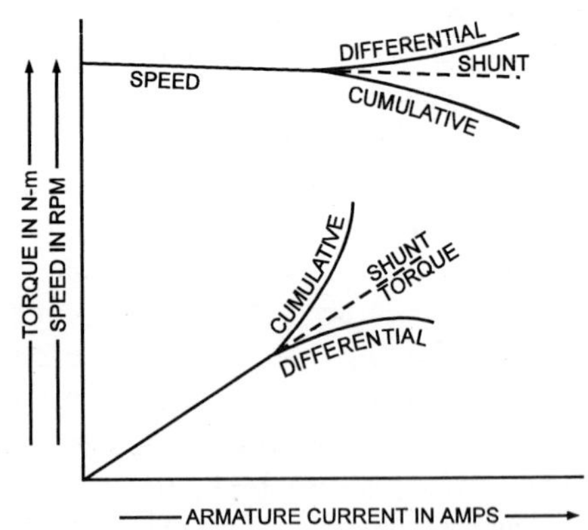

Fig. 14.14 *Characteristics of Compound Wound Motors*

By varying the proportions of shunt field mmf and series field mmf, compound wound motors may be designed to give speed and torque characteristics anywhere intermediate between those for shunt and series motors. In many cases only enough shunt field is provided to guarantee a safe no-load speed, in which case the motor be called a *stabilised series motor*. Compound wound motors are usually designed so that there will be a drop in speed of about 20% between no load to full load.

Cumulative compound wound motors are used in driving machines which are subject to sudden applications of heavy loads, such as occur in rolling mills, punching and shearing machines, lifts, mine-hoist etc. This type of motor is used also where a large starting torque is required but series motor cannot be employed conveniently such as in cranes and elevators.

Another advantage of the cumulative compound wound motor for suddenly applied loads is that the motor automatically undergoes a substantial drop in speed when the load is applied. Accordingly much of its stored kinetic energy becomes available for supplying a part of the increased load thus reducing the electrical load on the motor as well as the peaks on the power system. The available kinetic energy is frequently increased by the use of a flywheel, particularly with rolling mill motors.

Cumulative compound wound motors are not practical for applications requiring adjustable speed for field control. With a weakened shunt field, the series field becomes dominating hence the changes in load may produce unstable speeds.

14.16.2. Operating Characteristics of Differential Compound Wound DC Motors. In differential compound wound motor, since the series field opposes the shunt field, the resultant flux decreases with the increase in load; thus the machine runs at a higher speed than it would do as a shunt motor. The decrease in flux with the increase in load causes the torque to be less than that of a shunt motor, as illustrated in Fig. 14.14. Such motors are rarely employed in practice as the differential arrangement causes difficulties during overloads and starting.

14.17 SPEED CONTROL OF DC MOTORS

Speed control means intentional change of the drive speed to a value required for performing the specific work process.

One of the attractive features the dc motor offers over all other types is the relative ease with which speed control can be achieved and, therefore, dc motors are indispensable for many adjustable speed drives. The various schemes available for speed control can be deduced from the expression of speed for a dc motor which is repeated here with one modification

$$N = K\frac{V - I_a(R+R_a)}{\Phi} \qquad \text{...(14.34)}$$

The modification involves the inclusion of an external resistance in the armature circuit. The above expression reveals that the speed can be controlled by adjusting any one of the three factors appearing on the right hand side of the expression: (*i*) applied voltage to the armature terminals, V (*ii*) external resistance in the armature circuit, R and (*iii*) flux per pole, Φ. The first two possibilities involve adjustment affecting the armature circuit, whereas the third involves change in the magnetic field. Therefore, speed control methods are broadly classified as *armature control methods* and *field control methods.* Sometimes a combination of the two methods is employed. With armature control the speed decreases as the voltage applied to the armature terminals is reduced, whereas with field control the speed increases as the flux is reduced.

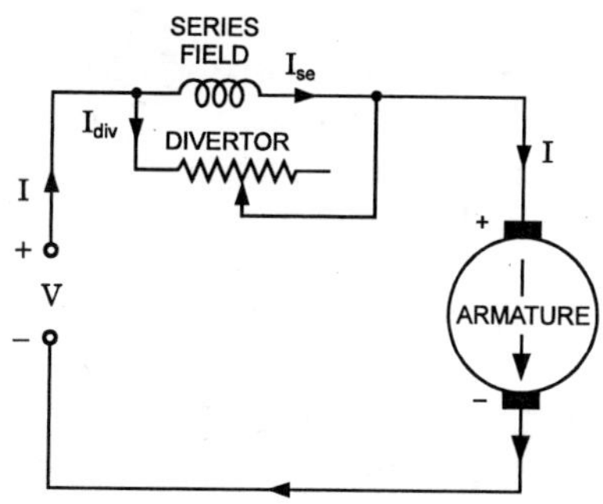

Fig. 14.15 (*a*) *Field Divertor Method of Speed Control For DC Series Motors*

1. *Field Control Methods.* In case of *dc series motors,* the flux can be varied by any one of the following methods.

(*i*) ***Field Divertor Method.*** The field flux can be reduced by shunting a portion of motor current around the series field, thus reducing the excitation mmf and weakening of field. This method is illustrated in Fig. 14.15 (*a*).

This method provides speeds above normal because flux is reduced by this method. Lesser the divertor resistance, less the field current, less flux and, therefore, more the speed. This method is convenient as well as economical and provides the speed control range usually not exceeding 2 : 1. This method of speed control is used in electric drives in which the speed should rise sharply as soon as the load falls.

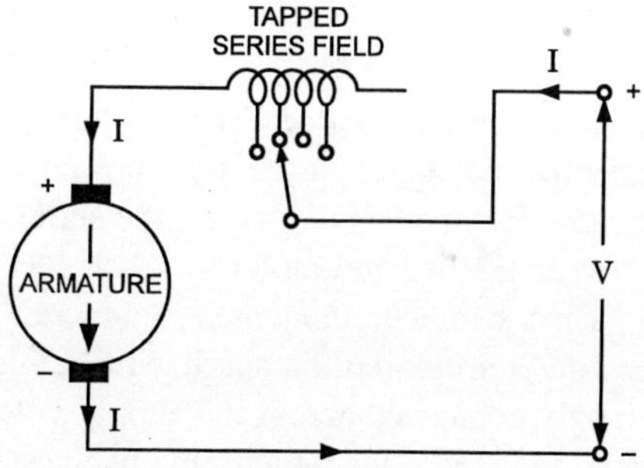

Fig. 14.15 (*b*) *Tapped Field Control For DC Series Motors*

(ii) Tapped Field Control. This is another method of increasing the speed by reducing the flux and it is accomplised by reducing the number of turns of the field winding through which the current flows. In this method of speed control of dc series motors a number of tappings from the field winding are brought outside, as illustrated in Fig. 14.15 (*b*). A number of series field turns can be taken out according to the requirement. When all field turns are in circuit, the motor runs at the lowest speed and speed increases with cutting out some of the series field turns. This method is usually employed in electric traction.

(iii) Paralleling Field Coils Method. In this method, used in fan motors, several speeds can be obtained by regrouping of field coils, as illustrated in Fig. 14.15 (*c*).

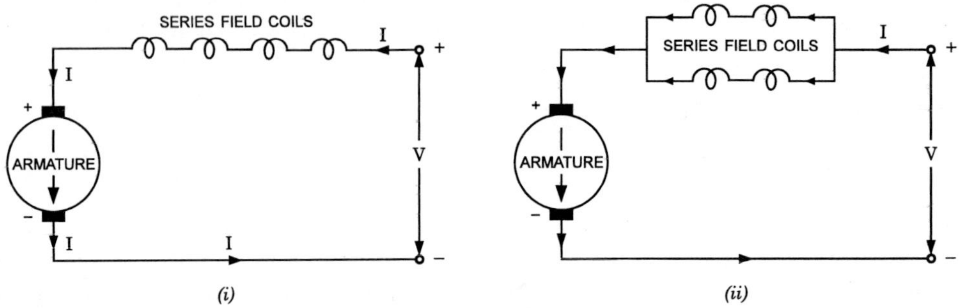

Fig. 14.15 (*c*)

In case of dc **shunt motors** the flux can be varied by inserting a variable resistance in series with the field winding (refer to Fig. 14.16). Since this resistance has to carry only a small current, it is made up of slide-wire type of resistor to have continuously variable speed over the range.

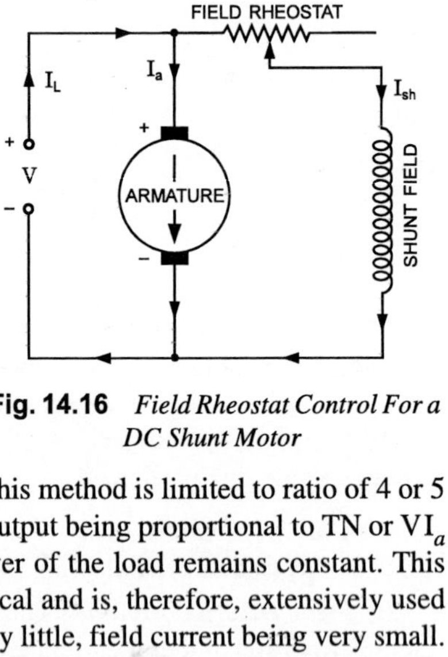

Fig. 14.16 *Field Rheostat Control For a DC Shunt Motor*

Since the flux can be only decreased (not increased), the speeds only above normal one can be obtained by this method. The speed is minimum at the maximum value of flux and depends upon the design of the field and its saturation point. The speed is maximum at the minimum value of flux, which is governed by the demagnetizing effect of armature reaction on the field as at higher speeds the motor tends to be unstable and difficulties in commutation arise. The high speed is also restricted due to mechanical considerations as the centrifugal forces are set up at high speeds. Speed variation by this method is limited to ratio of 4 or 5 to 1. Creeping speeds cannot be obtained by this method. The power output being proportional to TN or VI_a remains constant. This method is, therefore, suitable only where power of the load remains constant. This method of speed control is very simple, convenient and most economical and is, therefore, extensively used in electric drives. The power wasted in the controlling resistance is very little, field current being very small. This method of speed control is independent of load on the motor and permits the remote control of speed.

2. *Armature Resistance Control Methods.* In this method of speed control, reduced speeds are obtained by inserting resistance in the armature circuit. It may be employed with series, shunt and compound wound motors. For the last two types, the series resistance is connected between the shunt field and armature, not between the line and motor.

In case of a dc *shunt motor* an increase in armature circuit resistance will cause more voltage drop in the armature circuit, so the speed will be reduced. Field current will remain unaffected, as the shunt field is directly connected across the supply voltage. For a constant torque load, the armature current remains the same so input to the motor remains the same but the output decreases in proportion to the speeds. Operating costs are, therefore, comparatively high for long time running at reduced speed. In case of fans and centrifugal pumps where the load torque decreases with the decrease in speed, the losses are considerably low and because of its low initial cost and simplicity this method may be quite convenient and economical for short-time or intermittent slowdowns. Wide range of speed (below normal one) can be obtained by this method and at the same time motor will develop any desired torque over its operating range. The main advantage of this method is that speeds below base speed down to creeping speeds of only a few rpm are easily available. This method of speed control, therefore, is employed where speeds lower than rated one are required for a short period only and also occasionally as in printing machines, cranes and hoists where the motor is frequently started and stopped. This method of speed control is also employed where the load drops off rapidly with the decrease in speed, as in fans and blowers.

In case of a dc *series motor* an increase in armature circuit resistance will cause more voltage drop in the armature circuit, so speed will be reduced. The torque which is proportional to the product of the flux and armature current will also be reduced with the increase in armature circuit resistance as increase in armature circuit resistance will reduce the armature current and the flux which also depends upon the armature current. In order to obtain different speeds for constant load torque, the armature current is kept constant so the flux. The resistance in the armature circuit is increased to reduce the speed. The drawbacks of armature resistance control for machines with shunt fields are not as important in the speed control of dc series motors.

The poor speed regulation that is inherent in this method has no significance for the control of dc series motors, since the speed characteristic of a dc series motor is a rapidly drooping curve. The power loss in the control resistance for many applications of dc series motors is not too serious, since in these applications the control is utilized for a large portion of time for reducing the speed under light-load conditions and is only employed intermittently when the motor is carrying full load. The maximum range of speed control of about 3 : 1 will be available depending on the load. This method of speed control is most economical for constant torque drives. This method of speed control is employed chiefly for dc series motors driving cranes, hoists, trains etc. because such drives operate on intermittent duty.

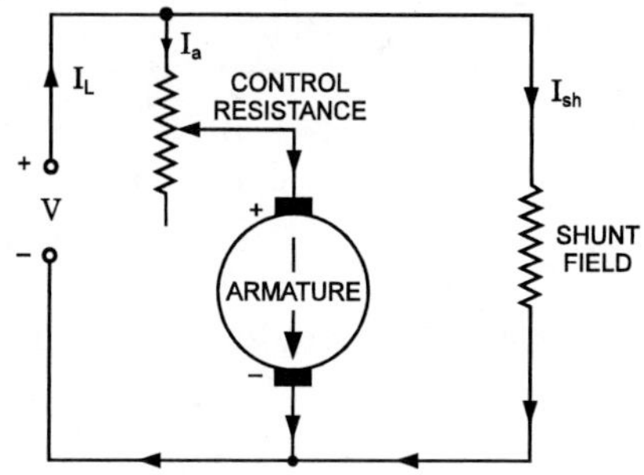

(*a*) *For DC Shunt Motor*

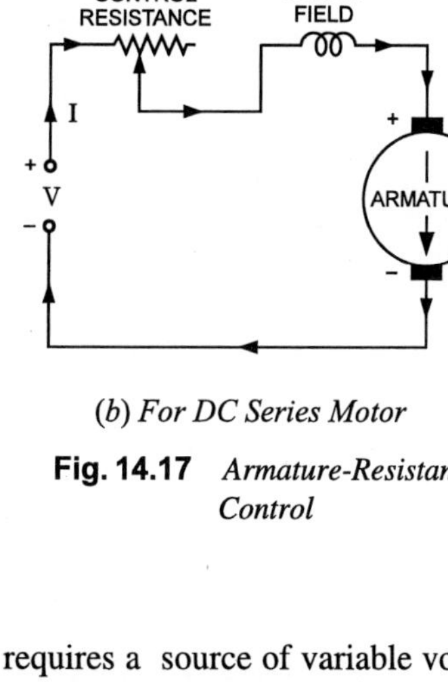

(*b*) *For DC Series Motor*

Fig. 14.17 *Armature-Resistance Control*

3. *Armature Voltage Control Method.* This method of speed control requires a source of variable voltage separate from the source supplying the field current. This method avoids the disadvantages of poor speed regulation and low efficiency which are the characteristics of the armature–resistance control method but it is more expensive in initial cost. The adjustable voltage for the armature is obtained from an adjustable voltge generator or from an adjustable electronic rectifier*. This method provides a large speed range with any desired number of speed points. It is essentially a constant–torque system, because the output delivered by the motor decreases with a decrease in applied voltage and a corresponding decrease in speed.

Example 14.19 **A 500 V shunt motor runs at its normal speed of 250 rpm when the armature current is 200 A. The resistance of armature is 0.12 Ω. Calculate the speed when a resistance is inserted in the field, reducing the shunt field to 80% of normal value and the armature current is 100 A.**

* For most applications dc motors are fed by rectification of the available ac supply. It is usually necessary to insert a transformer immediately ahead of the rectifier.

Solution:

$$\text{Back emf, } E_{b1} = V - I_{a1} R_a = 500 - 200 \times 0.12 = 476 \text{ V}$$

$$\text{Back emf, } E_{b2} = V - I_{a2} R_a = 500 - 100 \times 0.12 = 488 \text{ V}$$

$$\Phi_2 = 80\,\% \text{ of } \Phi_1 = 0.8\,\Phi_1$$

$$\text{Speed, } N_1 = 250 \text{ rpm}$$

$$\text{Speed, } N_2 = N_1 \times \frac{E_{b2}}{E_{b1}} \times \frac{\Phi_1}{\Phi_2} = 250 \times \frac{488}{476} \times \frac{\Phi_1}{0.8\Phi_1} = 320.4 \text{ rpm } \textbf{Ans.}$$

Example 14.20 **A 240 V shunt motor runs at 1,450 rpm on full load with an armature current of 11 A. The total resistance of the armature and brushes is 0.6 Ω. If the speed is to be reduced to 1,000 rpm with the same armature current, calculate the amount of resistance to be connected in series with the armature and power lost.**

Solution: In normal conditions

$$\text{Armature current, } I_{a2} = I_{a1} = 11 \text{ A}$$

$$\text{Back emf, } E_{b1} = V - I_{a1} R_a = 240 - 11 \times 0.6 = 233.4 \text{ V}$$

$$\text{Speed, } N_1 = 1{,}450 \text{ rpm}$$

With additional resistance (say of R ohms) in armature circuit

$$\text{Speed, } N_2 = 1{,}000 \text{ rpm}$$

$$\text{Back emf, } E_{b2} = V - I_{a2}(R + R_a) = 240 - 11(0.6 + R) = 233.4 - 11\,R$$

$$\frac{E_{b2}}{E_{b1}} = \frac{N_2}{N_1} \qquad \text{Assuming flux to remain constant}$$

$$\text{or} \quad \frac{233.4 - 11\,R}{233.4} = \frac{1000}{1{,}450}$$

$$\text{or } R = 6.585\ \Omega \textbf{ Ans.}$$

$$\text{Power lost} = I^2(R + R_a) = 11^2 \times (6.585 + 0.6) = 869.4 \text{ watts } \textbf{Ans.}$$

Example 14.21 **A 440 V shunt motor while running at 1,500 rpm takes an armature current of 30 A and delivers an output of 15 hp, the load torque varies as the square of speed. Calculate the value of resistance to be connected in series with the armature for reducing the motor speed to 1,300 rpm.**

Solution: In normal operation

$$\text{Armature current, } I_{a1} = 30 \text{ A}$$

$$\text{Back emf, } E_{b1} = V = 440 \text{ V neglecting armature resistance drop}$$

$$\text{Speed, } N_1 = 1{,}500 \text{ rpm}$$

Let the resistance of R ohms be connected in series with the armature circuit to reduce the speed to 1,300 rpm.

$$\text{Since } T \propto N^2$$

$$\therefore \quad T_2 = T_1 \times \left(\frac{N_2}{N_1}\right)^2 = T_1 \times \left(\frac{1{,}300}{1{,}500}\right)^2 = 0.75111$$

$$\text{or} \quad I_{a2}\,\Phi_2 = 0.75111\, I_{a1}\,\Phi_1 \qquad \because T \propto \Phi I_a$$

$$\text{or} \quad I_{a2} = 0.75111 \times 30 = 22.5333 \text{ A} \qquad \because \Phi_2 = \Phi_1$$

$$E_{b2} = V - I_{a2} R = 440 - 22.5333\,R$$

$$\text{Since} \quad \frac{E_{b2}}{E_{b1}} = \frac{N_2}{N_1}$$

$$\therefore \quad \frac{440 - 22.5333\,R}{440} = \frac{1{,}300}{1{,}500}$$

$$\text{or} \quad R = 2.604\ \Omega \textbf{ Ans.}$$

Example 14.22 **A 200 V dc series motor runs at 500 rpm when taking a current of 25 A. The resistance of the armature is 0.5 Ω and that of field is 0.3 Ω. If the current remains constant, calculate the resistance necessary to reduce the speed to 250 rpm.**

Solution: Motor input current, $I_1 = 25$ A

Back emf, $E_{b1} = V - I_1 (R_a + R_{se}) = 200 - 25 (0.5 + 0.3) = 180$ V

Speed, $N_1 = 500$ rpm

Let the resistance R be connected in series with the motor to reduce the speed to

$N_2 = 250$ rpm

Motor input current, $I_2 = I_1 = 25$ A

Back emf $E_{b2} = V - I_2 (R + R_a + R_{se}) = 200 - 25 (R + 0.5 + 0.3) = 180 - 25$ R

Since $N \propto \frac{E_b}{\Phi}$

$$\therefore \frac{N_2}{N_1} = \frac{E_{b2}}{E_{b1}} \times \frac{\Phi_1}{\Phi_2} = \frac{E_{b2}}{E_{b1}} \qquad \because \Phi_2 = \Phi_1 \text{ as field current remains the same}$$

$$\text{or} \quad \frac{250}{500} = \frac{180 - 25\text{ R}}{180}$$

or R = 3.6 Ω **Ans.**

14.18 POWER LOSSES

The basic function of a dynamo is the conversion of energy, either from mechanical to electrical form or from electrical to mechanical form. Whole of the input energy to the dynamo is not converted into useful output energy but a part of the input energy is converted into heat and is lost for the useful output purpose of the machine. The energy converted into heat is called the *energy loss.*

For steady-state operating conditions the average rate of energy conversion is constant, and the energy converted is equal to the product of average power and time duration of steady-state operation.

Therefore, it is more effective to deal with the power input, power losses and power output than to deal directly with the energy values. The basic power equation for a dynamo is

$$P_{\text{input}} = P_{\text{output}} + P_{\text{losses}}$$

The power losses in a dc machine consist of input power that is converted into heat. Power losses occurring in a dc machine are divided into (*i*) copper or electrical losses, (*ii*) iron or magnetic losses, and (*iii*) mechanical losses.

1. Copper or Electrical Losses. Copper or electrical losses include power lost in heating the shunt field, and also in heating armature circuit, which includes armature winding, brush contacts, series field and interpole field. These losses are due to power consumed in forcing a current against the resistance of windings and connections.

(a) Armature Copper Loss. This is given as $I_a^2 R_a$ where I_a is the armature current and R_a is the armature resistance at the operating temperature. Brush contact resistance loss is also taken into account by including the brush contact resistance in the armature resistance R_a. This loss is proportional to the square of armature current and is about 30 to 40% of full-load losses.

(b) Field Copper Loss. This is further sub divided into (*i*) shunt field loss, (*ii*) interpole loss, and (*iii*) series field loss. Shunt field loss is given as $I_{sh}^2 R_{sh}$ where I_{sh} is the shunt field current and R_{sh} is the shunt field resistance. The shunt field loss incurs in shunt and compound dc machines and is practically constant if applied voltage across the machine remains the same. Interpole loss is given as $I_a^2 R_i$ where I_a is the armature current and R_i is the interpole field resistance. Interpole loss incurs only in the machines provided with interpoles. The series field loss, which incurs in series and compound machines, is given as $I_{se}^2 R_{se}$ where I_{se}

is series field current and R_{se} is series field resistance. With the increase in load on machine armature current, therefore, series field current increases resulting increase in interpole and series field losses. Field copper loss is about 20 to 30% of the full-load losses.

2. Iron or Magnetic Losses. These losses are also called core losses and include the hysteresis and eddy current losses which are present in any part of the machine that is constructed of iron and subjected to variations of magnetic flux. The hysteresis loss depends upon the range of variations of the flux density and the frequency of this variation of flux. The hysteresis loss can be expressed with reasonable accuracy by following expression:

$$P_h = \eta \, (B_{max})^{1.6} f \, v \text{ joules/second or watts}$$

where η = Steimmetz hysteresis coefficient; v = volume of core in metres^3 and f is the frequency of the magnetic cycles per second.

The eddy current loss also depends upon the range of variation of flux density and the frequency of the variation of the flux. In addition it depends upon the thickness of the iron laminations. It can be determined with reasonable accuracy by the following expression:

$$P_e = K_e (B_{max})^2 f^2 \, v \, t^2$$

where K_e is a constant depending upon the electrical resistance of the magnetic material and the units employed in the expression, B_{max} is the maximum flux density attained throughout the magnetic cycle, f is the frequency of magnetic cycles per second, v is the volume of magnetic material and t is the thickness of the magnetic material laminations.

The core losses are approximately constant in a machine whose flux and speed remain constant as in case of shunt wound dc machines but are variable in case of series wound and compound wound dc machines. These losses are about 10 to 20% of full-load losses.

3. Mechanical Losses. These losses consist of power losses due to friction of the bearings, air friction or windage, as it is called, caused by the motion of the moving parts through the surrounding medium, and the friction between brushes and commutator or slip rings. These losses are constant in a machine operating at approximately constant speed and are independent of load. These losses are about 10 to 20% of full-load losses.

Usually magnetic and mechanical losses collectively termed as *stray losses* and shunt field copper losses are collectively known as *constant losses.*

The total losses occurring in a dc machine are summarized below:

- Total input
  - Useful output
  - Total losses
    - Copper losses
      - Armature copper loss
      - Field copper loss
    - Iron losses
      - Hysteresis loss
      - Eddy current loss
    - Mechanical losses
      - Friction loss
      - Windage loss

14.19 EFFICIENCY OF DC MACHINES

The ratio of useful output to the total input is called the efficiency of the machine. It is expressed as

$$\eta = \frac{\text{Output}}{\text{Input}} = \frac{\text{Input} - \text{total losses}}{\text{Input}} = \frac{\text{Output}}{\text{Output} + \text{total losses}} \quad ...(14.35)$$

The various efficiencies of dc machines are expressed as follows:

1. Generator: (*i*) Mechanical efficiency, $\eta_m = \dfrac{\text{Electrical power developed by armature}}{\text{Total mechanical power input}}$

$$= \frac{E_g I_a}{\text{BHP of prime mover} \times 735.5}$$

$$\text{(ii) Electrical efficiency, } \eta_e = \frac{\text{Useful electrical power output}}{\text{Electrical power developed}} = \frac{V I_L}{E_g I_a}$$

(*iii*) Commercial or overall efficiency,

$$\eta_G = \eta_m \eta_e = \frac{\text{Useful electrical power output}}{\text{Total mechanical power input}}$$

$$= \frac{V I_L}{\text{BHP of prime mover} \times 735.5} \quad \text{...(14.36)}$$

2. Motor: (*i*) Electrical efficiency,

$$\eta_e = \frac{\text{Mechanical power developed}}{\text{Total electrical power input}} = \frac{E_b I_a}{V I_L}$$

$$\text{(ii) Mechanical efficiency, } \eta_m = \frac{\text{Useful mechanical power output}}{\text{Mechanical power developed}} = \frac{\text{BHP of motor} \times 735.5}{E_b I_a}$$

(*iii*) Commercial or overall efficiency,

$$\eta_M = \eta_m \eta_e = \frac{\text{Useful mechanical power output}}{\text{Total electrical power input}}$$

$$= \frac{\text{BHP of motor} \times 735.5}{V I_L} \quad \text{...(14.37)}$$

The overall efficiency of dc generator and motor can also be expressed as follows:

$$\text{Overall efficiency of generator, } \eta_G = \frac{\text{Useful power output}}{\text{Total power input}} = \frac{\text{Useful power output}}{\text{Useful power output + total losses}}$$

$$= \frac{V I_L}{V I_L + \text{total losses}} \quad \text{...(14.38)}$$

$$\text{Overall efficiency of motor, } \eta_M = \frac{\text{Useful power output}}{\text{Total power input}} = \frac{\text{Total power input} - \text{total losses}}{\text{Total power input}}$$

$$= \frac{V I_L - \text{total losses}}{V I_L} \quad \text{...(14.39)}$$

14.20 CONDITION FOR MAXIMUM EFFICIENCY

Condition for maximum efficiency of a dc generator or of a dc motor is the same and for a dc generator it is derived as follows

Generator output = $V I_L$ watts where V is the terminal voltage and I_L is the output or load current

Generator input = Output + total losses

$$= V I_L + I_a^2 R_a + P_C \text{ (constant losses)}$$

$$= V I_L + (I_L + I_{sh})^2 R_a + P_C = V I_L + I_L^2 R_a + P_C$$

$\because$ Shunt field current I_{sh} is very small in comparison with load current I_L

where R_a is the total resistance of armature circuit (including brush contact resistance and resistance of series, interpole and compensating windings, if any).

$$\text{Generator efficiency, } \eta = \frac{\text{Output}}{\text{Input}} = \frac{V I_L}{V I_L + I_L^2 R_a + P_C} = \frac{1}{1 + \frac{I_L R_a}{V} + \frac{P_C}{V I_L}}$$

Efficiency will be maximum when denominator will be minimum *i.e.* when

$$\frac{d}{dI_L}\left(\frac{I_L R_a}{V}+\frac{P_C}{VI_L}\right) = 0$$

$$\text{or } \frac{R_a}{V}-\frac{P_C}{VI_L^2} = 0$$

$$\text{or } I_L^2 R_a = P_C$$

Hence efficiency will be maximum when variable losses are equal to constant losses.

The load current corresponding to maximum efficiency is given by the relation.

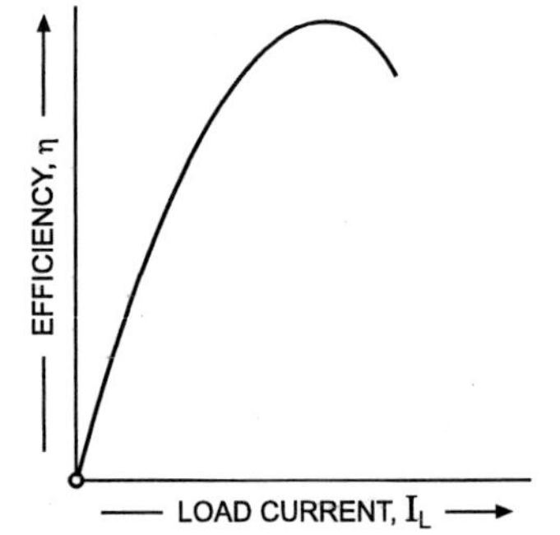

Fig. 14.18

$$I_L = \sqrt{P_C/R_a} \qquad \text{...(14.40)}$$

Variation of efficiency with load is illustrated in Fig 14.18 which indicates that the efficiency increases with the increase in load current, attains a maximum value when the load current equals the value given by equation (14.40) and then starts decreasing. The machines are usually designed to give maximum efficiency at or near the rated output of the machine.

Example 14.23 A dc shunt motor develops 10 bhp at 600 rpm when drawing 18 A at 500 V. Calculate efficiency and torque.

Solution:

$$\text{Motor output} = 10 \times 735.5 = 7{,}355 \text{ watts or J/s}$$

$$\text{Motor input} = V \times I = 500 \times 18 = 9{,}000 \text{ watts}$$

$$\text{Motor efficiency, } \eta = \frac{\text{Output}}{\text{Input}} \times 100 = \frac{7{,}355}{9{,}000} \times 100 = 81.72\ \% \quad \textbf{Ans.}$$

$$\text{Torque, T} = \text{Output} \times \frac{60}{2\pi N} = \frac{7{,}355 \times 60}{2\pi \times 600} = 117 \text{ newton-metres} \quad \textbf{Ans.}$$

Example 14.24 A 400 V dc motor takes 5 A at no load. Its armature and field resistances are 0.5 Ω and 200 Ω respectively. Calculate the efficiency when motor takes 50 A on full load.

Solution:

$$\text{Shunt field current, } I_{sh} = \frac{V}{R_{sh}} = \frac{400}{200} = 2 \text{ A}$$

$$\text{No-load line current, } I_{L_0} = 5 \text{ A}$$

$$\text{No-load input} = VI_{L_0} = 400 \times 5 = 2{,}000 \text{ W}$$

$$\text{No-load armature current, } I_{a_0} = I_{L_0} - I_{sh} = 5 - 2 = 3 \text{ A}$$

$$\text{No-load armature copper loss} = I_{a_0}^2 R_a = (3)^2 \times 0.5 = 4.5 \text{ W}$$

$$\text{Constant losses} = \text{No-load input} - \text{no-load copper losses} = 2{,}000 - 4.5 = 1{,}995.5 \text{ W}$$

On Full Load:

$$\text{Line current, } I_L = 50 \text{ A}$$

$$\text{Input} = VI_L = 400 \times 50 = 20{,}000 \text{ W}$$

$$\text{Armature current, } I_a = I_L - I_{sh} = 50 - 2 = 48 \text{ A}$$

$$\text{Armature copper losses} = I_a^2 R_a = 48^2 \times 0.5 = 1{,}152 \text{ W}$$

$$\text{Total losses} = \text{Constant losses} + \text{armature copper loss} = 1{,}995.5 + 1{,}152 = 3{,}147.5 \text{ W}$$

$$\text{Output} = \text{Input} - \text{total losses} = 20{,}000 - 3{,}147.5 = 16{,}852.5 \text{ W}$$

$$\text{Efficiency, } \eta = \frac{\text{Output}}{\text{Input}} \times 100 = \frac{16{,}852.5}{20{,}000} \times 100 = 84.26\ \% \quad \textbf{Ans.}$$

Example 14.25 **A 400-V shunt generator has a full-load current of 200 A, its armature resistance is 0.06 Ω field resistance 100 Ω; the stray losses are 2,000 W. Find the horse power of its prime mover when it is delivering full-load.**

Solution: **On full load**

$$\text{Output power, } P_{out} = V \times I_L = 400 \times 200 = 80{,}00 \text{ watts}$$

$$\text{Shunt field current, } I_{sh} = \frac{V}{R_{sh}} = \frac{400}{100} = 4 \text{ A}$$

$$\text{Armature current, } I_a = I_L + I_{sh} = 200 + 4 = 204 \text{ A}$$

$$\text{Armature copper loss} = I_a^2 R_a = (204)^2 \times 0.06 = 2{,}497 \text{ W}$$

$$\text{Shunt field copper loss} = V I_{sh} = 400 \times 4 = 1{,}600 \text{ W}$$

$$\text{Output power of prime mover} = \text{Input power to generator}$$

$$= V I_L + I_a^2 R_a + V I_{sh} + P_S = 80{,}000 + 2{,}497 + 1{,}600 + 2{,}000 = 86{,}097 \text{ W}$$

$$\text{Horse power of prime-mover} = \frac{\text{Output of prime mover in watts}}{735.5} = \frac{86{,}097}{735.5} = 117 \textbf{ Ans.}$$

Example 14.26 **A 17 hp, 230 V, 1200 rpm motor has 4 poles, four parallel armature paths, and 868 armature conductors. The armature circuit resistance is 0.18 Ω. At rated speed and rated output the armature current is 70 A and the field current is 1.5 A.**

Calculate (i) the electromagnetic torque (ii) the flux per pole (iii) the rotational losses, and (iv) efficiency.

Solution:

$$\text{Motor output, } P_{out} = 17 \times 735.5 = 12{,}503.5 \text{ W}$$

$$\text{Line current, } I_L = I_a + I_{sh} = 70 + 1.5 = 71.5 \text{ A}$$

$$\text{Input power to motor, } P_{in} = V I_L = 230 \times 71.5 = 16{,}445 \text{ W}$$

$$\text{Back emf, } E_b = V - I_a R_a = 230 - 70 \times 0.18 = 217.4 \text{ V}$$ neglecting brush contact drop

$$(i) \text{ Electromagnetic torque, } T_e = \frac{E_b I_a}{2\pi N/60} = \frac{217.4 \times 70}{2\pi \times 1{,}200/60} = 121 \text{ Nm} \quad \textbf{Ans.}$$

$$(ii) \text{ Flux per pole, } \Phi = \frac{E_b \times 60 \times A}{Z \times N \times P} = \frac{217.4 \times 60 \times 4}{868 \times 1200 \times 4} = 0.0125 \text{ Wb or } 12.5 \text{ mWb} \quad \textbf{Ans.}$$

$$\text{Copper losses, } P_c = I_a^2 R_a + V I_{sh} = (70)^2 \times 0.18 + 230 \times 1.5 = 882 + 345 = 1227 \text{ W}$$

$$\text{Total losses, } P_T = P_{in} - P_{out} = 16{,}445 - 12{,}503.5 = 3{,}941.5 \text{ W}$$

$$(iii) \text{ Rotation losses} = P_T - P_c = 3{,}941.5 - 1{,}227 = 2{,}714.5 \text{ W} \quad \textbf{Ans.}$$

$$(iv) \text{ Efficiency, } \eta = \frac{P_{out}}{P_{in}} \times 100 = \frac{12{,}503.5}{16{,}445} \times 100 = 76.03\ \% \quad \textbf{Ans.}$$

Example 14.27 **A 10 kW, 200 V, 1200 rpm series dc generator has armature resistance of 0.1 Ω, field winding resistance of 0.3 Ω. The frictional and windage loss of the machine is 200 W and brush contact drop is 1 volt per brush. Find the efficiency of the machine.**

Solution:

$$\text{Output power, } P_{out} = 10 \text{ kW} = 10{,}000 \text{ W}$$

$$\text{Load current, } I_1 = \frac{P_{out} \text{ in kW} \times 1{,}000}{V} = \frac{10 \times 1{,}000}{200} = 50 \text{ A}$$

$$\text{Copper power losses, } P_c = I^2 (R_a + R_{se}) + I \times \text{brush drop}$$

$$= 50^2 (0.1 + 0.3) + 50 \times 2 = 1{,}100 \text{ W}$$

$$\text{Friction and windage losses, } P_i = 200 \text{ W}$$

$$\text{Power input, } P_{in} = P_{out} + P_c + P_i = 10{,}000 + 1{,}100 + 200 = 11{,}300 \text{ W}$$

$$\text{Efficiency of generator, } \eta = \frac{P_{out}}{P_{in}} \times 100 = \frac{10{,}000 \times 100}{11{,}300} = 88.5\% \textbf{ Ans.}$$

14.21. APPLICATIONS OF DC MOTORS

The dc motor is often called upon to do the really tough jobs in the industry because of its high degree of flexibility and ease of control. These features cannot easily be matched by other electromechanical energy-conversion devices. The dc motor offers a wide range of control of speed and torque as well as acceleration and deceleration. The applications of the three types of dc motors are given below:

1. DC Series Motor. It is a variable speed motor i.e. very low speed at high torque and vice versa. However, at light or no load, the motor tends to attain dangerously high speed. The motor has a very high starting torque (up to 5 times of full-load torque). It is, therefore, used for drives requiring very high starting torque and where adjustable varying speed is satisfactory. Industrial uses are hoists, cranes, trolley cars, conveyors, electric locomotive, elevators, air-compressors, vacuum cleaners, sewing machines etc.

Loads must be positively connected, not belted. To prevent overspeed, lightest load should not be much less than 15 to 20% of full-load torque.

2. DC Shunt Motor. It is an approximately constant speed motor and has medium starting torque (usually limited to 2.5 times of full-load torque by a starting resistor).

DC shunt motor is, therefore, used essentially for constant speed applications requiring medium starting torque. May be used for adjustable speed not greater than 2 : 1 range.

Industrial applications are lathes, centrifugal pumps, reciprocating pumps, fans, blowers, conveyors, woodworking machines, machine tools, printing presses, spinning and weaving machines.

3. Compound Wound Motors. *Differential compound wound dc motors* are rarely used because of their poor torque characteristics and difficulties experienced during overloads and starting.

Cumulative compound wound dc motors are used in driving machines which are subject to sudden applications of heavy loads, such as occur in rolling mills, punching and shearing machines, lifts, mine-hoists etc. This type of motor is also used where a large starting torque is required but series motor cannot be conveniently used such as in cranes and elevators.

HIGHLIGHTS

1. As regards fundamental principles, the dc motors are identical with the dc generators which have the same type of excitation i.e., a machine that operates as a motor will also operate satisfactorily as a generator.
2. When a dc motor is connected to dc supply mains, the armature continues to rotate due to motor action, the armature conductors cut the magnetic flux and, therefore, emfs are induced in them. The direction of this induced emf, known as back emf or counter emf, is such that it opposes the applied voltage.

$$\text{Back emf in a dc motor, } E_b = \frac{\Phi Z N}{60} \times \frac{P}{A} \text{ volts}$$

3. Armature current of dc motor is given as

$$I_a = \frac{V - E_b}{R_a}$$

4. Voltage equation of the motor is

$$V = E_b + I_a R_a + \text{brush drop}$$

5. Power equation of a dc motor is given as

$$V I_a = E_b I_a + I_a^2 R_a \text{ in case of dc shunt motor}$$

$$\text{and } V I_a = E_b I_a + I_a^2 (R_a + R_{se}) \text{ in case of dc series motor}$$

6. Similar to dc generators, the dc motors can be classified as (*i*) permanent magnet (*ii*) separately excited (*iii*) series wound (*iv*) shunt wound and (*v*) compound wound dc motors.
7. Direction of rotation of a dc motor can be reversed by reversing the current through either the armature winding or field coils not in both.

8. From emf equation $E_b \; \alpha \; \Phi \, N$

$$\text{or } N \; \alpha \; \frac{E_b}{\Phi} = K \frac{V - I_a R_a}{\Phi}$$

$$\text{or } N = \frac{E_b}{\Phi} \times \frac{60A}{Z \times P} \text{ rpm}$$

9. Speed regulation of a motor is given by

$$\text{Percentage speed regulation} = \frac{\text{No-load speed} - \text{full-load speed}}{\text{Full-load speed}} \times 100$$

10. Torque developed by armature in a dc motor is given as

$$T_a = 0.159 \, \Phi \, Z \, P \times \frac{I_a}{A} \text{ newton-metres}$$

$$\text{Also } T_a = 9.55 \, \frac{E_b I_a}{N} \text{ N-m}$$

11. Shaft or useful torque is given as

$$T_{sh} = \frac{\text{Output in watts}}{2\pi N/60} \text{ N-m}$$

$$\text{Torque lost in iron and friction losses} = \frac{\text{Iron and friction losses in watts}}{2\pi N/60} \text{ Nm}$$

12. *Important Characteristics.*

Series motor develops high torque at load and has ability to pickup. Its speed increases with the decrease in load. DC series motors are never started on no load because their speed at no load is dangerously high. Series motors are best suited for electric traction work.

Shunt motor is a constant speed motor. Such a motor is used in applications requiring medium starting torque such as centrifugal pumps, blowers, fans, conveyors, boring mills, shapers, woodworking machines etc.

13. Speed of a dc motor is given as $N = K \dfrac{E_b}{\Phi} = K \dfrac{V - I_a(R_a + R)}{\Phi}$

Based on above equation, the speed of a dc motor can be controlled by field control, armature resistance control or armature voltage control.

14. *In field control of dc motors*

In case of dc shunt motors, speed variation is accomplished by means of a variable resistance inserted in series with the shunt field.

In case of dc series motors, speed variation is accomplished by using a divertor across the series field winding. Field control method is more economical and convenient and gives speeds above the normal ones.

15. *In armature control of dc motors*, a variable controlling resistance is connected in series with the armature circuit. Rheostatic control or armature control gives speeds below the normal, and is expensive and less efficient.

16. *In armature voltage control*, the speed of a dc motor can be varied by varying the voltage applied to the armature of the dc motor.

17. The total losses occurring in a dc machine are summarized below:

- Total input
  - Useful output
  - Total losses
    - Copper losses
      - Armature copper loss
      - Field copper loss
    - Iron losses
      - Hysteresis loss
      - Eddy current loss
    - Mechanical losses
      - Friction loss
      - Windage loss

18. Efficiency of electrical machine:

$$\eta = \frac{\text{Power output}}{\text{Power input}} = \frac{\text{Power input} - \text{total losses}}{\text{Power input}} = \frac{\text{Power output}}{\text{Power output} + \text{losses}}$$

19. Efficiency of an electrical machine will be maximum when variable losses = Constant losses.

EXERCISES

1. Explain back emf in a dc motor. Explain the effect of field current on the speed of a dc motor.
2. State and explain the expression of back emf (E_b) and torque (T_e) in dc machines. What are the basic roles of back emf in dc machines? **[U.P. Technical Univ. Electrical Engineering Odd Semester 2013-14]**
3. Derive an equation for generated torque in dc motor. **[G.B. Technical Univ. Electrical Engineering First Semester 2009-10]**
4. Derive an expression for torque in dc motor. Draw the load characteristics of dc series and shunt motors. **[U.P. Technical Univ. Electrical Engineering 2013-14]**
5. Describe the working principle of dc series motor and draw its various characteristics. **[G.B. Technical Univ. Elec. Engineering Second Semester 2010-11]**
6. Explain speed-torque characteristic of dc series motor. Also mention typical applications of it. **[Electrical Engineering G.B. Technical Univ. First Semester 2011-12; U.P. Technical Univ. Second Semester 2013-14]**
7. Derive the expression of torque for dc motor. Also discuss the characteristics of dc shunt motor. **[G.B. Technical Univ. Electrical Engineering Second Semester 2010-11]**
8. What are the different types of dc motors? Also write the applications of each. **[G.B. Technical Univ. Electrical Engineering Even Semester 2012-13]**
9. Explain the speed-torque characteristics of dc shunt and series motors. **[Mahamaya Technical Univ. Electrical Engineering First Semester 2011-12]**
10. Sketch and explain the speed-load characteristics of following dc motor:
    (*i*) Series motor (*ii*) Shunt motor
    (*iii*) Cumulatively compounded motor (*iv*) Differentially compound motor.
    [G.B. Technical Univ. Electrical Engineering Second Semester 2009-10]
11. Sketch and explain speed-torque characteristics of following dc motor:
    (*i*) Shunt motor
    (*ii*) Compounded motor. **[Mahamaya Technical Univ. Electrical Engineering First Semester 2011-12]**
12. What are the factors that affect the speed of a dc motor? Explain. Also, explain how the speed can be controlled above and below the normal speed. **[U.P. Technical Univ. Electrical Engineering, Second Semester 2003-04]**
13. How can the speed of dc motor be varied? Explain the method by which you can vary the speed above and below the base speed. **[U.P. Technical Univ. Basic Electrical Engineering Second Semester 2013-14]**
14. What are the various methods of speed control of dc shunt motor? Explain them in brief with relative advantages and disadvantages. **[U.P.Technical Univ. Elec. Engineering, First Semester 2004-05]**
15. Discuss the methods of speed control of dc shunt motor. **[G.B. Technical Univ. Electrical Engineering First Semester 2010-11]**
16. Derive efficiency equation of a dc motor. **[U.P. Technical Univ. Electrical Engineering January 2003]**
17. What are the losses taking place in a *dc* machine and how do they vary with load current and derive the condition for maximum efficiency ?
18. Draw speed-torque characteristics of (*i*) dc shunt motor, (*ii*) dc series motor (*iii*) dc cumulatively compounded motor. Mention one industrial application of each of these motors. **[U.P. Technical Univ. Elec. Engineering February 2001]**
19. Explain briefly typical application of each type of dc motor (stating the reason for choice). **[U.P. Technical Univ. Elec. Engineering June 2001]**
20. Write short notes on the following:
    (*i*) Working principle of a dc motor. (*ii*) Direction of rotation of dc motor.
    (*iii*) Characteristics and applications of different types of dc motors. (*iv*) Speed control of dc motors.
    (*v*) Losses in dc machines.

SHORT ANSWER TYPE QUESTIONS WITH ANSWERS

Q. 1. How does a dc motor differ from a dc generator in construction ?

Ans. The frame of the generator can, as a rule, be open but those of motors should be either partly or totally enclosed.

Q. 2. What is a commutator in a dc motor ? **[R.G.P.V. Bhopal December 2002]**

Ans. The commutator is a form of rotating switch placed between the armature and the supply circuit and so arranged that it will reverse the connections to the supply circuit at the instant of each reversal of current in the armature coils of the motor.

Q. 3. Why is it (commutator) needed ? **[R.G.P.V. Bhopal December 2002]**

Ans. In case of a dc motor, it is necessary that the current through the coils of the armature winding be reversed as a particular coil leaves one pole (say the N-pole), crosses the neutral line and comes under the influence of next pole which is of opposite polarity (i.e. S-pole). The commutator is required for this purpose.

Q. 4. Why is induced emf in a dc motor called the back or counter emf ?

Ans. Since induced emf in a dc motor opposes the voltage applied to the armature of a dc motor, it is called the back or counter emf.

Q. 5. What is the significance of back emf in dc motor? **[U.P. Technical Univ. 2013-14]**

Ans. The presence of back or counter emf makes the dc motor a self-regulating machine *i.e.* it makes the dc motor to draw as much armature current as is just sufficient to develop the required load torque. For details refer to Art. 14.3.

Q. 6. What do you understand by self excitation mode of dc machine ? Name two dc machines working in this mode. **[P.T.U. June 2003]**

Ans. The dc machine in which field winding(s) is/are excited by the current supplied by the machine itself is called the self-excitation mode of the machine. In such machines the field coils are interconnected with the armature winding.

The self excited machines may be (*i*) dc series machine (*ii*) dc shunt machine.

Q. 7. How can the direction of rotation of a dc motor be determined ?

Ans. By applying Fleming's left hand rule.

Q. 8. How will you change the direction of rotation of dc motor? **[U.P. Technical Univ. Odd Semester 2013-14]**

Ans. Direction of rotation of a dc motor can be reversed by reversing the current flow through either the armature winding or the field winding.

Q. 9. How can the direction of rotation of a dc series motor be changed ?

Ans. By reversing the supply terminals to either armature or series field.

Q. 10. What will happen if the back emf of dc motor vanishes? **[G.B.T.U. First Semester 2011-12]**

Ans. If back emf of a dc motor vanishes, heavy current will flow through the armature winding and the motor may get damaged.

Q. 11. What happens if the field winding of a running shunt motor suddenly breaks open? **[M.T.U. First Semester 2011-12]**

Ans. When the field winding of the dc shunt motor is disconnected while in normal operation, the flux will drop to practically zero value, and therefore, the speed will increase to an extremely high value and the motor may get damaged due to heavy centrifugal forces.

Proper use of overload devices should protect the motor. Back emf would be reduced due to reduction in flux and therefore there will be increase in armature current or current drawn from the supply mains by the motor. The excessive current would operate the overload device, disconnect from the supply, and thus protect the motor.

Q. 12. What is meant by the term speed regulation ?

Ans. The speed regulation of a dc motor is defined as the change in speed when the load on the motor is reduced from full load to zero load and is expressed in percentage of rated full-load speed.

Q. 13. Why the shaft torque is always less than that developed by the armature of a motor ?

Ans. Because of loss of certain percentage of torque developed by the armature in overcoming the iron and friction losses.

Q. 14. Why a series motor should not be fitted to a load through belt ?

Ans. Because the mishap to the belt would cause the motor to run on no load and on no load the series motor will attain dangerously high speed and the motor may get damaged due to heavy centrifugal forces set up in the rotating parts.

Q. 15. Why dc series motor is never started on no load? **[U.P. Technical Univ. 2013-14]**

Ans. Since on no load the speed is dangerously high, the machine may get damaged due to heavy centrifugal forces set up in the rotating parts. *This is the reason that series motors are never started on no load, which is explained below.*

When the motor is connected across the supply mains without load, it draws a small current from supply mains flowing through the series field and armature, the speed tends to increase so that back emf may approach the applied voltage in magnitude. The increase in back emf weakens the armature current and hence the field current. This causes again increase in speed so in back emf. Thus the field continues to weaken and speed continues to increase until the armature gets damaged.

Q. 16. Why a dc series motor is ideally suited for traction purposes ?

Ans. Because dc series motor develops large starting torque and slows down with the increase in load and so gets automatically relieved from heavy excessive load.

Q. 17. For what type of load dc shunt motor is suitable and why ?

Ans. DC shunt motors are suitable for loads requiring medium starting torque, approximately constant speed and which are totally and suddenly thrown off because dc shunt motors develop modern starting torque and run almost on constant speed from no load to full load.

Q. 18. Why the dc motors are mostly gear motors ?

Ans. Because the low-speed motors are much more expensive than high-speed motors of the same kW output rating.

Q. 19. What are the factors that affect the speed of a dc motor? **[U.P. Technical Univ. Second Semester 2003-04]**

Ans. The factors influencing the speed of a dc motor are:

(*i*) Applied or terminal voltage (*ii*) Flux per pole Φ (*iii*) Resistance of armature circuit and (*iv*) Load or armature current.

Q. 20. What factors do limit the increase of speed of dc shunt motors in field rheostat control?

Ans. The factors limiting the increase of speed of dc shunt motors in field rheostat control are demagnetizing effect of armature reaction on the field, commutation problems at high speed and large centrifugal forces developed at high speed.

Q. 21. Why is field control considered superior to armature resistance control for dc shunt motors?

Ans. Because power wasted in controlling resistance is very small and field control method of speed control is independent of load on motor and permits remote control of speed.

Q. 22. How are slow speed obtained at light load in case of dc series motors?

Ans. By connecting a control resistance directly in series with the supply to the complete motor.

Q. 23. How the speed above normal can be obtained in case of dc series motors?

Ans. By weakening the field which can be accomplished by

(*i*) shunting a portion of motor current around the series field *i.e.*, field divertor method.

(*ii*) reducing the number of turns of the field winding through which the current flows *i.e.*, tapped field control.

(*iii*) regrouping of field coils *i.e.*, paralleling field coils.

Q. 24. In the armature circuit of dc motor, an external resistance is introduced. How shall it affect the speed of the motor? **[P.T.U. May 2002]**

Ans. The speed will decrease.

Q. 25. What will be the effect on the speed of a dc shunt motor if an additional resistance is introduced in the field circuit?

Ans. Speed will increase.

Q. 26. Which method we must adopt to control the speed of a dc shunt motor above the base speed?

Ans. Field control method.

Q. 27. While controlling the speed of a dc shunt motor what should be done to achieve a constant torque drive?

Ans. Applied voltage should be maintained constant so as to maintain field strength constant.

Q. 28. Sketch speed-torque characteristics of different types of dc motors in one figure.

Ans. Speed-torque characteristics of dc series, shunt, separately-excited, cumulatively compounded and differentially compounded motors are shown in Fig. 1.

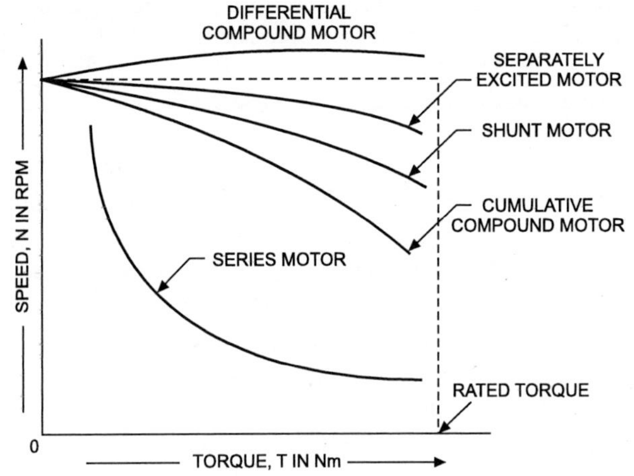

Fig. 1 *Speed-Torque Characteristics of Different Types of DC Motors*

Q. 29. Which losses occur in a dc machine? **[P.T.U. 1st Semester 2005-06]**

Ans. The different types of losses that occur in a dc machine are (*i*) copper or electrical losses (*ii*) iron or magnetic losses and (*iii*) mechanical or friction and windage losses.

Copper or electrical losses include power lost in heating the shunt field, and also in heating the armature circuit, which includes armature winding, brush contacts, series field and interpole field.

Iron or magnetic losses include the hysteresis and eddy current losses which are present in any part of the machine that is constructed of iron and subjected to variations of magnetic flux.

Mechanical losses consist of power losses due to friction of the bearings, air friction or windage, as it is called, caused by the motion of the moving parts through the surrounding medium, and the friction between brushes and commutator.

Q. 30. What is the field of application of dc shunt motor and dc series motor. **[P.T.U. January 2000]**

Ans. *DC shunt motors* are almost constant speed motors, develop medium starting torque and their speeds can be adjusted by using speed control methods. So the shunt motors can be used for the loads which are totally and suddenly thrownoff without resulting in excessive speed. Being constant speed motors, they are best suited for driving of line shafts, machine lathes, milling machines, conveyors, fans and for all purposes where constant speed is required.

DC series motors are variable speed motors, develop large starting torque but attain dangerously high speed on no load. So these motors cannot be operated on no load. Series motors are best suited for electric traction, elevators, cranes, vacuum cleaners, sewing machines etc.

Q. 31. State four major applications of dc motors. **[G.G.S.I.P. Univ. Delhi May/June 2006]**

Ans. Four major applications of dc motors are: Hoists, cranes, conveyors, electric locomotives.

Q. 32. Why is dc series motor preferred in elevators? **[U.P. Technical Univ. Second Semester 2013-14]**

Ans. The series motor has a very high starting torque (up to 5 times) of full-load torque. Therefore, dc series motor is preferred in elevators.

Q. 33. Give two applications of series motor and shunt motor each. **[P.T.U. December 2002]**

Ans. Applications of dc series motors. Electric locomotives, hoists and cranes.

Applications of dc shunt motors. Lathes, reciprocating and centrifugal pumps.

Q. 34. Give at least one application of dc series motor and dc shunt motor and draw their speed-torque characteristics. **[M.T.U. Second Semester 2011-12]**

Ans. Refer to Answer of Question 33 and Figs 14.11 and 14.13.

Q. 35. What are the applications of dc series and dc shunt motors? **[U.P. Technical Univ. 2013-14]**

Ans. *DC series motors* are variable speed motors, develop large starting torque but attain dangerously high speed on no load. So these motors cannot be operated on no load. Series motors are best suited for electric traction, elevators, cranes, vacuum cleaners, sewing machines etc.

DC shunt motors are almost constant speed motors, develop medium starting torque and their speeds can be adjusted by using speed control methods. So the shunt motors can be used for the loads which are totally and suddenly thrownoff without resulting in excessive speed. Being constant speed motors, they are best suited for driving of line shafts, machine lathes, milling machines, conveyors, fans and for all purposes where constant speed is required.

PROBLEMS

1. A 250 V dc shunt motor takes 41 A at full load. Resistance of motor armature and shunt field windings are 0.1 Ω and 250 Ω respectively. Find the back emf on full load. What will be its generated emf, if working as generator and supplying 41 A to load at terminal voltage of 250 V. [**Ans.** 246 V, 254.2 V]
2. A 230 V dc shunt motor takes 51 A at full load. Resistances of armature and field windings are 0.1 Ω and 230 Ω respectively. Determine (*i*) field current (*ii*) armature current and (*iii*) back emf developed at full load? [**Ans.** (*i*) 1 A (*ii*) 50 A (*iii*) 225 V]
3. A dc series motor has an armature resistance of 0.12 Ω and field resistance of 0.08 Ω. The supply voltage across the motor terminals is 230 V. Determine the back emf and power developed by motor when line current is drawn by the motor is 30 A. [**Ans.** 224 V; 6.72 kW]
4. A 250 V dc shunt machine has line current of 80 A. It has armature and field resistances of 0.1 Ω and 125 Ω respectively. Calculate power developed in armature when running as (*a*) generator (*b*) motor. [**Ans.** (*a*) 21.172 kW (*b*) 18.89 kW]
5. A 220 V shunt motor takes 5 A on no load and is running at 800 rpm. The resistances of armature and shunt field are 0.25 Ω and 110 Ω respectively. Determine the speed of the motor when loaded and taking 40 A from the line. [**Ans.** 768 rpm]
6. A dc machine develops an open-circuit emf of 250 V at 1,500 rpm. Find the developed torque for an armature current of 20 A. [**Ans.** 31.833 N-m]
7. A 250 V, 4-pole wave wound dc series motor has 782 conductors on its armature. It has armature and series field resistance of 0.75 Ω. The motor takes a current of 40 A. Estimate its speed and gross torque developed if it has a flux per pole of 25 mWb. [**Ans.** 337.6 rpm; 248.9 Nm]
8. A 240 V shunt motor has a field resistance of 400 ohms and armature resistance 0.1 ohm. The armature current is 50 A and the speed is 1,350 rpm. Assuming a straight line magnetisation curve determine the additional resistance in the field to increase the speed to 1,500 rpm for the same armature current. [**Ans.** 44.4 ohms]
9. A 230 V dc shunt motor runs at 800 rpm and takes armature current of 50 A. Find the resistance to be added to the field circuit to increase speed to 1,000 rpm at an armature current of 80 A. Assume flux proportional to field current. Armature resistance 0.15 Ω and field winding resistance = 250 Ω. [**Ans.** 68.95 Ω]
10. A 200 V dc shunt motor running at 1,000 rpm takes an armature current of 17.5 A. It is required to reduce the speed to 600 rpm. What must be the value of resistance to be inserted in the armature circuit if the original armature resistance is 0.4 Ω ? Take armature current to be constant during this process. [**Ans.** 4.41 Ω]
11. What value of resistance should be inserted in the armature circuit of a 200 V, 1,000 rpm shunt motor to reduce its speed to 800 rpm when drawing 20 A. Assume resistance to be 0.5 Ω. [**Ans.** 1.9 Ω]

12. A series motor with series field and armature resistance of 0.06 ohm and 0.04 ohm respectively is connected across 220 V mains. The armature takes 40 A and its speed is 900 rpm. Calculate the speed when the armature takes 75 A from the same supply mains and the excitation is increased by 15%. [**Ans.** 0.1867 Ω]

13. A long shunt dynamo running at 1,000 rpm delivers a load of 22 kW at 220 V. Its armature, shunt field and series field resistances are 0.05, 110 and 0.06 ohms respectively. Determine (*i*) copper loss (*ii*) iron and friction losses (*iii*) output of the prime mover and (*iv*) torque exerted by the prime mover. Overall efficiency of the dynamo at full load may be taken 88%. [**Ans.** (*i*) 1,584.44 W (*ii*) 1,415.56 W (*iii*) 25 kW (*iv*) 238.73 Nm]

14. A 110 V dc shunt generator has full-load current of 100 A. Its shunt field resistance is 55 Ω and it has constant loss of 500 W. If the full-load efficiency is 88%, find armature resistance. [**Ans.** 0.096 Ω]

15. A 4-pole lap wound dc shunt motor has armature and field resistances as 1 Ω and 200 Ω respectively. The motor is drawing a load current of 20 A from a 220 V dc supply source. Draw electric circuit for the motor and find the value of (*i*) back emf induced in the motor (*ii*) current in each path of the armature and (*iii*) total copper losses.

[**Ans.** (*i*) 201.1 V (*ii*) 4.725 A (*iii*) 539 watts]

❑❑❑

15

CHAPTER

Three-Phase Induction Motor

15.1. INTRODUCTION

The polyphase induction motor is, by a very considerable margin, the most widely used ac motor (almost more than 90 per cent of the mechanical power used in industry is provided by 3-phase induction motors). The reasons are its low cost, simple and rugged construction, absence of commutator, good operating characteristic (reasonably good power factor, sufficiently high efficiency and good speed regulation). An induction motor of a medium size may have an efficiency as high as 90 per cent and a power factor of 0.89. It is substantially a constant speed motor with a shunt characteristic; a few per cent speed drop from no load to full load. The physical size of such a motor for a given output rating is relatively small as compared with other types of motors.

The distinguishing feature of such a motor is that it is a singly-excited machine, although such machines are equipped with both field and armature windings. In such a machine the field* (or stator) winding is connected to an ac supply and there is no electrical connection from the armature* (or rotor) to any source of supply. Currents are made to flow in the armature (or rotor) conductors by induction which interact with the field produced by the field (or stator) winding and thereby produce a net unidirectional torque. Such motors are also called *asynchronous motors* as they run at a speed other than the synchronous speed of the rotating field developed by the stator currents.

Like other electrical machines, the asynchronous machine is reversible, i.e., it can operate as both a motor and a generator. The mode of operation of the machine is determined by the speed of the rotating field in relation to the rotor.

An asynchronous machine may be considered to be a transformer in the sense that the power is transferred from the stator (primary) to the rotor (secondary) winding only by mutual induction. For this reason such a machine is often called the *induction* machine.

There is no simple and cheaper method of controlling the speed of the induction motor as is possible in case of a dc shunt motor and the torque developed by a polyphase induction motor is also not so strong as that developed by a dc shunt motor. It still finds stiff competition from the dc shunt motor in industrial applications.

15.2. CONSTRUCTION

The 3-phase induction motor is very simple in construction compared to a dc motor or a synchronous motor. The essential features of a polyphase induction motor are: a laminated stator core carrying a polyphase winding; a laminated *rotor* core carrying either a cage or polyphase winding, the latter with shaft-mounted slip rings; a stiff *shaft* to preserve the very short air gap; a *frame* to form the stator housing and carry the end covers,

* The terminology of the field and armature windings is seldom used for polyphase induction motors.

bearings, and terminal box. It is not possible, as in the dc machines, to use the frame as part of the magnetic circuit. The end covers receive the ball- or roller-bearings with their clamping plates. Non-salient pole construction is used for all polyphase induction motors.

15.2.1. Frame. It is the outer body of the motor. Its functions are to support the stator core and winding, to protect the inner parts of the machine and serve as a ventilating housing or means of guiding the coolant into effective channels. The frame may be die cast or fabricated. Machines up to about 50 kW rating may have frame die cast in a strong silicon-aluminium alloy, sometimes with the stator core cast in. The frames for medium and large machines are almost exclusively fabricated.

15.2.2. Stator Core. The stator of an induction motor is quite similar in construction to that of a 3-phase synchronous alternator or motor. The stator core is to carry the alternating flux which produces hysteresis and eddy current losses. In order to reduce eddy currents and hysteresis loss in the stator core it is assembled of high grade, low electrical loss, silicon steel punchings. The thickness of punchings varies from 0.35 mm to 0.65 mm. In small motors, or if low core loss is not so important, slightly thicker laminations may be used. In the smaller sizes, the stator plates form complete annular rings, as shown in Fig. 15.1, but for larger machines they are prepared of segmental laminations to avoid wastage of steel. The punchings are insulated from one another by coating of varnish or oxide produced by heat treatment. The stampings are assembled under hydraulic pressure and are keyed to the frame. Careful alignment is required to prevent the rotor from rubbing against the stator, the air gap between the stator and rotor being very small. The radial ventilating ducts are provided along the length of the stator core, spaced every 5 or 7 cm. These are provided by the use of spacers placed between the laminations. The air gap between the stator and rotor is made as small as practicable (0.3 to 0.35 mm in small machines and 1.0 to 1.5 mm in high power machines) so as to make air-gap reluctance minimum. The slots are punched out on the inner periphery of the stator laminations, as shown in Fig. 15.1, to accommodate the stator winding, also sometimes known as *primary winding*. Some motors, especially of large ratings, have open type slots so that form-wound coils can be used but usually the motors are provided with semi-closed type slots to make the air-gap reluctance minimum and to prevent excessive tooth loss. In large, high-speed motors it is not essential to employ semi-closed slots to reduce the reluctance. The reason of it is that in large, high-speed motors, pole pitch is large, therefore, the number of conductors per pole is large, so the desired flux density in the air gap can be attained without an excessive magnetizing current, even if open type slots are employed.

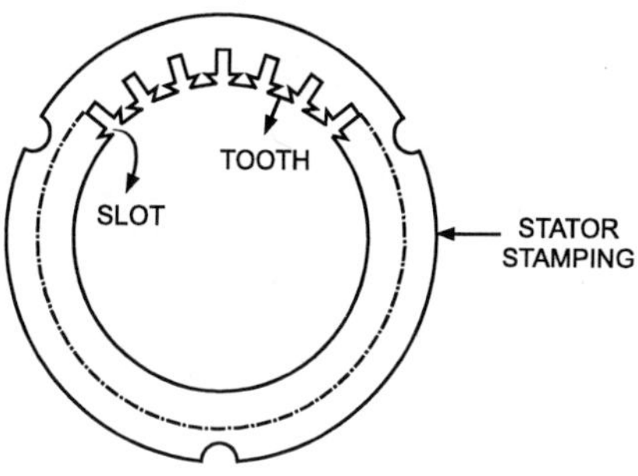

Fig. 15.1

15.2.3. Stator or Primary or Field Winding. Motors up to 250 kW operate usually at voltages in the range 380-440 V. For 3.3 kV the minimum economical rating is 250-300 kW, and for 11 kV about 750 kW.

In a polyphase induction motor the stator winding is usually 3-phase winding which is usually supplied from a 3-phase supply mains. The three phases of the winding can be connected in either star or delta depending upon the methods of starting used. The squirrel cage motors are usually started by star-delta starters and, therefore, their stators are designed for delta connections and six terminals (two from each phase) are brought out to be connected to the starter. The wound rotor motors are started by inserting resistance in the rotor circuit and, therefore, the stator winding can be connected in either star or delta as desired. It is wound for a definite number of poles as per requirement of speed. Greater the number of poles, lesser the speed or vice versa for a supply of given frequency.

In general, the same windings can be used on the stator of induction motor as were suitable for synchronous generators or motors. The double layer winding is mostly used in medium size motors because of its greater ease of manufacture, assembly, and repair. Moreover, stator windings are almost always short-pitched because of reduced copper weight and winding resistance as well as reduced leakage reactance and harmonic torque disturbances that result. Small motors, operating at ordinary supply voltages, with a small number of slots having a large number of turns per phase may use single layer mush winding. For large motors, or high-voltage motors the stator windings may be formed by single layer concentric coils. Double layer coils are placed in open slots so that form-wound coils can be used while mush winding is housed in semi-closed slots.

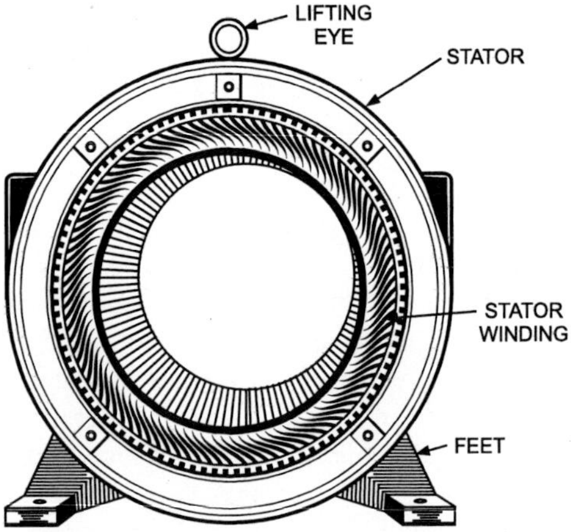

Fig. 15.2 *Stator of a 3-Phase Induction Motor*

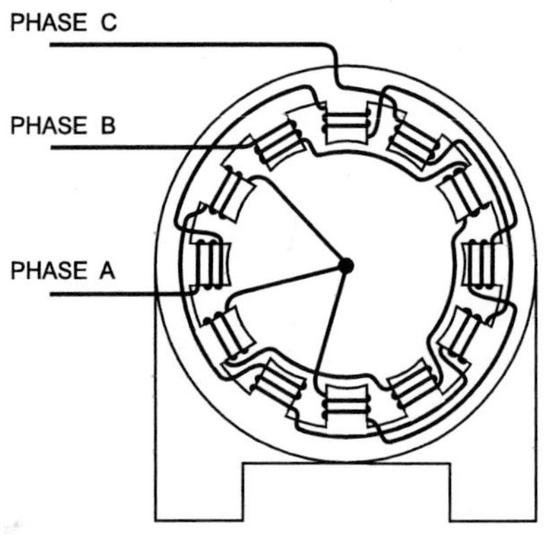

Fig. 15.3 *3-Phase, 4-Pole, Star-Connected Stator Winding*

Stator of a 3-phase induction motor and 3-phase, 4-pole, star-connected stator winding are illustrated in Figs. 15.2 and 15.3 respectively.

15.2.4. Rotor. The rotor comprises a cylindrical laminated iron core, much the same as that of a dc machine, with slots, around the core, carrying the rotor conductors. In general the same sheet steel laminations are employed for the rotor core as for the stator, but owing to the lower frequencies of the rotor flux, thicker laminations can be employed without excessive iron loss. The rotor stamping is shown in Fig. 15.4. Like stator, the rotor laminations are punched in one piece for small machines; in larger machines the laminations are segmented and dovetailed to a fabricated spider. If there are ventilating ducts on the stator core, an equal number of such ducts is provided on the rotor core also. Fan blades are usually employed on the ends of the rotor core to force circulating air through the motor for cooling.

The rotor has a smaller number of slots than the stator and must be a non-integral multiple of stator slots so as to prevent magnetic locking of rotor and stator teeth at the starting instant.

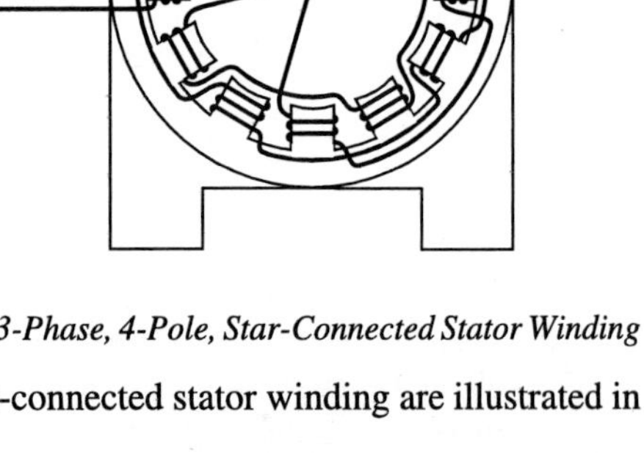

Fig. 15.4 *Rotor Stamping*

The rotors employed in 3-phase induction motors, according to the type of windings used, are of two types, viz; squirrel cage rotor and the wound rotor.

1. *Squirrel Cage Rotor. Almost* 90 per cent of induction motors are provided with squirrel cage rotor because of its very simple, robust and almost instructible construction. In cage construction, copper, brass, or aluminium bars are placed, as the rotor conductors, parallel or approximately parallel to the shaft (one bar in each slot) and close to the rotor surface. The conductors are not insulated from the core, since the rotor currents naturally follow the path of least resistance, *i.e.*, the rotor conductors. At both ends of the rotor, the rotor conductors are all short circuited by the continuous end rings of similar material to that of the rotor conductors. The rotor conductors and their end rings form a complete closed circuit in itself, resembling a squirrel cage, thus explaining the name. In motors with ratings up to 100 kW, the squirrel cage structure is formed

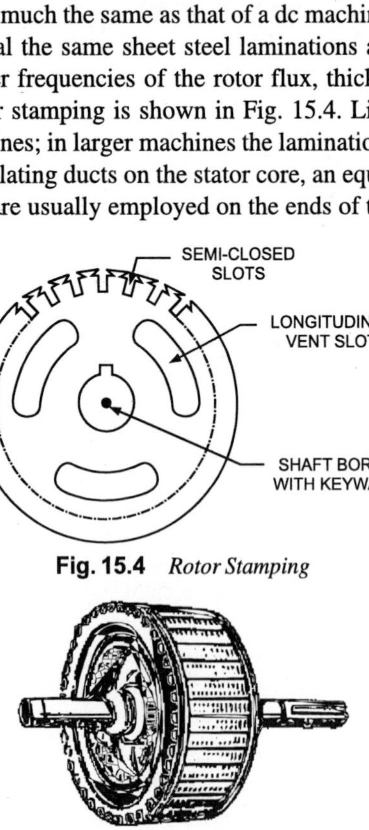

Fig. 15.5 *General View of Squirrel Cage Rotor*

by aluminium cast (under pressure) into the slots of the rotor (Fig. 15.6). In large motors the rotor bars, instead of being cast, are wedged into the rotor slots and are then welded securely to the end rings. The slots on the rotor are always not parallel to the motor shaft but are usually skewed in order to obtain a uniform torque, reduce the magnetic locking of the stator and rotor and reduce the magnetic humming noise while running.

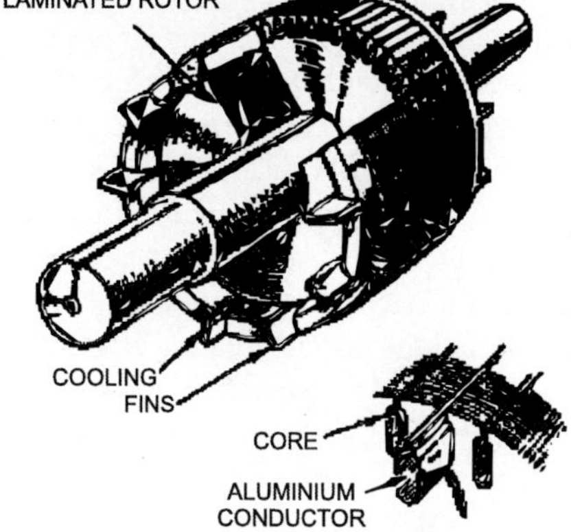

Fig. 15.6 *Cutaway View of Squirrel Cage Rotor*

The slots on rotor are either of semi-closed type or of totally closed type, because there is little difficulty in inserting the rotor bars in such slots. The advantage of semi–closed and totally closed slots is that the effective cross-sectional area of the air gap is increased, therefore, magnetizing current is reduced. Such slots also reduce the pulsations of flux in the individual teeth, therefore, tooth loss is reduced. The disadvantage of such slots is that these give higher slot inductance than the open slots, which lowers the power factor and reduces the starting and breakdown torques of the motor.

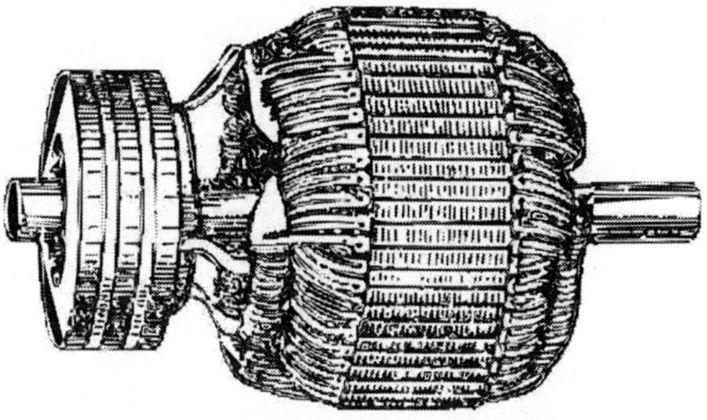

Fig. 15.7 *Rotor With Slip Rings*

The squirrel cage rotor windings are perfectly symmetrical and have the advantage of being adaptable to any number of pole pairs. The distribution of current due to electromagnetic induction in the rotor bars varies from bar to bar sinusoidally and depends upon the position and time, assuming sinusoidal distribution of radial flux density in space and also the applied voltage to be varying sinusoidally with time.

Since the rotor winding is permanently short circuited in cage construction, there is no possibility of adding any external resistance in the rotor circuit.

2. *Wound Rotor. As* the name implies, such a rotor is wound with an insulated winding similar to that of the stator except that the number of slots is smaller and fewer turns per phase of a heavier conductor are used. Bar, strap, or wire is used for rotor windings, the last being used where many turns are desired. A large number of rotor turns increases the secondary voltage and reduces the current that flows through the slip rings. The secondary voltage determines the insulation that must be provided; furthermore, the voltage and current influence the value of the resistance to be employed across the slip rings. The motor operation is not influenced by the number of rotor turns, but the ratio of transformation is determined by consideration of secondary current, danger of high secondary emf at starting, and distance to secondary resistors. The standstill open-circuit slip-ring voltage is usually 100 to 400 V for small machines using hand-operated gear and maximum up to 1 kV for large machines.

The rotor is wound for the same number of poles as that of the stator. The rotor winding is always 3-phase* winding even when the stator is wound for two phases. The rotor winding may be star or delta connected but star connection is usually preferred. The three finish terminals are connected together to form star point and the three start terminals are connected to three phosphor-bronze (or brass) slip rings mounted on but insulated from the rotor shaft. The brushes, which carry the current from and to the rotor windings, are held in box type holders mounted on insulated steel rods and securely bolted to the end shield. Each brush is

*The flux set up by the stator windings is a rotating flux with a definite number of poles; operation is only slightly influenced by a change in the number of phases provided on the rotor winding so long as it is greater than one, but the number of poles must be the same on both (stator and rotor). Three-phase rotors are standard for both three-and two-phase motors and are sometimes employed in single-phase motors even.

fed forward by a lever held in tension by an adjustable spring. These brushes are further externally connected to a 3-phase star-connected rheostat for the purpose of starting and speed control (Fig. 15.8). At the time of starting, the entire resistance is included in the rotor circuit and this resistance is gradually cut out as the rotor picks up the speed. For the normal running condition the entire external resistance is cut out and the rotor windings are short circuited automatically through the slip rings by means of a metal collar which is pushed along the shaft and connects all the rings together. Some machines are provided with brush lifting and slip ring short-circuiting arrangement for running condition. However, brush lifting and short-circuiting gear is usually not used in modern machines. The rotor is skewed in this case also. Since the connection of the wound secondary to the external terminals is made through slip rings and brushes, wound secondary motors often are called *slip-ring induction* motors. A sectional diagram of a slip-ring induction motor is shown in Fig. 15.9.

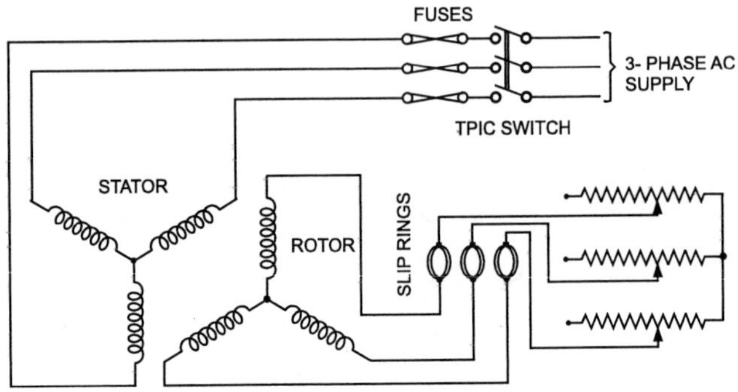

Fig. 15.8 *Slip-Ring Induction Motor With Starting Rheostat*

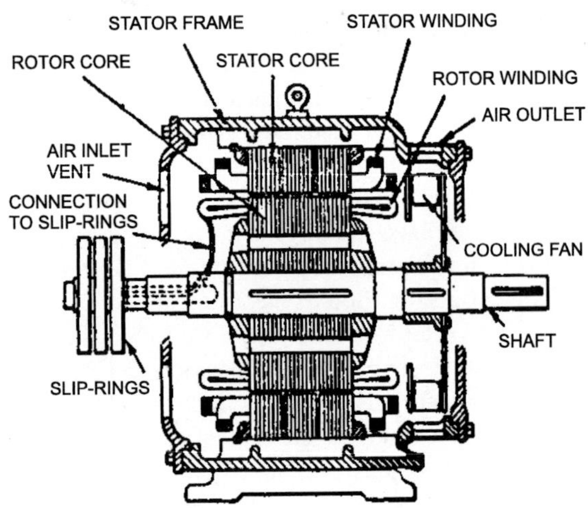

Fig. 15.9 *Section of a Slip-Ring Induction Motor*

15.2.5. Shafts and Bearings. The rotor shaft is supported by bearings housed in the end shield. The air gap of an induction motor is kept necessarily small, therefore, the rotor shaft must be made short and stiff so that the rotor may not have any significant deflection. Even a minor deflection in the shaft would develop large irregularities in the air gap which would lead to production of an *unbalanced magnetic pull.* There is also a possibility of rotor and stator fouling with each other.

Ball- and roller-bearings are generally employed as their use makes accurate centring much simpler than with journal bearings. Also the overall length of the machine is reduced. In small motors, a roller-bearing may be employed at the driving end and a ball-bearing at the other end. For large and heavy rotors journal bearings of self aligning spherical seated type are useful.

15.3. TYPES OF 3-PHASE INDUCTION MOTORS

The 3-phase induction motors are of two types namely (*i*) squirrel cage and (*ii*) wound rotor or slip-ring induction motors. Both motors operate on the same principle and have the same stator construction but differ in rotor construction. The rotor of the squirrel cage induction motor is of squirrel cage type while slip-ring induction motor employs wound rotor. The slip-ring induction motors are less extensively used than squirrel cage type because of their higher first cost and greater maintenance cost. The slip-ring induction motors are employed only when speed control or high starting torque is required.

Induction motor with squirrel cage rotor and that with wound rotor, in disassembled form are shown in Fig. 15.10 (*a*) and 15.10 (*b*) respectively.

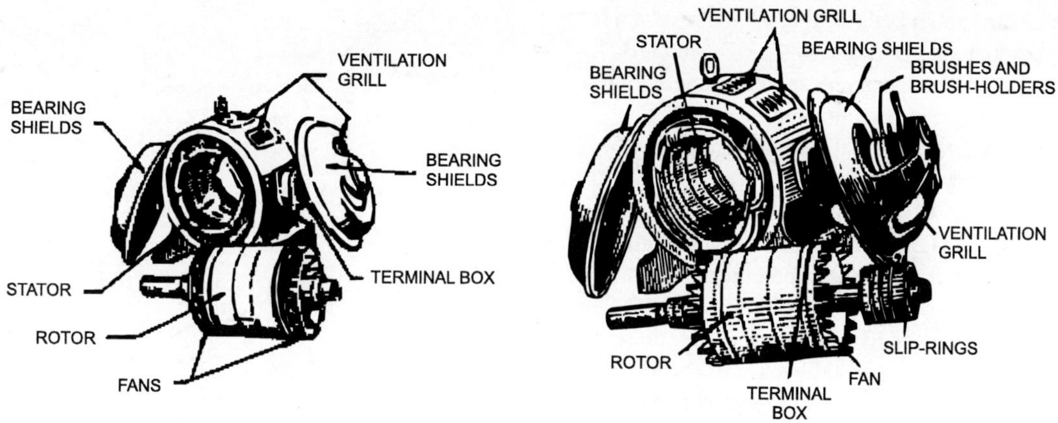

(a) Induction Motor With Squirrel Cage Rotor (Disassembled)

(b) Induction Motor With Wound Rotor (Disassembled)

Fig. 15.10

15.4. ROTATING MAGNETIC FIELDS

A rotating magnetic field is that field which is constant in magnitude but whose axis of direction rotates in space as field system of a dc machine. With a stationary field system the magnetic field is stationary in space, but if the field system is rotated at a certain speed, its magnetic field will rotate with it at the same speed.

The magnetic field produced by single-phase alternating current is an *alternating magnetic field*, the field acting along a fixed axis, varying in magnitude periodically and changing its direction alternately positive and negative.

15.4.1. Ferraris Principle. The magnetic field produced by single-phase alternating current is an alternating magnetic field which, by Ferraris principle, can be resolved into two rotating fields of half its amplitude and rotating in opposite directions at synchronous speed.

Consider two rotating magnetic fields OA_1 and OA_2, each having magnitude of H units and travelling in opposite directions with angular velocity ω, as shown in Fig. 15.11.

Let both the fields start travelling from axis OX at time $t = 0$.

After time t seconds the angle through which fields OA_1 and OA_2 have rotated, $\theta = \omega t$ radians.

i.e. Field represented by vector OA_1 has moved in counter-clockwise direction through an angle $\theta = \omega t$ from axis OX and field represented by vector OA_2 has moved in clockwise direction through same angle θ, because both fields are travelling with same angular velocity ω.

Resolving the magnetic fields represented by vectors OA_1 and OA_2 along axis X and axis Y, we get

$$\text{X-component, } H_x = OA_1 \cos\theta + OA_2 \cos\theta$$

$$= H\cos\theta + H\cos\theta = 2H\cos\theta = 2H\cos\omega t$$

$$\text{Y-component, } H_y = OA_1 \sin\theta - OA_2 \sin\theta = H\sin\theta - H\sin\theta = 0$$

Hence resultant magnetic field is 2H cos ω*t* along X-axis. It is, therefore, obvious that two rotating magnetic fields, travelling in opposite directions with the same angular velocity, result in an alternating field of twice their amplitude. *Conversely an alternating field can be replaced exactly by two rotating fields of half its amplitude travelling in opposite directions at synchronous speed.*

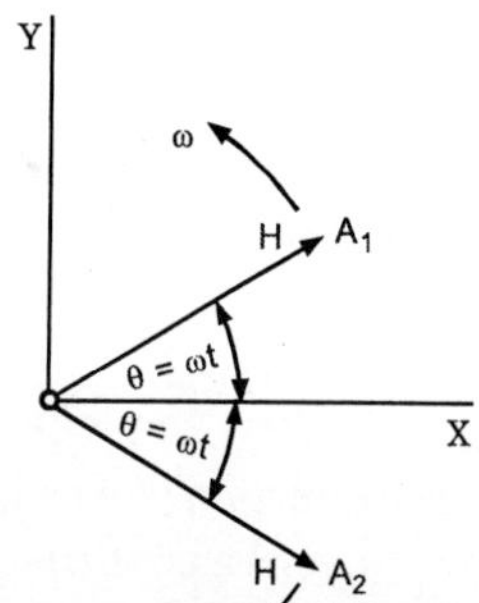

Fig. 15.11

15.4.2. Production of 3-Phase Rotating Magnetic Field. Consider three similar coils A, B and C displaced in space by 120°, as shown in Fig. 15.12, and connected to 3-phase ac supply. Let each of alternating fields due to currents in coils A, B and C be resolved into two components OA_1 and OA_2, OB_1 and OB_2 and OC_1 and OC_2 respectively travelling with angular velocity ω in the directions shown in Fig. 15.12, where $\omega = 2\pi f$, f being the supply frequency.

Consider the instant when the current in coil A is maximum. At this instant both of the components OA_1 and OA_2 of the alternating field produced by the current in coil A are along the axis of the coil A *i.e.* along OA. The current in coil B is 120° behind the instant of its maximum value, therefore, each of its components OB_1 and OB_2 will have to rotate through 120° in order to reach the axis of the coil B *i.e.* along OB. Hence components OB_1 and OB_2 are along OC and OA respectively. The current in coil C is 240° being the instant of its maximum value, therefore, each of components OC_1 and OC_2 of the field produced by current in coil C will have to rotate through 240° in order to reach the axis of coil C *i.e.* along OC. Hence at this instant components OC_1 and OC_2 are along OB and OA respectively.

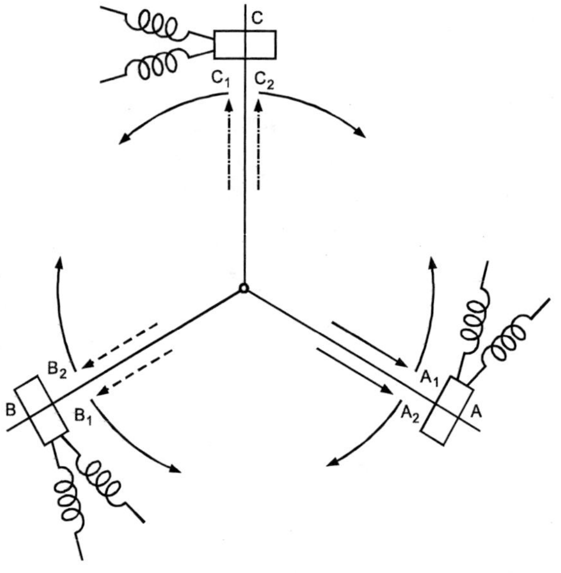

Fig. 15.12

From Fig. 15.13. it is clear that components OA_1, OB_1 and OC_1 are rotating in counter-clockwise direction with same angular velocity ω but are 120° apart, therefore, their resultant is always zero. Components OA_2, OB_2 and OC_2 are rotating in clockwise direction with same angular velocity ω and also points in same direction, therefore, resultant gives a pure rotating field rotating at synchronous speed.

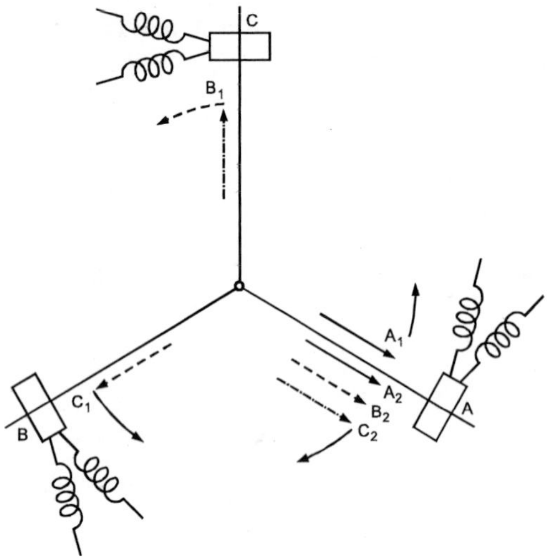

Fig. 15.13

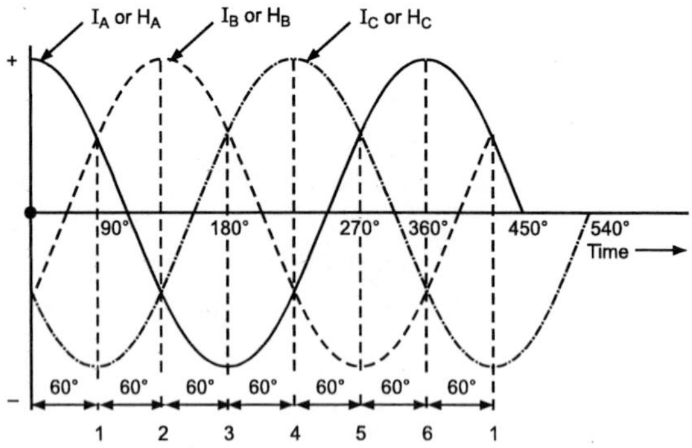

Fig. 15.14

The direction of rotation is the same as the direction of phase sequence *i.e.* first current in coil A attains its maximum value, then current in coil B attains its maximum value and then current in coil C attains its maximum value, as shown in Fig. 15.14. Hence to reverse the direction of rotation of the magnetic field, it is necessary to change the phase sequence which can be done by interchanging any two of the terminals.

15.5. PRINCIPLE OF OPERATION

In a direct current motor, current is drawn from the supply and conducted into the armature conductors through the brushes and commutator. When the armature conductors carry current in the magnetic field established by the field, a force is exerted on the conductors which tends to move them at right angles to the field.

Though in an induction motor, there is no electrical connection to the rotor, but currents are induced in the rotor circuit, and therefore, the same condition exists as in the dc motors i.e., the rotor conductors carry current in the stator magnetic field and thereby have a force exerted upon them tending to move them at right angles to the field.

When the stator or primary winding of a 3-phase induction motor is connected to a 3-phase ac supply, a rotating magnetic field is established which rotates at synchronous speed. The direction of rotation of this field will depend upon the phase sequence of the primary currents and, therefore, will depend upon the order of connection of the primary terminals to the supply. The direction of rotation of the field can be reversed by interchanging the connection to the supply of any two leads of a 3-phase induction motor. The number of magnetic poles of the revolving field will be the same as the number of poles for which each phase of the primary or stator winding is wound. The speed at which the field produced by the primary currents will revolve is called the *synchronous speed* of the motor and is given by an expression, $N_s = 120\ f/p$ where f is supply frequency and P is the number of poles on stator.

As the rotating magnetic field produced by the primary currents sweeps across the rotor conductors, an emf is induced in these conductors just as an emf is induced in the secondary winding of a transformer by the flux set up by the primary current. Since the rotor winding is either directly shorted or closed through some external resistance, the emf induced in the secondary by the revolving field causes a current to flow in the rotor conductors.

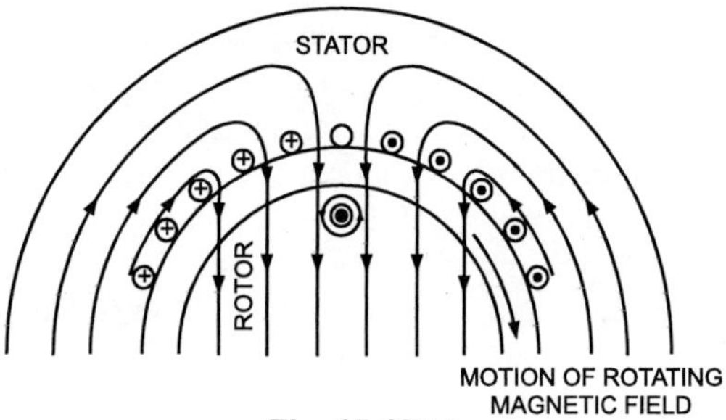

Fig. 15.15 (*a*)

The setting up of the torque for causing the rotor to rotate is explained below:

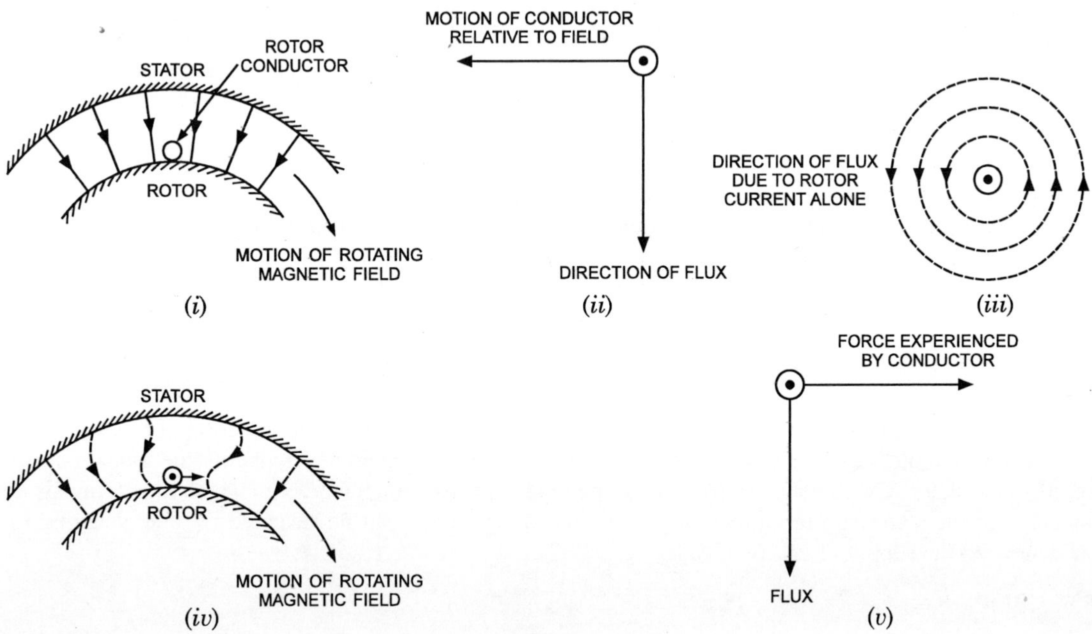

(*b*) *Production of Torque*

Fig. 15.15

A section of an induction motor stator and rotor, with the magnetic field assumed to be rotating in a clockwise direction and with the rotor stationary, as at starting, is shown in Fig. 15.15 (*a*). The relative motion of the rotor with respect to the stator field is anticlockwise. By applying right hand rule, the direction of induced emf or current in the rotor conductor shown is found outward. Hence the direction of the flux due to rotor current alone is anticlockwise, as shown in the Fig. Now, by applying the left hand rule, or by the effect of combined field, it is clear that the rotor conductor shown experiences a force tending to move the conductor to the right. For simplicity, only one rotor conductor is shown. However, other adjacent rotor conductors in the stator field likewise carry current in the same direction as the conductor shown and also have a force exerted upon them tending to move them towards the right. One half cycle later, the stator field direction will have reversed, but the rotor currents will have also reversed, so that the force on the rotor is still the same. Likewise rotor conductors under other stator field poles will have a force exerted upon them all tending to turn the rotor in the clockwise direction. If the developed torque is great enough to overcome the resisting torque of the load, the rotor will accelerate in the clockwise direction or in the same direction as the rotation of the stator field.

The above facts have been illustrated diagrammatically in Fig. 15.15 (*b*).

This may also be explained in terms of Lenz's law. According to Lenz's law we know that the direction of induced emf would be in such a direction that it would try to oppose the very cause for which it is due. Because the cause producing the induced currents is the relative speed between the rotating magnetic field and stationary rotor conductors, therefore, they circulate in such a way that a torque is produced in the rotor tending to cause it to follow the rotating magnetic field and thus reducing the relative speed.

When the rotor is stationary and about to start, the frequency of induced emf in the rotor is equal to that of the supply fed to the stator, because the relative motion is at synchronous speed. As the motor picks up the speed, the relative motion between the rotor and the synchronously rotating magnetic field becomes less and the frequency of emf induced in the rotor decreases. The magnitude of rotor induced emf, induced rotor current and so that of torque developed depends upon the relative motion. In case the relative motion is zero, *i.e.*, the rotor runs at synchronous speed, there would be no induced emf and no current in the rotor conductors, no rotor field and hence no torque. Thus we see that an *induction motor cannot run at synchronous speed.*

An induction motor running at no load will have a speed very close to the synchronous speed and, therefore, emfs in the rotor winding will be very small. This small emf gives a small current producing a torque just sufficient to overcome the losses such as due to friction and maintain the rotor in motion. As the mechanical load is applied on the motor shaft, it must slow down because the torque developed at no load will not be sufficient to keep the rotor revolving at the no-load speed against the additional opposing torque of load. As the motor slows down, the relative motion between the magnetic field and the rotor is increased. This results in greater rotor currents and greater developed torque. Thus, as the load is increased, the motor slows down until the relative motion between the rotor and the rotating magnetic field is just sufficient to result in the development of the torque necessary for that particular load. The decrease in speed from no load to full load is usually 4 to 5 per cent in case of small and medium size induction motors and in case of large size motors it varies from 2 to 2.25 per cent. In respect of speed- load characteristics, a 3-phase induction motor is quite similar to a dc shunt motor.

15.5.1. Reversal of Direction of Rotation of a 3-Phase Induction Motor. The direction of rotation of a 3-phase induction motor depends upon the direction of rotation of the field developed by the 3-phase supply applied to the stator. The direction of rotation of the field depends upon the order in which line terminals are connected to the stator. So the direction of rotation of the motor can be reversed by interchanging the connections to the supply of any two leads.

15.6. SLIP

In Art. 15.5, it has been determined that the speed of a polyphase induction motor must always be less than the synchronous speed and that, as the load is increased, the speed of the motor will decrease. The difference between the speed of the stator field, known as *synchronous speed* (N_s), and the actual speed of the rotor (N) is known as the *slip* and is denoted by s. Though the slip can be expressed in rpm or in radians per second, but usually it is expressed as a fraction or percentage of synchronous speed.

$$\text{Fraction slip, } s = \frac{\text{Synchronous speed} - \text{rotor speed}}{\text{Synchronous speed}} = \frac{N_s - N}{N_s} \quad ...(15.1)$$

$$\text{and percentage slip} = \frac{N_s - N}{N_s} \times 100 \quad ...(15.2)$$

At normal load the slip of an induction motor is usually between 2 and 5 per cent. At no load the slip is as small as 0.5 per cent and for many purposes the machine may then be considered to be running at synchronous speed. As the load is applied, the natural effect of the load torque is to cause the machine to slow down. As it does so, slip increases, and with it current and torque until the driving torque of the machine balances the retarding torque of the load. This determines the speed at which the machine runs on load.

The induction motor is thus a motor with substantially constant speed and plays the same role as a dc shunt motor.

Example 15.1 **An induction motor has a rated speed of 715 rpm. How many poles have its rotating magnetic field?** [*G.B. Technical Univ. Electrical Engineering Even Semester 2012-13*]

Solution: The number of poles, even in number and to provide a synchronous speed nearest higher than actual speed of motor (*i.e.* 715 rpm), that the motor must have is 8 **Ans.**

$$\because P_m = \frac{120 f}{N_m} \simeq \frac{120 \times 50}{715} \simeq 8 \text{ assuming } f = 50 \text{ Hz}$$

Example 15.2 **A 60 Hz induction motor has 2 poles and runs at 3,510 rpm. Calculate (*a*) the synchronous speed and (*b*) the per cent slip.** [*U.P. Technical Univ. Electrical Engineering Second Semester 2008-09*]

Solution:

(*a*) $$\text{Synchronous speed, } N_S = \frac{120 \times f}{P} = \frac{120 \times 60}{2} = 3{,}600 \text{ rpm} \quad \textbf{Ans.}$$

(*b*) $$\text{Per cent slip} = \frac{N_S - N}{N_S} \times 100 = \frac{3{,}600 - 3{,}510}{3{,}600} \times 100 = 2.5\% \quad \textbf{Ans.}$$

Example 15.3 **A 4-pole, 3-phase, 50 Hz, star-connected induction motor has a full-load slip of 4%. Calculate full-load speed of motor.** [*U.P. Technical Univ. Electrical Engineering Odd Semester 2013-14*]

Solution:

$$\text{Synchronous speed, } N_s = \frac{120 f}{P} = \frac{120 \times 50}{4} = 1{,}500 \text{ rpm.}$$

Actual speed of induction motor, $N = N_s (1 - s) = 1{,}500 (1 - 0.04) = 1{,}440$ rpm **Ans.**

Example 15.4 **A-12 pole alternator is coupled to an engine running at 500 rpm. It supplies a 3-phase induction motor having full-load speed at 1,440 rpm. Find percentage slip and number of poles of the motor.**

[*Mahamaya Technical Univ. Electrical Engineering Second Semester 2011-12*]

Solution:

$$\text{Frequency of generated emf, } f = \frac{P_a \times N_a}{120} = \frac{12 \times 500}{120} = 50 \text{ Hz}$$

The number of poles, even in number and to give a synchronous speed nearest higher than actual speed of motor (i.e. 1,440 rpm), that the motor must have is 4. **Ans.**

$$\because P_m = \frac{120 f}{N_m} = \frac{120 \times 50}{1{,}450} \simeq 4$$

$$\text{Synchronous speed of motor, } N_s = \frac{120 f}{P} = \frac{120 \times 50}{4} = 1{,}500 \text{ rpm}$$

$$\text{Slip, } s = \frac{N_s - N}{N_s} = \frac{1{,}500 - 1{,}440}{1{,}500} = 0.04 \text{ or } 4\ \% \quad \textbf{Ans.}$$

Example 15.5 **A motor-generator set used for providing variable frequency ac supply consists of a 3-phase, 10-pole synchronous motor and a 24-pole, 3-phase synchronous generator. The motor-generator set is fed from a 25 Hz, 3-phase ac supply. A 6-pole, 3-phase induction motor is electrically connected to the terminals of the synchronous generator and runs at a slip of 5%. Determine (i) the frequency of the generated voltage of the synchronous generator, (ii) the speed at which the induction motor is running.**

[*U.P. Technical Univ. Electrical Engineering February 2001*]

Solution: Speed of motor-generator set, $N_s = \dfrac{120 \times \text{supply frequency}, f_1}{\text{Number of poles on synchronous motor}} = \dfrac{120 \times 25}{10} = 300$ rpm

(*i*) Frequency of generated voltage, $f_2 = \dfrac{\text{Speed of motor-generator set} \times \text{number of poles on synchronous generator}}{120}$

$$= \frac{300 \times 24}{120} = 60 \text{ Hz} \quad \textbf{Ans.}$$

(*ii*) Speed of induction motor, $N_m = N_s(1-s) = \dfrac{120 f_2}{P_m}(1-s) = \dfrac{120 \times 60}{6}(1-0.05) = 1{,}140$ rpm **Ans.**

15.7. FREQUENCY OF ROTOR CURRENT (OR EMF)

It has been shown in Art 15.5 that the rotor runs in the direction of rotating magnetic field. When the rotor is stationary, rotor conductors are being cut by the rotating flux at synchronous speed $N_s\left(N_s \text{ being equal to } \dfrac{120 f}{P}\right)$, therefore, the frequency of rotor current (or emf) is the same as that of supply frequency. But when the rotor starts revolving, the rate at which the rotor conductors are being cut by the rotating flux depends upon the relative speed between the rotor and the stator revolving magnetic field, called the *slip speed.* The frequency of rotor current (or emf) induced by the relative motion between rotor conductors and the stator revolving magnetic field is given by the equation:

$$\text{Rotor emf frequency}, f' = \frac{\text{Relative speed in rpm}}{120/P} = \frac{N_s - N}{120/P} \quad \text{...(15.3)}$$

$$\text{Since slip}, s = \frac{N_s - N}{N_s}$$

$$\text{or } N_s - N = s N_s = s \times \frac{120 f}{P}$$

Substituting, $N_s - N = s \times \dfrac{120 f}{P}$ in Eq. (15.3) we have

$$\text{Rotor emf frequency}, f' = s \times \frac{120 f}{P} \times \frac{P}{120} = sf \quad \text{...(15.4)}$$

Thus the frequency of rotor current (or emf) in an induction motor is given by the product of slip s and the supply frequency f. That is why, it is also called the *slip frequency.*

Example 15.6 **A three-phase, 50 Hz, induction motor has a full-load speed of 1,460 rpm. Calculate slip, number of poles and frequency of rotor induced emf.**

[*Electrical Engineering G.B.T.U. First Semester 2011-12; U.P. Technical Univ. Second Semester 2013-14*]

Solution: The number of poles, even in number and to give a synchronous speed nearest higher than actual speed of motor (i.e., 1,460 rpm), that the motor must have is 4. $\because P_m = \dfrac{120 f}{N_m} \approx \dfrac{120 \times 50}{1{,}460} = 4.$

$$\text{Synchronous speed}, N_s = \frac{120 f}{P} = \frac{120 \times 50}{4} = 1{,}500 \text{ rpm}$$

$$\text{Slip, } s = \frac{N_s - N}{N_s} = \frac{1{,}500 - 1{,}460}{1{,}500} = 0.02667 \text{ or } 2.667\% \textbf{ Ans.}$$

Frequency of rotor induced emf, $f' = sf = 0.02667 \times 50 = 1.333$ Hz **Ans.**

Example 15.7 **A 3-phase, 4-pole induction motor is supplied from 3-phase, 50 Hz ac supply. Calculate:**

(*i*) the synchronous speed.

(*ii*) the rotor speed when slip is 4%.

(*iii*) the rotor frequency when rotor runs at 600 rpm.

[*U.P. Technical Univ. Electrical Engineering Second Semester 2005-06; Basic Electrical Engineering Second Semester 2013-14*]

Solution: (*i*) Synchronous speed, $N_s = \frac{120 \times f}{P} = \frac{120 \times 50}{4} = 1{,}500$ rpm

(*ii*) Speed when slip is 4% or 0.04, $N = N_s(1 - s) = 1{,}500(1 - 0.04) = 1{,}440$ rpm **Ans.**

(*iii*) Slip when the motor runs at 600 rpm $s' = \frac{1{,}500 - 600}{1{,}500} = 0.6$

Rotor frequency, $f' = s'f = 0.6 \times 50 = 3$ Hz **Ans.**

Example 15.8 **A 12-pole, 3-phase alternator is coupled to an engine running at 500 rpm. It supplied a 3-phase induction motor having a full-load speed of 1,440 rpm. Find the percentage slip, frequency of rotor current and number of poles of the motor.** [*U.P. Technical Univ. Elec. Engineering First Semester 2005-06*]

Solution: Frequency of supply from alternator, $f = \frac{P_a \times N_a}{120} = \frac{12 \times 500}{120} = 50$ Hz

Full-load speed, $N_f = 1{,}440$ rpm

The number of poles, even in number, to give a synchronous speed nearest higher than full-load speed of motor (i.e. 1,440 rpm) that the motor must have is 4 $\left[P = \frac{120f}{N} = \frac{120 \times 50}{1{,}440} \simeq 4\right]$

$$\text{Synchronous speed, } N_s = \frac{120f}{P_m} = \frac{120 \times 50}{4} = 1{,}500 \text{ rpm.}$$

$$\text{Percentage slip, } s = \frac{N_s - N}{N_s} \times 100 = \frac{1{,}500 - 1{,}440}{1500} \times 100 = 4\% \textbf{ Ans.}$$

Rotor current frequency, $f' = sf = 0.04 \times 50 = 2$ Hz **Ans.**

Number of poles of the motor = 4 **Ans.**

Example 15.9 **A 3-phase slip-ring, 4-pole induction motor has rotor frequency of 2 Hz while connected to 400 V, 3-phase, 50 Hz supply. Determine slip and rotor speed.**

[*U.P. Technical Univ. Electrical Engineering Second Semester 2006-07*]

Solution: Synchronous speed, $N_s = \frac{120f}{P} = \frac{120 \times 50}{4} = 1{,}500$ rpm

$$\text{Slip, } s = \frac{\text{Frequency of rotor emf, } f'}{\text{Supply frequency, } f} = \frac{2}{50} = 0.04$$

Speed of motor on load, $N = N_s(1 - s) = 1{,}500(1 - 0.04) = 1{,}440$ rpm **Ans.**

Example 15.10 **A 4-pole, 3-phase induction motor is energized from a 60 Hz supply, and is running at a load condition for which the slip is 0.03. Determine : (*a*) Rotor speed, in rpm (*b*) Rotor current frequency, in Hz (*c*) Speed of the rotor's rotating magnetic field with respect to the stator frame, in rpm.**

[*U.P. Technical Univ. Electrical Engineering First Semester 2008-09*]

Solution: Synchronous speed, $N_s = \frac{120f}{P} = \frac{120 \times 60}{4} = 1{,}800$ rpm

Slip, $s = 0.03$

(*a*) Rotor speed, $N = N_s(1 - s) = 1{,}800\,(1 - 0.03) = 1{,}746$ rpm **Ans.**

(*b*) Rotor current frequency, $f' = sf = 0.03 \times 60 = 1.8$ Hz **Ans.**

(*c*) Speed of rotor's rotating magnetic field with respect to stator frame $= N_s = 1{,}800$ rpm **Ans.**

Example 15.11 **A 3-phase, 50 Hz induction motor has 6 poles and operates with a slip of 5% at a certain load. Determine**

- **(i) the speed of the rotor with respect to the stator.**
- **(ii) the frequency of rotor current.**
- **(iii) the speed of the rotor magnetic field with respect to rotor.**
- **(iv) the speed of the rotor magnetic field with respect to stator and**
- **(v) the speed of the rotor magnetic field with respect to the stator magnetic field.**

[U.P. Technical Univ. Electrical Engineering February 2002; Odd Semester 2013-14]

Solution: Supply frequency, $f = 50$ Hz

Number of poles, $P = 6$

Slip, $s = 5\%$ or 0.05

Synchronous speed, $N_s = \frac{120\,f}{P} = \frac{120 \times 50}{6} = 1{,}000$ rpm

(*i*) Speed of rotor w.r.t. the stator, $N = N_s(1 - s) = 1{,}000\,(1 - 0.05) = 950$ rpm **Ans.**

(*ii*) Frequency of rotor current, $f' = sf = 0.05 \times 50 = 2.5$ Hz **Ans.**

(*iii*) Speed of rotor magnetic field with respect to the rotor

$$= \frac{120\,f'}{P} = \frac{120 \times 2.5}{6} = 50 \text{ rpm}$$ **Ans.**

(*iv*) Speed of rotor magnetic field with respect to the stator $= N_s = 1{,}000$ rpm **Ans.**

(*v*) Rotor field and stator field are revolving at the same speed of 1,000 rpm, therefore, speed of rotor field w.r.t. stator field is zero.

Example 15.12 **A three-phase, 50 Hz induction motor has a full-load speed of 960 rpm. Calculate:**

- ***(i)* slip**
- ***(ii)* frequency of rotor induced emf**
- ***(iii)* Number of poles**
- ***(iv)* Speed of rotor field with respect to rotor structure**
- ***(v)* Speed of rotor field with respect to stator field.**

[G.B. Technical Univ. Electrical Engineering Second Semester 2009-10]

Solution: Actual speed of motor, N = 960 rpm

The number of poles even in number and to provide a synchronous speed nearest higher than actual speed of motor (i.e. 960 rpm) that the motor must have is 6. $\because \; P_m \simeq \frac{120f}{N_m} \simeq \frac{120 \times 50}{960} = 6$

Synchronous speed, $N_s = \frac{120f}{P} = \frac{120 \times 50}{6} = 1{,}000$ rpm

(*i*) Slip, $s = \frac{N_s - N}{N_s} = \frac{1{,}000 - 960}{1{,}000} = 0.04$ **Ans.**

(*ii*) Frequency of rotor induced emf, $f' = sf = 0.04 \times 50 = 2$ Hz **Ans.**

(*iii*) Number of poles, $P = 6$ **Ans.**

(*iv*) Speed of rotor field with respect to rotor structure $= \frac{120\,f'}{P} = \frac{120 \times 2}{6} = 40$ rpm.

(*v*) Speed of rotor field with respect to stator field = zero, because both the fields are revolving at the same speed.

Example 15.13 **A three-phase, 50 Hz induction motor has a full-load speed of 1,440 rpm. Calculate:**

(*i*) **Slip** (*ii*) **Number of poles.**

(*iii*) **Frequency of rotor induced emf.** (*iv*) **Speed of rotor field with respect to rotor structure.**

(*v*) **Speed of rotor field with respect to stator field.**

[*Mahamaya Technical Univ. Electrical Engineering Second Semester 2010-11*]

Solution: The number of poles even in number and to provide synchronous speed nearest higher than actual speed of motor (i.e. 1,440 rpm) that the motor must have is 4. **Ans.** $\because \; P_m = \dfrac{120\,f}{N_m} = \dfrac{120 \times 50}{1{,}440} \simeq 4$

Synchronous speed of motor, $N_s = \dfrac{120 \times f}{P} = \dfrac{120 \times 50}{4} = 1{,}500$ rpm.

(*i*) Slip, $s = \dfrac{N_s - N}{N_s} = \dfrac{1{,}500 - 1{,}440}{1{,}500} = 0.04$ or 4% **Ans.**

(*iii*) Frequency of rotor induced emf, $f' = sf = 0.04 \times 50 = 2$ Hz **Ans.**

(*iv*) Speed of rotor field with respect to rotor structure $= \dfrac{120\,f'}{P} = \dfrac{120 \times 2}{4} = 60$ rpm **Ans.**

(*v*) Rotor field and stator field are revolving at the same speed of 1,500 rpm, therefore, speed of rotor field w.r.t. stator field is zero. **Ans.**

15.8. ROTOR EMF

When the rotor is at standstill, the motor is equivalent to a 3-phase transformer with secondary short circuited. So induced emf per phase E_2 in the rotor, when it is at standstill i.e. at the instant of starting is given by

$$E_2 = E_1 \times \frac{N_2}{N_1} \qquad \text{...(15.5)}$$

where E_1 is applied voltage per phase to primary *i.e* stator winding, N_2 and N_1 are the numbers of turns per phase on rotor and stator respectively.

When the rotor starts running, the relative speed of the rotor with respect to stator flux i.e. slip s drops. Hence, the rotor induced emf which varies in direct proportion with the relative speed or slip s and is given by sE_2.

Hence for slip s, the induced emf in the rotor is s times the induced emf in the rotor at standstill.

15.9. ROTOR CURRENT AND POWER FACTOR

The circuit diagram of an induction motor rotor is given in Fig. 15.16.

Let the resistance and inductance per phase of rotor be R_2 and L_2 respectively and induced emf per phase in rotor at standstill be E_2.

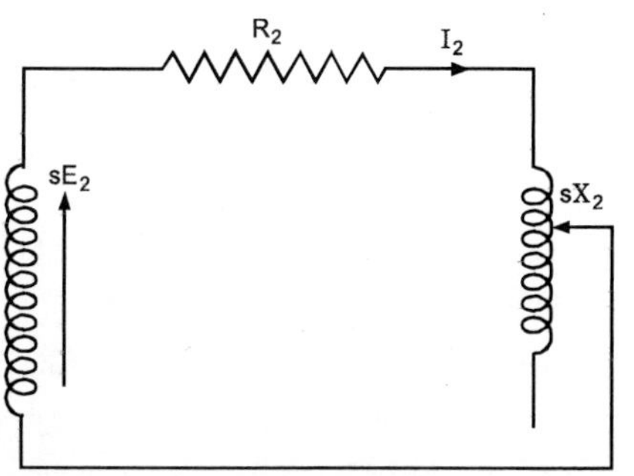

Fig. 15.16

At Standstill Induced emf per phase in rotor $= E_2$

Rotor winding resistance / phase $= R_2$

Rotor winding reactance / phase, $X_2 = 2\,\pi f L_2$

where f is supply frequency

$$\text{Rotor winding impedance / phase, } Z_2 = \sqrt{R_2^2 + X_2^2} \qquad \text{...(15.6)}$$

$$\text{Rotor current, } I_2 = \frac{E_2}{Z_2} = \frac{E_2}{\sqrt{R_2^2 + X_2^2}} \qquad \text{...(15.7)}$$

$$\text{Power factor of rotor current, } \cos \phi_2 = \frac{R_2}{\sqrt{R_2^2 + X_2^2}} \qquad \text{...(15.8)}$$

At Slip s

Induced emf per phase in rotor winding $= sE_2$

Rotor winding resistance / phase $= R_2$

Rotor winding reactance / phase $= 2\pi s f L_2 = s(2\pi f L_2) = s X_2$

$$\text{Rotor winding impedance / phase} = \sqrt{R_2^2 + (sX_2)^2} \quad ...(15.9)$$

$$\text{Rotor current per phase, } I_2 = \frac{sE_2}{\sqrt{R_2^2+(sX_2)^2}} = \frac{sE_2}{\sqrt{R_2^2+s^2X_2^2}} \quad ...(15.10)$$

$$\text{Power factor of rotor current, } \cos\phi_2 = \frac{R_2}{\sqrt{R_2^2+s^2X_2^2}} \quad ...(15.11)$$

The Eq. (15.10) for rotor current I_2 may be rewritten as

$$I_2 = \frac{E_2}{\sqrt{\left(\frac{R_2}{s}\right)^2 + X_2^2}} \quad ...(15.12)$$

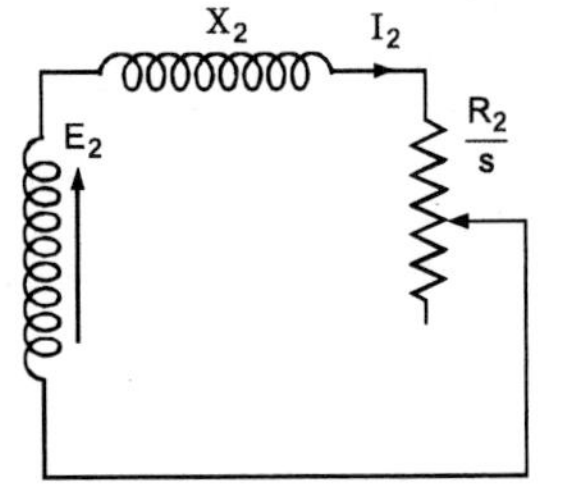

Fig. 15.17 *Equivalent Circuit of Rotor*

The above equation gives convenient form of rotor equivalent circuit, as shown in Fig. 15.17.

The resistance $\frac{R_2}{s}$ is a function of slip and can be splitted into two parts R_2 and $R_2\frac{(1-s)}{s}$ where R_2 is rotor resistance and $R_2\frac{(1-s)}{s}$ is the fictitious resistance representing electrical load on the rotor. The final simplified equivalent circuit of an induction motor rotor is given in Fig. 15.18.

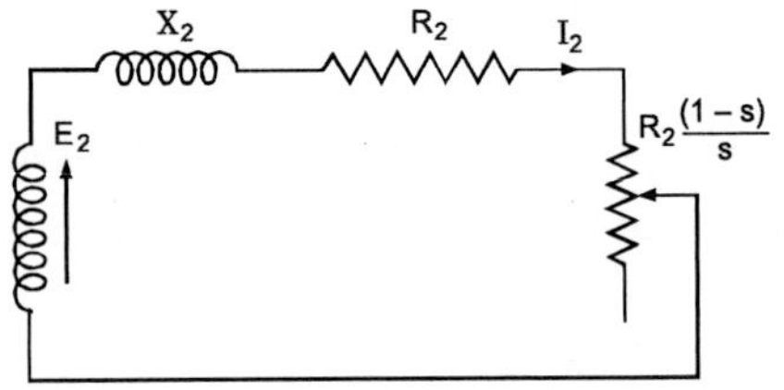

Fig. 15.18

The power consumed by the fictitious resistance representing the rotor electrical load is given as

$I_2^2 R_2 \frac{(1-s)}{s}$ and is converted into mechanical power.

Example 15.14 **A 3-phase, 440 V, 50 hp, 50 Hz induction motor runs at 1,450 rpm when it delivers rated output power. Determine:**

(*i*) **Number of poles in the machine.**

(*ii*) **Speed of rotating air gap field.**

(*iii*) **Rotor induced voltage if stator to rotor turns ratio is 1 : 0.80. Assume the winding factors are the same.**

(*iv*) **Frequency of rotor current.** [*U.P. Technical Univ. Electrical Engineering Second Semester 2004-05*]

Solution: (*i*) The number of poles even in number and to give synchronous speed nearest higher than actual speed of motor (i.e. 1,450 rpm) that the motor must have is 4 **Ans.**

(*ii*) Speed of rotating air gap field. $N_s = \frac{120 \times f}{P} = \frac{120 \times 50}{4} = 1{,}500$ rpm **Ans.**

(*iii*) Induced emf in rotor at standstill $= KE_1 = KV_1 = 0.8 \times 440 = 352$ V

$$\text{Slip, } s = \frac{N_s - N}{N_s} = \frac{1{,}500 - 1{,}450}{1{,}500} = 0.03333$$

Rotor induced voltage $= sKV_1 = 0.03333 \times 352 = 11.733$ V **Ans.**

(*iv*) Frequency of rotor current, $f' = sf = 0.03333 \times 50 = 1.6666$ Hz **Ans.**

Example 15.15 **A three-phase delta-connected 440 volts, 50 Hz, 4-pole induction motor has a rotor standstill emf per phase of 130 volts. If the motor is running at 1440 rpm, calculate slip, frequency of rotor induced emf, the value of the rotor induced emf per phase and stator to rotor turn ratio for same speed.**

[*Mahamaya Technical Univ. Electrical Engineering First Semester 2011-12*]

Solution: Synchronous speed of induction motor, $N_s = \dfrac{120 \times \text{supply frequency}}{\text{Number of poles}} = \dfrac{120 \times 50}{4} = 1{,}500 \text{ rpm}$

Motor speed, N = 1,440 rpm

(*i*) Slip, $s = \dfrac{N_s - N}{N_s} = \dfrac{1{,}500 - 1{,}440}{1{,}500} = 0.04$ or 4% **Ans.**

(*ii*) Frequency of rotor induced emf = Slip × supply frequency = $sf = 0.04 \times 50 = 2$ Hz **Ans.**

(*iii*) Rotor induced emf per phase at the speed of 1,440 rpm or at slip of 0.04

= Slip × induced emf per phase in rotor winding at standstill

= $s \times E_2 = 0.04 \times 130 = 5.2$ V **Ans.**

(*iv*) Stator-to-rotor turn-ratio $= \dfrac{\text{Induced emf per phase in stator}}{\text{Induced emf per phase in rotor at standstill}}$

$= \dfrac{E_{P_1}}{E_{P_2}} = \dfrac{440}{130} = \dfrac{44}{13}$ **Ans.**

Example 15.16 **A 5 hp, 230 V, 50 Hz induction motor has a rated full-load speed of 950 rpm. The induced voltage per phase of rotor at standstill is 100 V. Calculate: (*i*) Number of poles and % full-load slip. (*ii*) Rotor induced voltage and its frequency at full-load.** [*U.P. Technical Univ. Electrical Engineering Odd Semester 2013-14*]

Solution: (*i*) The number of poles, even in number and to provide a synchronous speed nearest higher than actual speed of motor (*i.e.* 950 rpm), the motor must have is 6. **Ans.** $\because P_m = \dfrac{120f}{N_m} = \dfrac{120 \times 50}{950} \simeq 6$

Synchronous speed, $N_s = \dfrac{120f}{P} = \dfrac{120 \times 50}{6} = 1{,}000$ rpm.

% Full-load slip, $s = \dfrac{N_s - N}{N_s} \times 100 = \dfrac{1{,}000 - 950}{1{,}000} \times 100 = 5\%$ **Ans.**

(*ii*) Rotor induced voltage at full-load = Slip × induced voltage per phase in rotor winding at standstill

$= \dfrac{5}{100} \times 100 = 5$ V **Ans.**

Frequency of induced emf at full-load = Slip × supply frequency $= \dfrac{5}{100} \times 50 = 2.5$ Hz **Ans.**

15.10. ROTOR TORQUE

The torque of an induction motor, being due to interaction of the rotor and stator fields, depends on the strength of those fields and the phase relation between them. Thus the torque developed by the rotor of an induction motor is directly proportional to (*i*) rotor current, I_2 (*ii*) stator flux per pole, Φ and (*iii*) power factor of the rotor circuit, $\cos \phi_2$

i.e. $T \propto \Phi I_2 \cos \phi_2$

Since rotor emf per phase at standstill, $E_2 \propto \Phi$

∴ Torque developed, $T \propto E_2 I_2 \cos \phi_2$

or $T = K E_2 I_2 \cos \phi_2$ ...(15.13)

where K is any constant

15.10.1. Torque Under Running Conditions

Substituting $I_2 = \dfrac{sE_2}{\sqrt{R_2^2 + s^2 X_2^2}}$ from Eq. (15.10) and $\cos \phi_2 = \dfrac{R_2}{\sqrt{R_2^2 + s^2 X_2^2}}$ from Eq. (15.11) in Eq. (15.13) we have torque developed by an induction motor rotor,

$$T = K E_2 \frac{sE_2}{\sqrt{R_2^2 + s^2 X_2^2}} \times \frac{R_2}{\sqrt{R_2^2 + s^2 X_2^2}} = \frac{K s R_2 E_2^2}{R_2^2 + s^2 X_2^2} \qquad ...(15.14)$$

It may be seen from Eq. (15.14) that the running torque is proportional to the square of supply voltage because rotor induced emf E_2 is proportional to supply voltage.

15.10.2. Starting Torque. At start, the rotor is stationary and so the slip $s = 1.0$ and the starting torque may be obtained by substituting $s = 1.0$ in Eq. (15.14).

$$\text{Starting torque, } T_{st} = \frac{K R_2 E_2^2}{R_2^2 + X_2^2} \qquad ...(15.15)$$

Generally, the stator supply voltage V is constant, so the rotating flux Φ set up by the stator currents and rotor induced emf E_2 are also constant.

$$\therefore T_{st} = \frac{K_1 R_2}{R_2^2 + X_2^2} = \frac{K_1 R_2}{Z_2^2} \quad \text{where } K_1 \text{ is another constant.} \qquad ...(15.16)$$

15.10.3. Condition For Maximum Running Torque. Torque under running condition given by Eq. (15.14) is reproduced here

$$T = \frac{K R_2 s E_2^2}{R_2^2 + s^2 X_2^2}$$

Torque T for fixed supply voltage will be maximum when $\dfrac{s R_2}{R_2^2 + s^2 X_2^2}$ or $\dfrac{R_2}{\dfrac{R_2^2}{s} + s X_2^2}$ or

$\dfrac{R_2}{\left(\dfrac{R_2}{\sqrt{s}} - X_2 \sqrt{s}\right)^2 + 2 R_2 X_2}$ is maximum viz $\dfrac{R_2}{\sqrt{s}} - X_2 \sqrt{s} = 0$ or $s = \dfrac{R_2}{X_2}$

Substituting $s = \dfrac{R_2}{X_2}$ in Eq. (15.14) we have

$$T_{max} = \frac{K R_2^2 E_2^2 / X_2}{R_2^2 + R_2^2} = \frac{K E_2^2}{2 X_2} \qquad ...(15.17)$$

From Eq. (15.17) it is obvious that

(*i*) Maximum torque is independent of rotor circuit resistance.

(*ii*) The slip at which maximum torque occurs is governed by the rotor resistance. As seen from above, the torque becomes maximum when rotor reactance equals the rotor resistance. So by varying the rotor circuit resistance, which is possible with wound rotors, maximum torque can be made to occur at any desired slip or motor speed.

(*iii*) Maximum torque varies inversely as standstill reactance of the rotor, hence to have maximum torque, standstill reactance *i.e.* inductance of the rotor should be kept as small as possible. This is achieved by placing the rotor conductors very close to the surface of the rotor and reducing the air gap between stator and rotor to the smallest possible value.

(*iv*) Maximum torque varies directly as the square of supply voltage.

15.10.4. Condition For Maximum Starting Torque

Starting torque T_{st} will be maximum when $\frac{R_2}{X_2} = s = 1$ or $R_2 = X_2$

Since the rotor resistance is not more than 1 or 2 per cent of leakage reactance otherwise efficiency will be low, so to increase the starting torque, extra resistance is required to be inserted in the rotor circuit at start and cut out gradually as the motor picks up speed. This is possible in case of wound rotor motors. This is the reason that wound rotor motors are used where heavy loads are to be accelerated at start such as cranes, lifts, elevators etc.

15.11. TORQUE-SLIP AND TORQUE-SPEED CURVES

Torque developed by an induction motor rotor is given by the equation $T = \frac{K s R_2 E_2^2}{R_2^2 + s^2 X_2^2}$.

From the above equation it is revealed that

(*i*) When speed is synchronous *i.e.* when slip s is zero, the torque T is zero so the torque-slip curve starts from origin O.

(*ii*) When the speed is very near to synchronous speed N_s *i.e.* when slip s is very low the value of term sX_2 is very small and is negligible in comparison with rotor resistance R_2, therefore, torque T is approximately proportional to slip s, if rotor resistance R_2 is constant *i.e.* at speeds near synchronous speed the torque-speed and torque-slip curves are approximately straight lines, as shown in Fig. 15.19 (*a*) and 15.19 (*b*) respectively.

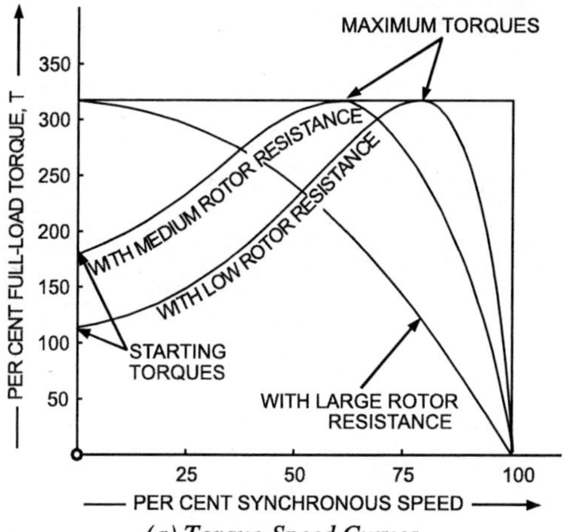

(a) Torque-Speed Curves

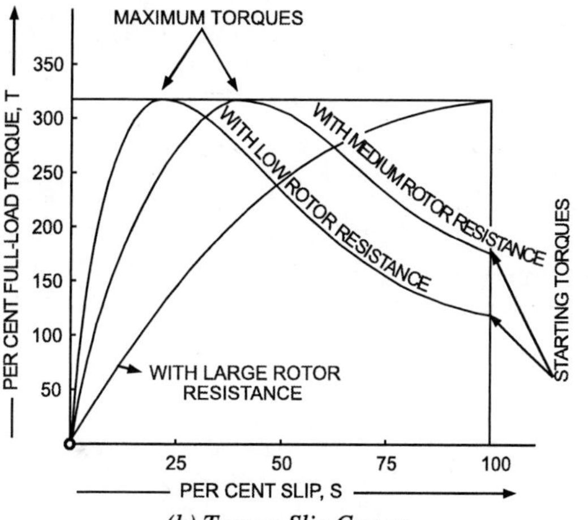

(b) Torque-Slip Curves

Fig. 15.19

(*iii*) As the slip increases , *i.e.* as the speed drops with the increase in load, torque increases, reaches its maximum value when slip $s = \frac{R_2}{X_2}$. The maximum torque is also known as *breakdown or pull-out torque.* The slip corrresponding to breakdown torque is called the breakdown slip s_b.

(*iv*) With the further increase in slip or drop in speed due to increase in load beyond the point of maximum torque the torque begins to decrease. The result is that the motor slows down and eventually stops. The motor operates for the value of slip between zero and that corresponding to breakdown or pull-out torque.

With higher values of slip, R_2 becomes negligible as compared to $s X_2$ and the torque varies according to the relation $T \propto \frac{s}{s^2 X_2^2} \propto \frac{1}{s}$ if standstill reactance X_2 is constant. It means that torque-speed curves are rectangular hyperbola with the speed or slip beyond that corresponding to breakdown or pull-out torque.

Torque-speed and torque-slip curves are shown in Figs. 15.19 (*a*) and 15.19 (*b*) respectively for different values of rotor resistance R_2.

It is seen that although maximum torque is independent of rotor resistance R_2, yet the exact location of T_{max} is dependent on it. Greater the R_2, greater is the value of slip at which the maximum torque occurs.

15.11.1. Effect of Rotor Resistance Upon Torque-Slip or Torque-Speed Relationship. From the torque equation it is obvious that, for a constant supply voltage., when rotor resistance R_2 is very small compared with $s\,X_2$, the torque for a given slip is directly proportional to R_2; but when rotor resistance R_2 is large compared with $s\,X_2$, the torque for a given slip is inversely proportional to R_2.

From the Fig. 15.19 (*b*) it is seen that variation of rotor resistance does not change the magnitude of maximum torque T_{max}; merely changes the value of slip at which it occurs. Larger the rotor resistance the greater is the slip at which the maximum torque occurs.

For a given value of torque T, slip *s* is proportional to rotor resistance R_2, so addition of external resistance in the rotor circuit does not lower the torque curve but merely stretches it so that the same torque values occur at lower speeds (or higher slips) as shown in Fig. 15.19 (*a*) and 15.19 (*b*).

15.12. EFFECT OF CHANGE IN SUPPLY VOLTAGE ON TORQUE AND SLIP (OR SPEED)

When the motor is running with slip *s*, then torque acting on the rotor is given by

$$T = \frac{K\,s\,R_2\,E_2^2}{R_2^2 + s^2\,X_2^2} \quad \text{... Refer to Eq. (15.14)}$$

Since emf induced in the rotor at standstill, $E_2 \propto V$,

$$\text{torque developed, } T = \frac{K' s\,R_2 V^2}{R_2^2 + s^2 X_2^2}$$

where K′ is another constant

Since at full load, slip *s* is very low, $s^2 X_2^2$ can be neglected in comparison with R_2^2 and equation for torque may be written

$$\text{as } T = \frac{K' s\,R_2\,V^2}{R_2^2} = \frac{K'\,s\,V^2}{R_2} \quad \text{or } T \propto sV^2$$

If the supply voltage is changed, it changes the torque under running conditions also. With the decrease in supply voltage, torque under running condition decreases. As the motor now develops a reduced torque, the motor slows down (*i.e.* slip increases) so as to cause increase in torque to meet the load torque and the motor may come to a standstill (if the load torque exceeds the pull-out torque of the motor corresponding to reduced supply voltage) unless the load is removed. Fig. 15.20 shows the variation of torque with slip for three different values of line voltage.

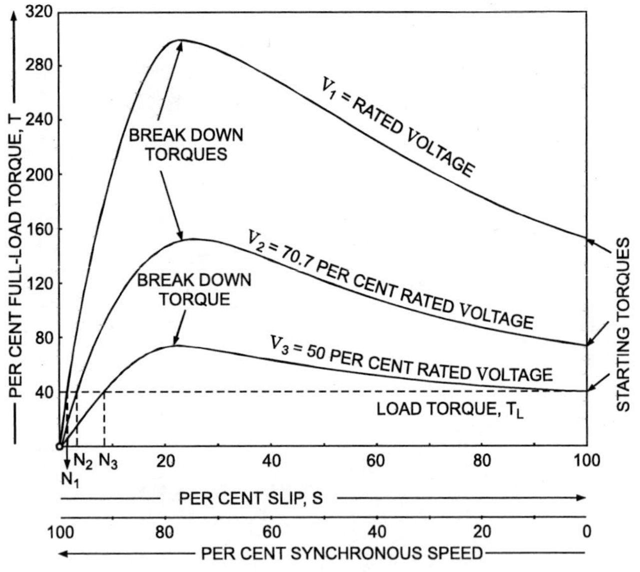

Fig. 15.20

15.13. TORQUE-SPEED CURVE AND OPERATING REGION

The torque-speed curve for an induction motor is given in Fig. 15.21. From Fig. 15.21 it is evident that for any load torque, there are two operating points (say B and D). But at position B, the operation is unstable because if there is tendency of speed rise, the developed torque also increases than the load torque causing further rise in speed. Thus at position B the operation is unstable. At position D operation is stable because a tendency to rise in speed will be opposed by the decrease in developed torque. Similarly when there is a tendency to fall in speed, there will be increase in developed torque to bring the motor to operating position D. Thus the region AC of the torque-speed characteristic is *unstable region* of operation while the region CF is stable region of operation.

At full load, the motor runs at a speed of N rpm. If the mechanical load applied to the motor is increased, motor speed drops till the torque developed by the motor equals the load torque. As long as the two torques are in balance, the motor will run at constant speed. However, if the load torque exceeds the breakdown torque of the motor, the motor will immediately stop.

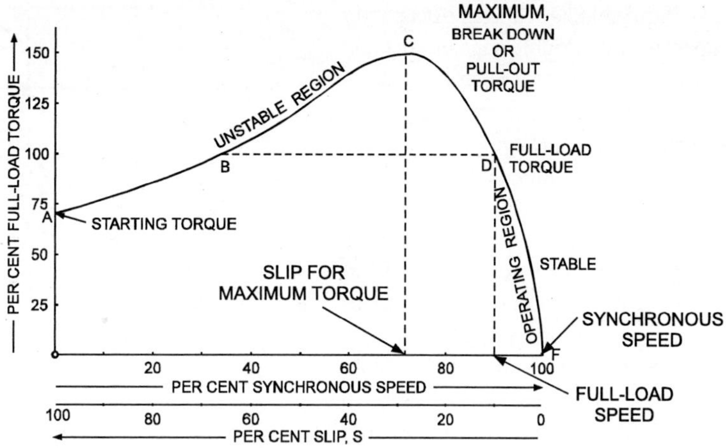

Fig. 15.21

The value of breakdown torque varies with the design of the motor but ranges from 200 to 300 per cent of full-load torque in standard squirrel cage motors.

The torque-speed characteristics of an induction motor in the operating region are quite similar to those of a dc shunt motor. Even at full load, the drop in speed from no load does not exceed 5 per cent.

15.14. STARTING OF 3-PHASE SQUIRREL CAGE INDUCTION MOTORS

Small induction motors up to 2 kW capacity can be directly put on the supply mains without any disturbance to the supply system. Since at the instant of starting, 3-phase induction motor acts as a polyphase transformer with a short-circuited rotating secondary, if motor of higher capacity is connected directly to the mains, then it will draw heavy current from the supply mains, which will cause large voltage drop in the distribution network and thus affect the operation of other electrical equipment connected to the same distribution network.

Therefore, to control the starting current, some starting device known as *starter* must be used. The principle of all types of starters is to impress the lower voltage across stator winding at the starting instant.

There are several methods of starting squirrel cage induction motors. The most common methods are described below.

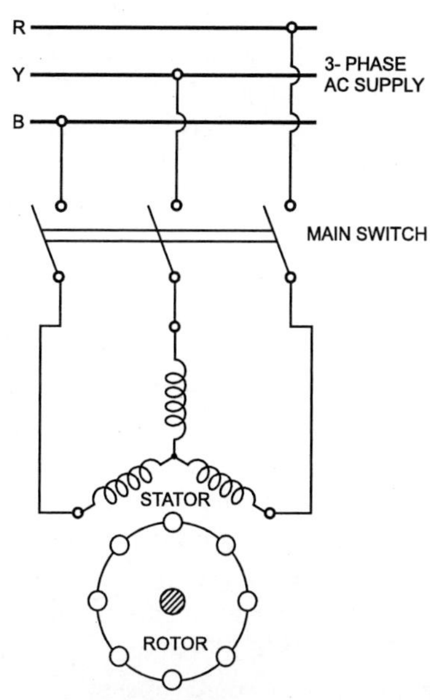

Fig. 15.22

1. Direct-on -Line Starting Method (or DOL Starter). Although there is no limitation on the size of the motor that may be started by this method, it should be understood that objectionable line voltage drops will usually occur, especially if large motors are started frequently. Whether or not normal voltage is employed will, therefore, depend upon the following factors :

(*i*) the size and design of the motor
(*ii*) the kind of application
(*ii*) the location of the motor in the distribution system and
(*iv*) the capacity of the power system and the rules governing such installations as established by power supply companies.

Squirrel cage motors of capacity up to 1.5 kW, double cage motors and squirrel cage motors of large capacity having a large rotor resistance are started by this method.

The push button type direct-on-line starter, which is very common in use, is shown in Fig. 15.23. It is simple, inexpensive and easy to instal and maintain. It consists of a set of 'start' and 'stop' push buttons, a contactor (an electromagnet) with its associated contacts and usually an overload and undervoltage protection devices. The start button is a momentary contact switch that is held normally open by a spring. The stop button is held normally closed by a spring. When the start button is pressed the operating coil (or main contactor) gets energized through the overload relay contacts OL (normally closed). This closes the three main contacts M that connects the motor to the supply. At the same time a set of auxiliary or maintaining contacts MC are closed. When the maintaining contacts MC close, a new circuit is established through stop button, maintaining contacts MC and operating coil. Since operating coil circuit is now maintained by the auxiliary contacts MC, the start button may be released. When the stop button is pressed, the operating coil gets de-energized, thereby opening all main contacts and auxiliary contacts.

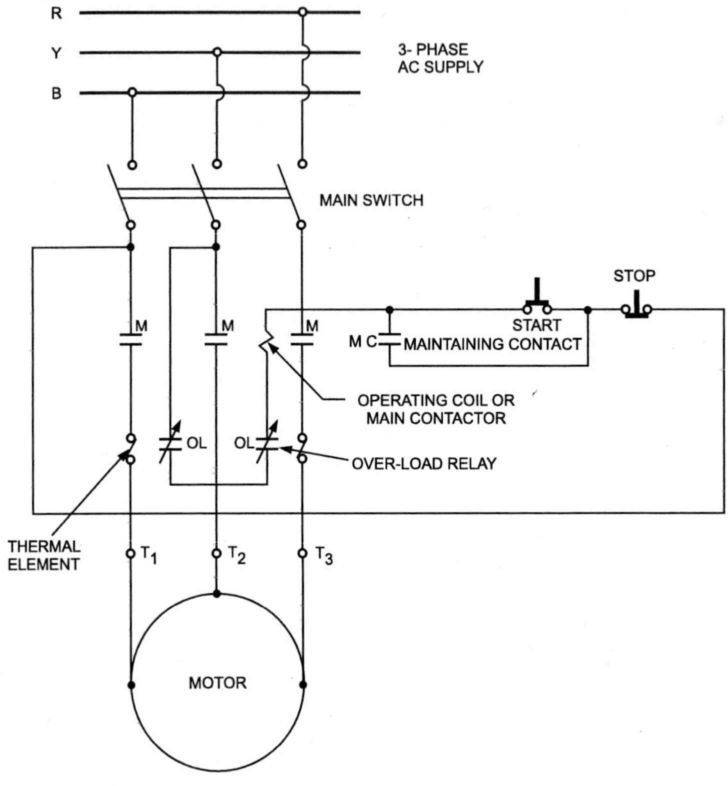

Fig. 15.23 *Wiring Diagram of a Direct-on-Line Starter With Protective Devices*

If the supply fails or line voltage drops below a certain value, the main contacts and the maintaining contacts are both opened. Upon return of the supply, the contactor cannot close until the start button is again closed. Because a contactor that is controlled by a three-wire control circuit maintains the interruption of the circuit even after the supply is restored, it is said to provide *undervoltage protection* for the motor. This protection is employed when it is desired to prevent the unexpected starting of a motor.

Overload protection is employed for protecting the motor and control apparatus from excessive heating due to overloads on motor. Thermal overload relays are commonly used for motor overload protection. Thermal overload relays are of two types. Both types are operated from the heat generated in a thermal element through which the motor current flows. In one type, the heat bends a bimetallic strip, and in the other the heat melts a film of solder. Both act to open the motor control circuit and, therefore, to disconnect the motor from the source of supply.

A bimetallic strip is made of two different metals whose surface has been welded together. One of the metal expands rapidly as compared to the other, when heated. When heat is applied, the strip is caused to deflect or curl up due to unequal expansion of the two metals. When the motor current attains a predetermined value, the heat generated deflects the strip far enough to trip a latch that opens the motor control circuit. When the strip cools down sufficiently, the relay may be reset and the motor re-started.

In the solder-film type of overload relay the heat generated at a given overload current melts the film, which releases a latch arrangement and opens the motor control circuit. When the solder has cooled enough to hold the latch, the relay may be reset.

In either type of the thermal relay, the time required to operate is determined by the magnitude of current flowing in the heater. Thus, the relay operates slowly for light overloads but disconnects the motor in a short time for dangerously heavy overloads. Relays in which the time of operation is inversely proportional to the amount of current flowing are known as the *inverse time relays.*

Torque Developed On Starting of Induction Motor By Direct Switching.

$$\text{Input power to rotor} = T\omega$$

$$\text{Rotor copper loss} = s \times \text{input power to rotor}$$

$$\text{or } 3 I^2 R_2 = s T \omega$$

where I_2 is the rotor current per phase, R_2 is rotor resistance per phase, T is the torque developed, s is the slip and ω is the supply angular velocity.

$$\text{or Torque developed, } T = \frac{3 I^2 R_2}{s \omega}$$

$$\text{i.e. } T \propto \frac{I_2^2}{s} \quad \text{if rotor resistance is constant}$$

Since rotor current is proportional to stator current I_1,

$$T \propto \frac{I_1^2}{s}$$

$$\text{or } T = K \frac{I_1^2}{s} \qquad \text{where K is any constant}$$

$$\text{At starting slip, } s = 1$$

$$\text{So starting torque, } T_{st} = K (I_{st})^2$$

where I_{st} is the starting current

$$\text{Full-load torque, } T_f = \frac{K I_f^2}{s_f}$$

where I_f is the full-load motor current and s_f is the full-load slip

$$\text{So } \frac{T_{st}}{T_f} = \frac{K I_{st}^2}{K I_f^2 / s_f} = \left(\frac{I_{st}}{I_f}\right)^2 s_f$$

$$\text{or } T_{st} = T_f \times \left(\frac{I_{st}}{I_f}\right)^2 s_f$$

When the motor is connected directly across the supply mains, the starting current is equal to the short-circuit current. I_{SC}

$$\therefore \text{Starting torque, } T_{st} = T_f \left(\frac{I_{SC}}{I_f}\right) s_f \qquad (15.18)$$

2. Primary Resistor (or Reactor) Starting. This method of starting of a polyphase cage motor is very simple and provides smooth acceleration of the motor. In this method of starting of a 3-phase squirrel cage induction motor reduced voltage is obtained by means of resistors (or reactors) that are connected in series with each stator lead, as shown in Fig. 15.24, during the starting period. The voltage drop in resistors (or reactors) causes a reduced voltage across the motor terminals. As the motor picks up the speed, the resistors (or reactors) are cut out in steps and finally short circuited when the motor attains the operating speed.

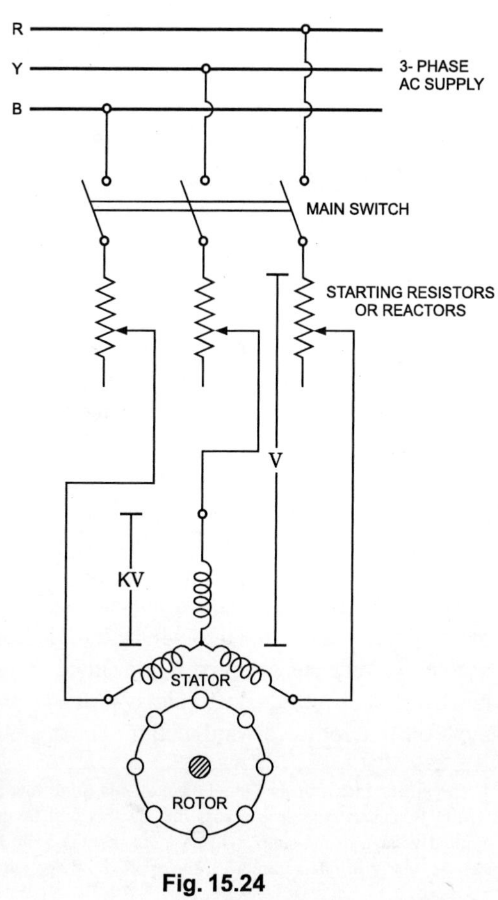

Fig. 15.24

Although the initial cost of reactors is high in comparison to that of resistors, reactor starting is preferred because this method incurs small power losses and is more effective* in reducing the voltage applied to the stator at starting.

The advantages and disadvantages of this method of starting of cage motors are given below:

Advantages. 1. Smooth acceleration.

2. High power factor during start.
3. Less expensive than auto-transformer starter in lower output ratings.
4. Closed transition starting.
5. Available with as many as 7 accelerating points.

Disadvantages 1. Resistors give off heat.

2. Low torque efficiency.
3. Starting duration usually exceeds 5 seconds, so needs expensive resistors.
4. Starting voltage is difficult to adjust to meet varying conditions.

If the normal supply phase voltage = V volts and by using line resistance starter the voltage is reduced to KV volts, then starting current is also reduced in the same ratio *i.e.* starting current $I_{st} = KI_{SC}$ where I_{SC} is the short-circuit current.

$$\text{Starting torque, } T_{st} = T_f \times \left(\frac{I_{st}}{I_f}\right)^2 s_f$$

$$= T_f \times K^2 \left(\frac{I_{SC}}{I_f}\right)^2 s_f \qquad \text{...(15.19)}$$

$= K^2 \times$ Torque obtained by switching the motor directly.

3. Auto-Transformer Starting. In auto-transformer starting method the reduced voltage is obtained by taking tappings at suitable points from a three phase auto-transformer, as shown in Fig. 15.25. The auto-transformers are generally tapped at the 50, 60 and 80 per cent points, so that adjustment at these voltages may be made for proper starting torque requirements. Since the contacts frequently break large values of current, arcing is sometimes quenched effectively by having them assembled to operate in an oil bath.

Auto-transformer starters may be either manually or magneti-cally operated.

The manual auto-transformer starter is essentially a multi-pole double-throw switch. It consists of three sets of contacts starting, running (both stationary) and the movable contacts (the contacts attached to the operating handle). The operating handle is spring loaded lever mounted on the outside of a steel cabinet to stand vertically in the "off" position. When the operating handle is moved to the start position, the movable contacts are moved against the starting contacts. This energizes the star-connected auto-transformer and impresses reduced voltage (50, 60 or 80 per cent according to the setting on auto-transformer) across the motor. After the motor has accelerated to about full speed, the operating handle is moved to the run position. This instantly opens the starting contacts (or disconnects the auto-transformer from the line) and connects the motor directly to the line through the running contacts. A latch, fixed to the mechanism, is made to drop into a notch so that the operator is prevented from throwing the handle accidently to the run position first, however, when the handle is quickly pushed from the start to the 'run' position, the latch is kicked up to make the lever free for its forward motion. The operating handle is held in the run position by an electromagnet or

* The power factor of an induction motor is quite low at start and if reactors are used for reducing voltage across the motor terminals, the voltage drop across the reactors will be nearly in phase with the supply voltage and so the net reduction in the applied voltage to the stator will be more. On the other hand in case of resistance starting, the voltage drop across the resistors will be nearly in quadrature with the applied voltage and so the net reduction in the applied voltage to the stator terminals will be less.

undervoltage (latch) coil until the stop button is pressed. If the supply fails or supply voltage drops to low value, the electromagnet will release and trip the holding mechanism. Overload protection is provided by thermal overload relays.

Let the motor be started by an auto-transformer having transformation ratio, K. If I_{SC} is the starting current when normal voltage is applied, and applied voltage to stator winding at starting = KV

$$\text{then motor input current, } I_{st} = K\, I_{SC} \quad \text{....(15.20)}$$

$$\text{Supply current} = \text{Primary current of auto-transformer}$$

$$= K \times \text{secondary current of auto-transformer} = K^2 I_{SC}$$

$$\text{Starting torque, } T_{st} = T_f \left(\frac{I_{st}}{I_f}\right)^2 \times s_f = T_f K^2 \left(\frac{I_{SC}}{I_f}\right)^2 \times s_f \quad \text{...(15.21)}$$

$$= K^2 \times \text{Torque obtained by direct switching.}$$

Hence the line current and the starting torque are reduced in the square ratio.

The advantages of this method of starting of cage induction motors lies in the fact that the voltage is reduced by transformation and not by dropping the voltage in the resistors (or reactors) and, therefore, the current and power drawn from the supply mains are also reduced in comparison to primary resistor (or reactor) starting. The internal losses of the starter itself are small during long starting periods. Other advantages of this method are (*i*) availability of highest torque per ampere of supply current (*ii*) adjustment of starting voltage by selection of proper tap on the auto-transformer (*iii*) suitability for long starting periods (*iv*) closed transition starting and (*v*) motor current larger than supply current.

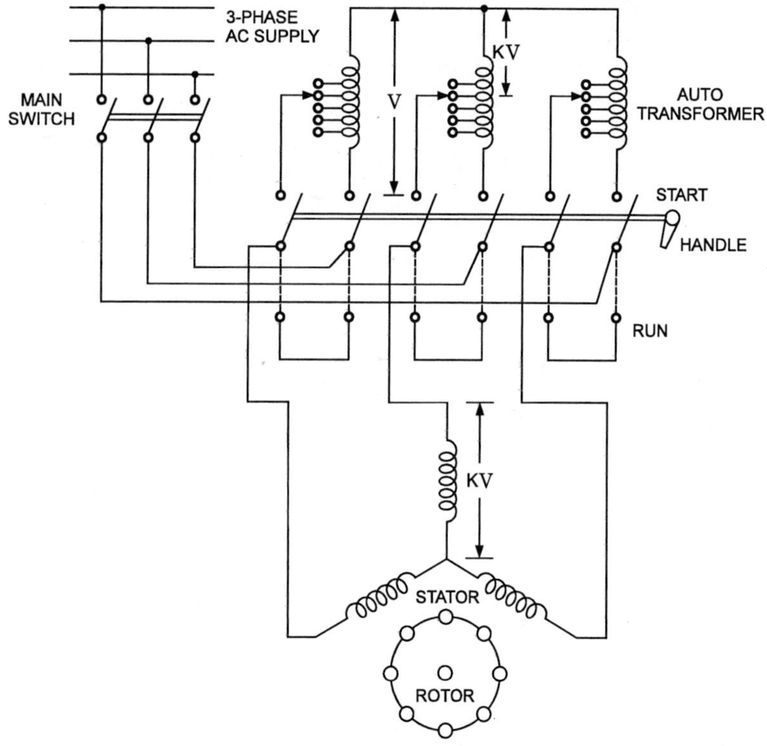

Fig. 15.25 *Auto-Transformer Starter*

The drawbacks of such starters are low power factor and higher cost in case of lower output rating motors.

This method can be employed for starting of star-connected as well as delta-connected motors.

For starting of large cage motors (of output rating exceeding 20 kW) this method of starting is often used.

4. Star-Delta Starting. This method of starting of cage induction motors is based upon the principle that with 3 windings connected in star, the voltage across each winding is $1/\sqrt{3}$ *i.e.* 57.7 % of the line to line voltage whereas the same winding connected in delta will have full line-to-line voltage across each.

The star-delta starter connects the three stator windings in star across the rated supply voltage at the starting instant. After the motor attains speed the same windings, through a change-over switch, are reconnected in delta across the same supply voltage.

The basic diagram of connection is shown in Fig. 15.26. Air break star-delta starter with single phasing preventive and overload relay is shown in Fig. 15.27. The starter is also provided with a mechanical interlocking device to prevent the handle from being put in the "Run" position first.

Before putting the starter into service, the starter is mounted in vertical position and magnet armature, which is tied up to avoid damage in transit, is released.

For starting the handle is moved to START position and quickly to RUN position after motor has accelerated.

For stopping of motor, STOP push-button is used.

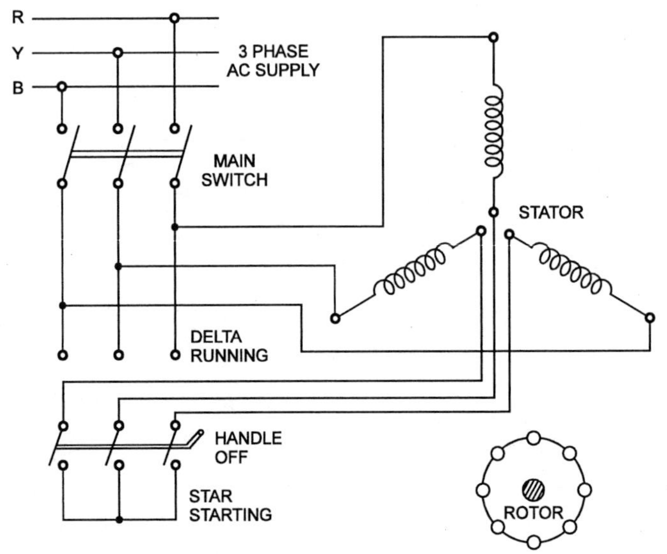

Fig. 15.26 *Star-Delta Starter*

The load indicator may be set at full-load rated current of motor when thermal overload trips will be operated at approximately 20 to 25 % overload.

Since at starting instant, the stator windings are connected in star, voltage across each phase winding is reduced to $1/\sqrt{3}$ of line voltage and, therefore, starting current per phase becomes equal to $I_{SC}/\sqrt{3}$.

Starting line current by connecting the stator windings in star at the starting instant

$$= \text{Starting motor current per phase} = I_{SC}/\sqrt{3}.$$

Starting line current by direct switching with stator windings connected in delta = $\sqrt{3}\ I_{SC}$.

$$\therefore \quad \frac{\text{Line current with star-delta starting}}{\text{Line current with direct switching}} = \frac{I_{SC}/\sqrt{3}}{\sqrt{3}\, I_{SC}} = \frac{1}{3}$$

Hence by star-delta starting line current is reduced to one-third of line current with direct switching.

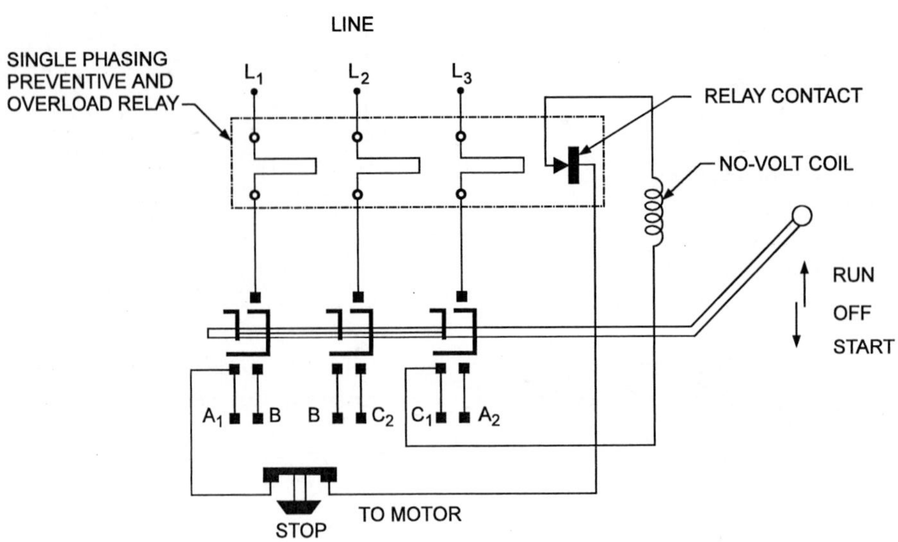

Fig. 15.27 *Air Break Star-Delta Starter With Single Phasing Preventive and Overload Relay*

$$\text{Starting torque, } T_{st} = T_f\left(\frac{I_{st}}{I_f}\right)s_f = T_f\left[\frac{I_{SC}/\sqrt{3}}{I_f}\right]^2 s_f = \frac{1}{3}T_f\left(\frac{I_{SC}}{I_f}\right)^2 s_f = \frac{1}{3} \times \text{torque obtained by direct switching} \quad ...(15.22)$$

Hence with star-delta switching, the starting torque is also reduced to one-third of starting torque obtained with direct switching.

This method of starting of cage motors is simple, cheap, effective and efficient since no power is lost in auxiliary components. This method is also suitable for high inertia and long acceleration loads.

This method needs a motor to be delta-connected for normal operation and all the six terminals of the 3-phase stator windings are to be brought out. The reduction in voltage is fixed and starting torque is also low. So this method is limited to application where high starting torque is not the essential requirement e.g. machine tools, pumps, motor-generator sets etc. This method is unsuitable for line voltage exceeding 3,000 V, because of excess number of stator turns required for delta connection.

Such starters are employed for starting 3-phase squirrel cage induction motors of rating between 4 and 20 kW.

Precaution With Star-Delta Starting. The initial current flowing when the motor is started in star is 57.7 % of the short-circuit current in delta together with a transient in each phase. The transient currents decay rapidly but the steady state is not reached until the motor has attained 70 per cent of its synchronous speed. The change-over from star to delta connection should not be made until the motor attains about 90 per cent of synchronous speed, otherwise there will be a current surge considerably greater than full-load current which may even be greater than the standstill current with star connection.

Example 15.17 A small 3-ϕ induction motor has a short-circuit current equal to 4 times the full-load current. Determine the starting torque as a percentage of full-load torque if full-load slip is 2.5 %.

Solution: Short-circuit current, $I_{SC} = 4\ I_f$

Full-load slip, $s_f = 0.025$

Starting torque, $T_{st} = T_f\left(\frac{I_{SC}}{I_f}\right)^2 s_f = T_f \times (4)^2 \times 0.025 = 0.4\ T_f$ or 40 % of full-load torque. **Ans.**

Example 15.18 A small 3-phase induction motor has a short-circuit current 5 times of full-load current and full-load slip 5 %. Determine the starting torque and starting current if starting resistance starter is used to reduce the impressed voltage to 60 % of normal voltage.

Solution: Starting current, $I_{st} = 0.6 \times 5\ I_f = 3\ I_f$ **Ans.**

Starting torque, $T_{st} = T_f\left(\frac{I_{st}}{I_f}\right)^2 \times s_f = T_f \times (3)^2 \times 0.05 = 0.45\ T_f$ or 45 % of full-load torque. **Ans.**

Example 15.19 A 3-phase induction motor takes a starting current which is 5 times full-load current at normal voltage. Its full-load slip is 4 per cent. What auto-transformer ratio would enable the motor to be started with not more than twice the full-load current drawn from the supply ? What would be the starting torque under this condition ?

Solution: Let the transformation ratio of auto-transformer be K

Now the starting current, $I_{st} = K^2\ I_{SC}$

or $\frac{I_{st}}{I_f} = K^2 \frac{I_{SC}}{I_f}$

or $2 = K^2 \times 5$ $\quad \because I_{SC} = 5\ I_f$ and $I_{st} = 2\ I_f$

or transformation ratio, $K = \sqrt{\frac{2}{5}} = 0.6325$ or 63.25 % **Ans.**

$$\text{Starting torque, } T_{st} = K^2\left(\frac{I_{SC}}{I_f}\right)^2 s_f\, T_f = 0.6325^2 \times 5^2 \times 0.04\, T_f$$

$$= 0.4\, T_f \text{ or } 40\% \text{ of full-load torque } \textbf{Ans.}$$

Example 15.20 **Determine approximately the starting torque of a 3-phase induction motor in terms of full-load torque when started by (*i*) star-delta starter (*ii*) an auto-transformer starter with 50 per cent tapping. The short-circuit current of motor is 5 times the full-load current and the full-load slip is 5 per cent.**

Solution:

Short circuit current, $I_{SC} = 5\, I_f$

Full-load slip, s = 5 per cent or 0.05

$$(i) \text{ Starting torque with star-delta starter} = \frac{1}{3}\left(\frac{I_{SC}}{I_f}\right)^2 \times s_f\, T_f = \frac{1}{3}(5)^2 \times 0.05 = 0.4167\, T_f$$

or 41.67% of full-load torque **Ans.**

(*ii*) Starting torque with an auto-transformer starter with 50 % tapping (*i.e.* k = 0.5)

$$= k^2\, \frac{1}{3}(5)^2 \times 0.05 = (0.5)^2 \times (5)^2 \times 0.05\, T_f = 0.3125\, T_f$$

or 31.25 % of full-load torque. **Ans.**

15.15. STARTING OF 3-PHASE SLIP-RING INDUCTION MOTORS

Though the methods described in the preceding section can also be employed for starting of 3-phase slip-ring induction motors but this is not done in general because then the advantages of wound rotor cannot be realized fully. *Rotor resistance starter* cannot be employed in case of squirrel cage type motors because external resistance cannot be introduced in the cage rotor.

Slip-ring induction motors are usually started with full line voltage across the stator terminals and by introducing variable resistance in each phase of the rotor circuit. The external resistance introduced in each phase of the rotor circuit not only reduces the current at the starting instant but increases the starting torque also due to improvement in power factor. The external resistance at the starting instant is introduced in the rotor circuit by taking the rotor winding terminals out to the slip rings mounted on the shaft with brushes resting on them, as shown in Fig. 15.8. As the motor accelerates, the external resistance is cut out in steps so that the available electromagnetic torque remains maximum during the accelerating period. Ultimately when the motor attains the normal speed, the rotor winding is short circuited automatically through the slip-rings by means of a metal collar which is pushed along the shaft under the slip rings touching them all there by eliminating the resistance of brush contacts on the slip rings. The brushes are then lifted to reduce the wear and frictional losses.

For small and medium size motors face-plate starters are employed but for large size motors liquid starters are employed.

A starter is also provided with no-voltage (or under voltage) and overcurrent protective devices. It is also provided with some type of interlocking to ensure proper sequential operation of the line contactor and starter. This interlocking arrangement prevents the closing of stator contactor unless the starter is 'all in'.

Since slip-ring induction motor develops high starting torque due to introduction of external resistance in the rotor circuit at the starting instant with reasonable starting current, such motors can be started under load also.

15.16. ADVANTAGES, DISADVANTAGES AND APPLICATIONS OF INDUCTION MOTORS

Squirrel cage induction motors are comparatively *cheaper* than wound rotor induction motors and are *simple* and *rugged* in construction.

A cage rotor needs considerably less conductor material than a wound rotor, so copper loss in cage rotor is low and *efficiency* of a cage motor is a *little more* than that of a wound rotor motor.

Cage motor is more rugged in construction and needs no slip rings, brushes etc., therefore, its *maintenance cost is low.*

Cage motor is *explosion-proof,* because the absence of slip rings and brushes eliminates risk of sparking.

Cage motor can be *cooled better* because of its bare end rings. Also there is more space for rotor fans.

A squirrel cage induction motor has very small length of overhang, so it has low rotor overhang leakage flux. This causes standstill reactance X_2 low for a cage rotor than wound rotor. So a cage motor has *more pull-out torque, greater maximum power output* and *better operating power factor* as compared to a wound rotor motor.

Other advantages of squirrel cage motors are (*i*) nearly constant speed, high overload capacity and simple starting arrangement which may be made automatic.

The disadvantages of squirrel cage induction motors are enumerated below :

(*i*) Since in cage rotors, rotor bars are permanently short circuited, it is not possible to insert any external resistance in the rotor circuit for the purpose of increasing starting torque. Consequently its *starting torque is low* with *large starting current :* — starting torque varying between 1.5 and 2 times the full-load torque with starting current 5 to 9 times the full-load current. *Power factor* is also *low* at start in case of cage motors.

(*ii*) No possibility of speed control.

(*iii*) Squirrel cage induction motors are very sensitive to fluctuations in supply voltage.

(*iv*) The total energy lost during starting of cage motor is much more than with the slip-ring induction motor and this fact is very important where frequent starting is required.

Squirrel cage induction motor is suitable for constant speed industrial drives of small power where speed control is not required and where starting torque requirements are of medium or low value, such as for printing machinery, flour mills and other shaft drives of small power.

When the squirrel cage motors are used for crane or hoist work where large starting torque is of more importance than high efficiency owing to the intermittent nature of load, the rotor resistance is increased by employing the end rings, and sometimes the bars as well, of high resistance metal such as German silver.

Advantages of wound rotor (or slip-ring) induction motors are given below :

(*i*) *High starting torque* with low starting current. By inserting an external resistance of proper value (so that rotor resistance R_2 becomes equal to its standstill reactance X_2) in each phase of the rotor circuit, maximum torque can be obtained at start.

(*ii*) *Adjustable speed.* Speed can be controlled easily in case of a wound rotor motor simply by adding external resistors in the rotor circuit. This is the reason that wound rotor induction motors are considered to be variable speed motors.

Other advantages are high overload capacity, nearly constant speed, smooth acceleration under heavy loads, no abnormal heating during starting, adoptability to automatic starting and good running characteristics after the external resistances are taken out of the rotor circuit.

The disadvantages of wound rotor induction motors are :

(*i*) Initial and maintenance costs are more than those of a squirrel cage motor, because a slip-ring induction motor needs slip rings, brushes, short-circuiting devices etc.

(*ii*) Speed regulation is poor when operated with external resistances in the rotor circuit.

(*iii*) Lower efficiency and power factor as compared with cage rotor motor.

(*iv*) Low power factor at light loads.

(*v*) Sensitivity to fluctuations in supply voltage.

Wound rotor (or slip-ring) induction motors are used for loads requiring severe starting conditions or for loads requiring speed control such as for driving line shafts, lifts, pumps, generators, winding machines, cranes, hoists, elevators, compressors, small electric excavators, printing presses, turn tables, strokers, large ventilating fans, crushers etc.

15.17. COMPARISON OF SQUIRREL CAGE AND SLIP-RING (PHASE WOUND) INDUCTION MOTORS

S.No.	*Characteristics*	*Squirrel Cage Induction Motor*	*Slip-Ring Induction Motor*
1.	Construction	Simple and rugged	Needs slip rings, brushes, short-circuiting device etc.
2.	Overhang	Less	Large
3.	Space factor in slots	Better	Poor
4.	Cost (initial as well as maintenance)	Less	More
5.	Copper losses	Small	More
6.	Efficiency	High (only for machines, not designed for high starting torque)	Low
7.	Power factor	Better	Poor
8.	Cooling	Cooling better because of its bare end rings and availability of more space for rotor fans	Not quite efficient
9.	Speed regulation	Better	Poor when operated with external resistances in the rotor circuit
10.	Starting	Simple	The motor needs slip rings, brushgear, short-circuiting device and starting resistors etc.
11.	Starting torque	Starting torque is low with large starting current	Possibility of increasing starting torque by insertion of external resistances in the rotor circuit
12.	Power factor at start	Poor	Can be improved
13.	Speed control	No possibility	Possible by insertion of external resistor in the rotor circuit
14.	Protection against explosion	Explosion-proof	Not explosion-proof

HIGHLIGHTS

1. An induction motor is simply a transformer whose magnetic circuit is separated by an air gap into two relatively movable portions, one carrying the primary winding and the other the secondary winding.

It essentially consists of (*i*) a stator and (*ii*) a rotor.

Rotors are of two types (*i*) squirrel-cage rotor and (*ii*) phase-wound or wound rotor.

2. An alternating field can be replaced exactly by two rotating fields of half its amplitude travelling in opposite directions at synchronous speed. This is the Ferraris principle.

3. When a 3-ϕ supply is given to a 3-ϕ winding, a resultant field is set up which rotates in space at a constant speed called the synchronous speed given as

$$N_s = \frac{120 f}{P}$$

4. In a polyphase induction motor, slip-frequency currents are induced in the rotor windings as the rotor slips past the synchronously rotating stator flux wave. These rotor currents in turn produce a flux wave which rotates in synchronism with the stator flux wave, the torque is produced from the interaction of these two flux waves. For increased load on the motor, the rotor speed falls, resulting in larger slip and increased induced rotor currents, and greater torque.

Examination of the flux-mmf interactions in a polyphase induction motor shows that, electrically, the machine is a generalized transformer. The synchronously rotating air-gap flux wave in the induction machine is the counterpart of the mutual core flux in the transformer. The rotating field induces emfs of stator frequency in stator windings and of slip frequency in the rotor windings for all rotor speeds other than synchronous speed. Thus, the induction machine transforms voltages and at the same time changes frequency. When viewed from the stator, all rotor electrical and magnetic phenomena are transformed to stator frequency. The rotor mmf reacts on the stator windings in the same way that the mmf of secondary current in a transformer reacts on the primary.

5. *Working principle.* When a 3-ϕ supply is provided to a 3-phase wound stator of an induction motor, a rotating magnetic field is established rotating at synchronous speed. At start, stationary rotor conductors cut across the revolving magnetic field and an emf is induced in them by the electromagnetic phenomenon. Current flows through the rotor conductors, as they form a closed path, and so rotor field is developed. By interaction of stator magnetic field and rotor field, torque develops and causes the rotation of rotor in the same direction as that of revolving magnetic field.
6. The direction of rotation of a 3-phase induction motor can be reversed by interchanging the connections to the supply of any two terminals.
7. *Synchronous speed,* $N_s = \dfrac{120\,f}{P}$, f is supply frequency and P is the number of poles on stator.
8. **Slip.** The rotor never attains the synchronous speed because at that speed there would be no relative speed between rotor conductors and stator revolving field and induction phenomenon is not possible. Its speed is always less than synchronous speed.

   The difference between the speed of the stator field, known as synchronous speed (N_s), and the actual speed of the rotor (N) is known as the slip and is denoted by s

$$s = \frac{N_s - N}{N_s}.$$

9. Both the stator and rotor fields rotate synchronously, which means that they are stationary w.r.t. each other at all possible rotor speeds.
10. Frequency of rotor current or emf, $f' = sf$.
11. Rotor emf at start, $E_2 = E_1 \times \dfrac{N_2}{N_1}$.
12. Rotor current at start, $I_2 = \dfrac{E_2}{\sqrt{R_2^2 + X_2^2}}$.

    Rotor current at slip s, $I_2' = \dfrac{sE_2}{\sqrt{R_2^2 + (sX_2)^2}}$.
13. Starting torque, $T_{st} = \dfrac{KR_2E^2}{R_2^2 + X_2^2} = \dfrac{K_1R_2}{R_2^2 + X_2^2}$, if applied voltage is constant.

    Torque under running condition, $T = \dfrac{K_1 sR_2E_2^2}{R_2^2 + s^2X_2^2} = \dfrac{K_2 sR_2}{R_2^2 + s^2X_2^2}$, if applied voltage is constant.
14. Condition for maximum torque under running condition

$$R_2 = s\,X_2.$$

15. Slip corresponding to maximum torque, $s = \dfrac{R_2}{X_2}$.
16. Condition for maximum starting torque, $R_2 = X_2$.
17. Maximum torque, $T_{max} \propto \dfrac{1}{X_2}$.

18. One of the salient facts affecting induction motor applications is that the slip at which maximum torque occurs can be controlled by varying the rotor resistance. A high rotor resistance gives optimum starting conditions but poor running performance. A low rotor resistance, on the other hand, may result in unsatisfactory starting conditions.

19. Starting torque of an induction motor is proportional to the square of applied voltage.

20. When the induction motor is running with slip *s*, the torque acting on the rotor,

$$T \propto s \times (\text{supply voltage})^2 \propto s\, V^2.$$

21. The rotor current of a 3-phase induction motor reduces slightly with the slight increase in speed but with large increase in speed the current drops rapidly (almost nearly) until at synchronous speed the rotor emf and rotor current are both zero. This is because at high speed, the slip *s* is low and so the rotor reactance (sX_2) becomes small as compared to rotor resistance R_2 and the rotor current becomes proportional to rotor induced emf which is proportional to slip.

22. The value of breakdown torque varies with the design of motor but ranges from 200 to 300% of full-load torque in standard squirrel cage motors.

23. The starter is employed for starting 3-phase induction motor in order to (i) limit the initial rush of current to a predetermined value at the starting instant and (ii) provide overload and undervoltage protection.

24. The most commonly used starters for cage motors are :

(*a*) Direct-on-line (D-O-L) starter.

(*b*) Primary resistor or reactor starter.

(*c*) Auto-transformer starter.

(*d*) Star-delta starter.

(*a*) **Direct-on-line starting** is the cheapest method of starting of a 3-phase induction motor. Squirrel cage motors of capacity up to 1.5 kW capacity can be started by this method.

In D-O-L starting Starting current, $I_{st} = I_{sc}$

$$\text{and starting torque, } T_{st} = T_f \left(\frac{I_{sc}}{I_f}\right)^2 s_f$$

(*b*) **Primary resistor or reactor starting method** is very simple and provides smooth acceleration of the motor. In this method reduced voltage is obtained by means of resistors (or reactors) connected in series with each stator lead. The drawbacks of this method are requirement of expensive resistors and low torque efficiency.

$$\text{Starting current, } I_{st} = K\, I_{sc}$$

$$\text{Starting torque, } T_{st} = K^2\, T_f \left(\frac{I_{sc}}{I_f}\right)^2 s_f = K^2 T_f$$

(*c*) **Auto-transformer starting** is often employed for starting of large cage motors (of output rating exceeding 20 kW). In this method, the reduced voltage is obtained by taking tappings at suitable points from a 3-phase auto-transformer. The advantages of auto-transformer starting are (*i*) availability of highest torque per ampere of supply current (*ii*) suitability for long starting periods (*iii*) closed transition starting and (*iv*) motor current larger than supply current. The drawbacks of this method are poor power factor and higher cost in case of lower output rating motors.

$$\text{Starting current, } I_{st} = K^2\, I_{sc}$$

$$\text{Starting torque, } T_{st} = K^2\, T_f \left(\frac{I_{sc}}{I_f}\right)^2 \times s_f = K^2 \times \text{Torque obtained by direct switching}$$

(*d*) **Star-delta starting method** is simple, cheap, effective and efficient method of starting cage motors as no power is lost in auxiliary equipment. The star-delta starter connects the three stator windings in star across the rated supply voltage at the starting instant. After attainment of normal speed, the same windings are re-connected in delta. Star-delta starter reduces line current at starting and starting torque to one-third of those with direct switching.

25. Slip-ring induction motors are usually started with full-line voltage across the stator terminals and by introducing variable resistor in each phase of the rotor circuit. The external resistances introduced in each phase of the rotor circuit not only reduces the current at the starting instant but increases the starting torque also due to improvement in *pf.*

26. For applications requiring a substantially constant speed without excessive severe starting conditions, the squirrel cage induction motor usually is unrivaled because of its ruggedness, simplicity, and relatively low cost. Its only disadvantage is its relatively low pf (about 0.85 to 0.90 at full load and considerably lower at light loads and for lower-speed motors). A wound-rotor motor can be used for very severe starting conditions or when speed control by rotor resistance is required. A wound-rotor motor is more expensive than a squirrel cage motor.

EXERCISES

1. Discuss the constructional details and working principle of 3-phase induction motor.
[U.P. Technical Univ. Electrical Engineering 2013-14]

2. Explain the principle of operation of three-phase induction motor.
[U.P.Technical Univ. Elec. Engineering Second Semester 2005-06]

3. Explain with the help of diagrams, how a rotating field is produced in the air gap of a three-phase induction motor.
[G.B. Technical Univ. Electrical Engineering Even Semester 2012-13]

4. Explain why 3-ϕ induction motor cannot run at synchronous speed $\left(N_s = \dfrac{120f}{P}\right)$, where symbols having their usual meaning. **[U.P. Technical Univ. Electrical Engineering Odd Semester 2013-14]**

5. Show that the frequency of the rotor induced emf in an induction motor is slip times its stator frequency.
[Mahamaya Technical Univ. Electrical Engineering Second Semester 2011-12

6. Derive an expression for the frequency of rotor current in a three-phase induction motor.
[U.P. Technical Univ. Elec. Engineering Second Semester 2003-04]

7. Draw and explain the slip-torque characteristic of 3-ϕ induction motor and mention the various operating regions on same. **[U.P. Technical Univ. Electrical Engineering Odd Semester 2013-14]**

8. Define slip in 3-phase induction motor. What is its value at starting and at the synchronous speed? Draw torque-slip characteristics of 3-phase induction motor. **[U.P. Technical Univ. Elec. Engineering Second Semester 2005-06]**

9. Induction motor cannot operate between slip 1 and slip at maximum torque. Explain it.
[U.P. Technical Univ. Electrical Engineering Second Semester 2004-05]

10. (*i*) Draw torque-slip characteristics of a 3-phase induction motor indicating therein the starting torque, the maximum torque and the operating region.
(*ii*) Discuss the effect of variation of rotor resistance and supply voltage on the torque-slip characteristics.
(*iii*) State the operating conditions of the motor when the slip is either positive, greater than one or negative.
[U.P. Technical Univ. Electrical Engineering First Semester 2005-06]

11. Draw torque-speed characteristics of a 3-phase induction motor. Show the different operating regions. What will happen if rotor resistance of motor changes ? **[U.P. Technical Univ. Electrical Engineering Second Semester 2006-07]**

12. What is the torque-speed characteristic of a three-phase squirrel-cage induction motor ? If it is driving a constant torque load, show the running speed on the torque-speed curve. Mark salient regions on the torque-speed characteristics.
[U.P. Technical Univ. Elec. Engineering First Semester 2003-04]

13. Explain the slip-torque characteristics of three-phase induction motor.
[G.B. Technical Univ. Electrical Engineering Second Semester 2010-11]

14. Draw slip-torque characteristics of a three-phase induction motor and explain its various regions of operation.
[G.B. Technical Univ. Electrical Engineering First Semester 2009-10]

15. Draw torque-slip characteristics of a 3-phase induction motor indicating therein the starting torque, the maximum torque and the operating region. **[U.P. Technical Univ. Elec. Engineering First Semester 2005-06]**

16. Explain working of 3-phase induction motor. Also draw torque-slip characteristic showing operating regions.
[U.P. Technical Univ. Basic Electrical Engineering Odd Semester 2013-14]

17. Name various methods of starting the polyphase induction motors and describe auto-transformer method of starting, in detail. **[U.P. Technical Univ. 2002-03]**
18. Why starters are necessary for starting of three-phase induction motors. Describe with the help of diagram, the rotor resistance starter for 3-phase slip-ring induction motor. Mention its merits in comparison to other types of starters. **[U.P. Technical Univ. Elec. Engineering First Semester 2005-06]**
19. What are the different types of induction motors ? Enumerate the merits and demerits of each of these. **[U.P. Technical Univ. Elec. Engineering Second Semester 2003-04]**
20. Differentiate the constructional details of squirrel cage and slip ring (wound rotor) 3-phase induction motors and also give their relative merits and demerits. **[U.P. Technical Univ. Elec. Engineering First Semester 2004-05]**
21. What is the use of starter? Explain various methods of starting of 3-phase induction motors. **[G.B. Technical Univ. Electrical Engineering Odd Semester 2013-14]**
22. Write short notes on the following:
    (*i*) Rotating magnetic field in a 3-phase induction motor.
    (*ii*) Induction motor as a transformer.
    (*iii*) Direct-on-line starter
    (*iv*) Star-delta starter

SHORT ANSWER TYPE QUESTIONS WITH ANSWERS

Q. 1. What is meant by an induction motor ?

Ans. An induction motor derives its name from the fact that the current in the rotor conductors is induced by the motion of rotor conductors relative to the magnetic field developed by the stator currents.

Q. 2. What are the two types of 3-phase induction motors ? Which type is generally preferred ?

Ans. Two types of 3-phase induction motors are (*i*) squirrel cage and (*ii*) wound rotor or slip-ring induction motors. Squirrel cage induction motor is generally preferred.

Q. 3. Why are induction motors called ''asynchronous'' ?

Ans. Because the rotor does not turn in synchronism with the rotating field developed by the stator currents.

Q. 4. Explain why the number of poles of the stator and of the rotor of an ac motor should be equal. **[U.P. Technical Univ. 2001-02]**

Ans. The flux set up by the stator windings is a rotating flux with a definite number of poles; operation is only possible when the number of poles is same on both (stator and rotor).

Q. 5. Which type of induction motor develops higher starting torque ?

Ans. Slip-ring induction motor.

Q. 6. What is the necessity of short-circuited rotor conductors in a squirrel cage induction motor ?

Ans. If the rotor conductors in a squirrel cage motor are not short circuited, the rotor winding will be open circuited and when the stator is connected to 3-ϕ ac supply, the rotating magnetic field will be set up in the air gap and an emf will be induced in the rotor but no current will flow through the rotor conductors and, therefore, no torque will be developed and the rotor will remain stationary.

Q. 7. Explain what is rotating magnetic field and how this is made use of in electrical machinery ? **[P.T.U. December 2001]**

Ans. A rotating magnetic field is that which is constant in magnitude but whose axis of direction rotates in space as field system of a dc machine. When a 3-phase winding is energized by 3-phase supply, a rotating magnetic field is developed that tends the rotor of the machine to move in the direction of rotation.

The operation of ac rotating machines (such as synchronous machines and induction machines) is based on the production of rotating magnetic field.

Q. 8. How can the direction of rotation of the 3-phase induction motor be reversed ?

Ans. The direction of rotation of a 3-phase induction motor may be reversed by interchanging the connection to the supply of any two leads of the motor.

Q. 9. Give the expression of speed in terms of poles and frequency of supply. **[U.P. Technical Univ. 2002-03]**

Ans. Speed, $N = \dfrac{120 \times \text{frequency of supply}}{\text{Number of poles}}$.

Q. 10. Define slip in 3-phase induction motor. What is its value at starting and at synchronous speed? **[U.P.Technical Univ. Elec. Engineering Second Semester 2005-06; 2008-09]**

Ans. The difference between the speed of the stator field and the actual speed of the rotor expressed as a fraction of synchronous speed is known as slip of an induction motor.

i.e. Fraction slip, $s = \dfrac{N_s - N}{N_s}$

At starting, slip, $s = 1.0$ **Ans.**

At synchronous speed, $s = 0$ **Ans.**

Q. 11. Explain the term "slip" in induction motor. **[G.B.T.U. First Semester 2011-12]**

Ans. The difference between the speed of the stator field and the actual speed of the rotor expressed as a fraction of synchronous speed is known as slip of an induction motor.

i.e. Fraction slip, $s = \dfrac{N_s - N}{N_s}$.

Q. 12. Define the term "slip". **[G.B.T.U. Second Semester 2011-12; U.P. Technical Univ. Second Semester 2013-14]**

Ans. Refer to Answer of Question 11.

Q. 13. State the operating conditions of the motor when the slip is either positive greater than one or negative. **[U.P. Technical Univ. Elec. Engineering First Semester 2005-06; 2008-09]**

Ans. When the slip is positive greater than 1, the induction motor acts as a brake and it absorbs kinetic energy from the still revolving field causing its speed to fall. When the slip is negative (i.e. when the speed is more than synchronous speed), the induction machine operates as an induction generator and feeds power back to the supply line.

Q. 14. What is the speed of rotor mmf of a 3-phase induction motor w.r.t. its stator mmf ?

Ans. zero.

Q. 15. In case of a 3-phase induction motor why is slip always positive ? **[P.T.U. 1999]**

Ans. The induction motor cannot run at synchronous speed because at synchronous speed, there will not be relative speed between the revolving field and the rotor conductors and so the basic phenomenon will not occur to develop torque. It means that a 3-phase induction motor always run at a speed lesser than that of the field. The slip being the difference of synchronous speed, N_s and actual speed N ($N_s > N$) is, therefore, always positive.

Q. 16. Explain why rotor of an induction motor can never attain synchronous speed. **[P.T.U. June 2003]**

Ans. An induction motor cannot run at synchronous speed. If it were possible, by some means, for the rotor to attain synchronous speed, the rotor would then be standstill with respect to rotating flux. Then no emf would be induced in the rotor, no rotor current would flow, and therefore, there would be no torque developed.

Q. 17. Show that in induction motor, the rotor always run in the direction of stator field. **[P.T.U. December 2002]**

Ans. According to Lenz's law we know that the direction of induced emf would be in such a direction that it would try to oppose the very cause for which it is due. Because the cause producing the induced currents in the rotor conductors is the relative speed between the rotating magnetic field and stationary rotor conductors, therefore, they circulate in such a way that a torque is produced in the rotor tending to cause it to follow the rotating magnetic field and thus reducing the relative speed.

Q. 18. How does the slip vary with load ? **[P.T.U. May 2001]**

Ans. With the increase in load, the motor slows down until the relative motion between the rotor and the rotating magnetic field is just sufficient to result in the development of torque required for a particular load. Thus slip increases with the increase in load.

Q. 19. If the full-load speed of an induction motor is 960 rpm, what will be the synchronous speed ? Give an intuitive answer.

Ans. 1,000 rpm.

Q. 20. At what slip the torque developed in an induction motor will be maximum ?

Ans. Torque developed in an induction motor will be maximum at a slip given by the ratio of rotor resistance to rotor standstill reactance. i.e. $s = \frac{R_2}{X_2}$.

Q. 21. What is the condition for maximum torque at starting in a 3-phase induction motor ?

Ans. Torque developed in an induction motor at start will be maximum when rotor resistance equals rotor standstill reactance i.e. $R_2 = X_2$.

Q. 22. Is the maximum torque of a 3-phase induction motor dependent on the rotor resistance ?

Ans. No, it is independent of rotor resistance.

Q. 23. What is the relationship of developed torque of a 3-phase induction motor with the supply voltage?

Ans. $T \alpha sV^2$.

Q. 24. When the applied rated voltage per phase is reduced to one-half, what will be the starting torque of a squirrel cage induction motor in terms of its starting torque with full voltage ?

Ans. One-fourth of starting torque with full rated voltage.

Q. 25. How the starting torque and maximum torque vary with rotor circuit resistance ? **[U.P. Technical Univ. 2002-03]**

Ans. As the rotor resistance R_2 is increased from zero, power factor is improved but rotor impedance is increased and so the rotor current is reduced. At first, the effect of improved power factor predominates over the effect of decrease in rotor current due to increase in impedance and the starting torque, therefore, increases. But after a certain value of rotor resistance, the effect of increased impedance predominates over the effect of improved power factor and so the torque starts decreasing.

Maximum torque is independent of rotor circuit resistance.

Q. 26. Why are starters used for starting of induction motors (3-phase)? **[U.P. Technical Univ. December, 2003]**

Ans. At the starting instant, a 3-phase induction motor acts as a polyphase transformer having an air gap with a short-circuited secondary. If the motor of higher capacity (larger than 2 kW) is connected directly to the supply mains, it will draw heavy current (about 5-8 times the rated full-load current) at the starting instant which is objectionable because it will cause large voltage drop in the distribution network and thus affect adversely the operation of other electrical equipment connected to the same distribution network. Hence, to control the starting current, use of starter is essential for starting of 3-phase induction motors of larger size (rating exceeding 2 kW).

Q. 27. What is typical use of starter in motor? **[U.P. Technical Univ. Second Semester 2013-14]**

Ans. Refer to Answer of Q. 26.

Q. 28. What are the various types of starters used for squirrel cage motors ?

Ans. Starters used for starting of a squirrel cage induction motor are

(a) Direct-on-line starter

(b) Line resistance starter

(c) Star-delta starter and

(d) Auto-transformer starter

Q. 29. Which is the cheapest method of starting a 3-phase induction motor.

Ans. Direct-on-line starting.

Q. 30. How many terminals do you expect to find on the terminal box of a squirrel cage induction motor to be started by star-delta starter ?

Ans. Six.

Q. 31. In what ratio line current and starting torque are reduced with star-delta starting ?

Ans. Line current and starting torque are reduced to one-third with star-delta starting.

Q. 32. What is the serious objection to the practice of employing reduced voltage for the starting of a squirrel cage induction motor ?

Ans. Large reduction in staring torque because the starting torque varies as the square of voltage applied to the stator.

Q. 33. What type of protection is provided in the starter meant for 3-phase induction motors ?

Ans. Overload and undervoltage protection.

Q. 34. A 3-phase induction motor is to be started first using an auto-transformer with 80 per cent tapping and next direct-on-line. What will be the ratio of the starting torques ?

Ans. Starting torque with auto-transformer with 80 % tapping/starting torque on direct on line = 0.64.

Q. 35. What are the main advantages of a cage motor ?

Ans. The main advantages of a cage motor are simple and rugged construction, low initial as well as maintenance cost, better efficiency and power factor, better cooling, more pull-out torque, greater maximum power output, nearly constant speed and simple starting arrangement.

Q. 36. What are the main disadvantages of a cage motor ?

Ans. The main drawbacks of a cage motor are low starting torque with large starting current and no possibility of a speed control.

Q. 37. What are the advantages and disadvantages of wound rotor over squirrel cage rotor in case of induction motor ?

[P.T.U. December 2002]

Ans. In wound rotor external resistance can be inserted and, therefore,

(*i*) high starting torque can be obtained by inserting an external resistance of proper value in each phase of the rotor circuit and

(*ii*) adjustable speed can be obtained simply by adding external resistors in the rotor circuit.

The drawback of wound rotor over squirrel cage rotor are initial and maintenance costs are more because wound rotor needs slip rings, brushes, short-circuiting devices etc.

Q. 38. What is meant by single phasing ?

Ans. The term 'single phasing' refers to operation with one stator terminal accidently disconnected (as by a blown fuse or a faulty contact on the run side of a star-delta switch).

PROBLEMS

1. A 12-pole, 3-phase alternator driven at a speed of 500 rpm supplies power to an 8-pole, 3-phase induction motor. If the slip of the motor is 0.03 pu, calculate the speed. **[U.P. Technical Univ. Electrical Engineering July 2002]**

   [**Ans.** 727.5 rpm]

2. A three-phase 4-pole induction motor is running with 4% slip. The supply frequency is 50 Hz. Find out the speed of induction motor. **[G.B. Technical Univ. Electrical Engineering First Semester 2010-11]**

   [**Ans.** 1,440 rpm]

3. A 50 Hz, 4-pole, 3-phase induction motor has rotor current frequency of 2 Hz. Obtain slip and speed of motor.

   [Pb. Technical Univ. Basic Electrical Engineering and Electronics Second Semester 2004-05]

   [**Ans.** 4%, 1,440 rpm]

4. A 3-ϕ induction motor is wound for 4 poles and is supplied from 50 Hz system. Calculate:

   (*i*) synchronous speed, (*ii*) rotor speed when slip is 4%, and (*iii*) rotor frequency when rotor runs at 600 rpm.

   [**Ans.** (*i*) 1,500 rpm (*ii*) 1,440 rpm (*iii*) 30 Hz]

5. A 4-pole, 3-phase induction motor operates from a supply whose frequency is 50 Hz, calculate (*i*) the speed at which the magnetic field of the stator is rotating (*ii*) the speed of the rotor when the slip is 0.04 (*iii*) the frequency of rotor currents when the slip is 0.03 (iv) the frequency of the rotor currents at standstill.

   [**Ans.** (*i*) 1,500 rpm (*ii*) 1,440 rpm (*iii*) 1.5 Hz (*iv*) 50 Hz]

6. A 3-ϕ, 6-pole, 50 Hz, induction motor has a slip of 1% at no-load and 3% at full load. Determine synchronous speed, no-load speed, full-load speed, frequency of rotor current at standstill and frequency of rotor current at full load.

   [**Ans.** 1,000 rpm, 990 rpm, 970 rpm, 50 Hz, 1.5 Hz]

7. A 3-phase, 50 Hz induction motor has 6 poles and operates with a slip of 5% at a certain load. Determine

   (*i*) the speed of the rotor with respect to the stator.

   (*ii*) the frequency of rotor current.

   (*iii*) the speed of the rotor magnetic field with respect to rotor.

(*iv*) the speed of the rotor magnetic field with respect to stator and

(*v*) the speed of the rotor magnetic field with respect to the stator magnetic field.

[U.P. Technical Univ. Electrical Engineering February 2002]

[**Ans.** (*i*) 950 rpm (*ii*) 2.5 Hz (*iii*) 50 rpm (*iv*) 1,000 rpm (*v*) zero]

8. A 3-phase delta-connected 440 volts, 50 Hz, 4-pole induction motor has a rotor standstill emf per phase of 130 volts. If the motor is running at 1,440 rpm, calculate for this speed: (*i*) the slip, (*ii*) the frequency of rotor induced emf, (*iii*) the value of the rotor induced emf per phase, and (*iv*) stator to rotor turn ratio.

[U.P. Technical Univ. Electrical Engineering Second Semester 2002-03]

$$\left[\textbf{Ans.}\ 1{,}500 \text{ rpm } (i)\ 4\%, (ii)\ 2 \text{ Hz}, (iii)\ 5.2 \text{ V}, (iv) \frac{44}{13}\right]$$

9. Determine approximately the starting torque of an induction motor in terms of its full-load torque when started by means of (i) star-delta starter and (ii) by an auto-transformer starter with 50 % taps. The motor draws 5 times full-load current when switched directly on line and has a full-load slip of 4%. [**Ans.** One-third and one-fourth of full-load torque]

10. A 3-phase squirrel cage type induction motor, when started by means of a star-delta starter, takes 200 % of full-load current (line) and develops 44 % of full-load torque at starting. Calculate the starting torque and current if an autotransformer with 75 per cent tapping was employed. **[U.P. Technical Univ. 2002-03]**

[**Ans.** 74.2% of T_f; 337.5% of I_f]

❑❑❑

16

CHAPTER

Single-Phase Induction Motor

16.1. INTRODUCTION

There are two basic reasons for the use of single-phase rather than 3-phase motors.

First, for reasons of economy, most houses, offices and also rural areas are supplied with single-phase ac, as power requirements of individual load items are rather small.

The second factor is the economics of the motor and its branch circuit. Fixed loads requiring not more than 0.5 kW can generally be served most economically with single-phase power. Single-phase motors are simple in construction, reliable, easy to repair and comparatively cheaper in cost and, therefore, find wide use in fans, refrigerators, vacuum cleaners, washing machines, other kitchen equipment, tools, blowers, centrifugal pumps, small farming appliances etc. To avoid the need for special branch circuits, portable tools are also usually single phase.

Because of above reasons motors of comparatively small ratings (mostly in fractional kW ratings) are manufactured in large number to operate on single-phase ac at standard frequencies.

Single-phase ac motors may be divided in three general classes namely (*i*) induction motors (*ii*) commutator motors and (*iii*) synchronous motors.

Induction motors are further classified as split-phase motors, shaded-pole motors and repulsion– start induction motors according to the method of producing starting torque.

Single-phase induction motors in very small sizes $\left(\frac{1}{400} \text{kW to } \frac{1}{25} \text{kW}\right)$ are used in toys, hairdryers, vending machines etc.

The main drawbacks of single-phase motors are low overload capacity, low efficiency, low power factor and low output as compared to that of a 3-phase motor of the given frame size.

16.2. PRINCIPLE OF OPERATION OF SINGLE-PHASE INDUCTION MOTORS

If one line of a three-phase induction motor is opened while the motor is running with moderate or light load, it is found that the motor continues to run though at slightly lower speed. This condition is known as *single-phase operation.* Opening of one line (phase) leaves the equivalent of a single phase connected to the stator. In other words a 3-phase induction motor has become a single-phase motor.

A single-phase induction motor is similar to a 3-phase squirrel cage induction motor in physical appearance. The rotor of a single-phase squirrel cage motor is essentially the same as that employed in 3-phase induction motors and needs no further description. There is uniform air gap between stator and rotor but no electrical connection between them (stator and rotor). Except for shaded-pole types, the stator core is also very similar. A single-phase motor can be wound for any even number of poles, two, four, and six being most common. Like three-phase machines, adjacent poles have opposite magnetic polarity and synchronous speed equation ($N_s = \frac{120f}{P}$) also applies.

An elementary single-phase induction motor is shown in Fig. 16.1. Instead of being a concentrated coil, the actual stator winding is distributed in slots to develop an approximately sinusoidal space distribution of mmf.

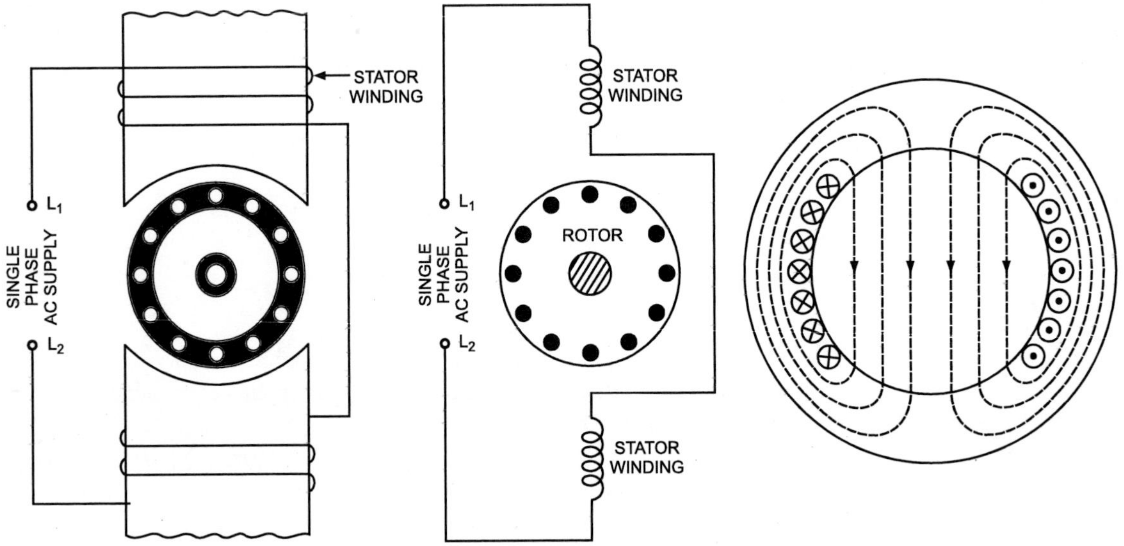

Fig. 16.1 *Elementary Single-Phase Induction Motor*

When the stator winding of a single-phase induction motor is connected to single-phase ac supply, a magnetic field is developed, whose axis is always along the axis of stator coils. With alternating current in the fixed stator coil the mmf wave is stationary in space but pulsates in magnitude and varies sinusoidally with time. Currents are induced in the rotor conductors by transformer action, these currents being in such a direction as to oppose the stator mmf. Thus the axis of the rotor mmf wave coincides with that of the stator field, the torque angle is, therefore, zero, and no torque is developed at starting. However, if the rotor of such a motor is given a push by hand or by another means in either direction, it will pick up the speed and continue to rotate in the same direction developing operating torque. Thus a single-phase induction motor is not self starting and needs special starting means.

The peculiar behaviour, mentioned above, of a single-phase induction motor may be explained by any one of the following two theories

1. Double revolving field theory
2. Cross-field theory.

Both theories have their own advantages and fields of application. They ultimately lead to the result that are approximately same. Here we shall discuss the double revolving field theory in brief.

Double Revolving Field Theory. Though double revolving field theory is not simpler in application than the cross-field theory, but once this theory is understood, the fundamental concepts are more readily grasped. This theory is based on the idea that pulsating field produced in a single phase motor can be resolved into two components of half its amplitude and rotating in opposite directions with synchronous speed (Ferraris principle—Art 15.4.1).

Under stationary rotor condition (i.e. when speed $N = 0$ or slip $s = 1$), the two rotating fields slip past the rotor at the same slip, $s = 1$ and inducing equal currents in the squirrel cage rotor. The two rotating fields are of the same strength and develop equal and opposite electromagnetic torques resulting in net torque of zero value. Thus the starting torque is zero and the single-phase induction motor is non-self starting. Further, the two rotating fields induce a resultant emf in the stator which balances the applied voltage assuming low leakage impedance of the stator winding.

If, however, the rotor is made to run at speed N by some external means in any direction, say in the direction of forward field, the two slips are now s and $(2 - s)$, as shown below:

The slip of the rotor with respect to the forward rotating field F_f,

$$s_f = \frac{N_s - N}{N_s} = s \qquad ...(16.1)$$

The slip of the rotor with respect to the backward rotating field F_b,

$$s_b = \frac{N_s - (-N)}{N_s} = \frac{2N_s - (N_s - N)}{N_s} = (2 - s) \qquad ...(16.2)$$

For normal operation $(2 - s) >> s$ and as a consequence the backward field rotor currents are much larger than at standstill and have a low power factor. The corresponding opposing rotor mmf, owing to stator impedance, causes the backward field to be greatly reduced in strength. On the other hand, the low-slip forward rotating field induces smaller currents of a high power factor in the rotor than at standstill. This leads to greatly strengthening of the forward field. Thus weakening of backward field and strengthening of forward field depends upon the slip (or speed of rotor) and the difference increases with the decrease in slip with respect to the forward field or with the increase in rotor speed in forward direction. In fact, at near about the synchronous speed, the forward field may be several times stronger than the backward field. As a result there is a net running torque.

Let us assume that the torque developed by forward field in the direction of rotor rotation is positive. Then the torque developed by the backward field will be negative and, evidently, be a braking torque.

Fig. 16.2 shows torques developed by the two revolving fields and also the resultant (or net) torque developed by the motor.

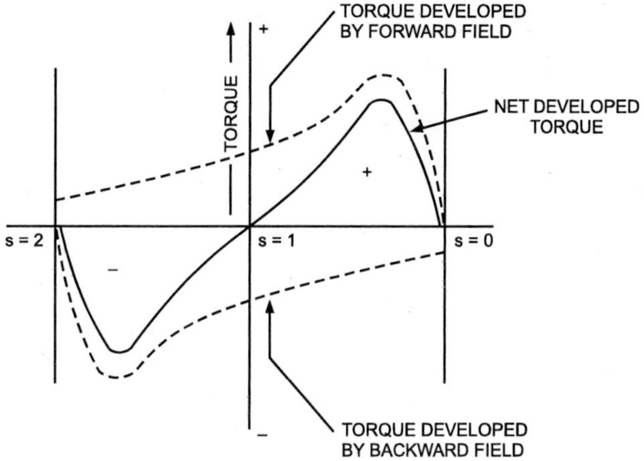

Fig. 16.2 *Torque-Slip Curves For a Single-Phase Induction Motor*

From Fig. 16.2 it is noted that the curve of the resultant torque passes through zero at a slip of unity (or at standstill), thus showing that no starting torque is developed and the motor cannot start rotating independently. However, for other values of slip, the motor develops a net torque in the direction of rotation and the operating conditions with the rotor running in either direction at a certain speed are identical. The breakdown or maximum torque is also less than that what is expected in a 3-phase induction motor. A further point is that the resultant developed torque drops to zero at a speed slightly below synchronous speed whereas in a polyphase induction motor the torque does not drop to zero until synchronous speed is attained. The fact is that the torque developed at syschronous speed is negative. This is due to the torque developed by the backward field.

The effect of increasing rotor resistance is rather different in a single-phase induction motor than in a 3-phase induction motor. In the case of a 3-phase induction motor the value of the maximum torque is not affected by the change in rotor resistance, the only change being the value of slip at which maximum torque occurs. In a single-phase machine, the increase in rotor resistance increases the effectiveness of the backward field, which reduces the breakdown torque, lowers the efficiency and increases the slip corresponding to maximum torque.

The performance characteristics of a single-phase induction motor, in general, are of the same nature as those for polyphase induction motor. However, the performance of a single-phase induction motor, in every respect, is somewhat inferior to that of a 3-phase induction motor for the given frame size. This is due to the presence of backward rotating field. It has a lower breakdown torque at larger slip and increased power losses. Further, it has greater power (true as well as apparent) input because of their consumption in the backward rotating field. Even the copper losses occuring in the stator winding of a single-phase induction motor are higher because all the current flows through a single winding. As a consequence, efficiency is lowered and temperature rise is increased. Further, the speed regulation of a single-phase induction motor tends to be poorer than that for a polyphase motor. The normal slip of a single-phase induction motor under load conditions is rather greater than that of the corresponding 3-phase motor and, therefore, power factor of a single-phase motor tends to be lower. For a given power and speed rating, a single-phase motor must therefore, have a larger frame size than that of a 3-phase motor.

The forward field and the rotor's backward reaction field and also the backward field and the rotor's forward reaction field move in opposite directions with relative speeds of double the synchronous speed and develop second harmonic pulsating torques with zero average value. As a consequence single-phase motors tend to be somewhat noisier than three-phase ones, which have no such pulsating torque.

16.3. STARTING METHODS AND TYPES OF SINGLE-PHASE INDUCTION MOTORS

As already discussed, the single-phase induction motor is not self starting and, therefore, it is necessary to employ some means for making it self starting. Commercial single-phase induction motors are classified in accordance with the methods of starting employed.

It is known that when two windings, spaced 90 electrical degrees apart on the stator of a motor, are excited by two alternating emfs that are 90° displaced in time phase, a rotating magnetic field is produced. If two windings so placed are connected in parallel to a single-phase source, the field produced will alternate but will not revolve since the two windings are equivalent to but one phase winding. If, however, an impedance is connected in series with one of these windings, the currents may be made to differ in time phase. By proper selection of such an impedance the currents may be made to differ by as much as 90°, thereby producing a rotating field very much like the field of a two-phase motor. This is the principle of phase-splitting and the single-phase induction motors employing this principle for starting are known as split-phase motors. Split-phase motors are of the following types.

(*a*) Resistance-start motor (*b*) Capacitor-start motor (*c*) Capacitor-start-capacitor run motor (*d*) Permanent capacitor motor (*e*) Shaded pole motor.

16.4. RESISTANCE-START SINGLE-PHASE INDUCTION MOTOR

It is a form of split-phase motor having a high resistance connected in series with the starting winding, as illustrated in Fig. 16.3 (*a*). Because of low resistance and high inductive reactance of main winding and high

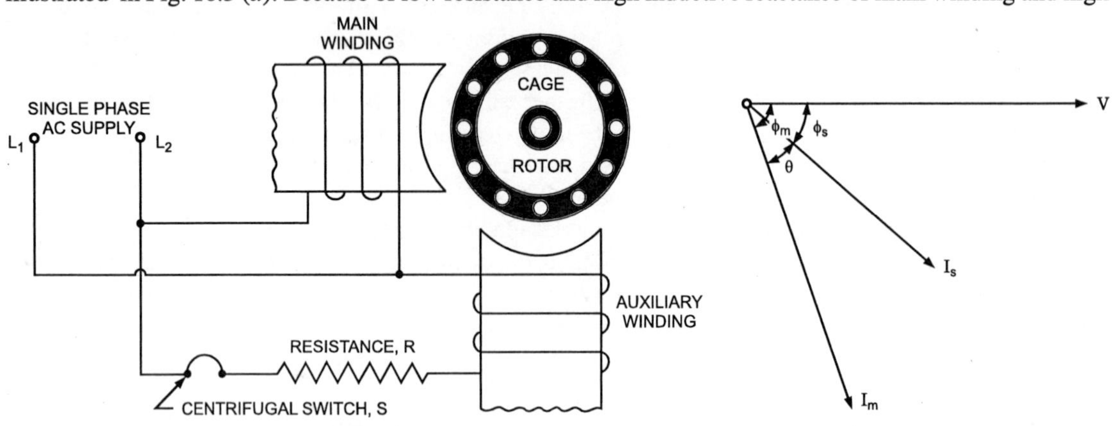

(a) Schematic Diagram *(b) Phasor Diagram*

Fig. 16.3

resistance and low inductive reactance of starting winding, the starting winding current I_s does not lag so much as the main winding current I_m. At starting the main winding current I_m lags behind the applied voltage V by 65-75 degrees, whereas the starting winding current I_s lags by 35-45 degrees. Thus the phase displacement between main winding current and starting winding current is about 20-30 degrees, as shown in phasor diagram [Fig. 16.3 (*b*)]. Since the currents in the two windings are different in magnitude and constants of two windings are different, the arrangement forms imperfect two-phase motor, that produces a rotating magnetic field that is not uniform in either time or space, but is sufficient for starting the motor. The best starting condition will be obtained if I_s and I_m differ in time by about 90°, but this condition is not possible in the motor of this type.

When the motor picks up the speed 75 or 80 per cent of synchronous speed, the starting winding (or auxiliary winding) is taken out of circuit by a transfer or centrifugal switch. It is important that this should be done because otherwise it will result in overheating and burning out of auxiliary winding owing to its low current carrying capacity and inefficient and noisy performance.

The speed regulation of a resistance–start single-phase induction motor is very good. The slip in most fractional kilowatt split-phase motors is about 4–6 per cent. Such a motor may operate with a power factor of 0.55–0.65 and efficiency of 60–65 per cent.

Owing to their low cost, resistance start split–phase motors are most popular single-phase motors. These motors are made in fractional kilowatt $\left(\frac{1}{20} \text{ to } \frac{1}{4} \text{ kW}\right)$ ratings with speed ranging from 2,875 to 700 rpm. Such motors find wide applications for low inertia loads, continuous operating loads and applications requiring moderate starting torque such as for driving washing machines, fans, blowers, centrifugal pumps, domestic refrigerators, duplicating machines, office appliances, food processing machines, woodworking tools, grinders, oil burners etc. Because of low starting torque these motors are seldom employed in sizes larger than $\frac{1}{4}$ kW.

The direction of rotation of such a motor can be reversed by interchanging the connections to the supply of either starting (auxiliary) or running (main) winding. The noteworthy point is that reversal of a split-phase induction motor takes place only when it is at rest and the starting switch is in its normally closed de-energized position.

16.5. CAPACITOR- START SINGLE-PHASE INDUCTION MOTOR

The schematic diagram of a capacitor–start single-phase induction motor is given in Fig. 16.4 (*a*). It is also known as *capacitor-start, induction-run motor.* This is an improved form of a resistance-start split-phase motor. Fundamentally this motor is very similar to the resistance- start split -phase motor, except that the

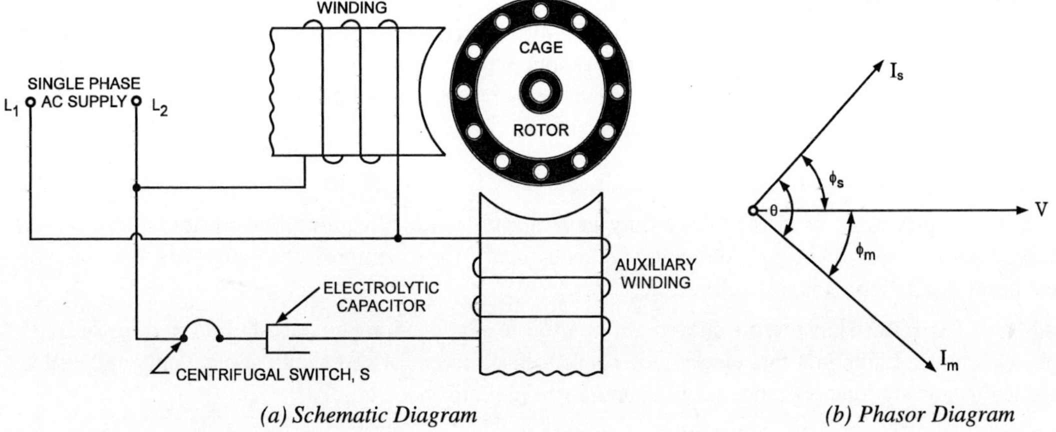

(a) Schematic Diagram *(b) Phasor Diagram*

Fig. 16.4

starting winding has a few more turns and consists of a heavier wire than the starting winding of a resistance–start motor; there is also a large electrolytic capacitor in place of resistor, connected in series with the starting winding. By using a capacitor of proper value, the starting winding current I_s at standstill, can be made to lead the main winding current I_m by 90° (i.e $\theta = 90°$) . The phasor diagram of a capacitor–start motor is given in Fig. 16.4 (*b*).

The capacitor is usually mounted in a metal casing on the top of the motor although it may be mounted in any convenient external position on the motor frame. In some cases, the capacitor is mounted inside the motor housing. The capacitor used in the capacitor-start motor is of the electrolytic type and is designed for extremely short - duty service and is guaranteed for not more than 20 periods of operation per hour, each period not to exceed 3 seconds. The size of the capacitor depends upon the rating of the motor and has a value around 50 μF to a few hundred microfarads. It is primarily due to the development of the inexpensive yet reliable ac electrolytic* capacitor that the capacitor- start motor has largely replaced all other single-phase motors where high starting torque is the essential requirement.

The capacitor-start motor, like resistance–start motor, has the starting winding disconnected by means of a centrifugal switch as the motor picks up speed. Because of short-duty service rating of electrolytic capacitor used in capacitor-start motor, proper operation of the starting (or centrifugal) switch is doubly important. A faulty switch may keep the capacitor in the circuit for too long time during each starting period, and thereby appreciably shorten its life. Of course, the starting winding itself is affected by the switch in the same manner as in the resistance–start motor.

The direction of rotation of a capacitor-start induction motor may be reversed by interchanging the connections to the supply of either the main or starting winding, as in case of a resistance-start single-phase induction motor but it is subject to the same limitations.

Since the torque developed by a phase- split motor is proportional to the sine of phase displacement angle θ between the starting winding current I_s and main winding current I_m, it is obvious that the increase in phase displacement angle θ (from 30° to 80°) alone increases the starting torque twice the value developed by a resistance-start motor. The increase in starting winding turns causes decrease in starting current on one hand and produces a large rotating flux, and therefore a further increase in the starting torque on the other hand. Finally it is seen that, for the same magnitudes of two stator winding currents, I_s and I_m the line current I_L is less in the capacitor-start motor, because of the greater angle between them. In addition, the starting winding power factor is also improved. In short it can be said that because of the closer approximation to a true two-phase motor, a greater number of turns on the starting winding, and more allowable current in both of the windings due to lower line current, the capacitor-start motor develops a greater starting torque than does an equally rated resistance-start motor. The line current of a resistance-start motor was found to be two-third higher than line current of a corresponding capacitor-start motor, while the starting torque of the capacitor–start motor was twice that of resistance-start motor.

It is to be noteworthy that the capacitor-start motor differs from the resistance-start motor only during the starting period, for it develops more starting torque (about 3.5-4.5 times full-load torque) with a lower inrush current. After the machine attains normal speed and the starting winding is cut out, its operating characteristics are identical to those of a corresponding resistance-start motor.

Capacitor-start motors are manufactured in rating ranging from $\frac{1}{10}$ kW to $\frac{3}{4}$ kW, but larger sizes are also available. They are widely used for heavy duty general purpose applications like refrigerator units, air-conditioners, compressors, jet pumps, sump pumps, centrifugal pumps, compressors, conveyors, fans, blowers, farm and home workshop tools, oil burners etc.

Cautions : 1. Capacitor-start motors should not be used for applications having starting periods of longer duration. This is due to the fact that electrolytic capacitors are rated for short-duty service only and will get damaged if the starting load does not allow the motor to pick up speed quickly.

* AC electrolytic capacitor is not the same as the dc electrolytic capacitor often employed in a filter in radios, amplifiers, and TV sets etc.

2. The capacitor-start motor should not be started more than 20 times an hour. If the motor is started too frequently in a short time interval, the dielectric of the capacitor will get damaged by the frequent current surges. As a result, the dielectric will rupture or breakdown and short the capacitor plates and cause the fuse protection for the branch motor circuit to blow. If the fuse ratings are quite high and do not interrupt the power supply to the motor, the starting winding may burn out.

16.6. CAPACITOR-START, CAPACITOR-RUN SINGLE-PHASE INDUCTION MOTOR

In order to have the advantages that result from the use of a capacitor during the normal running operation as well as starting period, the capacitor-start, capacitor-run motors have been developed. The capacitor-start, capacitor-run motor is similar to capacitor-start, induction-run motor, except that the auxiliary winding and a capacitor remain connected in the circuit at all times. Since the value of the capacitance which will result in optimum running performance is not the same as the value which will give the best starting torque conditions and also the capacitor employed for normal running should be of continuous duty rating and that employed for starting should be of short-duty rating, therefore, in such a motor two different capacitors — one small value oil impregnated paper continuous rating capacitor C_R and another much larger value electrolytic short-duty capacitor C_S are employed. It is also necessary to provide a centrifugal switch to take out the electrolytic capacitor out of auxiliary winding circuit when the motor has attained a predetermined speed. The schematic diagram of a capacitor-start, capacitor-run motor is given in Fig. 16.5. When the motor attains 75 per cent of synchronous speed, the starting capacitor C_S is taken out of the circuit by the operation of centrifugal or transfer switch, which is normally closed.

It is to be noted that when the centrifugal switch is closed during the starting period, the two capacitors C_S and C_R are in parallel, so that total capacitance in the auxiliary winding circuit is sum of their values; after the switch opens, only the continuous oil type capacitor C_R remains in the auxiliary winding circuit. In practice starting capacitor C_S is about 10-15 times as large as the running capacitor C_R.

Such a motor operates as a two-phase motor from single-phase supply, thereby, producing a constant torque and not a pulsating torque, as in other single-phase motors. Besides their ability to start heavy loads they are extremely quiet in operation, have better efficiency (55-65%) and power factor (0.8-0.95) when loaded and develop up to 25% greater overload capacities. They are indeed splendid machines where the load requirements are severe as in case of compressors, refrigerators and fire strokers, pumps, conveyors etc. The disadvantage of such machines is only high cost. These motors are often employed requiring a quiet operating motor such as in hospitals, studios etc.

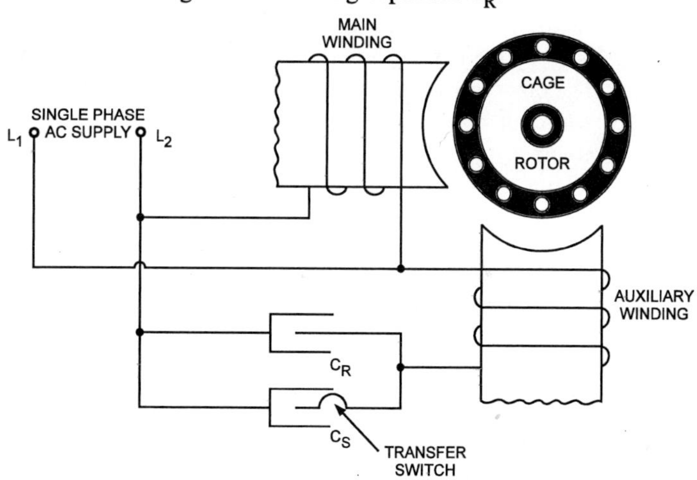

Fig. 16.5 *Schematic Diagram of Capacitor-Start, Capacitor-Run Single-Phase Induction Motor*

In recent years, home air-conditioners have become extremely popular. In many cases, the electrical system in the home may not have the capacity to handle the electrical load of the air-conditioner(s). For this reason, capacitor-start capacitor-run motor have found more wide spread use in the compressors. Both the starting and running currents are low — full-load current may be $\frac{3}{4}$ th of that for an equivalent capacitor-start motor.

Such motors have become increasingly popular and are now manufactured as the larger sized single-phase motors where previously only repulsion-start motors were available. The direction of rotation of motor may be reversed by interchanging the connections to the supply of either the main or auxiliary winding.

16.7. PERMANENT CAPACITOR SINGLE-PHASE INDUCTION MOTOR

Permanent capacitor single-phase induction motor, also sometimes called the *single-value capacitor–run motor*, has two stator windings placed mutually 90 electrical degrees apart. The main or running winding is connected directly across the single-phase ac supply. A capacitor in series with the auxiliary winding is also connected across the supply lines. There is no centrifugal switch, since the auxiliary winding is energized at all times when the motor is in operation. The auxiliary winding in such motors usually has almost the same size of wire and almost as many turns as the main winding; in fact, in some motors the two windings are identical. The schematic diagram of a permanent capacitor motor is given is Fig. 16.6.

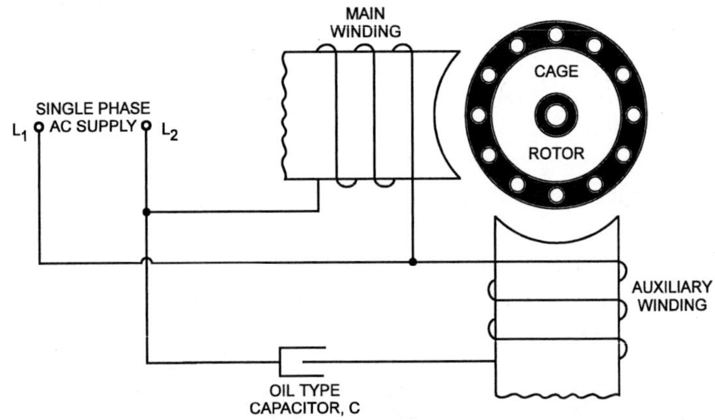

Fig. 16.6 *Schematic Diagram of Permanent Capacitor Single-Phase Induction Motor*

The auxiliary winding is always in the motor circuit, and therefore, the operation of the motor when loaded resembles more closely to that of a 2-phase motor. One advantage of such operation is that the rotating field developed by the two stator windings is a more nearly uniform, and therefore, the motor is quieter in operation. The other advantages of such a motor are (*i*) increased pull-out torque (*ii*) a higher power factor (0.8-0.95), (*iii*) higher operation efficiency (55-65 per cent) and (*iv*) smaller full-load line current.

Such motors have some drawbacks also. Since in such motors the same capacitor is used for starting and running, therefore, neither optimum starting nor running performance can be obtained. Generally capacitors of 2-20 μF capacity are employed. The electrolytic capacitor cannot be employed owing to the requirement of continuous operation, and therefore, more expensive oil or Pyranol insulated foil-paper capacitor is to be used. Under starting conditions, the capacitor is too small and so the starting torque tends to be low (50 to 80 per cent of full-load torque); but the manufacturers can partly compensate for this by using a high resistance squirrel cage. The permanent capacitor induction motor is more expensive than the equivalent resistance-start or capacitor -start induction motor mainly owing to the increased cost of auxiliary winding which is now wound with thicker wire and consists of as many turns as the main winding.

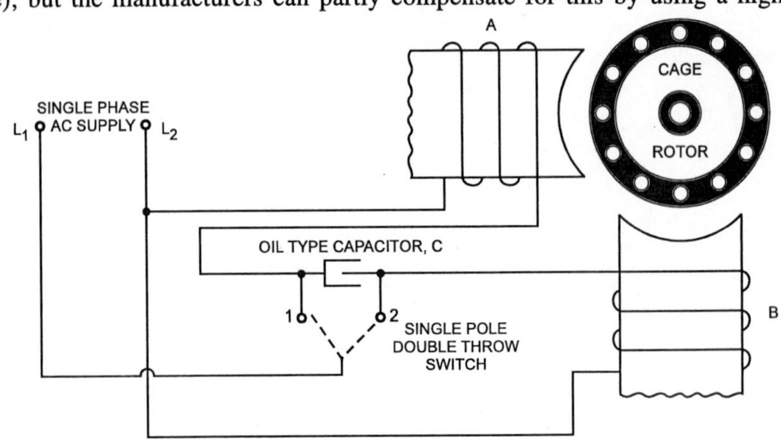

Fig. 16.7 *Reversing Permament Capacitor Motor*

A unique feature of such a motor is that it can be very easily reversed by an external switch, provided the two windings are identical. Because of absence of starting switch it is possible to reverse the leads of either winding while the motor is running. The torque is thus reversed, the motor comes to a standstill, and then starts in the opposite direction. If the two stator windings are identical, it is possible to place the capacitor in series with either of the two windings, making that particular winding the auxiliary or leading one. Circuit for such an operation is given in Fig. 16.7. When the SPDT switch is in position 1 winding A serves as a main winding and winding B as an auxiliary winding. When the SPDT switch is moved in position 2, winding A becomes the auxiliary winding and winding B the main one.

Such motors are widely used in table and ceiling fans, blowers, oil burners, room coolers, portable tools, other domestic and commercial electrical appliances, where low starting torque is required.

Because of the simplicity of reversal of motor, such motors are often used to operate devices that must be moved back and forth frequently such as induction regulators, rheostats, furnace controls, valves and arc-welding controls.

Because of the absence of starting switch, it is possible to run a permanent (or single value) capacitor run motor over a wide range of speeds, and such a motor has many applications where adjustable speed is required. Such a motor can be arranged for an adjustable speed by the use of a tapped winding or an auto-transformer regulator. It is known that the torque developed in an induction motor is proportional to the square of the applied voltage. For a given load, therefore, if the applied voltage is reduced, the motor will operate with increased slip to develop the required torque. Thus the speed can be controlled by varying the voltage applied to the motor.

16.8. SHADED-POLE MOTOR

Shaded-pole motor is a split-phase type single-phase induction motor. The stator has salient poles, each provided with its own exciting coil. A part $\left(\text{about } \frac{1}{4} \text{ to } \frac{1}{3}\right)$ of each pole is wrapped by a copper strap forming a closed loop, known as a *shading coil,* as shown in Fig. 16.8. Stator usually has 2 or 4 poles giving synchronous speed of 3,000 or 1,500 rpm for 50 Hz power supply. The rotor is a simple squirrel cage type with a 60° skew for optimum starting torque and for limiting the torque dip during run-up. A low cage resistance impairs both the starting torque and the dip, and usually the resistance is greater than for any other type of single-phase cage motor, giving a starting torque in the range of 40-80 per cent of full-load torque and pull-out torque of about 110-140 per cent of full-load torque. The gap length is of the order of 0.25-0.5 mm. Too short air gap may result in starting-torque variations due to rotor slotting. It has no commutator, brushes, collector rings, contactors, capacitors or moving switch parts, so it is relatively cheaper, simpler and extremely rugged in construction and reliable (absence of centrifugal switch or starting mechanism auxiliary eliminates the possibility of motor failure due to faulty centrifugal switch mechanisms).

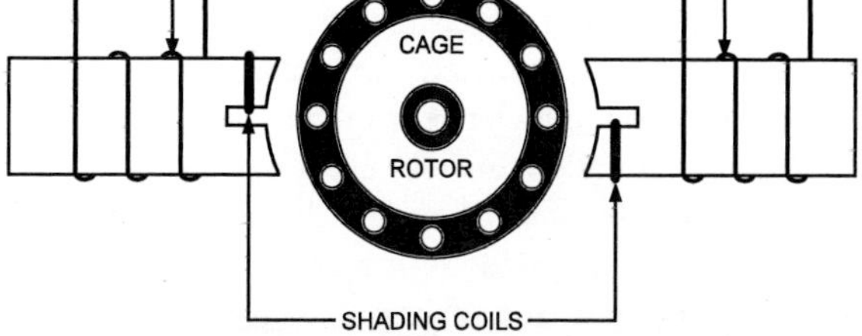

Fig. 16.8 *Schematic Diagram of Shaded-Pole Single-Phase Induction Motor*

Action. When the exciting winding is connected to a single-phase ac supply source an alternating current flows through it and sinusoidally varying flux is set up in the core. This change of flux induces short-circuit currents in the shading coil in such a direction as to oppose the core flux. The flux in the shaded part of the pole thus lags behind the flux in the main part of the pole. At the same time, the main flux and shaded-pole flux are obviously displaced in space, although by less than 90°. Since two fields have both space and time displacement, the conditions for setting up a rotating field are present, no matter how imperfectly. Torque is thus developed in the squirrel cage rotor.

The effect of shading coil is to cause the field flux to shift across the pole face from the unshaded portion to shaded portion. This shifting flux is like a rotating weak field moving in the direction from unshaded portion to the shaded portion of the pole. Consequently a small starting torque is developed. As soon as the rotor starts rotating under the influence of starting torque, an additional torque is developed by single-phase induction motor action. The motor ultimately attains a speed slightly less than the synchronous speed and runs as a single-phase induction motor.

Because of their poor starting torque, low power factor and poor efficiency, such motors are only suitable for low power applications such as for toys, small fans, electric clocks, hairdryers, time phonographs, slide projectors, humidifiers, small business machines (photocopying machines, vending machines, advertising displays etc.) and other similar applications. They are manufactured in small sizes ranging from below 5 W to about 100–150 W. They are available with built-in gear reducers to give almost any speed (as low as below 1 rpm.)

16.9. UNIVERSAL MOTORS

For some applications, it is desirable to employ a motor that operates on either dc or ac. It is possible to build, by a compromise design, small series motors up to about $\frac{1}{2}$ kW rating to operate satisfactorily on dc or ac at 115 or 230 V. Such motors are called the *universal motors.*

A universal motor is a specifically designed series wound motor, that operates at approximately the same speed and output on either dc or ac of approximately same voltage.

Owing to reactance voltage drop, which is present on ac but not on dc, the motor speed is somewhat lower for the same load in ac operation, especially at high loads. On ac, however, the increased saturation of the magnetic circuit at the peak of the sine wave of current may materially reduce the flux below the dc value, and this tends to increase the speed in ac operation. It is possible, therefore, to design small series motors, called the *universal motors*, to have approximately the same torque-speed performance over the operating range, for all frequencies from 0 to 50 Hz.

Because of the difficulty in obtaining like performance on dc and ac at low speeds, most universal motors are designed to operate at speeds exceeding 3,500 rpm. Motors operating at speeds 8,000 to 10,000 rpm are common. High speed results in small size and weight of the motor.

16.9.1. Construction. Basically, a universal motor is similar to a dc series motor in construction. However, it is constructed with few series field turns, laminated armature and field circuits, low-reluctance magnetic path, increased armature conductors and commutator segments and using low flux densities. This is done to minimize the adverse effects caused by high field reactance, eddy current and hysteresis losses.

The commutation on ac is much poorer than on dc, owing to current induced in the short-circuited armature coils, and this provides a definite limitation on their size and utility. If the wide brushes are used, the short-circuit currents are excessive, and the motor starting torque is reduced; while if narrow

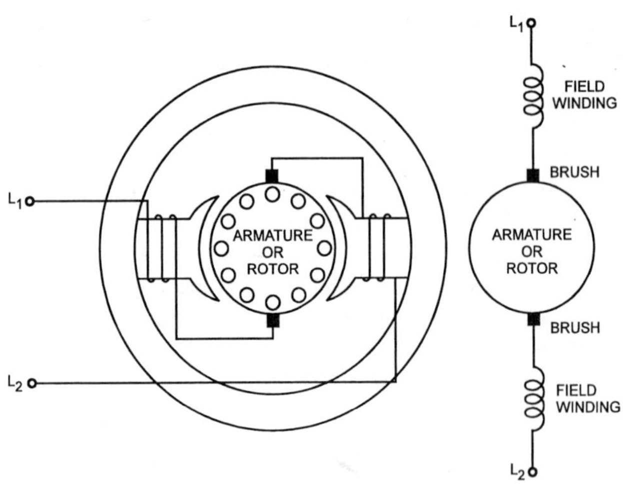

Fig. 16.9 *Universal Motor.*

brushes are used, there may be excessive brush chatter at high speeds, causing short brush life. Good design, therefore, needs careful proportioning of commutator and brush rigging to meet conflicting electrical, mechanical and thermal requirements. Brushes employed are high resistance carbon ones so as to aid commutation.

Compensating winding is sometimes used in large rating motors to improve commutation but in a small rating and high speed universal motor this winding needs not to be provided because in such a motor armature current is so small to cause serious commutation troubles.

Universal motors may be either compensated (distributed field) or uncompensated (concentrated field) type, the latter type being used for the higher speeds and smaller output ratings (usually not exceeding 200 W) only.

A motor of the concentrated field type is usually a salient pole machine with two poles and a winding of relatively few turns to give opposite magnetic polarity. It consists essentially of a laminated core having either straight or skewed slots and a commutator to which the leads of the windings are connected.

Distributed field universal motors are of two types viz; single-field compensated motor and two-field compensated motor. The field windings of a 2-pole, single-field compensated motor resemble the stator winding of a 2-pole, split-phase ac motor. A two-field compensated motor has a stator winding consisting of a main winding and a compensating winding spaced 90 electrical degrees apart.

The frames of universal motors are usually made of aluminium, cast iron, and rolled steel.

The uncompensated motor is simpler and less expensive than the compensated motor but gives poorer speed regulation than a motor with compensation. The distributed-field compensated motor is employed for larger output ratings at high speeds, such as $\frac{3}{4}$ kW at 7,500 rpm, or for lower output ratings at low speeds, such as 200 watts at 3,000 rpm.

16.9.2. Operation. In a series wound motor, the same current flows through field windings and armature, being connected in series with each other, when motor is connected to either dc or ac supply. The magnetic fields developed by the series field winding and armature currents react with each other and hence develop unidirectional torque.

16.9.3. Characteristics. The torque-speed characteristic of a universal motor is quite similar to that of a series wound dc motor i.e., high starting torque and high no-load speed. In a very small series motor, the losses are usually large enough at no load to restrict the speed to a definite value. Where there is a danger of high no-load speed, a governor may be employed to keep speed within specified limits. The governor consists of a centrifugal switch mounted on the motor shaft. The tension of the springs of the switch is adjusted so that the switch opens at a predetermined speed and thus place an external resistor R in series with the armature, thereby reducing the speed. When the speed drops because of increase in load, the switch contacts close, thereby shorting out the series resistor R, thus again raising the speed. The connections are shown in Fig. 16.10. The capacitor C is placed across the governor to reduce sparking.

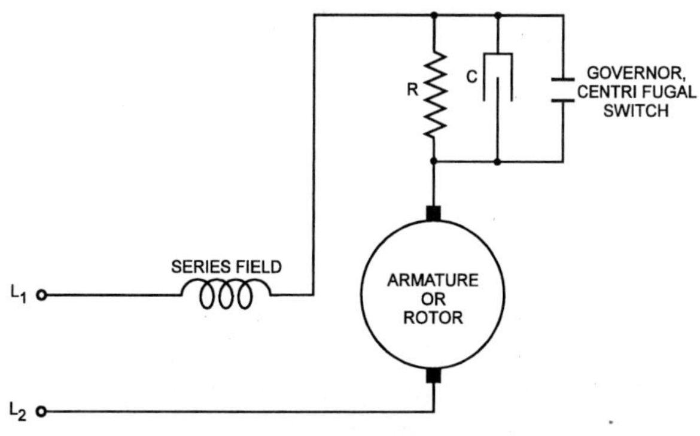

Fig. 16.10

Universal motors being high-speed motors, are smaller in size and lighter in weight as compared to other motors of the same output. Full-load power factor is high (approximately 0.9) but it is poor at start and on overloads.

16.9.4. Reversal of Direction of Rotation. The direction of rotation of any series wound motor can be reversed by reversing the direction of the current flow in either the field or armature circuit (but not through both).

Universal motors are sensitive to brush position and severe arcing at the brushes occur if the direction of rotation is reversed without shifting the brushes to the neutral plane. This is the reason that the universal motors are usually wound for operation in only one direction.

16.9.5. Applications. The very small power output rating universal motors, which usually does not exceed 5 or 10 watts, are employed in such equipment as sewing machines, fans, portable hand tools, hairdryers, motion-picture projectors, and electric shavers. The higher rating (5-500 W) universal motors are used in vacuum cleaners, electric typewriters, food mixers and blenders, motion-picture projectors, cameras, adding machines, and calculating machines.

The small series motors are frequently furnished as motor parts, i.e., consisting of bare stators and rotors (with shaft) but without bearings or feet. They can then be compactly "built in" to the power-using devices.

16.10. STEPPER MOTORS

There are many industrial situations which call for precise positioning of an object or precise control of speed without having to resort to closed-loop feedback. Stepper motors are ideally suited to such tasks. A stepper motor rotates a precise angular distance, one step, for each pulse that is delivered to its drive circuit. For example, if a particular stepper motor has a step angle of 7.5° per pulse and the shaft is to be turned through 225° in order to place an object in a certain target position, then it is only necessary to supply 225/7.5 i.e. 30 pulses to the motor's drive circuit. Because the step pulses can be counted digitally and stopped when the desired number have been delivered, stepping motors lend themselves to digital control by a programmable controller or on-line microcomputer. These motors can also provide precise open-loop speed control, since their rotational speed is determined solely by the step pulse frequency, independent of load.

The stepper motor is a form of synchronous motor which is designed to rotate through a specific number of degrees for each electrical pulse received by its control unit. Typical steps are 2,2.5,5,7.5 and 15° per pulse. Stepper motors are built to follow signals as rapid as 1,200 pulses per second and with equivalent power ratings up to several kilowatts.

A stepper motor is an *incremental motion machine* i.e. the motor which turns in discrete movement (called the steps) is known as the stepper motor. Stepper motor does not rotate continuously, as a conventional motor does.

16.10.1. Operating Principle. The principle of operation of a stepper motor can be easily explained by considering a series of solenoids or electromagnets arranged in a circle and an iron bar (rotor), as illustrated in Fig. 16.11. When these solenoids are energized in sequence, the mmf developed in them interact with the iron bar (or rotor) and causes it to turn either in clockwise or counter-clockwise direction, depending upon the switching sequence.

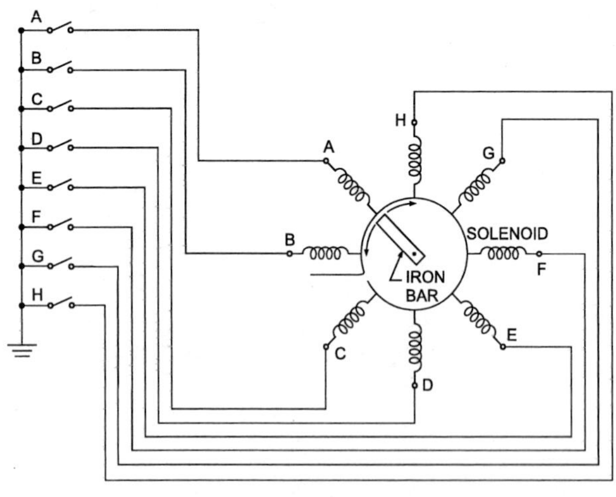

Fig. 16.11

16.10.2. Constructional Features and Working. The construction of the stepper motor is quite simple and consists of a slotted stator equipped with a multi-pole, multi-phase winding that is not unlike the windings of the conventional machines and a rotor structure that carries no winding. They typically use three and four phase windings, the number of poles being determined by the required angular change per input pulse. The rotors are either of the *permanent magnet type* or the *variable reluctance* type. Stepper motors operate with an external drive logic circuit; as a train of pulse is applied to the input of the drive circuit, the circuit delivers appropriate currents to the stator windings of the motor to make the axis of the air-gap field around in coincidence with the input pulses. Depending upon the pulse rate and load torque, including inertia effects, the rotor follows the axis of air-gap magnetic field by virtue of the permanent magnet torque and/or the reluctance torque. Both types i.e. permanent magnet and the variable reluctance type stepper motors operate on the same basic magnetic principle.

1. Permanent Magnet Stepper Motor. The representative structure of a permanent magnet type stepper motor having 4 stator poles and 6 rotor poles producing a full step angle of 30° is shown in Fig. 16.12. The end terminals of all the windings are brought out to the terminal box for dc excitations. The rotor of such a motor has even number of poles made of high retentivity steel alloy (Alnico). Both rotor and stator may employ salient or non-salient pole construction. Usually the stepper motors having small stepping angles are of non-salient pole construction. In the motor illustrated in Fig. 16.12, the stator pole windings are wrapped in such a way that closure of any winding's control switch causes that winding's pole to become magnetically north. If the six pole rotor is permanently magnetized as shown, and if only a single stator winding is energized at one time, then the rotor will always orient so that the nearest rotor south pole is aligned with the energized stator pole.

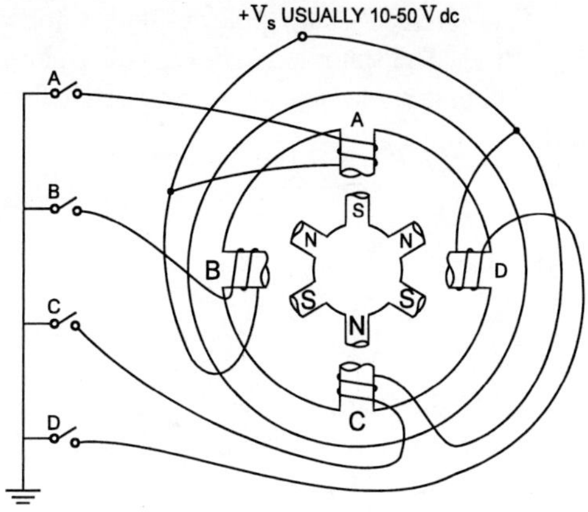

Fig. 16.12 *4-Stator-Pole, 6 Rotor-Pole Permanent Magnet Stepper Motor*

For instance if switch A is closed, energizing stator pole winding A, the rotor will align as shown in the figure. If switch A is opened and the switch B is then closed, so that only winding B is energized, the rotor will step 30° clockwise. This step will incur due to the tendency of the south pole nearest the winding B to align with the north stator pole.

If the switch B is then opened and switch C subsequently closed, the rotor will make a second 30° clockwise step, due to attractive tendency of the north stator pole (C) for the rotor pole nearest it. It is not hard to see that when the switching sequence ABCD is repeated continually, the rotor shaft will continue to turn clockwise, one 30° step per switch closure. If the switching sequence is reversed, to ADCB, the rotor shaft would turn in opposite i.e. counter-clockwise direction.

Many stepping motor designs allow the selection of full steps or half steps. This is accomplished by altering the switching sequence to simultaneously energize two stator coils during every second timing interval. Simultaneous energization of two stator windings creates the effects of moving the stator's magnetic north to a position between the two poles.

In reality, stator winding switching is accomplished by transistors rather than mechanical switching. A typical stepper motor needs 1 to 10 A of winding current in order to develop its rated torque. Therefore, the switching transistors must be high power devices.

2. Variable-Reluctance Stepper Motor. The stator of a variable reluctance stepper motor is similar to that of a permanent magnet stepper motor. It has no permanent magnet on the rotor and the rotor employed is a ferromagnetic multi-toothed one.

The torque is developed due to large differences in magnetic reluctances that exist between the direct and quadrature axis. The stationary field developed by the direct current in some stator coil tends to develop a torque which causes the rotor to move to the position where the reluctance of the flux path is minimum.

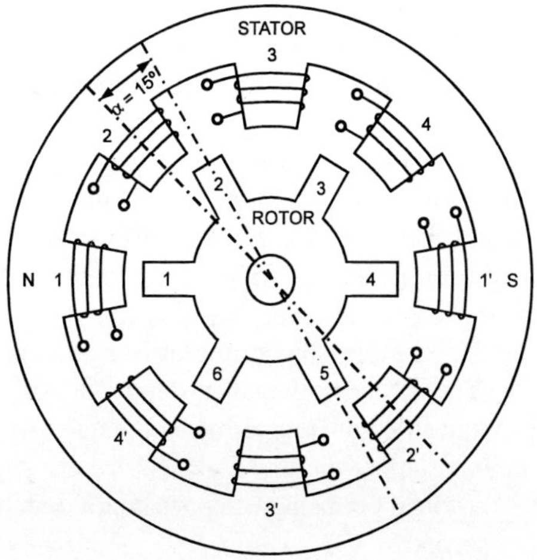

Fig. 16.13 *4-Phase, 8-Pole Stator, 6-Tooth Rotor Variable Reluctance Stepper Motor*

A 4-phase, 8-stator pole, 6-tooth rotor variable reluctance stepper motor is shown schematically in Fig. 16.13. The stator is provided with 4-phase winding and the rotor gives the effects of 6-poles.

Stepping angle is given as, irrespective the type of stepper motor (i.e. permanent magnet type or reluctance type)

$$\alpha = \frac{360^\circ}{\text{Number of phases} \times \text{number of poles}} = \frac{360}{np}$$

In the above case

$$\alpha = \frac{360^\circ}{4 \times 6} = 15^\circ$$

When the phase 1-1' is energized, keeping other coils unexcited, the rotor aligns as shown in the figure. If now the phase 1-1' is switched off and phase 2-2' is switched on to develop north pole at coil 2 and south pole at coil 2', a reluctance torque is produced to align the axes of rotor poles 2-5 with the axis of stator poles of phase 2-2'. The rotor advances 15° counter-clockwise. Successive excitation of phases 3, 4, 1' yields a total advance of 60° i.e. on rotor pole pitch. Six-times repetition of this switching sequence completes one revolution of the shaft.

Rotor can be made to rotate in clockwise direction by reversing the switching sequence i.e. in sequence 1, 4', 3', 2', 1'.

3. Hybrid Stepper Motor. This is in fact a permanent magnet stepper motor with constructional features of toothed rotor used in variable reluctance stepper motor. A 2-phase, 4-pole stator, 6-tooth rotor hybrid stepper motor is shown in Fig. 16.14. Stepping angle is $\frac{360}{2 \times 6} = 30^\circ$.

When phase 1-1' is switched on, keeping phase 2-2' unexcited, the rotor aligns as shown in Fig. 16.14. If now the phase 1-1 is switched off and phase 2-2' is closed such that lower pole becomes south pole and upper pole becomes north, this would cause rotor to turn in counter-clockwise direction by 30° and the rotor will get locked into the new position. The rotor will turn in opposite i.e. clockwise direction if the phase 2-2' were oppositely excited.

The main advantage of the hybrid motor is that if the stator excitation is removed, the rotor continues to remain locked into the same position, as before removal of excitation. This is due to the reason that the rotor is prevented to move in either direction by the torque because of the permanent magnet excitation.

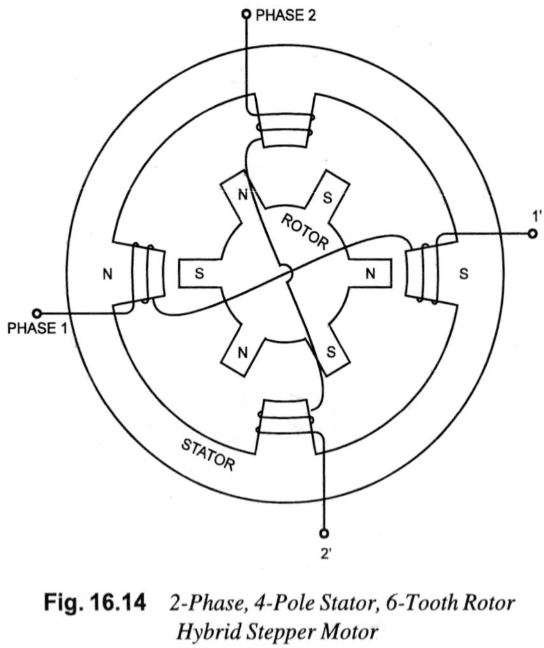

Fig. 16.14 *2-Phase, 4-Pole Stator, 6-Tooth Rotor Hybrid Stepper Motor*

Hybrid type stepper motor is used in applications where stepping angle is small (e.g.1.5, 2.5 degrees).

The variable reluctance stepper motor's step angles are small, making possible finer resolution than that is obtainable with the permanent magnet type. Maximum stepping rates usually are higher than in permanent magnet stepper motor.

Typical permanent stepper motor operates at larger steps, up to 60°, and at maximum rates of 300 pulses per second.

16.10.3. Operational Features. The stepper motor's performance can be visualised from Fig. 16.15. Every time a pulse is supplied to the motor, the rotor steps forward or backward by a definite angle, known as the step angle. It is seen that the rotor overshoots its final position and comes to rest after series of oscillations. This performance is true for low stepping rate. With the increase in pulse rate, the average velocity of the rotor increases and when the pulse rate becomes very high, i.e. the interval between pulses falls below that the rotor setting time, then the rotor behaves, as illustrated in Fig. 16.16. At very high pulse rate, the motor is said to slew. When the safe slew rate is exceeded, the rotor may pull out of synchronism. Similarly the rotor may pull out of synchronism when there are abrupt changes in the pulse rate.

For applications such as machine-tool control, it is necessary to achieve a high slewing rate to start successive machining cycles, and the tendency to error or stalling can be eliminated by incorporating the stepper motor into a closed-loop control system.

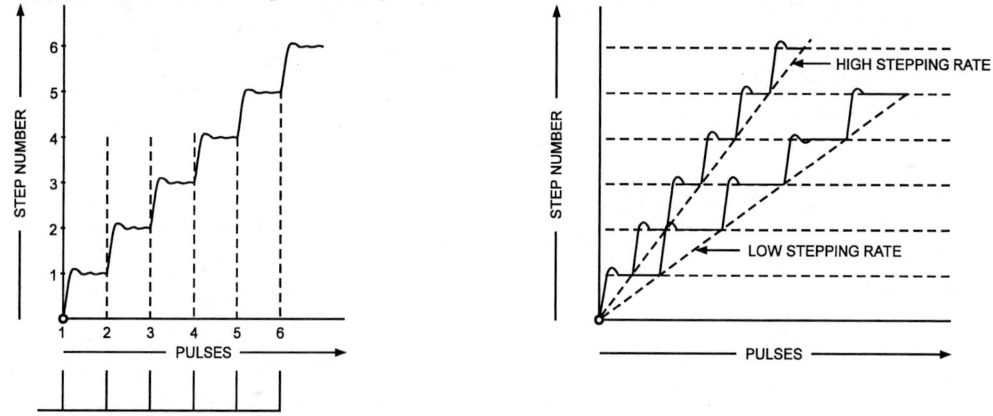

(a) (b)

Fig. 16.15

16.10.4. Applications. The stepper motor is essentially a position control device and has the following advantages over a conventional machine :

1. The angular displacement can be precisely controlled without any feedback arrangement.

2. It can be readily interfaced with microprocessor / computer based controller.

Stepper motors have a wide range of applications — paper feed motors in typewriters and teleprinters, positioning of print heads, pens in X-Y graphical plotters, recording heads in computer disk drives and in positioning of work tables and tools in numerically controlled machining equipment.

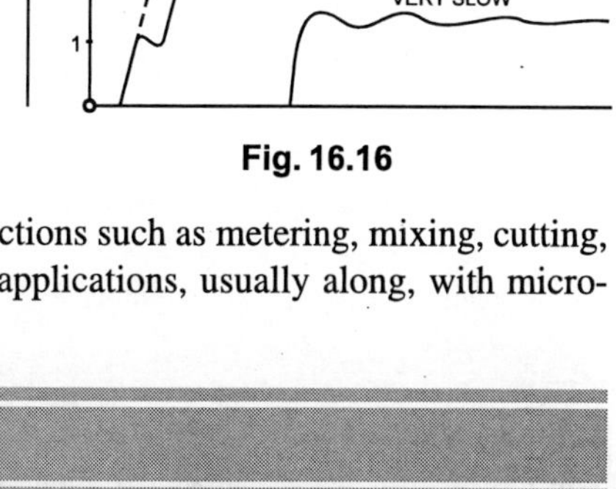

Fig. 16.16

The stepper motor is also employed to perform many other functions such as metering, mixing, cutting, blending, stirring etc. in many commercial, military and medical applications, usually along, with microprocessor and controlled switches.

HIGHLIGHTS

1. A single-phase split-induction motor has a squirrel cage rotor and a stator which has two windings. One is running winding or main winding and the other is the starting or auxiliary winding which is disconnected from the supply by a centrifugal switch when rotor attains 70 or 80 per cent of its rated speed.
2. *Operating Principle.* When a single-phase induction motor is connected to a single-phase supply, two fields of magnitude $\Phi_{max}/2$ are developed which rotate in opposite directions. An equal and opposite torque is set up by the two fields and the resultant torque is zero at start. Thus single-phase induction motor is inherently not a self-starting motor.

However, if the rotor is rotated in either direction by some external means, torque will be developed and the rotor will pick up the speed in that direction.

To obtain starting torque in a single-phase induction motor, the single winding is splitted into two windings displaced in space by 90° and having different resistance and inductance. They carry currents of different phase angle which develop a resultant field revolving in space at synchronous speed.

3. Single-phase split-phase induction motors are of two types viz resistance-start induction motor and capacitor-start induction motor.

   The *resistance-start induction motor* has a relatively poor starting torque because the phase angle between the two winding currents is only about 30–50 electrical degrees.

   The capacitor-start induction motor has a comparatively good starting torque because phase angle between the currents flowing in the two windings (running and starting winding) can be as large as 90°.

4. *Capacitor motors* may be (*i*) capacitor-start (*ii*) capacitor-start, capacitor-run and (*iii*) permanent capacitor motors.

   The capacitor-start capacitor-run motor is similar to capacitor-start motor except that the auxiliary winding and a capacitor remain connected in the circuit at all times.

   The permanent capacitor motor has no centrifugal switch and the auxiliary winding remains energized at all times when the motor is in operation.

5. *Shaded-pole motors* have projected poles with one-third pole part wrapped with a copper strip. This portion of the pole is called shaded part of the pole and because of this a revolving field is produced in the stator and starting torque is developed.

   The shaded-pole motors have poor starting torque, poor efficiency (20-50%), very poor power factor (0.5–0.6 lagging) and little overload capacity.

6. A universal motor is a specifically designed, small series wound motor up to about $\frac{1}{2}$ kW rating to operate satisfactorily on dc or ac at 115 or 230 V. These motors have approximately the same torque-speed performance over the operating range, for all frequencies from 0 to 50 Hz.

   Universal motors being high-speed motors, are smaller in size and lighter in weight as compared to other motors of the same output. Full-load power factor is high (approximately 0.9) but it is poor at start and on overloads.

   The very small power output rating universal motors, which usually does not exceed 5 or 10 watts are employed in such equipment as sewing machines, fans, portable hand tools, hairdryers, motion-picture projectors, and electric shavers. The higher rating (5–500 W) universal motors are used in vacuum cleaners, electric typewriters, food mixers and blenders, motion-picture projectors, cameras, adding machines, and calculating machines.

7. The stepper motor is a form of synchronous motor which is designed to rotate through a specific number of degrees for each electrical pulse received by its control unit. Typical steps are 2,2.5,5,7.5 and 15° per pulse. Stepper motors are built to follow signals as rapid as 1,200 pulses per second and with equivalent power ratings up to several kilowatts.

EXERCISES

1. Give applications of 3-phase induction motor and single-phase induction motor. Why is 3-phase induction motor used for medium and large load and single-phase induction motor for small load applications ?

   [U.P. Technical Univ. Elec. Engineering Second Semester 2005-06]

2. Explain why 1-ϕ induction motors are not self-starting. Discuss various methods of starting of 1-ϕ induction motor.

   [U.P.Technical Univ. Electrical Engineering Odd Semester 2013-14]

3. Why single phase induction motor is not self-starting? What are the methods of starting? Explain any one of them which is used in fan. **[U.P.Technical Univ. Electrical Engineering Odd Semester 2013-14]**

4. Explain principle of operation of single-phase induction motor using two revolving field theory. List various methods of starting. **[U.P.Technical Univ. Elec. Engineering First Semester 2005-06]**

5. Using double-revolving field theory, explain the principle of operation of a single phase induction motor.

   [G.B. Technical Univ. Electrical Engineering Odd Semester 2012-13]

6. Discuss the principle of operation of single-phase induction motor.

   [G.B. Technical Univ. Electrical Engineering First Semester 2011-12]

7. Why there is problem of starting the single-phase motor? Describe its one commonly used method of starting.

[Mahamaya Technical Univ. Electrical Engineering Second Semester 2011-12]

8. Explain why do pure, 1-phase induction motors have no starting torque ? What modifications are being done to have starting torque in normal 1-phase fan motors ? Give brief explanations.

[U.P.Technical Univ. Elec. Engineering First Semester 2004-05]

9. Explain the starting methods of single-phase induction motor.

[G.B. Technical Univ. Electrical Engineering First Semester 2010-11]

10. Discuss the principle of operation of a single-phase induction motor. How the motor is started? Explain any one method of starting of it. **[Mahamaya Technical Univ. Electrical Engineering Second Semester 2010-11]**

11. Discuss the principle of operation of a single-phase induction motor. How the motor is started? Explain any method of starting. **[U.P. Technical Univ. Electrical Engineering First Semester 2008-09]**

12. Discuss why single-phase induction motors do not have starting torque. Explain its principle of operation and list various methods of starting. **[U.P. Technical Univ. Electrical Engineering First Semester 2006-07]**

13. Why single-phase induction motor is not self starting ? What are different methods to make it self starting ? Explain one of them. **[U.P. Technical Univ. Electrical Engineering First Semester 2007-08]**

14. How the motors generally used in ceiling fans of daily use, are made self starting ? Write a short note on it.

[U.P. Technical Univ. Electrical Engineering Second Semester 2007-08]

15. Why should the auxiliary winding be disconnected in a capacitor-start motor after the motor has picked up speed ?

[U.P. Technical Univ. Electrical Engineering Second Semester 2003-04]

16. What do you know about splitting of phase in a single-phase induction motor? How does it help in starting the capacitor start single-phase induction motor? **[G.B. Technical Univ. Electrical Engineering Second Semester 2011-12]**

17. Discuss the principle of operation of a single-phase induction motor. Also write its applications.

[G.B. Technical Univ. Electrical Engineering Second Semester 2010-11]

18. Describe any one method of starting single phase induction motor with neat diagram.

[U.P. Technical Univ. Basic Electrical Engineering First Semester 2013-14]

19. Explain the working principle and applications of single-phase shaded-pole motors. Is it possible to reverse the direction of rotation of such a motor ? If yes, how ? If not, why ?

20. Describe the constructional features and operating characteristics of a shaded-pole motor. Give its uses.

21. Describe shaded-pole method of starting a single-phase induction motor.

[U.P. Technical Univ. Electrical Engineering First Semester 2003-04]

22. Describe a series (universal) motor. Give its x-sectional view when connected to ac supply. Describe its performance characteristics.

23. Name three types of stepper motors and comment on their constructional differences.

24. Describe the constructional features of 'variable-reluctance' type and 'permanent magnet' type stepper motors. Compare them. Explain the working principle of any one. Discuss the effects of the stepping rate of pulses on the stability of the stepper motor.

25. Explain the working principle of permanent magnet stepper motor with constructional diagram.

[U.P. Technical Univ. 2006-07]

SHORT ANSWER TYPE QUESTIONS WITH ANSWERS

Q. 1. How will you resolve a single pulsating field ?

Ans. A single pulsating field can be resolved into two components of half its amplitude and rotating in opposite directions with equal speed.

Q. 2. State the principle that the double revolving field theory makes use of.

Ans. The double revolving field theory makes use of the principle that pulsating field produced in a single-phase motor can be resolved into two components of half its amplitude and rotating in opposite directions with synchronous speed.

Q. 3. Is the starting torque of a single-phase induction motor zero ?

Ans. Yes, the starting torque of a single-phase induction motor is zero.

Q. 4. Is the torque developed by a single-phase induction motor at synchronous speed zero ?

Ans. No, it is negative.

Q. 5. What is the effect of increasing rotor resistance in a single-phase induction motor ?

Ans. Increase in rotor resistance of a single-phase induction motor reduces its breakdown torque, lowers the efficiency and increases the slip corresponding to the maximum torque.

Q. 6. Why single-phase induction motors are usually set on rubber spring mounts ?

Ans. In a single-phase induction motor, pulsating torque is developed as a direct consequence of the pulsating power in a single-phase circuit.

Q. 7. What is a split-phase motor ?

Ans. This is a single-phase induction motor equipped with an auxiliary winding, also called the starting winding, connected in parallel with the main winding on stator and differing from it in both phase and space. The starting winding is usually opened by a centrifugally operated device when the motor attains a predetermined speed.

Q. 8. A single-phase induction motor is provided with a main winding and an auxiliary winding. Which of the two windings should be more resistive ?

Ans. Auxiliary winding should be more resistive.

Q. 9. In the design of resistance-start single-phase induction motor the starting winding is disconnected once the slip reaches 0.25 or less. Explain.

Ans. The starting winding of a resistance-start motor is not designed for continuous operation and so it is necessary to disconnect it from the motor circuit when the motor attains a definite speed. The starting winding is usually taken out of circuit at a speed of about 75% of synchronous speed because at this speed the torque developed by the main winding alone is greater than that developed by both windings.

Q. 10. How is the direction of rotation of a single-phase induction motor reversed ?

Ans. The direction of rotation of a single-phase induction motor (split-phase type) is reversed by reversing the leads to the main or starting winding, but not both.

Q. 11. How does a capacitor-start motor differ from a resistance-start motor ?

Ans. Fundamentally the capacitor-start motor is quite similar to a resistance-start motor, except that its starting winding has a few more turns and is wound with comparatively thick wire than the starting winding of a resistance-start motor and also the external resistance, connected in series with the starting winding, is replaced by a large electrolytic capacitor.

Q. 12. How the direction of rotation of a capacitor-start induction motor is changed ?

Ans. The direction of rotation of a capacitor-start, induction run motor is reversed by reversing either the starting or running winding leads, not both, to the line.

Q. 13. What happens to a capacitor-start induction motor if the auxiliary winding is left in the circuit after it has been used to start the motor ?

Ans. The auxiliary winding of a capacitor-start induction motor is not designed for continuous operation and if left in the circuit after it has been used to start the motor, it will burn out. In addition to it, the short-duty rated electrolytic capacitor connected in series with the starting winding will also get damaged.

Q. 14. Are capacitor-start induction-run motors usually dual voltage or single voltage motors ?

Ans. Capacitor-start, induction-run motors are usually dual voltage motors.

Q. 15. If a capacitor-start motor is rated as 115/230-V motor (the capacitor is usually rated at 115 V), how is it connected, when the motor is used on 230 V ?

Ans. The main or running winding of a dual voltage motor is in two sections so that they may be connected in parallel for 115-V operation and in series for 230-V operation. The capacitor is connected in series with the starting winding; therefore, when operating at 115 V, the winding leads are connected across the line at start, but when operating on 230 V, one lead is connected to the one side of the line and the other lead to the mid point of the running windings.

Q. 16. What is the advantage of using a capacitor-start motor over a resistance-start split-phase motor ?

Ans. The capacitor-start motor develops more starting torque than does an equally rated resistance-start split-phase motor with a lower inrush current.

Q. 17. What is a capacitor-start, capacitor-run motor ?

Ans. This is the same motor as the capacitor-start motor, except that it has an extra oil type capacitor connected in series with the starting winding, which always remains in the circuit.

Q. 18. In a 2-phase capacitor motor which capacitor is of fairly high value and to what order ?

Ans. In a 2-phase capacitor motor the starting capacitor is about 10 – 15 times as large as the running capacitor.

Q. 19. State the advantage of capacitor-run over capacitor-start motor.

Ans. Capacitor-start, capacitor-run motors are quieter in operation, develop more pull-out torque, operate with higher power factor and efficiency, and consequently draw less current than the capacitor-start motors.

Q. 20. What happens when the auxiliary winding of a capacitor motor is disconnected during running condition ?

Ans. The motor will continue running but will develop small torque. Once it stops, it will not start again.

Q. 21. Name the motor being used in a ceiling fan ?

Ans. Split-phase capacitor type single-phase induction motor.

Q. 22. What is the use of condenser in single-phase ac motor?

[Mahamaya Technical Univ. First Semester 2011-12; G.B. Technical Univ. Odd Semester 2012-13]

Ans. Single-phase induction motors are not self starting. Single-phase induction motors can be made self-starting by using the principle of phase splitting. The capacitor is used in single-phase capacitor motors (capacitor-start motor, capacitor start-capacitor-run motor and permanent capacitor motor) to cause phase splitting. By using a capacitor of suitable value in starting or auxiliary winding, the starting winding current at standstill can be made to lead the main winding current by 90°.

Q. 23. What is a shaded-pole motor ?

Ans. Shaded-pole motor is a salient-pole, split-phase type, squirrel cage rotor single-phase induction motor provided with one or more auxiliary short-circuited stator windings displaced in magnetic position from the main winding.

Q. 24. Does a shaded-pole motor have much starting torque ?

Ans. No, it has very little starting torque.

Q. 25. In what direction do shaded-pole motors turn ?

Ans. Shaded-pole motor rotates in a direction from the unshaded part of the pole to the shaded part.

Q. 26. What role do 'magnetic bridges' play in the operation of a shaded-pole motor ?

Ans. The magnetic bridges cause the flux in the shaded portion of the pole to lag behind the flux in the unshaded portion. This gives, in effect a motion of flux across the pole face and under the influence of this moving flux a starting torque is developed.

Q. 27. State the applications of shaded-pole single-phase induction motors.

Ans. Shaded-pole motor develops very little starting torque and, therefore, it is used only for small fans, electric clocks, hairdryers, time phonographs, advertising displays, toys, instruments and such other applications.

Q. 28. How can a shaded-pole motor be reversed in direction of rotation ?

Ans. Unless the machine is constructed so that the shading coil can be shifted to the other side of the pole, the direction of rotation of such a machine cannot be reversed.

Q. 29. What is a universal motor ?

Ans. A universal motor is a series wound motor, compensated or uncompensated, specifically designed to operate on direct or single-phase alternating current of the same voltage at approximately the same speed and output.

Q. 30. Why should a motor be named as universal motor ?

Ans. Because the motor runs on dc as well as on ac, it is named as universal motor.

Q. 31. What is the principle of operation of universal motor ?

Ans. DC motors rotate in the same direction regardless polarity of the supply i.e. if the polarity of the line terminals of any dc motor are reversed, both the direction of the flux and direction of the current in the armature conductors is reversed and, therefore, torque developed remains in the same direction, and the motor continues to rotate in the direction in which it was rotating before the line terminals were reversed.

Q. 32. Why compensating winding is provided in a universal motor for ac operation ?

Ans. The compensating winding is provided in a universal motor for ac operation to neutralize reactance voltage set up by the alternating flux due to transformer action.

Q. 33. How can the direction of rotation of a universal motor be reversed ?

Ans. The direction of rotation of universal motor can be reversed by reversing either the field or armature leads, but not both.

Q. 34. Why are high speeds often desirable in operation of universal motors ?

Ans. Universal motors are designed for operation at high speeds (exceeding 3,500 rpm) owing to the difficulty in obtaining like performance on dc and ac at lower speeds.

Q. 35. What is a stepper motor ?

Ans. A motor in which the rotor turns in discrete movements is called *a stepper motor.*

Q. 36. What are the two types of stepper motors ?

Ans. Permanent-magnet type and variable-reluctance type motors.

Q. 37. For the same number of poles which stepper motor produces (*i*) largest stepping angle, (*ii*) smallest stepping angle ?

Ans. (*i*) Permanent magnet type—largest stepping angle

(*ii*) Variable reluctance type—smallest stepping angle.

Q. 38. Where is stepper motor used ?

Ans. Stepper motors are used in industrial situations which call for precise positioning of an object or precise control of speed without having to resort to closed-loop feedback.

❑❑❑

17

CHAPTER

Three-Phase Synchronous Machines

17.1. INTRODUCTION

A synchronous machine is an ac machine in which the rotor moves at a speed which bears a constant relationship to the frequency of currents, in the armature winding. A synchronous machine is one of the important types of electric machines. Large ac networks operating at constant frequency of 50 Hz rely almost exclusively on *synchronous generators,* also called the *alternators*, for the supply of electrical energy. Private, stand-by and peak-load plants with diesel or gas turbine prime movers also have synchronous generators. Synchronous motors provide constant speed industrial drives with the possibility of power factor correction. Synchronous machines are generally constructed in larger sizes. Small size alternators are not economical. The modern trend is to build alternators of very large sizes capable of generating 500 MVA or even more. The synchronous motor is rarely built in small sizes owing to superior performance characteristics and economical construction of induction motors.

17.1.1. Operating Principle. The operating principle of a synchronous machine is fundamentally the same as that of a dc machine, but, unlike the latter, in the synchronous machine there is no need to rectify the time varying emf which is induced in the armature winding. Consequently a synchronous machine does not require a commutator. It is, in fact, quite possible to use a dc generator as an alternator by placing a set of collector rings on the shaft and connecting these rings to the proper points on the armature winding; brushes riding on the rings can then be connected to the load. But unlike dc generator, they are to be driven at a very definite constant speed as the frequency of generated emf is determined by that speed. The latter is usually referred to as the *synchronous speed,* for which reason these machines are frequently called the *synchronous generators*.

Synchronous generators, because of absence of commutator, are comparatively simple and possess several important advantages over the dc generators.

17.1.2. Classification of Synchronous Machines. Synchronous machines, according to their applications, may be *synchronous generators* or *synchronous motors*. A synchronous generator is a synchronous machine which receives mechanical energy from a prime mover (steam turbine, hydraulic turbine or diesel engine) to which it is mechanically coupled and delivers electrical energy. A synchronous motor receives electrical energy from ac supply main and drives mechanical load.

The synchronous machines may be *single-*, *two-* or 3-*phase* types. Three windings may be connected either in star or delta. Large ac machines are invariably 3-phase type.

Based on the construction of the machines, the synchronous machines may be classified as (*i*) rotating armature type and (*ii*) rotating field type.

Since it is immaterial for generation of an induced emf whether a conductor moves across a magnetic field or vice versa, synchronous generators may be constructed with either the armature or the field structure as the revolving member.

Rotating armature type alternator looks very much like a dc generator except that there are 3 slip rings in place of commutator (or 4 slip rings if it is desired to provide a connection to the generator neutral). In such generators the required magnetic field is produced by dc electromagnets placed on the stationary member called the *stator*, and the current generated is collected by means of brushes and slip rings on the revolving member, called the *rotor*. Such an arrangement is economical for the small low-voltage generators. Rotating armature type alternators are built only in small ratings up to about 200 or 250 kVA because the voltage generated is comparatively low and the current to be collected by the brushes small, no difficulty being experienced in collecting such a current. Such machines are suitable for small power plants, isolated lighting plants, where medium or small size machines are required.

Practically all medium and large machines are always constructed with revolving field. The advantages of stationary armature and revolving field system are given below:

1. It is easier to insulate stationary armature winding for very high voltage e.g. as high as 33,000 volts.
2. The load circuit can be connected directly with the fixed terminals of the stator (or armature winding) without passing through slip rings and brushes.
3. The armature winding can be more easily braced in a rigid frame to prevent any deformation which could be developed by the mechanical stresses set up due to short-circuit currents and the high centrifugal forces brought into play.
4. The armature winding is cooled more readily because the stator core can be made large enough with many air passages or cooling ducts for forced air circulation.
5. Only two slip rings are required for the supply of direct current to the rotor and since the exciting current is to be supplied at low voltage of 125 or 250 V, there is no difficulty in insulating them.
6. Since the exciting current is relatively small, the slip rings and the brush gear need be of only light construction.
7. Due to simple, light and robust construction of the rotor, higher speed of rotating dc field is possible.

Since in dc machines, the commutator makes necessary that either the armature should be rotating one, or brushes should revolve with the field, so it is convenient to have armature as a rotating member in dc machines.

Revolving field alternator with essential parts is shown in Fig. 17.1.

The synchronous machines may be classified as (*i*) salient pole machines and (*ii*) cylindrical rotor machines depending upon the type of construction used for the rotor. The salient pole construction is used for generators and motors of all ranges of output and up to all but the higher speeds. Medium-and large-sized generators for the highest speeds are of the cylindrical rotor type.

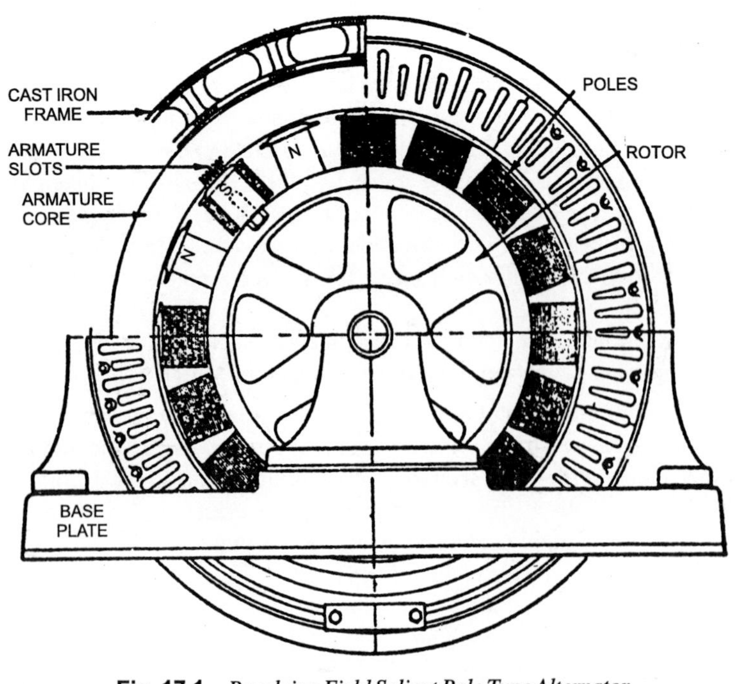

Fig. 17.1 *Revolving Field Salient Pole Type Alternator*

The synchronous generators, based on the type of prime movers to which they are mechanically coupled, may be classified as (*i*) turbo-generators, (*ii*) hydrogenerators and (*iii*) diesel engine driven generators.

17.2. CONSTRUCTION

Synchronous machine consists essentially of two parts namely the armature (or stator) and the field magnet system (or rotor).

1. Stator. The armature is an iron ring, formed of laminations of special magnetic iron or steel alloy (silicon steel) having slots on its inner periphery to accommodate armature conductors and is known as *stator*. The whole structure is held in a frame which may be of cast iron or welded steel plates. The field rotates in between the stator, so that flux of the rotating field cuts the core of the stator continuously and therefore causes eddy current loss in the stator core. To minimize the eddy current loss, the stator core is laminated. The laminations are stamped out in complete rings (for smaller machines) or in segments (for larger machines) and insulated from each other with paper or varnish. The stampings also have openings which make axial and radial ventilating ducts to provide efficient cooling. A general view of stator and frame is shown in Fig. 17.2.

Slots provided on the stator core are mainly of two types (*i*) open slots and (*ii*) semi-closed slots, as shown is Figs 17.3 (*a*) and 17.3 (*b*) respectively.

The open slots are more commonly used because the coils can be form-wound and insulated prior to being placed in the slots giving least expenditure and more satisfactory winding method. This type of slots also facilitate in removal and replacement of defective coils. But this type of slots have disadvantage of distributing the air gap flux into branches or tufts which tends to produce ripples in the emf wave. The semi-closed type slots are better in this respect but do not permit the use of form-wound coils. Totally closed slots are rarely used.

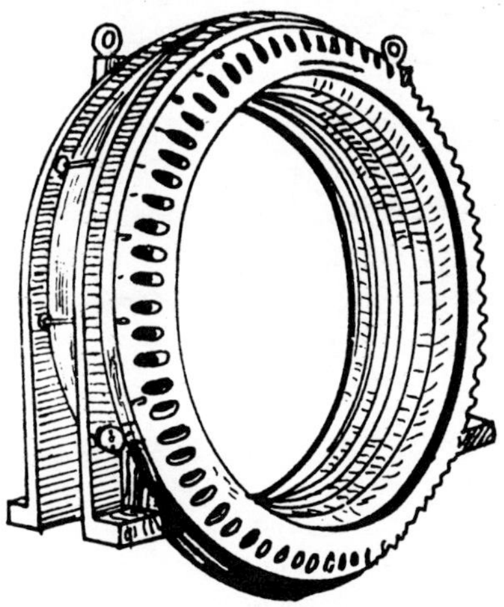

Fig. 17.2

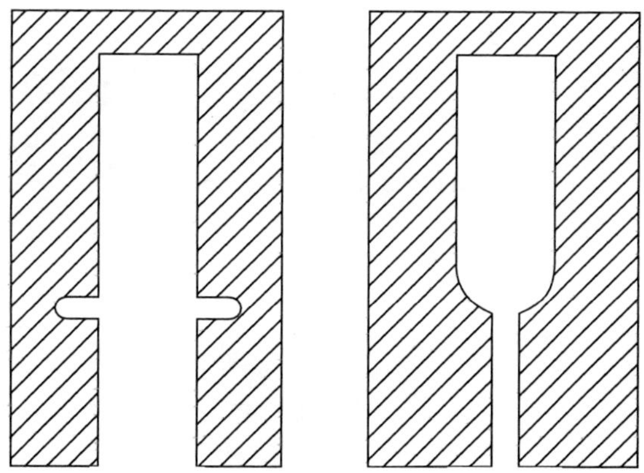

(*a*) *Open Slot* (*b*) *Semi-Closed Slot*

Fig. 17.3

2. Rotor. The field system is just like that of a dc generator which is excited from a separate source of 125 or 250 V dc supply. The excitation is usually provided from a small dc shunt or compound generator, *known as an exciter*, mounted on the shaft of the alternator itself. The field system of the alternator is rotated within the armature ring and is known *as rotor*. The exciting current is supplied to the rotor through two slip rings and brushes. The polarities of the field produced are alternately North and South.

The power rating of the exciter is ordinarily 0.3 to 1% of power rating of synchronous-generator. The rated voltage of the exciter is usually between 125 and 250 volts.

Rotors are of two types namely (*i*) salient-pole type and (*ii*) smooth cylindrical type or non-salient pole type. Both types are shown in Fig. 17.4.

Salient Pole Type. The rotor of this type is used almost entirely for low- and moderate-speed alternators, since it is least expensive and provides ample space for the field ampere-turns. Salient poles cannot be employed in high speed generators on account of very high peripheral speed (100 to 170 metres per second) and the difficulty of obtaining sufficient mechanical strength.

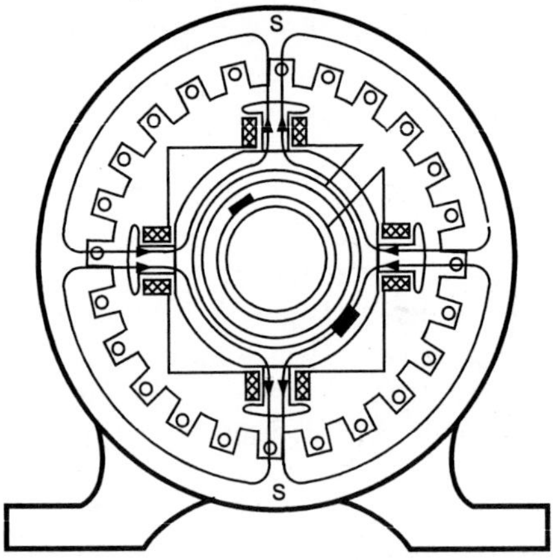

(a) Salient Pole Type Field Structure

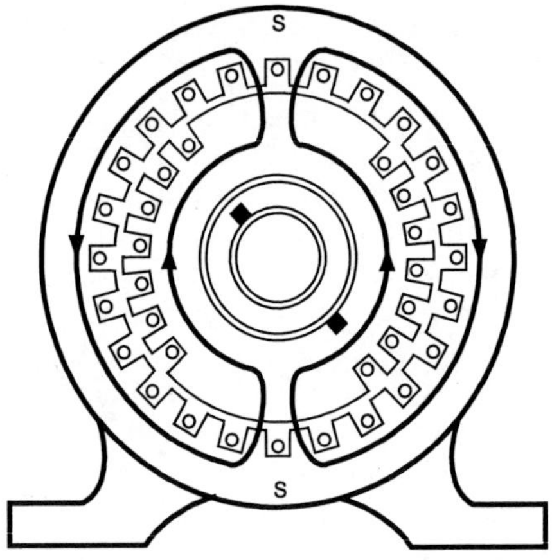

(b) Non-Salient Pole Type Field Structure

Fig. 17.4

The salient poles are made of thick steel laminations riveted together and are fixed to rotor by a dovetail joints. The pole faces are usually provided with slots for damper windings. These dampers are useful in preventing the hunting. The pole faces are so shaped that the radial air gap length increases from the pole centre to the pole tips so that the flux distribution over the armature is sinusoidal and waveform of generated emf is sinusoidal. The field coils are placed on the pole-pieces and connected in series. The ends of the field windings are connected to a dc source through slip rings carrying brushes and mounted on the shaft of the field structure.

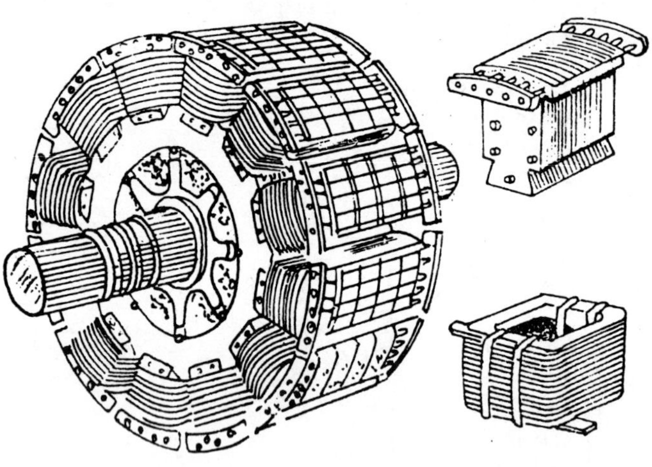

Fig. 17.5 *Salient Pole Field Structure*

The salient pole field structure has the following special features :

(*i*) They have large diameter and short axial length.

(*ii*) The pole shoes cover about 2/3 of pole pitch.

(*iii*) These are employed with hydraulic turbines or diesel engines. The speed is 100 to 375 rpm.

Smooth Cylindrical or Non-Salient Pole Type. The rotors of this type are used in very high-speed alternators (driven by steam turbines). To reduce the peripheral velocity, the diameter of the rotor is reduced and axial length is increased. Such rotors have two or four poles.

It consists of a cylindrical steel forging which is suitably worked mechanically and treated thermally. The forging has radial slots in which the field copper, usually in strip form, is placed. The coils are held in place by steel or bronze wedges and coil ends are fastened by metal rings. The slots over certain portions of the core are omitted to form pole faces, as shown in Fig. 17.7. Each slot is provided with a ventilation hole at the bottom. General view of an assembled smooth cylindrical rotor is shown in Fig. 17.6.

The non-salient field structure has the following special features :

(*i*) They are of small diameter and of very long axial length.

(*ii*) Less windage loss.

(*iii*) High operating speed (3,000 rpm).

(*iv*) Robust construction and noiseless operation.

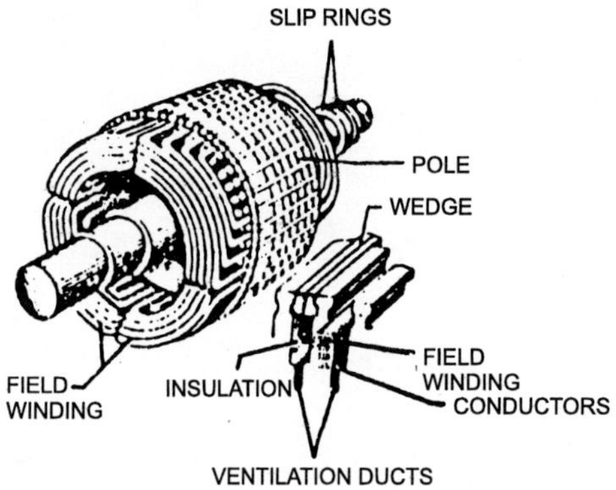

Fig. 17.6 *Non-salient Pole (Cylindrical) Field Structure*

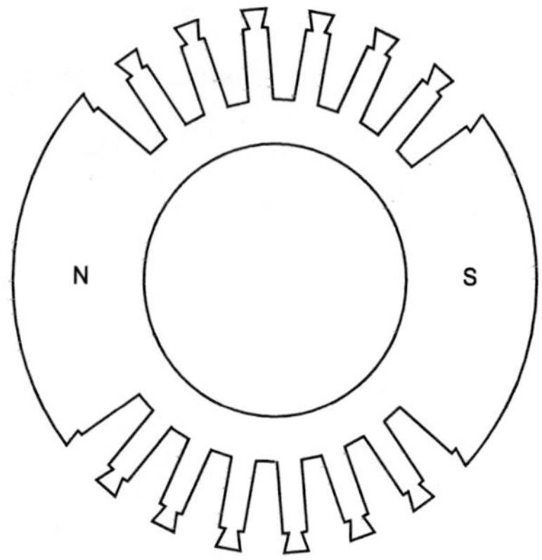

Fig. 17.7

Air Gap. A very small air gap increases the stray-load loss; also the eccentricity in the air gap can result from mechanical difficulties. This also increases the synchronous reactance. A large air gap needs larger excitation. A compromise has to be made. Generally the ratio of air gap to the pole pitch is between 0.008 to 0.02.

17.3. PRODUCTION OF SINUSOIDAL ALTERNATING EMF

When the rotor is rotated by means of some prime mover, the armature conductors cut the magnetic flux, therefore, an emf is induced in the armature conductors, due to electromagnetic induction effect.

When the conductor is opposite the neutral planes, as at A, C and E, induced emf in it is minimum because flux density is minimum there. When the conductor is opposite the middle of the poles, as at B, D and F, induced emf in it is maximum, the direction of induced emf depends upon the name of the pole influencing the conductor at any given instant. Thus an alternating emf is induced in the conductors which goes through one complete cycle in an angular distance equal to twice of the pole pitch. Though the shape of the wave of alternating induced emf is not exactly the sinusoidal but taken as sinusoidal.

17.4. FREQUENCY OF INDUCED EMF

As mentioned above, an emf induced in conductor goes through one complete cycle in an angular distance equal to twice the pole pitch.

If the number of poles on rotor of an alternator is P, then P/2 cycles of emf are completed in one revolution. The number of cycles per second, *known as frequency f,* will be equal to the product of number of cycles of emf per revolution and number of revolutions made per second by the rotor.

i.e. Frequency, $f = \frac{P}{2} \times n$

where n is the number of revolutions made per second by rotor

$$\text{or } f = \frac{P}{2} \times \frac{N}{60} = \frac{PN}{120} \quad \text{...(17.1)}$$

where N is the number of revolutions made per minute by the rotor.

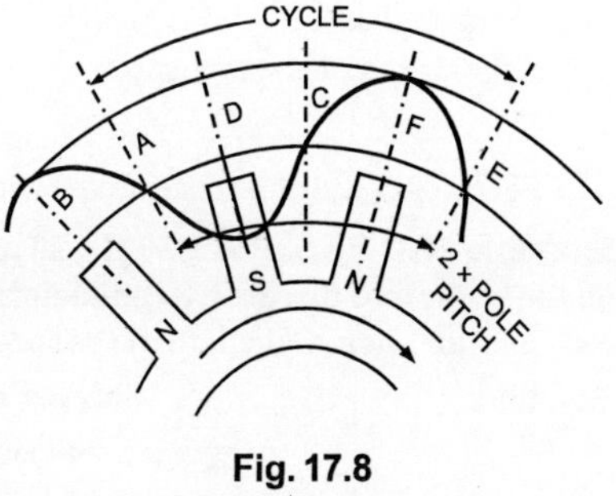

Fig. 17.8

Hence frequency of induced emf or current induced in stator conductors depends upon the number of poles and speed of rotor.

Example 17.1 **A 6-pole ac generator is running and producing the frequency of 60 Hz. Calculate the revolutions per minute of the generator. If the frequency is reduced to 20 Hz, how many number of poles will be required if the generator is to be run at the same speed.**

Solution:

$$\text{Frequency generated, } f = 60 \text{ Hz}$$

$$\text{Number of poles, P} = 6$$

$$\text{Speed, N} = \frac{120\, f}{\text{P}} = \frac{120 \times 60}{6} = 1{,}200 \text{ rpm } \textbf{Ans.}$$

$$\text{When frequency, } f' = 20 \text{ Hz}$$

$$\text{Number of poles required, P}' = \frac{120\, f'}{\text{N}} = \frac{120 \times 20}{1{,}200} = 2 \textbf{ Ans.}$$

17.5. EMF EQUATION

Let Z_p = Number of conductors in series per phase

P = Number of poles

Φ = Useful flux per pole in webers

N = Rotational speed in rpm

and f = Frequency in hertz and is equal to $\frac{NP}{120}$

The flux cut by any conductor while passing from the centre of one inter polar gap to the centre of the next is Φ webers and since during the movement, the emf wave completes half cycle *i.e.* the time taken is $\frac{1}{2f}$ seconds, therefore, the average rate of cutting the flux,

$$\frac{d\Phi}{dt} = \frac{\Phi}{1/2f} = 2f\Phi \text{ Wb/s}$$

Hence average emf induced in each conductor = $2f\Phi$ volts

Average emf per phase, E_{av}/phase = Number of conductors in series/phase arranged in one slot/pole × average emf induced per conductor

$$= Z_p \times 2f\Phi = 2\text{T} \times 2f\Phi = 4\,\Phi f \text{T volts} \qquad \because Z_p = 2\text{T (turns)}$$

For distributed winding the average value of emf per phase will be K_d times the above value

$$i.e.\ E_{av}/\text{phase} = 4\,K_d\,\Phi f \text{T volts}$$

For short-pitched winding the true average value of emf per phase will be K_p times the above value

$$i.e.\ E_{av}/\text{phase} = 4\,K_d\,K_p\,\Phi f \text{T volts}$$

$$\text{and } E_{rms}/\text{phase} = \text{Form factor, } K_f \times E_{av}/\text{phase} = 4\,K_f\,K_d\,K_p\,\Phi f \text{T volts}$$

For sinusoidal wave of emf, K_f = 1.11

$$E_{rms}/\text{phase} = 4.44\,K_d\,K_p\,\Phi f \text{T volts} \qquad ...(17.2)$$

For full-pitched and concentrated windings, $K_p = 1$ and $K_d = 1$

If the alternator is star-connected, as usually,

$$\text{Line voltage, } E_L = \sqrt{3}\ E_{rms}/\text{phase} = \sqrt{3} \times 4.44\,K_d\,K_p\,\Phi f \text{T volts} \qquad ...(17.3)$$

For full-pitched and concentrated windings, $K_p = 1$ and $K_d = 1$

Example 17.2 **A 3-phase, 50 Hz, 20 pole salient pole alternator with star-connected stator winding has 180 slots on the stator and there are 8 conductors per slot. The flux per pole is 25 mWb and is sinusoidally distributed. The coils are full pitch. Calculate (i) the speed, (ii) the generated emf per phase and (iii) line emf.**

Solution:

$$\text{Flux per pole, } \Phi = 25 \text{ mWb} = 0.025 \text{ Wb}$$

$$\text{Frequency, } f = 50 \text{ Hz}$$

$$\text{Number of armature conductors, } Z = 180 \times 8 = 1{,}440$$

$$\text{Number of armature conductors per phase} = \frac{1{,}440}{3} = 480$$

$$\text{Number of turns per phase, } T = \frac{480}{2} = 240$$

$$\text{Number of poles, } P = 20$$

(*i*) Speed, $N = \dfrac{120\,f}{P} = \dfrac{120 \times 50}{20} = 300$ rpm **Ans.**

(*ii*) Generated emf per phase $= 4.44\,\Phi f T$ volts $= 4.44 \times 0.025 \times 50 \times 240 = 1{,}332$ V **Ans.**

(*iii*) Line emf $= \sqrt{3} \times 1{,}332 = 2{,}307$ V **Ans.**

17.6. SYNCHRONOUS MOTORS

A synchronous motor has the same relationship to an alternator as a dc motor has to a dc generator *i.e.* if an alternator is supplied ac power it is capable of rotating as a motor and doing mechanical work. If the mechanical power supplied to a rotating alternator is removed while the dc field remains energized, and an ac supply is then connected across the armature terminals, torque will be developed and the alternator will continue to rotate at a speed determined by the ac supply frequency and number of poles on the synchronous machine. Changes in mechanical load within the machine's rating will not cause a change in speed.

In case of a dc motor the field and armature, both, are supplied from dc supply mains, since both require direct current. But in case of a synchronous motor the field structure is to be energized by direct current, as in case of an alternator, whereas the armature winding is connected to a 3-phase ac supply mains. Such a machine, therefore, requires two sources of supply, viz., an ac source to supply the power for driving the armature and a dc source to excite the field. DC excitation may be provided either from an exciting plant dc system or, if no direct current is available, from separate dc generators. Direct-connected exciters are frequently found on high-speed motors. These are overhung from specially constructed motor brackets with the exciter armature mounted on an extension of the motor shaft. On low-speed machines, belted exciters are sometimes employed, notably on air compressor drives. Separate motor-generator sets are widely used for excitation of low-speed synchronous motors. Where a group of synchronous motors is installed, it is sometimes economical to use a single motor-generator set for exciting the entire group, although greater flexibility is obtained when each motor is provided with its own exciter. Synchronous motor with a directly connected exciter is shown in Fig. 17.9.

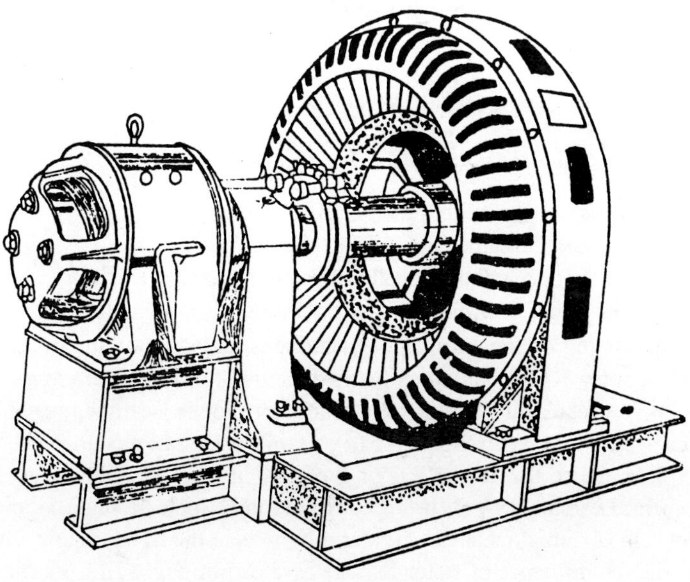

Fig. 17.9 *Synchronous Motor With Directly Connected Exciter*

Essentially, the construction of a polyphase synchronous motor is the same as that of an alternator. Certain features of its design may be varied to suit the particular requirements. A synchronous motor may have either a revolving armature or a revolving field, but the revolving field construction is the more common.

The essential parts of a three-phase synchronous motor are (*i*) laminated stator core with 3-phase armature winding (*ii*) rotating field structure complete with damper winding and slip rings (*iii*) brushes and brush holders and (*iv*) two end shields to house the bearings that support the rotor shaft.

The stator core and windings of a synchronous motor are similar to those of a 3-phase alternator. The leads from the stator windings terminate in a terminal box usually mounted on the side of the motor frame. The rotor of the synchronous motor usually has salient field poles connected to give alternate polarity. The number of rotor field poles must be the same as that of stator field poles. The field circuit leads are brought out to two slip rings mounted on the shaft. Carbon brushes mounted in brush-holders make contact with the two slip rings. The terminals of the field circuit are brought out from the brush holders to a second terminal box mounted on the motor frame.

In the most common type of field construction, the poles are somewhat different from those of the alternator. In addition to the exciting winding, a synchronous motor often is provided with a squirrel cage winding on the rotor. The slots of this winding are in the pole faces and are parallel to the shaft. The ends of the copper or bronze bars embedded in the slots are short circuited at both ends by conducting rings. This short-circuited winding on the rotor is known as the *cage* winding, the *damper* winding, the *amortisseur* winding, or sometimes, to describe its function, the *starting* winding.

17.7. PRINCIPLE OF OPERATION OF SYNCHRONOUS MOTOR

The action of the motor may be followed by referring to Fig. 17.10, which illustrates a 3-phase, 2-pole synchronous machine. The rotor has two poles and the stator has two poles per phase. In practice it is usual to have salient poles on the rotor, but the armature winding is housed in slots in the concave periphery of the stator. In the figure salient type stator poles have been shown for convenience only.

As the rotor is excited from dc supply, so the poles of the rotor retain the same polarity throughout but the polarity of the stator poles changes as it is connected to an ac supply mains.

First of all let us consider the rotor as stationary and in the position shown in Fig. 17.11 (*a*), At this instant the rotor S-pole is attracted to the stator N-pole and, therefore, the rotor tends to rotate in clockwise direction. After half a period $\left(i.e. \text{ after } \frac{T}{2} \text{ second whereas } T = \frac{1}{f}\right)$ the polarity of stator poles is reversed but the polarity of the rotor poles remains the same [Fig. 17.11 (*b*)]. At this instant, the rotor S-pole is repelled by the stator S-pole, being similar in nature and, therefore, rotor tends to rotate in counter-clockwise direction. Thus the torque acting on the rotor of a synchronous motor is not unidirectional but pulsating one and due to inertia of the rotor, it will not move in any direction. So the *synchronous motor has got no self-starting torque.*

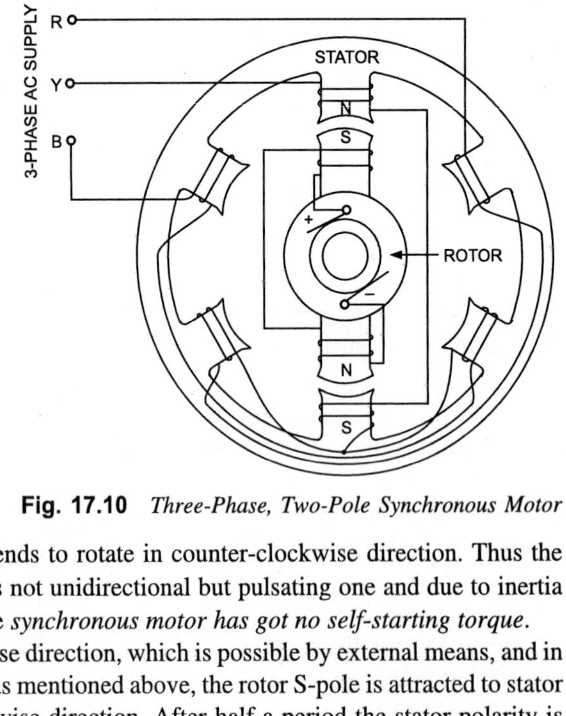

Fig. 17.10 *Three-Phase, Two-Pole Synchronous Motor*

Now let us consider the rotor rotating in clockwise direction, which is possible by external means, and in the position shown in Fig. 17.11 (*a*). At this instant, as mentioned above, the rotor S-pole is attracted to stator N-pole and so the torque acts on the rotor in clockwise direction. After half a period the stator polarity is reversed *i.e.* stator S-pole becomes N-pole and N-pole becomes S-pole but if the rotor is rotated at such a speed by some external means at the starting moment that rotor S-pole advances by a pole pitch so that it is again under the influence of stator N-pole, as shown in Fig. 17.11 (*c*), the torque acting on the rotor will be again clockwise. Hence a continuous (unidirectional) torque will be obtained. Now if the external means is removed

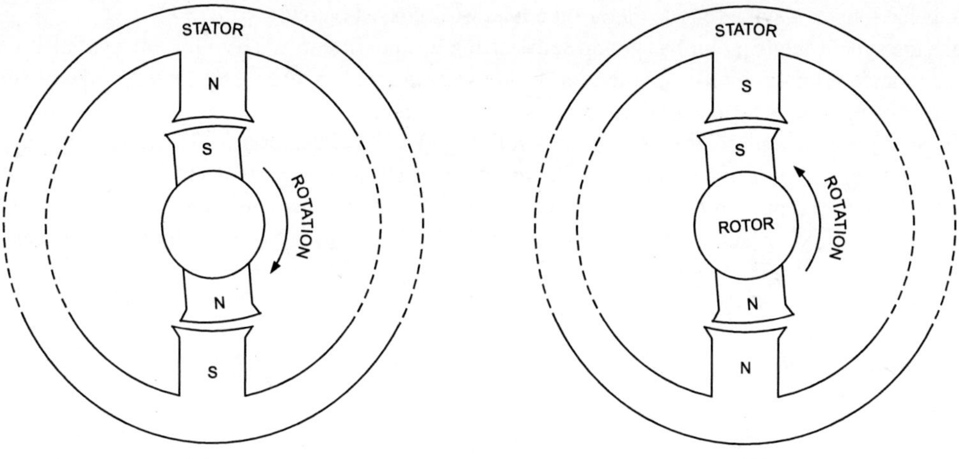

(a) (b)

Fig. 17.11

the rotor will continue to rotate in clockwise direction under the influence of clockwise continuous torque acting on the rotor.

Hence to obtain a continuous torque it is necessary that the rotor rotates at such a speed that it moves through the distance equal to pole pitch in half the period i.e. in $\frac{T}{2}$ or $\frac{1}{2f}$ second where f is the supply frequency. If there are P poles then to complete one revolution it should take $\frac{1}{2f}$ second or the rotor should rotate at a speed of $\frac{2f}{P} \times 60$ revolutions per minute.

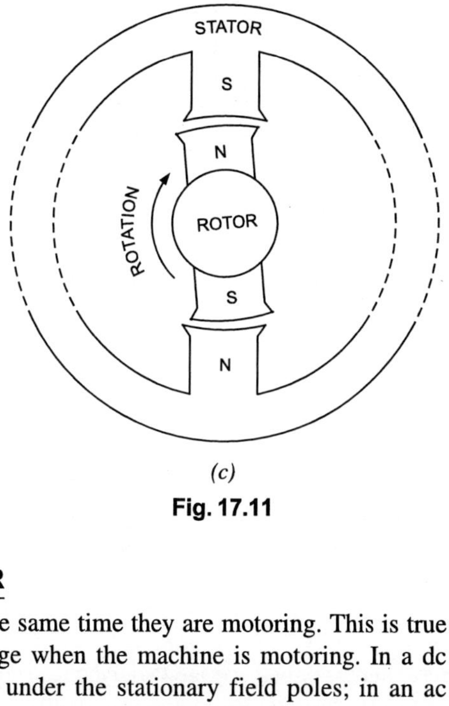

(c)

Fig. 17.11

Hence to obtain a continuous torque the rotor should rotate at synchronous speed given by the expression $N_s = \frac{120f}{P}$.

17.8. EFFECT OF LOAD ON A SYNCHRONOUS MOTOR

All the electric motors, whether dc or ac act as generators at the same time they are motoring. This is true because an emf is always induced opposing the applied voltage when the machine is motoring. In a dc machine, the rotating armature conductors cut magnetic flux under the stationary field poles; in an ac machine, as in an induction motor or a synchronous motor, the rotating magnetic field cuts the stationary armature conductors of the stator. In fact, it is this very opposing induced emf that limits the armature current to a value that is just enough to take care of the power requirements of the motor for a particular load.

When the load on a dc shunt motor is increased, speed of motor falls slightly. This reduces the induced emf, called the back emf, so that resultant voltage across the armature $(V - E_b)$ increases, which in turn increases the armature current, since armature current is given by the expression $I_a = \left(\frac{V - E_b}{R_a}\right)$, to supply extra torque and power required by the increased load. In the case of synchronous motor, however, the speed, as we have already discussed, remains constant irrespective of magnitude of load, therefore, method of drawing increased current from supply mains on synchronous motor cannot be same as that of a dc shunt motor.

The behaviour of a synchronous motor on a load is discussed below:

Since the synchronous motor has got no self-starting torque, first it is driven by some external means as an alternator and is synchronised with the supply. At the instant of synchronising the induced emf across the stator circuit of the motor (also called the back or counter emf or excitation voltage) is equal to the applied line voltage V but opposite in direction, as shown in Fig. 17.13 (*a*) vectorially. The relative positions of a stator pole and dc field pole of a synchronous motor at the synchronising instant are shown in Fig. 17.12 (*a*). The centre lines of the two poles coincide. Since V = E, hence the resultant voltage is zero and so is the armature current. Motor intake is zero, as there is neither load nor losses to be met by it. Under these conditions the motor is said to be *floating*.

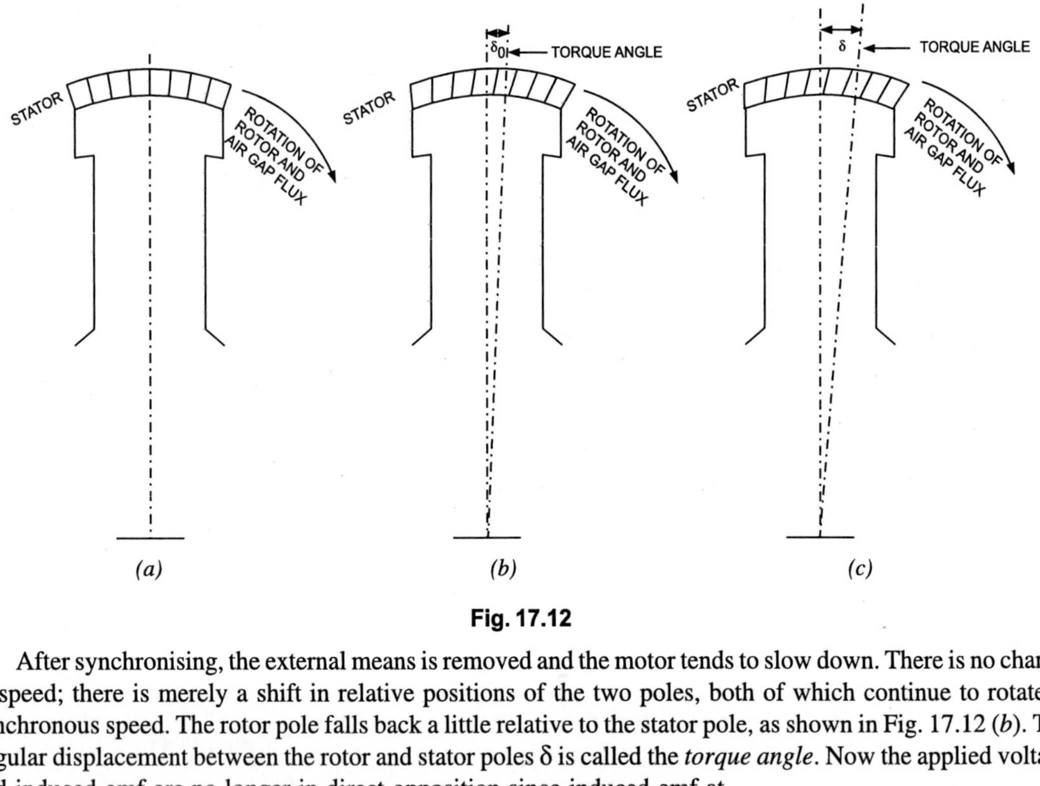

Fig. 17.12

After synchronising, the external means is removed and the motor tends to slow down. There is no change in speed; there is merely a shift in relative positions of the two poles, both of which continue to rotate at synchronous speed. The rotor pole falls back a little relative to the stator pole, as shown in Fig. 17.12 (*b*). The angular displacement between the rotor and stator poles δ is called the *torque angle*. Now the applied voltage and induced emf are no longer in direct opposition since induced emf at this instant lags behind the induced emf at the synchronising instant, say by an angle δ_0, as shown in Fig. 17.13 (*b*). The resultant voltage $\mathbf{E_R}$ (the resultant of applied voltage **V** and induced emf **E**) causes a current $I_0 = \frac{\mathbf{E_R}}{\mathbf{Z}_s}$ to flow in the stator winding; where Z_s is the synchronous impedance per phase of the armature circuit. I_0 lags E_R by an angle θ, called the *internal angle*, where $\theta = \tan^{-1} \frac{X_s}{R_e}$; X_s and R_e being the stator synchronous reactance and effective resistance respectively. Hence on no load motor draws power equal to $VI_0 \cos \phi_0$ (for one phase) from the supply mains which is sufficient to make the motor running continuously at synchronous speed.

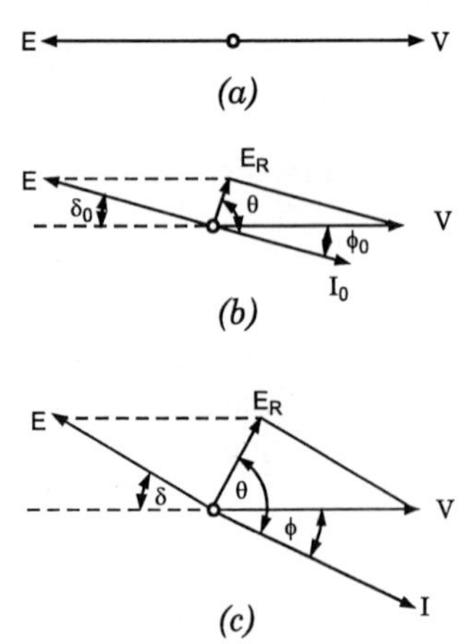

Fig. 17.13

At no load the torque angle δ_0 is very small, so the resultant voltage E_R acting across the armature circuit is small, due to which a small current is drawn from the supply mains. At no load, the torque angle δ_0 is so small that it can be assumed zero (neglecting the motor internal losses).

Now when the motor is loaded, it slows down momentarily to adjust itself to the change in loading conditions, so the rotor pole falls back a little more relative to the stator pole, as shown in Fig 17.12 (*c*). Hence the torque angle increases with the increase in load. Due to increase in load (or torque) angle δ, the resultant voltage E_R across the armature circuit increases, and, therefore, current drawn from the supply mains increases, as illustrated in Fig 17.13 (*c*).

Thus a synchronous motor is able to supply increased mechanical load, not by reduction in speed, but by shift in relative positions of the rotor and rotating magnetic field. From Fig. 17.13 it is obvious that for increasing load with a constant value of induced emf E the phase angle ϕ increases in lagging direction.

If too great mechanical load is applied to a synchronous motor, the rotor will pull out of synchronism, after which it will come to standstill. The maximum value of torque that a motor can develop without losing synchronism is called the *pull-out torque*, like pull-in torque, may be designed for widely varying values, depending on the applications of the motor. These values range from 125% to 350% of full-load torque values.

17.9. EFFECTS OF VARYING EXCITATION ON ARMATURE CURRENT AND POWER FACTOR

The change in field excitation neither affects the speed of the motor nor the output of the motor but does affect the power factor and consequently armature current for constant supply voltage and constant input power.

The operation of a synchronous motor at a constant supply voltage demands nearly constant flux, as mentioned in the foregoing Art. Both the dc source and ac source cooperate in establishing this resultant flux. When the dc excitation is weak, the flux produced by it is weak and needs stengthening, so the stator draws the current lagging behind the supply voltage by 90° so as to magnetise the air gap to the extent needed to provide the necessary resultant flux. The magnetising component of current being a large part of the total input current results in low lagging power factor. For constant applied voltage and constant input power, the active component of current drawn from the supply mains should be constant and so because of large wattless (magnetising) component of current armature current I is much more. With the increase in excitation to 100%, the magnetising current drawn from the 3-phase ac supply mains is reduced *i.e.* the magnetising component becomes a smaller part of the total input current and so the power factor improves. This results in reduction of armature current I drawn from 3-ϕ ac supply mains. If the excitation is further increased such that power factor becomes unity, the stator draws only energy or active current, and the dc field circuit supplies all the current necessary to magnetize the rotor field. For unity power factor, the armature current drawn will be minimum. On increasing the dc excitation further, the field produced by it becomes stronger than it is required and needs demagnetization. So the motor draws the leading current so as to demagnetise the main field. This makes the power factor leading and lesser than unity and so the armature current increases.

The effect of variation in field excitation on armature current and power factor can also be explained as below.

Consider a synchronous motor loaded with a constant mechanical load and operating at a constant supply voltage. When the excitation is 100% *i.e.* when E = V, armature current I lags behind the resultant voltage E_R (phasor sum of V and E) by a fixed angle θ, given as $\theta = \tan^{-1}\dfrac{X_s}{R_e}$ and lags behind the applied voltage by an angle ϕ, power factor angle [Fig. 17.14 (*a*)]. X_s is the synchronous reactance per phase, R_e is the effective resistance of armature per phase, θ is the *internal angle* and δ is the load or torque angle. The power drawn per phase from the 3-ϕ ac supply mains is V I cos ϕ. Since applied voltage and input power or load is constant, the active component of the current drawn from the supply mains, I cos ϕ remains constant for constant mechanical load (neglecting internal losses). As the speed of the motor remains constant so increase

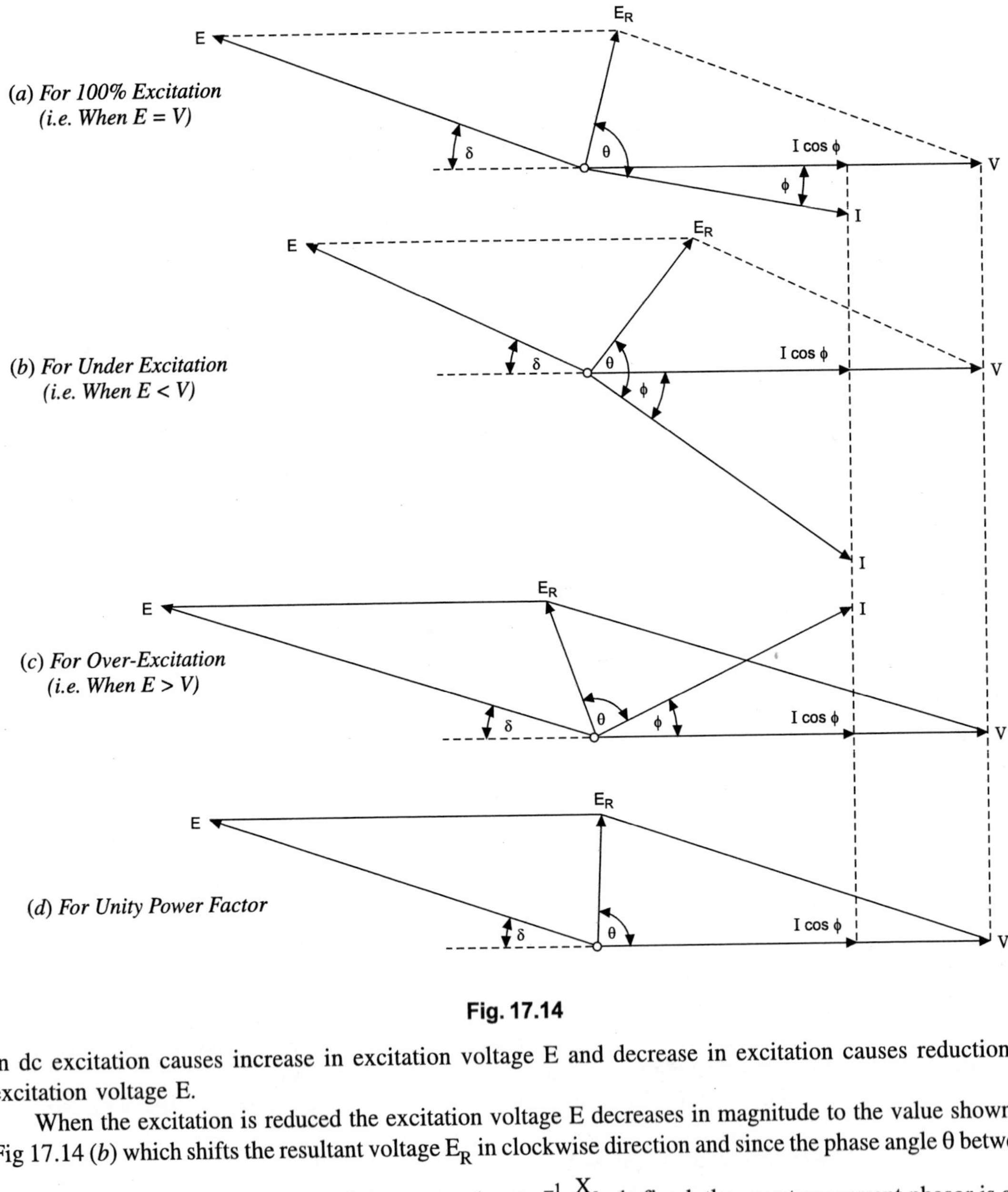

Fig. 17.14

in dc excitation causes increase in excitation voltage E and decrease in excitation causes reduction in excitation voltage E.

When the excitation is reduced the excitation voltage E decreases in magnitude to the value shown in Fig 17.14 (*b*) which shifts the resultant voltage E_R in clockwise direction and since the phase angle θ between the resultant voltage E_R and current I, being equal to $\tan^{-1}\dfrac{X_s}{R_e}$, is fixed, the armature current phasor is also shifted in clockwise direction. Thus power factor angle φ increases, power factor decreases and armature current increases because I cos φ is constant. *Hence with reduction in dc excitation synchronous motor draws more current from the supply mains at lower lagging power factor.*

When the excitation is increased, the excitation voltage E increases in magnitude to a value shown in Fig. 17.14 (*c*) which shifts the resultant voltage E_R in counter-clockwise direction. This results in change in position of phasor of current I with respect to terminal voltage so that the motor then operates at a leading power factor. For a certain value of excitation phase angle φ between voltage V and current I is zero *i.e.* the current I will be in phase with voltage V and at this instant the value of current drawn from supply mains, I will be minimum, as shown in Fig. 17.14 (*d*). Now if the excitation is further increased, it draws current with leading power factor, as shown in Fig. 17.14 (*c*).

Hence when the excitation of a synchronous motor is increased above 100 per cent, first power factor improves until it becomes unity; at this instant the current drawn from the supply mains is minimum and is in phase with supply voltage. With further increase in field current the power factor becomes leading one and decreases and current drawn from supply main increases.

The variation of current and power factor of a synchronous motor with a variation of field current (excitation) and for a constant load is shown in Fig. 17.15. Because of their shape, excitation-armature current curves are called the *V-curves* of a synchronous motor.

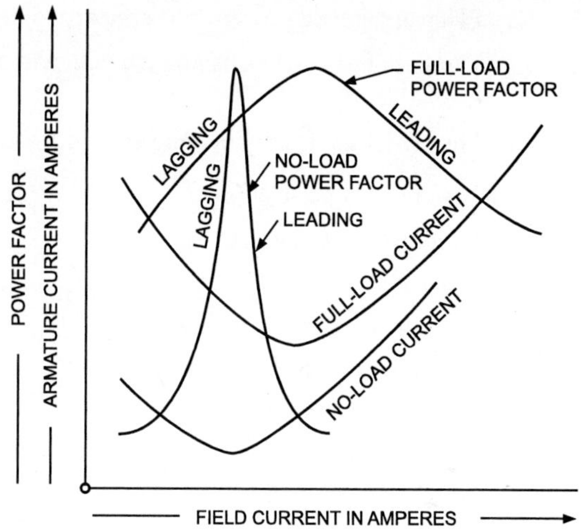

Fig. 17.15

17.10. MAIN CHARACTERISTICS OF SYNCHRONOUS MOTORS

1. Synchronous motor operates only at one speed, known as *synchronous speed*, given by the expression $N_s = \frac{120f}{P}$. The speed is independent of load and can be varied either by varying the supply frequency or by changing the number of poles.
2. It is, inherently, not a self-starting motor and, therefore, before connecting it to an ac supply it is to be run up to the synchronous speed by some external means.
3. It can be operated under a wide range of power factor both leading and lagging by controlling its excitation current.
4. On no load synchronous motor draws very little current from the supply mains to meet the internal losses. With fixed excitation the input current increases with the increase in load. After the input current reaches maximum no further increase in load is possible. If the motor is further loaded, the motor will stop.
5. If field excitation is small, the stator inductance will be more so the motor draws lagging current and, therefore, power factor is low. If the excitation is increased the stator flux is neutralized by the rotor flux and, therefore, power factor becomes unity. In this position, the current and voltage are in phase and motor current will be minimum.

If excitation is further increased rotor flux increases. It neutralizes the stator flux and causes leading current. This property makes the motor extremely useful as a power factor corrector. By taking leading current it compensates the lagging current drawn by other equipment from the supply mains and thus brings the total current drawn from the supply mains more nearly in phase with the voltage.

17.11. MERITS AND DEMERITS OF SYNCHRONOUS MOTORS

Merits. (*i*) The ease with which the power factor can be controlled. An overexcited synchronous motor having a leading power factor can be operated in parallel with induction motors and other power apparatus operating at lagging power factors, thereby improving the power factor of the supply system.

(*ii*) The speed is constant and independent of load. This characteristic is mainly of use when the motor is required to drive another alternator to generate a supply at a different frequency as in frequency changers.

(*iii*) Electromagnetic power varies linearly with the voltage.

(*iv*) These motors can be constructed with wider air gaps than induction motors, which make them better mechanically.

(*v*) These motors usually operate at higher efficiencies, especially in the low-speed unity-pf ranges.

Demerits (*i*) The cost per kW output is generaly higher than that of an induction motor.

(*ii*) It requires dc excitation which must be supplied from external source.

(*iii*) The synchronous motor is inherently not self-starting motor and needs some arrangement for its starting and synchronizing.

(*iv*) It cannot be used for variable speed jobs as there is no possibility of speed adjustment.

(*v*) It cannot be started under load. Its starting torque is zero.

(*vi*) It has a tendency to hunt.

(*vii*) It may fall out of synchronism and stop when overloaded.

(*viii*) Collector rings and brushes are required.

(*ix*) For some purposes synchronous motors are not desirable as for driving shafts in small workshops having no other power available for starting and in cases where frequent starting or strong starting torque is required.

17.12. SYNCHRONOUS CONDENSER

An outstanding advantage of the synchronous motors is that they can operate at unity or leading power factor. When operated in an electrical system in parallel with induction motors, or other power apparatus operating at lagging power factors, the leading reactive volt-amperes (or kVAR) supplied by the synchronous motors compensate for the lagging reactive volt-amperes (or kVAR) of other power apparatus and improve the overall power factor of the system.

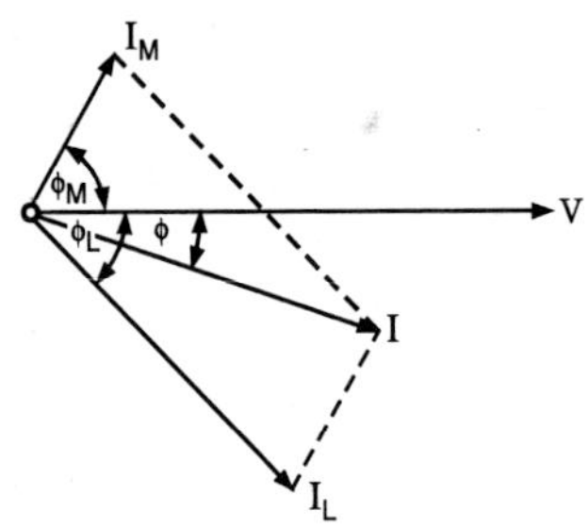

Fig. 17.16

In phasor diagram for a synchronous condenser connected in parallel with an industrial load drawing current I_L lagging behind the supply voltage by large angle ϕ_L shown in Fig. 17.16, I_M represents the current drawn by the synchronous condenser leading the supply voltage V by an angle ϕ_M. The resultant current I is the phasor sum of I_L and I_M and we see that its angle of lag ϕ is much smaller than ϕ_L. Thus overall power factor is improved from cos ϕ_L to cos ϕ by the use of synchronous condenser. In this way the power factor can be made unity even.

In large industrial plants, which have a low lagging power-factor load, it is often found economical to install a synchronous motor, even though the motor is not required to drive a load. The motor is operated overexcited at no load so that the current drawn by it leads the voltage by nearly 90°. A synchronous motor used in this way is said to "float"on the line because it has no mechanical output. Since the motor operating at no load acts in the same manner as a static capacitor and when operated in this manner it is called the *synchronous condenser* or a *synchronous capacitor*.

When a synchronous motor is operated on no-load, it still takes power for its own rotational losses (*i.e.* friction, windage and iron losses). Although the inphase component of armature current is usually comparatively small under no-load operation, the total current may be rather large, as it depends upon dc excitation. When a synchronous condenser is carrying full-rated leading kVA, the current leads the supply voltage by 88 to 89°, depending on the size of the machine. The angles correspond to power factors of 3.5 and 1.7 per cent respectively.

The synchronous condenser is especially designed so that practically all its rated kVA are available for power factor correction. The design of a synchronous capacitor is characterized by its large synchronous reactance and field windings, because its function is to develop zero-power factor currents with the least expenditure of power in losses. In comparison to a synchronous motor with equal armature voltage and current ratings, a synchronous condenser needs more copper in the field winding to carry the increased field current. The synchronous condenser does not require so large shaft and bearings as the synchronous motor because no shaft torque is required.

The advantage of synchronous condensers over static capacitors is that power factor can very easily be controlled by variation of the field excitation. The disadvantages of synchronous condensors are (*i*) except in size above 500 kVA, the cost is higher than that of static capacitors of the same rating, (*ii*) the possibility of their falling out of synchronism with resulting interruption of supply, and (*iii*) increase of short-circuit currents when the fault occurs near the synchronous condensers.

So the synchronous condensers are employed only when power factor correction is required on a large scale.

17.13. COMPARISON OF 3-PHASE SYNCHRONOUS AND INDUCTION MOTORS

S. No.	*Synchronous Motors*	*Induction Motors*
1.	It has got no self starting torque and some external means is required for its starting.	It has got self starting torque and no special means is required for its starting.
2.	Its average speed is constant and independent of load.	Its speed falls with the increase in load and is always less than synchronous speed.
3.	It can be operated under a wide range of power factor, both lagging and leading.	It operates at only lagging power factor, which becomes very poor at light loads.
4.	It requires dc excitation so it is a doubly excited machine.	It requires no dc excitation so it is a singly excited machine.
5.	No speed control is possible.	Speed can be controlled but to small extent.
6.	It is used for supplying mechanical load as well as for power factor improvement.	It is used for supplying mechanical load only.
7.	Its torque is less sensitive to change in supply voltage.	Its torque is more sensitive to change in supply voltage.
8.	Breakdown torque is proportional to the supply voltage.	Breakdown torque is proportional to the square of the supply voltage.
9.	It is more complicated and costs more comparatively. However, synchronous motors with speeds below 500 rpm and ratings exceeding about 40 kW or with medium speeds from 500 to 1,000 rpm and ratings exceeding about 500 kW cost less than induction motors.	It is more simple and costs less comparatively.

17.14. APPLICATIONS OF SYNCHRONOUS MOTORS

Synchronous motors are rarely used below 40 kW output in the medium speed range because of their much higher initial cost in comparison to that of induction motors. In addition they need a dc excitation source, and the starting and control devices are usually more expensive—especially where automatic operation is required. However, there is a kW output and speed range where the disadvantage of higher initial cost vanishes, even to the point of putting the synchronous motors to the advantage. Where low speeds and high kW outputs are involved the induction motors are no longer cheaper because they need large amounts of iron in order to restrict the air-gap flux density to 0.7 T. In the synchronous machines, on the other hand, a value twice this Figure is permissible because of the separate excitation.

The various classes of service for which synchronous motors are employed may be classified as (*i*) power factor correction (*ii*) voltage regulation and (*iii*) constant-speed, constant-load drives.

1. They are used in powerhouses and substations in parallel to the bus-bars to improve the power factor. For this purpose they are run without mechanical load on them and overexcited.
2. In factories having a large number of induction motors, or other power apparatus operating at lagging power, they are employed to improve the power factor.
3. Such motors are also used to regulate the voltage at the end of transmission lines.
4. Because of the higher efficiency possible with synchronous motors, they can be employed advantageously for the loads where constant speed is required. Typical applications of high-speed synchronous motors (above 500 rpm) are drives such as fans, blowers, dc generators, line shafts, centrifugal pumps and compressors, reciprocating pumps and compressors, constant speed frequency changers, rubber and paper mills etc.

The fields of applications of low-speed synchronous motors (below 500 rpm) are drives such as reciprocating compressors when started unloaded, dc generators, centrifugal and screw type pumps, vacuum pumps, electroplating generators, line shafts, rubber and band mills, ball and tube mills, chippers, metal rolling mills etc. Flywheel is used for pulsating loads.

High power electronic convertors generating very low frequencies made the operation of synchronous motors at ultra low speeds. Synchronous motors in very large sizes (as high as 10 MW size) operating at ultra low speeds are employed to drive crushers, rotary kilns and variable speed ball mills.

HIGHLIGHTS

1. A *synchronous machine* is an ac machine in which the rotor moves at a speed which bears a constant relationship to the frequency of currents in the armature winding.
2. *Alternator* is that machine which converts mechanical power into ac electrical power at a desired frequency. This is also known as a synchronous generator.
3. An *alternator* consists of a stator and a rotor. The stator provides the armature windings whereas the rotor provides the rotating magnetic field.

   The rotors are of two types :

   (*i*) Salient pole type used almost entirely for low- and moderate-speed alternators (hydrogenerators or diesel-engine driven generators) being least expensive and providing ample space for field ampere-turns.

   (*ii*) Smooth cylindrical or non-salient pole type used in very high speed alternators (turbo-generators), being of smaller diameter and robust construction.
4. The alternators may be rotating armature type or rotating field type but practically all medium and large machines are always constructed with revolving field because of its superiority over the rotating armature type.
5. The frequency of the alternating currents produced is

$$f = \frac{PN}{120} \text{ Hz}$$

6. EMF equation is E_{rms} per phase = 4.44 $K_d K_p \Phi f T$ volts.
7. *Synchronous motor* is identical in construction with the alternator and converts ac electrical power into mechanical power.
8. When 3-phase supply is given to the stator of a 3-phase synchronous motor, a revolving field is produced. This field tries to drag the rotor with it but could not do so owing to rotor inertia. Hence starting torque is not developed. Thus a synchronous motor is inherently not self-starting and, therefore, it is necessary to provide some external means to start it and bring its speed near synchronous speed.
9. Synchronous motor operates only at one speed known as synchronous speed given as $N_s = \frac{120 f}{P}$. It can be operated under a wide range of power factor both leading and lagging by controlling its excitation current.
10. Synchronous motors are rarely used in small ratings (below 40 kW output) in the medium speed range because of being expensive in comparison to induction motors. Where low speeds and high kW outputs are involved the synchronous motors prove economical.

    Synchronous motors are used for (*i*) constant-speed, constant-load drives (*ii*) power factor correction and (*iii*) voltage regulation.

EXERCISES

1. The field of a dc machine is usually kept on stator but in synchronous machines it is on rotor. Justify.
[U. P. Technical Univ. Tutorial Question Bank]
2. Give constructional differences between salient pole and cylindrical rotor synchronous machines. Why is cylindrical rotor machine used in thermal power plant and salient pole machine in hydroelectric plant ?
[U.P. Technical Univ. Tutorial Question Bank]
3. Explain the principle of operation of alternator.
Deduce the relationship between the number of poles, the frequency and the speed of the alternator.
4. Derive the emf equation for a 3-phase synchronous generator.
5. Explain construction and principle of operation of synchronous alternator.
[Mahamaya Technical Univ. Electrical Engineering Second Semester 2010-11]
6. Explain the operation of alternator. **[G.B. Technical Univ. Electrical Engineering First Semester 2010-11]**
7. Describe briefly the construction and working of a synchronous generator.
[Electrical Engineering G.B. Technical Univ. Second Semester 2011-12; U.P. Technical Univ. Second Semester 2013-14]
8. Discuss principle of operation of synchronous motor with diagram.
[U.P. Technical Univ. Basic Electrical Engineering Odd Semester 2013-14]
9. Why synchronous motor is not self-starting? Explain the various methods of starting these motors.
[G.B. Technical Univ. Electrical Engineering Even Semester 2012-13]
10. Discuss the working principle of a three-phase synchronous machine. Also differentiate synchronous motor from induction motor. **[G.B. Technical Univ. Electrical Engineering Second Semester 2010-11]**
11. Three-phase induction motor is self-starting but 3-phase synchronous motor is not self-starting. Explain, why? Enlist applications of a 3-phase synchronous motor.
[Mahamaya Technical Univ. Electrical Engineering First Semester 2011-12]
12. Briefly discuss the principle of operation of alternator and also give its applications. Draw V-curve for synchronous motor. **[U.P. Technical Univ. Electrical Engineering First Semester 2008-09]**
13. The field excitation of a 3-phase synchronous motor operating on a fixed load is increased from a very low value. With the help of a V-curve or otherwise explain how the magnitude and power factor of the armature current will vary.
[U.P. Technical Univ. Electrical Engineering July 2002]
14. Give the constructional details of salient pole synchronous machine. Explain 'V' curve and give the applications of synchronous motor. **[U.P. Technical Univ. Electrical Engineering Second Semester 2004-05]**
15. Explain the working principle of synchronous motor. Draw V-curves and give its applications.
[U.P. Technical Univ. Electrical Engineering Second Semester 2006-07]
16. Explain principle of operation of synchronous motor. Give its applications.
[U.P. Technical Univ. Electrical Engineering First Semester 2007-08]
17. Three-phase induction motor is self starting but three phase synchronous motor is not self starting. Explain why ? Give applications of 3-phase synchronous motor. **[U.P. Technical Univ. Electrical Engineering First Semester 2006-07]**
18. Explain the following:
(*i*) Synchronous motor always runs at synchronous speed.
(*ii*) Field winding of synchronous generators is placed on rotor.
[U.P. Technical Univ. Electrical Engineering Second Semester 2004-05]
19. Discuss the principle of operation of a three-phase synchronous machine. Give the various applications of a 3-phase synchronous motor. **[U.P. Technical Univ. Electrical Engineering Second Semester 2008-09]**
20. Explain principle of operation of a synchronous motor and give its applications.
[G.B. Technical Univ. Electrical Engineering First Semester 2009-10]
21. Write down the principle of operation of 3-phase synchronous motor. Also write its applications.
[G.B. Technical Univ. Electrical Engineering First Semester 2010-11]

22. Explain working principle of synchronous motor and two applications.

[U.P. Technical Univ. Electrical Engineering Odd Semester 2013-14]

23. Discuss the working principle of a three-phase synchronous machine. Also differentiate synchronous motor from induction motor. **[G.B. Technical Univ. Electrical Engineering Second Semester 2010-11]**

24. In a large industrial set-up, largest capacity three-phase induction motor is being replaced by equal capacity synchronous motor. What advantages and disadvantages would be seen?

[Mahamaya Technical Univ. Electrical Engineering Second Semester 2011-12]

25. Write short notes on the following :

(*i*) Constructional features and principle of operation of an alternator.

(*ii*) Construction and working principle of a synchronous motor.

(*iii*) Main characteristics of a synchronous motor.

(*iv*) Synchronous condenser. **[U.P. Technical Univ. Second Semester 2013-14]**

(*v*) Applications of synchronous motors.

SHORT ANSWER TYPE QUESTIONS WITH ANSWERS

Q. 1. Write the name of different types of rotor of an alternator. **[G.B.T.U. Odd Semester 2012-13]**

OR

Enlist types of rotor of an alternator. **[G.B. Technical Univ. Even Semester 2012-13]**

Ans. Rotor are of two types namely (*i*) salient pole type and (*ii*) smooth cylindrical type or non-salient pole type. [Fig. 17.4(*a*) (*b*)]

Q. 2. Under no-load running condition of a synchronous machine, what will be the angle between the induced voltage and supply voltage? **[G.B.T.U. Odd Semester 2012-13]**

Ans. Negligibly small.

Q. 3. What factors determine the number of poles of a synchronous motor ?

Ans. The number of poles of a synchronous motor is determined by supply frequency and speed.

Q. 4. A synchronous motor runs at N_S rpm at full load. What will be its speed if the load is reduced to half the full-load ?

Ans. N_s rpm.

Q. 5. Why a 3-phase synchronous motor will always run at synchronous speed ?

Ans. The synchronous motor always runs at synchronous speed because there is an interlocking action between stator and rotor fields of this motor.

Q. 6. How the power factor of a synchronous motor is changed keeping the shaft load undisturbed ?

Ans. By varying the field excitation.

Q. 7. Does the change in excitation affect the synchronous motor speed and power factor ?

Ans. Change in excitation will affect the power factor (not the speed) of the synchronous motor.

Q. 8. What is the effect of increase in excitation of a synchronous motor ?

Ans. When the excitation of a synchronous motor is increased, first power factor improves until it becomes unity and with the further increase in excitation the power factor becomes leading one and decreases.

Q. 9. What is meant by V-curves of a synchronous motor ?

Ans. The curves drawn between armature current and field current for different constant loads are known as V-curves.

Q. 10. What are the advantages of synchronous motor ?

Ans. The advantages of synchronous motor are (*i*) constant-operation speed, (*ii*) higher efficiency, especially in the low-speed unity-pf ranges and (*iii*) capability of being operated under a wide range of power factor both lagging and leading.

❑❑❑

Experiments

EXPERIMENT-1

Objective: Verification of Kirchhoff's current law.

Apparatus Required: 24 V battery or regulated power supply–1, PMMC ammeters 0–1 A — 3 nos., Rheostats 10 Ω, 1 A — 3 nos. and connecting wires.

Theory: According to Kirchhoff's current law, in any network of wires carrying currents, the algebraic sum of all currents meeting at a point (or junction) is zero or the sum of incoming currents towards any point is equal to the sum of outgoing currents away from that point.

Connection Diagram:

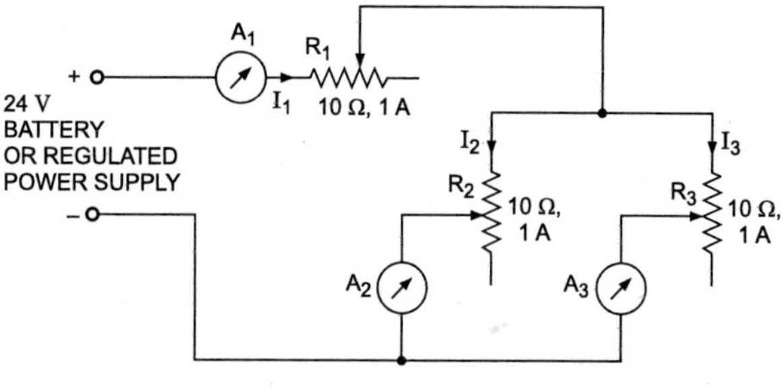

Fig. 1

Procedure: Three rheostats R_1, R_2 and R_3 and ammeters A_1, A_2 and A_3 are connected to 24 V battery or regulated power supply as shown in Fig. 1. The three rheostats are set to their maximum values, supply is switched on, and the readings of ammeters A_1, A_2 and A_3 are noted. The process may be repeated by varying either of rheostats R_1, R_2 or R_3.

Observations:

S. No.	*Readings of Ammeter* A_1, I_1	*Readings of Ammeter* A_2, I_2	*Readings of Ammeter* A_3, I_3	$I_2 + I_3$
1.				
2.				
3.				
4.				
5.				

Conclusion: It will be found that current I_1 is equal to the sum of currents I_2 and I_3. Hence Kirchhoff's current law is verified.

EXPERIMENT-2

Objective: Verification of Kirchhoff's voltage law.

Apparatus Required: 24 V battery or regulated power supply–1, PMMC voltmeters 0–24 V — 3 nos., Rheostats 10 Ω, 2.5 A — 2 nos. and connecting wires.

Theory: According to Kirchhoff's voltage law, in any closed circuit or mesh, the algebraic sum of emfs acting in the circuit or mesh is equal to the algebraic sum of the products of the currents and resistances of each part of the circuit or mesh.

Connection Diagram:

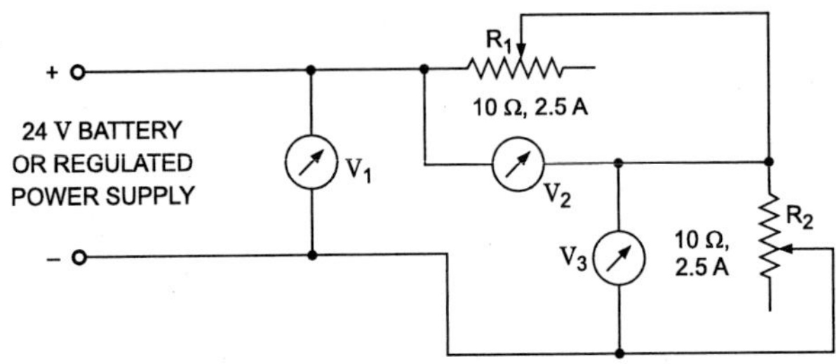

Fig. 2

Procedure: Two rheostats R_1, R_2 and three voltmeters V_1, V_2 and V_3 are connected to 24 V battery or regulated power supply as shown in Fig. 2. Two rheostats are set to their maximum values, supply is switched on, and the readings of voltmeters V_1, V_2 and V_3 are noted. The process may be repeated by varying either of the rheostats R_1 or R_2.

Observations:

S. No.	*Readings of Voltmeter V_1 in Volts*	*Readings of Voltmeter V_2 in Volts*	*Readings of Voltmeter V_3 in Volts*	$V_2 + V_3$
1.				
2.				
3.				
4.				
5.				

Conclusion: It will be found that voltage V_1 is equal to the sum of voltages V_2 and V_3. Hence Kirchhoff's voltage law is verified.

EXPERIMENT-3

Objective: Verification of superposition theorem.

Apparatus Required: Two batteries of 12 V and 6 V, three PMMC ammeters 0–2 A, Three rheostats of 10 Ω, 2.5 A, Two keys and connecting wires.

Theory: The superposition theorem may be stated as below:

In a linear resistive network containing two or more voltage sources, the current through any element (resistance or source) may be determined by adding together algebraically the currents produced by each source acting alone, when all other voltage sources are replaced by their internal resistances. If a voltage source has no internal resistance, the terminals to which it was connected are joined together. If there are current sources present they are removed and the network terminals to which they were connected are left open.

Connection Diagram:

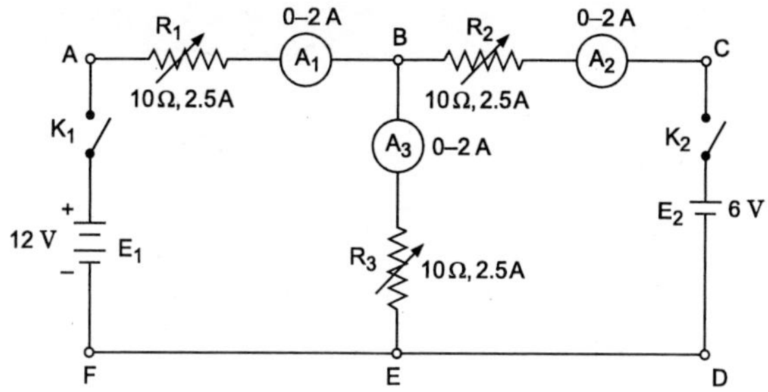

Fig. 3

Procedure: Two batteries E_1 of 12 V and E_2 of 6 V, three ammeters A_1, A_2 and A_3, three rheostats R_1, R_2 and R_3 and two keys K_1 and K_2 are connected as shown in Fig. 3. Rheostats are placed at the position so as to make the readings in all the three ammeters within the range. Keys K_1 and K_2 are closed and readings of all the three ammeters are noted. 6 V battery is removed the points C and D are shorted key K_1 is closed and the readings of the three ammeters are noted. Now 6 V dc source and key K_2 are reinserted in the circuit as shown in Fig. 3 but 12 V battery and key K_1 are replaced by a short-circuit, key K_2 is closed and the readings of all the three ammeters are noted. The settings of the rheostats are kept undisturbed throughout.

The process is repeated by varying either of the rheostats R_1, R_2 or R_3.

Observations:

S.No.	*Ammeter Readings With Both Sources in Amperes*			*Ammeter Readings With 12 V Source Only, in Amperes*			*Ammeter Readings With 6 V Source Only, in Amperes*			$I_1' + I_1''$	$I_2' + I_2''$	$I_3' + I_3''$
	I_1	I_2	I_3	I_1'	I_2'	I_3'	I_1''	I_2''	I_3''			
1.												
2.												
3.												
4.												
5.												

Conclusion: It will be found that $I_1 = I_1' + I_1''$; $I_2 = I_2' + I_2''$ and $I_3 = I_3' + I_3''$. Thus superposition theorem is verified.
Note: There may be small error in result due to internal resistance of batteries, which has been considered negligible.

EXPERIMENT-4

Objective: Verification of Thevenin's theorem in dc circuits.

Apparatus Required: Two batteries of 6 V and 12 V, DC ammeter (0 – 5 A)-one, 3 rheostats, One resistor, Four keys, DC voltmeter (0 – 10 V)-one, and Connecting wires.

Theory: According to this theorem if a resistor of R_L ohms be connected between any two terminals of a linear bilateral network (the circuit which has the same characteristics in either direction such as transmission line), then resulting steady-state current through resistor will be equal to $\frac{V_{TH}}{R_L + R_{TH}}$ where V_{TH} is the pd across these two points and R_{TH} is the resistance of network measured between these two points.

For details Art. 2.16 may be referred to.

Connection Diagram:

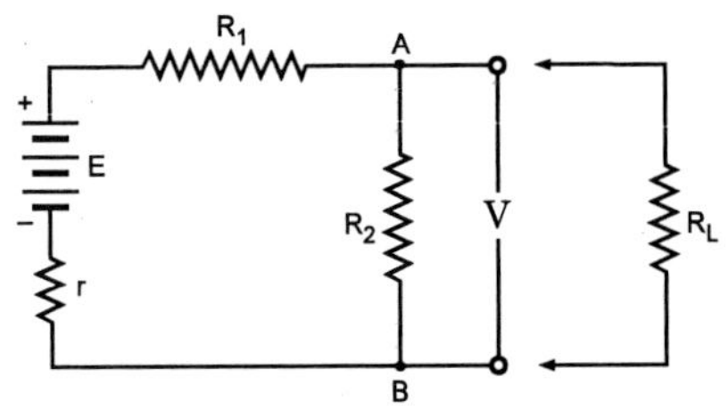

Fig. 4

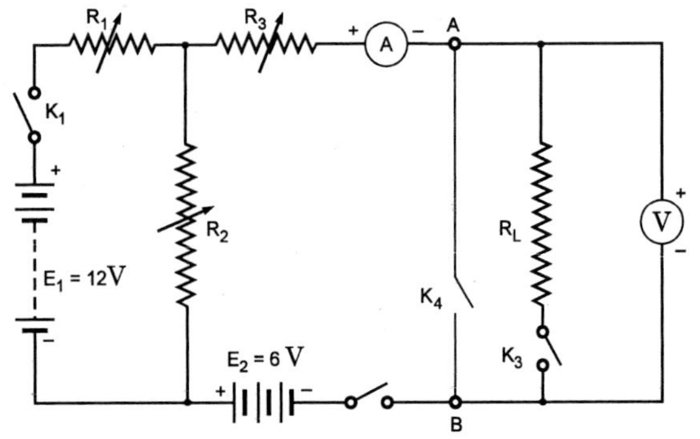

Fig. 5

Procedure: The batteries E_1 of 12 V and E_2 of 6 V, rheostats R_1 , R_2 and R_3, load resistance R_L, voltmeter V and ammeter A are connected, as shown in the Fig. 4. Keys K_1 , K_2 and K_3 are closed, reading of ammeter A giving current flowing through load resistance R_L, I_L is noted and also voltage across load resistance R_L is measured by voltmeter.

Now load resistance R_L is removed by opening key K_3 and voltage across terminals A and B is measured by a dc voltmeter. This is open-circuit voltage V_{OC} (or Thevenin's voltage, V_{TH}).

Now terminals A and B are short-circuited by closing key K_4 and reading of ammeter is noted down. This is the short-circuit current, I_{SC}. The Thevenin resistance R_{TH} is given as $\frac{V_{OC}}{I_{SC}}$.

The load current is determined by Thevenin's theorem *i.e.* from the relation $I_L = \frac{V_{TH}}{R_{TH} + R_L}$.

R_L is determined from the ratio of measured values of voltage across the load resistance and current flowing through it.

The process is repeated by varying either of the rheostats R_1, R_2 or R_3 or by varying load resistance R_L.

Observations:

S.No.	*Load Current in Amperes,* I_L	*Voltage Across Load in Volts,* V_L	*Load Resis-tance in Ohms,* $R_L = V_L / I_L$	*Open-Circuit Voltage Across Points A & B in Volts,* V_{TH}	*Short-Circuit Current Through Points A & B Under Short-Circuit Condition in Amperes,* I_{SC}	$R_{TH} = \frac{V_{TH}}{I_{SC}}$	*Load Current by Thevenin's Theorem,* $I_L = \frac{V_{TH}}{R_{TH} + R_L}$
1.							
2.							
3.							
4.							

Conclusion: It will be found that measured values of current flowing through the load, I_L are the same as determined by Thevenin's theorem.

EXPERIMENT-5

Objective: Verification of maximum power transfer theorem in a dc circuit.

Apparatus Required: 12 V accumulator–1, DC voltmeter (0–12 V)–1, DC ammeter (0–2.5 A)–1, Resistance box–1, Rheostat of suitable rating–1 and Connecting wires.

Theory: According to maximum power transfer theorem a resistive load, served through a resistive network, will abstract maximum power when the load resistance R_L is equal to the internal resistance of the network, as looked back into the network. For details Art. 2.19 may be referred to.

Circuit Diagram:

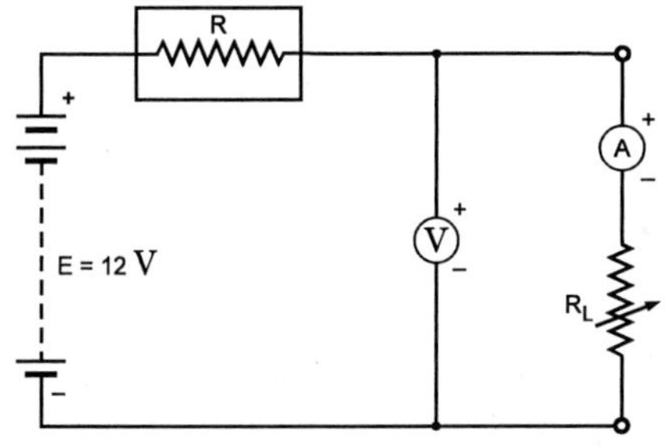

Fig. 6

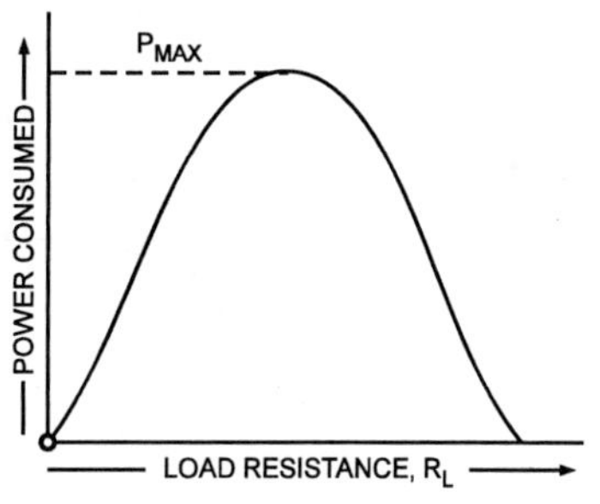

Fig. 7

Procedure: Connection diagram is shown in Fig. 6. R is fixed at some suitable value, load resistance R_L is kept maximum and ammeter and voltmeter readings are noted. Load resistance R_L is reduced to a very low value in steps and each time ammeter and voltmeter readings are noted.

The process is repeated for different values of R.

For every value of R, curve is plotted between power consumed in the load resistance and load resistance R_L and from the curve so drawn the value of R_L for maximum power drawn is determined.

Observations:

S.No.	*Resistance R in Ohms*		*Current I_L in Amperes*	*Voltage, V_L in Volts*	*Power Consumed $P = V_L I_L$ watts*	*Load Resistance $R_L = V_L/I_L$ Ohms*
1		a				
		b				
		c				
		d				
2		a				
		b				
		c				
		d				

Conclusion: It will be found that power consumed ($V_L I_L$) will be maximum when R_L becomes equal to R. This verifies the maximum power transfer theorem.

EXPERIMENT-6

Objective: Measurement of power and power factor in a single phase ac series inductive circuit and study of improvement of power factor using capacitor.

Appartus Required: AC ammeter 10 A, 5A, 5A - one each, AC voltmeter 0 – 300 V - 1, Single phase inductive load (lamps and choke coils), Variac 10 A, 250 V-1. Non-inductive resistor of known value -1, Capacitor - 1 and connecting wires.

Procedure: Keeping the variac at its minimum position, supply is switched on. Variac position is gradually varied to increase the voltage applied to the circuit so that the readings of ammeter A_3 and voltmeter are appreciable. Readings of ammeters A_1, A_2 and A_3 are noted down.

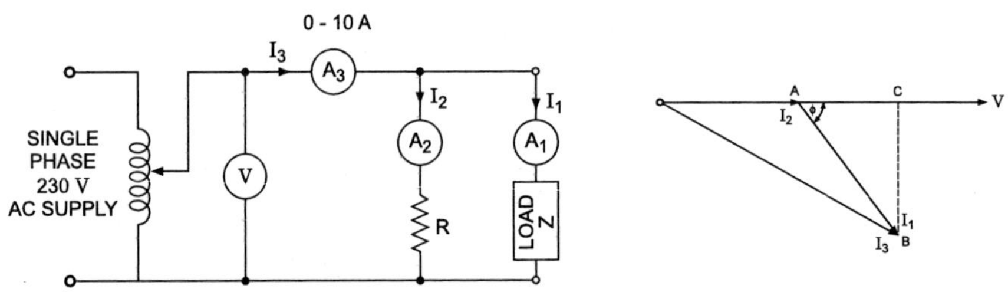

(*a*) *Connection Diagram* (*b*) *Phasor Diagram*

Fig. 8 *Measurement of Power By 3-Ammeter Method*

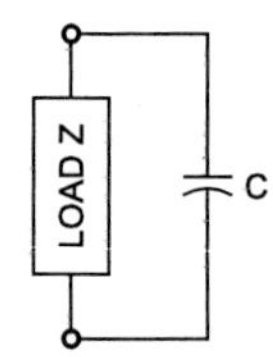

Fig. 9

Supply is *switched off,* capacitor C is connected across the load as shown in Fig. 9 and supply is *switched on.* Readings of ammeters A_1, A_2 and A_3 are noted down.

Process may be repeated with different inductive loads or different non-inductive resistors R.

S.No.	Reading of Ammeters in Amperes						Power in Watts		Power factor of load	
	Without Capacitor			With Capacitor			Without Capacitor in Circuit	With Capacitor Connected Across Load	Without Capacitor in Circuit	With Capacitor Connected Across Load
	I_1	I_2	I_3	I'_1	I'_2	I'_3	$P = (I_3^2 - I_1^2 - I_2^2) \times \frac{R}{2}$	$P = (I_3'^2 - I_1'^2 - I_2'^2) \times \frac{R}{2}$	$\cos\phi = \frac{I_3^2 - I_1^2 - I_2^2}{2I_1 I_2}$	$\cos\phi' = \frac{I_3'^2 - I_1'^2 - I_2'^2}{2I'_1 I'_2}$
1										
2										
3										
4										
5										

Conclusion: Power consumed by the circuit remains unchanged on connecting capacitor C across the load but the power factor improves.

The above experiment can also be performed by three-voltmeter method but three-ammeter method is more accurate and normal supply voltage is required.

EXPERIMENT-7

Objective: Study of phenomenon of resonance in R-L-C series circuit and obtain resonant frequency.

Apparatus Required: AC voltmeters–2, AC ammeter–1, Power factor meter–1, Adjustable non-inductor resistor–1, Inductor–1, Variable capacitor–1, DPIC switch–1 and connecting wires.

Theory: Refer to Art. 7.2.

Connection Diagram: Connection diagram is shown in Fig. 10.

Procedure: An adjustable non-inductive resistor R, an inductor L and a variable capacitor C are connected in series; a power factor meter is also included, as well as voltmeters and an ammeter, as shown in Fig. 10. With R adjusted to a value, such that the current I = V/R does not exceed the safe value and with a small capacitor, the supply is switched on and all readings are recorded. The capacitance is gradually increased, readings being noted at intervals; if, on approaching resonance, the capacitor voltage tends to reach a dangerous value, the switch must be opened and the capacitance increased beyond the resonant point, the switch may be closed, and the observations continued. The results are tabulated, and curves are plotted for current, condenser voltage and power factor versus capacitance (Fig. 11).

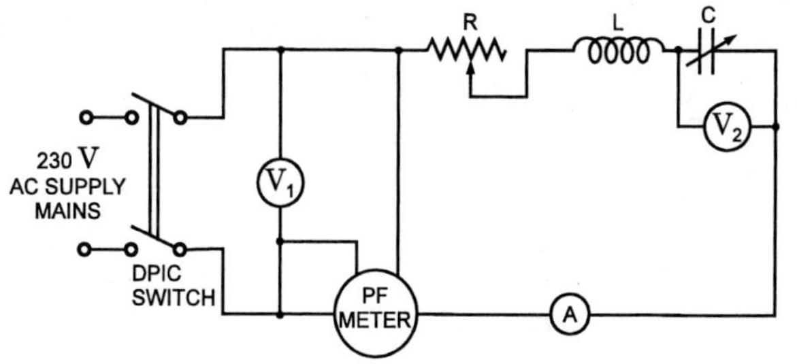

Fig. 10

Observations: R = ... ohms; Supply voltage, V_1 ...volts (ac); L = ... henries

Capacitance C in μF	*V_2 in Volts*	*I in Amperes*	*Power Factor in Fraction*

$$\text{Resonant frequency, } f_r = \frac{1}{2\pi\sqrt{LC}}$$

$$\text{and Q-factor} = \frac{1}{R}\sqrt{\frac{L}{C}}$$

where C is the value of capacitance corresponding to maximum current.

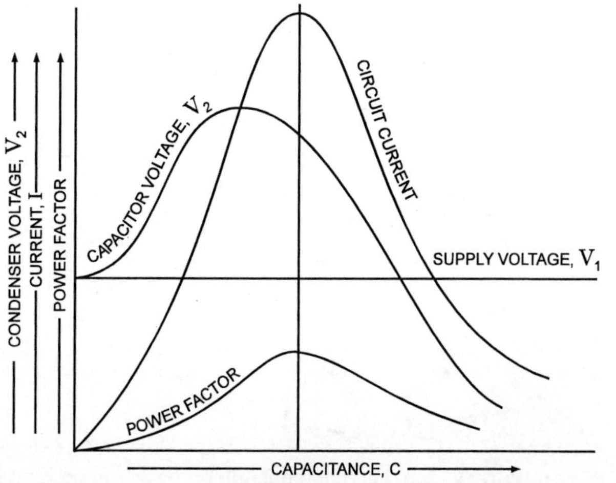

Fig. 11

Conclusions: 1. The circuit current rises from zero, when the capacitance is zero (reactance is infinite), to a maximum value V/R at resonance, beyond which it decreases, tending to a constant value as capacitance C becomes very large.

2. The capacitor voltage equals the supply voltage when its value is zero. It is because of zero current and no potential drop across either R or L. This increases at first to a maximum value near resonance, and then falls gradually to zero as capacitance C tends to infinity.

3. The power factor varies in a similar way, as the circuit current, *i.e.* it rises first to a maximum value of unity at resonance point and then falls. At resonance, the circuit behaves like a pure resistor.

EXPERIMENT-8

Objective: Measurement of power in 3-phase balanced circuit by two wattmeter method and determination of its power factor.

Apparatus Required: Triple pole iron clad (TPIC) switch-1, Dynamometer type wattmeters (500 V, 10 A)-2, 3-phase balanced load (say a 3-phase, 440 V, 50 Hz, 7.5 kW induction motor-1). Moving iron voltmeter (0-500 V)-1, Moving iron ammeter (0 – 10 A)-1, Connecting leads.

Theory: The connection diagram is given in fig. 12.

The total power is given as the sum of two wattmeter readings W_1 and W_2

i.e. Total power of load, $P = W_1 + W_2$

$$\text{and power factor of the load, } \cos\phi = \text{Cos } \tan^{-1} \sqrt{3}\,\frac{(W_1 - W_2)}{W_1 + W_2}$$

$$\text{Also power factor, } \cos\phi = \frac{P}{\sqrt{3}VI} = \frac{W_1 + W_2}{\sqrt{3}VI}$$

Connection Diagram:

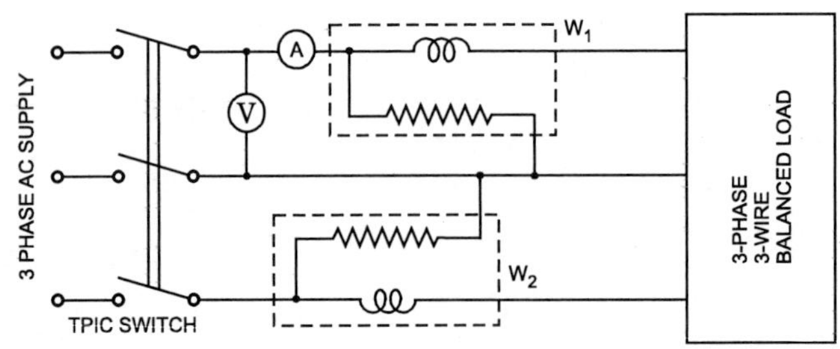

Fig. 12

Procedure: The load, wattmeters W_1 and W_2, ammeter A and voltmeter V are connected to the 3-phase ac supply through a TPIC switch, as shown in Fig. 12. The readings of wattmeters W_1 and W_2, ammeter A and voltmeter V are noted with different loads (in case of an induction motor, the load can be varied by changing the mechanical load on it). If any one of the wattmeters gives down scale reading (it happens when the load power factor is below 0.5), the supply is switched off and the connections of current coil (CC) or potential coil (PC) are reversed and the reading is considered as negative.

Observations:

S.No.	Ammeter Readings in Amperes, I	Voltmeter Readings in Volts, V	Wattmeter Readings in Watts		Total Power $= W_1 + W_2$	Power Factor	
			W_1	W_2		$\text{Cos } \tan^{-1} \sqrt{3}\frac{(W_1 - W_2)}{W_1 + W_2}$	$\frac{W_1 + W_2}{\sqrt{3}VI}$

Conclusions: 1. When the induction motor is run on light load (or on no load), one of the wattmeter gives down scale reading *i.e.* power factor is less than 0.5.

2. Power factor computed from the two relations given in theory is the same.

EXPERIMENT-9

Objective: Determination of parameters of ac single phase series R-L-C circuit.

Apparatus Required: Variac or 1-phase auto-transformer-1, Non-inductive resistor (or lamp load)-1, Inductor (or choke)-1, Capacitor-1, Moving-iron voltmeter (0-250V)-1, Moving-iron voltmeter (0-150V)-2, Moving iron-ammeter (0-5A)-1, Dynamometer type wattmeter (250V, 5A)-1, Double pole, single throw switch (DPST) switch-1, Connecting leads.

Theory: Current flowing through an ac circuit is given as $I = \frac{V}{Z}$ where V is the ac supply voltage (voltage applied to the circuit) and Z is the impedance of circuit in ohms.

Power factor of an ac circuit is given as $\cos\phi = \frac{P}{VI}$ where P is the power of the given circuit in watts, V is the voltage applied to the circuit in volts and I is the current flowing through the circuit.

$$\text{Resistance of the circuit, } R = Z\cos\phi$$

$$\text{Reactance of the circuit, } X_L \text{ or } X_C = \sqrt{Z^2 - R^2} \text{ or } = Z\sin\phi$$

In case of R-L-C circuit,

$$X = X_L \sim X_C = \sqrt{Z^2 - R^2} \text{ or } = Z\sin\phi$$

$$X_C = \frac{V_C}{I} \text{ and } C = \frac{1}{2\pi f X_C} \text{ and } L = \frac{X_L}{2\pi f} = \frac{X_C \pm X}{2\pi f}$$

+ sign for lagging power factor *i.e.* when $X_L > X_C$

– sign for leading power factor *i.e.* when $X_C > X_L$

Inductance can also be determined by measuring voltage drop across inductor directly.

$$\text{In this case } L = \frac{X_L}{2\pi f} = \frac{V_L}{I \times 2\pi f}$$

Similarly capacitance can be determined by measuring voltage drop across capacitor directly and $C = \frac{I}{2\pi f V}$.

Connection Diagram:

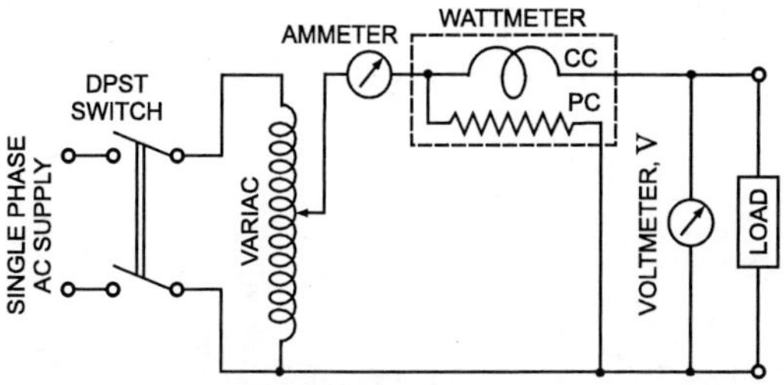

Fig. 13

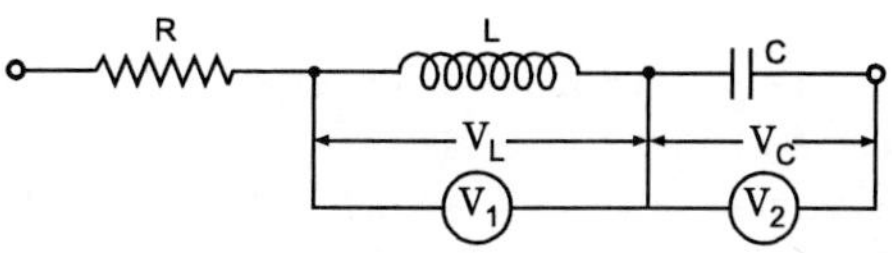

Fig. 14

Procedure: Variac, ammeter, voltmeters, wattmeter, load circuit consisting of resistor R, inductor L and capacitor C are connected through DPST switch to a single phase ac supply mains as shown in Fig. 13. The readings of ammeter, voltmeters and wattmeter are noted for various settings of variac.

Observations: Supply frequency f = 50 Hz

S.No.	V in Volts	V_1 in Volts	V_2 in Volts	I in Amperes	P in Watts	$\text{Cos } \phi = \frac{P}{VI}$	$Z = \frac{V}{I}$ in ohms	$X_L = \frac{V_1}{I}$ in ohms	$X_C = \frac{V_2}{I}$ in ohms	$L = \frac{X_L}{2\pi f}$ in Henry	$C = \frac{1}{2\pi f X_C}$ in Farads
1.											
2.											
3.											
4.											

Conclusions: 1. Current I increases in proportion to applied voltage V.
2. Load impedance Z remains constant for a given combination.
3. Power factor cos ϕ also remains constant for a given combination.

EXPERIMENT-10

Objective: Determination of polarity and voltage ratio of a single phase transformer.

Apparatus Required: Small transformer, Voltmeter, Connecting leads.

Theory:

Polarity Test: On the primary side of a two winding transformer, one terminal is positive with respect to the other one at any instant. At the same instant, one terminal of the secondary winding is positive with respect to the other one. *Polarity test* is performed to determine the terminals having the same instantaneous polarity (assuming that the terminals are not marked). The relative polarities of the primary and secondary terminals at any instant must be known for connecting windings of the same transformer in parallel, or series, or for interconnecting two or more transformers in parallel, or for connecting single-phase transformers for polyphase transformation of voltages.

In subtractive polarity, the voltage between A_1 and a_1 (or A_2 and a_2) is reduced. The leads connected to these terminals and the two windings are, therefore, not subjected to high voltage stress. On the other hand in additive polarity the two windings and leads connected to A_1, A_2, a_1 and a_2 are subjected to high voltage stresses. This is the reason that subtractive polarity is preferred over additive polarity.

Voltage-Ratio Test: The true ratio is based on turn ratio, as explained in Art 12.6. If the secondary and primary voltages are measured on no load, their ratio is very nearly equal to the true value. Measurement of primary and secondary currents in short-circuit test also gives fairly accurate results $\left(\text{voltage ratio, } \frac{V_2}{V_1} = \frac{I_1}{I_2}\right)$, especially if the transformer has little leakage flux and low core reluctance.

Procedure: For determination of relative polarity of the two windings of a transformer the two windings are connected in series across a voltmeter, while one of the windings (preferably hv winding) is excited from a suitable ac voltage source, as illustrated in Fig. 15. If the polarities of the windings are as marked on the diagram i.e. when the transformer has a subtractive polarity, the voltmeter will read the difference of E_1 and E_2. If the voltmeter reads $E_1 + E_2$, the polarity markings of one of the windings must be interchanged.

The voltage ratio of a transformer is obtained from the readings of two voltmeters, one on the primary, and the other on the secondary side, on open-circuit.

$$\text{Voltage ratio} = \frac{\text{Voltage on secondary side, } V_2}{\text{Voltage on primary side, } V_1} \text{ on open-circuit.}$$

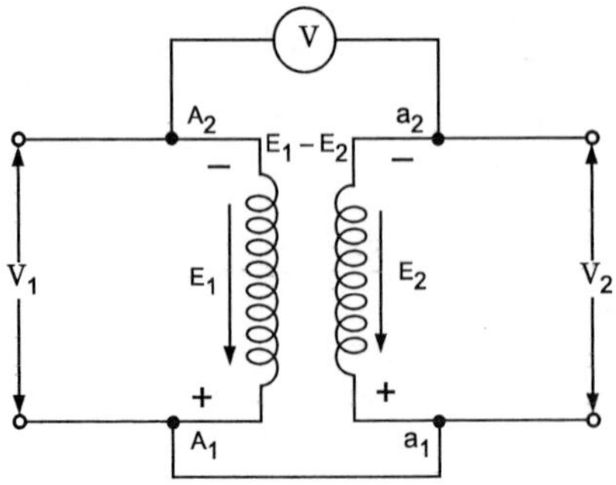

Polarity Test

Fig. 15

EXPERIMENT-11

Objective: Measurement of efficiency of a single phase transformer by load test.

Apparatus Required: AC ammeters-2, AC voltmeters-2, Wattmeters-2, Single phase transformer-1, Lamp load in series with an inductor-DPIC Switch-1, Connecting Wires.

Theory: The efficiency of a transformer is given, as in case of other machines, by the expression

$$\eta = \frac{\text{Output power}}{\text{Input power}} \times 100 = \frac{V_2 I_2 \cos \phi_2}{V_1 I_1 \cos \phi_1} \times 100$$

The method of determination of transformer efficiency by direct measurement of output and input does not give accurate result, as the power losses are quite low (of the order of 1-4%). The difference between the readings of output and input instruments is then so small that an instrument error as low as 0.5% would cause an error of the order of 15-20% in the power losses. Further it is inconvenient and costly to have necessary loading devices of the correct current and voltage ratings and power factor to load the transformer. There is also a wastage of large amount of power and no information is available from such a test about the proportion of copper and iron losses.

Connection Diagram:

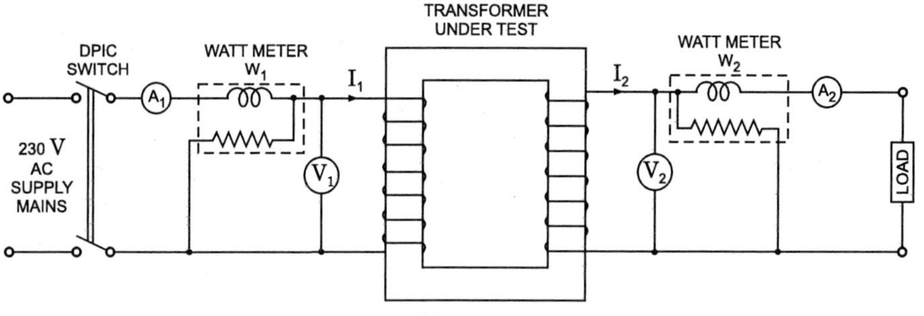

Fig. 16

Procedure: The connection diagram is shown in fig. 16. The load is made by connecting a lamp board consisting of several lamps (may be of different or equal wattage) and an inductor (or choke coil). The load, wattmeters, ammeters and voltmeters are connected to the primary and secondary of the transformer under test, as shown in Fig. 16. The supply is switched on to the circuit through DPIC switch. The readings of voltmeters, ammeters and wattmeters on input (primary) and output (secondary) are noted for different loads and tabulated as below:

S.No.	Input or Primary Side				Output or Secondary side				$\eta =$
	V_1 in Volts	I_1 in Amperes	W_1 in Watts	PF $\frac{W_1}{V_1 I_1}$	V_2 in Volts	I_2 in Amperes	W_2 in Watts	PF $\frac{W_2}{V_2 I_2}$	$\frac{W_2}{W_1} \times 100$
1									
2									
3									
4									
5									

Conclusions:

1. Transformer efficiency is quite high (of the order of — per cent).
2. Power factor on primary and secondary sides are almost same.

EXPERIMENT-12

Objective: To perform the OC and SC test on 1-phase transformer and establish equivalent circuit and full-load efficiency.

Apparatus Required: Low range ammeter–1, Low range voltmeter–1, Low range low power factor wattmeter–1, Wattmeter, Ammeter and Voltmeter of normal ranges (depending upon the rating of transformer under test) one each, Transformer and Connecting wires.

Theory: Refer to Arts. 12.16.3, 12.16.4, 12.17, 12.17.1.

Connection Diagrams: Refer to Figs. 12.22 and 12.23.

Procedure:

1. Open-circuit Tests: The transformer is connected, as shown in Fig. 12.22, on the low voltage side. The ammeter A is low-range instrument and wattmeter W is low-range and low-power factor instrument. The high voltage side is kept open circuited. Maintaining the supply frequency constant, the applied voltage is adjusted to normal voltage and instrument readings are noted and tabulated as below.

2. Short-circuit Tests: The transformer is connected as shown in Fig. 12.23, the low-voltage winding is short-circuited by connection having negligible resistance and good contacts. The applied voltage V_s to the primary is gradually increased till the ammeter A indicates the high-voltage side rated current. The instrument readings are noted and tabulated as below.

Observations:

Open-circuit Test

Ammeter reading, I_0 = ...

Voltmeter reading, V_1 = ...

Input power on no-load, W_0 = ...

Short-circuit Test

Ammeter reading, I_s = ...

Voltmeter reading, V_s = ...

Wattmeter reading, W_s = ...

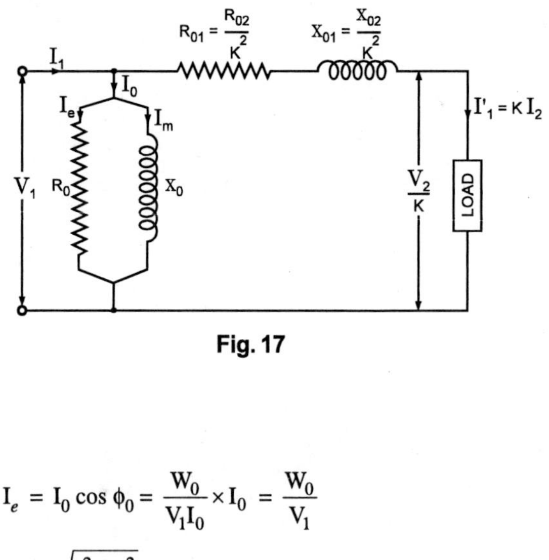

Fig. 17

Calculation:

$$\text{Iron loss, } P_i = W_0$$

$$\text{No-load current} = I_0$$

$$\text{No-load power factor, } \cos\phi_0 = \frac{W_0}{V_1 I_0}$$

$$\text{No-load current energy component, } I_e = I_0 \cos\phi_0 = \frac{W_0}{V_1 I_0} \times I_0 = \frac{W_0}{V_1}$$

$$\text{No-load current magnetising component, } I_m = \sqrt{I_0^2 - I_e^2}$$

$$\text{No-load resistance, } R_0 = \frac{V_1}{I_e} = \frac{V_1^2}{W_0}$$

$$\text{No-load reactance, } X_0 = \frac{V_1}{I_m} = \frac{V_1}{\sqrt{I_0^2 - I_e^2}}$$

$$\text{Full load copper losses, } P_c = W_s$$

$$\text{Equivalent resistance referred to high voltage side, } R_{02} = \frac{W_s}{I_s^2}$$

$$\text{Equivalent impedance referred to high voltage side, } Z_{02} = \frac{V_s}{I_s}$$

$$\text{Equivalent reactance referred to high voltage side, } X_{02} = \sqrt{(Z_{02})^2 - (R_{02})^2}$$

Let the transformation ratio $\frac{V_2}{V_1} = K$, then equivalent resistance and reactance referred to primary (low voltage) side will be $\frac{R_{02}}{K^2}$ and $\frac{X_{02}}{K^2}$ respectively.

From the values of R_0, X_0, R_{02}, X_{02} and K determined above the equivalent circuit is drawn, as shown in Fig. 17.

$$\text{Percentage regulation} = \frac{I_2R_{02}\cos\phi \pm I_2X_{02}\sin\phi}{E_2} \times 100$$

$$\text{Full-load efficiency, } \eta = \frac{V_2I_2\cos\phi}{V_2I_2\cos\phi + W_0 + W_s} \times 100$$

Precautions:

1. For performing open-circuit test (*i*) the wattmeter used should be of low range and low power factor because no-load current and no-load power factor are very low and (*ii*) the ammeter used should also be of low range.
2. Since no-load current is very small, pressure coils of the wattmeter and the voltmeter should be connected such that currents taken by them should not flow through the current coil of the wattmeter while performing open-circuit test.
3. For performing the short-circuit test the voltmeter used should be of low range.

Conclusions: 1. Transformer efficiency is quite high (of the order of _____ per cent).

2. Power factor on primary and secondary sides are almost the same.

EXPERIMENT-13

Objective: To study speed control of dc shunt motor using (i) armature voltage control (ii) field flux control.

Apparatus Required: 250 V dc shunt motor of capacity, say, 2 kW–1, PMMC ammeter (0-2.5 A)–1 and (0–10 A)–1, Rheostats-high resistance, low current–1 and low resistance, high current–1, Tachometer–1, DPIC switch–1 and Connecting leads.

Theory: The speed of a dc shunt motor is given as

$$N = K\frac{E_b}{\phi} = K\frac{V - I_aR_a}{I_{sh}}$$

Hence the speed of a dc shunt motor can be changed by either changing the shunt field current I_{sh} (by inserting an external resistance in the field circuit) or changing the back emf, E_b, (*i.e.* $V - I_aR_a$) by inserting an external resistance in the armature circuit.

Connection Diagram:

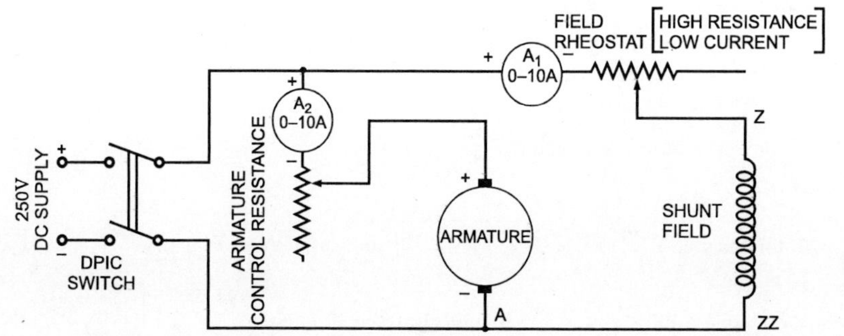

Fig. 18

Procedure: The connections are made as shown in Fig. 18 and the rheostats are put in zero positions. The dc supply is switched on by closing the DPIC switch.

Field Control: The reading of ammeter A_1 is noted and the speed of the motor is measured by tachometer. Now the field rheostat setting is changed in steps (increasing the resistance) and ammeter A_1 readings are noted and the motor speed is measured and recorded.

Now the field rheostat setting is brought back to zero.

Armature Control: Resistance is increased in steps in the armature circuit for armature resistance control. The ammeter A_2 readings are noted and the motor speeds are measured and recorded.

Armature control rheostat is brought back to zero and the supply is switched off through DPIC switch.

Observations:

S.No.	*Field Control*		*Armature Control*	
	Ammeter A_1 Readings in Amperes, I_{sh}	*Speed of Motor in RPM*	*Ammeter A_2 Readings in Amperes, I_a*	*Speed of Motor in RPM*
1.				
2.				
3.				
4.				
5.				

Results: 1. With the increase in resistance in the field circuit, the field current decreases *i.e.* field becomes weaker and, therefore, speed increases.

2. With the increase in resistance in the armature circuit, voltage drop in armature increases *i.e.* back emf E_b decreases and, therefore, speed decreases.

EXPERIMENT-14

Objective: Determination of efficiency of a dc shunt motor by load test.

Apparatus Required: Small dc shunt motor-1, DC voltmeter-1, DC ammeter-1, DPIC switch-1, Brake arrangement and connecting wires.

Procedure: DC shunt motor is connected electrically, as shown in Fig. 19. The brake is applied to a water cooled pulley mounted on the motor shaft, as illustrated in Fig. 19. The brake band is fixed with the help of wooden blocks gripping the pulley. One end of the band is fixed to earth through spring and the other is connected to the suspended weight W_1.

The motor is run with rated voltage applied across the motor terminal on no-load and the speed of the motor is measured. The readings of voltmeter and ammeter are also noted. Then the mechanical load is applied to the motor and increased in steps and corresponding motor speed is measured and the readings of voltmeter and ammeter are read and recorded. The load on the motor is increased upto about 20 per cent of overload. The results are tabulated.

Calculations:

$$\text{Power input} = \text{V I watts} = \frac{\text{VI}}{1{,}000}\ \text{kW}$$

$$\text{Motor torque, T} = (W_1 - W_2)\left(r + \frac{t}{2}\right) \text{kg m/s or } (W_1 - W_2)\left(r + \frac{t}{2}\right) \times 9.81 \text{ Nm}$$

where W_1 is the suspended weight in kg, W_2 is the spring balance reading, r is the radius of pulley in metres and t is the thickness of belt in metre.

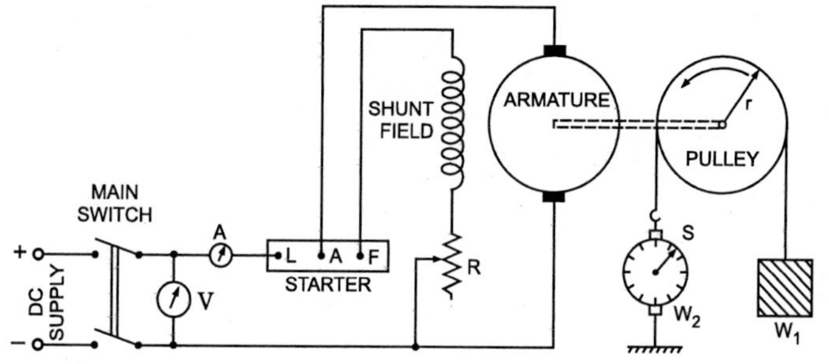

Fig. 19

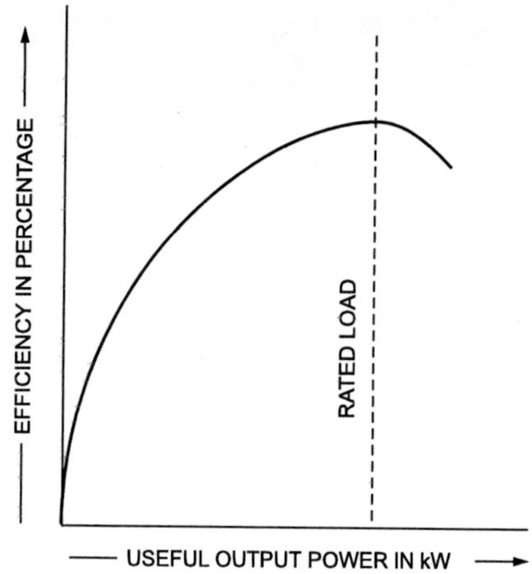

Fig. 20 *Efficiency – Output Characteristic of a DC Shunt Motor*

$$\text{Motor output} = T \times \frac{2\pi N}{60} \text{ watts} = \frac{T \times 2\pi N}{60{,}000} \text{ kW}$$

$$\text{Motor efficiency, } \eta = \frac{\text{Output}}{\text{Input}} \times 100$$

Observations: Radius of pulley, r = ... metre

Thickness of belt, t = ... metre

S.No.	Suspended Weight, W_1 in kg	Spring Balance Reading, W_2 in kg	Voltmeter Reading, V in Volts	Ammeter Reading, I in Amperes	Speed, N in RPM	Input Power to Motor $= \frac{VI}{1,000}$ kW	Torque Applied to Motor in N-m, $T = (W_1 - W_2)\left(r + \frac{t}{2}\right) \times 9.81$	Motor Output $= \frac{T \times 2\pi N}{60,000}$ kW	Motor Efficiency, $\eta = \frac{\text{Output}}{\text{Input}} \times 100$
1.									
2.									
3.									
4.									
5.									
6.									

Conclusion: The efficiency rises from zero, attains maximum value near full load and then decreases as shown in Fig. 20.

EXPERIMENT-15

Objective: To study running and speed reversal of a 3-phase induction motor and record speed in both directions.

Apparatus Required: Three-phase induction motor-1, Star-delta starter-1, TPIC switch-1, Speedometer-1, Connecting leads etc.

Theory: Motor is connected to the 3-phase ac supply mains through star-delta starter and TPIC switch, as shown in Fig. 21. The direction of rotation of a 3-phase induction motor can be reversed by interchanging any two terminals at the TPIC switch, as shown in Fig. 22. Speed can be measured by a speedometer.

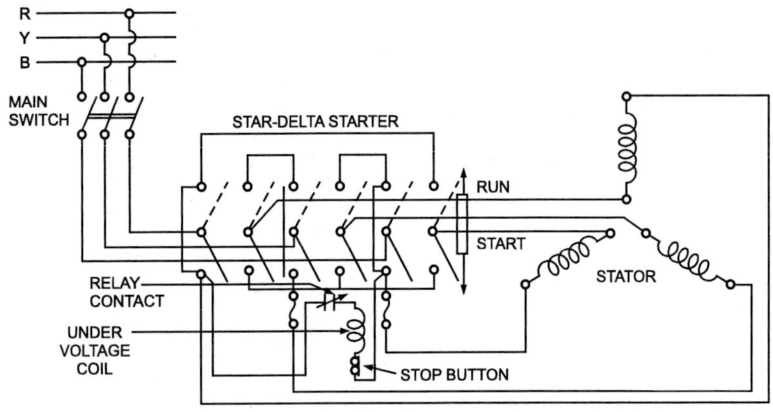

Fig. 21

Procedure: The connections of a 3-phase induction motor are made to the star-delta starter and to the TPIC switch, as shown in Fig. 21. The TPIC switch is closed and the motor is started by taking the lever of the starter to the start (star) position and then with a jerk to the run position (or delta connections). The direction of the rotation of the motor is observed. Say, it is in clockwise direction. Speed of the motor is measured by a speedometer.

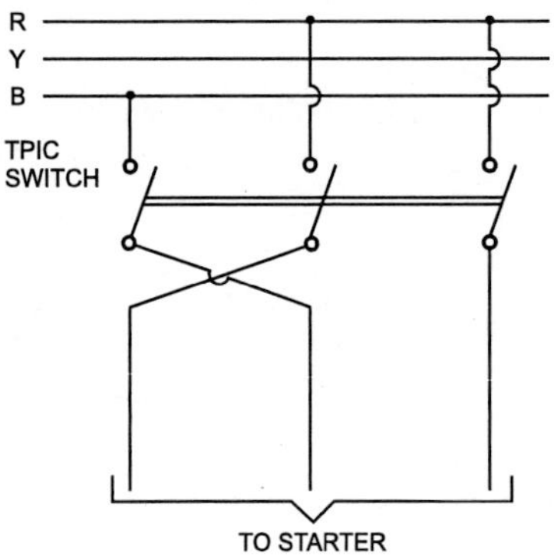

Fig. 22

Now the motor is stopped by pushing the stop button and supply to the motor is removed by opening the TPIC switch. The two leads of the motor are interchanged to the TPIC switch, as shown in Fig. 22. TPIC switch is closed and the motor is started again. The direction of rotation of the motor is observed. Speed of motor is again measured by a speedometer.

The push button is pushed and the TPIC switch is made off.

Observations and Conclusions:

1. The direction of rotation of the motor in second case is found opposite to that in first case.
2. The speed of motor is same in both cases.

EXPERIMENT-16

Objective: To measure energy by a single phase energy meter and determine error.

Apparatus Required: Energy meter-5 A, 240 V-1, Moving -iron voltmeter (0–300 V)-1, Moving iron ammeter (0–5 A)-1, Dynamometer type wattmeter 5 A, 240 V-1, Auto-transformer or variac-1, Lamp load, Stop watch-1, DPIC switch-1 and connecting leads.

Theory: Energy meter is an integrating instrument which is used to record the energy consumed by the load during a given time period. The datas mentioned on the name plate of the energy meter are as follows:

Number of phases — single
Volts — 230/240/250 V
Frequency — 50 Hz
Current rating — 5/10/25 A
Energy meter constant — ....revolution/kW*h*

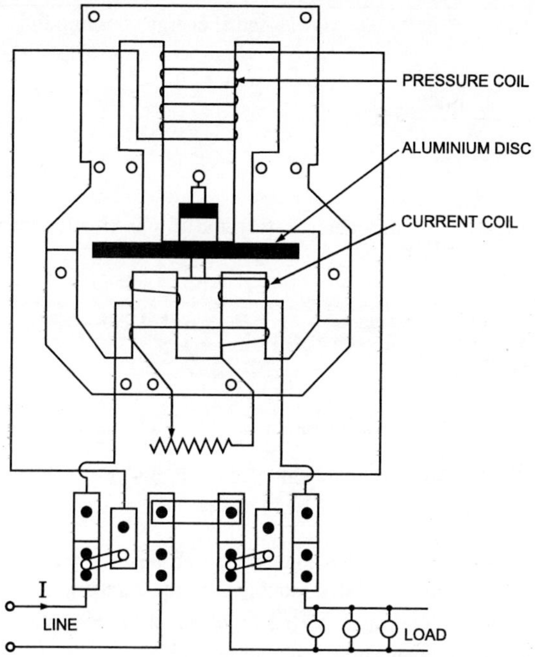

Fig. 23 *(a) Induction Type Single-Phase Energy Meter*

Connection Diagram:

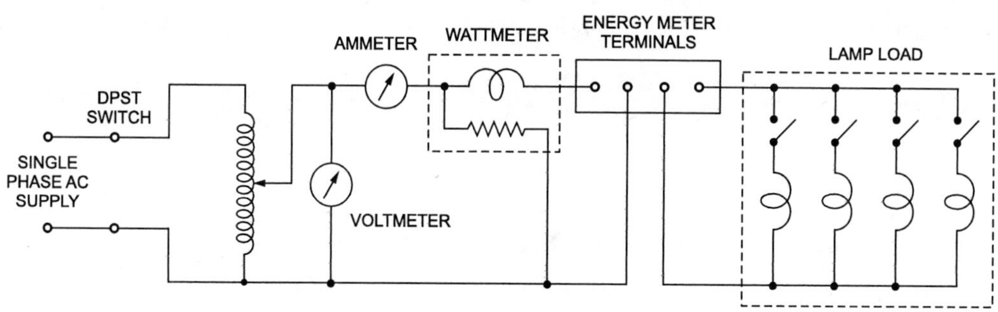

Fig. 23 *(b)*

Procedure: Ammeter, voltmeter, wattmeter and energy meter and load are connected to single phase ac supply through an auto-transformer or variac and a DPST switch as shown in Fig. 23 (*b*). Supply is *switched on*, autotransformer is adjusted to provide rated voltage and reading of energy meter is noted. Now the load is *switched* on and adjusted so that the ammeter reading is about 1 A and stop watch is started. After just 15 minutes, reading of energy meter, wattmeter are noted and load is switched off.

The process is repeated for load currents of 2, 3, 4 and 5 A.

Calculation: Energy consumption indicated by energy meter

$$= \text{Final reading} - \text{initial reading} = \text{kWh}_2 - \text{kWh}_1 = x \text{ kWh}$$

$$\text{Actual energy consumption} = \frac{\text{Reading of wattmeter in watts} \times \text{time duration in hours}}{1{,}000} = \frac{W \times t}{1{,}000} \text{kWh}$$

$$\text{Percentage error} = \frac{\text{Energy consumption indicated by energy meter} - \text{actual energy consumption}}{\text{Actual energy consumption}} \times 100$$

$$= \frac{x - \dfrac{W \times t}{1{,}000}}{\dfrac{W \times t}{1{,}000}} \times 100$$

Conclusion: If x is higher than $\dfrac{W \times t}{1{,}000}$, the meter is fast and if x is smaller than $\dfrac{W \times t}{1{,}000}$ the meter is slow.

EXPERIMENT-17

Objective: To start and reverse single-phase induction motor.

Apparatus Required: Small single-phase induction motor (say, 1 hp)-1, 2-wire cord-2 m, 15 A plug-1.

Theory: Refer to Arts 16.2 and 16.5.

Procedure: Plug is connected to 2-wire cord, which is connected to terminal box provided on the motor. Insert the plug in the socket and switch on the supply. The motor will start rotating in normal direction of rotation.

For reversing the direction of rotation of the motor, switch off the supply, take out the plug from the socket, let the motor come to standstill and then interchange the connection to the supply of either starting (auxiliary) or running (main) winding of the motor – the terminals of auxiliary and main windings are provided on the terminal box. Now insert the plug in the socket and switch on the supply. The motor will start rotating in the direction opposite to that of previous one (normal one).

Conclusion: The direction of rotation of such a motor can be reversed by interchanging the connections to the supply of either starting (auxiliary) or running (main) winding. The noteworthy point is that reversal of a split-phase induction motor takes place only when it is at rest and the starting switch is in its normally closed de-energized position.

EXPERIMENT-18

Objective: To study P-N diode characteristics.

Apparatus Required: Regulated power supply-1, Resistor-1, DC Voltmeter (0-1V)-1, DC Milliammeter-1, DC Micro-ammeter-1, DC Voltmeter (0-10V)-1 and Connecting wires.

Theory: When the external voltage is zero *i.e.* when the circuit is open, the potential barrier at the junction does not allow the flow of current and, therefore, the circuit current is zero, as shown in Fig. 24 (*a*).

With forward bias to the P-N junction (*i.e.* P-region connected to + ve terminal and N-region connected to –ve terminal of the battery) very little current, called the *forward current,* flows until the *forward voltage exceeds* the junction barrier potential (0.3 V for Ge and 0.7 V for Si). The characteristic follows an exponential law. As the forward voltage is increased to the knee of the characteristic, the barrier potential is progressively reduced to zero, allowing more and more majority charge carriers to flow across the junction. Beyond the knee of the characteristics, the potential barrier is completely eliminated, forward current increases almost linearly with the increase in forward voltage and the P-N junction starts behaving as a resistor. If the forward voltage is increased beyond a certain value, extremely large current will flow and the P-N junction may get destroyed due to over heating.

Because of the nonlinear knee of the curve, if a very small signal voltage is applied to the diode so that it only operates around the knee, the signal will be distorted. The signal must, therefore, be large enough to operate the diode over the linear part of the characteristic. Forward resistance of P-N junction is low (of the order of 75 Ω in case of Ge and 150 Ω in case of Si).

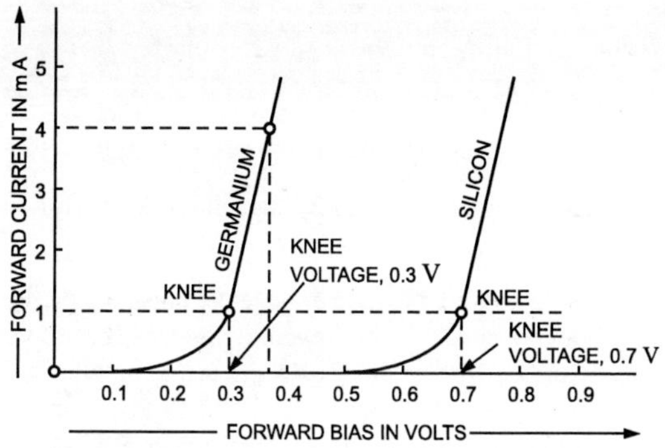

Fig. 24 *(a) Forward Characteristics For P-N Junction*

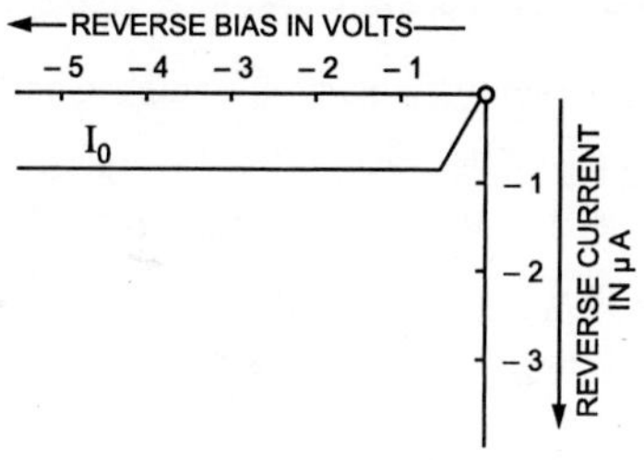

Fig. 24 *(b) Reverse Characteristics For P-N Junction*

2. ***Reverse Characteristics.*** When the reverse bias is applied (*i.e.* when P-region is connected to the negative bias terminal and N-region is connected to the positive bias terminal), the potential barrier at junction is increased, therefore, the junction resistance becomes very high and there is no possibility of a majority carriers flowing across a reverse biased junction. But still minority carriers generated on each side can cross the junction. Electrons on the P-side are attracted across the junction to the positive potential on the N-side and holes on the N-side may travel across the junction to the negative potential on the P-side. This results in a very small current, as illustrated in Fig. 24 (*b*). This current is called *reverse current* and is due to minority carriers.

For silicon the reverse current is typically less than 1 μA and may be as low as 1 *n*A. In case of germanium, the reverse current may exceed 10 μA. Only a very small reverse voltage is necessary to direct all minority carriers across the junction, and when all minority carriers are flowing across, further increase in bias voltage will not increase the current. This current is referred to as a *reverse saturation current* and is designated as I_0. Reverse resistance is very high (of the order of MΩ).

As discussed above, the reverse current I_0 is made up of minority charge carriers crossing the junction. When the temperature of the semiconductor material is increased, the additional thermal energy causes more electrons to break away from atoms. This creates more hole-electron pairs and generates more minority charge carriers. So reverse saturation current I_0 increases with the increase in the junction temperature. It has been found that I_0 approximately doubles for each 10° C rise in temperature in case of germanium and for each 8° C rise in temperature in case of silicon.

Procedure: Silicon or germanium diode, milliammeter A, voltmeter V and regulated power supply is connected, as shown in Fig. 25. Reading of voltmeter V and ammeter A are noted for different applied voltages.

Now the connection of diode to supply are reversed, milliammeter is replaced by microammeter and voltmeter (0-1V) is replaced by a voltmeter of high range (0-10V). Readings of voltmeter V and microammeter are noted for different input voltages.

The results are tabulated, and curves are plotted between voltage V and current I.

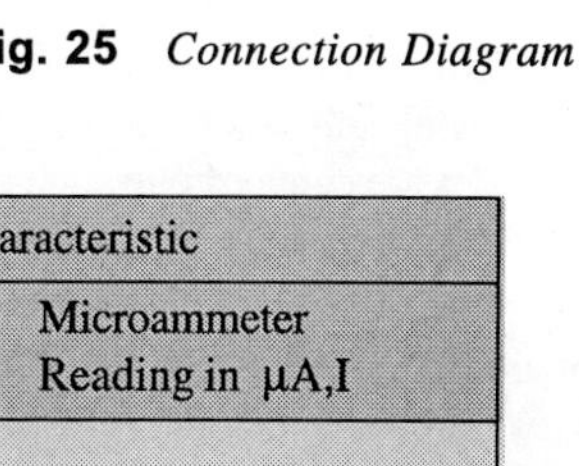

Fig. 25 *Connection Diagram*

Observations:

Forward Characteristic		Reverse Characteristic	
Voltmeter Reading in Volts, V	Milliammeter Reading in m A, I	Voltmeter Reading in Volts, V	Microammeter Reading in μA,I
0.0		0	
0.1		1.0	
0.2		2.0	
0.5		3.0	
0.8		4.0	
1.0		5.0	

Conclusion: The plotted V-I characteristics resemble to those shown in Figs. 25 (*a*) and 25 (*b*) respectively.

EXPERIMENT-19

Objective: To study full wave and half wave rectifier circuits with and without capacitor and determine ripple factors.

Apparatus Required: Semiconductor diodes-2; Centre-tapped transformer-1 Switches-2, Capacitor (10 μF)-1, Load resistor (1.2 kΩ)-1, Multimeter-1, Connecting wires.

Theory: *Rectifier* is a device which converts the sinusoidal ac voltage into either positive or negative pulsating dc. P-N junction diode, which conducts when forward biased and practically does not conduct when reverse biased, can be used for rectification i.e. for conversion of ac into dc. The rectifier typically needs one, two or four diodes. Rectifiers may be either half-wave or full-wave (centre-tap or bridge) type.

When a single rectifier unit is placed in series with the load across an ac supply, as illustrated in Fig. 26, it converts alternating voltage into unidirectional pulsating voltage, using one half cycles of the applied voltage, the other half cycles being suppressed because it conducts only in one direction. Unless there is an inductance or battery in the circuit, the current will be zero, therefore, for half the time. This is called *half–wave rectification.*

As already discussed, diode is an electronic device consisting of two elements known as cathode and anode. Since in a diode electrons can flow in one direction only *i.e.* from cathode to anode so the diode provides the unilateral conduction necessary for rectification. This is true for diodes of all types–vacuum, gas-filled, crystal or semiconductor, metallic (copper oxide and selenium types) diodes. Semiconductor diodes, because of their inherent advantages are usually used as a rectifying device. However, for very high voltages, vacuum diodes may be employed.

The half-wave rectifier circuit using a semiconductor diode with a load resistance R_L but no smoothing filter is given in Fig. 26. The diode is connected in series with the secondary of the transformer and the load resistance R_L, the primary of the transformer is being connected to the ac supply mains.

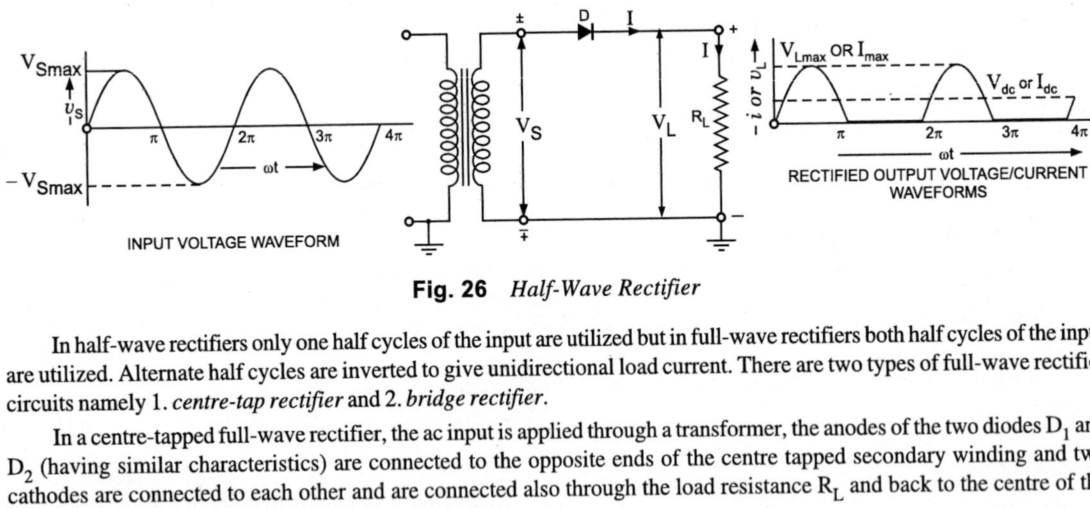

Fig. 26 *Half-Wave Rectifier*

In half-wave rectifiers only one half cycles of the input are utilized but in full-wave rectifiers both half cycles of the input are utilized. Alternate half cycles are inverted to give unidirectional load current. There are two types of full-wave rectifier circuits namely 1. *centre-tap rectifier* and 2. *bridge rectifier.*

In a centre-tapped full-wave rectifier, the ac input is applied through a transformer, the anodes of the two diodes D_1 and D_2 (having similar characteristics) are connected to the opposite ends of the centre tapped secondary winding and two cathodes are connected to each other and are connected also through the load resistance R_L and back to the centre of the transformer, as shown in Fig. 27.

When the top of the transformer secondary winding is positive, say during the first half-cycle of the supply, the anode of diode D_1 is positive w.r.t. cathode, and anode of diode D_2 is negative w.r.t. cathode. Thus only diode D_1 conducts, being forward biased and current flows from cathode to anode of diode D_1, through load resistance R_L and top half of the transformer secondary making cathode end of load resistance R_L positive. During the second half-cycle of the input voltage the polarity is reversed, making the bottom of the secondary winding positive w.r.t. centre tap and thus diode D_2 is forward biased and diode D_1 is reverse biased. Consequently during this half-cycle of the input only the diode D_2 conducts and current flows through the load resistance R_L and bottom of the transformer secondary making the cathode end of the load resistance R_L positive. Thus the direction of flow of current through the load resistance R_L remains the same during both halves of the input supply voltage. Thus the circuit shown in Fig. 27 acts as a full-wave rectifier.

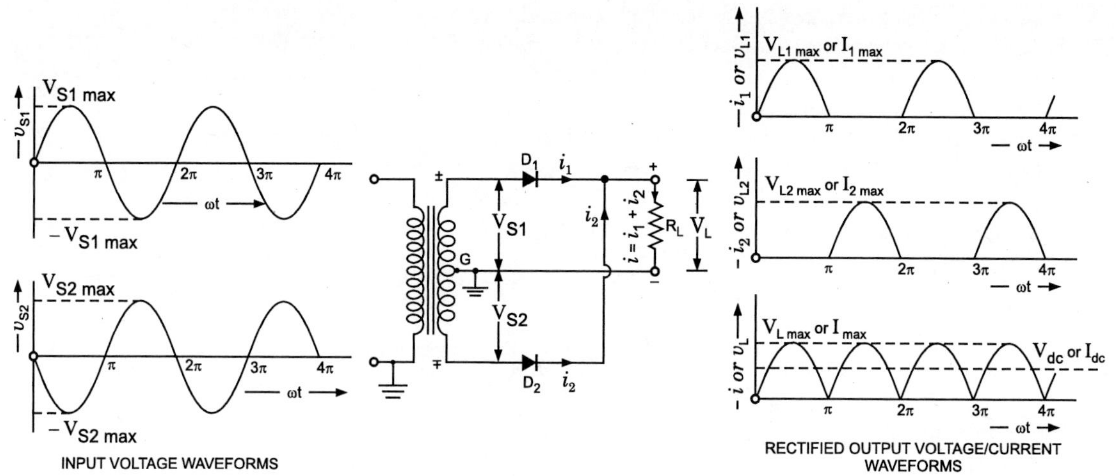

Fig. 27 *Centre-Tap Full-Wave Rectifier*

The output voltage from a rectifier circuit has a pulsating character *i.e.*, it contains unwanted ac components (components of supply frequency *f* and its harmonics) along with dc component. For most supply purposes, constant direct voltage is required than that furnished by a rectifier. To reduce ac components from the rectifier output voltage a *filter circuit* is required. Thus *filter* is a device which passes dc component to the load and blocks ac components of the rectifier output. Filter is typically constructed from reactive circuit elements such as capacitors and/or inductors and resistors.

Commonly used filter circuits are (*a*) series inductor filter (*b*) shunt capacitor filter (*c*) choke input filter (*d*) capacitor input or π filter and (*e*) R-C filter.

Shunt capacitor filter is the most simple form of the filter circuit and in this arrangement a high value capacitor C is placed directly across the output terminals, as shown in Fig. 28. During the conduction period it gets charged and stores up energy in the electrostatic field and discharges through the load resistance R_L delivering energy to it during non-conduction period. Through this process, the time duration during which current flows through the load resistance gets prolonged and ac components or ripples get considerably reduced. It is to be noted here that the capacitor C gets charged to the peak value of input voltage quickly because charging time constant is almost zero. It is so because there is no resistance (except the negligible forward resistance of diode) in the charging path. But the discharging time is quite large (roughly 100 times more than the charging time depending upon the value of R_L) because it discharges through load resistance R_L.

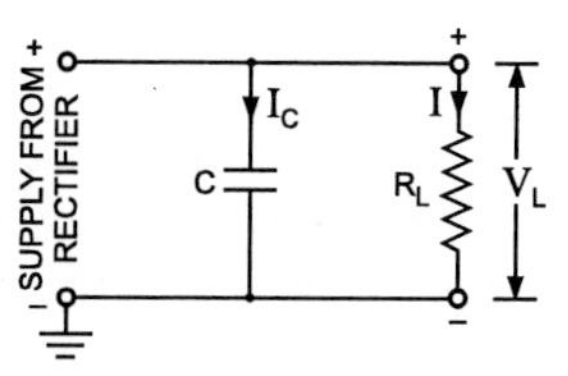

Fig. 28

The function of the capacitor filter may be viewed in terms of impedances. The large value capacitor C offers a low impedance shunt path to the ac components or ripples but offers high impedance to the dc component. Thus ripples get bypassed through capacitor C and only dc component flows through the load resistance R_L.

The pulsating output of a rectifier can be considered to contain a dc component and ac components called the *ripples*. The ripple current is undesirable and its value should be the smallest possible in order to make the rectifier effective.

The ripple voltage or current is measured in term of the ripple factor which is defined as the ratio of the effective value of the ac components of voltage (or current) present in the output from the rectifier to the direct or average value of the output voltage (or current).

$$i.e.\ \text{ripple factor},\ \gamma = \sqrt{\left(\frac{I_{rms}}{I_{dc}}\right)^2 - 1}$$

Circuit Diagram: Circuit diagram is shown in Fig. 29. When switch S_1 is open the circuit behaves as a half-wave rectifier and when switch S_1 is closed the circuit behaves like a full-wave rectifier. Shunt capacitor C can be connected across load resistor R_L by closing switch S_2 and taken out from the circuit by opening switch S_2.

Procedure: The switches S_1 and S_2 are kept in open positions. The circuit now behaves as a half-wave rectifier without capacitor filter. AC voltage across the secondary of the transformer in the circuit (*i.e.* half the secondary) at X and the dc voltage across load resistor R_L are measured with a multimeter.

Now switch S_2 is closed, keeping switch S_1 in open position. The circuit is now behaving as a half-wave rectifier circuit with capacitor filter.

AC voltage across the secondary of the transformer and the dc voltage across the load resistor R_L are measured with a multimeter.

The procedure is repeated with switch S_1 closed and switch S_2 open (*i.e.* full-wave rectifier without capacitor filter) and with both switches S_1 and S_2 closed (*i.e.* full-wave rectifier with capacitor filter).

The readings are noted in tabular form as below:

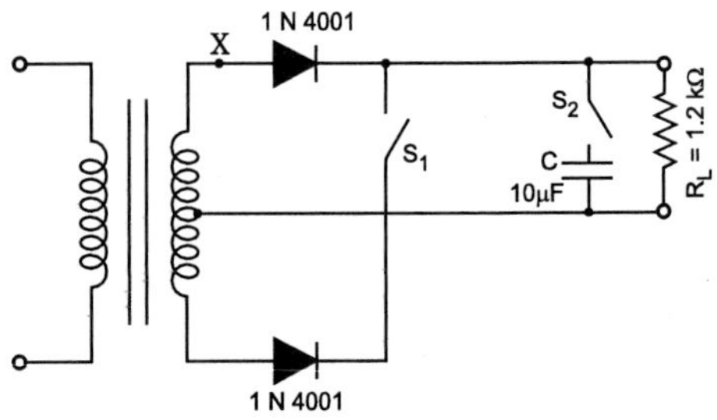

Fig. 29 *Circuit Diagram*

Observations:

Parameters	*HW Rectifier without capacitor filter*	*HW Rectifier with capacitor filter*	*FW Rectifier without capacitor filter*	*FW Rectifier with capacitor filter*
Voltage at the diode input, V_{rms} in volts				
DC output voltage measured V_{dc} in volts				
DC, output voltage calculated $V'_{dc} = \frac{\sqrt{2}V_{rms}}{\pi}$ volts				
Ripple factor $\gamma = \sqrt{\left(\frac{V_{rms}}{V_{dc}}\right)^2 - 1}$				

Conclusion: It is observed that the ripple factor is the maximum in the case of half-wave rectifier, decreases with the capacitor filter circuit and is minimum in case of full-wave rectifier with capacitor filter circuit.

EXPERIMENT-20

Objective: To study various logic gates.

Apparatus Required: Bread board, Regulated power supply (5 V) ICs 7408, 7432, 7404, 7400, 7402, 7486, Digital multimeter or LED indicator, Logic probe or Digital trainer practical kit.

Theory: A circuit, which performs the arithmetic and logic functions, is a combination of different logic circuits. Various types of gates are used in logic circuits. A gate is a digital circuit with one or more input voltages but with one output voltage. The most basic gates are:

OR (IC7432), AND (IC7408), NOT (IC7404), NOR (IC7402), NAND (IC7400), EXOR (IC7486) and EX-NOR (IC74266) [Fig. 30]

Gate	Symbol	Function	Truth Table		
			Inputs		Output
			A	B	Z
OR		Z = A OR B = A + B	0	0	0
			0	1	1
			1	0	1
			1	1	1

Contd...

AND

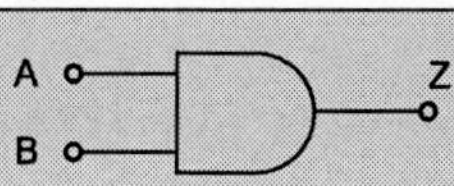

$Z = A \text{ AND } B$
$= A \cdot B$

Inputs		Output
A	B	Z
0	0	0
0	1	0
1	0	0
1	1	1

NOT

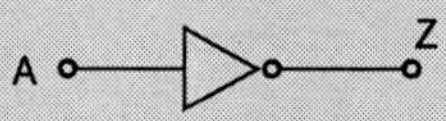

$Z = \text{NOT } A$
$= \overline{A}$

Input A	Output Z
0	1
1	0

NOR

$Z = A \text{ NOT OR } B$
$= A \text{ NOR } B$
$= \overline{A + B}$

Inputs		Output
A	B	Z
0	0	1
0	1	0
1	0	0
1	1	0

NAND

$Z = A \text{ NOT AND } B$
$= A \text{ NAND } B$
$= \overline{A \cdot B}$

Inputs		Output
A	B	Z
0	0	1
0	1	1
1	0	1
1	1	0

EX-OR

$Z = A \text{ EX-OR } B$
$= A \oplus B$
$= A\overline{B} + \overline{A}B$

Inputs		Output
A	B	Z
0	0	0
0	1	1
1	0	1
1	1	0

EX-NOR

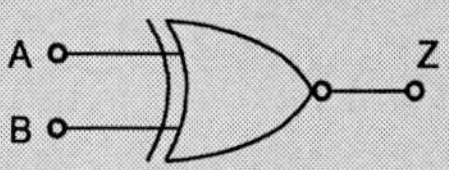

$Z = \overline{A \text{ EX-OR } B}$
$= A \text{ EX-NOR } B$
$= \overline{A \oplus B}$
$= AB + \overline{A}\,\overline{B}$

Inputs		Output
A	B	Z
0	0	1
0	1	0
1	0	0
1	1	1

Fig. 30

Procedure: 1. Place the different gates ICs on the bread board.

2. Apply power supply V_{CC} (+ 5V) to pin 14 and ground to pin 7 of the gates with the help of switches.

3. Verify the logic at output (*low* or *high*) with the help of indicator LED or measure output voltage with the help of multimeter. Logic probe can also be used for this purpose.

Conclusion: All the logic gates of 74 series verify the Truth table for given gate as mentioned in Fig. 30

EXPERIMENT-21

Objective: To study operational amplifier as adder and subtractor.

Apparatus Required: Regulated power supply (± 15 V), Experimental board of Op-Amp IC 741, Multimeter and Connecting wires.

Theory: An *operational amplifier*, abbreviated as *op-amp*, is basically a multistage, very high gain (typically 200,000), direct-coupled, negative feedback amplifier that uses voltage-shunt feedback to provide a stabilized voltage gain.

The most useful of the op-amp circuits employed in analog computers is the summing amplifier circuit. This circuit can be used to add ac or dc signals. This circuit provides an output voltage proportional to or equal to the algebraic sum of two or more input voltages each multiplied by a constant gain factor. Three-input summing circuit is shown in Fig. 31 (*a*).

Using the virtual equivalent circuit, as shown in Fig. 31 (*b*), the output voltage can be given in terms of inputs as

$$V_{out} = -R_f\left[\frac{V_1}{R_1}+\frac{V_2}{R_2}+\frac{V_3}{R_3}\right]$$

$$\text{If } R_1 = R_2 = R_3 = R_f$$

$$\text{then } V_{out} = -(V_1 + V_2 + V_3)$$

Op-amp summing amplifiers are also called *mixers*. One of the advantages of inverting op-amp mixers is that there is no interaction between the inputs. The inverting input is a virtual ground. This prevents one input signal from appearing at the other inputs.

The circuit shown in Fig. 31 (*a*) can be used as a *summing amplifier, scaling amplifier* or *averaging amplifier*.

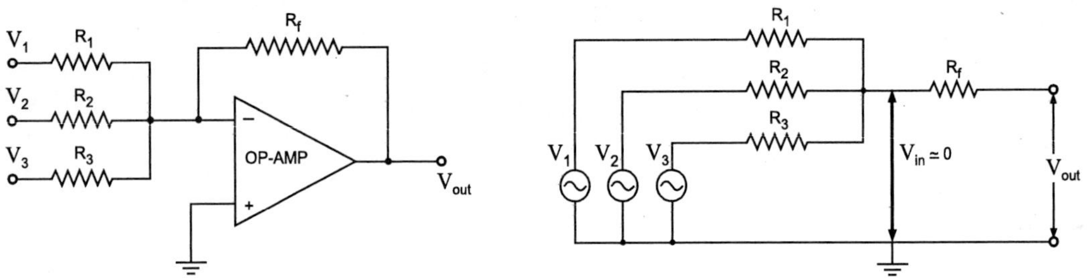

(*a*) *Summing Operational Amplifier* (*b*) *Virtual-Ground Equivalent Circuit*

Fig. 31

Since differential amplifier amplifies the difference of two input signals applied to the inverting and noninverting terminals of an op-amp, it can be used as subtractor circuit. The circuit is shown in Fig 32. This circuit provides an output equal to the difference of two input signals.

By selecting appropriate values for the external resistors, the input signals can be scaled (or attenuated) to the desired value. The circuit so obtained is called the *scaling amplifier*.

In circuit shown in Fig. 32 all external resistors are of equal value, and, therefore, gain of the amplifier is unity.

So output voltage, $V_{out} = \frac{-R}{R}[V_a - V_b] = V_b - V_a$

Thus the output voltage V_{out} is equal to the voltage applied to the noninverting input terminal less the voltage applied to the inverting input terminal. Hence the circuit is called a *subtractor*.

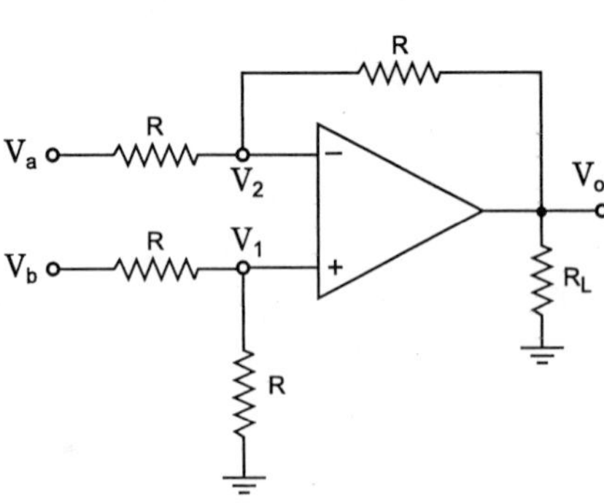

Fig. 32 *OP-Amp Subtractor*

Procedure: Using experiment board for IC 741 (or bread board to design the circuit), apply two or more input voltages from dc power supplies to the inverting terminal of the op-amp and measure them with the help of suitable meters. Then measure output voltage at pin 6 of op-amp 741. Change the inputs and repeat the process.

Now apply two input voltages V_a and V_b from dc power supplies to the op-amp noninverting and inverting input terminals respectively and measure them with the help of suitable meters. Then measure output voltage at pin 6 of the op-amp 741. Change the input and repeat the process.

Note: AC voltage can also be used for adder and subtractor. Procedure and result will be the same for both.

Observations: (1) Adder or summing Amplifier

$$V_1 = ...$$
$$V_2 = ...$$
$$V_3 = ...$$
$$V_{out} \text{ (measured)} = ...$$
$$V_{out} = -(V_1 + V_2 + V_3)$$

Result can also be verified using CRO traces.

(2) Subtractor:

$$V_a = ...$$
$$V_b = ...$$
$$v_{out} \text{ measured} = ...$$
$$v_b - v_a = ...$$

Conclusion: It is observed that both the op-amp circuits namely *adder* and *subtractor* follow the principle as laid down in theory.

EXPERIMENT-22

Objective: To study transistor as a switch.

Apparatus Used: Bread board, Regulated power supply, Resistors, LED, Ammeter and Voltmeters (or Multimeter).

Theory: Transistor is a three layer (PNP or NPN), three terminal (base, emitter and collector) bipolar junction device. On proper biasing it provides amplification to given input signal. Transistor has two junctions. For suitable amplification input junction should be forward biased and output junction should be reverse biased.

Output characteristics of a transistor has three main regions - cutoff, active, and saturation.

1. If the input is zero, output will also be zero and transistor is to be in cutoff region with no output.

2. If the transistor is in active region, it will amplify the input by a factor called amplification factor or gain factor

3. For a certain input signal, output attains a maximum value. If the input signal is further strengthened, the output will not increase and it will be saturated. This is called the *saturation region.*

Now if any transistor is operated in cutoff region and saturation region, with being stayed in active region it works as switch and its output will be either Zero or V_{CC} (i.e. either *low* or *high*).

Procedure: Design the circuit on bread board using the components of suitable values as shown, in Fig. 33.

1. Apply low voltage (< 0.3 V) to the input, transistor will be in cutoff and the output at the collector will be *high* (≈ V_{CC}), and LED will start glowing.
2. Apply high voltage (> 3.5 V) to the input base, transistor will be in saturation and the output at collector will be *low* (≈ 0.2 V) and LED will not glow.
3. Apply rectangular wave (or train of pulses) to the input, the transistor will oscillate between cutoff and saturation region. The LED will start blinking (i.e. *on* and *off*).

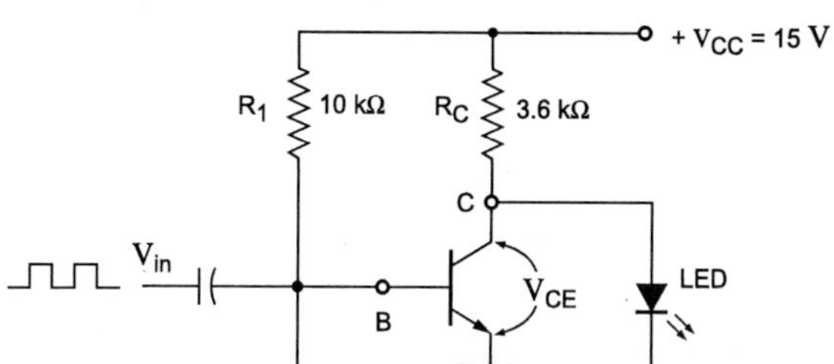

Fig. 33

Conclusion: If a transistor is given such an input that it is either in cutoff or saturation region, it will work as a switch and its output will oscillate between *high* and *low*.

❑❑❑

B.TECH.

(SEMESTER-I) THEORY EXAMINATION, 2011-12
ELECTRICAL ENGINEERING

Time : 3 Hours *Total Marks : 100*

Note: Attempt **all** Sections. Assume any missing data, if any.

SECTION A

1. Answer all parts with brief explanation: **(2 × 10 = 20)**

(*a*) Write the properties of ideal voltage source.

(*b*) Write an expression of resonance frequency for high Q-series RLC circuit.

(*c*) Why two-wattmeter method for power measurement is universal one?

(*d*) Explain the term "creep" in energy meter.

*(*e*) Write the abbreviation of ACSR in power system.

*(*f*) What is the concept of grid in power system?

(*g*) Classify the losses in power transformer.

(*h*) Explain the term "slip" in induction motor.

(*i*) What will happen if the back emf of dc motor vanishes?

(*j*) Why dc series motor is preferred in elevators?

SECTION B

2. Answer any **three** parts of the following: **(10 × 3 = 30)**

(*a*) State Norton's theorem in dc circuit. Also calculate Norton's equivalent of the network shown in Fig. 1 at terminal AB. Determine the current through 4 Ω resistor across AB.

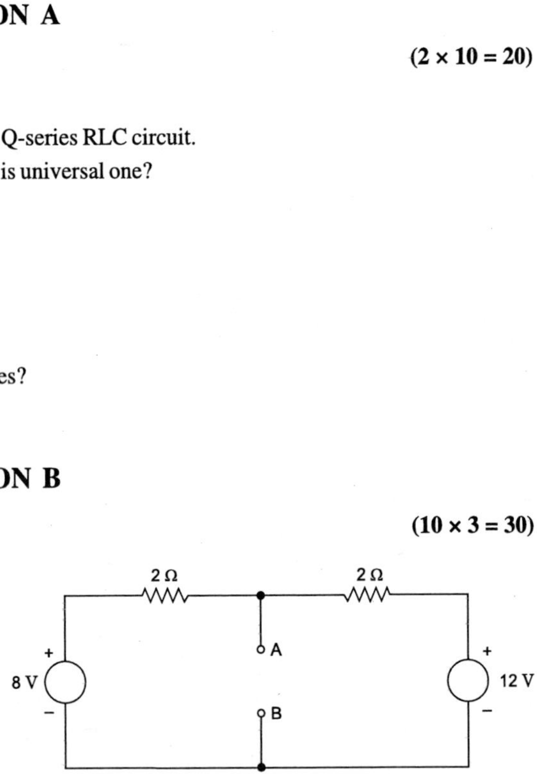

Fig. 1

(*b*) Show that power in 3-phase, balanced system is constant at every instant and is given by 3 $V_pI_p \cos\phi$, where V_p, I_p and ϕ have usual meanings.

(*c*) Describe the construction and principle of operating of attraction type moving iron instrument.

(*d*) Derive the Quality factor Q of the series RLC circuit at resonance. Define the bandwidth for the same.

(*e*) A three-phase 50 Hz, induction motor has a full-load speed of 1,460 rpm. Calculate slip, number of poles and frequency of rotor induced emf.

SECTION C

Note: Attempt **all** questions in this Section.

3. Answer any **two** parts of the following: **(5 × 2 = 10)**

(*a*) Explain voltage and current sources of a network with characteristics. Explain source-transformation principle in any circuit.

(*b*) State and prove maximum power transformer theorem.

(*c*) Using nodal analysis, find the current through 10 Ω resistor shown in Fig. 2.

* Not covered in latest syllabus.

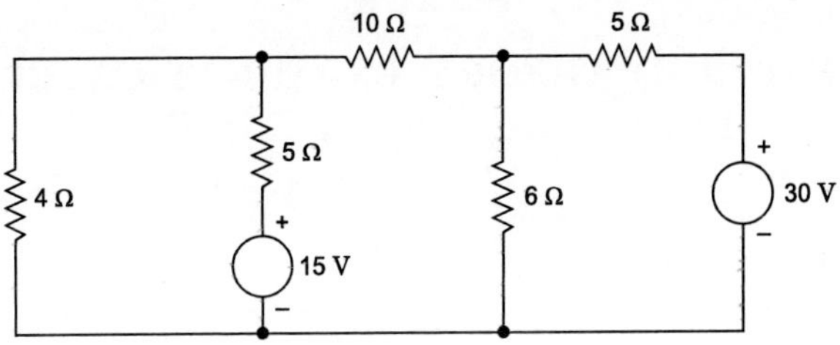

Fig. 2

4. Answer any **one** part of the following: **(10 × 1 = 10)**

(*a*) The voltage applied to a circuit is $v = 100 \sin (\omega t + 30°)$ and current flowing in the circuit is $i = 20 \sin (\omega t + 60°)$. Determine the impedance, resistance, reactance, power and power factor of the circuit.

(*b*) Calculate the resonance frequency of the circuit shown in Fig. 3.

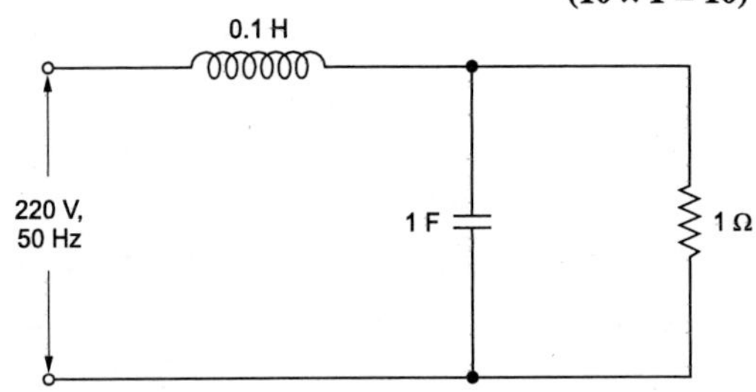

Fig. 3

5. Answer any **two** parts of the following: **(5 × 2 = 10)**

(*a*) Describe the construction details of a single-phase induction energy meter with neat diagram.

(*b*) A balanced delta connected load of $(12 + j9)$ Ω/phase is connected to 3-phase 400 V supply. Calculate line current, power factor and power drawn by it.

(*c*) Show that in a 3-phase star connected system, the line voltage is $\sqrt{3}$ times of the phase voltage.

6. Answer any **one** part of the following: **(10 × 1 = 10)**

(*a*) Derive an emf expression of power transformer. Also draw an equivalent circuit of it.

*(*b*) Discuss the classification of power system in terms of voltage level. Also draw line diagram of typical substation.

7. Answer any **two** parts of the following: **(5 × 2 = 10)**

(*a*) Discuss the principle of operation of single-phase induction motor.

(*b*) Explain speed-torque characteristic of dc series motor. Also mention typical applications of it.

(*c*) A dc shunt motor develops an open-circuit emf of 250 volt at 1,500 rpm. Find its developed torque for an armature current of 20 amperes.

❑❑❑

* Not covered in latest syllabus.

B.TECH.

(SEMESTER-II) THEORY EXAMINATION, 2011-12
ELECTRICAL ENGINEERING

Time : 3 Hours *Total Marks : 100*

Note: Attempt all the **three** sections.

SECTION A

1. Attempt the following **(2 × 10 = 20)**

(*i*) Is 'L' a linear element? Explain the answer.

(*ii*) If resistance of each branch is 3 ohms in a delta connected load, what would be the resistance of each branch in its star equivalent?

(*iii*) Draw suitable waveforms to show the following:

$$v = V_m \sin \omega t;\ i = I_m \sin (\omega t + \theta)$$

Label the quantities suitably.

(*iv*) For a series R-L-C circuit draw the phasor diagram and show the power factor angle in that. Assume the circuit draws inductive current.

(*v*) Give the relationship between quality factor, resonant frequency and bandwidth for a series R-L-C circuit.

(*vi*) What is the major difference between PMMC type and dynamometer type of instruments?

(*vii*) Show the residual magnetism in a B-H curve.

(*viii*) Name the constant losses taking place in a transformer.

(*ix*) Name two motors used for constant speed operation.

(*x*) Define the term "slip".

SECTION B

Attempt any **three** of the following: **(10 × 3 = 30)**

2. Apply mesh analysis, obtain the current through 5 ohm resistance in the following circuit:

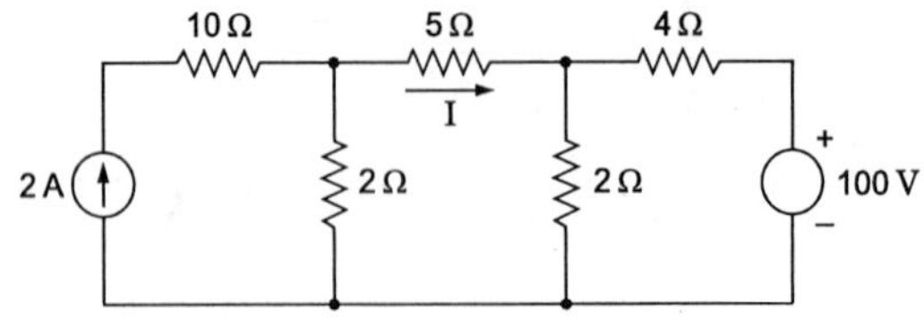

3. Voltages across R, L, C connected in series are 5, 8 and 10 volts respectively. Calculate the value of supply voltage at 50 Hz. Also find the frequency at which this circuit would resonate.

4. Find the current in phase R in the following circuit:

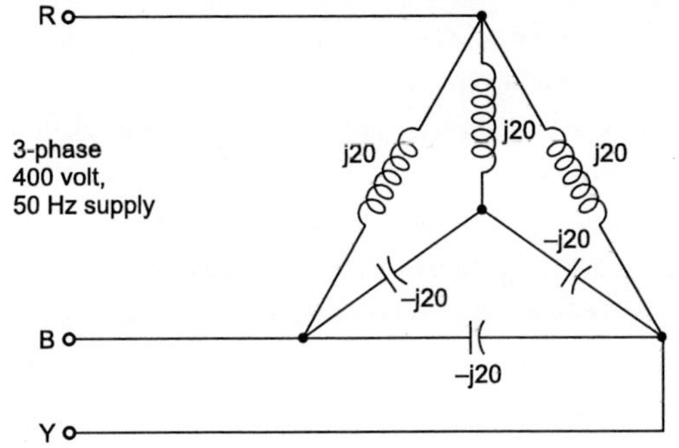

5. The maximum efficiency of a 100 kVA, 1100/440 volt, 50 Hz transformer is 96%. This occurs at 75% of full load at 0.8 pf lagging. Find the efficiency of transformer at 3/4 FL at 0.6 pf leading.

6. The armature resistance of a 200 V dc shunt motor is 0.12 ohm. It runs at 600 rpm at constant torque load and draws a current of 21 amperes. Calculate its new speed if the field current of the motor is reduced by 10%.

SECTION C

Attempt **all** questions. **(10 × 5 = 50)**

7. Attempt any **two** of the following:

(*a*) Find the equivalent resistance for the following circuit and hence calculate the current supplied by the source.

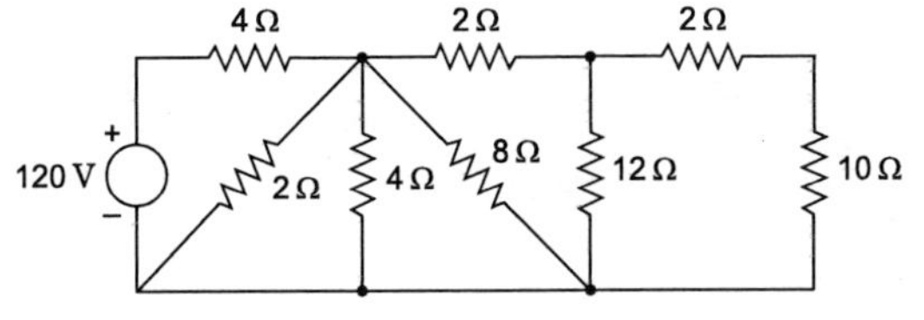

(*b*) Explain the duality between a Thevenin's and a Norton's equivalent circuits.

(*c*) Find the current flowing through 10 Ω resistance in the following circuit. Use Superposition theorem.

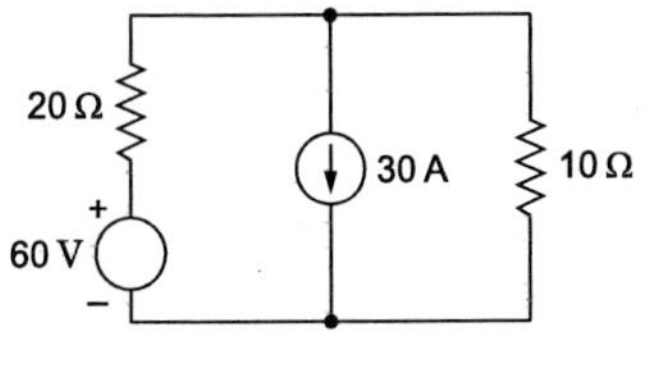

8. Attempt any **two** of the following:

(*a*) If load draws a current of 10 ampere at 0.8 pf lagging; when connected to 100 volt supply, calculate the values of real, reactive and apparent powers. Also find out the resistance of the load.

(*b*) Deduce the formula for half-power frequencies for a series R-L-C circuit under resonance. Why are they called 'half-power frequencies'?

(*c*) The two branches of a parallel circuit draw currents I_1 and I_2 such that $I_1 = 10\sqrt{2}\ \sin \omega t$ and $I_2 = 5\sqrt{2}\ \sin (\omega t - 60°)$. What is the total current drawn by them?

9. Attempt any **two** of the following:

(*a*) Discuss about induction type instrument in detail.

(*b*) With the help of connection diagram and the related phasor diagram show that $W_1 + W_2$ gives the total 3ϕ power in a two wattmeter method of power measurement.

(*c*) An ammeter has internal resistance of 2 ohms and it is designed to measure 1 A current. What would you do to make this meter read 15 A current?

10. Attempt any **two** of the following:

(*a*) Draw and explain B-H curve. What is meant by saturation, coercive force, residual magnetism? Show them in the diagram.

(*b*) Write down the expression of efficiency for a transformer. How is it affected by change in power factor?

(*c*) Discuss in brief about Auto-transformer.

11. Attempt any **two** of the following:

(*a*) Define 'Slip'. Discuss the principle of operation of $3\text{-}\phi$ induction motor.

(*b*) What do you know about splitting of phase in a single phase induction motor? How does it help in starting the capacitor start single phase induction motor?

(*c*) Describe briefly the construction and working of a synchronous generator.

❑❑❑

B.TECH.

FIRST SEMESTER EXAMINATION, 2012-13
ELECTRICAL ENGINEERING

Time : 3 Hours *Total Marks : 100*

Note: Attempt **all** sections. Assume missing data, if any.

SECTION A

1. Answer ***all*** parts of this section: **(2 × 10 = 20)**

(*i*) On what factors do the resistance of a conductor depends?

(*ii*) Calculate the form factor of a square wave.

(*iii*) What is the use of condenser in single-phase ac motor?

(*iv*) The dimension of L/CR is ________.

(*v*) The maximum and minimum values of power factor can be ________.

(*vi*) Write the name of different types of rotor of an alternator.

(*vii*) What is typical use of an autotransformer?

(*viii*) What happens if dc supply is given to a transformer?

(*ix*) Under no load running condition of synchronous machine, what will be the angle between the induced voltage and supply voltage?

(*x*) Enlist the types of moving iron instrument.

SECTION B

2. Answer any ***three*** parts of this section: **(10 × 3 = 30)**

(*a*) Using superposition theorem, calculate the current in the AB branch in the circuit shown in the below figure:

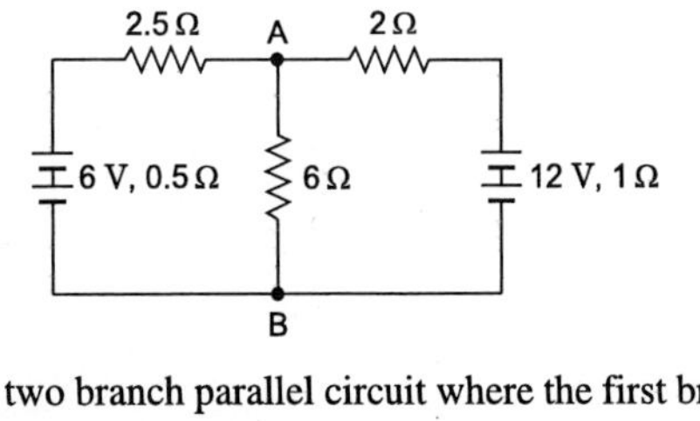

(*b*) Obtain the power factor of a two branch parallel circuit where the first branch has $\overline{Z}_1 = (2 + j4)\ \Omega$ and second $\overline{Z}_2 = (6 + j0)\ \Omega$. To what value must the 6 Ω resistor be changed to result in the overall power factor 0.9 lagging?

(*c*) A moving coil instrument having a resistance of 50 Ω has a full-scale deflection of 1 mA. Calculate:

(*i*) Shunt resistance to convert the instrument into an ammeter of 2 A range.

(*ii*) Net resistance of the meter.

(*d*) Prove that for a single-phase transformer:

$$(\text{KVA})_{\text{at max}} = (\text{KVA})_{\text{rated}} \sqrt{\frac{P_{\text{core}}}{P_{\text{ohmic (rated)}}}}$$

(*e*) A dc shunt motor runs at 600 rpm taking 60 A from a 230 V supply. Armature resistance is 0.2 ohm and field resistance is 115 ohms. Find the speed when the current through the armature is 30 A.

SECTION C

Note: Answer any **five** questions of the following: **(10 × 5 = 50)**

3. Why power factor improvement is important? What are causes and problems of low power factor? How power factor can be improved?

4. A series RLC circuit consisting of a resistance of 20 Ω, inductance 0.2 H and capacitance of 150 μF is connected across a 230 V, 50 Hz source. Calculate:

(*i*) the impedance

(*ii*) the current

(*iii*) power factor

(*iv*) the frequency of supply to be adjusted to make power factor unity.

5. Calculate current in a 1000 Ω resistor connected between terminals A and B, as shown in the below figure with the help of Thevenin's theorem.

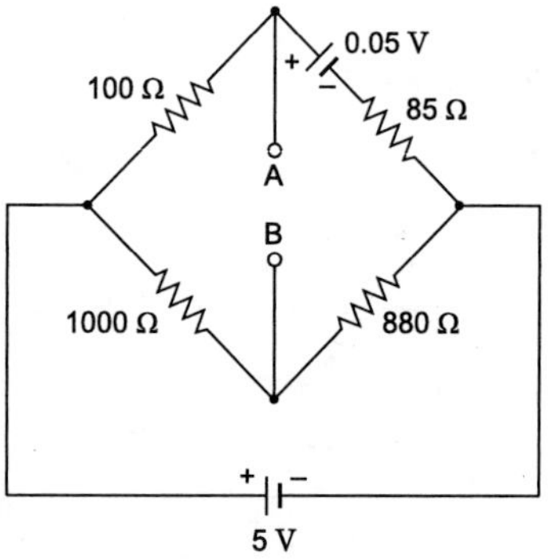

6. Explain the speed-torque characteristics of a dc shunt and series motors.

7. What is the use of starter? Explain various methods of starting of 3-phase induction motors.

8. Using double-revolving field theory, explain the principle of operation of a single-phase induction motor.

9. A balanced 3-phase star-connected load of 18 kW taking a leading current of 60 amperes when connected across a 3-phase 440 V, 50 Hz supply. Find the values and nature of load.

❑❑❑

B.TECH.

EVEN SEMESTER EXAMINATION, 2012-13
ELECTRICAL ENGINEERING

Time : 3 Hours *Total Marks : 100*

Note: Attempt **all** sections. Assume missing data, if any.

SECTION-A

1. Answer **all** parts of this section: **(2 × 10 = 20)**

(*a*) $V = \sqrt{2} \times 200 \cos 500t$, $P_{av} = 250$ W, power factor = 0.7 lagging. Calculate the reactive power of the system.

(*b*) What will be the current at resonance in series and parallel RLC circuit?

(*c*) Enlist types of rotor of an alternator.

(*d*) The full-load copper loss and iron loss of a transformer are 6,400 W and 5,000 W respectively. The above losses at half load are ....

(*e*) Write an expression of sensitivity of a voltmeter.

(*f*) An induction motor has a rated speed of 715 rpm. How many poles have its rotating magnetic field?

(*g*) A three-phase induction motor connected in star takes 5 A. When connected in delta, what will be current?

(*h*) A voltage source of 100 V has an internal impedance 2 Ω and supplies a load having that same impedance. How much power is absorbed by the load?

*(*i*) What is the voltage level range of power generated in a power station?

(*j*) Enlist the various types of moving iron instruments.

SECTION-B

2. Answer any **three** parts of this section: **(10 × 3 = 30)**

(*a*) Obtain the Norton's equivalent circuit at terminal AB of the network shown in Fig. 2(*a*).

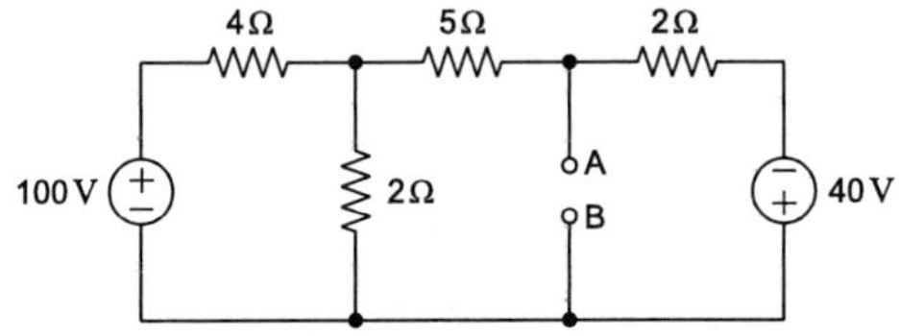

Fig. 2(*a*)

(*b*) A resistance and inductance are connected in series across a voltage $v = 283 \sin 314t$. An expression of current is found to be $i = 4 \sin(314t - 45°)$. Find the values of resistance, inductance and power factor.

(*c*) Explain the essential torques of an indicating instrument in detail.

(*d*) What do you understand by the term "ideal transformer"? Why the efficiency of transformer is high as compared to other machines?

(*e*) What are the different types of dc motors? Also write the applications of each.

SECTION-C

Note: Answer **all** questions of this ection: **(10 × 5 = 50)**

3. What is the criterion for star-delta transformation? Develop the relationship for delta-to-star and star-to-delta transformation.

4. Obtain the power factor of a two branch parallel circuit, where the first branch has $\mathbf{Z_1}$ = (2 + *j*4) Ω and second $\mathbf{Z_2}$ = (6 + *j*0) Ω. To what value must the 6 Ω resistor be changed to result in overall power factor 0.9 lagging?

OR

A 100 V, 80 W lamp is to be operated on 230 V, 50 Hz ac supply. Calculate the inductance to be connected in series with the lamp for the above operation. Lamp can be taken as pure resistance.

5. Describe working principle of PMMC instruments. Why is the scale linear?

OR

A 50 A, 230 V energy meter on full-load test makes 61 revolutions in 37 seconds. If the normal disc speed is 500 revolution per kWh, find the percentage error.

6. Prove that, for a single-phase transformer

$$(\text{kVA})_{\text{max}} = (\text{kVA})_{\text{rated}}\sqrt{\frac{P_{\text{core}}}{P_{\text{cu rated}}}}$$

OR

A balanced star connected load of $(8 + j6)\ \Omega$ per phase is connected to a 3-phase 400 V supply. Find the line current, power factor and three phase VAs.

7. Why synchronous motor is not self-starting? Explain the various methods of starting these motors.

OR

Explain with the help of diagrams, how a rotating field is produced in the air gap of a three-phase induction motor.

❑❑❑

B.TECH.

ODD SEMESTER EXAMINATION, 2013-14
ELECTRICAL ENGINEERING

Time : 3 Hours *Total Marks : 100*

Note: Attempt **all** sections.

SECTION-A

1. Answer **all** parts: **(10 × 2 = 20)**

(*a*) Define active and passive elements with examples.

(*b*) Define form factor and peak factor.

(*c*) A series circuit has R = 10 Ω, L = 0.05 H, C = 10 μF. Calculate Q-factor of the circuit.

(*d*) What is the significance of back emf in dc motor?

(*e*) Why dc series motor is never started on no load?

(*f*) Draw approximate equivalent circuit of transformer referred to primary side.

(*g*) A 4-pole, 3-phase, 50 Hz, star connected induction motor has a full-load slip of 4%. Calculate full-load speed of motor.

(*h*) Define mmf, reluctance, flux in magnetic circuit.

(*i*) What are the applications of dc series and dc shunt motors?

(*j*) Write an expression of hysteresis loss in a transformer. Why it decreases at higher frequencies with constant V?

SECTION-B

2. Attempt any **three** parts: **(10 × 3 = 30)**

(*a*) Derive emf equation for a single phase transformer. A 25 kVA, 2,200/220 V, 50 Hz, 1-phase transformer has following parameters:

$$R_1 = 1.75\ \Omega,\ R_2 = 0.0045\ \Omega,\ X_1 = 2.6\ \Omega,\ X_2 = 0.0075\ \Omega.$$

Calculate:

(*i*) Equivalent resistance referred to primary and secondary.

(*ii*) Equivalent reactance referred to primary and secondary.

(*b*) Derive an expression for torque in dc motor. Draw the load characteristics of dc series and shunt motors.

(*c*) Define series resonance and resonant frequency. Why in series circuit voltage across L and C is very high? Draw resonance curve.

(*d*) Explain two wattmeter method to measure three-phase power. If in a two wattmeter method, readings of two wattmeters are 1,200 W and 300 W. Find the power factor of the load.

(*e*) Describe the working principle and construction of single phase energy meter and its applications.

SECTION-C

Note: Attempt **all** parts. **(10 × 5 = 50)**

3. Attempt any **two** parts:

(*a*)

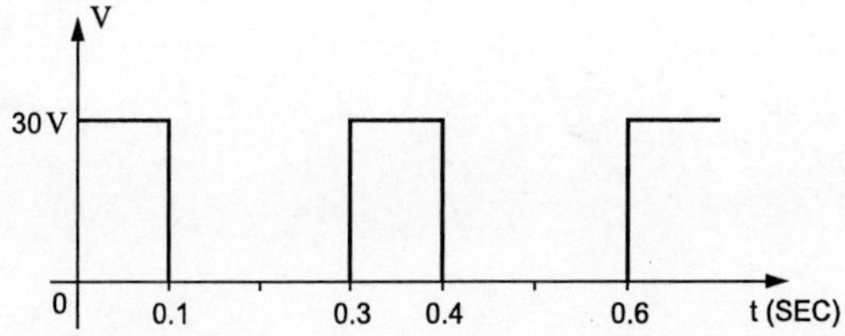

Find rms value, average value and form factor of the wave.

(*b*) A 120 V, 100 W lamp is to be connected to a 220 V, 50 Hz ac supply. What value of pure inductance should be connected in series in order that lamp is run on the rated voltage?

(*c*) Define resonance in parallel RLC circuit. Draw resonance curve.

4. Attempt any **two** parts:

(*a*) State and prove maximum power transfer theorem in dc circuit.

(*b*) Find the voltage V_1 across 6 Ω resistance using loop analysis method.

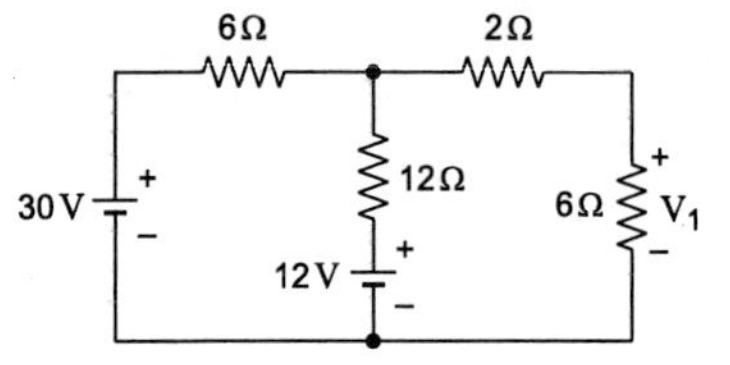

(*c*)

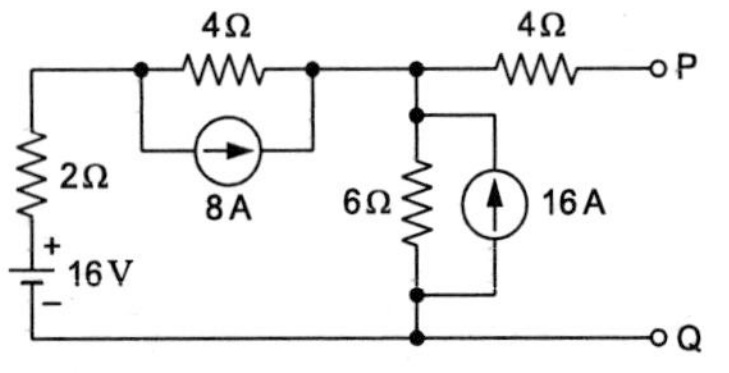

Find V_{th} and R_{th} for the circuit shown in above figure.

5. Attempt any **one** part:

(*a*) Prove that in a 3-phase delta-connected system $I_L = \sqrt{3}\, I_{ph}$. A 3-phase, 400 V supply is connected to a 3-phase star balanced load. The line current is 20 A and the power consumed by the load is 12 kW. Calculate the impedance of the load, phase current and power factor.

(*b*) (*i*) A moving coil instrument having internal resistance of 50 Ω indicates full-scale deflection with a current of 10 mA. How can it be made to work as:

(*a*) Voltmeter to read 100 volts.

(*b*) Ammeter to read 1 A on full scale?

(*ii*) Define analogy between electric and magnetic circuits.

6. Attempt any **two** parts:

*(*a*) Draw single line diagram of power system and explain.

(*b*) Explain working principle of autotransformer. What are its advantages and applications?

(*c*) What are different losses in transformer? Explain.

7. Attempt any **one** part:

(*a*) Why single phase induction motor is not self starting? What are the methods of starting? Explain any one of them which is used in fan.

(*b*) (*i*) A 5 hp, 230 V, 50 Hz induction motor has a rated full-load speed of 950 rpm. The induced voltage per phase of rotor at standstill is 100 V. Calculate:

(*a*) Number of poles and % full-load slip.

(*b*) Rotor induced voltage and its frequency at full load.

(*ii*) Explain working principle of synchronous motor and two applications.

❑❑❑

* Not covered in latest syllabus.

B.TECH.

ODD SEMESTER EXAMINATION, 2013-14
ELECTRICAL ENGINEERING

Time : 3 Hours *Total Marks : 100*

Note: Attempt **all** the questions. All questions carry equal marks.

1. Attempt any **two** parts of the following: **(10 × 2 = 20)**

(*a*) Explain the following:

(*i*) Active and passive elements.

(*ii*) Unilateral and bilateral elements.

(*iii*) Time variant and time invariant elements.

(*iv*) Distributed and lumped parameters.

(*b*) State and explain maximum power transfer theorem. Mention its significance in communication circuits. Consider the circuit diagram shown in Fig. 1.

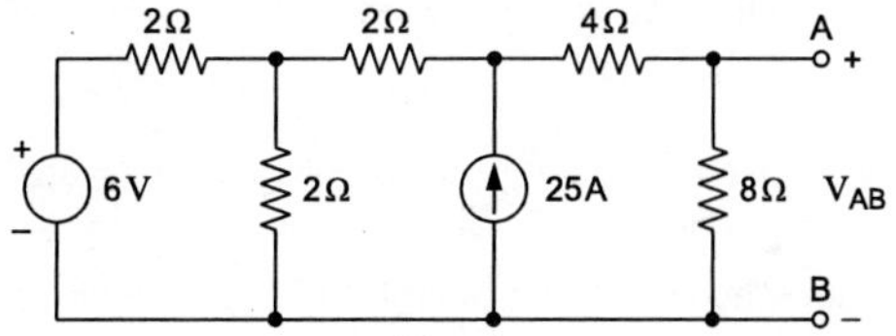

Fig. 1

Determine open-circuit voltage across AB terminal shown in Fig. 1 by Thevenin's Theorem.

(*c*) State and explain Kirchhoff's law. What are the limitations and applications of Kirchhoff's law in circuit theory? Determine i_1, i_2 and i_3 using Nodal analysis for a given circuit shown in Fig. 2.

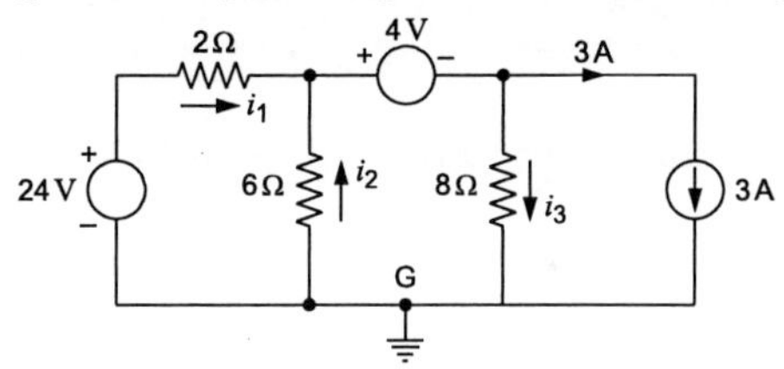

Fig. 2

2. Attempt any **two** parts of the following: **(10 × 2 = 20)**

(*a*) What do you understand by "series resonance" and "parallel resonance"? What are the applications of tank circuits?

(*b*) Explain the following:

(*i*) Series resonance circuits

(*ii*) Parallel resonance circuits.

(*c*) Consider series R-L-C circuit as shown in Fig. 3.

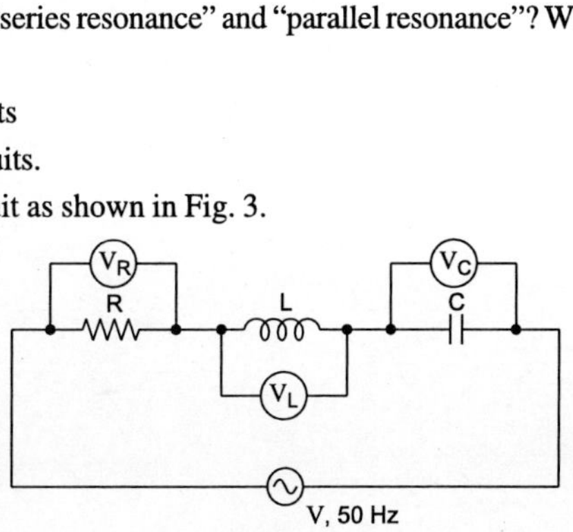

Fig. 3

When $V_R = 8$ V, $V_L = 18$ V, $V_C = 12$ V, find V in above circuit.

3. Attempt any **two** parts of the following: **(10 × 2 = 20)**

(*a*) Draw and explain the neat circuit diagram of 2-wattmeter method and 3-wattmeter method for measurement of power and power factor in 3-ϕ balanced circuit.

(*b*) A 3-phase balanced load connected across a 3-ϕ, 400 V ac supply draws a line current of 10 A. Two wattmeters are used to measure input power. The ratio of two wattmeter readings is 2 : 1. Find readings of two wattmeters.

(*c*) Discuss the constructional details and working principle of PMMC type of voltmeters. What are the advantages and disadvantages of PMMC type of voltmeter? Explain why PMMC type of instruments belongs to linear scale instrument.

4. Attempt any **two** parts of the following: **(10 × 2 = 20)**

(*a*) What are the similarities and dissimilarities between electric and magnetic circuits? Enlist any five applications of magnetic circuits in engineering fields.

(*b*) What are the performance parameters of 1-ϕ transformer? Differentiate between 2 winding transformers and autotransformers. Mention any two applications of shell type of 1-ϕ transformer.

(*c*) A ring of ferromagnetic material has rectangular cross section. The inner diameter is 7.4 inch, the outer diameter is 9 inch and the thickness is 0.8 inch. There is a coil of 600 turns wound on the ring. When coil carries a current of 2.5 A, the flux produced in the ring is 1.2×10^{-3} wb. Find:

(*i*) Magnetic field intensity (*ii*) Reluctance (*iii*) Permeability.

5. Attempt any **four** parts of the following: **(5 × 4 = 20)**

(*a*) State and explain the expression of back emf (E_b) and torque (T_e) in dc machines. What are the basic roles of back emf in dc machines?

(*b*) Discuss the constructional details and working principle of 3-phase induction motor.

(*c*) Explain why 3-ϕ induction motor cannot run at synchronous speed $\left(N_s = \frac{120f}{P}\right)$, where symbols having their usual meaning.

(*d*) Explain why 1-ϕ inductive motor are not self-starting. Discuss various methods of starting of 1-ϕ induction motor.

(*e*) Draw and explain the slip-torque characteristic of 3-ϕ induction motor. And mention the various operating regions on same.

(*f*) Discuss the principle of operation of synchronous machine. A 3-ϕ, 50 Hz, induction motor has 6 poles and operates with a slip of 5% at a certain load. Determine:

(*i*) the speed of rotor with respect to stator.

(*ii*) frequency of rotor current.

(*iii*) speed of rotor magnetic field with respect to rotor.

(*iv*) speed of rotor magnetic field with respect to stator.

(*v*) speed of the rotor magnetic field with respect to stator magnetic field.

❑❑❑

B.TECH.

SECOND SEMESTER EXAMINATION, 2013-14
ELECTRICAL ENGINEERING

Time : 3 Hours *Total Marks : 100*

Note: Attempt **all** questions. Assume any missing data, if any.

SECTION-A

1. Attempt **all** parts of the following: **(2 × 10 = 20)**

(*a*) Write the properties of ideal voltage source.

(*b*) Is L a linear element? Explain the answer.

(*c*) Classify the losses in power transformer.

(*d*) Define the term slip.

(*e*) Why is dc series motor preferred in elevators?

(*f*) Show the residual magnetism in a B-H curve.

(*g*) Explain the term creep in energy meter.

(*h*) Name two motors used for constant speed operation.

(*i*) Write an expression of resource frequency for high Q-series R-L-C circuit.

*(*j*) What is concept of grid in power system?

SECTION-B

2. Attempt any **three** parts of the following: **(3 × 10 = 30)**

(*a*) Describe the construction and principle of operation of attraction type moving iron instrument.

(*b*) Apply mesh analysis, obtain the current through 5 ohm resistance in the following circuit:

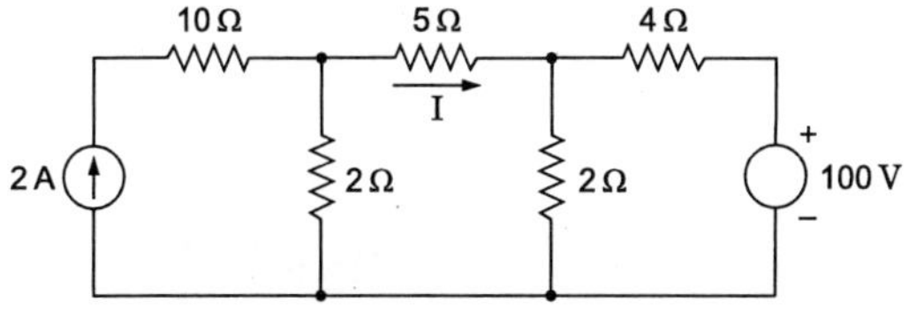

Fig. 1

(*c*) A three-phase 50 Hz, induction motor has a full-load speed of 1,460 rpm. Calculate slip, number of poles and frequency of rotor induced emf.

(*d*) Voltages across R, L, C connected in series are 5, 8 and 10 volts respectively. Calculate the value of supply voltage at 50 Hz. Also find the frequency at which this circuit would resonate.

(*c*) Derive the quality factor Q of the series R-L-C circuit at resonance. Define the bandwidth for the same.

SECTION-C

3. Attempt any **two** parts of the following: **(2 × 5 = 10)**

(*a*) State and prove maximum power transfer theorem.

(*b*) Find the current flowing through 10 Ω resistance in the following circuit. Use superposition theorem.

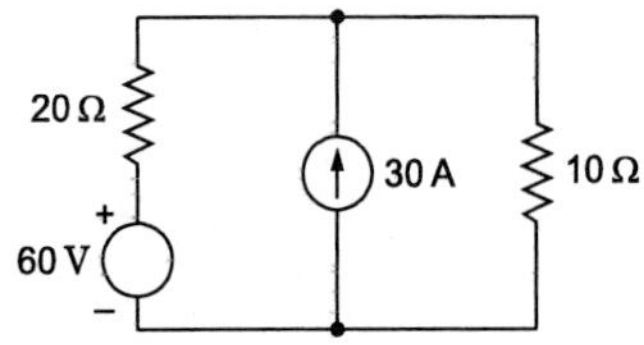

Fig. 2

(*c*) Explain voltage and current sources of a network with characteristics. Explain source-transformation principle in any circuit.

4. Attempt any **one** part of the following: **(1 × 10 = 10)**

(*a*) Calculate the resonance frequency of the circuit shown in figure.

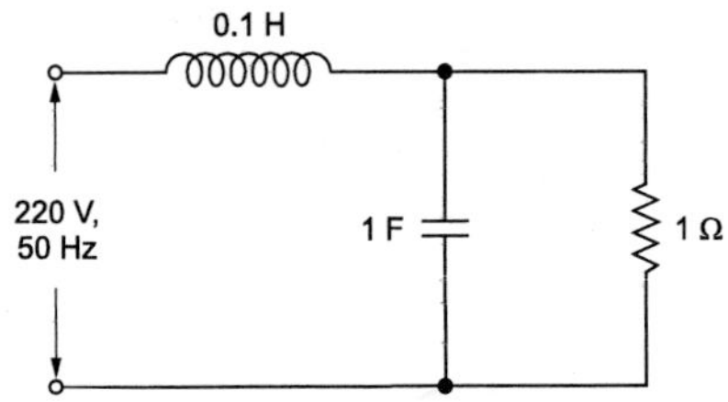

Fig. 3

(*b*) Derive an emf expression of power transformer. Also draw an equivalent circuit of it.

5. Attempt any **two** parts of the following: **(2 × 5 = 10)**

(*a*) Show that in a 3-ϕ star connected system, the line voltage is $\sqrt{3}$ times of the phase voltage.

(*b*) Discuss in brief about autotransformer.

(*c*) A balanced delta connected load of $(12 + j9)$ Ω/phase is connected to 3-phase 400 V supply. Calculate line current, power factor and power drawn by it.

6. Attempt **one** part of the following: **(1 × 10 = 10)**

*(*a*) Discuss the classification of power system in terms of voltage level. Also draw line diagram of typical substation.

(*b*) A dc shunt motor develops an open-circuit emf of 250 V at 1,500 rpm. Find its developed torque for an armature current of 20 A.

7. Attempt any **two** parts of the following: **(2 × 5 = 10)**

(*a*) Describe briefly the construction and working of a synchronous generator.

(*b*) Explain speed-torque characteristics of dc series motor. Also mention typical applications of it.

(*c*) A dc shunt motor develops an open-circuit emf of 250 V at 1,500 rpm. Find its developed torque for an armature current of 20 A.

❑❑❑

B.TECH.

(SEMESTER-I) ODD SEMESTER THEORY EXAMINATION, 2013-14
BASIC ELECTRICAL ENGINEERING

Time : 3 Hours *Total Marks : 100*

SECTION-A

1. Attempt **all** parts of this section: **(2 × 10 = 20)**

(*a*) What are the properties of ideal voltage and current sources?

(*b*) A moving coil instrument has resistance of 2 Ω, reads up to 250 V when a resistance of 5,000 Ω is connected in series with it. Find current range of instrument when it is used as ammeter with coil connected across a shunt resistance of 2 mΩ.

(*c*) A series R-L-C circuit has R = 10 Ω, L = 0.1 H, C = 8 μF. Calculate half power frequencies for the circuit.

(*d*) Find current in 2 Ω.

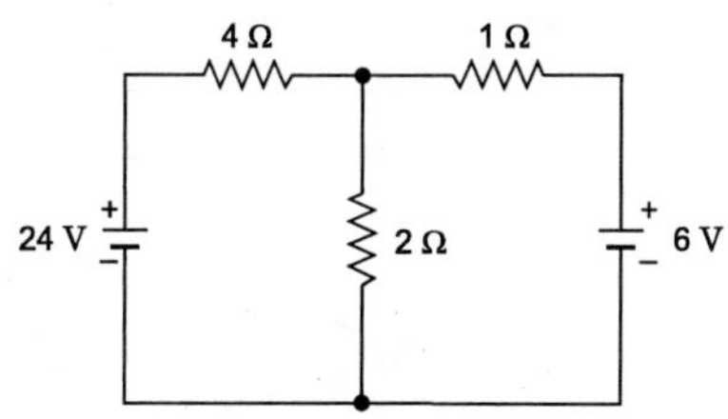

Fig. 1

(*e*) What is the significance of phasor and vector in circuit analysis?

(*f*) Define coefficient of coupling.

(*g*) If in a single phase transformer core, hysteresis and eddy current losses are 80 W and 50 W at normal voltage and frequency then calculate losses when voltage and frequency are increased by 20%.

(*h*) What is magnetomotive force and write its unit.

(*i*) How will you change the direction of rotation of dc motor?

(*j*) What do you understand by active and reactive powers? Why reactive power is not present in dc circuits?

SECTION-B

2. Attempt any **three** parts of this section: **(10 × 3 = 30)**

(*a*) State Norton's theorem. Find current across 15 Ω by using Norton's theorem.

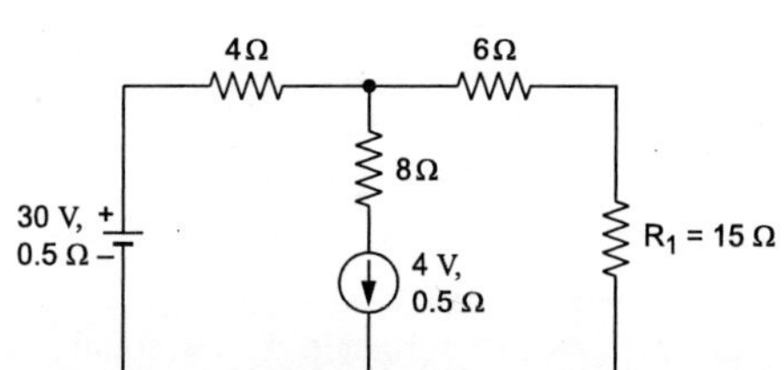

Fig. 2

(*b*) (*i*) Derive Bandwidth of series resonance.

(*ii*) If two alternating currents represented by $i_1 = 7 \sin \omega t$ and $i_2 = 10 \sin [\omega t + \pi/3]$ are fed into a common conductor, then find equation for resultant current and its rms value.

(*c*) A star-connected system has impedance as $\mathbf{Z} = 8 + j6\ \Omega$ in each phase. It is fed from 400 V 3-phase ac supply. Find active and reactive powers. If power measurement is done by two wattmeter method then calculate readings of each wattmeter. Also draw connection diagrams for this power measurement.

(*d*) Explain the need of earthing electrical devices. What are the important safety issues?

(*e*) Discuss principle of operation of synchronous motor with diagram. Draw characteristics for dc series motor.

SECTION-C

Note: Attempt **all** the questions.

3. Attempt any **two** parts: **(5 × 2 = 10)**

(*a*) Find current in 6 Ω using Thevenin's theorem.

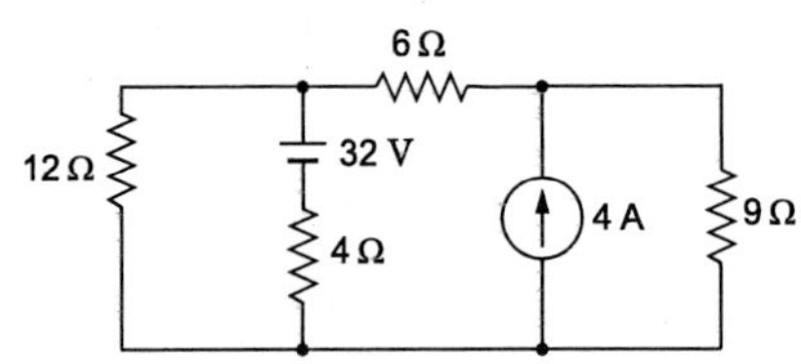

Fig. 3

(*b*) Find current in each branch by using nodal analysis. Also calculate total power loss.

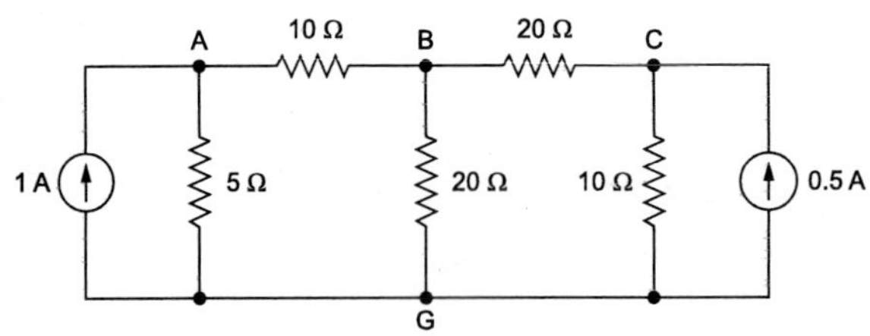

Fig. 4

(*c*) Deduce delta-connected system from star-connected system.

4. Attempt any **two** parts: **(5 × 2 = 10)**

(*a*) What are causes and disadvantages of low power factor?

(*b*) A metal filament lamp, rated at 750 W, 100 V, is to be connected in series with a capacitor across 230 V, 50 Hz supply. Calculate the value of capacitor.

(*c*) Two impedances $\mathbf{Z}_1 = 5 + j10\ \Omega$ and $\mathbf{Z}_2 = 10 - j15\ \Omega$ are connected in parallel. If total current is 20 A then find:

(*i*) current taken by each branch,

(*ii*) power factor,

(*iii*) power consumed in each branch.

5. Attempt any **two** parts: **(5 × 2 = 10)**

(*a*) Explain the construction and working of electrodynamic type instrument with diagram.

(*b*) Explain construction, working and torque in energy meter with neat sketch.

(*c*) A 3-phase load consisting of resistance 25 Ω, inductance of 0.15 H and capacitor of 100 μF is connected to 400 V, 50 Hz. Calculate line current, power factor, total power (*i*) when connected in star, (*ii*) when connected in delta.

6. Attempt any **two** parts: **(5 × 2 = 10)**

(*a*) Describe the analogies between electric and magnetic circuits in brief.

(*b*) Explain eddy current loss and its remedy in magnetic circuit.

(*c*) The efficiency of a 400 kVA, single phase transformer is 98.77% at full load 0.8 pf and 99.13% at half load unity pf. Find iron and cu losses at half load.

7. Attempt any **two** parts: **(5 × 2 = 10)**

(*a*) Explain any one method of starting single-phase induction motor with neat diagram.

(*b*) Explain working of 3-phase induction motor. Also draw torque-slip characteristic showing operating regions.

(*c*) A 4 pole dc generator with wave-connected armature has 41 slots and 12 conductors/slots. Armature resistance and shunt field resistance are 0.5 Ω and 200 Ω. Flux per pole is 125 mWb. Speed N = 1,000 rpm. Calculate the voltage drop across terminals. The load resistance is 10 Ω.

B.TECH.

SECOND SEMESTER EXAMINATION, 2013-14
BASIC ELECTRICAL ENGINEERING

Time : 3 Hours *Total Marks : 100*

Note: Attempt **all** sections.

SECTION-A

1. Answer **all** parts of this section: **(2 × 10 = 20)**

(*a*) Enlist the limitations of Ohm's law.

(*b*) Write the practical units of electrical energy and mechanical power.

(*c*) Determine the form factor of ac current $i = 400 \sin(157t + \pi/6)$.

(*d*) Enlist the main causes of a large kVAR in a system.

(*e*) Write two problems of low power factor in a system.

(*f*) What is selectivity?

(*g*) How is range extended in moving coil ammeter and voltmeter?

(*h*) What do you understand by the term "ideal transformer"?

(*i*) Write the emf equation of a dc generator.

(*j*) What is typical use of starter in motor?

SECTION-B

2. Answer any **three** parts of the following: **(10 × 3 = 30)**

(*a*) Find the value of resistance 'R' to have maximum power transfer in the circuit shown in Fig. 1. Also obtain the amount of maximum power.

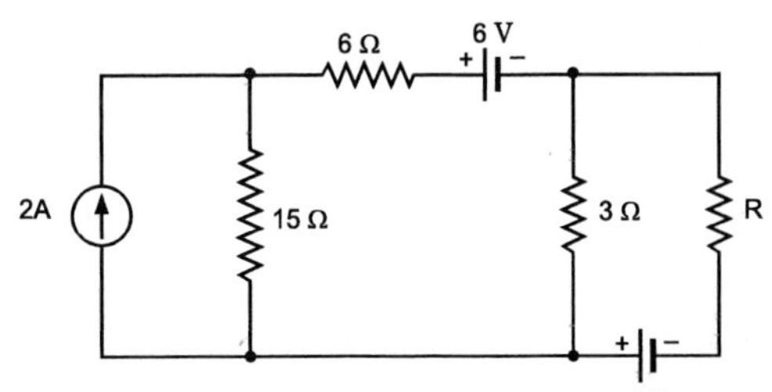

Fig. 1

(*b*) Draw a phasor diagram showing the following voltages:

$$v_1 = 100 \sin 500\,t$$
$$v_2 = 200 \sin (500t + 45°)$$
$$v_3 = \cos 500t$$

Find rms value of resultant voltage.

(*c*) Explain the principle and construction of attraction type moving iron instruments.

(*d*) Why earthing is needed in an electrical system? Enlist the various types of earthing systems.

(*e*) How can the speed of dc motor be varied? Explain the method by which you can vary speed above and below the base speed.

SECTION-C

Note: All questions are compulsory.

3. Attempt any **one** part of the following: **(10 × 1 = 10)**

(*a*) Derive an expression of star to delta transformation and vice versa.

(*b*) State Norton's theorem. Obtain the Norton's equivalent circuit at terminals AB of the network given in Fig. 2.

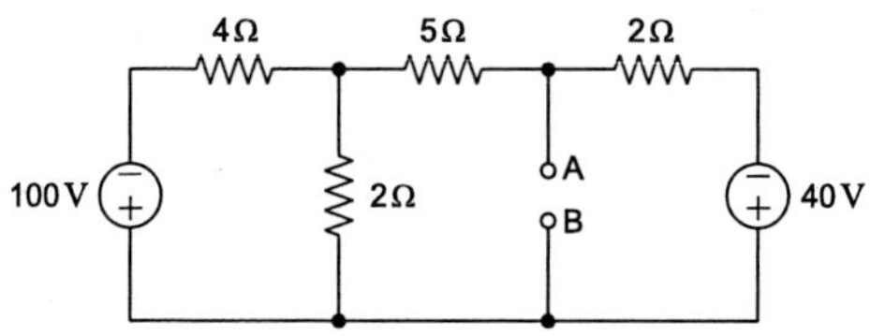

Fig. 2

4. Answer any **one** part of the following: **(10 × 1 = 10)**

(*a*) A resistance and inductance are connected in series with voltage $v = 283 \sin 314t$. The current expression is found to be $i = 4 \sin (314t - 45°)$. Find the value of resistance, inductance and power factor.

(*b*) Derive the condition for resonance in series RLC circuit. What are the different applications of resonance?

5. Answer any **one** part of the following: **(10 × 1 = 10)**

(*a*) Describe the following in case of measuring instruments.

(*i*) Deflecting torque

(*ii*) Controlling torque

(*iii*) Damping torque.

(*b*) Show that power in 3-phase balanced system is constant at every instant and is given by $3V_pI_p \cos \phi$.

6. Answer any **one** part of the following: **(10 × 1 = 10)**

(*a*) Define the following terms as applied to magnetic circuit.

(*i*) MMF

(*ii*) Flux density

(*iii*) Reluctance

(*iv*) Permeability

(*b*) A single phase 250 kVA transformer has an efficiency of 96% on full load at 0.8 power factor and on half load

(*i*) Iron loss

(*ii*) Full load copper loss.

7. Answer any **two** parts of the following: **(5 × 2 = 10)**

(*a*) A 230 V dc series motor is taking 50 A. Resistance of armature and series field winding is 0.2 Ω and 0.1 Ω respectively. Calculate:

(*i*) Brush voltage

(*ii*) Back emf.

(*b*) A 3-phase 4 pole induction motor is supplied from 3-phase 50 Hz supply. Calculate:

(*i*) N_s

(*ii*) Rotor speed when slip is 4%

(*iii*) Rotor frequency when rotor runs at 60 rpm.

(*c*) Write a short note on synchronous condenser.

❑❑❑

INDEX

A

B

C

D

E

F

G

H

I

J

K

L

M

N

O

P

Q

R

S

T

U

V

W

❑❑❑

Win Prizes!

ATTENTION READERS

We request you for your frank assessment regarding some of the aspects of the book given as under:

A Textbook of
ELECTRICAL ENGINEERING

By

J.B. Gupta

Please fill up the given space in neat capital letters.

(i) What topics of your syllabus that are important from you examination point of view are not covered in the book ?

....

....

....

....

(ii) What are the chapters wherein the treatment of the subject matter is not systematic or organised or updated ?

....

....

....

....

(iii) Have you come across misprints/mistakes/factual inaccuracies in the book? Please specify the chapters and the page numbers.

....

....

....

....

(iv) What other books on the same subject have you found/heard better that the present book? Please specify in terms of price and quality.

....

....

....

....

(v) Further suggestions and comments for the improvement of the book:

....

....

....

....

(PS: If need be please attach a separate sheet to write your views.)

Other Details :

(i) Who recommended you the book ? (Please tick in the box near the option relevant to you.)

☐ Teachers ☐ Friends ☐ Bookseller

(ii) Name of the recommending teacher, his designation and address.

....

....

....

....

(iii) Name and address of the bookseller you purchased the book form :

....

....

....

....

(iv) Name and address of your institution (Please mention the University or Board, as the case may be)

....

....

....

....

(v) Your name and complete postal address :

....

....

....

....

(vi) Course you are enrolled in :

....

....

....

....

The best assessment will be awarded each month. The award will be in the form of our publications as decided by the Editorial Board.

Please mail the filled up coupon at your earliest to:

S.K. Kataria & Sons®
4885/109, Prakash Mahal, Dr. Subhash Bhargav Lane,
Opposite Delhi Medical Association, Daryaganj, New Delhi-110002 (India)
Phone: +91-11-43551243; Mobile: +91-9871775858, 7678425209
e-mail: katariabook@gmail.com, katariabooks@yahoo.com
Website: www.skkatariaandsons.com